Environmental Science and Sustainability

Environmental Science and Sustainability

Daniel J. Sherman

UNIVERSITY OF PUGET SOUND

David R. Montgomery

UNIVERSITY OF WASHINGTON

100

W. W. NORTON & COMPANY

Celebrating a Century of Independent Publishing

To Sally and John, who taught me that one question always leads to another. —Dan Sherman

To those who seek to build a sustainable future. —Dave Montgomery

W. W. Norton & Company has been independent since its founding in 1923, when William Warder Norton and Mary D. Herter Norton first published lectures delivered at the People's Institute, the adult education division of New York City's Cooper Union. The firm soon expanded its program beyond the Institute, publishing books by celebrated academics from America and abroad. By midcentury, the two major pillars of Norton's publishing program—trade books and college texts— were firmly established. In the 1950s, the Norton family transferred control of the company to its employees, and today—with a staff of five hundred and hundreds of trade, college, and professional titles published each year—W. W. Norton & Company stands as the largest and oldest publishing house owned wholly by its employees.

Editor: Eric Svendsen
Senior Developmental Editor: Sunny Hwang
Associate Managing Editor, College: Melissa Atkin
Assistant Editor: Mia Davis
Copyeditor: Pat Wieland
Editorial Intern: Devon N. Rodgers
Associate Director of Production, College: Benjamin Reynolds
Managing Editor, College: Marian Johnson
Media Editor: Ariel Eaton
Media Editorial Assistant: Rachel Bass

Ebook Producer: Sophia Purut
Managing Editor, College Digital Media: Kim Yi
Marketing Research and Strategy Director: Stacy Loyal
Art Director: Jillian Burr
Director of College Permissions: Megan Schindel
College Permissions Specialist: Joshua Garvin
Photo Editor: Catherine Abelman
Composition: GraphicWorld / Sunil Kumar, Project Manager
Infographics and Cover by Hyperakt and Mattias Mackler
Manufacturing: Transcontinental—Beauceville, QC

Permission to use copyrighted material is included at the back of the book.

ISBN: 978-0-393-89350-2 (pbk.)

W. W. Norton & Company, Inc., 500 Fifth Avenue, New York, NY 10110
wwnorton.com
W. W. Norton & Company Ltd., 15 Carlisle Street, London W1D 3BS

1 2 3 4 5 6 7 8 9 0

About the Authors

Daniel J. Sherman is the Luce-Funded Professor of Environmental Policy and Decision Making and director of the Sound Policy Institute at the University of Puget Sound. He studies the role individuals and groups play in environmental politics, policy, and sustainability. He received BA degrees from Canisius College (1995, political science) and Victoria University of Wellington (1996, Maori studies), MA degrees from Colorado State University (1999, political science) and Cornell University (2002, government), and a PhD from Cornell University (2004, government). Sherman has written *Not Here, Not There, Not Anywhere: Politics, Social Movements, and the Disposal of Low-Level Radioactive Waste* with Resources for the Future Press. He is an award-winning teacher who loves to take his students outside and helps them to engage directly in environmental decision-making contexts. Sherman has also written articles about and conducted workshops for faculty on the integration of sustainability across the curriculum.

David R. Montgomery is a professor of Earth and space sciences at the University of Washington. He studies the evolution of topography and the influence of geomorphological processes on ecological systems and human societies. He received a BS from Stanford University (1984, geology) and a PhD from UC Berkeley (1991, geomorphology). His field studies have included projects in the Philippines, eastern Tibet, South America, California, and the Pacific Northwest of North America. In 2008 Montgomery received a MacArthur Fellowship. His books *Dirt: The Erosion of Civilizations*, *King of Fish: The Thousand-Year Run of Salmon*, and *The Rocks Don't Lie: A Geologist Investigates Noah's Flood* have all won the Washington State Book Award in General Nonfiction. Montgomery's *Growing a Revolution: Bringing Our Soil Back to Life*, was a finalist for the PEN/E. O. Wilson Award for Literary Science Writing. He also coauthored with Anne Biklé *The Hidden Half of Nature* and, most recently, *What Your Food Ate: How to Heal Our Land and Reclaim Our Health*. His books have been translated into ten languages.

Brief Contents

1   Environmental Science and Sustainability: What's the Big Idea?....................2

2   Ethics, Economics, and Policy: Who or What Do We Value?....................32

3   Matter and Energy: What Are the Building Blocks of Sustainability?....................62

4   Life: What Shapes Biodiversity?....................86

5   Conservation: Why Is It Important to Protect Biodiversity?....................118

6   Human Population: Can We Have Too Many People?....................150

7   Water: How Do We Use It and Affect Its Quality?....................180

8   Air: What Are We Breathing?....................212

9   Land: How Does It Shape Us?....................244

10   Systems and Cycles: Are We Harming Earth's Life-Support System?....................276

11   Climate: How Does Global Climate Change?....................302

12   Food: How Do We Feed Ourselves?....................336

13   Fossil Fuels: Energy of the Industrial Age....................368

14   Energy Alternatives: How Are Our Energy Decisions Changing?....................396

15   Waste: What Happens to All the Stuff We Use?....................426

16   Urbanization: Why Are Cities Growing?....................458

17   Environmental Health and Justice: How Do Environmental Factors Affect the Places People Live, Work, and Play?....................488

18   Decision Making: Why Do Our Choices Matter?....................520

19   Groups and Organizations: How Do We Work Together for Sustainability?....................544

20   Government: How Can Policy Influence Sustainability?....................570

Contents

To the Instructor xxix

To the Student xxxiv

Acknowledgments xxxvi

1 Environmental Science and Sustainability:
What's the Big Idea? 2

Chapter Objectives 3

1.1 What Is the Environment, and What Is an Ecosystem? 5
What Are Systems? 5
Where Do Humans Fit in the Environment? What's Our Impact? 6

1.2 What Is Sustainability? 7
Defining Sustainability 8
Ecological Resistance and Resilience 9

1.3 What Is Environmental Justice? 12
Working for Environmental Justice 14

1.4 What Is Science? 14
The Scientific Method 14
Ways of Observing and Testing 15
Models 16

1.5 What Are Challenges to Good Science? 18
Fraud and Pseudoscience 18
Bias and Misinformation 19
The Function of Peer Review 20

1.6 What Shapes Our Decisions on the Environment? 20
Factors Influencing Individual Decisions 20
STORIES OF DISCOVERY When Small Observations Lead to
Questions, Problems, and Solutions 21
Trade-Offs and Incentives 22
Making Sense of Our Environmental Effects 23
Groups and Organizations 24
What Are Some of the Big Decisions We Will Explore in This Book? ... 24
SUSTAINABILITY MATTERS Mostafa Tolba: Scientist as Diplomat 25

1.7 What Can I Do? 26

Consider Sustainability When You Make Purchases 26

Research and Participate in Sustainability Groups, Offices, or
Plans on Your Campus 26

Register and Exercise Your Right to Vote in Elections 27

Use What You Learn to Start Making Sustainable Decisions Every Day 27

Chapter 1 Review 27

Summary 27

Key Terms 28

Review Questions 28

For Further Thought 28

Use the News 29

Use the Data 30

Learn More 31

2 Ethics, Economics, and Policy:
Who or What Do We Value? 32

Chapter Objectives 33

2.1 How Does Ethics Influence Our Decisions? 35

Duties and Rights 35

The Greatest Good for the Greatest Number 37

Who or What Do We Care About? 37

2.2 How Are the Environment and Economy Connected? 38

Supply and Demand 38

SUSTAINABILITY MATTERS Is Eating a Vegetarian Diet More
Sustainable? 39

Government and Markets 40

Scarcity Inspires Innovation and Substitution 41

Efficiency Gains Can Inspire Increased Consumption 42

**2.3 How Can Economics Help Us Understand Environmental
Problems?** 43

Negative Externalities 43

Tragedy of the Commons 44

Positive Externalities, Ecosystem Services, and Public Goods 45

Placing Economic Values on Nature 46

2.4 Why Do We Have Environmental Policy? 47

What Is Policy? 47

What Is Politics, and How Are Policies Made? 47

Environmental Policy in the United States 48

STORIES OF DISCOVERY Harnessing Energy from Waves
in the Ocean 49

State and Local Policies 51

Global Environmental Politics 51

2.5 What Can I Do? 52
Exercise Your Economic Power 52
Influence Public Policy 53
Research and Influence Your Elected Officials 53

Chapter 2 Review 56
Summary 56
Key Terms 56
Review Questions 56
For Further Thought 57
Use the News 57
Use the Data 60
Learn More 61

3 Matter and Energy:
What Are the Building Blocks of Sustainability?

.... 62

Chapter Objectives 63

3.1 What Is Matter Anyway? 64
Elements and Atoms 64
What's Inside an Atom? 65

3.2 What Distinguishes Different Kinds of Matter? 66
Conservation of Mass 66
How Do Atoms Bond Together? 66
STORIES OF DISCOVERY Mining Garbage for Plastic: One
Way to Profit from the Different Properties of Matter 69
Acids and Bases 72

3.3 Can Matter Change? 72
SUSTAINABILITY MATTERS Can We Tinker with Earth's Chemistry
to Address the Effects of Climate Change? 73
Phase Changes 74
Chemical Reactions 74

3.4 What Is Energy? 75
Forms of Energy 76

3.5 What Happens to Energy When We Have Used It? 77
The First and Second Laws of Thermodynamics 77

3.6 How Does Energy Affect Life? 79
Food Chains and Trophic Levels 79
The 10% Law 79

3.7 What Can I Do? 80
Understand the Connections and Help Others to Do So Too ... 80
Contact Your School's Office of Sustainability 80
Learn to Follow the Matter 80

Understand That Everything Goes Somewhere 81

Find Your Opportunity to Innovate 81

Chapter 3 Review 81

Summary 81

Key Terms 81

Review Questions 82

For Further Thought 82

Use the News 82

Use the Data 84

Learn More 85

4 Life:
What Shapes Biodiversity?

.... 86

Chapter Objectives 87

4.1 What Do Living Things Have in Common? 88

4.2 How Does Life Evolve? 90

Mutations, Selection, and Extinction 91

Five Misconceptions about Evolution 92

4.3 What Causes Population Traits to Change? 93

Genetic Diversity 94

4.4 What Shapes Biodiversity? 95

How Many Species Are There? 95

Species Evenness 96

Ecosystem Diversity and Biogeography 98

Biomes 98

4.5 How Do Communities of Organisms Interact? 98

Competition 99

Predation 102

Interspecific Competition 103

Coevolution 103

Symbiotic Interactions 104

STORIES OF DISCOVERY Predators and Trees in Yellowstone 105

4.6 What Controls Population Size? 106

Population Growth and Responses to Limits 106

Population Density 107

Reproductive Strategies 108

4.7 What Can Cause Loss of Biodiversity? 109

Identifying Areas and Species Vital to Biodiversity 110

SUSTAINABILITY MATTERS Killing to Protect Biodiversity
in New Zealand 111

4.8 **What Can I Do?** 112

Create Wildlife-Friendly Habitats Where You Live 113

Reduce Some of the Negative Effects on Wildlife Where You Live 113

Minimize Your Impact on Wildlife When You Are out Enjoying Nature 113

Chapter 4 Review 114

Summary 114

Key Terms 114

Review Questions 114

For Further Thought 115

Use the News 115

Use the Data 116

Learn More 117

5 Conservation:

Why Is It Important to Protect Biodiversity? 118

Chapter Objectives 119

5.1 **What Is the Status of Earth's Biodiversity?** 121

Human Impact on Biodiversity 121

5.2 **Why Protect Biodiversity?** 123

What Are Ecosystem Services? 124

Placing Economic Value on Ecosystem Services 125

Criticisms of the Ecosystems Services Approach 126

5.3 **What Are Protected Areas?** 126

STORIES OF DISCOVERY Can Biodiversity Boost Soil Fertility
and Agricultural Productivity? 127

National Parks, Monuments, and Recreation Areas 130

National Wildlife Refuges 130

National Forests 131

Bureau of Land Management Lands 131

Wilderness Areas and Wild and Scenic Rivers 131

Aquatic and Marine Areas 131

Protected Areas throughout the World 131

5.4 **What Are the Limitations of Protected Areas, and
Are There Alternatives?** 132

The Ecological Island Effect 132

Protected Areas Do Not Necessarily Match Protection Priorities 133

Human–Wildlife Conflict 133

Alternative Protection Strategies 135

5.5 **Can Laws Protect Biodiversity?** 135

SUSTAINABILITY MATTERS How Should We Prioritize Our
Efforts to Conserve Biodiversity? 137

Role of the States 138

International Laws and Agreements 138

5.6 What Can Be Done to Reduce Biodiversity Loss? 139

Sustainable Forest Management 140

Grazing and Grassland Management 141

Agriculture and the Protection of Biodiversity 141

Urbanization and Land-Use Planning 142

5.7 What Can Be Done to Fix Damage That Has Already Occurred? 142

Species Reintroduction 142

Ecological Restoration 142

5.8 What Can I Do? 144

Get Involved with a Campus Restoration Project 144

Learn about Your Local Biodiversity and Get Others Involved:
Start or Join a BioBlitz 144

Join a Conservation Group 144

Buy Products That Have Less Impact on Threatened Species 144

Chapter 5 Review 145

Summary 145

Key Terms 145

Review Questions 145

For Further Thought 146

Use the News 146

Use the Data 148

Learn More 149

6 Human Population:
Can We Have Too Many People?

.... 150

Chapter Objectives 151

6.1 What Is the History of Human Population Change? 152

Historical Population Crashes 153

**6.2 What Has Driven Recent Population Growth? Can
Earth Support It?** 156

Rates of Birth and Death 156

Agricultural Development and Technology 157

Advances in Health 158

How Much Human Activity Can Earth Support? 158

6.3 How Have Societies Attempted to Control Population? 159

Population-Control Policies 159

Social Stratification and Inequality 160

STORIES OF DISCOVERY Soap Operas and Fertility 161

6.4 How Is Global Population Changing? 162

Demographic Transition 163

Examples of Demographic Transition 164

Waiting for or Passing by the Stability Transition 164

Consequences of Low Birth Rates 165

6.5 What Factors Facilitate a Demographic Transition? 166

Drops in Infant Mortality 166

Urbanization 166

Health and Education 167

6.6 How Does Empowering Women Change Populations? 168

Access to Birth Control 168

Education for Girls and Women 169

International Response 170

6.7 How Is Population Connected to Consumption? 170

Impact = Population × Affluence × Technology 170

Global Comparisons 170

SUSTAINABILITY MATTERS Overpopulation or Rising
Consumption? 173

6.8 What Can I Do? 174

Know Your Personal Ecological Footprint 174

Get Involved with Education and Rights Issues 174

Explore How Technological Solutions Might Help 174

Chapter 6 Review 175

Summary 175

Key Terms 175

Review Questions 175

For Further Thought 176

Use the News 176

Use the Data 178

Learn More 179

7 Water:

How Do We Use It and Affect Its Quality?

.... 180

Chapter Objectives 181

7.1 Where on Earth Is All the Water? 183

7.2 How Does Fresh Water Support Life? 186

Lakes and Ponds 186

Rivers and Streams 186

Life in Wetlands and Estuaries 187

7.3 How Do Humans Impact Fresh Water? 189

Increased Withdrawals from Groundwater Systems 190

Disruptions to Surface Water Systems 191

Water Pollution 193

7.4 Can We Improve Access to Fresh Water? 194

Where Does the Water Go? 196

Water Shortages 196

Water Conservation 196

Water Scarcity 197

7.5 How Do We Keep Fresh Water Clean? 198

SUSTAINABILITY MATTERS Cow Milk or Almond Milk—What Uses More Water? 199

STORIES OF DISCOVERY From Toilet to Tap? 201

7.6 Why Is Frozen Water Important? 202

7.7 How Does the Ocean Support Life? 204

7.8 How Can Humans Impact the Ocean? 204

7.9 What Can I Do? 207

Learn Where Your Water Comes From 207

Figure Out How Much Water You Use 207

Consume Less Water for Personal Use 207

Use Less Water for Landscaping 207

Learn More about How Others Are Facing Water-Conservation Challenges and Sustaining Reforms 207

Chapter 7 Review 208

Summary 208

Key Terms 208

Review Questions 209

For Further Thought 209

Use the News 209

Use the Data 210

Learn More 211

8 Air:
What Are We Breathing?

.... 212

Chapter Objectives 213

8.1 What Makes Up the Air We Breathe and Our Atmosphere? 215

Pressure 215

How Is Earth's Atmosphere Structured? 216

8.2 What Is Weather and How Does It Change? 217

Temperature 218

Water Vapor 219

Barometric Pressure 219

Wind, Fronts, and Storms 221

Lake and Land Effects 221

8.3 How Does the Atmosphere Circulate? 222

Patterns of Atmospheric Pressure and Prevailing Winds 222

STORIES OF DISCOVERY What's in the Clouds? 223

Atmospheric Dust Transport 224

8.4 How Does the Atmosphere Become Polluted? 225

Primary Pollutants 226

Secondary Pollutants: Smog and Ground-Level Ozone 227

8.5 **What Is Happening to the Ozone Layer?** 229

8.6 **What Is Acid Rain?** 230

SUSTAINABILITY MATTERS Air Pollution: How Will the Chinese Government Respond? 235

8.7 **How Are We Responding to Pollution in the Atmosphere?** 236

8.8 **What Can I Do?** 237

Choose Your Transportation with Pollution in Mind 237

Look for Ways to Limit Pollution at Home and on Campus 237

Explore the Impacts of New Technology 238

Learn More about Power Generation in Your Community in Order to Influence Local Policy 238

Chapter 8 Review 238

Summary 238

Key Terms 239

Review Questions 239

For Further Thought 239

Use the News 240

Use the Data 242

Learn More 243

9 Land:
How Does It Shape Us? 244

Chapter Objectives 245

9.1 **How Do Mountains Rise and Animals Find Their Way Home?** 246

Earth's Internal Structure 247

Plate Tectonics 248

9.2 **How Does Earth Recycle and Renew Its Surface?** 250

Weathering 250

Erosion and Lithification 251

9.3 **Why Are Minerals Important, and How Do We Get Them?** 253

Minerals and Mineral Resources 253

9.4 **What Shapes Earth's Surface?** 255

Landforms 255

Processes That Shape the Land 257

9.5 **What Is Soil, and Where Does It Come From?** 259

Soil Horizons 259

9.6 **What Makes Soil Alive?** 260

Organic Matter and Soil Structure 260

STORIES OF DISCOVERY Worms Make Soil 261

Soil Classification and Factors Affecting Soil Formation 262

Topsoil and Plant Growth 262

Water in Topsoil 263

9.7 How Can Humans Repair and Sustain the Land? 263
Agriculture and Grazing 264
SUSTAINABILITY MATTERS The Erosion of Civilizations 265
Mining 266
Forestry 266
Urbanization 267

9.8 What Can I Do? 270
Conserve and Recycle Rare Earth Minerals 270
Know How Your Food Affects Soil Health 270
Get Informed about Development Decisions in Your Area ... 270
Be Prepared for Natural Hazards 270

Chapter 9 Review 271
Summary 271
Key Terms 271
Review Questions 271
For Further Thought 272
Use the News 272
Use the Data 273
Learn More 275

10 Systems and Cycles:
Are We Harming Earth's Life-Support System? 276

Chapter Objectives 277

10.1 What Are Systems? 278
Parts, Interactions, and Emergent Properties 279

10.2 How Can We Show How a System Works? 280
Stocks and Flows 280
Feedback 281

10.3 How Do Earth's Life-Support Systems Work? 283
The Story of Oxygen 283
Harnessing the Phosphorus Cycle 284

10.4 How Do Living Things Change the Nitrogen Cycle? 286
SUSTAINABILITY MATTERS We Are Flushing Phosphorus
Down the Drain, but Can We Reclaim It? 287
STORIES OF DISCOVERY Biological Nitrogen Fixation 289
Human Impacts on the Nitrogen Cycle 290

10.5 How Are We Changing Earth's Carbon Cycle? 291
Natural Cycling of Carbon between the Lithosphere, Atmosphere,
and Hydrosphere 291
Carbon Cycling through the Biosphere 291
Carbon Cycling from the Biosphere to the Lithosphere ... 293
Humans and the Carbon Cycle 293

10.6 What Can I Do? 296
Know Your Footprints for Carbon and Nitrogen 296
Practice Systems Thinking I: Stock and Flow Models 296
Practice Systems Thinking II: The Iceberg Model 297
Chapter 10 Review 298
Summary 298
Key Terms 298
Review Questions 298
For Further Thought 299
Use the News 299
Use the Data 300
Learn More 301

11 Climate:
How Does Global Climate Change?

.... 302

Chapter Objectives 303

11.1 How Do We Measure Global Climate? 304
Temperature Records 305
Temperature Measurements from the Ocean 306
Geothermal Gradients 306
Proxies 306

11.2 What Caused Climate Change in the Past? 307
The Greenhouse Effect 308
Continental Drift 309
The Composition of the Biosphere 310
Earth's Orbit 310
Eruptions and Asteroids 310

11.3 How Do Oceans Influence the Climate? 311
Ocean Currents 311
Countercurrents and El Niño 312

11.4 What's Happening with Our Climate Now? 313
The Concentration of Carbon Dioxide in the Atmosphere
Is Increasing 316
Concentrations of Other Human-Caused Greenhouse Gases
Are Also Increasing 316
STORIES OF DISCOVERY The Development of Global Climate Models 317
How Much Do Human-Caused Greenhouse Gas Emissions
Affect the Climate? 318

11.5 What Are Some Effects of Climate Change? 319
Precipitation and the Water Cycle 320
Hurricanes and Severe Weather 321
Sea-Level Rise 322
Ecosystem Effects 322

11.6 What about the Future? 325
Modeling Alternative Scenarios for Future Human Response 325
SUSTAINABILITY MATTERS Fuel Efficiency and the Carbon
Footprint of Your Vehicle 327

11.7 What Can I Do? 328
Know Your Carbon Footprint 328
Consider the Impact of Your Transportation 328
Reduce Home Energy Use and Choose Green Power 328
Consider the Whole Impact of What You Consume 328
Learn about Climate Justice 329
Make Changes Visible to Others 329

Chapter 11 Review 330
Summary 330
Key Terms 330
Review Questions 331
For Further Thought 331
Use the News 331
Use the Data 333
Learn More 335

12 Food:
How Do We Feed Ourselves? 336

Chapter Objectives 337

12.1 How Did Our Modern Agriculture Develop? 338
Historical Agricultural Practices 338
Modern Agricultural Practices: The Green Revolution 339

12.2 How Does Modern Agriculture Impact the Environment? 341
The Loss of Agrobiodiversity 342
Pesticides 342
Genetically Modified Organisms 343
Soil Loss and Degradation 343
Water Quality and Air Pollution 344
Water Availability 344

12.3 How Is Meat Production Changing What We Consume? 345
Changing Livestock Production Practices 345
Chicken Production 348
Pork Production 348
SUSTAINABILITY MATTERS Genetically Modified Organisms 349
Beef Production 350
Catching and Raising Seafood 351

12.4 Is Conventional Meat Production Sustainable? 352
Impacts of Livestock Production 352
Impacts of Seafood Production 353

12.5 How Have Our Food Systems Changed? 353

Agricultural Production: Increasing Scale and Shrinking Diversity 354

Food Security 355

Food Consumption: Increasing Role of Meat and Processed Foods 355

Shifting Food Culture 356

Food Policy 357

12.6 How Do Our Food Choices Link to Sustainability? 358

The Efficiency Argument: Eating Low on the Food Chain 358

Concern for Animal Welfare 359

Organic and Sustainable Agricultural Practices 359

Concern for Global Equity and Community: Fair Trade and
Buying Local 359

Reducing Food Waste 360

STORIES OF DISCOVERY Sir Albert Howard, Grandfather
of the Organic Farming Movement 361

12.7 What Can I Do? 362

Make Sustainable Choices about the Meat You Eat 362

Support Sustainable Fishing Practices 362

Reduce Your Food Waste 362

Support a Community Garden, Farmers' Market, or Food Bank 362

Chapter 12 Review 363

Summary 363

Key Terms 363

Review Questions 363

For Further Thought 364

Use the News 364

Use the Data 366

Learn More 367

13 Fossil Fuels:
Energy of the Industrial Age 368

Chapter Objectives 369

13.1 What Are Fossil Fuels, and How Important Are They Today? 370

13.2 Have We Always Used Fossil Fuels? 372

Energy and Early Human History 372

Power and Productivity 373

13.3 What Drove the Rise of Fossil Fuels? 374

Economic Forces and the Rise of Fossil Fuels 375

Government Actions and the Rise of Fossil Fuels 377

13.4 **What Are the Environmental Impacts of Obtaining and Using Fossil Fuels?** 378

Coal 378

Oil 380

Natural Gas 382

13.5 **What Factors Will Impact the Future of Fossil Fuels?** 383

The Costs of Fossil Fuel Dominance 386

SUSTAINABILITY MATTERS What Is Fracking? 387

Policy Responses 388

STORIES OF DISCOVERY Capture, Use, and Storage of Carbon Emissions 389

13.6 **What Can I Do?** 391

Choose Transportation That Uses Less Fossil Fuel 391

Opt for Products and Services That Use Less Fossil Fuel 391

Find Out How Familiar Organizations Are Committed to Reducing Fossil Fuel Use 392

Be Informed about the Policies and Economics of Fossil Fuel Use 392

Chapter 13 Review 392

Summary 392

Key Terms 393

Review Questions 393

For Further Thought 393

Use the News 394

Use the Data 395

Learn More 395

14 Energy Alternatives:
How Are Our Energy Decisions Changing? 396

Chapter Objectives 397

14.1 **Why Are Wind and Solar Power Use Growing?** 398

Wind Power 398

Solar Power 400

14.2 **What Are Other Energy Alternatives?** 403

Hydropower 403

Nuclear Power 404

Geothermal Power 406

14.3 **How Is Alternative Energy Used for Transportation?** 407

Ethanol and Other Biofuels 407

Electric Vehicles 408

Fuel Cells 409

Adoption of Technologies 410

14.4 What Role Does Energy Conservation Play in Energy Transitions? 410

STORIES OF DISCOVERY The Solutions Project 411

Cogeneration 412

14.5 What Could Drive a Transition away from Fossil Fuels? 412

Economic Forces and Employment 413

Innovation and Infrastructure for Energy Storage and Distribution 414

Government Actions 416

SUSTAINABILITY MATTERS Self-Driving Cars 417

14.6 What Can I Do? 420

Explore Options and Benefits for Trading in Your Less Fuel-Efficient Car 420

Support Alternative Energy Initiatives on Campus 420

Join in the Debate about Local Alternative Energy Projects 421

Learn More about Global Energy Policies 421

Chapter 14 Review 422

Summary 422

Key Terms 422

Review Questions 422

For Further Thought 423

Use the News 423

Use the Data 425

Learn More 425

15 Waste:
What Happens to All the Stuff We Use? 426

Chapter Objectives 427

15.1 What Is Waste? 429

Municipal Solid Waste 429

Solid Waste 430

Life-Cycle Assessment 430

15.2 What Happens When Waste Is Dumped? 430

Polluting Gases 434

Leachate 434

Ocean Dumping 434

Waste Trade 435

15.3 How Do We Manage Waste? 435

Isolation 436

Incineration 438

SUSTAINABILITY MATTERS Who Bears the Impacts of Waste? 439

Conversion 440

15.4 How Do We Recycle and Reuse Waste? 441

Recycling 442

The Economics of Recycling 444

The Advantages and Limitations of Recycling 445

Reuse, Refurbish, Repurpose, and Upcycle 445

15.5 How Does Our Culture Affect Consumption? 446

Culture and Consumption in the United States 447

15.6 Can We Reduce Our Waste? 448

Business and Consumer Strategies 448

STORIES OF DISCOVERY What Can We Learn from Our Garbage?
The Archaeology of Us 449

Government Action 450

Community Initiatives and Individual Actions 450

15.7 What Can I Do? 451

Reduce Your Food Waste 451

Stop Hitting the "Buy" Button So Often 451

Track How Much Waste You Produce 451

Take Action against Wasteful Practices 451

Help Others Use Less 452

Chapter 15 Review 452

Summary 452

Key Terms 452

Review Questions 453

For Further Thought 453

Use the News 454

Use the Data 456

Learn More 457

16 Urbanization:
Why Are Cities Growing?

...... 458

Chapter Objectives 459

16.1 How Are Urban Areas Changing? 460

Putting the Urban Transition into Historical Perspective 460

Urbanization and Environmental Impacts 460

The Opportunity of Urban Areas 464

16.2 What Are Slums? 465

History and Development of Slums 466

Opportunities in Slums and Cities in the Developing World 466

16.3 What Is Suburban Sprawl? 467

Characteristics of Sprawl 468

What Are the Problems of Suburban Sprawl? 469

16.4 How Did Suburban Sprawl Develop in the United States? 471
Building Policies 471
Transportation Infrastructure 473
Sprawl and Consumption 473

16.5 What Is Urban Planning? 474
Urban Planning 474
Challenges of Planning 474
SUSTAINABILITY MATTERS How Can Cities Respond When Disaster Strikes? 475

16.6 Why Is Transportation Important? 476
The Challenge of Induced Traffic 476
Transportation Alternatives 477
The Relationship between Transit and Density 477

16.7 How Are Cities Changing? 478
Walkability 478
Transit-Oriented and Pedestrian-Oriented Developments 478
Green Buildings and Infrastructure 480
STORIES OF DISCOVERY The Bullitt Center—Meeting the Living Building Challenge 481

16.8 What Can I Do? 482
What Is Transportation Like Where You Live? 482
Learn More about Your Campus Plan 482
Attend Planning Board Meetings in Your City or Town 483
Chapter 16 Review 483
Summary 483
Key Terms 483
Review Questions 484
For Further Thought 484
Use the News 484
Use the Data 486
Learn More 487

17 Environmental Health and Justice:
How Do Environmental Factors Affect the Places People Live, Work, and Play?

..... 488

Chapter Objectives 489

17.1 How Are Environmental Benefits and Burdens Distributed? 491
Responding to Issues of Environmental Justice 492

17.2 What Is the Study of Environmental Health? 496
Epidemiology 496
Toxicology 496
STORIES OF DISCOVERY Alice Hamilton 497
Public Health 498

17.3   How Do Microorganisms Make Us Sick? 498

What Are the Major Types of Biological Hazards? 498

Where Do These Pathogens Strike? 499

Future Risks 500

SUSTAINABILITY MATTERS DDT: Trading a Biological Hazard
for a Chemical Hazard? 501

17.4   What Does It Mean When Something Is Toxic? 502

Risk Factors 502

17.5   How Can I Be Exposed to Something Toxic? 503

Where Exposure Occurs 504

17.6   How Do We Manage Risks Associated with Toxins? 506

Addressing Disproportionate Impacts 507

17.7   What Are the Most Common Physical Hazards in the Environment? 507

Radiation Hazards 509

17.8   How Is Climate Change Affecting Environmental Health Issues? 510

Extreme Weather 511

Expanded Ranges for Diseases 512

17.9   Where Are Vulnerable Communities Located? 512

What Actions Promote Environmental Justice? 512

17.10   What Can I Do? 514

Follow Good Sanitation and Hygiene Practices 514

Avoid Chemical and Radiation Hazards When Possible 514

Prepare a Personal and Community Disaster Plan 514

Become Aware of Environmental Injustices in Your Area 515

Understand Your Rights to a Safe Workplace 515

Chapter 17 Review 515

Summary 515

Key Terms 516

Review Questions 516

For Further Thought 516

Use the News 517

Use the Data 518

Learn More 519

18 Decision Making:

Why Do Our Choices Matter? 520

Chapter Objectives 521

18.1   What Are the Key Factors Influencing Our Decisions? 522

Hierarchy of Needs 522

Automatic Thinking 522

Emotional Defense Mechanisms 524

Social Acceptance and Conformity 524

What Do You Believe? 525

18.2 What Are Some Successful Strategies to Influence Behavior? 526

Prompts 526

SUSTAINABILITY MATTERS Addressing Product Obsolescence:
How Can We Extend the Useful Life of Our Stuff? 527

Feedback 528

Commitments 528

STORIES OF DISCOVERY Applying a Nudge to Reduce Food Waste 529

18.3 How Can Incentives Motivate Behavior? 530

How Do Incentives Work? 530

Incentives as a Strategy to Influence Choice 531

Social Incentives 531

18.4 How Important Are the Words We Use? 534

Using Frames to Design Narratives 535

Labels and Guides 536

18.5 What Can I Do? 538

Make Commitments, Use Prompts, and Use Tools to Keep
Track of Your Progress 538

Communicate More Effectively to Influence Changes You Want to See 538

Find Places Where Prompts Might Influence Behavior on Campus 538

Chapter 18 Review 539

Summary 539

Key Terms 539

Review Questions 539

For Further Thought 540

Use the News 540

Use the Data 542

Learn More 543

19 Groups and Organizations:
How Do We Work Together for Sustainability?

.... 544

Chapter Objectives 545

19.1 How Are We Connected? 546

Social Groups and Organizations 546

Social Networks 547

Structure and Influence in Social Networks 547

19.2 Why Are Organizations Important for Change? 548

For-Profit Businesses 549

Nonprofit Organizations 550

Benefit Corporations 551

19.3 How Do Organizations Integrate the Ideal of Sustainability? 551
Mission Statement and Strategy 552
STORIES OF DISCOVERY B Corp: Inventing a New Kind
of Corporation 553
Organizational Culture and Structure 557

19.4 How Can Members within an Organization Facilitate Change? 557
SUSTAINABILITY MATTERS Should Organizations Divest
from Fossil Fuel Companies? 559

19.5 How Can Those Outside an Organization Facilitate Change? 560
Environmental Nonprofits Exerting Influence on Businesses 560
Forces Changing Environmental Groups 561
Media and Social Media as a Force for Change 562
Partnerships and Economic Incentives 563

19.6 What Can I Do? 564
Examine Your Sphere of Influence 564
Get Involved with What's Happening on Your Campus 564
Bring Sustainability Groups Together on Campus 564
Connect to Other Campuses 564
Learn More about Financial Factors That Drive Business Decisions 564

Chapter 19 Review 565
Summary 565
Key Terms 565
Review Questions 565
For Further Thought 566
Use the News 566
Use the Data 568
Learn More 569

20 Government:
How Can Policy Influence Sustainability? 570

Chapter Objectives 571

20.1 How Are We All Subject to Government Decisions? 573
How Do Governments Maintain Authority? 573

20.2 Why Is the Way a Government Is Organized Important? 574
How the Type of Government Impacts Environmental Policy 574

20.3 How Do Elections Shape Government? 576
Organizing Election Systems 576
Political Parties 577

20.4 How Are Environmental Policies Made? 577
Politics and the Arctic National Wildlife Refuge 578

20.5 **How Are Policies Designed to Influence Behavior?** 580

Prescriptive Regulations 580

Payments 580

Penalties 581

Property Rights 581

Persuasion 581

20.6 **How Did Important Environmental Policies in the United States Develop?** 582

Pollution Control, the EPA, and *Silent Spring* 582

STORIES OF DISCOVERY Rachel Carson: Speaking Out about *Silent Spring* 583

Natural Resource Policy 584

20.7 **How Are Environmental Issues Addressed on the International Scale?** 584

Intergovernmental Organizations 584

International Agreements 586

The Montreal Protocol on Substances That Deplete the Ozone Layer 586

The UN Framework Convention on Climate Change 587

US Participation in International Agreements 588

20.8 **What Can I Do?** 588

SUSTAINABILITY MATTERS Can the World's Nations Work Together to Confront Climate Change? 589

Get Registered and Vote 592

Support a Campaign 592

Work with a Group You Support 592

Tune In to Social Media 592

Chapter 20 Review 592

Summary 592

Key Terms 593

Review Questions 593

For Further Thought 593

Use the News 594

Use the Data 596

Learn More 597

Appendix AP-1

Selected Answers SA-1

Selected Sources SS-1

Glossary G-1

Credits C-1

Index I-1

At a Glance

Major Human Actions Affecting the Environment...10

Environmental Policies in the United States..54

What about Water Makes It So Important?...70

The Earth's Biomes...100

Protected Areas..128

Global Population: Past, Present, and Future...154

The Water Cycle..184

Air Pollution..232

Mining Impacts...268

The Carbon Cycle: It's a Carbon-Based World..294

Indicators of a Warming World...314

Environmental Impacts of Conventional Agriculture..346

Fossil Fuel Energy Disasters...384

An Energy Transition Is Under Way..418

The Life Cycle of One Pair of Jeans...432

The Urban Transition..462

The Environmental Justice Movement in the United States: A Timeline...............494

Strategies for Influencing Behavior..532

Why Are Organizations Addressing Sustainability?..554

Opportunities to Influence Policy Making...590

To the Instructor

How This Book Can Help Your Students Engage with Environmental Science and Sustainability

It's not hard for an instructor to argue for the relevance of an environmental science course, but that doesn't mean teaching and learning about the environment is easy. Whether or not environmental issues are among the top stories on their news feeds, most students have at least a general sense that problems like pollution, endangered species, and climate change are among the current challenges faced by society and that competing responses to these challenges feed into policy debates. And many colleges and universities now promote "green" initiatives that can range from recycling programs and lower-impact transportation and dining options to energy and water conservation in buildings and grounds, and even most recently (often due to pressure from student activists) to divestment from fossil fuels.

But familiarity with environmental issues and actions isn't enough to cultivate an understanding of how Earth works, how this planet supports us, and how our actions can either disrupt and degrade or conserve and restore the systems and resources on which we depend. And deeper understanding can help equip our students to critically assess, decide, and act in response to environmental problems. In writing this book, we recognize that an environmental science course is an opportunity to do more than just build knowledge of key concepts; it is also a chance to develop an appreciation for scientific inquiry, to cultivate a curiosity about the complex ways our actions affect our home planet, and to develop a critical approach to environmental problem solving.

Solving environmental problems also requires an understanding of the ways in which environmental harms and benefits are inequitably distributed across communities. In this second edition we have expanded our coverage of and sharpened our emphasis on the concept of environmental justice, the many ways environmental injustices persist, and new policies and initiatives designed to address these concerns. We hope this new emphasis will help students think about environmental justice issues where they live, and what they can do to help alleviate them. On a global scale we also consider the ways in which environmental justice concerns are exacerbated by climate change. In particular, we have updated this edition to incorporate this and many other findings of the Intergovernmental Panel on Climate Change's (IPCC) Sixth Assessment Report (AR6), as well as many other sources of the most up-to-date science.

Sustainability as a Way of Thinking

Most fundamentally, this book engages the concept of sustainability as a way of thinking—a way of interrogating our individual and collective impacts on the planet, considering what kind of future we want for our world and then planning, deciding, and acting to realize that future. When we employ sustainability as a way of thinking, we ask questions like, How can we satisfy our needs and wants in a way that does not diminish or degrade Earth's ability to provide for future generations? How are the human harms resulting from our environmental impacts distributed across different groups in society? How can we ensure that the places we live, depend on, and cherish can recover from disruption and adverse environmental effects?

While exploring questions like these requires that we establish and build on a foundation of environmental science concepts and relationships that help us understand how Earth works, it is also essential that we approach science as an unfolding method of inquiry rather than a list of facts. To do this, we created, organized, researched, and wrote this book as a set of questions posed in the chapter titles, sections, subsections, and text itself. In this way, the book models the scientific and environmental curiosity we hope our students develop themselves. Every chapter in the book also includes mini case boxes (Stories of Discovery and Sustainability Matters) and sidebars on science-in-context, sustainability, and environmental justice that provide more insight into how scientific understanding of certain environmental issues has developed, as well as the challenges and opportunities of addressing environmental problems. We have benefited greatly from our reviewers, who helped us develop and refine these features and link them effectively to the core concepts in the text.

Building a foundation of environmental science knowledge and skills also necessitates a serious and in-depth examination of climate change. This is the overarching environmental issue of our times. In this book, not only do we dedicate a chapter (Chapter 11) to Earth's changing climate and the science that helps us understand the recent warming trend and the human factors contributing to it, but we also have climate-relevant cases and decision contexts throughout the book, along with supporting environmental science material. For example, Chapter 8 on air includes detailed coverage of weather systems and climate trends, and Chapter 10 on biogeochemical systems includes the climate effects of a changing carbon cycle. In addition, the chapters in the second half of the book that focus on human actions—like the way we grow our food, supply our energy, dispose of waste, and build our cities—each call out both the human causes of climate change and the impacts of climate change that will affect our lives. Many proposed responses to climate changes are highlighted throughout the book.

Engaging deeply with sustainability also takes us beyond environmental science into other fields of inquiry. Science alone cannot tell us what kind of future we ought to work toward, what ought to be sustained, or how environmental benefits and harms ought to be distributed in society. Questions like these, and the decisions and actions that stem from them, invoke values. So in this book we integrate tools and concepts from other academic fields—such as ethics, economics, psychology, political science, sociology, and anthropology—to critically assess how individuals, groups, and governments form and act on values. When coupled with a scientific understanding of the consequences of various decisions, a critical approach to values will help students see that making a sustainable choice is rarely simple or without controversy. It will also help students look for and weigh the trade-offs of various courses of action and engage more thoughtfully with and as decision makers. We emphasize the complexity of sustainable choices with mini case boxes (Sustainability Matters) that both outline the challenges of an environmental problem-solving context and ask the students to weigh in as decision makers on that issue.

Helping students assume roles as decision makers is a key feature of this book. While students may be familiar with environmental issues, "green" actions, and even basic environmental science concepts, wading into this arena from a problem-solving perspective can be daunting. For this reason, we begin each chapter with an introduction that directly connects to day-to-day student experiences. We make no assumptions about what the reader knows about environmental issues, science, or other academic fields. Instead, we build questions and concepts from the ground up in each chapter, while developing contexts in which students are asked to apply this knowledge to environmental problems—either from their own individual perspectives or by taking on the role of a decision maker in a more official capacity. These opportunities to consider "what would you do?" linked to mini case boxes, end-of-chapter reviews, and data and news analysis exercises can serve to empower, excite, and motivate students to be more than just consumers of information—to become active participants in the teaching and learning process and decision makers regarding their planet's future.

How the Book Is Organized

We recognize that environmental science instructors rarely assign the chapters of a text in order. So, although there is a logic to the organization of this book, we also wrote the chapters so that they could be assigned in a modular way to suit the needs of each instructor and course. The themes of science, sustainability, and decision making are interwoven throughout each chapter.

The first two chapters outline and define key aspects of the study and application of environmental science and sustainability. Chapter 1 lays out the key themes of the book, including Earth systems, human environmental impacts, and the definition and use of the term "sustainability." It also delves into the characteristics and methods of scientific inquiry while identifying problems such as bias, misinformation, and pseudoscience. Finally, this chapter introduces readers to the role that values and other human factors play in decisions affecting the environment. Chapter 2 expands on the concept of values by exploring key tools and concepts from the fields of ethics, economics, and political science to understand human decision-making processes that shape the environment.

Chapters 3 through 11, the largest section of the book, engage the readers in an exploration of how Earth works and human impacts on our planet. This includes basic background on matter and energy (Chapter 3); life on Earth with a focus on biodiversity (Chapter 4); efforts to conserve, protect, and restore biodiversity (Chapter 5); and the population dynamics of humans as a species (Chapter 6). Earth's water, air, and land each have a chapter devoted to them (Chapters 7, 8, and 9), as do the biogeochemical cycles that work among these aspects of the planet—with special emphasis here on systems and systems thinking (Chapter 10). Chapter 11 on Earth's climate teaches students how climate science has helped us understand climate change over Earth's history, the current role that human actions are playing in global warming, and the consequences of this warming and those of alternative emission scenarios in the future.

Chapters 12 through 17 focus on how humans make a living on Earth, environmental impacts of human actions, and alternatives for the future. This includes chapters devoted to food and agriculture, energy from fossil fuels, alternative energy, consumption and waste, and urbanization. These chapters each include a historical approach, so students can get a sense of the changing ways in which humans have provided for themselves over time—and the impacts of those changes on the environment. Each chapter also provides details on how these areas of human activity currently affect the planet and what the prospects are for the future. Chapter 17 provides an in-depth focus on environmental threats to human health and well-being and the way environmental benefits and harms are unequally distributed around the globe and among racial, ethnic, and socioeconomic groups. This focus on environmental justice is woven into environmental health concerns including biological, chemical, and physical hazards.

Chapters 18 through 20 introduce the reader to factors affecting our individual and collective decision making on the environment. Chapter 18 focuses on the ways that individual choices affect the environment and factors that affect our decisions and actions. This chapter reviews research on both the psychology and social psychology affecting our choices and strategies to promote sustainable behaviors. Chapter 19 takes a closer look at the role groups and organizations—including businesses and nonprofit organizations—play in sustainability. Chapter 20 provides the student with an understanding of the environmental policy-making process, different levels of government policy (from the local to the international), and different policy strategies for responding to environmental problems. Each of these chapters emphasizes many ways in which the reader can engage in environmental decision-making processes that make a difference.

A Thoroughly Revised Second Edition

Working with feedback from 42 reviewers, our own teaching experiences, and new research and important issues in the field, we worked to create a significant and meaningful new edition. **Changes have been made to every chapter, and a comprehensive list of these can be obtained from your local Norton representative, or by visiting wwnorton. com/environsci2 and choosing the Instructor Resources tab.** Selected highlights include:

>> As noted earlier, we have increased our coverage of Environmental Justice, adding it as a core theme to the text, including a new section on this topic in Chapter 1, expanding and reorganizing this coverage in Chapter 17 on Environmental Health and Justice, and adding Environmental Justice sidebars to every chapter.

>> We have updated the text with new examples and lessons from current events, including the pandemic. For example, Chapter 8 includes a new Use the Data activity on how decreased pollution levels from COVID lockdowns created a natural experiment showing how factors like driving less would change CO_2 levels.

>> We have updated the data in the text, using the IPCC Sixth Assessment and other sources like NOAA, NASA, the EPA, and the EIA.

>> Additional Science coverage has been added based on reviewer feedback. For example, Chapter 3 now has increased coverage and a dedicated figure illustrating photosynthesis. We also show/depict more of the chemistry of ozone formation in Chapter 8 and the basic science of how PVC panels work in Chapter 14.

>> We increased coverage of how to evaluate science and information in the news starting with an expanded section on this topic in Chapter 1.

>> Approximately one-third of the Use the News and Use the Data activities and the end of each chapter are new.

We have made important updates and improvements in our media and support package with our text too. Our data show that in the first edition over 75% of students used this text in an electronic format; important features have been added to make this format more useful to them and to you. These include

>> The new Norton Illumine Ebook features Dynamic Data figures and low stakes Check Your Understanding questions (written and edited by the text authors). Check Your Understanding figures contain answer-specific feedback to motivate students and build confidence in their learning progress. Students' progress is tracked as they work toward completion through easy-to-use assignment tools and LMS integration.

>> An expanded set of What Would You Do? questions is now in an easier-to-use, web-ready interface and assignable with premade questions via an instructor's LMS. There are two new activities, and existing activities were reedited to tighten the stories and decrease their overall length.

>> Norton's Inquizitive assessment system is still included with every new textbook purchase, electronic or print, or is available for purchase on its own. Inquizitive has been updated with new questions written for this edition. All new questions were written by Dan Sherman.

How We Organized Each Chapter

Always Speak to the Students When we wrote, we tried to speak directly to the student. Since this is an active discipline, we tried to use an active voice whenever we could, with short and easy-to-digest sentences. We also tried to consider the background and experiences of our student audience. When framing examples, instead of starting with far-flung and lengthy cases, we used short examples of common experiences. For example, our chapter openers address and engage students in day-to-day experiences and then build connections to broader environmental issues and challenges. Overall, our goal was to have every student understand the role that they play in sustaining the environment in their daily experiences.

Asking and Answering Questions and Making Decisions As mentioned earlier, another key goal of the text is giving students the background needed to be real decision makers. So every chapter title and section head is written in the form of a question, to give students practice in framing ideas this way. Mini case boxes and Use the Data and Use the News sections ask students to consider, answer, and defend responses to What Would You Do? questions in real-life situations. Every chapter ends with a dedicated What Can I Do? section with suggestions and ideas.

Further, a unique set of chapters probes the science of decision making and influencing at individual, group, and government levels. These are designed to be integrated either singly or as a group, whenever you choose. We assume other instructors may also consider only topics within them or use them as student reference chapters.

Just the Right Amount of Science Environmental science is a broad discipline, with a lot to potentially cover. Through many drafts and the help and input of reviewers, we sought to strike the right balance of what students can successfully digest. This includes every topic we considered, the depth of scientific coverage, the art and figure program, the end-of-chapter

material, and even the key terms we selected. Our goal was to not cover too much but also to not present too little: the perfect balance. In addition, we included chapters on matter and energy and on systems and cycles. While we know many instructors skip at least one of these chapters in their syllabi, we tried to carefully construct them to make these sometimes difficult topics accessible. We thought many instructors would find these useful in teaching the interconnectedness of our environment.

Providing the Whole Story: Science, Context, Sustainability, and Environmental Justice In each chapter, while covering the science, we try to provide the context and perspective needed to understand the issue at hand. For example, in our chapter on energy alternatives, we not only describe the science behind different energy technologies but also describe the history of past energy transitions. Similarly, our chapter on urbanization describes not only the characteristics of urban and suburban places but also factors like past housing policies that drove choices and decisions to move. While these ideas flow through every chapter, they are easiest to see in the short, topical sidebars that highlight the concepts. Two-page infographic spreads called At a Glance also try to capture the complete picture of one core idea in each chapter.

Sourced and Supported In writing this book, through several drafts and now two editions, we used many sources to support each chapter. A list is provided at the end of the text, sorted by chapter. Sources for figure data are also provided with the figures.

Features of Every Chapter With these broad goals in mind, you will find each chapter features the following:

>> *Chapter subtitles and section heads in the form of questions.* These are designed to engage students and stimulate key skills of framing questions and considering solutions.

>> *Modular chapters.* We understand that most instructors teach these chapters in their own preferred order. We have designed the chapters to be flexible and work with this teaching style.

>> *Engaging openers based on day-to-day experiences.* These frame the beginning of chapters and help students understand that this discipline is about what they are doing now. These openers provide bridges for students to cases and examples in the rest of that section.

>> *Written for students.* We use an active writing style as much as possible. We frame chapter examples in common experiences.

>> *What Would You Do?* One of the core ideas of this text is to get students thinking about solutions. The mini case boxes (Sustainability Matters and Stories of Discovery) and Use the News and Use the Data sections at the end of each chapter all finish with a brief question asking students to make and support a decision.

>> *What Can I Do?* Many students want to know actions they can take now. Unique, dedicated sections at the end of every chapter (called What Can I Do?) provide some ideas. These sections flow within the main text. The range of suggestions starts with simple individual steps while also including larger ideas, especially those that might take place within the setting of where students are studying.

>> *Sustainability Matters.* These mini case boxes focus on ideas, topics, challenges, and innovations related to sustainability. One of these generally page-length mini case boxes appears in each chapter.

>> *Stories of Discovery.* These are short profiles and cases focusing on how a scientist or scientists innovated in the area of environmental science. These are also approximately one page and appear in each chapter.

>> *Use the News questions.* Each chapter contains an actual news article from a real news source. Students are first asked simple comprehension questions on this article and then are asked short-answer questions that progressively require more consideration.

>> *Use the Data questions.* Learning to understand and make use of data is an important skill in this course. The end of each chapter contains a Use the Data feature with questions on an actual scientific data source (when possible, we tried to obtain the actual graphic). Each set of questions starts with basic questions that help students understand the data source and then progresses to short-answer questions that require more thought.

>> *At a Glance.* Each chapter has a carefully crafted two-page spread designed to highlight and teach a key point in the chapter. Spreads were created from art, text, and photos, as appropriate, and with the expertise of a design firm that specializes in these graphics.

>> *Sidebars: Science, Context, Sustainability, and Environmental Justice.* Each chapter contains a selection of these sidebars that highlight the big themes of the text. These are short and topical, contain illustrations, and are designed to be engaging "quick reads."

>> *Graphs and data.* Representations of graphs and data were all drawn with simplicity, consistency, and accuracy in mind. A simple color scheme has been used so as not to confuse students, important information is highlighted, titles are provided when possible, and bubble captions are used to provide further explanation. In all cases, our goal is to create accurate graphs while also considering the skills of students using the text.

>> *Chapter Objectives.* Each chapter begins with objectives written with Bloom's taxonomy of learning.

>> *Take-Home Message.* Each major section has a brief takeaway message to help students understand the major points of that section.

>> *Key terms in the margin.* Carefully selected key terms with definitions appear as they are defined in the text. Some of these terms appear in multiple chapters, and the text notes when this happens.

>> *Review material.* Each chapter ends with a review section including a bulleted summary, list of key terms, simple Review Questions, and more thought-provoking For Further Thought short-answer questions for assignment, as well as Use the News and Use the Data question features, each with What Would You Do? elements. Learn More sections provide additional readings and websites for students to explore.

>> *Sources.* As noted earlier, this book has been heavily researched in its preparation. An extensive list of selected sources is included at the end of the text.

>> *Appendices.* In several places in the text, we refer students to online calculators that allow them to calculate measurements like their individual carbon footprints. Appendix A provides additional reference material to help locate and use different calculators. In addition, Appendix B provides a listing of carbon footprints by country (the most recent data available at the printing of this text).

>> *Glossary.* A complete glossary of all key terms in the text is provided at the end of this book.

Tools for Teaching and Learning

The teaching and learning package provides all the tools needed for students to understand broad group concepts and learn key vocabulary and for instructors to test knowledge. **To obtain and learn more about these tools, visit digital.wwnorton.com/environsci2.**

Norton Illumine Ebook

Whether purchased through an inclusive access program or through our Norton website, ebooks are far and away the most-used format of this text. Norton's high-quality content shines brighter through engaging and motivational features that illuminate core concepts for all students in a supportive, accessible, and low-stakes environment. In the second edition, Embedded Dynamic Data figures engage students with applications and explorations of important course content. Check Your Understanding questions with rich feedback motivate students and build confidence in their learning progress. The active reading experience also includes the ability to highlight, take notes, search, read offline, and more. Instructors can promote student accountability by adding their own content and notes and through easy-to-use assignment tools in their LMS.

INQUIZITIVE

InQuizitive is Norton's award-winning, easy-to-use adaptive learning tool that personalizes the learning experience for students and helps them master—and retain—key learning objectives. Through a variety of question types, answer-specific feedback, and game-like elements such as the ability to wager points, students are motivated to keep working until they've mastered the concepts. Students then come to class better prepared, giving you more time for meaningful discussions and activities. InQuizitive can also improve students' test grades when assigned consistently for points as a part of the overall course grade. The convenience of Learning Management System (LMS) integration saves you time by allowing InQuizitive scores to report right to your LMS gradebook.

InQuizitive for *Environmental Science and Sustainability* offers a range of questions aimed both at helping students understand the science and at ensuring they understand the processes laid out in the text. Questions incorporate artwork from the book, helping reinforce students' understanding. Videos and animations in the feedback give students a visual framework for challenging science concepts. Questions have been written by a team of classroom instructors, including the book's authors.

Student access codes for InQuizitive are included at no charge with all new books in any format (paperback, ebook, or custom). If students need to purchase a stand-alone access code for InQuizitive, they can do so at an affordable price at wwnorton.com.

What Would You Do? Activities

Available from Norton, these interactive, web-based activities help students understand the impact of personal and policy decisions. For the second edition, stories have been streamlined for clarity, two new ones have been added, and they now use an improved web-based interface. These decision-making cases let students role-play making choices and considering the environmental consequences in a risk-free online environment. They help students better understand the choices, options, and consequences

Based on your decision, you receive a grant to help farmers prepare for long-term variations in regional climate patterns. Part of the grant is for irrigation systems that will benefit these farmers even if precipitation doesn't change. The farmers are happy about implementing this part of the grant. But they're uncertain about the grant's initial plan for addressing temperature shifts. This plan requires that farmers adjust established planting times and pest control practices. Given these concerns, two alternative ideas are proposed to help address temperature shifts.

Photo Credit: United States Department of Agriculture

The digital What Would You Do? activities challenge students to engage as decision makers in a variety of complex environmental issues, including debates around GMO's.

available when faced with a situation impacting the environment. The decision they make informs their outcomes, as well as their subsequent decisions. Each activity includes a summary helping the student to understand the consequences of their actions, an LMS connection to prebuilt Check Your Understanding exercises, and an essay prompt encouraging students to reflect on and apply their decisions to other situations.

Instructor's Manual

This text continues to provide an innovative Instructor's Manual. In particular, this manual includes a point/counterpoint feature that helps instructors discuss issues in class, guided by quotes from real individuals.

Further, the Instructor's Manual offers helpful and creative resources for lecture planning and classroom activities. Point/counterpoint activities present quotes from actual current public figures and provide instructors with guidelines on how to discuss and debate the quotes in class, from context on the speakers and the issue to guided discussion questions on the quotations.

Test Bank

Norton uses an evidence-based model to deliver high-quality and pedagogically effective quizzes and testing materials. The framework to develop our test banks, quizzes, and support materials is the result of a collaboration with leading academic researchers and advisers. Questions are classified by section and difficulty, making it easy to construct tests and quizzes that are meaningful and diagnostic. With now approximately 50 questions per chapter, the revised *Environmental Science and Sustainability* test bank provides a valuable array of questions to test student learning. The Test Bank is available from the Instructor Resources section of the website.

Lecture PowerPoint Slides and Art Slides

Designed for instant classroom use, these slides using art from the text are a great resource for your lectures. All art from the book is also available in PowerPoint and JPEG formats. Download these resources from the Instructor Resources section of the website.

Curated Selection of Videos

Text chapters are supported by a curated selection of videos and animations. Distributed across chapters, available to both instructors and students, and close captioned, these multimedia elements help support the science and decision-making pedagogy of the book.

To the Student

Environmental Science and Sustainability: What's It All About?

You've probably heard the word "sustainability" before. If you're like most people, you associate it primarily with "environmentally friendly" practices like recycling, carpooling, using a reusable water bottle, or eating organic foods. But whatever you think about these and other practices, it is worth asking: what is it that makes something more or less sustainable? Your actions certainly do matter (we'll get to that), but in this book we want to consider sustainability not just as a list of practices but as a way of thinking that engages big questions about your role in the world. In what ways does your well-being depend on certain natural resources and environmental conditions? How do your actions affect the environment? How can we satisfy our needs and wants in a way that does not diminish or degrade Earth's ability to provide for future generations? How can we ensure that the places we live, depend on, and cherish can recover from disruption and adverse environmental effects?

To engage with questions like these, you'll need to explore environmental science. Science is our way of asking and answering questions and testing ideas about the natural world by using evidence gathered from the natural world. It is best understood as a process of inquiry rather than a collection of known facts, and in this book you'll gain insight on the scientific curiosity that has driven key discoveries and understanding. But digging deep into sustainability also entails examining the factors that drive human decisions and behaviors—from individuals to groups and governments. So you'll gather insights from social science fields outside the natural sciences as well. Let's take a closer look at what the study of environmental science and sustainability entails and what you'll experience and learn as you read this book.

It's about This Planet

Earth, of course is not just any planet; it's your home. Like all humans, you make your living on this planet by drawing on its resources. What kind of resources? Earth's natural systems provide essentials like oxygen, water, food, energy, and building materials that we all need to survive and carry out our various pursuits. But your planet is more than just a source of essential "stuff": you rely on Earth's systems to do things like assimilate your waste, cycle nutrients, and provide a climate and conditions favorable for your survival. In short, you depend on Earth. So in this book you'll explore how Earth's processes support your survival. How do energy and matter provide the building blocks for Earth's processes? How does water on Earth cycle between the ocean, atmosphere, and freshwater resources we depend on? How does soil form, and how do landscapes change over time? What are the dynamics in the atmosphere shaping short-term weather and longer-term climate trends? How are the many species of plants, animals, and other life-forms on Earth situated in interrelated communities that, like us, are also dependent on Earth's resources to survive? As you engage with questions like this, especially in the first half of this book, we hope you develop a better understanding of, and a curiosity about, your home planet.

It's about People

But as an environmental science student, you'll do more than just study how Earth works; you'll learn about the ways in which humans are transforming the world in which we live. Now that there are more than 7.5 billion people on Earth, the magnitude of our impact is considerably greater than that of other species. Some scientists have even suggested that we live in a new time in Earth's history called the Anthropocene—one marked by conspicuous human effects on the planet. In addition to learning about the dynamics of human population growth, you'll also develop a sharper understanding of the way human activities (including your own) are linked to troubling environmental effects like air and water pollution, climate change, and declining biodiversity. In particular, chapters in the second half of the book on food, energy, waste, and urbanization detail the link between human actions and environmental impacts. And you'll learn that the adverse impacts of these problems on human well-being are not distributed equally across the population, raising issues of equity and justice. But what's behind the human actions causing these environmental effects? Throughout this book you'll go beyond the natural sciences to engage insights from fields like history, economics, political science, sociology, and anthropology that clarify the larger economic, political, and social forces that influence our collective human impacts on the planet.

It's about the Future

While the environmental impacts of human actions are serious, as humans we have the ability to consider and pursue alternative scenarios in the future. While collectively our actions may be causing environmental problems, we are also the source of potential solutions. Throughout this book we'll introduce you to individuals and groups working to foster sustainability and devise solutions to environmental problems. But a future that reduces our environmental impacts is not a foregone conclusion. It requires that individuals and groups prioritize sustainability as a value—an assertion of how we want things to be—and set specific goals and plans to meet those ends. While science can help us understand the environmental consequences of past human actions and the current state of environmental conditions, and can even model alternative future trajectories based on key variables, it cannot tell us what our values should be or set our goals for us. And competing interests among individuals and groups can make agreeing on values, goals, and plans challenging and contentious work. However, the study of ethics, which you'll encounter early in this book, can help us develop an awareness of how we form and apply values. Throughout this book you'll read about groups like businesses, cities, states, nations, and even the international community of the United Nations that have agreed to and are pursuing sustainability plans.

It's about Choices

The extent to which our future becomes sustainable will come down to individual and collective choices that affect the environment. This includes a multitude of decisions that affect where and how we live, what we consume, how we dispose of our waste, and how we draw on energy to transport ourselves and power our pursuits. You'll learn in this book that making sustainable choices is not always easy or straightforward. From individual consumer choices (like which brand of sneakers to purchase) to competing energy policy proposals, making sustainable decisions often involves trade-offs. And, while science can't make your mind up for you, in this book you'll see how environmental science can help assess the environmental impacts of alternative courses of action. You'll also learn from the fields of psychology and economics about the factors that go into influencing decision-making processes. The final three chapters of this book focus on the factors influencing choices at the individual, group, and governmental levels. This will make you more aware of your own thinking and behavior and give you some ideas about what types of prompts, incentives, and communication strategies motivate others as well.

It's about Action

Most of the students reading this book will not go on to become environmental scientists (although many of you may!). But each person reading this book—by virtue of being alive on Earth—will be a decision maker affecting the future of the planet. This could be as an individual commuter deciding how to travel to work, a consumer making purchasing decisions, or a voter supporting a candidate for election; or it could also be in a role that influences others as well, like a leader within a business or other organization, a teacher, a community activist, or even a parent planning for a family's future. Each chapter in this book concludes with a What Can I Do? section that outlines several options for taking actions that will raise your sustainability awareness, help you gain important experience, and make a difference. But it's best not to think of sustainability actions as a preset list of things to do. Your impact will be much greater if you also start to adopt a critical way of thinking about your role in the environment, the impact of your actions, and the influence you have on your community. We hope this book will help you get to know your home, the Earth, better and will help you see yourself as a decision maker whose choices and actions will help shape Earth's and humanity's shared future.

Acknowledgments

In developing this new edition, we continue to see how much of a team effort it takes to produce a new text and supplement package. This includes the experts at W. W. Norton, our design studio, the authors and reviewers that helped create the supplements, and reviewers and focus group attendees. There are a lot of people to thank! To start, we thank the professionals at W. W. Norton. Your hard work, care, commitment, and careful eyes were instrumental in producing what we see. Our thanks go to Eric Svendsen, who has been with us from day one. From just a few ideas, Eric has worked with us for years to shape this text. Working with him has been our developmental editor, Sunny Hwang. Sunny is not only a great and thoughtful editor, but a scientist and educator experienced in making difficult concepts easy to understand. Many others dedicated their time and made very important contributions to this effort. Carla Talmadge expertly orchestrated and managed the complexities of getting all these ideas together into the first edition, and Melissa Atkin has continued with more excellent work in the second edition; Lissi Sigillo created the stunning design for the book along with directing the look and usability of the two-page spreads; and Jillian Burr developed the second-edition improvements. Cat Abelman and Melinda Patelli have continued to go the extra mile to find the perfect photo to illustrate a concept; Debra Morton-Hoyt led the design of our innovative cover; Benjamin Reynolds has expertly guided the production process and coordinated with a variety of vendors for two editions now; and Joshua Garvin helped us immeasurably with the permissions and figures in this update. We'd also like to thank Stacy Loyal, our marketing manager, whose ideas have been driving the text's direction since she joined the team. Mia Davis has worked closely with us as the assistant editor on the second edition, including recruiting and managing reviewers, updating all the sources, and meticulously tracking and preparing the final manuscript including all of the art and photos; Pat Wieland and Laura Sewell brought keen eyes and insights as copyeditor and proofreader; and intern Devon N. Rodgers provided valuable support, especially with compiling new sources we used in the second edition. This book also benefited from a dedicated media team now led by Ariel Eaton. She has worked hard to make major improvements to this program, especially with the Norton Illumine Edition, and the revised What Would You Do? interface and activities. They were helped by dedicated associate editors and assistants, including the focused attention of Alex Park and Rachel Bass. Projects in media production were led by Sophia Purut. We also thank the professionals at Hyperakt Studio, Imagineering, and GraphicWorld, who partnered with Norton to create the first edition, and Mattias Mackler who developed new At a Glance activities and a new cover for the second edition. Finally, we thank Stephen Marshak and Clark Larsen. In creating our book, we often referred to yours as examples of successful texts. We only hope we can be as successful as the two of you.

Many others from outside Norton helped develop this text and its supplement package. In particular, we thank Heidi Marcum, who has provided feedback on every chapter of this text through several drafts of the first edition. Heidi then used this experience to create the Instructor's Manual for this text, complete with an innovative Point/Counterpoint feature for two editions now. In addition, Brian Shmaefsky embraced the themes of this text and worked with Norton to create the What Would You Do? online activities for both editions too. We think you will enjoy the results of his hard work. All text supplements take a lot of hard work from their authors and reviewers, as well as a lot of creativity. *We want give our heartfelt thanks to the team of authors who have worked on these supplements for the first or second editions, or both!*

Matthew Allen Abbott, *Des Moines Area Community College*: Test Bank

Joni Thomas Backstrom, *The University of North Carolina at Wilmington*: InQuizitive reviewer

Matthew Paul Badtke, *Jackson College*: Test Bank

Sudipta Biswas, *South Mountain Community College*: InQuizitive

Michelle Cawthorn, *Georgia Southern University*: InQuizitive

Melinda Coogan, *Baylor University*: PowerPoint slides

April Ann Fong, *Portland Community College*: InQuizitive reviewer and lecture PowerPoint author

Kim Largen, *George Mason University*: InQuizitive

Heidi Marcum, *Baylor University*: Author of the Instructor's Manual, Norton Teaching Tools, and What Would You Do? questions

Terri Matiella, *The University of Texas at San Antonio*: InQuizitive, lecture PowerPoint author, and Test Bank reviewer

Brian Shmaefsky, *Lone Star College–Kingwood*: What Would You Do? online activities

Keith Summerville, *Drake University*: Test Bank reviewer

Reviewers

We thank the following reviewers and focus group attendees for the first and second editions. Their feedback, ideas, and advice were instrumental in creating this text. This text has evolved over several drafts, and there is no way it could have done so without the help of these people:

Matthew Allen Abbott, *Des Moines Area Community College*

Shamili Ajgaonkar, *College of DuPage*

Jennifer Braswell Alford, *California State University, San Bernardino*

Aaron S. Allen, *The University of North Carolina at Greensboro*

Douglas E. Allen, *Salem State University*

Mark W. Anderson, *University of Maine*

Scott Applebaum, *University of Southern California*

Joni Thomas Backstrom, *The University of North Carolina at Wilmington*

Matthew Paul Badtke, *Jackson College*

Julie K. Bartley, *Gustavus Adolphus College*

Christy N. Bazan, *Illinois State University*

Andrew Reid Bell, *New York University*

Jennifer Bell, *Daytona State College*

Joressia Beyer, *John Tyler Community College*

Sudipta Biswas, *South Mountain Community College*

Susan Blas, *Augusta University*

Laura Brentner, *Loyola University Chicago*

John Buschek, *Carleton University*

Susan Caplow, *University of Montevallo*

Kyle Cavanaugh, *University of Los Angeles*

Rebecca Carter, *Coastal Carolina University*

Scott Connelly, *University of Georgia*

Diane Irene Doser, *The University of Texas at El Paso*

Jonathan Fingerut, *Saint Joseph's University*

Chris Fowler, *High Point University*

Robert Friberg, *University of British Columbia*

Rico M. Gazal, *Glenville State College*

David Gillette, *University of North Carolina, Asheville*

Curtis Greene, *Wayne County Community College District*

Ted Greenhalgh, *University of Nevada, Las Vegas*

Jillian Gregg, *Oregon State University*

Dierdre Hall, *Pikes Peak Community College*

Jacquelyn E. Hams, *Los Angeles Valley College*

Timothy W. Hawkins, *Shippensburg University*

Jameson Henkle, *Fresno City College*

Kelley Hodges, *Gulf Coast State College*

Matthew Julius, *St. Cloud State University*

Richard Jurin, *University of Northern Colorado*

Jihoon Kang, *University of Texas, Rio Grande Valley*

Megan Kelly, *Loyola University*

Erica Kipp, *Pace University*

Karen Knee, *American University* (Washington, DC)

Bess G. Koffman, *Colby College*

John Korstad, *Oral Roberts University*

George Kraemer, *Purchase College, State University of New York*

James D. Kubicki, *The University of Texas at El Paso*

Kate Lajtha, *Oregon State University*

Kim Largen, *George Mason University*

Elizabeth M. Larson, *Arizona State University*

Jennifer Latimer, *Indiana State University*

Kara Lazdinis, *John A. Logan College*

Hugh Lefcort, *Gonzaga University*

Libby Lunstrum, *Boise State University*

James Maki, *Marquette University*

Smita Malpani, *Washtenaw Community College*

Heidi Marcum, *Baylor University*

Sylvain Masclin, *University of California, Merced*

Terri Matiella, *The University of Texas at San Antonio*

J. Vaun McArthur, *University of Georgia*

Katherine McCarville, *Upper Iowa University*

Neusa H. McWilliams, *The University of Toledo*

Anna Mitterling, *Lansing Community College and Alpena Community College*

John Murphy, *Humboldt State University*

Tad Mutersbaugh, *University of Kentucky*

Michelle Nathan, *Honolulu Community College*

Taryn Oakley, *Portland Community College*

Brian Olechnowski, *Old Dominion University*

Natalie Osterhoudt, *Broward College*

Shirley Papuga, *Wayne State University*

Craig D. Phelps, *Rutgers, The State University of New Jersey*

Thomas Pliske, *Florida International University*

Lauren Elaine Smith Roberts, *Maricopa Community Colleges*

David Roon, *University of Idaho*

John G. Rueter, *Portland State University*

Amanda Rugenski, *University of Georgia*

Dork Sahagian, *Lehigh University*

Richard L. Shearman, *Rochester Institute of Technology*

Roger Shew, *The University of North Carolina at Wilmington*

Sheldon Skaggs, *Bronx Community College, City University of New York*

Robin Socci, *Georgia State University*

Benjamin W. Stanley, *Arizona State University*

Meredith Steele, *Virginia Tech*

Michelle Stevens, *California State University, Sacramento*

Keith Summerville, *Drake University*

Timothy Thomas, *California State University*

Craig Tinus, *Bethune-Cookman University*

Marta Toran, *Appalachian State University*

Shannon Wells, *Old Dominion University*

William Winner, *North Carolina State University*

Shuang-ye Wu, *University of Dayton*

James Robert Yount, *Eastern Florida State College*

Yan Zheng, *Queens College, City University of New York*

1

Environmental Science and Sustainability
What's the Big Idea?

Sustainability means managing natural resources in ways that do not diminish or degrade Earth's ability to provide them in the future. Using less-polluting forms of energy (such as solar power) and harvesting fish at levels that allow for stable populations in the future are two examples of sustainable practices.

Good morning. Your alarm is going off, and it is time to wake up! Like it or not, even people who are groggy in the morning face a series of choices that affect the environment. What's for breakfast? How much time to spend in the shower? What kind of coffee to drink? How to get to work or school? What to wear? While you may not be all that mindful of these choices on any particular morning, you might be more familiar with environmental effects linked to how much water you use during a long shower or to the amount of gas you burn while your car idles in traffic. Maybe you look for labels such as "organic" or "fair trade" to guide your coffee, tea, or food choices.

But what about clothing? When you get dressed in the morning, you wrap yourself in cotton, polyester, wool, silk, or other textiles. We might think of clothing as serving a fundamental need for protection *from* the environment—a wool sweater providing warmth in the winter or a nylon raincoat treated to repel water during a storm. But like many other things we consume, our clothing and shoes also have a significant impact *on* the environment.

Think about the materials our clothes are made of. Some textiles are derived from crops such as cotton that require large amounts of water and often rely on large inputs of fertilizer and pesticides. Other materials such as leather and wool come from animals that need land and feed and also generate waste. Synthetic fabrics such as polyester and spandex are made from petroleum products, which are associated with the extraction and refinement of oil. And most textiles are dyed and treated with various chemicals to add color or make them soft, waterproof, or wrinkle and stain resistant. All of these processes generate waste and consume energy, as does the transport of the goods throughout the world. If you looked at the tags on your clothes, you would probably find that their journeys to your closet frequently began in China, Southeast Asia, or Central America (**FIGURE 1.1**).

This connection between clothing and environmental effects came as a revelation for scientist and engineer Linda Greer

Chapter Objectives

By the end of this chapter, you should be able to . . .

A. describe the range of ways in which humans affect the environment.

B. explain the meaning of sustainability and ways it relates to human development and environmental justice.

C. summarize how the scientific method works.

D. describe the various methods of observation and testing involved in scientific inquiry.

E. identify ways in which values influence our individual and collective responses to environmental issues.

What is the good of having a nice house without a decent planet to put it on?

—Henry David Thoreau

(**FIGURE 1.2**). She was studying the release of toxic chemicals and water pollution in China in the early 2000s when she realized that many of the most-polluting industrial facilities were producing textiles used by major clothing brands in the United States. Environmental scientists such as Greer study the ways humans transform the world in which we live. They use scientific inquiry to explore the consequences of our effects on the environment and to evaluate ways to address them. Once Greer identified the environmental effects of the textile industry, she led a team of scientists in an initiative called Clean by Design, which developed best practices for factories in China to reduce these effects. By insulating and reducing leaks from piping systems and introducing water-recycling systems, the factories that worked with the initiative not only reduced pollution but also collectively saved more than 800 million gallons of water, reduced their use of coal for energy by about 61,000 tons, and thus decreased operating costs by nearly $15 million every year.

Reduction of environmental effects involves more than just using science to identify and address them. With respect to the apparel industry, it also entails influencing the choices of designers who first select various textiles for products, of companies that make and sell the products, and of consumers who ultimately choose whether to buy them. This can lead to some interesting collaborations.

For example, a team of Nike scientists partnered with students at the London College of Fashion to help make the environmental effects of various materials easier to see and understand. The Nike team catalogued the impact of more than 75,000 materials used in their apparel and shoes on things such as energy and water use, waste generation, recyclability, and toxicity. The student team developed an app that would help designers and consumers add environmental values to the other factors (cost, durability, performance, and fashion) that already shape their decisions. Nike has shared the results with the Sustainable Apparel Coalition, which includes other big brands (such as Walmart, Target, and Levi Strauss) that together account for more than

FIGURE 1.1 Red Dyes Can Mean Red Pollution The Jian River in China's Henan Province flows red with pollution from a factory that dyes textiles for clothing. The clothes we wear have environmental effects—such as those resulting from the chemicals used to dye the clothes. Studies by the World Bank have identified 72 toxic chemicals from dyeing practices that have polluted water supplies near textile factories.

FIGURE 1.2 Linda Greer The findings of environmental scientists such as Linda Greer can help us make changes that reduce our impact on the environment. These changes could include encouraging textile manufacturers to reduce pollution and water use and guiding consumers to purchase clothing that is less harmful to the environment.

40% of apparel sold worldwide. Similar apps help consumers incorporate environmental costs and benefits into their clothing purchasing decisions. The Good on You app was developed to make sustainable shopping easy, providing sustainability ratings for almost 3,000 brands (**FIGURE 1.3**).

So you see, even small decisions such as what clothes we choose to buy and wear each day are linked to global processes that affect the environment. And many forward-thinking people are working to understand and reduce these effects. In this book, we explore the science of our environment and examine the factors that drive our decisions. As we introduce the core ideas of environmental science, we do so in light of the decisions that underlie sustainable actions. To start, in this chapter we will look a little more closely at some core environmental and scientific concepts and how we think about them. We then explore some of the challenges, choices, and debates that humanity will need to address in the years ahead.

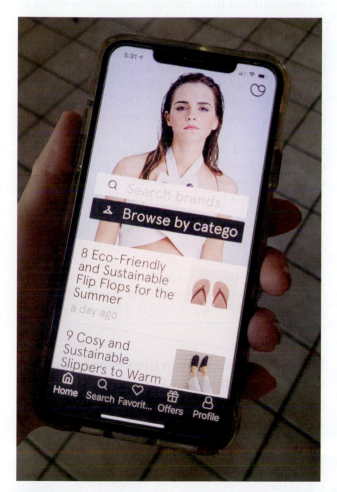

FIGURE 1.3 Greening Your Closet The Good on You app (pictured) helps you compare the environmental impacts of various clothing brands. Similarly, the Save Your Wardrobe app helps you learn how to make your clothes last longer, and connects you to repair services and secondhand clothing markets.

1.1 What Is the Environment, and What Is an Ecosystem?

We have already used the word *environment* several times. It is in the title of this book, and you probably encounter the word in your everyday life as well. It often comes up in news stories and publicity campaigns that describe how actions are either preserving or damaging the environment. So let's define this term.

On a planetary scale, the **environment** is the multitude of living and nonliving things and conditions on Earth that sustain life, including our own. Throughout this book, we will see that components of the environment interact with one another in different ways that shape, favor, and sustain life. The environment is a collection of systems—an important scientific concept that will be the subject of an entire chapter in this book. So let's briefly introduce this concept before we start to examine how humans play a role in shaping our environment.

What Are Systems?

Although the term *system* may sound technical, most of us are very familiar with the idea of how systems work in our everyday lives. Let's consider a house. Compared to biological systems, a house is a very basic system—but it's one we can easily observe and relate to. First, the structure of a house can be broken down into many parts, including building materials, pipes and plumbing fixtures, electrical wiring and appliances, a furnace and heating ducts, walls and windows, and much more. If these parts were piled together in a heap, they would not form the system we call a house; they would be just a pile of building materials (**FIGURE 1.4**). But assembled

FIGURE 1.4 A House Is a System It's not a house until all of the parts are assembled and functioning. This photo shows construction staging for the mass production of a suburban house by the developer Levitt & Sons in the 1950s—a trend in land use we address in Chapter 16.

FIGURE 1.5 The Forest Ecosystem When we look at a forest from a distance like this, we may see only trees. But a forest is much more than just trees: it is an ecosystem with many interacting parts. A multitude of plants, animals, and microscopic organisms make their home in the forest, and the overall conditions in the forest influence life both above- and belowground.

system a collection of components interacting with each other to produce outcomes that each component could not achieve on its own.

emergent property an outcome arising from the function of the system as a whole.

ecosystem a community of life and the physical environment with which it interacts.

ecosystem services a concept that assigns a value to human benefits derived from naturally functioning ecosystems.

Anthropocene a new, current epoch of Earth history recognized by many scientists, which is marked by conspicuous human effects on the planet.

in the right way, they can form a plumbing system, an electrical system, a heating system, and an overall structure that functions as the whole system we recognize as a house. Now imagine considering not just the physical house but the entire household, which includes the people who live there. They too are part of the system, carrying out various functions big and small, such as adjusting the thermostat, paying the utility bills, cooking, cleaning, and so forth. Thus, a household is an example of a system with living and nonliving factors interacting to carry out certain ongoing functions.

A **system** consists of components that interact to produce patterns of behavior over a period of time. In other words, the whole is greater than the sum of its parts because the interactions among the components of a system produce results that each component could not achieve on its own. These outcomes arising from the function of the system as a whole are known as **emergent properties**.

The environment is made up of many systems, though instead of pipes, electrical wires, and concrete, it is made up of complex and dynamic natural systems termed **ecosystems**, a word that describes communities of life and the physical environments with which they interact. (*Oikos*, the Greek word for "household," is the root of the prefix "eco" in words like *ecosystem*, *ecology*, and *economics*.) For example, a forest that may be the source of lumber used to build a house is also an ecosystem that includes populations of plants, animals, and microbes as well as the soil, water, and local

or regional climate. A coral reef is an ecosystem that forms primarily in shallow tropical ocean waters. It includes not only coral—the colonies of tiny marine organisms whose skeletons give a reef its structure—but also other animals such as sponges, sea anemones, and a variety of worms, shrimp, and fish. This ecosystem also includes plants, microscopic organisms, and nonliving factors such as sunlight, water, and minerals. The relationships in these systems are complex because each living thing responds to the environment in which it is situated and draws on environmental resources to survive, while also modifying the environment in ways that can then affect other parts of the system. So in forests, trees rely on their environment for sunlight, nutrients, and water. While doing so, they also modify the environment by sending down roots that disturb the soil and even break up the rock below, by producing shade that lowers air temperature below the canopy, and by dropping leaves and branches that accumulate as organic matter on the forest floor. Shade, air temperature, and organic matter are aspects of the environment that then affect the community of life in the forest (**FIGURE 1.5**).

In a similar way, we humans make our living on Earth by drawing on the environment around us. Our planet's natural systems provide resources, such as oxygen, water, food, energy, and building materials, that we need to survive and carry out our various pursuits. Ecologists use the term **ecosystem services** to describe a concept that assigns a value to the human benefits derived from naturally functioning ecosystems. For example, coastal wetlands can protect communities in these areas from storm surges and flooding. The term also captures the many ways that natural systems provide the conditions on which our well-being depends. We rely on the environment to do things such as assimilate our waste, cycle nutrients, and provide a climate and the conditions favorable for our survival. Ecosystem services are discussed in more detail in Chapter 5, and throughout the book we highlight the ways in which humans depend on natural systems.

Where Do Humans Fit in the Environment? What's Our Impact?

Now that there are more than 7.75 billion people on Earth, the magnitude of our impact is considerably greater than that of other species. Scientists have suggested that we live in a new epoch in Earth's history, which they call the **Anthropocene**. Just as some past periods of time are identified with certain species—such as the Mesozoic Era, when dinosaurs roamed Earth—the Anthropocene Epoch is marked by conspicuous human effects on the planet. A simple example of human dominance is revealed in the night view of the United States from space, in which lighted areas mark centers of human activity (**FIGURE 1.6**). While most of us take this nocturnal activity for granted, consider that about 125 years ago this entire area would have been pitch black at night. Now, think about the land that is occupied by a big city near you. What was that land like before it became a major population center? It was home to more natural features, such as forests, wetlands, and free-flowing creeks and rivers, than it has now. Even rural areas far from the city that are now used for crop production have replaced large areas of forest and prairie. In other words, as we use the land and water to build and expand our own habitat, we also transform it. And these transformations generally result in habitat loss for other species. **FIGURE 1.7** illustrates global habitat loss on land and impacts on the oceans due to human activity. Looking at these figures, it is easy to see how humans dominate the planet. We have done so in a relatively short amount of time and have only recently learned—and are still learning—about some of the indirect effects of our actions on our environment.

Today, very few if any areas on Earth are unaffected by human influences.

⬡ **TAKE-HOME MESSAGE** The environment is made up of the living and nonliving things that sustain life on Earth. Humans draw on Earth's natural systems, and as we do so, we affect the environment. Our impact on the environment is now much greater than that of other species.

FIGURE 1.6 The United States at Night Cities light up the sky at night in the United States.

1.2 What Is Sustainability?

When scientists characterize the time period in which we are living as the Anthropocene, they are highlighting the fact that environmental effects resulting from everyday human activity may be the most important factor now affecting life on Earth—including human life. As we will explore throughout this book,

FIGURE 1.7 Percentage of Terrestrial Habitat Loss and Human Impact on Ocean Habitat

Adapted from (a) Hoekstra et al. (2010) and (b) United Nations Environment Programme (2012).

Terrestrial Habitat Loss

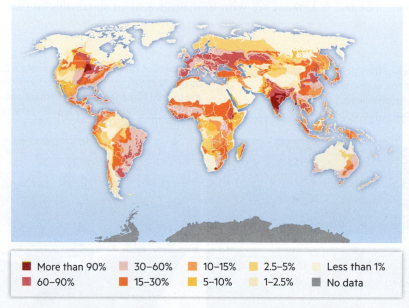

■ More than 90%	■ 30–60%	■ 10–15%	■ 2.5–5%	□ Less than 1%
■ 60–90%	■ 15–30%	■ 5–10%	■ 1–2.5%	■ No data

(a) One way that human environmental effects are becoming apparent is through tracking human activities, such as urbanization, agriculture, and forestry, that have transformed the landscape—causing the loss of habitat for other species.

Human Impacts on Ocean

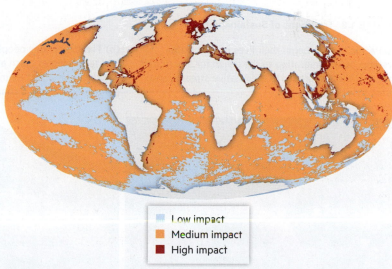

■ Low impact
■ Medium impact
■ High impact

(b) Human environmental effects are not confined to land. This map shows that most areas of the oceans have also experienced at least medium impact due to human effects—such as pollution and the harvest of sea creatures—on marine ecosystems. In fact, the researchers that produced this study concluded that no areas of the oceans were unaffected by human activity.

these effects include those on natural systems such as nutrient cycles and climate and on the availability of resources such as forests and fish. Human impact on the environment both raises concerns about the future and sets the stage for understanding how adaptation of our own behaviors and practices can reduce such effects.

Acting in a way to protect Earth's resources is not a new idea. There are many historical examples of human practices developed to address our long-term impact. Consider the Indigenous peoples of the Pacific Northwest who relied on Pacific salmon as their primary source of protein. Although the various tribes had technologies such as large nets and traps capable of overharvesting salmon swimming up their native rivers and streams to reproduce, the tribes also had rituals and rules in place to allow a substantial share of the salmon to swim upstream unhindered. This ensured that successive generations of fish would continue to be available for future generations of people in the area (**FIGURE 1.8**).

Or in a completely different cultural context, consider the way in which construction was planned in 1379 for the dining hall of New College in Oxford, England. The ceiling of the dining hall was built from massive oak beams that were each 2 feet square and 45 feet long. After about 500 years, the beams had begun to rot, were infested with beetles, and needed to be replaced. The college owns and maintains forests of mixed trees for whatever needs it might encounter, so in 1862 the senior warden had the largest oaks in these forests cut down to replace the beams (**FIGURE 1.9**). These forests, which the college acquired in 1441, are still maintained to this day. Similarly, when the famous Notre Dame cathedral

in Paris was badly damaged by fire in 2019, some of the new beams for the construction came from 1,000 oak trees in France's Villefermoy Forest. Here, as well as other places in France, centuries-old trees are regularly cut down and replaced with new saplings as part of a forest management program. This time, some of the cut trees are being used to rebuild Notre Dame. As one forester explained, "It's about making the best forest for the future, not about making money from trees."

Defining Sustainability

Today, when we hear about practices like these, we admire the foresight and describe them as sustainable because they enabled people to continue their reliance on a natural resource over long periods of time. We say the decisions were made with sustainability in mind. **Sustainability** is the management of natural resources in ways that do not diminish or degrade Earth's ability to provide them in the future. When we consider sustainability in terms of a natural resource, such as fish or trees in the earlier examples, then we can think of it as humans using or consuming the resource in a way that ensures it will regenerate in time for future use. But on a planetary scale, we can also think of sustainability in terms of maintaining ecosystems and vital Earth processes such as nutrient cycles and the global climate system that we will learn about later in this book.

Although the definition of sustainability is short, it is not simple. It requires that we select and think ahead to a specific goal (what we aim to sustain) and also consider the factors that will affect our ability to achieve that goal. Let's say we want to manage a forest sustainably so that

FIGURE 1.8 Traditions of Sustainability

(a) Pacific salmon begin their lives in freshwater systems before migrating to the ocean for a period of time and then returning to their native streams and rivers to reproduce.

(b) The Indigenous peoples of the Pacific Northwest had technologies (such as this traditional fish trap) that could have enabled them to overharvest the salmon before the fish had a chance to reproduce.

(c) However, they had traditions in place (such as this First Salmon ceremony performed by the Lummi Nation) that allowed many salmon to swim upstream, thereby ensuring salmon harvests in the future.

current and future generations can continue to use and benefit from it. To make decisions that will help us meet our goal, we will need to answer questions such as the following:

» What kind of uses and benefits are we trying to sustain?

» What sorts of factors affect the continuation of those uses and benefits?

» At what rate should we allow trees to be cut?

» How much forest should we leave standing?

» How should we manage other uses of the forest?

Measuring performance against goals or benchmarks that we set can foster sustainability. For example, thousands of companies, including Apple and Walmart, follow sustainability guidelines created by the Global Reporting Initiative (GRI), an independent agency that creates standards for businesses, governments, and citizens. These guidelines help companies track and be accountable for the environmental and social impact of their business practices, such as their energy and water consumption, greenhouse gas emissions, and waste production. Similarly, many US cities, such as Dallas, Cleveland, and Atlanta, have sustainability plans that establish community goals for water and land use, air quality, transportation systems, public health, and education.

Many of these sustainability plans for cities, companies, and even college campuses have adopted the broad idea of **sustainable development**. The term comes from a 1987 United Nations (UN) report titled *Our Common Future* that defined it as "development that meets the needs of the present without compromising the ability of future generations to meet their own needs." This concept refers to positive changes in social and environmental conditions and not necessarily economic expansion or growth. The UN report proposed a framework that entailed careful consideration of not only the environmental effects of human actions but also the economic and social effects related to human development. These became known as the **3Es—environment, economy, and equity**—and have served as key considerations for sustainability efforts.

In 2016, the United Nations released 17 sustainable development goals designed for nations to use as guidelines for and measurements of achievement in the coming years (**FIGURE 1.10**). The goals cover the 3Es and include environmental objectives such as combating climate change and halting the loss of biodiversity; economic objectives such as providing decent work and economic growth; and equity objectives such as eliminating poverty and hunger, reducing inequality, and promoting gender equity. Importantly, they also

FIGURE 1.9   New College, Oxford   The founders of New College acquired forests in 1441 to provide wood for whatever uses the college might need, such as new beams for the dining hall centuries after its building. Today's foresters continue to plant trees to regenerate the forest and provide wood for generations to use in the future.

provide a framework for addressing changes that we are seeing due to Earth's changing climate. **Climate change** and its impacts lie at the heart of many environmental problems, and understanding what is driving this change is an important goal of this text. Although climate change has its own chapter, important ideas about this phenomenon and how we experience it are woven into most chapters.

For each goal, specific indicators were identified to measure progress over time. For example, progress toward the goal of conserving and sustainably using ocean ecosystems is tracked through indicators assessing marine pollution of various kinds, the percentage of coastal areas that are protected from human use, and the policies in place to regulate fishing. Although the United Nations as a whole developed the goals, the manner and extent to which member nations decide to pursue them is up to each country's government. Although setting goals for sustainability is an idea that has made its way into American language and culture, the debate about how, when, and why to act on the goals remains lively in the United States.

Ecological Resistance and Resilience

Despite the best planning, damaging events can occur that disrupt an environment and confound sustainability efforts. These events can be natural, such as hurricanes or wildfires, or human caused, such as overfishing or mining. For this reason, it's also important to consider factors that help systems resist degradation and then recover. In natural systems, **ecological resistance** is the ability of an ecosystem to remain unchanged in the face of a disturbance. **Ecological resilience** is the ability of an ecosystem to recover from damage and

sustainable development
development that meets the needs of the present without compromising the ability of future generations to meet their own needs.

3Es (environment, economy, and equity)
a framework that considers the environmental effects as well as the economic and social effects of human actions.

climate change   long-term change in climate conditions, such as temperature and precipitation.

ecological resistance
the ability of an ecosystem to remain unchanged in the face of a disturbance.

ecological resilience
the ability of an ecosystem to recover from damage suffered in a disturbance and return to its pre-disturbed state.

Major Human Actions Affecting the Environment

These pages highlight some of the more important impacts that the ever-increasing human population is having on Earth. As we explore these impacts in this text, we will discuss how we can modify our actions to be more sustainable.

Increasing Carbon Dioxide in the Atmosphere

Fossil fuel use and other human activities have dramatically increased the concentration of carbon dioxide (CO_2) in the atmosphere.

This and emissions of other greenhouse gases that trap heat in the lower atmosphere have led to a recent warming of Earth's climate.

Harvesting the Ocean

Humans take more than one-third of the ocean life from the ocean areas nearest land.

This area holds the most easily accessible of Earth's fisheries, and nearly one-third of these are considered overharvested.

Decreasing the Variety of Life

Humans are causing a rapid decline in biological diversity—the variety of life-forms on Earth—such that many biologists estimate that species extinction rates are 100 to 1,000 times more than the background rate that has persisted over much of Earth's history.

Freshwater Usage

Humans use more than half of the freshwater that falls on land each year. This leaves less water to be used or stored by Earth's natural systems.

Increasing Acidity of Oceans

The surface water in the oceans absorbs human-caused CO_2 emissions, which makes them more acidic.

This phenomenon affects the survival of marine organisms, especially coral reefs and the life systems dependent on them and any sea animal that has a shell.

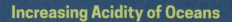

Consumption of Plants

Each year, humans use more than 40% of Earth's net primary productivity—the green plant matter produced on Earth—and the amount we use is increasing.

Nitrogen Conversion

Humans now convert more free nitrogen from the atmosphere into other compounds than that converted by all other processes on Earth.

Most of this is to produce synthetic fertilizers for agriculture. This use of nitrogen contributes to air pollution and low-oxygen marine "dead zones."

What Can I Do?

A challenge for all of us is to find ways to use these resources more sustainably. We will help you explore practical changes that can make a positive impact on Earth as you read this book.

SUSTAINABLE DEVELOPMENT G**O**ALS

FIGURE 1.10 The United Nations Sustainable Development Goals These guidelines were created in 2016 to help countries measure progress toward goals in 17 areas.

1 NO POVERTY
2 ZERO HUNGER
3 GOOD HEALTH AND WELL-BEING
4 QUALITY EDUCATION
5 GENDER EQUALITY
6 CLEAN WATER AND SANITATION

7 AFFORDABLE AND CLEAN ENERGY
8 DECENT WORK AND ECONOMIC GROWTH
9 INDUSTRY, INNOVATION AND INFRASTRUCTURE
10 REDUCED INEQUALITIES
11 SUSTAINABLE CITIES AND COMMUNITIES
12 RESPONSIBLE CONSUMPTION AND PRODUCTION

13 CLIMATE ACTION
14 LIFE BELOW WATER
15 LIFE ON LAND
16 PEACE, JUSTICE AND STRONG INSTITUTIONS
17 PARTNERSHIPS FOR THE GOALS

SUSTAINABLE DEVELOPMENT G**O**ALS

environmental justice the principle that no community should bear more environmental burdens or enjoy fewer environmental benefits than others.

return to its pre-disturbed state. Planning for sustainability means considering factors that promote resistance and resilience to predicted changes and risks in an environment.

Consider New York City, which sustained tremendous damage to its more than 520 miles of coastline during Superstorm Sandy in 2012. As a result, city planners are evaluating coastal areas and designing new construction with an eye toward their preservation. One area that held up particularly well during the storm was Brooklyn Bridge Park, which was designed with resilience in mind and features elevated landforms and a saltwater wetland (**FIGURE 1.11**). Learning from this lesson, designers of a newly expanded recycling facility built on the waterfront in a low-lying area of Brooklyn the year after the storm intentionally elevated the structure 4 feet above peak flood levels in anticipation of future storms and rising seas. Analysis of similar successes and failures became part of the city's 2013 Coastal Climate Resilience plan.

⬡ **TAKE-HOME MESSAGE** Sustainability is the management of natural resources in ways that do not diminish or degrade Earth's ability to provide them in the future. It aims to meet the needs of the present without compromising the needs of future generations. The ecological resistance and resilience to tolerate and recover from damage is a key factor in sustainability.

1.3 What Is Environmental Justice?

One of the 3Es of sustainability is equity, which conveys a concern for fairness. Fairness is at the heart of **environmental justice**, the principle that no community

FIGURE 1.11 Coastal Resilience After New York City sustained considerable damage during Superstorm Sandy in 2012, the city developed a coastal resilience plan to preserve its shorelines and related construction. Strategies such as elevated landforms and restored wetlands helped Brooklyn Bridge Park resist damage during the storm and have been integrated into resilience planning for other areas.

should bear more environmental burdens or enjoy fewer environmental benefits than others. With roots in the modern civil rights movement, civil rights and environmental concerns coalesced in the 1970s and 1980s as communities of color protested the disproportionate location of landfills, incinerators, and hazardous waste sites in their neighborhoods.

Warren County, North Carolina, is widely recognized as the birthplace of the environmental justice movement. The state decided to dispose of more than 30,000 cubic yards of soil contaminated with the toxin polychlorinated biphenyl (PCB) in this predominantly Black community in 1982. Residents from the area joined civil rights activists from across the region in protest at the site, resulting in more than 500 arrests. The protest did not stop the building of the landfill, but it spotlighted disproportionate environmental harm as a serious issue (**FIGURE 1.12A**).

Problems of environmental justice are very real across the United States and are rooted in historically discriminatory policies shaping housing, employment, and the location of polluting industries. For example, a 2018 study by the US Environmental Protection Agency (EPA) found that Black Americans are 75% more likely than other Americans to live near hazardous waste facilities. It also found that Black Americans suffer higher exposure to air pollution than White Americans regardless of income levels, with people of color exposed to more air pollution than are White people in 46 US states. Further studies show that communities of color are disproportionately located in proximity to major sources of air pollution, such as highways, industrial facilities, and oil and gas refineries. Disproportionate exposure to air pollution is also linked to health disparities including elevated incidence of asthma, other types of lung disease, heart disease, certain kinds of cancer, and elevated death rates from COVID-19 (**FIGURE 1.12B**).

On a different scale, there is also a mismatch between the people enjoying the greatest benefits of consumption and those bearing the brunt of the environmental impacts from the production of goods and services. Black and Latino Americans are exposed to 56% and 63% more air pollution, respectively, than they generate through consumption, while white and other Americans are exposed to about 17% less air pollution than they produce through the goods and services they consume (**FIGURE 1.13**). High-income countries have much higher per capita consumption rates than low-income countries, yet low-income countries bear more of the environmental impacts associated with the manufacture, distribution, and disposal of consumer goods (see more on these impacts in Section 1.6)

Further, the National Park Service has reported that just 23% of the visitors to its natural spaces were people of color, even though they make up more than 40% of the US population. Spatial analysis of urban parks and green spaces in US metropolitan areas has found that wealthier and whiter communities tend to have more access to these environmental amenities.

FIGURE 1.12 Environmental Injustices

(a) In 1982, civil rights protesters attempted to block a hazardous waste landfill from being sited in Warren County, North Carolina—a predominantly Black area of the state. This struggle over disproportionate exposure to environmental harms is widely recognized as the origin of the environmental justice movement.

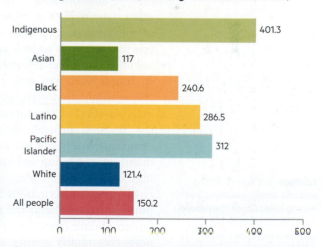

COVID-19 Deaths per 100,000 People through March 2, 2021, USA (Age-Standardized Rates)

Group	Deaths per 100,000
Indigenous	401.3
Asian	117
Black	240.6
Latino	286.5
Pacific Islander	312
White	121.4
All people	150.2

(b) In the midst of the COVID-19 pandemic, researchers learned that many historically marginalized racial groups were suffering higher death rates from the disease. This stemmed from long-term inequities in exposure to air pollutants (linked to health conditions associated with more severe cases of COVID-19), the inability to socially distance at work and home, and access to health care.

Adapted from APM Research Lab (2021).

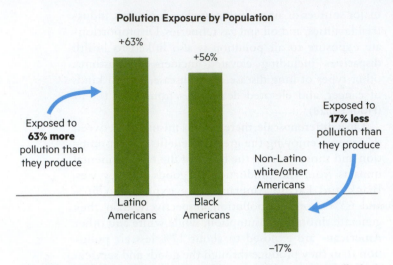

Pollution Exposure by Population

+63%

+56%

Latino
Americans

Black
Americans

Non-Latino
white/other
Americans

−17%

Exposed to
63% more
pollution than
they produce

Exposed to
17% less
pollution than
they produce

FIGURE 1.13 The Imbalance of Consumption and Exposure to Pollution The consumption of goods and services produces pollution. But the benefits enjoyed through consumption are not balanced with equal exposure to the harms of pollution across societal groups. Latino and Black Americans are exposed to a much larger share of pollution than they produce with their consumption.

Adapted from Tessum et al. (2019).

Working for Environmental Justice

Today, an understanding of environmental justice issues and efforts toward remedying injustice are growing with new research tools and methods. This includes environmental justice mapping tools from the EPA that can help researchers and affected communities identify and understand the cumulative impact of various environmental factors. Some states have recently passed new laws promoting environmental justice. New Jersey and several other states require environmental justice analysis as part of the permitting process for private and public projects that are likely to have environmental impacts. Many scientists and other researchers who study environmental justice concerns now follow a method known as community-based participatory research. This approach involves affected community members in all phases of the research process, including identifying potential environmental harms and health issues, forming research questions, collecting and interpreting data, sharing findings, and recommending solutions to environmental challenges. Collaborations between scientists and community partners helps the researchers gain insights and understanding from community perspectives and supports community members in raising issues of concern, proposing solutions, and pressing policy makers for needed changes.

This book examines all three Es, as the environment, economy, and equity are linked through how we make our living on Earth. To understand the full dimensions of sustainability and how to achieve it, we need to address the physical, biological, and social processes that shape our individual and collective impacts on the world—and thereby on ourselves.

⬡ **TAKE-HOME MESSAGE** Environmental justice is a principle that strives for the fair treatment of all people when it comes to environmental harms and benefits. It requires meaningful involvement of all people in decisions affecting their environment. The environmental justice movement was born out of struggles by communities disproportionately affected by environmental harms. Environmental injustice is an ongoing problem, though many are working to address it through social movements, policies, and research.

1.4 What Is Science?

If sustainability calls attention to the fact that our decisions affect the ecosystems around us, how do we gain an understanding of how these systems work and the ways they are and can be affected? **Science** is our way of asking and answering questions and testing ideas about the natural world by using evidence gathered from the natural world.

The Scientific Method

Science is best understood as a process of inquiry rather than a collection of known facts. This process is often summarized as following a series of steps, called the **scientific method** (**FIGURE 1.14**), designed to examine and test problems and ideas:

1. Observe a problem.
2. Formulate a **hypothesis**—a statement that attempts to explain a phenomenon or answer a scientific question.
3. Conduct experiments and gather data.
4. Evaluate your data.
5. Refine, alter, expand, or reject your hypothesis.
6. Repeat the process.

Let's see how this process worked for a real environmental problem. Often, the observation of a problem happens by accident while scientists are examining something else. This was the case in the 1950s when Caltech geochemist Clair Patterson was attempting to establish our planet's age. Certain elements on Earth are radioactive and decay at fixed rates. Uranium is one of these elements, and it decays into lead. Patterson was attempting to date Earth's antiquity by measuring the lead content in rocks. However, as he was gathering and analyzing his samples, he noticed that they were contaminated by lead that was in the air, the water, and the dust on Earth's surface. Patterson wondered where all this lead was coming from (**FIGURE 1.15**).

science our way of asking and answering questions and testing ideas about the natural world by using evidence gathered from the natural world.

scientific method a formal process of inquiry designed to test problems and ideas.

hypothesis a proposed explanation for a phenomenon or answer to a scientific question.

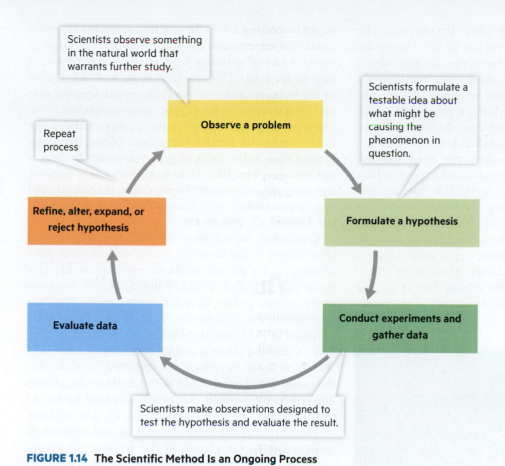

FIGURE 1.14 The Scientific Method Is an Ongoing Process

The following text boxes appear in the diagram:

- Scientists observe something in the natural world that warrants further study.
- Scientists formulate a testable idea about what might be causing the phenomenon in question.
- Repeat process
- Scientists make observations designed to test the hypothesis and evaluate the result.

Cycle boxes:
- **Observe a problem**
- **Formulate a hypothesis**
- **Conduct experiments and gather data**
- **Evaluate data**
- **Refine, alter, expand, or reject hypothesis**

FIGURE 1.15 Leaded Gasoline Geochemist Clair Patterson stumbled on the problem of lead in the environment and hypothesized that additives in gasoline were depositing fine lead particles on land and in water via air pollution generated by automobile emissions.

After testing water samples from the Pacific Ocean, Patterson noticed that the presence of lead particles diminished as samples were taken at increasing depths. This led him to hypothesize that the lead was being deposited from the air. He knew that a lead-containing compound (tetraethyl lead) had been added to gasoline since the 1920s to improve engine performance and reduce wear. He then reasoned that very fine lead particles from automobile emissions were being deposited and accumulating over Earth's surface.

To test this hypothesis, Patterson took samples in a variety of places, including urban areas where automobile use was high and regions in which prevailing weather patterns would likely deposit air pollution from these urban areas, such as in the coastal waters and mountain snow near Los Angeles. Not only did he find higher concentrations of lead particles in these regions than in areas not affected by urban air pollution, but he was able to determine, for instance, that the type of lead in the regions near Los Angeles matched the type of lead in air pollution in Los Angeles itself. After repeated sample gathering and analysis, he concluded that automobile pollution had raised the lead levels in the atmosphere to 1,000 times the natural amount. His findings and conclusions were challenged by automobile and gasoline companies for years, but they held up to repeated observations and experiments by the scientific community. In response, the United States banned lead additives in gasoline in 1986, which is why when you pull into a gas station today, the fuel is labeled "unleaded."

Ways of Observing and Testing

Patterson's research on lead levels is an example of the way scientific research is designed so observations can be used to repeatedly test and evaluate alternative explanations. Ideally, research is designed in such a way that observations either meet or clearly fail to meet the expectations associated with a proposed explanation and thus provide evidence to support or refute it. If, for example, Patterson had found that the concentrations of lead particles were lower in the regions where pollution from automobiles was typically deposited than in other regions where it was not deposited and that the type of lead in the polluted regions did not match that of the polluted air in the respective urban areas, it would have sent him searching for other explanations.

There are many ways observations can be set up to test alternative explanations. Sometimes a **controlled experiment** can be used. This is a test in which researchers intentionally manipulate some specific aspect of a

controlled experiment a test in which researchers intentionally manipulate some specific aspect of a system to see how this change affects the outcome.

system to see how this change affects the outcome. The parts or participants that are subject to this change are known as the *test group*, and those that are not subject to the change are called the *control group*. For example, many coral reef ecosystems are undergoing a phenomenon known as coral bleaching, a process in which colorful living organisms called algae that normally live together with the coral are expelled because of stresses in the environment. This weakens the coral and also turns it white. We know that coral-bleaching events have become much more common worldwide over the past 40 years because scientists have conducted careful surveys on the location, scale, and characteristics of reefs throughout the world. One hypothesis for the cause of coral bleaching is that increasing ocean temperature is the responsible environmental stress. Another hypothesis is that increasing acidity of the oceans due to the rise in carbon dioxide (CO_2) concentrations in the water

might be playing a role. To test this, scientists created a controlled experiment by putting samples of coral in several tanks of seawater kept at a constant temperature in their lab. They then manipulated the acidity of the water in some of the tanks, while keeping one tank of water as a control group with no manipulation. They found that the greater the acidity of the water in the tanks, the greater the extent of the bleaching beyond that in the control group. But one finding is not enough, regardless of how convincing it is. After repeating the experiment many times, they concluded that ocean acidification was in fact playing a role in coral bleaching (**FIGURE 1.16**).

Controlled experiments are just one way to use observations to test ideas. We cannot subject many phenomena in the natural world to experiments like this because it is either difficult, impossible, or inappropriate to manipulate variables. For example, Nicolaus Copernicus was a mathematician and astronomer who proposed in 1514 that the Sun was the center of the universe, a startling claim at a time when convention held that Earth lay at the center of everything. To make this discovery, Copernicus could not manipulate the planets and conduct controlled experiments. Instead, he had to test the expectations stemming from his new theories of planetary motion using meticulous and methodical observations of the planets in the night sky. Similarly, geologists cannot go back in time to observe what led to the extinction of the dinosaurs. Instead, they tested their hypotheses—such as the prevailing theory that an asteroid colliding with Earth started this process—by predicting the evidence we would likely find on Earth today if their hypothesis was correct.

Ethical concerns lead researchers away from experimental designs that would intentionally harm research subjects in order to observe various health effects. So public health researchers often rely instead on natural experiments or observational studies, opportunities that arise for observation and testing because of an unplanned event or occurrence. For example, a group of public health researchers had a hypothesis that the increasing frequency of childhood asthma attacks requiring emergency room visits or hospitalization was linked to increases in air pollution from automobile emissions. The 1996 Summer Olympics in Atlanta provided a natural experiment. During the weeks that Atlanta hosted this event, the city mandated restrictions on car use, and researchers found that air pollution declined significantly and that area hospitals saw a 40% decline in childhood asthma attacks compared to the incidences in the time periods just before and just after the Olympics.

Models

Experiments and many other kinds of research designs are set up to help us understand how certain variables can affect outcomes. But natural phenomena are rarely

(a)

(b)

FIGURE 1.16 Coral Bleaching
(a) Coral turns white (a process known as coral bleaching) because it expels colorful microscopic algae when under environmental stress.
(b) The coral in the lab setting is being tested for its responses to different acidity levels to gauge the effect of different levels of ocean acidity on these organisms.

explained by a single factor, especially when multiple factors play a role in a system. To examine such a system, we often create a **model**, a simplified concept or representation of a complex process. Models are designed to assemble observations and data to try to simulate aspects of what actually happens in the world, so that we can better understand the interactions among many different factors.

Models are inherently simplified because it is impossible to incorporate and compute the relationships between every aspect of the real system. But this does not mean that models are simple. Models employed in environmental science are often mathematical representations of reality. Consider the models that provide you with a weather report each day (**FIGURE 1.17**). These

Grid spacing
6–12 miles

Variables at the surface, including temperature, humidity, pressure, changes in moisture, heat, and radiation, are added to the model.

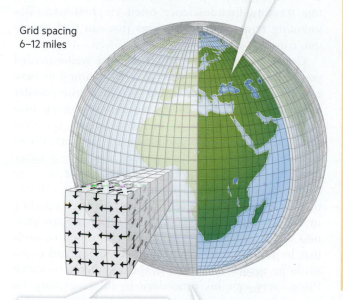

Arrows indicate how cells can interact with other cells next to, below, or above them.

Variables in the atmosphere, including measurements of wind vectors, humidity, clouds, temperature, height, precipitation, and aerosols, are also added to the model. Supercomputers simulate interactions between cells using both surface and atmospheric variables.

FIGURE 1.17 Weather Forecast Modeling Modern weather forecast models break Earth's atmosphere into vertical stacks of cube-like units called grid cells. Using the world's most powerful supercomputers, researchers enter data on various conditions thought to affect weather into programs that simulate interactions among grid cells of the atmosphere. The results of the simulations can be checked against what actually happens, so that the modeling programs are constantly updated and improved.

SCIENCE

Can We Predict the Past?

In 1980, physicist Luis Alvarez and his son Walter Alvarez, a geologist, found a thin layer of clay in rock formations around the world that held what they thought might be debris from an asteroid strike. They hypothesized that a large asteroid struck Earth, raising a huge dust cloud that blocked sunlight for several years, killing most plant life and causing the dinosaurs to become extinct. But how can you test a hypothesis about something that already happened? What you can do is develop a better understanding of the type of evidence you would likely find if an asteroid impact occurred that was large enough to cause this huge cloud of dust. So over the next decade, scientists collected and analyzed debris found at similar locations around Earth and predicted the characteristics and location of this impact. Later, during oil exploration in the Gulf of Mexico, scientists identified Chicxulub Crater (on the Yucatán Peninsula in Mexico) as having the expected age and size of such an asteroid impact, a discovery that drove general acceptance of the Alvarezes' hypothesis.

models are run on the most powerful computers in the world and use millions of lines of code to program the numerical representations of Earth processes on which they are based. These weather models are widely shared and constantly revised. Weather models are not merely updated moment to moment with new observations; their formulas are also tweaked as actual weather outcomes are observed. For models simulating short-term processes, such as models for tomorrow's forecast, this refinement comes as forecasts are compared to what actually happens. When models simulate longer-term processes, such as the climate models we will learn about in Chapter 11, they are tested and refined through a process called *hindcasting*. In these cases, the initial conditions from a known starting point in the past can be plugged into a model to see how closely the model results match up

model a simplified concept or representation of a complex process that is designed to help understand interactions among different factors.

to what actually happened in a given historical period. Such exercises help researchers estimate the uncertainty associated with a model and make changes to improve its performance. Like the other types of observations described in this section, models are a tool of scientific inquiry that enables testing and refinement.

⬠ **TAKE-HOME MESSAGE** Science asks and answers questions about the natural world by testing hypotheses against evidence from the natural world. The scientific method uses observations to test and sort through alternative explanations. Models assemble observations to simulate what happens in the world, helping to show interactions of different factors.

fraud an attempt to deceive people by communicating findings that are simply false.

pseudoscience claims that are not the result of scientific inquiry or are derived by a process that is not open to scientific scrutiny.

1.5 What Are Challenges to Good Science?

Scientific inquiry is a cautious and skeptical approach. Imagine one of your friends shares a story on Facebook about a newly invented serum that contains a powerful fragrance that makes you irresistible and boosts your romantic life if you just dab a little bit behind your ears. The story is linked to a video that shows the inventor in a white lab coat and includes interviews with satisfied customers. Your friend is excited about the product and thinks you should both buy some. What do you think? While we all want to trust the people we know, should you be skeptical and seek out some more information before you commit?

Fraud and Pseudoscience

Unfortunately, it is not uncommon to hear of claims that have the appearance of scientific information but lack the full integrity of the scientific process. Claims like these are not new. Sometimes they are the result of **fraud**, an attempt to deceive people by communicating findings that are simply false. In other cases, the person or group may make claims that are based in **pseudoscience**; that is, the claims are not the result of scientific inquiry or are derived by a process that is not open to scientific scrutiny. Pseudoscience often employs scientific-sounding terms and descriptions that can fool people into trusting certain claims.

Consider the case of James Price, a well-regarded young English chemist who in 1781 claimed to have experimented with something he called "the powder of production" that could transmute mercury into gold. Price was a fellow of the Royal Society, the most prestigious scientific organization of his time. When other Royal Society fellows asked Price to demonstrate his process, he at first claimed that it was too costly and dangerous to reproduce. Then he claimed that his reputation as a scientist and his high standing in society ought to be enough to verify the truthfulness of his claim. Other scientists continued to insist that he replicate his experiment for them, and eventually he agreed. As several of them gathered to watch Price carry out his procedure in his laboratory, he instead ingested poison and committed suicide rather than admit to his fraudulent results.

One of the things that this extreme (and tragic) example shows is that a simple claim of a finding from an observation (even by a well-established expert) is not enough on its own to convince the scientific community. The process and results used to arrive at the finding must be methodically documented, shared, and demonstrated in such a way that they can be replicated by others. This makes scientific inquiry a collaborative and continual process of repeatedly testing findings and conclusions. In other words, science must be reproducible and open to revision.

So we should be wary of claims about the natural world that are not subjected to the scrutiny of the scientific community. For example, you might have heard of the *Old Farmer's Almanac*, an annual publication founded in 1792 that provides information about planting and

→ **CONTEXT**

The Truth about "Cold Fusion"

In 1989 two chemists, Martin Fleischmann and Stanley Pons, gained widespread media attention when they held a press conference to describe experiments they had run that appeared to demonstrate cold fusion, using a setup like the one shown. Nuclear fusion—the joining together of nuclei from different atoms—releases tremendous amounts of energy and powers stars, like the Sun. If humans could harness the energy of nuclear fusion at "cold" temperatures (i.e., at or near room temperature), it could usher in a new, cheap, and efficient source of energy. However, a press conference is no substitute for the vetting of the scientific community. Many teams of scientists tried and failed to reproduce Fleischmann and Pons's results. The prestigious science journal *Nature* declined to publish their work both because similar experiments did not support the findings and because the two chemists had no theory to explain their results.

FIGURE 1.18 **Reliable Sources?** What source do you check for your weather forecast? Not only do these two sources represent different technologies (print vs. electronic media); they also represent different approaches to inquiry and knowledge. The *Old Farmer's Almanac* is based on a centuries-old formula that has been kept secret in a locked box. This formula is used to make long-range weather predictions 18 months in advance for each week of the year. In contrast, the forecast on your phone is based on observations and on models from the National Weather Service that are constantly tested, updated, and modified by meteorologists and other members of the scientific community to create short-term (7- to 10-day) weather forecasts.

makes predictions about the coming year's weather. The basis for these predictions is a secret formula that is kept in a locked box in Dublin, New Hampshire, and has never been subjected to testing or revision by meteorologists or other scientists. So scientists would recommend that it is probably better to rely on the forecasts you get on your phone, which are based on National Weather Service models (**FIGURE 1.18**).

Bias and Misinformation

Scientific inquiry also guards against errors in our thinking that are unintentional. For example, all humans are prone to **bias**, an unreasonable weighting, inclination, or prejudice of one's thinking that leads to misunderstandings. Bias can be the result of generalizations we make from personal experiences, and we can pick up biased perspectives and ideas from the people we interact with. In Chapter 18, we will learn in more detail how various biases play important roles in shaping our decisions. For example, psychologists have found that humans are prone to certain cognitive biases, which are common errors in our intuition. One of these is known as the "gambler's fallacy." Imagine you are repeatedly flipping a coin, and you've just observed five consecutive coin flips that landed tails. Intuition leads many people to erroneously predict an increase in the likelihood that the next coin toss will land heads. Because there have been so many tails on prior flips, our intuition tells us that heads is due to turn up the next time. But the odds of a coin coming up heads on any one flip—including those after a string of five tails—are still 50/50. We know this thanks to scientific inquiry. Experiments recording the results of thousands of coin flips and identifying all of the instances where there were five consecutive tails have found that in these cases, the odds are still 50/50 that the very next coin lands tails. So we can conclude that despite the bias of our intuition, there is no good reason to believe that those first five coin flips have any bearing on the next one.

Sometimes our understanding of phenomena in the natural world is just based on **misinformation**. For example, there is a common but misinformed belief about lightning that is captured in the saying, "Lightning never strikes the same place twice." But we could test this by methodically observing and keeping track of where lightning strikes. In fact, there is good reason to believe that lightning would repeatedly strike places that are particularly prominent and conductive. If we set up our test of lightning strikes at the top of the Empire State Building, we would find that the building often absorbs multiple strikes during the same storm event and that it is struck more than 100 times each year (**FIGURE 1.19**). Scientific inquiry is a kind of organized skepticism that uses empirical evidence to continually test and refine our understanding of phenomena, though it may not always produce definitive answers.

Unfortunately, the spread of misinformation and **disinformation**—deliberately misleading, incorrect, or false information—has increased with our growing reliance on social media for news. In 2020, the flood of false claims on social media about cures for COVID-19, ranging from eating sea lettuce to injecting bleach, made it clear that even areas of scientific research like public health can be besieged by mis- and disinformation. The Pew Research Center has found that more than half of US adults get their news from social media sites like Facebook, Twitter, YouTube, and Instagram either "sometimes" or "often." While social media is the primary news source for adults between the ages of 18 and 29, most social media users also recognize that inaccurate and one-sided news stories are "very big problems" on social media. Research tracking social media engagement with mis- and disinformation news sources demonstrates that these concerns are warranted. Between 2019 and 2020, social media engagement with "unreliable news" websites and sources more than doubled from 8% to 17% of all news engagement on social media. In response to this, librarians at California State University, Chico, developed the "CRAAP" guide to vetting sources and making sure you're not spreading

bias an unreasonable weighting, inclination, or prejudice of one's thinking that leads to misunderstandings.

misinformation false or incorrect information that may be spread intentionally or unintentionally.

disinformation deliberately misleading, incorrect, or false information.

FIGURE 1.19 **Lightning Does Strike Twice—or More!** Lightning strikes the Empire State Building more than 100 times per year.

Is it Current?
When was it published? Are their references current?
Is currency important for your topic?

Is it Relevant?
Does the info relate to my topic? What audience is it written for?
Is it at an appropriate level for my needs?

Is it Authoritative?
Who is the author/organization? Are they qualified? Is it edited or peer-reviewed? If a website, does the URL tell you anything?

Is it Accurate?
Where does the information come from? Are there references?
Are there errors, broken links, etc.?

What is its Purpose?
What's the purpose of the information? Advertising? Scholarly work?
Opinion? Is there bias?

FIGURE 1.20 Is This Story CRAAP? Librarians at California State University, Chico, developed a quick guide to vetting digital information sources that has been adopted widely by college libraries in the United States. Asking these questions can help you assess the news sources you engage with on social media and avoid inadvertently contributing to the spread of mis- or disinformation.

false or misleading claims. These questions are designed to help you decide if internet and social media content is valid and reliable (**FIGURE 1.20**).

The Function of Peer Review

To help make certain that scientific work meets the standards of the field of knowledge to which it is contributing, it goes through **peer review**—a process for refining research design and ensuring that conclusions can indeed be drawn from the evidence presented. This is a much more rigorous and thorough process than the journalistic editing and fact-checking employed by reputable newspapers and magazines. Peer review occurs when the authors of a scientific study write up their question, methods, findings, and conclusions in an article and submit it to an academic journal in their field. The journal editor sends the article out for review by several experts in the field. The reviewers submit feedback to the editor and the authors that identifies areas for improvement, suggests changes to the procedures followed or the conclusions drawn, and/or points the authors toward additional related research to consider and address. The reviewers also make a recommendation as to whether or not the research is suitable for publication. The bar for publication is typically high, with far more candidate articles rejected than accepted by prestigious journals. It is not unusual for this process to take a year or more and to involve one or more rounds of revision and resubmission before the work is refined and finalized. Publication in a peer-reviewed journal does not mean that the findings are correct, let alone conclusive. But it does provide a measure of confidence that the procedures followed and conclusions drawn meet the standards of science. Journal articles serve to inform others in the scientific community who can then attempt to replicate the studies, challenge the findings, develop competing ideas, or otherwise refine our understanding.

peer review a process for refining research design and ensuring that conclusions can indeed be drawn from the evidence by subjecting work to assessment by experts in the field of study in question.

values reflections of our understanding of how we want things to be—what we desire, aim for, or demand.

TAKE-HOME MESSAGE Scientific inquiry is a collaborative and continual process of testing findings and conclusions. It must be reproducible and open to revision. This guards against fraud, pseudoscience, bias, misinformation, and disinformation. The process of peer review helps to ensure that the procedures followed and conclusions drawn from a study meet the standards of science.

1.6 What Shapes Our Decisions on the Environment?

Sustainability involves making decisions and taking actions in response to environmental issues. This requires more than just scientific inquiry. Although science is a powerful tool for developing, testing, and refining our understanding of the natural world—and our environmental problems—it cannot determine what ought to be done about a particular issue. Any such determination requires the assertion of values. **Values** reflect our understanding of how we want things to be—what we desire, aim for, or demand. They are what we use to assign relative worth, merit, or importance to different things. Acting on values is human, and all of us use values to make decisions.

This applies at the level of governmental decision making and to the decisions you make each day. Consider the question about the clothes you wear that was discussed in the introduction to this chapter. Scientific inquiry can help you compare the relative environmental effects of different clothes or shoes. But your decision on what clothes to buy is based on far more than this scientific understanding of relative effects. Not only do we make ethical judgments about what is more or less "good" or "bad" (by, for example, considering that reducing our environmental impact is good), but we also assess our alternatives according to other personal needs and desires. In the case of clothing, most of us will think about appearance, overall style, comfort, and cost, and not only about the environmental effects. So to understand and explore sustainability we must consider a wide range of factors that influence the choices we make as individuals and groups, choices that in turn affect the environment.

Factors Influencing Individual Decisions

The individual choices we make are influenced by a variety of factors, including common biases in our thinking (like those mentioned earlier in this chapter), the way we prioritize our needs and desires, and the pressure to conform to those around us. Consider a choice between riding a bicycle across campus or driving a car instead. Let's

When Small Observations Lead to Questions, Problems, and Solutions

When we think of science applied to environmental problems, we tend to assume that the problem is identified first and then scientists work to explain why the problem is occurring and what can be done about it. But it doesn't always work this way. Often, scientists are not initially looking to explain an environmental problem, but they identify a potential problem while studying something else. This was the case with the scientific discoveries linking chlorofluorocarbons (CFCs)—chemicals widely used in the 1950s and 1960s in air conditioners, refrigerators, and aerosol cans—to the damage to a protective part of Earth's atmosphere called the ozone layer.

Chlorofluorocarbons were not found in the atmosphere until the 1970s, when British scientist James Lovelock began using new equipment to understand the composition of Earth's atmosphere. Lovelock identified CFCs in every air sample he took over England, Ireland, and vast reaches of the Atlantic Ocean. These findings demonstrated the circulation of these substances in the atmosphere. However, at about 60 parts per trillion (the equivalent of just a couple of drops in an Olympic-size swimming pool), the concentration seemed insignificant. Lovelock himself concluded that CFCs in the atmosphere were not an environmental hazard.

Nevertheless, the finding piqued the curiosity of F. Sherwood Rowland and Mario J. Molina, two chemists at the University of California, Irvine. Initially interested in how these chemicals might behave in the atmosphere, they found that more than 18 miles (29 km) above Earth's surface, ultraviolet (UV) radiation from the Sun began to break the CFCs apart, releasing atoms of the element chlorine. Rowland and Molina determined that chlorine in the atmosphere could react with a protective atmospheric gas called ozone that filters out many of the Sun's harmful UV rays. They had discovered an emerging environmental problem. In 1974, the two scientists published a paper predicting

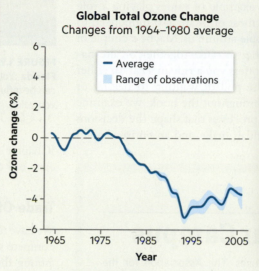

Global Total Ozone Change
Changes from 1964–1980 average

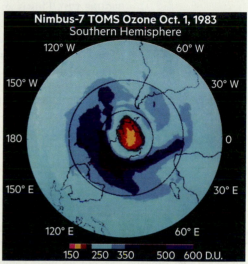

Nimbus-7 TOMS Ozone Oct. 1, 1983
Southern Hemisphere

Chemists F. Sherwood Rowland and Mario J. Molina helped direct attention toward the damaging effect of CFCs on Earth's ozone layer. It was first confirmed by NASA satellites in 1983, which recorded a growing "hole" over the Antarctic. The graph shows how quickly the amount of total atmospheric ozone dropped in the late 1970s and the 1980s before it was stabilized and started to recover after CFC use began to diminish in the 1990s. D.U. = Dobson unit.

Graph adapted from Chipperfield and Fioletov (2007).

that if CFC emissions continued at their then current rate, total atmospheric ozone would drop by as much as 13% by 2050—removing a significant amount of Earth's protection from UV radiation. This would slow plant growth and increase the incidence of skin cancer.

In subsequent years, further tests supported this finding, including 1983 NASA satellite data that provided confirmation of a growing ozone hole over Antarctica. Samples taken from aircraft flying over the Southern Hemisphere showed that as the concentration of chlorine in the air increased, ozone concentrations decreased.

These scientific findings led to a global policy response that eventually phased out the use of CFCs. Scientists now project that recovery to 1980 levels of protective ozone will be achieved by 2050. When Rowland and Molina received the Nobel Prize in Chemistry for their work in this area, they spoke to the importance of letting scientists explore questions and pursue research even when applications to real-world problems are not immediately obvious. Our understanding of problems often emerges from explorations that begin with general curiosity about how the world works.

What Would You Do?

After CFCs were phased out, different chemicals that do not interact with ozone were put to use in refrigerators and air conditioners. What if these newer chemicals are found to damage some other aspect of the environment? If you were the CEO of a refrigeration company, what would you do? For example, would you try to make changes immediately, run more studies, or wait for the government to tell you the steps to take? Would you look for alternative chemicals or try to invent a different technology? Briefly explain what actions you would take and why.

trade-offs pros and cons, benefits and costs of alternative courses of action.

communication strategy verbal or nonverbal ways of connecting with others to influence or inform.

incentive positive or negative signal that pulls us toward or pushes us away from a certain choice or behavior.

assume that in this case, once you factor in the traffic, low speed limit, stop signs, and time it takes to park the car, it is actually quicker to bike. But maybe you've never made the trip by bike before, and you simply assume that it is faster to drive because cars can move faster than bikes. This misunderstanding would be an example of bias playing a role in your decision. But other factors might shape your choice as well. Perhaps in addition to preferring the fastest option, you also care about reducing your environmental impact, so you decide to bike (**FIGURE 1.21**). This is an example of values playing a role in your choice. Finally, most of us are also influenced to some extent by the people around us. Maybe a couple of friends are suggesting that you bike together, or, alternatively, perhaps a friend offers you a ride in a car. In either case, you might feel the pull of wanting to belong or of social acceptance. Throughout the book, we examine the behind-the-scenes processes that shape the decisions that groups and individuals make, and an entire chapter

FIGURE 1.21 Getting across Campus in Gainesville, Florida Your decision on how to get across campus can be influenced by many factors, including what you value.

(Chapter 18) takes a closer look at what factors influence our decisions and actions.

Trade-Offs and Incentives

We have many needs and desires, so sometimes our values compete with each other and we are forced to prioritize among them as we make up our minds. In these cases, we are considering the **trade-offs**—or pros and cons, benefits and costs—of alternative courses of action. For some major decisions (such as choosing a college), we might even formalize the consideration of trade-offs by making a two-column list of the pros and cons of each alternative. But we also quickly sort among our needs and desires as we make smaller choices, such as deciding whether to bike or drive, as mentioned above. In this case, you might have to sort among an ethical desire to do what you consider is "good" for the environment, practical desires for the cheapest and quickest option, and even social values such as the relationships with your friends. Sometimes, public strategies are used to alter the trade-offs you consider as you make decisions. These can include **communication strategies** that strive to make you aware of certain costs and benefits, such as a campus campaign that publicizes the environmental cost of cars driven on campus. These can also include **incentives**, which are positive or negative signals that pull us toward or push us away from certain choices or behaviors. For example, expensive parking meters might act as a negative signal discouraging us from driving, and a free and easy-to-use bike-sharing program might function as a positive signal encouraging us to bike. When we become aware of incentives, we factor them into the trade-offs we consider when making decisions. And institutions (such as universities or governments) often make use of incentives as they design policies to shape behavior. We consider trade-offs and incentives throughout this book as we explore how our choices are linked to

Sustainability on Your Campus

Sustainability is a growing concern on most campuses. The Association for the Advancement of Sustainability in Higher Education (AASHE) lists almost 2,500 academic programs in areas related to these fields, from professional certificates to doctoral degrees. You may even be pursuing or considering one. On a broader scale, many colleges and universities are making formal declarations about their sustainability goals. The Talloires Declaration is a commitment to sustainability signed by more than 500 colleges and universities around the world that outlines values and actions to reduce environmental impacts. Other sustainability commitments have followed the model of the Talloires Declaration, cultivating support from the leadership of colleges and universities. The AASHE even has a Sustainability Tracking and Rating System (STARS), a sustainability auditing tool that measures progress toward sustainability goals.

environmental effects and study the ways in which economic and social forces together with policies can reshape the context in which we make our decisions.

Making Sense of Our Environmental Effects

The choices we make and actions we take as individuals can add up to significant environmental effects, both in terms of one person's cumulative impact over time and the impact of many people collectively. To get a better sense of this, we often use the tool of **footprint analysis**. Just as your foot makes an impression in the soft sand of a beach, your actions make an impression on the environment, and footprint analysis attempts to quantify this impact. For example, ecological footprint analysis translates individual actions, such as energy use, food consumption, and material use over the course of a year, into an estimate of the total natural resources needed to support these actions, expressed in terms of land area. You could learn more about your own ecological footprint by answering a series of questions about things such as your transportation use, food consumption, and other material uses at a website such as the one featured in **FIGURE 1.22A**. Chances are you will find that your footprint is more than a dozen acres; the average US footprint is just over 17.2 acres (about 13 football fields). And footprints can also be calculated for countries. The United States and Canada join China and India as the countries with the largest total footprints (**FIGURE 1.22B**). Finally, footprints can also be used to make sense of a particular environmental impact. For example, carbon footprints estimate the greenhouse gas emissions that are generated by various actions, such as driving certain vehicles or eating certain foods, and are linked to global warming. Carbon footprints are covered in more detail in Chapter 10 and ecological footprints in Chapter 6. All footprint analyses strive to make clear the environmental trade-offs of the various choices we make.

footprint analysis a method to understand the magnitude of the impact of choices and actions individuals make, both over time and collectively.

(a)

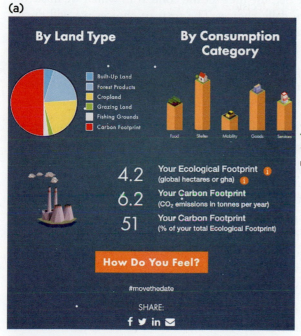

FIGURE 1.22 Ecological Footprints
(a) Conducting an ecological footprint analysis using an online calculator can help you learn about how your own choices affect the environment. See Appendix A to learn more about using online assessment tools. **(b)** Overall ecological footprints of entire nations vary greatly. The bar graph shows the totals for the entire country. The total area of the United States is 983 million hectares, but its ecological footprint is just over 2.6 billion hectares. **(c)** The ecological footprints of the average person in some countries, measured in standard units of land area called global hectares (gha), are larger than the actual areas of those countries. (All ecological footprint data are for 2018.)

Adapted from (b) and (c) Global Footprint Network (2022).

(b)

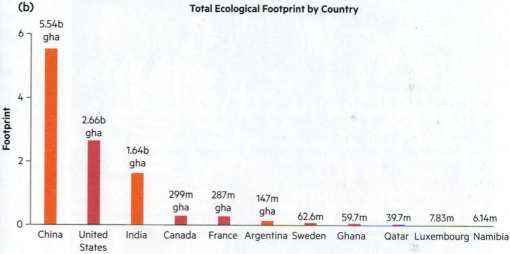

(c)

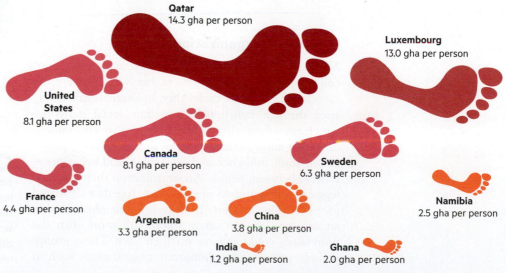

Impacts of Fast Fashion

We live in an unequal world, with much disparity rooted in consumption differences. This disparity is particularly evident in fast fashion—a trend whereby retailers sell large quantities of clothing at cheap prices while rapidly shifting designs. Fast fashion increases clothing consumption and encourages consumers to regard clothing as a disposable good. While the United States is the world's leading consumer of clothing, more than 90% of all clothing is produced in middle- and low-income countries where communities bear the costs of water scarcity and pollution while workers are often subjected to hazardous conditions for low pay. Yet the average US consumer disposes of 80 pounds of clothing each year. The UN Sustainable Development Goal #12—Responsible Consumption and Production—seeks to address this challenge by influencing consumer behavior, business practices, and national policies. Its #ACTNOW Fashion Challenge is a push to develop zero-waste fashion.

Groups and Organizations

Not only do our individual actions accumulate into collective environmental effects, but we also have the capacity to influence each other. Individuals gain influence through collaborations within social networks, which are systems that include people as well as connections that allow certain kinds of things—money, goods, information, influence, and even values and behaviors—to flow between people. An organization such as a college can link networks of individuals together for shared purposes. The concerted efforts of individuals within an organization can often accomplish more than the individuals could achieve on their own. These groups include governments, nonprofit organizations such as environmental advocacy groups, and for-profit businesses as well.

What Are Some of the Big Decisions We Will Explore in This Book?

In this chapter, we have introduced the idea that as we make a living on Earth, we have an impact on this planet that supports us. In the chapters that follow, we will learn about many of Earth's processes that provide the environmental conditions on which we depend. We will also consider the extent to which our choices and actions affect environmental conditions in ways that could jeopardize (or improve) the capacity of humans to survive and thrive on Earth. This includes not only our individual choices, such as what we consume and how we use energy to power our various pursuits, but also the ways in which our individual choices accumulate to produce action across societies. Our governments, other aspects of social organization, and economic forces also play a role because they set the context in which we make our own daily choices. The sorts of choices we will discuss include the following:

>> *Population.* The human population continues to grow. What sorts of factors have affected this growth rate, and how could various policies affect it in the future?

>> *Food and agriculture.* As we grow, harvest, process, and distribute food for more than seven billion people through various practices, what impact are we having on the land, air, water, and biodiversity of Earth?

>> *Energy.* Where do we get the energy to fuel our transportation, heat our homes, and power our many pursuits? What are trade-offs associated with various energy sources and conservation strategies?

>> *Consumption and waste.* We use a multitude of material goods every day. What sorts of environmental effects are associated with the production of these things? What do we do when we are done using them?

>> *Urbanization and land use.* Most of the world's people now live in cities. How do the urban habitats we construct for ourselves shape our environmental impact and prospects for sustainable development?

⬡ **TAKE-HOME MESSAGE** What we value often determines any actions taken as the result of scientific inquiry, especially when it comes to decisions about the environment. Values reflect what we desire or demand. They are what we use to assign relative worth, merit, or importance to different things.

Mostafa Tolba: Scientist as Diplomat

One of the biggest sustainability challenges at the global level is how to reduce environmental effects linked to industrialization—such as climate change and air and water pollution—while also enabling less industrialized developing countries to attain the same social and economic measures of progress already enjoyed in more developed countries.

For example, you probably live or work in a building with air-conditioning. Nearly 90% of US households have this technology. In contrast, air-conditioning is rarely used in many poor countries whose economies are just developing. In India, only 5% of households have air-conditioning units. In hot climates, air-conditioning is not just for comfort; it is essential for economic productivity, and it can even save lives during heat waves. So this technology will become even more essential in the wake of global warming.

Yet many of the chemicals used as refrigerants have negative effects on our environment. Banning these gases and shifting to other technologies would be expensive, and it would slow the growing availability of air-conditioning in areas where it may help spur economic growth. Is it fair to require all countries to ban the troublesome refrigerants, even as some nations are just starting to benefit from their use?

This fairness issue is often a sticking point in international environmental negotiations. Mostafa Tolba, an accomplished Egyptian biologist, devoted his life to addressing diplomatic challenges like this. In 1972, Tolba led the Egyptian delegation to the United Nations Conference on the Human Environment, which was held in Stockholm, Sweden, and was the world's

The Egyptian biologist Mostafa Tolba is known as "the father of the Montreal Protocol" for his work spearheading and negotiating the agreement.

first attempt to reach comprehensive agreement on how to address the impact of human actions on the environment. At the conference, leaders of what are known as "developing countries," such as Indian prime minister Indira Gandhi, asserted that addressing environmental problems must be carried out together with poverty alleviation. Tolba, for his part, argued that the global community needed to support "development without destruction," a vision that would later come to be known as sustainable development.

In the 1980s, Tolba got his chance to act on that vision. Although he was the author of several books and nearly 100 scientific articles, Tolba is now best remembered as "the father of the Montreal Protocol." The Montreal Protocol on Substances That Deplete the Ozone Layer is a 1987 international agreement phasing out use of the chlorofluorocarbon (CFC) refrigerants that were depleting Earth's protective ozone layer. Tolba devised a framework by which all countries agreed on plans to curtail use of CFCs, but developing countries would be granted additional support in this transition in the form of funding along with technological knowledge, training, and equipment. Tolba also recognized the important role that private industry would play in developing these alternatives.

The Montreal Protocol is commonly seen as the most successful international agreement on the environment, and Tolba's model of balancing the needs of richer and poorer nations was followed in future agreements, including the 2016 Kigali Amendment to the Montreal Protocol, which will phase out hydrofluorocarbon (HFC) refrigerants that contribute to global warming. Tolba's leadership highlights the role that scientists can play not only as researchers in their respective fields but also as diplomats who can forge agreements across a variety of interests. Addressing our most difficult sustainability challenges requires both skill sets.

❓ What Would You Do?

Imagine that you are the environmental minister for a developing country that is a signatory to the Kigali Amendment. After your country receives some initial support from more developed countries, would you recommend that your country try to phase out HFCs at a faster pace to improve environmental conditions? Or would you continue to phase out HFCs at a slower pace so that your country can receive support for a longer period of time? Which do you think would be better for your country?

1.7 What Can I Do?

In this book, you are invited to think about the future you want and consider the actions that you can take both individually and with others. A thoughtful understanding of how we want the future on Earth to be for ourselves, others, and our descendants requires a grounding in environmental science and an understanding of the many social dynamics that shape our individual and collective decisions and actions. In several chapters, you will learn how to assess your impact (or "footprint") or that of a local group in a specific environmental area. But even armed with this knowledge, it can be difficult to understand the actions to take. Especially in a world in which the actions of groups such as businesses and governments have so much impact on the environment, where can one person start? So each chapter will end with ideas, suggestions, or examples to show what you can do as an individual or as part of a larger group.

Consider Sustainability When You Make Purchases

For example, we began this chapter by considering the environmental effects associated with the apparel industry. The choices we make as consumers when we buy clothing and shoes can play a role in shaping these effects. Some companies, such as TOMS shoes, have missions that directly allocate company resources toward sustainability. In this case, the company not only strives to reduce the adverse environmental effects of its production processes but also invests one-third of its profits to help communities in need. This involves not only providing shoes in these communities but also establishing partnerships to provide jobs, clean water, medical care, and other social services.

In addition to purchasing decisions, you can exercise your power as a consumer by questioning retailers about

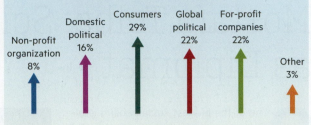

WHO WILL DRIVE SUSTAINABILITY?

Non-profit organization 8%
Domestic political 16%
Consumers 29%
Global political 22%
For-profit companies 22%
Other 3%

Consumers believe their actions are most likely to drive changes, followed by the actions of for-profit companies and international groups.

Adapted from Simon-Kucher & Partners (2021).

the impacts of the products they sell. Consumer pressure for product details can encourage more careful accounting of environmental impacts and affect decision making in the production process. Some apparel companies, such as Patagonia, now provide a detailed footprint analysis for each of the products they sell. By following this company's "footprint chronicles," you can see where the factories, farms, and mills are located that produce each product and track environmental and social effects. Patagonia also encourages waste reduction by buying back used clothing from its customers. It then fixes, cleans, and resells these products through its Worn Wear store. Of course, most towns throughout the United States also have consignment shops and other outlets for used clothing where you can sell or donate your clothes and add some vintage apparel to your wardrobe. Online consignment tools like Poshmark and ThredUp also facilitate clothing reuse. Many large companies of all types, including Walmart, Apple, and General Electric, now issue annual sustainability reports. If you visit their websites, you can assess how they define sustainability, the goals they are setting, and their strategies to meet these goals, as well as their progress.

Research and Participate in Sustainability Groups, Offices, or Plans on Your Campus

While we interact with companies as individual consumers, in our daily lives we are also members of many groups and organizations that can play a role in making sustainable choices. For example, as a student you are a member of a campus community, and most colleges and universities have student-led environmental groups as well as campus committees dedicated to sustainability. See which groups and committees are working on your campus and how you can get involved in setting sustainability goals and working toward them. If you want to see how your campus compares to others or get ideas from what other schools are doing, you can check out the Association for the Advancement of Sustainability in Higher Education (AASHE) website to see case studies, comparative sustainability ratings, and ideas for action. Throughout this book, individual choices and actions you can take to influence others around you are highlighted to encourage you to look for ways that you can make a difference.

Register and Exercise Your Right to Vote in Elections

To recognize our full potential to exert leverage over decision making, we need to think of ourselves as more than just consumers making purchasing decisions at the end of a process. The most powerful decisions upstream of us are policies—decisions made by authorities designed to influence the actions of government, companies, institutions, and individuals. As voters, we have the ability to influence policies by supporting candidates and ballot measures whose goals align with ours. When you are voting, you are not merely a customer; you are a boss making a hiring decision for a very influential position. Although voting might seem like a simple action to take, only half of US adults ages 18–29 cast a ballot in the 2020 general election—the highest-turnout election in over a century. Casting a ballot is not necessarily easy, and it requires planning. State and local voting policies and processes differ widely, so getting registered in the right place and understanding how the voting process works in your location is the first step. A website like FairElectionsCenter.org that connects you

to registration and voting details by location is a good place to start. Once you're registered, make sure you check the election dates and contact your elected officials about issues that matter to you to make your voice heard.

Use What You Learn to Start Making Sustainable Decisions Every Day

Throughout this book, you will be provided with skills, tools, and examples to help you make sustainable decisions. An important step, and a goal of this text, is for you to start using them.

Chapter 1 **Review**

SUMMARY

- Earth is a system—a set of living and nonliving things that interact to sustain life.
- Human actions have environmental effects as we draw on Earth's natural systems to make a living.
- Sustainability is the management of natural resources in ways that do not diminish or degrade Earth's ability to provide them in the future. Acting sustainably requires establishing what we aim to sustain and considering and acting on the factors that will affect our ability to achieve that goal.
- Sustainable development is that which meets the needs of the present without compromising the ability of future generations to meet their own needs.
- Ecological resistance and resilience are, respectively, the ability of an environment to tolerate disturbance and its ability to recover from damage.
- Environmental justice is the principle that no community should bear more environmental burdens or enjoy fewer environmental benefits than others.

- Science is a process of inquiry about the natural world that uses the scientific method to test hypotheses and sort through alternative explanations by making observations and gathering evidence from the natural world.
- Models are designed to assemble observations into simulations of what actually happens in the world, so that we can better understand the interactions among different factors.
- Science must be reproducible and open to revision to guard against fraud, pseudoscience, bias, misinformation, and disinformation.
- Scientific inquiry cannot aid determination of what we ought to do about a problem without the assertion of values.
- A thoughtful consideration of our future on this planet requires a grounding in environmental science and an understanding of the factors influencing our decisions as individuals and as members of larger groups and organizations.

KEY TERMS

3Es (environment, economy, and equity)

Anthropocene

bias

climate change

communication strategy

controlled experiment

disinformation

ecological resilience

ecological resistance

ecosystem

ecosystem services

emergent properties

environment

environmental justice

footprint analysis

fraud

hypothesis

incentive

misinformation

model

peer review

pseudoscience

science

scientific method

sustainability

sustainable development

system

trade-offs

values

REVIEW QUESTIONS

The letters following each Review Question refer to the Chapter Objectives.

1. What's the difference between a system and a collection of things that do not function as a system? **(A)**

2. Give two examples of ecosystem services. **(A)**

3. Why are some scientists using the term *Anthropocene* to refer to the current period of Earth's history? **(A)**

4. Give two examples of human activities that have environmental effects. **(A)**

5. What are the 3Es of concern for sustainable development? **(B)**

6. What is the principle of environmental justice, and what are two examples of disproportionate exposures to environmental harms across societal groups? **(B)**

7. Identify the steps of the scientific method. **(C)**

8. Name two challenges to scientific thinking that can be met by ensuring that scientific inquiry is reproducible and open to revision. **(D)**

9. Give two examples of values that influence human decisions and of actions that affect the environment. **(E)**

FOR FURTHER THOUGHT

The letters following each item refer to the Chapter Objectives.

10. In this chapter, we defined sustainability and stated that acting sustainably often requires us to think ahead to a specific goal (what we aim to sustain) and consider the factors that will affect our ability to achieve this goal. Earlier in the chapter, an example of this type of planning used a series of questions related to sustainable forest management. List a similar series of questions and how you might answer them for something you would like to sustain. **(B)**

11. In this chapter, we identified several problems of environmental justice and highlighted recent policies in states like New Jersey that require environmental justice analysis in the permitting of new projects. We also gave examples of some tools used for environmental justice analysis and decision making. If you were responsible for the permitting of a new highway with several possible routes, how would you ensure that the decision-making process incorporated the principle of environmental justice? **(B)**

12. Find a dubious claim made on social media or another outlet that could be subjected to scientific inquiry. Identify whether the claim may be the result of fraud, pseudoscience, bias, misinformation, disinformation, or a combination of these factors, and describe the steps by which scientific inquiry could be used to test the claim. **(D)**

13. Choose and describe one individual decision or action that has a significant environmental impact. Identify and describe two factors that you think influence this decision or action and one step that could be taken to change the impact associated with this individual behavior. **(E)**

Make Your Case?

The following exercises use real-world data and news sources. Check your understanding of the material and then practice crafting well-supported responses.

Use the News

The following article is a University of Cambridge news release. It describes social scientific research on a game that might help people get better at spotting misinformation. Use this article to answer the questions that follow. The first three questions test your understanding of the article. Question 4 is short answer, requiring you to apply what you have learned in this chapter and cite information in the article. Answers to Questions 1–3 are provided at the back of the book. Question 5 asks you to make your case and defend it.

"Cambridge Game 'pre-bunks' coronavirus conspiracies," Fred Lewsey, University of Cambridge, 2021.

A new online game that puts players in the shoes of a purveyor of fake pandemic news is the latest tactic in efforts to tackle the deluge of coronavirus misinformation costing lives across the world.

The Go Viral! game has been developed by the University of Cambridge's Social Decision-Making Lab in collaboration with the UK Cabinet Office and media collective DROG.

It builds on research from Cambridge psychologists that found by giving people a taste of the techniques used to spread fake news on social media, it increases their ability to identify and disregard misinformation in the future.

Go Viral! is launched on the heels of a new study from the team behind it, just published in the *Journal of Experimental Psychology: Applied*. The latest findings show that a single play of a similar game can reduce susceptibility to false information for at least three months.

"Fake news can travel faster and lodge itself deeper than the truth," said Dr Sander van der Linden, who leads the project and the Social Decision-Making Lab at Cambridge.

"Fact-checking is vital, but it comes too late and lies have already spread like the virus. We are aiming to pre-emptively debunk, or pre-bunk, misinformation by exposing people to a mild dose of the methods used to disseminate fake news. It's what social psychologists call 'inoculation theory'."

The new 5–7 minute game introduces players to the basics of online manipulation in the era of coronavirus. It acts as a simple guide to common techniques: using emotionally charged language to stoke outrage and fear, deploying fake experts to sow doubt, and mining conspiracies for social media Likes.

"By using a simulated environment to show people how misinformation is produced, we can demystify it," said Dr Jon Roozenbeek, co-developer of Go Viral! and researcher at Cambridge's Department of Psychology. "The game empowers people with the tools they need to discern fact from fiction."

Go Viral! is based on a pre-COVID iteration, Bad News, which has been played over a million times since its 2018 launch. Cambridge researchers developed and tested Bad News, and found that just one play reduced perceived reliability of fake news by an average of 21% compared to a control group.

The research team, including DROG and designers Gusmanson (who also worked on Go Viral!), argue that this neutralising effect can contribute to a societal resistance to fake news when played by many thousands of people.

These initial results were confirmed in an even more rigorous replication study published in January this year. "Our pre-bunk game not only improved people's ability to spot fake news but also their confidence in judging what is true or false," said Melisa Basol, a Cambridge Gates Scholar who led the study.

"This confidence boost only occurred for those who got better at accurately identifying misinformation. By exposing people to the tactics behind fake news we can help create a general 'inoculation', rather than trying to counter each specific falsehood."

Intervention effects in social psychology often dissipate within days. However, the team's latest findings show that - when paired with added testing - the 'inoculation' of a single Bad News play lasts at least three months (the time limit of the study).

"We were very encouraged by the new results on longevity," said Rakoen Maertens, lead author and Cambridge PhD candidate. "In a society with ever-changing manipulation threats, the unique approach of interventions such as Bad News and Go Viral! can offer long-lasting effects not found when using a simple fact-check."

Go Viral! is a leaner, COVID-focused experience. The team used research on the current surge in coronavirus conspiracies - called an 'infodemic' by the WHO - to hone the game, creating a more direct version that is faster to complete and easier to adapt for different languages and cultures.

Recent research suggests that close to six thousand people around the world were hospitalised in just the first three months of this year due to coronavirus misinformation, with many dying after consuming cleaning products.

The game exposes the most pervasive 'infodemic' tactics. Players find out how real news gets discredited by exploiting fake doctors and remedies, and how false rumours such as the notorious 5G conspiracy get promoted.

It also touches on how out-of-context video gets used to add credibility to fake news - and ends with conspiracies slipping beyond your control, even seeping into the mainstream. Players are provided with a sharable score and connected to the WHO's COVID 'mythbusters'.

1. Per the article, which response correctly describes what "pre-bunking" is?

   a. Checking all the facts in a news article before it is published

   b. Exposing people to the methods that individuals use to produce fake news

   c. Researching a topic to gain some understanding of it before going to a news site

   d. Exposing people to the poor-quality sources in an article before they read it

2. What is the purpose of the *Go Viral!* game developed at the University of Cambridge?

   a. To teach people how others use different techniques to gain traction with fake news stories

   b. To help students understand the way that viral news stories spread from one news agency to another

   c. To teach the science behind how viruses spread across different communities and populations

   d. To show students how big data is often used to push trends across news sites

3. What were the findings of the study involving *Go Viral!* described in the article?

   a. People better understood the techniques that political parties use to influence them.

   b. People became better at influencing others and selling more of the products they were asked to market.

   c. People better understood how viruses spread across community boundaries.

   d. People who played this game became better at identifying fake news than those using other techniques.

4. By playing *Go Viral!* students become better at recognizing the techniques used to create influential fake news articles. In doing so, students also see how they can use these techniques to possibly create their own fake material to influence articles. Explain briefly if you think the potential benefit is worth the risk.

5. **Make Your Case** Imagine by playing *Go Viral!* you were able to identify a fake news item that came across a social media feed. While you know what it is saying is fake, you agree with the broad point or the cause it is trying to promote. Since you feel strongly about this issue, do you think it is OK to forward the link to your friends to help influence them? Make your case for or against forwarding the information. Use what you learned in this article, in this chapter, or through information from outside sources to help back up your claim.

Use the Data

Asthma is a condition that is a leading cause of childhood death in the United States, and it has been linked to air pollution. Because exhaust from cars is a major cause of air pollution in urban areas, public health researchers have long wondered what effect changes to urban transportation systems might have on pollution levels and asthma. In 1996, there was a 17-day window of time when Atlanta limited and thus decreased automobile use in the city as it hosted the Summer Olympics. Researchers at the Centers for Disease Control and Prevention compared the reported frequency of asthma attacks during this period to the reported frequencies during baseline (typical) periods of 4 weeks before and 4 weeks after the Summer Olympics period. The data in the table and the graph shown here are from this research (M. S. Friedman et al., "Impact of Changes in Transportation and Commuting Behaviors during the 1996 Summer Olympic Games in Atlanta on Air Quality and Childhood Asthma," *JAMA* 285 [2001], 897–905). Part (a) summarizes the percent change in the mean number of insurance claims submitted to HMOs and to Medicaid for asthma–related events and for non-asthma events each day during the 1996 Summer Olympics compared to the mean number of those each day during the baseline periods. Part (b) graphs the change in different air pollutants before, during, and after the Olympics.

Study the table and graph, and use them to answer the questions that follow. The first three questions test your understanding of the table and graph. Questions 4 and 5 are short answer, requiring you to apply what you have learned in this chapter. Answers to Questions 1–3 are provided at the back of the book. Question 6 asks you to make your case and defend it.

(a)

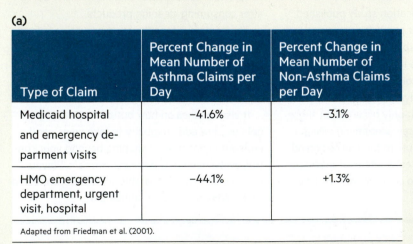

Type of Claim	Percent Change in Mean Number of Asthma Claims per Day	Percent Change in Mean Number of Non-Asthma Claims per Day
Medicaid hospital and emergency department visits	−41.6%	−3.1%
HMO emergency department, urgent visit, hospital	−44.1%	+1.3%

Adapted from Friedman et al. (2001).

(b)

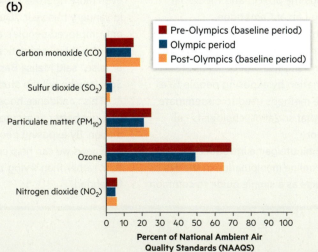

1. Which air pollutant shows the greatest reduction during the Olympic period?
   a. PM$_{10}$
   b. SO$_2$
   c. NO$_2$
   d. Ozone

2. Which statement most accurately describes the findings on asthma events?
   a. Asthma events occurred more frequently during the Olympic period than during the baseline periods.
   b. Between the Olympic period and the baseline periods, the percent change in the mean number of HMO asthma claims per day was about the same as the percent change in the mean number of HMO non–asthma claims per day.
   c. Between the Olympic period and the baseline periods, the percent change in the mean number of HMO asthma claims per day was less than the percent change in the mean number of Medicaid asthma claims per day.
   d. Between the Olympic period and the baseline periods, the percent change in the mean number of all types of asthma claims per day was much greater than the percent change in the mean number of all types of non–asthma claims per day.

3. On the basis of the data, which statement is most accurate?
   a. CO levels decreased during the Olympic period compared to that during either baseline period.
   b. CO levels increased during the Olympic period compared to that during either baseline period.
   c. CO levels were unchanged during the Olympic period compared to that during either baseline period.
   d. Data on CO levels are not reported.

4. Considering the material presented in this chapter about observations and experiments, what type of research design did these public health officials use?

5. What are two conclusions you can draw from the data in the table and the graph? In making your conclusions, cite specific data to support your claim.

6. **Make Your Case** Vehicle counts during rush-hour traffic on weekdays demonstrated a 22.5% decrease in number of vehicles during the Olympic period compared to that during either baseline period. If you were an elected official in Atlanta serving on a transportation planning committee, what sort of policy recommendations would you make? Use the data in the table and the graph to raise questions and/or make your points. How might personal values (in addition to the results of this scientific inquiry) factor into your decision?

LEARN MORE

- UC Berkeley Understanding Science 101: undsci.berkeley.edu
- UN Sustainable Development Goals—17 Goals to Transform Our World: www.un.org/sustainabledevelopment/
- Video: A Year in the Life of Earth's CO$_2$: digital.wwnorton.com/environsci
- Anthropocene website from the Stockholm Center for Resilience: www.anthropocene.info/anthropocene-timeline.php
- Global Footprint Network: www.footprintnetwork.org/
- Environmental Justice at the EPA: www.epa.gov/environmentaljustice

2

Ethics, Economics, and Policy

Who or What Do We Value?

Imagine that your car is slowly leaking oil. You know this because you can see oil stains on the driveway where you park, and when it rains, you see a rainbow-colored sheen of oil in the puddles. You can see that the oil is flowing with the water into the storm drain, which will ultimately empty into a nearby river and bay. The cumulative effect of small leaks like this from millions of vehicles is a major source of water pollution. Do you feel any moral obligation to fix your car in order to prevent the oil pollution? How much would it cost to fix the leak? Do you ignore the problem and just make sure to add oil frequently enough to keep your engine lubricated? How do you respond?

On a far grander scale, in April 2010 Americans followed the news of an explosion on the *Deepwater Horizon* drilling rig in the Gulf of Mexico, which killed 11 workers and led to a spill of more than 180 million gallons of oil into the Gulf over the next five months. Oil from the spill polluted the Gulf and reached the shores of all five US Gulf Coast states, killing fish, crabs, and birds and causing staggering economic losses to the region (**FIGURE 2.1**). In the wake of the spill, the fishing industry lost an estimated $2.5 billion, and the economic impact of lost tourism on the Gulf Coast was estimated at $23 billion. The US government placed a moratorium on offshore oil drilling in the Gulf during the spill response, affecting more than 300,000 people employed directly or indirectly by this industry. While many people saw their livelihoods threatened, tens of thousands volunteered in the response—to help rescue animals covered in oil, clean up beaches, and provide support for people affected by the spill. Four years after the spill, a federal judge found British Petroleum, the owner of the drilling rig, guilty of gross negligence and willful misconduct. The company had failed to carry out safety tests before the accident and had intentionally underestimated the flow rate from the well in the days following the spill. These seem like actions that could and should have been avoided: Why weren't they?

When you see oil running down a drain, have you ever thought about where it might end up? All these small leaks are a major source of water pollution.

Chapter Objectives

By the end of this chapter, you should be able to . . .

A. define ethics, and describe different ways that people apply its principles.

B. understand what the terms *supply* and *demand* mean in economics.

C. describe the relationship between efficiency and consumption.

D. understand how markets often fail to account for all costs and benefits associated with a good or service, leading to environmental harm.

E. describe how policies influence our behavior and affect our environmental impact, and name some important environmental policies in the United States.

F. name some daily decisions you make that are affected by ethics, economics, and policies.

> **All ethics so far evolved rest upon a single premise: that the individual is a member of a community of interdependent parts.**
>
> —Aldo Leopold

(a)

FIGURE 2.1 **Deepwater Horizon Oil Spill** (a) The extent of the 2010 *Deepwater Horizon* oil spill. (b) Fire engulfs the *Deepwater Horizon* drilling rig.

Houston • Baton Rouge • New Orleans • Mobile • Pensacola • Panama City

Deepwater Horizon accident site

Area affected by *Deepwater Horizon* spill

Gulf of Mexico

(b)

When we consider environmental challenges like this one, we might be curious not only about what is happening but also about who is affected, who is to blame, what ought to be done about it—or what should have been done to avoid it in the first place—and how we can help. Answering such questions can prove complicated because the answers involve ethics, economics, and policy. Deciding whether to help in a troubling situation and judging who is responsible and what ought to be done are influenced by our notions of ethics: of what is right, wrong, good, bad, just, or unjust. An understanding of our economic system helps us to trace human effects on the environment, both good and bad. And an understanding of how our government works (or doesn't work) reveals how political institutions, elections, parties, public involvement, and other factors influence policy. In this chapter, we will introduce and briefly explore the topics of ethics, economics, and policy, as they often influence the choices we make—such as fixing that oil leak in your car.

2.1 How Does Ethics Influence Our Decisions?

To determine whether a particular action is right or wrong, good or bad, our personal **ethics**—a set of moral principles—provide guidelines for our behavior. We make decisions on the basis of ethical judgments every day by combining our personal beliefs and values to guide our own actions, sometimes with powerful effects. For example, in 2010 the Food and Drug Administration (FDA) issued a report that identified possible hazards to fetuses, infants, and young children from an additive to plastic products known as bisphenol A (BPA). As consumers learned of this potential harm, many shifted toward the use of glass, metal, or BPA-free plastic baby bottles and sippy cups. In doing so, most parents exercised the almost instinctive ethical choice of protecting the well-being of their children and let this choice drive their purchasing decisions. Individual consumer values for health and safety coalesced into public campaigns to ban the use of BPA. Several states passed legislation to this effect, and the FDA ultimately passed a regulation banning the use of BPA in baby bottles and sippy cups (**FIGURE 2.2**).

When you make decisions—like whether to fix your leaky car, how to dispose of your trash, or what to eat—you may apply a similar thought process. Formal ethical principles of right and wrong can apply to many environmental decisions extending beyond the well-being of ourselves and those close to us. Let's take a closer look at two common systems of ethics: one focuses on duties and rights, and the other focuses on the consequences of the actions we take.

Duties and Rights

Consider a parent and child relationship, where a general moral and legal understanding is that parents have a *duty* to care for their children and the children have a *right* to an adequate standard of care. This approach is known as **deontological ethics** (from the Greek word *deontos* for "duty"). It assesses whether actions are right or wrong by establishing general rules in the form of *duties* that ought to be upheld and *rights* that ought to be protected. We can see this approach in laws and agreements. For example, the National Environmental Policy Act of 1970 established a duty for the federal government to "assure for all Americans safe, healthful, productive, and aesthetically and culturally pleasing surroundings" (**FIGURE 2.3**). Rights and duties often emphasize the **intrinsic value**, or worth, of whatever is deemed worthy of protection. If something has intrinsic value, it is valuable in and of itself apart from its usefulness or value to others. For example, ascribing intrinsic value to salmon would be holding that salmon are valuable in their own right, apart from values that humans place on them for food, recreation (through fishing), or cultural practices. The United Nations (UN) initiated the **Earth Charter**, a document completed in 2000 that outlines environmental duties for the people of Earth, including the protection and restoration of ecological systems. The Earth Charter declares that "every form of life has value regardless of its worth to human beings" (**FIGURE 2.4**).

ethics a set of moral principles that provides guidelines for our behavior.

deontological ethics an approach to ethics that establishes duties that ought to be upheld and rights that ought to be protected.

intrinsic value the value of something in and of itself and apart from its usefulness to others.

Earth Charter a United Nations initiative completed in 2000 that outlines duties for the people of Earth, including protecting and restoring Earth's ecological systems.

FIGURE 2.2 BPA Free Many parents, out of a sense of duty to protect their children's health, began choosing baby bottles and sippy cups made without the chemical bisphenol A (BPA). In 2012, the FDA banned the use of BPA in these products.

FIGURE 2.3 Signing the National Environmental Policy Act In 1970, President Richard Nixon signed the National Environmental Policy Act. This law acknowledged that the government's actions could affect the environment and that these effects needed to be assessed (through a public process) before decisions were made.

United Nations Earth Charter

I. Respect and Care for the Community of Life

1. Respect Earth and life in all its diversity. Recognize that all beings are interdependent and every form of life has value regardless of its worth to human beings.

2. Care for the community of life with understanding, compassion, and love.

3. Build democratic societies that are just, participatory, sustainable, and peaceful.

4. Secure Earth's bounty and beauty for present and future generations.

II. Ecological Integrity

5. Protect and restore the integrity of Earth's ecological systems, with special concern for biological diversity and the natural processes that sustain life.

6. Prevent harm as the best method of environmental protection and, when knowledge is limited, apply a precautionary approach.

7. Adopt patterns of production, consumption, and reproduction that safeguard Earth's regenerative capacities, human rights, and community well-being.

8. Advance the study of ecological sustainability and promote the open exchange and wide application of the knowledge acquired.

III. Social and Economic Justice

9. Eradicate poverty as an ethical, social, and environmental imperative.

10. Ensure that economic activities and institutions at all levels promote human development in an equitable and sustainable manner.

11. Affirm gender equality and equity as prerequisites to sustainable development and ensure universal access to education, health care, and economic opportunity.

12. Uphold the right of all, without discrimination, to a natural and social environment supportive of human dignity, bodily health, and spiritual well-being, with special attention to the rights of indigenous peoples and minorities.

IV. Democracy, Nonviolence, and Peace

13. Strengthen democratic institutions at all levels, and provide transparency and accountability in governance, inclusive participation in decision making, and access to justice.

14. Integrate into formal education and lifelong learning the knowledge, values, and skills needed for a sustainable way of life.

15. Treat all living beings with respect and consideration.

16. Promote a culture of tolerance, nonviolence, and peace.

FIGURE 2.4 Earth Charter In 2000, the UN Earth Charter applied deontological ethics to establish duties for protection of the environment according to the four pillars and 16 principles outlined here.

Adapted from the Earth Charter Initiative (n.d.).

The Greatest Good for the Greatest Number

Imagine having to decide whether to build a large dam to provide clean hydroelectric power when doing so would flood an area that is home to several communities. How might you make this decision? You might use an approach known as **utilitarianism**, which defines what is right in a given situation by determining what actions would bring as much good (or as little harm) as possible. Utilitarianism is often expressed in terms of providing the greatest good for the greatest number and is informed by two important ideas. First, the rightness or wrongness of an act is determined by the balance of good and bad consequences that result from it. In other words, the end is given more moral weight than the means in decision making. Second, utilitarianism typically assesses the consequences by attempting to calculate the total benefits and the total harms experienced by all those affected. When comparing alternative actions, the one that produces the greatest net benefit is the one that should be pursued. So, using this approach for the dam, you might assess whether the broad benefits of clean power outweigh the hardship caused to the adversely affected communities and environments. Calculations like these are not always simple or straightforward. Conflicts often arise over different interpretations of who is affected by an action and what counts as a good or bad consequence. Further, we will see in the next subsection that both deontological ethics and utilitarianism can be applied in assessing impact on future generations and on nonhumans and nonliving things as well. The National Environmental Policy Act (discussed earlier) directs the government to "attain the widest range of beneficial uses of the environment without degradation, risk to health or safety, or other undesirable and unintended consequences." In other words, the government should balance overall benefits and harms to consider which actions it should take.

Who or What Do We Care About?

When we apply ethics, we are considering how we ought to behave in relation to others. But who are the "others" we consider? In the earlier example of health and safety concerns about the use of BPA in baby bottles and sippy cups, the ethical concern is clearly focused on children. But there are many different ways we can draw ethical boundaries to identify the "others" about whom we are concerned. One approach is to draw ethical boundaries only around the interests of humans. This is known as **anthropocentrism** (**FIGURE 2.5**). When anthropocentrism is used as an ethical approach, it holds that there are no moral aspects to environmental decisions apart from how actions that affect the environment will affect people. The

FIGURE 2.5 How Far Do Our Ethical Concerns Extend? Our sphere of concern determines whose interests "count" when we make ethical judgments. Anthropocentric ethics extends to human concerns, biocentric ethics extends to all living things, and ecocentric ethics extends to the living and nonliving aspects of the environment that enable ecosystems to function in a natural state.

smallest sphere of concern would be drawn around those closest to us, such as family and friends. In this case, our food ethics might center on how our choices affect these people when we prepare, provide, or otherwise share food with them. This is often the way in which efforts to address human health concerns linked to diet are targeted. Consumers are warned of obesity, diabetes, heart disease, and other effects that unhealthy food choices could have on themselves and those closest to them. A significantly larger sphere of concern could include the health of communities—for example, our city or town, our college or university, or, at the largest scale, all of humanity—as when we focus on global problems such as world hunger or access to clean drinking water. Anthropocentrism could also be extended more broadly to include future generations of humans. This is at the heart of the UN definition of sustainability, now the cornerstone of the 17 sustainable development goals the international community is striving to achieve by 2030 (see Chapter 1, Figure 1.10).

For most people, what we truly care about does extend beyond humans—to pets, for example, or to the land or neighborhoods we call home—though often these things are still connected to our anthropocentric interests. But others take broader positions. Advocates of animal rights might assert that sentient

utilitarianism an approach to ethics that defines what is right by determining what actions would bring as much good (or as little harm) as possible.

anthropocentrism an approach to ethics where moral concerns are focused on the interests of humans.

biocentrism an approach to ethics where the interests of all living things are considered.

ecocentrism an approach to ethics concerned with all the living and nonliving components of ecosystems.

economic system a chain of exchange that helps shape the production, distribution, and consumption of things we use.

beings (such as pigs and chickens) are deserving of protection because they have the capacity to experience pain. Others may extend the boundaries to include all living things (regardless of whether they experience pain), an approach known as **biocentrism**. This approach holds that we have a duty to protect all species of life and that we ought not to take actions that jeopardize species survival because each life-form has intrinsic value. Biocentrism is at the heart of many arguments to protect endangered species and can include the entire living world, part of which is plants (see Figure 2.5). Finally, **ecocentrism** goes even further, drawing boundaries around the living and non-living components of ecosystems. Wildlife biologist Aldo Leopold popularized this idea in the 1940s by proposing a "land ethic" that would extend our sphere of concern to include soils and water bodies as well as plants and animals. He was not arguing against human management of these aspects of the world but for their right to continued existence in a natural state (**FIGURE 2.6**).

FIGURE 2.6 Aldo Leopold In 1935, wildlife biologist Aldo Leopold and his family began an ecological restoration project on an old farm along the Wisconsin River. Leopold documented his observations of ecological recovery in *A Sand County Almanac*. In this book, he argued for a "land ethic," an ecocentric ethics that would extend not only to other forms of life but to functioning ecosystems as well.

⬡ **TAKE-HOME MESSAGE** Ethics provides guidelines for how we ought to act on our notions of right and wrong. Deontological approaches focus on duties and rights as general principles. Utilitarian approaches focus on the good or bad consequences for all those affected. Anthropocentrism limits concern to the interests of humans. Other approaches stretch beyond human interests to include other sentient beings, all living beings, or ecosystems.

2.2 How Are the Environment and Economy Connected?

We all consume stuff. All of this stuff comes from somewhere and has an environmental impact. And if you think about how you got a particular item, chances are you will realize that you engaged in some sort of exchange or transaction with someone else to get it. Now consider this common system of exchange: You perform some work for which you are paid, and you use the money to buy something you need or want. Your payment goes to a variety of sources that helped create what you bought—whether by cultivating, making, processing, distributing, or finally selling it—and also to investors in the product, who invested with the hope of future profits. Throughout this chain of exchange, raw materials might be extracted, cultivated, processed, and distributed. Equipment and infrastructure—such as roads, buildings, and power systems—are used, and many workers are involved in the various stages of these processes. In other words, you aren't just buying something but are engaging in an **economic system**, and your participation helps shape the production, distribution, and consumption of things we use. Choices made in economic systems can have significant effects on the environment, and while we often hear more about negative effects, positive environmental outcomes can happen too. So, in this section, we will explore the basics of how economic systems work. Because the development of energy is central to some of the most significant environmental effects, we will focus our exploration on this topic.

Supply and Demand

Most days if you read the news, you will find someone reporting the price of a barrel of oil, which in general is the cost that a buyer (such as an oil refinery) pays to a producer (the company that extracted the oil from the ground). Very broadly, if many buyers want each barrel of oil, the producer can charge more for it. The buyers bid for the barrels, and the producer takes the highest price. However, if there is a significant increase in the

Is Eating a Vegetarian Diet More Sustainable?

The answer to this question may be more complicated than you think. The raising of animals to produce animal products generally requires more land, water, and energy than does the raising of crops. Most of the animal products now consumed in the United States come from animals that are raised on grains, which require land, water, and energy of their own to grow. Vegetarians and vegans often argue that by avoiding meat—and even dairy and eggs in the case of vegans—their food consumption has a smaller environmental footprint because it leads to less soil degradation and loss, less water scarcity, and lower greenhouse gas emissions. The precise numbers vary depending on how livestock are raised, but 1 pound of feedlot-raised beef requires at least 25 pounds of grain and more than 200 gallons of water to produce; a pound of conventional pork requires about half as much grain, and a pound of chicken requires 2 pounds of grain. Most cattle are initially raised on grass and do not enter feedlots until they've already put on about half their adult weight, but some cattle remain on pastures until slaughter (grass-fed beef). So, while it takes about 25 pounds of grain to produce a pound of beef in a North American feedlot, it does not require any grain to produce a pound of 100% grass-fed beef.

The efficiency argument of those favoring a plant-based diet maintains that fewer resources—less land, water, and energy—are required for humans to obtain their calories and nutrients from eating plants than from eating animals. Add to this the additional emissions of often harmful gases from conventional livestock production, and a strong case can be made that less meat consumption is more sustainable. One UN estimate states that almost a fifth of global emissions of carbon dioxide (a gas involved in climate change) comes from livestock production. And all told, conventional meat-rich diets produce about twice the gas emissions as vegetarian and piscivorous (fish-eating) diets.

While grain-fed cattle emit (primarily by burping) somewhat less methane (another potent climate-affecting gas) than do grass-fed

Feed	**6.7** Pounds of grains and forage
Water	**52.8** Gallons for drinking water and irrigating feed crops
Land	**74.5** Square feet for grazing and growing feed crops
Fossil fuel energy	**1,036** BTU for feed production and transport. That's enough to power a typical microwave for 18 minutes.
Carbon footprint	**14.4** Pounds of CO_2 equivalents from feed production and transport, as well as emissions and manure from cattle.

What It Takes to Make a Quarter-Pound Hamburger Before the food you eat reaches your table, it has already made a significant environmental impact through the resources that are required to produce and transport it. Red meat from cattle raised with grain on conventional feedlots has a particularly big impact (as shown here) through the large amount of grain, water, land, and energy required to produce beef and the greenhouse gas emissions directly from the cattle. BTU = British thermal unit.

Adapted from Capper (2011), Barclay et al. (2012), and Pelletier et al. (2010).

cattle, feeding cattle on grains that people could consume is an issue. Some studies have concluded that about 4 billion people could be fed with the crops currently devoted to fattening livestock. This would prove more than enough to feed the expected increase in the human population over the course of the 21st century.

Still, humans cannot eat grass, and some of the world's semi-arid grasslands are not suitable for reliable crop production without additional irrigation that is unsustainable. Grass-fed cattle, however, eat something that people cannot, converting cellulose into beef and milk that people can consume. Some studies have shown that grazing cattle and their manure can remove a lot of harmful gases from the atmosphere by building up more than enough soil organic matter to offset the methane produced by livestock. And it is known that plowing up grasslands to plant grains reduces soil organic matter and sends a lot of carbon dioxide skyward. It is also necessary to consider the climate-changing gases produced by the diesel vehicles and fertilizer used to grow the conventionally raised corn that animals are fed. So grazing practices tailored to build up soil organic matter can offer a more sustainable way to produce food, especially in grasslands ill-suited to support long-term crop production.

There are strong ethical arguments that can be made in favor of a vegetarian diet because of issues such as animal rights or the cruelty of confined feeding operations. But while consuming less meat is generally more sustainable, consuming grass-fed beef is not necessarily unsustainable. So, although reducing consumption of conventionally raised feedlot beef is one of the easiest ways to reduce one's carbon footprint, the overall environmental impact of meat consumption depends on where—and how—livestock are raised.

What Would You Do?

Imagine you had a long-lost relative who gave you a working 10,000 acre ranch in the western United States with 1,000 head of cattle on it. What would you do with the ranch? Given just this information, is there additional information about the ranch you would want to know to help you make your decision?

demand in economics, a concept reflecting how much someone (the consumer) desires and is willing to pay for a good.

supply in economics, the total quantity of a good or service that is available.

market in economics, a system that brings buyers and sellers together to exchange goods.

number of barrels of oil for sale on the market or there is a decrease in interested buyers, prices will go down as producers seek to attract buyers (**FIGURE 2.7**).

This example illustrates the basic economic relationship in a market economy, where the pressures of supply and demand influence the distribution of goods under competitive conditions. **Demand** measures the amount people are willing and able to pay for a good, and **supply** measures how much of that good is available. In reality, each day the supply and the demand for oil change depending on many factors, such as the number and types of cars that people drive, the number of people who want more oil for power or are using alternative sources of energy (e.g., wind or solar), or even the predicted weather patterns. The relationship between supply and demand introduces you to the core relationship in economic systems. Next, let's turn to the **market**, the system that brings buyers and sellers together to exchange goods, and learn how this system can be affected by governments.

Government and Markets

In 1776, Scottish economist Adam Smith published *An Inquiry into the Nature and Causes of the Wealth of Nations* (commonly called *The Wealth of Nations*), in which he famously described an ideal market as an "invisible hand." Smith argued that a market worked best when it simply allowed a multitude of self-interested buyers and sellers to adjust production and prices on their own. Prices and economic factors, such as how many of an item

to produce, would naturally form from the day-to-day interactions of this "free market." While many of Smith's ideas now structure the core of modern economic thought, prices created by these markets are not necessarily to everyone's liking. For example, if hurricanes on the Gulf Coast shut down or destroy oil refineries there, the supply of gasoline and diesel produced by the refineries can dip, and fuel prices can rise rapidly. Because gasoline and diesel are still the primary fuels for personal travel and for the trucks that distribute goods, these effects can disrupt the entire economy. For this reason, the Department of Energy maintains the Strategic Petroleum Reserve: a supply of crude oil stored in Louisiana and Texas that can be released during emergencies to increase the supply (and lower the price) of fuel. The governments of many other oil-producing nations—such as Saudi Arabia and Mexico—play an even more active role in controlling the supply of oil in an attempt to influence global oil prices for their own benefit. It is not unusual for governments to intervene in markets in a variety of ways on behalf of buyers who want lower prices or sellers who want higher prices: such interventions include setting prices, regulating the provision of certain goods and services, providing financial support or subsidies for producers or industries, and collecting taxes. For example, the US government intervenes in the agricultural market by providing many types of subsidies. Nearly 40% of US farms receive some form of federal subsidy, such as direct government payments, subsidized loans, or crop insurance. These subsidies influence both

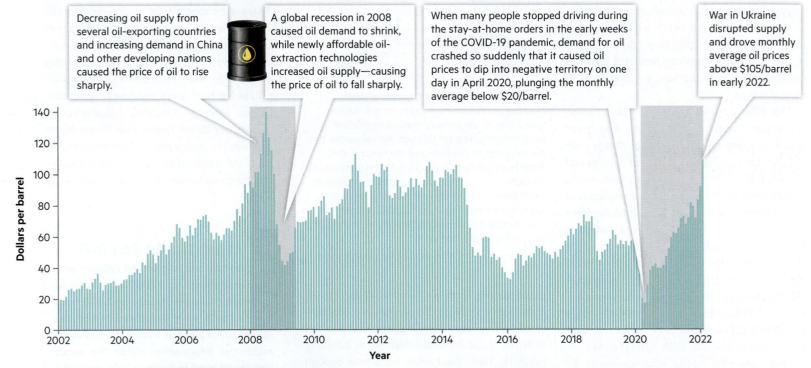

FIGURE 2.7 Oil Prices In a market, prices are shaped by supply and demand. Increasing demand and decreasing supply cause prices to rise, while decreasing demand and increasing supply cause prices to fall.

Adapted from US Energy Information Administration (n.d.).

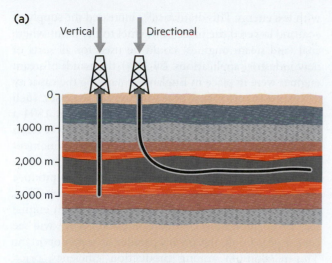

(a)

Vertical   Directional

0
1,000 m
2,000 m
3,000 m

FIGURE 2.8 **What Happens When a Good Becomes Scarce?**
When a good becomes scarce, the price increases, which
motivates innovations in the production of the good (such as
new ways to extract oil and gas). **(a)** Directional drilling greatly
expands the area of a deposit that can be opened for extraction.
A technology known as hydraulic fracturing ("fracking";
discussed in Chapter 13) has been developed to access deposits
like this. **(b)** Deepwater drilling provides access to oil that was
previously inaccessible beneath the seafloor. This depicts an
actual well drilled off the coast of Brazil. m = meters.

Adapted from (a) Marshak (2016) and (b) Stillman (2010).

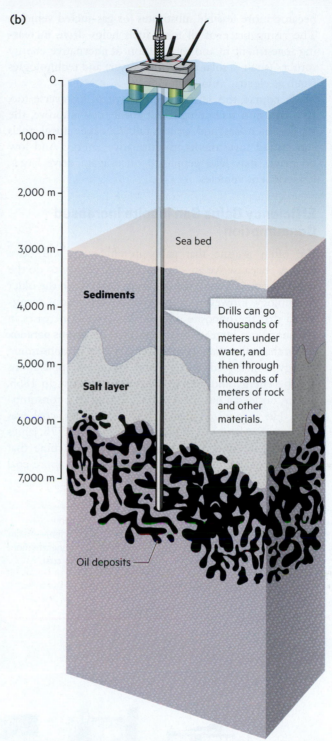

(b)

0
1,000 m
2,000 m
3,000 m — Sea bed
Sediments
4,000 m
5,000 m — **Salt layer**
6,000 m
7,000 m

Drills can go
thousands of
meters under
water, and
then through
thousands of
meters of rock
and other
materials.

Oil deposits

the cost of agricultural production and the prices that
consumers pay for food.

How, when, or whether interventions happen is normally a major source of political debate, and the level
of government intervention in markets is used to categorize national economies. Broadly, we think of *market
economies* as having relatively little government interference, while *command* or *planned economies* are more
government directed. Countries often cited for having the most market-driven economies are Australia,
Switzerland, and the United States, while command
economies include Cuba, North Korea, and Venezuela.

Scarcity Inspires Innovation and Substitution

Another basic economic relationship is that markets
will spur creative efforts to fill unmet demand. This was
observed in the first decade of the 21st century when
demand for oil increased, causing prices to rise dramatically to well over $140 a barrel as oil became more
scarce. But this scarcity and expense made the industry
try harder and harder to find oil. New exploration technologies and extraction techniques let producers obtain
huge quantities of oil from more difficult sources, such
as tar sands, oil locked in rocks in the Bakken Formation of North Dakota and Montana, and oil located
thousands of meters deep beneath the ocean floor off
the coast of Brazil. Similarly, as natural gas prices rose,
a new process called hydraulic fracturing ("fracking")

was developed to get more gas from sources once
thought to be used up (we will discuss emerging fossil fuel technologies in Chapter 13). In this case, high
prices ultimately spurred innovations: technologies that
were once too expensive became economically feasible
when the high price of oil and gas covered the costs of
extraction (**FIGURE 2.8**).

High gasoline prices also led to a greater demand for
more energy-efficient vehicles that require less money
to fill up. As oil prices rose, hybrid and electric vehicles

became more feasible substitutes for gas-fueled vehicles. The rising price of all fossil fuels helps drive increasing investment in and substitution of alternative energy sources (such as solar and wind power) and technologies (such as electric vehicles).

Of course, this relationship can work in reverse too. As extraction technologies for fossil fuels improve, the supply increases and prices fall. Cheaper fossil fuels can curtail investments in alternative energy. And low oil prices may lead people to once again drive larger, less-efficient vehicles.

Efficiency Gains Can Inspire Increased Consumption

We often assume that gains in efficiency will "save energy" because we are consuming less energy to do the same amount of work that we were doing with the older technology. However, economist William Jevons recognized that the opposite often proves true as increases in efficiency can increase consumption. The **Jevons paradox** holds that efficiency gains lower the cost of consuming energy, which leads to new and expanded applications for the energy and then greater demand for it. In 1865, Jevons noticed a historical trend affecting coal consumption. He recognized that in the mid-1700s, coal was scarce in England and its price was rising. By 1775, James Watt had developed the steam engine, a machine that enabled coal-mining operations to extract far more coal

with less energy. This dramatically increased the supply of coal and lowered the price of the fuel to the point where coal (and steam engines) could be used for all sorts of new industrial applications. By 1800, thousands of steam engines were in place in England, expanding the capacity of textile factories and other industrial processes, such as use of blast furnaces for iron production. In 1804 a high-pressure steam engine was designed to power trains and ships with coal, and in 1812 gas from coal refinement was used for streetlights and later fueled early automobiles. By 1882, steam engines were using coal to produce electricity, which is still coal's primary use today. Thus, cheap coal drove more applications of the steam engine that led to increasing coal consumption. As we will see later in this book, the Jevons paradox is very important. This relationship among production efficiency, price, innovation, and expanding consumption helps explain many environmental effects (**FIGURE 2.9**).

⬡ **TAKE-HOME MESSAGE** Economic systems shape the production, distribution, and consumption of goods and services. Markets bring together buyers and sellers, and supply and demand interactions generally determine prices. Governments sometimes intervene to adjust prices, regulate production, or distribute goods. Scarcity can inspire innovation to meet the same need. Efficiency gains that lower the price of a good often inspire new uses of that good in ways that increase consumption (the Jevons paradox).

Jevons paradox a finding that efficiency gains in the use of a resource can lower the cost of that resource, which can cause consumption of the resource to rise.

FIGURE 2.9 Jevons Paradox
British economist William Jevons found that gains in energy efficiency did not necessarily lead to less overall use of energy resources (such as coal). Efficiency gains tend to lower the price of energy, which motivates new and expanded uses for the energy resource, thereby increasing consumption.

Adapted from Jevons (1865).

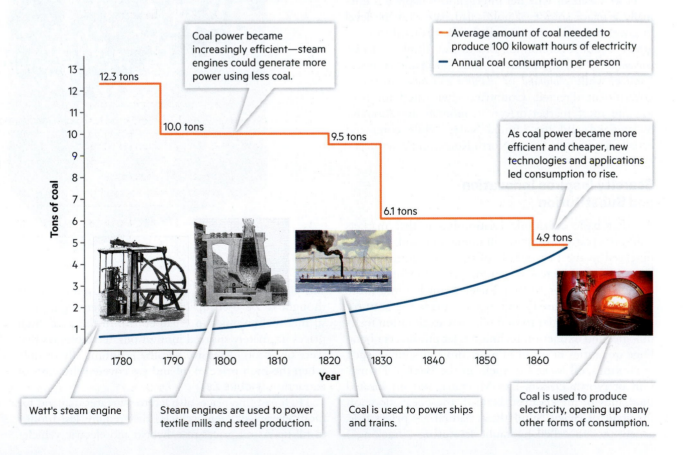

Coal power became increasingly efficient—steam engines could generate more power using less coal.

— Average amount of coal needed to produce 100 kilowatt hours of electricity
— Annual coal consumption per person

As coal power became more efficient and cheaper, new technologies and applications led consumption to rise.

12.3 tons
10.0 tons
9.5 tons
6.1 tons
4.9 tons

Tons of coal

1780 1790 1800 1810 1820 1830 1840 1850 1860
Year

Watt's steam engine

Steam engines are used to power textile mills and steel production.

Coal is used to power ships and trains.

Coal is used to produce electricity, opening up many other forms of consumption.

2.3 How Can Economics Help Us Understand Environmental Problems?

Economists recognize that markets can fail to account for all of the costs and benefits associated with goods or services. These market failures are often the result of **externalities**. To see how this works, let's first consider a simple example involving noise pollution. Imagine that you and some neighbors get together and decide to host a block party with a live band. You close off the street, and each of you chips in to pay the band. You also invite your friends and ask every guest to contribute money to help pay for the party. Everyone who attends agrees that the party is a great success—until the police arrive. It seems that some of the people in the surrounding area experienced the music as noise pollution and were angry that they had difficulty parking in front of their houses because of the congestion. The noise and congestion that angered people not present at the party were *negative externalities*: costs associated with the block party that did not fall on those directly involved in the party. However, some nearby residents enjoyed the music. Even though they did not contribute money to the party, these residents were still able to enjoy the band from the comfort of their own porches. These residents enjoyed *positive externalities*: benefits associated with the party that they did not have to pay for. Either way, some of the costs and benefits of the party were borne by those external to it.

Negative Externalities

What does this have to do with the environment? Consider the negative externalities associated with coal-fired power plants. Relative to other forms of available energy, especially in the United States, coal is inexpensive to obtain, and it is used to power a large share of US energy needs. The prices for energy that most of us pay reflect the relatively low costs associated with coal's mining, production, transportation, and burning. But these prices do not reflect other costs of coal use, including the health harms to the people who mine it, carbon emissions from burning of coal, diminished land value of abandoned mines, pollution from mining operations, and accidents caused by rail transportation. A 2011 study by the Harvard Medical School Center for Health and the Global Environment estimated that the costs of negative externalities for coal amount to as much as $500 billion per year. Because the market price does not reflect these costs, it is failing to supply coal at what economists would consider the right price. If these costs were included in the price, this study estimated the "true cost" of coal power could be 4 times higher—far higher than the current price of energy alternatives (**FIGURE 2.10**). In fact, goods associated with negative externalities are *overproduced* because the buyers and sellers of these goods aren't forced to take their full costs into consideration. The artificially low prices of

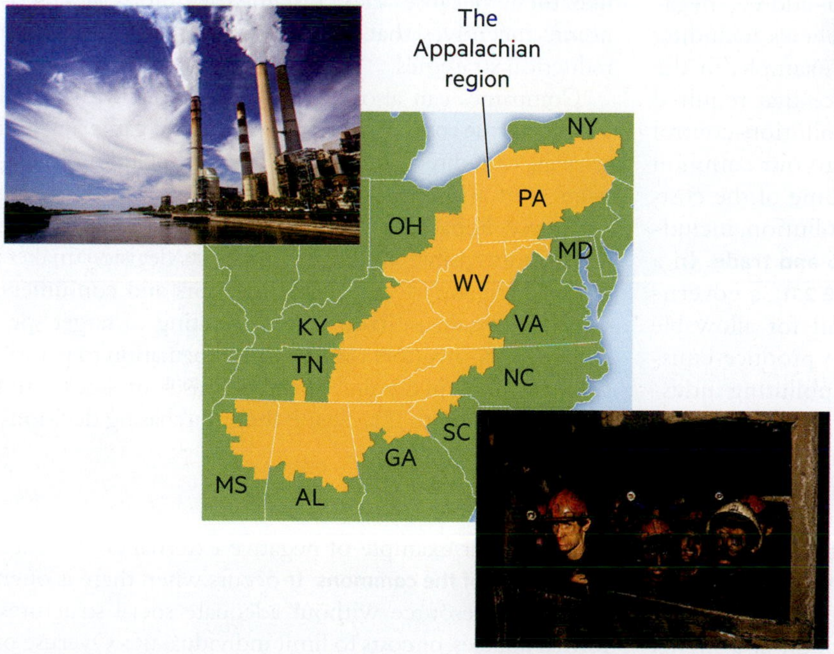

> **externality** failure of markets to account for all the costs and benefits of goods and services.

The Appalachian region

Health and environmental impacts	Costs (per year)
Land disturbance	$54 million–$3.3 billion
Methane emissions from mines	$684 million–$6.8 billion
Fatalities in the public due to coal transport	$1.8 billion
Public health burden of communities in Appalachia	$74.6 billion
Emissions of air pollutants from combustion	$65 billion–$187 billion
Excess cardiovascular disease from mercury emissions	$246 million–$17.9 billion
Climate damage from combustion emissions of CO_2 and N_2O	$20 billion–$205 billion
Climate damage from combustion emissions of black carbon	$12 million–$161 million

FIGURE 2.10 The True Cost of Coal Coal power has many costs that are not accounted for in the market exchange between coal plants and consumers. These costs, known as externalities, are felt acutely in communities where the coal is extracted. This chart shows estimates of some of these costs in the Appalachian region of the United States. Environmental damage associated with coal in this region is estimated to total $330 billion to $970 billion per year.

Adapted from Epstein et al. (2011).

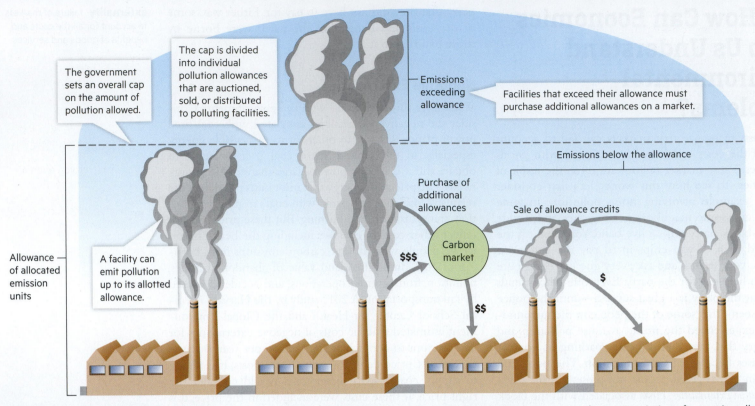

The government sets an overall cap on the amount of pollution allowed.

The cap is divided into individual pollution allowances that are auctioned, sold, or distributed to polluting facilities.

Emissions exceeding allowance

Facilities that exceed their allowance must purchase additional allowances on a market.

Emissions below the allowance

Allowance of allocated emission units

A facility can emit pollution up to its allotted allowance.

Purchase of additional allowances

Sale of allowance credits

$$$

Carbon market

$

$$

FIGURE 2.11 How Cap and Trade Works In a cap-and-trade system, the "cap" is the overall limit that the government places on emission of a certain pollutant. The total emission of the pollutant stipulated by the cap is then divided into allowances that are distributed among polluting facilities. A market is then established so that facilities that do not use their full allowance (e.g., by taking measures to reduce their pollution) can sell the balance to facilities that exceed their allowance. Over time, the government can reduce the overall cap to encourage more pollution-reduction measures.

such goods help them outcompete alternative goods that could otherwise replace them.

Governments often take action to address negative externalities, using a variety of policies including regulations, taxes, and exchanges. For example, in the 1990s, US regulations were put in place that required new coal-fired power plants to install pollution-control devices to reduce harmful emissions into our common air. Other policies tried to recognize some of the costs of pollution by creating a market for pollution, including an exchange system known as **cap and trade**. In a cap-and-trade exchange system (**FIGURE 2.11**), a government sets an overall maximum amount for allowable emissions (a cap), and then the right to produce emissions up to this cap is given to firms in polluting industries in the form of allowances. While the total number of allowances won't change, they can be bought, sold, traded, or banked for the future. If firms pollute less than their allowance, they can sell the remainder to firms that pollute more and pocket the money they make as profit. This market creates an incentive for firms to reduce pollution and allows firms that pollute more to maintain their operations, though at a higher cost. Cap-and-trade systems typically do not cover all of the externalities associated with the pollution, and they can be quite complicated to administer effectively. Yet they do provide a high degree of certainty in

the amount of overall emissions reductions over time through the establishment of the cap. Many economists like them because they provide flexibility and economic incentives that encourage innovative pollution reduction strategies.

Companies can also address externalities with a tool known as **true cost accounting**, a method of gathering and assessing the direct and indirect costs (including externalities) of a product. This tool puts the environmental and social impacts into a bookkeeping format, making them more visible and measurable for decision makers in the company, as well as for investors and consumers. Companies can use true cost accounting to target specific impacts, investors can use this information to prioritize environmentally and socially conscious stocks, and consumers can use it to shape their purchasing decisions.

Tragedy of the Commons

An important example of negative externalities is called the **tragedy of the commons**. It occurs when there is open access to a resource without adequate social structures, formal policies, or costs to limit individual use. Overuse of this common area eventually degrades it until it is no longer usable, eliminating the shared benefit provided by the resource. The tragedy of the commons can help explain many types of environmental problems, including

cap and trade a system where a government sets an overall maximum allowable emissions standard (cap) and then creates a market that enables pollution allowances to be bought, sold, traded, or saved for the future.

true cost accounting a bookkeeping method that incorporates direct and indirect costs (including externalities) throughout the life cycle of a product.

tragedy of the commons a situation in which there is open access to a resource without adequate social structures, formal policies, or costs to limit individual use, leading to the degradation of the common benefit provided by the resource.

land degradation, air and water pollution, and overfishing. But let's first consider a less serious open-access resource—a shared refrigerator in an office kitchen or college residence hall (**FIGURE 2.12**). The fridge provides a common benefit, but it can easily become a disgusting, overcrowded mess over time if users fail to clean up spills or remove spoiling food. There may be little incentive to take personal responsibility for one's own leftovers or to mop up spills. And it's not in any one user's interest to put in the time to deal with the unpleasantness of cleaning the entire fridge. Many communal refrigerators suffer from a tragedy of the commons because although each user benefits from the resource, incentives are not in place to constrain individual behavior in a way that keeps it tidy. The shared fridge may even become so full of spoiled food that it ceases to be of use to anyone.

Several solutions to the tragedy of the commons have been recognized. First, strongly enforced government regulations can constrain individuals from overexploiting or degrading a resource. For example, governments can establish marine protected areas that restrict or ban fishing in regions where overharvesting has decimated certain species. The success of such strategies hinges on effective enforcement and the penalties placed on violators. Second, in some cases—such as land use—privatizing resources could also constrain use by giving property owners a long-term interest in sustaining their private resource over time. Third, there are many instances where common ownership or shared resource use has proven sustainable. Elinor Ostrom won the Nobel Prize in Economics in 2009 for her work challenging the notion that common, shared resources will inevitably lead to the "tragedy" of degradation (**FIGURE 2.13**).

Local communities in many areas of the world have devised systems for sustainable resource use of pastures, fisheries, water, and forests without private ownership or government regulation. These communities distributed, limited, and monitored resource use through informal, face-to-face meetings and other small-scale social interactions. In many cases local shared control protected resources better over the long term than did private ownership or government control that was detached from local interests. Consider how face-to-face interactions and other informal tools could help maintain a clean refrigerator. Users could agree to clean up their own spills and empty old food, while holding each other accountable with direct conversations or reminder notes. Or maybe a chore wheel could be used to give everyone a turn cleaning the entire fridge.

Positive Externalities, Ecosystem Services, and Public Goods

In our earlier example of a block party, the people experiencing a positive externality were the neighbors who did not pay for the party but who still enjoyed listening

FIGURE 2.12 Tragedy of the Commons In 1968, biologist Garrett Hardin borrowed the tragedy of the commons concept from 19th-century British economist William Lloyd to help draw attention to the social and economic dimensions of a wide range of environmental problems. You might experience this whenever you open a shared fridge in your dorm, workplace, or apartment. Whose turn is it to clean the refrigerator?

FIGURE 2.13 Elinor Ostrom Elinor Ostrom won the Nobel Prize for Economics for her insights on tragedy of the commons and the environment.

to the music. They got a benefit they didn't have to pay for. Now consider the environmental example of a farmer who relies on pollinators—such as insects, birds, and bats—for crop production. The farmer does not pay to support these species. Instead, neighbors who have kept their land undeveloped are providing habitat for the pollinators—and in turn providing an uncompensated benefit or positive externality to the farmer. There are many positive externalities in functioning ecosystems, including clean water and air, fertile soils

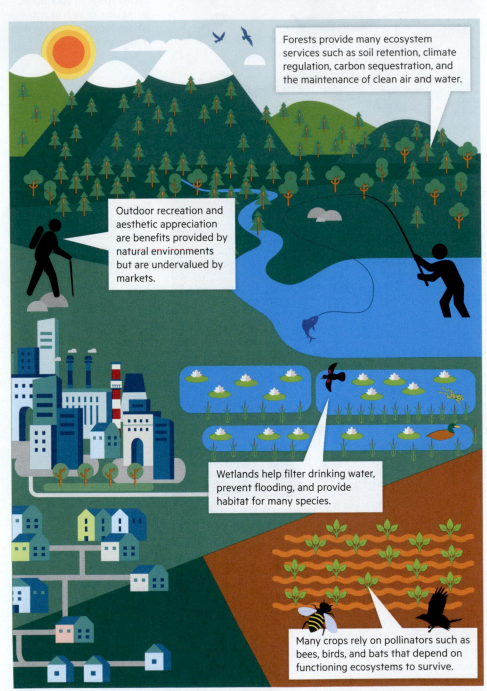

Forests provide many ecosystem services such as soil retention, climate regulation, carbon sequestration, and the maintenance of clean air and water.

Outdoor recreation and aesthetic appreciation are benefits provided by natural environments but are undervalued by markets.

Wetlands help filter drinking water, prevent flooding, and provide habitat for many species.

Many crops rely on pollinators such as bees, birds, and bats that depend on functioning ecosystems to survive.

FIGURE 2.15 Ecosystem Services Ecosystems provide many services for us that are not exchanged on a market and are therefore undervalued. Environmental economists have estimated values for many of these services in an effort to raise awareness of their importance.

for growing crops, and better opportunities for outdoor recreation and aesthetic appreciation. But economists have observed that markets normally underproduce positive externalities. This is because the would-be producers of these goods and services are not being fully compensated for all of the benefits that flow from them. In the case of the pollinators, landowners willing to preserve habitat that protects these species may be in short supply.

Positive externalities often overlap with what economists call **public goods**: things that cannot be profitably produced because it is difficult to keep nonpaying customers from receiving the benefits. These include many things we take for granted in our everyday lives, such as roads and parks. Governments often step in to ensure the provision of public goods. For example, in the early 19th century, coal gas was a new technology that could be piped to streetlights to illuminate a city. There was a great demand for this service, as it would enhance both safety and commerce. But because firms could not easily profit from the investment in manufacturing gas and then providing the infrastructure for public lighting, city governments often provided this public good. Later, other public goods, such as government-supported electrical infrastructure, natural gas and oil pipelines, and modern highways, helped kick-start fossil fuel use for electric power, home heating, and transportation. Some governments are taking similar approaches to stimulate alternative energy industries by providing infrastructure that accommodates charging stations for electric vehicles along major roads (**FIGURE 2.14**).

Placing Economic Values on Nature

Economists look for ways to determine the value associated with benefits people derive from goods and services, including the benefits that nature provides. Some benefits are direct uses that have a clear market value—like the price of timber harvested from a forest. But economists also try to put a price on aspects of nature that are not so obviously valued in market transactions. These include the soil, water, and other aspects of a forest ecosystem that support the growth of trees harvested as timber. Economists also look for ways to value nonconsumptive uses of a forest, such as bird-watching, hiking, and even the satisfaction individuals get from simply knowing that a certain forest exists to be enjoyed in the future. As we read in Chapter 1, economists and conservation biologists created the concept of **ecosystem services**, which is a way to assign these various types of values to the human benefits derived from naturally functioning ecosystems (see also Chapter 5). Placing economic value on ecosystem services can increase recognition of these benefits and inspire additional policy steps to protect, sustain, and restore them (**FIGURE 2.15**).

TAKE-HOME MESSAGE Markets often fail to consider all costs and benefits associated with a good or service. These costs or benefits are externalized to people other than the buyers and sellers. Goods with negative externalities tend to be oversupplied because they are artificially cheap, while those with positive externalities tend to be undersupplied. Unmanaged open access to resources can cause a phenomenon known as the tragedy of the commons. Governments sometimes address both with policies—for instance, assigning prices to negative externalities or producing public goods such as parks and transportation infrastructure.

2.4 Why Do We Have Environmental Policy?

Every morning when you walk out your door to head to class, work, breakfast, or the gym, your transportation choices are being shaped by the effects of government policies. Although you choose how to get from one place to another, policies have a lot to do with the range of choices available and how appealing those choices are relative to each other. For example, most public transportation options, such as buses and commuter trains, are provided by local or regional governments with the help of funding established through policies at the state and federal levels. Cars are built and sold by businesses, but government sets their safety and fuel economy standards. The cost of the vehicles is affected by trade policy, and the cost of fuel is affected by various energy policies, including federal, state, and local taxes on gas. Funding for infrastructure, such as sidewalks, bike paths, and roads, and the zoning ordinances governing the location of residential and commercial areas are also determined by policies established at the local and state levels (**FIGURE 2.16**).

What Is Policy?

So your choices are significantly influenced by **policies**: authoritative decisions, such as laws, regulations, and court rulings, that guide the behavior of the people and organizations subject to a government. Governments are particularly influential because they create and use policies to direct the behavior of those they govern toward particular goals. Think of the range of goals influencing transportation policies. They might include developing the local economy, adding manufacturing jobs in the automotive sector, reducing pollution, promoting public health and safety, and preserving open spaces. Changing these goals or any related rules or regulations enforced by government would likely change the context of your decisions about your daily transportation.

Policy changes can also have a major effect on the environment. In the 1960s and 1970s, the United States developed a series of environmental laws (e.g., the Wilderness Act, Clean Air Act, Clean Water Act, and Endangered Species Act) that established new national goals, rules, and standards for environmental protection to influence actions. These policy changes have led to improved air and water quality in many parts of the country, the protection of millions of acres of federal lands, and recovery plans for hundreds of threatened and endangered species. Nevertheless, human environmental impacts still present serious challenges, and policy making is a powerful way to influence them. Policy making involves political struggles among competing interests, and there are contentious processes under way right now—from local town land-use planning commissions to international treaty negotiations—that will influence our impact on the environment.

What Is Politics, and How Are Policies Made?

The term *politics* refers to the processes where decisions are made for and applied to a group of people. Governments are charged with making policies that have serious implications for high-stakes matters, such as the distribution and use of resources, provision of public goods, and maintenance of order. In these contexts, political processes are power struggles—contests among competing interests with different perspectives on what ought to be done. Political scientists—people who study systems of government and political behavior—describe their work as the study of "who gets what, when, and how." In this context, they define political power as the capacity of individuals and groups to make their concerns or interests count in decision-making processes shaping society.

Because of the range of interests that compete for many decision outcomes, it is difficult or impossible for each interest to achieve its initial preference. For example, if one group of citizens wants to protect a forested area as a nature preserve and another group wants to use the area for residential development, it is unlikely that a decision will fully satisfy the desires of both groups. If a decision is reached, it is more likely that one group's desire wins out over the other's or that a compromise is made. People and organizations with shared interests often form interest groups to influence government. They attempt to directly persuade a government decision maker through a practice known as lobbying. They can also try to influence the electoral process by supporting candidates for office and mobilizing their members to vote for certain candidates. Interest groups may also try to influence decision makers indirectly by mobilizing public opinion behind their cause. This can involve media and public awareness campaigns as well as protest events designed to focus attention on a particular issue.

public good a thing that can't be profitably produced because it is difficult to exclude nonpaying customers from receiving the benefits.

ecosystem services a concept that assigns a value to human benefits derived from naturally functioning ecosystems (Chapter 1).

policy authoritative decisions such as laws, regulations, and court rulings guiding the behavior of people and organizations subject to a government.

FIGURE 2.16 Transportation Choices In many cases, policies shape our choices. Government decisions on transportation influence our decisions on how to get around.

FIGURE 2.17 Stages of the Policy Process The policy-making process is often summarized as a series of steps (explored in greater detail in Chapter 20). Although these steps are useful for understanding what happens as policy decisions are made, it is important to remember that in the political arena, this process is typically quite complicated and contentious as competing groups struggle to influence the outcome.

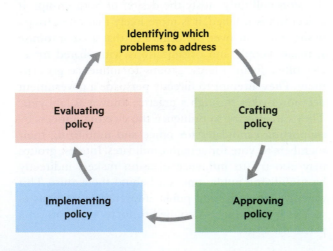

The political activities that compose the process of policy making are summarized in **FIGURE 2.17**, but real-world events do not unfold in a neat step-by-step process. Each step in the process is an opportunity for influence and is likely to be hotly contested by groups with competing views on the issue at hand. Moreover, the steps of the process do not necessarily follow sequentially. It is common for many of these steps to be occurring simultaneously.

Environmental Policy in the United States

Environmental policies fall into two broad issue areas: those that regulate pollution and those that manage natural resources. In the United States, natural resource

Harnessing Energy from Waves in the Ocean

The energy generated by the motion of ocean waves is the third largest source of renewable energy on Earth behind wind and solar. Yet wave-energy conversion technology is still in the early stages of development and has yet to be commercialized on a large scale.

The Department of Energy recognized that new technological developments to harness energy from ocean waves were often promising on a conceptual level but lacked prototype testing because funding from private investors was difficult to secure. So the agency created a contest that funds testing of wave-energy conversion technologies at a small scale in indoor tanks.

One of the finalists in 2015 was CalWave Power, a diverse team of engineers, physicists, and business development specialists from the United States, Europe, and Africa. Much of the team's work is based out of the Theoretical and Applied Fluid Dynamics Laboratory at the University of California, Berkeley, where graduate student Marcus Lehmann and Professor Mohammed-Reza Alam led development of a "wave carpet."

This device, placed 30–60 feet (9–18 meters) beneath the water near the shore, covers an area of the seafloor with a highly elastic synthetic fabric that is laid over double-action pistons attached to pumps. The motion of the waves over the fabric moves the pistons up and down, absorbing the wave energy in the same way that muddy seafloors near shore can absorb the energy and dampen the impact of waves passing over them. The wave carpet can be used to generate electricity by pumping high-pressure seawater to a turbine onshore. It can also be used to pump seawater at high pressure through a desalination membrane to produce fresh water. And because the carpet is submerged well below the surface, it can withstand stormy seas.

The support from the Wave Energy Prize will surely help speed the research and development process. As a finalist, the CalWave Power team received $500,000 and support to test their device on an increasingly large scale in a variety of wave conditions at several indoor tank laboratories.

What Would You Do?

Many environmental groups see technological innovation as a key component of solutions to environmental problems and feel that governments should contribute to their funding. However, these funds could be bestowed on other, more proven environmental initiatives. Others question whether governments should bother cultivating these technologies at all. If you worked for the US government and oversaw the funds of a large grant program, would you fund technology development or prizes like those described in this box? Or would you make another decision?

(a)

(b)

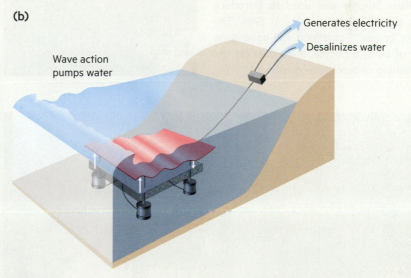

Generates electricity

Desalinizes water

Wave action pumps water

(a) The CalWave Power team with their $500,000 award check from the Department of Energy. (b) A wave carpet uses pistons powered by the motion of waves to pump water and generate electricity.

Who Were Some Early Leaders of Environmental Policy?

Gifford Pinchot (a) and John Muir (b) are considered leaders of the movement to set aside public lands. Both worked alongside President Theodore Roosevelt (pictured at bottom in Yosemite Valley) to establish protected status for various lands in the early 20th century. Pinchot is known for championing *conservation*, a belief that public lands should be managed for current and future human uses, such as forestry, grazing, hydropower, and irrigation. His legacy is seen in the National Forest System, which was originally established to provide a "continuous supply of timber." In contrast, Muir is known for advocating *preservation* of certain lands to be protected from development and thereby provide places where people can gain recreational and spiritual benefits in nature. Muir wrote, "When we try to pick out anything by itself, we find it hitched to everything else in the Universe." His legacy is seen in the National Park System. Preservation was taken a step further in 1964 with the Wilderness Act, which sets aside certain federal lands to remain "untrammeled by man." A letter from author and activist Wallace Stegner was used to introduce this legislation in Congress in 1964. The letter began, "Something will have gone out of us as a people, if we ever let the remaining wilderness be destroyed."

(a)

(b)

policy at the federal level dates to the late 19th century and thus has a longer history than that of pollution-control policy. At that time, the effects of rapid urbanization and industrialization inspired a new ethic of resource conservation and a movement to preserve some areas still considered natural and scenic. New policies directed the government to reserve some public lands and even acquire some from private ownership for special protection.

In 1872, Congress set aside the world's first national park at Yellowstone in Montana and Wyoming, reserving 2.2 million acres of federal land to be retained in its "natural condition" and managed as a public park "for the benefit and enjoyment of the people." The US government then began reserving large tracts of land for national forests in 1891 to "furnish a continuous supply of timber for the use and necessities of citizens of the United States." Lands that were neither forested nor deemed scenic or suitable for farming often remained in federal ownership, even while private interests used them for ranching and mining. In the 20th century, Congress established four main executive land-use agencies to manage these public lands: the National Park Service, US Forest Service, Bureau of Land Management, and US Fish and Wildlife Service. These agencies have different missions that have undergone significant changes over the past century, and they are each authorized to make and enforce rules and regulations for the land they manage.

Then, some very important laws passed in the 1960s and 1970s established protections for certain kinds of public lands and new processes for land management agencies. For example, the Wilderness Act of 1964 established the highest level of environmental protection for federal lands designated as wilderness areas, "where the earth and community of life are untrammeled by man, where man himself is a visitor who does not remain." Two other important environmental laws—the National Environmental Policy Act of 1970 and the Endangered Species Act of 1973—have overarching influence on the actions taken by federal agencies. The National Environmental Policy Act requires that an Environmental Impact Statement (EIS)—lengthy documents compiled over many months or even years with a high level of public involvement—be prepared when actions taken by the federal government are likely to have a significant environmental impact. The Endangered Species Act extends federal protection over species deemed to be in danger of extinction.

Policies concerning pollution control are more recent. In the 1970s, the federal government responded to a groundswell of public support for more expansive environmental laws and a growing environmental movement. With bipartisan majorities, Congress passed the Clean Water Act governing water pollution, the Clean Air Act governing air pollution, and the Resource Conservation and Recovery Act governing solid and hazardous waste disposal. Also created in 1970, the Environmental Protection Agency (EPA) sets minimum national standards for pollution levels and the procedures governing polluting activities. The EPA then works with state governments to enforce the standards by granting permits to various categories of polluters and administering penalties to those who fail to comply with the standards. Other pollution-control laws include the Comprehensive Environmental Response, Compensation, and Liability Act (better known as Superfund) under which the EPA identifies and prioritizes hazardous waste sites for cleanup, and the Toxic Substances Control Act, which gives the EPA authority to inventory, test, and regulate chemical substances currently manufactured or processed within the United States.

State and Local Policies

While there are many important environmental policies set at the national level, state and local governments in the United States also set policies that have far-reaching environmental effects. For example, while the federal government sets minimum environmental standards for things such as air and water pollution, toxins and waste management, and workplace health and safety, states are often responsible for meeting these standards and have the authority to enact and enforce their own policies with stricter standards. For example, California is well known as a state that has more stringent standards than the federal government in many areas, including air quality, automobile emissions, water quality, and energy efficiency in products and buildings. State governments also have primary jurisdiction over laws governing the use and distribution of water, regulations on energy utilities, and the use of natural resources (e.g., forestry, mining) on state and private land. In Texas, state agencies such as the Texas Commission on Environmental Quality, the Public Utility Commission, and the Texas Parks and Wildlife Department oversee a wide range of policies affecting the environment. The governments of counties, cities, and towns often can pass zoning ordinances that regulate where and how certain areas of land are used. Together with building codes, these policies influence the ways we occupy and affect the environment where we live. This means that politics—and especially elections—at the state and local levels has a big environmental impact. See **AT A GLANCE: ENVIRONMENTAL POLICIES IN THE UNITED STATES** on pp. 54–55 for more information about federal, state, and local environmental policies.

Global Environmental Politics

Many human effects on the environment extend beyond national borders. Air pollution from coal-fired power plants in the United States causes acid rain in Canada. Hazardous electronic waste from developed countries gets shipped to developing countries where it can cause environmental and health hazards during disassembly and disposal. Water quality and

→ ENVIRONMENTAL JUSTICE

New Voices in the Environmental Movement

Much early environmental policy was based on the idea of preserving wilderness in an untouched state. For early leaders, this often meant preserving the land for people like themselves, while discriminating against others. For example, the Sierra Club's executive director in 2020 acknowledged his group's past discriminatory membership practices: Some club chapters screened applicants by class and race until at least the 1960s. The club's founder John Muir's writings often reinforced racial stereotypes of Black people and supported the forced removal of the Miwok people from their ancestral lands in the newly established Yosemite National Park. Theodore Roosevelt, who as president established many national parks, refuges, and national forests, also championed the removal of indigenous peoples from their lands in his book *The Winning of the West*.

Today, while groups besides the Sierra Club are reconsidering their past, new ones are working to bring new voices to the table. For example, WE ACT for Environmental Justice in New York City works to "expand political participation among communities of color and low-income residents in order to impact environmental planning and decision making." Founded in the 1980s to decrease pollution from bus idling in West Harlem, this pioneering environmental justice group has expanded to take on many issues throughout the community and state and at the federal level.

quantity is affected as rivers flow through multiple countries. Demand for ivory or other products derived from endangered species in one place can affect the survival of that species in another (**FIGURE 2.18**). Dealing comprehensively with these and other issues presents a political challenge because no single country has authority over all the human actions that contribute to them. Each country has sovereignty: the ability to make policies and carry out actions within its territory. But sovereignty ends at the national border.

Countries can coordinate actions to address environmental problems in a variety of ways. Most nations are members of several intergovernmental organizations (IGOs), which provide venues for different countries to collaborate. With 193 member countries, the

(a)

(b)

FIGURE 2.18 Global Environmental Problems Many environmental problems extend across national borders: **(a)** on the outskirts of Accra, Ghana, locals recycle e-waste, mostly from Western nations, in toxic conditions; **(b)** smuggled ivory tusks and shark fins seized in Hong Kong.

United Nations is the largest IGO. Sometimes this collaboration can lead to international agreements that commit signatories to take various actions and abide by rules to meet a common goal. As **FIGURE 2.19** shows, since the 1970s, international agreements have been reached on a range of environmental issues. But reaching international agreements is challenging, and it is often a slow and incremental process (and is considered in more detail in Chapter 20.)

1970s
● 1972 UN Conference on the Human Environment (UNCHE; "Stockholm Declaration")
● 1972 UN Environment Programme (UNEP) established
● 1973 Convention on International Trade in Endangered Species (CITES)
● 1979 Convention on Long-Range Transboundary Air Pollution (CLRTAP)

1980s
● 1982 UN Convention on the Law of the Sea (UNCLOS)
● 1987 Montreal Protocol on Substances That Deplete the Ozone Layer
● 1987 World Commission on Environment and Development (WCED)
● 1989 Basel Convention on the Control of Transboundary Movements of Hazardous Wastes and Their Disposal

1990s
● 1992 UN Framework Convention on Climate Change (UNFCCC)
● 1992 UN Convention on Biological Diversity
● 1994 International Convention to Combat Desertification
● 1997 Kyoto Protocol on Climate Change

2000s
● 2001 Stockholm Convention on Persistent Organic Pollutants
● 2002 World Summit on Sustainable Development (WSSD)

2010s
● 2016 Paris Agreement on Climate Change

FIGURE 2.19 Timeline of Major International Environmental Agreements

 TAKE-HOME MESSAGE Policies affect large numbers of people on important issues such as the distribution and use of resources, provision of public goods, and maintenance of order. Policy making involves power struggles among competing groups. Modern US environmental policy has shifted toward increased federal regulation of natural resource management and pollution control. Human effects on the environment extend beyond national borders and must be addressed through international organizations and agreements.

2.5 What Can I Do?

So how do your choices affect the environment? Once your own ethical considerations help you clarify the issues and concerns important to you, what can you do about them? Thinking more about the role that economics and politics play in our collective decisions and in environmental issues can help us identify ways in which we as individuals can make a difference.

Exercise Your Economic Power

It may not seem like your daily choices as a consumer add up to much, but over time they can. If you fix the oil leaking from your car, you're not only saving your engine from potential damage; you're also keeping oil from pouring into the environment. One quart of leaked motor oil can contaminate 250,000 gallons of water. You can also see the cumulative nature of daily choices if you fill up a reusable bottle at a fountain with a digital counter. If each refill replaces the purchase of water in a disposable bottle, the environmental benefit of avoided waste adds up quickly. In 2015, Duke University installed 50 stations like this and avoided the use of more than 400,000 disposable bottles in a single year. And because bottled water costs hundreds of times more than tap water, when you avoid purchasing it, you also save money. You can also think of your consumer choices as support for various business practices. In Chapter 19, we'll learn more about steps many businesses are taking to reduce their environmental impact and even provide environmental benefits.

As a consumer, you can support businesses that support issues of concern to you by choosing their products over those of their competitors. You can also withhold your buying power from companies in response to negative effects or business practices you disagree with. When this is done on a large scale, it is known as a **boycott**, which sometimes can motivate a company to change its practices. For example, consumer boycotts played an important role in motivating apparel companies such as Nike to improve their labor standards in factories overseas and in influencing tuna companies to adopt fishing practices that lessen adverse effects on dolphins.

Influence Public Policy

Elections determine the officials who make policy, so voting for candidates who support issues important to you is an important way to participate, as we noted in the previous chapter. This is true for campus government as well as local, state, and national offices. Candidates for office also need support during their campaigns, and individuals can help by raising funds and mobilizing other people to vote for their preferred candidate. At the state and local levels, there are often ballot measures on policy that people vote on directly. For example, ballot initiatives in various states have enacted policies ranging from a ban on plastic bags in grocery stores, to requirements for renewable energy use by utilities, to the provision of public money for parks and wildlife protection.

Individuals can join the ranks of decision makers by running for office themselves. This is what Mary-Pat Hector chose to do in her second year at Spelman College. After engaging in a number of political efforts, including one to bring clean water to residents of Flint, Michigan, after their water supply was contaminated, Hector decided at the age of 19 to run for a city council position in her home of Stonecrest, Georgia. Although she narrowly lost the election, she continues to work as a community advocate and for organizations that help young people vote.

Research and Influence Your Elected Officials

Once the winning candidates are in office, they become representatives of their constituents: the people living in the districts they represent. It is important to know who your representatives are at the national, state, and local levels and the ways you can most effectively communicate your interests to them. The USA.gov website lists elected officials and can help you find contact information for your federal, state, and local officials (www.usa.gov/elected-officials). Voting guides like the VOTE411 website organized by the League of Women Voters (vote411.org) can help you learn what's on your local ballot when it comes time to vote, and voting scorecards like the ones compiled by the League of Conservation Voters can help you see how your representatives voted on environmental issues (scorecard.lcv.org).

Staff working for members of Congress are employed to respond to constituent communications and have noted that the more personal and direct the communication, the more effective it is in influencing the member. It is often more effective to call a member's office, attend and comment at a town hall meeting, or visit a member's office. It is also important to realize that all forms of communication on a particular issue are typically tallied to give the member an idea of how many constituents are concerned, so the number of e-mails, letters, phone calls, and public comments matters. Members of Congress can also be influenced indirectly with highly visible rallies, protests, and other events that draw attention to issues in their districts.

While elected officials establish the laws governing the environment, career employees of public agencies are charged with establishing a wide range of rules and regulations that determine how the laws will be implemented. The agency decision-making processes at the local, state, and federal levels often require a period of public comment. These are also opportunities for individuals and groups to engage in the policy-making process.

boycott the act of withholding your buying power (as an individual or group) from a company to motivate change.

Environmental Policies in the United States

In the United States, environmental policies fall into two broad categories—pollution regulation and natural resource management—and are administered by federal, state, and local governments.

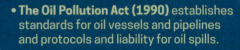

- **Offshore oil-drilling permits and leases** are managed by the Department of the Interior.

- **The Ocean Dumping Ban Act (1988)** prohibits dumping waste into the ocean.

- **The Oil Pollution Act (1990)** establishes standards for oil vessels and pipelines and protocols and liability for oil spills.

- **The Marine Mammal Protection Act (1972)** prohibits the taking and importation of marine mammals.

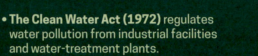

- **The Clean Water Act (1972)** regulates water pollution from industrial facilities and water-treatment plants.

- **The Safe Drinking Water Act (1974)** sets standards for public drinking water supplies.

- **State and local governments** make and enforce land-use laws, zoning laws, and building codes and determine transportation planning.

- **The Clean Air Act (1970) and the Energy Policy and Conservation Act (1975)** set emissions and fuel-economy standards for vehicles.

- **The Clean Air Act (1970)** regulates emissions from sources such as factories and power plants.

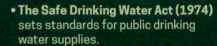

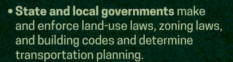

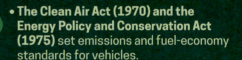

- **The Resource Conservation and Recovery Act (1976)** establishes standards for solid and hazardous waste disposal.

- **The Comprehensive Environmental Response, Compensation, and Liability Act (1980),** commonly known as **Superfund,** enforces cleanup of hazardous waste sites.

- **The Toxic Substances Control Act (1976, updated 2016)** requires premarket testing of new chemical compounds.

- **The Bureau of Land Management** issues permits and leases for grazing and oil, gas, and coal leases on federal lands.

- **The Clean Water Act** also regulates development that affects wetlands.

- **State water laws** govern surface and groundwater use as well as water pollution from sources such as agricultural water runoff.

- **The National Environmental Policy Act (1970)** requires federal agencies to prepare Environmental Impact Statements.

- **The Food Quality Protection Act (1996)** establishes safety standards for pesticide use.

- **The National Wildlife Refuge System** is managed by the US Fish and Wildlife Service.

- **The National Forest System** is managed for timber harvest, wildlife habitat, and recreation by the US Forest Service.

- **The Wilderness Act (1964)** designates certain federal lands as wilderness areas.

- **The Federal Energy Regulatory Commission** regulates and issues permits for hydroelectric dams.

- **The Endangered Species Act (1973)** protects species that are threatened or in danger of extinction.

- **The National Park System** is managed by the National Park Service to ensure "unimpaired enjoyment of future generations."

- **National Monuments** are established by the president under the **Antiquities Act (1906)** "to protect significant natural, cultural or scientific features."

Chapter 2 **Review**

SUMMARY

- When we seek to determine whether a particular action is right or wrong, good or bad, we use ethics—a set of moral principles that provides guidelines for our behavior. Two common systems of ethics include one focused on duties and rights (deontological) and another focused on the amount of benefit or harm that results from a course of action (utilitarian).

- Ethical considerations concern how we ought to behave in relation to others, and our range of concern on environmental issues can vary from a strictly human focus (anthropocentrism) to wider concerns for all living things (biocentrism) and whole ecosystems (ecocentrism).

- Markets are economic systems that bring buyers and sellers together to exchange goods and services, and the interaction of supply and demand in a market determines the price. Scarcity of certain goods can often inspire innovation in the use of that good or the substitution of other goods to meet the same need. Efficiency gains that lower the price of a good often inspire new uses of that good in ways that increase consumption (Jevons paradox).

- Governments sometimes intervene in markets to adjust the price, regulate production, or distribute goods.

- Externalities result when markets fail to encompass all of the costs and benefits associated with a good or service: this causes some things to be oversupplied (negative externalities, such as economic activities that cause pollution) and other things to be undersupplied (such as parks and wilderness areas that provide benefits outside of the market).

- The tragedy of the commons is a situation in which inadequate management of an open-access resource leads individuals to degrade the resource over time.

- Governments sometimes attempt to address externalities with rules and regulations or by providing public goods such as parks and transportation infrastructure.

- Policies are decisions made by authorities to guide the behavior of those subject to them. Policy making involves struggles among competing groups attempting to influence the outcome.

KEY TERMS

anthropocentrism	Earth Charter	intrinsic value	tragedy of the commons
biocentrism	ecocentrism	Jevons paradox	true cost accounting
boycott	economic system	market	utilitarianism
cap and trade	ecosystem services	policy	
demand	ethics	public good	
deontological ethics	externality	supply	

REVIEW QUESTIONS

The letters following each Review Question refer to the Chapter Objectives.

1. What is the difference in focus between a utilitarian approach to ethics and a deontological approach? **(A)**

2. Which ethical approach extends concern about environmental issues to the broadest community of interests: anthropocentrism, biocentrism, or ecocentrism? **(A)**

3. Describe how the scarcity of a good (such as oil) can lead to rising prices and to innovation and substitution in a market economy. **(B, C)**

4. What are two examples of government interventions in markets? **(B)**

5. What is an externality? What is an example of an externality that is associated with adverse environmental effects? **(D)**

6. What is an example of an environmental problem that fits the definition of the tragedy of the commons? What would one possible solution be to address the incentives behind this tragedy of the commons problem? **(E, F)**

7. Why will governments sometimes intervene in markets to provide public goods (such as streetlights and transportation systems) instead of waiting for the market to provide these benefits that many enjoy? **(D)**

8. What are three examples of policies affecting the environment? **(E)**

9. What are some factors that often make the policy-making process contentious? **(E)**

10. Over the past century, has the role of the US government in environmental issues been increasing or decreasing? Provide examples that support your answer. **(E)**

11. What are two ways countries address environmental problems that cross national boundaries? **(E)**

FOR FURTHER THOUGHT

The letters following each item refer to the Chapter Objectives.

11. If the price of oil rises dramatically, would you expect to see more investment in alternative energy and vehicles (such as electric cars), more investment in expensive technologies that can extract oil from places that have been difficult to access, both of these developments, or neither? Explain your answer using economic terms with help from Section 2.2 in this chapter. **(B, C)**

12. Plastic is a big component of our waste stream. Past innovations have made thinner and stronger plastic resins, which make it possible to use 50% less material for plastic containers such as water bottles, 75% less for plastic wrap, and 85% less for hard molded plastic used in car bumpers. Explain why these innovations have likely led to less or more plastic in the waste stream. Use the concept of the Jevons paradox to explain your answer. **(C)**

13. Imagine that a group of environmental economists used the ecosystem services approach to determine that each tree in the parks and greenspaces of a major US city has an average dollar value of about $1,000 (for benefits such as shade, erosion control, carbon sequestration, and scenic and recreational enjoyment).

Then suppose this is about $500 more on average than the value of each tree as lumber at that time. When news of the study is published, a group of environmental ethicists argues that this approach is wrongheaded because it devalues the intrinsic value of trees: as living things they have a right to protection regardless of their dollar value, and we have a duty to protect them. Both the economists and the ethicists claim they are trying to protect the trees. Which approach do you find more compelling from an ethical perspective? Explain why. Which approach do you think would be more effective and why? **(D, F)**

14. Consider a place that is familiar to you (perhaps someplace near your hometown or someplace you visit often). Identify at least three federal policies that likely influence the environment in that place, and briefly describe the specific influences. Next, using the same place or a new one of your choosing, identify at least two state or local policies that likely influence the environment in that place, and briefly describe the specific influences. **(E, F)**

Make Your Case ?

The following exercises use real-world data and news sources. Check your understanding of the material and then practice crafting well-supported esponses.

Use the News

The following article is from sciencemag.org/news, the news site of the journal *Science*. It describes a decision in 2016 to change a US safety regulation policy. Use this article to answer the questions that follow. The first three questions test your understanding of the article.

Questions 4–6 are short answer, requiring you to apply what you have learned in this chapter and cite information in the article. Answers to Questions 1–3 are provided at the back of the book. Question 7 asks you to make your case and defend it.

"Updated: United States Adopts Major Chemical Safety Overhaul," Puneet Kollipara, *Science*, June 8, 2016

The U.S. Senate yesterday unanimously approved a major overhaul of the nation's primary chemical safety law—marking one of the last steps in a decades-long reform effort. The House of Representatives on 24 May overwhelmingly approved the rewrite of the Toxic Substances Control Act (TSCA), which governs how industrial chemicals are tested and regulated. The legislation now moves to President Barack Obama for signing.

The measure—H.R. 2576, named for the late Senator Frank Lautenberg (D–NJ), a long-time TSCA reform champion—is perhaps the most far-reaching and influential environmental statute passed by Congress since the body updated the

Clean Air Act in 1990. The measure aims to make chemical safety reviews more science-based, and includes provisions designed to reduce the use of animals in chemical testing and promote the study of so-called cancer clusters.

"The end result . . . is a vast improvement over current law," said Representative John Shimkus (R–IL), who co-sponsored the House bill, on the House floor. The bill, he added, is "a careful compromise that's good for consumers, good for jobs, and good for the environment."

"While this is a compromise bill, it is a long overdue step forward in protecting families and communities from toxic chemicals," said Representative Frank Pallone Jr. (D–NJ), top Democrat on the House Committee on Energy and Commerce.

Numerous Fixes

Both environmentalists and industry have long agreed that the TSCA, originally passed in 1976, has numerous flaws. It includes legal barriers, for example, that essentially prevent the Environmental Protection Agency (EPA) from acquiring toxicity data on chemicals and imposing new restrictions on them—even on highly toxic substances such as asbestos. Critics say the current legislation also favors economic concerns over scientific findings, and has led to thousands of chemicals entering the market without adequate health and safety oversight.

The reform bill seeks to fix a number of these flaws. It aims to make chemical safety reviews purely science-based, by eliminating a longtime requirement that EPA weigh regulatory costs in the safety review process. It also repeals a long-time requirement that EPA select the "least burdensome" method of regulating a toxic substance. And the bill would require EPA to deem a new chemical safe before it could enter the marketplace; under current law, a chemical can enter the marketplace unless EPA deems it unsafe within a certain time period.

The bill would also make it easier for EPA to order chemical companies to generate any toxicity data that the agency needs to inform its reviews; under current law, EPA can only order these data by going through a lengthy rulemaking process that often ends up mired in litigation. And the bill would require EPA to take tougher action on persistent, bioaccumulative, and toxic chemicals, and ensure that chemicals are safe for vulnerable groups such as infants, seniors, and chemical workers.

Animal protection and animal rights groups hailed another provision that aims to reduce EPA and chemical companies' use of animal-based toxicity testing methods. It would task EPA with using non–animal-based methods "to the extent practicable," and the agency would have to devise a plan to research, develop, and eventually use more nonanimal methods—including computational modeling, high-throughput screening, and cell-culture testing.

The bill also includes a measure known as Trevor's law that encourages federal agencies to study "cancer clusters"—areas that appear to have unusually high numbers of cancer cases that may be linked to a shared environmental cause. The Society of Toxicology in Reston, Virginia, although praising the bill, expressed some concern about including the cancer-cluster measure and other topic- or chemical-specific language in the bill. Doing so "detracts from the wider range of priority chemical-specific or analytical issues that, as toxicologists, we address every day," society President John Morris said in a 23 May letter.

Rocky History

The TSCA reform bill is the result of years of negotiations involving lawmakers in both parties and a wide range of stakeholders. Many previous efforts to overhaul the TSCA failed after lawmakers couldn't strike a consensus among competing interest groups, such as chemical companies and environmental groups. The current effort succeeded, however, despite the toxic political climate in Washington, D.C., and a government divided between a Democratic-held White House and Republican-held Congress.

To arrive at the current bill, the House and Senate first approved their own bipartisan—but widely different—versions of TSCA reform. Then, lawmakers spent months negotiating a compromise between the chambers.

It wasn't clear for instance, whether the animal-testing provisions—which were in the Senate bill, but not the House's—would ultimately survive. "But the fact that we are now going to severely restrict the unnecessary cruelty to animals is something that I'm very proud that the leadership helped preserve," Senator Cory Booker (D–NJ), a proponent of the language, told reporters outside the U.S. Capitol on 19 May in announcing his support of the bill.

A much bigger sticking point was concern, voiced by many liberal Democrats and environmental groups, that the legislation would weaken states' ability to issue their own chemical regulations. Senator Barbara Boxer (D–CA), the top Democrat on the Senate environment panel, had argued especially forcefully against language in the Senate bill that would have kept existing state chemical regulations on the books, but reduced the states' ability to issue new regulations in the future.

But Boxer ultimately supported the final compromise. The final bill is far from perfect on that issue, but it's better than current law, she said in announcing she would support the reform measure. "What a battle that was," she said. "Well, we no longer have that battle."

1. The article describes a change to the nation's primary chemical safety law. On the basis of the article, how would you describe the support of this final bill? Select the response that best describes the support given.

   a. Support was divided in both the House and the Senate. Most Republicans supported it, and most Democrats opposed it.

   b. Support was divided in both the House and the Senate. Most Democrats supported it, and most Republicans opposed it.

   c. Support was bipartisan. While compromise was needed on both sides, the Senate in particular passed it unanimously.

   d. Support was divided in the Senate but not in the House. The House passed the bill unanimously, but it failed to pass in the Senate.

2. In the "Numerous Fixes" section, the article describes the many changes that the reform bill made. From the following list, select two changes that are described in this section.

   a. Review and approval by the EPA can now be halted and skipped by an executive order from the president approving or rejecting its use.

   b. The bill requires the EPA to ensure chemicals are safe for groups such as infants, seniors, and chemical workers that are particularly susceptible.

   c. The bill eliminates all testing on animals, permitting only use of computer models and other technical screening methods.

   d. The bill eliminated a provision that made companies generate data on the toxicity of a chemical on demand from the EPA and without legal review.

   e. Review and approval of a new chemical by the EPA will be based purely on results of scientific reviews and will not take into account the costs of regulating it.

   f. The bill leaves the study of cancer clusters to state agencies rather than federal agencies.

3. In the "Rocky History" section, several compromises are described that were made by lawmakers and competing interest groups. Choose the two statements from the following list that describe actual compromises made.

   a. Companies agreed to a provision to not market chemicals in areas where states issued laws that were more stringent than federal laws.

   b. Environmental groups gave up on trying to limit animal testing in the bill because of pressure from corporations.

   c. More animal-testing prohibitions were included in the final bill than were included in many versions of early bills.

   d. The bill was made temporary so it could easily be overturned or changed in the future by executive orders from the president.

   e. States were given less ability to issue their own regulations than environmental leaders in states such as California wanted.

4. This law affects many different interests. Name at least three broad categories of interest groups that were likely engaged in this policy-making process. For each group, choose an outcome of the bill that the group may have favored and an outcome it may have been less happy about.

5. Find at least two examples in this article of ethical concerns arising from a duty to protect the interests of someone or something. Give the example, and name the ethical approach that was used.

6. Under the old law, a new chemical could enter the marketplace if the EPA failed to deem it unsafe within 90 days, while the new law requires the EPA to deem a new chemical safe before it can enter the marketplace. What is the significance of this change?

7. **Make Your Case** In the last paragraph of this article, Senator Barbara Boxer says that this compromise is "better than the current law," having ceded a point she thought was important. Others in this article mention compromise as well. On the basis of this article, what is your opinion of this bill? Do you agree with its provisions? Do you think the lawmakers (both Republican and Democrat) were right to compromise? In answering these questions, answer as if you were a US senator from your state, defending your decision to your local voters.

Use the Data

Gasoline is the dominant transportation fuel, so gas prices have a big effect on our driving habits and vehicle-purchasing decisions.

Study the graphs shown here, and use them to answer the questions that follow. The first three questions test your understanding of the graphs.

Questions 4 and 5 are short answer, requiring you to apply what you have learned in this chapter and cite data in the graphs. Answers to Questions 1–3 are provided at the back of the book. Question 6 asks you to make your case and defend it.

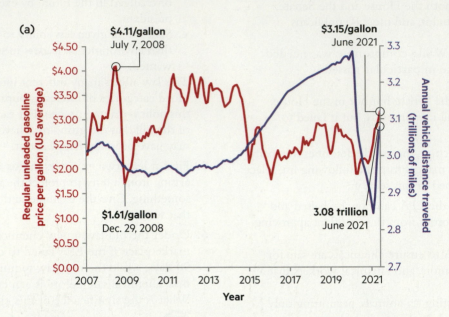

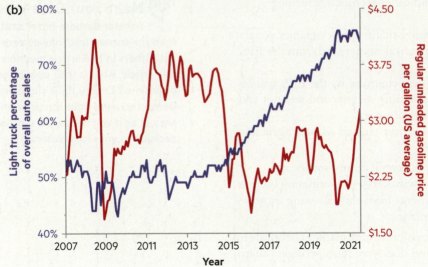

Adapted from **(a)** US Federal Highway Administration (2021), US Bureau of Labor Statistics (2021) and **(b)** US Bureau of Economic Analysis (2021), US Bureau of Labor Statistics (2021)

1. The red line in graph (a) shows that gas prices rose steeply and peaked in 2008. Turning to the blue line on the same graph, what does it show about the amount of driving consumers of vehicles did between 2008 and 2009? Select the answer that best describes the change.

   a. The peak number of annual miles driven occurred at the same time as peak gas prices.

   b. The number of annual miles driven began to climb immediately after gas prices decreased from the peak.

   c. As gas prices peaked, consumer demand dropped, and consumers drove fewer annual miles.

   d. As gas prices peaked, consumer demand increased, and consumers drove more annual miles.

2. Graph (b) shows the relationship between gas prices and light truck sales between 2007 and 2021.

   (i) Looking at the period between 2008 and 2010, what relationship can you see between gas prices and light truck sales?

   a. As gas prices increase, light truck sales decrease, and as gas prices decrease, light truck sales increase.

   b. As gas prices increase, light truck sales increase, and as gas prices decrease, light truck sales decrease.

   c. Gas prices and truck sales do not show an observable relationship.

   (ii) Looking at the entire graph, how does the relationship between gas prices and light truck sales between 2014 and 2015 compare to what happened between 2008 and 2010?

   a. A different relationship is seen: a large change in gas prices did not dramatically affect light truck sales.

   b. The same relationship is seen: a large change in gas prices caused a large change in light truck sales.

   c. Gas prices and truck sales do not show an observable relationship.

3. At the start of 2011, gas prices climbed almost as high as the last peak price of 2008. How did consumers respond in the later part of 2011?

   a. Light truck sales increased to a high of 45% by the middle of 2011.

   b. Light truck sales decreased to a low of 40% by the middle of 2011.

   c. Annual miles driven increased to a high of 3.1 trillion miles by the end of 2011.

   d. Annual miles driven decreased to a low of 2.95 trillion miles (5 trillion km) by the end of 2011.

4. Gas prices fell sharply at the beginning of the COVID-19 pandemic in early 2020. However, miles driven dropped too. Why did this happen? In the six months after prices hit bottom, did light truck sales respond as expected? Briefly describe why or why not.

5. Gas prices plummeted between July 7, 2008, and December 29, 2008. Using the language of an economist, how would you describe the change in terms of supply and demand? What do you think might have caused this price drop? Hint: Consider or research the overall state of the economy during this time period or use the internet to research possible causes.

6. **Make Your Case** Imagine you are working for an investment firm and are asked to predict the relationship between light truck sales and gas prices in the year 2025. Citing data from graph (b), what relationship would you predict and why? To make a better prediction, what is some other information that would be helpful that is not on this graph?

LEARN MORE

- Aldo Leopold. 1949. *A Sand County Almanac.* New York: Oxford University Press.
- Peter Singer. 1975. *Animal Liberation: A New Ethics for Our Treatment of Animals.* New York: Random House.
- EPA, "Environmental Economics": www.epa.gov/environmental-economics
- Khan Academy, "Market Failure and the Role of Government": www.khanacademy.org/economics-finance-domain /microeconomics/market-failure-and-the-role-of-government

- EPA, "Summaries of Environmental Laws and Executive Orders": www.epa.gov/laws-regulations/laws-and-executive-orders #majorlaws
- EPA, "The Basics of the Regulatory Process": www.epa.gov /laws-regulations/basics-regulatory-process
- Video: NOAA, *Deepwater Horizon*/BP Oil Spill: digital.wwnorton .com/environsci

3

Matter and Energy

What Are the Building Blocks of Sustainability?

Imagine you are having a cookout in a park, backyard, or campground with your friends and have lit a grill to start barbecuing. Whether you build your fire with charcoal purchased from your local store or with wood you gathered, the fuel you are using will seem to disappear as it burns: your stack of wood or pile of charcoal briquettes will get smaller and smaller as your fuel turns to ash. It would seem in this burning that you have destroyed something, right? A lot was there, and now most of it is gone.

In 1772 a French nobleman, lawyer, and tax collector named Antoine Lavoisier thought the same thing when he began a series of experiments that are often regarded as the foundation of modern chemistry. Lavoisier placed different substances in separate, sealed glass containers and used these to conduct various transformation experiments, including burning fuels (combustion), as well as allowing metal to rust and food to rot. He precisely weighed each container and its contents before and after the experiment to see if anything changed. He found that the total amount of the substances present before and after the reactions remained the same. Lavoisier summarized his results as showing that "nothing is lost, nothing is created, everything is transformed."

What Lavoisier discovered was that just because the wood, charcoal, or other fuel for our fire seems lost to us, this does not mean that it has been destroyed. If we collected all of the gases produced along with the remaining ash and the tiny burning particles and substances in our fire, this basic *matter* would weigh the same as the original stack of wood or pile of charcoal. And what about the energy of the fire, which produced its heat and light? Scientists in the 19th century discovered that this too was conserved, although it can change forms.

But how do matter and energy affect the choices we make about the environment? Let's think about a fire of a different type and its potential consequences. Each summer, large forests in Indonesia are purposely burned, clearing the land for plantations of oil palm

Forests such as this one pictured in Bali, Indonesia, are beautiful, but they also store tremendous amounts of energy. If these forests are destroyed, where does this energy go?

Chapter Objectives

By the end of this chapter, you should be able to . . .

A. explain what matter, elements, and atoms are and how they relate to each other.

B. name the types of chemical bonds and describe how they can affect the properties of matter.

C. discuss the states of matter and how chemical reactions can change molecules.

D. define what energy is and differentiate between different forms of energy.

E. describe the first and second laws of thermodynamics and recognize why these are important for energy efficiency.

F. understand what a trophic level is and how it relates to the concept of energy flow.

Atoms are what make us all matter.

—Anonymous

trees, often at the expense of habitats for orangutans, elephants, rhinoceroses, and tigers (**FIGURE 3.1**). In addition, the matter fueling these fires fills the air as smoke and burned particles, causing hundreds of thousands of people in the region to suffer respiratory ailments. Burning the trees and a common ground material called *peat* releases both energy and large amounts of the gas carbon dioxide. As the latter moves into the air, it contributes to climate change, as we'll see in Chapter 11.

Like Lavoisier and earlier scientists, we will explore the basics of matter and energy: the stuff making up Earth and the universe. To make decisions on sustainability, we need to understand and try to trace where things come from and go to and how they change. Following the paths of matter and energy can reveal both the sources of environmental problems (such as air pollution, water pollution, and climate change) and ways to address these problems (such as the reuse and recycling of materials, energy conservation strategies, and innovations that enable us to produce energy and materials with less harmful side effects). Grasping the basic science of matter and energy is the first step in understanding and addressing the environmental impacts of the choices we make.

FIGURE 3.1 Forests Burned for Agricultural Development When tropical forests are burned to convert them to agricultural land, carbon dioxide is released into the air and wildlife habitats are destroyed. In past years, 2.4 million acres of Indonesian forests were destroyed each year through burning or other methods. By 2020, however, more recent government policies and economic trends had decreased this total to 370,000 acres per year.

3.1 What Is Matter Anyway?

You are made of matter, as are all other tangible things such as plants, animals, soil, air, water, buildings, cars, plastic, and all the planets, moons, and stars in the universe. **Matter** is defined as anything that takes up space, and the amount of matter in a particular object is its **mass**.

Matter comes in many forms—solids, liquids, and gases; living and nonliving things; things that are human-built and things that only occur in nature. We have a vast but finite amount of tangible matter on the planet, and aside from a relatively tiny amount of outer-space debris that arrives on Earth through collisions with meteorites and comets, matter on Earth stays within a closed system: little lands on or gets off our planet (**FIGURE 3.2**). But what is matter made of? The ancient Greek philosopher Democritus (460–370 BCE) asked this question more than 2,400 years ago and hypothesized that tangible matter is composed of tiny units that can be assembled and disassembled in various combinations. He called these fundamental units "atoms"—from the Greek word *atomos* meaning "indivisible"—and we use the same word today. Let's take a closer look at the basic units of matter.

Elements and Atoms

The ancient Greeks believed that every living thing on the planet was made of some mixture of four substances: earth, air, fire, and water. The philosopher Empedocles (490–430 BCE) called these basic substances "elements," another term scientists still use. However, today we use

→ CONTEXT

The Republic Has No Need of Scientists

In his experiments, Antoine Lavoisier made other important discoveries: he identified and named oxygen as one of the elements involved in his combustion reactions and observed that these basic substances can exist in different forms, as solid, gas, or liquid. Unfortunately, Lavoisier's scientific career ended abruptly in 1794, during the French Revolution, when he was executed by guillotine along with many other members of the nobility. When he and his supporters pled for mercy so he could continue his experiments, the judge reportedly declared, "the Republic has no need of scientists." At the time, Joseph-Louis Lagrange, an Italian mathematician and scientist, mourned, "It took but a moment to cut off that head: perhaps a hundred years will be required to produce another like it."

the term **element** to describe substances such as gold, calcium, oxygen, and silicon that cannot be broken down into other substances. Chemists now identify 118 of these different elements on the periodic table. More than 90 of these have been observed to occur in nature, and the rest have been created in laboratories (**FIGURE 3.3**).

While the Greeks first proposed that elements are the building blocks of nature, our modern understanding of this concept did not arise until the early 19th century. Today we know that an **atom** is the smallest unit of an element that still has all the characteristics of that element. But what do we see if we look even closer?

What's Inside an Atom?

While atoms of an element can't be broken down into other elements, they are all made up of three even smaller particles: protons, neutrons, and electrons. Every atom of the same element always has the same number of **protons**, which determines the element's *atomic number* on the periodic table. Each proton has a positive electric charge. **Electrons** carry a negative electric charge. If the number of electrons and protons in an atom is equal, the atom

FIGURE 3.2 Matter Incoming from Space Earth receives only small additions of matter, such as from meteor showers from space.

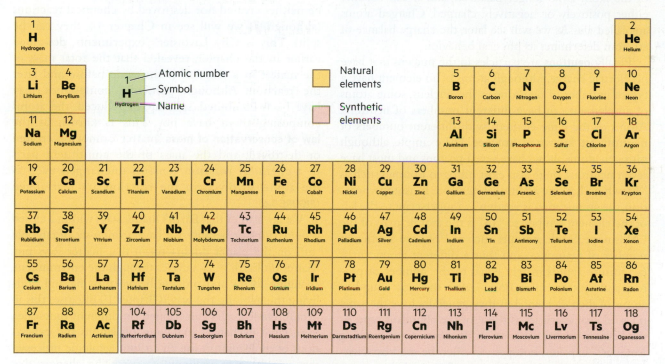

matter anything in the universe that takes up space.

mass a measure of the amount of matter in something.

element a substance that cannot be broken down into other substances.

atom the smallest unit of an element that retains all the characteristics of that element.

proton a particle within atoms that has a positive charge.

electron a particle within atoms that carries a negative charge.

FIGURE 3.3 Periodic Table The periodic table categorizes elements by the number of protons in each element (its atomic number) and by the physical and chemical properties of the elements.

neutron a particle within atoms that has no electric charge.

isotope an atom of an element with a particular number of neutrons.

molecule what is formed when two or more atoms (either the same or different) bond together.

compound a substance made of atoms of two or more elements bonded to each other.

law of conservation of mass the physical law that matter cannot be created or destroyed, and that the mass of the constituent parts in a chemical reaction remains unchanged even as the atoms involved in the reaction are rearranged.

chemical bond the force that holds the atoms in molecules together.

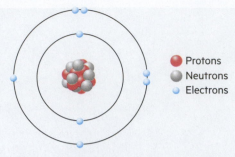

● Protons
● Neutrons
● Electrons

FIGURE 3.4 **The Bohr Atomic Model** Danish physicist Niels Bohr devised a model of the atom with a relatively small, positively charged nucleus containing protons and neutrons and a larger area containing orbiting, negatively charged electrons. These orbits are called shells or electron energy levels. They can hold a defined set of electrons at different distances from the nucleus.

remains electrically neutral. But unlike protons, which reside in an atom's nucleus, electrons orbit the nucleus in specifically defined paths called shells (as **FIGURE 3.4** illustrates). Electrons weigh almost 2,000 times less than protons, and atoms sometimes lose or gain electrons in the outer shell. When this happens, the electric charge of the atom is no longer balanced, and it can become either positively or negatively charged. Charged atoms are called *ions*. As we will see later, the charge balance of an atom determines its physical behavior.

Finally, **neutrons** are particles in the nucleus that have no electric charge. Although atoms of an element have a typical number of neutrons in the nucleus, some atoms of an element can have different numbers of neutrons. Atoms of the same element with different numbers of neutrons are known as **isotopes**. For example, although nitrogen atoms always have seven protons and most have seven neutrons, there also exist nitrogen isotopes such as ^{15}N, which has one additional neutron. In isotope notation, the atomic mass is written as a superscript to the left of the element symbol, so "15" is the atomic mass of the ^{15}N isotope. Stable isotopes such as ^{15}N have a different number of neutrons—and thus a different atomic mass (a measure approximately equal to the number of protons and neutrons in an atom)—than the most abundant atoms of their element, but they generally have the same physical characteristics as those of the abundant type. As the Science sidebar "What Can We Learn from Tracking Isotopes?" explains, isotopes can often be used by scientists to trace where material comes from and goes to.

🏠 **TAKE-HOME MESSAGE** Matter is anything that takes up space and has mass. It is made of elements: substances that cannot be broken down into other substances. Atoms are the smallest subunits of elements. Protons, neutrons, and electrons are particles within each atom that determine properties such as atomic mass and electric charge.

3.2 What Distinguishes Different Kinds of Matter?

The physical characteristics, or properties, of any material on Earth depend on three things: (1) the kind of atoms in the material, (2) the way these atoms are connected or bonded to each other, and (3) the way the atoms are arranged. In this section, we'll look at the kinds of atoms and the way atoms connect or bond.

When two or more atoms join or bond together, they form a **molecule**. Molecules can be combinations of atoms of the same element (such as oxygen gas, O_2, made of two oxygen atoms) or atoms from more than one type of element (such as water, H_2O, or carbon dioxide, CO_2). When the atoms of two or more elements are bonded to each other, this substance is called a **compound**. Molecules can range in size from two atoms, as in oxygen gas, to thousands, as in proteins and plastics (**FIGURE 3.5**).

Conservation of Mass

While chemical processes can break down a compound into its component atoms, these individual atoms can be neither created nor destroyed by chemical reactions (although, as we will see in Chapter 14, they can be split). This is why Lavoisier's experiments, described earlier in the chapter, revealed that the total mass of the matter in a reaction is the same before and after the reaction. Although particular compounds may have been combined, split up, or reduced, the atoms composing them have not. This is known as the **law of conservation of mass**: matter cannot be created or destroyed, and the mass of the constituent parts in a chemical reaction remains unchanged even as the atoms involved in the reaction are rearranged (**FIGURE 3.6**).

We saw earlier that Earth's matter stays within a closed system. But just because we have a finite set of tangible stuff on the planet doesn't mean that matter stays put. Matter on Earth in fact cycles through the environment, changing forms and moving between living and non-living things along the way (Chapter 10 discusses this in detail). These cycles can happen on different timescales, from rapid exchanges like your inhalations and exhalations of molecules in the air to long geologic processes that operate over millions of years. The changes that happen during these cycles use the same set of atoms over and over again.

How Do Atoms Bond Together?

The precise ordering and joining of atoms are fundamental to the properties of larger molecules and the materials of which they are a part. In fact, just 50 of the 118 known elements account for nearly

FIGURE 3.5 Chemical Formulas and Structures

(a) Chemical formulas are used to represent what makes up a molecule. A chemical formula includes symbols of the elements (such as "H" for hydrogen, "O" for oxygen, and "C" for carbon) and subscripts that denote the proportions of the elements. So the chemical formula for a molecule of carbon dioxide, which has one carbon atom and two oxygen atoms, is written as CO_2. Another way to depict this is with a ball-and-stick model, in which atoms are represented by spheres, while sticks represent the bonds holding them together.

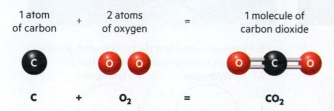

1 atom of carbon	+	2 atoms of oxygen	=	1 molecule of carbon dioxide
C	+	O_2	=	CO_2

(b) Some compounds are complex combinations of thousands of atoms—like this polyvinyl chloride (PVC) molecule.

everything we interact with in our lives. How do so few atomic "ingredients" account for the great variety of materials we see on Earth? Part of the reason comes from the **chemical bonds** that hold the atoms in molecules together. Different types of bonds work like the different types of glue you might have in your utility drawer. The different glues confer stronger and weaker bonding abilities to different types of materials.

How does this work with atoms? Atoms bond together because of the interaction of the electrons in their outer shells.

» *Ionic bond.* These bonds form when one atom "donates" an electron from its outer shell to another atom (**FIGURE 3.7A**). Depending on how many electrons are donated, these bonds can be very strong and result in very high melting points for ionic

What Can We Learn from Tracking Isotopes?

Many isotopes are relatively rare, so by closely examining where they occur, we can understand more about relationships in nature. For example, scientists have found significant amounts of the ^{15}N isotope in trees in the Pacific Northwest, though this isotope is almost always otherwise found in oceans. So what happened? Scientists have determined that much of this nitrogen comes from Pacific salmon. Salmon hatch in fresh water and then migrate to the ocean, where they grow to adult size and feed on marine organisms rich in ^{15}N. To reproduce, salmon return to their home rivers, swimming upstream to spawn farther inland. The salmon die soon after spawning, and if you visit one of these spawning areas at the right time, you'll see their carcasses littering the stream and surrounding forest. These dead fish are food for large predators such as bears and for much smaller creatures such as insects that feed on the rotting flesh. The bears and other creatures then transfer the nitrogen from the fish into the soil, either when they die and decompose or in their waste. The trees use this nitrogen as an important source of nutrition. By tracing this isotope, we see how spawning salmon fertilize the forest.

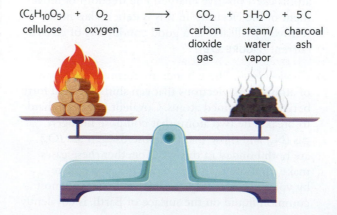

$(C_6H_{10}O_5)$	+	O_2	$\longrightarrow$	CO_2	+	$5\,H_2O$	+	$5\,C$
cellulose		oxygen	=	carbon dioxide gas		steam/ water vapor		charcoal ash

FIGURE 3.6 Conservation of Mass The law of conservation of mass means that when wood (which is mostly cellulose, or $C_6H_{10}O_5$) reacts with oxygen in a fire, it yields an equal number (and mass) of atoms—although they are now in the form of carbon dioxide gas, water vapor, and charcoal or ash.

(a)

Sodium donates an electron from its outer shell and shares it with the outer shell of chlorine.

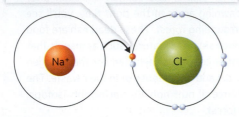

Salt bonds form in a three-dimensional lattice of ionic bonds. It takes a lot of energy to separate them.

(b)

Electrons from outer shells of metal atoms are shared.

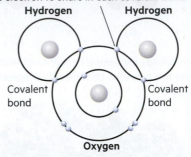

Metal ions

(c)

Each hydrogen atom and the oxygen atom provide one electron to share in each covalent bond.

Hydrogen   **Hydrogen**

Covalent bond   Covalent bond

Oxygen

FIGURE 3.7 Types of Chemical Bonds **(a)** Ionic bonds, such as those that hold sodium chloride (table salt) together, form when one atom "donates" an electron from its outer shell to another atom. **(b)** Metallic bonds, such as those in aluminum or copper, occur when groups of positively charged metal atoms (metal ions) share one another's outer electrons. This enables metals to be relatively flexible and good conductors of electricity. **(c)** Covalent bonds, such as those in water, share electrons that can shift back and forth between the joined atoms.

solids: common table salt (NaCl) melts at 801°C (1,474°F), which is the temperature at which the ions start losing their attraction to each other.

» *Metallic bond.* The bonds that connect metal atoms such as those in gold, iron, and copper are not permanent "donations" of electrons from one atom to another but rather are a system of outer electrons shared between metal atoms. In metallic bonds, negatively charged electrons flow freely among the metal atoms while the unmoving nuclei of the atoms exert positive charges. The freedom of movement of the electrons in this system makes metals relatively malleable and good conductors of electric current (**FIGURE 3.7B**).

» *Covalent bond.* These bonds are formed when pairs of atoms share electrons that can shift back and forth between the joined atoms. Covalent bonds are hard to break. The two atoms that compose nitrogen gas (N_2) and those that compose oxygen gas (O_2) are both bonded in this way. Together these gases make up more than 99% of Earth's atmosphere by volume. Water (H_2O), which is by far the most common liquid on the surface of Earth, is covalently

bonded because electrons are shared between an oxygen atom and two hydrogen atoms (**FIGURE 3.7C**).

» *Hydrogen bond.* These relatively weak bonds are formed when a hydrogen atom with a slight positive charge is attracted to an atom with a negative charge. Unlike in the chemical bonds described above that hold atoms together in molecules and metals, with hydrogen bonds no electrons are shared between atoms. A web of hydrogen bonds enables water molecules to "stick" to each other. As we learned above, each water molecule itself is formed by covalent bonds between oxygen and the two hydrogen atoms. However, the oxygen atom in water attracts electrons slightly more than do the hydrogen atoms, causing the oxygen atom to carry a slight negative charge while the hydrogen atoms carry a slight positive charge. This imbalance is called **polarity**, and it leads the hydrogen atoms of one water molecule to be attracted to the oxygen atoms of other water molecules like tiny magnets. The attraction created by these hydrogen bonds is why liquid water forms into droplets and can be "pulled up" through the porous structures of plant roots and blood vessels.

polarity an imbalance of positive and negative charge within an atom.

Mining Garbage for Plastic: One Way to Profit from the Different Properties of Matter

Today, most of us automatically toss our glass bottles and aluminum cans into a recycling bin, as recycling programs have become widespread. We are very successful at recycling some materials; for example, close to 90% of the world's steel is recycled each year. But not all materials lend themselves to easy recycling processes. Plastic has been a difficult material to recycle because plastic objects are difficult to sort from each other. Piles and piles of plastics of all different types arrive at recycling centers but are often discarded because so many different types are mixed together. And even when plastics are recycled, the end product is most often a less pure, lower-quality material that has more limited uses than newly created plastic. Manufacturers are less interested in buying it to make their products, and then it is more difficult to recycle in the future.

In 1994, chemical engineer Mike Biddle left his job making plastics at Dow Chemical to launch his own company, MBA Polymers, to focus on solving how to better recover plastics from the waste stream. He began the work out of his garage but soon gained investors and research grants from others who saw the economic and environmental potential of this project: buying recycled plastic would be cheaper and greener than making brand-new materials.

Eventually, Biddle succeeded in creating a new plastic-recycling process. While this process is a closely guarded secret forming the core of his business, Biddle does say that his discoveries all had to do with identifying essential properties of matter that aid in sorting. For example, after all the plastics are ground into confetti-like flakes, they are floated in various liquids for sorting because different types of plastics respond differently to certain liquids as a result of their chemical makeup. Flakes can also be sorted by color and by other means, such as scanning the appearance of their broken edges. Biddle's process involves more than 30 sorting steps.

Once sorted, each type of plastic can be cleaned, mixed, melted, molded, and chopped into pellets that can serve as the raw materials for new products. You can now find MBA Polymers' recycled plastics in many consumer goods, such as Nespresso coffee machines and Electrolux vacuum cleaners. After 20 years, $150 million of investments, and more than 60 patents, Biddle's company opened three international processing plants that can recycle plastics from some of the most difficult mixed-material sources. Plants in China and Austria process plastics from shredded e-waste such as computers and cell phones, while a plant in England processes plastics from shredded automobile components. In each case, the processing yields plastic pellets categorized by type, grade, and color that match the quality of newly produced plastic while requiring 80% to 90% less energy to make.

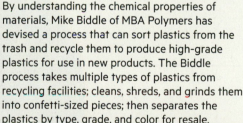

By understanding the chemical properties of materials, Mike Biddle of MBA Polymers has devised a process that can sort plastics from the trash and recycle them to produce high-grade plastics for use in new products. The Biddle process takes multiple types of plastics from recycling facilities; cleans, shreds, and grinds them into confetti-sized pieces; then separates the plastics by type, grade, and color for resale.

What Would You Do?

MBA Polymers keeps its process secret to give it a business advantage and make profits for investors. Biddle realizes his competitors are trying to discover the techniques and eliminate his advantage. If Biddle made these secrets freely available, even more plastic might be recycled using this process. However, Biddle might never have gotten the money to develop his process if he hadn't agreed to keep the process secret: if the process were known, larger corporations would have taken advantage of his research to earn money for themselves. If you were Biddle, what decision would you have made when you were getting started? Would you have tried to develop this on your own, partnered with larger players, or possibly even given the idea away?

What about Water Makes It So Important?

All life, including your own, depends on water. Not only does water cover more than two-thirds of Earth's surface, it is also present in the air we breathe and in the ground beneath us. And although less than 1% of Earth's water is accessible to us, the living cells of all organisms (including us) are largely composed of water. We humans are about 60% water by weight on average.

1 Water is a polar molecule.

The hydrogen atoms in a water molecule are in a "mouse ear" arrangement, with a 104.5° angle between them. Each hydrogen region of the molecule has a positive charge, like the positive pole of a magnet. In contrast, the oxygen region has a negative charge, like the negative pole of a magnet. This concentration of opposite charges into different poles makes water a polar molecule.

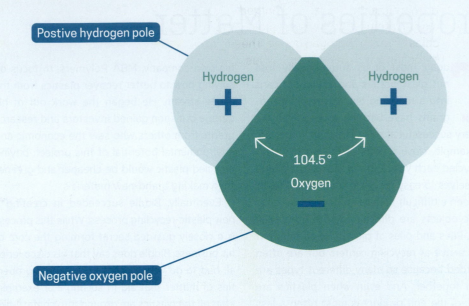

Postive hydrogen pole

Hydrogen +

Hydrogen +

104.5°

Oxygen −

Negative oxygen pole

2 Water is known as the "universal solvent."

It dissolves more substances than any other liquid. Water's positive and negative charges help it break down many kinds of molecules.

Table salt is made of alternating ions of sodium (positive charge, Na⁺) and chlorine (negative charge, Cl⁻).

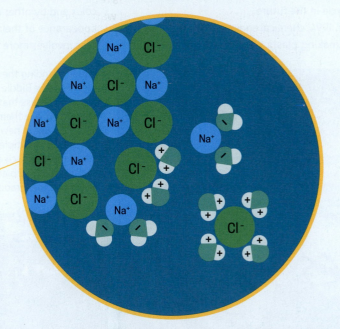

In water, the positive ends of multiple water molecules attach to and pull off negative chlorine ions. Negative ends of water molecules attach to and pull off positive sodium ions. This dissolves the salt.

3 Water is cohesive.

Water molecules "stick" to each other as the positively charged hydrogen atoms of one water molecule are attracted to the negatively charged oxygen atoms of other water molecules, forming hydrogen bonds. These hydrogen-bonded water molecules form a cohesive network.

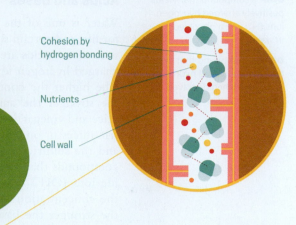

Cohesion by hydrogen bonding

Nutrients

Cell wall

Cohesion enables plants to draw up columns of water from their roots as water evaporates from the leaves. And because it is a solvent, as the water moves up the plant it also conveys the essential nutrients of life.

4 Water can absorb a lot of heat.

Water's cohesive hydrogen bonds require a lot of energy to break. This means that water can absorb a lot of heat before it turns into water vapor, its gaseous form. Water on the surface of Earth and within organisms can thus moderate temperature fluctuations by absorbing additional heat and distributing it to other places as it flows.

Liquid water is more dense than frozen water. It is the only commonly found compound on Earth with this property.

Ice floats in water, and water freezes from the top down rather than from the bottom up, which enables aquatic life to exist beneath a layer of ice.

acid a compound that yields positively charged hydrogen ions (H⁺) when dissolved in water.

base a compound that yields negatively charged hydroxide ions (OH⁻) when dissolved in water.

Acids and Bases

Water is one of the most important types of matter for life, and certain substances differ in how they react with water. **Acids** are compounds that yield positively charged hydrogen ions (H^+) when dissolved in water. The higher the concentration of H^+ in the solution, the stronger the acid. Weaker acids include orange juice and vinegar, while stronger acids include the gastric juices in our stomachs that aid in food digestion and the sulfuric acid used in car batteries. **Bases** are compounds that produce negatively charged hydroxide ions (OH^-) when dissolved in water. The higher the concentration of OH^- molecules in the solution, the stronger the base. Weak bases include blood and baking soda, while stronger bases include bleach and ammonia. The pH scale is used to express the strength of acid and base solutions. A pH of 7 is neutral. As pH decreases, the solution becomes more acidic, and as the pH increases, the solution becomes more basic (**FIGURE 3.8**).

The pH of a substance is very important as even slight changes can have large effects, especially on living things. The human body normally regulates the pH of blood to stay between 7.35 and 7.45. When these regulatory functions fail and blood pH is outside of this range, serious health problems ensue as proteins are destroyed and enzymes lose their ability to function. Similarly, when CO_2 present in the air dissolves in oceans, lakes, and rivers or when acid rain falls into these bodies of water, these habitats become more acidic. Fish and other aquatic organisms are very sensitive to changes in the pH of the water in which they live, so these changes can have major environmental impacts.

⬡ **TAKE-HOME MESSAGE** Matter cannot be created or destroyed, and matter on Earth cycles through the environment. The kinds of atoms that compose matter and the way these atoms bond to each other help determine the properties of the matter. Electric charges account for the variety of chemical bonds that hold compounds of atoms together. Acids and bases are substances that when dissolved in water produce H⁺ and OH⁻ ions, respectively.

3.3 Can Matter Change?

At the beginning of this chapter, we asked you to imagine a cookout. Although the burning pile of wood or charcoal shrinks—or a canister of propane empties—we learned that the matter continues to exist in the emissions and ash. But it is not as though the matter is merely moving from one place to another; it is also undergoing physical and chemical transformations. For example, in the heat of a campfire, some of the water within the wood will turn into steam. And some of the carbon in the fuel will react with oxygen to

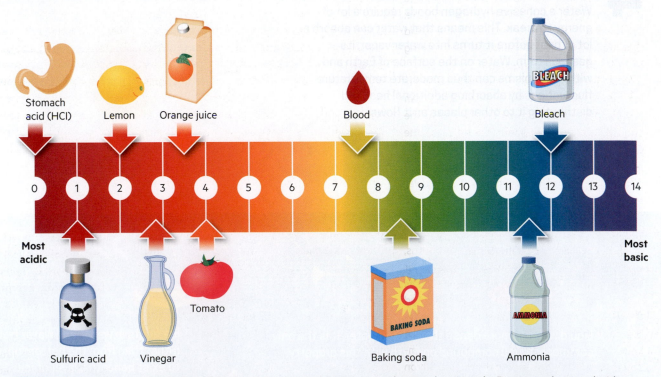

FIGURE 3.8 The pH Scale How acidic or basic a substance is determines its place on the pH scale. Pure water is neutral with a pH of 7. Substances with a pH greater than 7 are basic, while substances with a pH less than 7 are acidic.

Can We Tinker with Earth's Chemistry to Address the Effects of Climate Change?

Catastrophic climate change has often been the triggering event in postapocalyptic science-fiction movies. In the film and TV series *Snowpiercer*, the last remaining humans struggle to survive after a new ice age was triggered by scientists trying to engineer Earth's climate to combat global warming.

Today, some scientists are actually exploring the possibility of large-scale human interventions to redesign Earth's climate. These climate intervention approaches are known as *geoengineering*. Rather than addressing the source of climate change, they attempt to directly address its symptoms—such as rising temperatures and increasing acidity of ocean water—by altering the chemistry of the atmosphere or ocean. For example, Nobel Prize–winning chemist Paul Crutzen has proposed releasing sunlight-blocking particles known as sulfates into Earth's atmosphere via giant guns or balloons. This approach mimics the cooling effect very large volcanic eruptions can have on Earth's climate. Scientists considering this proposal have found that it could theoretically limit global warming to 2°C (3.6°F) above preindustrial levels.

Other geoengineering ideas propose altering the composition of the ocean so that it can absorb larger quantities of CO_2 gas from the atmosphere. As it is, about one-third of the CO_2 humans release into the atmosphere is absorbed by the oceans. This has slowed global warming, but when CO_2 is absorbed by seawater, chemical reactions make the water more acidic. One team of scientists has suggested countering this by dumping massive amounts of calcium oxide (quicklime) into the oceans, which would raise the pH and enable seawater to absorb more CO_2 from the air. Other chemists have found that similar results could be attained by sprinkling dust made of the mineral olivine throughout the ocean. Another approach known as iron fertilization would add iron compounds to the ocean to boost the population of phytoplankton. These microscopic organisms live in the ocean and, like plants, have the

Iron fertilization would use iron compounds to boost the growth of photosynthesizing phytoplankton—consuming more CO_2.

Injecting sulfate particles into the sky could block sunlight from Earth's surface.

Adding quicklime to the ocean could enable it to absorb more CO_2.

ability to consume CO_2. Under this scheme, large blooms of phytoplankton would absorb CO_2 and then sink to the ocean floor when they die, trapping the carbon in their bodies.

But each of these proposals has serious drawbacks. Large releases of sunlight-blocking particles in the past—such as those from the emissions of unregulated European factories and power plants in the 1970s and 1980s—caused droughts and famine in north-central Africa by disrupting weather patterns. Further, these particles would likely thin the protective ozone layer in Earth's atmosphere that limits the amount of harmful solar radiation reaching Earth's surface. Adding iron to the water could also stimulate toxic phytoplankton blooms that may not sink to the bottom.

Opponents of ocean geoengineering plans note that extracting, processing, and transporting the materials to be deposited in the ocean would require enormous amounts of energy. Similarly, the addition of olivine would entail large energy inputs because it needs to be ground to a fine powder and at least 100 large ships would need to operate continually to distribute the 1 billion tons per year needed to have the desired effect.

The geoengineering proposals described above would be very costly and would likely require government funding and complex international agreements—just like current global initiatives to cut greenhouse gas emissions. Because of the risks, impracticalities, and uncertainties of geoengineering proposals, most experts agree that reducing emissions is the better bet to combat climate change and that geoengineering should be considered as a last resort.

What Would You Do?

Imagine you are part of the next National Academy of Sciences review panel. Would you support US government funding for a geoengineering project? Consider the science, risks, ethics, role of government, or anything else that you feel is important. Use the internet to find more information to help make your decision, searching terms such as "geoengineering," "climate engineering," and "carbon sequestration."

phase change a change in matter from one state (solid or liquid or gas) to another without a change in its chemical composition.

pressure the force exerted on or acting against something.

chemical reaction the process by which one or more substances are converted into different substances.

energy the capacity to do work.

work in science, the act of applying force to an object over some distance.

kinetic energy energy embodied in something due to its being in motion.

potential energy energy in something that has yet to be released.

form CO_2 gas. In this section, we'll look at the two general ways matter changes: through phase changes and chemical reactions.

Phase Changes

We know that we can heat molecules of liquid water to change them into steam (a gas) or freeze them into ice (a solid). While these changes are familiar to anyone with a stove and a teakettle or a freezer and an ice cube tray, what is happening—and how? In science, these are termed **phase changes**, as the organization of atoms and molecules in a substance is altered by various conditions of heat and pressure that determine its *state of matter*—whether it is solid, liquid, or gas.

First, it is important to understand that all atoms and molecules move and vibrate constantly and are never completely still. Changes from one state to another occur because of changes in the vibration of the atoms and molecules. In *solids*, atoms and molecules vibrate relatively slowly and are fixed in a rigid arrangement by relatively strong bonds. This arrangement gives solids a fixed volume and shape. When you put an ice cube in a larger container and keep it frozen, it remains a cube, with its molecules maintaining a relatively fixed relationship with each other.

Atoms or molecules in *liquids* vibrate more rapidly than those in solids and are held together loosely by attractive forces that enable them to freely slide over each other. Liquids have a fixed volume but changeable shape: if you put a quantity of liquid water into a container, it will slosh around, changing shape as the molecules bang into and slide past one another. However, it will retain the same volume in any shape of container. At the molecular level, this is like beans in a beanbag chair: the molecules, or "beans," are able to move past one another, rearrange themselves, and change the shape of the beanbag while staying within its overall boundaries. Atoms and molecules in *gases* vibrate even more rapidly and randomly fly about, bouncing off everything they contact. In this state, molecules are not bonded to each other, and so gases have neither a fixed volume nor a fixed shape. Gases will expand to fill whatever volume contains them (**FIGURE 3.9**). If you put steam in a closed container, it will expand to fill the volume of the container as the molecules vibrate rapidly. If the container is a teakettle, the steam will eventually push out through any opening, such as the spout, and cause the kettle to whistle. The **pressure** of a gas—for example, the air pressure in car tires—is a result of the force of these rapidly bouncing and colliding molecules acting against something.

Chemical Reactions

So we see that the form or the state of matter can change depending on how the atoms or molecules vibrate and are physically arranged. However, molecules can be broken apart entirely and combine into new molecular arrangements through processes called **chemical reactions**. What are some common chemical reactions? A very familiar one occurs when we turn on a gas stove or fire up an outdoor grill. The burning, or combustion, of natural gas or other fuels is an *oxidation–reduction* (or redox) reaction involving the transfer of electrons between atoms. In these reactions, some substances lose electrons (this is called oxidation) and some substances gain electrons (this is called reduction). Oxidation and reduction always occur together. For example, when natural gas (methane, or CH_4) reacts with oxygen (O_2) on your stove, it produces carbon dioxide (CO_2) and water (H_2O). The carbon in the reaction loses eight electrons in the process, so it has undergone oxidation. The hydrogen atoms maintain the same number of electrons in the process, but the oxygen atoms end up gaining eight electrons, so they have undergone reduction. Oxidation–reduction reactions also happen on a cellular level when we eat. Carbon-rich food molecules (usually sugars) are oxidized in the cells of our body, reacting with oxygen to produce energy for our cells and the waste products CO_2 and H_2O.

Polymerization—the process by which many smaller molecules link up to create complex new chain-like

FIGURE 3.9 Properties of Gas Gases have neither a fixed volume nor a fixed shape. They will expand to fill whatever volume contains them, be it a bicycle tire or a blimp.

structures—is also a kind of chemical reaction. For example, the main protein in hair and fingernails is created by this process, as is the nylon in backpacks; the latter is manufactured through a synthetic polymerization process that joins six-carbon molecules together. Conversely, *depolymerization* breaks down biopolymers. If you have ever marinated steaks or vegetables and then cooked them, you have been depolymerizing their tough, stringy animal proteins or plant fibers (**FIGURE 3.10**).

⬡ **TAKE-HOME MESSAGE** Differences in the organization of atoms and molecules under various conditions of heat and pressure determine the particular states of matter: solid, liquid, and gas. Phase changes between these states leave the molecules themselves unchanged. But various kinds of chemical reactions can break molecules apart and form new molecular arrangements.

FIGURE 3.10 Polymerization and Depolymerization Nylon—like the material in this backpack—is made through a chemical reaction known as polymerization whereby many smaller molecules join up to build complex chain-like structures (polymers). The reverse of this process is the chemical reaction known as depolymerization—as with a marinade working to break apart the biopolymers in the animal tissue of a steak.

3.4 What Is Energy?

Heading back to the cookout we described at the beginning of the chapter, as our briquettes start to turn gray, we know they are ready to start cooking whatever is on our menu. If you put your hand over the grill, you can feel the heat emanating from the coals. Of course, if it's a sunny day, you can feel the heat from the Sun above you too. While we have spent the first part of the chapter discussing matter, our senses tell us that the universe is more than just a collection of stuff: there is also heat, light, and motion.

In the 19th century, physicists began using the concept of energy to describe these less tangible aspects of reality. We all use the term frequently throughout the day to describe how we feel. Are you full of energy, low energy, or somewhere in between? We even refer to some beverages as energy drinks. Formally, **energy** is defined as the capacity to do work. While most people associate the word *work* with a job or a task, in science the definition differs from our everyday use of this word. **Work** is the act of applying force to an object over some distance. Work accomplishes movement, and it can occur on a range of scales, acting over a few centimeters to flip a hamburger, many millions of kilometers to produce planetary motion, or a few billionths of a meter to cause the vibration of atoms. We often see energy as motion, and the energy possessed by something in motion is termed **kinetic energy**. A bird in flight, a moving car, and a falling rock all have kinetic energy. But energy can also be stored. The bird and the car each have chemical energy stored in the bonds of organic compounds they use as fuel, and this is called **potential energy** because it has yet to be released. Similarly, a rock perched on a cliff or water in a reservoir behind a dam has potential energy due to the pull of gravity. As the water flows through the dam, its

→ **ENVIRONMENTAL JUSTICE**

Toxins and Fish Consumption

Some substances, known as persistent bioaccumulative toxins (PBTs), are particularly harmful because they can cycle through food chains. As bigger animals consume smaller animals and plants containing the contaminants, those at the top of the food chain (including humans) accumulate higher concentrations of the toxins in their bodies. PBTs often enter the food chain when industrial and agricultural discharges and stormwater pollution flow into bodies of water. People living near the water may have seen signs by a dock, pier, or beach warning them not to eat the fish in the area because of PBTs.

Although the sources of PBTs are all around us, the EPA has found that communities of color, low-income communities, certain immigrant groups, and Indigenous peoples depend on fish to a much greater extent than the general population. Fishing is also a central facet of the cultural practices in many of these communities. These communities ultimately are at greater risk from PBTs.

Until 2000, the EPA used just one-half pound of fish per month per person to guide its policy decisions—a number far below the rate consumed in many communities. Since that time the agency has raised the fish consumption rate and set a distinct consumption rate for "subsistence fishers" at roughly 10 pounds (4.5 kilograms) per month per person, and it encourages states to develop criteria "to protect highly exposed population groups." Still, many states have been slow to adopt new standards, and federal court cases have ensued over which fish consumption rates should be used to set policies.

temperature in common terms, a measurement of the hotness or coldness of something; formally, a measure of how vigorously the atoms in a material are moving and colliding with each other.

photosynthesis a series of chemical reactions involving water and CO_2 through which plants and some other organisms store the Sun's energy in simple sugars.

potential energy becomes kinetic energy when it turns a turbine to produce electricity. If this water flows down the river and then is held up behind a second dam, it turns into a source of potential energy again (**FIGURE 3.11**).

Forms of Energy

Just as energy changes from potential to kinetic energy, its form can change too. In fact, you can trace the conversion of energy into its different forms and back again. Let's look at the different forms of energy we use every day. In our daily life we use our bodies—our physical labor—to push, pull, lift, throw, or otherwise exert force on some object that moves it some distance. We also use many labor-saving devices, machines that exert force on some object for us: snowblowers that move heavy loads of snow, cordless drills that turn screws, and electric mixers that blend ingredients together. The energy used to perform these tasks is referred to as *mechanical energy*. But most of these examples of mechanical energy result from a transformation of *chemical energy*: energy stored in the bonds of atoms and molecules and released with chemical reactions. All organisms store the energy required for their mechanical tasks in the bonds of molecules. Carbohydrates, such as the sugar glucose ($C_6H_{12}O_6$), are a common source of energy for many organisms. When bonds in glucose are broken, energy is released. Similarly, a snowblower is powered by energy stored in the bonds of carbon-based molecules in the petroleum fuel filling its gas tank. A cordless drill uses the chemical energy stored in the bonds of lithium cobalt oxide ($LiCoO_2$) in its lithium-ion battery, the same type of battery that powers most tablets, laptops, and cell phones. Almost every day, we plug these batteries into a wall socket to recharge them, so they are powered by a transformation

of *electrical energy*. The electric mixer we use in cooking and baking is also powered by electrical energy, but in its case the electrical energy is transformed directly into mechanical energy rather than acting through a chemical intermediary like a battery. Electrical energy is created by moving electrons that are carried along a wire or other conductive material. If we follow those wires far enough, back through your wall, through power lines, and across a regional electric grid, we would arrive at some sort of generator that is transforming another form of energy into electricity. In Chapters 13 and 14, we will see that there are many ways to generate electricity, but currently, most of the world's electricity is generated by coal-fired or natural gas–fired power plants—which utilize another form of chemical energy.

However, the coal or natural gas burned at the power plant, the petroleum burned by the snowblower, and the carbohydrates powering your body can all be traced back to the common source of all energy: the light, or *radiant energy*, of the Sun. You can feel thermal energy radiating directly from the Sun when sunlight hits your skin. *Thermal energy* comes from the movement of molecules, and **temperature** is not just a measure of how "hot" something is; it is a measure of how vigorously the atoms in a material are moving and colliding with each other. Light from our Sun is made up of tiny elementary particles called photons that travel at the speed of light in waves through space. When they reach Earth, their energy is used by plants, algae, and some bacteria to power a series of chemical reactions that uses water and CO_2 and is called **photosynthesis** (**FIGURE 3.12**). Photosynthesis stores this energy in molecules such as simple sugars that are based on the element carbon. As we will study in more detail later, the incomplete decomposition of these types of organisms over millions of years formed the coal, petroleum, and natural gas that we use today.

The Sun itself is powered by *nuclear energy*, energy contained within the nuclei of atoms. Within the Sun, the force of gravity is so strong that it can smash two hydrogen atoms together with enough force that their nuclei fuse, forming atoms of helium. This reaction, called *nuclear fusion*, releases tremendous amounts of energy, some of which reaches Earth as the photons we described. Our nuclear power plants harness the energy within the nuclei of atoms in a different way, by splitting the nucleus of an atom and producing two or more different types of atoms in the process. This is *nuclear fission*, which we'll learn about more in Chapter 14.

Although sunlight provides Earth with a massive external source of energy, heat from the Earth's core—*geothermal energy*—also powers global processes. Geothermal energy powers huge natural events such as earthquakes and volcanic eruptions. But it can also be harnessed for more practical purposes, such as powering the electric grid of the country of Iceland or, more locally, the heating and cooling systems in homes equipped with heat pumps.

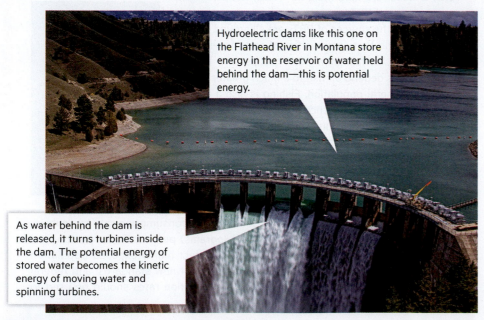

Hydroelectric dams like this one on the Flathead River in Montana store energy in the reservoir of water held behind the dam—this is potential energy.

As water behind the dam is released, it turns turbines inside the dam. The potential energy of stored water becomes the kinetic energy of moving water and spinning turbines.

FIGURE 3.11 **Using Dams to Store and Release Energy**

FIGURE 3.12 Photosynthesis
The ultimate source of all energy in living and formerly living things is the Sun. Plants, algae, and some bacteria store the radiant energy of the Sun in carbon-based molecules through the process of photosynthesis. Photosynthetic organisms like this tree use the energy of the Sun to break apart water molecules and combine its hydrogen atoms with carbon dioxide to form simple sugars.

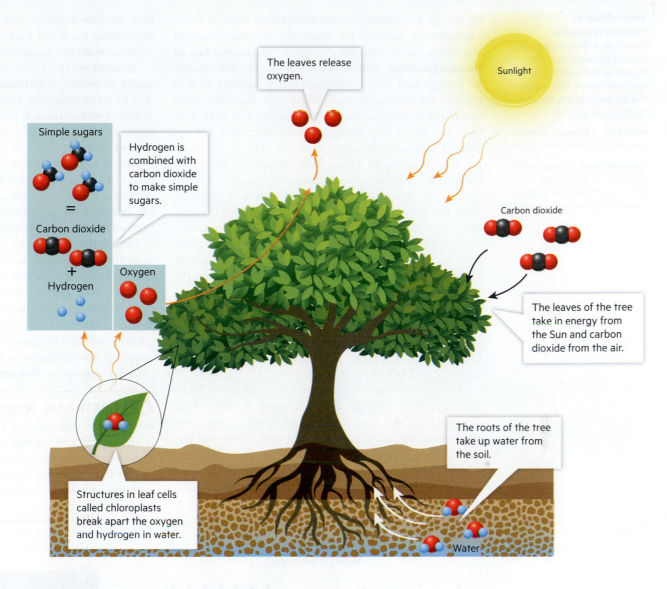

The leaves release oxygen.

Sunlight

Simple sugars

Hydrogen is combined with carbon dioxide to make simple sugars.

=

Carbon dioxide

+

Hydrogen

Oxygen

Carbon dioxide

The leaves of the tree take in energy from the Sun and carbon dioxide from the air.

Structures in leaf cells called chloroplasts break apart the oxygen and hydrogen in water.

The roots of the tree take up water from the soil.

Water

⬡ **TAKE-HOME MESSAGE** Energy is the capacity to do work. There is energy in motion, called kinetic energy, but energy can also be stored as potential energy. Forms of energy include mechanical, chemical, electrical, radiant, nuclear, and thermal.

3.5 What Happens to Energy When We Have Used It?

When we use a cell phone or laptop to send texts or photos or to stream music or videos, the device will gradually get hotter and the battery power level will go down. If we don't recharge the device, the battery will die. It seems as though the energy in the battery is gone and that we'll need to refill it from a source that will produce some more. However, this is not exactly how things work. Energy, like matter, is constant in the universe: more is not being created, and what we have is not being destroyed. The energy that was in your device's battery is not gone; it has just changed into another form. Underlying this basic concept are two important relationships in the universe, the first and second laws of thermodynamics. In the early 19th century military engineers in Britain and Germany developed these laws as they observed the heat generated from working steam engines and the boring of cannon barrels.

The First and Second Laws of Thermodynamics

The **first law of thermodynamics** tells us that energy is conserved: it cannot be created or destroyed; it merely changes form. Because of this, we can follow a series

first law of thermodynamics
the physical law that energy is conserved: it cannot be created or destroyed; it merely changes form.

second law of thermodynamics the physical law that with each transformation or transfer of energy, some energy is degraded or wasted, and that the tendency of any isolated system is to disorder.

entropy a measure of the degree of disorder in a system.

trophic level an organism's position in a food chain indicating its number of steps away from primary producers.

of energy exchanges or transformations such as those illustrated in Section 3.4. However, this succession of transformations also demonstrates the **second law of thermodynamics**, which considers the overall quality of energy and how it degrades or is wasted as it is transferred or transformed. With each transformation of energy, a certain amount is dissipated as heat. This does not mean that energy is destroyed, but it is considered degraded because it is less able to do work for us. Think again about your phone or computer battery. With each transformation of energy, a certain amount is dissipated as heat that you feel radiating from the device. Remember, heat results from the random motion of vibrating molecules. At a molecular level, when faster-moving (warmer) molecules collide with slower-moving (cooler) molecules, the slower molecules speed up when they are bumped, while the faster molecules are slowed down a bit by the contact. Because of this, heat dissipates, moving spontaneously from warmer regions to colder ones. Heat is considered a degraded, or lower-quality, form of energy because as it dissipates, it eventually escapes into the universe and is lost to us.

This means that energy conversions are never 100% efficient. Some energy always escapes as waste heat. Think about the way heat escapes from energy conversions in our daily lives. The hot coffee that you buy before class gradually cools. When we drive our cars, friction from the engine, the application of brakes, and the wheels on the road produce some heat. For example, even the most efficient internal combustion automobile engines still lose more than 60% of the energy they produce in forms such as heat rather than in power generation. Even our bodies radiate waste heat into the universe from the metabolic processes that keep us alive.

Sometimes the second law of thermodynamics is stated in terms of **entropy**. This is a measure of the degree of disorder in a system. "Order" implies that things are positioned in a regular and predictable pattern, like the neatly arranged bricks of a building or the carefully arranged cells of your body. The loss of useful energy to heat means that without energy inputs, things tend to become more disordered: entropy increases. One implication is that all things tend to break down over time due to increasing entropy. Nonliving and living things alike eventually wear out. So energy flows in one direction not only toward the dissipation of heat into space but also toward disorder. A second implication is that highly organized systems—cells, organisms, ecosystems, and even our cities—depend on constant inputs of energy. In the absence of energy inputs, the amount of useful energy available to do the work required to maintain order can only decline, and disorder will ensue (**FIGURE 3.13**).

⬡ **TAKE-HOME MESSAGE** The first law of thermodynamics tells us that energy cannot be created or destroyed but can change from one form to another. The second law of thermodynamics tells us that no conversion of one form of energy to another is 100% efficient. With each energy transformation, a certain amount of energy is dissipated as heat.

What Is Entropy?

High entropy

Low entropy

Bricks thrown in a pile will likely fall in a disordered pile. This pile has high entropy.

It takes energy to order these in a chimney. This finished chimney has low entropy.

Over time, without upkeep, the chimney will crumble. It will tend back toward a state of high entropy.

FIGURE 3.13 Entropy in a System The second law of thermodynamics can also be cast in terms of entropy—or disorder. Entropy increases in any system left to itself. In other words, it will become increasingly disordered. It takes inputs of energy to organize and form a load of bricks into a chimney, and without upkeep and maintenance the highly ordered chimney will break down over time.

3.6 How Does Energy Affect Life?

Unlike matter, which cycles through living and non-living things on Earth, energy degrades in each transformation. You can think of energy as following a one-directional rather than cyclical flow that diminishes with each use. How does the one-directional flow of energy affect life? To answer this question, let's look at a food chain—a model that shows how organisms get their energy from food.

Food Chains and Trophic Levels

Ecologists who study living animals and plants categorize living things by their **trophic level**, which is their place in the *food chain*. Food chains basically show who eats what, so a trophic level represents the level at which an organism feeds in an ecosystem. Thus, a great white shark in the ocean is categorized at a higher trophic level than the fish that it eats. In this scheme, plants that transform sunlight into chemical energy held within the molecules of their structures through photosynthesis are called **primary producers**. Organisms that consume plants directly are **primary consumers**. Herbivores such as deer, rabbits, elephants, and grasshoppers all belong to this trophic level. Organisms that consume primary consumers are called *secondary consumers*. These include meat eaters, or carnivores, such as wolves and lions, and omnivores that eat both meat and plants, such as bears, raccoons, and of course humans. And these trophic levels could continue to extend upward; if an organism (like a lion) were to eat a secondary consumer (like a meat-eating human), it would be functioning as a *tertiary consumer*. Food chains also include organisms called *decomposers* that feed on dead organisms.

FIGURE 3.14A shows a simple food chain and the energy available at each trophic level. We will see in Chapter 4 that most feeding relationships in ecosystems are more complicated and web-like. But this simplified depiction illustrates the way energy is available in diminishing amounts at each trophic level. The feeding relationships between the trophic levels also represent conversions of energy. Plants transform sunlight into the chemical energy held in their cells via photosynthesis. Primary consumers get their energy from eating the plants and so on up the food chain.

The 10% Law

Organisms at each trophic level transform some of the energy from the biomass they eat into biomass of their own. However, the second law of thermodynamics tells us that these transformations are never 100% efficient: some amount of energy flows from a more usable to a less usable form. Notice how the further removed an organism is from the energy of the Sun, the less

FIGURE 3.14 Trophic Levels and the Second Law of Thermodynamics

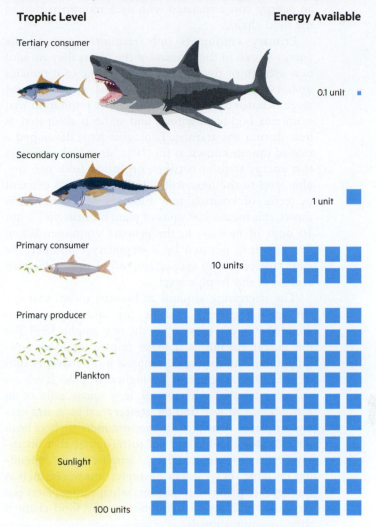

Trophic Level	Energy Available
Tertiary consumer	0.1 unit
Secondary consumer	1 unit
Primary consumer	10 units
Primary producer (Plankton)	100 units

Sunlight

(a) In a food pyramid, organisms are arranged in trophic levels on the basis of how they obtain energy. Because of the second law of thermodynamics, less energy is available at each step up the pyramid as energy is transferred between trophic levels. The "10% law" holds that for each step up the trophic pyramid, only about 10% of the energy from consumed organic matter is put to use creating biomass. In other words, there is a significant energy cost associated with each transformation up the food chain.

(b) For this reason, organisms at higher trophic levels (such as lions) are less abundant and compose far less mass per unit area than those at lower levels (such as grass).

primary producer
an organism that transforms sunlight into chemical energy held within the molecules of its structure.

primary consumer
an organism that consumes plants directly.

energy there is available at that trophic level. There is an energy cost associated with each transformation in the food chain.

Primary consumers only transform part of the energy stored in the biomass of the plants they eat into new growth or tissue replacement. Some of it remains undigested and is expelled untapped as waste; even more is burned off during the cellular respiration that maintains bodily functions; and some is dissipated as heat during the transfer. Ecologists have developed a rule of thumb known as the *10% law*, which estimates that energy transfer between organisms from one trophic level to the next will only be about 10% efficient in terms of creating new biomass. As Figure 3.15a shows, this means 100 units of plant biomass yields just 10 units of biomass in the primary consumer. When this organism is eaten by a secondary consumer, the same rule of thumb applies, and only 1 unit of biomass accrues at this trophic level.

The decreasing amount of biomass means that the amount of energy available for consumption also diminishes as you move up to the next trophic level. For this reason, organisms at lower trophic levels are typically more abundant and comprise more mass per unit of area than organisms at higher trophic levels in an ecosystem. In other words, in a square mile of an ecosystem, you will find far greater total mass and species of plants than of tertiary consumers such as lions (**FIGURE 3.14B**). It is important to recognize that organisms can consume at different trophic levels depending on what they eat. For example, omnivores, such as humans, are primary consumers when eating plants but secondary consumers or higher on the food chain if they consume animals.

⬡ **TAKE-HOME MESSAGE** Trophic levels show where living things feed in a food chain. There is less energy available at higher levels of a food chain. Primary producers are organisms such as plants that transform energy from sunlight into stored chemical energy in the form of simple sugars. Primary consumers eat the primary producers, and higher-level consumers eat lower-level consumers. The 10% law holds that each energy transfer between trophic levels will only be about 10% efficient.

🌍 ❓ 3.7 What Can I Do?

Understanding the basic science of matter and energy is fundamental to understanding environmental problems and knowing how to make sustainable choices. Getting a broader sense of all the linked causes and effects helps identify ways to tackle environmental problems. At the start of this chapter, we asked you to imagine a fire at a cookout, and we learned that every bit of the matter and energy in that fire goes somewhere. Now consider the importance of following the pathways of matter when we consider fires on a much larger scale.

Understand the Connections and Help Others to Do So Too

At the beginning of the chapter, tropical forest fires were linked to the production of palm oil. So what is this oil used for? Who uses it? Chances are you do—because palm oil is the world's most commonly used vegetable oil. By some estimates, it is an ingredient in more than half of the products commonly found on grocery store shelves, including peanut butter, cookies, pizza, soap, and lipstick. The Roundtable on Sustainable Palm Oil (RSPO) standard is a program that encourages palm oil producers and processors, along with food product companies, to support only oil production that avoids fires, deforestation, and other negative effects. Companies and products that meet these standards are allowed to put the RSPO label on their products. Students can also lobby for more sustainable practices in campus dining. One group that has helped do so is the Real Food Challenge organization, which supports student efforts in sustainable campus dining.

Contact Your School's Office of Sustainability

Many institutions today have staffed offices charged with leading sustainability efforts on campus. Search for yours on the Web (for instance, see if your school has an "office of sustainability" or search with the word "sustainability" along with your campus's name) and reach out to them. New York University, for example, provides opportunities to become an EcoRep who works with residence halls to promote sustainable practices.

Learn to Follow the Matter

Consider the materials that make a product and what happens to those materials when you are done with it. Trying to use materials that are easily recycled takes into account that whatever you consume ends up somewhere. And this saves energy too. Consider that

›› recycling steel and aluminum cans saves 74% and 95%, respectively, of the energy needed to produce them.

›› paper makes up 29% of the municipal waste stream.

›› Americans recycle only 9% of the plastic they use.

›› circuit boards in electronics can be recycled for valuable elements such as gold, silver, and platinum.

Understand That Everything Goes Somewhere

For years, tiny spheres of plastic called microbeads were used in skin care products and toothpastes. However, because these don't break down, most ended up in waterways, damaging that habitat. An environmentalist named Stiv Wilson noticed this and began a campaign to eliminate the use of microbeads in these products, ultimately leading to a bill banning microbeads that passed with bipartisan support and became law in 2015. You too can become more aware of products that contain unnecessary polluting materials and choose to use more sustainable alternatives.

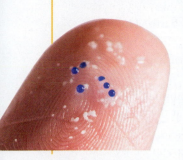

Find Your Opportunity to Innovate

The MBA Polymers case (see the Stories of Discovery feature) showed that there are interesting business opportunities in the life cycle of waste and energy. Could one of the next innovators be you? Others have taken the plunge. When he was 27, David Amster-Olszewski moved to Colorado Springs to start a company because of a state incentive to provide solar power. He founded a company called SunShare where local citizens own parts of solar panels. When Inna Braverman was 29, she founded Eco Wave Power, which harvests energy from ocean waves. Sanwal Muneer started Capture Mobility, a company that places windmills by the side of freeways to turn the turbulent air from passing vehicles into usable energy. Where do you see opportunities to innovate?

Chapter 3 **Review**

SUMMARY

- Matter is made up of atoms, and the properties of any kind of matter are determined by the kinds of atoms that compose it, the way these atoms bond to each other, and the way the atoms are arranged.
- The properties within atoms—such as atomic mass and electric charge—are determined by subatomic particles known as protons, neutrons, and electrons.
- Molecules form when two or more atoms bond together.
- Electric charges account for the variety of chemical bonds that hold atoms together in molecules.
- Differences in the organization of atoms and the molecules they compose under various conditions of heat and pressure determine the various states of matter: solid, liquid, and gas. Phase changes between these states leave the molecules themselves unchanged.

- Chemical reactions can break molecules apart and form new molecular arrangements.
- Energy is the capacity to do work, and it can come in many forms, including mechanical, chemical, electrical, radiant, nuclear, and thermal.
- The first law of thermodynamics tells us that energy cannot be created or destroyed but can only change from one form to another.
- The second law of thermodynamics tells us that no conversion of one form of energy to another is 100% efficient. With each energy transformation, a certain amount of energy is lost to us as dissipated heat.
- The feeding relationships between different life-forms involve conversions of energy, and because of the second law of thermodynamics, there is less energy available at each level up a food chain.

KEY TERMS

acid	energy	molecule	primary producer
atom	entropy	neutron	proton
base	first law of thermodynamics	phase change	second law of thermodynamics
chemical bond	isotope	photosynthesis	temperature
chemical reaction	kinetic energy	polarity	trophic level
compound	law of conservation of mass	potential energy	work
electron	mass	pressure	
element	matter	primary consumer	

REVIEW QUESTIONS

The letters following each Review Question refer to the Chapter Objectives.

1. What is the definition of matter? Which of the following items can be called matter, which cannot, and why? (a) brick, (b) helium gas, (c) oil, (d) fire, (e) bees **(A)**

2. At the most fundamental level, describe what all matter is made of. **(A)**

3. Explain what determines the electric charge of an atom. **(A)**

4. Explain how different atoms of the same element could have different atomic mass. **(A)**

5. What three qualities determine the properties of any material on Earth? **(B)**

6. What does the law of conservation of mass say? Imagine you watch a muddy puddle of water evaporate during a hot day. Explain what happened in terms of the law of conservation of mass. **(C)**

7. Name the four types of chemical bonds, and describe how each bond holds atoms together. **(B)**

8. What are two examples of matter undergoing a phase change? **(C)**

9. Give an example of a chemical reaction. **(C)**

10. What is the ultimate source of all energy on Earth? How do energy and work differ? **(D)**

11. If energy cannot be created or destroyed, what happens to it as we use it? **(E)**

12. What does the term *trophic level* describe? Consider a shark, a goldfish, and seaweed. Which of the three is at the highest and which is at the lowest trophic level? **(F)**

FOR FURTHER THOUGHT

The letters following each item refer to the Chapter Objectives.

13. As we drive a gas-powered car, the fuel tank empties and eventually we'll need to fill it up again. How can the law of conservation of mass explain what is happening to the matter that was recently in our fuel tank? How might it explain why the air we breathe is changing? **(C)**

14. Natural gas water heaters that provide hot water for showers and taps in many households work by igniting a flow of natural gas into a flame that heats a metal tank filled with water. A simple way to make this system more efficient is to wrap the metal tank with foam or another insulating material. Starting with the natural gas, identify the energy in this system as it changes forms, and use the first and second laws of thermodynamics to explain why the insulation can make the system more efficient. Use the same reasoning to describe two other home heating/cooling energy-conservation measures that could be taken in a typical house. Can a building ever be perfectly efficient with regard to energy for heating and cooling? Why or why not? **(E)**

15. Consider two life-forms in a food chain: grass and a herd of cows in a meadow that eat the grass. Identify which of these two beings is at a higher trophic level. Explain which of the two requires more of the Sun's energy to produce its body mass and why. **(F)**

Make Your Case ?

The following exercises use real-world data and news sources. Check your understanding of the material and then practice crafting well-supported esponses.

Use the News

The following article appeared on the website ScienceDaily.com. It describes how a study using isotopes is helping us better understand the food web of sharks. Use this article to answer the questions that follow. The first two questions test your understanding of the article.

Questions 3 and 4 are short answer, requiring you to apply what you have learned in this chapter and cite information in the article. Answers to Questions 1–2 are provided at the back of the book. Question 5 asks you to make your case and defend it.

"Global Analysis Reveals How Sharks Travel the Oceans to Find Food," University of Southampton, January 18, 2018

You've heard of "you are what you eat"—this research shows that for sharks, the more relevant phrase is "you are where you ate."

A major international collaboration led by the University of Southampton could help global efforts to overturn recent declines in the world's shark population by providing greater insight into the feeding habits of the world's most misunderstood fish.

Led by Dr. Christopher Bird during his PhD at Southampton, the study published in *Nature Ecology & Evolution* used chemical markers in the form of carbon isotopes found in sharks to investigate where in the world they have been feeding—an unresolved question for many shark species. Knowing which parts of the global ocean are important shark feeding areas may help to design more effective conservation measures to protect declining shark populations.

All life depends on carbon at the bottom of the food chain. Carbon comes in three forms or isotopes, and the proportions of two of the most common isotopes vary across the world's ocean. In the study, 73 scientists from 21 countries compared the carbon isotopes from more than 5,000 sharks from 114 species across the globe with those from phytoplankton at the bottom of the food web.

"If an animal feeds in the same place where it was caught, the carbon isotope signals in the shark and phytoplankton will match," says Christopher, whose PhD research was focused on deep-sea sharks. "However, if the shark has moved between feeding and where it was caught, then the signals will be different."

"You've heard of 'you are what you eat'—well this is more 'you are where you ate,'" Bird continued. "We were able to show that sharks living close to land and those that live in the open ocean have very different ways of feeding."

The results show that sharks living near to the coast feed locally across a range of different food webs—this is like people living in a city with access to lots of different restaurants in the neighbourhood and no need to travel far to find the food they want. On the other hand, oceanic sharks that are found throughout the world's oceans appear to get most of their food from specific areas of cooler water in the northern and southern hemispheres. This is more like travelling long distances from rural areas to spend lots of time eating in a few restaurants in a distant city.

"With over 500 known species around the world, sharks are certainly amongst our most diverse and misunderstood group of fish but we still have limited knowledge of their habits and behaviours, particularly relating to feeding and movement," said Dr. Christopher Bird. "Over the last 50 years, the pressures of fishing and habitat degradation have resulted in declines amongst some of the world's shark populations, the effects of which are also not fully understood."

Senior author Dr. Clive Trueman, Associate Professor in Marine Ecology also from the University of Southampton, added, "The results have important implications for conservation. Globally, sharks are not doing well. Many shark populations have declined in the last few decades, particularly in the wide-ranging oceanic sharks that are targeted by fishing boats and caught accidentally in tuna fisheries as 'by-catch.' Governments are now creating large marine protected areas around the globe, which help to reduce fishing, but most of these protected areas are in tropical waters, and may not provide effective protection for oceanic sharks."

"Sharks urgently need our help, but to help them we also need to understand them. Our study has helped by identifying important shark feeding grounds. New technologies like satellite and isotope tracking are giving us the information we need to turn the tide on these beautiful and fascinating animals."

---

1. Which of the following statements accurately describes how scientists used isotopes to discover more about the feeding patterns of sharks?

   a. Scientists measured the amount of carbon isotopes floating in the water in different areas and compared this to the isotopes in sharks.

   b. Scientists compared carbon isotopes found in sharks with the carbon isotopes in plankton in different areas to find matches.

   c. Scientists inserted carbon isotopes under sharks' skin to track them as they moved to different parts of the globe.

   d. Scientists fed special carbon isotopes to fish they knew sharks liked to consume to see how much of those types of fish each type of shark normally ate.

2. Which of the following statements most accurately describes the feeding habits of coastal sharks versus oceanic sharks?

   a. Coastal and oceanic sharks travel frequently and eat whatever they can find, but they prefer to eat in areas of cooler water.

   b. Coastal and oceanic sharks both only travel short distances to find food, so both eat from a small number of local food webs.

   c. Coastal sharks normally eat from locations near them across many food webs; oceanic sharks travel long distances to eat from food webs in specific places.

   d. Coastal sharks normally eat from just a few food webs located in different oceans; oceanic sharks eat from many food webs across all the oceans.

3. Reread the Science sidebar "What Can We Learn from Tracking Isotopes?" in this chapter. Briefly describe how the process for tracking salmon with isotopes compares with the process for tracking sharks with isotopes.

4. The article implies that governments should be creating more areas protected from fishing in important oceanic feeding zones for sharks. Using ideas that you have learned in this chapter, discuss how protecting these zones may have an impact on ocean ecosystems in terms of the 10% law.

5. **Make Your Case** If the phytoplankton in an important cold-water feeding area for sharks died off because they were susceptible to rising ocean temperatures, what would happen to the sharks that normally fed there? If you were put in control of this feeding area, would you advocate using geoengineering techniques to restore historical ocean temperatures? Explain why or why not using information from the chapter to support your answer.

Use the Data

The following graphic shows information about the energy use (represented by wattage, where W = watts), energy savings, life span, and cost of different types of lightbulbs.

Study the figure, and use it to answer the questions that follow. The first two questions test your understanding of the figure. Questions 3–5 are short answer, requiring you to apply what you have learned in this chapter and cite information in the figure. Answers to Questions 1–3 are provided at the back of the book. Question 6 asks you to make your case and defend it.

1. According to the figure, how much less energy input is required for a light-emitting diode (LED) bulb than for a traditional incandescent bulb to produce the same 800 lumens of light?

   a. 0%
   b. ~25%
   c. ~75%
   d. 85%

2. Per the prices in the figure, which type of lightbulb is cheapest in terms of cost per hour of light?

   a. LED bulb
   b. traditional incandescent
   c. halogen incandescent
   d. compact fluorescent (CFL)

   What is the cost per hour of the option you chose?

3. The figure shows bulbs that provide light using several forms of energy. Each of the items listed below involves a particular form of energy mentioned in the chapter (chemical, mechanical, nuclear, electrical, radiant, thermal, or geothermal). For each item, provide the form of energy that is being used or produced. Some may have more than one form of energy in use.

   a. the halogen gas in the halogen incandescent
   b. the LED as voltage is applied to it
   c. the heat from a lit traditional incandescent bulb
   d. the electricity powering all the bulbs

4. Each bulb puts out 800 lumens of light, but which bulb wastes the most energy doing so? What information on the figure gives you this answer? Considering what you learned in the chapter, what is this wasted energy an example of?

5. Which type of lightbulb do you think produces the least heat? Explain your answer using information from the figure and reasoning about the loss of heat as discussed in the chapter.

6. **Make Your Case** Imagine you were opening a clothing store and needed to purchase lights for your business. What factors might influence your decision, including data from the figure? What lightbulbs would you choose if you owned an unattended parking lot that must be lit 24 hours a day? Explain your answer citing data from the figure and/or information from the chapter to support your choice.

Comparisons of Types of Lightbulbs
Each bulb produces the same amount of light, about 800 lumens.

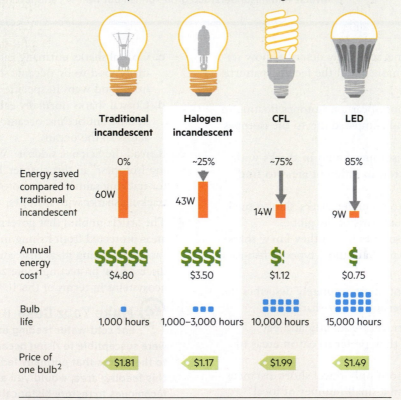

	Traditional incandescent	Halogen incandescent	CFL	LED
Energy saved compared to traditional incandescent	0% / 60W	~25% / 43W	~75% / 14W	85% / 9W
Annual energy cost[1]	$$$$$ $4.80	$$$$ $3.50	$ $1.12	$ $0.75
Bulb life	1,000 hours	1,000–3,000 hours	10,000 hours	15,000 hours
Price of one bulb[2]	$1.81	$1.17	$1.99	$1.49

[1]Based on 2 hours per day of usage and electricity rate of 11 cents per kilowatt hour, in US dollars.
[2]Based on typical per-bulb cost when purchased in a package of four bulbs.

LEARN MORE

- Robert W. Sterner, Gaston E. Small, and James M. Hood. 2011. "The Conservation of Mass." Nature Education Knowledge Project. www.nature.com/scitable/knowledge/library/the -conservation-of-mass-17395478

- US Energy Information Administration. 2017. "What Is Energy? Explained." www.eia.gov/energyexplained/index.cfm?page=about _forms_of_energy

- Jim Lucas. 2015. "What Is the Second Law of Thermodynamics?" Live Science. www.livescience.com/50941-second-law -thermodynamics.html

- Video: An Atomic View, Parts of an Atom: digital.wwnorton.com /environsci

- Video: NSF, Turning CO_2 Emissions into Fuel and More: digital .wwnorton.com/environsci

4

Life
What Shapes Biodiversity?

I t might surprise you to learn that apple trees have sex. It might surprise you even more that college students like you can play a role in it. Let's start with the trees themselves. To create another tree, an egg from one apple tree needs to connect with a sperm from the pollen of another apple tree. But the trees are rooted in the ground, so how does this work? If you pay close attention to apple blossoms in bloom in the springtime, you will likely see bees feeding on their nectar. As the bees feed, pollen particles from these blossoms stick to their bodies. When the bees fly to other trees, some of this pollen rubs off on the apple blossoms of those trees, fertilizing the egg cells there. These fertilized eggs develop into fruits with seeds that can grow into a new tree.

But how are college students engaging in this process? They are finding ways to help bees. Scientists are seeing a decline of bee populations, which some are even calling a "collapse." This problem is seen in not only wild bees but also bees that are "commercially managed" to pollinate plants on large farms. Bees pollinate not just apples of course but 70 of the top 100 crops we rely on for food. Without pollinators like them, we humans would have a hard time feeding ourselves. Bees are losing their habitat through land use and urbanization and are being affected by the use of pesticides called neonicotinoids that are broadly applied to crops and garden plants. Without these bees, the apple trees would never become fertilized: their pollen would never meet their eggs, and the trees would produce no apples.

To address this crisis, some students are initiating changes in how their campus grounds and landscaping are managed. Students at Auburn University in Alabama worked with their grounds-keeping staff first to identify all of the landscaping beds on campus. Then they worked together with faculty to create a locally native, pollinator-friendly plant list along with an integrated pest management plan that reduces the use of toxic chemicals. Signs around campus and in the dining halls raise awareness of pollinator issues and the way our landscaping and food choices affect the health of these important animals. Another way to address the crisis is to partner with local beekeepers, farmers, and gardeners to promote plantings that support healthy bee populations, as students have done at Eastern Washington University. This work also supports biology faculty research examining bacteria, fungi, and parasites that contribute to bee population collapse. In both cases, students are proactively taking steps to help plant fertilization occur (**FIGURE 4.1**).

Many flowering plants we rely on for food, including apple trees, depend on animals such as bees for pollination. Owing to declining bee populations, commercial beekeeping is on the rise.

Chapter Objectives

By the end of this chapter, you should be able to . . .

A. name the characteristics that distinguish living things from nonliving things.

B. explain how life evolves via natural selection and how the genetic makeup of a population can change, leading to new species.

C. identify environmental factors that shape biodiversity and list the criteria used to define a biome.

D. explain why biodiversity is important and list different ways to measure it.

E. compare and contrast the ways that communities of organisms interact.

F. describe ways that populations grow and respond to limits.

In nature nothing exists alone.

—**Rachel Carson**

This apple–bee–human–landscape relationship is not unique. In this chapter, we will explore what it means to be alive and how various life-forms develop and survive on Earth. We will see that interactions among living things are often far-reaching and complex. Why? One reason is that a lot of living things share this planet. So far, scientists have already identified and described more than 2 million distinct **species** of animals, defined as groups of living organisms breeding together. And thousands more are discovered each year. Many adopt seemingly extreme behaviors to live on a crowded planet, ranging from dracula ants in Madagascar that drink the blood of their young to survive, to glow-in-the-dark ninja lanternsharks living nearly a mile beneath the surface of the Pacific Ocean (**FIGURE 4.2**). While human interactions can help a process, as with apple trees and bees, scientists are learning that human choices and their effects are reducing the overall number of different species—the overall **biodiversity** of the planet—and sometimes threatening their survival. One of the big challenges of sustainability is finding ways to support the diversity of life on this planet, especially as effects caused by humans increase. But before we examine the diversity of life on

Earth and the way humans are affecting it, let's develop an understanding of the key characteristics of life itself—characteristics all organisms on Earth share.

4.1 What Do Living Things Have in Common?

An apple tree, a bee, and a human being seem as different from each other as they could possibly be. A tree, a flying insect, and you and I walking down the street differ not only in size and shape but also in how we make a living on this planet. However, all living things share several characteristics that, taken together, set them apart from inanimate objects. Let's consider the distinguishing characteristics of life.

On every scale from single-celled organisms to giant sequoia trees, living things maintain their physical form and respond to changes in their environment. What does this mean? A living thing is more precisely ordered to maintain itself than is something that is nonliving (**FIGURE 4.3**). Consider the sea star on a rock in Figure 4.3. Both the rock and the sea star are made of matter, and

species a group of organisms that are closely related to each other and are usually able to breed with each other to produce viable offspring.

biodiversity the variety of species and life in the world or in a particular ecosystem.

population in biology, a collection of the same species living in a given area.

FIGURE 4.1 Supporting Pollinators Is your campus bee friendly? Students at colleges are stepping in to help bees and other pollinators by raising awareness and planting pollinator gardens. Many campuses, including Occidental College and the University of North Texas, and cities, such as Washington, DC, and Albuquerque, New Mexico, are now certified as "Bee Friendly."

(a)

(b)

FIGURE 4.2 Interesting Species **(a)** Dracula ants, first described in 1994, chew holes in their larvae to feed on their blood, a form of nondestructive cannibalism. **(b)** About 20% of all known shark species, including the ninja lanternshark shown here, have been discovered in the past 15 years.

both absorb radiant energy. But in order to maintain itself the sea star must use energy to precisely order and reorder hundreds of components, from the smallest parts of cells to its body as a whole.

To maintain their physical form, living things can also adapt to environmental changes, such as temperature, the availability of food or water, and injuries or threats from predators. When sea stars are injured and lose an arm or even an organ outside of their central ring, they have the ability to regrow these parts and continue functioning. Sea stars also have the ability to reproduce. Living things come from other living things because the living can generate new organisms like themselves; this is the most common way to distinguish members of the same species.

When we examine living things, we can study them at different scales, as **FIGURE 4.4** shows. Using wolves as an example, we can zoom in on an individual wolf to its tissue or cellular level to study it at a smaller scale. Or we could zoom out by examining a pack of wolves as a **population**: a collection of the same species living in a given area. This approach considers the way in which the number of individuals, the genetic makeup, and other characteristics of the population respond to various environmental conditions.

At larger scales, populations of different species living in a particular place are called a **community**, and

we think of wolves and bison as having a predator–prey community interaction. Zooming out further, entire **ecosystems** encompass all the life as well as the nonliving physical environment of a particular area. By studying how ecosystems function, we can understand the ways in which energy and matter are transformed as they move through the living and nonliving environment. Particular regions of Earth that have distinctive types of

community in biology, populations of different species living and interacting in a particular place.

ecosystem a community of life and the physical environment with which it interacts (Chapter 1).

Ecosystem

Community

Population/Species

Organism

Tissue/Organ

Cell

FIGURE 4.4 Scales of Life Life on Earth can be understood as systems operating at different scales, as shown here for wolves.

biome a particular region of Earth that has a distinctive type of climate, organisms, vegetation, and overall ecosystem.

biosphere the entirety of the regions of Earth occupied by living things.

evolution the process of genetic change in populations over generations.

habitat in ecology, the place(s) an organism inhabits.

climate, organisms, and overall ecosystems, such as temperate rain forests or grasslands, are called **biomes** (see At a Glance: The Earth's Biomes), and all of the world's ecosystems together constitute the **biosphere**.

What happens in ecosystems is particularly important as we move to the next big topic of the chapter: the fact that over many generations of offspring, species can change, or evolve. The discovery of **evolution**, the process of genetic change in populations over generations, transformed our understanding of the living world from the snapshot view of the species alive today to how life has changed over geologic time.

⬡ **TAKE-HOME MESSAGE** Living things are characterized by their complexity, their ability to maintain themselves in response to a changing environment, and their capacity to reproduce and evolve. Living systems operate at different scales, from the cellular level up to the entire biosphere. A common way to distinguish species is to categorize organisms by their ability to successfully interbreed to produce fertile offspring.

4.2 How Does Life Evolve?

In 1831, at the age of 22, the young English scientist Charles Darwin accepted an invitation to join what would become a 5-year voyage throughout the Southern Hemisphere aboard HMS *Beagle*. Darwin's father judged this opportunity to be a waste of time and advised against it, but Darwin nevertheless joined the crew as a naturalist, which enabled him to observe and collect organisms and fossils in far-flung places. He also made observations of each specimen's **habitat**: the place an organism inhabits. As he described and categorized species, he recognized that their physical traits were remarkably well adapted to the nature of their surroundings. He noticed there were more than a dozen species of finches endemic to (native or restricted to) the Galápagos Islands, with different species inhabiting different islands. These species had a variety of beak types that were well adapted to the food sources available in their habitats. For example, large-beaked finches had an ability to crack large seeds, the woodpecker finch had the ability to use its beak to retrieve insects from dead wood, and the sharp-beaked ground finch, or "vampire finch," could peck nesting seabirds for their blood.

Darwin reasoned that individual variations in physical traits within a population are passed on to the following generations through reproduction. If more individuals are born than can survive to breed successfully, then those with the traits best adapted to a particular environment would be more likely to obtain essential resources, avoid threats to survival, and reproduce than those without such traits. These individuals would pass their traits to the next generation. Over many generations, individuals with the well-adapted traits would increase as a share of the population, and those without those traits would decrease. This is the process of **natural selection**, in which organisms better adapted to their environment survive and tend to produce more offspring. Darwin further reasoned that one ancestral species of finch on the Galápagos Islands had branched into different, related species as each adapted to different environmental conditions through a process now known as **adaptive radiation**. Even on this small island chain, over many generations an ancestral species could diversify, producing a variety of new species adapted to specific sets of conditions, or *niches*, of an ecosystem (**FIGURE 4.5**). We will discuss ecological niches in more detail later in the chapter.

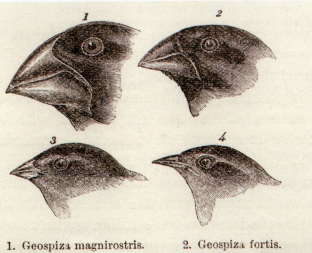

1. Geospiza magnirostris.
2. Geospiza fortis.
3. Geospiza parvula.
4. Certhidea olivacea.

Pacific Ocean

South America

Pacific Ocean

Galápagos Islands

FIGURE 4.5 Darwin's Finches Charles Darwin made careful observations of the different species of finches he encountered in the Galápagos Islands. The photos and their numbers correspond to the numbered species shown in the drawing. Among the many attributes he studied were differences in beak shape, which he reasoned were related to the food each type of finch ate. This led Darwin to conclude that although the species of finches were related through a common ancestor, each had adapted to exploit the food resources of a particular niche on the islands. In his own words, Darwin proclaimed, "One species had been taken and modified for different ends."

Mutations, Selection, and Extinction

While Darwin recognized the process of natural selection, he did not identify the basic mechanism of inheritance—the way in which traits are passed to successive generations through reproduction. We now know that an organism's **genes** determine its traits. Genes are composed of molecules called DNA (short for deoxyribonucleic acid). DNA is located within all cells and holds the instructions for building the cell's structure and performance of the cell's functions. DNA contains chemical sequences that direct the production of proteins, which in turn govern the function of all the cells in the body of an organism.

Individuals within a species exchange genetic information through breeding. In doing so, they pass some of their genes on to the next generation. Individuals inherit different forms of a gene, called **alleles**, from each parent. An individual can inherit two alleles of the same type (an identical allele from each parent), or the parents could each contribute a distinct allele to the individual. **FIGURE 4.6** shows how the pairing of alleles during reproduction can affect various traits, as some alleles dominate others.

natural selection the process where organisms better adapted to their environment survive and tend to produce more offspring.

adaptive radiation an evolutionary process where over many generations an ancestral species diversifies, producing a variety of new species adapted to specific sets of conditions, or niches, of an ecosystem.

gene an organism's basic unit of inheritance between a parent and its offspring.

FIGURE 4.6 Where'd You Get Your Genes?
(a) Genes are inherited from our parents and are contained within chromosomes: of the 46 chromosomes (23 pairs) in humans, one chromosome of each pair comes from the mother's egg and the other chromosome of each pair comes from the father's sperm, giving us two copies of each gene at fertilization. The combination of these genes produces our genetic traits.

(b) Some observable human genetic traits are pictured here: cleft chin (dominant), freckles (dominant), and attached earlobes (recessive). And still observable, though more gross, wet earwax is dominant over dry earwax!

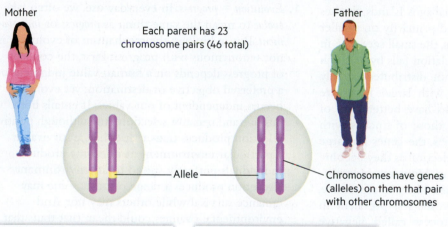
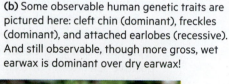
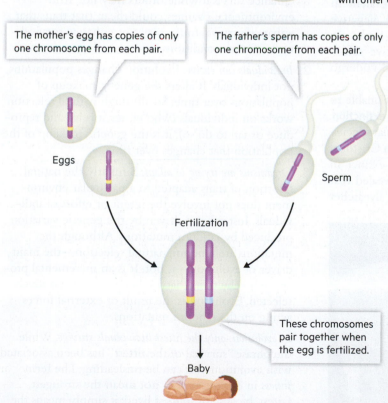

Mother

Each parent has 23 chromosome pairs (46 total)

Father

Allele

Chromosomes have genes (alleles) on them that pair with other chromosomes

The mother's egg has copies of only one chromosome from each pair.

The father's sperm has copies of only one chromosome from each pair.

Eggs

Sperm

Fertilization

These chromosomes pair together when the egg is fertilized.

Baby

allele alternate form of a gene.

genotype the genetic makeup of an individual.

phenotype an individual's observable characteristics, or traits.

mutation a random change to DNA that can produce an altered trait in an organism.

extinction a complete and permanent loss of that particular species.

The genetic makeup of an individual is called a **genotype**, and an individual's observable characteristics, or traits, are known as the **phenotype**.

But if natural selection works because of variations, how do these variations emerge? In other words, what is the mechanism through which completely new characteristics arise in a population? Gene **mutations** are the fundamental source of new genetic material. These are random changes to DNA that can produce altered traits, such as the short legs on the sheep in **FIGURE 4.7**. Mutations can be caused by errors made during the replication of DNA or by outside agents such as chemicals or radiation. Mutations can be harmful, neutral, or beneficial to the individual depending on the way the resulting trait interacts with environmental conditions.

Some mutations benefit individuals, making them better suited for their environment and more likely to reproduce and pass on these mutations. This is called positive selection. For example, researchers have found that during wet years in the Galápagos Islands, finches with large or small beaks all feed primarily on smaller seeds. However, in drought years, the small seeds are in short supply, and the finch population falls by as much as 85%. But these deaths are not distributed equally across the population. Finches with larger beaks are able to eat larger seeds and thus have better rates of survival and reproduction than those of finches with smaller beaks. During these times, the genes for large beaks are undergoing positive selection as they become more prevalent in the population. However, individuals with harmful mutations are less likely to reproduce, and their genes (and traits) become less prevalent in a population over time, a process called negative selection. So while the birds with larger beaks do better during droughts and undergo positive selection, the ones with smaller beaks die off and undergo negative selection.

When the traits of an entire species make it unable to adapt to a changing environment, it undergoes **extinction**: a complete and permanent loss of that particular species. For example, on the Galápagos Islands, the San Cristóbal vermilion flycatcher was officially declared extinct in 2016 (just one month after DNA analysis revealed that it was a distinct species). Rats that prey on flycatcher

eggs and parasitic flies that afflict newly hatched chicks are thought to have contributed to the demise of this species, which hasn't been spotted since 1987. Because environmental conditions are always changing on Earth and have changed dramatically over the billions of years that life has existed on this planet, extinction has been the fate of more than 99% of the species that have ever existed. As we will see later in the chapter, the pace of extinctions can be large-scale and sudden or steady and gradual. And as human activity decreases biodiversity, it is causing one of the greatest extinction events in Earth's long history. In Section 4.3, we'll look at the forces that influence biodiversity in general.

Five Misconceptions about Evolution

Now that we have introduced what evolution is, let's take a moment to highlight what it isn't. Here are five common misconceptions about evolution.

1. *Evolution = progress.* In everyday use, we often take *evolve* to mean the same thing as *progress* or *improvement*. But in biology, the mechanism of evolution is not synonymous with progress. First, the concept of progress depends on a human value judgment of a preferred objective or destination, yet evolution occurs independent of our values. It entails both positive and negative selection. And although positive selection produces traits that enhance survival in a particular environment, it does not produce individuals perfectly suited to an environment. Evolution produces a range of traits: some may enhance survival while others may not. And environmental changes could mean that traits that enhance survival today may not enhance survival under future conditions.

2. *Individuals can evolve.* Evolution changes populations, not individuals. It alters the genetic makeup of populations over time. So although natural selection works on individuals (who either survive and reproduce or fail to do so), it is the genetic makeup of the population that changes over time.

3. *Organisms are trying to adapt.* Similarly, the natural selection of traits adapted to a particular environment does not involve the intent or effort of individuals. Instead, it is driven by the genetic variation produced by random mutations. Although the mutations are random, natural selection—the main driver of evolution—is not. It is an incremental process whereby traits that are beneficial are positively selected. Evolution is the result of external forces acting on the random mutations.

4. *In evolution, only the fittest individuals survive.* While the phrase "survival of the fittest" has been associated with evolution, this can be misleading. The term *fitness* in evolution does not mean the strongest, fastest, biggest, or longest lived; it simply means the

FIGURE 4.7 Mutations and Selection The short legs on these Ancon sheep resulted from one individual with a mutation in 1791.

ability to pass one's genes to the next generation. Individuals have a range of traits—some may make them more likely to pass on their genes, and some may make them less likely to pass on their genes—and they need not be the fittest in every trait, just fit enough to survive and produce offspring.

5. *Evolution takes a long time, so human actions do not have an impact.* There are many examples of human actions influencing the evolution of species. Two examples are the development of insects resistant to pesticides and of bacteria resistant to antibiotics. In each of these cases, human action causes the share of the population with the resistant traits to grow rapidly. People also change environmental conditions, such as through climate change, deforestation, or river alterations, which will then promote selection for different traits in species and shift the course of their evolution. And, of course, in some cases humans can directly cause extinctions.

⬠ **TAKE-HOME MESSAGE** Life evolves, or changes over time, through natural selection. Individuals with the traits best adapted to a particular environment are most likely to survive and reproduce, passing their genetic material to the next generation. Over generations, those with well-adapted traits increase as a share of the population. Random and heritable changes to genetic material, known as mutations, cause completely new characteristics to arise in a population.

4.3 What Causes Population Traits to Change?

Did you know that birds and crocodiles are relatives? They are not close relatives but do share ancestors that long ago diverged because of **speciation**, the process by which subsets of a population diverge enough genetically to no longer produce fertile offspring when they interbreed. **FIGURE 4.8** shows how related but distinct types of species, such as birds and crocodiles, are at the tips of branches in a family tree, and their last common ancestor is at the node, or junction, from which they branch. Most speciation begins with allopatric (or "different place") factors, which involve the geographic separation of populations. This can happen as some individuals are dispersed to distant places (such as islands) or separated by geographic barriers. For example, populations of snapping shrimp were separated as the Isthmus of Panama formed about 4.5 million years ago, blocking the passage between the Caribbean Sea and the Pacific Ocean. Over time, separated populations like these can become distinct species as each evolves in a distinct environment and its genetic makeup changes.

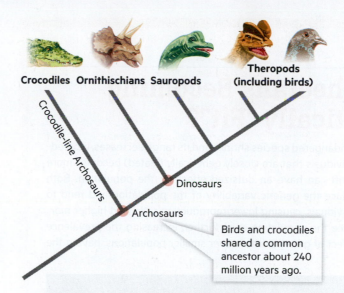

FIGURE 4.8 **Are Birds and Crocodiles Related?** Evolutionary family trees such as this one can reveal some unusual relatives—for example, birds and crocodiles.

The resulting incompatibility due to factors such as geographic barriers between separated populations is called **reproductive isolation**. In many cases, morphological differences make interbreeding impossible. The males of some fruit flies have genitalia that fit like a lock and key only with females of their own species, prohibiting interbreeding with other species of fruit flies. Closely related flowering plants may be reproductively isolated because the pollen from one will not fertilize the flowers of the other. Differences in the timing of reproductive behavior can also prohibit interbreeding. For example, two species of fruit fly, *Drosophila persimilis* and *Drosophila pseudoobscura*, breed at different times of the day, with the former preferring the morning and the latter preferring the afternoon. Plants may flower at different times of the year, making cross-pollination impossible. In other cases, genetic differences between the parents prevent the normal development of offspring after the egg is fertilized. One commonplace example of this is a horse and donkey interbreeding to produce a sterile *hybrid* known as a mule. Another example occurs on the island of Plaza Sur in the Galápagos Islands. Here, the different mating seasons of marine iguanas and terrestrial iguanas overlap slightly, and these two species, which diverged some 4.5 million years ago, can produce sterile hybrids.

However, these processes can also work in reverse. If two previously isolated populations of a species mix—as they do when individuals migrate from one island to another—their allele frequencies (the proportion of different versions of a gene present in the population) can change. The isolated populations had genetic differences because they each had adapted to the conditions on their respective islands. Once they begin to interbreed, **gene flow**—the transfer of genetic material from one population to another—will cause genetic differences between

speciation the process by which subsets of a population diverge enough genetically to no longer produce fertile offspring when they interbreed.

reproductive isolation the inability of populations to successfully interbreed due to factors such as geographic isolation; this can lead to evolutionary divergence into distinct species.

gene flow a transfer of genetic material from one population to another.

Why Are Cheetahs Becoming Less Genetically "Fit"?

As the population of an endangered species shrinks and its range decreases, inbreeding (breeding among individuals that are closely genetically related) becomes more common, and genetic drift can have an outsized effect on the population. Both of these phenomena reduce the genetic variability of the population and tend to reduce the fitness of individuals, causing lower reproduction rates and higher mortality. These trends reduce the population even further, increasing the prevalence of inbreeding and the effect of genetic drift in ever smaller populations, paving the way to extinction.

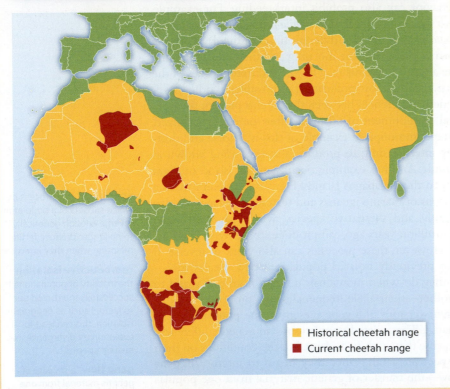

🟨 Historical cheetah range
🟥 Current cheetah range

individuals to decrease but will increase the genetic diversity of the overall population.

Genetic Diversity

The DNA of any two humans on Earth is about 99.9% identical. Although we are all much more similar than we are different, we also know that even a small difference in the sequence of our nearly identical genes can lead to a wide variety in such traits as hair color, eye color, facial structure, and much, much more. Variations in these traits are indicators of **genetic diversity**, the number of different kinds of genetic characteristics present within a population or species. Populations with high genetic diversity exhibit a wide range of characteristics. Populations with relatively low genetic diversity will have more shared characteristics, and individuals will differ less. Why does this matter? As a general rule, higher genetic diversity within a population tends to increase its productivity and make it more resilient to disease and environmental stress. For example, honeybee colonies with greater genetic diversity are more resistant to disease, achieve better foraging and food storage rates, produce larger populations, and are more likely to survive the winter than those with less genetic diversity.

What factors influence genetic diversity? Gene flow after the migration of individuals of a species from one population to another can increase genetic variation relatively quickly. If separate populations have distinct characteristics that are adaptations to local environmental conditions, the migrating individual removes its genetic material from the population it leaves and adds it to the population it joins. The impact of a small amount of migration can be profound.

This was the case with wolves on the Scandinavian Peninsula in northern Europe. In 1990, the wolf population across the entire peninsula consisted of a single, roaming pack of just 10 individuals with low genetic diversity. Each individual, including the breeding female and male in the pack, had descended from a pair of wolves who moved into the area in the early 1980s. When populations are this small, the number of potential mates is also small, and breeding between closely related individuals, or **inbreeding**, becomes more likely. Inbreeding reduces genetic diversity and increases the likelihood of reproductive failure and genetic diseases in offspring. However, in 1991 a single new breeding male migrated into the area, increasing genetic diversity and producing new genetic combinations in offspring. This migration enabled breeding among individuals that were more genetically distinct, known as *outbreeding*. Migration and outbreeding greatly enhanced the reproductive success of the wolf population, which climbed to more than 100 individuals organized into 9–10 packs over the next 10 years.

It is also possible for allele frequencies within a population to change simply through the random combination of alleles during sexual reproduction. Some individuals in a population will leave behind a few more offspring with certain alleles simply by chance. These alleles can change in frequency within the gene pool of a population even if they are not necessarily beneficial. This is known as **genetic drift**. The impact of this phenomenon depends on population size, and **FIGURE 4.9** shows how it can have significant effects on extremely small populations. Low genetic diversity can be a contributing factor to the extinction of a species. For example, the population of cheetahs in East Africa has very little genetic variation and is considered to be at especially high risk of extinction.

⬡ **TAKE-HOME MESSAGE** As genetic differences and their resulting traits cause reproductive isolation among different populations, they diverge into distinct species. Genetic variation within a population can increase or decrease over time. The genetic differences between isolated populations decrease when they meet and interbreed. Genetic drift reduces diversity over time simply by chance because of the random exchange of genetic material during sexual reproduction.

4.4 What Shapes Biodiversity?

When you travel far away from the places that are familiar to you, you notice significant differences. If you've spent most of your life in the Pacific Northwest (where the authors live) and you travel to Costa Rica, for example, you'll notice some obvious cultural distinctions during your visit, such as the use of Spanish as the official language, the custom of eating bigger meals in the middle of the day rather than in the evening, and the prevalence of black beans, rice, and plantains as dietary staples. You'll also notice differences in the natural world. Even if you didn't know the names of trees common to northwestern US forests, such as Douglas fir, western red cedar, western hemlock, and red alder, you would surely recognize visual differences between these species and the trees in the tropical rain forests of Costa Rica. You might also be lucky enough to see some creatures you'd never observe at home, such as a scarlet macaw, mantled howler monkey, or giant anteater (**FIGURE 4.10**). If you guessed that environmental conditions are determining what sorts of living things are present at a location, you'd be right. Let's discuss how we count species and measure biodiversity, and then we'll move on to how the nature of a place dictates the species that live there.

How Many Species Are There?

When early naturalists visited tropical climates, they recognized that they were not just observing new and different species but that these places hosted a greater

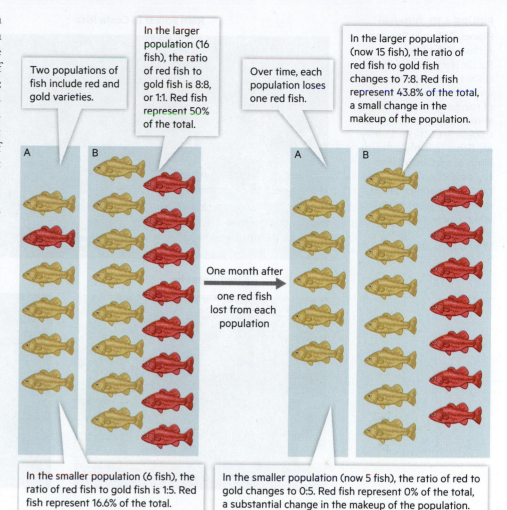

Two populations of fish include red and gold varieties.

In the larger population (16 fish), the ratio of red fish to gold fish is 8:8, or 1:1. Red fish represent 50% of the total.

Over time, each population loses one red fish.

In the larger population (now 15 fish), the ratio of red fish to gold fish changes to 7:8. Red fish represent 43.8% of the total, a small change in the makeup of the population.

One month after one red fish lost from each population

In the smaller population (6 fish), the ratio of red fish to gold fish is 1:5. Red fish represent 16.6% of the total.

In the smaller population (now 5 fish), the ratio of red to gold changes to 0:5. Red fish represent 0% of the total, a substantial change in the makeup of the population.

FIGURE 4.9 Genetic Drift The impact of genetic drift depends on population size. In extremely small populations (such as population A in this example), the loss of individuals with certain alleles can dramatically alter the genetic diversity of the population.
Adapted from Larsen (2017).

diversity of species than did temperate regions. Darwin's contemporary Alfred Russel Wallace commented on the "great diversity of details" he observed in tropical rain-forest trees. In contrasting tropical forests to those of his English homeland, Wallace wrote, "Instead of endless repetitions of the same forms of trunk such as are to be seen in our pine, or oak, or beech woods, the eye wanders from one tree to another and rarely detects two of the same species." These scientists were observing differences in **species richness**, the number of different kinds of species in an area.

You might think we have identified most of the living things on our home planet and that newly discovered species are rare. After all, scientists have identified and described over 2 million species (**FIGURE 4.11A**). But that is not the case. According to the International Institute for Species Exploration (IISE), on average more than 17,000 new species are identified every year. In fact, most estimates of Earth's species richness suggest that the number of species not yet discovered far

genetic diversity the number of different kinds of genetic characteristics present within a population or species.

inbreeding breeding between closely related individuals, often occurring when populations are small.

genetic drift a change in frequency of a trait within the gene pool of a population, caused by chance.

species richness the number of different kinds of species in an area.

Rain Forest in Costa Rica

Howler Monkey

Giant Anteater

Rain Forest in Washington State

exceeds the number we have catalogued. One recent study estimated that as many as 86% of all species on land and 91% of all species in the ocean have yet to be discovered. Moreover, genetic analysis of known species is revealing more variety. For example, genetic analysis of wasps collected in Costa Rica thought to fall into 171 described species revealed an additional 142 previously undescribed species.

Estimates of Earth's total number of species range widely from 5 million to 15 million, though these estimates are put forth with a high degree of uncertainty. The first species that come to mind for many of us are vertebrates: animals with backbones, such as various kinds of fish, birds, and mammals. But although more than 90% of all existing vertebrates are thought to have been identified, they make up a relatively small share of the total number of species on Earth. Other taxonomic categories, such as nematodes, protists, fungi, and insects, have far larger numbers of distinct species, and the share of existing species that have been identified and described within many of these groups is less than 15%. Bacteria and archaea are even less thoroughly catalogued, and their species likely number in the millions (**FIGURE 4.11B**). However, the concept of distinct species is coming into question for the microbial world, where horizontal gene transfer helps explain things like the rapid development of antibiotic resistance in microbes. In the visible world of nature that we are familiar with, organisms pass their genes on to their progeny through the reproductive mechanics of sex. But bacteria can swap genes the way that humans swap stories. In horizontal gene transfer, bacteria transfer genes to other bacteria that are not their offspring, and even to other species.

Species Evenness

Imagine walking through the two different forests depicted in **FIGURE 4.12**. Each forest has 100 total trees and 10 distinct species of trees, each represented by a different color and shape. The two forests have identical species richness. But notice the variation in the relative abundance of the 10 species across the two forests. In forest A, there are far more red trees than any other species, but in forest B each color of tree is equally abundant. If you walk in a straight line at random through each forest, you would likely observe that forest B is more diverse than forest A despite the fact that they have identical species richness. Put another way, your odds of seeing any of the non-red trees in forest A would be considerably lower than they would be in your walk

(a) Number of Described Species

Species	Estimated number of described species	Species	Estimated number of described species
VERTEBRATES		**PLANTS**	
Mammals	6,578	Mosses	21,925
Birds	11,162	Ferns and allies	11,800
Reptiles	11,690	Gymnosperms	1,113
Amphibians	8,395	Flowering plants	369,000
Fishes	36,058	Green algae	12,090
Subtotal	73,883	Red algae	7,445
INVERTEBRATES		*Subtotal*	423,373
Insects	1,053,578	**FUNGI & PROTISTS**	
Mollusks	83,706	Lichens	17,000
Crustaceans	80,122	Mushrooms, etc.	120,000
Corals	5,610	Brown algae	4,381
Arachnids	110,615	*Subtotal*	141,381
Velvet worms	208	**TOTAL**	2,130,023
Horseshoe crabs	4		
Others	157,543		
Subtotal	1,491,386		

(b)

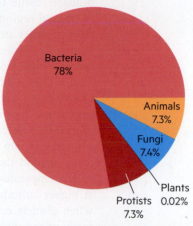

Estimated Composition of Earth's Species

Bacteria 78%
Animals 7.3%
Fungi 7.4%
Plants 0.02%
Protists 7.3%

FIGURE 4.11 Described Species (a) Scientists have discovered and described (catalogued) over 2 million species, but plants and animals are likely drastically overrepresented in our growing catalogue of described species. **(b)** The pie chart shows the likely percentage of total species that each group represents. Bacteria are by far the most abundant.
Adapted from (a) IUCN (2021) and (b) Larsen et al. (2021).

through forest B. We say that forest B has a higher measure of **species evenness** than forest A, a measure of the relative abundance of each species.

This example illustrates the important point that if we move from considering Earth as a whole to focusing on communities of life in particular areas, species diversity entails not only the number of species but also how many individuals of each species live in the area. When a single or small number of species are far more abundant than other species in a community—as with the red trees in forest A—the abundant species are dominant. **Dominance** is the opposite of diversity. So simply tallying up the distinct number of species is not enough to fully describe species diversity in an area.

species evenness a measure of the relative abundance of each species in a certain area.

dominance in biology, the opposite of diversity; when a single or small number of species is far more abundant than other species.

Forest A: Lower Species Evenness

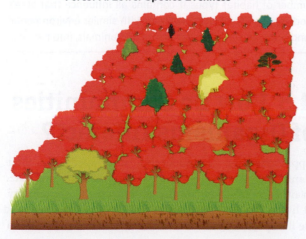

Forest B: Higher Species Evenness

FIGURE 4.12 Differences in Species Evenness Forest A and forest B each have the same total number of trees and the same number of different tree species. However, they differ in species evenness because forest A is dominated by one species of tree (constituting more than 80% of the trees), whereas forest B has an equal number of each tree species.

Ecosystem Diversity and Biogeography

When you hear the words *jungle* or *rain forest*, images of dense landscapes teeming with animals, insects, and fish probably come to mind. In fact, this image is accurate, as tropical regions tend to have greater species richness than do other areas of the globe, a pattern known as the latitudinal gradient (**FIGURE 4.13**). Why are the tropics so diverse? First, tropical conditions have existed on Earth longer than colder climates and at times have encompassed much larger areas of the planet. This longer time period has meant more opportunity for the evolution of tropical species. Second, temperatures within the large tropical belt are also fairly uniform, so organisms can move around the area without facing the challenge of severe temperature fluctuations. In contrast, parts of Earth at higher latitudes have experienced very cold ice ages when glaciers covered major habitat zones, eliminating many of the species living there. And large tropical areas of the planet—which now span Earth's waistline—have at certain times in Earth's history provided more opportunities for groups within species to spread out and later become separated by geographic barriers, such as mountains, canyons, and large rivers. Third, large areas also typically encompass more diverse environmental conditions and habitat types. Places with a greater diversity of habitats support more species diversity as more species can evolve to use each unique set of conditions.

Ecologists have found that the diversity of bird and bat species in the tropics is linked to the greater availability of fruit, nectar, and insects. These many different food types let species specialize in what they consume, and would-be competitors can coexist because they draw on different food niches in the area. Such niche specialization increases diversity because interspecies competition is weaker.

Biomes

Ecologists have developed global categories of ecosystems called biomes that are organized by environmental conditions and the types of plants and animals living there. Biomes typically refer to land-based ecosystems, while aquatic life zones (covered more in Chapter 7) apply the same organizational principles to ecosystems in freshwater and saltwater bodies. **AT A GLANCE: THE EARTH'S BIOMES** shows how ecologists have organized terrestrial regions of the world into various biomes identified by general climatic conditions (temperature and precipitation) and dominant plant species. North American biomes range from tundra and taiga in Alaska and Canada to temperate grasslands in the American Midwest, Mediterranean chaparral and desert in the American Southwest, temperate deciduous forest in the American East, and tropical rain forest in parts of Mexico and Cuba. It also shows the global distribution of biomes.

Notice how particular biomes are found in widely different places around the planet. This means that although the types of species will be different in African and South American rain forests, many will share similarities in form and function. The species in **FIGURE 4.14** are not closely related. Despite their superficial resemblance, they have very distinct evolutionary histories, yet they adapted to similar environmental conditions. When unrelated species resemble one another because of evolution under similar environmental conditions, it is known as **convergence**. In the next section, we will see that the way different organisms interact within a community also has a profound influence on each species and the overall diversity of life.

⬡ **TAKE-HOME MESSAGE** Biodiversity is measured using both species richness (number of different species in an area) and species evenness (relative abundance of each species). In addition to the climatic conditions present in an area, the amount of time it has existed, its size, and the number of habitat types it contains all influence that area's biodiversity. Ecologists sort areas with similar environmental conditions, and thus similar plants and animals, into biomes.

4.5 How Do Communities of Organisms Interact?

Apple trees are dependent on other species in their community—such as bees or even humans—to propagate. In other words, they are dependent on the community of living things around them to survive. In this section, we'll explore several different ways that interactions in

➜ **CONTEXT**

Why Are Certain Species of Bacteria Important for Our Health?

Plants and animals have populations of microorganisms that inhabit different portions of their bodies—from roots and leaves to skin and gut. These communities of microorganisms, or *microbiota*, interact with their hosts in various ways. Some, we all know, are pathogens that cause maladies, from the common cold to infectious diseases that can ravage human populations. But others are beneficial in their relationships with their host organism. The human gut microbiota, for example, helps us by digesting complex carbohydrates to produce vitamins B and K, thereby benefiting our immune systems.

(a)

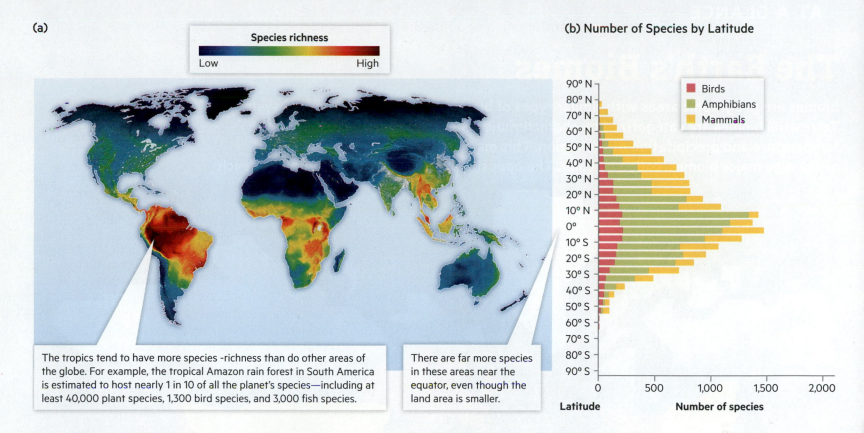

Species richness

Low   High

The tropics tend to have more species -richness than do other areas of the globe. For example, the tropical Amazon rain forest in South America is estimated to host nearly 1 in 10 of all the planet's species—including at least 40,000 plant species, 1,300 bird species, and 3,000 fish species.

(b) Number of Species by Latitude

- Birds
- Amphibians
- Mammals

There are far more species in these areas near the equator, even though the land area is smaller.

Latitude   Number of species

(c)

FIGURE 4.13 Where on Earth Is the Biodiversity?
(a) This map shows the areas of greatest species richness for mammals, amphibians, and birds, per BirdLife International and the International Union for Conservation of Nature.
(b) Species richness by latitude for certain species classifications.
(c) One of the most biodiverse places on Earth is the Raja Ampat archipelago, home to more than 1,400 fish species and 75% of known coral species. This chain of more than 1,500 islands in Indonesia is sometimes nicknamed the "species factory."

Adapted from (a) BiodiversityMapping.org (2012) and (b) Millennium Ecosystem Assessment (2005).

communities can affect the species involved and even the ecosystem as a whole.

Competition

If there is one piece of pizza left at dinner and five people at the table, there is a good chance there will be some competition for it. **Competition** occurs when one individual reduces for others the availability of a resource—such as food, water, potential mates, or even available sites for living, nesting, or

competition in biology, what occurs when one individual reduces for others the availability of a resource—such as food, water, or potential mates.

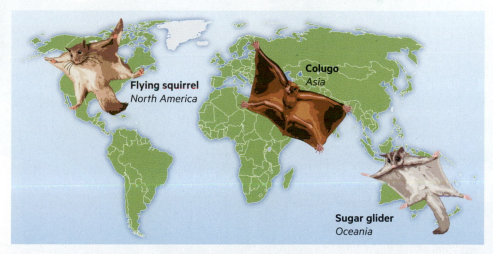

Flying squirrel
North America

Colugo
Asia

Sugar glider
Oceania

FIGURE 4.14 Similar Traits Shared by Unrelated Organisms Convergent evolution of unrelated organisms adapting to similar environments results in development of similar traits. These small, tree-dwelling mammals have all evolved flaps of skin between their front and hind limbs that catch the air and allow them to glide from tree to tree.

The Earth's Biomes

Biomes are geographic areas with similar types of biological communities and ecosystems. Terrestrial biomes are categorized by characteristic vegetation, which is related to the temperature and precipitation of the region. This map of Earth shows the locations of the nine major biomes, coded by color. The opposite page provides brief profiles of each.

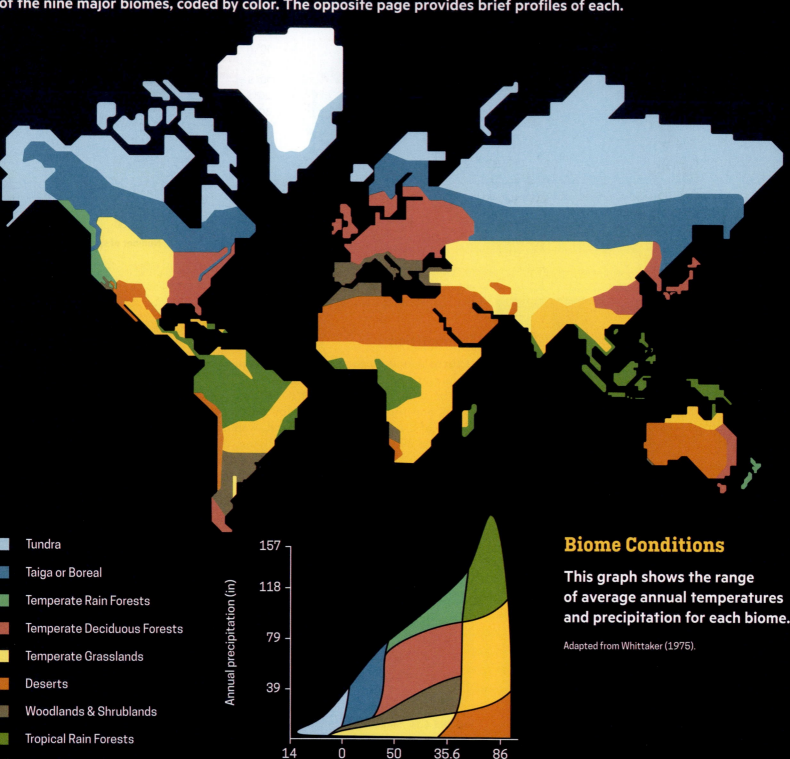

Tundra

Taiga or Boreal

Temperate Rain Forests

Temperate Deciduous Forests

Temperate Grasslands

Deserts

Woodlands & Shrublands

Tropical Rain Forests

Tropical Deciduous
Forests & Savannas

Annual precipitation (in)

157

118

79

39

14   0   50   35.6   86

Average annual temperature (°F)

Biome Conditions

This graph shows the range of average annual temperatures and precipitation for each biome.

Adapted from Whittaker (1975).

Tundra

The coldest biome, it receives very little precipitation. It is characterized by permanently frozen soil, or permafrost, and treeless expanses populated by mosses, lichens, and low shrubs.

Taiga or Boreal

These cool, wet forests located between 50° N and 60° N contain stands of relatively short coniferous trees, like spruces and firs.

Temperate Rain Forests

Common along the Pacific Northwest coast of North America, they receive lots of precipitation and support large trees like Douglas firs and redwoods, with abundant ferns below.

Temperate Deciduous Forests

Covering much of eastern North America and western Europe, they are dominated by trees that lose their leaves seasonally, like maples, oaks, and birches.

Temperate Grasslands

Often called prairies, they sweep across the midwestern United States. Low levels of precipitation limit tree growth, so grasses dominate.

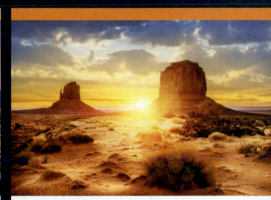

Deserts

Plants here, like short juniper or cacti, are adapted to the low precipitation and high evaporation rates by deep roots and/or the ability to store water.

Woodlands & Shrublands

Called chaparral in North America, these experience mild, moist winters and summer droughts. Drought-resistant shrubs and small trees, like eucalyptus and acacia, characterize them.

Tropical Rain Forests

These span the equator, where high temperatures and heavy rainfall support a high diversity of plant species. Taller trees form a canopy, and shorter plants tangle in an understory below.

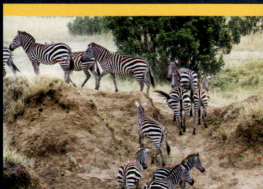

Tropical Deciduous Forests & Savannas

These are located beyond 10° N to 10° S and experience a dry season. Trees typically lose their leaves during this season of low precipitation. Savannas are grasslands with scattered trees and shrubs.

FIGURE 4.15 Intraspecific Competition Antarctic fur seals engage in intraspecific competition (competition within a species) by battling each other for access to resources and mates. This form of intraspecific competition—direct and aggressive behavior between organisms—is known as interference competition.

(a)

(b)

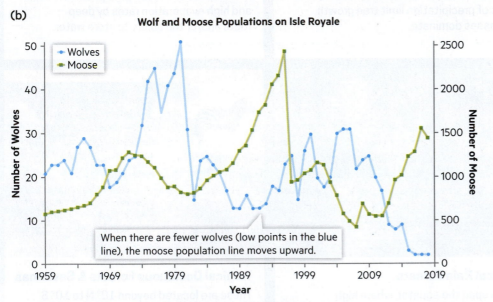

FIGURE 4.16 Predators, Prey, and Parasites (a) On Isle Royale in Lake Superior, moose are prey for wolves and hosts for a parasite known as the moose tick that causes them to lose fur. Moose themselves prey on the plants they eat, such as the balsam fir. (b) Scientists have tracked the relationship between wolf and moose populations here for over 60 years.

hiding. Charles Darwin recognized that competition is fundamental to natural selection, noticing that "more individuals are produced than can possibly survive," so that "there must in every case be a struggle for existence." While the competing individuals may be harmed to different extents, the one that suffers the least harm wins out over the one that suffers more. Individuals that outcompete others can survive, reproduce, and pass on their traits to successive generations.

Competition ensues when there are not enough resources to meet each individual's needs in a population. Competition among members of the same species is called intraspecific competition and can take two forms. Most commonly, the competition occurs without direct contact; individuals are simply drawing from the same total supply of resources available. This is called exploitation competition. The roots of each tree in a particular area are exploiting the soil water resource in a way that limits the water available for other trees. Similarly, grazing zebras reduce the grass available for each other on the savanna. Other times, competing individuals interact directly and aggressively with each other through interference competition, as one organism directly prevents another organism from accessing the resource. Animals that will battle each other for territory or access to mates, such as the Antarctic fur seals in **FIGURE 4.15**, are engaged in interference competition.

Predation

An interaction that is plain to see and present in all ecosystems is **predation**, where one organism, called the predator, feeds on another, known as the prey. When we think of predators, we might first think of carnivores, such as wolves, preying on herbivores, such as moose, but contrary to the popular perception of predation, it does not have to be violent or between two animals. Thus, herbivores are also predators—of plants. So we can think of cows as predators of grass and baleen whales as predators of plankton. Even parasitic plants, fungi, bacteria, and animals that feed on their living hosts are predators.

Predation affects the populations of both predators and prey, a fact clearly seen in the relationship between wolves and moose at Isle Royale National Park, where in 1958 the world's longest continual study of a predator–prey interaction began (**FIGURE 4.16A**). Isle Royale encompasses just over 200 square miles in area in northwestern Lake Superior. In 1900, moose were observed arriving on the island by crossing the lake when it was frozen. Later, in 1949, a pack of wolves made the crossing. Since then, scientists have observed that each significant rise in the moose population is followed by a rise in the wolf population. This is because wolves prey on moose: more moose means more food for the wolves. Each significant rise in wolf population is then followed by a decline in the moose population as they face more predation. Soon after the moose population falls, the wolf population also declines because of the decrease in available food. Because the remaining moose population consists primarily of the strongest and healthiest individuals that can escape attack, there are few easy targets for the wolves, such as calves and adult moose weakened by old age, disease, or injury. With a lower wolf population, the small but relatively strong and healthy population of moose again begins to rise. In this way, the island populations of predator and prey regulate each other.

But moose are also prey for a different sort of predator, the moose tick (*Dermacentor albipictus*), which is about one-millionth the weight of the moose and survives exclusively on moose blood. In this **parasitic relationship**, one organism known as a parasite lives off a host organism without immediately killing it. In the winter, as many as 80,000 ticks can live on a single moose, ingesting gallons of blood and causing the loss of the moose's insulating fur. Infestation weakens the moose, making it increasingly vulnerable to wolf predation and to starvation. A steep decline of moose on Isle Royale in 1996 is attributed to proliferation of the moose tick. Similarly, in the 1980s the population of wolves on the island was decimated by a deadly pathogen known as parvovirus that was introduced by a domestic dog (**FIGURE 4.16B**). As with moose and wolves, we find the population dynamics of parasites and hosts can follow the patterns of predators and prey. When the population of the host species is high, parasites thrive and infection rates increase. When the population of the host species declines, the parasite population also declines, leaving hosts who have avoided or resisted the parasites to once again begin growing their population.

Herbivores, such as moose, are predators themselves of the plants they consume. Before moose arrived on Isle Royale, balsam fir covered much of the island, composing close to 50% of the forest canopy. But moose depend on balsam fir for nearly 60% of their winter diet, and the tree now accounts for only about 5% of the island's forests. Moreover, there is evidence that annual balsam fir growth responds to the population dynamics of moose and to the wolves and ticks that are affecting moose abundance. Tree growth declines with growth of the moose population and increases when moose populations are lowered by increasing tick infestations and wolf predation.

Interspecific Competition

Isle Royale illustrates clear ways in which the predator and prey relationships of multiple species affect the community. But these interactions are often much more complex and diffuse because most predators prey on more than one species, and most prey are preyed upon by multiple predators. Earlier we saw that

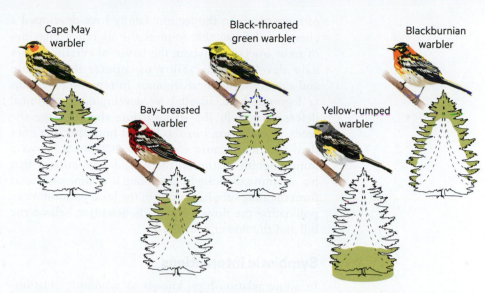

FIGURE 4.17 Resource Partitioning Each of these warbler species can rely on the resources of the same spruce tree habitat by feeding in different areas (indicated in green for each warbler). While some areas may overlap, none of them are exactly the same. In this way, each species occupies its own niche and avoids competing with others.
Adapted from MacArthur (1958).

intraspecific competition occurs between individuals of the same species. On a broader scale, interspecific competition occurs when individuals of different species are each working to obtain something necessary for survival, either through exploitation or interference. For example, the tree roots of white pines in New England forests exploit water and nutrients so efficiently that they suppress the growth of other plants in the area. Interference competition has occurred between wolves and coyotes in Yellowstone National Park as they fight—sometimes to the death—over carcasses.

If two species compete directly for the same resources, the species that better exploits those resources will eventually eliminate the other. But different species can and do exist together if they each exploit a different **ecological niche**, or particular role in a community. For example, birders in North America sometimes see five different species of warbler all inhabiting the same tree and all eating insects from that tree (**FIGURE 4.17**). In the 1950s, ecologist Robert MacArthur made careful observations of the wood warblers on the branches of individual spruce, fir, and pine trees and saw that different species divided themselves among various parts of the tree. Moreover, the warblers also varied in their nesting times, which caused significant variations in each species' season of peak food demand.

Coevolution

Two or more species can evolve together in a relationship known as **coevolution**. Here, the adaptations of one species cause a second species to adapt, which in turn may effect another adaptation in the first species in a type of "evolutionary arms race." For example, the seeds

predation an interaction where one organism, called the predator, feeds on another, known as the prey.

parasitic relationship an interaction where one organism (known as a parasite) lives off a host organism without immediately killing it.

ecological niche a particular role or position of a species in a community.

coevolution where two species evolve together, and the adaptations of one species cause a second species to adapt too.

of most plants in the legume family have developed a chemical defense that inhibits the digestive enzymes of most insects. However, the larvae of certain beetles have developed the ability to counter this defense and can consume most legumes. In response, soybeans (a legume) have adapted by developing a chemical defense that kills the larvae before they damage the seeds. Coevolution can also benefit both species rather than set them against each other. **FIGURE 4.18** shows a hummingbird in the northern Andes mountains that has an unusually long bill, enabling it to dine on nectar from the long tubular flowers of the local passionflower, pollinating the flower as it feeds. Scientists believe the bill and the flower coevolved.

Symbiotic Interactions

In some relationships, known as symbiotic relationships, two organisms are closely associated with each other in a long-term relationship. As we have seen, a *parasitic* relationship is one type, in which one organism is harmed and the other is helped. *Commensalism* benefits one organism while having no effect on the other (**FIGURE 4.19**). For example, some orchids use tree limbs as a surface on which to grow but do not draw resources away from the tree or harm the tree. Smaller animals sometimes hitch a ride with larger animals; the smaller animals gain transportation, protection, and access to sources of food around the larger animal. This is the case for barnacles that attach themselves to whales and tiny pseudoscorpions that hitch a ride underneath the wings of some beetles. *Mutualism* benefits both organisms: a win–win situation. Many flowers and their pollinators are mutualists. The plant benefits from the pollination, and the pollinator benefits from the nectar. Mutualism often facilitates nutrient transfer. This is the case for vertebrate animals and the microbes

FIGURE 4.18 Coevolution This sword-billed hummingbird has a tongue that spans the length of its tube-shaped bill. It can extract nectar from passionflowers that is inaccessible to most other pollinating species. The shape of the hummingbird's bill and tongue is thought to have coevolved with the flowers that it pollinates.

living in their intestines. The microbes get nutrients from the food the animal has eaten, and in turn the animal relies on the microbes to break down certain large food molecules to aid in digestion. Consider, for example, that although cows eat grass, they can't digest it and so depend on their gut microbiota to convert the cellulose in the grass to fatty acids that *can* nourish them. This is also the case in the human digestive tract; recent research estimates put the number of

FIGURE 4.19 Community Interactions There are many types of community interactions among organisms. We can think of the interactions of organisms as some combination of benefit, harm, or no significant impact for the species involved. For example, for the orchids growing on the tree, the orchid benefits from the structure but has no significant impact on the tree. The blue-tailed day gecko is an example of a vertebrate pollinating a flower—and both organisms benefit. The predation example in the final image shows a relationship where the predator benefits and the prey is harmed.

Predators and Trees in Yellowstone

Sometimes scientific insights are sparked by observations and ideas that are not all that scientific. When ecologist Bill Ripple first began his vegetation study of declining tree species in Yellowstone National Park in 1996, he was struck by a picture for sale in the visitors' center depicting wolves seemingly standing guard over a grove of aspen trees. This image led him to wonder what the ecological relationship might be between these two seemingly distant species.

That year, Ripple's colleague Robert Beschta, a hydrologist, began investigating the rapidly widening Lamar River that flows through the park. He found that the erosion causing the widening was linked to the loss of trees that used to line and stabilize the riverbank. Fewer trees also meant less habitat for beaver and songbirds in the area. The decline in aspen trees was particularly striking. What could be causing the decline of these trees? Climate change was initially considered but later ruled out because aspen trees just outside of the park were thriving in the same climatic conditions. Ripple arrived to drill cores and count the growth rings in more than 100 aspen trees in the park. This analysis revealed that most trees were at least 70 years old and that the growth of new aspen trees in Yellowstone had dramatically fallen off in the 1930s.

Remembering the image of the wolves and the aspens in the visitors' center, Ripple focused the research team on the fact that wolves, which had been plentiful in Yellowstone, were eradicated from the park in 1926 by hunting and predator elimination programs. Why would this matter for the trees? Wolves in the park had preyed on elk. So in the absence of wolves, the elk population, which is protected within the park boundaries, soared. Larger numbers of elk could now graze on young aspen shoots, cottonwoods, and willows without fear of predation. So unchecked elk browsing drastically reduced new tree growth.

As this research was in process, Yellowstone began reintroducing wolves, which set up a natural experiment on how reestablished wolf populations would affect not only elk in the park but also less directly related species, such as trees. As wolf populations were reestablished over the next two decades, researchers found that the elk population declined by more than two-thirds. Since 1995, studies of aspen, cottonwoods, and willows have shown a decrease in browsing and increases in plant height, stem diameter, and canopy cover. Populations of species that rely on trees along the riverbank, such

Ecological relationships are often more complicated than they seem. Research at Yellowstone National Park has documented how its wolf population's role in controlling the elk population also impacts songbirds, beavers, and even groves of aspen trees.

Similar complexities have been discovered in other ecosystems—for instance, the work of ecologist Arian Wallach in Australia uncovering the effect that dingoes have on small mammals and vegetation.

(continued)

as beaver and songbirds, have also shown signs of recovery.

The species interactions in Yellowstone show how feeding relationships in ecosystems are often more complex than the simple food chains described in the previous chapter. One species often eats more than another species, and a single species is often food for several other species. These interwoven relationships are more properly described as *food webs*. The impact that top predators, such as wolves, can have directly and indirectly across several levels of a food web is known as a trophic cascade. After their initial findings in Yellowstone, Ripple and Beschta founded the Global Trophic Cascades Program at Oregon State University to bring together a community of scientists throughout the world who examine the ecological impact of top predators. For example, ecologist Arian Wallach's work in Australia has found that dingoes there play a role similar to that of wolves in Yellowstone. Dingoes prey

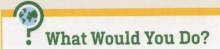
on carnivores, such as foxes, and herbivores, such as kangaroos. Her research has found that declining dingo populations reduce the diversity of small mammals and vegetation, while dingo reintroduction helps recover this biodiversity. Research on trophic cascades reminds us that relationships between members of an ecological community do not need to be direct or immediately obvious to be consequential for the ecosystem as a whole.

bacteria in a typical human large intestine at more than 100 trillion!

🛑 **TAKE-HOME MESSAGE** Populations of different species in a given space interact as a community. These interactions take many forms, including predation—where one organism feeds on another—and competition for resources among species. If two species compete directly for the same resources in a particular place, the species that exploits those resources most efficiently will eventually eliminate the other species. Long-term interactions among organisms are called symbiotic and can involve any combination of benefit or harm for the species involved. When species interact over time, they can stimulate evolutionary responses in each other.

While this presents an extreme example of rapid population growth in the animal kingdom, populations change over time as individuals of particular species are added and lost and (as we saw in Section 4.5) compete for the resources available to them. These changes are known as population dynamics. Additions occur when individuals reproduce or immigrate from different areas into the population. Loss occurs when individuals die or emigrate away from the area. If the number of births and increases from immigration equals the number of deaths and decreases from emigration over a period of time, the population will be in equilibrium, neither increasing nor decreasing. If births and immigration outpace deaths and emigration, the population will increase. If the reverse is true, the population will decrease.

4.6 What Controls Population Size?

In the early 1800s the population of rabbits in Australia was near zero: they essentially did not exist on this continent except as pets. Then around 1850, 24 rabbits were released into the wild as something for the locals to hunt. By 1920 an estimated 10 billion rabbits hopped about Australia because there were no natural predators to keep them in check. The presence of rabbits in Australia has been termed an "ecological nightmare" because they have contributed to the decline or loss of native animal species and also eat seedlings before the plants can reproduce.

Population Growth and Responses to Limits

Under ideal conditions, populations can increase very quickly, as conditions allow for a constant increase of individuals over time. This pattern is known as **exponential growth** (or *geometric growth*). Some species of bacteria can divide and produce a new generation in as short a time as 20 minutes. Imagine examining a population of bacteria like this under a microscope. After 20 minutes, one bacterium would become 2, then 2 would become 4, and then 8, 16, 32, 64, 128, and so on at 20-minute intervals (**FIGURE 4.20**). At this rate of growth, the population would exceed 2 million after just 7 hours and more than 4 sextillion (4 billion trillion) by the end of a day.

exponential growth rapid population growth that occurs when environmental conditions allow for a constant ratio in the increase of individuals over time, resulting in population growth proportional to population size.

carrying capacity the maximum number of individuals of a species that a habitat can sustainably support.

So what stops species from overrunning our planet? Conditions are not always favorable in the real world for the growth of populations of living things. The **carrying capacity** is the maximum number of individuals of a species that a habitat can sustainably support. If a population grows beyond the carrying capacity, it will exhaust the resources needed to survive, and there will be a population crash. In the bacteria example, if we do not provide food and fail to give the bacteria more and more space to expand, the population will soon crash.

Another classic example involves the population of reindeer on St. Matthew Island, Alaska (**FIGURE 4.21**). In 1944, the US military introduced 29 reindeer to the island as an emergency food source for troops in the area. The reindeer were never used in this way and flourished there for nearly 20 years. The reindeer population was about 6,000 by 1963, with about 47 of the animals per square mile. However, in 1964 the reindeer faced a harsh winter after having eaten almost all the lichen that served as their primary food. Many reindeer starved, and the population on the island crashed. To survive, populations of animals need to have a minimum number of individuals. In this case, the reindeer population fell below this threshold, was unable to recover, and died out completely in the 1980s.

Exponential growth is not always followed by a population crash. Often it will slow before it overshoots the carrying capacity. **FIGURE 4.22** depicts a typical population growth pattern that follows a logistic curve (an *S-shaped curve*). Notice how the rate of growth gradually slows as the curve flattens out, reflecting the onset of factors that limit population growth. As the growth rate slows, populations following this pattern will often fluctuate around the carrying capacity. This pattern was documented in a study of the fur seals on St. Paul Island, Alaska. Their population grew exponentially after hunting was banned in 1911. Between about 1930 and 1960, the population began to oscillate around a population of 10,000. After hunting female seals for fur was permitted again in 1956, the population crashed in the 1960s.

Population Density

As the size of a population in an area increases, the availability of resources for each individual decreases, and the population becomes more susceptible to certain environmental harms. These resources and harms are termed density-dependent factors and can cause population trends such as increased death rates and/or decreased birth rates. For example, the space available for the growth of barnacles on a rock outcrop is a density-dependent factor: as population increases, less unoccupied space is available for new barnacles to colonize (**FIGURE 4.23**). Food availability in an area can also be a density-dependent factor. When the population density of harp seals increases and the availability of food decreases, the average size of the seals then decreases as well. In turn, the number of females giving birth decreases because

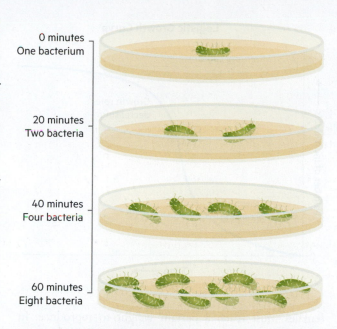

FIGURE 4.20 **Exponential Growth** If we provide ideal conditions for bacteria that divide every 20 minutes, their population can expand exponentially to more than 2 million in just 7 hours.

0 minutes
One bacterium

20 minutes
Two bacteria

40 minutes
Four bacteria

60 minutes
Eight bacteria

FIGURE 4.21 **Population Crash** Population growth is subject to limits in the environment. This graph shows the population crash of the reindeer on St. Matthew Island. The number of reindeer grew from 29 to 6,000 and then rapidly dropped, as indicated by the vertical line, around 1965.

Graph adapted from Klein (1968).

As the population grew and consumed more resources, its food became more scarce, leading to a population crash.

Introduced reindeer thrived and their population grew exponentially.

The reindeer population never recovered from the crash—eventually dying out completely in the 1980s.

FIGURE 4.22 **Logistic Growth, or S-Curves**

Many populations, like the fur seals of St. Paul Island, approximate a logistic growth model, in which the rate of population growth slows as it approaches the carrying capacity.

Count graph adapted from NOAA Fisheries (n.d.).

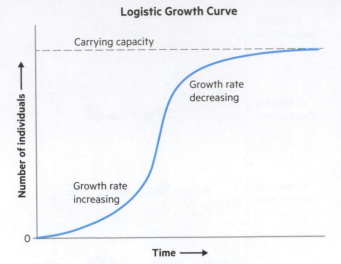

Logistic Growth Curve

Carrying capacity

Growth rate decreasing

Growth rate increasing

Number of individuals

Time

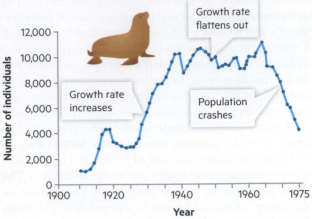

Growth Curve of St. Paul Island Seals

Growth rate flattens out

Growth rate increases

Population crashes

Number of individuals

Year

females must reach a certain weight to reproduce. In some cases, increasing population density leads to an increase in waste products that regulate population. For example, one waste product generated by yeast is alcohol, a substance toxic to yeast. As the density of yeast in a space increases, their death rates increase because of the higher concentrations of alcohol (something very useful to winemakers). In other cases, higher population densities may make individuals more susceptible to predators or more vulnerable to the spread of disease.

Other factors affecting population growth have no relationship to density. These density-independent factors include certain environmental conditions, such as temperature, precipitation, pH, extreme weather, and geologic events. For example, the 1980 eruption of Mount St. Helens in Washington State dramatically lowered the population of many plant and animal species in the area, but this event was not related to the density of those populations; it was related to the heat from the eruption and the 540 million tons of ash that blanketed the area.

Reproductive Strategies

Consider the dramatic difference between the number of offspring dandelions produce each year and the number of offspring bears produce each year. Bears and dandelions are similar in that under ideal conditions, each dandelion seed and each bear cub can produce a mature adult. However, you don't need to know the precise numbers of offspring for each species to recognize that there are considerably more potential dandelions than there are baby bears. What's going on?

We use the term *biotic potential* to describe the reproductive capacity of a species under ideal conditions, and dandelions and bears represent two different reproductive strategies. Dandelions are *r-strategists*: the "r" connotes a rapid growth rate. They produce a lot of offspring but do not invest resources in the care of those offspring. The strategy here is to produce high numbers quickly. Even though most of the offspring will not survive to reproduce, the odds are that enough will survive to fill your local park or yard with yellow blossoms. Bears are

FIGURE 4.23 **Density-Dependent Factors**
The available space for barnacles to attach to and make their living is a density-dependent factor influencing population.

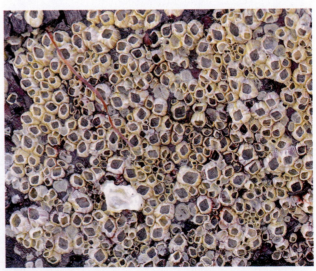

r- and K-Strategists: How Many Offspring and How Often?

	r-strategist "quick and many"	K-strategist "slower and fewer"
Age of maturation	Young—before the next group of offspring	Older—usually many seasons after birth
Number of offspring	Many	Few
Breeding frequency	Very frequent, and with many eggs produced each time	Infrequent, and with only a few eggs at most each time
Physical size of offspring	Usually small	Larger
Mortality rates	High—many offspring do not live to sexual maturity	Low—offspring generally survive
Examples	Mice, rabbits, most insects, octopuses, mass spawning organisms such as dandelions	Whales, elephants, humans, some birds

FIGURE 4.24 r-Strategists and K-Strategists Dandelions are r-strategists because they produce a lot of offspring rapidly, each with a low probability of survival. In contrast, bears are K-strategists, which have much lower reproductive rates but invest heavily to ensure the survival of their offspring.

K-strategists: the "K" comes from *Kapazitätsgrenze*, the German word for "carrying capacity." They have a much lower reproductive rate, producing just two or three cubs in a litter. Once a cub is born, the mother will invest 2 years of resources in caring for the cub until it can support itself. During this time, she does not produce more offspring. The strategy here is to invest in a small number of offspring and ensure that they have a very high chance of surviving to reproduce (**FIGURE 4.24**).

Not all species fit neatly into these two categories, but ecologists have made useful generalizations about population dynamics associated with each strategy. The r-strategists tend to be relatively small organisms with short life spans that are equipped to exploit resources present only for short periods in unstable environments. This is why they are sometimes called weedy or opportunistic species. Their populations tend to fluctuate in boom and bust cycles: increasing rapidly while resources are available, then crashing once they reach or exceed the carrying capacity. By contrast, K-strategists tend to be relatively large organisms with long life spans that are specialists in more stable environments where resource availability is relatively predictable. Their populations tend to follow logistic growth curves that keep them more or less in equilibrium near the carrying capacity.

🛑 **TAKE-HOME MESSAGE** Population dynamics explores how the size of populations changes over time as individuals are added and lost. Each population also has a carrying capacity, the maximum population size that a habitat can support. Organisms can be divided into two main groups on the basis of their reproductive strategies: r-strategists that reproduce rapidly with large quantities of offspring and K-strategists that invest more in caring for their far fewer offspring.

4.7 What Can Cause Loss of Biodiversity?

Even though increasing biodiversity and speciation has been the norm on Earth, extinction has also always been a fact of life. The permanent loss of particular species occurs both as a relatively steady, slow process and occasionally through rapid mass extinction events. Five such events in Earth's history resulted in the loss of an extremely large number of species over a relatively short geologic time span (tens to hundreds of thousands of years). The largest of these events took place about 252 million years ago with the loss of more than 90% of terrestrial and marine species. This mass extinction was perhaps the result of an extraordinary volume of volcanic activity that clouded the atmosphere, changed the climate, and acidified the oceans (**FIGURE 4.25**). Biodiversity does rebound after mass extinctions, but this process takes millions of years, and the new species that evolve are different from those that have been lost. Prominent biologists have proposed that human effects on Earth's species have placed us in the midst of a global decline in biodiversity that is on track to constitute a sixth mass extinction. In the next chapter, we take a closer look at the rapid pace of biodiversity decline and at conservation efforts to counter this.

Extinction is an absolute term that indicates the death of the last existing individual of a species. However, it is important to remember that a species may cease to play a major role in its ecosystem long before it goes extinct. For example, prior to the 19th century, tens of millions of American bison roamed the Great Plains and significantly shaped the landscape and influenced the plant and animal communities in the region. American bison

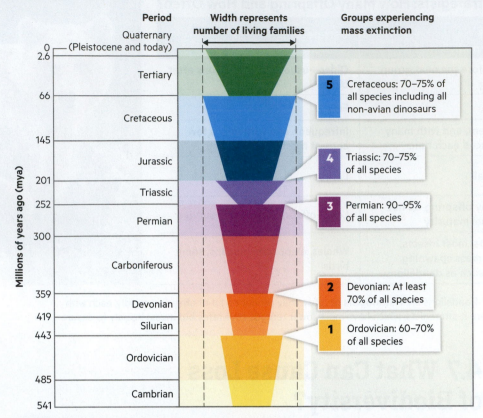

FIGURE 4.25 Mass Extinction Events Five mass extinction events have occurred in Earth's history. Each extinction is marked by a sharp decrease in the width of the colored area showing the number of living families.

Period | Width represents number of living families | Groups experiencing mass extinction

5 Cretaceous: 70–75% of all species including all non-avian dinosaurs

4 Triassic: 70–75% of all species

3 Permian: 90–95% of all species

2 Devonian: At least 70% of all species

1 Ordovician: 60–70% of all species

still exist in the thousands and therefore are not extinct, but they are considered to be *functionally extinct* because their abundance and range have diminished to such an extent that they no longer play the significant role in the ecosystem that they once did.

Identifying Areas and Species Vital to Biodiversity

Places where large numbers of species are particularly vulnerable to extinction are termed **biodiversity hot spots** (**FIGURE 4.26**). Biodiversity hot spots have a relatively high number of *endemic species*: species that specialize in a particular set of environmental conditions that effectively restrict their location to a single place where those conditions exist. By studying fossils as well as modern extinctions, we see that some kinds of organisms are more prone to extinction than others. Specialist species that live in a relatively narrow geographic range are much more likely to go extinct than are generalists such as rats and cockroaches that can make their living almost anywhere.

What factors lead to endemism? Unique environmental conditions along with barriers to migration lead to these isolated pockets of highly specialized species. Many of the hot spots are on islands (such as New Zealand, Madagascar, the Hawaiian Islands, and the islands of the Caribbean) as opposed to continental landmasses.

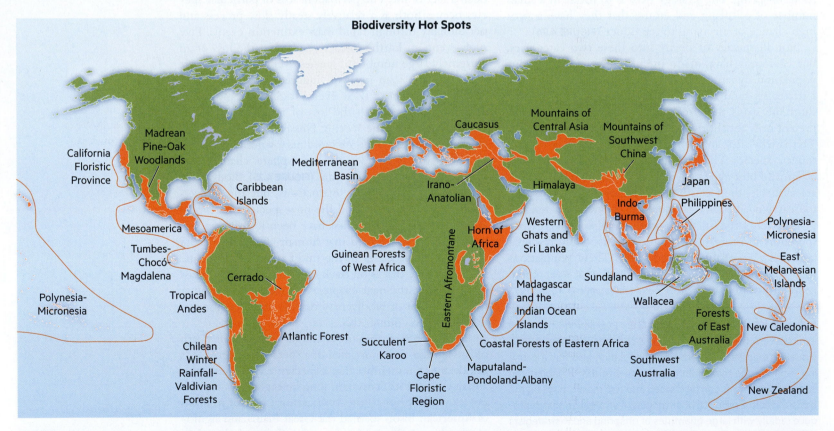

Biodiversity Hot Spots

FIGURE 4.26 Hot Spots Biodiversity hot spots are places where large numbers of species are particularly vulnerable to extinction. These places tend to have a high number of endemic species: those that only exist in a particular geographic region. Adapted from Critical Ecosystem Partnership Fund (n.d.).

Killing to Protect Biodiversity in New Zealand

In New Zealand, the most significant threat to biodiversity is introduced species such as rats, stoats (short-tailed weasels), and Australian brushtail possums. As part of their government's ambitious agenda, they plan to kill hundreds of millions or even billions of the creatures in an effort to make the country "predator free" by 2050. It's even part of a post–COVID-19 Labour Party "Jobs for Nature" agenda, which aims to "supercharge" this goal.

When Polynesians arrived on the islands, they introduced Pacific rats as a food source. The rats quickly overran the islands, preying on the eggs of native birds. When Europeans colonized New Zealand in the middle of the 19th century, they caused more extinctions by introducing many more species. These included more species of rats that stowed away on the arriving ships, stoats used to control an exploding population of European rabbits, and brushtail possums that were introduced to create a fur industry. All of these species compete with native birds for habitat and food and prey on bird eggs and chicks. Today, introduced species like these kill an estimated 25 million native birds a year. Many of New Zealand's remaining native species of birds are threatened with extinction, including the country's national symbol: the flightless kiwi.

New Zealand's recovery efforts, which include killing the predators of its native birds, began on

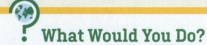

Predator Free New Zealand is a government effort to rid the nation of invasive predators of native birds by 2050. This effort includes the use of lethal means such as traps and the application of poison via air drop.

small islands offshore of the main North Island and South Island. Once these islands were cleared of predators, native birds were reintroduced to build up populations that can then be reintroduced on the main islands. Efforts on the mainland include dividing neighborhoods into 100-meter-square grids to ensure that traps are

set, cleaned, and reset uniformly throughout the area. On an even larger scale, a poison known as compound 1080 (sodium fluoroacetate)—which induces cardiac arrest—is applied via air drop to millions of acres of forestland.

Although this government-led effort has widespread support in New Zealand, not everyone approves of the extermination effort. For example, an animal rights group named SAFE (Save Animals from Exploitation) is opposed to the effort because it judges the methods of killing to be inhumane. Hunting groups oppose the poison drops because of their impact on deer, which are also an introduced species. Other commentators have questioned whether the effort is worth the billions of dollars it will cost and if it is practical to expect that a human-dominated landscape can ever be returned to a state approximating what it was like before human arrival.

What Would You Do?

Take the role of a member of the animal rights group, a hunter opposed to the poison drops, or an ecologist working on the eradication plan and explain your reasons for either opposing or supporting the mass poisoning of rats and stoats in New Zealand.

But water is just one type of isolating factor for land-based species. High levels of endemism also occur when landscape features such as mountain ranges create stark differences in environmental conditions between neighboring places. For this reason, mountainous areas surrounded by low lying deserts, like those in the southwestern United States or the Andean region in South America, are sometimes called "islands in the sky" and host a relatively large number of endemic species. Biodiversity hot spots contain a large percentage of species within certain taxonomic groups relative to the land area they occupy. For example, although hot spots occupy less than 3% of Earth's land area, they are home to more than 35% of all terrestrial vertebrate species (land animals with a spine) and more

than 44% of all plant species. These are also places where human populations are growing at a rapid rate and human effects are causing rapid habitat loss.

Some species also exhibit a particularly strong influence over the abundance and diversity of other organisms in their ecosystem. These are known as **keystone species**, and the loss of these species can cause the loss of many other species as well (**FIGURE 4.27**). For example, when sea stars are removed from their intertidal habitats, populations of the species of mussels and barnacles that they preyed on grow out of control. They then outcompete and exclude the wider variety of mussels, barnacles, limpets, and chitons that existed prior to removal of the sea stars. Thus, the biodiversity of the overall area

biodiversity hot spot place on Earth where a large number of species are particularly vulnerable to extinction.

keystone species species that exhibit a particularly strong influence over the abundance and diversity of other organisms in their ecosystem.

Biodiversity Hot Spots and Overlapping Vulnerabilities

Indigenous peoples and local communities living in biodiversity hot spots feel biodiversity losses most acutely. Although biodiversity hot spots comprise less than 3% of Earth's landmass, they host about 20% of the human population. Moreover, the average human population density and growth rate in hot spots is higher than the global average. Environmental economists have found that hot spots overlap significantly with areas of severe poverty where people struggle to access clean water and adequate nutrition. Researchers have also documented unique cultural vulnerabilities in biological hot spots; for example, they account for 70% of all the languages spoken on Earth—many of which are endemic to particular locales and are in danger of dying out. Further, their peoples draw directly on the ecosystem services of biodiversity hot spots for their livelihoods and cultural practices. Sometimes measures put in place to protect hot spots can harm the people who rely most on these places, as when conservation policies and the establishment of protected areas evict, exclude, displace, and/or disempower local communities.

The knowledge accrued by local people and communities can be critical in conserving these ecosystems. A study of more than 100 protected areas found that positive conservation and socioeconomic outcomes were more likely when policies empowered Indigenous peoples and local communities in decision making and maintained the benefits people gained from the protected area.

declines beyond just the reduced abundance of sea stars. Identifying keystone species is another way to prioritize conservation efforts (**FIGURE 4.28**).

🛑 **TAKE-HOME MESSAGE** Although the history of life on Earth is one of increasing biodiversity, extinction is a natural process that has occurred both at a relatively steady background rate and in large and rapid mass extinction events. Biodiversity hot spots are areas with relatively high numbers of endemic species that are also subject to rapid rates of habitat loss due to human impact.

The sea star *Pisaster ochraceus* living in Mukkaw Bay, Washington, feeds on mussels.

1

| Mussels | *Pisaster ochraceus* | 15 other species |

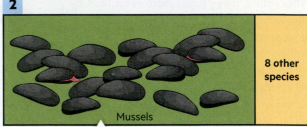

2

| Mussels | 8 other species |

When *Pisaster ochraceus* was removed, the mussel population increased and the number of other species decreased.

FIGURE 4.27 Species That Have a Pronounced Effect on Diversity The keystone species concept was discovered through research on *Pisaster ochraceus* sea stars. They have a pronounced effect on the diversity of organisms in their ecosystem. (1) Sea stars feed on mussels. When sea stars are present, 15 other species live in the ecosystem. (2) When sea stars are removed, the mussel population increases, forcing out other species that had existed when the sea stars were present.

Art adapted from Harris (2010).

4.8 What Can I Do?

We are just individual members of one species on a planet teeming with diverse life-forms. What can you do to support biodiversity? In the next chapter, we'll take a closer look at many different conservation strategies—efforts to protect biodiversity ranging from endangered species laws to protected areas to restoration efforts. In

FIGURE 4.28 **Keystone Species** A keystone species is an organism that helps the entire ecosystem and is often a predator. Sea otters, for example, eat sea urchins, thus keeping the populations of sea urchins in check. Without the otters, the urchins would overfeed on kelp and destroy the complex ecosystem. Similarly, grizzly bears keep large herbivores such as moose and elk in check, while also distributing seeds of berry bushes and other plants in their feces. Beavers build natural dams that create pools in rivers and wetland habitats, which benefit many other species.

the meantime, here are some ideas on how to help, either by improving habitat for other species or by minimizing your negative impact.

Create Wildlife-Friendly Habitats Where You Live

At the beginning of this chapter, we learned about efforts at some colleges and universities to provide habitat for

pollinators such as bees and butterflies. You can do similar things at your residence. The National Wildlife Federation offers a Wildlife Habitat Certification program that offers tips on how to create wildlife-friendly habitat even in spaces as small as an outdoor balcony or patio area (www .nwf.org/Garden-for-Wildlife/ Certify). Your action could be as simple as establishing a small container garden that is pollinator friendly, and this small action could be another step toward the Million Pollinator Garden Challenge—an effort to register one million public and private gardens that support pollinators. There are also more ambitious projects such as constructing nesting boxes for certain kinds of bats or birds.

Reduce Some of the Negative Effects on Wildlife Where You Live

Throughout this book we'll see that our environmental effects—including those affecting biodiversity—are linked to our consumption of material goods, food, and energy. So conserving energy, avoiding wasteful consumption, and eating organic foods that don't use synthetic fertilizers and pesticides are all good steps to take no matter where you live. There are also numerous certifications for products that minimize habitat impact, including Forest

Stewardship Council certification of wood and paper products, fair trade certification of coffee and chocolate, and Marine Stewardship Council certification of seafood. You also can help to reduce the impact of some lesser-known effects:

» Reduce light pollution by minimizing the use of outdoor light in the evening. Artificial lights can disrupt migrating birds and impede nocturnal predators such as owls.

» Keep track of your pet. Pets, particularly house cats, can have a big impact on species in your neighborhood. Free-roaming house cats are estimated to kill more than one billion birds each year.

» Be careful what you put in the water. Certain medicines can have a negative impact on aquatic species—even in very small amounts—so avoid flushing unused pharmaceuticals down the drain. Similarly, certain active ingredients in sunscreen, such as oxybenzone and octinoxate, are causing damage to coral reefs, so look for sunscreen that is mineral based, with titanium dioxide or zinc oxide as the active ingredient.

Minimize Your Impact on Wildlife When You Are out Enjoying Nature

Follow the seven steps of "leave no trace" ethics when you are out hiking, camping, biking, or otherwise enjoying nature. These include properly disposing of your waste, leaving in place natural items you find (taking only photographs), walking and camping on durable surfaces to avoid trampling vegetation, and avoiding actions that would disturb wildlife. You can find a complete list of the leave-no-trace principles at the lnt.org website.

Enjoy Your World.
Leave No Trace.

Plan Ahead and Prepare
Travel and Camp on Durable Surfaces
Dispose of Waste Properly
Leave What You Find
Minimize Campfire Impacts
Respect Wildlife
Be Considerate of Other Visitors

Leave No Trace™
Center for Outdoor Ethics

To learn more and get involved in
your community, visit www.LNT.org

Chapter 4 **Review**

SUMMARY

- Living things have an ability to maintain themselves in response to a changing environment and the capacity to reproduce and evolve.

- The components of life and living organisms interact in systems of various scales from the cellular level up to ecosystems and the entire biosphere.

- The process of natural selection enables life to evolve over time.

- Random heritable mutations to genetic material can give rise to new characteristics in a population.

- Genetic diversity within a population can increase or decrease over time.

- Extinction can occur at a steady background rate or in sudden mass extinction events.

- Geographic areas with similar environmental conditions are called biomes, and these conditions along with the amount of time a biome has existed, its size, and its number of habitat types influence the biome's biodiversity.

- Populations of species in an area interact in various ways, sometimes harming and sometimes benefiting individuals in one or more of the interacting species.

- Each population has a carrying capacity—a maximum size beyond which the habitat can no longer support it.

KEY TERMS

adaptive radiation	competition	gene flow	parasitic relationship
allele	convergence	genetic diversity	phenotype
biodiversity	dominance	genetic drift	population
biodiversity hot spot	ecological niche	genotype	predation
biome	ecosystem	habitat	reproductive isolation
biosphere	evolution	inbreeding	speciation
carrying capacity	exponential growth	keystone species	species
coevolution	extinction	mutation	species evenness
community	gene	natural selection	species richness

REVIEW QUESTIONS

The letters following each Review Question refer to the Chapter Objectives.

1. What characteristics distinguish living things from nonliving things? **(A)**

2. Name three different scales at which biologists study life within the hierarchy of living organisms. **(A)**

3. Define and describe the process of natural selection. **(B)**

4. What is the basic biological mechanism of inheritance that determines an organism's traits? **(B)**

5. What role do gene mutations play in natural selection? **(B)**

6. Distinguish between positive selection and negative selection. **(B)**

7. Describe one way in which speciation occurs. **(B)**

8. Define and describe gene flow. Does it increase or decrease the genetic diversity of a population? **(B)**

9. Describe two different ways to measure biodiversity. **(D)**

10. What does the latitudinal gradient have to do with the global pattern of biodiversity? **(C)**

11. Name two ways in which members of the same species can compete with each other. **(E)**

12. Describe two ways in which populations of different species in a given space interact. **(E)**

13. Name two factors that limit population size. **(F)**

14. Distinguish between "r" and "K" reproductive strategies and provide examples. **(F)**

15. What is a mass extinction event? Provide an example. **(F)**

16. What types of areas are particularly vulnerable to extinction and declining biodiversity? **(F)**

The letters following each item refer to the Chapter Objectives.

17. Factors that limit population size include both density-dependent and density-independent factors. Provide two examples of ways that human environmental effects are causing density-independent factors to affect population growth of other species. **(F)**

18. The concept of "biodiversity hot spots" has generated some controversy at times because it is often used to direct limited conservation resources toward these areas in an effort to protect as many species (and as much biodiversity) as possible. Some critics would argue that species in danger of extinction that live outside

of these areas are just as deserving of protection and should not be overlooked. What do you think of this ethical dilemma? How should we prioritize efforts to protect biodiversity? **(D)**

19. Imagine one of your friends has expressed one of the misconceptions about evolution listed in the "Five Misconceptions about Evolution" section. In your own words, write a paragraph that responds to the misconception. **(B)**

20. From what you have read so far, what reasons would you find compelling for the protection of biodiversity on Earth? **(D)**

Make Your Case?

The following exercises use real-world data and news sources. Check your understanding of the material and then practice crafting well-supported responses.

Use the News

Before the COVID-19 pandemic, other viral diseases like Zika and West Nile were often in the news. These diseases were spread to humans by mosquitoes, so one way to control them was to reduce the populations of species of mosquitoes that carried these diseases. This technique engages issues of population dynamics and genetics. The following article appeared in *Sierra*, the magazine of the Sierra Club.

Use this article to answer the questions that follow. The first three questions test your understanding of the article. Questions 4 and 5 are short answer, requiring you to apply what you have learned in this chapter and cite information in the article. Answers to Questions 1–3 are provided at the back of the book. Question 6 asks you to make your case and defend it.

"Scientists Race to Kill Mosquitoes Before They Kill Us," Sara Novak, *Sierra*, January 9, 2018

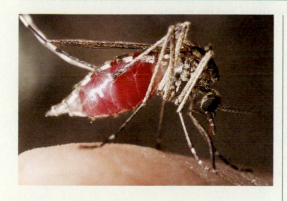

Since West Nile virus made its debut in New York City over a decade ago, outbreaks of mosquito borne diseases, especially West Nile virus, have become increasingly commonplace. As temperatures reach new highs as a result of global climate change, mosquitoes that once called the tropics home find the United States just as habitable. A recent report from the Centers for Disease Control and Prevention

found that *Aedes aegypti*—which is capable of transmitting the Zika, dengue, and chikungunya viruses—could find suitable breeding habitats in 75 percent of the contiguous United States.

Efforts to deal with the unwelcome vectors, however, are already running into trouble. In a poorly executed plan to suppress mosquitoes in South Carolina last year, officials in Dorchester County misted an insecticide called Naled—deadly to both honeybees and mosquitoes—through the air over Summerville without warning local beekeepers. The subsequent deaths of millions of bees served as a wake-up call that weapons beyond pesticides are needed to fight the spread of mosquito-borne diseases.

According to Jose Luis Ramirez, an entomologist with the USDA's Agricultural Research Service in Peoria, Illinois, many pesticides can not only cause adverse reactions in animals and humans, but they also become ineffective over time. "It's an arms race," he says. "At the same

time we're trying to kill them, mosquitoes are evolving to evade death and developing resistance to pesticides."

In order to prevent mosquitoes from developing resistance, Ramirez and his team of scientists are developing bio-pesticides made from fungi that kill mosquitoes but do not impact other insects or humans. The fungi are entomopathogenic—that is, they act like parasites—and are particularly effective at killing mosquitoes.

The magic of the fungi is in their ability to morph. When fungal spores land on a mosquito, they awaken from a resting phase and burrow into the insect. Once inside, the fungus changes into a blastopore [*sic*], a simpler structure that lacks molecules that could alert the insect's immune system. This allows them to multiply in great numbers, eventually killing the mosquito. Soon after the insect is dead, the fungal blastopores [*sic*] continue to grow and re-emerge from the corpse as threadlike filaments called hyphae.

"We're trying to pinpoint the best fungal candidates and the compounds found in the fungi that allow the fungi to kill or evade the mosquito's immune system," Ramirez says. If his team can pinpoint such compounds, they could be extracted from the spores and made into targeted bio-pesticides designed to kill mosquitoes without harming anything else.

Scientists from the biotech company Oxitec are working on a more controversial approach. They have genetically modified mosquitoes to produce a protein that kills them in the absence of the antibiotic tetracycline, which they are given in the lab to survive. Once released, however, the mosquitoes die, but not before passing this trait on to their offspring. Field trials in other countries such as Brazil have reported an 86-to-90 percent decrease in the number of mosquitoes after the introduction of this modified variety. U.S. field tests were to have taken place in Key Haven, Florida, but residents fought hard to oppose it, concerned about unforeseen repercussions from the genetically modified insects.

The issue is complex, and scientists need to carefully weigh the risk of disease with that of GMO mosquitoes, according to Stephen Mahoney, conservation chair of the Sierra Club Miami Chapter. "The Sierra Club recognizes that action is needed now using the best available techniques that are reasonably safe for the environment, which, at the moment, appear to include the use of genetically modified sterile mosquitoes in Zika-infected areas of Florida," Mahoney said in a statement.

Last year, the EPA also approved the use of mosquitoes that are not genetically modified but are infected with the *Wolbachia pipientis* bacterium in the lab. The males are then separated from the females and released into the wild. The males then mate with wild female *Aedes albopictus* mosquitoes, which can transmit a number of tropical diseases. The resulting fertilized eggs fail to hatch.

---

1. Why have mosquito-borne diseases become more common in the United States?

   a. New species of mosquitoes have evolved in the past decade.
   b. Rising temperatures have made more of the United States into suitable habitat for disease-bearing mosquitoes.
   c. Genetically modified mosquitoes are introducing more diseases into the United States.
   d. New strains of fungi have infected humans and made them more susceptible to mosquito-borne disease.

2. Which statement does *not* describe a drawback to conventional pesticides?

   a. They can harm humans.
   b. They can harm helpful insects.
   c. Mosquitoes can develop resistance to them.
   d. They can act as parasites.

3. How would introducing genetically modified (GMO) mosquitoes combat the spread of diseases such as dengue and Zika?

   a. The GMO mosquitoes would prey on the disease-spreading mosquitoes.
   b. The GMO mosquitoes would outcompete the disease-spreading mosquitoes for resources.
   c. The GMO mosquitoes would mate with the disease-spreading mosquitoes and produce offspring with very short life spans.
   d. The GMO mosquitoes would mate with the disease-spreading mosquitoes and produce offspring that do not carry diseases.

4. How does the development of pesticide resistance in mosquitoes relate to the concept of positive and negative selection in evolution?

5. What relationship in the article would be described as predator–prey? Explain why.

6. **Make Your Case** As Stephen Mahoney, conservation chair of the Sierra Club Miami Chapter, stated elsewhere, "Scientists need to carefully weigh the risk of disease with that of GMO mosquitoes." If you were the health commissioner for the state of Florida, which mosquito-control measure described in the article would you choose? Describe the benefits and risks of your decision using information from the article and the chapter.

Use the Data

These graphs show the results of research on the species richness of islands in and around Indonesia. Graph (a) plots the number of bird species contained on the large island of Timor and different islands nearby. These 15 islands are in the South Pacific, and most are part of Indonesia. Graph (b) shows the number of bird species found on small islands at various distances from New Guinea, the second largest island in Indonesia.

Study the graphs and use them to answer the questions that follow. The first two questions test your understanding of the graphs. Questions 3–5 are short answer, requiring you to apply what you have learned in this chapter and cite data in the graphs. Answers to Questions 1–2 are provided at the back of the book. Question 6 asks you to make your case and defend it.

(a)

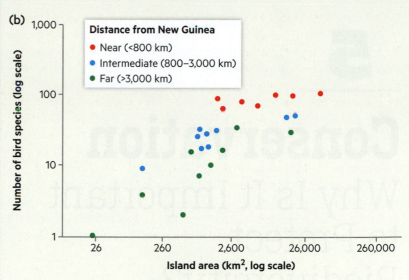

(b)

Adapted from (a) Trainor and Leitão (2007) and (b) MacArthur and Wilson (1963).

1. Using the data in graph (a), match up each island with the number of species living on it.

Island	Number of bird species
Terbang Utara	50
Kisar	100
Damar	20
Atauro	170
Roti	38
Timor	70

2. On the basis of information in graph (b), about how far away from the main island of New Guinea is the island with the greatest number of species? How far away from New Guinea is the island with the fewest number of species?

3. Without looking at a map, do you think Atauro or Sermata is closer to Timor, the largest island in the group of graph (a)? Explain the reason for this assumption.

4. The islands plotted on graph (b) are volcanic islands, which are bare of living things when they first form. In this graph, how does the species richness of the near islands compare overall to that of the far islands and intermediate islands? Considering what you have read about biodiversity, why do you think this relationship exists?

5. Consider another set of islands not represented on these graphs. Trinidad and Tobago is an island nation that is just over 5,100 square kilometers (km²) in area and is 11 kilometers (km) away from the mainland nation of Venezuela. Barbados is an island nation that is 432 km² in area and about 1,100 km away from Venezuela. How do you think the species richness of Trinidad and Tobago compares with that of Barbados? Explain your answer.

6. **Make Your Case** The main island of the nation of Mauritius is just over 2,000 km² in area and about 2,000 km off the southeastern coast of Africa. Imagine you are the president of this country, which is becoming more and more developed to accommodate a growing number of tourists, many of whom like to experience the unique species of life on your island. If you were presented with a development proposal to double the tourist traffic, how would you respond? Would you endorse or oppose this proposal and why? What sorts of development projects would you pursue?

LEARN MORE

- Millennium Ecosystem Assessment: www.millenniumassessment.org/en/index.html
- United Nations Decade on Biodiversity, 2011–2020: www.cbd.int/2011-2020/
- US Fish and Wildlife Service Endangered Species page: www.fws.gov/endangered/
- Video: Darwin's Theory of Natural Selection: digital.wwnorton.com/environsci

- Darwin 200 special in the journal *Nature*: www.nature.com/collections/qjtwzbswgh
- University of California, Berkeley, resource on biomes: www.ucmp.berkeley.edu/glossary/gloss5/biome/
- Video: DNA Replication: digital.wwnorton.com/environsci
- Video: Human Variation: digital.wwnorton.com/environsci

5

Conservation
Why Is It Important to Protect Biodiversity?

Pacific salmon return to their native streams to spawn. Many populations of Pacific salmon are now listed as threatened under the Endangered Species Act because of human impact.

Wild Pacific salmon. It is one of the highest-priced items at your local fish market or the seafood counter of your grocery store. It is also often one of the highest-priced entrees at fine restaurants and near the top of the list of the most valuable commercial fishery resources in the United States. Each year, the value the fish bring when they're caught and sold totals nearly half a billion dollars. Yet you may have heard that salmon populations are in decline and that the US government has issued protections for some species, restricted harvest in certain places, and funded projects to help the populations recover. Pacific salmon provide a good example of human efforts to conserve biodiversity.

What do we know about Pacific salmon? Although they spend most of their lives in the ocean, they begin and end their lives in freshwater systems. They hatch in rivers and streams and journey downstream to the sea, where they put on most of their body weight. Near the end of their lives, salmon return hundreds of miles through the ocean back to the freshwater systems from which they came, swimming upstream to spawn and dying after leaving their eggs to hatch another generation in the river gravel. The nutrients in the dead salmon are then released into the freshwater and forest ecosystems, where they play an important role in supporting the communities of life in these places—including the insects that the next generation of young salmon depend on for food.

While this might sound like a straightforward process involving some very long swims, it in fact is a very delicate one. While humans can casually swim in both salt and fresh water, salmon live and breathe in the water. This means their bodies must change each time they migrate from fresh to salt water and then back to fresh water again (**FIGURE 5.1**). Salmon also rely on highly specialized navigational abilities to travel long distances back to a precise location. Because returning salmon rarely stray from their home streams and spawn there during particular seasons, there is little interbreeding among the populations from stream to stream. This isolation means that different populations of salmon have accumulated genetic distinctions and developed adaptations specific to their

Chapter Objectives

By the end of this chapter, you should be able to . . .

A. explain how human actions are contributing to the loss of global biodiversity.

B. identify how humans depend on biodiversity and ecosystem services.

C. evaluate the range of policy tools we can use to protect biodiversity.

D. summarize what steps can be taken in areas of significant human impact to reduce biodiversity loss.

E. understand how species reintroduction and ecological restoration can be used to address biodiversity loss.

> **Nothing can survive on Earth unless it is a cooperative part of a larger global life.**
>
> —**Barry Commoner**

home stream. In other words, the salmon migrating up the Copper River in southeastern Alaska are different—genetically, morphologically, and behaviorally—than those migrating up the Columbia River in Washington and Oregon.

How are the salmon doing in our changing world? The abundance of any particular species of Pacific salmon varies by population. For example, the Chinook salmon spawning in the Copper River watershed have met or exceeded the sustainability targets set by state fish and wildlife officials nearly every year for the past 15 years. In contrast, during this time Chinook in the Columbia River and Puget Sound systems along the same Pacific coast have been listed as threatened under the Endangered Species Act (**FIGURE 5.2**).

In this chapter, we will begin by considering the status of Earth's biodiversity and the reasons for protecting it. Why should we care about the survival of particular populations of Pacific salmon in the first place? One reason is that humans often directly benefit from abundant, healthy populations of certain species. We already mentioned the commercial importance of the salmon fishery, but these fish are also central to the cultural traditions and identities of Indigenous peoples and other communities in the Pacific Northwest. A second reason can be less obvious: the decline of a species (including those largely "unseen" by humans) often indicates a decline in the ecosystems on which we all depend. A third reason might be to simply prevent the loss of wild species such as salmon because of their intrinsic value: their value apart from human use. Next, we will examine efforts to protect biodiversity, broadly known as conservation.

FIGURE 5.2 **Same Species, but One Population Is Threatened** Although the Chinook salmon of the Copper River basin and Columbia River basin are part of the same species, they are genetically and morphologically distinct populations due to their geographic separation. The health of these populations also varies significantly due in large part to different human habitat impacts in these two regions.

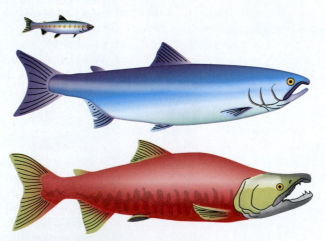

FIGURE 5.1 **Life Stages of Pacific Salmon** Pacific salmon (like these sockeye salmon) undergo physiologic changes as they move from fresh water to the ocean and then back to fresh water again to spawn. Juvenile sockeye salmon (top) begin life in fresh water. Then in their adult years (middle), the salmon adapt to the ocean before changing again (red body, green head) to return to fresh water to spawn.

This includes species-protection laws, the establishment of protected areas, and recovery and restoration projects. For example, many populations of Pacific salmon species are listed under and receive protection from the Endangered Species Act. Efforts to restore these populations involve restoration projects of various scales. These can include local efforts to plant trees to provide shade along the edges of streams as well as more ambitious projects such as restoring straightened river channels or even removing large dams. We will learn that Earth's biodiversity affects all of us, and protecting it requires our concerted effort.

5.1 What Is the Status of Earth's Biodiversity?

When biologists assess our planet's biodiversity, often the focal point is human impact on the extinction rate of life-forms. Every species possesses some genetic traits that are not shared with others, and each extinction permanently subtracts from Earth's biodiversity. Though biodiversity does rebound after mass extinctions (we saw in Chapter 4 that Earth has had five so far), it takes millions of years for species to evolve that are different from those that were lost.

Human Impact on Biodiversity

Extinction Rates Today, most biologists have concluded that a sixth mass extinction is under way, driven by human actions. Estimates of extinction rates range from hundreds to more than 10,000 species per year. One United Nations finding estimated that since the dawn of industrialization, humans have increased the species extinction rate by at least 1,000 times and that 10% to 30% of existing bird, mammal, and amphibian species are now threatened with extinction. Because we have not yet catalogued a majority of all the species currently in existence, many species go extinct before they have ever been identified and described (**FIGURE 5.3**). Conservation groups often prioritize their species protection and recovery efforts on the biodiversity hot spots we discussed in Chapter 4. Because these are places where

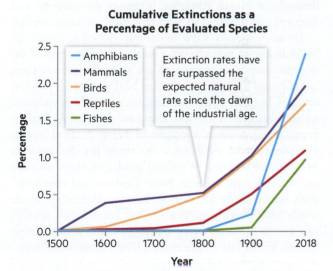

Cumulative Extinctions as a Percentage of Evaluated Species

- Amphibians
- Mammals
- Birds
- Reptiles
- Fishes

Extinction rates have far surpassed the expected natural rate since the dawn of the industrial age.

FIGURE 5.3 Cumulative Vertebrate Extinctions This graph shows how cumulative extinctions for different types of animals have been mounting over the centuries. Evaluated species are those that are known and have been subject to monitoring by the International Union for Conservation of Nature—a network of scientists who monitor and evaluate the status of Earth's species.

Adapted from Intergovernmental Science-Policy Platform on Biodiversity and Ecosystem Services (2019).

→ CONTEXT

Extinction of the Passenger Pigeon

The passenger pigeon was once the most abundant bird in North America and possibly the world. With a population in the billions, these birds migrated in massive flocks throughout the midwestern and eastern regions of the United States and Canada, devouring acorns and beechnuts along the way. Accounts from the 19th century report that the flocks were so large they made a noise like thunder as they approached and shaded the sunlight as they passed. As they nested in trees in large numbers, they would break branches and even tree trunks. How could a species numbering in the billions fall to zero? Overexploitation by humans who hunted them for food and sport drove them to extinction. Their concentration in large flocks made the pigeons vulnerable to being killed on a large scale by methods such as capturing them in large nets, poisoning them with corn soaked in whiskey, suffocating them by burning sulfur beneath the trees where they roosted, and simply shooting them with guns. Advances in communication with the invention of the telegraph and in transportation with the railroad also played a role, enabling hunters to receive word of arriving flocks in the spring and to travel quickly to their nesting grounds.

large numbers of species occur, they are areas particularly vulnerable to extinctions caused by human actions.

Alteration of Habitats The greatest effect humans have on biodiversity is through our ability to physically alter the habitats of other species as we pursue our own needs and desires. Large areas are cleared for agricultural use, paved for roads and residential development, excavated for mines, or altered in other ways. While habitats may be converted entirely for human development, in many other cases they are fragmented—carved up by roads and other developments into disconnected chunks that make it difficult or impossible for the species dependent on the habitat to survive (**FIGURE 5.4**).

FIGURE 5.4 Deforestation and Habitat Loss (a) This satellite photo of the border region between Guatemala and Belize in 2007 shows a stark difference in forest habitat loss: Guatemala has lost far more habitat because of human actions such as logging, farming, settlements, and road building. **(b)** This graph compares the area of land in various biomes that has been converted to human uses with the area that is protected. In most cases, the percentage of a biome's area that is protected is far less than the percentage that has already been converted.

Adapted from (b) Hoekstra et al. (2005).

(a)

(b)

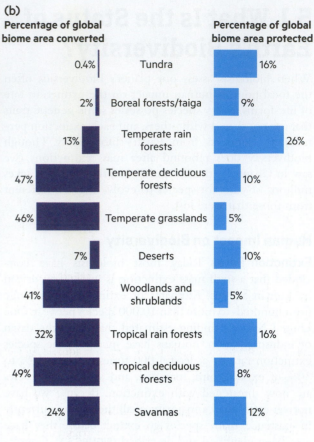

Percentage of global biome area converted	Biome	Percentage of global biome area protected
0.4%	Tundra	16%
2%	Boreal forests/taiga	9%
13%	Temperate rain forests	26%
47%	Temperate deciduous forests	10%
46%	Temperate grasslands	5%
7%	Deserts	10%
41%	Woodlands and shrublands	5%
32%	Tropical rain forests	16%
49%	Tropical deciduous forests	8%
24%	Savannas	12%

Overexploitation of Resources Humans also overexploit species directly through activities such as hunting and fishing. This was the case for the now-extinct passenger pigeon, which once numbered in the billions in North America. These birds were hunted aggressively for food and sport, and some individual hunters in the 19th century were reported to have killed tens of thousands of passenger pigeons in a single day. We risk behaving similarly with seafood today: more than 3 billion people rely on seafood for at least 20% of their animal protein intake, and modern fishing technologies have driven serious overharvest of many of the world's fisheries.

Introduction of Invasive Species When humans introduce an organism that is not native to a particular area, the predators that naturally restrict its growth are not present. As a result, the organism can become overabundant, harming native species and reducing biodiversity. Invasive plants such as kudzu vines in the southeastern United States and English ivy on the East and West Coasts spread rapidly to smother trees and other native plants (**FIGURE 5.5**). Some invasive plants cause more fundamental changes to ecosystems. The evergreen fire tree that has invaded Hawaii and Scotch broom shrub that has invaded much of the western United States have altered the soil chemistry in these areas to the detriment of native plant communities. Cheatgrass, an invasive species throughout North America, increases the severity and frequency of wildfires in ways that displace native shrubs.

Invasive species have been introduced accidentally by humans, as was the case with rats that arrived on oceanic islands as stowaways on ships. These rats then contributed to the extinction of bird species by preying on eggs. In other cases, humans intentionally introduced invasive species. This was the case for the Australian brush-tailed possum, brought to New Zealand in the 19th century to establish a fur trade for the animal's pelts. In the absence of natural predators, these animals decimated native trees in New Zealand and competed with native birds for insects, berries, and other foods while also devouring their eggs. Introduced species can also sometimes carry pathogens that attack native species. In the late 19th century, when cattle were introduced to eastern Africa, they carried a virus that spread disease among native wildebeest and buffalo. Population declines of up to 95% in the native animals caused human famines in the region.

Effects on Water Humans impact water sources in many ways. Dams and other water diversions on rivers can disrupt the life cycles of migrating fish, such as the Pacific salmon described earlier. Freshwater withdrawals for irrigation

FIGURE 5.5 Invasive Species Kudzu (left) and English ivy are examples of invasive species in the United States that smother native species. The evergreen fire tree (center) outcompetes native plant species in Hawaii and alters soil chemistry. Australian brush-tailed possums (right) are an example of an invasive mammal that has decimated tree canopies and native bird populations in New Zealand.

and municipal water supplies can threaten aquatic species by reducing the quantity of water they live in. Water pollution of many types also has numerous negative effects. In Chapter 10, we will see how dead zones—areas of the ocean that are devoid of most organisms—form close to shore due to oxygen depletion triggered by the overuse of fertilizers that eventually wash into the ocean.

Climate Change We will see in Chapter 11 that the effects of climate change on species are wide ranging. For example, significant habitat alteration due to climate change directly affects species such as polar bears and walruses that depend on polar habitats, calcium carbonate–shelled marine organisms that are damaged by ocean acidification, and corals that can tolerate only a narrow range of temperature. As **FIGURE 5.6** illustrates, when stressed by higher water temperatures, corals turn white, which leaves them weakened and increases their mortality rate.

⬡ **TAKE-HOME MESSAGE** Human actions are contributing to a sixth mass extinction event. Habitat destruction is the primary human-based cause of declining biodiversity. Other human actions and effects include the overuse of species, the introduction of invasive species, water withdrawals, pollution, and climate change.

5.2 Why Protect Biodiversity?

If we are in fact driving a mass extinction, what motivations do we have to take action? Thinking back to the fish counter at the grocery store, some people might question whether it is important to ensure that pricey salmon

fillets remain available. There are other, cheaper fish to buy, so why protect salmon? One motivation focuses on **instrumental value**: the usefulness of particular species for human purposes. In the salmon example, this includes factors such as the commercial availability of a particular fish and the livelihoods of the people who catch the fish. Another motivation might focus on **intrinsic value**: the

instrumental value the usefulness of particular species for human purposes.

intrinsic value the value of something in and of itself apart from its usefulness to others (Chapter 2).

HEALTHY CORAL
1 Coral and algae depend on each other to survive.

STRESSED CORAL
2 If stressed, algae leave the coral.

BLEACHED CORAL
3 Coral is left bleached and vulnerable.

Corals have a symbiotic relationship with microscopic algae called zooxanthellae that live in their tissues. These algae are the coral's primary food source and give them their color.

When the symbiotic relationship becomes stressed due to increased ocean temperature or pollution, the algae leave the coral's tissue.

Without the algae, the coral loses its major source of food, turns white or very pale, and is more susceptible to disease.

FIGURE 5.6 Coral Bleaching Coral bleaching is a whitening that occurs when the symbiotic relationship between coral and their colorful resident algae is disrupted. It is caused by warming ocean conditions and can weaken or kill coral.

ecosystem services
a concept that assigns a value to human benefits derived from naturally functioning ecosystems (Chapters 1 and 2).

provisioning services
goods humans consume that are directly provided by ecosystems.

regulating services natural processes of ecosystems that provide favorable conditions for humans.

supporting services
the fundamental natural conditions on which many other ecosystem services depend.

cultural services benefits to humans from the environment that may not be essential to survival but enhance our quality of life.

value of something in and of itself apart from its usefulness to others. From this perspective, humans have a duty to protect biodiversity from our adverse effects simply because species have a right to survive and flourish.

But more broadly, we might consider how salmon are part of a diverse ecosystem and how their diversity improves human lives. In 2005, the United Nations set out to do just this for ecosystems in general. They assembled more than 1,000 scientists from nearly 100 different countries to synthesize research on the consequences of ecosystem change for human well-being. The resulting 2005 Millennium Ecosystem Assessment found that loss of species diversity reduces the productivity and resilience of ecosystems and compromises their function. This is troubling for humans because we depend on functioning ecosystems in so many different ways, such as for food, clean air and water, medicine, materials for various goods and energy production, recreational pursuits, educational opportunities, aesthetic appreciation, and culture. The Millennium Ecosystem Assessment notes that although human well-being has increased since the 1950s, these gains have come at a cost to Earth's ecosystems, costs that may well contribute to future declines in human well-being.

What Are Ecosystem Services?

We all depend on certain services every day. For example, local governments provide fire protection, a police force, roads, sewers, water, and education. As we have seen in previous chapters, ecologists use the term **ecosystem services** to assign a value to human benefits derived from naturally functioning ecosystems. How do you rely on ecosystem services? **Provisioning services** are the first type that comes to mind for most people: these are goods we consume that are directly provided by ecosystems. So the Pacific salmon described earlier and other kinds

of seafood caught in the wild for human consumption provide a provisioning service. Fresh water—something on which every member of the human species depends—provides another ecosystem service, whether we use it to keep us alive or use it to provide hydropower. Other examples of provisioning services include wood from forests and wild plants that provide pharmaceutical products (**FIGURE 5.7**).

Regulating services are another type of ecosystem service: these are natural processes that provide favorable conditions for humans. Examples include the water filtration provided by soil, the erosion reduction provided by vegetation, the waste decomposition provided by microbes, the carbon sequestration and climate regulation provided by forests, and the storm surge and flood protection provided by coral reefs, wetlands, and mangroves. Crop pollination and pest control are other familiar services provided by insects, birds, bats, and spiders. Often, the importance of these systems becomes most apparent when they are absent, and we see the problems that result: landslides in an area of forest that has been clear-cut, flooding and hurricane damage in areas that have lost their natural protection, and declines in agricultural production when areas have a shortage of bees to pollinate crops.

Scientists typically see the most essential ecosystem services as **supporting services**, which provide fundamental conditions on which many other ecosystem services depend. Examples of supporting services are soil formation, photosynthesis, and the processes governing nutrient and water cycling.

In addition, there are ecosystem services that may not be essential to survival but that greatly enhance our quality of life. These are grouped into a category known as **cultural services**. Recreation is one such service, and the millions of dollars that hunting and fishing groups, such as Ducks Unlimited and Trout Unlimited, have devoted

(a)

(b)

(c)

FIGURE 5.7 Biodiversity as a Source of Medicine Wild plants are a resource for new pharmaceutical products. The search for species that can yield commercially viable drugs is called bioprospecting. **(a)** The rosy periwinkle flower from Madagascar and **(b)** the Pacific yew from the Pacific Northwest in the United States have each been the source of breakthrough cancer treatments, and **(c)** Chinese star anise is the source of the active ingredient in the antiviral drug Tamiflu.

FIGURE 5.8 **Natural Beauty** Artistic images such as photographs and landscape paintings remind us that ecosystems also provide intangible values such as aesthetic appreciation for places we find beautiful.

to conservation show its value. Education and research (like field-based lab projects in high school and college classes), aesthetic benefits, and even spiritual needs depend on healthy ecosystems (**FIGURE 5.8**). When environmentalist John Muir, the founder of the Sierra Club, was arguing against the construction of a dam inside Yosemite National Park in 1912, he wrote, "Everybody needs beauty as well as bread, places to play in and pray in, where Nature may heal and cheer and give strength to body and soul alike." The Jémez Pueblo people have argued for The Jemez Pueblo people successfully argued for restricted public access to Redondo Peak, a mountain in northern New Mexico, because it is central to their religious beliefs and practices. (**FIGURE 5.9**).

Placing Economic Value on Ecosystem Services

Ecosystem services are obviously valuable to humans: we are all happy when there are bees to pollinate so that we can eat fruit, and scenic property usually costs more to buy. But how are dollar values placed on ecosystem services? Provisioning services often have a market value for their goods. In these cases, we can estimate **value chains**, economic assessments that integrate the value of the good itself and the market value of things linked to the production of that good. For example, in 2007 the World Bank estimated that the world's fisheries contributed $624 billion to the global economy. This figure includes both commercial and recreational fisheries; preharvest activities, such as boatbuilding and equipment sales; and postharvest activities, such as food processing. In other cases, economists can use **replacement cost**: an estimate of how much it would cost to replace an ecosystem service with an artificial substitute. The supply of clean drinking water to New York City from watersheds in the Catskill Mountains provides one example of this approach. The replacement cost for this service is

equivalent to the cost of constructing and maintaining a water-filtration plant serving the same number of people. In 1996, the replacement cost of this service was more than $8 billion.

In still other cases, economists can estimate value from people's *stated preferences* or *revealed preferences*. Stated preferences are typically gathered by surveying large numbers of people and asking questions about how much they would be willing to pay to realize one environmental condition over another. The US Forest Service used surveys like this to determine that US citizens' willingness to pay for carbon storage provided by the forests of the Pacific Northwest was $65 per ton. Revealed preferences estimate the value people place on something

value chain economic assessment that integrates the market value of a good with the market value of things linked to its production.

replacement cost in economics, an estimate of how much it would cost to replace something with a substitute good.

FIGURE 5.9 **Cultural Practices in Natural Environments** Ecosystems can also be a source of cultural values. The Jemez Pueblo people have worked to protect Redondo Peak because it is a sacred site where residents conduct religious practices.

national park an area of scenic, historical, or scientific importance protected and maintained by the federal government through the National Park Service.

based on their actions. Travel expenses people incurred to visit Australia's Great Barrier Reef have revealed that the value people put on this place for recreation is more than $1 billion each year. Economists can also combine valuations of individual locations to tally the total ecosystem services provided by a particular area. An analysis of the St. Louis River in northeastern Minnesota in 2015 placed a value of between $5 billion and $14 billion on the total ecosystem services of the water and land in the watershed area. On a much larger scale, in 1997 a team of economists compiled analyses like these for 17 types of ecosystem services across 16 of Earth's biomes to estimate an economic value of $33 trillion for Earth's entire biosphere: nearly twice the total of the gross national product for all the countries in the world at that time.

Criticisms of the Ecosystem Services Approach

Many ecologists have raised concerns that economic valuation of ecosystem services as described above is not adequate. First, many would argue that humans have a duty to protect endangered species and to preserve biodiversity because life has intrinsic value. Similarly, ecosystem functions do not always benefit humans, and there may even be ecosystem "disservices": organisms that spread disease (like mosquitoes) or systems that increase the risk of disease (wetlands). Critics ask if it is right to destroy these elements just because their economic value may not be obvious. Second, because many ecosystems can maintain their functions and continue to provide services to humans even with the loss of some species, some ecologists believe that the ecosystem services viewpoint can be used to wrongly justify the loss of "expendable species." Third, humans may develop artificial substitutes for natural

services, and critics worry that this then may reduce the motivation to protect the original service. For example, when the cancer drug Taxol was first isolated from the Pacific yew tree, laws were passed to preserve and manage forests of Pacific yew. Taxol can now be synthesized in the laboratory, but ecologists would argue that this does not mean we should start clearing these forests again.

⬠ TAKE-HOME MESSAGE Reasons for protecting biodiversity include the ways humans use and benefit from ecosystems and a sense that species have intrinsic value and humans have a duty to protect against species loss. The concept of ecosystem services places economic values on the ways humans depend on functioning ecosystems.

5.3 What Are Protected Areas?

One way to address adverse human effects on biodiversity is for government authorities to create areas where ecosystems are provided with protection from certain kinds of human activities (**AT A GLANCE: PROTECTED AREAS**). A familiar example of a protected area is a **national park**, an area of scenic, historical, or scientific importance protected and maintained by the National Park Service. Yellowstone National Park is perhaps the most famous national park and one of the last remaining places in the continental United States with a diversity of mammals similar to what it had in the 1800s (when it opened). The park provides critical habitat to species now largely absent over most of their historical range, such as American bison and grizzly bears (**FIGURE 5.10**).

FIGURE 5.10 The First National Park Millions of visitors flock to Yellowstone National Park each year to see iconic species such as American bison.

Can Biodiversity Boost Soil Fertility and Agricultural Productivity?

At the dawn of the 20th century, the German agronomist and plant pathologist Lorenz Hiltner made several key—and long-neglected—discoveries about the role of soil life in promoting the health of life aboveground and that of plants in particular. As the founding director of the Bavarian Agricultural–Botanical Institute in Munich, Hiltner tested the then radical idea that microbes could benefit plant health. He also found that special soil additives that boosted the population of beneficial bacteria in the soil promoted plant growth and could cure ailing plants. Hiltner's work showed that plants in life-filled soils were better able to resist illness-causing microorganisms (*pathogens*) and diseases. He concluded that soils with high densities of these beneficial microbes benefit the growth and health of plants and that soils with low densities of soil life promote plant diseases.

His classic experiments are among the most replicated botanical experiments. This effect is so general that scientists now consider life-filled soils "disease suppressive." One of Hiltner's most famous experiments involved growing similar plants in sterilized and unsterilized soils. He would then introduce a pathogen into each type of soil. Time and again, plants grown in sterilized soil succumbed to the pathogen, while plants grown in unsterilized soil thrived. In other experiments, Hiltner demonstrated that this ability to suppress plant diseases results from microbial action. Sterilizing of soil destroyed its disease suppression ability, and mixing even a tiny amount of microbe-rich soil with sterile soil conferred disease suppression.

Hiltner noticed that soil microorganisms were far more abundant in proximity to plant roots and named this zone the rhizosphere. Scientists have discovered that when beneficial microbes are present in the soil around roots, they send chemical messages to plants. These chemicals can trigger an immune-like response that helps plants to fend off pests and pathogens. Thus, plants in life-filled soils are better equipped to combat pests and pathogens and fare better than plants unprepared to face an attack.

That more organisms in the soil makes for a more resilient system is one reason why diverse crop systems are more sustainable over the long haul. And in the aboveground world too, experiments that manipulate the number of species in sample areas have shown that diverse ecosystems are more productive. Diversity also helps maintain ecosystem functions and the ability of ecosystems to resist or recover from certain kinds of disturbances. The general lesson is that greater biodiversity translates into greater ecosystem productivity and resilience.

A recent example comes from the work of ecologist David Tilman at the University of Minnesota. Tilman studies prairie ecosystems and has found that mixtures of prairie grasses discharge a lot of carbon out of their roots and into the soil. In fact, this amount is more than enough to offset the amount of carbon dioxide (CO_2) that would be produced if the aboveground biomass the grasses produce was burned as a fuel. This means that using biofuels produced from such crop mixtures would actually reduce greenhouse gases in the atmosphere. Most biofuels research has focused on growing single-species plots (monocultures) of high-biomass plants (such as poplar trees or switchgrass). But Tilman's work suggests that planting mixed prairie grasses could produce twice the bioenergy per acre. And planting prairie grasses that squirrel away carbon in root systems and soil organic matter could improve the fertility of degraded and marginal lands. Tilman has even estimated that planting diverse prairie grass mixtures on the world's degraded grasslands could produce enough energy to replace about 13% of global oil use and store enough carbon in the soil to cancel out an additional 15% of global fossil fuel emissions. Tilman's research suggests that it would make more ecological and economic sense for biofuel producers to plant a diverse community of plants instead of monocultures.

Prairie grass root systems are much larger than those of plants like wheat, and they can hold a lot of carbon. Compare the extensive root system of the perennial prairie grass, being held aloft and seen growing in the left part of photo, to the shallow root system of the annually cultivated wheat on the right.

What Would You Do?

Many college campuses and cities dedicate a significant amount of space and money to landscaping, using extensive plantings of trees, bushes, flowers, lawns, and sometimes even vegetable gardens. Knowing what you know about your campus and/or city landscape and the information in this box, do you think this is a wise investment? How do you think grounds-keeping decisions influence biodiversity, plant and soil health, and carbon sequestration?

Protected Areas

There are many protected areas in the United States, as indicated on this map. National parks are some of the oldest and best known of these areas and are labeled. Although four federal agencies (National Park Service, US Forest Service, Bureau of Land Management, and US Fish and Wildlife Service) manage most of the protected lands shown, tribal, state, and local governments also play an important role in managing protected areas.

National Park

North Cascades
Olympic
Glacier
Mount Rainier
Theodore Roosevelt
Crater Lake
Yellowstone
Redwood
Grand Teton
Badlands
Wind Cave
Lassen Volcanic
Great Basin
Rocky Mountain
Haleakalā
Yosemite
Capitol Reef
Arches
Black Canyon
Kings Canyon
Bryce Canyon
Canyonlands
Zion
Pinnacles
Sequoia
Mesa Verde
Great Sand Dunes
Hawai'i Volcanoes
Death Valley
Grand Canyon
Channel Islands
Petrified Forest
Joshua Tree
White Sands
Kobuk Valley
Saguaro
Carlsbad Caverns
Gates of the Arctic
Guadalupe Mountains
Denali
Lake Clark
Kenai Fjords
Wrangell–St. Elias
Big Bend
Katmai
Glacier Bay

Protected Areas and Managing Agencies

- Bureau of Land Management (BLM)
- US Forest Service (USFS)
- National Park Service (NPS)
- National Oceanic and Atmospheric Administration (NOAA)
- US Fish and Wildlife Service (USFWS)
- Bureau of Reclamation (BOR)
- Department of Defense (DOD)
- Department of Energy (DOE)
- National Resources Conservation Service (NRCS)
- American Indian Lands (BIA)
- State fish and wildlife
- State parks and recreation
- Other state (NHP, DOT, HS, etc.)
- County land/ regional agency land
- Private conservation; private corporation
- US Army Corps of Engineers (USACE)
- Other federal (TVA, ARS, BPA, etc.)
- Nongovernmental organization
- Territorial land
- City land
- Joint, other
- Unknown
- State trust land

Map labels: Voyageurs, Isle Royale, Acadia, Indiana Dunes, Cuyahoga Valley, Shenandoah, Gateway Arch, New River Gorge, Mammoth Cave, Great Smoky Mountains, Hot Springs, Congaree, Everglades, Biscayne, Dry Tortugas

Total Land under Federal Government Management

The federal government owns and manages roughly 650 million acres of land. This graph shows which agencies manage this land, the acres they manage, and the percentage this is of the total.

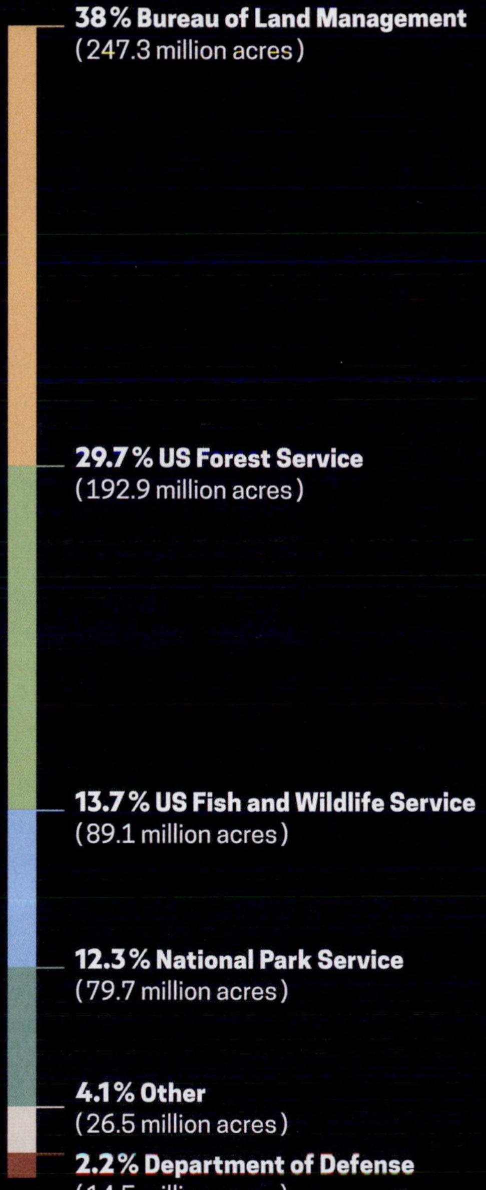

38 % Bureau of Land Management
(247.3 million acres)

29.7 % US Forest Service
(192.9 million acres)

13.7 % US Fish and Wildlife Service
(89.1 million acres)

12.3 % National Park Service
(79.7 million acres)

4.1 % Other
(26.5 million acres)

2.2 % Department of Defense
(14.5 million acres)

FIGURE 5.11 National Monuments Under the Antiquities Act, national monuments can be created by the president without congressional approval. Many national parks, such as Grand Teton National Park and Grand Canyon National Park, began as national monuments.

national recreation area an area that conserves and provides recreation, typically around the reservoirs created by large dams.

National Wildlife Refuge System a system of public lands and waters set aside and managed by the US Fish and Wildlife Service to conserve America's fish, wildlife, and plants.

At the same time, the law establishing Yellowstone National Park also provided for a certain level of protection of the area. It mandates "preservation" of the area "from injury or spoliation," maintenance of its "natural condition," and protection of the fish and wildlife from "wanton destruction." In the United States, the federal government owns and manages more than 620 million acres of land (about 28% of the nation's land surface) with varying levels of protection (see At A Glance: Protected Areas). About 90% of this land is managed and protected by one of four land-use agencies: the National Park Service, US Fish and Wildlife Service, US Forest Service, and Bureau of Land Management. In addition, there are also protected rivers, aquatic and marine areas, and a variety of international designations of protected status. Let's take a closer look at each of these.

National Parks, Monuments, and Recreation Areas

The National Park Service, established by Congress in 1916, administers more than 400 different parks, monuments, and other designated lands covering more than 80 million acres of public land. Establishing a national park requires congressional approval, and there are 63 in total. Congress also passed the Antiquities Act of 1906 that empowered the US president to set aside certain federal lands as national monuments without congressional approval. Many areas that are now national parks, including the Grand Canyon and the Tetons, first gained protection through national monument designation (**FIGURE 5.11**). **National recreation areas**, such as the Lake Mead and Glen Canyon areas on the Colorado River, were established to conserve areas and provide recreation, typically around the reservoirs created by large dams. In the 1960s, a greater priority on protecting ecosystems and biodiversity led to largely roadless parks, such as North Cascades National Park in Washington State and Canyonlands National Park in Utah, which prioritize preserving ecosystems over providing public access to natural landmarks.

National Wildlife Refuges

The US Fish and Wildlife Service, established in 1940, administers an even larger acreage of public lands and waters than does the National Park Service through the **National Wildlife Refuge System**. There are more than 560 national wildlife refuges covering more than 100 million acres of public land. President Theodore Roosevelt established the first national wildlife refuge at Pelican Island, Florida, "as a preserve and breeding ground for native birds" by executive order in 1903 (Roosevelt established 55 bird reservations and game preserves during his presidency). Although wildlife refuges allow hunting and fishing, since 1966 federal law has mandated that any use of refuge lands must be "compatible" with the purpose for which each refuge was established.

Yet when Yellowstone became the first national park in 1872, the primary management mission was recreation for humans, not biodiversity protection. Congress set aside the 2.2 million acres in Yellowstone as a "public park or pleasuring ground for the benefit and enjoyment of the people." The boundaries of Yellowstone, like most national parks, were drawn primarily to feature scenic vistas rather than to preserve ecosystems. And until the 1960s, the National Park Service was authorized to carry out predator-control programs, killing large numbers of wolves, coyotes, cougars, and other animals thought to have a "detrimental" effect on human recreation.

Often this means that hunting, fishing, and other forms of recreation are strictly limited.

National Forests

The US Forest Service was established in 1905 to manage the nation's forest reserves. The original policies governing **national forests**—lands owned, managed, and preserved by the US Forest Service for multiple purposes, including timber harvests, recreation, and fish and wildlife conservation—focused only on the provisioning services of timber and water. Reserves were created to "improve and protect forests" for the purpose of "securing favorable conditions of water flows, and to furnish a continuous supply of timber." However, new forest laws in the 1960s and 1970s required the agency to manage the lands for recreation and fish and wildlife conservation as well as natural resource extraction. Timber is still harvested in national forests, but the practice is now subject to long-term decisions about the size of harvests and how and where they will be conducted.

Bureau of Land Management Lands

The Bureau of Land Management (BLM) oversees more land than any other federal agency. It was originally established to manage grazing and mining on federal lands. The BLM collects fees and sets the conditions for more than 18,000 grazing permits issued to ranchers who lease these public lands. The BLM also issues permits and leases for more than 300 coal-mining operations and more than 63,000 oil and gas wells on public lands. In return, these companies pay royalties—a percentage of the profit they make on the commodities they extract—to the federal government. But as with the US Forest Service, new laws in the 1970s required the BLM to administer its lands for multiple uses beyond resource extraction and to conduct long-term planning with public involvement. Some have said the BLM is undergoing a transformation from the "Bureau of Livestock and Mines" to the "Bureau of Landscapes and Monuments."

Wilderness Areas and Wild and Scenic Rivers

The United States also has a National Wilderness Preservation System that offers special protection to certain areas. The **Wilderness Act of 1964** required land-use agencies to review their lands as possible wilderness areas and Congress to then determine which lands receive this designation. Road construction and motorized transportation are prohibited in these areas, as are permanent structures, such as cabins, bathrooms, and picnic shelters. There are now more than 750 wilderness areas totaling more than 100 million acres. Similarly, the federal government also manages a National Wild and Scenic Rivers System that includes parts of more than 200 rivers and totals more than 2,500 miles (4,000 kilometers)—or about one-quarter of 1% of the river miles in the United States.

FIGURE 5.12 Collecting Data in the Chesapeake Bay National Estuarine Reserve Marine protected areas have been documented to improve fishing conditions near them, as fish grow larger in the protected areas.

Aquatic and Marine Areas

National lakeshores and national seashores focus on the preservation of shorelines and islands, and individual states and local governments also have land designations for protecting these areas. The federal government coordinates with state, tribal, and local governments to provide protection in the ocean through more than 1,600 marine protected areas (MPAs). MPAs are situated in a variety of aquatic life zones, including the open ocean, coastal areas, intertidal zones, and estuaries (**FIGURE 5.12**).

Protected Areas throughout the World

Although our focus has been on protected areas in the United States, nearly every country in the world employs some form of protected area status. It is estimated that about 14% of Earth's land surface has been granted some level of protected status, while less than 1% of the planet's oceans enjoy protection. In particular, the United Nations Educational, Scientific and Cultural Organization (UNESCO) has encouraged national governments to protect areas with "outstanding value to humanity" as **World Heritage sites**. The United Nations Environment Programme (UNEP) has also assembled a World Database on Protected Areas that includes more than 160,000 marine and terrestrial protected areas. As is the case with the designations in the United States described earlier, the human uses permitted in these areas vary considerably.

⬡ **TAKE-HOME MESSAGE** Government-designated protected areas are an important tool for conserving biodiversity. In the United States, four agencies manage most of the federal protected areas. Each agency has a mission that privileges certain human uses, but each also makes and enforces rules that offer certain protections against human use.

national forest lands owned and managed by the US Forest Service for multiple purposes, including timber harvests, recreation, and fish and wildlife conservation.

Wilderness Act of 1964 an act creating areas where road construction and motorized transportation are prohibited, as are permanent structures such as cabins, bathrooms, and picnic shelters.

World Heritage site an area with "outstanding value to humanity" established by a national government and recognized by the United Nations Educational, Scientific and Cultural Organization (UNESCO).

Indigenous Peoples and the 30 by 30 Initiative

The 30 by 30 initiative is a commitment by more than 60 nations to put more than 30% of Earth's land and water into protected status, in an effort to address the crises of climate change and biodiversity loss. In June 2021, the leaders of the G7 nations—a group of the world's wealthiest democracies, including the United States—pledged to do just that by 2030.

These announcements have worried some Indigenous communities because past conservation efforts—such as the creation of national parks—removed their ancestors from their lands and livelihoods. A recent report by the United Nations found that while 75% of land and 66% of oceans have been significantly altered by humans, those areas held or managed by Indigenous peoples have suffered far less severe human impacts. In addition, Indigenous peoples hold or manage 20% to 25% of Earth's land, including more than one-third of the world's remaining intact forest ecosystems, an area greater than that protected by the world's national parks.

The UN report concluded that the full and effective participation of Indigenous peoples in creating and governing new conservation measures would best preserve biodiversity. Similarly, the Biden administration's plan for implementing its 30 by 30 commitment (called "America the Beautiful") acknowledged that Indigenous land was taken by past US conservation actions and recognized the sovereignty of Native American tribes over their lands and waters. The success of the 30 by 30 initiative hinges on the support of the global community for biodiversity and for the Indigenous peoples continuing to steward the intact ecosystems they control.

US Interior Secretary Deb Haaland, the first Native American to serve in this role, speaks at a White House Tribal Nations Summit. It was convened to receive input and recommendations from Tribal leaders for their lands and communities.

FIGURE 5.13 The Limits of Protected Areas Wolf #10 was brought to Yellowstone National Park from Canada in 1995 as part of a species-reintroduction program. Shortly after reintroduction, this wolf was shot and killed when it strayed outside the park boundary.

of the first wolves brought to Yellowstone National Park in a 1995 program to reintroduce the species into the wild. Officials chose Yellowstone in part because its large size would help protect the wolf, but 1 month after his release into the wild, Wolf #10 was shot and killed just outside the park boundary. When protection stops at the park boundary, preserving species in protected areas, even those as big as Yellowstone, can fail if the area's size does not meet the needs of the species.

The Ecological Island Effect

Wide-ranging species such as wolves often suffer from the **ecological island effect**: negative effects on a population when its protected habitat is isolated amid wider unprotected areas. If the boundaries of the protected area do not align well with the scale of habitat certain species require, these species can migrate outside of a protected area, where they face threats to their survival. Other negative effects associated with this isolation include an increased vulnerability to stresses such as disease and an increased likelihood of inbreeding as the population within the area shrinks. In the 1980s, biologist William Newmark studied populations of mammalian carnivores, hoofed animals, and rabbits in 14 national parks in the western United States. He not only found that more than 40% of his study species had become extirpated (locally extinct) in these areas but also discovered that the smaller the protected area, the greater the number of mammal species that was lost.

This finding also fits a widely recognized ecological pattern known as **island biogeography**, in which

---

ecological island effect
negative effects on a population when its protected habitat is isolated amid wider unprotected areas.

island biogeography
an ecological pattern in which large isolated areas tend to have more species richness than smaller isolated areas of the same habitat type.

wildlife corridor a protected strip of land that enables migration from one habitat to another.

5.4 What Are the Limitations of Protected Areas, and Are There Alternatives?

Protected areas are an important tool in the effort to preserve biodiversity, but they have limitations. For example, how do we protect salmon that swim across vast expanses of ocean? Or consider the case of "Wolf #10" (**FIGURE 5.13**). Wolf #10 was the designation given to one larger areas tend to host more species richness than smaller areas of the same habitat type. This pattern was first observed by E. O. Wilson and Robert MacArthur in the 1960s on oceanic islands, but subsequent studies have found that it also applies to patches of land cut off from the surrounding area by deforestation or development. Areas that are isolated like this will have lower rates of new colonizers entering the population. Smaller areas also cannot support as many individuals as large areas. So, when a habitat area is reduced in size and cut off from intact ecosystems around it (for example, by deforestation or development), the extinction rate will increase and species diversity will decline.

Conservation biologists have learned that biodiversity protection can be enhanced if protected areas are surrounded by buffer zones that limit the most destructive human effects and if they are connected by **wildlife corridors**: protected strips of land that enable migration from one habitat to another (**FIGURE 5.14**). And corridors are not just for animals. Experiments demonstrate that connected areas harbor more native plant species over time than do unconnected areas.

Protected Areas Do Not Necessarily Match Protection Priorities

Another limitation arises when the areas with restricted human uses do not align with the type of habitats most needing protection. In the United States, for example, most of the largest national parks were established before critical habitats for endangered species were identified. Also, the percentage of land area protected is not equally distributed by biome (see Figure 5.4b). A closer look at Figure 5.4b reveals that areas with the least percentage of protected land tend to be those most favorable to human crop production and settlement. So grasslands have a relatively low percentage of protected land area and have largely been converted to agricultural uses. Similarly, the world's coastal areas see little protection and have largely been converted for human settlement. The early focus on tourism and scenic natural wonders when establishing protected areas favored high-elevation regions. According to one analysis, nearly half of the most-protected areas in the United States are above 3,000 meters (about 10,000 feet) in elevation. Finally, establishment of protected areas on land has far outpaced that in the marine environment.

Human–Wildlife Conflict

Human livelihoods near or within protected areas are often in conflict with wildlife. For example, the reintroduction of wolves in Yellowstone National Park led to the recovery of wolf packs throughout the Rocky Mountain west. Wolves have come in

→ **SCIENCE**

Repopulating Lynx in Spain's Donana National Park

Donana National Park

In 1969, the Spanish government created Donana National Park, largely to protect the endangered lynx in that area. However, the lynx living in this ecological island soon faced extinction because of disease, automobiles on the roads surrounding the park, and inbreeding. Lynx diversity and fitness declined, leading to an ever-shrinking population. Although researchers and conservationists were aware of this problem, there was little they could do, especially due to the small size of the remaining population and the lack of adequate habitat connectivity outside of the park boundaries. Ultimately, lynx bred in captivity had to be reintroduced to the park after the original population died off.

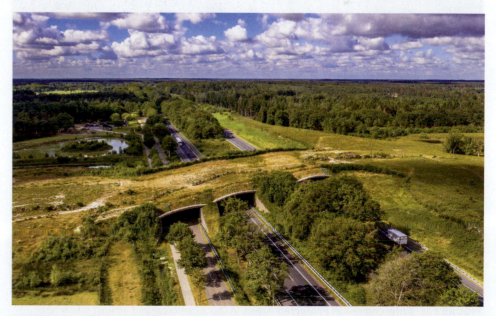

FIGURE 5.14 Wildlife Corridors Wildlife corridors are connections among protected areas through areas with significant human uses. They include wildlife bridges like this one in Dwingelderveld National Park in the Netherlands that enable species to safely cross highways.

Sierra del Divisor National Park

In November 2015, the government of Peru attempted to address many of the common limitations of protected areas when they established Sierra del Divisor National Park. This new protected area in the Amazon rain forest encompasses 3.3 million acres, more than Yellowstone and Yosemite National Parks combined. This area is home to some of the Amazon's highest estimates of species richness for mammals, plants, and fish. Not only is this land allotment large on its own, but Sierra del Divisor protects against habitat fragmentation by connecting to a larger 67-million-acre contiguous swath of protected areas known as the Andes–Amazon Conservation Corridor that stretches more than 1,000 miles (1,600 kilometers). Moreover, the new national park will serve as an extractive reserve for 21 different Indigenous communities who will continue to make their livelihoods within its boundaries.

FIGURE 5.15 Enforcement in Protected Areas A lack of enforcement against unwanted human uses can weaken the effectiveness of protected areas. Poaching as well as illegal mining and logging—like this unpermitted timber harvest in Papua New Guinea—are all threats to protected areas.

conflict with ranchers across this region by preying on cattle and sheep. This has led wildlife agencies in some western states to kill packs that have preyed on livestock and issue permits for wolf hunting. Similarly, Maasai peoples in Kenya and Tanzania who live near game parks and make a living raising cattle often come into conflict with lions that prey on their herds. Lion hunting plays an important role in Maasai culture and is often carried out against lions that have caused a loss of livestock. In each of these cases, managing the recovery of large predators while balancing the impacts on local culture and livelihoods has been challenging. One approach that is being implemented in both cases is paying ranchers or herders for any loss of livestock confirmed to have been caused by a predator.

Other human–wildlife conflicts are caused more directly by human harvest through activities such as hunting and fishing. Even when laws prohibit such activities to protect endangered species, the laws must be enforced to be effective. Many of the world's protected areas still experience frequent poaching (illegal hunting) as well as illegal logging and mining operations (**FIGURE 5.15**). But enforcement is a challenge because protected areas are often difficult to access, enforcement actions are likely to provoke conflicts, and training and employing armed rangers and guards are expensive tasks.

Important habitats are also home to some of the world's poorest and most marginalized people. Until the 1970s, most protected areas exercised enforcement by prohibiting human habitation within the area, often forcibly displacing Indigenous and local peoples. For example, the legislation that created Yellowstone National Park in 1872 authorized the removal of all persons trespassing in the area. The US Army then removed the Indigenous peoples who had used the area for millennia, prohibiting them from entering and occupying the national park. Scholars' conservative estimates of such displacements range from 10 million to 20 million people globally.

Conservation groups now advocate engaging the local and Indigenous peoples who depend on these areas through community-based conservation. This can be done in a number of ways. The national government and Indigenous peoples can decide to comanage the protected area. For example, the government of Australia and Australian Indigenous peoples comanage Booderee National Park on the southeastern coast. This arrangement exempts the Indigenous community from some use restrictions in the park and gives them a share of responsibility for managing the area. So the people who have traditionally depended on protected areas are allowed to continue using them in a way that is deemed sustainable. These areas are often called extractive reserves. In the Terai Arc Landscape region of Nepal, deemed vital to tiger conservation, people living close to protected areas are assembled in community forest user groups and manage and harvest forest resources for household use at sustainable levels.

In other cases, governments can support local ecotourism businesses in the protected area or provide compensation for not extracting certain resources from the park. In Tortuguero, Costa Rica, the government has helped local people promote tourism centered on green sea turtle nesting. The revenue generated from the lodging, food, transportation, and guide services provided to tourists has exceeded $6 million per year in this area (**FIGURE 5.16**). But ecotourism itself can cause ecological damage. Ecotourist destinations in Thailand, India, and Kenya have all struggled with water scarcity due to an influx of tourist lodges, and animals can even become habituated to interactions with tourists.

Alternative Protection Strategies

Still, a lot of critical habitat remains on private land. What can be done to protect these areas? **TABLE 5.1** highlights some options.

⬡ **TAKE-HOME MESSAGE** The benefits of protected areas can be limited if their boundaries do not match the habitats most in need of protection. Wildlife corridors that connect protected areas can help reduce the ecological island effect. Indigenous and traditional human uses of protected areas can be maintained in ways that reduce adverse effects through community-based conservation and extractive reserves.

5.5 Can Laws Protect Biodiversity?

In the 1800s, hunters throughout the United States targeted spectacular and beautiful birds for their feathers, often for particular styles of ladies' hats. The demand for these products was so great that it threatened the survival of some species, which inspired the formation of the Audubon Society—an environmental group devoted to the protection of birds—and ratification of the **Migratory Bird Treaty Act** by the United States and Canada in 1918. This act made it unlawful to collect the feathers, eggs, or nests of these birds (**FIGURE 5.17**). It is still in effect today, with more than 800 types of birds protected under this law, including many iconic species, such as the national bird of the United States: the bald eagle.

Another example of a specific and familiar law targeted at conservation is the **Marine Mammal Protection Act** of 1972, which protects species such as whales, dolphins, seals, and manatees. A very important law

FIGURE 5.16 Ecotourism Lodges that promote opportunities to observe and engage in conservation work to protect sea turtles have flourished in Costa Rica. Tourists flock to the northern Caribbean coast of Costa Rica in the summer and early fall to see green sea turtles nesting on the beach.

FIGURE 5.17 Fashion as a Threat to Biodiversity A fashion trend in the late 1800s saw ladies' hats festooned with bird feathers (or in extreme cases, entire stuffed birds).

Migratory Bird Treaty Act a treaty between the United States and Canada protecting more than 800 types of birds, including the bald eagle.

Marine Mammal Protection Act a law protecting marine mammal species such as whales, dolphins, seals, and manatees.

TABLE 5.1 Other Ways to Establish Protected Areas

Method	Description	Example	
Eminent domain	A legal tool by which governments acquire private property. In the United States, the federal government has used this power, while offering landowners "just compensation," to obtain large portions of popular protected areas.		Eminent domain was used to acquire substantial parts of Rock Creek Park, Shenandoah National Park, and Great Smoky Mountains National Park in the late 19th and early 20th centuries.
Land trusts	Nonprofit organizations that purchase lands for conservation purposes. They can retain and manage the lands they acquire or transfer them to the government.		One of the earliest land trusts was the Save the Redwoods League in California. Since its founding in 1918, this group has purchased and protected more than 1,000 groves of redwoods covering nearly 200,000 acres of land.
Conservation easements	Contracts that limit a private property owner's right to certain uses—such as housing developments or other changes that could degrade or destroy ecosystems—often in exchange for payments or tax incentives.		Between 1995 and 2006, New York State and a land trust secured conservation easements on more than 500,000 acres of forest land adjacent to Adirondack State Park.
Tradable development rights (TDR)	A legal arrangement ensuring that the right to develop land that is deemed of high conservation value is forfeited by the owner in exchange for a payment. The money to purchase these rights comes from permission fees to develop properties in urban or urbanizing areas.		Just north of Milwaukee, Wisconsin, the city of Mequon has a TDR program through which developers who purchase land and then forfeit rights to it are given rights to develop higher-density properties in areas with suitable transportation options.
Payments for ecosystem services (PES)	A system whereby landowners receive payments for a share of the services the ecosystem on their land provides.		In Uganda, the national government and the Chimpanzee Sanctuary and Wildlife Conservation Trust have partnered to provide PES for forest landowners situated between two protected areas deemed critical to chimpanzee population.

Endangered Species Act an act making it illegal to directly and intentionally harm endangered species through activities such as hunting, commercial development, and trapping.

with a broader umbrella of protection is the **Endangered Species Act** of 1973, which was passed during the Nixon administration with broad bipartisan support. It is the nation's most far-reaching species-protection law, protecting both plants and animals. The Endangered Species Act (ESA) makes it illegal to directly and intentionally harm endangered species through activities such as hunting, commercial development, and trapping. Species and distinct populations are added to the ESA protection list by the US Fish and Wildlife Service or the National Marine Fisheries Service. When a species is listed, a recovery plan is devised that designates a critical habitat for survival. More than 1,600 species have been listed as threatened or endangered under the ESA.

How Should We Prioritize Our Efforts to Conserve Biodiversity?

When you think of endangered species, what organisms come to mind? Most people think first of animals, particularly mammals and birds (such as the giant panda or California condor), or perhaps well-known fish or reptiles (such as salmon or sea turtles). In the conservation world, these are all known as charismatic megafauna: large animals with widespread popular and symbolic appeal. These species are often used by environmental groups to draw public attention to and support for efforts to preserve biodiversity. For example, the Endangered Species brand makes chocolate candy bars with wrappers sporting beautiful photos and descriptions of "at-risk" species, such as cheetahs, puffins, and orangutans, and donates

10% of the net profits from their products to conservation groups. The logo of the World Wildlife Fund is in fact a panda, and its gift shop sells mugs, T-shirts, and plush toys depicting popular species of animals, such as tigers, rhinos, dolphins, and polar bears.

But what about endangered plants, fungi, snails, insects, and other species that are not the center of publicity campaigns? And what about the larger fate of functioning ecosystems that support biodiversity more broadly? Many conservation biologists question the wisdom of the traditional focus on charismatic megafauna. First, popular symbolic species can raise public awareness and support for protecting biodiversity, but focusing on single species can overlook the broader need to protect the ecosystems

supporting biodiversity. Second, while some charismatic megafauna are keystone species (such as wolves and salmon) whose protection and recovery play a major role in preserving and restoring ecosystems, others (such as pandas and rhinos) are not. Third, studies have found that prioritizing areas using only the habitat range of mammalian charismatic megafauna tends to miss the places with the greatest species richness of invertebrates, reptiles, amphibians, and plants. Finally, not all ecosystems have the good fortune of being home to a species that draws widespread popular support.

Since there is a limit to resources such as money and human effort that can be devoted to the research, protection, and recovery of biodiversity at any given time, advocates are

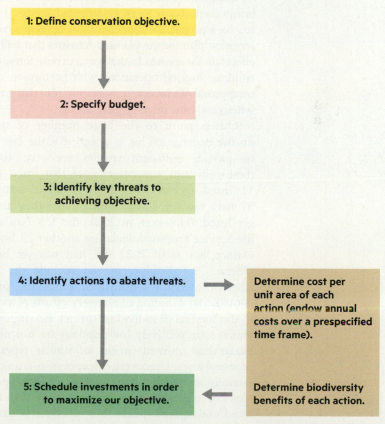

1: Define conservation objective.

↓

2: Specify budget.

↓

3: Identify key threats to achieving objective.

↓

4: Identify actions to abate threats. → Determine cost per unit area of each action (endow annual costs over a prespecified time frame).

↓

5: Schedule investments in order to maximize our objective. ← Determine biodiversity benefits of each action.

Efforts to protect endangered species (including products such as these sold to support conservation) often focus on the most charismatic species.

(continued)

looking for other ways to prioritize conservation work throughout the world. Identifying entire areas for conservation work is one way to ensure that efforts extend beyond single species. Maps identifying biodiversity hot spots, such as Figure 4.26, are one way to do this. The criteria used to identify hot spots typically include the number of endemic species—those that do not exist anywhere else—and the amount of primary vegetation that has been lost in the area. So focusing conservation on hot spots prioritizes those places with unique species that have experienced a relatively high degree of habitat loss.

Although the biodiversity hot-spot approach is now widely accepted, critics warn against placing too much emphasis on this method of prioritizing conservation actions. First, hot spots tend to be concentrated in particular regions and particular kinds of ecosystems, such as tropical rain forests. So relying solely on this approach would not necessarily protect

a wide range of functioning ecosystems throughout the world. Second, some biodiversity "cold spots" have little species richness but provide important ecosystem services to humans and other species. For example, many types of wetlands have relatively few plant species but function to maintain water quality and provide flood control in the surrounding areas. Third, only one type of threat to biodiversity, habitat loss, is used to identify hot spots. Not considered are other threats such as invasive species or overexploitation by humans. Finally, hot spots do not factor in the relative cost of effectively implementing conservation efforts.

For these reasons, some biologists are developing more-complex models to compare similar types of ecosystems throughout the world. These models compare not only the number of species in each place but also the specific local threats to biodiversity and the estimated cost of addressing those threats. The result is

a kind of balance sheet for each category of ecosystem. Conservation leaders can use these to direct conservation funds to the place where the most species can be saved for the least amount of investment. Although many biologists and conservationists are uncomfortable putting these decisions in such stark financial terms, this "return on investment" approach is increasingly used to prioritize the efforts of international conservation organizations.

What Would You Do?

If you were the leader of a conservation organization, how would you handle a choice of either devoting your organization's funding to a habitat recovery plan for polar bears in the Arctic or protecting a tropical rain forest in Indonesia? What sort of criteria would you use to make your decision?

The ESA can be a source of controversy. These regulations sometimes are criticized as being bad for business, particularly agricultural, construction, mining, and logging interests. The ESA was amended in 1982 to allow incidental take permits, which authorize development activity that brings some harm to an endangered species. In order to receive a permit, the applicant must develop a habitat conservation plan for the planned activities that will minimize effects on the species. In this way, a private forest owner can continue logging operations with permission contingent on approval of a habitat conservation plan that governs how, where, and how often logging occurs.

Others point to the large number of species still on the endangered list as a sign that the law has failed to provide sufficient species recovery. To counter these criticisms, supporters note that since 1973 only 11 listed species have been declared extinct, while 51 have recovered well enough that they are no longer listed. (However, in 2021, the US Fish and Wildlife Service proposed declaring another 23 listed species extinct, but as of 2022 this had not yet been made official.) Successful recoveries include the American alligator, Steller sea lion, bald eagle, and gray whale (**FIGURE 5.18**). Funding of recovery efforts plays a key role in the survival of individual species. Endangered species that receive relatively low funding are not much more likely than unlisted species of similar types to make progress toward recovery. Moreover, funding and protections are not equitable across various types of species. For example, plants make up more than half of the listed endangered species but receive less than 10% of the funds for recovery and don't receive protection on

private property. A further concern is that the ESA protects only species whose populations are so low that they are likely to become extinct. A species in this dire situation can suffer significant losses while it awaits protected status, and as the species approaches extinction, recovery efforts become much more costly.

Role of the States

Regulations for species of plants and animals not listed as endangered or threatened generally fall to the states. Forty-six states have some version of species-protection laws. These include laws that only establish a listing process and prohibitions on import, sale, and harm toward listed species and those that empower state agencies to protect critical habitat and undertake species-reintroduction programs. State laws can play an important role in recovering species that are not listed under the ESA. Colorado listed the Canada lynx as endangered in 1976 under its own species-protection laws, enabling the Colorado Division of Wildlife to develop and launch a successful lynx-reintroduction program within its boundaries. This occurred well before the federal government could list and develop a recovery plan for this animal in the continental United States.

International Laws and Agreements

Many countries besides the United States have species-protection laws that vary considerably. Australia's Environment Protection and Biodiversity Conservation Act of 1999 enables the government to protect threatened and endangered ecosystems as well as individual

FIGURE 5.18 **Delisted Species** More than 50 species have been delisted from Endangered Species Act protections because their populations have been deemed sufficiently recovered. These include the gray whale, bald eagle, American alligator, and Steller sea lion.

species. Ecuador adopted a new constitution in 2008 that provided rights for nature that extend beyond anthropocentric interests. Under this provision, "nature in all its life forms" has "the right to exist, persist, maintain and regenerate its vital cycles."

There are also several important international agreements concerning the protection of species. Two of the most important follow:

>> **Convention on International Trade in Endangered Species of Wild Fauna and Flora (CITES).** One hundred eighty-three countries (including the United States) are party to this agreement, which regulates international trade for roughly 25,000 species of plants and about 5,000 species of animals. This agreement bans the hunting, capture, and selling of endangered and threatened species and has done much to reduce the trade in ivory, rhino horns, and various products derived from such species. For example, the CITES member countries agreed to a suspension of ivory trading in 2007.

>> **Convention on Biological Diversity.** Ratified by 196 countries, this agreement drafted in 1992 commits member countries to develop national conservation strategies for habitats as well as individual species. It commits countries to pass laws to expand protected

areas, restore degraded ecosystems, and ensure the sustainable and equitable use of ecosystem services. The US Senate has yet to garner enough votes to ratify this agreement; therefore the United States is not a member of this convention.

⬠ **TAKE-HOME MESSAGE** Species-protection laws such as the Endangered Species Act extend protection over plants and animals deemed to be in danger of extinction. The ESA has succeeded in preventing extinctions and in recovering many species as well. While international species-protection agreements can protect endangered species, they rely on the enforcement of each member country.

5.6 What Can Be Done to Reduce Biodiversity Loss?

Much of the land in Earth's biomes has already been converted to human use. What steps can be taken to reduce the effects caused by human action? In this section, we'll take a closer look at possible restorative actions, starting with forests, which are the most diverse terrestrial ecosystems on Earth.

Convention on International Trade in Endangered Species of Wild Fauna and Flora (CITES) an agreement banning the hunting, capture, and selling of endangered and threatened species.

Convention on Biological Diversity an agreement committing countries to pass laws to expand protected areas, restore degraded ecosystems, and ensure the sustainable and equitable human use of ecosystem services.

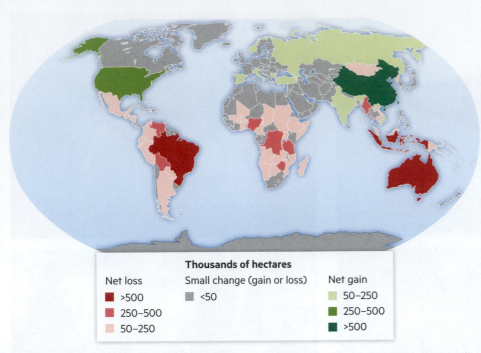

Thousands of hectares

Net loss	Small change (gain or loss)	Net gain
■ >500	■ <50	■ 50–250
■ 250–500		■ 250–500
■ 50–250		■ >500

FIGURE 5.19 Annual Change in Forest Area Tropical areas shaded deep red (Brazil, Australia, and Indonesia) are seeing the most deforestation. Some areas shown in darker green shades are replacing past forest loss with some net gain from reforestation.

Adapted from Food and Agriculture Organization of the United Nations (2010).

deforestation the clearing of large areas of forested land.

edge effect the distinctive environmental conditions and community of life that occur in a place where two or more types of habitats come together.

Sustainable Forest Management

Forest ecosystems are home to the majority of Earth's land-based plants and animals, yet they are under the constant threat of **deforestation**, the clearing of large areas of forested land. About 32 million acres of land per year are deforested—an area slightly larger than that of New York State. **FIGURE 5.19** shows that tropical areas, which host the highest levels of biodiversity, are experiencing the biggest net losses per year in forest area. This threatens tens of thousands of species—many of which have yet to be described—with extinction. The cause of deforestation varies by region. In temperate and boreal forests, such as those in Russia, deforestation is primarily the result of logging for timber. Forested lands are also harvested or converted to produce energy. In Africa, more than 90% of the harvested wood is used directly for fuel in the form of firewood. In Southeast Asia, deforestation is primarily due to land conversion for oil palm plantations. Palm oil can be used to produce biodiesel fuel, and it is also an ingredient in many processed foods. Purchasing products grown on deforested land provides economic incentive for deforestation to continue. In Australia, most of the deforestation over the past decade has occurred due to forest fires that have blazed throughout a period of prolonged drought.

It is also important to recognize that 7% of the world's remaining forests are managed as plantations for growing and harvesting single species of trees that may not be native to the area (**FIGURE 5.20A**). These industrial forests of cloned trees planted at the same time differ from more resilient native forests, which have more biodiversity. While plantation forests can provide some environmental benefits such as removal of CO_2 from the atmosphere and protection against erosion, they do not fulfill all of the ecosystem functions of native forests.

When areas of forest are lost, often a patchwork of fragmented habitat remains, which adversely affects wildlife populations throughout the remaining intact forest. To understand just how much of the world's forest cover is cut up into fragments, consider that a recent study of forest habitat across five continents found that about 70% of the forested area was located within a kilometer (0.6 miles) of a forest edge. Moreover, the ecologists conducting the study found that once a forest is fragmented, it loses an average of about half its plant and animal species within 20 years. This loss of biodiversity is due to **edge effects**, the distinctive environmental conditions and community of life that occur in places where two or more types of habitat come together. In this case, small, fragmented patches of forests have more edge per unit of area than do large, contiguous forests. This means the habitat near nonforested areas, such as meadows or human developments (e.g., roads, farmed fields, and residential areas), is also larger (**FIGURE 5.20B**). Typical edge effects in these areas include increased exposure to wind and sunlight, decreased humidity, and higher risk of forest fire. Many species in these areas are also more vulnerable to predation in the open areas and to death at the hand of humans, whether by hunting, roadkill, or other means.

FIGURE 5.20 Industrial Forests (a) Trees in plantations like this tend to be a monoculture (just one species) and lack understory (small trees, shrubs, and ground-level plants) growing between them. As a result, they do not provide as healthy a habitat for other forest-dwelling organisms. (b) When large areas of forest are harvested, fragmented habitat often remains.

(a)

(b)

Further, deforestation disrupts other services we receive. Forests help maintain soil fertility, prevent erosion, and regulate the movement of water through the Earth system. They also increase relative humidity, cloud cover, and precipitation of the regions in which they are situated. Loss of these functions affects not only the species that depend on forest habitat but also other species—including humans—who benefit from clean water and flood protection. When high rates of erosion in deforested areas cause high rates of runoff into waterways, the habitat of aquatic species (such as that of the salmon described earlier in the chapter) is damaged by the large amounts of sediment entering the water. Deforestation also increases water temperature because of loss of shade. Finally, forests play an important global role by taking in and locking up large quantities of carbon, thereby reducing the atmospheric levels of the greenhouse gas CO_2. Deforestation means more CO_2 stays in the atmosphere, where it contributes to global warming.

One strategy to address deforestation is through **sustainable forest management**. Sustainable forest management attempts to manage forests not only for harvest but also as ecosystems that maintain their biodiversity. Some important forestry practices associated with sustainable forest management include cutting small areas of forest to leave patches of living trees to reseed the area along with tall, standing dead trees, known as "snags," as habitat for birds of prey. These areas can be harvested in patterns that resemble natural disturbances such as fires or windstorms, and logging can also be done selectively to ensure that the trees remaining represent a diversity of age, size, and species. Sustainable forest management minimizes road building and typically prohibits logging on steep slopes that could erode rapidly or generate landslides and alongside bodies of water where aquatic species could be affected.

Sustainable forest management practices can be incorporated into certification programs. The Forest Stewardship Council (FSC) administers the most widely used certification. To date, forests covering more than 450 million acres in over 100 different countries have attained FSC certification. Home Depot, the largest retail vendor of lumber in the United States, and IKEA, the world's single largest consumer of wood, have both set goals based on FSC standards. IKEA reached its goal to be "forest positive" by 2020, meaning it restores more forests than it uses, and set new goals for 2030 to make forest positivity a global standard. Of course, business interests lie at the heart of these programs, as they are normally seen as driving the bottom line. IKEA lost its FSC certification temporarily in 2015 for using wood from old growth forests in Russia. In 2021 the company was once again linked to illegal logging practices in eastern Europe. FSC certifications are ultimately dependent on consumers demanding that companies have and abide by them.

Grazing and Grassland Management

Left to their own devices, grassland ecosystems are populated with native grazing animals that feed on the grasses growing there. Grasses grow from the base of the blade rather than the tip and can withstand periodic grazing that would kill most other plants. Their extensive root systems regrow rapidly—if allowed time to recover. Grazing even can stimulate roots to exude sugars and proteins that feed soil microbes, which in turn produce metabolites that promote plant regrowth and build soil carbon and thereby enhance soil fertility. Grazing animals also spread seeds in grasslands and help to fertilize the soil with their droppings.

Modern livestock production can alter this balance, though it need not if carefully managed. Rotating confined herds of livestock from area to area can ensure that each grazed area has time to recover before the livestock return. Rotation can also prevent herds that prefer a select few species of plants from overgrazing these preferred species over a wide area. While the herd is confined to a particular area over a period of time, it will be forced to feed on other species after it has consumed its favorites. Innovative ranchers have even adopted intensive rotational grazing as a means to restore native prairie vegetation. This style of grazing emulates that of native bison herds that would intensively graze an area and then move on in an endless migration, returning the following year after the grass recovered.

Agriculture and the Protection of Biodiversity

Even land that has been converted for crop production can play a role in protecting biodiversity in the surrounding region. First, some types of crops, such as nuts and fruits harvested from trees or shrubs, can serve as habitat for particular species. Research has shown that areca nut palm plantations in the Western Ghats region of India have retained 90% of the bird species richness found in nearby native forests. Similarly, coffee can be grown in the shade of a canopy of native trees that are left in place, which helps maintain diversity of insects, mammals, and birds almost as well as do nearby native forests (**FIGURE 5.21A**).

In other cases, crops can be grown while retaining strips of the original ecosystem. In Europe and the United States, hedgerows served this purpose. While they marked boundaries between properties or crops, restricted livestock movement, and protected against wind erosion, they also provided habitat corridors for many species. In the 1950s, many hedgerows were eliminated to maximize the acreage under cultivation, but more recent research in the United Kingdom demonstrated that more than 100 threatened species depend to some extent on hedgerows. In

sustainable forest management a strategy to manage forests not only for harvest but also as ecosystems that maintain their biodiversity.

FIGURE 5.21 Agriculture and Habitat Conservation **(a)** Shade-grown coffee plantations leave native trees to grow over the coffee plants to maintain biodiversity. **(b)** Hedgerows of native plants grown around cultivated fields in England provide a network of habitat corridors across the rural landscape while also protecting against wind erosion.

(a)

(b)

1997, the United Kingdom prohibited the destruction of hedgerows without a permit, while also providing funds to plant new hedgerows consisting of a diverse mix of native plants (**FIGURE 5.21B**). In the United States, the Conservation Reserve Program encourages farmers to remove environmentally sensitive land from agricultural production through direct payments from the federal government. Farmers instead plant cover species on those former farmlands that help reduce soil erosion, provide wildlife habitat, and improve water quality. More than 4 million acres of land are protected in this way.

Urbanization and Land-Use Planning

How we plan where we live affects biodiversity. Most land-use planning in the United States is done at the state and local levels, guided by zoning ordinances and building codes. **Zoning ordinances** are regulations that mandate the types of development, land uses, and human activities that are allowed (or not allowed) in particular places. Strategic zoning often attempts to manage growth, while preventing and limiting *suburban sprawl* (covered in Chapter 16) and its patchwork of relatively low-density construction spread widely over an area.

Growth management laws plan for population growth by concentrating development in particular areas and by requiring local governments to protect other areas from development within their boundaries. For example, in Washington State—where the recovery of Pacific salmon described earlier is a priority—the Growth Management Act requires cities and counties to designate fish and wildlife habitat conservation areas. But growth management laws and other kinds of comprehensive land-use planning measures are also controversial. Although all states have the authority to pass comprehensive planning laws, most have not done so because there is considerable pressure against constraining the practices of private landowners and developers. And even states that have well-received planning laws, such as Oregon and Washington, have faced challenges. For example, some

landowners argue that the government should compensate them for the property value they lose because of restrictions on the way they can develop their land.

⬡ **TAKE-HOME MESSAGE** Deforestation results in habitat loss and a decline in many vital ecosystem services. Sustainable forest management can reduce the adverse effects of deforestation. Even in cases where humans have completely and deliberately converted an ecosystem to another type of land use, steps can be taken to minimize the impact on biodiversity.

5.7 What Can Be Done to Fix Damage That Has Already Occurred?

In the *Jurassic Park* movies, dinosaurs have been reintroduced to the planet through genetic engineering. While this remains a fantasy tale even with significant advances in technology, what can we do to reverse or undo human effects on biodiversity? Although it is easiest and cheapest to prevent biodiversity decline before it happens, there are measures that we can take after the fact.

Species Reintroduction

Species can sometimes be reintroduced to an area where they once lived. This approach can be applied to large animals such as wolves (as we saw earlier) and to plants and even microbes, as in the restoration of soil communities. A famous reintroduction is that of the California condor—the largest bird in North America—which once had a range stretching from Baja California in Mexico to British Columbia in Canada. But its population was reduced to fewer than two dozen individuals by the 1980s. The condors were wiped out by lead poisoning as they consumed large quantities of animal remains left by hunters that were riddled with lead ammunition (**FIGURE 5.22**). So in 1986, the US Fish and Wildlife Service captured the remaining birds and launched a **captive breeding** program. In these cases, surviving individuals are removed from the wild and bred in captivity to produce more individuals. The hope is that captive breeding will restore the population to a level that allows its eventual reintroduction to its original home. This is an expensive undertaking, often with low probability of success. The captive population is always small, so a lack of genetic diversity and inbreeding can decrease the fitness of the offspring. Generations of species bred in confinement can adapt to their environment in captivity and become less fit for life in the wild. However, the condor was successfully reintroduced to a portion of its original range with a captively bred population of hundreds of birds.

Ecological Restoration

On a bigger scale, **ecological restoration** is the process of assisting the recovery of an entire ecosystem. The Comprehensive

(a)

(b)

FIGURE 5.22 Toxic Ammunition (a) Lead shot (left) caused lead poisoning in California condors, who consumed shot fragment–riddled carcasses left behind by hunters. State policies encouraged the use of non-lead ammunition (on the right side of the circle). **(b)** US Fish and Wildlife recovery efforts for the California condor have included breeding the birds in captivity to increase the population.

Everglades Restoration Plan (CERP) is one example of a very large-scale ecological restoration project in the United States. Prior to European settlement in the region, Florida's Everglades were a vast system of lakes, streams, and wetlands that moved in a slow, shallow flow toward the ocean over an area larger than 5 million acres (**FIGURE 5.23**). It was one of the most extensive wetland ecosystems in the world. In the 1950s and 1960s, the US Army Corps of Engineers built nearly 1,000 miles (1,600 kilometers) of canals and levees, as well as hundreds of tide gates and dozens of pumping stations to dramatically alter the flow of the Everglades and open the land for farming and residential development. By the 1980s, it became clear that this transformation was causing significant loss of biodiversity, as more than 50 species of plants and animals had become threatened or endangered because of habitat loss. In 1996, Congress authorized the US Army Corps of Engineers to work with the South Florida Water Management District to develop CERP, which entails removing more than 240 miles (385 kilometers) of levees and canals and restoring about 70,000 acres of wetlands. The cost of the project and the impact on sugar producers and other large water users in the region have caused both funding and legal challenges to the restoration efforts. Complications like this are not uncommon for ambitious restoration projects of this scale. The entire project will take decades to complete and may cost well over $10 billion.

A project like the Everglades restoration described above is massive in scale and is authorized and funded at a national level, though there are other means of funding restoration projects. In some cases, nongovernmental groups such as land trusts raise money to put toward restoration projects, and some laws in the United States draw money for restoration from those who have damaged ecosystems in some way. For example, the **Natural Resource Damage Assessment** legal process levies fines on polluters who create oil spills or hazardous waste sites, with the money collected then used for restoration projects.

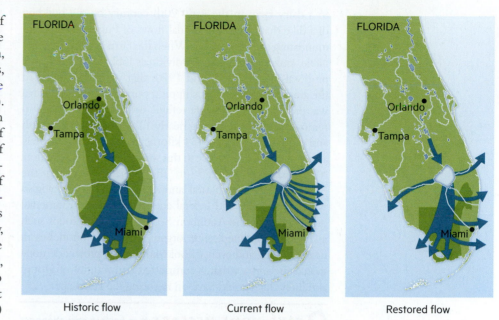

Historic flow Current flow Restored flow

FIGURE 5.23 Restoring the Everglades This series of illustrations shows the historical wide flow of the Everglades system (left) that moved north to south feeding the expansive wetlands. During the 1950s and 1960s, much of this flow in South Florida was diverted via construction of canals, levees, and pumping stations to its current flow (center). The Comprehensive Everglades Restoration Plan aims to restore much of the historic flow southward (right).

Adapted from LoSchiavo et al. (2013).

Restoration projects may not seem like something an individual can play much of a role in. But most restoration projects are smaller and rely heavily on the support of individuals and groups in the surrounding area. Even larger projects are often collections of smaller restoration pieces that come together over time to transform a larger landscape.

This was the case at the Nisqually River delta just north of Olympia, Washington (**FIGURE 5.24**). In 2009, this was the site of the largest estuary restoration in the Pacific Northwest—a key habitat improvement for the Pacific salmon we discussed earlier in the chapter. The site had been converted

Natural Resource Damage Assessment a legal process that fines polluters who create oil spills or hazardous waste sites and then uses the revenue for restoration projects.

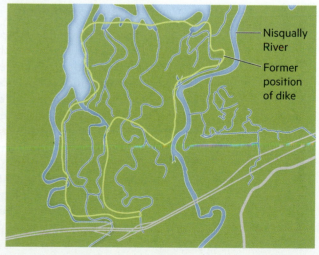

Nisqually River

Former position of dike

FIGURE 5.24 Estuary Restoration Restoring the Nisqually River delta in Washington State involved the removal of a dike that encircled the salt marsh (indicated by the yellow line) and the work of numerous individuals and groups in the region.

to a dairy farm in the early 20th century by the building of a dike that blocked the tidal flows of Puget Sound from inundating the salt marshes. When the farm became unprofitable, citizen groups and individuals lobbied and defeated proposals in the 1960s and 1970s that would have turned the area into a landfill or shipping port. Eventually, the area was protected as a wildlife refuge, and plans were made to remove the dikes, restoring nearly 1,000 acres of tidal marsh and more than 21 miles (34 kilometers) of stream channels, an estimated 50% increase in this sort of habitat in the region. Many partners worked on the project, including the Nisqually Tribe, agricultural landowners in the area, ecologists working for state and federal agencies, local environmental groups such as the Audubon Society, hunting groups such as Ducks Unlimited, civic organizations and businesses in nearby towns, and even schoolchildren. The project demonstrated the impact that a committed community of groups and individuals could have on restoring the landscape.

⬡ **TAKE-HOME MESSAGE** It is easiest and cheapest to preserve biodiversity before it declines. However, extraordinary measures such as captive breeding and reintroduction can sometimes recover severely depleted species or populations. Large- and small-scale ecological restoration projects can help recover degraded, damaged, or destroyed ecosystems.

5.8 What Can I Do?

For many people, getting involved in conservation projects is very rewarding because of the hands-on nature of these projects. You can often see the results of your efforts within a short period of time. Great activities to promote conservation and biodiversity include the following ideas.

Get Involved with a Campus Restoration Project

There are restoration projects large and small under way throughout the country: there are probably even some near you. The success of these projects often relies on the collaboration of volunteers, nonprofit groups, and local government agencies and elected officials who together raise the funds, undertake the planning, and in some cases contribute the physical labor required to restore a site. For example, in 2018, students at Eastern Carolina University could sign up to spend their spring break working to restore the Atlantic Beach area in North Carolina. Efforts in this long-term project include restoration of oyster habitats, building rainwater gardens, and other work with the local community. In some cases,

you may even earn credit for your work. This is the case with the University of Washington's Restoration Ecology Capstone. Students take courses in which they are put into multidisciplinary teams to complete projects in the local community. For example, in 2016 and 2017 a group of students worked to restore the Yesler Swamp in Seattle, an area that was being overtaken by nonnative plants. They also created a plan for future and continued restoration of the area with other teams.

Learn about Your Local Biodiversity and Get Others Involved: Start or Join a BioBlitz

A BioBlitz is an event that gathers detailed information about the species in a certain area. This data collection blitz creates useful information, while also being an event designed to engage members of the community. Events typically involve an introduction by a naturalist guide and instructions on how to photograph, document, and upload the species you find on your expedition. You can learn more about starting or joining one through the National Geographic website.

Join a Conservation Group

Conservation projects are everywhere, and you can donate money to and/or volunteer for them. Some of the largest conservation groups include the World Wildlife Fund, Nature Conservancy, Sierra Club, Wildlife Conservation Society, and National Audubon Society. But you can often find local opportunities to volunteer for smaller conservation groups.

Buy Products That Have Less Impact on Threatened Species

You might have heard of dolphin-safe tuna, but what about bee-safe plants? A class of pesticides known as neonicotinoids has been linked to bee colony collapses in the United States and parts of Europe. The latest Environmental Protection Agency review of neonicotinoids found that they have adverse effects on birds and aquatic invertebrates as well. By buying organic, you can avoid purchasing produce that has been sprayed with neonicotinoids (or other pesticides) and reduce risk to pollinators and other organisms. You can also choose products that avoid damage to endangered organisms and natural habitats: shade-grown coffee, products containing sustainable palm oil, avocados from plantations that do not displace monarch butterfly habitat, and yes, dolphin-safe tuna.

Chapter 5 **Review**

SUMMARY

- Most scientists agree that we are in the midst of a mass extinction event, and the most significant human contribution to this loss of biodiversity is habitat destruction. Other human effects include the overuse of species, introduction of invasive species, disruption to freshwater ecosystems, pollution, and climate change.

- One reason to protect biodiversity is that humans rely on and benefit from functioning ecosystems, and the concept of ecosystem services attempts to quantify these human uses in economic terms.

- Another reason to protect biodiversity stems from a belief that species have intrinsic value and humans have a duty to protect against extinction.

- Protected areas conserve biodiversity by establishing places designated by government authorities that limit certain kinds of human activities.

- There are a variety of protected-area designations managed by various government agencies, each with its own rules and levels of protection.

- The benefits of protected areas can be limited if they do not match the habitats most in need of protection and if they are too small and isolated, resulting in the ecological island effect. Habitat corridors and strategies for managing human livelihoods in and around protected areas can help conservation efforts.

- Species-protection laws, such as the Endangered Species Act, aim to protect plants and animals facing extinction and to develop recovery plans. There are also international agreements to protect against the overexploitation of endangered species and to conserve biodiversity.

- Sustainable forest management and certain grazing and agricultural practices can reduce habitat damage associated with these human land uses.

- Although it is easiest and cheapest to prevent biodiversity loss before it happens, captive breeding, species reintroduction, and ecological restoration can help to recover species and/or ecosystems that have been degraded, damaged, or destroyed.

KEY TERMS

captive breeding
Convention on Biological Diversity
Convention on International Trade in Endangered Species of Wild Fauna and Flora (CITES)
cultural services
deforestation
ecological island effect

ecological restoration
ecosystem services
edge effect
Endangered Species Act
instrumental value
intrinsic value
island biogeography
Marine Mammal Protection Act

Migratory Bird Treaty Act
national forest
national park
national recreation area
National Wildlife Refuge System
Natural Resource Damage Assessment
provisioning services

regulating services
replacement cost
supporting services
sustainable forest management
value chain
Wilderness Act of 1964
wildlife corridor
World Heritage site
zoning ordinances

REVIEW QUESTIONS

The letters following each Review Question refer to the Chapter Objectives.

1. What is the most significant human-based cause of biodiversity loss? **(A)**

2. What is the difference between the intrinsic and instrumental values used as reasons for protecting biodiversity? **(B)**

3. Name the four types of ecosystem services, and provide an example of each. **(B)**

4. Describe one method by which economists put dollar values on an ecosystem service. **(B)**

5. What are two types of protected areas in the United States? What are the government agencies that manage each one? **(C)**

6. Explain how the ecological island effect can limit the benefits of a protected area, and offer one way to address this challenge. **(C)**

7. Identify and describe two laws or agreements designed to protect endangered species. **(C)**

8. Explain how sustainable forest management can address the effects of deforestation. How can consumers of wood products also support these strategies? **(D)**

9. Identify two actions that can be taken to restore biodiversity after a species or ecosystem begins to decline. **(E)**

FOR FURTHER THOUGHT

The letters following each item refer to the Chapter Objectives.

10. In this chapter, we introduced two kinds of values that can motivate people to protect biodiversity: instrumental value and intrinsic value. Which type of value do you find most compelling? If it was your job to argue for protection for a species, how would you use these values to make your case to the public? **(B)**

11. Imagine you are a landowner in a rural area and your property has significant acreage of both forest and grazing land. You rely on harvesting timber and grazing cattle for a portion of your income. What steps could you take to maintain this income while also conserving biodiversity on your property? **(D)**

12. Imagine that you are the leader of an environmental organization charged with gaining public support for habitat protection of endangered species in the grasslands and forests of a rural area. The area is ripe for real estate development, and many farmers and ranchers have already sold their properties to developers who are converting it to residential housing. Which of the following conservation strategies mentioned in this chapter would you advocate for and why: eminent domain used to establish a new protected area, conservation easements, or land trust acquisition of the property? **(C)**

13. Military bases in the United States cover many acres of land in areas that are otherwise urbanized and developed. As a result, the military often finds itself the de facto steward of some of the last remaining habitat for certain threatened and endangered species. Although the Department of Defense is subject to the Endangered Species Act like all federal agencies, some lawmakers have argued that this requirement does not properly prioritize our national security. For example, some argue that the need to practice tank maneuvers or test land mines in a grassland should not be obstructed or delayed due to the ESA. Similarly, some in Congress have introduced bills that would exempt areas near our northern and southern borders from complying with the ESA when border walls and other defenses are constructed. What do you think? When national security conflicts with protection of biodiversity, how should we prioritize these values? Explain your thinking. **(C)**

Make Your Case

The following exercises use real-world data and news sources. Check your understanding of the material and then practice crafting well-supported responses.

Use the News

Species that are listed as threatened or endangered under the Endangered Species Act are often imperiled by human actions and by predators. Government officials usually first try nonlethal methods to protect a species from predators, but sometimes the predators have to be killed, even when they are charismatic species themselves. The article that follows details a recent case of lethal predator control in the Columbia River in Oregon. Use this article to answer the questions that follow. The first three questions test your understanding of the article. Questions 4 and 5 are short answer, requiring you to apply what you have learned in this chapter and cite information in the article. Answers to Questions 1–3 are provided at the back of the book. Question 6 asks you to make your case and defend it.

"Sea Lion-Salmon Cycle Starts Again in Columbia, This Time with Expanded Ability to Kill," Karina Brown, *Courthouse News Service*, March 25, 2021

The Columbia River was misty and grey an hour from the coast on a recent spring morning, as clouds mingled with the Douglas Fir forests atop the hill above a small town and rolled down to the city's marina, the water choppy from passing ships. The hoarse barks of 100 sea lions echoed over the water, voices of animals unlikely to survive the year, doomed by their appetite for endangered fish.

Highway 30 runs through Rainier, parallel to the aquatic thoroughfare of the Columbia River. Perched just below the mouth of the Cowlitz River,

the waters in Rainier's front yard are a mingling point for numerous species of fish. In early March, tens of millions of eulachon smelt leave the Pacific and swim up the Columbia River. Eulachon, a threatened species, were a mainstay in the diets of previous generations of Oregonians and the tribes that have lived at this spot for thousands of years. Some peel off the Columbia at Rainier, heading north up the Cowlitz. Following close behind the smelt are spring Chinook, headed up the Columbia to spawn in the rivers and streams of their birth.

The fish are struggling. The various runs of spring Chinook, listed as either threatened or endangered, are in turn the main food source for endangered Southern Resident killer whales. But the fish face habitat loss, warming oceans, inaccessible culverts and dozens of dams, in addition to natural predators. State and federal regulators tightly control fishing here and everywhere in the Columbia River basin. But sport fishing is still an economic driver, and boats crowd the water here each fall, anchoring in groups called hog lines.

And all that fatty traffic has attracted another visitor: sea lions. They're a major predator here, and a new one that fish didn't face before the last couple of decades. Sea lions themselves are a success story. California sea lions rebounded from below 90,000 animals in 1975 to a high of 306,000 in 2012—above carrying capacity for sustainability in the ecosystem. Since then, their numbers have declined to slightly below carrying capacity because of unusually warm ocean conditions known as "the blob." Stellar [sic] sea lions were removed from the Endangered Species List in 2013.

The animals have turned the docks at Rainier into a predation hot spot, similar to feasting grounds the sea lions have established further upriver, on the rocks below Bonneville Dam and at Willamette Falls. In those two spots, tribes and state fishery managers pushed for and won federal approval to kill sea lions in order to prevent them from pushing endangered fish into extinction. They began killing individual sea lions shown to have been eating salmon at Bonneville Dam in 2007.

Just over 500 winter steelhead returned from the ocean to Willamette Falls in 2017, spurring the drive toward more widespread lethal removal of sea lions. Before that, state wildlife managers tried various nonlethal methods, including capturing, tagging and relocating the animals to the southern Oregon coast. Determined swimmers, they reappeared at the falls days later.

In 2019, wildlife managers removed 33 sea lions from Willamette Falls that had been returning there for years, according to Shaun Clements, senior policy analyst for the Oregon Department of Fish and Wildlife. Clements said lethal removal of individual sea lions has reduced numbers at the area's two biggest predation hotspots from around 100 animals each year at Bonneville Dam and about 40 at Willamette Falls down to 40 at Bonneville and 20 at Willamette Falls.

A new permit issued by the National Oceanic and Atmospheric Administration last August allows three states and six tribes to kill as many as 540 California sea lions and 176 Steller [sic] sea lions along the 180 miles between Portland and the McNary Dam. The idea is to halt the sea lions' annual tradition of following fish up the Columbia.

"They're very social and take cues from other animals," Clements said. "So by removing those individuals you break that cycle."

Despite these efforts, the animals' appearance now at Rainier marina suggests that this year will be no different. Rainier sits outside the lethal management zone, though managers can cull animals that leave the Columbia to follow salmon up smaller tributaries to their spawning grounds. But it's a near certainty that the animals congregating here will soon make their way to Bonneville or Willamette Falls—some likely already have.

In past years, the sea lions had caused significant damage to the city-owned docks at the Rainier public marina, according to Rainier Mayor Jerry Cole. They gnawed sections of wood and shoved the underwater balloons that keep the docks afloat out from underneath the walkways.

"That's not an easy fix," Cole said.

Since then, the city installed electric fencing to keep the animals off the public docks. Even that won't stop the occasional determined bull, Cole said. On Wednesday, about 100 sea lions squeezed onto a metal dock owned by a local tugboat company. Arranged in a haphazard order, their hoarse barks erupted each time a newcomer hoisted himself up into the pile, disrupting the slumber of those already plopped atop the metal rails. In recent weeks, Cole said, there were hundreds more dotting the river, perched atop shifting sandbar islands.

Cole said local opinion on the sea lions' seasonal residency was mixed.

"We've got some people who hate them for what they do to the salmon," Cole said. "Others love them like puppy dogs."

In a healthy ecosystem, where salmon and smelt runs were thriving, there would be plenty of fish for sea lions, as well as birds and humans. But that's not the current reality in the Columbia River Basin.

Chuck Hudson, now retired from his position as governmental affairs director with Columbia River Intertribal Fish Commission, which represents four of the six tribes that proposed the plan approved by NOAA alongside Oregon, Washington and Idaho. Hudson told Courthouse News at the time that controlling predators—both mammalian and avian—would be necessary until salmon no longer struggled to survive.

"The long-term pathway is not perpetual killing of sea lions, but it is a necessary measure now and until we can accomplish restoration of ecological abundance—not only in the Columbia, but in other salmon-producing systems," Hudson said. "We simply do our very best, with the authorities we have, to preserve a tenuous balance in a drastically altered environment."

---

1. Which of the following threatened or endangered animals are state and tribal officials trying to protect in the Columbia River?
   a. Chinook salmon
   b. pelicans
   c. Steller sea lions
   d. California sea lions

2. Which of the following predators are wildlife managers attempting to capture and kill?
   a. killer whales
   b. eulachon
   c. sea lions
   d. steelhead

3. Which of the following nonlethal methods to discourage predators was mentioned in the article?
   a. spraying docks with strong-smelling repellents
   b. capturing, tagging, and relocating predators
   c. removing underwater balloons from docks
   d. stretching netting over parts of the river

4. Describe why wildlife managers think killing individual predators will stop the predator population from following prey up the Columbia River.

5. Explain why the prey populations in the Columbia River are in danger of extinction.

6. **Make Your Case** If you were a NOAA official deciding whether to approve a new permit to kill predators in the Columbia River, how would you make your decision? Based on the information in the article, how would you decide if the current predator-control program is successful and if it is still necessary? What additional information would you want to see before making your decision and why?

Use the Data

The US Forest Service produces reports on the ownership of forest land in the United States. The table shown here breaks down the forest land in the United States into different ownership categories.

US Forest Land, by Ownership Class, 2012

Owner Class/Land Class	US	REGION North	South	West
		Million acres		
All owners	**766**	**176**	**244**	**346**
Timber land	521	167	210	144
Reserved forest	74	7	4	63
Other forest	172	2	31	139
National Forest	**145**	**12**	**13**	**120**
Timber land	98	10	12	75
Reserved forest	27	1	1	24
Other forest	20	0	0	20
Other public	**176**	**35**	**20**	**122**
Timber land	63	29	15	19
Reserved forest	47	5	3	39
Other forest	67	0	2	65
Private corporate	**147**	**29**	**65**	**53**
Timber land	111	29	61	21
Reserved forest	0	—	0	0
Other forest	36	0	4	32
Private noncorporate	**298**	**100**	**147**	**51**
Timber land	249	99	121	28
Reserved forest	0	0	0	0
Other forest	48	1	25	22

Adapted from USDA (2014).

Study the table, and use it to answer the questions that follow. The first three questions test your understanding of the table. Questions 4 and 5 are short answer, requiring you to apply what you have learned in this chapter and cite information in the table. Answers to Questions 1–3 are provided at the back of the book. Question 6 asks you to make your case and defend it.

1. According to the table, how many acres of forest are owned by the federal government (National Forest) in the United States?
   a. 120 million acres
   b. 145 million acres
   c. 176 million acres
   d. 766 million acres

2. The US Forest Service considers "reserved forest" to be those areas where timber is prohibited from being harvested. What is the total amount of reserved forest in the United States?
   a. 27 million acres
   b. 47 million acres
   c. 63 million acres
   d. 74 million acres

3. The Forest Service divides the United States into three regions: North, South, and West. Which region has the most acres of reserved forest?
   a. North
   b. South
   c. West

   How many more acres does it have than the other two regions combined?

4. The Forest Service classifies "timber land" as land that can be harvested. How does the amount of timber land owned by private corporate and noncorporate owners compare to the total amount of timber land in the United States? Based on what you have learned in this chapter, briefly explain why you think this ownership pattern exists.

5. Elsewhere in this report, the Forest Service refers to the South as the nation's "woodbasket" for timber. Cite data from this table to explain why it deserves this nickname.

6. **Make Your Case** This table shows that the US Forest Service manages almost 100 million acres of timber land spread out over all regions, substantially more area than the reserved forest. Do you feel that this meets the mission of the organization as you understand it from reading this text? Using the data from this table, information from this text, and other sources such as www.fs.fed.us, explain if you think the US Forest Service is meeting its mission and why. Or if you think the US Forest Service's overall mission should change, explain why.

LEARN MORE

- Millennium Ecosystem Assessment synthesis report, *Ecosystems and Human Well-Being*: www.millenniumassessment.org/documents /document.356.aspx.pdf
- Conservation International guide to biodiversity hot spots: www .conservation.org/How/Pages/Hotspots.aspx
- US Fish and Wildlife Service, "ESA Basics": www.fs.fed.us/biology /resources/pubs/tes/esa20basics.pdf
- *Science* special issue, "Ecosystem Earth": science.sciencemag.org /content/356/6335
- David R. Montgomery. 2003. *King of Fish: The Thousand-Year Run of Salmon*. Boulder, CO: Westview Press.

- Video: NASA, The Sting of Climate Change: digital.wwnorton .com/environsci
- Video: NPS, Monitoring California's Mediterranean Ecosystem: digital.wwnorton.com/environsci
- Video: Science360, Bumblebees in Peril: digital.wwnorton.com /environsci
- Video: NSF, Can an Ecosystem Recover from Damage?: digital .wwnorton.com/environsci

6

Human Population
Can We Have Too Many People?

Your phone buzzes with news of a big family event: relatives of yours have just had their first child! Another human being has been added to the nearly 8 billion people living on Earth. But unlike this new baby, you were here before the global population surpassed 7 billion on October 31, 2011. Earth had far fewer people when you came into the world than it has now. You can use **FIGURE 6.1** to match your birth year to global population growth over recent decades. Consider an individual born in 1995; this person is now sharing the planet with 33% more people than at the time of his or her birth. Someone you know born around 1980 now lives in a world with 71% more people than in the year he or she was born. Those born in 1970 or earlier are among the only humans in history to have experienced a doubling of the global population within their lifetime. What is going on here, and how long can it go on?

Two centuries ago, the vicar of a rural church south of London named Thomas Malthus asked the same question. During a 3-year period, he had presided over 5 times as many baptisms as funerals and saw the size of his village increase. Malthus recognized that our collective birth rates and death rates determine population growth. Simply put, when births outpace deaths, the population grows. In 1798, after seeing the population of England double from 5 million to 10 million people since 1750, Malthus began to write a series of influential essays on population dynamics. He determined that England's population was experiencing exponential growth (as described in Chapter 4), increasing at a rate proportional to the total number of people, and thus accelerating over time. Importantly, Malthus argued that food production typically does not grow at this rate (**FIGURE 6.2**). He held that food production grew at a constant rate and thus could not keep up with exponential population growth. In Chapter 4 we learned about population ecology and the way exponential population growth of a species is met by limits in the environment. Malthus feared that humans were approaching such limits and warned of human "extermination" from disease, famine, and war as food production failed to keep pace with population growth.

Hundreds of thousands of babies are born each day, and births outpace deaths on Earth by about 2 to 1. As the human population expands, so does the strain our species places on the environmental resources that provide for our well-being.

Chapter Objectives

By the end of this chapter, you should be able to . . .

A. describe the history of human population growth and provide examples of population collapses.

B. understand the ways in which human innovations have increased Earth's capacity to support our species.

C. consider the strategies societies use to influence their populations and the cultural factors that influence these strategies.

D. describe a demographic transition, including its associated economic and social factors.

E. explain how empowering women affects population growth.

F. compare how environmental impact, as estimated by ecological footprint analysis, varies greatly across individuals and countries.

The more of us there are, the less planet there is per person.

—Headline from Earth Overshoot Day

Yet the human population has continued to grow over the past 200 years. Malthus's predictions of large-scale population collapse failed to materialize. What happened? While most species have very little capability to change their surrounding environment to support an increase in their numbers, humans are different. Malthus underestimated the ways in which people can innovate and adapt, expanding Earth's ability to support a larger population through technological advances, such as increased agricultural production, improved medicine, and disease prevention.

However, Malthus was correct in his understanding that Earth's resources are not endless. We have a fixed reserve of matter that makes up everything on the planet, and the way we live, what we create, and what we consume together determine how many people Earth can sustain. While in Chapter 15 we will focus more on consumption and waste, in this chapter we'll look at the factors driving the number of people on this planet and various ways in which societies and governments have attempted to influence population dynamics. We will also see how **demographers**, social scientists who study the characteristics and consequences of human population

growth, examine both current and historical trends to look for lessons. Then, we will see how changing population patterns are intertwined with environmental trade-offs of increased wealth and consumption. When you finish this chapter, there will be more people on Earth than when you started, and it is important to see how this impacts the planet and its ability to support us.

6.1 What Is the History of Human Population Change?

For most of our species' existence, there were far, far fewer of us scattered around the world than there are now. About 10,000 years ago, the world's population is estimated to have been about 2 million to 3 million people, and it probably had been at that level for many thousands of years. But at that time humans were just starting to domesticate plants and animals, no longer relying solely on "hunting and gathering." In particular, the domestication of wheat, corn, and rice powered

demographer a social scientist who studies the characteristics and consequences of human population growth.

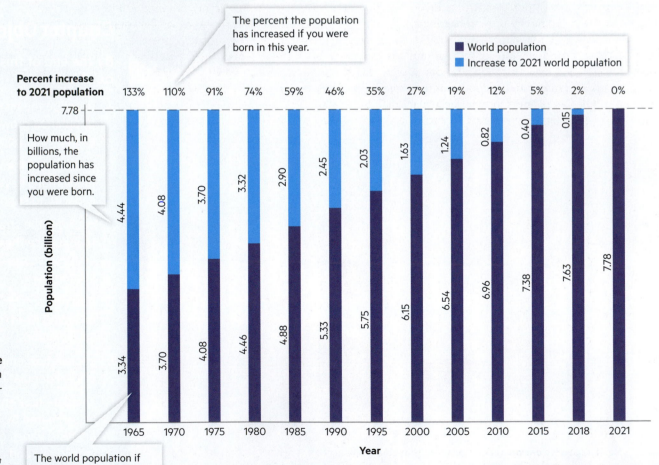

FIGURE 6.1 How Many More People? The dark blue bar in this figure shows world population for each year since 1960. What is the percentage of growth in population since the year you were born?

Adapted from United Nations, Department of Economic and Social Affairs, Population Division (2021).

population growth. Agricultural society made communities more stable and led to the development of villages and towns, especially around the Fertile Crescent region in the Middle East, as well as in southwestern Asia and China. Agricultural development enabled the human population to expand rapidly over the ensuing millennia. Just over 2,000 years ago, worldwide population had grown to approximately 250 million to 300 million. By the year 1804, it reached approximately 1 billion.

Historical Population Crashes

In the opening days of the 19th century, Thomas Malthus thought humanity was headed for disaster. Why did he think this? In his lifetime, Malthus saw increasingly dense industrial cities struggle with waste accumulation, poor sanitation, and rapidly spreading outbreaks of diseases such as typhoid and cholera. And up to this point in history, population collapses had not been uncommon. A closer look at population growth in Europe prior to the 17th century shows that growth was bumpy and checked by periodic increases in death rates from epidemics, conflict, and famine. For example, after 250 years of prosperity and population growth, 14th-century Europe was struck by the Great Famine of 1315–1317, which killed 10% to 25% of the population. Then just a generation later, between 1347 and 1352, another third of Europe's population died of bubonic plague in an event known as the Black Death.

Researchers have found that population collapses have at times checked population growth around the globe. In particular, societies that degrade once-fertile land can eventually cause drastic reductions in their own populations. A classic example of this occurred on Easter Island, a fragile island environment in the South Pacific

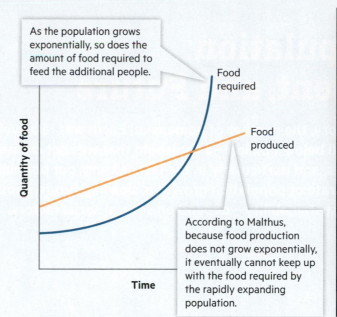

As the population grows exponentially, so does the amount of food required to feed the additional people.

Food required

Food produced

According to Malthus, because food production does not grow exponentially, it eventually cannot keep up with the food required by the rapidly expanding population.

Quantity of food

Time

FIGURE 6.2 Malthus's Prediction In the late 18th century, English vicar Thomas Malthus predicted that if the human population grew exponentially and food production did not, the amount of food produced would not keep up with the amount of food required by the expanding population.

Graph adapted from Malthus (1798).

(**FIGURE 6.3A**). Studying soil samples and ancient pollen preserved in lake beds, researchers found that for 700 years this remote island supported a small but healthy community of people—people who famously left enormous statues across the island. However, as the inhabitants cleared forests and intensified agricultural practices, their actions had stripped fertile soil from much of the island by the mid-17th century. Rats that humans introduced

(a)

(b)

(c)

FIGURE 6.3 Examples of Population Collapse There are many examples of human populations that collapsed because of the overuse of resources and changing environmental conditions. These include the **(a)** Easter Islanders in the Pacific, **(b)** Ancestral Puebloans of Chaco Canyon in the state of New Mexico, and **(c)** Viking settlements on Greenland.

Global Population: Past, Present, and Future

For most of human history, the number of humans on Earth was relatively small and steady, totaling well below 1 billion—a threshold that was not crossed until 1804. In recent centuries, and particularly in the last century, our population has skyrocketed. While the rate of population growth is slowing, future population projections—and their sustainability—depend on several social factors.

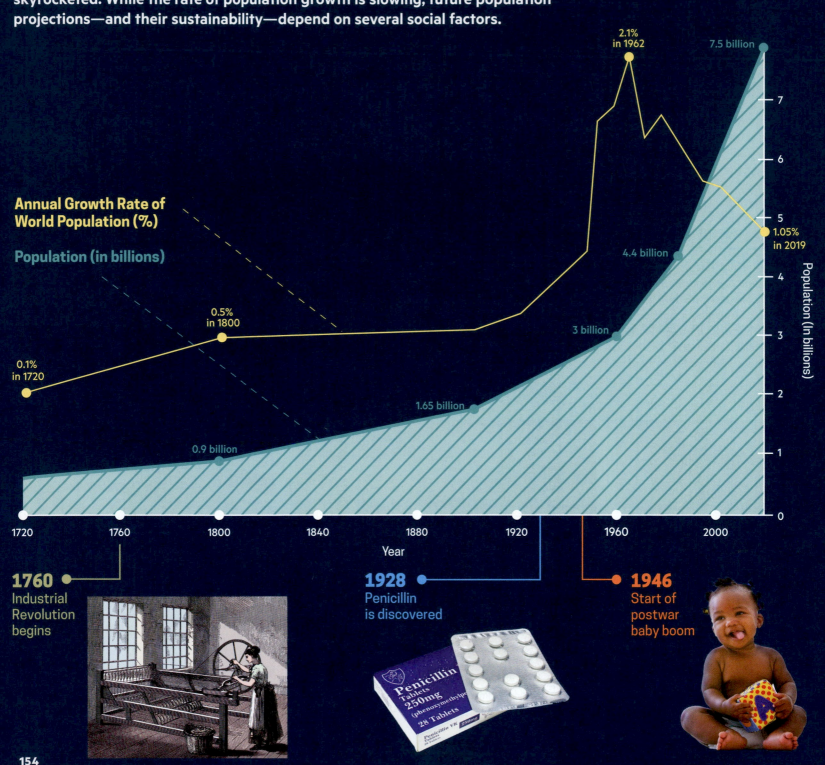

Annual Growth Rate of World Population (%)

Population (in billions)

2.1% in 1962

7.5 billion

4.4 billion

0.5% in 1800

1.05% in 2019

3 billion

0.1% in 1720

1.65 billion

0.9 billion

Population (in billions)

Year
1720 1760 1800 1840 1880 1920 1960 2000

1760
Industrial Revolution begins

1928
Penicillin is discovered

1946
Start of postwar baby boom

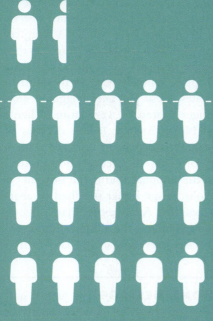

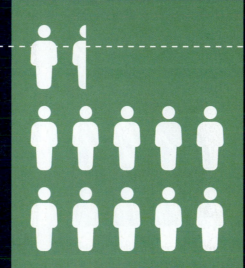

United Nations Population Projections for 2100

UN estimates of future population growth (high, middle, and low) depend on assumptions made about the empowerment and health of women and girls that will affect fertility rates.

Current Population
8 billion

High: 16.5 billion

Decreased access to birth control and decline in status of women are coupled with improved maternal and infant health.

Middle: 11.5 billion

Current trends in access to birth control, education, employment, and health care for women remain unchanged.

Low: 7 billion

Significant improvement in access to birth control and education and employment opportunities for women occur.

Historical Predictions of Population Collapse

Thomas Malthus was not the only prominent scholar to predict population decline. Well-known 20th-century examples include ecologist William Vogt and biologist Paul R. Ehrlich (pictured here in 1974). In 1948, Vogt predicted in his book *Road to Survival* that humans had already met Earth's carrying capacity and that 75% of the human race would be "wiped out" once the population rose to 3 billion people; however, humanity moved past this threshold in 1960 without a collapse. Ehrlich's 1968 book, *The Population Bomb*, issued predictions of dire population collapse by the 1980s that also failed to materialize. These failed predictions demonstrate how difficult it is to make accurate predictions that tie catastrophe to specific human population thresholds. Time and again, humans develop new ways to support a growing population. Yet despite these achievements, the fact remains that there are physical resource limits on human population growth.

They had deforested nearby food-bearing trees and bushes and lost enough irrigation water that they had to import nearly all of their raw materials and food. Ultimately, these peoples left their more densely populated areas to form and settle pueblos—smaller villages that relied on subsistence farming and were located closer to major waterways—that still persist throughout the American Southwest as examples of adaptability and resilience.

Farther north, a third example comes in the cautionary tale of Viking colonies on the southwestern coast of Greenland in the late 10th century. Viking colonists from Scandinavia cleared the coastal forest so their livestock could graze grass-covered tundra. Foreshadowing what would happen centuries later in the Dust Bowl in the United States, these practices triggered erosion of topsoil no longer rooted by vegetation. When a cold period in the North Atlantic led to decades of poor harvests, the colonists depended on trade with Iceland and Norway for what they could not produce. But as ice clogged the shipping routes, the isolated colonists ran out of resources and either perished or abandoned Greenland altogether. After more than 400 years of agricultural development, the Viking colony collapsed due to disease and starvation in the late 14th century (**FIGURE 6.3C**).

⬢ **TAKE-HOME MESSAGE** The human population has grown at an exponential rate in recent history. Historical population collapses demonstrate that humans are subject to limiting factors in the environment.

to the island played an important role in destroying trees, other plants, and many species of birds. As wood resources disappeared, the islanders' ability to build tools, canoes, and cooking fires declined, and native bird populations that inhabited the trees plummeted. These human-caused problems were then greatly amplified at several points in time when changes to the climate affected rainfall on the island. Finally, the island's population was further diminished when islanders were exposed to diseases introduced by European explorers in 1722.

Consider a second example in the abandonment of a major cultural center in the American Southwest by the Ancestral Puebloan people around the 12th century. After domesticated maize arrived at Chaco Canyon in present-day New Mexico about 1500 BCE (**FIGURE 6.3B**), farming gradually expanded to all the suitable nearby land over the next 2,600-plus years, to around 1150 CE. But a 50-year drought that began in the year 1130 eventually forced the Ancestral Puebloan people to abandon Chaco Canyon after centuries of occupation. Excavations of the spectacular ruins have revealed that during prolonged droughts, the Chacoan civilization was unable to sustain itself in its population-dense settlements.

6.2 What Has Driven Recent Population Growth? Can Earth Support It?

Despite historical regional collapses, many more societies have found ways to adapt to environmental challenges, and the human population as a whole has continued to grow, surpassing the levels that Malthus and many others after him warned would bring catastrophe. The population of Great Britain that so worried Malthus 200 years ago is now more than 60 million people. What is causing this growth?

Rates of Birth and Death

Malthus saw that the simple relationship between the number of births and the number of deaths drives growth. Let's look at births first. Population growth is driven by increases in the fertility rate, or number of births per woman. Today, **total fertility rate (TFR)** is a tool

total fertility rate (TFR) the average number of children a woman would have in her reproductive years in a given population.

demographers use to understand changes in population growth over time. The TFR is the average number of children a woman in a given population would have in her reproductive years (roughly between 15 and 40 years of age). Demographers use TFR rather than a birth rate (usually figured as births per 1,000 people) because it is not influenced by the age structure of a population and instead focuses exclusively on birth rate among those women in the population who are able to produce offspring. A TFR of 2.1 children is known as **replacement fertility**, where the population does not grow or decline. This occurs because two children will take the place of their parents and the extra fraction accounts for children who will not survive very long after their birth or who die without themselves producing offspring. The higher the TFR is above 2.1, the more the population will grow, and the lower it falls below 2.1, the more the population will shrink.

Total fertility rates are not uniform across populations. The TFR in the United States recently fell below 2 children per woman of childbearing age, a number just below replacement fertility, and a similar rate is seen in many developed nations, such as New Zealand, Chile, and Sweden. Other nations have a much lower TFR, including China and Italy at around 1.5, Ukraine and South Korea around 1.3, and the island city-state of Singapore at 0.8. However, the TFRs in developing countries, such as the Philippines and Guatemala, are often above 3, while many African countries, such as Ethiopia and Nigeria, have TFRs between 5 and 7. Global TFR is just under 2.5 (**FIGURE 6.4**).

Declines in the **crude death rate**, the total number of deaths per year per 1,000 people, have also driven population increases. While there have certainly been famines, wars, and pandemics (including COVID-19 most recently), the crude death rate has fallen from more than 25 people per year per 1,000 during Malthus's time to just 7.5 people per year per 1,000 at present. This combination of TFRs above 2.1 and lower crude death rates has caused global human population to rise steeply. Let's take a closer look at some of the factors affecting these important trends.

Agricultural Development and Technology

Malthus's prediction failed to account for the agricultural and resource development of previously underdeveloped areas such as North America. For example, he had not anticipated that the United States would become the world's largest agricultural producer or that this production could help supply less agriculturally productive places overseas.

By the middle of the 20th century, predictions of population collapse resurfaced because of the belief that humans had already cultivated much of the best arable land on the planet. Ecologist William Vogt (see Historical Predictions of Population Collapse) thought

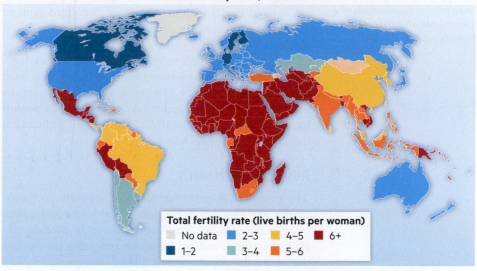

Total Fertility Rate, 1965–1970

Total fertility rate (live births per woman)
- No data
- 1–2
- 2–3
- 3–4
- 4–5
- 5–6
- 6+

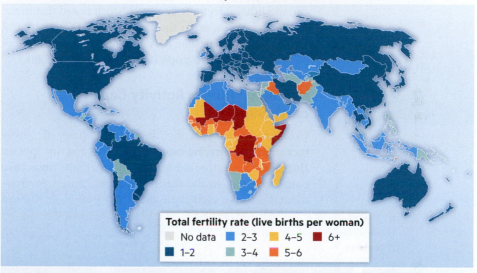

Total Fertility Rate, 2010–2015

Total fertility rate (live births per woman)
- No data
- 1–2
- 2–3
- 3–4
- 4–5
- 5–6
- 6+

FIGURE 6.4 **Changes in Total Fertility Rate** The total fertility rate varies by country and has declined in many places over recent decades. These two maps show rates between 1965 and 1970 and 2010 and 2015. Notice the significant changes from reds and oranges to shades of blue in many countries.

Adapted from United Nations, Department of Economic and Social Affairs, Population Division (2017).

that places with high populations but less agriculture, such as England, were acting as "parasites" siphoning off resources from the rest of the world and that this would trigger a collapse. And in the 1960s, biologist Paul R. Ehrlich warned that India's population would increase by 200 million people between 1967 and 1980. Ehrlich believed India could not possibly feed such numbers, let alone achieve its goal of self-sufficient food production. The population of India did in fact increase from 500 million to more than 700 million by 1980, but by this time India was also more than self-sufficient in food production and had become a net exporter of agricultural products. India's population is now over 1.2 billion people, and the country

replacement fertility a TFR of 2.1 children, which is the rate at which the population does not grow or decline.

crude death rate the total number of deaths per year per 1,000 people.

continues to be a major rice exporter. Observers such as Vogt and Ehrlich failed to imagine the rapid gains in yield per acre that new crop varieties and agricultural methods would reap through technological advances of the Green Revolution (which we discuss in Chapter 12). From 1967 to the present, grain yields have increased at a greater rate than the population. In short, the amount of land it takes to support a person has declined as dramatically as the human population has increased (**FIGURE 6.5**).

Advances in Health

The discovery, mass production, and application of antibiotics and vaccines in the past century saved hundreds of millions of lives. If not for advances in medical science, sanitation, and public health programs throughout the world, the population would not have risen as rapidly. Medical and human health advances have dramatically reduced deaths from malaria, yellow fever, smallpox, cholera, and other infectious diseases. Ehrlich noted that in the 1940s and 1950s, using the insecticide DDT to control malaria-carrying mosquitoes rapidly decreased death rates by as much as 40% in tropical places such as Sri Lanka.

How Much Human Activity Can Earth Support?

Contrary to many predictions, humans have effectively increased the **carrying capacity** of Earth for our species. But as we'll see throughout this book, these gains have come at a cost in the form of human-caused environmental impacts, such as air and water pollution, soil loss

and degradation, shrinking biodiversity, and climate change. So are we still bound by environmental limits? How many of us can Earth now support? This is a difficult question to answer for a number of reasons. Not only does it require us to foretell other human innovations that might expand the planet's carrying capacity for our species; it also requires that we estimate the typical lifestyle—and corresponding environmental impacts—for the people of the future.

If the objective is simply basic survival, some experts have estimated that Earth could support as many as 15 billion or 20 billion people. But bare survival in such a crowded world would probably not be an existence desirable to most of us. These numbers represent more than double our current population, which at present requires the consumption of more than 40% of all the plant matter produced on land. In our world today, more than 80% of the population is concentrated in countries that use more resources than they can gather from their own territory. Packing more than twice as many people onto the planet would amplify these impacts.

If the objective is a more comfortable lifestyle, such as that enjoyed by many Americans, most experts agree that we are already pushing past planetary limits. In fact, as we will see later in this chapter, if the nearly 8 billion of us adopted the average US level of consumption, it would take more than four Earths to provide the resources we need and to absorb the wastes we produce.

So the question of how many people Earth can support is not only a function of our population. It also depends on our ability to adapt and expand Earth's carrying capacity and the environmental impacts associated with our consumption patterns. When questioned about failed predictions, Ehrlich would say that dismissing his and others' concerns would be like someone jumping off a tall building and thinking everything will be fine because they have not yet hit the pavement below. Global population is projected to approach 10 billion people by the end of this century. The well-being of our species in this context will likely require we slow or reverse population growth, find new ways to expand our carrying capacity, and make changes to our lifestyles and patterns of consumption. In the sections that follow, we'll first explore human population dynamics and then turn our attention to our lifestyle and patterns of consumption.

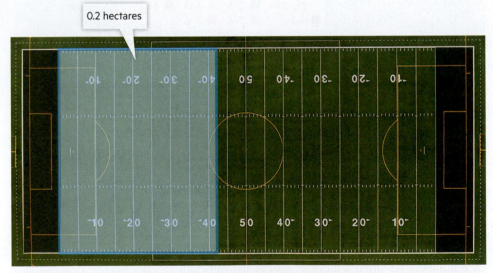

FIGURE 6.5 How Much Land Do We Need to Grow Our Food? Increases in agricultural productivity have led to dramatic reductions in the area of land required to support the food needs of a person. Today, we require 0.2 hectares (0.5 acres) of land to support each person on the planet. The area of a typical 100-yard-long athletic field is about 0.45 hectares (1 acre), so just under half of this area supports one person, whereas ancient preagricultural societies (10,000 or more years ago) required 100 hectares (250 acres) to support one person, or about 222 athletic fields. Researchers project the amount needed in the future will drop even further to 0.1 hectares (0.25 acres).

🛑 **TAKE-HOME MESSAGE** Humans have the ability to innovate and adapt in ways that have expanded the carrying capacity of the planet for our species. The question of how many people Earth can support is a function of human population, our ability to adapt and expand Earth's carrying capacity, and the environmental impacts associated with our consumption patterns.

6.3 How Have Societies Attempted to Control Population?

Families and individuals have long taken steps to prevent pregnancy. We are now accustomed to highly successful and safe **contraception** technologies, such as condoms and birth-control pills, that greatly reduce the probability of impregnation. When these technologies are used, fertility rates drop dramatically. For example, after gaining independence in 1966, Botswana quickly established a wide-ranging family planning program that distributed contraceptives to almost every part of the country and coupled it with maternal and child health services. Its TFR dropped from greater than 6 in 1966 to less than 3 today and is the fastest-declining TFR in sub-Saharan Africa. Demographers now estimate that we have reached a point in time globally where birth-control methods dramatically limit the global pregnancy rate. Reduced fertility rates have enormous benefits for families, especially for women in developing countries, as the responsibilities of childbirth and child care fall disproportionately on their shoulders. The health, education level, and workforce participation of women and girls all increase as TFR drops. We will discuss these benefits further in Sections 6.4, 6.5, and 6.6.

But it is important to recognize that even before modern contraception, humans pursued various strategies to reduce the number of pregnancies during the reproductive years. While these methods were not as successful as modern methods, they could reduce the fertility rate by half, from 6 or 8 births per woman to 3 or 4. For example, ovulation is suppressed during breast-feeding by release of the hormone prolactin. Women in most hunter-gatherer societies tended to breast-feed 3 to 5 years, leaving significant gaps between pregnancies and reducing the TFR. Abortion has been practiced historically as well, and today the World Health Organization of the United Nations (UN) estimates that more than 20% of pregnancies end in abortion. Roughly half of these abortions are conducted legally, and half are conducted in places where abortion is against the law. Sexual practices such as periodic or total abstinence have long been favored methods for many religious groups, including the modern Roman Catholic Church. Other groups such as the Shakers, a Christian sect in late 18th- and early 19th-century England and New England, required all male and female members of the community to practice celibacy. Children were only brought into the community by adoption or conversion.

Population-Control Policies

At various times and in different societies, authorities have structured policies designed to either promote

FIGURE 6.6 A One-Child Policy Chinese propaganda posters, such as this one depicting a single happy, healthy baby, promoted China's one-child policy as an effort to slow population growth. China has since relaxed this policy, now allowing women to have up to three children.

or reduce population growth. One of the best-known population policies occurred in China. As fertility rates rose in the 1960s, the Chinese government became concerned about overpopulation and in the 1970s instituted policies to limit childbirth, eventually enforcing a one-child rule. Towns, workplaces, neighborhoods, and couples all had state-controlled quotas for babies (**FIGURE 6.6**). It was thus possible to be illegally pregnant, and in many such cases abortions and sterilization were coerced. Due to traditional cultural preference for baby boys over baby

contraception technology, such as condoms or birth-control pills, that greatly reduces the probability of impregnation.

> ### SUSTAINABILITY

Family Planning Success Story

The TFR in Bangladesh fell from more than 7 average births per woman of childbearing age in the 1970s to 2.2 today—the lowest in South Asia. This successful reduction is largely credited to an approach that employed tens of thousands of female family planning workers who met regularly with women to educate them on birth-control methods and to distribute contraceptives if the women wanted them. These workers also connected women to other health-care services, leading to steep declines in both maternal and infant mortality.

The Eugenics Movement

Government efforts to control population have at times moved into the realm of *eugenics*, a word with Greek roots meaning "well born." The eugenics movement was founded in England in the late 1800s by Francis Galton, who believed that the human species could benefit from selective breeding that encouraged the "genetically superior upper classes" to multiply while instituting population-control measures on others. The eugenics movement flourished in the United States during the 1920s and 1930s when many states had forced-sterilization policies focusing on the elimination of "undesirable traits" believed to be hereditary, including criminality, promiscuity, poverty, and mental disability. These policies were often used to subjugate people of color and women. In the 1930s, the Nazi Party in Germany consulted with California eugenicists when designing their own forced-sterilization policies. These of course then expanded horrifically into the Holocaust, in which more than 10 million people were killed. In the decades after World War II, the eugenics movement in the United States was discredited, and policies allowing forced sterilizations were eliminated.

FIGURE 6.7 Policies Encouraging Reproduction When a country's population growth slows, governments sometimes create policies to encourage people to have more children. This couple is passing a strawberry mouth-to-mouth during a game at a 2008 Valentine's Day celebration in Singapore. The government organized this celebration to promote the holiday because of concerns about low fertility.

girls and sex-selective abortions, this policy also led to a skewed sex ratio in the population that rose as high as 117 males for every 100 females—about 10% higher than the natural ratio. It is estimated that China now has more than 30 million more males than would naturally be expected. After reaching a high of nearly 6 children per woman of childbearing age in the 1960s, the TFR in China is now well below the replacement rate at 1.6. The Chinese government has recently changed its policy and now allows women to have up to three children.

Significantly, even in areas of China where the one-child rule did not apply, fertility rates fell below the replacement rate. Demographers have concluded that simply making birth control readily available to people lowers the TFR. Iran provides another example of a government population-control policy. In the late 1980s, the country's fertility rate rose above 7 children per woman. The national government responded by establishing clinics to distribute free contraception. Without quotas or coercion, Iran experienced the largest and fastest drop in fertility on record—falling below the replacement rate in just 10 years.

When population growth is the goal, authorities may institute policies to push the TFR above the replacement rate. For example, Singapore has the lowest TFR in the world, which has fallen below 1. In response, the government attempted to encourage women to have more children by setting up online matchmaking services, relationship classes, and speed dating sessions and even introducing cash bonuses for childbirth (**FIGURE 6.7**). Other countries with TFRs below 2 that now offer bonuses or tax incentives to promote childbirth include Russia, Australia, Spain, and Germany. Germany, which has one of Europe's lowest TFRs (1.54 in 2019), offers parents of newborn babies a payment of up to €1,800 ($2,000) per month for a year. Most countries in Europe and North America offer income tax credits to families with children. However, these and other similar governmental efforts have not increased the TFR above 2.1. Once a country's TFR has been reduced below the replacement level, it becomes difficult to raise it back above that level, as a lower TFR is typically associated with a rising standard of living and more investment of resources per child.

Social Stratification and Inequality

Regulations designed to control population often include attempts to influence reproductive decisions. This can lead to ethically fraught policies, by which the question "Are there too many of us?" changes to "Are there too many of them?" Historically this has produced what are widely considered humanity's worst breaches of ethical conduct. In the early 20th century, population-control arguments were often linked to such ideas and promoted *eugenics*: attempts to "improve" the human race by selecting for some types of people over others. Policies in many countries, including the United States,

Soap Operas and Fertility

Of the many factors associated with falling fertility rates, including improved living conditions, health-care advances, and access to education and birth control for women, here's one you probably didn't expect: soap operas. In 2008, economist Eliana La Ferrara studied the effect of Brazilian *telenovelas* on fertility rates between the 1970s and 1990s, a period when broadcasting of these gradually spread geographically across Brazil.

Why look for a soap opera effect on fertility rates? In 1977 Miguel Sabido, the vice president of Televisa (Mexico's national TV network at the time), used a popular telenovela to convey a positive message about birth control. The main character on the show was a poor young woman who was born to a large family and trying to find a better life for herself—which she did by choosing contraception and limiting her own family size. The show even concluded with advice about how women could access family planning services. The Mexican government reported that while the show was on the air, the number of women seeking family planning services increased by more than 500,000, and contraceptive sales increased by nearly 25%. Other shows were soon released injecting a similar social message, and over the next decade Mexico's TFR fell from more than

5 births per woman of childbearing age to about 3.5. Although there was no good research done on the link between the soap operas and the fertility rate, the phenomenon became known as the "Sabido effect," and similar shows were programmed in other developing countries as well.

In Brazil, La Ferrara and researchers at the Inter-American Development Bank took advantage of a natural experiment. The Brazilian TV network Rede Globo had a near monopoly on the broadcasting of telenovelas in the country, and as access to its programming spread, Brazil's TFR was cut in half, dropping from more than 5 to 2.5 births per woman between 1970 and 1995. Moreover, for much of that time the government was not taking an active role in promoting family planning. The researchers could track the local fertility impact of the television shows as they entered various markets across the country, while controlling for various demographic and economic factors. They found that gaining access to the television network was associated with a fertility decrease of 6% overall and 11% among women over 35 years of age. This might not sound like much, but it is equivalent to the fertility decrease demographers have found is associated with an increase of 2 years of education on average for women.

La Ferrara's research team also conducted content analysis on more than 115 telenovelas that aired between 1965 and 1999 to gain insight into the messages the shows were conveying. Many of the shows did not have a direct and overt family planning lesson like that associated with the Sabido effect. Instead, the researchers hypothesized that the shows exerted an influence simply by representing delayed childbearing, smaller families, and associated lifestyles that differed from the average Brazilian reality when they aired. In other words, the soap operas presented role models for family planning decisions to which viewers seemed to aspire.

What Would You Do?

Imagine you are the public relations director for an environmental issues–based organization. If you wanted to influence behavior on an issue important to you, who would your target audience be, and how might you utilize popular culture outlets (e.g., movies, music, TV, social media) to influence behavior? What message would you try to convey, and how would you present it?

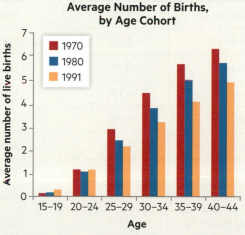

Average Number of Births, by Age Cohort

Economist Eliana La Ferrara found that Brazilian *telenovelas* (soap operas) can have a big cultural impact on public perceptions about family size and the age at which women have children. From 1970 to 1991 the number of children born to Brazilian women aged 25–44 dropped significantly, and La Ferrara's research demonstrated a connection between this decline and the messages delivered in *telenovelas*. Although the start of childbearing was not affected, as shown by the lack of change in fertility rates for the youngest age groups (women aged 15–24), the spacing and stopping of childbirth was dramatically affected in older women.

Graph adapted from La Ferrara et al. (2012).

FIGURE 6.8 Population Dynamics Vary by Region This map shows that between 2010 and 2020 some counties in the United States increased (green shades) in population while others decreased (brown shades). The cities with the highest rates of growth and decline are indicated on the map and in the tables below.

Adapted from US Census Bureau (2021).

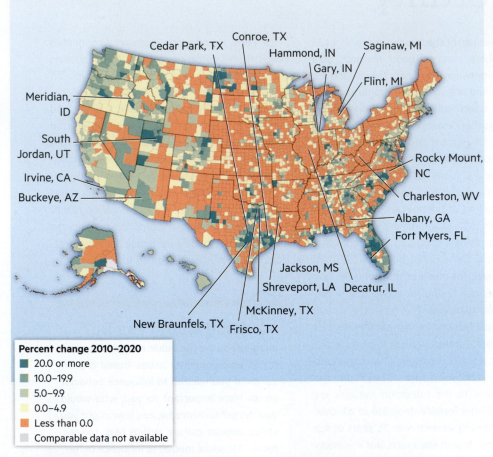

Percent change 2010–2020
- ■ 20.0 or more
- ■ 10.0–19.9
- ■ 5.0–9.9
- □ 0.0–4.9
- ■ Less than 0.0
- □ Comparable data not available

Fastest-declining Cities

City	Percent change
Charleston, WV	−9.4%
Jackson, MS	−7.4%
Decatur, IL	−7.1%
Shreveport, LA	−6.9%
Albany, GA	−6.9%
Gary, IN	−6.7%
Flint, MI	−6.6%
Hammond, IN	−6.6%
Rocky Mount, NC	−6.5%
Saginaw, MI	−6.5%

Fastest-growing Cities

City	Percent change
Frisco, TX	71.1%
Buckeye, AZ	56.6%
New Braunfels, TX	56.4%
McKinney, TX	51.9%
South Jordan, UT	51.8%
Meridian, ID	48.3%
Cedar Park, TX	44.2%
Fort Myers, FL	39.8%
Conroe, TX	39.3%
Irvine, CA	35.5%

linked the goals of population control with eugenics. Often these policies took the form of sterilization of the "feeble-minded," which could include the mentally ill, criminals, the poor, unmarried mothers, or other groups authorities found undesirable. Practices adopted in the United States in the 1920s included sterilization of the deaf and blind, people with epilepsy, and Native Americans and African Americans. Nazi Germany of the 1930s and 1940s represents the most horrendous extreme, as the Nazis attempted to exterminate Jewish, Romani, gay, and other marginalized people during the Holocaust, while also offering cash payments to "pure-bred" Aryan women to have children with Aryan soldiers in hopes of producing a "master race."

More recent and less extreme examples of population control targeted at social groups range from financial incentives to coercion. In the 1960s and early 1970s, India initiated government programs to pay men who underwent sterilization. However, what began as a voluntary program sometimes slid into a targeted and coercive tool. In some cities, poor couples had to produce a certificate of sterilization to apply for new government housing. In other cases, sterilization was a condition for obtaining rickshaw licenses, medical care, irrigation water, or ration cards. Population-control policies were largely funded by international organizations such as the United Nations and the World Bank, which made aid to developing countries conditional on the adoption of such policies. In response to concerns over human rights in India, the World Bank at the time stated that it had no formal policy for or against forced sterilization.

Yet development agencies found that coercive measures backfired. Such policies inspired widespread opposition and mistrust for the authorities seeking to implement them. India's annual population growth rate actually underwent a slight but steady increase from 2% to 2.3% when these policies were in place—adding more than 200 million people to the nation.

🛑 **TAKE-HOME MESSAGE** Population-control policies attempt to either increase or decrease the growth rate of human populations in particular places. When these policies follow systems of social stratification in societies, they can amplify inequalities. Experience in many countries has shown that coercive population-control policies are not required to reduce fertility rates; safe and effective birth-control methods need only to be widely accessible.

6.4 How Is Global Population Changing?

Think about where you live. Does it seem like there are more or fewer people than you remember from when you were younger? In the United States, over the past

20 years our population has grown from 282 million to 328 million. But depending on where you live, the experience of population growth or decline may be very different. Population in southern, coastal, and urban areas of the United States has grown, often at the expense of rural areas and areas in the industrial Midwest. If you live in many parts of central Texas or southern Florida, you may feel like population is booming. But if you live in many parts of Appalachia in the eastern United States, you may have seen your population decline significantly (**FIGURE 6.8**).

In a way, the experience of global population growth is similar. Although the number of people on the planet continues to increase, over the past several decades the rate of human population growth has slowed, and there is a wide variation in the TFR of different countries. As we discussed earlier, many women of childbearing age in Japan, Germany, and other relatively high-income countries are having just one child or no children at all, while in countries such as Burundi and Somalia it is not uncommon for childbearing women to have more than five children. What accounts for these differences?

Demographic Transition

Today, global population increase is slowing, and demographers believe it is not being caused by any extreme measures but by a phenomenon known as **demographic transition**—a decrease in the birth and death rates of a population linked to improvements in basic human living conditions, the availability of modern birth-control technologies, and economic growth. What are some signs of slowing population growth?

The rate of global population growth peaked in the 1960s at more than 2% per year but has since fallen to 1% per year. The number of additional people added annually to the planet in absolute terms peaked in the late 1980s at 87 million. It has since fallen to about 70 million per year and will likely keep declining. The TFR for the world peaked long ago in the 1950s at 5 children per woman. The global TFR is now near 2.3, less than half of what it was in the 1970s (**FIGURE 6.9**). In fact, 40% of the world population now lives in countries with TFRs below 2. Despite these trends, countries such as Niger, Ethiopia, Yemen, Afghanistan, and Pakistan still have high TFRs, meaning their populations will continue to grow rapidly for some time.

Demographers have found that while the populations of some countries are still rising, those of many other countries are stabilizing. This stabilization effect of a demographic transition occurs in three phases: a mortality transition, a fertility transition, and a stability transition—the end result of which is low death and birth rates (**FIGURE 6.10**). The **mortality transition** occurs as access to food, clean water, and medical care improves and the country's death rate declines. However, during this phase the birth rate remains high, so the population grows quickly. Most significant, infant mortality drops and more children survive into adulthood. The rapid growth in this phase transforms the age structure of the population, creating a large proportion of young people. As young people in these expanding countries approach

demographic transition decrease in the birth and death rates of a population linked to improvements in basic human living conditions, the availability of modern birth-control technologies, and economic growth.

mortality transition a period that occurs as access to food, clean water, and medical care improves and the country's death rate declines.

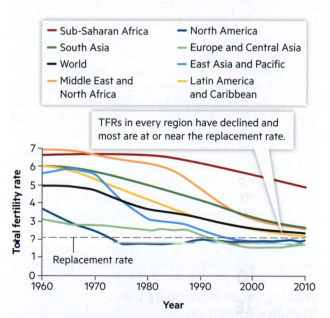

FIGURE 6.9 Global Total Fertility Rate The world TFR and the TFR on each continent have been declining in recent decades. Much of the world is now at, near, or even below the replacement TFR of 2.1.

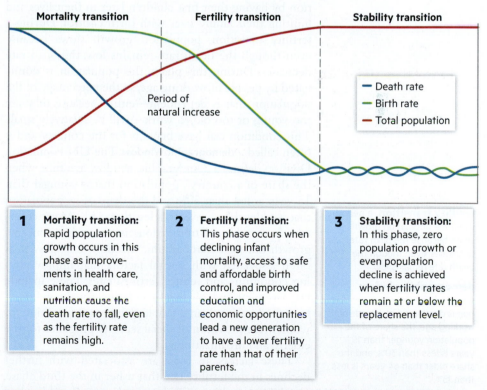

1	**Mortality transition:** Rapid population growth occurs in this phase as improvements in health care, sanitation, and nutrition cause the death rate to fall, even as the fertility rate remains high.
2	**Fertility transition:** This phase occurs when declining infant mortality, access to safe and affordable birth control, and improved education and economic opportunities lead a new generation to have a lower fertility rate than that of their parents.
3	**Stability transition:** In this phase, zero population growth or even population decline is achieved when fertility rates remain at or below the replacement level.

FIGURE 6.10 Demographic Transition A demographic transition occurs in three phases—a mortality transition, a fertility transition, and a stability transition—and results in both low death rates and low birth rates.

Nations with the highest percentage of people aged 65+
Japan 28%
Italy 23%
Finland 23%
Portugal 23%
Germany 22%
Greece 22%
France 21%
Bulgaria 21%
Croatia 21%
Slovenia 21%

FIGURE 6.11 Japan's Aging Population Japan entered a stability transition as the birth rate and death rate approached each other through the 1980s and 1990s. In 2007, the death rate began to exceed the birth rate, leading to a declining population. And with fewer births, today Japan has the highest percentage of people 65 or older among all nations.

their reproductive years, they are poised to raise the population to unprecedented levels.

However, history has shown that this situation can then follow a different course known as a **fertility transition**. This can happen if the younger generation has a lower fertility rate than that of their parents' generation by having their first children later in their lives and limiting their family sizes with birth control. During a fertility transition, population growth slows because even though the death rate remains low, the birth rate decreases. During this phase, the population is dominated by people of working age, so the percentage of the population that is dependent—either because they are too young or too old to work—will be relatively small. This condition can have benefits for the country and is often called a **demographic window**. The UN Population Division defines a demographic window as a time when the share of a country's population that is younger than 15 years is less than 30% and the share of its population that is older than 64 years is less than 15%. Many countries have achieved tremendous increases in economic growth during a demographic window. This growth is often associated with several factors that improve living conditions and reduce fertility rates. These include improvements in basic health and nutritional conditions, better educational and employment opportunities for women, the provision of social security programs for the elderly, and urbanization.

These factors, in turn, are associated with further declines in the fertility rate that usher in the third phase, a **stability transition** with low birth rates matching low death rates (**FIGURE 6.11**). During this phase of transition, there is zero population growth or even population

decline if death rates start to exceed birth rates. Japan and many European countries, such as Germany and Italy, are experiencing this phase of the transition.

Examples of Demographic Transition

Let's examine some specific examples of this pattern at work. We'll start with the United States. Between 1935 and 1955, the death rate in the United States fell from 11 people per year per 1,000 to 9.3, and infant mortality declined from 58 deaths per 1,000 live births to 28. During the same period, the birth rate increased from 18.7 births per year per 1,000 to 25, and the TFR ticked up from 2.1 to 3.7 children per woman of childbearing age, starting a period known as the "baby boom." Then, when the resulting "baby boomers" (those born between 1946 and 1964) began reaching reproductive age in the mid-1960s, they started to exhibit a lower fertility rate than that of their parents. The TFR in the United States fell steadily through the 1960s back to the replacement rate of 2.1 by the beginning of the 1970s. During the 1960s and 1970s, the baby boom generation also entered the workforce in large numbers. The working-age population rose steadily and eventually surpassed 75% of the US population before the first baby boomers began to reach retirement age in 2011. **FIGURE 6.12** uses graphs known as population pyramids to show how the baby boom generation rose through ages on the pyramid as the United States also underwent a demographic transition.

Perhaps the best example of how economic development can lead to a demographic transition occurred in Japan. Economists in the 1950s and 1960s argued that the island country was overpopulated and would not be able to feed or support itself. The TFR in Japan was as high as 4.4 during a post–World War II baby boom. So in the 1950s, with US support, Japan pursued an extensive voluntary birth-control program offering free contraception at every health clinic in the country. By the time that entire generation was of reproductive age, the TFR had fallen well below the replacement rate. In the decades that followed, Japan emerged from postwar chaos and became one of the world's leading economic powers, with rapid industrialization and urbanization and one of the world's highest literacy rates. Taiwan, Singapore, and South Korea followed Japan's course as well, and between 1965 and 1990 these countries had the world's fastest-growing economies. Each of these countries achieved a fertility transition by the 1980s when its baby boom generation had a lower fertility rate than that of the previous generation and saw its working-age population grow 4 times faster than its nonworking young and elderly populations.

Waiting for or Passing by the Stability Transition

Over the past 200 years, most countries appear either to have already experienced or to be in the midst of a demographic transition, with a mortality transition

fertility transition a period when the population growth slows because the birth rate decreases, even though the death rate remains low.

demographic window a time when a country's population is dominated by people of working age; the share of the population younger than 15 years is less than 30%, and the share older than 64 years is less than 15%.

stability transition a period when low birth rates match low death rates.

Population by Age Group in the United States

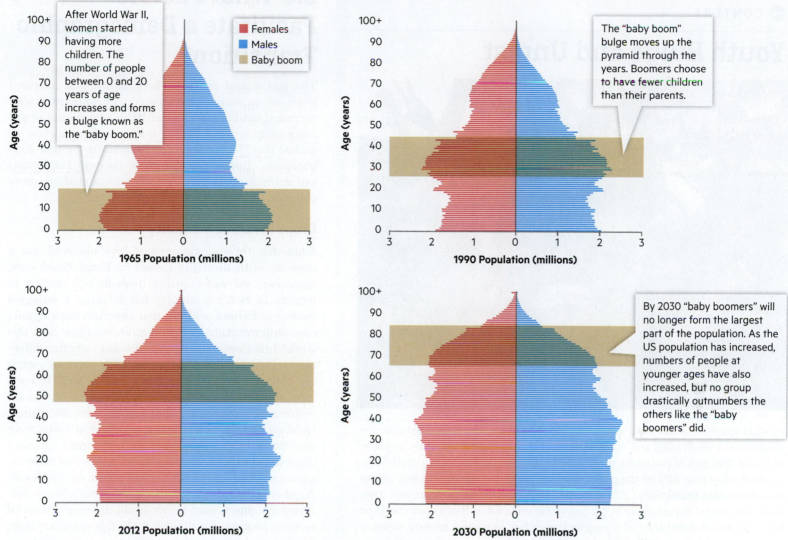

After World War II, women started having more children. The number of people between 0 and 20 years of age increases and forms a bulge known as the "baby boom."

The "baby boom" bulge moves up the pyramid through the years. Boomers choose to have fewer children than their parents.

By 2030 "baby boomers" will no longer form the largest part of the population. As the US population has increased, numbers of people at younger ages have also increased, but no group drastically outnumbers the others like the "baby boomers" did.

FIGURE 6.12 The Baby Boom: The United States' Demographic Transition These graphs, known as population pyramids, show the numbers of females and males in the United States by age group in different years. The tan areas highlight members of the "baby boom" generation, with a median birth year of 1955, and illustrate how a demographic transition took place in the United States.

Adapted from Colby and Ortman (2014).

followed by a fertility transition. But countries such as India and many African nations are still on the cusp of or still anticipating the demographic window. And if there is no fertility transition, a country can experience prolonged periods of low mortality with high fertility and explosive population growth. Some social scientists warn that in these cases, a **youth bulge** can occur, often associated with increasing incidences of civil unrest, crime, and terrorism as a large, undereducated, unemployed generation of young people becomes alienated and prone to radicalism and violent conflict.

Consequences of Low Birth Rates

While many nations have derived benefits from a declining fertility rate, having a TFR far below the replacement rate for an extended period of time can cause

problems. In these cases, the share of the population that is older increases. In 1950, the TFR of Japan was 4.3, and the median age of its population was 22 years. After decades of fertility rates below replacement, the median age of a Japanese citizen is now 45 years, and the TFR is 1.4. As the swelling older generation retires, such countries will be faced with significant costs for their pension and public health systems that are currently borne by working-age people who will soon constitute a shrinking share of the total population. However, there are also social and economic benefits of an aging population. Crime rates decrease and fewer services, such as education and health care, are needed for children, which saves money that could be used on the elderly. Extending the work years of older adults can also raise more money and keep them healthier and more active. Finally, if a

youth bulge a prolonged period of low mortality with high fertility, leading to explosive population growth and a very large population of young people.

Youth Bulge and Unrest

Political scientists and demographers have found a strong correlation between the presence of a youth bulge and the development of civil unrest in countries. Between 1970 and 1999, 80% of civil conflicts occurred in countries experiencing a youth bulge in which more than 60% of the population was under the age of 30. Here we see antigovernment protesters in 2011 in Yemen, where a youth bulge is present. They eventually forced the president out of power. Experts caution that a youth bulge on its own does not determine civil unrest; factors such as corruption, poverty, ethnic or religious tension, weakening political institutions, and adverse environmental conditions also play a role (as was the case in Yemen). But a youth bulge does seem to increase the likelihood that such conflict will erupt.

country has a population too small to support its citizens, it could increase immigration levels, which would provide younger, more productive workers (and relieve population pressures in the country the workers come from). In the face of declining population, Germany and Russia have both actively encouraged immigration, though not without controversy.

🛑 **TAKE-HOME MESSAGE** Demographers have identified a recent pattern of human population growth in many countries known as the demographic transition, in which a rise in population precedes a transition to a relatively stable national population. The pattern begins with a falling mortality rate and a large population of young people. However, when these people reach reproductive age, the fertility rate falls, and economic growth increases as they enter the workforce.

6.5 What Factors Facilitate a Demographic Transition?

The past several centuries have shown that increased levels of economic development are associated with decreased fertility rates. In the 1970s, recognition of this relationship led the Indian minister of population to declare that "[economic] development is the best contraceptive." But is this statement correct? Economists and demographers have actually seen something more complex.

Drops in Infant Mortality

Since the 1940s, demographers have observed that a drop in infant mortality caused by better health care, sanitation, and water quality helps drive a decline in fertility. In fact, no country has achieved a sustained decline in fertility without first experiencing a significant drop in infant and child mortality. How does this work? It is important to consider not only the different ways that children are expected to serve in a society but a society's desired family size as well. In rural areas of developing countries, families often depend on children as a source of labor and income. Even small children on a farm can plant and harvest crops, help care for animals, gather firewood, and tend to other domestic chores. As children mature, they can contribute more labor and be a source of wealth for the family. As parents look ahead to old age, they may rely on children for support, particularly in the absence of a social security program for the elderly. In this context, a large family is an asset.

But if in these developing areas health care is lacking and infant mortality is high, parents will likely conceive a large number of children just to ensure that enough babies will grow to adulthood to support the family. So high death rates among babies and children are normally associated with high fertility rates. For this reason, health-care improvements that increase infant survival are one of the most critical factors associated with fertility declines over time. If parents believe their children are likely to survive into adulthood, they tend to have fewer.

Urbanization

The process of development and industrialization is also associated with urbanization. People are drawn from rural areas to cities for employment opportunities. But even though we often see horrendous scenes of overpopulated urban areas in the developing world either on TV or in movies, urbanization is associated with lower fertility rates (**FIGURE 6.13**). Globally, the average TFR in cities is 2.2—near the replacement

FIGURE 6.13 Cities and Population Growth Cities in the developing world are growing rapidly, and TV and movie representations, such as *Slumdog Millionaire* and *City of God*, often focus on crowded and poor living conditions. But in most cases, these cities have lower fertility rates than those of the surrounding rural areas, and much of the urban population growth is caused by people moving into the cities from the outlying areas.

level—while TFR in the world's rural areas is 3. This may come as a surprise because, as we will see in Chapter 16, city populations are swelling rapidly, and nearly 1 billion people now live in places where city infrastructure and services have failed to keep pace with population growth. But we must remember that cities grow in two ways. Cities can grow by **natural increase**, where their birth rates exceed their death rates. But they can also grow as people migrate into cities from rural areas. The most rapidly growing cities in the world today—such as Shenzhen, China; Karachi, Pakistan; and Lagos, Nigeria—are in developing regions undergoing a massive urban migration.

Children in urban areas do not constitute as much of a source of labor or income for the family as they do in rural areas. Although child labor is still practiced in some places, cultural norms and policies that forbid the practice are increasingly common around the globe.

Children may be expected to provide support for the family as they mature, but until that point they are dependents. Survival, advancement, and success in urban areas hinges on education, which requires substantial investments on the part of both families and the public. Having a large family in the city is not an economic advantage. Families are more likely to focus time, energy, and financial resources on a smaller number of children to ensure success over a longer time frame.

Health and Education

A measure called **literate life expectancy**, the average number of years a person has the ability to read and write, attempts to capture the important factors of improved health and education. In countries that have already undergone a demographic transition in North America, Europe, and Asia, literate life expectancy is 65 years or more. For example, Spain now has a literate life expectancy of 70 years. This means that the average person has a life span that exceeds 70 by several years and is literate for nearly all of those years. However, a lower number tends to reflect shorter life spans and less literacy—perhaps due to less access to health care and education. The literate life expectancy in India is close to 30 years. In Afghanistan it is just 14 years (**TABLE 6.1**). Literate life expectancy can also reveal important differences between males and females in a given population. For example, in Afghanistan (with a TFR of 6.28), the literate life expectancy for males, 19 years, is more than double that of females, 9 years. We will see that this difference between the sexes can play an important role in the demographic transition. A high level of female literacy is associated with better family health and nutrition, enhanced life expectancy, and lower infant mortality—all factors associated with lower TFR.

natural increase a process in which population growth due to birth rates exceeds death rates.

literate life expectancy the average number of years in one's life a person has the ability to read and write.

TABLE 6.1 Literate Life Expectancy (LLE) for Selected Countries

Country	LLE
Italy	69.6
Israel	66.8
Philippines	56.5
Mexico	55.9
Brazil	48.5
India	28.4
Bangladesh	18.6
Senegal	14.5
Afghanistan	14.0

Adapted from Lutz and Goujon (2004).

The Importance of Secondary Education

According to research by the UN Educational, Scientific and Cultural Organization (UNESCO), women with a secondary education are far less likely to marry and have children at an early age. They are also more likely to provide adequate nutrition for their children, and their offspring are far less likely to die during childhood. As we learned in this chapter, improvements in infant and child health and survival are linked to lower TFRs. In terms of population, sub-Saharan Africa is an example where women without any education have 6.7 births on average compared to those with a secondary education who have 3.9 births on average. The impact of empowering girls and women has led many large philanthropic organizations (such as the Bill and Melinda Gates Foundation) to focus considerable resources on aid programs that provide education to girls.

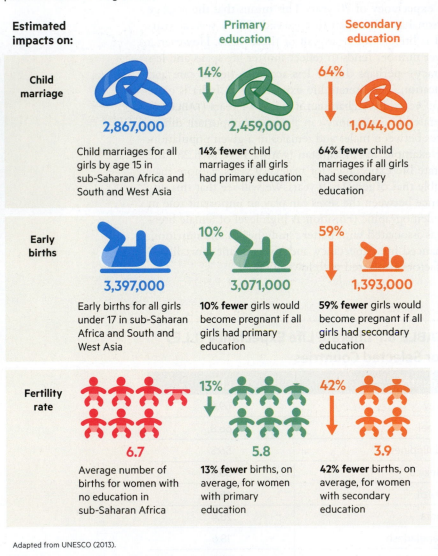

Estimated impacts on:		Primary education	Secondary education
Child marriage	2,867,000	14% ↓ 2,459,000	64% ↓ 1,044,000
	Child marriages for all girls by age 15 in sub-Saharan Africa and South and West Asia	**14% fewer** child marriages if all girls had primary education	**64% fewer** child marriages if all girls had secondary education
Early births	3,397,000	10% ↓ 3,071,000	59% ↓ 1,393,000
	Early births for all girls under 17 in sub-Saharan Africa and South and West Asia	**10% fewer** girls would become pregnant if all girls had primary education	**59% fewer** girls would become pregnant if all girls had secondary education
Fertility rate	6.7	13% ↓ 5.8	42% ↓ 3.9
	Average number of births for women with no education in sub-Saharan Africa	**13% fewer** births, on average, for women with primary education	**42% fewer** births, on average, for women with secondary education

Adapted from UNESCO (2013).

TAKE-HOME MESSAGE When a country's economic growth is associated with increased production of social goods and services, fertility rates tend to fall. Demographic transition is associated with a decline in infant mortality, urbanization, and closing of the gap between women and men on measures of education and health.

6.6 How Does Empowering Women Change Populations?

Over the past three decades researchers have reached widespread agreement on the link between women's empowerment and declining fertility rates. International aid groups and development agencies are promoting women's empowerment in lower-income countries, including access to reproductive services as well as educational and employment opportunities, to reduce the rate of population growth while fostering broad socioeconomic gains. Dozens of recent studies on fertility rates found positive associations between women's empowerment and lower fertility, age at first birth, longer intervals between births, and lower rates of unintended pregnancy.

FIGURE 6.14 shows how addressing early and unintended pregnancies can affect a region's overall population growth. The United Nations projects that the population of sub-Saharan Africa will grow from just over 1 billion people today to 4 billion by the year 2100. However, if the average age that women in the region have their first child increases just 2 years, from 20 to 22 years old, the projected population in 2100 dips by 9%. More significantly, if women in the region do not have any unintended pregnancies, the projected population will be 30% lower, and there will be 1.2 billion fewer people in 2100. Overall, research shows that access to birth control and high levels of education are the two factors most likely to increase a women's age at her first childbirth and decrease unintended pregnancies.

Access to Birth Control

The African country of Rwanda provides an example of how improving access to birth control can lead to an overall decline in the fertility rate. Since the early 2000s, the government invested in a network of community health workers and communication campaigns to make and promote access to birth control for women. Between 2000 and 2020, contraceptive use among women of reproductive age (or by their partners) increased from less than 10% to more than 50%. During this time the average age of first birth increased by 2 years (from 21 to 23 years old), and the average number of unintended pregnancies per woman decreased from nearly 2 to 1. The country's TFR declined from 5.64 in 2000 to 4.04 in 2020. While a TFR above 4 will still lead to a significant rate of population growth in the coming years, it is significantly lower than those of other countries in the region.

Education for Girls and Women

While access to reproductive services is seen as essential to lowering fertility rates, it works best in tandem with improved educational access for girls and women. Statistics show that populations decline when significant numbers of women in developing countries are able (or in many cases are allowed) to obtain a secondary (high school) education (**FIGURE 6.15**). Women who complete their secondary education tend to marry and start having children later. They are also more likely to use birth control. More highly educated women also tend to work and earn more. As they become employed and earn a salary, the money they earn outside the home becomes an essential part of the family budget, and anything that might disrupt that income, such as an unplanned pregnancy or additional children, becomes less desirable. More educated women also tend to invest more money in each of the children they do have—improving opportunities for the successive generation. Then these children tend to follow similar or reduced fertility patterns of their parents, helping to reinforce a declining TFR.

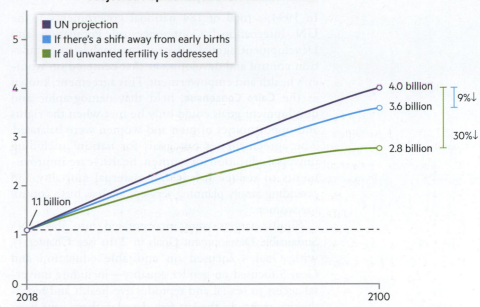

Projected Population in Sub-Saharan Africa

FIGURE 6.14 Empowering Women Decreases Population Forecasts Sub-Saharan Africa has one of the fastest growth rates in the world and is projected to add nearly 3 billion people by 2100 (dark blue line). The light blue line shows how this projection would change if women in the region had their first childbirth two years later on average than they do now (a shift from age 21 to age 23). The green line shows how the projection changes if women in the region no longer have unintended pregnancies. Empowering women with access to birth control and higher levels of education are two factors that facilitate later first childbirths and fewer unintended pregnancies—leading to a lower total fertility rate overall.

Adapted from The Track20 Project (n.d.).

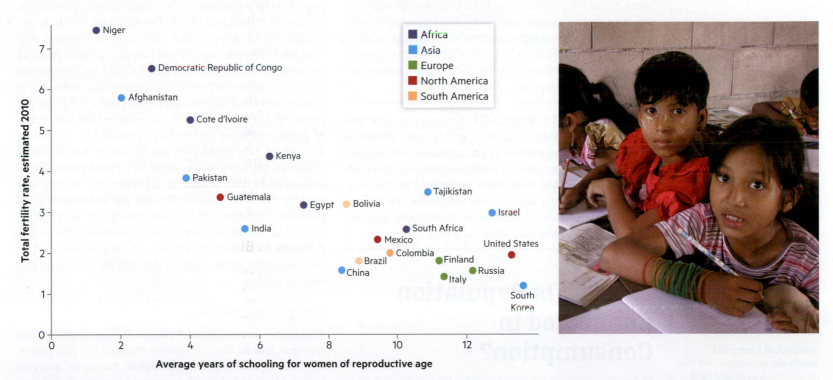

FIGURE 6.15 Education and Total Fertility Rate Research has found a strong correlation between the percentage of girls in a country that are enrolled in secondary school and the TFR. The dots on this graph represent various countries. In general, the greater the average number of years of schooling for women of reproductive age, the lower the TFR.

Graph adapted from Roser (2018).

International Response

In 1994, a total of 184 national governments at the UN International Conference on Population and Development agreed on new approaches to population control and development that centered on women's health and empowerment. This agreement, known as the **Cairo Consensus**, held that demographic and development goals could only be met when the rights and opportunities of men and women were balanced. The agreement set out goals for nations including universal education for women, health-care improvements to reduce infant and maternal mortality, and providing family planning services, such as birth control for women.

Similar targets were incorporated into the UN's Sustainable Development Goals in 2016 (see Chapter 1), with Goal 4 focused on equitable education and Goal 5 focused on gender equality—including universal access to sexual and reproductive health and reproductive rights. In the last few decades, there have been significant gains in the education of girls and women. Since 1995, the global enrollment rate in secondary school for girls increased from 53% to 75%. Contraceptive use by women of reproductive age (or by their partners) in developing countries more than doubled from less than 15% to more than 37% in 2020. Still, girls and women are deprived of schooling at much higher rates than boys and men, and the United Nations estimates that hundreds of millions of women who want to avoid pregnancy still lack access to safe and effective birth-control methods. International development experts warn that success on other sustainable development goals, including those concerning poverty reduction and environmental quality, depend on greater advances in empowering girls and women.

⬡ **TAKE-HOME MESSAGE** Empowering girls and women is linked to declining total fertility rates. Access to birth control and high levels of education are the two factors most likely to increase the average age women give birth to their first child and to decrease unintended pregnancies. Empowering girls and women helps reduce the rate of population growth and meet sustainable development goals for poverty reduction and environmental quality.

6.7 How Is Population Connected to Consumption?

If the sheer number of people on Earth was the only factor to consider when assessing our environmental impact, current trends might lead to a stable or even declining population by the end of this century. But earlier in the chapter we saw that human population is not the only factor to consider.

The declining fertility rates that many developed and developing countries have experienced in the past several decades came with dramatic increases in economic growth and standard of living. These trends reflect growing affluence and are directly associated with rising consumption and rising environmental impact. So while the rate of population increase may be slowing, what people are consuming and discarding in these countries is growing rapidly, especially in wealthier nations. The average resource consumption and pollution generation of the richest billion people on Earth today is more than 32 times that of the average for the other 6 billion people on the planet. In contrast, the world's highest fertility rates and most rapid population growth are in poorer developing countries with far less consumption and environmental impact. But if attaining lower fertility rates in those countries is accompanied by rising rates of consumption per person similar to those experienced in developed countries, the negative effects on the environment could expand even as the global population stabilizes or declines.

Impact = Population × Affluence × Technology

In the 1970s, scientists John Holdren and Paul Ehrlich devised a heuristic ("rule of thumb") known as **I=PAT**. I=PAT is shorthand for the idea that environmental impact (I) is a function of not only human population (P) but also affluence (A) and technology (T). It is not a mathematical formula but an attempt to represent and compare different aspects of human activity. Because we've spent most of the chapter focused on population, let's now explore affluence and technology. Affluence is usually measured as gross domestic product (GDP) per person in a country. GDP is a measure of the total value of goods produced and services provided in a country in a given year—and that are changing hands—so it is also an indicator of the level of consumption. Technology in I=PAT is used to represent the energy and resource intensity associated with the consumption of goods and services. I=PAT is too simplistic to produce numerical values for actual environmental effects, but as **FIGURE 6.16** shows, it can help us see that the level of human environmental impact depends on more than just our population.

Global Comparisons

Taking the environmental effects of consumption into account reveals that not every human on this planet has an equal impact. **Ecological footprint analysis** is a technique that makes the impacts associated with human consumption more visible. This analysis attempts to tally the area of land (and water) required for different categories of consumption and waste

Cairo Consensus an agreement saying that demographic and development goals can only be met when the rights and opportunities of men and women are balanced.

I=PAT shorthand for the idea that environmental impact (I) is a function of not only human population (P) but also affluence (A) and technology (T).

ecological footprint analysis an analysis that tries to tally the area of land (and water) required for each category of consumption and waste discharge to make human consumption impacts more visible.

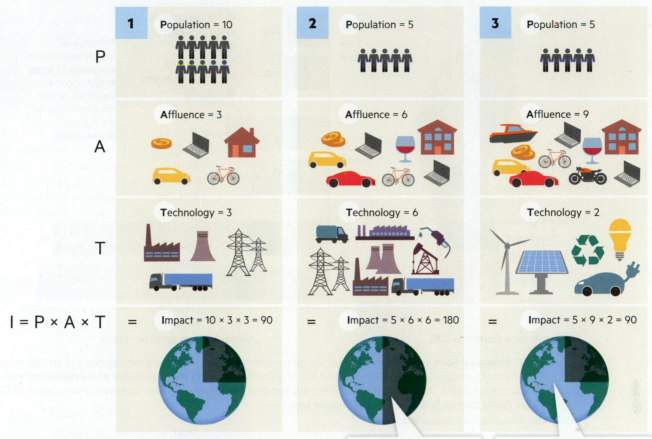

I = PAT: Three Scenarios

1 Population = 10
P

Affluence = 3
A

Technology = 3
T

I = P × A × T = Impact = 10 × 3 × 3 = 90

2 Population = 5

Affluence = 6

Technology = 6

= Impact = 5 × 6 × 6 = 180

3 Population = 5

Affluence = 9

Technology = 2

= Impact = 5 × 9 × 2 = 90

Although population decreased by half, the impact doubled because of a doubling of affluence and technology.

Although population decreased by half and new, less resource-intensive technology has developed, impact remained the same as that of scenario 1 because of a tripling of affluence.

FIGURE 6.16 Impact = Population × Affluence × Technology The I=PAT heuristic is used to emphasize the role that population (P) plays together with affluence (A) and technology (T) in determining environmental impact (I). Consider these three scenarios where we rate P, A, and T on scales of 1 to 10, where 10 = very high and 1 = very low. While the population decreases from a level of 10 to 5 in scenarios 2 and 3, changes in the ratings of technology and affluence can mean the overall impact stays similar or even increases.

(**FIGURE 6.17A**). More resource-intensive consumption creates a larger footprint area. If the resources and services Earth provides were put in terms of area and divided equally among the human population, it would amount to a little more than 1.5 hectares (3.7 acres) per person.

How do different countries compare? **TABLE 6.2** shows that Qatar has the largest footprint at more than 14.7 global hectares per person, while Eritrea has a footprint of half a global hectare per person. The footprint of the United States is more than 8 global hectares per person. Using this calculation, it would take more than 4 Earths to support 7 billion people all living at the current US consumption level. Globally, at current consumption levels, humans are using about 1.6 Earths' worth of resources and waste-absorption capabilities. How is it possible to use more than Earth can provide? According to this kind of footprint analysis, if we were using 1 Earth's worth of resources or less, we would be using resources (and waste-absorption capabilities) at a rate equal to or greater

than the planet's ability to renew these resources. However, we are currently using resources at a faster rate than they are being renewed. We can do this for some period of time without feeling all of the consequences of resource depletion, but if we maintain these rates of consumption, eventually the resources will be drawn down, and we will face the consequences.

One way to understand the long-term consequences of this unsustainable resource use is the concept of the tragedy of the commons introduced in Chapter 2. Earth's resources, or "commons," are being depleted at a rate that will degrade and diminish their availability for future generations. Countries that consume these resources at the highest rates per capita are contributing more to this destruction than others. Another way to represent this overall impact is with "Earth Overshoot Day," an analysis that tracks the environmental impact related to global human consumption each year and then compares that with Earth's ability to regenerate resources and absorb wastes. The analysis

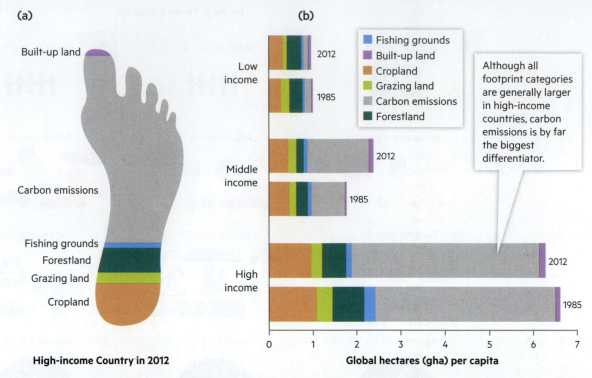

(a)

Built-up land

Carbon emissions

Fishing grounds
Forestland
Grazing land
Cropland

High-income Country in 2012

(b)

Legend:
- Fishing grounds
- Built-up land
- Cropland
- Grazing land
- Carbon emissions
- Forestland

Low income: 2012, 1985
Middle income: 2012, 1985
High income: 2012, 1985

Although all footprint categories are generally larger in high-income countries, carbon emissions is by far the biggest differentiator.

Global hectares (gha) per capita

FIGURE 6.17 Ecological Footprint Analysis (a) Ecological footprint analysis uses factors such as demand on fishing resources, land for development, cropland, and forestland in addition to greenhouse gas emissions to derive an overall estimate of a country's draw on resources per capita. **(b)** Footprint analysis like this has shown that higher-income countries draw on more of Earth's resources per capita than do middle- and lower-income countries.

Adapted from (b) World Wildlife Federation (2016).

TABLE 6.2 Countries Ranked by Ecological Footprint per Capita in 2018

Country		Ecological Footprint (global hectares)	Country		Ecological Footprint (global hectares)
10 LARGEST			**10 SMALLEST**		
1	Qatar	14.3	179	Burundi	0.8
2	Luxembourg	13.0	180	Eritrea	0.8
3	Bahrain	8.2	181	Mozambique	0.8
4	United States	8.1	182	Democratic Republic of Congo	0.8
5	United Arab Emirates	8.1	183	Pakistan	0.8
6	Canada	8.1	184	Rwanda	0.8
7	Estonia	8.0	185	Afghanistan	0.7
8	Kuwait	7.9	186	Haiti	0.6
9	Belize	7.9	187	Timor-Leste	0.6
10	Trinidad and Tobago	7.4	188	Yemen	0.5

Note: Ecological footprints are measured in global hectares. This is a measure of land and sea area required to provide the resources for the goods humans consume and the waste that must be absorbed from human activity.

Adapted from Global Footprint Network (2022).

Overpopulation or Rising Consumption?

The question about how best to address our sustainability challenges has often been framed as a debate over whether population growth or rising consumption is the primary threat to our continued well-being. Biologist Garrett Hardin developed a well-known and dramatic metaphor that framed the debate in terms of shipwrecked people lost at sea. Imagine yourself as one of the lucky ones. You managed to find a spot in a lifeboat, but you are also surrounded by a multitude of people in the water desperate to climb aboard. For Hardin, this story was about overpopulation. He feared that the lifeboats—Earth's resources—simply could not accommodate all of the people who needed to survive, and if too many people climbed aboard, the boats would be swamped and everyone would drown.

But what if we tell the story a different way? Imagine that instead of taking up just one seat, each person in the lifeboat is actually taking up several spaces and weighing down the boat with massive amounts of luggage and personal possessions? When told this way, the story is about how overconsumption in the developed world is threatening to sink us.

Although these two versions of the story do well to highlight population and consumption as key challenges to sustainability, they miss the way in which slowing the rate of population growth is associated with rising consumption. In this chapter, we learned that improvements in living conditions have been a common ingredient in countries that have successfully lowered and stabilized their TFRs. These improvements include gains in health care, nutrition, and education. But they also include a rising standard of living in terms of material goods, which on a national level is reflected in gross domestic product (GDP), the value of all the goods produced and services provided in a country.

As a country lowers its TFR and slows its population growth, it also tends to experience rising GDP per capita, a measure of affluence. And a rise in GDP per capita is associated with a rise in total resource use per capita. When GDP and total resource use per capita continue to climb even after the TFR has fallen significantly, great disparities emerge between the overall resource use of an individual in a developed country and that of someone in a less developed country.

So what is the way forward? A UN working group addressing this challenge concluded that our hope rests on reducing the amount of total resources needed for production of a good or service (or the resource intensity) and any typical environmental effects seen with economic growth. In other words, we need to devise ways to use fewer resources and reduce negative environmental effects while still achieving the economic output required to meet our development goals. This is no easy task of course; however, there have been some successful efforts to reduce the resource intensity of production. For instance, aluminum beverage cans are 38% lighter today and are typically made of 70% recycled aluminum. In 2015 in the United States, 91 million tons (or more than 34%) of all municipal solid waste was diverted from landfills and incinerators. In the coming years, reductions like these to the environmental impacts of consumption will need to be magnified and spread across all aspects of the economy in order to make real and lasting progress.

> **? What Would You Do?** Imagine you became the head of a large foundation and were tasked with using large sums of money to improve the welfare of families around the globe. Broadly, and given what you have learned so far, how would you prioritize your efforts and resources? To what extent and where would you focus on TFRs, for example? To what extent and where would you focus more on consumption issues? Why would you make this decision?

(a)

How Much Will Each Child Use?
The amount of resources used by each new person varies from country to country.

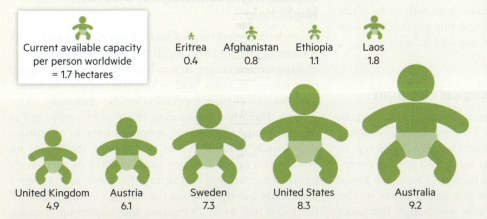

Current available capacity per person worldwide = 1.7 hectares

Eritrea 0.4
Afghanistan 0.8
Ethiopia 1.1
Laos 1.8

United Kingdom 4.9
Austria 6.1
Sweden 7.3
United States 8.3
Australia 9.2

(b)

The World's Fastest-Growing Economies

Country	Forecast GDP Growth, 2019
Uganda	7.61%
Egypt	6.81%
Myanmar	6.36%
China	6.09%
Vietnam	6.05%
Indonesia	5.80%
Cambodia	5.63%
India	5.50%
Tanzania	5.61%
Mali	5.37%

Adapted from The Growth Lab at Harvard University (n.d.).

(a) These values indicate each country's average ecological footprint per person measured in hectares (1 hectare is 2.5 acres). (b) The world's fastest-growing economies are developing nations in Asia and Africa.

estimates the day each year that consumption outpaces the ability of Earth to regenerate resources and absorb wastes. Each year, Earth Overshoot Day moves a few days earlier in the calendar year. In 2021, it fell on July 29.

Perhaps what's more revealing is the comparison of average ecological footprints across high-, middle-, and low-income countries (**FIGURE 6.17B**). This clearly shows the role of affluence and consumption on environmental impact, as the ecological footprint of high-income countries is several times that of low-income countries. Moreover, when the footprints are broken down into different consumption activities, we see that for low-income countries, nearly all of the footprint is created by the provision of food and shelter. In contrast, carbon emissions related to transportation and the production of consumer goods account for more than 50% of the ecological footprint in high-income countries such as the United States.

🛑 **TAKE-HOME MESSAGE** Demographic transitions come with dramatic increases in economic growth, consumption, and environmental impact. The I=PAT expression helps represent how consumption is factored into human environmental impact by accounting for affluence and technology as well as overall population. Resource consumption per person is much greater in wealthier countries than in poorer ones. Ecological footprint analysis estimates the environmental impact per person at particular levels of affluence and technology use.

🌿 6.8 What Can I Do?

It is clear that the human population cannot continue to grow forever if we want a sustainable future for our planet. A common theme in the chapter is that enabling women to control the size of their families (and enhancing their rights overall) leads to a decrease in fertility rate, which stabilizes population growth and generally improves the lives of women and their families. We've also seen that decreasing population growth alone is not enough to reduce environmental impact: consumption must decrease as well. What is our role in encouraging a sustainable population?

Know Your Personal Ecological Footprint

You can visit www.footprintcalculator.org to take a quiz that will assess your own personal ecological footprint and Earth Overshoot Day. You'll also see an estimate of how many planets we would need if everyone on Earth was living your lifestyle. Once you know your footprint, consider how you might lower it. The easiest way for many people is to change behaviors

related to transportation, food, energy, and water use at home. You could make a list of some behaviors you could change, try to live these changes for a week, and then retake the footprint quiz to see the difference you made.

Global Footprint Network®
Advancing the Science of Sustainability

Get Involved with Education and Rights Issues

In 2008, Nike launched the "Girl Effect," a public relations campaign designed to draw attention to the positive social, economic, and environmental changes that stem from empowering girls and women worldwide. Since then, many organizations and government agencies have redoubled their efforts to promote women's health issues and to improve women's access to education and economic opportunities throughout the world. You too could get involved with these initiatives. You can raise  your awareness of the status of women worldwide and efforts to address inequality by checking out the Global Women's Project and the World Bank's Global Partnership for Education. You can support and get involved with organizations like these and others that promote women's health and education globally, such as the Women's Health Initiative or the UN International Children's Emergency Fund (UNICEF), which focuses on the well-being of mothers and children. You could also get involved with groups that work closer to where you live, such as the Girl Scouts, Boys and Girls Clubs of America, or other local organizations. What do you think are the most effective steps to improving the status of women locally and in the wider world?

Explore How Technological Solutions Might Help

An economic challenge of smaller and older populations is figuring out who will do all the work. The economies of most countries are dependent on a constant supply of younger workers, and when this supply decreases, societies can change, especially for the remaining older citizens. Japan is experiencing this problem now and is experimenting with use of robots to help. In a solution that sounds straight out of the movies, the Henn na Hotel chain opened its first hotel in 2015 in Japan with minimal human staff and almost 250 robot staff. Despite some early hiccups, such as the "firing" of almost half of the robot staff in 2019 because they ended up creating more work for the human staff and did not reduce costs or workload, the Henn na Hotel has since expanded to 18 other locations in Japan, and 2 international locations, one in Seoul and one in New York City, opened in 2021. While these hotels are currently more focused on novelty (*Henn na* translates to "strange"), some experts see this increased "robotization" as the wave of the future. Do you agree? Do you see other practical opportunities for technology to replace human workers?

Chapter 6 **Review**

SUMMARY

- After growing slowly for many thousands of years, the human population has grown at an exponential rate in recent history.

- Although humans have demonstrated an ability to innovate and adapt to our environment in ways that expand the carrying capacity of the planet for our species, humans are also subject to limiting factors in the environment and have suffered population crashes.

- Advances in agriculture and in health are primarily responsible for the increase in fertility rates and decrease in death rates that have led to our recent boom in population growth.

- The number of people Earth can support depends on global population size, our ability to adapt and expand Earth's carrying capacity, and the environmental impact associated with our consumption patterns.

- Humans have the ability to alter reproductive practices in ways that either increase or decrease population growth. Although there are historical examples of coercive uses of population control, recent experiences in many countries show that birth-control methods need only be safe, effective, and widely accessible to bring down fertility rates.

- Many countries have experienced a demographic transition in which improvements in health care, nutrition, education, and employment lead initially to a falling mortality rate and a growing population of young people before ultimately achieving a relatively stable national population.

- Expanding rights and improving the health, education, and career opportunities for women in a population are important factors in lowering fertility rates and stabilizing population growth.

- Because consumption varies considerably by country and individual, not every human on the planet has an equal environmental impact. Resource consumption and pollution per person is much greater in wealthier countries than in poorer ones.

- The shorthand expression I=PAT summarizes how environmental impact (I) is a function of human population (P), affluence (A), and technology (T).

KEY TERMS

Cairo Consensus	demographic transition	literate life expectancy	total fertility rate (TFR)
carrying capacity	demographic window	mortality transition	youth bulge
contraception	ecological footprint analysis	natural increase	
crude death rate	fertility transition	replacement fertility	
demographer	I=PAT	stability transition	

REVIEW QUESTIONS

The letters following each Review Question refer to the Chapter Objectives.

1. Name two historical population crashes and their likely causes. **(A)**

2. Explain how two human developments have enabled the expansion of carrying capacity on Earth for our species. **(B)**

3. Explain what the total fertility rate (TFR) measures. **(C)**

4. What is replacement fertility? Provide an example of a country that is above replacement fertility and an example of a country that is below. **(C)**

5. Provide an example of a government policy designed to reduce population growth and an example of a government policy designed to encourage population growth. **(C)**

6. Explain what a demographic transition is, and provide an example of one country that has undergone this process. **(D)**

7. Identify two ways in which women in a country can gain empowerment that influences the total fertility rate. **(E)**

8. Identify four factors that facilitate a demographic transition. **(D, E)**

9. Explain how population, affluence, and technology contribute to human environmental impact (I=PAT). **(F)**

FOR FURTHER THOUGHT

The letters following each item refer to the Chapter Objectives.

10. In terms of the I=PAT concept, you count as just one person among all the other people that compose the human population. But can you identify two or three ways in which consumption in your daily life might differentiate your contribution to environmental impact from those of other people on Earth? Describe some steps you could take to reduce your impact. **(E)**

11. The Cairo Consensus on human population and sustainable development holds that development goals can only be met when the rights and opportunities of men and women are balanced. What do you think are the most important steps required to raise the status of women throughout the world? What obstacles do you think would have to be overcome to achieve these steps? **(D)**

12. While population growth affects all of us, government policies designed to control population can touch upon our most personal decisions and raise ethical issues related to individual rights. Can you identify one or more population-control approaches that raise ethical issues for you and describe what aspects make a policy objectionable to you? Conversely, can you identify and describe any policies designed to address population growth that you could support? **(C)**

Make Your Case?

The following exercises use real-world data and news sources. Check your understanding of the material and then practice crafting well-supported responses.

Use the News

The United Nations issues periodic reports and projections about population growth. The following article describes basic findings from one of these reports. Use this article to answer the questions that follow. The first three questions test your understanding of the article.

Questions 4 and 5 are short answer, requiring you to apply what you have learned in this chapter and cite information in the article. Answers to Questions 1–3 are provided at the back of the book. Question 6 asks you to make your case and defend it.

"9.7 Billion on Earth by 2050, but Growth Rate Slowing, Says New UN Population Report," UN News, June 17, 2019

"The World Population Prospects 2019: Highlights" estimates that the next 30 years will see the global population add an extra 2 billion people to today's figure of 7.7 billion, and, by the end of the century, the planet will have to sustain around 11 billion.

India will overtake China, sub-Saharan Africa population to double

India is expected to show the highest population increase between now and 2050, overtaking China as the world's most populous country, by around 2027. India, along with eight other countries, will make up over half of the estimated population growth between now and 2050.

The nine countries expected to show the biggest increase are India, Nigeria and Pakistan,

followed by the Democratic Republic of the Congo, Ethiopia, Tanzania, Indonesia, Egypt and the United States of America. In all, the population of sub-Saharan Africa is expected to practically double by 2050.

However, growth in these countries comes against the backdrop of a slowing global fertility rate. In 1990, the average number of births per woman was 3.2. By 2019 this had fallen to 2.5 births per woman and, by 2050, this is projected to decline further to 2.2 births: a fertility level of 2.1 births per woman is necessary to avoid national population decline over the long run (in the absence of immigration).

The population size of more and more countries is actually falling. Since 2010, 27 countries or areas have seen a drop of at least one per cent,

because of persistently low fertility rates. Between now and 2050, that is expected to expand to 55 countries which will see a population decrease of one per cent or more, and almost half of these will experience a drop of at least 10 per cent.

In some cases, the falling population size is reinforced by high rates of emigration, and migration flows have become a major reason for population change in certain regions. Bangladesh, Nepal and the Philippines are seeing the largest migratory outflows resulting from the demand for migrant workers; and Myanmar, Syria and Venezuela are the countries where the largest numbers are leaving because of violence, insecurity or armed conflict. For those countries where population is falling, immigration is expected to plug the gaps, particularly in Belarus, Estonia and Germany.

Population report a "roadmap to action and intervention"

"Many of the fastest growing populations are in the poorest countries, where population growth brings additional challenges," said Liu Zhenmin, United Nations Under-Secretary-General for Economic and Social Affairs (DESA). These challenges include the fight to eradicate poverty, and combat hunger and malnutrition; greater equality; and improved healthcare and education. The report, he said, offers a "roadmap" indicating where to target action and interventions.

At the same time, growth is providing opportunities in many developing economies: recent reductions in fertility mean that the working-age population (25 to 64) is growing faster than other age ranges, which could improve the possibilities for faster economic growth. The report recommends that governments make use of this "demographic dividend" to invest in education in health.

Proportion of older people increasing, life expectancy still lower in poorer countries

There will be one in six people over 65 by 2050, up from the current figure of one in 11. Some regions will see the share of older people double in the next 30 years, including Northern Africa, Asia and Latin America.

By 2050, a quarter of the population in [Europe and North America] could be 65 or over. The higher proportion and number of older people is expected to put increased financial pressure on countries in the coming decades, with the higher cost of public health, pensions and social protections systems.

Although overall life expectancy will increase (from 64.2 years in 1990 to 77.1 years in 2050), life expectancy in poorer countries will continue to lag behind. Today, the average lifespan of a baby born in one of the least developed countries will be some 7 years shorter than one born in a developed country. The main reasons are high child and maternal mortality rates, violence, and the continuing impact of the HIV epidemic.

---

1. According to the article, (i) which country will likely be the most populous country in 2027, and (ii) which three of these countries are part of a group of nine expected to have the biggest increase in population between now and 2050?

   a. United States
   b. Nigeria
   c. India
   d. China

2. Globally, which statement about projected overall fertility rate is true?

   a. It will decline globally from a rate of 2.5 births per woman in 2019 to 2.2 births per woman in 2050.
   b. It will decline in Africa and Asia from 2.5 births per woman in 2019 to 2.2 births per woman in 2050.
   c. It will increase globally from 2.2 births per woman in 2019 to 3.5 births per woman in 2050.
   d. It will increase in areas especially affected by emigration, such as the Philippines, from 2.2 births per woman in 2019 to 2.5 births per woman in 2050.

3. Which of the following statements is true?

   a. Life expectancy at birth has decreased significantly in the least developed countries in recent years.
   b. A significant aging of the population in the next several decades is projected for North Africa, Asia, and Latin America.

   c. The concentration of population growth is in the wealthiest countries.
   d. Life expectancy at birth will be almost the same in developed and developing countries in 2050.

4. Use material in the chapter to identify at least one advantage and one challenge of an aging population.

5. Describe the connection between an increasing life expectancy at birth in the world's poorest countries and the fact that these countries are where population growth is projected to be most pronounced.

6. **Make Your Case** The article identifies population growth being concentrated in the poorest countries as a challenge. If you were the manager of an international aid agency charged with addressing poverty and population growth, what sorts of programs and strategies would you strive to implement and why? Use the information in this article and in the chapter to support your position.

Use the Data

The population pyramids shown here depict the numbers of females and males as a percentage of a country's overall population by age. The shape of a population pyramid can be used to project future population growth. These three graphs each represent a different country (A, B, and C). Each has a different shape that portends a different population growth trend in the coming years.

Study the graphs, and use them to answer the questions that follow. The first three questions test your understanding of the graphs. Questions 4–6 require you to apply what you have learned in this chapter and cite data in the graphs. Answers to Questions 1–3 are provided at the back of the book. Question 7 asks you to make your case and defend it.

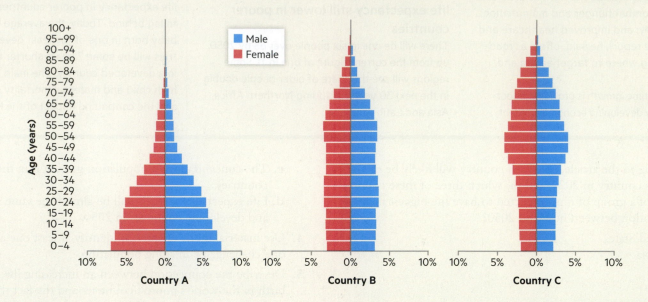

1. Which country (A, B, or C) has the largest share of people 60 years or older?

2. Which country (A, B, or C) has a larger share of its population in the 0–4 years age range than in any other age group?

3. Do women or men have a higher life expectancy across these three countries?

4. Based on the data in the three pyramids and what you read in the chapter about population transitions, which country (A, B, or C) is most likely to experience a decrease in population in the coming decades? Name the country and explain why.

5. Based on the data in the three pyramids and what you read in the chapter about population transitions, which country (A, B, or C) is most likely to experience rapid population growth in the coming decades? Name the country and explain why.

6. In the chapter, we learned that demographic transitions that result in declining population growth take several decades to occur. How can population pyramids like this help explain why these transitions take so long?

7. **Make Your Case** You are a government official working for an agency in charge of social services such as education, health care, and retirement support. If you worked for country C, what do you think would be the biggest challenges facing your agency in the coming decades on the basis of the population pyramid? If you worked for country A, how would the challenges facing your agency likely differ? Then, considering your answers, describe at least one idea you have to address each challenge.

LEARN MORE

- UN World Population Prospects, 2019: https://population.un.org/wpp/Publications/Files/WPP2019_Highlights.pdf
- Fred Pearce. 2009. "Consumption Dwarfs Population as Main Environmental Threat." Yale Environment 360. https://e360.yale.edu/features/consumption_dwarfs_population_as_main_environmental_threat
- International Conference on Population and Development: www.unfpa.org/icpd
- Philip Camill. 2010. "Global Change: An Overview." Nature Education Knowledge Project. www.nature.com/scitable/knowledge/library/global-change-an-overview-13255365

7

Water

How Do We Use It and Affect Its Quality?

Let's start this chapter with two questions about water. First, do you ever worry about the quality and accessibility of your water? Polling by the market research firm Morning Consult in 2020 found that although most people in the United States rate their own drinking water as excellent or good, a majority also worry about the country's general water quality and reliability. High-profile lead contamination crises of the water in cities such as Flint, Michigan, and Newark, New Jersey, have certainly raised awareness of threats to community water supplies. Second, how much water do you think you use each day? If you are like most Americans, you underestimated your typical daily usage of 80–100 gallons (300–380 liters) by a factor of 2. For many of us, safe fresh water may seem like an unlimited resource. But although water is abundant on Earth, the United Nations estimates that more than 3.5 million people a year die as a result of contaminated water and that one-fifth of the world's population (more than a billion people) suffers from water scarcity. How is this possible?

One reason is that most of the water on Earth is not available in a form we can readily use. It may be in the ocean as salt water, frozen in large formations of ice, suspended in the air, or contained in our bodies and other living things. Other reasons hinge on how a changing climate is affecting the water we do have available. For example, in Cape Town, South Africa—a city of 4 million people (roughly the size of Los Angeles) (**FIGURE 7.1**)—a severe drought that lasted more than 3 years reduced the level of water in local reservoirs to just 25% of their capacity. In 2018, officials warned that if the water level in the reservoirs dropped to less than 14% of their capacity, the government would shut off residential taps and ration water at collection stations. Each person would be allotted just 6.5 gallons (25 liters) per day to meet all of their needs, including drinking, cooking, bathing, and even toilet flushing. Cape Town residents responded with dramatic reductions in their water use, slashing their comparatively low consumption of 21 gallons (80 liters) per person per day to less than 10 gallons (38 liters) per person per day between 2015 and 2018. While this response avoided the worst-case scenario, water conservation and supply management efforts still continue in the city. Globally, similar

How do you feel about your water supply? What would you do if it stopped or became unsafe to use?

Chapter Objectives

By the end of this chapter, you should be able to . . .

A. describe how water on Earth moves among all three phases of matter: solid, liquid, and gas.

B. understand how water cycles between the atmosphere, ocean, snow and ice pack, groundwater, and surface water.

C. understand that groundwater and surface water are our sources of liquid fresh water and that these two systems are linked.

D. identify ways that human activities impact both the availability and quality of fresh water.

F. recognize that most of Earth's fresh water is frozen in permafrost, glaciers, and sea ice.

F. describe the diversity of life in the oceans, and how humans are impacting it.

We know the value of water when the well runs dry.

—Benjamin Franklin

embedded water a concept that accounts for the water that is used to produce goods that we consume.

droughts and shortages are projected to be a more frequent occurrence, making water scarcity more common. In 2017, a research team from Michigan State University estimated that climate change and the costs of updating our water supply might make clean water difficult for one-third of American households to afford. And the magazine *Consumer Reports* found in 2020 that prices of water and sewer service in 12 cities they surveyed had increased by an average of 80% in less than a decade, with disproportionate impacts on the poor and working classes.

Modern societies use tremendous amounts of the water that is available—not only for basic daily needs but also to produce what we consume. It's estimated that more than 500 gallons (1,900 liters) of water are needed for the production of a pound (0.5 kilograms) of pork, more than 100 gallons (380 liters) to grow a pound of potatoes, and nearly 700 gallons (2,650 liters) to make a gallon of beer. This so-called **embedded water** accounts for "costs" such as the water used to raise the pig or potato or the barley for beer, as well as water used in activities such as bottling and processing (**FIGURE 7.2**). Industrial uses of water are significant as well: in 2015 it was estimated that industry in the United States used close to 15 billion gallons (57 billion liters) of water a day. In fact, our thirst for water is causing societies to become more and more dependent on sources such as ocean water and reclaimed or recycled water to meet their daily needs. This is especially true in Middle Eastern countries such as Israel, where 60% of its water needs are met with ocean water that has been desalinated.

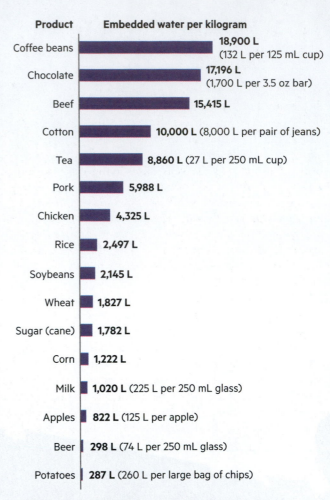

FIGURE 7.2 It Takes Water to Make Things We drink water to survive, and water goes into the production of nearly everything we use. This graph shows how many liters of embedded water are needed to produce a kilogram (2.2 pounds) of that item. Values are globally averaged estimates. L = liter (0.25 gallons); mL = milliliter (0.3 fluid ounces).

Adapted from Oleson (2014).

However, scientists are finding that all of these human activities have serious adverse impacts on aquatic ecosystems. Water provides the habitat for more than half of the species on Earth, but owing to human actions, biodiversity in aquatic environments is declining, with freshwater systems such as rivers and lakes declining at a faster pace than ocean habitats.

In this chapter, we take a closer look at water: where we find it, where it goes, the ecosystems it supports, and how it shapes our lives. While we will focus on fresh water because we rely on it to live, we will also discuss the oceans because they have a huge global impact on climate and biodiversity. Importantly, we will study how humans are affecting water resources, including issues of water supply (as seen in the Cape Town example) and water pollution. Although we will see the many ways our actions are making clean fresh water increasingly scarce, we will also consider opportunities to change and improve how we manage this essential resource.

FIGURE 7.1 Waiting for Water A severe drought in Cape Town, South Africa, between 2015 and 2018 caused water shortages and long lines to collect available water. Residents responded with conservation measures that cut their water usage by more than half.

7.1 Where on Earth Is All the Water?

We all know we need to drink water because we are constantly losing it; a typical person loses 2–3 liters of water per day through sweating, exhaling water vapor, and excreting waste. But this change is the norm with water. Water on Earth is also on the move, shifting not only in and out of the cells of all living creatures but also in and out of the oceans, the atmosphere, frozen ice and snow, surface water, and groundwater. We call each place in which water spends time a **reservoir**, and together all of the reservoirs constitute Earth's **hydrosphere**: a closed system that recycles water molecules over and over again from one reservoir to another (**AT A GLANCE: THE WATER CYCLE**).

We can see this cycling movement as water falls from the atmosphere as rain or snow. At any one time, the water held in the atmosphere is a tiny portion of Earth's water, about 0.001% of Earth's total water supply. But the atmosphere does not hold on to water vapor very long. Water molecules average only about 10 days of **residence time**—the time a molecule spends in a particular reservoir—in the air. For this reason, each year Earth's surface receives a tremendous volume of precipitation:

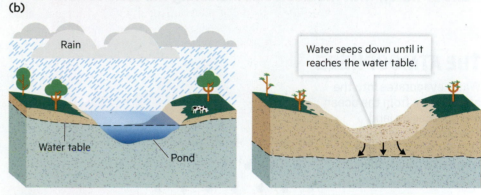

FIGURE 7.4 **How Groundwater Forms** **(a)** Groundwater forms when rainfall infiltrates into the ground and into an aquifer. The top of the groundwater reservoir is called the water table. It separates the saturated zone below from the unsaturated zone above. **(b)** Where the water table lies close to the surface, the water level in rivers, streams, lakes, or ponds is an extension of the water table. **(c)** In arid zones, the water table may lie deep below the ground surface, and ephemeral streams may flow only intermittently.

Adapted from Marshak and Rauber (2017).

about 500,000 cubic kilometers of water per year, more than 43 times the volume of Lake Superior.

About three-quarters of global precipitation falls directly into oceans. As Earth's largest reservoir of water, oceans hold about 97% of it with an average residence time of 3,000 years. Of course, the oceans are salty, so where is the fresh water? As **FIGURE 7.3** shows, the largest percentage (almost 80%) of Earth's fresh water is frozen. The largest reservoir of liquid fresh water is beneath the surface in groundwater (although some groundwater is also salt water). As precipitation falls to the Earth, some will sink into the ground and drain down until it reaches a layer of bedrock or dense clay that it cannot pass through. Groundwater will accumulate here, saturating the cracks and pores of the overlying material to form areas called **aquifers** (**FIGURE 7.4**). While groundwater can move through aquifers to feed into the ocean, lakes, and/or rivers, it moves at widely variable rates, often as slow as 1 foot or less per year. The average residence time of water molecules in an aquifer is long in human terms, often hundreds or many thousands of years. A much smaller amount of liquid fresh

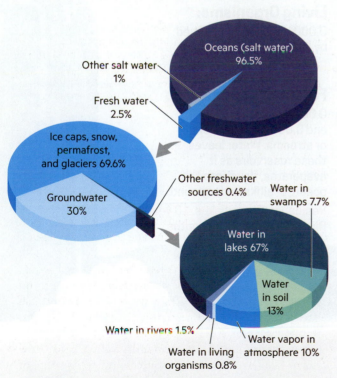

FIGURE 7.3 **Where's the Water?** Most of Earth's water is held as salt water in the oceans. Most of Earth's fresh water is held in ice and snow. Most of Earth's liquid fresh water is held as groundwater, and a much smaller fraction is available to us in freshwater lakes and streams.

reservoir a part of Earth where a material (such as water) remains for a period of time; also a term used for an artificial water body behind a dam.

hydrosphere all the places that hold water on Earth, including surface water, groundwater, glaciers, and water in the atmosphere.

residence time the time a molecule of a nutrient (like water) spends in a particular reservoir.

aquifer a subsurface area of rock or sediment where water can accumulate or pass through.

The Water Cycle

We see water appear and disappear in many forms. Rain falls and it trickles away. Ice melts and it seems to do the same. A puddle on the sidewalk disappears in the hot sun. Are these disappearances linked? Yes—these events are all part of the water cycle, which shifts molecules of water into different places or states of matter all around Earth. This includes the water inside you!

Here we show Earth's major water reservoirs and the average residence time of a water molecule in each. Arrows show typical flows of water, with the arrow size indicating the scale of flow.

Wind transportation of moisture

THE ATMOSPHERE

Water evaporates into the atmosphere from the ocean and sources on land. The atmosphere holds only about 0.001% of Earth's water at a given moment.

Residence time: 10 days

Evaporation of surface ocean water

Precipitation over oceans

THE LAND SURFACE

When water hits Earth's surface, it can go a lot of places:

Soil:
0.001% will be absorbed by soil.
Residence time: 1–2 months

Living Organisms:
0.0001% will be consumed by living things.
Residence time: A few hours to days

Freshwater Bodies:
0.007% falls into lakes and 0.0002% into rivers or streams. Water leaves these reservoirs as it evaporates, flows into the ocean, or sinks underground.
Residence times: Lakes, 10 years
Rivers/streams, 2–6 months

THE OCEAN

The oceans hold 96.5% of Earth's total water, and only about 0.03% of it evaporates in a year.

Residence time: 3,000 years

Spaces between rock grains are filled with water. This is groundwater. Underlying impermeable rock layers keep groundwater from moving deeper underground.

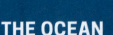

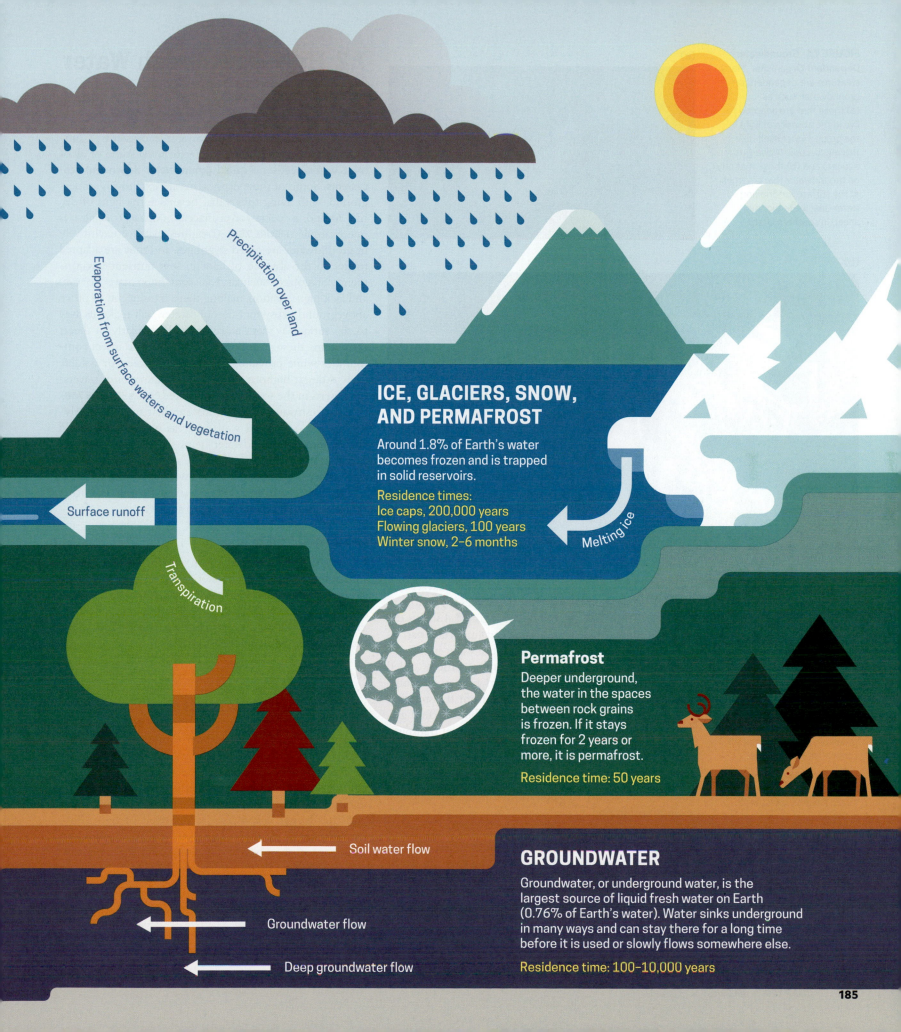

Evaporation from surface waters and vegetation

Precipitation over land

ICE, GLACIERS, SNOW, AND PERMAFROST

Around 1.8% of Earth's water becomes frozen and is trapped in solid reservoirs.

Residence times:
Ice caps, 200,000 years
Flowing glaciers, 100 years
Winter snow, 2–6 months

Melting ice

Surface runoff

Transpiration

Permafrost
Deeper underground, the water in the spaces between rock grains is frozen. If it stays frozen for 2 years or more, it is permafrost.

Residence time: 50 years

Soil water flow

GROUNDWATER

Groundwater, or underground water, is the largest source of liquid fresh water on Earth (0.76% of Earth's water). Water sinks underground in many ways and can stay there for a long time before it is used or slowly flows somewhere else.

Residence time: 100–10,000 years

Groundwater flow

Deep groundwater flow

FIGURE 7.5 Groundwater-Dependent Organisms Many creatures are dependent on groundwater, such as **(a)** those that are adapted to live in the absence of light in the groundwater itself (like this microscopic *Niphargus aquilex*) and **(b)** those that rely on groundwater that emerges at the surface for drinking water (like these bighorn sheep in southern Utah).

(a)

(b)

7.2 How Does Fresh Water Support Life?

When scientists see water, they expect to see life. In this section, we will examine freshwater ecosystems. Communities that require groundwater to meet some or all of their water needs are called **groundwater-dependent ecosystems**. Some of these ecosystems are exclusively underground, with living things that never see the light of day. In England, a species of crustacean, *Niphargus aquilex*, is one example (**FIGURE 7.5A**). Creatures that live exclusively in groundwater are called stygobites, and they are adapted to live in the absence of light in low-nutrient environments. Stygobites such as *Niphargus* often have no eyes and lack pigmentation, making them translucent. In other, aboveground ecosystems, groundwater will make its way to the surface as seeps and springs. Particularly in desert ecosystems, these areas host resident plant and insect communities while providing rare watering holes for wildlife such as desert bighorn sheep and migratory birds traversing deserts in the southwestern United States (**FIGURE 7.5B**).

The freshwater sources most familiar to us—the rivers, streams, lakes, and wetland areas we see—are called **surface water**. Groundwater helps shape these environments not only by providing a supply of fresh water during dry periods but also by lowering the temperature of the surface water, providing habitat for fish and other species that need cooler water to survive.

Lakes and Ponds

Let's start with lakes and ponds, which can form in various ways, such as when water fills areas cut out by the slow retreat of **glaciers**, massive ice sheets that last all year and flow over land surfaces (Great Lakes); volcanoes erupt (Crater Lake, Oregon); or rivers change course and abandon former channels (**FIGURE 7.6**).

Lakes and ponds host what are known as **lentic ecosystems**, which have relatively still water in zones defined by the distance from shore and the penetration of sunlight (**FIGURE 7.7**). The shallow-water *littoral zone* near the shore tends to host the greatest quantity and diversity of organisms. Here, plants that can root into the bottom, such as cattails, thrive. Well beyond the shore, the *pelagic zone* is further divided based on the extent to which sunlight penetrates the depths: the *photic zone* receives enough light to support photosynthesizing organisms, and the deeper *aphotic zone* is deprived of light. Photosynthesizing phytoplankton, such as green algae, are dominant in the photic zone along with the zooplankton that feed on them.

Rivers and Streams

Rivers and streams form as water flowing downhill collects along a common path, actively cutting channels in the ground. Rivers and streams host **lotic ecosystems**,

---

evapotranspiration the combined water vapor released from Earth's land and water surfaces along with transpiration from plants.

groundwater-dependent ecosystem a community of organisms that requires groundwater to meet at least some of its water needs.

surface water freshwater sources visible to us, such as rivers, streams, lakes, and wetlands.

glacier a mass of ice that flows over land surfaces and lasts all year.

lentic ecosystem an ecosystem in a lake or pond or other relatively still water.

lotic ecosystem an ecosystem in flowing water such as a river or stream.

---

water is in lakes, rivers, and streams, and in the next section, we will examine how these scarce resources support life.

Lastly, a thin sliver of the pie in Figure 7.4 shows the water in living organisms, a small but important part of the water cycle. For example, plants use water in the soil, then expel it as water vapor through a process called transpiration. Although it may seem like an insignificant amount, all of this water vapor from transpiration adds up. A large oak tree can give off about 40,000 gallons (150,000 liters) of water a year through transpiration, and on the whole, transpiration accounts for about 10% of the water vapor entering Earth's atmosphere. The release of water vapor from Earth's land and water surfaces along with transpiration from plants is called **evapotranspiration**.

⬡ **TAKE-HOME MESSAGE** Water cycles between Earth's major reservoirs in the atmosphere, in the ocean, on land, and beneath Earth's surface. Most of Earth's water is held in the ocean as salt water. All significant supplies of liquid fresh water on Earth exist as surface water or groundwater. But the largest reservoir of fresh water on the planet is frozen in ice and snow.

(a)

(b)

FIGURE 7.6 **Lake Formation** **(a)** Glacial lakes, such as the Great Lakes, were formed when continental-scale glaciers retreated, gouging and moving Earth to form basins that filled with water. **(b)** Volcanic lakes, such as Crater Lake, Oregon (the deepest lake in the United States), were formed when the crater at the summit of a collapsed volcano filled with water. **(c)** Oxbow lakes, such as this one on the Koyukuk River in central Alaska, form when a river changes course over time.

wetland an area where the ground is seasonally or permanently saturated with water.

(c)

Here, the oxbow is where the river once flowed before it changed course.

which have different habitats created by variations in their flows, ranging from turbulent rapids and riffles to relatively still pools (**FIGURE 7.8**). Food webs in streams and rivers rely on organic matter such as dead leaves and soil from the adjacent land, as well as tiny communities of photosynthesizing algae and blue-green bacteria, other microbes, and water moss clinging to debris such as rocks and logs. Larvae of insect species break up organic matter in the water as they consume other bacteria and fungi living on this debris. Then the finely shredded organic material, along with the feces of the larvae eating it, becomes food for mussels, worms, flies, and other filter feeders that strain their food from the water. Moving further out in the web, herbivores such

as snails feed directly on algae and moss attached to the rocks, and carnivorous creatures such as beetles, crayfish, and dragonfly larvae, as well as fish such as sculpin and trout, feed on the herbivores.

Life in Wetlands and Estuaries

Along the edges of streams and lakes, you'll often find **wetlands** (Figure 7.8) where the ground is seasonally or permanently saturated with water. Wetlands form along the flooded banks of rivers and streams during times of peak flow and along the coasts of large lakes when wind causes lake waves to flood the area. Others form in land depressions that fill with precipitation and/or groundwater

FIGURE 7.7 **Lentic Ecosystems** Lakes and ponds are known as lentic ecosystems, and they can be understood as a series of zones characterized by the distance from shore and the extent to which sunlight penetrates the water.

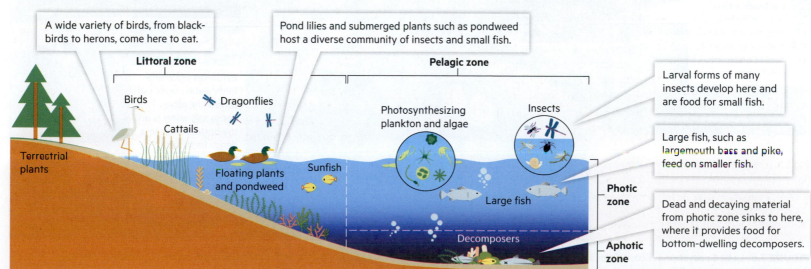

A wide variety of birds, from blackbirds to herons, come here to eat.

Pond lilies and submerged plants such as pondweed host a diverse community of insects and small fish.

Larval forms of many insects develop here and are food for small fish.

Large fish, such as largemouth bass and pike, feed on smaller fish.

Dead and decaying material from photic zone sinks to here, where it provides food for bottom-dwelling decomposers.

Littoral zone

Pelagic zone

Birds

Dragonflies

Cattails

Terrestrial plants

Floating plants and pondweed

Sunfish

Photosynthesizing plankton and algae

Insects

Large fish

Decomposers

Photic zone

Aphotic zone

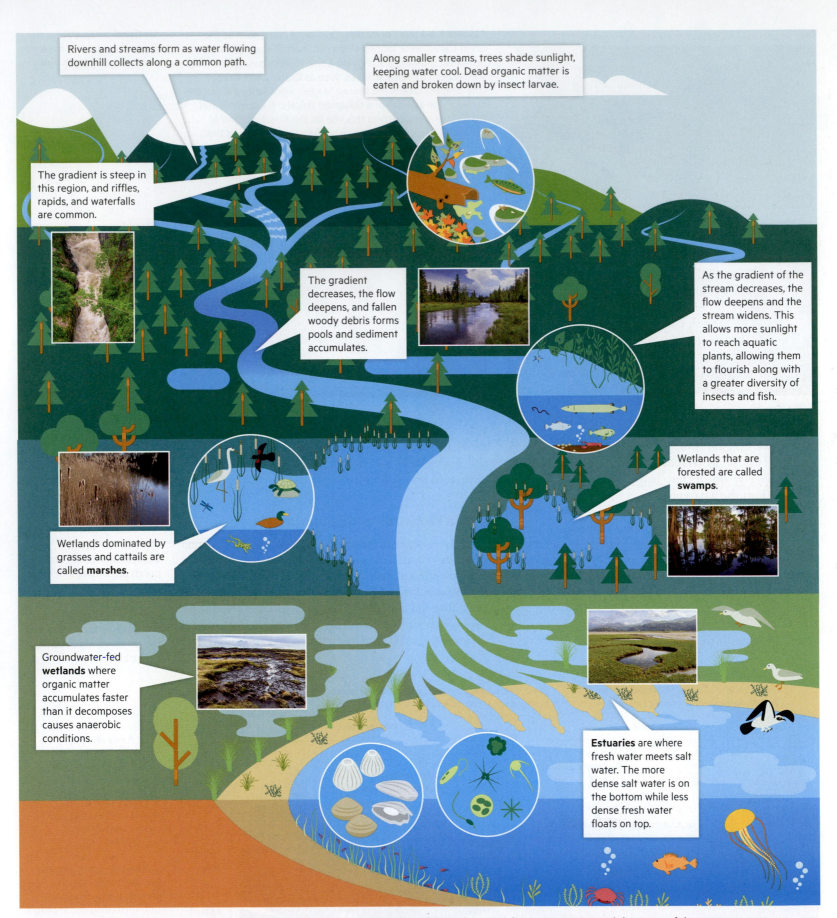

Rivers and streams form as water flowing downhill collects along a common path.

Along smaller streams, trees shade sunlight, keeping water cool. Dead organic matter is eaten and broken down by insect larvae.

The gradient is steep in this region, and riffles, rapids, and waterfalls are common.

The gradient decreases, the flow deepens, and fallen woody debris forms pools and sediment accumulates.

As the gradient of the stream decreases, the flow deepens and the stream widens. This allows more sunlight to reach aquatic plants, allowing them to flourish along with a greater diversity of insects and fish.

Wetlands that are forested are called **swamps**.

Wetlands dominated by grasses and cattails are called **marshes**.

Groundwater-fed **wetlands** where organic matter accumulates faster than it decomposes causes anaerobic conditions.

Estuaries are where fresh water meets salt water. The more dense salt water is on the bottom while less dense fresh water floats on top.

FIGURE 7.8 Lotic Ecosystems Ecosystems in flowing water such as rivers and streams are known as lotic ecosystems, and the zones of these systems are categorized primarily by differences in the flow of water.

flow. Forested wetlands, such as the woodlands adjacent to many rivers in North America, are called swamps, while areas dominated by grasses and cattails are called marshes. In some wetlands that are fed by groundwater, organic matter accumulates faster than it can be decomposed, resulting in anaerobic conditions. These bogs or peatlands have a sponge-like layer of *peat moss* that serves as habitat for other plants. Wetlands support not only plants but also birds such as herons, gulls, and ducks; amphibians and reptiles such as frogs and turtles; and mammals such as muskrats and beavers.

Estuaries are wetlands where fresh water meets salt water. These are among the most biologically rich areas on Earth due to abundant light and the nutrient-rich sediment delivered by the freshwater system. Salt marshes predominate in temperate climates, while mangrove swamps, such as those found along the Gulf Coast from the southern tip of Florida to Texas, are more common in tropical and subtropical climates (Figure 7.8). The plants in these areas can tolerate the higher salinity (salt content) of the water brought in by ocean tides. In salt marshes, low tides reveal large areas of muddy flats in which mussels and crabs are buried—food for herons, gulls, and raccoons. Mazes of meandering creeks carve channels among banks covered with plants such as cordgrass and glasswort. In the tropics, mangrove swamps support trees that grow up to 100 feet (30 meters) tall, which in turn support a wide variety of life forms. Oysters and barnacles cling to the bases of the roots, while snails live farther up the trunk. Insects and fish thrive in the protected water among the roots, and birds nest in the branches.

🏠 **TAKE-HOME MESSAGE** Freshwater ecosystems include those that are groundwater dependent, as well as those in lakes, ponds, rivers, streams, wetlands, and estuaries. Zones of life can be defined by distance from shore and depth in the water, as in lentic ecosystems, or by the speed of water flowing through them, as in lotic ecosystems. Wetlands are areas where the ground is seasonally or permanently saturated with water and include some of the most biologically diverse areas on Earth.

7.3 How Do Humans Impact Fresh Water?

In April 2014, the economically battered city of Flint, Michigan, switched its water supply to a cheaper source to address a budget shortfall. Though the new water source was more corrosive, officials neglected to treat it, causing lead in existing pipes and plumbing fixtures to be released into the drinking water. Gradually, citizens started to complain about the water, though they were assured it was safe. However, in September 2014, Mona Hanna-Attisha, a local pediatrician, studied her hospital's medical records

→ **CONTEXT**

What Happened to the Drinking Water in Flint?

For 50 years prior to 2014, Flint drew water from the Detroit River, but to reduce costs the city shifted its water intake to the Flint River while waiting for a new pipeline to be built that would deliver water from Lake Huron. The water from the Flint River was more corrosive than the water from the Detroit River, which caused the release of lead from pipes and plumbing fixtures. In 2015, scientists from Virginia Tech and the Environmental Protection Agency (EPA) found dangerous lead levels far in excess of EPA standards in water sampled from homes and schools in Flint.

Water treatment plant: The city draws and disinfects water from the Flint River.

Pipes in water mains and connections to homes can contain lead.

Flint River

Water mains

Lead solder: Copper pipe connections, especially in pre-1986 homes, can contain lead.

Corrosive water: Researchers have found Flint water to be more corrosive to pipes than water from the Detroit system, Flint's previous water source.

Lead into water: Some tap water samples are above the federal threshold for lead.

and publicized that the percentage of Flint children with high levels of lead in their blood had doubled after the change in water supply. What became known as the Flint water crisis soon gained national attention, revealing that changes in our water supply, distribution, and treatment can have serious public health impacts.

The situation in Flint illustrates just part of the complex relationship we have with our water. For many of us, our water comes from some distance away, and water resources are often brokered, sold, and transferred from

estuary a wetland where fresh water meets salt water.

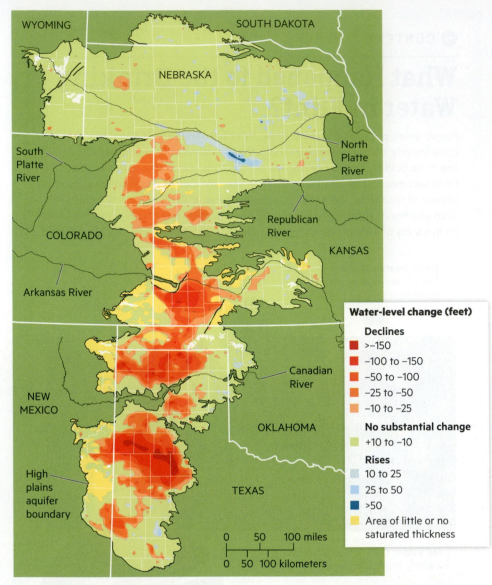

Water-level change (feet)

Declines
>−150
−100 to −150
−50 to −100
−25 to −50
−10 to −25

No substantial change
+10 to −10

Rises
10 to 25
25 to 50
>50

Area of little or no
saturated thickness

0   50   100 miles

0   50 100 kilometers

FIGURE 7.9 Drawing Down the Ogallala Aquifer Many western states rely on the Ogallala Aquifer, outlined in this figure, to provide water for agriculture. However, this aquifer was fed by glaciers more than 10,000 years ago and has a very slow recharge rate—just 0.02–6 inches (0.5–15 centimeters) a year. Currently, human activities are drawing this resource down faster than it can be replenished.

Adapted from McGuire (2004).

recharge process of adding water to a groundwater system.

water mining withdrawals from water reservoirs that are not renewable.

subsidence what occurs when land settles as formerly water-filled spaces collapse under the weight of overlying rock and soil.

one place to another as a commodity. Even in a relatively wealthy country such as the United States, clean water is not guaranteed, and access to resources can go to the highest bidder. Drinking water needs to be treated, as pollution is a constant threat. In this section, we will start to explore these two broad impacts: those resulting from how we choose to change freshwater systems by withdrawing and/or diverting water, and those resulting from pollutants that we add to these systems.

Increased Withdrawals from Groundwater Systems

Let's first consider the amount of fresh water we are using, starting with resources underground. Advances in technology, in particular better pumps and wells, have allowed

the global use of groundwater to expand 10-fold since 1950. But although groundwater is a renewable resource, the replacement, or **recharge**, time for groundwater often exceeds 10,000 years—which means that resupply can lag far behind the rate of human withdrawals. Moreover, few groundwater withdrawals are returned to the original recharge area after the water is put to use, which means that these withdrawals are not feeding and refilling the aquifer from which they came. Some aquifers are even sealed off by the surrounding rock and can no longer recharge from precipitation infiltrating through the ground. This "fossil water" can have a residence time of millions of years. These aquifers are considered nonrenewable, and withdrawals from them are considered **water mining**.

To get a sense of the scale and rate of groundwater withdrawals, let's consider the Ogallala Aquifer, located in an area that lies beneath parts of eight states in the western United States (**FIGURE 7.9**). It was filled with water from glaciers 10,000–25,000 years ago and can recharge only 0.02–6.0 inches (0.05–15 centimeters) annually. By the 1950s, tens of thousands of wells in the region were withdrawing more than 7 trillion gallons (26 trillion liters) of water per year, primarily for irrigation. Today, this aquifer irrigates more than 12 million acres of land and provides the drinking water for more than 80% of the people living in the region. But recently, the US Geological Survey found withdrawals have reduced the water level by an average of more than 11 feet (3.5 meters), and in some areas by more than 150 feet (46 meters). At the maximum recharge rate for the aquifer, it would take 300 years to refill the aquifer in the most depleted areas. The US Geological Survey estimates that the volume of water in the aquifer has been reduced by about 10% since extensive pumping began. Researchers at Kansas State University have found that the area of the Ogallala Aquifer on which Kansas agriculture depends has dropped by 30%, and at current levels of pumping it will fall another 39% over the next 50 years. At that point, the researchers estimate that agricultural productivity in the region will drop off sharply due to the diminished groundwater supply.

Excessive groundwater withdrawals can also damage groundwater-dependent ecosystems. For example, California pumps more groundwater than any other state, primarily from the Central Valley and Coastal Basin aquifer systems. Over the past century, this pumping shrank river and wetland habitats that depend on groundwater recharge during the dry summer and fall months. In many river systems, the loss of groundwater has caused low flows and shallow, warmer water that harms native species of salmon and trout. In coastal areas, the local groundwater is sometimes subject to *saltwater intrusion* from the ocean, which can contaminate freshwater wells (**FIGURE 7.10A**). Overpumping can cause the ground surface to sink, tilt, or even collapse through a process called **subsidence**, as the pore spaces belowground that once held water are drained and collapse under the weight of rock and soil above it (**FIGURE 7.10B**). Subsidence permanently reduces the ability

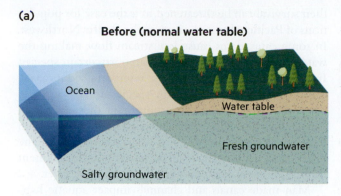

(a)

Before (normal water table)

Ocean

Water table

Fresh groundwater

Salty groundwater

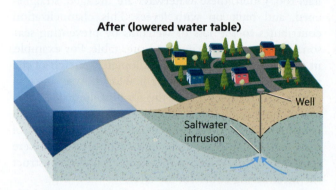

After (lowered water table)

Well

Saltwater
intrusion

(b)

1965

48 Years: 7.3 feet

SAN JOAQUIN VALLEY
CALIFORNIA
BM W 990 CADWR
SUBSIDENCE 7.3FT
1965-2013

2013

FIGURE 7.10 **Saltwater
Intrusion and Subsidence**
(a) Overpumping can lower
the water table and allow
salt water to seep landward
and intrude into freshwater
aquifers, where it can then
be drawn up into wells.
(b) This US Geological Survey
employee is demonstrating
how the land surface in the
San Joaquin Valley, California,
has been lowered by more
than 7 feet (2 meters) over
48 years due to subsidence
from pumping groundwater.

of an aquifer to store—and supply—water. Imagine the aquifer is like a wet sponge; while groundwater withdrawal drains the sponge, subsidence collapses it irrecoverably under the immense weight of the material above it.

Disruptions to Surface Water Systems

Humans also divert water from where it naturally flows on the surface. On small scales, houses divert rain into drainpipes and sewers instead of into the ground below. Impermeable surfaces (those that do not let water seep through them) such as paved driveways, roads, and parking lots send water to storm drains, which quickly channel the water to nearby waterways where the drains empty (see Chapter 9). During storm events, runoff from impermeable surfaces—also called impervious surfaces—can increase the peak flows of rivers, causing flooding and scouring fish eggs from streambeds. At the same time, this runoff prevents water from infiltrating into the land where it falls, thereby diminishing groundwater recharge.

On a larger scale, humans have long used technologies such as dams, reservoirs, canals, and pipelines to control water (**FIGURE 7.11**). Some of the first canal systems were invented by the Egyptians around 2000 BCE as they used water from the Nile River for irrigation. Fast-forward

(a)

(b)

(c)

FIGURE 7.11 **Surface Water Diversions** **(a)** Lake Powell is an artificial reservoir created behind the Glen Canyon Dam on the Colorado River. **(b)** The All-American Canal conveys water from the Colorado River on the Arizona border to the Imperial Valley in Southern California. **(c)** The Moffat Tunnel and Pipeline distributes water across the Rocky Mountains.

sediment eroded material that is transported and accumulates in different places.

more than 4,000 years, and we arrive at the monumental California State Water Project, which collects and diverts water from Northern California (which holds 75% of the project's water supply) and distributes it to water-scarce and heavily populated Southern California (which is responsible for 75% of the water demand for the project). These diversions, big and small, have impacts. Trapping water behind dams, while serving practical purposes for agriculture and power generation, increases the rate of freshwater evaporation as the water behind the dam is constantly exposed to the air and heating by the Sun.

Dams can be barriers to fish migrating upstream or downstream. When dams block fish from spawning areas,

their survival can be threatened, as is the case for populations of Pacific salmon in rivers in the Pacific Northwest. In some cases, dams cause low stream flow, making the water too shallow and warm to support certain species. In others, releasing water from reservoirs scours away the streambed gravel that many species need. Dams and reservoirs also influence the movement of **sediment** in stream systems. They typically prevent sediment from traveling downstream, forcing it to accumulate in the reservoir. This blockage keeps nutrient-rich sediment from reaching ecosystems at the mouth of the river.

Man-made canals and channels impact aquatic habitats too. Often, these waterways are dredged, straightened, and built up with levees. This channelization contributes to the loss of wetlands by preventing seasonal floods and lowering the water table. For example, in the 1960s, a 103-mile (166-kilometer) stretch of the meandering Kissimmee River in Florida was straightened and channelized into the 56-mile-long (90-kilometer-long) C-38 Canal (**FIGURE 7.12**). This project dried up about 50,000 acres of wetlands. Within 20 years of the canal's completion, new construction projects were started to restore some of the channels and reconstruct the meandering course of the river.

Surface water diversions diminish the water available to downstream users, often with ecological effects. A case in point is the mouth of the Colorado River in the Gulf of California. Here, the area of estuarine habitat has been reduced by more than half (**FIGURE 7.13**). Similarly, the area of the Macquarie Marshes of New South Wales, Australia, has been reduced by about 50% since the Macquarie River was dammed for irrigation.

FIGURE 7.12 Straightening Rivers Humans often alter rivers and streams by straightening or channelizing them in an attempt to control their course and flow. This disrupts, degrades, and in some cases destroys habitat in the river and along its banks. A straightened stretch of the Kissimmee River is pictured at the top, and a section of the river with its curves restored is shown at the bottom.

Small area of remaining estuarine wetland

Colorado River

Salt flat (former wetland)

Gulf of California

FIGURE 7.13 The River Runs Dry Estuarine habitat at the mouth of the Colorado River where it meets the Gulf of California has been reduced by more than half because of upstream diversions of river water.

(a)

(b)

point source a clearly identifiable source of pollution, such as a drainpipe, channel, or ditch.

nonpoint source a broad or diffuse source of pollution, such as agriculture or residential runoff.

eutrophication a process where marine and freshwater environments are enriched with nutrients, such as phosphorus or nitrogen from agricultural runoff or other sources, causing rapid growth, death, and decomposition of algae and phytoplankton. The decomposition process consumes and depletes oxygen levels in the water, harming oxygen-dependent organisms.

FIGURE 7.14 **Point-Source and Nonpoint-Source Water Pollution** Water pollution is often categorized by whether it comes from **(a)** a point source (a single identifiable source such as a pipe) or from **(b)** a nonpoint source (such as runoff) that enters water in a more diffuse way.

Water Pollution

The Flint water crisis, detailed earlier in this section, is a reminder not only that we need fresh water to survive but also that water must be clean and healthy. Unfortunately, freshwater pollution is a common occurrence. Sources of water pollution can be divided into two broad categories: **point sources** that discharge contaminants from a clearly identifiable conduit such as a pipe, ditch, channel, or well, and **nonpoint sources** that discharge pollutants in a broader, more diffuse, and less readily identifiable way. Point sources include drainpipes from factories and sewer systems that discharge directly into bodies of water. In contrast, nonpoint sources include contaminants from agricultural and residential activities that enter water through runoff, as well as airborne pollutants from vehicles and industry that rain out and settle into bodies of water (**FIGURE 7.14**).

Of the many different types of water pollution, sediment pollution is the most common. This occurs when large amounts of loose soil particles are swept into waterways by erosion from agriculture, forestry, and urban development. In rivers and streams, this causes cloudy or turbid water that can restrict the growth of vegetation and smother gravel beds where many fish lay their eggs. In lakes and ponds, it can similarly affect the clarity of the water, how far light penetrates beneath the surface, and nutrient levels in the water.

Chemical pollution also degrades water quality. In the case of Flint, Michigan, the pollutant was lead and the source was pipes and plumbing fixtures—a common risk in cities with older houses and water systems. But more common impacts of chemical pollutants are

linked to agriculture and energy production. For example, when agricultural fertilizers, whether synthetic or animal manure, run off fields into water systems, they can dramatically alter marine and freshwater ecosystems through a process known as **eutrophication** (**FIGURE 7.15**). The process starts when the fertilizer in the water causes phytoplankton and algae to grow and reproduce rapidly. With the right conditions, populations of these organisms can even double in size every 24 hours. When the density of the phytoplankton and algae becomes so high that the fertilizer in the water can no longer feed it, they die off rapidly, and this dead organic matter settles in the

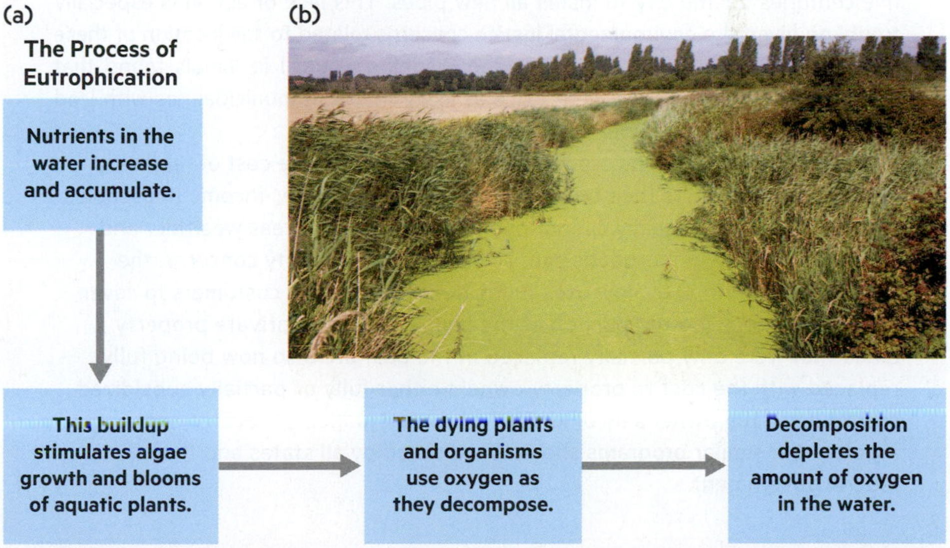

(a)
The Process of Eutrophication

| Nutrients in the water increase and accumulate. |

↓

(b)

| This buildup stimulates algae growth and blooms of aquatic plants. | → | The dying plants and organisms use oxygen as they decompose. | → | Decomposition depletes the amount of oxygen in the water. |

FIGURE 7.15 **Eutrophication** **(a)** Eutrophication is a process often triggered by excessive inputs of fertilizers into aquatic environments. **(b)** The rapid growth of plant matter in the water eventually reduces the amount of dissolved oxygen available to organisms in the water.

ENVIRONMENTAL JUSTICE

Who Should Pay to Get the Lead Out?

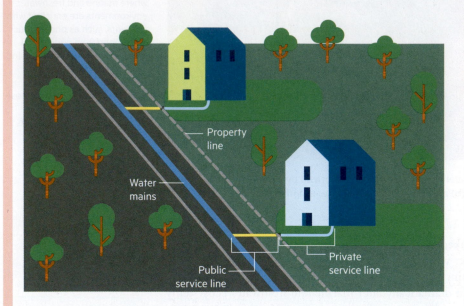

Lead is a toxin that impairs the physical and mental development of children and can cause kidney and heart problems for adults. Old lead service lines still deliver tap water to an estimated 10 million homes in thousands of communities across the United States and are the biggest source of lead exposure for children from birth to their first birthday—a critical developmental period. Although the installation of lead pipes was banned in the United States in 1986, cities have been slow to remove old service lines due to the high cost of locating, digging out, and replacing the pipes. For example, at the current rate of replacement in Chicago, it would take five centuries for the city to install all new pipes. This lack of action is especially troubling given the environmental justice concerns related to the location of these lines. A 2020 study by the Metropolitan Planning Council in Illinois found that people of color in this state were twice as likely to live in municipalities with lead service lines as white people.

Private property owners are typically responsible for the cost of replacing the pipe that connects their buildings to public lines. Lower-income households and neighborhoods usually cannot pay the high cost, whereas wealthier and typically whiter neighborhoods can. To address these equity concerns, the city of Washington, D.C., now uses funds from water utility customers to cover the cost of replacing the portion of the lines that are on private property. Lines that were only partially replaced in the past are also now being fully replaced with the cost to property owners either fully or partially subsidized according to the owner's income. Environmental justice advocates are arguing that similar programs should be adopted by all states and the federal government.

water. The organic matter is decomposed by bacteria, whose populations also explode in the presence of such abundant food. The bacteria consume oxygen in the water as they feed, depleting the available oxygen and suffocating other organisms like fish and other creatures that cannot move out of the area.

Agriculture also causes large increases in freshwater salt concentrations that can disrupt the particular water chemistry that aquatic species require. When water evaporates from irrigated fields, it leaves behind salts that concentrate in the soil. These salts will then move into freshwater systems when water runs off the fields. Salts also concentrate in river systems when water evaporates behind dams. The salt concentration of the Colorado River more than doubles as it passes through the heavily dammed and irrigated region of north-central Colorado and southwestern Arizona.

Chemical pollution from the energy industry includes common leaks from vehicles, fueling stations, and pipelines. But it also includes acid rain (covered more in Chapter 8), which occurs when air pollutants (normally from coal-fired plants) react with water vapor, fall as rain, and alter the chemistry of lakes and ponds. Similarly, when water runs through coal mines and other types of mines, this water can become contaminated and further affect other surface water and groundwater sources.

Finally, biological pollution occurs when disease-causing microorganisms such as certain bacteria, viruses, and protozoa are carried into freshwater sources by runoff tainted with manure, untreated sewage, or food-processing wastes. This type of pollution is of particular concern for human health, causing diseases such as cholera, hepatitis, and dysentery. The World Health Organization estimates that about 2 million people die from these diseases each year. We will discuss biological pollution in greater detail in Chapter 17.

🛑 **TAKE-HOME MESSAGE** Humans impact freshwater systems by physically disrupting and polluting them. The slow recharge rate of groundwater systems often lags behind the human withdrawals. Surface water withdrawals and diversions also cause ecological harm by diminishing and obstructing stream flows. Water pollution comes in the form of added chemicals, disease-causing organisms, and/or physical changes and can result from agriculture, mining, industry, and biological contaminants like sewage.

7.4 Can We Improve Access to Fresh Water?

In 2007, ultra–long distance runners Charlie Engle, Ray Zahab, and Kevin Lin became the first people to run the 4,000 miles (6,440 kilometers) across the Sahara Desert,

FIGURE 7.17 **Carrying Water** When water must be carried by hand long distances each day (a job that typically falls to women), it costs not only physical effort but also significant amounts of time simply to obtain this essential resource.

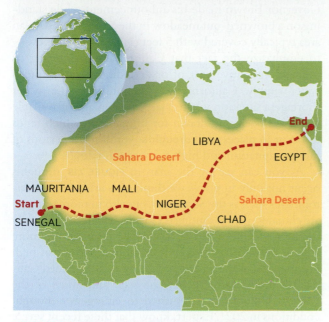

FIGURE 7.16 **Running for Water** Ultra-marathoners Charlie Engle, Ray Zahab, and Kevin Lin made their way 4,000 miles (6,440 kilometers) across the Sahara Desert over 111 days to bring attention to the need for adequate water supplies in Africa.

covering the equivalent of two marathons a day for 111 days (**FIGURE 7.16**). The distance, temperature extremes, and wind and dust storms all contributed to the challenges they faced, but if their crew had not transported ample water for them along the way, they never would have made it. Engle, Zahab, and Lin ran to support a nonprofit organization, H_2O Africa, and to raise awareness about the difficulties many people in the world face accessing adequate water supplies. Today, more than a billion people suffer from water scarcity. In many African countries, the problem is particularly acute. In some of the countries the team ran through, such as Mali and Niger, more than half of the population lacks access to safe drinking water. And often, the water that is available must be carried several miles in containers that weigh 40 pounds (18 kilograms) or more when full—a task that typically falls to women and can take 3 hours or more a day (**FIGURE 7.17**).

Ensuring reliable access to safe and affordable water— or **water security** —for everyone is one of the sustainable development goals of the United Nations (UN). Unfortunately, there are still grave inequalities in terms of access to water throughout the world. This is evident in both the Cape Town and Flint water crises covered earlier in this chapter. In the case of Cape Town, wealthier residents were able to construct tanks to stockpile water at their homes and to purchase additional supplies, while poorer residents were forced to deal with long lines at public distribution centers. And as the crisis unfolded in Flint, many observers questioned the judgment that led officials to approve the new water supply in order to save money—wondering whether this would have happened in a whiter or more affluent community. Globally, a UN report predicts that by 2025, as many as 48 countries (encompassing about 3 billion people) will likely be considered water scarce or water stressed. And because many water supplies span national and ethnic boundaries and divisions, strategic military and defense organizations such as NATO and the US Department of Defense have warned of increasing conflicts within and between countries over access to water, as well as an increase in refugee crises. As evidence, they point to water scarcity and inequitable access as precipitating factors in recent conflicts and humanitarian crises consuming Syria and Yemen. Because of the serious threat posed by water scarcity, the United States has a long history of bipartisan support for international aid efforts targeting water security. In recent years, both Republican and Democratic administrations have made improving global water security a national security priority. In this section, we look briefly at causes and possible solutions to the problem of water shortages.

water security reliable access to adequate safe and affordable water.

Bottle versus Tap

More than 90% of the people in the United States have access to cheap and highly regulated public drinking water systems, but Americans spend more than $30 billion each year on bottled water. This may be due to effective advertising and the widespread availability of bottled water, but polling also indicates that 40% of Americans believe bottled water is safer than tap water.

Is it safer? For one thing, the source of more than half of the bottled water sold in the United States is public tap water. And public tap water is governed by the Safe Drinking Water Act, which requires testing for more than 90 contaminants, making test results public, and supplying consumer confidence reports. In contrast, bottled water is regulated by the Food and Drug Administration (FDA). The FDA requires that bottled water companies conduct their own tests, but they are not conducted as frequently as those on public water supplies, and test results are not required to be made public.

You can save money (bottled water costs about 2,000 times more than tap water) and reduce waste (we'll learn more about waste associated with water bottles in Chapter 15) by choosing tap water. If you're curious about the source and quality of your tap water, you can look up its consumer confidence report at https://www.epa.gov/ccr (click "Find your local CCR").

Where Does the Water Go?

The supply of water in any area depends on the availability of surface water and groundwater, as well as the rate of precipitation, evaporation, and transpiration (the "exhalation" of water vapor by plants). But human water use often distributes water far from its source. Globally, agriculture uses the largest share of water (70%), followed by industry (20%) and public water supplies for residential use (10%). These categories are considered consumptive uses because they remove water in a way that does not return it to the same source. Instead, much of it is lost to the atmosphere through evaporation or is

embedded in various products. The most direct example of this is of course bottled water, which gets consumed far from its source. In 2019, the International Bottled Water Association reported that US bottled water consumption alone increased to more than 14.4 billion gallons (55 billion liters) a year: a volume equal to 21,800 Olympic-sized swimming pools and more than the amount of soda Americans consume.

Water Shortages

In the spring of 2015, Governor Jerry Brown of California ordered cities and towns across the state to cut water use by 25%. This was the first time in state history California had instituted mandatory water-use restrictions. Governor Brown made his announcement while standing on a browned-out meadow in the Sierra Nevada—an area typically covered with 5 feet (1.5 meters) of snow—to emphasize the fact that California was entering its fourth consecutive year of dry conditions and reduced water supply. California was responding to a water shortage, a situation where a large percentage of available water has been withdrawn for human use, which tends to occur in places with **droughts**—prolonged periods of low precipitation and high evaporation rates.

As you can see by the map in **FIGURE 7.18**, droughts occur not only in places we think of as arid, desert environments such as the Southwest but also in more temperate climates such as southern Florida, which we normally think of as quite wet. The longer the drought, the more serious the shortages become. Although periods of high precipitation in the winter and spring of 2017 ended California's drought emergency, the state and much of the western United States returned to emergency drought conditions in 2021. Experts know that these recent severe dry periods are part of a historic megadrought—a period of extreme dryness that lasts for decades. A growing population in California, heavy groundwater use, and hotter forecasted temperatures mean these water shortages will occur regularly. The first way to manage these problems is with conservation—using less.

Water Conservation

When water shortages occur, much of our attention focuses on ways to reduce water use in residential areas. Cape Town, South Africa, made infrastructure improvements, cutting the water lost to leaky pipes by about half. California set conservation goals for each local water district, and the districts implemented outreach campaigns for residents, employing social media and even door-to-door visits for high water users. The conservation measures with the biggest impact were those associated with landscaping. These included residents and businesses who stopped watering their lawns, allowing them to turn a "California golden" color through the summer, as well as many who removed their lawns altogether and planted drought-tolerant plants instead. The

drought a prolonged period of low precipitation and high evaporation rates that can lead to water shortages.

U.S. Drought Monitor

May 31, 2022
(Released Thursday, Jun. 2, 2022)
Valid 8 a.m. EDT

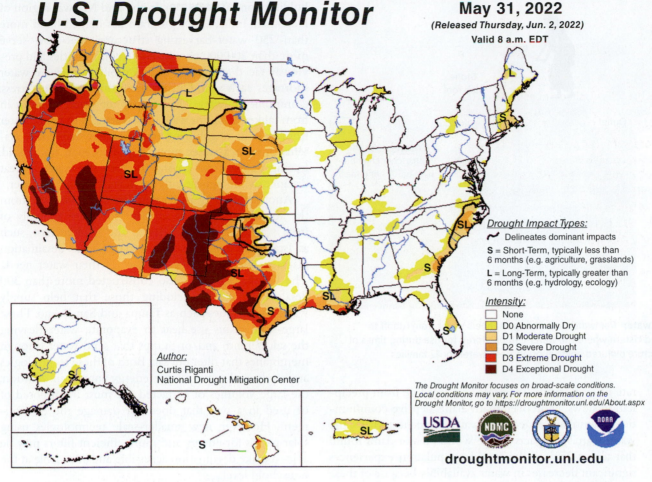

Author:
Curtis Riganti
National Drought Mitigation Center

Drought Impact Types:
∿ Delineates dominant impacts
S = Short-Term, typically less than 6 months (e.g. agriculture, grasslands)
L = Long-Term, typically greater than 6 months (e.g. hydrology, ecology)

Intensity:
☐ None
☐ D0 Abnormally Dry
☐ D1 Moderate Drought
☐ D2 Severe Drought
☐ D3 Extreme Drought
☐ D4 Exceptional Drought

The Drought Monitor focuses on broad-scale conditions. Local conditions may vary. For more information on the Drought Monitor, go to https://droughtmonitor.unl.edu/About.aspx

USDA   NDMC

droughtmonitor.unl.edu

FIGURE 7.18 Keeping Track of Drought On any given day, the US Drought Monitor identifies areas of the country facing prolonged periods of dryness. Note that on this day, portions of the Southwest, Oregon, California, and even Hawaii have some drought conditions.

state also exercised its authority to fine districts that were not meeting their goals, and the districts in turn levied fines on high water consumers who did not reduce their use.

But because globally most water consumption goes to agriculture, conservation measures outside of the cities can play an even bigger role in addressing water scarcity. This is particularly true in California where approximately 9 million acres of farmland are irrigated, accounting for about 80% of human water consumption in the state. Just repairing irrigation pipes and conduits can improve efficiency. Older methods of irrigation simply flood fields or send water running down channels. Estimates are that 50% of this water evaporates or flows off the fields. Newer systems are much more targeted, delivering water to precise locations and in small drips. Smart irrigation systems release water in small targeted amounts via drip and micro sprinkler units to fit local soil and weather conditions. On-site weather stations and instruments in the ground that constantly measure soil moisture to target irrigation only when needed can also help. Farmers can also switch to crops that require less water. For example, during the recent California droughts, many farmers turned away from water-intensive citrus and avocados to less thirsty crops, such as grapes, pomegranates, and even dragonfruit.

Another water-conservation measure involves **water recycling**. This strategy typically takes residential wastewater, treats it, and then applies it for various uses in agriculture, industry, or landscape irrigation. The Groundwater Replenishment System in Southern California gathers and treats wastewater and then injects it into the local aquifers to reduce pumping from an overdrawn, seawater-intruded local aquifer (see **STORIES OF DISCOVERY**).

Water Scarcity

As highlighted earlier, the biggest water-related problem in many areas of the world is access to affordable fresh water. It is a fact of life in many parts of the world that people (almost always women) make long walks to obtain scarce water from central sources every day. In the past, dams and other projects that control water (and that can also create power) have been seen as a solution. They typically provide year-round water supplies in reservoirs that can serve a wide area. In fact, dams and reservoirs were set up to provide an easily accessible water supply for Cape Town, South Africa. However, we have already learned that dams can have many adverse environmental impacts as well. As we saw for Cape Town, reservoirs can run dry during periods of drought as the

water recycling a strategy of taking residential water, treating it, and then using it typically for other agricultural or industrial uses.

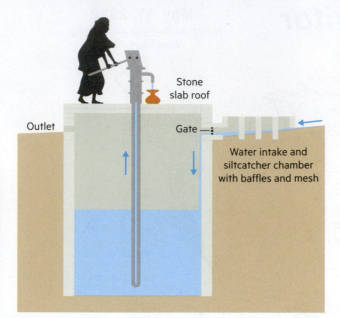

FIGURE 7.19 Harvesting Rainwater Rainwater harvesting channels water from runoff to structures like this underground cistern where the water can be stored for use during times of water scarcity. The simple structure pictured holds enough fresh water for 12 families.

Labels in figure: Stone slab roof · Outlet · Gate · Water intake and siltcatcher chamber with baffles and mesh · Photo © Niklas Hallen / UNDP India

rate of evaporation increases and recharge from precipitation slows. Moreover, many dams deprive communities downstream of access to water. A recent study on the water scarcity impacts of the world's major dams found that about 25% of the world's population experiences significant decreases in water availability because of these diversions.

Another concern is the growing trend of privatizing drinking water supplies. In these cases, a private company rather than a municipality—like a city or county—owns and manages the water supply. Price surveys in the United States have shown that on average, private companies charge more for the water than municipalities do—sometimes as much as 50% more. In the early 2000s, the city of Atlanta decided to retake control of its water supply after 4 years of operation by a private company led to high rates and declining water quality. Sacramento, California, provides an example of another type of privatization arrangement, through which Nestlé purchased the right to bottle and sell some of the city's water supply. This led critics to charge that Nestlé was profiting from shipping and selling California's water even while the state was facing a severe water shortage.

On a smaller scale, a practice known as water harvesting can provide a low-tech solution to water shortages. Water harvesting is simply the practice of catching water from precipitation and storing it for local use during times of water scarcity. This is not a new practice; water-harvesting systems as old as 4,000 years have been discovered in the Middle East. With new materials and technologies available today, runoff can be captured from rooftops or at ground level, and pipes and channels direct the water to a catchment such as a cistern or

holding pond (**FIGURE 7.19**). In the arid Marwar region of Rajasthan, India, a UN program has helped build more than 250 water-harvesting structures that now serve more than 200,000 people. These communities had previously faced an average of 3–6 months of severe water shortages each year. During these times of water stress, community members had to pay for water brought in from tanker trucks or walk long distances to bring water from wells in neighboring areas. The new harvesting systems provide a water supply close to each household, freeing people from having to carry the water a great distance, while significantly reducing the cost of water.

Another strategy is **desalination**: removing salt from the abundant supply of seawater to create a supply of fresh water. Many countries in the Middle East, such as Israel and Saudi Arabia, use large-scale desalination plants to meet a significant share of their water needs. The United States has also constructed more than 300 desalination plants, including those that help supply water to big cities such as Tampa and San Diego. These large-scale plants use heat to evaporate water (leaving the salt behind) and/or pumps that send water through membranes that filter out salt. Both of these technologies are costly and require large energy inputs; in addition, the large volumes of extracted salt must be disposed of or used in a way that does not damage the environment. However, new smaller-scale technologies using solar panels for energy and more efficient filters may be able to make desalination accessible and affordable at the household level.

⬡ **TAKE-HOME MESSAGE** Water security requires access to safe and affordable water. Unfortunately, many people around the world live in areas suffering from water scarcity. When rates of outflow exceed rates of inflow, the supply of available water in an area diminishes. Measures to improve access to water can include conservation measures such as water harvesting or converting seawater to fresh water through desalination.

7.5 How Do We Keep Fresh Water Clean?

People need access to fresh water that is safe for them to drink. Yet even in the United States, where more than 90% of the population is served by a water system with no reported violations, communities can suffer from an unsafe water supply—as the Flint water crisis illustrates. Moreover, the World Health Organization reports that 2.5 billion people lack access to sanitary wastewater systems, and that 1.8 billion people use a source of drinking water that is contaminated with sewage. More than 1 million liters (260,000 gallons) of raw sewage are dumped into India's Ganges River every minute.

desalination a process of removing salt from seawater to create fresh water.

Cow Milk or Almond Milk— What Uses More Water?

What do you pour on your breakfast cereal? For decades, the most popular addition to breakfast cereal has been cow milk. Although this remains the dominant milk product in the United States, its sales have been slipping as milk substitutes have increased in popularity. Grocery store dairy shelves now share space with almond milk, the most successful milk substitute. Since 2010, almond milk sales have grown more than 250%, with annual sales now exceeding $1 billion, while annual dairy milk sales have shrunk by about $1 billion. Almond milk production, however, uses a lot of water. Ninety-nine percent of the almonds consumed in the United States (and more than 80% of the almonds consumed globally) are grown in a 400-mile (645-kilometer) stretch of California's Central Valley, a region that since 2011 has been experiencing its worst drought conditions in more than a century. Because many of the state's water-conservation measures, such as restrictions on lawn watering and campaigns to reduce shower times, are focused on residential supplies, it is easy to overlook the fact that 80% of the state's water use is devoted to agriculture. And 10% of California's total water supply is devoted to almond orchards, which have doubled in acreage over the past two decades as demand for the nut has risen.

In this context, some have argued that milk from cows might be a better choice. But as we will see in later chapters, modern conventional beef and dairy production are also very resource intensive, especially in the amount of energy used to feed, process, and distribute the cows and the milk and in the amount of climate-damaging gases released during these processes. By this standard, conventional production of cow milk has an estimated 40 times the impact of production of almond milk. Water pollution from the fertilizer used to grow cow feed and manure from the cows also count against conventional dairy. And producing cow milk requires water not only for the cows but for their feed as well, which amounts to 640 gallons (2,400 liters) of water for each gallon of milk produced—more than for almond milk, which requires inputs of about 490 gallons (1,850 liters) of water for each gallon of product.

Almond growers point out that the expansion of almond orchards has occurred on lands that were being cultivated and irrigated for other crops. They maintain that switching to almond orchards has not caused an increase in the total amount of irrigated farmland in the state. Moreover, they note that the two most common crops replaced by almond orchards are cotton (produced for textiles) and alfalfa (produced for animal feed), both of which are more water intensive than almonds. In other words, the demand for almonds may be reducing overall agricultural water use.

But what about pasture-raised dairy that uses much less energy and water? Or soy milk? This case demonstrates that comparing competing products can involve considering environmental impacts (such as climate impacts and water use), impacts specific to the region of production, and what the production of a particular good is replacing. So the next time you reach for a product in the grocery store, you might pause to think about all that went into making it.

How Many Gallons of Water?

One glass of soy milk — 9 gallons

One glass of almond milk — 23 gallons

One glass of milk — 30 gallons

One yogurt — 35 gallons

One Greek yogurt — 90 gallons

Each of these items requires many gallons of water to produce, some more than others.

Note: Serving size is 6 ounces (180 milliliters). Figures do not include water for manufacturing or packaging.

Adapted from California Department of Food and Agriculture (2012–2013), Mekonnen and Hoekstra (2010), and Park and Lurie (2014).

? **What Would You Do?** Given the information above, which of these competing beverages do you believe is the best choice for daily consumption? And while you might not put the following on your cereal, other common beverage options are juices, soda, and bottled and tap water. Imagine you were assigned to choose beverages to serve at a local school program. What might you want to learn about some of these products to help you understand environmental impacts associated with them? Which would you recommend to put on the menu?

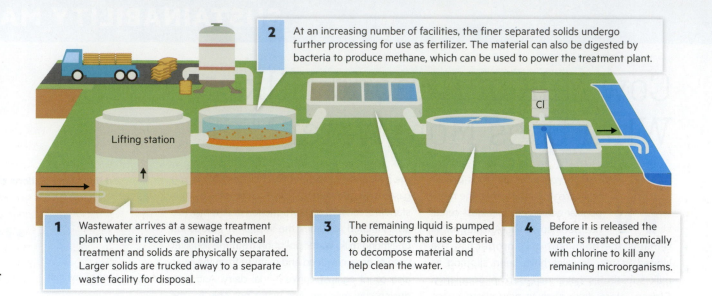

2 At an increasing number of facilities, the finer separated solids undergo further processing for use as fertilizer. The material can also be digested by bacteria to produce methane, which can be used to power the treatment plant.

Lifting station

Cl

FIGURE 7.20 Wastewater Treatment: An Overview

1 Wastewater arrives at a sewage treatment plant where it receives an initial chemical treatment and solids are physically separated. Larger solids are trucked away to a separate waste facility for disposal.

3 The remaining liquid is pumped to bioreactors that use bacteria to decompose material and help clean the water.

4 Before it is released the water is treated chemically with chlorine to kill any remaining microorganisms.

The task of improving water quality involves many things—for instance, good infrastructure. In most of the developed world, treated wastewater is the norm. Sewers empty into sophisticated plants that filter and treat wastes and pump out water that is often cleaner than that in the river or stream where it ends up (**FIGURE 7.20**). Septic tanks manage waste where sewers are not present.

The story can be different in developing nations. The United Nations invests more than $10 billion per year to help improve water quality and provide universal access to safe and affordable water and sanitation. This was part of the Millennium Development Goal to Ensure Environmental Sustainability, and according to the United Nations' assessment of these efforts, about 2.5 billion people have gained access to improved drinking water supplies since 1990. Strategies to improve water quality include not only large-scale sewage treatment and water-delivery systems but also smaller-scale water-purification technologies that can be used in households. For example, ceramic water filters can be manufactured by local potters. These work by filtering water through an inner porous ceramic pot into an outer pot that collects the clean water. Another effective household-scale water purifier is the solar Watercone (**FIGURE 7.21**), invented by German industrial designer Stephan Augustin, which uses the heat of the Sun to evaporate water, condense it, and collect it. This process kills pathogens, removes particulates and toxins, and can even be used to desalinate seawater. Wherever you are, infrastructure improvement can include strategies that reduce the pollution caused by runoff. Technologies such as green roofs and permeable pavement can reduce runoff by facilitating infiltration and uptake by plants (**FIGURE 7.22**) (see the discussion of low-impact development in Chapter 16).

Regulation has a role too. In the United States, the federal government regulates water quality primarily through the **Safe Drinking Water Act** and the **Clean Water**

FIGURE 7.21 Using the Sun for Desalination These plastic Watercones evaporate salty or brackish water placed in a tray, collecting clean droplets in the plastic shield. The shield is then turned upside down so that it can act as a funnel to collect the clean water.

Act. Under the Safe Drinking Water Act, the EPA identifies contaminants and sets maximum allowable levels for each in the water we drink, based on the agency's analysis of the health risks. State governments then typically enforce these standards. The Clean Water Act focuses on

Safe Drinking Water Act a federal law that sets the maximum allowable levels of contaminants in drinking water.

Clean Water Act a 1972 federal law that regulates the discharge of pollution in water.

From Toilet to Tap?

Can engineers devise a way to rapidly recycle wastewater from a major city's toilets into tap water that is safe to drink? Actually, more than 4 million Americans in several major cities, including Atlanta, Phoenix, and Dallas, already get at least some of their drinking water from recycled wastewater. Much of the pioneering work for water recycling has been carried out over the past five decades by the Orange County Water and Sanitation Districts in Southern California. Most recently, the Orange County Water District expanded its Groundwater Replenishment System to treat more than 100 million gallons (380 million liters) of wastewater a day that would otherwise be discharged into the Pacific Ocean. This water combats seawater intrusion and replenishes the groundwater basin that serves as the drinking water supply for more than 2.5 million people. This is the largest and most advanced water recycling project in the world, and it provides enough clean water each day to supply nearly 1 million people.

Water recycling addresses several environmental challenges in Southern California. First, water recycling provides a renewable source of fresh water in a region that faces water shortages due to prolonged drought. Second, injected fresh water fills the aquifers and effectively blocks saltwater intrusion.

Adding recycled fresh water to the groundwater has reduced Orange County's reliance on imported water from Northern California and the Colorado River. In the 1970s, when the plans for water recycling in the county began, imported water was thought to be cheaper than constructing new facilities to recycle water. However, drought and population growth over the ensuing decades have made recycled water a cost-effective option, while helping the county maintain more local control over its water supply. In addition, the new facility reduces the amount of treated wastewater that is discharged into the ocean. In the early 2000s, the county learned that population increases were leading to wastewater increases that would necessitate constructing a new system of pipes to drain wastewater into the ocean at a cost of more than $200 million. Instead, the county reduced the need for more waste disposal in the ocean by constructing the new water recycling plant.

But how can you turn raw sewage into potable water? The process starts with conventional sewage treatment that screens out solids and adds disinfectants (see Figure 7.20), but the treated wastewater still has a long way to go before it exceeds safe drinking water standards. The Groundwater Replenishment System uses polypropylene fibers with tiny holes just 1/300 the diameter of a human hair that can remove microscopic solids, bacteria, and many viruses through a microfiltration process. Then the water runs through a reverse osmosis plant that uses 1,000-horsepower engines to force water through the molecular structure of plastic sheets called reverse osmosis membranes. Under high pressure, the water will still move through the sheets, but contaminants are filtered out.

The Southern California research community has a long history with this technology. The first commercial reverse osmosis membrane was developed by chemical engineers at the University of California, Los Angeles, in the 1950s and was first put to use in the 1970s at an Orange County treatment plant known as Water Factory 21. Almost 40 years later and on a much larger scale, the reverse osmosis setup built in 2008 at the Groundwater Replenishment System can remove contaminants including pharmaceuticals, viruses, and salts at the molecular level. As a final step, the water is treated with ultraviolet light and tested for more than 400 contaminants to ensure that it exceeds state and federal drinking water standards.

What Would You Do?

Although the technology now exists to recycle wastewater into potable drinking water on a large scale, there may still be community resistance in many places. If you were a community outreach specialist working for a water district that wanted to pursue a water recycling project in your town, what types of opposing arguments do you think you would face? How would you design a public outreach strategy to build support?

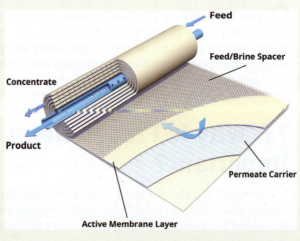

Feed

Feed/Brine Spacer

Concentrate

Product

Permeate Carrier

Active Membrane Layer

In the Orange County Water District's Groundwater Replenishment System, racks of reverse osmosis membranes filter microscopic contaminants from already-treated wastewater. Unwanted materials are caught in the membrane, leaving behind a salty brine ("Concentrate" in the diagram), and ultra-pure water ("Product") emerges.

FIGURE 7.22 Low-Impact Development The green roof of the Tianfu International Finance Center in Chengdu, China, absorbs rainwater and reduces runoff.

stopping pollution in surface water. This law makes it illegal and punishable to discharge pollutants from point sources into the water without a permit. Under this system, the EPA sets allowable pollution levels for each industry or other category of water polluter, and these rules are typically administered by state governments.

However, the most common source of surface water pollution in the United States comes from unregulated, nonpoint sources such as runoff from agricultural activities and urban development. The EPA estimates that about one-third of the stream miles in the United States are affected by nonpoint sources of water pollution, as are many groundwater supplies. Authority over nonpoint sources and the quality of groundwater supplies is largely left to state governments. While a few states have regulations for nonpoint sources, most rely on voluntary measures. For example, farmers might limit the amount of fertilizers and pesticides they use and leave permanent vegetated buffer zones around crops to slow surface runoff. Public awareness about the challenges of pollution can play a very important role in all cases. Whether through public awareness campaigns as seen in Flint or through a local community outreach campaign to combat pollution, public involvement plays an important role.

⬡ **TAKE-HOME MESSAGE** Addressing water pollution involves setting and enforcing standards for water quality, investing in and encouraging infrastructure and technology, and establishing policies that promote and preserve water quality.

7.6 Why Is Frozen Water Important?

For most of us, the everyday importance of frozen water—which is just solid fresh water—might focus on the ice to cool our drinks or on winter recreation.

cryosphere the frozen part of Earth's surface and crust.

permafrost water perpetually frozen in soil or rock.

sea ice frozen ocean water that is usually seasonal.

albedo a measure of the reflectivity of a surface.

ice age a time when the climate was much colder, glaciers at times covered larger areas of continents, and mountain glaciers grew.

But the ice on Earth plays a much more important role. One aspect involves glaciers, masses of ice formed by snowfall that accumulates over thousands of years without fully melting. Researchers have found that overall, Earth's glaciers have been losing mass since at least the 1970s (**FIGURE 7.23**). Why is this important?

(a)

Grinnell Glacier, 1936

Grinnell Glacier, 2013

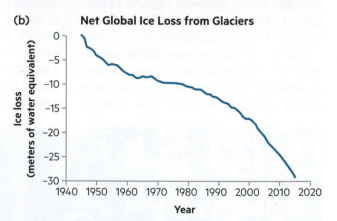

(b)

FIGURE 7.23 Glacial Retreat (a) The US Geological Survey's Repeat Photography Project has identified historical photos of glaciers in Glacier National Park and paired them with recent photos from the same locations to document glacial retreat—for example, the Grinnell Glacier shown here in 1936 and 2013. **(b)** Using a reference level taken in 1945, this graph shows a net loss of thickness (negative values on the *y*-axis) in the ice and snow of reference glaciers around the world.

Adapted from (b) US EPA (2016) and WGMS (2016).

Glaciers hold water for thousands of years. Loss of glaciers or other land ice causes a rise in sea level as the meltwater flows into the sea. Today, more than 95% of the world's glaciers are found near the poles, in Greenland and Antarctica. Greenland is about 3 times the size of Texas (656,000 square miles), and it is almost fully covered by an ice sheet. If all this ice melted, it would raise sea level worldwide by 20 feet (4 feet above the average elevation of Florida). If the entire Antarctic Ice Sheet melted (5.4 million square miles—about half the size of North America), it would raise sea level by about 200 feet (**FIGURE 7.24**). Although sea-level rise poses an immediate danger to coastal communities, a large input of fresh glacial meltwater into the oceans will also disrupt ocean currents. We will see in Chapter 11 that ocean currents are a major influence on global climate.

All the places on Earth where water is frozen make up the **cryosphere**, which includes seasonal snow and ice on land or sea, glaciers, and **permafrost**, water perpetually frozen in soil or rock. The residence time of water in ice and snow varies greatly from less than 1 year to more than 100,000 years depending on location and the depth of accumulation. Land itself can compose part of the cryosphere when it freezes. Permafrost can extend as deep as several thousand feet belowground, and some permafrost has remained frozen for hundreds of thousands of years. Like melting glaciers and ice sheets, thawing permafrost contributes to sea-level rise—although to a lesser extent—as much of the water that was previously locked up on land eventually finds its way into the ocean. Melting permafrost also releases methane, a potent greenhouse gas that can contribute to an increasing rate of global warming.

A final part of the cryosphere is **sea ice**. This frozen ocean water covers about 15% of the ocean for at least part of the year. Most sea ice is seasonal, growing in the winter months and melting again in the summer, although much of the sea ice near the North Pole persists for years. As sea ice is ocean water to begin with, its melting does not add a new stock of water to the system. Yet sea ice has a significant impact on global climate. Its reflectivity, or **albedo**, deflects the Sun's rays and has a cooling effect on the ocean water below it. So a decrease in sea ice means the ocean is warming; we will also see in Chapter 11 how the temperature of ocean water impacts the entire climate.

⬠ **TAKE-HOME MESSAGE** Earth's frozen water constitutes the cryosphere, including sea ice, glaciers, seasonal ice and snow, and permafrost. Glacial ice and permafrost are formed from fresh water frozen on land, and they raise the sea level when they melt or thaw, while sea ice does not.

⊙ **SCIENCE**

What Is an Ice Age?

At times, the temperature has been colder on Earth, causing glaciers to cover large areas of the continents and mountains. We commonly refer to these periods of time as **ice ages**. In the most recent of these glacial periods (from about 26,500–20,000 years ago), ice covered about one-third of Earth's land surface. This ice included sheets more than 1 mile (1.6 kilometers) thick that covered Canada, much of the midwestern United States, and even Manhattan. Far from stable, glaciers are rivers of ice that flow slowly downhill, grinding into rock and carving the landscape. Many features large and small, from the Great Lakes to the Alps to prairie potholes (pictured) in the midwestern United States, were carved or shaped by glaciers during ice ages.

FIGURE 7.24 Greenland's Ice Sheets Are Melting Measurements from satellites indicate that Greenland loses, on average, 260 billion tons of ice annually through icebergs breaking off and water flowing from the interior.

7.7 How Does the Ocean Support Life?

The oceans are the Earth's largest reservoir of water, teeming with diverse forms of life. But rather than thinking of oceans as a single place—like an aquarium with an array of colorful fish—it's better to understand them as a collection of many different ecosystems. These ecosystems are heavily influenced by the light available at different depths, ranging from shallow waters with relatively abundant light to the ocean floor where sunlight cannot penetrate (**FIGURE 7.25A**). Scientists have found some truly amazing animals deep below the surface where light does not penetrate. Creatures such as the fangtooth fish (found at a depth of 16,500 feet), the vampire squid (depth of 10,000 feet) (**FIGURE 7.25B**), and giant tube worms (depth of about 8,000 feet) give a small sense of the diversity of life in the sea.

Let's start with a place that is relatively easy for people to observe: the seashore, otherwise known as the intertidal zone (**FIGURE 7.25C**), where the ocean meets the land. During high tides these places are completely submerged, but during low tides the water retreats, leaving dry areas and small pools of water. If you visit the intertidal zone during low tide, you can get a sense of the wide range of species here, such as starfish, sea anemones, and barnacles clinging to rocks; crabs burrowing into holes in the sand or mud; and a range of seaweeds. The species that live in the intertidal habitat must be able to deal with both dry and wet periods.

If you swim a little farther out with a snorkel and a mask, you observe the shallow water that is just above the continental shelf, known as coastal waters, or the neritic zone. The seabeds here are part of the continental crust, so they are in relatively shallow water compared to the floor of the open ocean. Sunlight is plentiful here, and there is an abundance of tiny photosynthetic organisms such as phytoplankton. Phytoplankton is particularly important as it is the base of the food web for many other organisms, such as shrimp, jellyfish, snails, fish, and even other forms of plankton. This is also the zone where tropical coral reefs occur, complex ecosystems that support not only a variety of coral species but also algae, mollusks (such as giant clams), crustaceans (such as crabs), worms, sponges, and many species of fish.

Farther out still is the vast open ocean, which ecologists divide into a series of vertical zones. Close to the surface, algae, phytoplankton, and other photosynthetic organisms thrive, riding the currents and providing food for plant-eating animals big and small. These animals, in turn, feed a wide range of carnivores, including the largest in the world: whales. Fish, sharks, and whales can swim up and down through the ocean and travel between zones (**FIGURE 7.25D**). Some organisms in the dark deeper zones, such as lanternfish and squid,

produce their own light through chemical processes in their bodies, an ability known as *bioluminescence*, a trait that enables them to lure food toward them or to recognize individuals of the same species. On the ocean floor, also known as the benthic zone, many organisms live off of organic matter that sinks from dead and decaying creatures above. Other organisms, such as tube worms, thrive near hydrothermal vents where certain bacteria get energy from sulfur compounds emitted from the vents. The discovery of this community of life in the deepest parts of the ocean, in complete darkness, with vents heating water to temperatures in excess of 700°F (370°C), demonstrates the adaptability of life to extreme environments.

⬠ **TAKE-HOME MESSAGE** The ocean contains many types of ecosystems that can be understood as zones organized by depth and the penetration of sunlight, ranging from the intertidal zone on the seashore to the deep open ocean.

7.8 How Can Humans Impact the Ocean?

Although most people have little direct contact with the vast marine ecosystems that cover much of Earth's surface, human actions have profound impacts on the ocean. One way to think about these impacts is to consider the ways human food and energy production and waste disposal affect the ocean.

The most direct way human food production impacts ocean life is through fishing. Since 1950 the global annual catch for oceanic fisheries has increased from less than 20 million tons to nearly 100 million tons. Many fisheries that were once plentiful, like the Atlantic cod stocks off the coast of eastern Canada, have become so depleted that governments have closed areas to fishing to enable populations to recover. Commercial fishing also impacts marine mammals, seabirds, and other species caught up in fishing nets and lines. But food production on land also affects the ocean. The Baltic Sea, part of the Atlantic Ocean, hosts the world's largest dead zone—an area of water with low oxygen levels that is uninhabitable for most sea life. The main culprit for the lower oxygen levels is runoff from agricultural areas that contains nitrogen and phosphorus from fertilizers and livestock waste. The nitrogen- and phosphorus-rich runoff enters the ocean and starts the eutrophication process, which depletes the water of oxygen.

Energy production also affects oceans directly and indirectly through the use of fossil fuels. Since 2010, six large oil spills (those spilling 7 tons or more) a year, on average, have occurred in the ocean. Most of these are caused by ships that collide or run aground, such as the

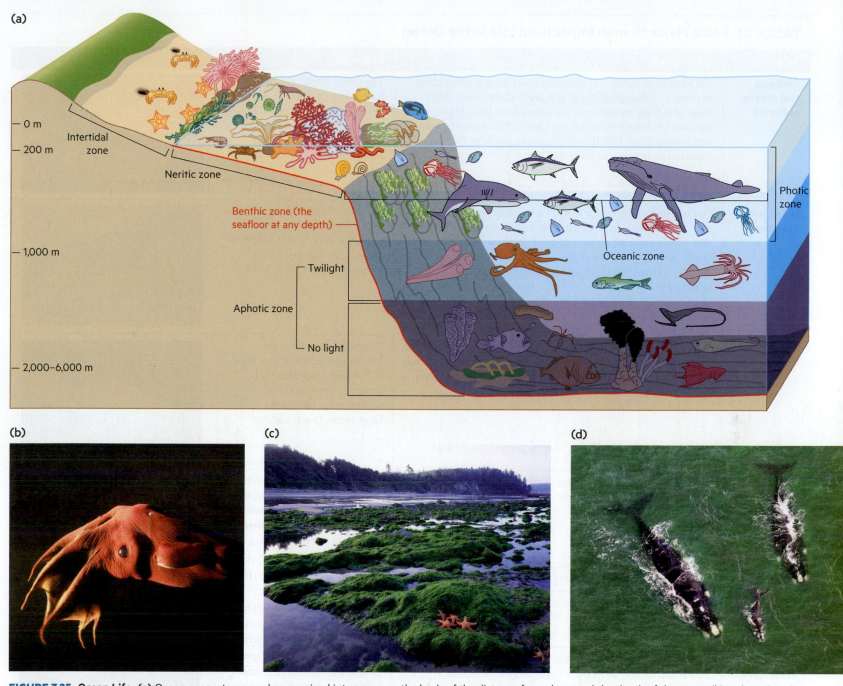

(a)

- 0 m
- 200 m
- 1,000 m
- 2,000–6,000 m

Intertidal zone

Neritic zone

Benthic zone (the seafloor at any depth)

Twilight

Aphotic zone

No light

Photic zone

Oceanic zone

FIGURE 7.25 Ocean Life **(a)** Ocean ecosystems can be organized into zones on the basis of the distance from shore and the depth of the water. **(b)** Aphotic benthic zones host creatures such as this vampire squid. **(c)** Organisms living in the intertidal zone must be able to live both in and out of water. **(d)** Large animals living in the open ocean, such as these southern right whales, can travel between vertical zones.

vessel that spilled more than 1,000 tons of oil near an environmentally sensitive coral garden off the coast of the island of Mauritius in the Indian Ocean in 2020. Far larger spills are caused by offshore oil production, such as the *Deepwater Horizon* disaster in the Gulf of Mexico in 2010. But fossil fuels also affect the ocean by contributing to climate change, which, as we will see in Chapter 11, is causing the water to warm and to become more acidic. Ocean acidification inhibits the formation of shells by many marine species that are at the base of the food chain. We'll also see in Chapter 11

that the oceans themselves play a major role in Earth's climate system, and changing ocean temperatures can affect major atmospheric circulation patterns that alter climate on land and at sea.

Finally, waste—particularly plastic—has become a major environmental problem in ocean ecosystems. Plastic waste not only from oceangoing vessels but also from bottles, bags, and other materials used and discarded on land have made their way into the ocean, where they have accumulated in large, swirling currents known as *gyres*. Large plastics can entangle sea life, but plastics also

TABLE 7.1 Some Major Human Impacts on Life in the Ocean

Type	Description	Example
Garbage and plastic pollution	Garbage of all kinds, but especially plastic, is accumulating in oceans. Although some of this trash comes from ships, most of it finds its way into the ocean from land. Plastic is a concern because it is long lived, can contain toxic chemicals, and breaks down into tiny microplastic particles that can be ingested by organisms large and small.	The Great Pacific Garbage Patch (the problem of plastics in the ocean is discussed in Chapter 15, where we learn about how we manage our waste)
Fossil fuel pollution	Our use of fossil fuels for energy also causes pollution at sea, most notably from oil spills and other commercial disasters such as sinking ships. But pollution from fossil fuel also comes in the form of mercury, a toxic heavy metal that accumulates in the food chain and is deposited in the ocean when air pollution from coal-fired power plants settles.	The 1989 *Exxon Valdez* oil tanker spill in Prince William Sound, Alaska, and the 2010 *Deepwater Horizon* oil rig blowout in the Gulf of Mexico
Agricultural runoff	When agricultural runoff containing synthetic fertilizers and animal waste makes its way to the ocean, it can cause "dead zones" due to eutrophication, which reduces the dissolved oxygen in the water.	The dead zone near the mouth of the Mississippi River in the Gulf of Mexico that grows as large as 8,700 square miles (about the size of New Jersey) each summer
Overfishing	The percentage of fishing stocks deemed fully exploited and overexploited has risen as fish consumption has grown. Adapted from FAO (2018).	Falling stocks of Atlantic cod and of other species such as dolphins, turtles, and seabirds that are incidentally taken up by fishing lines and nets
Ocean acidification	Much of the increase in carbon dioxide (CO_2) emissions from power plants and vehicles is absorbed by the ocean. When CO_2 dissolves in seawater it forms carbonic acid, which has increased the acidity of the ocean by more than 25% over the past 200 years. This change in the pH of ocean water threatens many marine organisms because it can weaken shells and skeletons made of calcium carbonate.	Shells of sea butterflies, a type of sea snail that lives in Antarctic waters, develop weak patches where acidic ocean water dissolves them.
Climate change	Ocean ecosystems are sensitive to even small changes in temperature, and global warming has caused ocean temperatures to rise 0.2°F on average over the past 100 years.	Increasing ocean temperature slows the reproduction of krill, a tiny crustacean at the base of the food chain that is consumed by penguins, seals, whales, and many other sea creatures

break down into smaller and smaller particles that are ingested by marine organisms.

In recognition of the magnitude of human impacts on ocean ecosystems, the United Nations has declared the years 2021 to 2030 as the Decade of Ocean Science for Sustainable Development. Efforts to address these impacts are wide ranging, from establishing marine protected areas that limit or reduce fishing and other human contact, to the European Union's effort to ban the single-use plastic items most commonly found in the ocean, to global efforts to adopt alternatives to fossil fuels.

TABLE 7.1 provides examples of some of the most significant impacts, including pollution, overfishing, and increased acidity. Most of these impacts will be discussed in more detail in later chapters.

⬠ **TAKE-HOME MESSAGE** Humans impact life in the ocean through various forms of pollution, by altering water chemistry and temperature, and by overharvesting seafood.

7.9 What Can I Do?

Life on Earth depends on water. When we damage this resource, it can take thousands of years to renew itself. As we said at the start of the chapter, water is a shared resource, so strategies to preserve it almost always involve working with others.

Learn Where Your Water Comes From

In the United States, the Safe Drinking Water Act requires that water providers issue a consumer confidence report to their customers that presents information about where their water comes from and the water quality. You can check out your own water supply at the EPA's website by looking for the "Consumer Confidence Report" (CCR) for your state or by searching the website of your water provider. What sources of water are you relying on? On the basis of what you learned in this chapter, what are some of the likely water issues of concern in your area?

Figure Out How Much Water You Use

The University of Arizona runs a water conservation program called Arizona Project WET in which K–12 and college students perform water audits in their school and college restrooms and then replace the existing faucet aerators with low-flow aerators. Students observe water-use patterns in their restrooms, measure how much water is used before and after the new aerators are installed, and then brainstorm methods to save water. On a more personal scale, you can use an online water footprint calculator (Appendix A) to determine which of your activities uses the most water and get personalized recommendations for reducing your water use.

Consume Less Water for Personal Use

Americans use a lot of water each day. The US Geological Survey estimates that the average American uses 80–100 gallons (300–380 liters) of water a day. This compares to 53 gallons (200 liters) for the average European and 3–5 gallons (11–19 liters) for the typical person in sub-Saharan Africa. As was demonstrated during the latest water shortage in California, limiting water use won't solve a shortage, but it is an important strategy to make the best use of what is available. Personal water-conservation efforts include taking small actions that can lead to cumulative savings over time, such as installing low-flow showerheads and aerators on faucets, taking shorter showers, turning off the water when you brush your teeth, and only running the dishwasher when it is full. And as we learned that water is embedded in the production of food and material goods, we know that reducing waste and consuming less stuff overall also conserves water.

Use Less Water for Landscaping

State experts in California found that during their recent drought, about 60% of residential water use was going toward watering landscaping. If you're not a homeowner, you might not face decisions about when to water the lawn or what plants you have in the yard. But college campuses make a significant impact on local water resources through the way they manage their grounds. Let your  college administration know that it should take steps to use less water in landscaping. For example, you might ask if your school is taking steps to reduce landscape irrigation, or to irrigate when water is less likely to evaporate, or to use plants suitable for your local climate.

Also, how does your school monitor problems that cause water loss on campus? The University of California, Santa Cruz, has a "Fix-it System" where individuals can make an online request for fixes to water losses they have observed and expect a high-priority response. Larger initiatives on campus include installing smart irrigation systems, using drought-tolerant plantings, and installing pervious pavement, especially for new construction. Some campuses such as the University of California, Irvine, and Arizona State University have also developed ways to recycle drinking water on campus for other uses such as irrigation and cooling systems.

Learn More about How Others Are Facing Water-Conservation Challenges and Sustaining Reforms

From 1996 to 2011, the city of Melbourne, Australia, suffered through a severe "Millennium Drought" that threatened the city's water supply. Melbourne made it through the crisis using both conservation measures and innovations, such as much greater use of recycled stormwater. Another innovation was electronic billboards that flashed reservoir levels. But just a few years after the drought ended, a report noted that Melbourne was already seeing some resistance to maintaining the practices it had put in place. As citizens felt more confident in their water supply, they started to adopt old habits again and needed to be reminded of the importance of continuing to conserve. One big lesson from the Millennium Drought was that it lasted far longer than anyone expected a drought could last. Communities need to plan for new scenarios and build resilient water-management plans to match them.

Chapter 7 **Review**

SUMMARY

- Earth's hydrosphere is a closed system that recycles water molecules over and over again through the water cycle and includes major planetary reservoirs—the atmosphere, ocean, water on land, and water beneath Earth's surface.

- Most of Earth's water is held in the ocean as salt water. The planet's largest reservoir of fresh water is frozen in ice and snow, and most of the liquid fresh water on Earth exists as surface water or groundwater.

- There is a variety of freshwater ecosystems—groundwater, lakes and ponds, rivers and streams, wetlands, and estuaries—each with unique characteristics that support life.

- The slow recharge rate of groundwater systems often lags behind the rate of human withdrawals of groundwater.

- Surface water withdrawals and diversions such as dams, reservoirs, canals, and pipelines cause ecological harm by diminishing or obstructing stream flows.

- Human activities that add chemicals, disease-causing organisms, sewage, and sediment into the water result in pollution that diminishes water quality.

- Addressing water pollution involves laws and regulations that set and enforce water-quality and water-treatment standards as well as development strategies that promote and preserve water quality.

- To achieve water security, everyone needs access to safe and affordable water. However, when rates of outflow exceed rates of inflow, the supply of water on hand in an area shrinks, threatening this security.

- Glaciers, ice sheets, ice shelves, and permafrost are formed from fresh water frozen on land, and they raise sea level when they melt or thaw.

- Marine ecosystems are influenced by the depth of penetration of sunlight.

- Humans impact life in the ocean through various forms of pollution, by altering water chemistry and temperature, and by overharvesting seafood.

KEY TERMS

albedo
aquifer
Clean Water Act
cryosphere
desalination
drought
embedded water
estuary

eutrophication
evapotranspiration
glacier
groundwater-dependent
  ecosystem
hydrosphere
ice age
lentic ecosystem

lotic ecosystem
nonpoint source
permafrost
point source
recharge
reservoir
residence time
Safe Drinking Water Act

sea ice
sediment
subsidence
surface water
water mining
water recycling
water security
wetland

REVIEW QUESTIONS

The letters following each Review Question refer to the Chapter Objectives.

1. Identify at least three of Earth's major "reservoirs" of water. **(A, B)**

2. Where is most water on Earth held? **(A, B)**

3. In what form is most of Earth's fresh water held? **(A, E)**

4. Identify at least three sources of liquid fresh water on Earth. **(C)**

5. Name at least two negative impacts of excessive groundwater withdrawals. **(D)**

6. Describe at least two structures that humans build to divert and distribute surface water. **(D)**

7. Explain at least two ways that human disruptions to freshwater systems impact other species. **(D)**

8. What's the difference between point sources and nonpoint sources of pollution? Give an example of each. **(D)**

9. List at least three sources of water pollution, and give an example of each. **(D)**

10. Explain how consumptive uses of water can cause water scarcity in some areas. **(D)**

11. Describe the characteristics of a drought. **(D)**

12. Among human uses of water, what use accounts for the biggest draw on freshwater resources? **(D)**

13. Describe two strategies and/or technologies for dealing with water scarcity. **(D)**

14. Identify the two major US laws that govern water policy. **(D)**

15. What's the difference between glaciers and sea ice, and how does this difference affect sea-level rise? **(E)**

16. Describe at least three human impacts on life in the ocean. **(F)**

FOR FURTHER THOUGHT

The letters following each item refer to the Chapter Objectives.

17. Imagine you are a state official charged with developing a water-conservation plan. You are given three categories of water use that can be reduced on your orders: agriculture, residential, and industrial. You are also given three ways in which the water supply can be augmented: increasing the rate of aquifer withdrawals, piping in more water from outside your state, and building desalination plants along the coast. What combination of these measures would you take and why? **(D)**

18. You are a public health worker, and you just learned that your local water source is suffering from biological pollution. What types of places would you look to find the likely source of this pollution? **(D)**

19. Consider the general area where you live. Where does the water go when it rains? What kind of low-impact development improvements could be made to the area to decrease stormwater runoff, increase groundwater recharge, and prevent flooding? **(D)**

20. Identify and explain what you think are the three most significant changes you could make in your personal life to conserve water. **(D)**

Make Your Case?

The following exercises use real-world data and news sources. Check your understanding of the material, and then practice crafting well-supported responses.

Use the News

The following article is excerpted from *The Hill*, a news website that focuses on political issues. In it the author discusses the role of the Environmental Protection Agency (EPA) in the contamination of a river by a gold mine in Colorado. Use this article to answer the questions that follow. The first four questions test your understanding of the article. Questions 5 and 6 are short answer, requiring you to apply what you have learned in this chapter and cite information in the article. Answers to Questions 1–4 are provided at the back of the book. Question 7 asks you to make your case and defend it.

"Animas River Spill: Root Causes and Continuing Threats," Joel A. Mintz, *The Hill*, September 2, 2015

The recent spill of gold-mine wastewater into the Animas River near Silverton, Colorado, has given EPA's die-hard critics ammunition to last the summer. Without question, EPA bears responsibility for the spill, since its contractor was at fault. But . . . it's important to put the incident into perspective.

First, although the mistakes by EPA's contractor were the immediate cause of the discharge of some 3 million gallons (11.4 million liters) of wastewater, the actual source of the pollution was an unremediated mineral mine, one of many such abandoned mines across Colorado and the American West. In fact, some 20,000 abandoned mine sites in southwestern Colorado are the sad legacy of an archaic—yet still unreformed—1874 federal mining statute that encourages excessive mining on sensitive public lands by permitting private mining companies to reap vast profits through mineral extraction while paying absurdly low royalties to the public treasury. In all, there are roughly 500,000 abandoned mine sites across the nation. They constitute a ticking time bomb that could cause additional, massive and spontaneous toxic discharges into rivers at any time.

Second, it's worth recalling that long before the recent spill, the Animas River was contaminated with mining wastes and other pollutants. Pre-spill testing revealed excessive levels of zinc, copper, cadmium and other metals, along with high concentrations of human fecal coliform bacteria. Moreover, still more damaging pollution incidents have taken place in the same vicinity in past years, including a disastrous 1975 discharge to the Animas of metal-loaded mine tailings and a discharge of roughly 500 million gallons (1.9 billion liters) of wastewater in 1978. That's no excuse for the mistakes made in this incident. Nevertheless, it's an example of why

EPA was in the vicinity to begin with: to clean up a huge mess of someone else's creation.

That task would have been made much easier if EPA had been able to designate the site as a Superfund site, because it would have allowed the agency to fund a wastewater treatment facility needed to clean up the mine discharge contamination. However, despite prodding from EPA and state officials, local business interests repeatedly obstructed the Superfund designation of the area. Their short-sighted resistance surely helped set the stage for the calamitous events of early August.

The Animas River accident happened when an EPA contractor drilled into the side of the now-closed Gold King Mine, as part of an effort to remedy a steady, 176 gallon per minute leak of toxic materials from the mine into the river. The torrent of wastewater that escaped turned the waters of the Animas to a mustardy yellow before coursing through Durango, Colorado, and traveling far downstream into other Western states.... Along with other EPA officials, the regional administrator of the agency's Denver office, Shaun McGrath, and its administrator, Gina McCarthy, both visited local

areas affected by the spill.... Administrator McCarthy announced an immediate suspension of all EPA mine reclamation work pending a full internal review of relevant EPA procedures and practices, as well as a fully independent outside review of the Animas River spill. McCarthy and her staff have also given regular public updates on the level of contaminants in the river revealed by post-spill sampling. Those tests initially indicated that the spilled wastewater plume, at its peak, contained elevated levels of certain heavy metals, and that the Animas River's acidity was at approximately the same pH level as black coffee. More recent tests indicate that, just below the point of the spill, water quality in the Animas has already returned to pre-spill levels.

The downstream impact of the spill is not yet clear. It does not appear to have killed many fish, or caused any direct human physical injury, and the unsightly discoloration of the affected water has already significantly faded. However, the incident is still taking a toll on businesses that rely on summer outdoor tourism, and the Navajo nation is understandably upset at the harm the spill could do to its farmers and ranchers.

1. What was the source of contaminants spilling into the Animas River?
   a. A train transporting mining waste away from the site derailed in the mountains and spilled its contents.
   b. A barge transporting chemicals used for mining activities leaked its contents into the river.
   c. Drilling activities from exploratory mining activities spilled waste into the areas downstream.
   d. An accident caused wastewater to spill from an abandoned mine site.

2. Which of the following groups does the article say was the immediate cause of the spill?
   a. contractors working for the Gold King mining company
   b. EPA officials
   c. contractors working for the EPA
   d. scientists working for the US Geological Survey

3. The article says the "vicinity" of the spill was a "huge mess" even before it happened. Which of the following two statements are given as evidence of this "huge mess"?
   a. In 1978, 500 million gallons of wastewater had spilled into the river.
   b. The mine had experienced a large oil spill in previous years when mining equipment became damaged.

   c. The river was contaminated with acid rain caused by industrialization in Mexico.
   d. Before the spill happened, the river was tested and had excess levels of zinc, copper, cadmium, and other metals.

4. Which of the following describes a known impact of this spill?
   a. closing the Gold King Mine
   b. extinction of several species of fish that live only in the river
   c. human fatalities of mineworkers
   d. economic losses suffered by the tourism industry

5. The article discusses downstream impacts of the spill. Given what you have read in this chapter, what are some characteristics of a typical river area that could be particularly impacted by a spill?

6. In this article, the author goes on to express a specific opinion about the EPA's role. Given what you have read, do you think the author believes any action should be taken against the EPA? In your answer, provide details that support your conclusion.

7. **Make Your Case** Imagine a lawsuit was filed by the state of Colorado against the EPA for the damage caused by this spill. If you were an attorney, who would you want to represent? Choose the party involved that you would like to represent, and give reasons why.

Use the Data

The figure shown here depicts water moving through the water cycle in the state of Illinois and is based on averaging data collected between 1971 and 2000. The values for each part of the figure are in billions of gallons of water per day.

Study the figure, and use it to answer the questions that follow. The first three questions test your understanding of the figure. Question 4 is short answer, requiring you to apply what you have learned in this chapter and cite data in the figure. Answers to Questions 1–3 are provided at the back of the book. Question 5 asks you to make your case and defend it.

1. Approximately what percentage of the atmospheric moisture passing overhead in Illinois falls as precipitation?

   a. 5%
   b. 10%
   c. 50%
   d. 104%

2. The figure shows how much water per day is coming from sources outside of the state. Match each source with the amount of water per day (in billions of gallons) it supplies.

   Wisconsin 1.5

   Lake Michigan 1.7

   Indiana 3.3

   Upper Mississippi River 34

   Upper Ohio River 95

3. The bending arrows labeled "runoff" and "stream flow" are meant to show that these pathways feed into surface rivers and streams relatively quickly. In contrast, the arrows labeled "sand aquifers," "bedrock aquifers," and "deep bedrock" indicate water that recharges aquifers where water has a much longer residence time. With this in mind, approximately how much more water per day runs off the land as stream flow than seeps into an aquifer?

   a. just under 3 times more
   b. just under 9 times more
   c. just over 5 times more
   d. just over 20 times more

4. The estimates in this figure are based on long-term averages, but the amount of actual usage is different every day. Thinking about what you've learned in the chapter and this text, what are two reasons that might cause daily evapotranspiration to be higher

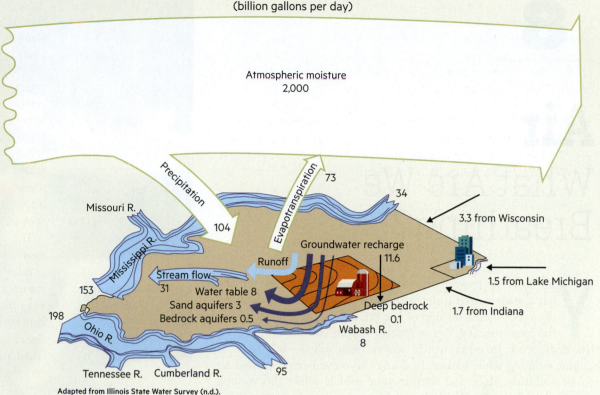

Illinois Water Budget, 1971–2000
(billion gallons per day)

Atmospheric moisture
2,000

Precipitation 104

Evapotranspiration 73

34

Missouri R.

Mississippi R.

153

198

Ohio R.

Tennessee R. Cumberland R. 95

Runoff

Stream flow
31

Water table 8
Sand aquifers 3
Bedrock aquifers 0.5

Groundwater recharge
11.6

Deep bedrock
0.1

Wabash R.
8

3.3 from Wisconsin

1.5 from Lake Michigan

1.7 from Indiana

Adapted from Illinois State Water Survey (n.d.).

than the average shown? In what months would you expect evapotranspiration to be highest?

5. **Make Your Case** As the figure suggests, agriculture is important to Illinois. In fact, as of 2017, the US Department of Agriculture calculates that approximately 75% of the state is farmland. However, most of the population in Illinois is centered in the north, upstream of farmland. Given this, how important is it for farmers to manage pollution and runoff that comes from farms? Should the map be capturing more relationships with and impacts on other states? Given what you have learned and the information in this figure, explain the scope of concern that you think farmers should have for their impacts on water.

LEARN MORE

- National Ocean Service: https://oceanservice.noaa.gov/facts/ocean_weather.html
- NASA's Garbage Patch Visualization Experiment: https://svs.gsfc.nasa.gov/4174
- US Geological Survey's Repeat Photography Project: www.usgs.gov/centers/norock/science/repeat-photography-project?qt-science_center_objects=0#qt-science_center_objects
- Video: NASA, Tracking Freshwater Movements: digital.wwnorton.com/environsci
- Video: NOAA, 25 Years after the *Valdez* Oil Spill: digital.wwnorton.com/environsci

- State Water Project of California: https://water.ca.gov/Programs/State-Water-Project
- Water rights and the western water shortage: https://projects.propublica.org/killing-the-colorado/story/wasting-water-out-west-use-it-or-lose-it
- The Flint water crisis: www.npr.org/sections/thetwo-way/2016/04/20/465545378/lead-laced-water-in-flint-a-step-by-step-look-at-the-makings-of-a-crisis

8

Air
What Are We Breathing?

You drove up to your friend's house 10 minutes ago, and she has already texted to let you know she's on her way out the door; but there's no sign of her yet, and all the while your car has been running. You wonder if maybe you should just turn the engine off. In other words, you've been idling—leaving your engine running while parked. In doing so, you've been contributing to air pollution inside and outside your vehicle while also wasting gas. If you idle more than 10 seconds, you begin to consume more gas than it would take to restart the engine. Depending on the vehicle you're driving, you may have burned as much as one-tenth of a gallon of gas in those 10 minutes. Moreover, vehicle engines operate less efficiently when they're idling, which means they pollute more than when you're driving. So while you've been waiting for your friend, you've also been adding harmful substances to the air around you, such as carbon monoxide (CO), particulate matter (PM), and nitrogen oxides (NO_x).

On a large scale, air pollution from idling adds up. A study of idling in New York City found that each year, curbside idling in the city generates as much air pollution as 9 million large trucks traveling the 30 miles (50 kilometers) between the Bronx and Staten Island. This is why many cities and schools have campaigns and policies to encourage drivers to turn off their engines while parked. For example, drivers that idle longer than 1 minute near schools in New York City are subject to fines. In 2008, The Ohio State University made it a policy that university vehicles and equipment should limit idling time as much as possible and also reduce warm-up time (Ohio State won an award for excellence in this area in 2017). Commercial vehicles, such as long-distance trucks and police cruisers, are now commonly equipped with devices called auxiliary power units that can run heating, air-conditioning, and other functions while the vehicle is parked and not burn gas. And many new vehicles have "stop-start" systems that cut off the engine automatically when it starts to idle, then starting it back up when the driver is ready to move again.

Most of us understand that air pollution is a problem. But to better understand how we contribute to air pollution and how it affects us, we need to learn more about the **atmosphere**: a dynamic envelope of gases that extends about 62 miles (100 kilometers) up from Earth's surface and clings to our planet because of gravitational pull.

Does this look like your commute? Exhaust from gas and diesel cars is a major source of air pollution in the United States.

Chapter Objectives

By the end of this chapter, you should be able to . . .

A. name the most prominent gases found in the atmosphere.

B. describe how the four layers of Earth's atmosphere are distinguished by varying characteristics.

C. explain what an air mass is and how it rises, falls, and circulates with changes in temperature, moisture, and wind conditions.

D. define and provide examples of primary and secondary pollutants.

E. describe how secondary pollutants in the atmosphere react to cause smog, ozone layer depletion, and acid deposition (acid rain).

F. compare and contrast different policies aimed at improving air quality.

Air, I should explain, becomes wind when it is agitated.

—Lucretius, 50 BCE

atmosphere a dynamic envelope of gases extending up from Earth's surface that clings to our planet because of gravitational pull.

This is the air we breathe, and although we often take it for granted when the air quality is good, the workings of the atmosphere seize our attention when air quality declines, either when haze obscures our view or foul air stings our nostrils. In extreme cases, a settling mass of polluted air can be deadly. This was the case on a fall day in Donora, Pennsylvania, in 1948 when a front of warm air moved over the town and trapped a cooler layer of air thick with pollution from the town's steel mills near the surface. Twenty people died, and one-third of the town's population was sickened before the weather pattern shifted 5 days later. Although we've made progress since this time in regard to pollution-control technologies and policies, air-quality challenges persist in cities around the world (**FIGURE 8.1**). The World Health Organization has found that air pollution is responsible for more than 4 million premature deaths worldwide each year.

Today, most scientists see that humans are changing the fundamental mix of what is in the air we breathe, no matter where we are in the world. As our air and atmosphere change, so does our quality of life, through increased pollution, increased exposure to ultraviolet radiation, and more extreme weather, among other impacts. To learn why this is happening, we will first need to understand what "air" actually is, how it forms our atmosphere, and the ways in which the composition and circulation of gases in the atmosphere shape weather. Next, we will explore the causes and consequences of human impacts on air quality. Understanding the chemistry and circulation of the atmosphere gives us a better understanding of what we are breathing in and out of our lungs every day. And finally, we will look at how we are trying to improve and restore air quality.

(a)

(b)

(c)

FIGURE 8.1 Polluted Air Air pollution is a major problem in many big cities such as **(a)** Seoul, South Korea and **(b)** Los Angeles, California and it has even been deadly, as it was in 1948 in **(c)** Donora, Pennsylvania.

8.1 What Makes Up the Air We Breathe and Our Atmosphere?

The air around us is made up of a mixture of many gases, including those produced by our idling cars, air-conditioning and heating systems, and other machines processing the air around us. On the whole, though, Earth's atmosphere is mostly composed of just two gases: 78% nitrogen (N_2) and 20% oxygen (O_2), with much smaller amounts of argon, carbon dioxide (CO_2), water vapor, and other compounds (**FIGURE 8.2**).

By definition, gases are constantly moving. To get a sense of this, imagine that you could shrink to become a microscopic version of yourself and slip inside an inflated balloon. You would find a busy place, with atoms and molecules bouncing like rubber balls, wildly and continually colliding within the space contained inside. This is what the atoms and molecules that form gases do in filling the available space or volume they occupy, be it a balloon or other enclosed area such as our lungs.

Pressure

The firmness you feel in a fully inflated balloon is not provided by a solid but by the many, many impact forces of those vigorously bouncing gas molecules trapped inside. This causes what we know as a gas sample's **pressure**, or the force exerted on or acting against something (**FIGURE 8.3**). If you put more air (more bouncing gas molecules) into the balloon, the pressure inside the balloon increases as more molecules press against the balloon walls. Or if you squeeze the balloon in both hands, decreasing its volume, the gas molecules bump into the walls more frequently, also increasing the pressure inside. All gases, including air, move from areas of high pressure to areas of low pressure. This is how we breathe: after we exhale and empty our lungs, relatively high-pressure air outside our bodies automatically flows into our low-pressure lungs. We will see later in the chapter that this fundamental behavior of gases is responsible for wind and weather.

Now, imagine a cylinder of air extending from Earth's surface up to the top of the atmosphere, roughly 60 miles (about 100 kilometers) above our heads (to help you picture this concept, consider that this is the height of a theoretical 15,840-story office building, or about 150 Empire State Buildings stacked on top of each other). The weight of the air at the top of the cylinder is pressing down on the lower layers of air, forcing the air molecules in the lower layers closer together. This causes the air

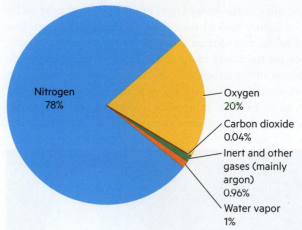

FIGURE 8.2 **What Is Earth's Air Made Of?** Earth's atmosphere is composed primarily of nitrogen and oxygen. Other gases including water vapor and carbon dioxide make up a smaller share of the air.

Nitrogen 78%
Oxygen 20%
Carbon dioxide 0.04%
Inert and other gases (mainly argon) 0.96%
Water vapor 1%

pressure the force exerted on or acting against something (Chapter 3).

(a)

(b)

(c)

FIGURE 8.3 **Air Pressure** **(a)** Blow into a balloon and you are adding to the air molecules inside it. Especially as the balloon expands near its breaking point, you can feel the increase in pressure as the molecules fill up the space of the balloon. **(b)** Squeezing a balloon decreases its volume. There is less room for the air molecules inside, and the pressure increases. **(c)** Deflate a balloon (such as this hot air balloon), and there are so few molecules of air in it that it starts to sag inside itself. The pressure is low, and the balloon collapses.

troposphere the lowest area of the atmosphere from 5 to 10 miles (8 to 16 kilometers) above the surface where Earth's weather occurs.

molecules in the lower layers of air in the cylinder to be more tightly packed than those in the upper layers. Just like when you squeeze a balloon, weight exerts a greater force on the lower layers in the cylinder of air, so the air pressure of the atmosphere increases closer to the surface.

A common way we think about this is to describe the air as getting "thinner" at increasing altitudes as the gas molecules are more dispersed and the air is therefore

(a)

FIGURE 8.4 **Why Is Air "Thinner" at Higher Elevations?** **(a)** On October 14, 2012, Felix Baumgartner (pictured) jumped from a helium balloon approximately 24 miles (39 kilometers) above Earth in the stratosphere. Two years later, Alan Eustace jumped from even higher, more than 25 miles (40 kilometers) above Earth, setting a new record. Both wore special pressure suits with an oxygen supply (much like what astronauts wear) to compensate for the low pressure and "thin air" environment at such high altitudes. **(b)** The weight of the air at the top of the atmosphere presses down on the lower layers. This forces air molecules at lower levels to be more tightly packed together. So both air density and air pressure of the atmosphere increase closer to Earth's surface.

Adapted from (b) Marshak and Rauber (2017).

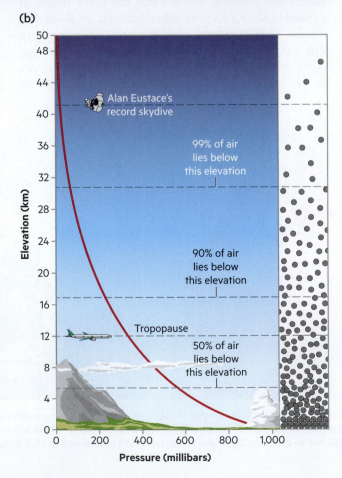

(b)

less dense (**FIGURE 8.4**). Even at altitudes relatively close to Earth's surface we can sense this. Mountaineers refer to altitudes above 8 kilometers (26,000 feet), such as the highest reaches of Mt. Everest and other Himalayan peaks, as the death zone because the human body cannot access enough oxygen in these thin-air environments to sustain vital functions for extended periods of time. Most commercial airplanes fly higher than this, reaching cruising altitudes between 9 and 12 kilometers (30,000 and 40,000 feet), to take advantage of the reduced resistance of thinner air. This is why airplanes are essentially sealed containers: air is pressurized within the cabin to create an artificial atmosphere similar to one at a much lower altitude (generally about 2.5 kilometers, or roughly 8,000 feet).

How Is Earth's Atmosphere Structured?

To better understand the different parts of the atmosphere, we divide it into four layers with different pressures, temperatures, and compositions: the troposphere, stratosphere, mesosphere, and thermosphere (**FIGURE 8.5**). Most of the atmosphere's mass (more than 75%) is concentrated in a 5- to 10-mile-thick (8- to 16-kilometer) layer closest to Earth's surface known as the **troposphere**, the lowest area of the atmosphere where Earth's weather occurs. The air here is very dynamic, rising, sinking, churning, and mixing as it warms and cools. It also holds most of the water vapor in the atmosphere. The temperatures within this layer range from those we are familiar with here on the ground to much colder temperatures. There is a constant temperature near −76°F (−60°C) in an area known as the tropopause at the top of the troposphere. Most water vapor condenses and forms clouds when it reaches the tropopause, which helps Earth retain its water.

The next layer, extending from the top of the troposphere to 30 miles (50 kilometers) above Earth's surface, is the **stratosphere**. Compared to the troposphere, the stratosphere has very little water vapor and is relatively calm. Materials such as pollution or ash from erupting volcanoes that make their way into the stratosphere can remain suspended there for years. The stratosphere also has an abundance of a molecule important for all life on Earth: ozone (O_3). Ozone forms when three oxygen atoms bond together and is a less stable form of the element oxygen than the O_2 that we breathe. The stratosphere holds a lot of ozone, 1,000 times more than the air does at ground level. As we will see in more detail later, the stratospheric layer of ozone absorbs ultraviolet solar radiation, shielding life below from damage and producing a warming effect on the stratosphere.

Next is the colder **mesosphere**, which lies above the stratosphere. Earth's coldest temperatures, falling as low as −148°F (−100°C), occur here, just above the maximum altitude of aircraft and just below the orbits of satellites. This is also the layer in which most meteors that reach Earth's atmosphere burn up due to the intense friction caused by their high-velocity passage through air molecules.

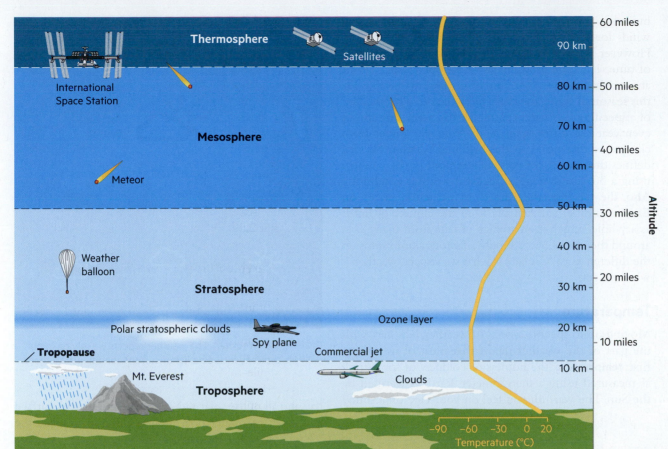

FIGURE 8.5 Layers of Earth's Atmosphere Earth's atmosphere varies in temperature, air pressure, and gas composition at various altitudes.

Further up still is the **thermosphere**, the top layer of the atmosphere that is warmer because of the influx of solar and cosmic radiation. Air molecules are much more diffuse (scattered) at this level: 99.9997% of the atmosphere's mass lies below the 62-mile (100-kilometer) height of the upper thermosphere. Here, air molecules are so diffuse that they offer little frictional resistance to spacecraft. The International Space Station and most satellites orbit within the thermosphere with little resistance from air.

⬡ **TAKE-HOME MESSAGE** The air in our atmosphere is a combination of gases that envelop Earth. Both the density and pressure of air increase closer to Earth's surface because the weight of the air squeezes air molecules near the surface. The four main layers of the atmosphere support life in different ways, including producing weather, containing the air we breathe, and filtering harmful ultraviolet radiation.

8.2 What Is Weather and How Does It Change?

Almost everyone loves a clear summer day, but we tend to hear more about extreme weather. Extreme weather events such as hurricanes, tornadoes, and droughts are some of the most damaging and disruptive ways in which we experience the natural world. Hurricanes and floods can force evacuations of major cities, cause fatalities, and leave billions of dollars in property damage in their wake. They also cause numerous environmental problems, as when floodwaters spread pollution from overflowing sewers and industrial sites. Similarly, prolonged periods of drought destroy crops, livestock, and natural habitats. Researchers estimate that the 2012–2016 drought in California killed more than 100 million trees. Increasingly, certain extreme weather events are being tied to human activities. Changes in the atmosphere have been linked to increased storm intensity, and a September 2017 report of the World Economic Forum noted that countries producing the most CO_2 emissions "had experienced the highest number of hydrological, meteorological and climatological disasters in recent years." We will learn more about how climate change is affecting weather in Chapter 11, but for now, let's answer a basic question: How does weather happen?

To start, we need to understand several core concepts, beginning with some commonly used words. Scientifically speaking, **weather** refers to short-term variations in temperature, moisture, and wind conditions in a specific place. It is important to distinguish weather from **climate**, the long-term average of weather conditions for a given region. For example, a weather report for your hometown typically includes current and expected temperature,

stratosphere the layer of the atmosphere above the troposphere. It holds a layer of ozone that protects and warms Earth.

mesosphere the layer of atmosphere above the stratosphere. It has Earth's coldest temperatures and is where most meteors burn up.

thermosphere the top layer of the atmosphere. It is very thin and where most man-made satellites orbit.

weather short-term variations in conditions such as temperature, moisture, and wind in a specific place.

climate the long-term average of weather conditions for a given region.

temperature in common terms, a measurement of the hotness or coldness of something; formally, a measure of how vigorously the atoms in a material are moving and colliding with each other (Chapter 3).

humidity and precipitation, air pressure, cloudiness, and winds for today and the rest of the week (**FIGURE 8.6**). However, if you have lived in a place for a longer period of time, you have probably also developed expectations about what the weather conditions tend to be like for this season in this place: the typical climate. The weather of a specific place over a particular day, week, month, or even year can differ significantly from the longer-term climate of that place, so the National Weather Service defines the climate for particular regions of the country using a 30-year average of regional weather conditions. Also, the weather and climate of particular places on Earth can differ substantially from the climate of Earth as a whole, which is an average of data from many sites around the planet. More simply, a famous axiom sums up the difference as, "Climate is what we expect, weather is what we get."

Temperature

Moving on to specific aspects of weather, let's start with the part of the weather report that most of us look at first: **temperature**, the hotness or coldness of something as measured using a thermometer. Earth is warmed by the Sun. The warmth you feel on your face on a sunny

FIGURE 8.6 **Weather** The weather reports we rely on in our daily life typically include a list of current conditions like these.

day, and all of the solar radiation reaching Earth's surface at a given moment, took about 8 minutes to make the 93-million-mile (150-million-kilometer) trip from the Sun to us. But not all places on Earth receive the same amount of the Sun's energy at a given time. For one thing, as our planet rotates, half of it is facing away from the Sun and experiencing lower temperatures during its nighttime phase.

Another key temperature variation across Earth's surface depends on the amount of solar radiation received at various latitudes. As **FIGURE 8.7** shows, the angle of the Sun's rays near the equator is more direct and closer to perpendicular than the angle at higher latitudes and the poles. This straighter path to the Sun is why the equator receives more of the Sun's energy per unit of surface area than do other latitudes. As you move toward the higher latitudes of the poles, the Sun's rays hit Earth at a lower, steeper angle that spreads the energy over a larger area of both the atmosphere and the planet's surface.

Local variations in heat gain and loss are also affected by bodies of water and land features. Temperatures in coastal areas fluctuate less than in areas farther inland because of the heat storage capacity of the ocean. A large body of water heats up and cools down slowly like a large pot of water on your stove does, and this affects the land around it, causing coastal areas to heat up and cool down more gradually than does land far from the sea. And one finds that at a given latitude, the higher that land is above sea level, the lower the air temperature. This is because air pressure decreases as altitude increases (as we learned in the last section), so the gas molecules are less tightly packed and are expanding into a bigger space. When air expands like this, it cools. As a rule of thumb,

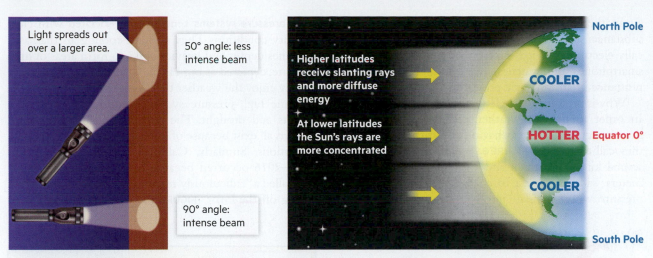

Light spreads out over a larger area.

50° angle: less intense beam

90° angle: intense beam

Higher latitudes receive slanting rays and more diffuse energy

At lower latitudes the Sun's rays are more concentrated

North Pole

COOLER

HOTTER   Equator 0°

COOLER

South Pole

FIGURE 8.7 Unequal Warming on Earth's Surface As the flashlight illustrates, the angle at which light reaches a surface affects the surface area that is warmed by it. The Sun's energy per unit of surface area is highest at the equator where it hits the surface most directly. This means that the energy is more concentrated, and the warming effect is greater here. Because of the curvature of Earth's surface, the Sun's rays hit at more of an angle at higher latitudes, which spreads the energy over a larger surface area, leaving it colder than the equator.

for every 1,000 feet (300 meters) in elevation gain, the air temperature decreases by about 3.6°F (2°C).

Water Vapor

Precipitation, humidity, and cloud cover all reflect the amount of water vapor in the air. As we saw in Chapter 7, at any given time, the atmosphere as a whole contains about 1% water vapor. Water vapor is constantly moving into and out of the atmosphere through evaporation and condensation. Let's first look at how **humidity**, the amount of water in a given volume of air, affects evaporation. When the air feels stuffy, thick, or moist, we say it is humid. For people susceptible to heatstroke, high humidity is dangerous when coupled with high temperatures during heat waves. At the other end, extended periods of low humidity can cause droughts and are often a factor leading to wildfires. We'll see in Chapter 11 that climate change is increasing the frequency of droughts and heat waves.

Strictly speaking, meteorologists measure the **relative humidity** of air in order to understand weather conditions. But relative to what? Meteorologists compare it to the maximum amount of water that the air can hold at a given temperature (its capacity). When air cannot hold any more water vapor, we say it is saturated. If the relative humidity is 100%, the air is fully saturated with water vapor. The drier the air, the more capacity it has for additional water molecules. This is why we often describe a place with dry heat, such as Los Angeles, as feeling more comfortable than a humid place with the same air temperature. Evaporation from our bodies occurs more readily in dry heat with lower relative humidity than it does in humid heat of the same temperature, leaving us feeling cooler and more comfortable (**FIGURE 8.8**).

Evaporated water moves into the atmosphere as water vapor and can form clouds. Clouds are bundles of condensed water vapor floating in the atmosphere, and they consist of tiny water droplets or ice crystals that develop by the freezing of the water droplets. The water droplets and ice crystals that form clouds start out tiny. To give you a better sense of the size of these droplets, it takes more than a million of these water or ice particles to compose a single raindrop or snowflake. These raindrops or snowflakes fall to Earth when they become too heavy to remain suspended in the air.

Barometric Pressure

On television, meteorologists typically give the forecast while standing in front of maps with big Hs and Ls to indicate high-pressure and low-pressure systems, as well as large storms they may describe as "extreme lows." They might also talk about whether the **barometric pressure**—a measure of the pressure exerted by Earth's atmosphere at any given point—is rising (indicating increasing air pressure) or falling (indicating decreasing air pressure). This is measured using a barometer, historically an instrument that detects the weight of the

humidity the amount of water in a given volume of air.

relative humidity the amount of water in the air as a percentage of the maximum amount of water that the air can hold at a given temperature.

barometric pressure a measure of the pressure exerted by Earth's atmosphere at any given point.

FIGURE 8.8 What Humidity Feels Like Relative humidity at the 2020 Tokyo Olympics reached 84% at times. In a six-day period more than 8,000 people in the city were hospitalized because of the stifling weather.

air mass a large volume of air typically several kilometers thick and a thousand or more kilometers wide that has relatively uniform temperature and humidity.

high-pressure system a system of air formed by cooling air that becomes denser and heavier and then sinks to form an area of high pressure. It is usually associated with clear, dry conditions.

air in the atmosphere as it presses against a confined substance, such as mercury. Today's barometers are typically electronic devices (like the ones included in many smartphones and watches) that measure the electrical resistance of atmospheric pressure.

Why is pressure a key element of the weather report? In order to forecast weather, meteorologists examine the characteristics and behavior of local volumes of air they call **air masses** (**FIGURE 8.9**). Each air mass, typically several kilometers thick and a thousand or more kilometers wide, is subject to changes in temperature and pressure, which translate into big effects on the weather.

High-pressure systems tend to be associated with clear and dry conditions. They contain cooling air that, as it becomes denser and heavier, sinks to form areas of high pressure, depicted as big Hs on weather maps. While we mostly enjoy the weather that comes with these systems, extreme high-pressure systems that don't move cause deserts and drought. The Sahara, Kalahari, and Gobi Deserts all exist because of semipermanent high-pressure conditions. Similarly, California's 5-year drought of 2012–2016 occurred because of what one meteorologist called a "ridiculously resilient ridge" of high pressure sitting off the coast (**FIGURE 8.10**). In contrast, when an

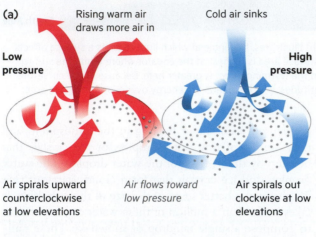

(a)

Rising warm air draws more air in

Cold air sinks

Low pressure

High pressure

Air spirals upward counterclockwise at low elevations

Air flows toward low pressure

Air spirals out clockwise at low elevations

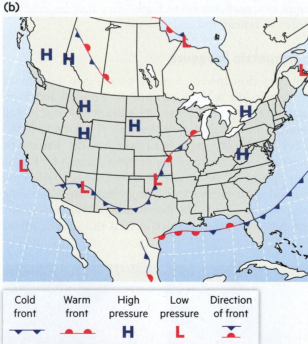

(b)

Cold front	Warm front	High pressure	Low pressure	Direction of front
▼▼▼	●●●	H	L	▼●▼

FIGURE 8.9 Moving Air Masses (a) Here we show how air masses move in the Northern Hemisphere. Low-pressure air masses have air molecules that are expanding and rising. As they spread out, the mass of air near Earth's surface weighs less and has lower pressure. In contrast, cooling air sinks, and the molecules become more tightly packed together, forming high-pressure areas. **(b)** Weather maps show areas of high and low pressure with Hs and Ls.

Adapted from New World Climate (2017).

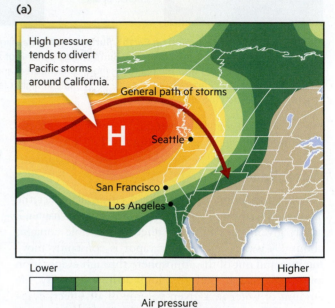

(a)

High pressure tends to divert Pacific storms around California.

General path of storms

H

Seattle

San Francisco

Los Angeles

Lower   Higher

Air pressure

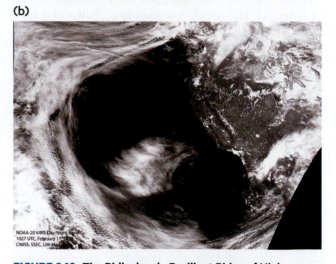

(b)

NOAA-20 VIIRS Day/Night Band
1027 UTC, February 1...
CIMSS, SSEC, UW-M...

FIGURE 8.10 The Ridiculously Resilient Ridge of High Pressure (a) From 2011–2016, a persistent high-pressure system off the West Coast of the United States was dubbed the "ridiculously resilient ridge." This weather system blocked North Pacific storms from reaching California, contributing to drought conditions in that state. **(b)** A similar event occurred in 2020. A high-pressure system right in the middle of the photograph is directing all the precipitation around it and toward the Pacific Northwest.

Adapted from Pierce (2015).

(a)

Precipitation occurs when warm air rises until it cools and condenses the water vapor it holds.

Precipitation occurs at the boundaries of high-pressure and low-pressure systems as the warm air in the low-pressure system is forced upward.

Snow occurs when cold air picks up moisture as it moves across large bodies of warmer water. The water vapor cools and falls as snow on inland areas.

Convective

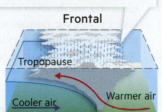

Cooler ground
Cooler ground
Warmer ground

Frontal

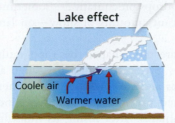

Tropopause
Cooler air
Warmer air

Lake effect

Cooler air
Warmer water

Precipitation occurs when air is forced upward over a topographical barrier such as a mountain; clouds form and precipitation falls where the weather system meets the mountain (often called the windward side).

Precipitation occurs when air flow comes together at Earth's surface and is forced upward, which can occur in the center of an area of low pressure.

Orographic

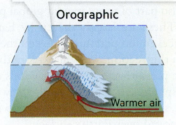

Warmer air

Convergent

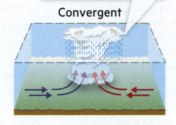

(b)

FIGURE 8.11 Precipitation **(a)** There are several processes by which air is lifted to higher altitudes, where it cools and creates clouds. Clouds produce precipitation such as rain and snow depending on the amount of moisture in the air. **(b)** Hurricanes form when warm, moist air converges on an oceanic low-pressure area, rises rapidly, and grows into a huge, swirling rainstorm.

air mass warms, its molecules expand to fill a greater volume, and it rises as it becomes less dense. Because this warm air is more spread out than cool air, it also weighs less and has fewer gas molecules per unit volume. So it forms a **low-pressure system**, indicated by a big L on a weather map.

A low-pressure system is often an indication of cloudiness and an increased chance of precipitation. This is because the warm, rising air cools as it expands into higher and colder altitudes. The cooling can cause the water vapor in the air mass to condense into clouds that may eventually produce precipitation. Whether or not it rains will also depend on the amount of moisture or humidity in the air. The more moisture in the air or in the area, the heavier the rains or snow will be.

Wind, Fronts, and Storms

Simply stated, **wind** is air in motion as it flows from high-pressure to low-pressure areas. On a larger scale, a cold air mass replacing a warm air mass is called a **cold front**, and a warm air mass replacing a cold air mass is called a **warm front**. Fronts form at the boundaries between high-pressure systems (cold air masses) and low-pressure systems (warm air masses). Sometimes a front is visible, with cloud formation marking the space where air masses with different temperatures meet. If the difference in

temperature and pressure between the air masses is large enough, there may be strong winds and even lightning. The anvil-shaped clouds known as thunderheads form when significantly warmer air masses collide with and rise above denser, cooler air, quickly condensing water vapor into clouds and rain (**FIGURE 8.11A**).

In extreme weather situations, low-pressure areas act almost like a powerful magnet pulling in air from other places (**FIGURE 8.11B**). In the case of hurricanes and typhoons, air is pulled into a low-pressure area over warm ocean water. When air from different locations arrives, it rises rapidly as it meets other incoming air masses and grows into a rapidly swirling storm hundreds of miles in diameter that can last for weeks. In the case of tornadoes, rotating horizontal rolls of air are drawn vertically into a low-pressure area over land that is caused by a rising column of air (an updraft). When sucked into an updraft, these rolls of air become the twisting funnels that we know as tornadoes. The strongest tornadoes are more powerful than the most severe hurricanes but only last a matter of hours or even minutes.

Lake and Land Effects

Continental landmasses and their topographic features also affect weather conditions. The interior of a continent typically receives less precipitation than

low-pressure system
a system of air in which warm, rising air cools as it expands into higher and colder altitudes. The cooling can cause the water vapor in the air mass to condense into clouds that may eventually produce precipitation.

wind air in motion as it flows from high-pressure to low-pressure areas.

cold front a cold air mass replacing a warm air mass.

warm front a warm air mass replacing a cold air mass.

Hadley cell a looping weather pattern forming circulation systems between the equator and latitude 30° N and between the equator and latitude 30° S.

coastal areas because it is a greater distance from the ocean, the major source of water evaporation. However, places near other large bodies of water, such as the Great Lakes, can experience "lake effect" rain and snow. When cold winds pick up water vapor as they move across the warmer water in the lakes below, this water vapor cools and drops precipitation—sometimes in the form of significant amounts of snow—on land in lake-effect areas.

Weather patterns are also shaped by topography. As Figure 8.11a shows, precipitation occurs when air masses meet mountains and are forced upward, where they cool and produce precipitation on the windward side of the mountains. After losing its moisture, the cooler, drier air mass then sinks and heats on the leeward side of the mountain, in a more arid region known as a rain shadow.

(a)

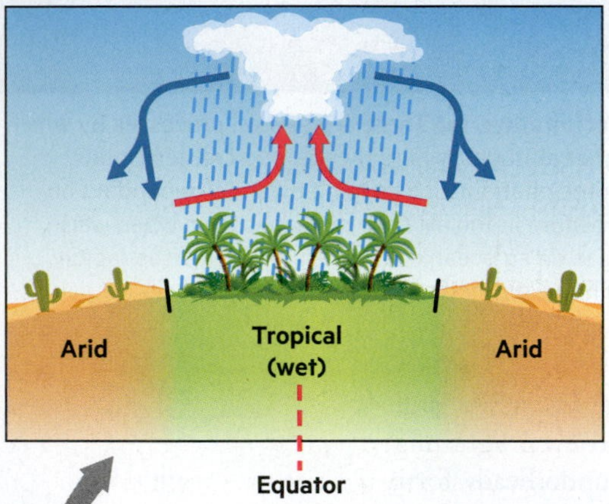

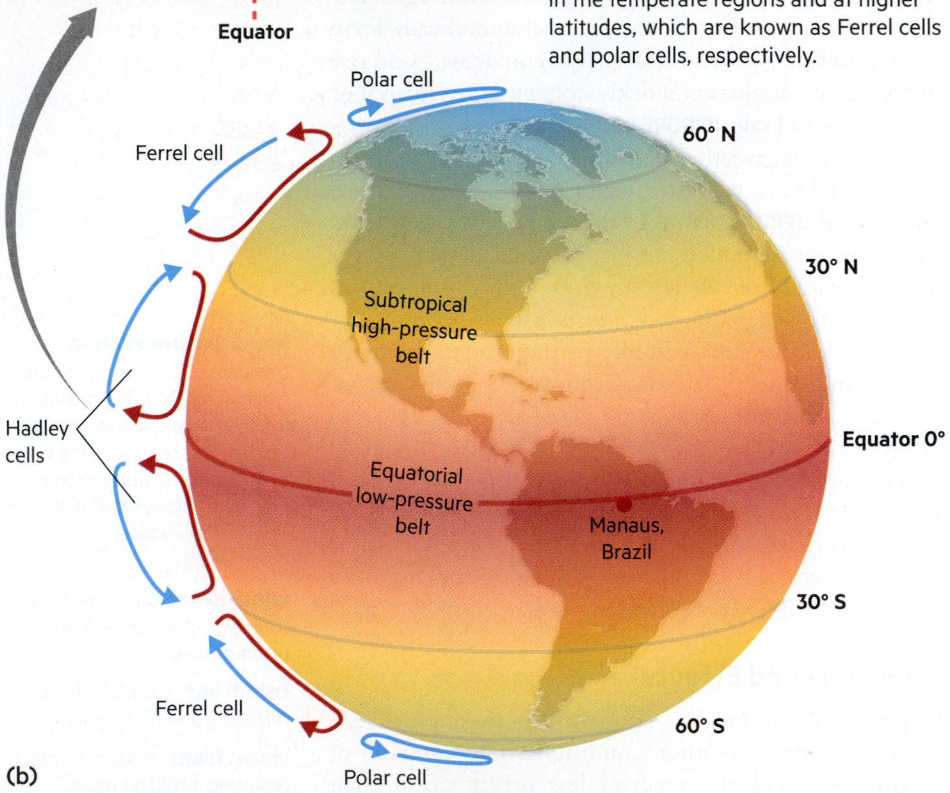

FIGURE 8.12 Air Circulation Cells Differential warming on Earth's surface creates air circulation cells. (a) Warm, moist air in the tropics rises and its water vapor condenses, resulting in abundant rainfall near the equator. The now cool and dry air descends back to the surface to the north and south, where arid conditions prevail. (b) This pattern of circulation is called a Hadley cell, and one occurs on either side of the equator. This style of circulation also drives less pronounced circulation cells in the temperate regions and at higher latitudes, which are known as Ferrel cells and polar cells, respectively.

(b)

○ **TAKE-HOME MESSAGE** Short-term variations in temperature, moisture, and wind conditions are known as weather. By contrast, climate refers to the long-term average of weather conditions for a given region. Weather is shaped by differences in the amount of solar energy received and by elevation, the location of landmasses, and proximity to bodies of water. Air masses rise, fall, and circulate horizontally with changes in temperature, pressure, and water vapor content, creating wind and local weather patterns.

8.3 How Does the Atmosphere Circulate?

Thinking back to the car idling as we waited for our friend, it's clear that exhaust was building up around us, probably in the general area of our vehicle. But how do these emissions impact other places? While we experience the weather effects of local air masses, our changing weather is also part of more regular global patterns of air circulation. These are largely driven by Earth's rotation and differences in temperature between the equator and the poles. In this section, we will explore the patterns of atmospheric pressure and prevailing winds that circulate in the atmosphere. We will also see the way that particles such as dust, soot, and ash can be swept up in this circulation and moved over Earth's surface.

Patterns of Atmospheric Pressure and Prevailing Winds

Let's start our examination of the circulation of Earth's weather around the middle of the globe: at the equator, in places like the Amazon jungle north of Manaus, Brazil, and Central Sumatra in Indonesia. Here, the Sun's rays strike Earth most directly, causing water to evaporate, rise into the air, and then fall as precipitation more rapidly than it does at other latitudes. This very large body of warm, humid, rising air at the equator is known as the *equatorial low-pressure belt*, and it supplies abundant precipitation to the tropical rain forests all around the equator (**FIGURE 8.12A**). But as this air continues to rise and approach the upper troposphere, it becomes cooler and drier as it spreads out to the north and south of the equator. These now cold air masses eventually descend, causing regular patterns of rising atmospheric pressure known as *subtropical high-pressure belts*. These high-pressure belts cause arid conditions at about 30 degrees of latitude to the north and south of the equator, in places such as the Sahara Desert and interior lowlands of Australia. Then, once this air reaches Earth's surface, it flows back toward the equator, where it warms and rises again. Taken together, these looping weather patterns form a circulation system between the equator and latitude 30° N and between the equator and latitude 30° S known as **Hadley cells** (**FIGURE 8.12B**). Differential

What's in the Clouds?

A cloudless blue sky means that, at least for the time being, there will be no precipitation—until the clouds roll in. Rain and snow fall from the clouds, but many clouds pass overhead without sprinkling us with precipitation. Why do clouds vary in their proclivity to drop rain or snow? Atmospheric chemist Kimberly Prather has been sampling dust particles in the air to answer this question. She flies planes through rain clouds, drives vehicles through dust storms, and sails research vessels through storms in the Pacific Ocean, all in pursuit of the dust they carry.

Why dust particles? One of the most mysterious aspects of precipitation has to do with the conditions that enable cloud-forming water droplets to start to freeze, eventually forming ever larger bits of ice heavy enough to fall back to Earth. Although we think of water freezing at 32°F (0°C), at a microscopic level, water droplets in clouds can remain liquid at temperatures as low as −40°F (−40°C). It seems they need a structure around which to form, and dust particles much smaller than a human cell can provide this necessary base.

But not just any kind of dust will do. Some particles trigger, or "seed," ice formation in clouds, while most do not. Prather works with a machine that uses vacuum tubes to collect dust that is then vaporized by a laser beam. Each particle that is blasted by the laser has an identifiable chemical pattern, or signature, that is recorded and analyzed to reveal the dust composition of clouds. While sampling dust within rain clouds over Wyoming, Prather found that the presence of icy droplets was strongly correlated with iron- and titanium-rich mineral particles. Cloud scientists such as Prather can help us understand not only what is seeding ice in clouds but also where these dust particles originated. In this case, the dust triggering rain and snow seems to be from the Taklamakan Desert in northwestern China or often even farther away in Africa. A separate study confirmed that a polluted air mass from Asia can reach the western United States in only 8 days.

In other words, precipitation in the western United States is facilitated by the circulation of a particular kind of dust that originates more than 8,000 miles (13,000 kilometers) away. There, desert winds in the spring send dust from the surface up more than 25,000 feet (7.5 kilometers) in the air into large weather systems that move over the western United States every year. These weather systems transport vast amounts of water evaporated from the equatorial Pacific and are responsible for most of the precipitation in the western United States. They have similar temperatures, water content, and wind patterns, yet they vary in how much precipitation they yield. The extent to which this dust from Asia is present in these systems helps to explain this variation. This research reminds us that the atmosphere is an intercontinental conveyor not only of air and water but also of particles like this rain- and snow-causing dust. Because of the effect of dust on weather, seemingly subtle changes in circulation patterns can lead to big environmental changes on the ground, such as extreme droughts or intense rain and snow events. And Prather's most recent research is finding microbes in the dust that affect the weather too.

What Would You Do?

Desertification, or the expansion of deserts into previously fertile land, is a problem for many regions, including the Taklamakan Desert area of China. The effects of desertification, such as dust storms and sandstorms, are not limited to desert countries, because (as we've just seen) winds can blow these sediments thousands of kilometers away. China has held meetings with the neighboring countries of South Korea and Japan to combat the desertification that affects all of them. Do you think it is worthwhile for countries that are farther away, such as the United States, but are nonetheless affected by desert dust, to participate too? If a policy was put forth for the US government to partner with China to help solve their desertification problems, explain whether you would be for or against it.

Atmospheric chemist Kimberly Prather and her team conduct research to identify and track the movement of dust particles in the atmosphere—including those on the move from the Taklamakan Desert in China to the western United States. This visualization from NASA shows the flight of different aerosols in the sky over Europe and Asia on the night of August 23, 2018. The purple area in the center is a large cloud of dust moving eastward away from this desert. (Blue swirls are huge amounts of salt in the air from two typhoons.)

Ferrel cell an air circulation pattern between latitudes 30° N and 60° N and between latitudes 30° S and 60° S.

jet stream a prevailing wind pattern near the top of the troposphere.

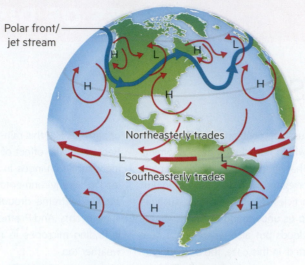

Polar front/ jet stream

Northeasterly trades

Southeasterly trades

FIGURE 8.13 Prevailing Wind Patterns Realistic representation of air circulation showing high-pressure and low-pressure centers near the surface and the polar-front jet stream.

Adapted from Marshak and Rauber (2017).

solar heating also drives similar circulation patterns known as **Ferrel cells** between latitudes 30° N and 60° N and between latitudes 30° S and 60° S and as polar cells at the higher latitudes.

All this movement and circulation of air causes winds. However, the prevailing wind patterns associated with these cells do not move in a straight line out from the equator to the north and south but rather are bent by Earth's rotation, a phenomenon called the Coriolis effect (see Chapter 11 for a more detailed discussion). As **FIGURE 8.13** shows, in the subtropical regions, Earth's rotation bends southward-flowing air to the right in the Northern Hemisphere and northward-flowing air to the left in the Southern Hemisphere, creating prevailing easterly winds that blow from east to west. This wind pattern is sometimes referred to as the trade winds, because sailing

ships used these winds to travel relatively swiftly from east to west across the Atlantic Ocean. At midlatitudes such as much of the continental United States, the Coriolis effect shapes prevailing winds that are westerly (blowing from west to east), and at polar latitudes the prevailing winds are once again easterly. At altitudes about 6 miles (10 kilometers) above Earth's surface, near the top of the troposphere, powerful **jet streams**, blowing as fast as 250 miles per hour (400 kilometers per hour), form prevailing wind patterns where the Hadley and Ferrel cells join and where the Ferrel and polar cells meet. These wind patterns explain why air travel across the United States is quicker when flying from west to east than the reverse.

Atmospheric Dust Transport

Maybe you have seen a kite or balloon flying high in the air after someone lost hold of the string or the fluffy white seeds of dandelions floating in the air after they were blown off their stems (and which can travel for miles). These examples make clear that the circulation patterns in the atmosphere and oceans carry more than just air and water. Wind at Earth's surface can sweep up dust and other small particles and transport them great distances. For example, the bedrock of the Colorado Plateau in the southwestern United States consists largely of nutrient-poor sandstone, which on its own would leave the area with relatively poor soil. Yet piñon–juniper woodlands thrive there in large part due to dust blown in from the Mojave Desert—from as far as 600 miles (965 kilometers) away. This dust accumulates over many years to form layers of soil as thick as 4 feet (1.2 meters) in some places. In fact, at elevations above 10,000 feet (3,000 meters) in Colorado's San Juan mountain range, it is estimated that more than half of the soil's mineral nutrients blow in from the Mojave Desert (**FIGURE 8.14**). But dust can migrate in the atmosphere for even greater distances.

Mojave Desert

Colorado Plateau

FIGURE 8.14 Dust Particles on the Move Blown dust from the distant Mojave Desert supports the piñon–juniper forests of the Colorado Plateau.

Research by cloud chemists and physicists has shown that dust from northwestern China's dune-filled Taklamakan Desert—sometimes called the Sea of Death—is forced up as high as 25,000 feet (7,600 meters) into the air by spring winds, entering a jet stream near the top of the troposphere that carries it thousands of miles east. Samples taken in the Pacific have revealed areas where more than half of the mud on the ocean floor is composed of dust from Asian deserts.

Other natural particles migrating in the atmosphere include soot from forest fires and ash from volcanoes. Forest fires send visible plumes of smoke into the air near the fire, often causing red, pink, and orange sunsets because of the color-filtering effects of the smoke-caused haze (**FIGURE 8.15**). Soot from forest fires in North America has been found to travel via the jet stream to Greenland, where it blackens the surface of ice. Soot-blackened ice absorbs more of the Sun's heat and accelerates melting. Volcanic ash particles can attract and collect enough water droplets to cause rain, thunder, and lightning in the area immediately surrounding an eruption. Very large eruptions can send material into the stratosphere and can affect global temperatures. In the stratosphere, larger volcanic particles will trap heat radiating from Earth's surface and have a warming effect. However, if the volcanic eruption produces tiny particles less than 2 micrometers in diameter (less than half the size of a human blood cell), they can have a cooling effect because they reflect incoming energy from the Sun. Such an event happened in 1991 after the eruption of Mount Pinatubo in the Philippines: scientists noticed a 10% reduction in the amount of sunlight reaching Earth's surface and a 0.5°F (0.28°C) drop in the average global temperature.

But just as dust, soot, and ash particles from natural processes circulate in the atmosphere, so too does pollution from human sources, which we'll explore in the next section.

⬡ **TAKE-HOME MESSAGE** Global patterns of atmospheric pressure and prevailing winds are shaped by temperature differences between the equator and the poles and also by Earth's rotation. Atmospheric circulation is a vehicle for transporting dust, soot, ash, and other particles around the globe.

8.4 How Does the Atmosphere Become Polluted?

In Chapter 3, we learned that matter is conserved. When we burn wood in a fire or fuel in our vehicles, these materials may go up in smoke, but they do not cease to

(a)

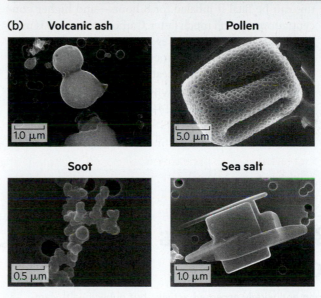

(b)

Volcanic ash — 1.0 μm
Pollen — 5.0 μm
Soot — 0.5 μm
Sea salt — 1.0 μm

FIGURE 8.15 Soot from Forest Fires (a) The color-filtering effects of soot in the atmosphere from forest fires or pollution can create brilliantly colored sunsets. **(b)** Common airborne particles (including soot particles) viewed through a scanning electron microscope. μm = micrometers.

exist. Although we may no longer be able to see them and they may undergo chemical changes, the matter that constituted the wood or fuel persists in emissions taken up by the atmosphere. When the addition of materials into the atmosphere adversely affects our health (or that of any organism), it is considered **air pollution**. We recognize air pollution and its effects almost immediately when it occurs indoors. For example, cigarette smoke was an ever-present source of indoor air pollution linked to an increased incidence of asthma, chronic bronchitis, and other ailments before smoking bans became common in public areas. Or you might notice a distinctive smell from new carpets, furniture, and paint as they emit volatile organic compounds, a category of pollutants we discuss in more detail later in this section.

air pollution what occurs when the addition of materials into the atmosphere adversely affects the health of humans and/or other organisms.

primary pollutant
a substance that is harmful in its directly emitted form.

particulate matter (PM) tiny particles and droplets less than 10 micrometers in size—about one-seventh the diameter of a strand of human hair—that are suspended in the air we breathe.

volatile organic compound (VOC) a gas released through the evaporation or incomplete combustion of fossil fuels and other organic chemicals.

secondary pollutant
a pollutant that is the product of reactions occurring in the atmosphere.

We know that because the atmosphere circulates, what is in the air will not necessarily stay put. The effects of air pollution are influenced to a large extent by atmospheric circulation patterns. In this section, we will focus on the broad categories of substances that often pollute the atmosphere and some of the ways they are transferred into our atmosphere.

Primary Pollutants

A **primary pollutant** is a substance that is harmful in its directly emitted form. One of the most prevalent primary pollutants that we are all exposed to is *carbon monoxide* (CO), a gas produced when organic materials such as wood, coal, gas, and oil are burned without abundant oxygen. But it is also emitted in car exhaust, and anyone who has been stuck in a long traffic jam may have experienced the headaches or sick feelings from CO exposure (**FIGURE 8.16**). Carbon monoxide gas is deadly to humans because it impairs the ability of our blood to carry oxygen. If the air has a concentration of just 150 parts per million of CO, this can be fatal if inhaled for 8 hours—and higher concentrations can kill much faster. Carbon monoxide is most dangerous in confined spaces—such as homes—which is why safety experts recommend that homeowners with furnaces, stoves, or water heaters fueled by wood, oil, or natural gas install carbon monoxide detectors.

Other primary pollutants that make their way into the atmosphere include pesticides applied in agricultural areas, benzene released from gasoline production

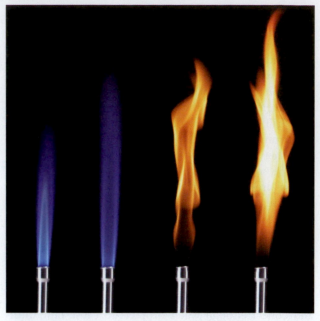

FIGURE 8.16 Carbon Monoxide from Incomplete Combustion When enough oxygen is present, organic materials combust and emit carbon dioxide (CO_2). When oxygen is scarcer, carbon monoxide (CO) is a by-product of the combustion. You can see this process in the color of the flame produced by a natural gas or propane appliance such as a stove, grill, or furnace. In the presence of abundant oxygen, the flame burns hot and blue with more complete combustion. Without enough oxygen—like when an air hole is closed or clogged—the flame will burn cooler and produce a yellowish-orange flame indicative of incomplete combustion and CO emissions.

or fueling stations, perchloroethylene used at dry cleaners, and asbestos dust from construction activities. These pollutants are primarily a risk to individuals who work or live in close proximity to their fumes and emissions. Other air pollutants can cause more far-reaching harm.

As we have seen, air currents carry dust around the globe; these same currents carry pollution too. For example, a form of pollution known as **particulate matter (PM)** is made up of tiny particles and droplets less than 10 micrometers in size—about one-seventh the diameter of a strand of human hair—that are suspended in the air we breathe (**FIGURE 8.17**). Although the particles are tiny, they can be composed of hundreds of different substances, such as dust and carbon from fires and industrial processes, dust and debris from mining and agricultural activities, or even just dust stirred up from unpaved road surfaces. We often perceive particulate matter as clouds of dust or smoke in the air or just a general haze reducing our visibility. When the particles settle at ground level, they form dark soot that coats surfaces we touch in our daily life. These particles can cause respiratory ailments such as asthma and chronic coughing when inhaled. The finest particulates are especially dangerous because they can be absorbed into the lungs and can even enter the bloodstream (through

Smoking Bans Address Indoor Air Pollution

Smoking was once a common and widely accepted activity, but public health campaigns and smoking bans changed public perception of this behavior relatively quickly. The Surgeon General's first report on the harmful effects of smoking was released in 1964, which was also the peak year of per capita cigarette consumption in the United States. The first smoking bans passed in the 1970s were mild by today's standards, mandating separate smoking sections in restaurants, workplaces, or airplanes, but by the early 2000s, total bans on indoor smoking were enacted in several states—including in bars and nightclubs. In 2011, New York City passed some of the strictest smoking bans in the country, prohibiting smoking outdoors in all public parks and beaches. Cigarette use and the percentage of Americans who smoke have both been cut by more than half since 1964, and the numbers continue to decrease.

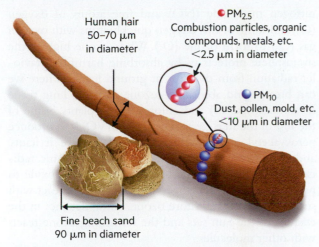

Human hair
50–70 μm
in diameter

● PM$_{2.5}$
Combustion particles, organic
compounds, metals, etc.
<2.5 μm in diameter

● PM$_{10}$
Dust, pollen, mold, etc.
<10 μm in diameter

Fine beach sand
90 μm in diameter

FIGURE 8.17 Particulate Matter Particulate matter (PM) is made up of many substances that take the form of tiny particles, much smaller than the diameter of a strand of human hair or fine beach sand. μm = micrometers.

the alveoli in the lungs); they have been linked to lung disease, irregular heartbeats, and heart attacks.

Lead and other toxic metals such as mercury, arsenic, nickel, and cadmium can also travel in airborne particles and fall on land and in water. Trace amounts of these chemicals are found in fuels such as coal and are emitted from industrial processes such as metal smelting. Lead, which is a neurotoxin that accumulates in the body and causes delays in mental and physical development, kidney damage, and a range of other ailments, was once a common ingredient in paints and a common additive to gasoline to improve fuel economy. Lead paint and leaded gasoline were banned in the United States in 1978 and 1986, respectively. Since that time, most countries have instituted similar bans.

Volatile organic compounds (VOCs), another category of primary pollutants, are gases released through the evaporation or incomplete combustion of fossil fuels and other organic chemicals. Often found in higher concentrations indoors than outdoors, VOCs are produced by a wide variety of paints, building supplies, cleaning supplies, pesticides, adhesives, and even nail polish: they are responsible for "new car smell." Outdoors, VOCs result from the evaporation or incomplete combustion of fossil fuels. Common VOCs include formaldehyde, toluene, and acetone. They can cause eye, nose, and throat irritation as well as headaches after only short periods of exposure in confined spaces. Chronic exposure—regularly breathing in VOCs over long periods of time—has been linked to damage to the liver, kidneys, and nervous system.

Secondary Pollutants: Smog and Ground-Level Ozone

Some pollutants are the product of reactions occurring in the atmosphere. We call these **secondary pollutants**, and the air pollution we call **smog** is an example.

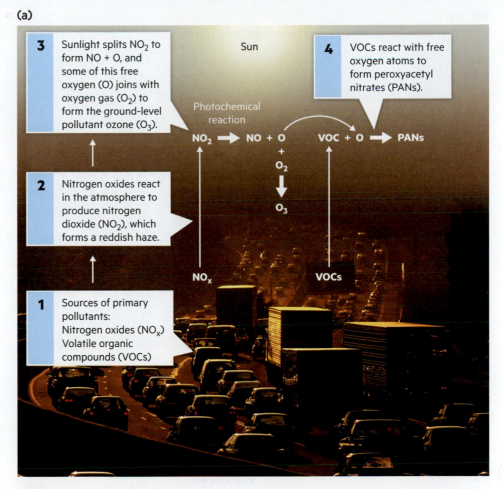

(a)

3 Sunlight splits NO$_2$ to form NO + O, and some of this free oxygen (O) joins with oxygen gas (O$_2$) to form the ground-level pollutant ozone (O$_3$).

4 VOCs react with free oxygen atoms to form peroxyacetyl nitrates (PANs).

Sun

Photochemical reaction

NO$_2$ → NO + O   VOC + O → PANs
+
O$_2$
↓
O$_3$

2 Nitrogen oxides react in the atmosphere to produce nitrogen dioxide (NO$_2$), which forms a reddish haze.

NO$_x$ VOCs

1 Sources of primary pollutants:
Nitrogen oxides (NO$_x$)
Volatile organic compounds (VOCs)

(b)

FIGURE 8.18 Smog Formation **(a)** The primary reactions driving photochemical smog formation. **(b)** Although smog forms during morning commutes when sunlight reacts with vehicle exhaust, regional weather patterns can cause smog to occur in low-traffic areas far from the source of the emissions. The red sky in this image of the Great Smoky Mountains is caused by pollution from sources such as traffic in Charlotte, NC.

Smog is a mixture of secondary pollutants resulting from reactions between sunlight and chemicals in the atmosphere. Smog forms when VOCs and nitrogen oxides (NO$_x$) from the combustion of fossil fuels interact in the presence of sunlight at warm temperatures (**FIGURE 8.18**). Morning commutes in crowded urban areas release large amounts of these chemicals, and as temperatures

smog a secondary pollutant that forms when chemicals from the combustion of fossil fuels interact in the presence of sunlight at warm temperatures.

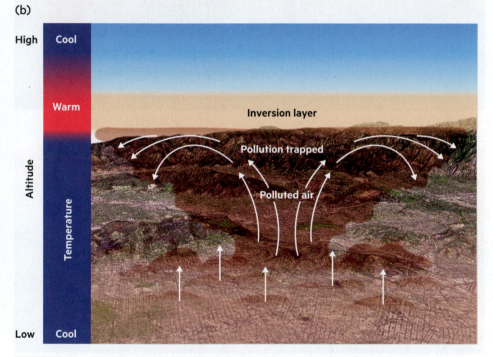

(a)

High — Cool

Altitude

Temperature

Low — Warm

Polluted air

Vertical mixing

Surface warmed by Sun

(b)

High — Cool

Warm

Inversion layer

Pollution trapped

Polluted air

Altitude

Temperature

Low — Cool

FIGURE 8.19 Thermal Inversion in Los Angeles **(a)** In normal conditions, air near the surface is relatively warm and gradually cools as it rises and mixes with cooler air at higher elevations above the mountains. **(b)** In a thermal inversion, cool air is trapped below a layer of warmer air above. This blockage in air movement can trap air pollution near the surface as well.

rise and sunlight intensifies through the day, a number of important photochemical reactions occur. Nitrogen oxides are highly reactive gases that form different compounds in the atmosphere, including nitrogen dioxide (NO_2), a chemical that forms a reddish haze and can cause airway inflammation. Sunlight splits NO_2 to form

nitrogen monoxide (NO) and free atoms of oxygen (O), and some of this oxygen quickly joins with oxygen gas (O_2) to form ozone (O_3). While ozone high in the stratosphere protects us by absorbing harmful ultraviolet radiation from the Sun, at ground level—where we can breathe and absorb it—ozone damages lung tissue in animals (like us) and chlorophyll in plants. Some of the rest of the free oxygen reacts with VOCs to produce peroxyacetyl nitrates (PANs), which are eye irritants, and highly reactive molecules known as organic radicals. Organic radicals react with nitrogen monoxide to produce even more nitrogen dioxide that can react with sunlight to yield still more ozone. Smog declines in the evening as the Sun sets and the remaining ozone reacts with other molecules.

Certain weather conditions can cause smog to accumulate a considerable distance from the source of the emissions. In the summer months, the air in Sequoia National Park in California often has ozone levels comparable to those of major cities such as Los Angeles and violates Environmental Protection Agency (EPA) air-quality standards. This is because regional winds draw distant, warm, polluted air from the San Joaquin Valley that is generated by its two major north–south highways, food-processing plants, and diesel tractors into the cooler mountains of the Sierra Nevada. Other national parks, such as Joshua Tree and the Great Smoky Mountains, have similar air-quality problems.

In some cases, local temperature and landforms can trap pollution over a particular area for an extended period of time. This happens when urban areas are adjacent to mountains or located in valleys, as is the case in Los Angeles and Mexico City. In normal conditions, air is relatively warm near the surface and gets cooler at high altitudes. But at certain times, a layer of warm air caps a cooler layer of air below, creating a *thermal inversion* (**FIGURE 8.19**). In Los Angeles, cool air from the ocean approaches the city from the west and sometimes slips below the warm air at ground level. Mountains to the east block air movement in that direction, preventing air circulation as emissions accumulate in the air near the surface. In Mexico City, warm polluted air is situated in a valley between mountain ranges. At night, the air over the mountain ranges cools, and this air can sink into the valley, pushing the warm air up until it is capped and kept in place by another cool layer above it.

⬡ **TAKE-HOME MESSAGE** Air pollution is the addition of materials into the atmosphere that affect the health of humans and/or other organisms. Primary pollutants are directly harmful in their emitted form. Secondary pollutants are the product of reactions that occur in the atmosphere. Landforms can worsen pollution by trapping it in a localized area.

8.5 What Is Happening to the Ozone Layer?

While ozone is harmful at ground level in smog, the ozone in the stratosphere is very important. Why is this? The Sun is constantly bathing Earth in ultraviolet (UV) radiation, and the ozone layer in the stratosphere prevents 99% of it from reaching Earth's surface. UV radiation damages the surface tissues of all life-forms. If Earth's surface were to absorb the full dose of UV radiation reaching the atmosphere, it is unlikely that the planet would support terrestrial life, or at least life as we know it. It is therefore vital that this protective part of the atmosphere be maintained. The ozone layer sustains itself through an ongoing process. As **FIGURE 8.20A** shows, ozone in the upper atmosphere is formed when UV light splits molecules of oxygen gas (O_2) into oxygen atoms (O), which then combine with unbroken oxygen gas molecules to yield ozone (O_3). UV radiation also splits these O_3 molecules, which then yield O_2 and an oxygen atom (O). This oxygen atom then combines with other O_2 molecules to re-form ozone.

However, other reactions can use up ozone faster than it is formed naturally, preventing the ozone layer from quickly repairing itself. One of these reactions was identified in the early 1970s. British scientist James Lovelock discovered that human-made chemicals known as chlorofluorocarbons (CFCs) were taking up residence in Earth's stratosphere. CFCs were developed in the late 1920s and were widely used in air conditioners, refrigerators, and aerosol spray cans. In 1974, chemists F. Sherwood Rowland and Mario J. Molina conducted lab experiments on CFCs that simulated solar radiation in the stratosphere. They found that although CFCs were very stable in the lower atmosphere, once the CFCs reached altitudes more than 18 miles (29 kilometers) above Earth's surface, UV radiation from the Sun began to break the CFCs apart, releasing chlorine atoms (**FIGURE 8.20B**). Using research on reaction rates of chlorine atoms in the stratosphere, Rowland and Molina found that the chlorine atoms would trigger a chain reaction with ozone molecules such that a single chlorine atom could remove as many as 100,000 molecules of ozone, thereby depleting the ozone layer. Worldwide measures of stratospheric ozone levels showed a steady and rapid decline throughout the 1970s and 1980s, and air samples taken from aircraft flying over the Southern Hemisphere showed that as the concentration of chlorine in the air increased, ozone concentrations decreased. Over the next decade, satellite images confirmed that the ozone layer was thinning.

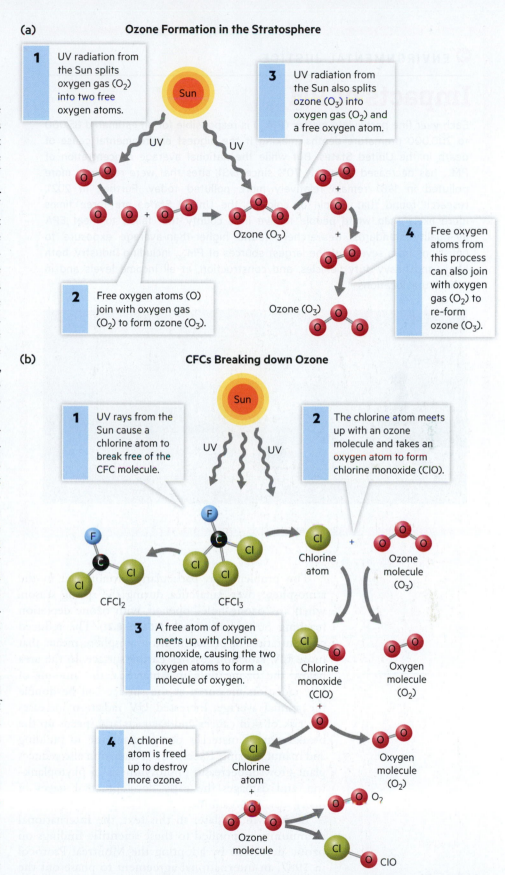

(a) Ozone Formation in the Stratosphere

1. UV radiation from the Sun splits oxygen gas (O_2) into two free oxygen atoms.

3. UV radiation from the Sun also splits ozone (O_3) into oxygen gas (O_2) and a free oxygen atom.

2. Free oxygen atoms (O) join with oxygen gas (O_2) to form ozone (O_3).

Ozone (O_3)

4. Free oxygen atoms from this process can also join with oxygen gas (O_2) to re-form ozone (O_3).

Ozone (O_3)

(b) CFCs Breaking down Ozone

1. UV rays from the Sun cause a chlorine atom to break free of the CFC molecule.

2. The chlorine atom meets up with an ozone molecule and takes an oxygen atom to form chlorine monoxide (ClO).

$CFCl_2$

$CFCl_3$

Chlorine atom

Ozone molecule (O_3)

3. A free atom of oxygen meets up with chlorine monoxide, causing the two oxygen atoms to form a molecule of oxygen.

Chlorine monoxide (ClO)

Oxygen molecule (O_2)

4. A chlorine atom is freed up to destroy more ozone.

Chlorine atom

Oxygen molecule (O_2)

Ozone molecule

O_2

ClO

FIGURE 8.20 Reactions in the Upper Atmosphere (a) Ozone in the stratosphere is replenished through natural processes. (b) However, chlorofluorocarbons (CFCs) can use up ozone faster than it is formed naturally.

Impacts of PM

Each year fine particulate matter ($PM_{2.5}$) is responsible for an estimated 85,000 to 200,000 premature deaths, making it the biggest environmental cause of death in the United States. But while the national average concentration of $PM_{2.5}$ has decreased by nearly 70% since 1981, sites that were relatively more polluted in 1981 remain relatively more polluted today. Further, in 2021, research found that people of color in the United States are three times more likely than white people to live in a county that failed to meet EPA air-quality standards. Researchers found higher-than-average exposure to emissions from several of the largest sources of $PM_{2.5}$, including industry, both light- and heavy-duty vehicles, and construction, at all income levels and in urban/rural locations.

The problem was particularly pronounced in the atmosphere over Antarctica during the spring season, when an "ozone hole" opened, with ozone depletion reaching 50% in this area (**FIGURE 8.21**). The reduced concentration of ozone in the stratosphere means that more UV radiation reaches Earth's surface. In the area below the ozone hole in Antarctica, the amount of UV radiation measured at the surface can be double the annual average. Increased UV radiation increases the risk of skin cancer and cataracts and speeds up the breakdown of materials that humans use in building and manufacturing. Excessive UV radiation also reduces plant growth, decreases photosynthesis in phytoplankton, and damages the early developmental stages of many marine organisms.

As we will see later in this text, the international community responded to these scientific findings on ozone depletion by adopting the Montreal Protocol in 1987, an international agreement to phase out the use of CFCs and other ozone-depleting substances. In one of the great success stories of environmental action, stratospheric ozone depletion has since leveled off, accompanied by the shrinking of the ozone hole over Antarctica. Scientists now project that we will achieve recovery to 1980 levels of protective ozone by 2050.

TAKE-HOME MESSAGE Chlorofluorocarbons are human-made chemicals that react with UV radiation in the stratosphere, triggering a chain reaction that destroys ozone molecules. This depletes Earth's ozone layer, which protects living things from the Sun's damaging UV radiation. The Montreal Protocol phased out the use of CFCs and other ozone-depleting substances beginning in 1989, and the ozone layer is now slowly recovering.

8.6 What Is Acid Rain?

Natural rainwater is slightly acidic, with a normal pH of 5.6. But in the 1970s, individual storms in New England were found to have precipitation with a pH as low as 2.1—nearly as acidic as lemon juice and vinegar. Researchers learned that rising sulfur dioxide (SO_2) and nitrogen oxides (NO_x) emitted from human sources such as automobile exhaust, coal-fired power plants, and a range of agricultural and industrial processes was reacting with water vapor in the atmosphere to form sulfuric acid (H_2SO_4) and nitric acid (HNO_3) (**FIGURE 8.22**). Then, prevailing wind and precipitation patterns in the continental United States would carry this pollution hundreds of miles from coal-fired power plants in the Midwest to the East Coast, especially the northeastern United States and Canada. This phenomenon became known as "acid rain"; more formally, scientists use the term **acid deposition**.

Persistent acid deposition changed the pH level of tens of thousands of lakes in eastern Canada and New England, eventually altering their ecology. Many organisms tolerate only a narrow range of variation in pH. Freshwater lakes have a natural pH range between 6 and 8 (a pH of 7 is considered neutral, neither acidic nor basic). Lower pH levels affect the developmental stages of both insects and fish and are related to malformed exoskeletons and skeletons. More acidic water also dissolves and releases elements from rocks and soil that are harmful in high concentrations. Few fish can survive at a pH of 5, and at a pH of 4.5, lakes are devoid of fish and often observed to be "dead" or "sterile," although acid-loving organisms such as mosses do populate these places.

Acid deposition can also change the soil composition and communities of microorganisms in ways that limit the growth of trees, particularly in high-elevation forests. Sharp increases of mortality in sugar maples and red spruce trees in New England have been linked to acid deposition. Acid deposition even wears away human structures and artwork, particularly those

acid deposition a phenomenon where air pollutants react with water, making an acid that precipitates from the sky, also known as "acid rain."

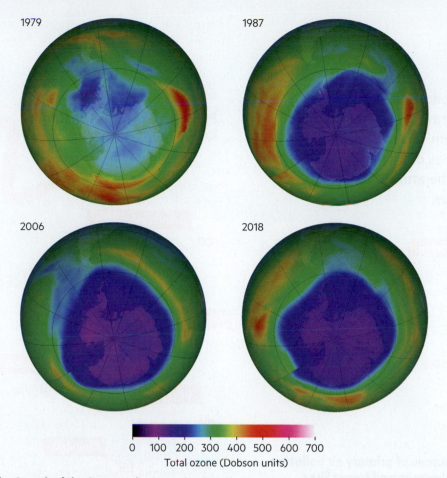

1979   1987

2006   2018

0   100   200   300   400   500   600   700
Total ozone (Dobson units)

FIGURE 8.21 **The Growth of the Ozone Hole** Ozone levels in the stratosphere dropped dramatically throughout the 1970s and 1980s due to increased atmospheric levels of CFCs, creating an "ozone hole" over Antarctica where ozone levels have been halved. Although the phaseout of CFCs has significantly decreased atmospheric CFC levels and caused the ozone hole to shrink slightly, it will still take several decades for ozone levels to recover and the hole to close.

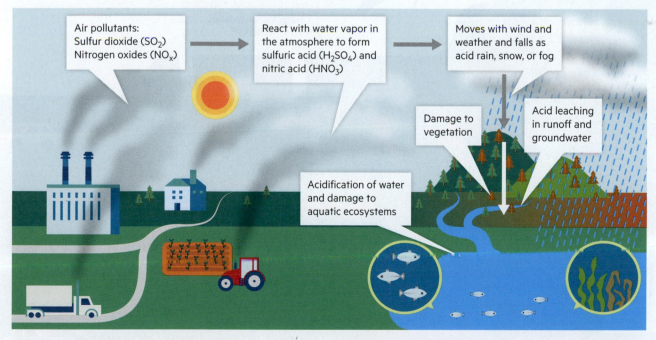

Air pollutants:
Sulfur dioxide (SO_2)
Nitrogen oxides (NO_x)

React with water vapor in the atmosphere to form sulfuric acid (H_2SO_4) and nitric acid (HNO_3)

Moves with wind and weather and falls as acid rain, snow, or fog

Damage to vegetation

Acid leaching in runoff and groundwater

Acidification of water and damage to aquatic ecosystems

FIGURE 8.22 **Acid Deposition** Acid deposition occurs when the air pollutants sulfur dioxide and nitrogen oxides react with water vapor in the atmosphere to form sulfuric acid and nitric acid—which can fall to the surface in precipitation, changing the pH of soil and water in ways that harm plants and animals.

Air Pollution

Pollution in the air can have a double impact. Substances that are primary pollutants in their natural forms can combine with others in the atmosphere or undergo chemical reactions to form new secondary pollutants. Both types of pollutants have numerous impacts on our environment, the atmosphere and air quality, and living things—including us.

PRIMARY POLLUTANTS

Primary pollutants are harmful in their emitted forms.

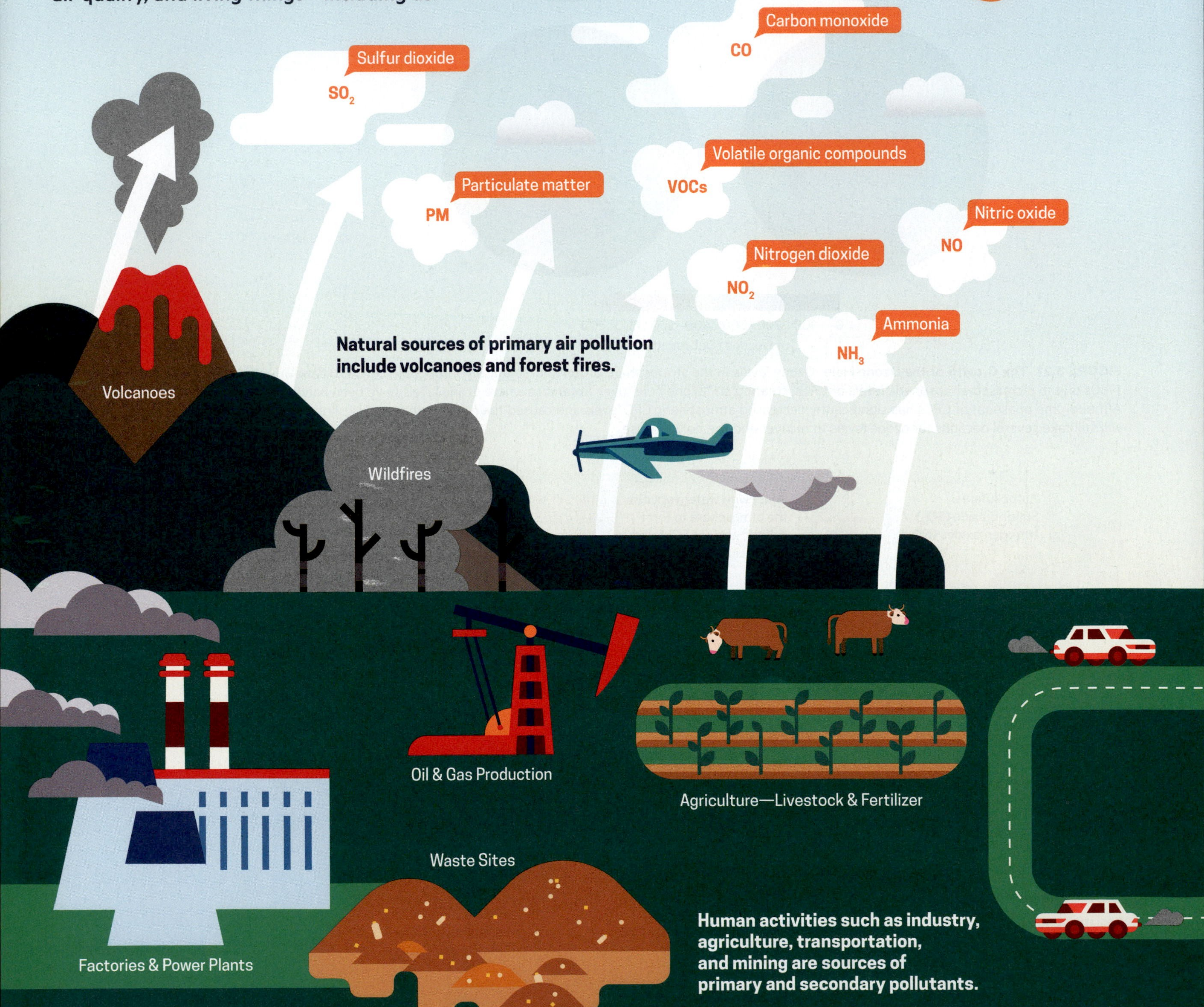

Carbon monoxide
CO

Sulfur dioxide
SO$_2$

Volatile organic compounds
VOCs

Particulate matter
PM

Nitric oxide
NO

Nitrogen dioxide
NO$_2$

Ammonia
NH$_3$

Natural sources of primary air pollution include volcanoes and forest fires.

Volcanoes

Wildfires

Oil & Gas Production

Agriculture—Livestock & Fertilizer

Waste Sites

Factories & Power Plants

Human activities such as industry, agriculture, transportation, and mining are sources of primary and secondary pollutants.

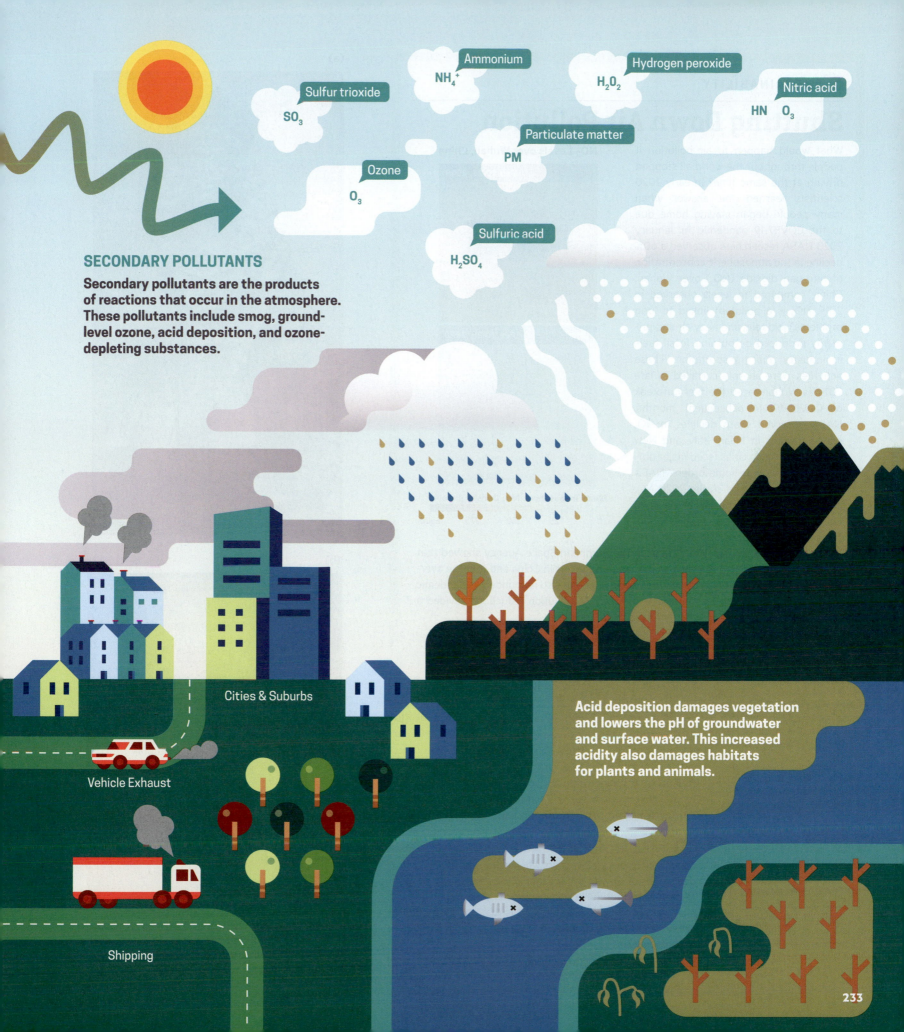

Sulfur trioxide

SO_3

Ammonium

NH_4^+

Hydrogen peroxide

H_2O_2

Nitric acid

HNO_3

Particulate matter

PM

Ozone

O_3

Sulfuric acid

H_2SO_4

SECONDARY POLLUTANTS

Secondary pollutants are the products of reactions that occur in the atmosphere. These pollutants include smog, ground-level ozone, acid deposition, and ozone-depleting substances.

Vehicle Exhaust

Cities & Suburbs

Shipping

Acid deposition damages vegetation and lowers the pH of groundwater and surface water. This increased acidity also damages habitats for plants and animals.

233

Shutting Down Air Pollution

What would happen to air pollution if people throughout the world all stopped driving at the same time? In early 2020 scientists learned the answer when many people began staying home due to the COVID-19 pandemic. In January 2020 NASA researchers observed a 60% decline in the atmospheric concentration of nitrogen dioxide (NO_2) over Wuhan, China, compared to levels observed in 2019. NO_2 is a pollutant linked to lung problems, acid rain, and hazy air and primarily produced by burning fossil fuels in industry and transportation. These activities were sharply curtailed in late 2019 as Wuhan faced the first outbreak of COVID-19. In subsequent months, NASA observed sharp declines in NO_2 concentrations in Milan, Italy, and New York City as these cities faced major outbreaks and stay-at-home orders. NASA researchers compared spring 2020 NO_2 concentrations with typical levels in 61 cities around the globe and found reductions between 20% and 50%. A year after the pandemic began, satellite data from the European Space Agency showed that NO_2 concentrations had returned to prepandemic levels in China and that the average level of air pollutants in other cities worldwide had also rebounded as pandemic restrictions were lifted. This natural experiment in pollution reduction provided a window into what a cleaner energy future might look like.

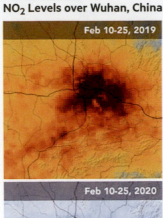

NO_2 Levels over Wuhan, China

Feb 10-25, 2019

Feb 10-25, 2020

100 km

Mean Tropospheric NO_2 Density ($\mu mol/m^2$)

0   125   250   375   ≥500

Clean Air Act of 1970 a law that established two important policy strategies: air-quality standards and regulations on the source of pollutants.

Energy Policy and Conservation Act a 1975 act that set minimum miles per gallon (mpg) fuel-economy standards for cars and light trucks.

tradable emission allowance a strategy where the EPA sets the total allowable annual emissions for a pollutant from utilities, then divides this total into tradable units called allowances that are bought and sold.

constructed of highly acid-soluble limestone and marble (**FIGURE 8.23**).

Of course, acid rain is not a phenomenon unique to the midwestern and eastern United States. It occurs wherever concentrations of these emissions can affect the water vapor in the air. Similar cases are found in Poland, Scandinavia, and the southeastern coast of China. In the next section, we will explore what can be done to combat this and other problems caused by air pollution.

⬡ **TAKE-HOME MESSAGE** Acid deposition, known commonly as acid rain, occurs when sulfur dioxide and nitrogen oxides react with water vapor in the atmosphere or water at ground level to form acidic solutions. Acid deposition can alter the pH of soil and water in ways that negatively impact ecosystems. These effects are often felt far from the source of the pollution.

(a)

(b)

FIGURE 8.23 Acid Deposition's Corrosive Effects on Human-Made Structures Acid rain is a global phenomenon. These affected stone carvings are on **(a)** Lichfield Cathedral in England and **(b)** a mountain in Cambodia.

Air Pollution: How Will the Chinese Government Respond?

What is it like to breathe seriously polluted air? In the final weeks of 2015 the Chinese government, in response to hazardous air pollution in Beijing, issued its first "red alert": a new policy to close schools, restrict traffic, halt the use of heavy machinery, and warn residents of declining air quality. During the red alert, residents described fatigue, difficulty breathing, and sore throats that made it painful to speak. While these symptoms were alarming, Beijing residents also reported that although the red alert was a new response, pollution events like this had become a fairly regular occurrence.

Recently, a World Health Organization study found that 1 million people die each year from polluted air in China—nearly one-third of all the air pollution deaths experienced worldwide. The report pinpointed the most dangerous pollutant as $PM_{2.5}$, tiny particulate matter less than 2.5 micrometers in size, which can penetrate the lungs and bloodstream, causing respiratory and heart conditions.

Statistics like this reveal that China's rapid industrial development and economic growth in recent decades is tied to significant environmental and public health costs. The majority of the $PM_{2.5}$ emissions in Beijing come from industrial production facilities and coal-fired power plants in the surrounding region. When setting up the current red alert system in 2013, Chinese premier Li Keqiang described air pollution as a warning against "inefficient and blind development." But air pollution in China is also about more than just the environmental and human health effects of development: it is also about political power. China's ruling Communist Party has announced that it believes environmental degradation could be a major source of social unrest in the near future.

Many Communist Party policy responses to air pollution in recent years have been rapid and ambitious, falling under the banner of a "war on pollution" announced by the national government. They included open-ended and ongoing fines and even plant closures for industrial

The difference in air quality in Beijing between a "good day" and a "bad day" is visible.

operations that fail to comply with stricter emissions and energy conservation standards. Further, citizen groups were empowered to file lawsuits against polluting industries, and the government has also set up public, real-time air monitoring stations. There is also a concerted push to increase energy production from alternative sources such as wind, solar, hydropower, and nuclear. The country has invested more than the United States and Europe combined in alternative energy sources, increasing their power generation from these sources from near zero to 25% over the past 10 years. However, the government also approved three new coal mines in 2022 after a three-year ban and has significantly increased its coal production every year since 2019. Furthermore, Beijing's air-quality problems continue.

Taken as a whole, there is evidence that these policy changes are having an effect. Chinese air-quality studies, as well as those conducted

China recently opened the world's largest floating solar plant on a lake formed over an abandoned coal-mining area.

by the US government and third-party research organizations, have all found significant annual decreases in $PM_{2.5}$ particles in Beijing and across much of eastern China. But while air-quality experts agree that the policy changes have played a significant role in reducing air pollution, declining industrial production and a slowing rate of economic growth also have been significant factors. These trends have led to job losses and declining incomes, which can be sources of social unrest. A key question facing China's government leaders today is whether and how they can meet their goals for increased economic growth while also continuing to reduce air pollution.

What Would You Do?

To improve air quality, the Chinese government is choosing to support some pollution regulations at least for now, though many economists are saying this will lead to different problems tied to job losses from these regulations. If you were in the Chinese government's position, what would you decide to do? Would you take the same policy and add pollution regulations to industries? Or would you recommend a different course of action? Faced with pressures to weaken regulations, how might you respond?

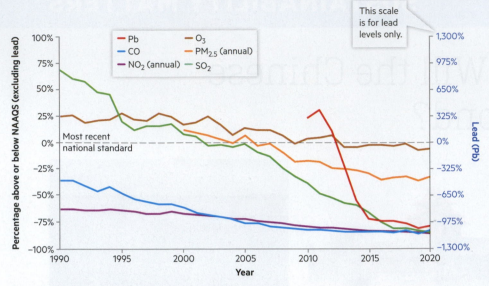

This scale is for lead levels only.

FIGURE 8.24 **National Air Quality Concentration Averages** The average national air quality in the United States has improved in recent years. The 0% line on the graph indicates that the national average air quality meets the standard (the allowable amount of pollution) for a particular pollutant. It is important to note that this graph does not show air-quality trends for particular cities or regions, which could differ considerably from the national average. NAAQS = National Ambient Air Quality Standards. Adapted from US Environmental Protection Agency (2021).

8.7 How Are We Responding to Pollution in the Atmosphere?

As the data in **FIGURE 8.24** demonstrate, overall air-quality trends in the United States for many pollutants show significant improvements since 1990. These improvements took place even as the population grew, the economy expanded, and vehicle miles and energy consumption increased. Although millions of Americans are still exposed to conditions that fail to meet federal air-quality standards, air-quality data also demonstrate that policy strategies addressing air pollution in the United States have been relatively successful.

The federal government launched a concerted attempt to address air pollution with the **Clean Air Act of 1970**, a law that established two important policy strategies: *air-quality standards* and *regulations on the source of pollutants*. Under this law, the EPA assesses the health risk for common pollutants in the outside air to determine maximum allowable concentrations of each. These form the National Ambient Air Quality Standards (NAAQS), which are measured and monitored for compliance across the country. Monitoring local conditions against these standards enables local, state, and federal government officials to identify air-quality problems and also helps make citizens aware of so-called "bad air days," when pollutants pose a risk to individuals with asthma or other respiratory problems. The Clean Air Act of 1970 also empowered the EPA to set emission standards for more than 70 categories of polluting industries such as agriculture, manufacturing, and utilities. These standards require new and upgraded facilities to use technologies that reduce smokestack pollution. In another example, in 1973, the EPA used its authority under the Clean Air Act of 1970 to phase out the use of lead additives in gasoline.

Similarly, in 1975 the **Energy Policy and Conservation Act**, among other regulations, empowered the Department of Transportation to set minimum miles-per-gallon (mpg) fuel-economy standards for cars and vehicles considered light trucks (such as vans, pickup trucks, and SUVs). These are known as corporate average fuel economy (CAFE) standards, and they are based on the size of the vehicle and the model year. The greater the mpg performance of a vehicle, the more efficiently it uses fuel (**FIGURE 8.25**).

In 1990, amendments to the Clean Air Act of 1970 added a new policy strategy known as **tradable emission allowances** to address sulfur dioxide (SO_2) emissions from coal-burning power plants. This is an example of the cap-and-trade policy strategies introduced in Chapter 2. Here's how it works. The EPA sets the total allowable annual emissions for this pollutant from utilities. It then divides this total into tradable units called allowances. Each facility's annual emissions must stay at or below the number of allowances it possesses to remain in operation. The only way a facility can legally exceed this level is by buying additional allowances from those who have not used their full allowance because of pollution-control measures they have taken. In other words, facilities can profit from taking pollution-control measures. Pollution-control technologies also compete with each other, prompting cost reductions and innovation. Over time, the EPA can reduce the total allowable annual emissions to reach national air-quality goals, placing more pressure on facilities to reduce their emissions.

FIGURE 8.25 **Comparison of CAFE Standards and Compliance** CAFE standards set the minimum miles per gallon that various models of vehicles must meet. This graph shows both the CAFE standard (darker lines) and actual performance (lighter lines) for cars and light-duty trucks (such as SUVs and pickup trucks).

Adapted from US Department of Energy (2021) and Congressional Research Service (2022).

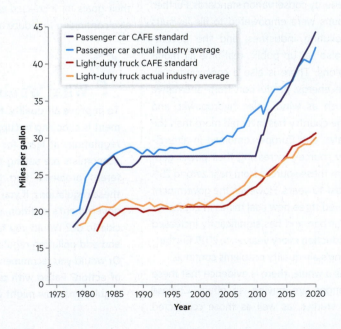

Assessments of this policy change are generally positive. The utilities enjoyed lower compliance costs with the change, and total SO_2 emissions from power plants declined from nearly 16 million tons in 1990 to just over 3 million tons in 2012. The EPA has estimated that the cumulative benefits of this policy from improved public health and fewer lost workdays exceeds $2 trillion. However, this policy has critics too. Some environmental groups say that it does not go far enough, because at its heart the policy does not eliminate "dirty energy" and does nothing to help communities suffering from pollution. And while some business groups support cap and trade as a market-based solution, others see it as a tax on their businesses.

While most air pollution policies focus on industries, some attempt to influence individuals such as commuters and consumers. For example, dedicated bus and carpool lanes on major commuter routes provide an incentive for public transportation and carpooling. The EPA also mandates fuel-efficiency labels for automobiles, hoping that more consumers will purchase more-efficient and thus less-polluting vehicles. Another strategy aimed at consumers is to offer buyback or rebate programs for technology that is more polluting. Examples include programs like the Car Allowance Rebate System, otherwise known as "cash-for-clunkers," whereby consumers got a generous rebate on a new car when they traded in an older, less fuel-efficient vehicle. While this federal program ended in 2009, a similar program is still in place in California.

🏠 **TAKE-HOME MESSAGE** Concerted federal action to address air pollution in the United States began with the passage of the Clean Air Act of 1970. Policy strategies to address air pollution include establishing and monitoring air-quality standards, regulating sources of air pollution, establishing tradable permits for pollutants, and designing incentives to influence commuter and consumer behavior.

8.8 What Can I Do?

We all share the air, so we all benefit from decisions to improve it. Most plans to improve air quality start with small decisions made at the individual level, then build toward broader impacts carried out by governments and institutions. Though the EPA reports that since 1990 the United States emits 1.5 million fewer tons of toxic air per year from industrial plants alone and that pollution-control technologies not common in 1970 are common today, few would argue that our air is *too* clean.

Choose Your Transportation with Pollution in Mind

Our transportation choices are often the most important way we can impact air quality. Riding a bicycle or walking is a powerful way to reduce your transportation impact. Sharing a ride or using public transportation greatly reduces the emissions you would have created traveling on your own. If you do drive your own vehicle, decisions that shorten your daily commute make a difference. And, as we saw at the beginning of the chapter, just cutting down on idling makes an impact. Since many auto manufacturers have built anti-idling features into new vehicles, a simple option is to use the feature if it's available to you. As we will learn in the chapter on fossil fuels (Chapter 13), choosing a vehicle that averages 30 miles per gallon over one that averages 20 miles per gallon reduces harmful emissions by a third. Maintaining your vehicle so that it is running properly, even if you just make sure that it has properly inflated tires, is an expense that pays off with better fuel economy.

Look for Ways to Limit Pollution at Home and on Campus

In Section 8.6, we learned that emissions from coal-fired plants cause a great deal of pollution. Because most of us draw energy from these plants in some way, choices made at home to reduce energy use will lead to lower emissions. Raising your thermostat when it is hot or lowering it when it is cold saves about 2% on your heating bill for every degree that you adjust it. Of course, this has more of an impact when it happens across your entire dorm or all the buildings on campus. Other easy savings you can both do yourself and request from your institution involve choosing replacement bulbs that are ENERGY STAR approved, which can use 75% less energy than traditional bulbs. And if you (or your school's maintenance staff) have a yard or garden to keep, limiting use of gas-powered tools and any supplies with VOCs adds up. For example, the University of Utah staff has switched to battery-powered motors and more efficient gas engines to limit pollution. Many hardware stores now clearly label paints and solvents that have low or zero VOC content, as well as cleaning and gardening supplies that are nontoxic.

Explore the Impacts of New Technology

It's probable that technology will continue to play a role in improving air quality. For example, a product called CityTree was created to "combine the natural abilities of air-purifying mosses with cutting-edge [information] technology." Installed in areas where trees are hard to plant (especially concrete areas in cities), this product uses mosses to remove particulate matter from the air and capture greenhouse gases: its creators estimate it can remove 240 grams of particulate matter a day and 240 metric tons of CO_2 a year. However, each CityTree currently costs $25,000. Portable air filters are a lower-tech and far more affordable option to clean indoor air; see the EPA website for advice on how to choose one.

Learn More about Power Generation in Your Community in Order to Influence Local Policy

An article in the October 13, 2017 issue of *Popular Science*, "There's No Such Thing as Clean Coal," states that "even though coal only made up 34 percent of electrical energy mix [in the United States] . . . as of 2015, it was responsible for 70 percent of emissions. The fastest way for a state to lower its emissions is to convert its coal to natural gas. . . . Reducing carbon emissions would be cheaper for customers . . . [and] would shave $7 off the average monthly electric bill." Is this an issue that you would support? While this argument seems powerful, recall the critics in this chapter who have noted that current pollution-reduction strategies are not going far enough. In contrast, politicians in coal-producing states argue that these decisions impact employment for miners, who have few other employment options in their area. Decisions on how we produce energy will have a great impact on air pollution but are almost always contentious. Understanding what you support and why is often the first step in making changes happen.

Chapter 8 **Review**

SUMMARY

- The atmosphere consists of a combination of gases that envelop Earth in four main layers that increase in density and pressure closer to Earth's surface.

- The troposphere is the atmospheric layer closest to Earth. This is the air we breathe and where weather—short-term variations in temperature, moisture, and wind conditions—occurs. The long-term patterns of weather determine a region's climate.

- The stratosphere is above the troposphere and contains the ozone layer that protects life on the surface from ultraviolet radiation.

- The mesosphere and thermosphere are the third and outermost atmospheric layers, respectively. The coldest temperatures occur in the mesosphere, while satellites and space stations orbit in the very diffuse air of the thermosphere.

- Factors affecting air temperature include the amount of solar energy received at a location, the elevation, and the location of landmasses and bodies of water. Air temperature in turn affects the exchange of water between the atmosphere and Earth's surface and the way air masses rise, fall, and circulate from areas of high to low pressure, causing wind and storms.

- The way that the Sun's rays strike Earth at different latitudes sets up predictable patterns of global circulation that include the Hadley, Ferrel, and polar cells, as well as the trade winds and jet streams.

- Atmospheric circulation also moves matter across Earth's surface, including air pollution in the form of primary pollutants that are directly harmful and secondary pollutants such as smog that are the product of reactions occurring in the atmosphere.

- Human-made chemicals such as chlorofluorocarbons can deplete Earth's ozone layer, increasing UV radiation at Earth's surface. An international agreement known as the Montreal Protocol has phased out the use of ozone-depleting substances, and the ozone layer is now slowly recovering.

- Acid deposition occurs when air pollutants from sources such as coal-fired power plants, vehicle emissions, and agricultural processes react with water to form acidic solutions that can negatively impact ecosystems far from the original source of the pollution.

- Policy strategies to address air pollution include air-quality standards and monitoring, regulation of emission sources, tradable permits for pollutants, and other incentives designed to influence behaviors linked to emissions.

KEY TERMS

acid deposition
air mass
air pollution
atmosphere
barometric pressure
Clean Air Act of 1970
climate
cold front

Energy Policy and
  Conservation Act
Ferrel cell
Hadley cell
high-pressure system
humidity
jet stream
low-pressure system

mesosphere
particulate matter (PM)
pressure
primary pollutant
relative humidity
secondary pollutant
smog
stratosphere

temperature
thermosphere
tradable emission allowance
troposphere
volatile organic compound (VOC)
warm front
weather
wind

REVIEW QUESTIONS

The letters following each Review Question refer to the Chapter Objectives.

1. What gases are most prominent in the atmosphere? Which layer of the atmosphere is closest to Earth's surface? **(A, B)**

2. Describe what causes air pressure on a molecular level. **(C)**

3. What is the difference between weather and climate? **(C)**

4. Which region of Earth receives more of the Sun's energy per unit of surface area than any other? **(C)**

5. Why does the air at higher elevations tend to be cooler than air at lower elevations? **(C)**

6. Compare and contrast the weather conditions associated with low-pressure and high-pressure weather systems. **(C)**

7. Name two sources of particles that migrate through the atmosphere. **(D)**

8. What is the difference between primary and secondary air pollutants? **(D, E)**

9. Name three types of air pollutants and at least one source for each. **(D, E)**

10. Smog forms from what atmospheric reaction? **(E)**

11. How does a thermal inversion form? **(E)**

12. What human-made chemical was primarily responsible for depleting the ozone layer? What was one intended use for this chemical? **(E)**

13. How did the international community respond to the environmental impact of ozone depletion? **(E)**

14. Name one source of pollution that leads to acid deposition (acid rain). **(E)**

15. Name two ecological impacts of acid deposition. **(E)**

16. What were the two main policy strategies established by the Clean Air Act of 1970? **(F)**

FOR FURTHER THOUGHT

The letters following each item refer to the Chapter Objectives.

17. While we all want to breathe clean air, the approaches we take to regulating pollution can be controversial because they affect the businesses and consumers linked to the polluting industries.

    a. In the United States, automobile manufacturers are subject to corporate average fuel economy (CAFE) standards set by the federal government that mandate fuel-efficiency improvements over time. Representatives of the auto industry traditionally resisted improving these standards because they argued that there is little consumer demand for more efficient vehicles and that the technology for these improvements either does not yet exist or that it is too expensive and will raise the cost of production and the price for consumers. Proponents of improving CAFE standards argue that auto companies and consumers will do little to improve fuel efficiency (and reduce pollution) without the incentives provided by improving the standards. What do you think? Should fuel efficiency be driven primarily by the voluntary actions of the companies, competition in the auto market, and consumer demand, or should the CAFE standards be advanced as an incentive to force these improvements by certain deadlines? Explain your reasoning.

    b. Under federal law, California has special status to set its own stricter CAFE standards because of the historically high levels of air pollution in its major cities. Other states can choose to follow California's CAFE standards or adopt the lesser federal standards. Representatives of the auto industry often argue against California's special status because it forces them to manufacture vehicles that meet the higher standards as well as their standard fleet, which is costly. They would prefer one uniform standard for the entire country. California argues that it should have the authority to adopt regulations tailored to its own air pollution problems. Should California have this special status? Explain your reasoning. **(F)**

18. What is the air quality like where you live? The American Lung Association assembles air-quality data for 220 metropolitan areas

in the United States and creates rankings of the most polluted and cleanest cities in the United States (www.lung.org/our-initiatives/healthy-air/sota/city-rankings/). What is the nearest metropolitan area to you, and how does it rank for high ozone days and annual particle pollution? What are the groups at risk from air pollution, and what percentage of the population does each group constitute? **(D, E)**

19. What are some of the sources of air pollution where you live? The EPA keeps a Toxics Release Inventory (TRI)

that you can search by zip code, city, county, and state (www.epaz.gov/toxics-release-inventory-tri-program). Search for your state report. Zoom in on the map of your state to identify the sites where toxic chemicals are released near you. How many pounds of toxic air pollutants were released in your county in the latest year measured? What are the top five chemicals released to the air in your state? What are the top five facilities releasing toxic pollutants in your state? How does your state rank in terms of total releases per square mile? **(D, E)**

Make Your Case

The following exercises use real-world data and news sources. Check your understanding of the material and then practice crafting well-supported responses.

Use the News

The following article describes how Volkswagen cheated emissions testing in the United States and how this was discovered. This huge scandal required Volkswagen to pay more than $3 billion in fines to the US government and to offer buybacks or repairs of almost half a million cars in the United States alone. Use this article to answer the

questions that follow. The first three questions test your understanding of the article. Questions 4 and 5 are short answer, requiring you to apply what you have learned in this chapter and cite information in the article. Answers to Questions 1–3 are provided at the back of the book. Question 6 asks you to make your case and defend it.

"VW Could Fool the EPA but It Couldn't Trick Chemistry," Eric Niiler, *Wired*, September 22, 2015

For decades, automakers have been caught between building an engine that squeezes a lot of energy out of the fuel it burns and one that has low emissions. It's not an easy tension to resolve. "Negotiating both fuel consumption and emissions is a hard tradeoff," says Anna Stefanopolou, professor of mechanical engineering at the University of Michigan.

When engineers at Volkswagen allegedly inserted a few lines of code into the diesel cars' electronic brains to circumvent emissions testing, they found a solution to this existential automotive conflict. Drivers got low emissions during the test, and high performance the rest of the time. The only problem: It's way outside of the rules. The company might have gotten away with it, too, if it hadn't been for those pesky engineers—and the basic chemistry of the diesel engine.

According to the US EPA, those lines of code hid the fact that nearly half a million diesel VWs in the US spewed up to 40 times more nitrogen oxide from their tailpipes than testing indicated. Volkswagen has now confirmed that

the problem actually affects approximately 11 million diesel cars worldwide. Diesel engines use a different mix of fuel than gasoline engines and don't use spark plugs to induce combustion—relying instead on highly compressed, heated air and fuel injected as droplets. If a diesel engine doesn't get enough oxygen to combust the fuel, it'll emit all kinds of gunk—nitrogen oxides, uncombusted fuel, and particulate matter (soot, basically).

All that gunk is a big problem. Exposed to sunlight, nitrogen oxides convert to ozone—making smog. How much depends on a bunch of variables, like sunlight exposure and what happens to the hydrocarbon emissions (the uncombusted fuel), plus the temperature and local winds.

However much extra crap came from the VWs, it won't be good. Exposure to nitrogen oxide and ozone is linked to increased asthma attacks, respiratory illnesses, and in some cases premature death. Ozone also worsens existing cardiovascular and lung disease.

To deal with those emissions, "you have a whole chemical factory at the tailpipe that traps

the oxides," Stefanopolou says. This bumps the sticker price for diesel cars by $5,000 to $8,000 per vehicle. (On the other hand, diesels get better mileage, especially in highway driving.)

For years, diesel trucks and buses were the biggest polluters on the highway. But carmakers adapted a relatively new technology called selective catalytic reduction—the same tech that scrubs pollutants from factory smokestacks—to the tailpipe of the diesel engine.

Here's how it works: Inside a honeycombed chamber, the scrubbing system sprays a liquid made of 30 percent urea and 70 percent water into the diesel exhaust. This sets off a chemical reaction that converts nitrogen oxides into nitrogen, oxygen, water, and small amounts of carbon dioxide—molecules that aren't as harmful to human health. Catalytic scrubbing was supposed to cut diesel NOx emissions up to 90 percent, according to the Diesel Technology Forum, an industry group based outside Washington. That made diesel engines clean enough to use in passenger cars, which have stricter emissions rules.

The scrubbing chemistry is also what gave away Volkswagen's alleged cover-up. In 2013, a small non-profit group decided to compare diesel emissions from European cars, which are notoriously high, with the US versions of the same vehicles. A team led by Drew Kodjak, executive director of the International Council on Clean Transportation, worked with emissions researchers at West Virginia University to test three four-cylinder 2.0-liter diesel cars in the Los Angeles area: a Jetta, a Passat, and a BMW. Only the BMW passed.

"We felt that it would be possible to get low emissions for diesels," Kodjak said. "You can imagine our surprise when we found two of the three vehicles had significant emissions."

The ICCT reported its findings to the EPA and the California Air Resources Board. Regulators met with VW officials in 2014 and the automaker agreed to fix the problem with a voluntary recall. But in July 2015, CARB did some follow-up testing and again the cars failed—the scrubber technology was present, but off most of the time.

How this happened is pretty neat. Michigan's Stefanopolou says computer sensors monitored the steering column. Under normal driving conditions, the column oscillates as the driver negotiates turns. But during emissions testing, the wheels of the car move, but the steering wheel doesn't. That seems to have been the signal for the "defeat device" to turn the catalytic scrubber up to full power, allowing the car to pass the test.

Stefanopolou believes the emissions testing trick that VW used probably isn't widespread in the automotive industry. Carmakers just don't have many diesels on the road. And now that number may go down even more.

---

1. How do diesel engines induce combustion differently than gasoline engines?
   a. Diesel engines use a bigger spark than gasoline engines.
   b. Diesel engines use spark plugs instead of compressed air and fuel injection.
   c. Diesel engines use compressed air and fuel injection instead of spark plugs.
   d. Diesel engines use the chemical urea to induce combustion.

2. What environmental problem results from an inadequate amount of oxygen combusting fuel in a diesel engine?
   a. Emissions of nitrogen oxides increase.
   b. Emissions contributing to ground-level ozone increase.
   c. Emissions contributing to smog increase.
   d. Emissions of particulate matter increase.
   e. All of the above

3. What technology did VW put in place on its diesel cars to reduce emissions?
   a. A honeycombed chamber captures and solidifies emissions for removal and disposal.
   b. A scrubbing system sprays water to cool the diesel exhaust before it leaves the vehicle.
   c. A scrubber system sets off a chemical reaction that converts nitrogen oxides into nitrogen, oxygen, and carbon dioxide.
   d. A governor is installed to regulate the speed of the car.

4. Briefly describe how the article says VW's technology to clean diesel exhaust worked and how other design aspects allowed them to cheat on the EPA emissions tests.

5. This fraud was first discovered not by a government agency but by an independent group working with researchers at West Virginia University. Given what you have learned about government regulation in this chapter, do you think current regulations are adequate, especially with private support as described here? Or do you think this example demonstrates that more regulation is needed? Explain your reasoning.

6. **Make Your Case** The end of the article notes an official who feels cheating like this is not widespread in the industry. Do you agree or disagree and why? Use information in this article and from other sources like chapters in this book to support your answer.

Use the Data

The COVID-19 pandemic caused many areas of the country to shelter in place and eliminate travel and a lot of business activity. One outcome was that researchers could observe how pollution and air quality change when transportation use—and use of the fossil fuels powering it—drops dramatically. The following two images were created by NASA from satellite data, showing (a) typical NO_2 levels in 2015–2019 and (b) those seen in early 2020 when lockdowns were in place across the Northeast.

Study the maps, and use them to answer the questions that follow. The first three questions test your understanding of the maps. Questions 4 and 5 are short answer, requiring you to apply what you have learned in this chapter and cite data in the maps. Answers to Questions 1–3 are provided at the back of the book. Question 6 asks you to make your case and defend it. All air pollution measures are in units of 10^{15} molecules/cm^2.

(a)

(b)

1. In map (a), which state shows the highest typical NO_2 pollution from 2015–2019?

   a. New Jersey (NJ)
   b. New York (NY)
   c. Maryland (MD)
   d. Massachusetts (MA)

2. Looking at maps (a) and (b), which statement is correct about NO_2 levels in and just outside of Boston (all measures are in units of 10^{15} molecules/cm^2)?

   a. NO_2 levels increased from almost zero to approximately 3.0.
   b. NO_2 levels increased from approximately 3.0 to over 5.0.
   c. NO_2 levels decreased from over 5.0 to approximately almost 0.
   d. NO_2 levels decreased from over 5.0 to approximately 3.0.

3. Looking at both maps, which of the following reasons most correctly describes the major factor(s) contributing to the changes in air quality across the two figures?

   a. Levels of NO_2 decreased in population centers but increased in most rural areas.
   b. Levels of NO_2 in population centers decreased as the pollution moved over ocean areas.
   c. NO_2 levels significantly decreased as transportation largely stopped around major population centers.

   d. NO_2 levels increased as transportation largely stopped around major population centers.

4. According to a recent study, New Jersey and Maryland are two of the five most congested states in the United States in terms of traffic. Comparing the two maps, what conclusion can you draw about how this congestion impacts NO_2 levels? Cite data from the map in your answer.

5. Transportation was much lower in the Northeast as states "closed down" for several months of the pandemic. Looking at the maps, note two areas that seem to see the most NO_2 pollution directly related to daily travel for work. Explain your choices.

6. **Make Your Case** As one of the most populous areas of the country, the Northeast corridor is a major driver of economic activity in the United States. Shutting down these states during the COVID-19 pandemic had a negative impact on the US economy but was seen as a trade-off for public health. Would drastically changing or limiting travel to decrease pollution be a similar trade-off? Or are there other alternatives that should be explored to impact pollution? Use the information in the maps, in the chapter, and in other sources to make your case.

LEARN MORE

- Weather and atmosphere with educational resources from NOAA: www.noaa.gov/resource-collections/weather-atmosphere-education-resources
- Air quality, observations, and forecasts: https://docs.airnowapi.org/aq101
- Video: How Air Pressure/Gas Pressure Works: digital.wwnorton.com/environsci

- Video: NASA, Seasonal Changes in Carbon Dioxide: digital.wwnorton.com/environsci
- UN Environment site on global air pollution: www.unenvironment.org/explore-topics/air/about-air
- EPA site on the basics of the Clean Air Act: www.epa.gov/clean-air-act-overview

9

Land

How Does It Shape Us?

What would you do if you were inside a building during an earthquake? If you grew up in an earthquake zone, you probably know the drill. Each year millions of students participate in the "Great Shakeout," the world's largest earthquake drill, where they are instructed to "drop, cover, and hold on." Sudden geologic events such as earthquakes remind us that the land beneath us does not stay put. Sometimes these reminders are terrifying—especially when they link up with environmental problems. In 2011, a massive earthquake off the coast of Japan sent a series of giant waves—a tsunami—crashing onto shore, killing thousands of people and destroying coastal towns. Damage from this tsunami released radioactive material from the Fukushima nuclear power plant, causing environmental impacts that remain today. Similarly, a 2010 earthquake in Haiti not only toppled buildings but also caused landslides that raced down steep slopes and across valleys deforested for agriculture (**FIGURE 9.1**).

Of course, the land changes in other, more gradual ways that we often don't notice. One is the loss of topsoil as it is carried away by water and wind. We depend on this outermost layer of Earth's surface to grow the crops we eat, yet in the United States we are losing topsoil at a rate at least 10 times faster than nature replenishes it. Other big agricultural producers such as India and China are experiencing topsoil loss at even faster rates. Similarly, beaches in coastal areas are disappearing as sand is washed away at a faster rate than it is deposited.

While these examples are caused by forces of nature, human actions increasingly influence processes that change the land. Agricultural practices such as tilling leave soil bare and hasten topsoil loss, as does cutting down or burning forests and any practice that removes plant cover holding soil in place. Damming rivers and constructing seawalls on coasts increase the rate at which beaches lose sand. Dams reduce the amount of sand flowing downstream to coastal areas, and seawalls impede sand's natural movement and distribution. Many tourist areas that are known for beaches, such as Miami Beach, Florida, are now importing hundreds of thousands of tons of sand each year at a cost of tens of millions of dollars. Further, rising sea level due to climate change is also accelerating the gradual erosion of land along shorelines.

In this chapter, we explore Earth's changing landscapes. We will begin deep underground, showing how rocks are created (and why this matters). Then we explore mineral resources that modern society depends on

Earth's surface is a dynamic place with forces both at the surface and below it that can shift the ground beneath our feet.

Chapter Objectives

By the end of this chapter, you should be able to . . .

A. explain how the processes shaping Earth's surface influence our lives.

B. describe Earth's internal structure and how its surface constantly changes and recycles itself.

C. describe how rocks and minerals form and some of their uses.

D. discuss how soils form and how soil-dwelling organisms and soil characteristics affect water infiltration, soil fertility, and plant growth.

E. explain how human activities can degrade or restore the land.

> **We know more about the movement of celestial bodies than about the soil underfoot.**
>
> —Leonardo da Vinci

which human activities impact the land. In the end, we will see that understanding how Earth's surface changes can help guide choices for building a more sustainable future.

and how erosion sculpts landscapes. We will also study soil, because humanity depends on fertile land to grow our food. As we go on, we will look at ways in

9.1 How Do Mountains Rise and Animals Find Their Way Home?

As you stand on Earth, you probably have the sense of standing on a solid mass of rock, or maybe you imagine it as a rocky ball with fire in the middle. Not surprisingly,

(a)

FIGURE 9.1 Earthquakes
(a) Earthquake awareness and preparation campaigns are reminders that the land beneath us is in motion.
(b) In 2011, a magnitude 9.0 earthquake off the coast of Japan triggered a massive tsunami that killed thousands of people and devastated towns (such as Otsuchi, pictured here) on the Pacific coast of Japan's northern islands.

(b)

→ SCIENCE

Migration, Magnetic Fields, and Monarch Butterflies

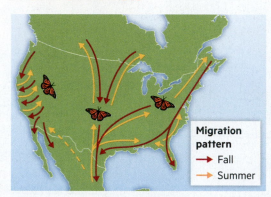

Adapted from US Forest Service (n.d.).

Recently, scientists discovered that monarch butterflies depend on Earth's magnetic field to guide their migrations. Magnetic sensors in their antennae guide them as they migrate thousands of miles from Canada and the United States to Mexico each fall and then back again in the spring and summer. Many other animals also use Earth's magnetic field for navigation, including fish such as rainbow trout, migrating birds such as the Arctic tern, and sea turtles. Experiments on creatures like these have found that even when they are displaced or confined in artificial environments with simulated magnetic fields, they can still use Earth's natural fields to chart a course and begin moving in the direction of their migration destinations.

FIGURE 9.2 Digging into Earth's Interior The drilling rig for Project Mohole (a model is shown here) was a major technological advancement for its time, but it had only dug 600 feet into the ocean floor when Congress stopped funding the project.

Earth's internal structure is much more complex. Understanding its complexity helps explain important aspects of our environment, such as what protects us from space radiation and how butterflies migrate. But our understanding of what's inside Earth is only decades old. At about the same time that the United States and Russia were competing to explore space, they also competed to explore Earth's interior. In 1961, the United States launched "Project Mohole" on a drilling rig in the Pacific Ocean. After plunging through 11,700 feet of water, the drill had only penetrated 600 feet into the ocean floor before Congress pulled funding from the project in 1966 (**FIGURE 9.2**). The Soviet Union, and later Russia, had more success. From 1970 to 1994, they drilled the world's deepest hole—the Kola Superdeep Borehole—9 inches wide and 7.5 miles down, more than 10,000 feet deeper than Mt. Everest rises above sea level. But as far as it went, even this deep borehole penetrated just one-third of the way through Earth's relatively thin, outermost layer and only two-hundredths of the way toward the center. It would have had to go more than 6,000 kilometers (3,700 miles) deeper to approach Earth's center.

Earth's Internal Structure

What would we find if we could dig deeper? Geologists like to use the analogy of a hard-boiled egg to illustrate the relationship between the three layers of material that

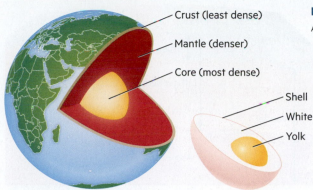

FIGURE 9.3 What's Inside Earth?
Adapted from Marshak (2018).

Crust (least dense)
Mantle (denser)
Core (most dense)
Shell
White
Yolk

make up Earth's core, mantle, and crust (**FIGURE 9.3**). We call the "yolk" at the center of Earth its **core**. The core consists of a solid inner section made primarily of the metals iron and nickel and a liquid outer section also made primarily of iron and nickel.

As Earth rotates, its liquid iron outer core flows and generates a magnetic field, the force that manipulates the needle of a compass. While many of us now navigate mostly by GPS, scientists have found that this magnetic field aids the migration of many animals (see Migration, Magnetic Fields, and Monarch Butterflies).

The magnetic field also establishes the **magnetosphere**, an area of space around Earth that shields the planet from highly charged particles emitted from the Sun. These particles would otherwise threaten life on this planet. The magnetosphere traps and then channels some of the particles toward Earth's poles. Here they collide with the atmosphere to produce spectacular colored auroras known as the northern and southern lights (**FIGURE 9.4**).

core the center of Earth, with a solid inner section and a liquid outer section both made primarily of iron and nickel.

magnetosphere an area of space around Earth that shields the planet from charged particles emitted from the Sun.

FIGURE 9.4 Magnetosphere Earth's magnetosphere produces the northern lights, pictured at top in Iceland, and the southern lights, pictured at bottom in Australia, when impacted by highly charged particles from the Sun.

FIGURE 9.5 The Lithosphere and Asthenosphere The crust and the uppermost part of the mantle together constitute the relatively rigid lithosphere. The asthenosphere is below it.

Adapted from Marshak (2018).

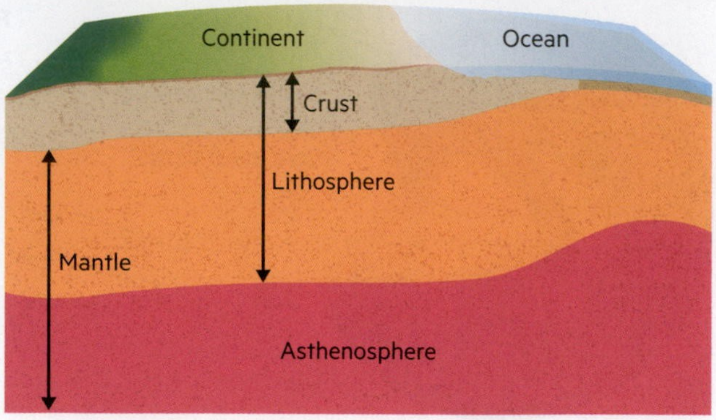

mantle the rock section of Earth above the core and below the crust.

crust Earth's outermost layer.

lithosphere the rigid outer portion of Earth, consisting of the crust and the very top of the mantle.

asthenosphere an area of Earth's mantle that is relatively pliable and is situated below the more rigid lithosphere.

tectonic plate a section of Earth's crust that rides on top of denser material below; interactions among these plates shape Earth's surface.

fault a fracture in Earth's crust, along which one body of rock slides past another.

The "egg white" section around Earth's core is called the **mantle**. It is less dense than the core and accounts for about two-thirds of Earth's mass and 84% of its volume. Most of the mantle is composed of solid but *ductile* rock; this means that although it is not liquid, it is malleable enough to flow and change shape under the intense heat and pressure this layer experiences. The mantle is overlain by a less dense and more brittle "shell" we call the **crust**, the outermost layer in which the United States and Soviet Union were drilling their boreholes. The crust accounts for less than 1% of the planet's volume, and its depth ranges from 7–70 kilometers (4 to more than 40 miles). But it creates the solid surface on which we live.

Another way to categorize Earth's layers is to distinguish between those that break or bend versus those that flow. As **FIGURE 9.5** shows, we refer to the rigid materials—the crust and the first few kilometers of the mantle—as the **lithosphere**, which extends down to roughly 100–150 kilometers (60–90 miles) belowground.

The denser and more malleable materials in the rest of the mantle beneath are the **asthenosphere**. While the materials in the asthenosphere can flow, the rigid materials in the lithosphere bend, break, and ride atop the asthenosphere, much like the crust on a hot cherry pie. Keeping this brittle crust in mind, let's take a closer look at how the lithosphere moves.

Plate Tectonics

The idea that the top layer of Earth moves is a newer discovery. As recently as 1960, it was widely assumed that Earth's lithosphere was stationary. German meteorologist Alfred Wegener challenged this idea in 1915, proposing that continents slowly wandered the globe. Wegener's idea was generally seen as outrageous for half of the 20th century. Widespread acceptance of the idea of moving continents did not happen until 1960 when oceanographer Harry Hess discovered that the seafloor was gradually spreading. By 1970, most geologists were embracing the concept of *plate tectonics*. This now accepted theory proposes that Earth's crust consists of seven major and a number of minor **tectonic plates** that ride on top of deeper, denser material below (**FIGURE 9.6**). The borders between tectonic plates are characterized by fractures called **faults**, and the movement of bodies of rock along these faults can produce earthquakes. Remember the Great Shakeout earthquake drill from the beginning of the chapter? The majority of participants in the Great Shakeout drill are in California, Mexico, and Taiwan, all areas located along active, moving faults. Tectonic plates can move at rates that average up to 6 inches per year. Such movements along California's San Andreas Fault caused both the great San Francisco earthquake of 1906

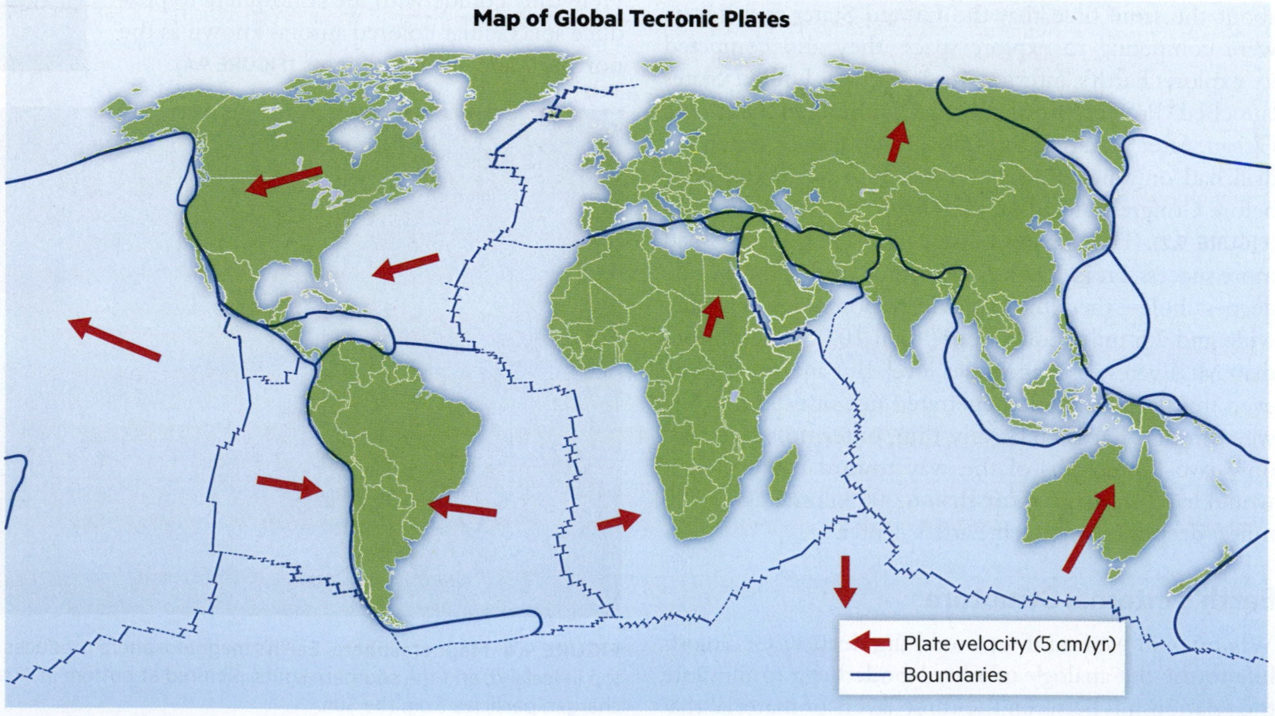

FIGURE 9.6 Tectonic Plates The lithosphere consists of tectonic plates that move relative to each other as they float on top of the denser asthenosphere below. Blue lines outline each plate, the arrows indicate the direction each plate is moving, and the length of an arrow indicates the velocity—with longer arrows showing plates moving relatively faster than those with shorter arrows.

Adapted from Marshak (2018).

Map of Global Tectonic Plates

← Plate velocity (5 cm/yr)
— Boundaries

and the Loma Prieta (Santa Cruz) earthquake of 1989. This pace may not seem very fast, but as **FIGURE 9.7** shows, over millions of years these movements added up, dramatically reconfiguring the continents over Earth's long history.

The different ways that plates move affect Earth's surface. Let's look at the three primary types of movement between plates. To help you picture this, you can think of the plates as broken halves of a crisp cookie pressed down on top of a creamy filling (**FIGURE 9.8**). For example, plates (the broken cookie halves) can spread apart from each other at a divergent plate boundary and form expanding areas such as rifts on land or at spreading sea-floors (Figure 9.8a). In other places, plates collide instead at a convergent plate boundary. If one colliding plate is made of denser oceanic crust, it is thrust under, or subducted, beneath the less dense plate. A chain of volcanoes forms as molten rock makes its way to the surface above the area where the sinking slab begins to melt. This activity is what formed areas such as the Andes mountain chain (Figure 9.8b). If both plates at a convergent plate boundary are made of similar low-density continental rocks, however, it's like pushing the cookie halves together and crushing them up into a high-standing pile. This kind of interaction is what formed the Himalayas, where India continues to crash into Tibet. A third kind of interaction occurs when tectonic plates move sideways relative to each other at a transform plate boundary. Imagine sliding the two halves of the cookie past each other, grinding the broken edges together. This is the type of boundary at the San Andreas Fault (Figure 9.8c). The global pattern of plate boundaries controls the distribution and character of earthquakes,

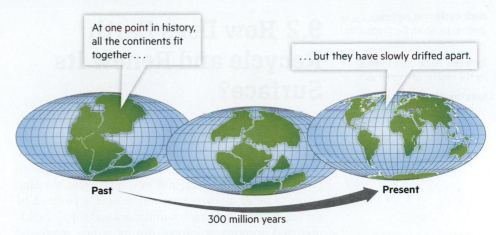

FIGURE 9.7 Slow Movements, Big Changes The arrangement of continents on Earth's surface slowly changes over time because of plate tectonics. Geologists have learned that all the continents were bunched together about 300 million years ago in a single supercontinent called Pangaea. Since then, they have drifted to their current placement.

Adapted from Marshak (2018).

volcanoes, mountains, plateaus, major rivers, and plains. In short, plate tectonics sets the geographic template that humanity lives on and uses.

⬡ **TAKE-HOME MESSAGE** Earth's structure includes the core, mantle, and crust. The lithosphere is the rigid outer portion that includes the crust and the upper reaches of the mantle. It is constantly in motion. The moving and shifting of tectonic plates produces phenomena and features at Earth's surface, such as earthquakes, volcanic activity, mountains, and plains. These dynamics have reconfigured continents throughout Earth history and created the geography we know today.

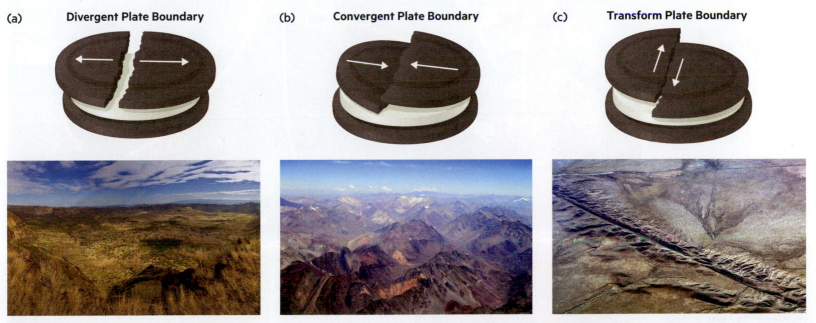

FIGURE 9.8 Tectonic Plate Movement (a) Divergent plate boundary, East African Rift. This flat-bottomed rift valley with a cliff on each side is a classic feature of divergent boundaries. **(b)** Convergent plate boundary, Andes Mountains of South America. **(c)** Transform plate boundary, San Andreas Fault in California.

rock cycle the geologic process by which Earth recycles and renews its surface.

igneous rock rock formed by the cooling of molten rock.

weathering physical and chemical processes that reduce rocks to smaller particles (such as gravel, sand, or silt) and alter minerals.

9.2 How Does Earth Recycle and Renew Its Surface?

Humans often find that the big and small events that change Earth's surface work in frustrating ways. On the plains, the wind whips dust in our faces, carrying away topsoil. On hillsides, houses often must be fortified to protect against landslides, and in coastal areas, the land seems to be disappearing around us as it is buffeted by the sea. Earthquakes crack foundations and topple buildings, and erupting volcanoes disrupt airline travel and threaten the areas surrounding them. But all these events are part of intertwined processes by which Earth recycles and renews its surface (**FIGURE 9.9**). Scientists call the process the **rock cycle**.

Weathering

In 2018, news feeds were dominated for months by images of molten lava spewing out of the ground in Hawaii, destroying neighborhoods as it flowed down to the ocean. As the lava cooled, hard rock formed, making the island a little bit bigger and changing the shape of its coastline. While the rock cycle has no true start or end, let's examine it beginning with these newly formed Hawaiian **igneous rocks**—rocks formed by the cooling of molten rock, whether deep belowground or at the surface (**FIGURE 9.10A**). Once exposed at Earth's surface, rocks such as the cooled Hawaiian lava are subject to different types of **weathering**: physical and chemical processes that alter minerals and reduce rocks to smaller particles (such as gravel, sand, or silt). When water freezes or salt crystals grow inside cracks in rocks, breaking the rocks apart over time, it's an example of physical weathering. Plant roots and animal burrowing can also cause physical weathering, as can extreme changes

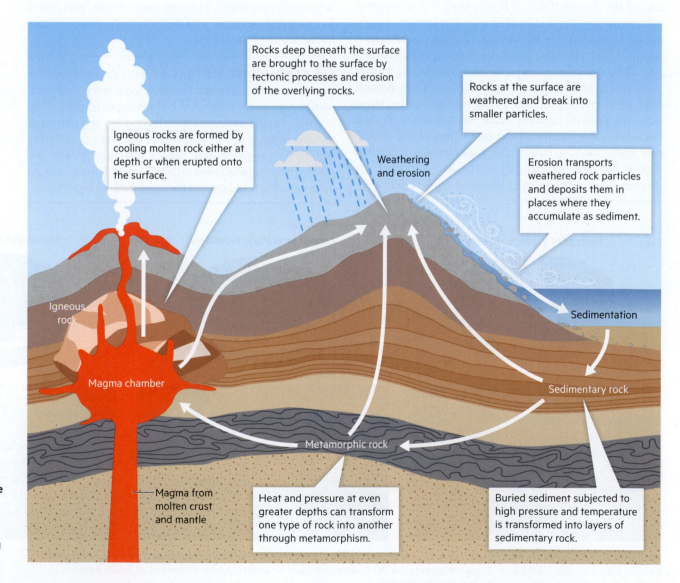

FIGURE 9.9 The Rock Cycle Sedimentary, igneous, and metamorphic rocks are formed, transformed, and recycled in the never-ending process of the rock cycle.

(a)

(b)

FIGURE 9.10 Rocks Forming and Weathering (a) Igneous rocks form from volcanic eruptions and when magma cools at depth. (b) Rusting is a familiar example of the chemical weathering process of oxidation.

in heat or pressure. As physical weathering breaks rocks into smaller pieces, it greatly increases the surface area of the material. This in turn increases the area exposed to chemical reactions that alter and further break down rocks through chemical weathering. Water in particular contributes to chemical weathering because it can dissolve some parts of rocks or react with and weaken them, creating new materials such as clay. Oxidation is another chemical weathering process: it occurs when the combination of minerals with oxygen causes the minerals to break down, as in the rusting of iron (**FIGURE 9.10B**).

Erosion and Lithification

We've all squinted to keep dust from blowing into our eyes on a windy day. Perhaps you've also experienced a strong current of water in a river or the power of waves pounding an ocean beach. Every day, natural forces—such as wind, water, ice, and gravity—pick up and move weathered rock particles, a process called **erosion**. Erosion is a powerful process that can carve dramatic landscapes (**FIGURE 9.11**). It also can be dramatically altered by humans. A famous example

erosion a process where natural forces—such as wind, water, ice, and gravity—move weathered rock particles.

FIGURE 9.11 Erosion Erosional processes driven by water and ice carve our landscapes, rivers, and valleys. The Grand Canyon (left) was carved through sedimentary rocks by the Colorado River. Mountain valleys in Norway (right) were carved by glaciers.

sediment eroded material that is transported and accumulates in different places (Chapter 7).

sedimentary rock rock formed when sediment buried under many layers of material cements together due to high pressures and temperatures.

metamorphic rock rock produced when one type of rock changes to another because of extreme heat and pressure.

mineral a natural solid from Earth's crust that forms rocks. Minerals also provide key materials for many practical human endeavors.

FIGURE 9.12 The Dust Bowl Human actions can facilitate erosion. In the 1930s, decades of plow-based agriculture and prolonged droughts in the midwestern United States left bare soil vulnerable to erosion, and winds carried dust storms of historic proportions.

⊙ **ENVIRONMENTAL JUSTICE**

Restoring Land Traditions

Haiti is a mountainous island country with steep, landslide-prone slopes that are highly susceptible to erosion. The island's Indigenous Taíno people conserved soil by shifting their plantings regularly to restore fertility, planting multiple crops in their fields to add nutrients, and forming planting mounds that retained moisture.

However, colonialism—the practice of occupying and acquiring control over another country and exploiting it economically—dramatically altered the social and physical landscape. Europeans first arrived in 1492, and by the early 1500s, their diseases and violent oppression decimated the Indigenous population by as much as 85%. By the 1700s French colonizers had cleared the forested hillsides to establish coffee, sugar, indigo, and cotton crops for export and imported nearly 800,000 enslaved people from Africa to labor on their plantations. The destruction of Indigenous populations—and their land conservation methods—by colonial powers, coupled with the deforestation of the original landscape, led to a massive loss of topsoil. Even after a successful revolt by enslaved people led to Haiti's independence in 1804, deforestation continued as the population grew, leading to the removal of ground cover and tree roots that hold soil in place from more than 70% of the land. Today efforts are under way to promote reforestation with trees and shrubs, using drought-tolerant cover crops between harvests to avoid bare soil and composting to enrich and restore topsoil.

is the Dust Bowl of the 1930s (**FIGURE 9.12**), where the introduction of plow-based farming in the mid-western United States in the previous decades broke up the prairie grass that held the fragile soil in place. When the next drought occurred, soil lofted into the air as dust, forming large black clouds that traveled across the continent to darken the skies of even New York City and Washington, DC. Recall that in Chapter 8 we learned that eroded dust particles from desertification in Central Asia can be carried by wind currents to the western United States in a little over a week.

As material erodes, it has to go somewhere. It often deposits and accumulates at the foot of a mountain, on plains where rivers flood, at the mouths of rivers and streams, or around reefs and on ocean floors. We call this transported and accumulating material **sediment**. Gradually, sediment buried under many layers of material experiences pressures and temperatures high enough to cement it together into solid **sedimentary rock**, a process known as lithification (**FIGURE 9.13**). An important aspect of lithification is that it allows dead matter from once-living organisms to become rock. For example, when the calcium carbonate ($CaCO_3$) shells of clams, oysters, and snails accumulate on the seafloor, they can eventually form limestone, one of the raw materials for making concrete. Similarly, plant matter in swampy environments can solidify over millions of years to produce coal. And when tiny floating organisms that make up plankton die and become buried, they can eventually transform into liquid oil or natural gas. In these ways, the carbon in organic materials gets worked into the rock cycle.

Where materials accumulate, they are gradually buried deeper and deeper as more material piles on top. The resulting heat and pressure cause metamorphism, the process in which one type of rock changes to another under extreme heat and pressure (more than 150°C and at least 1,500 times the atmospheric pressure we experience outside at sea level). The product of metamorphism is **metamorphic rock**. Metamorphism can change limestone into the Italian marble valued for sculptures, floor tiles, and countertops and can transform weak shale into hard slate used for floor and roofing material. Of course, the reason we can use marble and slate is because they have been pushed back to the surface by tectonic processes and exposed at the surface by erosion after their formation. Metamorphic rocks can be weathered, eroded, and deposited in sediments, perhaps to become sedimentary rocks once again. Or these same rocks could be pushed down deep enough to partially melt, producing fresh magma that can rise and possibly erupt in fresh lava flows.

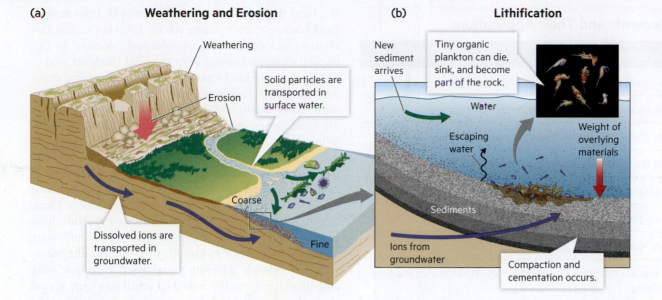

(a) **Weathering and Erosion**

Weathering

Erosion

Solid particles are transported in surface water.

Dissolved ions are transported in groundwater.

Coarse

Fine

(b) **Lithification**

New sediment arrives

Tiny organic plankton can die, sink, and become part of the rock.

Water

Escaping water

Weight of overlying materials

Sediments

Ions from groundwater

Compaction and cementation occurs.

FIGURE 9.13 **Weathering, Erosion, and Lithification** **(a)** Particles of weathered rock are eroded, moved, and deposited as sediment by gravity, wind, and water. **(b)** Lithification is the process by which heat, pressure, and chemical changes act on sediment that is buried under layers of material—transforming it into sedimentary rock. Even organic matter from dead organisms can become rock as it accumulates in sediment deposits.

Adapted from Marshak (2018).

⬡ **TAKE-HOME MESSAGE** The rock cycle is a process through which the material of Earth's crust is created, broken down, and recycled over time. At the planet's surface, rocks are reduced to smaller particles through physical and chemical weathering. Erosion moves weathered rock particles across Earth's surface and deposits them as sediment. Metamorphism transforms one type of rock into another through extreme heat and pressure.

9.3 Why Are Minerals Important, and How Do We Get Them?

Your phone vibrates, buzzes, chimes, or plays the latest ringtone. For most of us today, this phone is a mobile phone in our pocket, and almost always a smartphone, with a shiny glass face and powerful functionality. As you use this device, you are benefiting from the rare earth minerals necessary to create its parts, from the display to the speakers. These materials are mined from Earth, and the amount needed to construct the parts of a shopping bag full of devices can sell for tens of thousands of dollars. But this mining can be very intensive and dangerous, and it often occurs in remote parts of the world with little oversight. In this section, we will learn more about the importance of minerals, as well as some of the environmental impacts from obtaining them.

Minerals and Mineral Resources

All three types of rock we have studied—igneous, sedimentary, and metamorphic—are made of **minerals**, natural rock-forming solids from sources in Earth's crust. *Mineral*

➔ **CONTEXT**

Rare Earth Minerals: Baotou, Inner Mongolia

Baotou, a city in China's Inner Mongolia Autonomous Region, is one of the leading producers of rare earth metals for use in consumer electronics and alternative energy technologies such as wind turbines and electric cars. It is known as the "rare earth capital of the world." But these valuable minerals extract many costs. The refining process used to isolate the rare earth minerals produces waste sludge known as mill tailings that contain both hazardous waste and radioactive elements. In Baotou, this waste has been channeled into a large, toxic pond of sludge that is 11 square kilometers (4 square miles) and growing. The lake waste is a mixture of acids, heavy metals, carcinogens, and radioactive material derived from the refining process. As mining in the area grew, nearby crops began to fail, livestock sickened, and people in local communities fell ill.

TABLE 9.1 Rare Earth Elements and Their Applications

Element	Example Applications
Yttrium	Metal alloys, visual displays, lasers, lighting
Lanthanum	Optical glass, nickel-metal-hydride batteries
Cerium	Colored glass (flat-panel displays), automobile catalytic converters
Praseodymium	Super-strong magnets, metal alloys, specialty glass, lasers
Neodymium	Permanent magnets
Samarium	Permanent magnets, nuclear reactor control rods, lasers
Europium	Optical fibers, visual displays, lighting
Gadolinium	Shielding in nuclear reactors, X-ray and magnetic resonance imaging scanning systems
Terbium	Visual displays, fuel cells, lighting
Dysprosium	Permanent magnets, lighting
Holmium	Lasers, high-strength magnets, glass coloring
Erbium	Glass coloring, fiber-optic cables
Thulium	Lasers, portable X-ray machines
Ytterbium	Stainless steel, lasers
Lutetium	Petroleum refining

Adapted from Van Gosen et al. (2019).

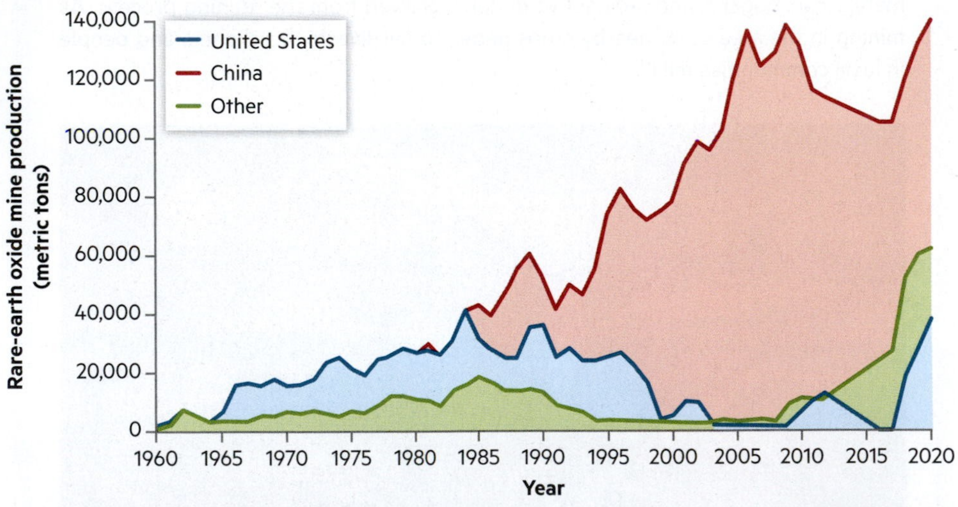

FIGURE 9.14 Rare Earth Elements This chart shows the history of rare earth element production. Some early demand came in the 1960s as new color-television technology became more popular. In the 1980s and 1990s, China began to dominate the rare earth market by selling at very low prices. Since then, rare earth mining has grown in other countries such as Australia, Russia, and Malaysia. Rare earth mining stopped entirely in the United States in 2016, only to restart in 2018.

Adapted from Van Gosen et al. (2019) and https://www.usgs.gov/centers/nmic/rare-earths-statistics-and-information.

is a term that many of us are familiar with from reading food labels—or from seeing ads for breakfast cereals that claim to include "essential vitamins and minerals." In fact, mineral elements coming from the solid Earth are vital to the chemistry of life. Humans need to consume five major mineral elements (calcium, magnesium, phosphorus, potassium, and sodium) and a host of trace minerals (including iron, copper, manganese, and zinc) to maintain health. While plants derive their mineral nutrients from rock and soil, we get ours from eating plants or animals that eat plants.

Similarly, minerals also provide key materials or mineral resources for human endeavors, from building materials to fertilizers and consumer electronics. These resources are often very valuable, and we pursue them in ways that are very intensive. Mineral resources are generally thought of as metallic and nonmetallic materials. For example, precious metals such as copper, gold, and silver are typically found in small amounts spread through a lot of rock—big nuggets and rich veins are rare. So a lot of metal-bearing rock needs to be dug up, crushed, and refined to produce the metals we know, usually in processes that use a lot of water. Minerals such as diamonds and metal ore can also form deep underground, with diamonds forming at depths more than 90 miles below the surface. A host of other important nonmetallic mineral resources such as decorative stone (for instance, marble) and even seemingly mundane resources such as sand, gravel, and salt are also mined.

Mining has a long history extending back before the dawn of agriculture. But most people have recently become very dependent on rare earth elements, which have unusual names and equally unusual properties that are useful for a wide variety of applications. In addition to their presence in your phone, rare earth elements are key components of electric car batteries and solar cells (**TABLE 9.1**). But rare earth elements are not distributed evenly around the globe because localized geologic processes form the deposits that concentrate them. Most of the world's supply comes from mines in China, and some observers are concerned that a restricted supply of strategically important rare earth elements could impact efforts to transition to alternative energy sources (**FIGURE 9.14**). Mining of rare earth minerals also comes with a heavy environmental cost, as 2,000 tons of toxic waste are produced to process just 1 ton of rare earths, and sometimes a human cost because of exploitation of mine workers.

⬡ **TAKE-HOME MESSAGE** Geologic processes govern the formation, preservation, and accessibility of mineral resources. We use mineral resources for a wide variety of items we rely on, from building materials to consumer electronics. Human extraction of these resources through mining causes significant adverse impacts. These include water pollution, disrupted groundwater flows, changes to the landscape, habitat destruction, and production of hazardous waste.

9.4 What Shapes Earth's Surface?

Artists in the mid-19th-century artistic movement known as the Hudson River School drew their inspiration from natural landscapes in the United States. Their paintings helped forge a sense of identity for the young nation and featured idyllic images of humans and nature peacefully coexisting (**FIGURE 9.15**). In fact, we still often closely identify ourselves with the world around us. Today you might spot license plates with slogans that tie state identity to specific **landforms**: surface features of our landscape. So Arizona is the Grand Canyon State, Minnesota is the Land of 10,000 Lakes, and Vermont is the Green Mountain State. Landforms and the natural processes that shaped them are of central importance to humans for the simple reason that we live on Earth's surface. The shape of the land, or its **topography**, has influenced events such as the migrations of our ancestors, the establishment of trade routes, and the settlement of major cities.

In ancient and modern times alike, when the accumulation of small day-to-day changes and occasional large events reshaped the land, humans were forced to adapt. Today, however, humans have become very effective at changing the land, often with unforeseen consequences. In this section, we will examine some basic landforms, how they are shaped, and how our attempts to manage or change them can go wrong.

Landforms

The Mississippi River, the longest river in North America, provides a good frame for our discussion of landforms. Flowing through the largest river system on this continent, the waters of the Mississippi River come from sources located in the 31 states and two Canadian provinces between the Rocky Mountains and the Appalachian Mountains, an area of land totaling nearly 1,250,000 square miles. This area is a grand-scale drainage basin, or **watershed**: land that drains into a particular point along a river or stream (**FIGURE 9.16**). Drainage basins come in all sizes, from areas draining into a single small stream, to small coastal basins that consist of mostly swampy land, to large continental-scale drainage basins such as the Mississippi River basin that extend across many states and over thousands of feet of elevation

landform a surface feature of the landscape.

topography the shape of the land.

watershed an area of land that drains to a particular point along a river or stream.

FIGURE 9.15 Hudson River School Art Artists of the Hudson River School, such as Thomas Cole, painted inspiring American landscapes in the mid-19th century.

→ **SUSTAINABILITY**

Development in Floodplains

Today, the Federal Emergency Management Agency (FEMA) estimates that more than 10 million residences are located in floodplains, along with many businesses and industrial facilities. However, FEMA also reports that development in floodplains increases the likelihood of flood damage. Not only does floodplain development put more people and property in harm's way, but the developed infrastructure can alter the dynamics of floods. As the land is filled and regraded (leveled or raised) for development and covered with buildings and pavement, the floodplain's capacity to store excess water diminishes—sending more water downstream and creating higher floodwater levels. Even levees and dikes designed to protect against flooding can cause problems during high-precipitation events because they confine the water in narrow channels that increase the velocity of floodwaters. When the water breaches levees or overflows the banks of these channels, it can do so in great volumes with tremendous energy and cause extensive damage.

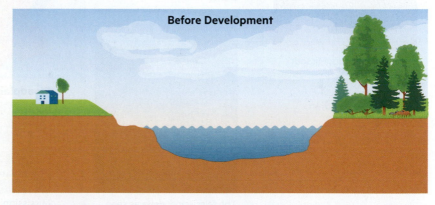

Before Development

After Development

Re-grading and filling

Adapted from US Government Accountability Office (2014).

Lowlands

Creek in northern Illinois prairie

Uplands

Hillslopes in the Great Smoky Mountains of Tennessee

Agricultural

Farm fields in northeast Iowa

Urban/Suburban

Skyline of St. Louis, Missouri

Wilderness

Protected swamp in Mississippi

Delta

The Mississippi delta as seen from the International Space Station

Floodplains

Mississippi River overflowing its banks in Davenport, Iowa

MISSOURI

UPPER MISSISSIPPI

Davenport, IA

St. Louis, MO

ARKANSAS–WHITE

RED

OHIO

LOWER MISSISSIPPI

FIGURE 9.16 The Mississippi Drainage Basin The Mississippi River basin drains about 40% of the area of the contiguous 48 states of the United States through the six river systems pictured. This basin includes diverse topography from mountains to lowlands and diverse human uses including highly urbanized areas and agricultural lands.

from the mountains to the ocean. Elevated areas change gradually as they are eroded and shaped by water, ice, and wind. Throughout a watershed, uplands (elevated areas such as hills) descend into lowland areas, as do the sediments, floodwaters, and pollutants produced in upstream areas.

One of the defining features of the Mississippi River, and a major factor driving settlement in this region, is its **floodplain**. As the name suggests, these are places where floods frequently send water over the banks of a river or stream channel, spreading the flow out, slowing it down, and thereby depositing sediment (**FIGURE 9.17**). Floodplains are very fertile areas for agriculture, which is why they have been home to some of the most ancient human civilizations, such as those along the Nile in Egypt or the Tigris and Euphrates Rivers in present-day Iraq and Syria. Because of their importance, floodplains are often highly modified to control their flooding. The Mississippi River now has 28 dams and locks designed to manage its flow, but they have caused significant changes

floodplain a place where floods frequently send water over the banks of a river or stream channel and deposit sediment.

(a)

(b)

FIGURE 9.17 **Flooding and Sediment Deposits** These photos taken (a) before and (b) during the 2018–2019 flood of the Mississippi River show how a flood sends water over the banks of river channels, where it spreads out and deposits sediment.

to the nature of the floodplain, which has allowed development to expand into naturally flood-prone areas. At the mouth of the Mississippi is its **delta**, a landform created where the river deposits its sediment load as it flows into the ocean.

Processes That Shape the Land

On March 22, 2014, at 10:37 AM, a tremendous landslide roared down a hillside near Oso, Washington, about 50 miles from Seattle. The debris from the event covered about a square mile of the valley floor, including a rural neighborhood in which 43 people lost their lives during this disaster. The scar left on the side of the hill clearly

Landslide scar

FIGURE 9.18 **The Deadly Oso Landslide** Landslides, such as this one that occurred in 2014 near Oso, Washington, are one of the dynamic processes that reshape landscapes.

showed a changed landscape (**FIGURE 9.18**). Landslides are one of the key processes that shape landscapes by moving sediment. Some landslides move slowly, gradually grinding downslope under the influence of gravity. Others fail rapidly and catastrophically like the Oso landslide, which launched more than 7 million cubic meters of sediment rapidly across the valley bottom. Though landslides are a natural process in many hilly environments, human actions such as forestry practices and excavation can influence their occurrence.

Not all landscape-shaping processes are dramatic and fast like some landslides. The power of flowing water (or glacial ice), wind, and more subtle influences such as the gradual pull of gravity can gradually reshape landscapes. In elevated terrain, rivers carve valleys over millennia by eroding material from the valley bottom and surrounding slopes. In steep and cold terrain, **glaciers**—rivers of ice—slowly grind away at and sculpt topography. The spectacular topography of the Swiss Alps, for example, was carved by glaciers that formerly extended across the landscape (**FIGURE 9.19A**). And the agricultural soils across much of the American Midwest developed from sediments that now-vanished glaciers carried south from Canada and that were redistributed across the landscape by strong glacial-era winds.

Of course, if you live near flowing water, it's easy to see sediment moving. Two rivers that flow into the Mississippi—the Big Muddy River in Illinois and the Missouri River (which is also sometimes nicknamed the Big Muddy)—flow brown with the sediment they carry downstream (**FIGURE 9.19B**). In watersheds, sediments generally originate in the elevated headwater region close to the source of a stream or river. The sediment is then transported by river networks, which connect the headwaters to *depositional environments*—the sediment storage areas such as major floodplains, river deltas, and estuaries—where sediments accumulate (**FIGURE 9.19C**).

Human activity can affect sediment movement; for example, floodplain development has diverted millions of tons of sediment from the Mississippi delta, making cities on the delta more vulnerable to flooding. Without this sediment, the land subsides. Runoff over the ground surface can carry all types of material, causing erosion if the flow becomes deep and fast enough. Smaller changes

delta a landform created where the river deposits its sediment load as it flows into the ocean.

glacier a mass of ice that flows over land surfaces and lasts all year (Chapter 7).

FIGURE 9.19 Sediment Transport (a) Glaciers are rivers of ice that grind and cut the topography as they flow, such as Aletsch Glacier in Switzerland—the largest valley glacier in the Alps. (b) Rivers, such as the Big Muddy in Illinois that joins the Mississippi River, transport sediment downstream as they flow. (c) Sediment is deposited and accumulates at river deltas such as the Mississippi River delta pictured here.

in elevated hillslope areas can cause new gullies to form downstream very quickly. You might observe *overland flow* happening during a rainstorm when the soil is fully saturated with water or when the rain is falling so hard that it accumulates more quickly than the soil can absorb it. Standing in a parking lot on a rainy day, you might notice overland flow streaming by your feet on its way to a storm drain because the surface on which you are standing is **impervious**: it does not allow water to pass through it. The proliferation of paved impervious surfaces sends much more of the water from storms flowing over the surface as runoff (**FIGURE 9.20**).

Finally, wind is also a major factor in shaping landforms, especially in arid and semi-arid regions. In many places, we may notice the effect of wind less because plant cover protects the ground from its force. But if you go to

impervious something that does not allow water to pass through it.

a place where the ground is exposed to the wind, you will feel and see the erosion during a windstorm. Some of that dust may travel well beyond where you are observing it. Fine-grained dust and silt can be lofted into the air and carried miles into Earth's atmosphere before it settles in a different region altogether. These plumes of dust (see Chapter 8) go where the wind carries them, sometimes crossing continents and oceans and carrying sediment (including pollution) from one region to another.

⬡ **TAKE-HOME MESSAGE** Landforms are shaped by a variety of processes that erode, transport, and deposit sediment. Erosion occurs from running water, landslides, flowing glaciers, blowing winds, and the gradual creep of material downslope under the influence of gravity.

FIGURE 9.20 Pervious versus Impervious Surfaces Impervious surfaces such as roads and parking lots in urban areas send much more of the water from storm events flowing over the surface as runoff. This graphic from the Philadelphia Water Department shows estimates of how impervious surfaces in an urban environment affect runoff as contrasted with runoff in natural habitats. Notice how the runoff from rain increases from 10% of the total to 55% of the total.

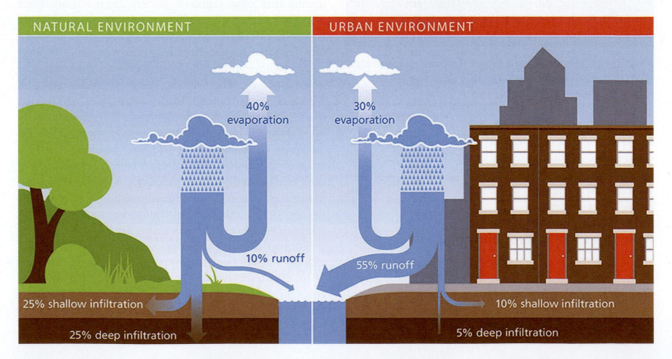

NATURAL ENVIRONMENT

URBAN ENVIRONMENT

40% evaporation

30% evaporation

10% runoff

55% runoff

25% shallow infiltration

10% shallow infiltration

25% deep infiltration

5% deep infiltration

9.5 What Is Soil, and Where Does It Come From?

Unlike many of the layers of Earth we have learned about thus far, soil is familiar to us. If we asked you to go out and find a scoop of soil right now, even if you live in a large city, you could quickly fulfill that assignment. We can see it, feel it underfoot, and dig it up. But this does not mean that it is well understood. Soil exists at the uppermost layer of Earth's surface, and it is sometimes referred to as Earth's skin, although there are some important ways in which this metaphor does not fit. Your skin is about one-tenth of an inch thick, about 0.14% (one seven-hundredth) of the height of the average person. In contrast, soil is much thinner relative to the size of Earth. The average depth of soil is less than a millionth of Earth's radius. And unlike skin, soil is not entirely a protective outer layer: soil generation depends on the destruction of the rocks below, which affects other parts of the environment. We are built of material that was once drawn out of the ground by plants, and eventually the material in our bodies will return to the soil, as related in the traditional phrase "ashes to ashes, dust to dust" that is read at many funeral services.

So what is **soil**? Taken as a whole, soil is a complex mixture of weathered rock and mineral particles (sediment), dead and decaying plant and animal matter, and the multitude of organisms that live within these materials. Soil is created by processes at Earth's surface, where the lithosphere (rocks) and the biosphere (life) interact, breaking down geologic source material, known as *parent material*, into different-sized particles. Air spaces within the soil allow water to move down into the soil, a process known as **infiltration**, and larger spaces of air in the soil promote *aeration*. The mixture of various sizes and types of parent material particles, together with the effects of soil-dwelling organisms and organic matter in various states of decomposition, determines the soil's overall structure and texture (**FIGURE 9.21**). The texture of soil is an important characteristic affecting plant growth. For agricultural purposes, *loam*—a mixture of roughly 40% sand, 40% slightly smaller silt particles, and 20% tiny clay particles—is considered the most desirable soil type. It allows water to infiltrate and drain yet retains enough moisture and nutrients to support vibrant plant growth.

Soil Horizons

Soil is much more than just weathered parent material or "dirt." Soil exists in large part due to the activities of living things. The action of living (biotic) and nonliving

(abiotic) factors in the soil creates layers known as **soil horizons** (**FIGURE 9.22**). The figure shows a **soil profile** in a cutaway view that is composed of up to four soil horizons labeled O, A, B, and C (Figure 9.22a). Soil profiles are typically less than 3 feet deep. The uppermost two horizons (O and A) constitute the **topsoil** and contain a complex mix of materials that includes organic matter derived from living things. Topsoil normally makes up the top 2–6 inches of soil. The O horizon is made of fallen leaves and other dead organic matter known as *detritus* that accumulates on the ground surface at the top of the soil profile. Detritus supports a vast and diverse community of life, including burrowing animals, insects, earthworms, fungi, and bacteria. The burrowing action of the largest soil fauna helps create air spaces and greatly influences overall

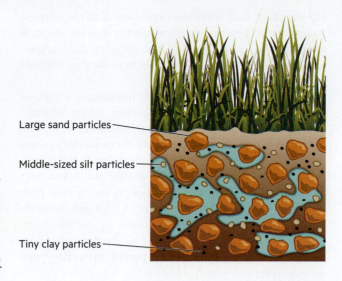

FIGURE 9.21 Soil Texture Soil texture is classified by the percentage of clay, silt, and sand in the mixture. Larger spaces can exist between larger particles, promoting aeration and infiltration.

Large sand particles

Middle-sized silt particles

Tiny clay particles

soil a complex mixture of weathered rock and mineral particles (sediment), dead and decaying plant and animal matter, and the multitude of organisms that live within these materials.

infiltration in science, a process where water (or other liquid) moves down into the soil.

soil horizon a layer in soil created by the action of living (biotic) and nonliving (abiotic) factors.

soil profile the collection of soil horizons at a location.

topsoil typically the first 2–6 inches of soil, encompassing the top two soil horizons.

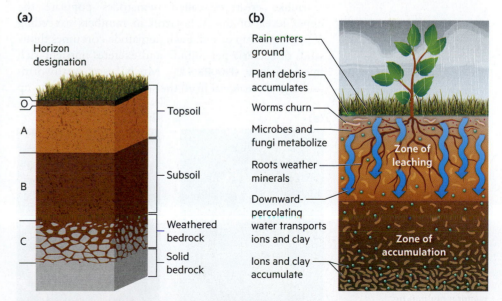

(a)

Horizon designation

Topsoil

Subsoil

Weathered bedrock

Solid bedrock

(b)

Rain enters ground

Plant debris accumulates

Worms churn

Microbes and fungi metabolize

Roots weather minerals

Downward-percolating water transports ions and clay

Ions and clay accumulate

Zone of leaching

Zone of accumulation

FIGURE 9.22 Soil Horizons (a) A typical soil profile. (b) A close-up of the processes occurring in the top layers of soil (O, A, and B horizons).

Adapted from Marshak (2016).

humus in soil, a complex, dark, sticky organic material that can remain relatively stable over time.

soil structure. Soil-dwelling organisms help decompose material and thereby release nutrients that are essential for plants. In this sense, soil life recycles dead organisms back into the raw materials necessary to support new life.

Below the O horizon, the A horizon is a mix of mineral fragments and organic matter. Water infiltrating through the A horizon leaches, or drains away, elements from there and delivers them to deeper soil horizons (Figure 9.22b). The underlying B horizon is composed primarily of mineral matter and constitutes what is typically called subsoil. It is less fertile than topsoil as it has less organic matter. Clays and elements dissolved in infiltrating water build up in this layer. Finally, the lowest horizon, the C horizon, is composed primarily of weathered parent material—rock or sediment.

⬡ **TAKE-HOME MESSAGE** Soil is a thin, complex mixture of materials and organic matter that forms on Earth's surface and is arranged into layers called horizons. It consists of mineral particles weathered from the rock cycle, dead and decaying plant and animal matter, and a multitude of organisms that live within this material.

9.6 What Makes Soil Alive?

Soil is rich with life and supports communities of animals, insects, worms, and microorganisms. Agricultural research shows that the most fertile soil supports abundant life. One study of New Zealand sheep pastures estimated that the weight of the worms underground equaled that of the flock of sheep grazing above. Tiny wormlike creatures called nematodes populate the upper levels of the A horizon in numbers exceeding 100,000 per cup of soil. Each nematode consumes thousands of bacteria per minute and excretes nutrient-rich micro-manure (**FIGURE 9.23**). Much larger earthworms pull organic material from the surface as they burrow up and down through the soil. In this way, worms not only digest material in the soil but also move and mix soil and organic matter as they tunnel. Over the course of a few centuries, the soil of an undisturbed field can come to consist almost entirely of nutrient-rich worm excrement known as *castings*. It is estimated that the digestive tracts of worms process as much as 15 tons of soil per acre each year in fertile ground.

But even more prevalent are much tinier microorganisms such as bacteria and fungi. If you scooped up a single cup of rich, fertile topsoil, it would contain more microorganisms than the number of people on Earth. Microorganisms speed the decomposition of organic matter, which releases carbon dioxide (CO_2) and produces vital molecules in forms that plants can take back up and reuse to spur new growth.

Organic Matter and Soil Structure

Not all of the organic matter working its way through the community of soil organisms decomposes quickly. Complex organic compounds can be quite resistant to breakdown, and this organic matter imparts the dark color most people associate with dirt or soil. A particularly important organic material known as **humus** is a complex, dark, sticky material that can remain relatively stable over time—decomposing at rates of less than 3% by weight per year. In fact, in some areas humus compounds have persisted for thousands of years (**FIGURE 9.24**). Humus is sticky and spongy and allows for a high degree of water and nutrient retention. A soil rich in humus contains an ideal arrangement of soil clumps interspersed with large open spaces that act as passages for the movement of air, water, and nutrients.

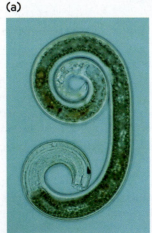

FIGURE 9.23 Nematodes and Worms (a) In the upper levels of the soil's A horizon, there can be more than 100,000 nematodes like these in a cup of soil—each consuming thousands of bacteria per minute. (b) Earthworms mix the soil and air as they tunnel beneath the ground.

(a)   (b)

FIGURE 9.24 Organic Matter in Soil These two images show soils in the Amazon basin. The photo on the right shows the lighter-colored, less fertile soil typical of the region. The photo on the left shows *terra preta*, or "dark earth," which was enriched over centuries by humans living in the area. They added charcoal, bone, and manure to increase the organic matter of the soil and boost fertility for crop production.

Worms Make Soil

Charles Darwin's best-known book, *The Origin of Species*, was not the one that sold the most copies in its first year. That honor goes to the last book he wrote, *The Formation of Vegetable Mould, Through the Action of Worms*, published in 1881 six months before he died. Darwin devoted the last years of his life to watching, experimenting with, and thinking about the effects of worms on the soil. While we know of Darwin mostly through his revolutionary influence on thinking about evolution, his last book helped radically reshape thinking about soil (what he called vegetable mould).

Darwin's interest started shortly after he returned to England to digest all he had seen on his fateful voyage aboard HMS *Beagle*. Darwin's father-in-law brought something odd

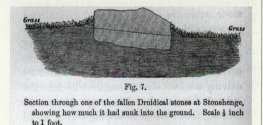

Fig. 7.

Section through one of the fallen Druidical stones at Stonehenge, showing how much it had sunk into the ground. Scale ½ inch to 1 foot.

Charles Darwin was curious about soil formation and the role of worms—which he explored in this work. One site Darwin studied was Stonehenge, where he saw the work of earthworms. He wrote, "The worms eject the earth which they have swallowed beyond the circumference of the stones; and thus the surface of the ground is raised around the stone. As the burrows excavated directly beneath the stone after a time collapse, the stone sinks a little."

to his attention. In several fields, his father-in-law had applied a layer of lime and burnt clay with cinders to cover the ground surface. Now, a few years later, this layer lay buried beneath fresh topsoil. What had done this?

It appeared as if the cinders had worked their way down into the soil. But Darwin thought it inconceivable that the pebbles and cinders could sink into the ground and still retain their initial arrangement as a well-defined layer. Close examination of the ground surface revealed that every other inch or so there was a worm burrow with a tiny heap of castings. Could these add up to build new topsoil given enough time? Did the collective action of a great number of worms move the layer of cinders downward, or did their activity build new soil on top of the layer?

After considering and ruling out other possible causes, Darwin concluded that worms had indeed built a new layer of soil atop the cinders, and he launched what became a lifelong investigation of the effects of worms on soil formation. He began observing their habits and measuring how much of their castings they pushed up and out of their burrows; he also noticed how the little piles retained their form until the first rain washed them away, spreading fresh soil across the ground surface.

Darwin published his first observations of worms in 1840. Several years later when he moved to the countryside, he spread a layer of chalk on one of his fields, intending to return in the future to see how far down it had been buried. Colleagues thought him daft for caring about, let alone studying, lowly worms. At the time, most scientists considered worms insignificant pests whose actions were incapable of amounting to anything significant.

But Darwin was not easily deterred from an intriguing idea. He entertained visitors with experiments on worms he kept in soil-filled pots that dotted his living room. He studied how worms ate, what foods they preferred, their sex

lives, and how they reacted to noise and light. He counted up the worm burrows in his fields and found there were more than 50,000 per acre. Each one of these creatures dragged vegetation down into its burrow, consumed it and mixed it with mineral soil, and then pushed the fertile result up and out of its burrow. Darwin was amazed at how much soil "God's ploughmen," as he came to call worms, could move. He calculated that over the course of a few centuries, every particle of English soil would pass through the intestines of worms.

Darwin even engaged his adult children to seek out churches abandoned at known dates in English history and far older Roman ruins so they could dig down to see how deeply the floors and foundations of these old buildings lay buried. From this he could determine how fast the overlying soil had formed. In the action of worms, Darwin saw the role of life in soil building. He went so far as to argue that worms helped shape the rolling hills of the English countryside. Small things, he realized, can add up to large effects if given enough time—a direct parallel to his more famous insights about evolution.

? What Would You Do?

Some species of earthworms are at high risk of becoming extinct. These include the endangered Oregon giant earthworm (*Driloleirus macelfreshi*) that lives in the forest soil of that state and the giant Palouse earthworm (*Driloleirus americanus*) native to the grasslands of Idaho and Washington. Both of these worms are threatened by habitat loss due to housing and agricultural development. What do you think would be required to protect these species? Would you support these kinds of conservation measures? Why or why not?

Soil Classification and Factors Affecting Soil Formation

Anyone who gardens can tell you that there can be big differences in the type of soil from one place to another and that this can have significant consequences for what can or can't grow. There are several hundred thousand specific soil types on Earth—more than 20,000 in the United States alone—and the characteristics of soil exert a significant influence on the community of life that survives aboveground. Although there is a great diversity of soil types on Earth, soil scientists classify various types by their physical features, which are a result of the environment in which they form. The *biomes* we learned about in Chapter 4—regions of Earth that host certain types of ecosystems—are thus linked to different types of soil. For example, temperate grasslands, such as mixed-grass prairie in the Great Plains of the United States, thrive atop soils with deep and dark A horizons rich in humus, while deserts such as the Mojave in the southwestern United States exist on exceedingly thin, lightly covered porous soils that often lack an O horizon (**FIGURE 9.25**).

Soil scientists sum up the factors that account for this diversity with the acronym CLORPT. The "CL" stands for climate, the single most important factor in determining soil characteristics. This includes the average precipitation and temperatures of an area. Wetter and warmer climates lead to more rapid leaching away of soluble materials, while drier and cooler climates slow the time it takes materials to leach nutrients away. The "O" stands for organisms, which highlights the role that biological communities play—something we emphasized earlier in this section. "R" stands for relief, or topography. In relatively flat areas such as wide valley bottoms, soil can accumulate to form thick layers, while on steep slopes, weathered material may wash or slide away before it can accumulate. "P" is for parent material, the type of rock or sediment the soil formed from. The parent material affects the size, texture, and chemical properties of the soil-forming particles that weather from it. Finally, soil formation takes time—signified by the "T." Areas that have been able to accumulate soil over long periods without disturbance will have thicker, better-developed soil profiles than those where soil formation is just beginning.

Topsoil and Plant Growth

Planting a bean plant in a cup of topsoil has long been a popular science activity for grade school students. After adding a little water and with a bit of patience, the students can watch the seedling unfurl and rise above the surface. But there are a few ways this activity can go wrong. The students can overwater, saturating the area around the roots, depriving them of oxygen. Or the students could pack the soil into the cup too tightly with their fingers, compressing the pores between the soil granules, again depriving the roots of oxygen that aerates these spaces. And if the students don't transplant their young, growing plants to a bigger container, the plants will not thrive because the tiny cup will soon confine root development. While getting their hands a little dirty and seeing seeds "in action" in this exercise, students are also seeing the delicate balance that must be maintained to get plants to grow. How saturated is the topsoil? Is it too compacted to facilitate aeration? A balance of these properties in topsoil

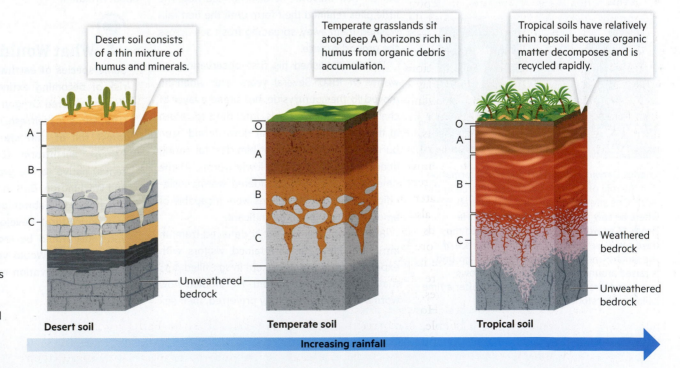

FIGURE 9.25 Biome Soil Types Biomes are categories of ecosystems, and each has a characteristic soil type. Example types for (a) normal deserts, (b) temperate areas, and (c) tropical areas are shown.

Desert soil consists of a thin mixture of humus and minerals.

Temperate grasslands sit atop deep A horizons rich in humus from organic debris accumulation.

Tropical soils have relatively thin topsoil because organic matter decomposes and is recycled rapidly.

Unweathered bedrock

Weathered bedrock

Unweathered bedrock

Desert soil

Temperate soil

Tropical soil

Increasing rainfall

characterizes what farmers call its **tilth**, the soil's overall structure and conditions that facilitate plant growth.

As noted earlier in the chapter, soil is subject to erosion. When topsoil is forming faster than it is eroding, it thickens, which allows more plant growth to be supported. When the reverse is true, the soil thins. In general, topsoil accumulates very slowly. The US Department of Agriculture estimates that it takes centuries to produce an inch of topsoil. Accelerated erosion can strip off topsoil far faster. And as we will see in the next section, human activities can have a big impact on the rate at which topsoil is eroded, degraded, or rebuilt.

Water in Topsoil

The amount of water that topsoil can hold is a very important characteristic for plant growth. Plants consume water in the soil by absorbing it and transporting it up through the plant. If there is not enough water available to the roots, the leaves may wilt to conserve water, an action that slows or completely halts photosynthesis. Or as we saw in our example with schoolchildren, if there is too much water, roots will effectively drown from the lack of oxygen. Obviously, topsoil's water supply depends on factors such as rainfall or irrigation (watering) systems provided by humans, but the characteristics of the soil—in particular its proportions of clay, silt, and sand particles, and the amount of soil organic matter—determine the ability of soil to retain water.

Water in topsoil also plays a role in two processes that inhibit plant growth. The first, **salinization**, is when mineral salts build up in the soil, sometimes enough to impact growth. All water contains dissolved salts, and when water evaporates, it deposits its formerly dissolved load of salts in the soil. However, if the soil drains poorly or is repeatedly irrigated, salts can accumulate, and soil salinity can rise to levels that inhibit plant growth. High salt concentrations limit the ability of roots to draw up water and may even draw water out of the plant roots and back into the soil (**FIGURE 9.26**). Second, while water can convey soil nutrients into plants through their roots, it also can carry nutrients away through **leaching**. In this process, driven by water, mineral elements move down to deeper soil layers, potentially beneath the reach of plant roots, to be carried away in groundwater or stream flow. Leaching and chemical weathering also affect soil pH. Decomposing organic matter tends to increase the acidity of soil, decreasing its pH. More weathered minerals decrease its acidity and increase its pH. While different plants and soil organisms are adapted to tolerate various, often narrow, pH ranges, most plants do best in neutral to slightly acidic soil. However, some plants thrive in extremes. For example, cranberries thrive in the highly acidic conditions of peat bogs full of decomposing matter.

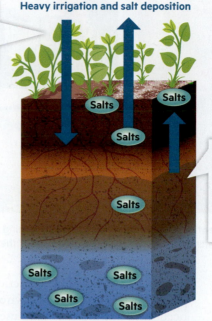

Heavy irrigation and salt deposition

When irrigation water evaporates, it leaves behind dissolved salts that accumulate in the soil.

Overirrigation and poor drainage can cause the water table to rise, dissolving mineral salts in the soil and carrying these back up toward the surface.

FIGURE 9.26 Salinization Excessive irrigation can cause salts to accumulate in the soil at levels that plants cannot tolerate. Salts are visible in the soil at the surface in this field.

🛑 **TAKE-HOME MESSAGE** A great variety of organisms live in soil and Earth hosts a great diversity of soils. Key factors shaping the characteristics of soil include climate, organisms, relief, parent material, and time (CLORPT). Topsoil refers to the uppermost layers of soil, and plant growth depends on soil structure and the balance of aeration, water retention, and nutrient availability. The amount of water in topsoil is a crucial factor affecting plant growth. Erosion can quickly remove topsoil that has taken centuries to build.

9.7 How Can Humans Repair and Sustain the Land?

Changes that are driven by Earth processes such as plate tectonics or erosion are inevitable, and at best humans can attempt to adapt to or mitigate their

tilth the soil's overall structure and conditions that facilitate plant growth.

salinization when mineral salts build up in the soil.

leaching a process driven by water where mineral elements move down to deeper soil layers, potentially beneath the reach of plant roots, to be carried away in groundwater or stream flow.

FIGURE 9.27 **Current Conventional Agriculture** Conventional farming practices such as tilling, irrigating the soil, and applying a lot of synthetic fertilizers and pesticides are carried out to increase crop yields in the short run, but they can also have negative impacts on the soil and its fertility over the long run.

tillage preparing the soil for planting by breaking it up and turning it over with a plow.

impacts. However, human actions can degrade the land, and many forms of environmental degradation occur as a side effect of land use—such as soil degradation and erosion by certain farming practices or the destruction of headwater streams as a result of mountaintop removal for mining operations. But degraded landscapes can also be rehabilitated or restored after human use. In this section, we will look at some of the options to restore and repair the land and use it sustainably, covering impacts from agriculture and

mining first, and then focusing on those from the loss of forests and growth of urban areas.

Agriculture and Grazing

While farming enabled humans to settle in places that grew to become the towns, cities, and nations in which we live, farming practices often accelerate soil erosion and degrade soil fertility (**FIGURE 9.27**). One of the largest impacts comes from preparing the soil for planting by breaking it up and turning it over with a plow, a process called **tillage**. This

Terracing
Slopes are cut to form stair-like ledges, which slow water as it flows downhill, reducing the amount of soil that is carried away by runoff.

No-Till Agriculture
Seeds are planted without plowing up the soil and disturbing beneficial soil communities. Dead plants from past seasons remain on the ground to help hold soil in place and to regenerate soil as they decompose.

Windbreaks and Hedgerows
Rows of trees and large bushes are planted between fields to reduce wind erosion and provide habitat for predators that reduce pest populations.

FIGURE 9.28 **Agricultural Methods That Reduce Soil Loss and Build Soil Health**

Cover Crops
Fields are planted continually throughout the year to avoid seasons of bare soil after harvests that make soil more vulnerable to erosion.

The Erosion of Civilizations

Erosive farming practices helped undermine the Roman Empire. How so? Plowing a field leaves the soil bare and exposed to erosion by wind or rain, leading to soil loss. Because of this effect, plow-based farming is only sustainable on flat-lying floodplains and deltas where sediment eroding from upland areas regularly replenishes soil. Not coincidentally, these are the environments where agricultural civilizations flourished for millennia on the floodplains of major rivers such as the Nile, Tigris and Euphrates, Indus and Brahmaputra, and those of lowland China. But in the Roman heartland in central Italy, important agricultural fields were on soil-covered upland slopes. Further, Roman tax records document the importance of agricultural regions in Syria and North Africa, where bountiful harvests were also from upland landscapes. These areas eventually eroded down to bedrock, and even today they have not recovered enough to support the existing populace. As these important fields feeding Rome and its empire eroded

The Romans suffered from the problem of diminishing soil fertility due to plow-based agriculture.

away, food shortages and riots repeatedly swept the capital.

So could the same thing happen again? Today there are regions, such as parts of the American Southeast, where the topsoil is gone and farmers plowing the subsoil bolster yields through chemical fertilization. And while modern conventional agriculture can enhance productivity in the short run, reliance on the plow and

intensive use of chemical fertilizers degrades soil organic matter and fertility over the long run. The accompanying disruption of ecosystems below- and aboveground also can lead to dependence on pesticides for controlling pests and pathogens. So far, conventional agriculture has led to the loss of about half of the soil organic matter originally in American agricultural soils. This has, in turn, increased reliance on chemical fertilizers to maintain harvests.

What Would You Do?

Some of the negative impacts of plow-based agriculture, which remains the dominant technology today, are described here. Do you think that farmers should be exploring new technologies? Or possibly only farmers in some areas? Given what you know, if you were a leader in your state's department of agriculture, what would you encourage farmers to do?

practice provides effective weed control but exposes the soil to wind and water erosion while also disrupting earthworms, beneficial fungi, and communities of soil-dwelling organisms. Other common practices, such as overuse of pesticides, herbicides, or fertilizers and even overirrigating crops, can damage the land too. Similarly, when ranchers graze too many livestock in an area for too long, it expands areas of bare earth that are eroded quickly by rain and wind. Carried out across generations, overgrazing can even create deserts, a process known as desertification.

Sustainable agricultural practices can have the opposite effect, enhancing soil production and building soil quality (**FIGURE 9.28**). Some of these methods are ancient. The thick, fertile black soils known as *terra preta* in the Amazon basin were created by people adding ground bone, charcoal, and manure to otherwise nutrient-poor soil near human settlements over thousands of years (see Figure 9.24). Another ancient farming method, terracing, is widespread in Asia and was also used by the Inca in the uplands of South America. This radical alteration of the landscape turns steep slopes into a series of flat

steps, dramatically reducing hillside soil erosion—as long as the terraces are maintained. Planting ground cover such as hedgerows between rows of crops can also greatly reduce erosion, as can planting cover crops once a field is harvested. Adding organic fertilizer such as compost, manure, and crop residues to soils can return nutrients that bolster fertility. Farmers in the United States and elsewhere around the world are increasingly adopting no-till methods that minimally disturb the soil and planting cover crops to reduce agricultural soil loss and build soil health (**FIGURE 9.29**).

Likewise, not all grazing practices degrade the soil. The fertilizing effects of livestock manure have long been recognized. Grazing can be carefully managed by rotating the livestock to new areas to integrate grazing (and manuring) into efforts to restore degraded soils. In some grassland habitats, such as in southern Africa and the American Midwest, intensive grazing followed by a longer recovery period between grazing events can promote native grasses, build up soil organic matter, and increase water retention in

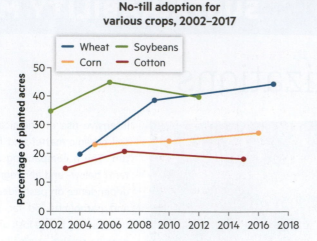

No-till adoption for various crops, 2002–2017

the soil. This style of intensive rotational grazing emulates the natural patterns of large herds of herbivores such as bison that long roamed native grasslands of North America.

Mining

There is no such thing as sustainable mining, and as also discussed in this chapter's earlier section on minerals, mining is associated with a long list of environmental impacts (see **AT A GLANCE: MINING IMPACTS**). However, abandoned mine sites can be reclaimed and rehabilitated to improve their ecological and economic value. Efforts to reclaim mine sites generally seek to return degraded land to an economically usable state and to reduce environmental impacts. Typical measures include importing topsoil and planting trees or reintroducing livestock on rangeland where grazing better emulates the natural ecosystem. For example, the remote Ereen Mine in northern Mongolia was restored to grazing land after its closure. In the European Union, more than half of the former mines that have been reclaimed are now forests or grasslands. And almost three-quarters of reclaimed mines in China are used for farmland. A particularly intriguing example of mine reclamation is the Eden Project, the world's largest greenhouse and a center of environmental education, located in an abandoned clay mine in Cornwall, England (**FIGURE 9.30**).

Forestry

Humans are causing *deforestation*—the loss of forestland—at a global annual rate of about 13 million hectares (32 million acres) per year. This means that each year, forest covering an area larger than Pennsylvania is lost to either logging or burning. Removing large numbers of trees in a forested area eliminates the primary source of organic material and the binding capacity of the trees' roots. It also exposes the soil to the eroding force of rain and wind. Harvesting trees from steep slopes

FIGURE 9.30 **The Reclaimed Mines of the Eden Project in Cornwall, England**

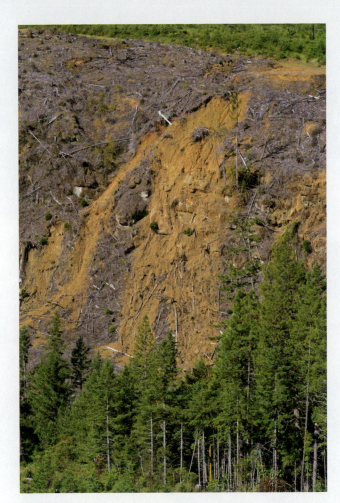

FIGURE 9.31 Deforestation and Erosion Deforestation, such as the clear-cutting timber harvest depicted in Oregon here, speeds the erosion of topsoil. On steep slopes, like the one shown here, it can also cause landslides.

increases the susceptibility of the latter to landslides during intense rainstorms (**FIGURE 9.31**). Tropical rain forests are particularly vulnerable to deforestation because in these high-rainfall areas, most of the available nutrients are held in the decomposing organic matter of the topsoil O horizon, as nutrients in the lower horizons have been leached away by infiltrating rainwater. So rain forests persist by efficiently recycling nutrients from the trees to the soil and back again. When the trees are removed, the loose topsoil rapidly erodes along with the nutrients it holds.

As we learned in Chapter 5, *sustainable forest management* strives to reduce the impacts of forestry, including soil loss. The Forest Stewardship Council, an organization that certifies forest products that "promote environmentally sound, socially beneficial and economically prosperous management of the world's forests," advocates for sustainable forest management practices. These methods include selective and shelterwood cutting, which ensure that enough mature trees remain to prevent erosion and provide seeds for new trees; reducing or eliminating road building that causes erosion in forested areas; and prohibiting logging beside streams and on landslide-prone slopes.

Urbanization

There are a lot more people on the planet than there used to be, with far more living in the "urban" cities and suburbs. Of course, building a new city block, subdivision, or shopping area changes the landscape considerably. So what can be done to ensure sustainability in the face of a growing human population? First, many cities have embarked on programs to reclaim contaminated industrial sites (often called brownfields) such as abandoned factories, railroad yards, and gas stations. Other cities have restored coastal and riverine waterfronts as community amenities, such as San Antonio's River Walk and San Francisco's redeveloped tourist-friendly waterfront.

Going forward, more and more cities are focusing on sustainable building practices that can lessen the impacts of urbanization. Constructing rain gardens or bioswales—landscaping techniques that help rainwater infiltrate the ground—offers alternatives to conventional stormwater systems and can greatly reduce urban runoff. Creating urban green spaces and adopting design standards that promote walking (such as attractive street trees and wide sidewalks) are recognized as efficient ways to enhance urban environments. A number of cities have embarked on municipal composting programs where organic wastes are collected, composted, and sold back to residents as a soil enrichment product. Urban planners understand they can increase an area's resilience to natural disasters by simply placing critical infrastructure more strategically and avoiding hazardous locations. We focus more on urban planning and green building practices in Chapter 16.

Sustainable land management typically involves economic and political choices about how our values influence land use. Over the long run, understanding how to minimize environmental impacts early on is the key to producing and obtaining a needed resource sustainably and cost effectively. For example, adopting soil conservation practices can be easier to manage in the long run because they involve less input-intensive farming practices. They can also lead to higher yields. But the motivation must be there to make the change. And sometimes, as in the case of mining, serious impacts cannot be avoided. The cost and methods of rehabilitating mined land must be considered at the onset.

🛑 **TAKE-HOME MESSAGE** Land uses such as farming and grazing, mining, forestry, and urban development can have an enormous negative impact on the land. However, regenerative agricultural practices can enhance rather than degrade soil fertility. While mining, forestry, and urbanization impact landscapes, their effects can be mitigated to varying degrees by changes in practices or restoration efforts.

Mountaintop removal mining, which is a common method of coal mining in the Appalachian Mountains of the eastern United States, fills in stream valleys and levels hilly terrain.

Mining of sand and gravel from beaches, riverbeds, and the seafloor for use in construction and fracking destroys habitats and disrupts groundwater flow to communities. This operation in Ontario, Canada, demonstrates just how much land can be disrupted for sand mining.

Water pollution by heavy metals and acidic drainage from mine waste can decimate rivers downstream. Failure of an earthen dam at the Ok Tedi Mine in Papua New Guinea sent copper-rich mine waste coursing down the Fly River, poisoning fish and contaminating surrounding farmlands.

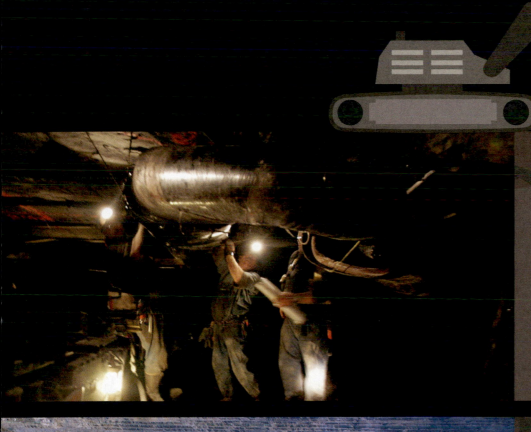

Subsidence and sinkholes can result when underground mining of resources such as salt and coal triggers collapses of the land above. Nearby residents have filed 10 lawsuits against the Pinnacle Mine in West Virginia, alleging that the coal-mining activities have caused subsidence and damage to their properties.

Open-pit mines create enormous artificial craters. The Bingham Canyon Mine in Utah, primarily producing copper, is the largest open-pit mine in the world, about 2.5 miles wide and a little more than half a mile deep.

Working in mines is one of the world's most dangerous trades. Hazards include respiratory ailments from dust and fumes (a constant presence at this sulfur mine in Indonesia), accidents from rock falls and heavy machinery, and exposure to hazardous chemicals.

9.8 What Can I Do?

The scale of Earth's surface and even that of individual landforms is enormous compared to that of a single person, but there are still steps you can take to protect the land and adapt to its changes. As we've seen in this chapter, human activity can alter the land, but it can also restore it.

Conserve and Recycle Rare Earth Minerals

Each of us is now carrying around small amounts of rare earth minerals in our consumer electronics. This ties us to the impacts of mining these resources. Unfortunately, recycling rare earth minerals is much more difficult than for most other materials, because they are often distributed in tiny amounts at the molecular level throughout device components. And there is no curbside pickup for these materials. But you can still reduce your impact.

First, as with all material use, you can avoid buying things that you will not use very much and replace electronics only when necessary: these practices reduce your waste production. Second, although studies show that most people hoard their unused electronics in drawers and closets or toss them in the trash, often the best thing to do is to send them back to the manufacturer. Many (though not all) companies producing phones, tablets, and computers have take-back programs for old devices. These companies are in the best position to refurbish and reuse the product or make use of its components. Many electronics companies are developing design processes that make it easier for them to recycle materials from their devices. If the company that made your device doesn't have a take-back program, see if your city or town has an e-waste recycling collection program. These programs seek to ensure that the devices will not be disassembled or disposed of in unsafe conditions.

Know How Your Food Affects Soil Health

We will learn in Chapter 12 that your decisions about what to eat connect you to the environmental impacts of the agricultural practices that produce your food. This includes topsoil loss and/or degradation caused by plowing, the overuse of pesticides and synthetic fertilizers, and overirrigation. Becoming aware of the agricultural practices that produce the food you buy is a first step in taking action to address topsoil loss and degradation.

The next time you buy groceries, do some research on the products you bring home. Try to learn more about how the food was produced, and use what you learned in this chapter to assess impacts on the land. You can also check where your school gets its food from. Does it have a student farm? Does its dining services source food from local farms? If so, what kind of farming practices do they use?

Get Informed about Development Decisions in Your Area

Most cities and counties in the United States have boards or committees governing things such as land use, development, planning, and flood management. Research what institutions oversee policies and decisions affecting floodplain development or sustainable building practices where you live. Attend a meeting or look at some recent decisions, and use what you learned in this chapter to determine the likely impact of these decisions on impervious surfaces, runoff, erosion, and flooding.

Be Prepared for Natural Hazards

Dynamic landscapes are a fact of life for many parts of the United States. Are you aware of and prepared for the natural hazards in your region? For example, if you live in an earthquake zone, it would be a good idea to have an emergency preparedness kit that includes a first aid kit, survival supplies, and emergency food and water. Similarly, an emergency preparedness kit would also be helpful if you live in a flood or fire danger zone, and it would also be wise to learn the safe evacuation route for your neighborhood. The Centers for Disease Control and Prevention provides advice on how to prepare for and be safe during natural disasters: see www.cdc.gov /disasters/index.html.

Chapter 9 **Review**

SUMMARY

- Earth consists of three layers: the core, mantle, and crust. The outer surface of our planet is broken up into rigid tectonic plates that ride on top of the more malleable asthenosphere.

- The shifting of Earth's tectonic plates causes surface events, including volcanic activity and earthquakes, and governs the formation of mountains and lowlands.

- The material in Earth's crust is continually broken down and recycled over time.

- Physical and chemical weathering break down rocks, and erosion moves weathered rock particles and deposits them in places where they accumulate as sediment. Intense heat and pressure metamorphose one type of rock into another.

- Geologic processes shape the mineral resources we rely on. Human mining of these resources causes adverse environmental impacts, including water pollution, disrupted groundwater flows, changes to the landscape, habitat destruction, and production of hazardous waste.

- The processes that erode, transport, and deposit sediment shape landforms such as hillslopes, floodplains, rivers, and streams. Soil consists of sediment, organic matter, and the abundant organisms that live within this material. It is arranged into layers called horizons.

- The main factors that determine soil type are climate, organisms, relief, parent material, and time (CLORPT). Plant growth depends on soil type, aeration, and water and nutrient content.

- Land uses such as farming and grazing, mining, forestry, and urban development can have an enormous impact on the land, hastening erosion that can quickly remove topsoil that has taken centuries to build.

- Soil loss and damage can be mitigated and restored with regenerative agricultural practices, sustainable forestry, and low-impact development.

KEY TERMS

asthenosphere	igneous rock	mineral	tillage
core	impervious	rock cycle	tilth
crust	infiltration	salinization	topography
delta	landform	sediment	topsoil
erosion	leaching	sedimentary rock	watershed
fault	lithosphere	soil	weathering
floodplain	magnetosphere	soil horizon	
glacier	mantle	soil profile	
humus	metamorphic rock	tectonic plate	

REVIEW QUESTIONS

The letters following each Review Question refer to the Chapter Objectives.

1. What three layers make up Earth's internal structure? **(B)**

2. What are the differences between the lithosphere and the asthenosphere? **(B)**

3. What causes the movement of Earth's tectonic plates? Name two types of geologic phenomena caused by this movement. **(B)**

4. Explain how the rock cycle recycles the materials of Earth's surface. **(B)**

5. Compare physical weathering and chemical weathering and provide an example of each. **(B)**

6. Name two forces that cause erosion and two types of places where eroded material accumulates as sediment. **(B)**

7. Describe how minerals are essential to human health and society. **(C)**

8. Describe two environmental impacts caused by mining. **(E)**

9. Match each part of a watershed to the erosional or sedimentary process that occurs there. **(A, B)**

10. What types of materials compose soil? **(D)**

11. Name three factors that characterize soil type. **(D)**

12. What properties of topsoil facilitate plant growth? **(D)**

13. Describe two strategies for restoring soil health. **(E)**

FOR FURTHER THOUGHT

The letters following each item refer to the Chapter Objectives.

14. Some of the major human activities responsible for soil loss and degradation covered in this chapter include agriculture and grazing, forestry, and urbanization. Do some research on your local area, and describe what you think are the most significant threats to soil in your area. What steps could be taken to address these threats? **(D, E)**

15. In this chapter, we outline some of the environmental impacts of mining and some of the ways mine sites can be reclaimed. Unfortunately, these sites, after they have been productively mined, are often abandoned by the companies that operated them. How would you design a policy to ensure that mining sites are reclaimed in a way that addresses their environmental impacts? In your description, be sure to explain how you would fund the work on these sites. **(C, E)**

16. Rare earth minerals are now an important part of many of the items we rely on, including our phones, other electronics, batteries, and even renewable energy technologies such as wind turbines. As we saw in this chapter, the people who work in these mines and processing facilities often face unsafe conditions and low pay. Identify two or three items in your own life that rely on rare earth minerals. Would you be willing to pay more for these items if the environmental and labor conditions involved in their production were improved? Why? And if so, how much more? What sort of process could be put in place that would give consumers confidence that such improvements were being made? **(C, E)**

Make Your Case?

The following exercises use real-world data and news sources. Check your understanding of the material and then practice crafting well-supported responses.

Use the News

The following article from the *Smithsonian* magazine website discusses issues with soil degradation and salinization. Use this article to answer the questions that follow. The first three questions test your understanding of the article. Questions 4 and 5 are short answer, requiring you to apply what you have learned in this chapter and cite information in the article. Answers to Questions 1–3 are provided at the back of the book. Question 6 asks you to make your case and defend it.

"Earth's Soil Is Getting Too Salty for Crops to Grow," Sarah Zielinski, Smithsonian.com, October 28, 2014

In the . . . film *Interstellar*, Earth's soil has become so degraded that only corn will grow, driving humans to travel through a wormhole in search of a planet with land fertile enough for other crops. In the real world things aren't quite so dire, but degraded soil is a big problem—and one that could be getting worse. According to a new estimate, one factor, the buildup of salt in soil, causes some $27.3 billion annually in lost crop production. "This trend is expected to continue unless concrete measures are planned and implemented to reverse such land degradation," says lead author Manzoor Qadir, assistant director of water and human development at the United Nations University Institute for Water, Environment and Health. Qadir and his colleagues published their findings October 28 in *Natural*

Resources Forum. Irrigation makes it possible to grow crops in regions where there is too little rainfall to meet the plants' water needs. But applying too much water can lead to salinization. That's because irrigated water contains dissolved salts that are left behind when water evaporates. Over time, concentrations of those salts can reach levels that make it more difficult for plants to take up water from the soil. Higher concentrations may become toxic, killing the crops.

Qadir and his colleagues estimated the cost of crop losses from salinization by reviewing more than 20 studies from Australia, India, Pakistan, Spain, Central Asia, and the United States, published over the last two decades. They found that about 7.7 square miles of land in arid and semi-arid parts of the world is lost to salinization

every day. Today some 240,000 square miles—an area about the size of France—have become degraded by salt. In some areas, salinization can affect half or more of irrigated farm fields.

Crop production is hit hard on these lands. In the Indus Valley of Pakistan, for instance, salinization causes an average decline in rice production of 48 percent, compared to normal soils in the same region. For wheat, that figure is 32 percent. Salty soils also cause losses of around $750 million annually in the Colorado River basin, an arid region of the US Southwest.

"In addition to economic cost from crop yield losses, there are other cost implications," Qadir says. These include employment losses, increases in human and animal health problems, and losses in property values of farms with degraded land.

There could be associated environmental costs as well, because degraded soils don't store as much atmospheric carbon dioxide, leaving more of the greenhouse gas to contribute to climate change. The total cost of salt degradation, therefore, could be quite a bit higher than the most recent estimate.

Salt damage can be reversed through measures such as tree planting, crop rotation using salt-tolerant plants, and implementing drainage around fields. Such activities can be expensive and take years, but the cost of doing nothing and letting lands continue to degrade is worse, the researchers argue. "With the need to provide more food, feed, and fiber to an expanding population, and little new productive land available, there will be a need for productivity enhancement of salt-affected lands in irrigated areas," they write.

On a cautiously hopeful note, Qadir adds that the issue is reaching the ears of policy makers: "Amid food security concerns, scarcity of new productive land close to irrigated areas and continued salt-induced land degradation have put productivity enhancement of salt-affected lands back on the political agenda," he says. "These lands are a valuable resource that cannot be neglected."

1. According to the article, what is causing the buildup of salt in the soil?

   a. irrigation
   b. the application of fertilizers containing salts
   c. runoff from the salting of roads in the wintertime
   d. salts rising up from the bedrock beneath the soil

2. How did the researchers estimate the cost of crop losses from salinization?

   a. They conducted soil salinity studies throughout the world.
   b. They conducted soil salinity studies in the United States.
   c. They reviewed studies of salinization conducted in the past year on cropland in the United States.
   d. They reviewed studies on salinization in several countries conducted over the past two decades.

3. What does the article say is the cost of soil salinization on wheat production in the Indus Valley?

   a. 48% decline in production
   b. 32% decline in production

   c. $750 million loss annually
   d. $27.3 billion loss annually

4. Briefly describe the human, animal, and overall environmental harms associated with soil salinization that are discussed in the article.

5. The article says that salt damage can be reversed, but it is expensive. Briefly, explain if you think damage reversal is worth the investment, citing information in the article or chapter in your answer.

6. **Make Your Case** The article identifies several practices that can help to reverse soil salinization and expresses a hope that the issue is reaching policy makers. Do you think this is an issue that should be addressed by policy, addressed by other measures, or not addressed at all? Explain what you think should be done (or possibly avoided), supporting your answer with information from the article, the text, and other sources if applicable.

Use the Data

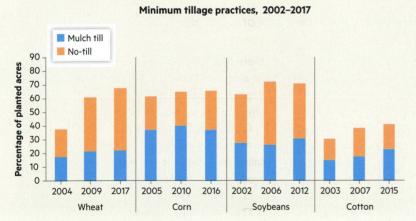

Minimum tillage practices, 2002–2017

The graph shown here is from a report produced by the US Department of Agriculture. The graph shows the number of acres by year planted for different crops in the United States that use either a no-till method or a "mulch till," which involves some periodic tilling but not nearly as much as conventional agricultural practices require. Use the graph to answer the questions that follow. The first three questions test your understanding of the graph. Questions 4

and 5 are short answer, requiring you to apply what you have learned in this chapter and cite data in the graph. Answers to Questions 1–3 are provided at the back of the book. Question 6 asks you to make your case and defend it.

1. On the basis of the data in this graph, which answer best describes the trend for combined no-till and mulch till practices?

   a. The total acres percentage is showing steady increases for wheat, soybeans, corn, and cotton.
   b. The total acres percentage is showing steady increases for wheat, corn, and cotton, but not soybeans.
   c. The total acres percentage has gone up and down over the years for all of the crops.
   d. The total acres percentage has gone up for wheat and cotton but is inconsistent for corn and soybeans.

2. On the basis of the graph, which of the following crops/year had the most acres planted using no-till methods?

   a. wheat in 2017
   b. wheat in 2009
   c. soybeans in 2012
   d. corn in 2010

3. Which of the following statements is true based on the information in the graph?

   a. In 2006, more than 70% of soybean crops were grown using either no-till or mulch till practices.

   b. For all three years reported for corn, it had the lowest percentage of no-till or mulch till acres planted out of all the crops listed.

   c. For all the crops reported, no-till practices always make up more of the total acres planted than mulch till.

   d. The graph reports data for the same years for each of the four crops.

4. On the basis of the graphs, do no-till and mulch till practices look like a success or a failure for farmers? Explain your answer.

5. Based on the information in the graph, if you looked at the acres planted with no-till or mulch till methods today, what do you think the approximate percentages would be, and how certain would you be of your answer and why?

6. **Make Your Case** Using the data in the graph, what recommendations would you make to a farmer who had opened up a new field for planting one of these four crops? Would you recommend these practices, and what else would you need to know to help make this recommendation?

LEARN MORE

- David R. Montgomery. 2007. *Dirt: The Erosion of Civilizations*. Berkeley: University of California Press.
- David R. Montgomery. 2017. *Growing a Revolution: Bringing Our Soil Back to Life*. New York: Norton.
- UN Food and Agriculture Organization. n.d. "The 10 Elements of Agroecology: Guiding the Transition to Sustainable Food and Agricultural Systems." Accessed January 19, 2019. www.fao.org/3/I9037EN/i9037en.pdf
- Video: How Plates Move/Plate Boundaries: digital.wwnorton.com/environsci
- Video: The Rock Cycle: digital.wwnorton.com/environsci
- Video: Forming Soil: digital.wwnorton.com/environsci
- US Department of Agriculture, Natural Resources Conservation Service "Soils" page: www.nrcs.usda.gov/wps/portal/nrcs/site/soils/home/
- US Geological Survey, Science Explorer "Natural Hazards" page: www.usgs.gov/science/science-explorer/Natural+Hazards
- Joel B. Gruver. 2013. "Prediction, Prevention and Remediation of Soil Degradation by Water Erosion." Nature Education Knowledge Project. www.nature.com/scitable/knowledge/library/prediction-prevention-and-remediation-of-soil-degradation-113130829

10

Systems and Cycles

Are We Harming Earth's Life Support System?

Imagine you're at home on a chilly winter evening and you want to warm things up, so you adjust the thermostat to raise the temperature a few degrees. But after a while you might notice that this warm air is also really dry, so you turn on a humidifier to bring more moisture into the air. You become more comfortable, but later when you're cooking something spicy in the kitchen, you find the smell overpowering, so you throw open a window to let some fresh air in. This causes the temperature to drop, so the furnace sends more hot, dry air into the house, and eventually you're adding more water to the humidifier so it can redouble its efforts. Each adjustment you make is impacting something else. Your heating system impacts your humidity system, and your cooking system impacts your air-quality control system (the window), which in turn affects the heating system. Even at a basic level, controlling the comfort of your home involves a series of interactions. And of course, in this example the climate control of your home is linked to larger power systems that are providing energy. So even on a small scale, the adjustments you're making are connected to sources well beyond the place you live.

What if instead of managing the comfort of a single room, you had to create your own environment to survive? You would soon become familiar with the idea that the environment functions as many nested **systems**: collections of components interacting with each other to produce outcomes that each component could not achieve on its own. This is the idea behind the 2015 movie *The Martian*, a fictional representation of someone stranded on Mars trying to create his own enclosed miniature world to survive. While the main character has initial success, at one point small amounts of oxygen escaping from a mask seal throw his habitat out of balance—causing an explosion that almost kills him. His attempt to create a safe place he can inhabit while he waits for help turns into an all-out push to escape the planet.

While *The Martian* is fantasy, a real attempt to re-create the systems that support our planet had been tried 30 years before the movie's

Can we make our own biosphere? In 1985, researchers attempted to make a self-sustaining enclosed environment that could support human life. The glass-enclosed structure in Arizona is called Biosphere 2.

Chapter Objectives

By the end of this chapter, you should be able to . . .

A. explain how systems can be open or closed and how their parts interact to produce outcomes.

B. describe the basic systems concepts of stocks and flows as well as reinforcing and balancing (positive and negative) feedback.

C. recognize that matter can cycle between the lithosphere, biosphere, hydrosphere, and atmosphere and that Earth is a closed system.

D. describe the biogeochemical cycles for oxygen, phosphorus, nitrogen, and carbon.

E. understand how human activities affect major biogeochemical cycles and the environmental consequences of these effects.

> You think that because you understand "one" that you must therefore understand "two" because one and one make two. But you forget that you must also understand "and."
>
> —Rumi

system a collection of components interacting with each other to produce outcomes that each component could not achieve on its own (Chapter 1).

open system a system that is affected by outside influences.

closed system a system that is self-contained, neither receiving inputs nor sending outputs beyond the system's borders.

release (though not on Mars). In 1985, construction began on Biosphere 2, a sealed steel, concrete, and glass structure covering more than 3 acres in the Arizona desert. Its builders attempted to create a self-sustaining habitat for humans by mimicking Earth features such as rain forests, rivers, and oceans in miniature and providing areas suitable for agricultural production (it was named Biosphere 2 because the builders considered Earth to be "Biosphere 1") (**FIGURE 10.1**). In 1991, eight people who called themselves "Biospherians" sealed themselves in Biosphere 2 for 2 years to see if it could sustain human survival. They survived but not without serious challenges and significant help from the outside. For example, Biosphere 2 failed to adequately provide oxygen (O_2), our most essential resource. Though Biosphere 2 was designed with lush communities of plants to provide O_2 for the humans to breathe, after several months the air in

Biosphere 2 had lost about 7 tons of O_2, dropping from 21% O_2 to just 14%. What happened? Unfortunately, the designers had failed to account for various effects. The oxygen was being used not only by humans and plants but also by the tiniest organisms in the soil, and it was even absorbed by the biosphere's structure itself. These unexpected reactions essentially caused Biosphere 2's life-support system to fail, and additional oxygen had to be added so that the Biospherians could breathe.

The problems of Biosphere 2 illustrate the complex workings of our planet and the importance of taking steps to ensure that we protect the life-support systems of "Biosphere 1." Earth is more than just a collection of resources: it is a multitude of interconnections and nested systems. In this chapter, we'll first examine what systems are and see how they can be represented or modeled using simple examples. Then we'll learn more about how some important elements—oxygen, phosphorus, nitrogen, and carbon—interact in ways crucial for life. In doing so we'll encounter some human impacts affecting outcomes on which we, and other life-forms, rely. Given the interconnectedness of Earth's systems, achieving sustainability will depend in large part on the way we address these impacts.

FIGURE 10.1 Biosphere 2 In Oracle, Arizona, an Earth systems research facility named Biosphere 2 was built to be a self-sustaining ecological system closed to outside material inputs and outputs.

Diagram adapted from MacCallum et al. (2004).

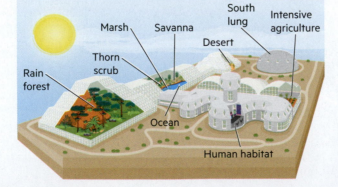

10.1 What Are Systems?

In the opening scenario of this chapter, we saw that simply attempting to maintain a comfortable climate in your home can make you aware of the way one change affects another. When scientists tried to solve the mystery of the missing oxygen in Biosphere 2, they needed to understand changes unfolding within this synthetic world as the product of a series of interactions. This meant they had to analyze how Biosphere 2 was operating as a system. Systems can be either closed or open to outside influence. In terms of energy flow, Earth is an **open system**—powered almost entirely by the rays of the Sun. But Earth is largely a **closed system** in regard to matter. A closed system is self-contained, neither receiving inputs nor sending outputs beyond the system's borders. Aside from a relatively tiny amount of space debris that enters Earth by collisions with meteorites and comets, matter on Earth is used and reused in different places, for different purposes, and in different states.

We already have described some natural systems in earlier chapters: ecosystems in Chapter 4, the water cycle in Chapter 7, weather systems in Chapter 8, and the rock cycle in Chapter 9. But systems can also be human creations. For example, plumbing operates as a system, as does the electrical grid and many of the other facilities operating within the cities and towns where we live. We have less tangible cultural, economic, social, and political systems too. So systems can involve a very broad array of objects, interactions, and concepts. To get started, let's first get a better understanding of systems in general. As an example, we will examine some of the most important systems to us all, the systems within our bodies. These biological systems enable us to live and make us who we are.

Parts, Interactions, and Emergent Properties

Your human body is composed of about 25 elements in various amounts and combinations. But what really distinguishes living things from a collection of elements is the way in which these elements form structures that work together (**FIGURE 10.2**). You are a living thing composed of trillions of cells of about 200 different varieties, with each cell being its own mini-system, with specialized parts that work together to process matter and energy, store and transmit genetic information, reproduce, and carry out many other functions. Further, in multicellular organisms like us, numerous cells work in a coordinated way to form organs or other parts of larger systems, such as the digestive, respiratory, and circulatory systems. In other words, systems are more than the sum of their parts. The stuff in your body interacts and forms a series of nested and coordinated systems that function to keep you alive and doing what you do.

Let's start with a simplified look at the system of breathing. Put your hand on your chest and take a breath. As your diaphragm and other muscles contract, the chest cavity expands and the pressure inside it decreases. This allows the pressure of the atmosphere to move about a half liter of air through your nose or mouth and into your larynx, trachea, bronchi, and lungs (**FIGURE 10.3**).

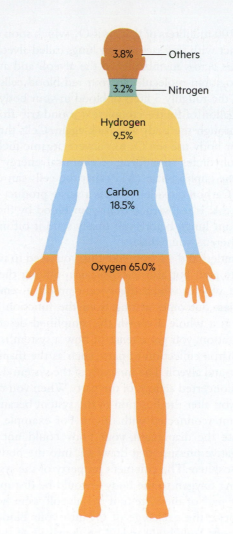

FIGURE 10.2 What Are We Made Of? The human body is a collection of elements. However, these elements work together to form a living organism—a type of system.

3.8% — Others
3.2% — Nitrogen
Hydrogen 9.5%
Carbon 18.5%
Oxygen 65.0%

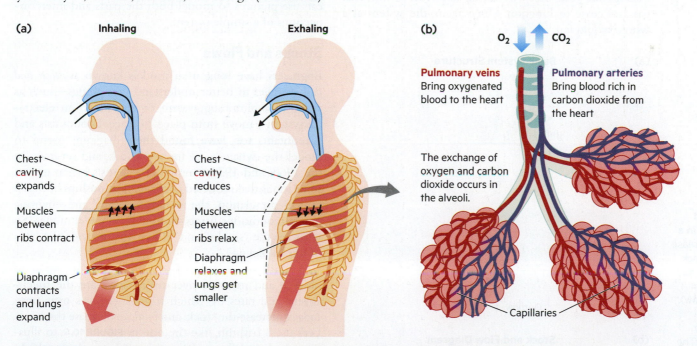

(a)

Inhaling

Exhaling

Chest cavity expands

Muscles between ribs contract

Diaphragm contracts and lungs expand

Chest cavity reduces

Muscles between ribs relax

Diaphragm relaxes and lungs get smaller

(b)

O_2 CO_2

Pulmonary veins
Bring oxygenated blood to the heart

Pulmonary arteries
Bring blood rich in carbon dioxide from the heart

The exchange of oxygen and carbon dioxide occurs in the alveoli.

Capillaries

FIGURE 10.3 Breathing as a System (a) One system each of us depends on to survive is the respiratory system, where specialized parts of our body, such as the diaphragm and the muscles between our ribs, work together to breathe in oxygen and exhale carbon dioxide. **(b)** Inside our lungs, specialized membranes on tiny sacs called alveoli exchange gases. This exchange provides the key element, oxygen, for many other systems throughout the body.

emergent property an outcome arising from the function of the system as a whole.

model a simplified concept or representation of a complex process that is designed to help understand interactions among different factors (Chapter 1).

stock a supply of something that we want to observe and measure over time.

flow the mechanism and rate by which a stock in a system changes over time.

About 100 milliliters of that air is O_2, which soon comes in contact with tiny sacs in your lungs called alveoli. The O_2 passes across membranes in the alveoli and attaches to hemoglobin molecules in your red blood cells. Each minute, the roughly 5 liters of blood in your body—just over a gallon of fluid—complete a round trip from the heart through the body and back again. Cells throughout your body use the O_2 to convert organic fuels such as carbohydrates to a form of chemical energy called adenosine triphosphate (ATP) that the cells can use for energy. Carbon dioxide (CO_2) is a waste product of this process that is reabsorbed from your blood by the alveoli in your lungs, before you finally send it off into the atmosphere as you exhale.

To understand a system like this, you need to identify not only *parts* but also *interconnections* among the parts, and the way in which the system produces **emergent properties**: outcomes arising from the function of the system as a whole. Even in this simplified description of respiration, you get a sense of how a system is much more than a collection of parts, such as the diaphragm, bronchi, and alveoli. The function of the system depends on the concerted actions of the parts. When you remove a part, you alter the function of the system because that part is interconnected with others. For example, if you eliminate the diaphragm, your body could not create the negative pressure that draws O_2 into the body from the atmosphere. The emergent property of the system—providing oxygen to the body—would be disrupted. In another part of the system, if the alveoli were not able to facilitate the exchange of O_2 into your blood, cells throughout your body would be deprived of this vital gas and cease to function. Once again, the system as a whole would fail.

Finally, consider also how the respiratory system is linked to other systems. Your circulatory system—heart, blood vessels, and blood—transports the O_2 and CO_2. Your nervous system uses receptors in your brain stem that detect CO_2 levels in your blood to regulate your breathing to maintain proper levels of O_2. And, as we'll see later in this chapter, planetary-scale Earth systems make the O_2 available for you to breathe in the first place.

⬡ **TAKE-HOME MESSAGE** Systems are sets of things that interact to produce an outcome distinct from the effect of each thing on its own. They can be either closed or open to outside influence. To understand a system, you need to identify its parts and their interconnections, as well as its emergent properties: outcomes arising from the function of the system as a whole.

10.2 How Can We Show How a System Works?

As you can tell from the example of your own respiratory system, systems are nested and interconnected in ways that make them quite complex. Scientists often use **models**, simplified concepts or representations of complex processes, to focus on the way certain aspects of a system function and interact. Models can take many forms, ranging from mathematical equations and computer simulations to simple diagrams. Let's see how we can use pictures to model both the parts and interconnections of a simple system.

Stocks and Flows

Engineers have long used models known as *stock and flow diagrams* to better understand how things—such as water in a plumbing system or current in an electrical system—move from place to place. Economists and accountants too have found these diagrams useful to model the dynamics of money, goods, and services. In the 1960s and 1970s, computer scientists began to run simulations that modeled complex relationships between human population, the world economy, and environmental conditions using stock and flow techniques. So what are the "stocks" and "flows" that provide the basis of this modeling?

A **stock** is a supply of something that we want to observe and measure over time. **Flows** are the mechanisms and rates by which a stock changes over time. *Inflows* increase the stock, and *outflows* decrease the stock. Let's use a bathtub, like the one in **FIGURE 10.4**, to illustrate stock and flow relationships. What is the stock? In this case it is the water in the tub. Where are the flows? There is an inflow from the faucet, and there is an outflow through the drain. We have designed this aspect of

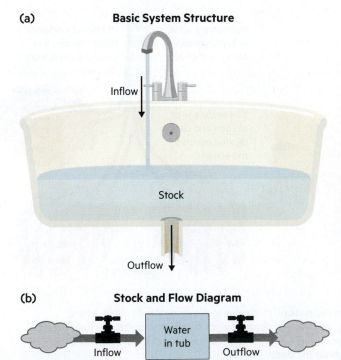

(a) **Basic System Structure**

Inflow

Stock

Outflow

(b) **Stock and Flow Diagram**

Inflow → Water in tub → Outflow

FIGURE 10.4 Stock and Flow Diagram of Water in a Bathtub (a) We can increase, decrease, or keep the stock of water in a bathtub steady by adjusting the faucets (inflow) and drain (outflow). **(b)** A simplified diagram represents this structure and these relationships. The stock of water in the tub can be increased by opening the faucet and decreased by opening the drain.

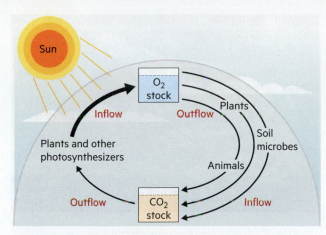

FIGURE 10.5 Biosphere 2 by Design This stock and flow diagram shows how Biosphere 2 was designed to function. Stocks of O_2 would be converted to balancing stocks of CO_2 by such sources as animals, plants, and soil microbes and then converted back to O_2 by photosynthesizers.

our plumbing system to be very easy to understand and operate. We manipulate the faucet and drain to regulate the amount of water in the tub. If you want the amount of water in the tub to increase, make sure the inflow outpaces the outflow. You can do this by plugging the drain (outflow) and turning on the faucet (inflow). If you want the amount of water in the tub to decrease, make sure the outflow outpaces the inflow: open the drain and turn off the faucet. If you want to keep the water level constant for your bath, you can either keep the drain plugged and turn off the faucet, or you can keep the faucet running at a rate that matches the outflow through the drain: either way the inflow equals the outflow, and the stock of water remains constant.

Similarly, we can use a simplified stock and flow diagram to help us understand where the O_2 was going in Biosphere 2. To do this, we illustrate a "two-stock" system that depicts the supplies of both O_2 and CO_2. **FIGURE 10.5** shows how Biosphere 2 was designed to function. We have a stock of O_2 within the facility when habitation begins. In this model, the outflows include the respiration of the humans as well as that of other animals, plants, and microbes. Notice how these outflows are an inflow for the stock of CO_2. First, recall that the process of **photosynthesis** (see Chapter 3) is a series of chemical reactions that uses water and CO_2 and stores this energy in molecules that are based on the element carbon, such as simple sugars. Plants and other *photosynthesizers* (organisms that perform photosynthesis) then provide an outflow for the CO_2 and an inflow for the stock of O_2 by producing more oxygen. The designers assumed that the inflows and outflows of O_2 would balance and the stock would remain constant.

But what was actually happening? **FIGURE 10.6** shows how several factors combined to lead the system to function differently than the designers assumed it would. First, the outflows from the O_2 stock were not all equal. The compost used in Biosphere 2 facilitated tremendous

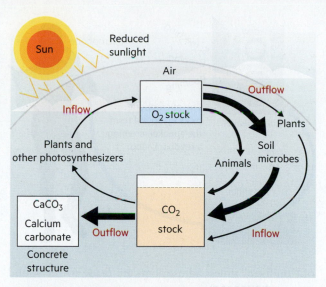

FIGURE 10.6 Biosphere 2 Actual Function Three factors contributed to the declining oxygen levels in Biosphere 2. Notice the increasing thickness of the O_2 outflow arrow moving through soil microbes. This represents the fact that the microbes were consuming much more O_2 (and expelling more CO_2) than the designers anticipated. Also notice the stock of calcium carbonate ($CaCO_3$). This shows that a significant amount of the oxygen in CO_2 was getting locked up in the concrete structure. Finally, notice the reflected sunbeams. Less sunlight than anticipated was penetrating the structure, which decreased O_2 production from photosynthesis.

growth in its soil microbes, which consumed far more O_2 and expelled far more CO_2 than the designers anticipated. Second, the glass structure of Biosphere 2 reduced available sunlight by about 50%, so the ability of plants and other photosynthesizers to draw in all of this CO_2 and produce O_2 was diminished and failed to match rising CO_2 supply. Finally, the excess CO_2 in the air was also reacting with the exposed concrete throughout the structure, which produced an unanticipated stock of calcium carbonate ($CaCO_3$). Samples from concrete on the inside and outside of the structure showed that concrete on the inside had 10 times as much calcium carbonate. This means that rather than continuously flowing through the air and the organisms in Biosphere 2, the missing tons of oxygen had been locked up inside the concrete structure itself.

Feedback

As oxygen was depleted in Biosphere 2, the people living in the system grew weaker and weaker The ever-worsening condition of the Biosphere 2 inhabitants is an example of a system concept known as **feedback**. Feedback is a loop in the system that responds to changes in levels of stocks by affecting the inflows or outflows of that stock. Musicians who use amplifiers and microphones encounter feedback when their microphones pick up sound from the amplifiers and run that sound back through the amplifiers, increasing the stock

photosynthesis a series of chemical reactions involving water and CO_2 through which plants and some other organisms store the Sun's energy in simple sugars (Chapter 3).

feedback a loop in a system that responds to and produces changes in levels of stocks that either amplify or counter a change.

FIGURE 10.7 Feedback for Musicians The loud screech of feedback at a concert is a reinforcing feedback loop in which the sound produced by the musician is amplified and then picked up by a microphone and amplified again (and again).

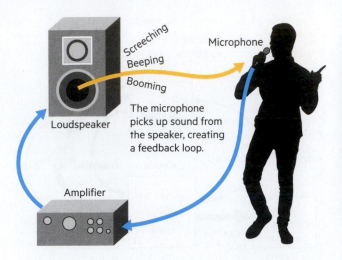

Screeching
Beeping
Booming

Microphone

Loudspeaker

The microphone picks up sound from the speaker, creating a feedback loop.

Amplifier

reinforcing feedback a loop that responds to the direction of change in the stock by enhancing that same direction of change.

balancing feedback in a system, a type of feedback that counteracts the direction in which a particular stock is changing.

of "loudness" unless adjustments are made to the sound system (**FIGURE 10.7**).

The particular kind of feedback experienced by the Biospherians is known as a **reinforcing feedback**: a loop that responds to the direction of change in the stock by enhancing that same direction of change. **FIGURE 10.8** shows how a reinforcing feedback loop in Biosphere 2 related to low air concentrations of O_2 affected the capacity of the scientists there to do the work required for survival, such as growing and preparing food. As their productivity slipped, they were less able to nourish themselves, which further diminished their capacity to do work, causing productivity to slip even more.

This sort of feedback goes by many other names as well, including *positive feedback*, *runaway* or *snowballing loops*, and *vicious* or *virtuous circles* (depending on whether it results in detrimental or beneficial outcomes). The important thing to remember is that reinforcing loops push the stock to change in the direction in which it is already heading. If the stock is falling, a reinforcing loop would cause the stock to fall further. If the stock is rising, a reinforcing loop would cause the stock to rise further.

Rapid growth of a population exemplifies the latter sort of reinforcing loop. In Chapter 4, we described how rabbits were introduced in Australia and gradually took over the landscape because of the lack of predators.

FIGURE 10.8 Feedback Loop Affecting Capacity for Biosphere 2 Scientists to Work Once oxygen levels declined in Biosphere 2, a reinforcing loop began to affect the scientists' capacity to do work. As oxygen levels in Biosphere 2 decreased further and further, the scientists were less and less able to do the necessary work to survive.

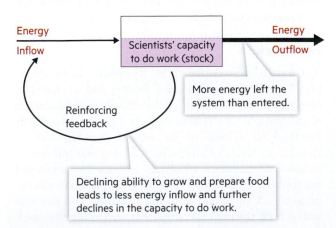

Energy
Inflow

Scientists' capacity to do work (stock)

Energy
Outflow

Reinforcing feedback

More energy left the system than entered.

Declining ability to grow and prepare food leads to less energy inflow and further declines in the capacity to do work.

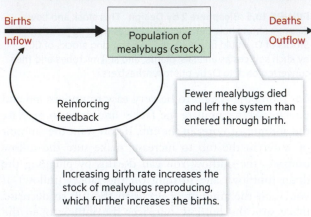

Births
Inflow

Population of mealybugs (stock)

Deaths
Outflow

Reinforcing feedback

Fewer mealybugs died and left the system than entered through birth.

Increasing birth rate increases the stock of mealybugs reproducing, which further increases the births.

FIGURE 10.9 Mealybugs in Biosphere 2 The rapid growth of mealybug populations over time in Biosphere 2 is an example of a reinforcing feedback loop. The thicker arrow for births relative to deaths represents how the birth rate was outpacing the death rate.

A similar loop happened with mealybugs (**FIGURE 10.9**), which had no predators in Biosphere 2. As their population increased, they produced more eggs, and their stock soon rose to such an extent that they covered the trunks, branches, and leaves of the trees.

Another type of feedback counteracts the direction in which a particular stock is changing. This is called **balancing feedback**, or a negative feedback loop, because it works against the direction in which the stock is changing (**FIGURE 10.10**). If the stock is falling, a balancing feedback will work to increase it. If the stock is rising, a balancing feedback would cause the stock to fall. When a musician's amplifier is screeching from the reinforcing feedback loop described earlier, an adjustment is made to break this cycle; that adjustment is a balancing feedback as it reduces the level of a rising stock of "loudness." You operate balancing feedback mechanisms in your everyday life. If you get uncomfortable when a room is too cold because the stock of heat has fallen, you might adjust a thermostat to heat the room back up, increasing the stock of heat. If you notice that the stock of money in your bank account is falling, you might reduce your spending and/or increase the hours you are working. In either case, you are operating a balancing feedback

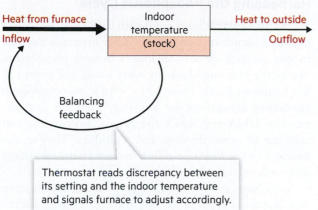

Heat from furnace
Inflow
→
Indoor
temperature
(stock)
Heat to outside
Outflow

Balancing
feedback

Thermostat reads discrepancy between
its setting and the indoor temperature
and signals furnace to adjust accordingly.

FIGURE 10.10 Thermostats as a Feedback Thermostats in
home heating and cooling systems act as a balancing feedback
to maintain a desired temperature.

mechanism to counteract the direction of change of the
stock.

⌂ **TAKE-HOME MESSAGE** Models are simplified
representations of systems in the real world. Using stock and
flow models helps us focus on the way certain aspects of a
system function. Feedback loops in a system respond to a
change in the level of a stock by changing the inflows or outflows of that stock. Feedback can either reinforce the direction of change in a stock or balance the direction of change
by counteracting it.

10.3 How Do Earth's Life-Support Systems Work?

As we learned at the beginning of the chapter, Earth
is a closed system in regard to matter. We have a vast
but finite set of tangible stuff on the planet, but matter
doesn't necessarily stay put.

So if we want to understand how the matter on which
we depend functions on Earth, we need to trace the path
that it follows through our environment. When we do

this work, we are tracing a **biogeochemical cycle**. The word
cycle means that we are identifying the complete path
of a particular element that returns to where it begins.
The "bio" in the term indicates that the element passes
through living things. And the "geo" in the term indicates
that the element also spends time in some of Earth's nonliving planetary components, such as rocks, water, and/or
the atmosphere. Another way to think about this is matter may spend time in any of Earth's four "spheres": the
lithosphere, biosphere, hydrosphere, or atmosphere. Each
sphere is a temporary storage compartment for a share of
a particular element's overall stock on Earth. And we can
look for paths by which the element flows into and out of
each storage compartment, or reservoir. To start, we will
compare and contrast the biogeochemical cycles of oxygen and phosphorus as they cycle through Earth's spheres.
These elements are both considered **macronutrients**: key
elements that organisms use in large amounts. Other macronutrients whose cycles we will trace in this chapter are
nitrogen and carbon. But let's begin with oxygen and see
how a shift in the balance of stocks and flows for this element billions of years ago led to life as we know it.

The Story of Oxygen

While we typically associate oxygen with the air we
breathe, Earth's atmosphere was not always so oxygen
rich. For about the first 1.3 billion years that life existed
on Earth, microorganisms existed anaerobically—without
relying on oxygen—by extracting energy from the
hydrogen sulfide of deep-sea vents or from other molecules dispersed in seawater. So where was the oxygen?
It was largely in the lithosphere—in rocks. Even today,
oxygen is the most abundant element by mass in the
lithosphere and makes up nearly half of Earth's crust,
primarily contained within silicate and oxide minerals.
We call a place where matter accumulates and is held

biogeochemical cycle
a path that shows how matter
on Earth flows through different
parts of the environment.

macronutrient an element
that organisms use in large
amounts.

⊘ **SCIENCE**

The Oxygenating Extinction

The oxygenation of the atmosphere
has clearly benefited humans, but this
change to the oxygen cycle was also
Earth's first mass extinction. Oxygen
was toxic to the once dominant species of anaerobic bacteria inhabiting
Earth. When cyanobacteria started
releasing oxygen into the atmosphere more than two billion years
ago, countless species of anaerobic
bacteria died off and were no longer
the dominant life-form on Earth.

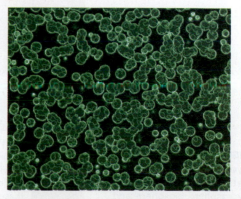

for a long period of time a **sink**, and most of the oxygen contained in the lithosphere stays in that sink for a very long time; the residence time of oxygen in the lithosphere is about 500 million years. Oxygen is also a major component of water (H_2O), composing 89% of the hydrosphere by mass both as a component of H_2O and as O_2 dissolved in water. Thus, the hydrosphere is a tremendous sink of oxygen.

Then about 2.7 billion years ago, the biosphere's evolving life-forms started changing the chemical composition of the air. The first photosynthesizers, called cyanobacteria, began taking in CO_2 and expelling O_2 into the atmosphere as a waste product. Eventually, this led to more oxygen in the atmosphere and the evolution of oxygen-breathing animals. Further, the accumulation of oxygen in the atmosphere yielded a thin layer of ozone (O_3) from the interaction of O_2 molecules with ultraviolet light. As we learned in Chapter 8, this ozone layer floats at the top of the atmosphere and filters out ultraviolet rays, thereby protecting life from the damaging effects of solar radiation.

Of course, these processes continue today (**FIGURE 10.11**). Oxygen moves between the atmosphere, the hydrosphere, and the biosphere, a process driven by photosynthesis, as roughly half of Earth's O_2 is still produced by single-celled plants in the ocean called phytoplankton. The other half comes from grasses, trees, shrubs, and other plants. Photosynthesis is, by far, the way in which most free oxygen is made available on Earth. This means

that the biosphere, which includes living organisms such as bacteria, algae, and plants, is responsible for the composition of the atmosphere on which we and so many other organisms depend.

To close the loop then, how can oxygen make it back into the lithosphere? Simple chemical processes such as rusting are one example of weathering processes that slowly return oxygen to the lithosphere. Oxygen can also move from the biosphere to the lithosphere when marine organisms that produce calcium carbonate ($CaCO_3$) shells die and sink to the seafloor, where they can eventually form limestone (**FIGURE 10.12**).

Harnessing the Phosphorus Cycle

All we have to do is hold our breath for a bit to recognize our dependence on oxygen, but other materials essential to our survival are less obvious. Consider phosphorus, one of the building blocks of every living cell within all life-forms on Earth. Phosphates, which are compounds containing the phosphate ion, PO_4^{3-}, are key components of DNA and RNA, the genetic instructions our cells use to grow, develop, and reproduce. They're also essential to the way our cells transport chemical energy and form cell membranes.

Sometimes the word *cycle* leads to the misconception that material circulates quickly through each of Earth's systems, but this is not always the case. Phosphorus has no gaseous phase on Earth, meaning that it does not cycle

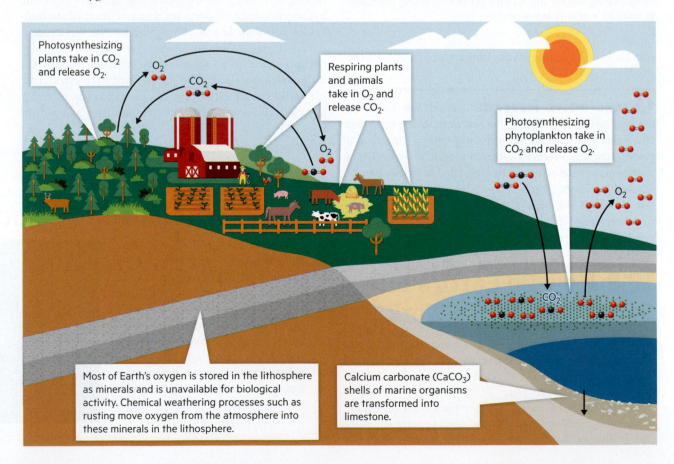

FIGURE 10.11 The Oxygen Cycle Oxygen cycles between Earth's atmosphere, hydrosphere, biosphere, and lithosphere.

FIGURE 10.12 **Limestone Formation Returns Oxygen to the Lithosphere** Limestone—pictured here in the Painted Cliffs of Maria Island, Australia—is made of the bodies of breathing organisms that incorporated oxygen into their calcium carbonate ($CaCO_3$) shells.

through the atmosphere at all. And while it does cycle in the lithosphere and hydrosphere, it does so very slowly (**FIGURE 10.13**). Slow weathering processes release phosphorus from rocks and minerals in the form of phosphate ions, which can be dissolved in water on land and in the ocean, thereby entering the hydrosphere. Plants then absorb phosphate and transform it into organic compounds that we can take in by eating plant and animal matter. Finally, when microorganisms break down dead matter and waste from plants and animals, phosphate is released back into the soil where plants can access it once again. But only a very tiny portion of this phosphorus will turn back to rock and become part of the lithosphere. Once in the biosphere, animals and plants instead

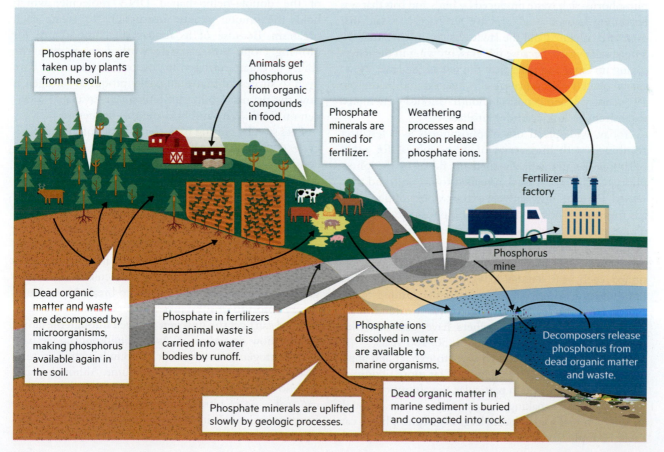

Phosphate ions are taken up by plants from the soil.

Animals get phosphorus from organic compounds in food.

Phosphate minerals are mined for fertilizer.

Weathering processes and erosion release phosphate ions.

Fertilizer factory

Phosphorus mine

Dead organic matter and waste are decomposed by microorganisms, making phosphorus available again in the soil.

Phosphate in fertilizers and animal waste is carried into water bodies by runoff.

Phosphate ions dissolved in water are available to marine organisms.

Decomposers release phosphorus from dead organic matter and waste.

Phosphate minerals are uplifted slowly by geologic processes.

Dead organic matter in marine sediment is buried and compacted into rock.

FIGURE 10.13

The Phosphorus Cycle Phosphorus cycles only between the lithosphere, hydrosphere, and biosphere, because on Earth it has no gaseous phase and therefore does not cycle through the atmosphere.

FIGURE 10.14 **Phosphorus Mines** Phosphorus for agricultural and industrial uses is extracted from mining phosphate minerals—as at this mining site in Bone Valley, Florida. Millions of years ago, Bone Valley was underwater and accumulated deposits that left it rich in nutrients. It is part of an area known as the Central Florida phosphate district.

(see Chapter 7), removing dissolved oxygen from the water and suffocating organisms in those ecosystems.

⬡ **TAKE-HOME MESSAGE** Matter on Earth can follow a biogeochemical cycle as it moves between the lithosphere, biosphere, hydrosphere, and atmosphere and back again. Each of these "spheres" can be thought of as a temporary storage compartment for a share of Earth's total stock of an element. Different elements can take very different paths through Earth, and the biosphere—and its life-forms—plays a vital role in biogeochemical cycles.

10.4 How Do Living Things Change the Nitrogen Cycle?

The element nitrogen is fundamental to creating and building life; it helps make you what you are. Nitrogen is one of the key components of proteins, which are the second most abundant type of molecule (behind water) in the human body. Proteins are the primary structural components of our muscle tissue, a source of energy in times of crisis, and important participants in countless bodily processes. Nitrogen-containing molecules also play an important role in forming the "rungs" of the double-helix-shaped DNA molecule, which holds the unique genetic instructions to create each organism. Because of its fundamental importance for growing crops, humans have worked to increase the ability to obtain nitrogen for use as a fertilizer. We have been so successful that industrially produced nitrogen is now the primary source of biologically available nitrogen on Earth. Why and how did this happen?

Let's start by getting back to the basics of the nitrogen cycle (**FIGURE 10.15**). About 80% of Earth's atmosphere consists of nitrogen. Except there is a catch. Nearly the entire stock of atmospheric nitrogen (99.9%) is in the form of nitrogen gas (N_2), often called *free nitrogen*, a highly stable, nonreactive form that is biologically unusable to humans and most other organisms. Although we breathe in great quantities of this gas that is all around us, our bodies cannot make use of it in this form. This is because the bond holding the molecule of free nitrogen together takes a lot of energy to break: N_2 is one of nature's most inert substances.

If we can't get the nitrogen we need through the air we breathe, how do we get it? As with phosphorus, we have to get nitrogen from organic compounds such as proteins in the plants and animals we consume. Although plants are also unable to make use of nitrogen in the air, they can make use of nitrogen compounds that are dissolved in water and taken up by their roots from the soil. This nitrogen is introduced to the soil through a process called

use and reuse phosphorus, as it rapidly and constantly cycles as organic matter decays and breaks down into forms that can be taken up and used by new life.

In the late 18th century, scientists in England recognized that fertilizing fields with phosphorus-rich materials would boost crop growth. In the 1790s, a student at Oxford named William Buckland used this information to play a college prank. He painted the large letters G-U-A-N-O on the Oxford lawn with bat guano—the name for the phosphorus-rich waste produced by birds and bats that was becoming known as an effective fertilizer. It is said that even after it rained and the lawn was cut many times, the letters were visible because the grass growing in the letters sprang up faster and taller than the rest of the lawn. News spread quickly about guano's effectiveness, and by the late 19th century, guano deposits in South America were mined extensively by Europeans for fertilizer.

Today, widespread use of phosphorus as a fertilizer combined with discoveries of its usefulness for other purposes has dramatically sped up flows of the phosphorus cycle, as it is used, returned to the soil, and then reused. Humans now actively extract stocks of phosphorus from the lithosphere by mining phosphate minerals. These minerals are used for industrial fertilizers, detergents, and a variety of industrial processes including steel production. Globally, 170 million tons of phosphate are mined each year (**FIGURE 10.14**), primarily from China, South Africa, Morocco, Jordan, and the United States, which are home to almost 90% of the world's known phosphorus reserves. Researchers have estimated that global reserves will be exhausted in 50–100 years.

As humans have developed ways to extract and apply phosphorus, we have created impacts far more significant than Buckland's prank. In particular, the amount of phosphorus gradually making its way to the oceans through runoff each year has tripled. In large doses, phosphorus can cause eutrophication in marine and freshwater ecosystems

We Are Flushing Phosphorus Down the Drain, but Can We Reclaim It?

Phosphorus is an essential nutrient. This mineral is in every cell of your body, with more than 85% of it in your bones and teeth. But you also depend on it to filter waste in your kidneys, to store and use energy at the cellular level, and to produce the DNA and RNA that are your genetic building blocks. You get phosphorus from the foods you eat, and a scarcity of phosphorus in the soil can limit plant growth, which in turn can affect crop yields. For decades now, modern agriculture has boosted crop yields by adding phosphorus fertilizer that is mined from phosphate minerals. But Earth's phosphorus reserves are a scarce resource that is being rapidly depleted. Natural resource experts estimate that at current levels of extraction, we may run out of phosphorus reserves within the next century. So where can we turn for phosphorus in the future?

To answer this question, let's follow the phosphorus. Phosphorus fertilizers are often overapplied to crops, used once, and then swept out to sea in runoff. A recent study of Chinese agriculture (the largest national consumer of phosphorus) found that nearly two-thirds of the phosphorus applied to crops was lost to the environment. This not only wastes a scarce resource but also causes damage to aquatic ecosystems through huge algal blooms that occur during the process of eutrophication.

In response, the College of Agricultural and Life Sciences at the University of Wisconsin has developed a software program known as the Wisconsin P Index. This software helps farmers determine exactly how much phosphorus they need on the basis of the crop they're growing, the soil conditions, the weather patterns, and other factors. Programs like this can reduce phosphorus waste by as much as 50%. Other practices can also conserve phosphorus and reduce runoff. These include leaving crop residues in place to return nutrients to the soil, avoiding tilling or plowing fields, and being careful not to overirrigate (see Chapter 12).

Humans excrete phosphorus in our waste. This valuable mineral can be captured in wastewater treatment plants and used as fertilizer.

Using animal waste as a fertilizer is yet another way to return phosphorus to the soil, as is recycling phosphorus in food waste as compost.

While fertilizer runoff from agriculture may be the largest source of phosphorus waste, there is also another large, untapped stream of phosphorus that remains mostly uncaptured: the excess phosphorus we excrete as human waste. Most of our wastewater treatment plants in the United States release the phosphorus contained in our sewage into rivers and out to sea. But it doesn't have to be this way. Although we are literally flushing this valuable mineral down the drain, it is possible for wastewater treatment plants to recapture phosphorus for agricultural use. One strategy is to harvest phosphorus-rich crystals known as struvite that form when bacteria break down urea in urine during waste treatment. Treatment plants have long been forced to shut down their operations to clear pipes clogged with these crystals, but now new filter systems have been designed to collect them because struvite is a rich, slow-release fertilizer that outperforms many other types of phosphorus fertilizer. Other methods of phosphorus recovery have also been devised, including the use of bacteria to concentrate the mineral from sewage in a removable sludge. Researchers have even tested phosphorus recovery systems to treat manure on large dairy farms. With technologies like these, it is now possible to recapture and reuse more than 90% of the phosphorus from human waste.

What Would You Do?

Although recapturing phosphorus in human waste is possible, it is expensive and time consuming. It typically requires infrastructure improvements to wastewater treatment systems that cost tens of millions of dollars. And the cost is not offset by the sale of the resulting phosphorus fertilizer alone. Some big cities, such as Chicago, Illinois, and Portland, Oregon, have made this investment and paid for it through increased rates for local utility customers. Others have argued that food companies or the fertilizer industry should pay for phosphorus treatment. If you were the mayor of a city and this opportunity was presented to you, what would you decide to do?

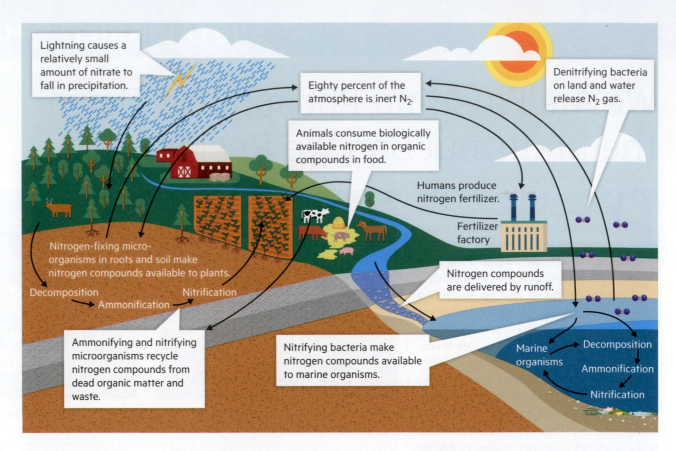

FIGURE 10.15 The Nitrogen Cycle Free nitrogen in the atmosphere must be converted to biologically available nitrogen compounds through nitrogen fixation before it can travel through the nitrogen cycle.

Lightning causes a relatively small amount of nitrate to fall in precipitation.

Eighty percent of the atmosphere is inert N_2.

Animals consume biologically available nitrogen in organic compounds in food.

Denitrifying bacteria on land and water release N_2 gas.

Humans produce nitrogen fertilizer.

Fertilizer factory

Nitrogen compounds are delivered by runoff.

Nitrogen-fixing micro-organisms in roots and soil make nitrogen compounds available to plants.

Decomposition
→ Ammonification
Nitrification

Ammonifying and nitrifying microorganisms recycle nitrogen compounds from dead organic matter and waste.

Nitrifying bacteria make nitrogen compounds available to marine organisms.

Marine organisms

Decomposition
Ammonification
Nitrification

nitrogen fixation a process where free nitrogen—inert nitrogen gas (N_2)—in the air is converted to soluble ammonium (NH_4^+) and nitrate (NO_3^-) ions that plants, algae, and bacteria can take in and use.

diazotroph a bacterium with an enzyme that uses free nitrogen from the air along with hydrogen to produce ammonia.

nitrogen fixation, whereby free nitrogen in the air is converted to soluble ammonium (NH_4^+) and nitrate (NO_3^-) ions that plants, algae, and bacteria can take in and use.

Nitrogen fixation occurs primarily in two ways. One way is very dramatic: through the power of lightning. The high energy of lightning triggers a series of reactions that result in nitrate ions falling to Earth's surface in precipitation. Observations by NASA satellites show that lightning strikes about 40 times each second on Earth—almost 3.5 million strikes per day—delivering about 13,000 tons of nitrate a day. While that may seem like a lot, it

accounts for less than 10% of nitrogen fixation on Earth. Most nitrogen fixation takes place in the biosphere. Bacteria known as **diazotrophs** possess an enzyme that uses free nitrogen from the air along with hydrogen to produce ammonia. This normally happens as part of mutualistic relationships with other organisms. For example, the *Rhizobium* genus of bacteria are diazotrophs that live in the roots of plants in the legume family, such as peanuts, lentils, beans, and alfalfa. They fix nitrogen, providing ammonia needed by their host plants, while utilizing sugars and the root structure provided by the plant (**FIGURE 10.16**).

(a)

(b)

(c)

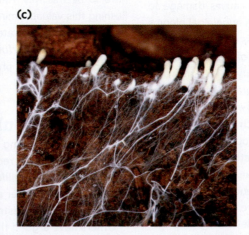

(d)

FIGURE 10.16 Nitrogen Conversion Processes (a) Lightning strikes trigger reactions that produce, or "fix," nitrate ions from N_2 molecules in the air. These fall to Earth in precipitation. (b) Bacteria-filled nodules often found in legumes such as these runner beans fix nitrogen from the atmosphere to produce ammonia, a chemical that can be used by plants. (c) Bacterial and fungal decomposers break down nitrogen compounds in animal waste and dead plant and animal matter. (d) Under the right conditions—often wet and muddy conditions in agricultural areas—nitrogen in the soil can be transformed by bacteria back to N_2 and released into the air.

Biological Nitrogen Fixation

It has been known since antiquity that planting legumes such as peas, beans, and clover helps restore soil fertility after grain crops are grown. But how this worked remained a mystery until 1888, when German chemists Hermann Hellriegel and Hermann Wilfarth discovered microbes living in nodules on the roots of peas. These nodules had been known since the 17th century, but they were generally thought to be growths caused by tiny insects. Hellriegel and Wilfarth found that the nodules actually convert inert atmospheric nitrogen (N_2) into ammonia (NH_3). This ammonia then dissolves in water to form ammonium (NH_4^+), which plants take up and use. Here was the secret to why the ancient practice of rotating crops of grains and legumes didn't rob the soil of nitrogen. Unlike grains such as wheat, legumes did not deplete nitrogen in the soil but rather built it up.

But what were the microbes that performed this biochemical feat, and how exactly did they do it? In the late 19th century, Dutch microbiologist Martinus Beijerinck isolated and cultured the microbes from a variety of legumes that carry out the process of nitrogen fixation, now named *Rhizobium* (from the Greek words for "root" and "life," *rhiza* and *bios*). Subsequent researchers found that different species of bacteria are responsible for nodulation and nitrogen fixation in different types of plants. The large, woody nodules on the roots of alder trees were also later found to host microbes that fixed nitrogen.

In the past several decades, increased interest has been focused on nitrogen fixation by microbes associated with plants other than legumes. Before the 1970s, little thought was given to the possibility that bacteria living in the soil outside of root nodules might fix nitrogen. The idea that nitrogen-fixing bacteria might be associated with non-legume crops, such as grains, gained credence through the

Johanna Döbereiner discovered nitrogen-fixing bacteria living in the soil around certain kinds of plant roots.

work of Johanna Döbereiner, a Czechoslovakian refugee who fled to Germany at the end of World War II and emigrated to Brazil in 1951 after completing her graduate studies. Working as a researcher for the Brazilian Ministry of Agriculture in the 1970s, Döbereiner discovered nitrogen-fixing bacteria in the soil surrounding the roots of cereal grasses and sugarcane. In the 1980s, she discovered nitrogen-fixing species of bacteria that colonized plant tissues. These *diazotrophic endophytes* (nitrogen-fixing symbiotic bacteria that live within a plant without causing harm or disease) were found to provide one-third to one-half of the nitrogen taken up by sugarcane. Introducing these nitrogen-fixing bacteria into plants has been proposed as a means for greatly reducing the need for chemical fertilizers in agriculture. And the idea of using processes known as "genetic engineering" (Chapter 12) to place nitrogen-fixing genes into corn and other grains is now being pursued as a strategy for more sustainable food production.

Some types of plants, such as peas and other legumes, host nitrogen-fixing bacteria in root nodules.

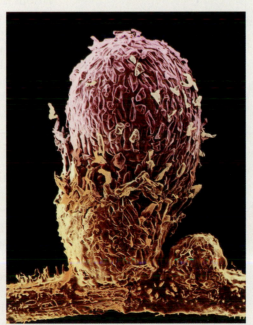

Rhizobium bacteria carry out the process of nitrogen fixation.

What Would You Do?

Current technology makes it possible to insert nitrogen-fixing genes directly into plants, thereby changing their essential genetics. While the final goal is to make food production more sustainable, do you agree with and support this practice? Briefly explain why you support or do not support it—or possibly if you think there is a middle ground.

ammonification a process where bacteria and fungi break down nitrogen compounds in animal waste products and in dead plant and animal matter and release ammonia.

nitrification a process in which microorganisms convert ammonia to nitrogen compounds.

denitrification a process in which microorganisms use nitrogen compounds for respiration and create N₂ gas, returning it to the atmosphere.

A variety of organisms make nitrogen available in soils and oceans in other ways. Some bacteria and fungi break down nitrogen compounds in animal waste products and in dead plant and animal matter, releasing ammonia in the process. This is called **ammonification**. Most of the ammonia from this process then undergoes **nitrification** as ammonia-oxidizing microorganisms convert ammonia to nitrogen compounds. When these compounds then dissolve in soil water, plants can take up the nitrogen through their roots. So, like the phosphorus cycle described earlier in this chapter, there is a loop within the biosphere that is circulating nitrogen between plants and animals at the surface and microbes in the soil. Finally, **denitrification** occurs when still other types of microorganisms underground or underwater use nitrogen compounds for respiration and convert them back into N₂ gas, returning it to the atmosphere. While we have focused on the relatively rapid cycling of nitrogen between the atmosphere and the biosphere, over longer timescales nitrogen is incorporated into the lithosphere. Nitrogen moves into the lithosphere as sedimentary rocks form and is released millions of years later when these rocks slowly weather.

Human Impacts on the Nitrogen Cycle

Over the past century, humans figured out how to fix nitrogen through a chemical reaction known as the Haber–Bosch process. This is an energy- and resource-intensive process that draws on 1% of the world's electricity output and consumes about 4% of the global natural gas supply. Synthetic nitrogen fertilizer from this process has dramatically increased the human contribution to nitrogen fixation on Earth. And globally, nearly half a million tons of synthetic fertilizers are now applied to croplands each day.

Yet only about half of the nitrogen in synthetic fertilizers is taken up by plants or remains in the soil; the rest is quickly lost to the water and air. Runoff and groundwater flows containing unused, dissolved ammonium and nitrate from these fertilizers are now the leading contaminants affecting public drinking water in the United States. Agricultural runoff also transports an abundance of industrially fixed nitrogen to aquatic and marine environments where, as with phosphorus described earlier, it can cause eutrophication and dead zones—areas where oxygen concentrations are so low that animals cannot survive. Each spring, fertilizer runoff creates a dead zone in the Gulf of Mexico as fertilizers are washed into the Mississippi River and then flushed into the gulf. This is one of the world's largest dead zones, with an average area of 5,000 square miles (roughly the size of Connecticut) and sometimes growing to more than 8,700 square miles in area (almost the size of New Jersey) (**FIGURE 10.17**).

Humans are also sending nitrogen back into the atmosphere. Ammonia from synthetic fertilizers that becomes airborne reacts with other compounds to form *fine particulate matter*, a kind of air pollution linked to a range of human lung problems, such as bronchitis and asthma, and to the formation of haze (see Chapter 8). Various nitrogen oxides from fertilizers, livestock, industrial processes, and the combustion of fossil fuels and biomass are also significant air pollutants, contributing to acid rain, urban smog, atmospheric ozone depletion, and climate change. This air pollution also sends nitrogen compounds back to Earth's surface in a process known as deposition. **FIGURE 10.18** sums up the various natural and human contributions to global nitrogen fixation and how they have grown over time. Human sources, including fertilizer production, fixation by crops, and air pollution, are now responsible for more than half of the nitrogen fixation occurring on Earth, with fertilizer production as the largest human source of nitrogen fixation.

⬡ **TAKE-HOME MESSAGE** Nitrogen is an essential component of proteins and DNA, and nearly 80% of Earth's atmosphere consists of nitrogen gas (N₂). Nitrogen compounds needed by plants and animals are made available primarily by nitrogen-fixing bacteria. In the past century, human activities have become the primary source of biologically available nitrogen on Earth because of the increasing use of synthetic fertilizers. About half of the nitrogen from these fertilizers is quickly lost to the water and air, where it can be a source of pollution.

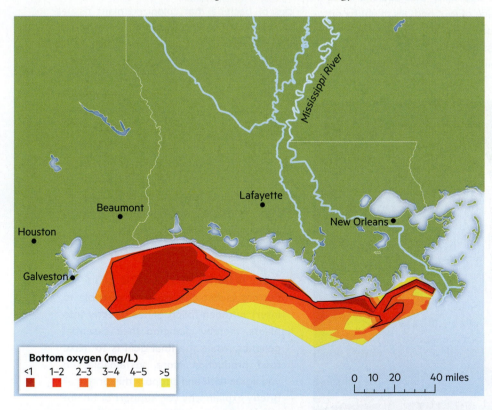

Bottom oxygen (mg/L)
<1   1–2   2–3   3–4   4–5   >5

0   10   20   40 miles

FIGURE 10.17 Dead Zone Eutrophication from fertilizer runoff creates a seasonal dead zone where the Mississippi River empties into the Gulf of Mexico. Bottom oxygen refers to oxygen levels in the water just above the seafloor. Levels less than 2 mg/L cannot support any animals, and levels from 2–5 mg/L cannot support most animals.

Adapted from NOAA (2021).

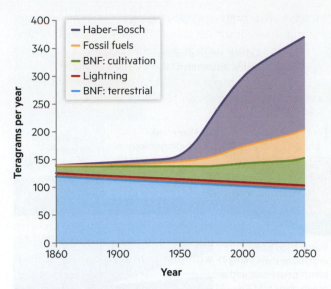

FIGURE 10.18 **Ways Reactive Nitrogen Is Produced over Time** This illustration shows the natural and human sources of nitrogen fixation over time and projected into the future, measured in teragrams. One teragram is equivalent to a million metric tons. Notice that nitrogen fixation from Haber–Bosch fertilizer production is more than double that of the natural sources of biological nitrogen fixation on land (BNF: terrestrial). Overall, human sources now contribute more than half of the nitrogen fixation on the planet.

Adapted from Nielsen (2005), Cowling et al. (2002), and Galloway et al. (2004).

10.5 How Are We Changing Earth's Carbon Cycle?

When scientists (and also science-fiction writers) discuss trying to locate life on other planets, they often describe searching for other "carbon-based life-forms" as a fundamental measure. Carbon is second only to oxygen as the most abundant element in our bodies by weight, and more significant, carbon is the primary component of organic molecules essential for life, such as carbohydrates, proteins, lipids, and nucleic acids. We depend on these organic molecules to form our cells and organs and to provide us with energy. But it's not just us. All living things on Earth are carbon-based life-forms, and many important materials we depend on use carbon. And, although carbon composes just a tiny fraction of a percent of Earth's crust by weight, its cycling from sinks in the lithosphere to the atmosphere and the hydrosphere are leading to adverse environmental impacts—most notably climate change (the subject of the next chapter). But before we discuss environmental impacts, let's learn more about the carbon cycle, starting far in the past.

Natural Cycling of Carbon between the Lithosphere, Atmosphere, and Hydrosphere

The natural cycling of carbon between the lithosphere, atmosphere, and hydrosphere takes a very long time, basically following the rock cycle introduced in Chapter 9.

Reducing Fertilizer Pollution

In addition to reducing or eliminating the use of synthetic fertilizers, several farming strategies reduce nitrogen pollution from agricultural runoff. Agronomists have found that most farmers tend to overapply nitrogen fertilizers, so one strategy is to simply apply less. Also, nitrogen pollution can be significantly reduced by applying fertilizer only at those times of year when the crops are most able to absorb the nutrients. Land-use practices along with the type and application method of fertilizer can also make a big difference. Minimizing how often fields are tilled reduces runoff and helps to build organic matter in soil. This reduces the need for fertilizer and holds nutrients in place longer. Maintaining large areas of vegetated land between crops and waterways forms a buffer that promotes infiltration and reduces runoff. Some synthetic fertilizers are formulated to release more slowly and last longer in the soil. We discuss the impacts of improved agricultural practices more in Chapter 12.

No-Till Farming

Vegetation Buffers

Most of Earth's carbon is stored in the lithosphere. It can move into the atmosphere as erupting volcanoes powered by tectonic movements emit CO_2 gas and to a lesser degree methane (CH_4). Once in the atmosphere, a significant share of carbon is dissolved directly into seawater at the surface to form bicarbonate (HCO_3^-) ions. Carbon also dissolves in water in the atmosphere to form a weak acid known as carbonic acid (H_2CO_3). This makes rain slightly acidic, which facilitates chemical weathering of rocks at and near Earth's surface and the release of calcium (Ca^{2+}) and HCO_3^- ions that are washed into the ocean.

Carbon Cycling through the Biosphere

But in the shorter term, the biosphere plays a very important role in the carbon cycle. Photosynthesis constantly works to pull CO_2 out of the atmosphere as plants use light energy and water to convert it into sugar

FIGURE 10.19 Carbon in the Biosphere Carbon moves through the biosphere as plants and other photosynthesizing organisms absorb and convert CO_2 into sugars and as plants, animals, and microorganisms conduct respiration that emits CO_2.

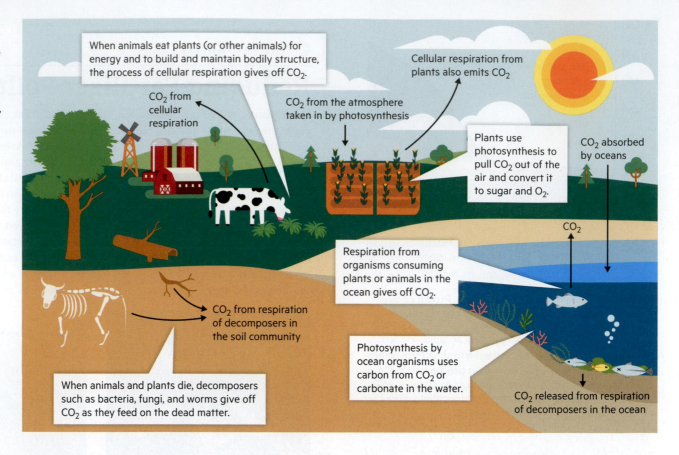

When animals eat plants (or other animals) for energy and to build and maintain bodily structure, the process of cellular respiration gives off CO_2.

CO_2 from cellular respiration

CO_2 from the atmosphere taken in by photosynthesis

Cellular respiration from plants also emits CO_2

Plants use photosynthesis to pull CO_2 out of the air and convert it to sugar and O_2.

CO_2 absorbed by oceans

CO_2

Respiration from organisms consuming plants or animals in the ocean gives off CO_2.

CO_2 from respiration of decomposers in the soil community

Photosynthesis by ocean organisms uses carbon from CO_2 or carbonate in the water.

When animals and plants die, decomposers such as bacteria, fungi, and worms give off CO_2 as they feed on the dead matter.

CO_2 released from respiration of decomposers in the ocean

and O_2. Various primary producers such as phytoplankton in the ocean utilize carbon either from CO_2 or from carbonates dissolved in the water for photosynthesis too. Then consumers of photosynthesizing organisms (like us) use the resulting organic molecules for energy and for building their own carbon-based structures, eventually returning CO_2 to the atmosphere through respiration and decomposition (**FIGURE 10.19**). As animals and plants die, their structures are consumed by decomposers such as bacteria, fungi, insects, and worms. These organisms also produce CO_2 through their respiration processes. This respiration by life in the soil community makes the concentration of CO_2 within soil 10–100 times greater than that of the atmosphere. This, in turn, leads to higher carbonic acid concentrations in soil water, which helps break down rocks and organic matter into soluble compounds, thereby making them available for plants to take up and use.

Carbon can also move from the biosphere to the atmosphere through methane-producing microorganisms known as **methanogens**. Methanogens cannot function in the presence of oxygen and instead live in anoxic (oxygen-free) environments. In these environments, they consume and convert (process) carbon compounds anaerobically (without oxygen), expelling CH_4 as a by-product. Where are these methanogens living? While some are in extreme environments such as hot springs and hydrothermal vents in the ocean, many are in places much more familiar to us. Methanogens live in and

play a key role in the function of animal digestive tracts, including those of humans. Cattle have multicompartment stomachs that host large quantities of methanogens, which they use to digest cellulose from plants. The cattle then release CH_4 from their guts through belching, flatulence, and defecation (**FIGURE 10.20**). Termites also host methanogens in their guts to aid the digestion of plant fibers. Warm, moist environments such as wetlands host

Mass of Methane Emitted Annually

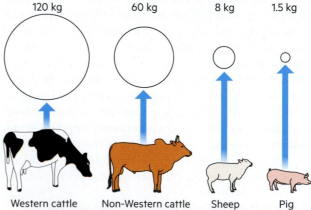

| 120 kg | 60 kg | 8 kg | 1.5 kg |

Western cattle   Non-Western cattle   Sheep   Pig

FIGURE 10.20 Methane Production by Common Farm Animals per Year Farm animals are major sources of methane, which is produced by methanogens in their guts that digest cellulose. The more farm animals there are, the more methane is released into the air. kg = kilograms.

methanogen a methane-producing microorganism that thrives in oxygen-poor environments.

large quantities of methanogens, as do anoxic underground environments where these microorganisms play a role in decomposing biomass. Carbon also moves from the biosphere to the atmosphere in CH_4 emissions from wildfires because of the incomplete combustion of trees and other biomass.

Carbon Cycling from the Biosphere to the Lithosphere

Take a walk on the beach and you will probably see evidence of carbon cycling from the biosphere into the lithosphere and back. Some marine organisms, such as clams, sand dollars, and a multitude of smaller floating organisms, combine calcium (Ca^{2+}) and bicarbonate (HCO_3^-) ions to construct the external skeletal material we call shells. When these creatures die, their shells eventually sink to the ocean floor, where they pile up to form carbonate sediment that can become limestone, the most abundant form of carbon in the lithosphere. Over long periods of time, carbonate sediment can be buried deep enough to become incorporated into magma, producing CO_2-rich magma, which can eventually erupt back onto the surface through volcanoes. Or carbonates can be exposed and weathered at the surface and returned to the ocean.

But the most significant impact on Earth's carbon cycle comes from humans using a product of long-term geologic processes. While much of the organic matter from the biosphere decays and flows into the soil, air, and water, the organic matter deposited in some environments resists decay because of low temperatures or lack of oxygen and accumulates over long periods of time.

Geologic processes can eventually bury this organic matter so deeply that it experiences very high temperature and pressure. Over many thousands or even millions of years, these conditions can produce coal, oil, and gas: fossil fuels.

Humans and the Carbon Cycle

Fossil fuel extraction and combustion add a lot of CO_2 to the atmosphere that would have remained in the lithosphere for a much longer period. In a way, the sink of carbon in the lithosphere has an unexpected hole in it that causes carbon to flow out (**FIGURE 10.21**). When we extract coal, oil, and gas from the ground, we transfer carbon that had been locked up beneath Earth's surface for millions of years into the atmosphere, where it is accumulating in increasing concentrations. Another significant way we are influencing the inflow of CO_2 to the atmosphere is through the production of cement. Cement is manufactured by a process known as calcination, which transforms the calcium carbonate in limestone into calcium oxide while releasing CO_2 into the atmosphere. Coal- or gas-fired furnaces heat the rock to temperatures reaching 1,400°C (2,550°F). Together, the direct emissions from calcination and the indirect emissions from the fossil fuel burned during calcination account for 5% of human-caused CO_2 emissions. Further, we are influencing how CO_2 flows out of the atmosphere through our land-use activities. As humans clear land, they reduce the quantity of trees and other primary producers that draw CO_2 out of the atmosphere. Conventional farms and urbanized landscapes store far less CO_2 than does forest or grassland.

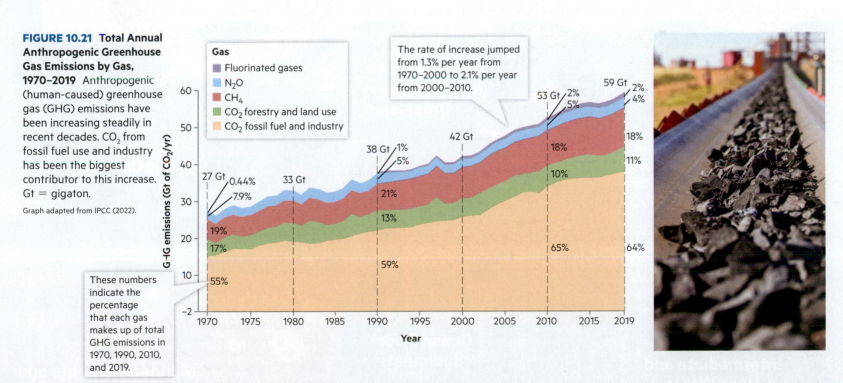

FIGURE 10.21 Total Annual Anthropogenic Greenhouse Gas Emissions by Gas, 1970–2019 Anthropogenic (human-caused) greenhouse gas (GHG) emissions have been increasing steadily in recent decades. CO_2 from fossil fuel use and industry has been the biggest contributor to this increase. Gt = gigaton.

Graph adapted from IPCC (2022).

Gas
- Fluorinated gases
- N_2O
- CH_4
- CO_2 forestry and land use
- CO_2 fossil fuel and industry

The rate of increase jumped from 1.3% per year from 1970–2000 to 2.1% per year from 2000–2010.

These numbers indicate the percentage that each gas makes up of total GHG emissions in 1970, 1990, 2010, and 2019.

GHG emissions (Gt of CO_2/yr)

The Carbon Cycle: It's a Carbon-Based World

Carbon seems to be everywhere. Carbon-based, or organic, molecules are in the food we eat, our DNA, and the air and materials around us. Wood, plastic, synthetic fibers, and, of course, the fuels that power our society are all primarily made up of organic molecules.

This diagram shows the major locations where carbon is stored on Earth, with the numbers indicating the amount of carbon stored in billions of metric tons. One-way arrows show flows of carbon through the carbon cycle in one direction, while circular arrows show paired exchanges of carbon. Larger/smaller arrows = larger/smaller exchanges. Arrow colors indicate the speed of exchange processes.

Speed of Exchange Processes

→ Very Fast (less than 1 year)

→ Fast (1–10 years)

→ Very Slow (more than 100 years)

Atmosphere
(871)

Exchange Atmosphere-Ocean

Surface Water
(900)

Marine Organisms
(3)

Exchange Surface Water-Deep Water

Dissolved Organic Carbon
(700)

Intermediate and Deep Water
(37,100)

Ocean Floor Sediment
(1,750)

Gas Hydrates
(500–2,500)

Marine Sediments and Sedimentary Rocks
(66,000,000–100,000,000)

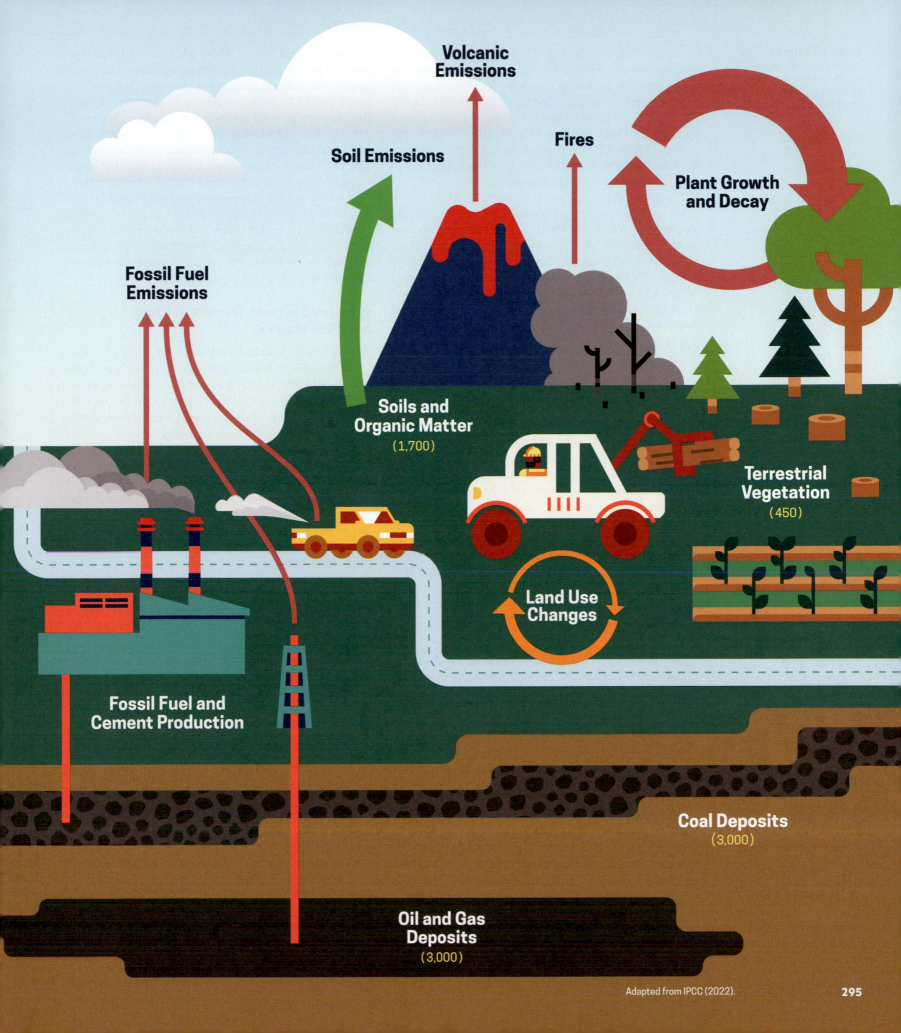

Volcanic Emissions

Soil Emissions

Fires

Plant Growth and Decay

Fossil Fuel Emissions

Soils and Organic Matter
(1,700)

Terrestrial Vegetation
(450)

Fossil Fuel and Cement Production

Land Use Changes

Coal Deposits
(3,000)

Oil and Gas Deposits
(3,000)

Adapted from IPCC (2022).

ocean acidification the ongoing decrease in the pH of Earth's oceans caused by absorption of carbon dioxide (CO_2) from the atmosphere.

Carbon dioxide is not the only climate-warming greenhouse gas (GHG) humans are moving into the atmosphere in increasing amounts. Methane can enter the atmosphere through human activities such as the production of natural gas for energy, a tremendous increase in cattle production, and other sources that include cultivated rice paddies, landfills, and wastewater treatment plants, all of which host large quantities of biomass-decomposing methanogens.

One important side effect is that the added atmospheric CO_2 is causing a phenomenon known as **ocean acidification**. About one-third of the CO_2 emissions from humans since the 18th century have been absorbed into the ocean (and if not for this, the atmospheric concentration of CO_2 would be even higher than it is now). When CO_2 is dissolved in water, it forms carbonic acid (H_2CO_3). As the water becomes more acidic, marine organisms have more difficulty building and maintaining their skeletons and shells, which significantly alters marine ecosystems.

Of course, the other side effect is climate change, and in the next chapter we will study how the CO_2 released by human activities impacts our planet.

⬠ **TAKE-HOME MESSAGE** Carbon is the primary component of organic molecules essential for life. Carbon can move out of the lithosphere through weathering and volcanic activity. It can move into the lithosphere when dead organisms form carbonate-rich rocks and fossil fuels. Carbon is removed from the atmosphere by photosynthesis and returned when products from this process decay or are used for fuel. Humans are releasing carbon long stored in the lithosphere through their activities, increasing the concentration of carbon in the atmosphere and ocean.

🌍 ❓ 10.6 What Can I Do?

Learning about systems is a broad and important skill, and much of what you learned about basic systems will be discussed further in later chapters. In the next chapter, we will take a closer look at the way human impacts on the carbon cycle influence the climate system. And later chapters will focus on ways to reduce impacts from energy use and transportation, agriculture, and consumption and waste. So for now, what are things you can do to build a better background in understanding some important systems concepts?

Know Your Footprints for Carbon and Nitrogen

You can see the ways your own decisions and actions influence essential systems and cycles by using nitrogen and carbon footprint calculators. If you have not done so already, go to Appendix A to learn how to calculate your carbon footprint. Seeing where you score high and score low will give you ideas about how to lessen your overall impact. Similarly, calculating your nitrogen footprint (also see Appendix A) gives you an idea of your overall use of this element. Some of the best ways to reduce your footprint in these calculations are by choosing more organic and sustainably grown foods that make less use of fertilizer. The National Agricultural Library at the US Department of Agriculture has a protein calculator to determine how much you should consume each day based on your age. Most people will find they need to eat far less than they consume currently. Substituting plant for animal proteins when possible is another way to reduce your impact, although as we'll see, the impact depends on how the animals were raised and the plants were grown.

Practice Systems Thinking I: Stock and Flow Models

We live in an interlinked world, so it's important to have a good idea of how systems work. Working with the models introduced in this chapter is a good place to start. Stock and flow models help you understand how systems fit together, but they take some practice to use properly. First, look at **FIGURE 10.22**. This is the bathtub model we used to introduce you to stock and flow models earlier in the chapter, and it is drawn in the standard stock and flow diagram format with more detailed labels. In a standard stock and flow diagram, clouds represent sources or sinks of a stock, rectangles represent stocks, and arrows represent flows. Valves on the flows indicate that there are ways to control the flows. On the basis of Figure 10.22, how would you fill in the blanks? (Answers are in the back of the book.)

Now let's look at another model. Here's a simplified stock and flow diagram of the carbon cycle (**FIGURE 10.23**). Study it, and then consider how you would answer the following questions. What are at least two different ways that the stock of carbon in the soil can be increased? What are at least two different ways that the stock of carbon in the atmosphere can be decreased? If the rate of photosynthesis doubled, what are different ways that the stock of carbon in terrestrial biomass could stay the same? (Answers are in the back of the book.) Think about what you read about this system to help you build an intuitive sense of how this stock and flow model works. Although the model is a simplification of the real world, it should help you see how different aspects of the system affect each other and how one change leads to a chain of other changes: this is systems thinking.

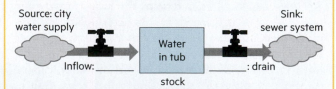

FIGURE 10.22 Stock and Flow Fill in the blanks of this stock and flow diagram.

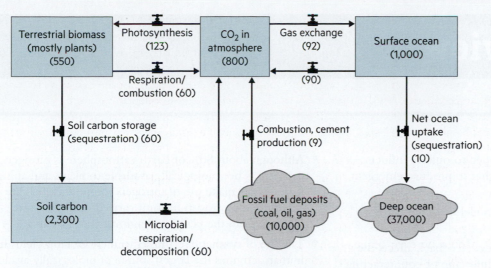

FIGURE 10.23 Stock and Flow Diagram of Carbon Cycle Numbers are in gigatons of carbon (GtC) in the reservoirs and GtC/year in the flows.

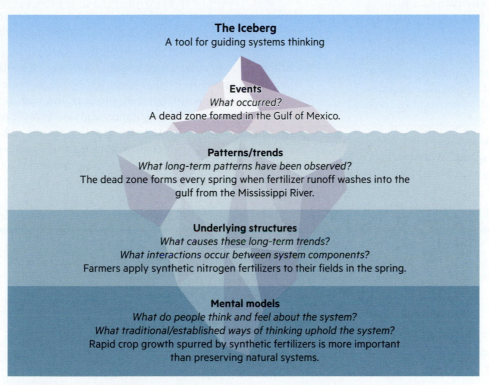

FIGURE 10.24 Iceberg Model for Systems Thinking What additional factors could be considered at each level below the "water"?

Adapted from Northwest Earth Institute (n.d.).

Practice Systems Thinking II: The Iceberg Model

Another way to practice systems thinking is to use the iceberg model developed by Michael Goodman. This model gets its name from the idea that a visible event represents only a small part of the system that sustains it, just like the visible, above-water portion of an iceberg. An example using the annual dead zone in the Gulf of Mexico (described earlier in the chapter) is shown in **FIGURE 10.24**.

See if you can add another item to each "underwater" level. For example, to the "Patterns/Trends" level, we could add, "The dead zone forms after explosive phytoplankton growth occurs." Thinking about the deepest level of the iceberg, the "Mental Models" level, helps us examine the roots of systems and the entrenched processes that will be most difficult to refine or reform. Overall, the iceberg model helps us think about the underlying causes of events and how we can work to correct or improve them. What is another event you could build an iceberg model for?

Chapter 10 **Review**

SUMMARY

- Systems, which can be either closed or open to outside influence, consist of parts that interact with each other to produce emergent properties.

- We can use stock and flow diagrams to model how aspects of a system function.

- Feedback is a response to changes in a stock; feedback affects the inflow or outflow of that stock, either reinforcing or counteracting the direction of change in the stock.

- Matter on Earth moves back and forth between the lithosphere, biosphere, hydrosphere, and atmosphere. The path of a particular kind of matter through its reservoirs and back again is its biogeochemical cycle.

- Oxygen cycles between all four of Earth's "spheres." Its movement between the atmosphere, hydrosphere, and biosphere is driven primarily by photosynthetic organisms. It moves into the lithosphere through weathering processes or the formation of rocks from oxygen-rich shells of dead organisms.

- Phosphorus does not cycle through the atmosphere. Because it moves very slowly through the lithosphere and hydrosphere, animals and plants continuously recycle this important macronutrient through the biosphere.

- Although about 80% of Earth's atmosphere is nitrogen, free nitrogen (N_2) is not biologically available to plants and animals in this form. Plants make use of nitrogen compounds that are made available through nitrogen fixation carried out by bacteria. Animals get nitrogen from the plants and animals they consume.

- The use of synthetic fertilizers to increase crop yields has made human activities the largest source of biologically available nitrogen on the planet.

- Agricultural applications of phosphorus and nitrogen as fertilizer are often washed away in runoff and can pollute aquatic systems, depleting oxygen levels and damaging these ecosystems through eutrophication. Synthetically produced nitrogen can also become airborne, contributing to air pollution.

- Carbon is the primary component of the organic molecules that make life possible on Earth. The biogeochemical cycle for carbon includes not only natural cycling mechanisms, such as weathering, volcanic activity, and the respiration and decomposition of organisms, but also human-driven processes that release carbon long stored in the lithosphere, such as burning fossil fuels and large-scale agriculture. These human processes are increasing the concentration of carbon in the atmosphere and ocean.

KEY TERMS

ammonification	emergent property	nitrification	sink
balancing feedback	feedback	nitrogen fixation	stock
biogeochemical cycle	flow	ocean acidification	system
closed system	macronutrient	open system	
denitrification	methanogen	photosynthesis	
diazotroph	model	reinforcing feedback	

REVIEW QUESTIONS

The letters following each Review Question refer to the Chapter Objectives.

1. How are systems classified with regard to outside influences? **(A)**

2. Explain how Earth is a different type of system with regard to energy than it is with regard to matter. **(A)**

3. Define an emergent property, and provide an example. **(A)**

4. Describe and provide an example of how a reinforcing feedback loop works. **(B)**

5. Describe and provide an example of how a balancing feedback loop works. **(B)**

6. Name three important biogeochemical cycles for life on Earth. Why are these processes considered cycles? **(C, D)**

7. Compare and contrast the oxygen and phosphorus cycles. **(D)**

8. Why can't we get the nitrogen we need through breathing in nitrogen in the air? How do we get the nitrogen we need instead? **(D)**

9. How do synthetic nitrogen fertilizers cause adverse environmental impacts? Why has this become a significant problem? **(D, E)**

10. Trace two different paths a molecule of carbon can take as it cycles through the biosphere. **(D)**

11. Describe how human activities are influencing the flow of carbon in and out of the atmosphere. **(D, E)**

12. Name two important environmental impacts of human interventions in the carbon cycle. **(D, E)**

FOR FURTHER THOUGHT

The letters following each item refer to the Chapter Objectives.

13. For most of us, our bank accounts have outflows and inflows. Draw a stock and flow diagram for your bank account or something else you know well enough to identify inflows and outflows. Identify at least one feedback loop (either reinforcing or balancing) that connects the stock to either the inflow or outflow. **(B)**

14. In theory, Earth's biogeochemical cycles recycle important elements and make them available again when the cycle repeats. In reality, the recycling of these elements is more complicated. Why are humans unable to count on a biogeochemical cycle to regenerate all elements when needed? **(C, D)**

15. First, identify at least three ways you are having an impact on the carbon cycle by increasing the inflow of carbon into the atmosphere. Then explain which changes in your life would be the easiest ones to make to reduce this impact and which changes would be the most difficult to make. Describe some ways in which you think government, businesses, personal connections, or other outside factors could make it easier to make these changes. **(D, E)**

Make Your Case ?

The following exercises use real-world data and news sources. Check your understanding of the material and then practice crafting well-supported responses.

Use the News

The following article appeared on the Michigan-based local news site Mlive.com and describes a new effort to recycle phosphorus, nitrogen, and other materials at the University of Michigan, Ann Arbor. Use this article to answer the questions that follow. The first four questions test your understanding of the article. Questions 5–7 are short answer, requiring you to apply what you have learned in this chapter and cite information in the article. Answers to Questions 1–4 are provided at the back of the book. Question 8 asks you to make your case and defend it.

"University of Michigan Adds Pee-Cycling to Its Recycling Efforts," Martin Slagter, Mlive.com, February 21, 2017

ANN ARBOR, MI—Three weeks after the launch of one of the University of Michigan's most intriguing research projects—"pee-cycling," as it's become

known—UM doctorate candidate Heather Goetsch estimated it has produced about a gallon of finished product. The UM research project is designed to find the most effective ways to convert urine into fertilizer that can be used to help plants grow. Project-coleader Krista Wiggington and Goetsch say the research also includes monitoring people's attitude toward this type of technology. A big portion of a $3 million grant is focused on understanding how to introduce this kind of technology to people and to educate them on its benefits and risks, Wiggington said. That's why tablet computers have been provided inside the G.G. Brown restrooms outside of stalls, asking

contributors to take a survey about their experiences using the new technologies and their broader views on the topic. "We are definitely trying to learn what the initial perception is, even with people just using a different kind of toilet," Wiggington said.

UM announced in September it would attempt to convert human urine into a safe fertilizer for agriculture crops as part of the grant, which is funded by the National Science Foundation. UM has since installed special demonstration toilets in the building on UM's north campus. The toilets, which include a waterless urinal and a "source separating" flush toilet, have been routing urine to a holding tank on the floor directly below it, where it is treated. The plan ultimately is for the fertilizer to be used on plants this spring at the Matthaei Botanical Gardens, Wiggington said. Wiggington is working with UM professor of civil and environmental engineering Nancy Love on the research, which is exploring the technology, systems requirements and social attitudes that go along with urine-derived fertilizers. . . .

"The reason we're interested in this is to make nutrient management more sustainable and also to make water management more sustainable," said Wiggington, an assistant professor of civil and environmental engineering at UM. "By diverting urine at the source and not mixing it and diluting it, you save a lot of energy trying to pull it back out before it goes into the environment, because if it goes into the environment, it's actually a pollutant because it leads to eutrophication and algal blooms, and things like that. But if you collect it and actually capture it, you can turn it into a resource," she added. . . .

Urine contains high levels of nitrogen, phosphorus and potassium, Wiggington said, because humans consume it in their diets, ending up in urine, rather than other wastes. The urinal

provided is a typical waterless variety, but it flushes periodically with a spray of vinegar to keep minerals from building up in the pipes. The toilet sends the solid waste and toilet paper that's collected to the wastewater treatment plant. The holding tank and treatment lab located in the floor below the restrooms then start the process of treatment, which involves concentrating the urine and distilling it to purify it of any microorganisms. . . . The process does require regular monitoring, Goetsch said, which includes watching pH levels of the lines coming into the treatment room. "If the pH level is high where we might see mineral precipitation, we add more acidic acid [sic; acetic acid] or vinegar into the system to prevent that from happening," Goetsch said. "We also have online tank sensors so we know how much we are generating throughout the day."

Wiggington understands there may be some apprehension from the public in embracing the idea, but explained that the final product is not offensive because it has been pasteurized, making it unrecognizable from normal human urine. "When it comes down to it, the nutrients

are the same, but the benefits are that it's more sustainable," Wiggington said. "It's recycled material, so it costs less and there's less energy turning it into a finished product."

The process also is a reminder of how wasteful the current system of going to the bathroom is, Wiggington said. Educating people about how this process is not conducive in helping preserve water, along with the phosphate mining involved in creating synthetic fertilizers is another advantage to the project, is all part of the teaching element that helps round out the research. "Right now we take a toilet full of drinking water, put waste in it and we flush it," Wiggington said. "In a place like Michigan, that might not be as big of a deal, because we have plenty of water, but in places like California and Arizona, to put that much treated drinking water down the drain every time someone goes to the bathroom, that's a big deal. Becoming comfortable with the idea of using fertilizer that came from human waste is a completely different social question, so we're trying to tackle all of those," she added. "The main thing is education, though, in helping people understand how wasteful the current system is."

1. Why are the researchers providing tablet computers in the bathrooms where the pee-cycling toilets are being tested?

   a. to distract bathroom users
   b. to monitor the pH levels in the new toilets
   c. to administer surveys to bathroom users
   d. to communicate with tank sensors in the toilet that signal when they are full

2. Why do the waterless urinals use vinegar spray?

   a. to counteract the smell of the urine
   b. to sterilize the urine
   c. to lower the pH of the urine
   d. to prevent minerals from building up in the pipes

3. What is the planned destination and use for urine from the pee-cycling toilets?

   a. It will be recycled into potable water.
   b. It will be used as a deterrent to deer that graze on rose bushes in the surrounding community.
   c. It will be a fertilizer in the botanical gardens.
   d. It will be sent to the sewage treatment plant where it requires less intensive treatment than typical sewage.

4. What is the environmental benefit of this practice?

   a. It keeps nutrient-rich urine from reaching the water where it can cause eutrophication.
   b. It reduces the use of treated fresh water for sewage disposal.
   c. It provides a recycled source of phosphorus that can reduce the use of mined phosphorus for fertilizers.
   d. all of the above

5. Describe what the article says are the environmental benefits of the program, in terms of both energy saved and management of the nutrients described in this chapter.

6. The article describes monitoring the pH level of the lines coming into the treatment room. Why does the article say this is being done? Explain this monitoring using the terminology of feedback loops.

7. What are some of the concerns that the article describes about this program? What is being done to monitor these concerns?

8. **Make Your Case** This program is funded by a National Science Foundation grant, with the goal of finding a cheaper and cost-saving source of fertilizers while reducing pollution. Do you think this money is being well spent? Make your case using information from the article and the chapter.

Use the Data

Chesapeake Bay suffers from excessive loads of nitrogen pollution from human activities on the land. The figures here produced by the Chesepeake Bay Foundation show (a) the primary sources of nitrogen pollution to Chesapeake Bay and (b) the relative cost per pound of several strategies to reduce nitrogen pollution.

Study the figures shown here and use them to answer the questions that follow. The first three questions test your understanding of the figures. Questions 3–6 are short answer, requiring you to apply what you have learned in this chapter and cite data in the figures. Answers to Questions 1–3 are provided at the back of the book. Question 7 asks you to make your case and defend it.

(a) Nitrogen pollution to the Chesapeake Bay by sector.

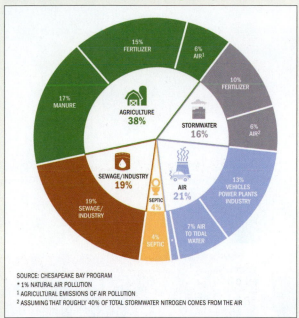

SOURCE: CHESAPEAKE BAY PROGRAM
* 1% NATURAL AIR POLLUTION
[1] AGRICULTURAL EMISSIONS OF AIR POLLUTION
[2] ASSUMING THAT ROUGHLY 40% OF TOTAL STORMWATER NITROGEN COMES FROM THE AIR

(b) Cost of nitrogen pollution reduction by sector and practice (per pound).

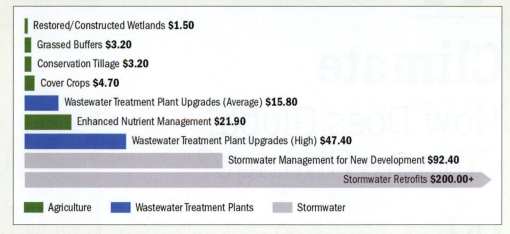

1. What share of the nitrogen pollution to Chesapeake Bay is contributed by agriculture?

   a. 6%
   b. 15%
   c. 17%
   d. 38%

2. What is the largest source of nitrogen in air pollution?

   a. agricultural emissions
   b. air to stormwater
   c. vehicles, power plants, and industry
   d. air to tidal water

3. According to image (b), changing practices in what sector provides the cheapest nitrogen-reduction strategies? What is the cost per pound of nitrogen saved?

4. How does the percentage of nitrogen coming from natural sources compare to that from agricultural sources? How does it compare to waste treatment and other urban/suburban sources?

5. Considering what you have read in the chapter, name at least one source of nitrogen pollution that is not addressed by the reduction strategies depicted in image (b).

6. Consider the cost per pound of the different strategies. Overall, do natural solutions or human-engineered solutions cost more? Support your answer with data from the figures.

7. **Make Your Case** Given the reduction strategies in image (b) and the sources in image (a), what strategies do you recommend communities in the area implement to reduce nitrogen in the bay? Imagine you are a consultant hired to prepare recommendations for what should be done (if anything) and who should do it. Make your case using these figures and information from the chapter to support your position.

LEARN MORE

• Fred Pearce. 2011. "Phosphate: A Critical Resource Misused and Now Running Low." Yale Environment 360. https://e360.yale.edu/features/phosphate_a_critical_resource_misused_and_now_running_out

• Christian Schwägerl. 2015. "With Too Much of a Good Thing, Europe Tackles Excess Nitrogen." Yale Environment 360. https://e360.yale.edu/features/with_too_much_of_a_good_thing_europe_tackles_excess_nitrogen

• Video: Rising CO_2 Levels Greening Earth: digital.wwnorton.com/environsci

• Allison M. Leach, James N. Galloway, Albert Bleeker, Jan Willem Erisman, Richard Kohn, and Justin Kitzes. 2012. "A Nitrogen Footprint Model to Help Consumers Understand Their Role in Nitrogen Losses to the Environment." *Environmental Development* 1 (1): 40–66. www.sciencedirect.com/science/article/pii/S221146451100008X

• NASA Carbon Cycle page: https://earthobservatory.nasa.gov/Features/CarbonCycle/

• Donella H. Meadows. 2008. *Thinking in Systems: A Primer*. Hartford, VT: Chelsea Green.

11

Climate
How Does Global Climate Change?

Where do you want to live? Imagine you are moving to a new city or town. Maybe it is for a promising new job, and you can see yourself settling in for a long time and even buying a place of your own. If you were in the market for a new home, what factors would shape your purchasing decision? You would probably consider the characteristics of the building itself, such as the size, condition, appearance, and amenities of the home. You might also consider factors related to location, such as proximity to work, schools, shops, and parks.

Yet all of these factors assess the current condition of the place and location. If your plan is to settle in for a while, you will be buying this home for years or decades of use. Shouldn't you also consider how this place and location are likely to change in the future and how that could influence your satisfaction and well-being in this home?

Individuals, businesses, and governments all confront this question as they consider a purchase or investment (including the mortgage or homeowner's insurance one may need) and try to predict factors that may affect any repayment obligations.

One type of information that institutions increasingly incorporate into their decision making is Earth's changing climate and its likely effects on the regions where they have money at risk. As one insurance company executive explained in a newspaper interview, "Insurance is heavily dependent on scientific thought."

But why does the concern of a big corporation matter to you? The answer relates to risk. When insurance companies determine that the risk of damage to a property is increasing, they raise the cost of the premium or even deny coverage. For example, in 2012, Superstorm Sandy damaged or destroyed more than 600,000 homes and businesses in New York and New Jersey, and insurers paid out more than $33 billion to policyholders in the wake of the storm (**FIGURE 11.1**). Because of this, in the first 2 years after Superstorm Sandy, the cost of flood insurance in the affected areas rose by 20% to 25%. Then in 2021, Hurricane Ida caused widespread damage from flooding and is expected to increase the cost of over 60% of flood insurance policies in New York City. In coastal areas of Florida, which are judged to be an even higher risk for insurance companies, hundreds of thousands of property owners have seen their premiums more than double in the past decade, and owners of waterfront property have had more difficulty finding private insurance companies willing to extend any coverage at all. Further, payouts from huge wildfires like the Camp Fire in California caused many insurers to

The city of Miami is seen as particularly susceptible to rising seas, as it already experiences impacts such as damage to low-lying buildings, salt intrusion, and widespread eutrophication events in Biscayne Bay.

Chapter Objectives

By the end of this chapter, you should be able to . . .

A. describe how Earth's climate is warmed by a greenhouse effect.

B. explain how scientists estimate historical changes in Earth's climate.

C. recognize that Earth's climate has changed dramatically in its history and that the recent temperature of air and water at Earth's surface has been warming.

D. explain how ocean currents influence Earth's climate.

E. describe how human activities are driving recent global warming.

F. discuss how climate models attempt to simulate Earth's climate system to forecast future climate.

G. summarize how global warming can alter precipitation patterns, the frequency and intensity of extreme weather, sea levels, and ocean acidity.

H. identify steps that individuals, businesses, and governments can take to address climate change.

> **Human beings are now carrying out a large scale geophysical experiment of a kind that could not have happened in the past nor be reproduced in the future.**
>
> —Roger Revelle

(a)

(b)

Top 10 Costliest Hurricanes In The United States

Rank	Year	Hurricane	Estimated insured loss in billions of 2020 dollars
1	2005	Hurricane Katrina	86.57
2	2012	Hurricane Sandy	33.93
3	2017	Hurricane Harvey	31.96
4	2017	Hurricane Irma	31.85
5	2017	Hurricane Maria	31.27
6	1992	Hurricane Andrew	29.70
7	2008	Hurricane Ike	21.76
8	2005	Hurricane Wilma	14.01
9	2018	Hurricane Michael	13.71
10	2004	Hurricane Ivan	12.06

FIGURE 11.1 Costly Storms (a) Areas of New Jersey and New York saw extensive flooding as Hurricane Ida traveled north. Flooding in the Bronx, New York City is pictured. Some estimates put damage from this storm at $31 billion. **(b)** According to the Insurance Information Institute, 9 of the 10 most costly storms for US insurance companies have occurred since 2004. Insurance companies are particularly sensitive to changing risks associated with climate change.

Adapted from (b) Insurance Information Institute (n.d.).

stop offering fire insurance entirely in some counties, until the state stepped in and required insurance companies to continue coverage.

Rising premiums and increasing risk are real harms being felt today by people living in the areas most affected by climate change. For example, the more than 100,000 residents of Kiribati, a nation of islands in the Pacific Ocean located between Hawaii and Australia, may need to resettle to a new country in the next 20 years because of a rising sea level that is already regularly flooding their homes, contaminating their water, and devastating their crops. In North America, residents of islands off the coast of Louisiana and of coastal communities in Alaska are seeking government aid to resettle their populations because their land has become uninhabitable due to the effects of climate change. Because so much of the world's population lives in coastal areas, some studies have estimated that by the year 2100, more than 2 billion people will likely become climate refugees.

Cities are now assessing the increasing risk of property damage associated with a changing climate. In summer 2013, shortly after suffering the effects of Superstorm Sandy, New York City unveiled a $20 billion plan to help the city adapt to the effects of climate change. The plan included proposals for new floodwalls and storm barriers, for an elevated electrical infrastructure, and for zoning regulations to protect against rising sea level and larger, more frequent storm surges. It also proposed building-code changes and infrastructure improvements to respond to hotter summer temperatures and heat waves. Similarly, states and cities in the Midwest are developing plans to improve their ability to respond to long-term droughts, which can have devastating effects on agricultural production. In 2011, for example, Texas suffered through its driest year on record, and estimated costs due to crop failure and livestock loss exceeded $8 billion for the year (**FIGURE 11.2**). A Texas A&M University study from 2020 projects that during this century the state will likely face the driest conditions of the past 1,000 years. Hundreds of cities around the world have begun work on climate action plans that outline not only adaptation strategies like the ones mentioned above but also mitigation strategies to reduce human influences spurring climate change.

While humans are adept at focusing on the present, decisions about where and how we live are increasingly driven by assessments of how our climate is changing. In this chapter, we will learn that Earth's climate has never been static. We will take a closer look at the methods scientists use to understand climate change and factors that have changed the climate during Earth's long history, including the oceans' influence on climate. Then we will examine what influences the climate today, what role humans have played in climate change, and what effects global warming is having. Finally, we will revisit the question of what the future holds by examining various scenarios of future climate change and its likely effects.

11.1 How Do We Measure Global Climate?

What is global climate anyway? In Chapter 8, we learned that climate is more than just the aspects of weather, such as air temperature, wind conditions, and precipitation at a given time and place. Climate is the average weather conditions over a significant period of time for a particular region of Earth. Just as we can average the weather conditions for a region of Earth, we can also examine the long-term trends in these conditions for the planet as a whole: the **global climate**.

So if you want to know what the temperature is where you are right now, you check a weather report or weather app or perhaps even consult a thermometer. But how do we characterize and measure global climate? And how do we estimate what the average temperature was on Earth hundreds, thousands, or even millions of years ago? In this section, we will look at the methods and the data that scientists use to measure climate change in the past and present.

Temperature Records

Scientists have been recording local temperatures since the invention of the thermometer. In 1714, Daniel Fahrenheit produced a sealed mercury thermometer that enabled weather stations to report and record the temperature according to a standard numerical scale. With this invention, countries such as England, Switzerland, and Sweden started to keep systematic temperature records at weather stations. By 1860, thermometers were both reliable and commonplace. Weather stations set up around the globe allowed scientists to keep careful temperature records. Thus, 1860 marks the beginning of what climatologists call the **instrumental period**—the era in which we have temperature records for Earth's surface air temperature that are based on readings taken directly with instruments such as thermometers. Since 1979, weather satellites have been estimating and recording the temperatures of the various layers of the atmosphere by measuring infrared radiation that emanates from vibrating gas molecules in the air (**FIGURE 11.3**).

FIGURE 11.3 Temperature Records (a) Since Daniel Fahrenheit invented the mercury thermometer in 1714, scientists have been able to report and record the temperature using a standard numerical scale. **(b)** Since 1979, weather satellites orbiting Earth have been estimating and recording the temperatures of the various layers of Earth's atmosphere. This polar orbiter (NOAA-20) gathers data on atmospheric variables for the entire planet as it rotates below.

FIGURE 11.2 More Droughts/Costly Fires (a) The increasing frequency of droughts, which cause crop and livestock losses, is yet another effect of climate change. This image shows a reservoir near Robert Lee, Texas, during a 2011 drought. **(b)** More frequent and intense wildfires, such as the River Fire of 2020 in Monterey County, California pictured here, can destroy entire communities.

FIGURE 11.4 shows the temperature record as measured during the instrumental period. In the earlier time periods, variations in the instruments used, the procedures governing the time and place the temperatures were taken, and the competence of people gathering and entering the data could make small but significant differences to the mean temperature records of particular stations. In more recent decades, as confidence in measurements increased, uncertainty about the trend line has narrowed. You might also notice that the graph, in addition

global climate the average weather conditions over a significant period of time for the planet as a whole.

instrumental period the current era in which we have access to temperature readings taken directly with instruments such as thermometers.

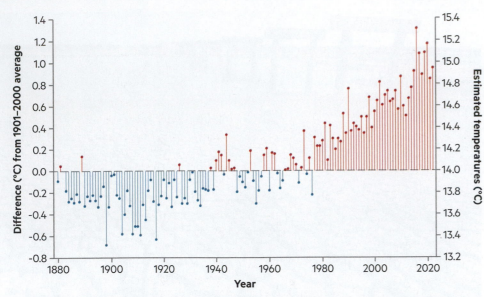

FIGURE 11.4 **Annual Global Average Temperature 1880–2022** The dot for each year shows both the global average temperature for that year and how it compares to the twentieth century (1901–2000) global average, indicated by the dashed line. Dots above the dashed line indicate warmer years, an increasingly common occurrence. The difference from the average in Celsius is indicated on the left-hand *y*-axis, and the actual value of a year's average global temperature is on the right-hand *y*-axis.

Adapted from NOAA National Centers for Environmental information (2022).

to plotting the annual average temperature, also plots the temperature difference, or anomaly, for a given year relative to a fixed base period. In this case, the base period is the average annual temperature for the 20th Century. Temperature records for each station are standardized by taking the average temperature for a given year and subtracting the average temperature over the century at that place. Why take this approach? Comparisons to a fixed base period reveal the extent to which climate is changing over time—which is what we are most interested in.

Temperature Measurements from the Ocean

While taking surface air temperature (SAT) measurements over land is very useful, oceans cover most of Earth's surface. So scientists also gather data to figure out the average marine air temperature (MAT) and sea-surface water temperature (SST). Collection of these latter two types of data began in

geothermal gradient
how temperature increases with depth in Earth's crust due to decaying radioactive elements.

proxy an observable and measurable phenomenon that serves as an indirect indicator of changes in climate.

the second half of the 19th century by crews aboard ocean-going vessels traveling along shipping routes. Today, in addition to ship recordings, specially equipped buoys gather data to calculate MAT. Since 2000, an array of more than 3,000 devices known as Argo floats now measures SST. These floats drift with the ocean currents and periodically descend to about 6,000 feet below the surface, then rise back to the surface to record temperatures at a range of depths.

Geothermal Gradients

For the years prior to the instrumental period (i.e., before reliable temperature records were kept), one way to measure and analyze surface temperatures stretching back hundreds of years is to drill bore holes deep into Earth's crust. The crust gets progressively hotter at deeper depths—an expected **geothermal gradient**—due to decaying radioactive elements that release energy and heat the crust. When there are sustained temperature changes at the Earth's surface, these slowly diffuse down through the crust and cause these measurements to change from what is normal. By studying these changes, researchers can reconstruct historical surface temperatures over the past several centuries.

Proxies

Another type of tool to help us understand climates of the past is called a **proxy**. Proxies are not direct records of temperature but rather observable and measurable phenomena that serve as indirect indicators of climate changes. Tree rings and coral skeletons are two examples of proxies that can be used to reconstruct past changes in climate. A core sample bored from a tree that is hundreds of years old will display light- and dark-colored bands that correspond to tree growth in the warm and cool seasons of each year, respectively. Similarly, the calcium carbonate skeletons of coral are constructed in annual layers that reflect annual changes in seawater temperature. Proxies such as these can be gathered across many regions of the world and then calibrated by checking them against instrumental records of temperature. Once calibrated, proxies can be used to estimate temperatures from pre-instrumental time periods (**FIGURE 11.5**).

(a)

(b)

(c)

FIGURE 11.5 **Proxies Help Indicate Past Temperatures** (a) Tree rings, (b) coral skeletons, and (c) ice cores are used to estimate temperatures from pre-instrumental time periods.

Other types of proxies can help us understand ancient climates, or **paleoclimates**, as well. For example, ancient glacial sediments are evidence of past glacial activity and an indicator of cooler climates. Existing glacial ice itself holds information about paleoclimate that we can obtain by examining ice cores from glaciers (see Figure 11.5) and measuring the ratio of oxygen isotopes in these ice cores. In these glacial samples, there is a heavier oxygen isotope (^{18}O) and a lighter oxygen isotope (^{16}O). Water vapor molecules containing the heavier ^{18}O condense more readily than do water vapor molecules with the lighter ^{16}O. The colder global temperatures are, the more quickly the heavy ^{18}O is depleted from air as the air moves toward the poles and condenses its water vapor. This means the ratio of the two oxygen isotopes changes with the climate. So in cores drilled from the polar regions, the finding of less ^{18}O in the ice means the temperatures were cooler. Glacial ice in Antarctica provides a climate record dating back more than 700,000 years.

There are many other ways to find signs of past temperatures. Ancient pollen grains preserved in lake or ocean floors can indicate where and when certain types of plant species formerly thrived. Fossilized deposits of warm-water species or cold-water species of plankton on the seafloor can serve to indicate past water temperatures. Coal deposits are indicative of warmer climates; glacial sediments are indicative of colder climates.

⬡ **TAKE-HOME MESSAGE** Scientists use a variety of methods to evaluate changes in global temperature over time. From the mid-19th century, an instrumental record of temperatures is available from stations, ships, and buoys across the planet. Since 1979, satellites have recorded the temperatures in the atmosphere. Proxies—such as tree rings, coral skeletons, and glacial ice—are indicators of past climates.

11.2 What Caused Climate Change in the Past?

Throughout Earth's history, global climate has changed naturally. Over geologic time, Earth's climate has been both significantly warmer and significantly cooler than it is today (**FIGURE 11.6**). We know from fossil evidence of warm-weather plants and animals that during the Cretaceous Period (145 million to 66 million years ago) polar regions were warm enough to be forested and free of ice. In contrast, 700-million-year-old glacial sediments at regions near the equator suggest that Earth was almost completely iced over for a long enough period to warrant the nickname "snowball Earth." Let's examine what causes these extreme differences in climate.

➔ CONTEXT

Earth's Early Climate Catastrophe

Geologists have found evidence that glaciers covered almost the entire Earth nearly 700 million years ago, even in regions that were at or near the equator at that time. This means that Earth went through a radical climate shift that left the entire planet cold enough to form what is often called a snowball Earth. Extensive ice or slush coverage in the oceans during this period decreased the amount of sunlight and dissolved oxygen available to life-forms, disrupting ocean ecosystems and causing many life-forms to become extinct.

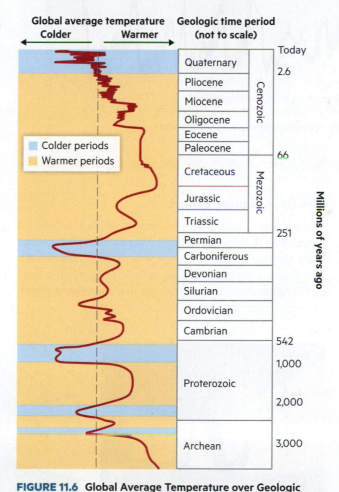

FIGURE 11.6 Global Average Temperature over Geologic Time Global average temperature, indicated above by the red line, has changed dramatically over Earth's long history, including fluctuations between extremely cold temperatures and extremely warm temperatures.

Adapted from Marshak and Rauber (2017).

paleoclimate ancient climate conditions understood through use of proxies.

How Are CO₂ and Temperature Related?

The history of global temperatures over the past million years shows a strong connection between CO₂ levels and temperature. Notice how the yellow line measuring CO₂ levels (on the left-hand axis) closely matches the blue line showing changes in temperature (on the right-hand axis). Researchers have found that CO₂ levels in the atmosphere are about 30% lower during glacial periods than during the warm periods between them. Why is this? A colder ocean can hold more CO₂. At the end of a glacial period, warming temperatures release CO₂ from the ocean. The warming ocean produces conditions that create a positive feedback loop, which enhances the pace of warming. When a new glacial period begins, Earth cools and atmospheric CO₂ levels fall, because a colder ocean can hold more dissolved CO₂.

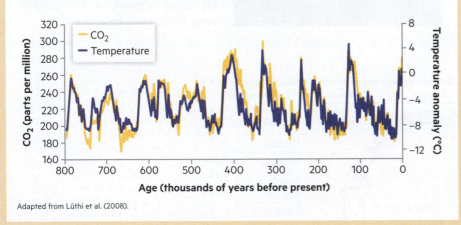

Adapted from Lüthi et al. (2008).

The Greenhouse Effect

Warming and cooling on a planetary scale can be thought of in terms of the relationship between incoming energy from the Sun and the energy that Earth radiates back into space. Central to this incoming and outgoing flow of energy is something known as the **greenhouse effect**.

If you have ever gotten into a car that has been parked in a sunny area with the windows shut on a warm day, you have experienced a warming system caused by a small-scale greenhouse effect (**FIGURE 11.7**). The car is receiving sunlight from the outside. Some of this energy is reflected off the exterior, but much of it is absorbed by the interior of the car, which then gives off, or radiates, energy in the form of heat. But you have probably observed that the interior of the car gets much, much warmer than the air outside. Although the sunlight can enter the car through the window glass, when this energy is converted into heat, it cannot readily escape back through the glass. For this reason, the interior of your car warms as sunlight continues to penetrate from the outside and heat is trapped in the interior. You can decrease the warming in the car somewhat by putting a reflective sunshade in the windshield. In this case, you are increasing the **albedo**, a term scientists use to describe the reflectivity of a surface. Actual greenhouses are designed to have low albedo because they are used to grow plants indoors when the weather is cold. Glass panels allow sunlight in but prevent much of the heat from escaping the facility, thereby creating an artificially warm climate. So the greenhouse effect refers to a system in which some sort of barrier enables the inflow of energy to outpace the outflow in a way that warms the interior.

Of course, there is no glass bubble around our planet, but as we learned in Chapter 7, Earth is encircled by a gaseous atmosphere. And water vapor, carbon dioxide (CO₂), and other gases in the atmosphere known as **greenhouse gases** (GHGs) play the role of the glass bubble. **FIGURE 11.8** shows a simplified diagram of how the global greenhouse effect works. Solar energy that reaches Earth's surface is absorbed and released as infrared radiation (a form of heat) that then rises back up into the atmosphere. Although some of the rising infrared radiation emitted from the warmed surface of Earth passes through the atmosphere and into space, much of it is absorbed by GHG molecules. The molecules scatter and bounce the infrared radiation around in the atmosphere

FIGURE 11.7 Examples of the Greenhouse Effect
(a) A car parked in sunlight will heat up as the sunlight penetrates the windows and the heat can only escape slowly from the inside. **(b)** A greenhouse creates a warm environment for plants. As sunlight passes through the glass, it warms the interior. Much of the heat that is reradiated from the ground, plants, and soil inside is reflected back by the glass and trapped inside.

(a)

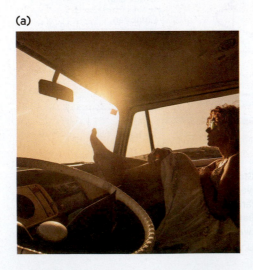

(b)

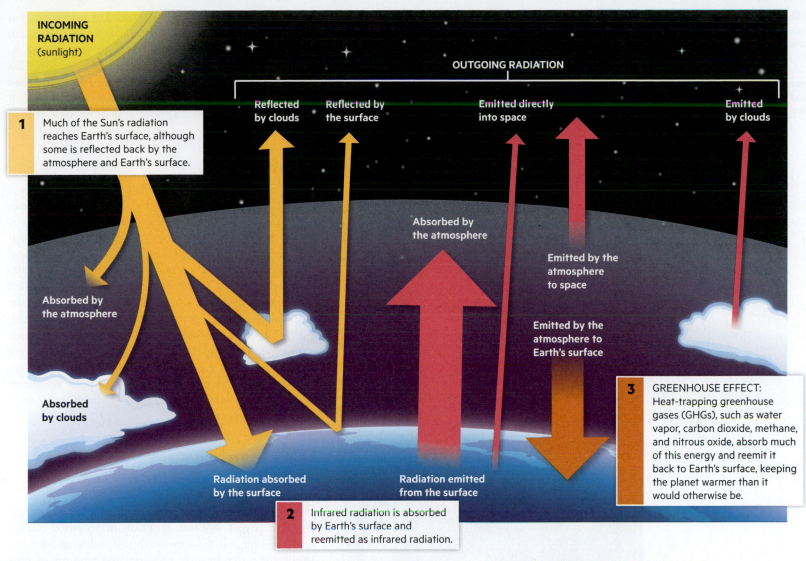

INCOMING RADIATION (sunlight)

OUTGOING RADIATION

Reflected by clouds

Reflected by the surface

Emitted directly into space

Emitted by clouds

1 Much of the Sun's radiation reaches Earth's surface, although some is reflected back by the atmosphere and Earth's surface.

Absorbed by the atmosphere

Absorbed by the atmosphere

Emitted by the atmosphere to space

Absorbed by clouds

Emitted by the atmosphere to Earth's surface

3 GREENHOUSE EFFECT: Heat-trapping greenhouse gases (GHGs), such as water vapor, carbon dioxide, methane, and nitrous oxide, absorb much of this energy and reemit it back to Earth's surface, keeping the planet warmer than it would otherwise be.

Radiation absorbed by the surface

Radiation emitted from the surface

2 Infrared radiation is absorbed by Earth's surface and reemitted as infrared radiation.

FIGURE 11.8 Earth's Energy Balance and the Greenhouse Effect Each arrow represents a transfer of energy and the relative size of that transfer.

Adapted from Marshak and Rauber (2017).

for a period of time before it is reradiated either upward into space or downward toward the planet's surface, where it has a warming effect. Water vapor is the dominant GHG in the atmosphere, and clouds can also absorb and reradiate infrared radiation back toward Earth's surface. Albedo on Earth is determined by the extent of light-colored reflective surfaces, such as snow and ice. Increasing albedo (caused by more extensive year-round coverage of snow and ice) decreases warming as more light is reflected from the surface and escapes into space. Decreasing albedo (from loss of snow and ice coverage) increases warming.

Without the greenhouse effect, Earth's average surface temperature would be about 60°F cooler than it is now, dropping to a chilly −2°F. We would live (or more likely never have evolved) in a climate more like that of Mars: one that would not support abundant liquid water or much of the life that now flourishes here. The geologic record shows that although Earth's average

surface temperature has stayed between the freezing and boiling points of water for the past 3 billion years, our global climate has changed quite a bit between these two extremes. Let's look at some of the factors associated with changes that can take place over millions to hundreds of millions of years.

Continental Drift

One factor associated with global climate changes is continental drift, or the movement of continents over geologic time. For example, when continental landmasses are clumped together, they can form large interior regions where extremely cold temperatures can develop in the winter. The positions of "drifting" continents relative to the equator affect the amount of solar radiation that can reach the land to warm it. The arrangement of the continents also influences ocean currents, which in turn affect the global distribution of heat.

greenhouse effect any system where a barrier causes the inflow of energy to outpace the outflow in a way that warms the interior. Specifically, the warming effect on Earth due to atmospheric greenhouse gases that prevent radiant heat emitted from the surface from escaping into space.

albedo a measure of the reflectivity of a surface (Chapter 7).

greenhouse gas a gas in the atmosphere that redirects heat rising from Earth's surface back toward the surface, causing a warming effect.

Importantly, the changes of elevation, or "uplift," caused by colliding landmasses also affect climate. This process starts because hills and mountains have more surface area than does flat land, hence more land in hills and mountains is exposed to natural acid in droplets of rain. This acid is formed when CO_2 in the atmosphere dissolves in droplets of water. When it reacts with rocks at the surface, carbon compounds are freed and flow to the ocean in groundwater and surface water. Once in the ocean, these dissolved compounds become part of the calcium carbonate shells and skeletons of various organisms. When these creatures die, their carbon-rich shells and skeletons sink to the ocean floor, where the carbon can become buried and locked up across geologic time. In this way CO_2 is removed very, very slowly from the atmosphere, which reduces the concentration of greenhouse gases and cools the climate.

The Composition of the Biosphere

The evolution of life in the biosphere also shapes climate over long time frames. The proliferation of certain forms of life can alter the composition of the atmosphere by either contributing or using CO_2 as part of the carbon cycle (see Chapter 10). For example, if the hot and humid conditions that promote the formation of swamps and marshes are prevalent, this organic matter (and the carbon in it) when it dies can be buried deep under layer upon layer of sediment (as we will see in Chapter 13) to eventually form coal. This process is thought to have contributed to an extended period of cooling in the Paleozoic Era, as abundant swamps gradually moved more and more CO_2 out of the atmosphere.

Earth's Orbit

Some factors significantly affect global climate over shorter time frames, though by shorter we still mean thousands of years. For example, the shape, or eccentricity, of Earth's orbit changes according to a 100,000-year cycle, and the tilt and direction of Earth's rotational axis also oscillate at intervals of 40,000 and 26,000 years, respectively. These systematic patterns are called **Milankovitch cycles**, and they alter the amount of solar radiation reaching Earth and the distribution of this energy at various latitudes. These cycles are associated with observed temperature changes in the atmosphere and oceans over these time frames, and their effects, particularly the waxing and waning of polar ice sheets and alpine glaciers, define glacial epochs popularly known as ice ages.

Eruptions and Asteroids

Volcanic eruptions and massive asteroid strikes can have a far more rapid effect on global climate. The 1883 eruption of Krakatoa in Indonesia—which is often considered the largest volcanic eruption in recorded human history—sent ash plumes nearly 50 miles up into the atmosphere, significantly reducing the amount of sunlight reaching Earth and reducing average global temperature by 1.2°C in the year of the eruption. On geologic timescales, even larger volcanic eruptions and dust from massive asteroid strikes have altered climate to such an extent that they initiated mass extinction events. Earth's most severe extinction event, the Permian–Triassic extinction (or "Great Dying") 252 million years ago, is linked to massive volcanic activity in Siberia. The Cretaceous–Tertiary extinction 66 million years ago that killed off most of the dinosaurs is linked to a giant meteorite (or comet) that struck Earth near the Yucatán Peninsula (**FIGURE 11.9**).

(a)

(b)

FIGURE 11.9 Past Extinctions (a) Scientists have discovered that the Cretaceous–Paleogene extinction was caused by a meteorite (or comet) impact. **(b)** Fossils in stone show evidence of a class of life known as trilobites that died off in the Permian–Triassic extinction.

⬡ **TAKE-HOME MESSAGE** Earth's climate is warmed by the greenhouse effect, when gases in the atmosphere, particularly water vapor and CO_2, reradiate heat toward the surface. Global climate has changed dramatically during Earth's history because of continental drift, rising landmasses, the life on the planet, Earth's orbit, volcanic eruptions, and asteroid strikes. Severe changes to Earth's climate have caused mass extinction events.

11.3 How Do Oceans Influence the Climate?

Saltwater oceans cover most of Earth, forming a huge reservoir of water that impacts global climate. Think about walking into an indoor pool area; if you've done this before, you probably are associating this thought with a sense of steamy, humid air. We intuit that the warm water in the pool is affecting the climate in the room. Now extend this pool to the size of the oceans that cover two-thirds of the surface of our planet, to an average depth of 2.3 miles, and you start to see why their immensity makes oceans major engines of our climate. Then consider that this water is always moving. Earth's oceans are not just humongous "pools" of water but a connected flowing mass that transports energy around the globe, affecting areas even thousands of miles from shore. In this section, we will explore how parts of the ocean system work together to influence global climate.

Ocean Currents

Currents are sometimes called the "rivers" of the oceans, in that they are sustained, large-scale streams of ocean water. Many scientists see the ocean currents as more like massive conveyor belts, moving cold and warm water around the world.

One type of ocean current, **surface currents**, affects the top 100–400 meters of water. Surface currents start because of the air moving above them. They do not travel in straight lines but are bent into large circular patterns called **gyres**. There are five main gyres in the oceans—in the North Atlantic, South Atlantic, North Pacific, South Pacific, and Indian Oceans—bending water toward their centers driven by two factors. The first is Earth's rotation, acting through a force known as the **Coriolis effect**. The Coriolis effect bends global airflow and creates winds that blow from east to west in equatorial latitudes, sometimes referred to as the trade winds. These winds also bend the surface ocean currents they generate in the same directions, to the right in the Northern Hemisphere and to the left in the Southern Hemisphere. The second factor governing the direction of ocean currents is the large continental landmasses, which redirect currents that press up against them (**FIGURE 11.10**).

But surface ocean currents also interact with currents in deeper waters. **Deep currents** are shaped by differences in water density, which are caused by variations in their temperature and **salinity** (the concentration of salt in the water). The colder and more saline the water, the greater the density. As less dense water rises and more

surface current an ocean current affecting the top 400 meters of water that starts because of air blowing across the water surface.

gyre a large circular ocean current.

Coriolis effect a force driven by Earth's rotation that deflects objects, winds, and currents on the surface of Earth and in the ocean or the atmosphere.

deep current the flow of ocean water below the surface caused by variations in density, temperature, and salinity.

salinity the concentration of salt in water, generally measured in parts per thousand.

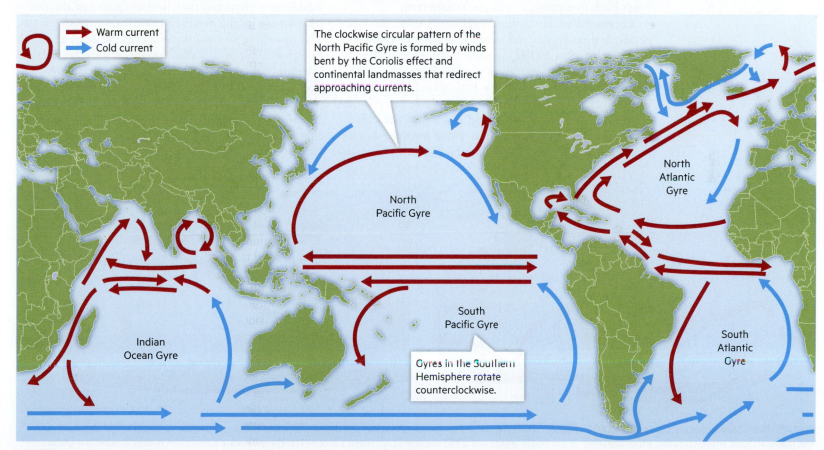

FIGURE 11.10 Ocean Gyres There are five ocean gyres formed by winds and landmasses. Notably, the North Pacific Gyre also spins large amounts of garbage in the ocean to its center, creating a huge "garbage patch" (learn more about this in Chapter 15).

Adapted from National Ocean Service (2018).

thermohaline conveyor
a large-scale ocean circulation driven by surface and deepwater ocean currents and changes in water temperature and salinity (density).

El Niño–Southern Oscillation (ENSO)
a change from normal ocean currents that significantly weakens or even shifts the direction of trade winds and ocean currents in the southern Pacific.

upwelling a place where ocean currents draw up colder water from the deep.

dense water sinks, this interaction of surface and deep-water currents mixes ocean waters, redistributing heat in ways that shape weather and regional climate while also distributing nutrients between the top and bottom of the ocean.

Let's use the Gulf Stream current as an example: it brings relatively warm water from the Gulf of Mexico all the way past Canada and northern Europe in a process of surface and deepwater circulation called the **thermohaline conveyor**. As warm water on the surface travels north away from the Gulf of Mexico, it cools. Some of the surface water in the ocean even freezes in winter near Iceland and Greenland. When the surface water freezes, most of the salt is left behind in the water under the ice. This cold and salty water below the ice sinks and forms a current moving cold water to the south called the North Atlantic Deep Water. The north-ward transfer of heat associated with currents in the North Atlantic helps moderate the climate of northern Europe, which is much warmer than it would otherwise be for its latitude. London, England, has a latitude similar to that of areas of central Newfoundland. But whereas London's average high and low January temperatures are 44°F and 34°F, respectively, those at a similar latitude in Newfoundland would be about 20 degrees colder. Eventually, the North Atlantic Deep Water delivers water back to equatorial regions (**FIGURE 11.11**).

Thermohaline circulation distributes heat around the planet, and ecologists and oceanographers have discovered that changes in this circulation are associated with major advances and retreats of glaciers. Recently, ocean-ographers at Duke University have found this current weakening by as much as 30%. One possible cause is the ongoing melting of icebergs and glaciers in the North Atlantic, which puts more fresh water into the ocean, thereby changing its salinity (and thus its density), which

weakens the current. Slower currents in the North Atlantic would lead to lower temperatures in Europe and likely stronger storms in this area, as well as far drier conditions in normally wet areas of India and Asia. In fact, toward the end of the last ice age about 13,000 years ago, an influx of fresh meltwater into the ocean stopped the density- and temperature-driven currents in the Atlantic, abruptly sending Europe back into a glacial deep-freeze for 1,300 years.

Countercurrents and El Niño

From time to time, countercurrents, or changes from normal ocean currents, take hold in the ocean and change the weather. Probably the best-known coun-tercurrent is the **El Niño–Southern Oscillation (ENSO)**, a cycle that significantly alters weather patterns in Pacific and Atlantic coastal areas every 2–10 years. Normally, the Pacific Ocean waters off the coast of Chile and Peru have cool **upwelling** currents where water is drawn up from deep in the ocean. Areas of upwelling are com-mon on the western coasts of continents where surface water currents flow away from the edge of the conti-nents (**FIGURE 11.12A**). In this case, the prevailing winds push warm Pacific surface waters westward. When the trade winds and ocean currents in the southern Pacific weaken or even reverse direction and push east, this rever-sal is known as the Southern Oscillation (**FIGURE 11.12B**). The upwelling of cool water along the coast of South America stops, and warm surface waters accumulate. Scientists are not sure what causes these shifts, but likely factors include cloud cover patterns over Asia and changes in deep ocean currents off the coast of South America. El Niño years bring heavier than normal pre-cipitation in Southern California and the southeastern United States. Other areas, such as Alaska, Washington,

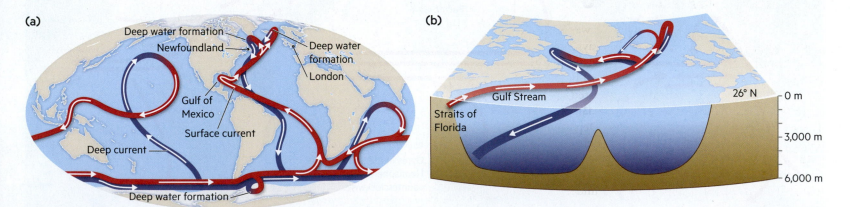

FIGURE 11.11 Thermohaline Conveyor (a) Differences in salinity and temperature affect the density of water and cause deepwater currents to circulate around the globe. In turn, these currents affect climates, in particular moderating the climate of northern Europe. The thermohaline circulation moves around the globe with warmer surface currents (red) and colder deepwater currents (blue). These currents only mix in certain areas. **(b)** This closeup of the North Atlantic shows how surface waters traveling north cool and then sink, forming deep currents traveling back to the south.

Adapted from (a) NASA (2009) and (b) Church (2007).

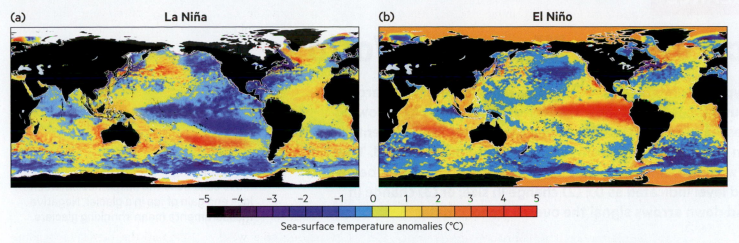

(a) La Niña   **(b)** El Niño

Sea-surface temperature anomalies (°C)

FIGURE 11.12 Upwelling and Countercurrents (a) Off the coast of South America, upwelling brings cold water to the surface. During times known as La Niña, this cold-water upwelling (in blue) is particularly strong. Upwelling is associated with tremendous biological productivity because nutrients from deeper levels of the ocean are carried to the surface. Some of the world's largest fisheries and seabird colonies are in upwelling zones. **(b)** El Niño is a countercurrent that occurs every 2–10 years. It is driven by a change in wind and currents that causes warm waters to stay off the coast of South and Central America (shown in red here) instead of moving west. El Niño years are associated with significant changes in weather patterns for many regions.

Adapted from (a) NOAA (n.d.) and (b) Marshak and Rauber (2017).

Oregon, and much of Canada, experience drier than normal weather.

> ⬡ **TAKE-HOME MESSAGE** Ocean currents play a big role in shaping Earth's climate. The ocean is a system in motion that distributes heat around the planet. Its surface currents are driven by Earth's rotation and continental landmasses. Wind-driven surface currents interact with currents deep below the ocean surface driven by differences in water temperature and salinity.

11.4 What's Happening with Our Climate Now?

Earth's climate is in a period of global warming that is starting to change our climate in ways that affect communities. For example, the population of Isle de Jean Charles, Louisiana, is being relocated because the island is sinking and becoming inundated with salt water from rising sea levels (**FIGURE 11.13**). The relocation is being financed by the US Department of Housing and Urban Development, which granted $1 billion to 13 states to help them adapt to effects of climate change. As one longtime resident put it, "This is our grandpa's land . . . but it's going under one way or another."

The following two pages sum up much of the data that indicate global warming (**AT A GLANCE: INDICATORS OF A WARMING WORLD**). These pages show increases in average global temperature measured over land and sea, in the air of the lower atmosphere, at the ocean's

surface, and in deep marine waters; they also show Earth system responses to warming expressed in diminishing Arctic sea ice, decreasing snow and glacier cover, and rising sea levels. In the rest of this section, we will examine the role that changing concentrations of GHGs in the atmosphere play in this warming and explore the evidence that these changes are driven primarily by human activities.

FIGURE 11.13 Effects of Rising Sea Levels Isle de Jean Charles, Louisiana, now floods so regularly that its residents need to be resettled.

Indicators of a Warming World

Different types of climate data recorded by many different research groups around the world all show the world is warming. The following data were assembled by NOAA's National Centers for Environmental Information and include data from researchers around the world. The graphs show measurements of (1) anomaly (changes above and below an expected level indicated as 0); (2) change in size; or (3) change in area. Up and down arrows signal the overall trend in data.

Glacier Mass

Most mountain glaciers are retreating. Mass balance is the difference between loss and gain of ice in a glacier. Negative measurements mean shrinking glaciers.

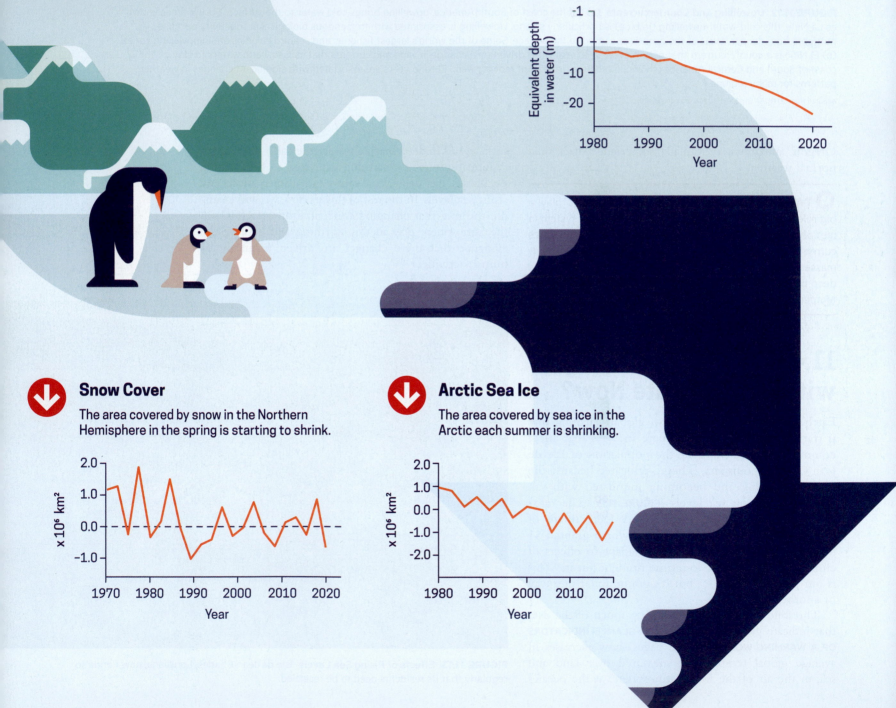

Snow Cover

The area covered by snow in the Northern Hemisphere in the spring is starting to shrink.

Arctic Sea Ice

The area covered by sea ice in the Arctic each summer is shrinking.

Air Temperature over Land

Average air temperature is rising.

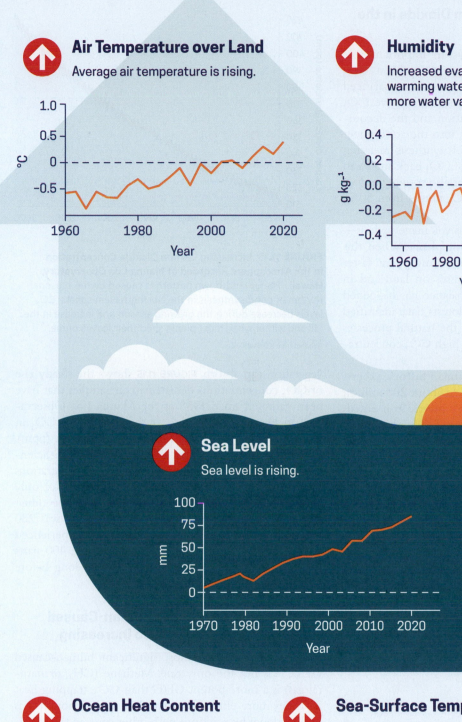

Humidity

Increased evaporation of warming waters leads to more water vapor in the air.

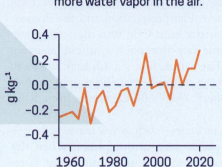

Temperature of the Lower Atmosphere

The air temperature of the lower atmosphere is getting warmer.

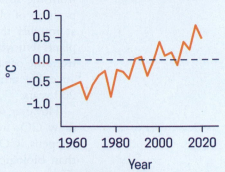

Sea Level

Sea level is rising.

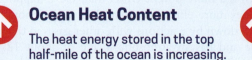

Air Temperature over Oceans

The air just above the ocean surface is warming, allowing more water to evaporate.

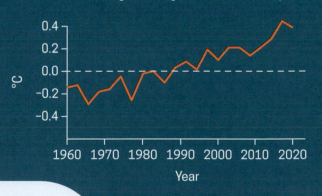

Ocean Heat Content

The heat energy stored in the top half-mile of the ocean is increasing.

Sea-Surface Temperature

The water on the surface of the ocean is getting warmer.

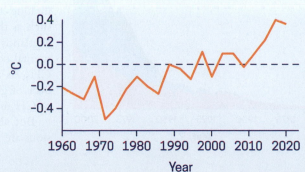

Adapted from Blunden and Boyer (2020)

The Concentration of Carbon Dioxide in the Atmosphere Is Increasing

Let's start with CO_2, which plays an important role in Earth's carbon cycle (see Chapter 10). Most CO_2 emissions are natural and are more or less balanced by physical and biological processes that remove CO_2 from the atmosphere. Plant respiration and the decomposition of plant matter emit CO_2 into the atmosphere at roughly the same rate that photosynthesis in living plants removes it. The natural flux of CO_2 back and forth from the ocean to the atmosphere is also roughly in balance. The atmospheric "life" of CO_2 ranges from 50 to 200 years.

However, some processes—such as volcanic eruptions—draw CO_2 from places deep in our planet. When this happens, CO_2 is emitted into the atmosphere faster than biological and physical processes on land and in the oceans can absorb it. This imbalance has happened before during Earth's history. Geologists have identified times in Earth's distant past when the natural processes of volcanic eruptions built up very high CO_2 concentrations in the atmosphere.

In the more recent period of human history, experts estimate that volcanoes have released about 200 million metric tons of CO_2 every year (1 metric ton equals about 2,205 pounds). This may sound like a big number, but it is 46 times smaller than a newer source: the 9.2 billion metric tons of CO_2 humans emit annually by combusting fossil fuels and through industrial processes (**FIGURE 11.14**). Together, the production of cement and the combustion of fossil fuels (such as coal, oil, and gas) account for about 80% of human-caused CO_2 emissions.

What is happening to all of this CO_2? While about 45% of the total of "new" CO_2 is absorbed by either the oceans or vegetation, much of the rest of the human-caused CO_2 accumulates in the atmosphere, increasing the concentration of CO_2 by about 0.5% per year. Then deforestation caused by land development for agriculture and urbanization, which we discuss more in Chapter 16, further diminishes the amount of CO_2 that Earth's

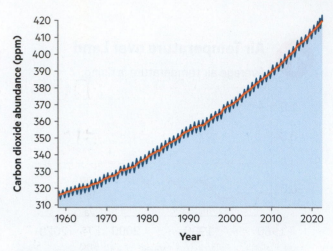

FIGURE 11.15 **Increasing Carbon Dioxide Concentration in the Atmosphere Recorded at Mauna Loa Observatory, Hawaii** The up-and-down pattern is caused by the seasonal rhythm of photosynthesis in the Northern Hemisphere. CO_2 levels decrease during the growing season and increase in the fall and winter when most plants shed their leaves or die.

Adapted from NOAA (2022).

vegetation can absorb. **FIGURE 11.15** shows the steady rise of CO_2 concentrations detected in air samples that have been gathered since 1958 at the Mauna Loa Observatory in Hawaii. In 2022, the concentration of CO_2 in the atmosphere surpassed 420 parts per million (ppm). How historically significant is this accumulation? Scientists have measured past carbon dioxide concentrations by sampling air bubbles trapped in ice cores. Not only have CO_2 levels increased by more than 40% since industrialization began in the late 18th century (it was 280 parts per million in 1750), but Earth has not experienced an atmospheric concentration of CO_2 above 400 parts per million in more than a million years—long before modern humans evolved.

Concentrations of Other Human-Caused Greenhouse Gases Are Also Increasing

Although CO_2 is the most significant human-caused GHG, it is not the only one. Methane (CH_4, or natural gas) is a more potent GHG than CO_2, trapping heat 25 times more effectively than does CO_2. Though it is broken down by reactions with other gases in the atmosphere and has an atmospheric life of about 12 years, CH_4—like CO_2—has been accumulating in the atmosphere. Natural biological sources of atmospheric CH_4 include the microorganisms known as methanogens, such as those that live in the digestive tracts of animals. Wetlands are also rich in methanogens, and the environmental conditions of these areas facilitate the release of CH_4. As a result, wetlands may account for as much as 30% of global CH_4 emissions. Natural geologic sources of CH_4 include volcanoes and forest fires.

But about two-thirds of Earth's current CH_4 emissions are from human sources. Methane is released from

FIGURE 11.14 **Major Sources of Carbon Emissions in the Industrial Age** Other significant human-caused emissions not represented in the figure come from land-use actions such as human-induced forest fires and agricultural practices like plowing that release CO_2 when soil organic matter is brought to the surface.

Adapted from Melillo et al. (2014).

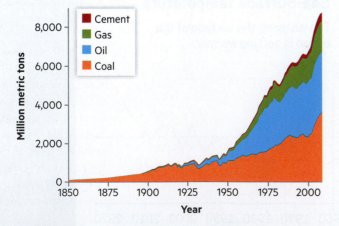

The Development of Global Climate Models

When we think of scientists working to understand global climate, often we picture them using observational tools: reading thermometers, looking at weather satellite data, and analyzing cores from trees or glaciers. But while gathering observational data is fundamental, making sense of changes in global climate also involves computational tools—mathematical equations that represent the many factors at work in the climate system.

During World War I, British mathematician and physicist Lewis Fry Richardson was working as an ambulance driver on the Western Front. For 6 weeks, when he wasn't tending to wounded soldiers, he did thousands of calculations with paper and pencil in an attempt to understand an 8-hour period of weather over a portion of Europe. Richardson reasoned that if he divided up the territory into a grid and gathered meteorological data such as air pressure, temperature, and humidity for each grid cell, he could then use a series of equations on the way heat, wind, and moisture function to forecast the weather. Although his forecast ultimately failed, he eventually published his methods, and this combination of grid-specific observational data and equations describing the behavior of air in the atmosphere became the basis many years later of the first global climate models.

Richardson's early efforts also revealed three of the major challenges to the development of such models. First, the computations required to simulate something as complicated as weather are so numerous and demanding that they are a challenge to complete. Even with the advent of computers that printed and sorted punch cards to break enemy codes during World War II, weather modelers using this technology in 1950 found that the calculation time required to produce a 24-hour forecast of a region exceeded 24 hours. It was not until the 1960s that increased digital computing power began to enable significant advances in climate modeling. Computational limitations also affected two other elements: the size of the cells in the grid and the number of

(a)

(b)

(a) In the 1960s, Akio Arakawa's work simplified and improved how air-flow simulations and global climate models were created. (b) In the 1980s, meteorologist Warren Washington coupled a model of the ocean to an atmospheric model to better represent the global climate system.

factors (such as data on clouds, oceans, vegetation, and much more) that could be plugged into a model. This is because the more cells there are in the grid and the more factors included, the more observational data there are to process. Adding factors into the model also increases the complexity of the computations because of interactions among the various factors.

So as computational power and scientific understanding of the climate system have increased, climate modelers have been able to make global grids with finer resolution and to add important new factors into the models. Two important early advances include Akio Arakawa's mathematical strategies that took the first steps in representing cloud cover in climate models in the late 1960s and Warren Washington's work coupling an atmospheric model with an ocean model in the early 1980s. Over time, factors linked to the land surface, ice and snow, the composition of gases in the atmosphere, the carbon cycle, and vegetation have been added to models as well. At the same time, the resolution of grid cells has increased. In 1990, the average size of a grid cell in climate models was 300 miles per side. Now some models can run simulations with grid cells as small as 15 miles per side.

Although some models are more successful than others in accurately hindcasting (reproducing) climate change across the past century, and different models vary to some extent in how they simulate future trends, all simulations of the future from these models predict that global temperatures will increase in this century in response to rising GHG emissions linked to human activities. There are more than 35 global climate models run by more than 20 modeling groups throughout the world, and these models continue to be refined as computational power increases and our understanding of the climate system expands. So observational tools help us refine our computational tools, and together these tools help us better predict changes in global climate.

What Would You Do?

In today's world, we often hear about the promise of "big data" and "artificial intelligence" because of increased computing power. In 2017, the World Bank issued a "Big Data Challenge" with $10,000 prizes to use modern computing to improve climate modeling. In your opinion, will advanced computing power help find solutions to global climate problems? If you were given access to a team of experienced computer engineers to develop a solution to try to win this prize, what are some areas you would ask them to explore and why?

coal mines and from natural gas leaks at production facilities and in transmission pipelines. The 2015 Aliso Canyon leak at a Southern California Gas Company underground storage facility emitted more than 88,000 metric tons of CH_4 over a 4-month period. While this is believed to be the worst natural gas leak in US history, smaller leaks at energy industry facilities are widespread and add up to significant CH_4 emissions. Agricultural activities such as rice farming and livestock production are also major sources of CH_4 emissions (as waste products of methanogens living in flooded rice fields and in the digestive tracts of cattle). Methane is also produced by the decomposition of garbage in landfills (primarily food waste) that is buried in the absence of oxygen. In the United States, landfills account for about 18% of human-caused CH_4 emissions. At 1.88 parts per million, the concentration of CH_4 in the atmosphere is now 250% greater than it was prior to the Industrial Revolution.

While less prominent than CO_2 and CH_4, two other gases also began significantly increasing in the industrial-age atmosphere. The atmospheric concentration of nitrous oxide (N_2O), a gas produced by bacteria, has increased 20% since the preindustrial era, and it has a particularly long atmospheric lifetime of about 120 years. Nitrous oxide is an even more potent GHG than CH_4, trapping heat 310 times more effectively than does CO_2. While tilling the soil, the use of nitrogen fertilizers and the production of livestock wastes all feed the microbial processes that yield N_2O emissions. N_2O is also a by-product of certain industrial processes, such as nylon production, and of fossil fuel combustion. Human activities now account for about one-third of global N_2O emissions. Finally, halocarbons such as hydrofluorocarbons are entirely a product of the industrial age: they have no natural source and are synthesized by humans for use as aerosol spray propellants, refrigerants, solvents, and fire retardants. These chemicals have residence times in the atmosphere ranging from 15 to 180 years, and their atmospheric concentrations are increasing (**TABLE 11.1**).

How Much Do Human-Caused Greenhouse Gas Emissions Affect the Climate?

Greenhouse gas emissions have increased more than 80% since 1970, and scientists have been observing and studying their effects during this period. In particular, the Intergovernmental Panel on Climate Change (IPCC) was created by the United Nations Environment Programme and the World Meteorological Association in 1988 to provide a scientific assessment every 5 years on the most recent research into climate change. In its first report in 1990, the IPCC reported that warming could be "largely due" to "natural variability" but that the "unequivocal detection" of an enhanced greenhouse effect due to human factors would not be likely for a decade or more. Then a little more than two decades later, the IPCC's Sixth Assessment Report in 2022 declared that "It is unequivocal that human influence has warmed the atmosphere, ocean and land." Similarly, the report of the 2018 National Climate Assessment, which was led by the National Oceanic and Atmospheric Administration, noted that "the warming trend observed over the past century can only be explained by the effects that human activities, especially emissions of greenhouse gases, have had on the climate."

What has driven this increasing certainty about cause and effect? First, Earth is responding in ways scientists predicted it would to increasing GHG concentrations. For example, there is now recorded evidence that over the past 50 years, the troposphere has been warming, while the stratosphere has been cooling. Why is this important? To begin with, this finding does not support the idea that recent global warming is due primarily to natural changes in the amount of solar radiation. If this were the case, increased solar output would increase average temperatures not only in the troposphere (at the lowest level of the atmosphere) but also in the stratosphere above it. Instead, the data show warming only in the troposphere—the warming is coming from the planet itself and not from an off-planet source—which is consistent with the idea that warming is caused primarily by increasing concentrations of GHGs.

A second reason is that climate scientists have developed increasingly sophisticated computer models to simulate the way various Earth processes work together to affect climate. These models are tested and refined through a process called **hindcasting**, in which simulations are run on a particular model to see how its predictions match up to actual or historical real-world observations. In other words, scientists assess confidence in various climate models by seeing how well the models can use actual collected data to reproduce the global temperature change that has already occurred. Scientists have found that when human factors are kept out of the models, leaving only natural factors such as solar variability, volcanic activity, and many other fluctuations in the climate system, the models do a good job simulating

TABLE 11.1 Important Greenhouse Gases

Greenhouse Gas	Average Lifetime in the Atmosphere (years)	Global Warming Potential of One Molecule of the Gas over 100 Years (relative to CO_2 = 1)
CO_2	50–200	1
CH_4	12	25
N_2O	114	298
CFCs	45–100	4,600–10,600
HFCs	270	14,800

CFC, chlorofluorocarbon; CH_4, methane; CO_2, carbon dioxide; HFC, hydrofluorocarbon; N_2O, nitrous oxide.

the actual global temperature changes until the mid-20th century. However, these same models then fail to simulate the warming actually experienced over the past 50 years unless both natural *and* human factors are considered. When both types of factor are included, the simulations closely match the actual warming in climate that has occurred. During the past 30 years, these models have made it increasingly clear that human factors are the culprit driving global warming.

> ⬡ **TAKE-HOME MESSAGE** Human activities that release greenhouse gases such as CO_2 and CH_4 are the dominant factor influencing the recent warming of Earth's climate. The atmosphere's CO_2 content is more than 40% higher than before industrialization began and higher than levels over the past million years. Climate models that include human and natural factors closely match actual warming, whereas those leaving out human factors do not.

11.5 What Are Some Effects of Climate Change?

Temperature variations of 10, 20, or sometimes 30 degrees in a single day are not unusual, but even a few degrees of change in global average temperature is significant. To put this in perspective, the average temperature change between the climate we enjoy now and that of the last ice age is about 6°C (11°F). Climate model simulations show there will be further significant increases in the global average temperature through the end of this century. So what are the likely consequences of this warming? In this section, we will take a closer look at some of the most notable likely effects.

A prime example is seen in the Maldives, a tropical paradise that is also the world's lowest-lying country. Each year more than 1 million tourists scuba dive, snorkel, and enjoy the white sand beaches on this 600-mile-long string of 26 coral atolls in the Indian Ocean. But with an average elevation of just 5 feet above sea level, the Maldives is out front in feeling the effects of climate change (**FIGURE 11.16A**). Because of global warming, the Maldives is experiencing rising sea levels, higher tidal surges, the destruction of coral reefs by an increasingly acidic ocean, and increasingly intense rains and other extreme weather events. These changes may render the islands uninhabitable or even completely submerged by the end of this century. In 2009, the government of the Maldives held a cabinet meeting underwater to publicize its call for global reductions in GHG emissions (**FIGURE 11.16B**). Although the effects of climate change may be felt earliest and most significantly in this small island nation, the Maldives contributes only a minuscule amount to human-caused GHG emissions. This is a global problem with local effects that vary from region to region and are not distributed equitably.

Of course, the problems facing the Maldives are not limited to islands: three-quarters of the world's largest cities are located in low-lying coastal areas. In the United States, 23 of the 25 most populous counties are located in coastal areas. Miami, New York, New Orleans, Tampa, and Boston are some of the large cities that will face the greatest risk of flooding due to climate change, but many smaller communities are at risk as well. And the effects of climate change are not limited to the coasts, with consequences varying by region. These effects include alterations to precipitation patterns, changes in the frequency and intensity of extreme weather events, rising sea levels, and ocean

(a)

(b)

FIGURE 11.16 **The Maldives** **(a)** Very low-lying island nations such as the Maldives are particularly vulnerable to sea-level rise. Malé, the capital city of the Maldives, is shown. **(b)** Cabinet ministers of the Maldives government held an underwater meeting to highlight future problems for this nation.

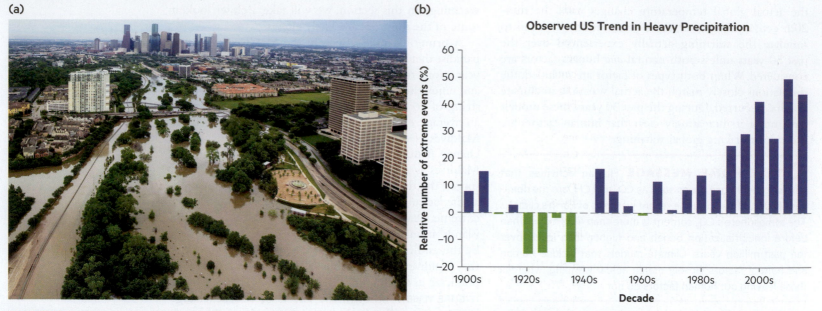

(a)

(b)

Observed US Trend in Heavy Precipitation

FIGURE 11.17 **Extreme Precipitation** **(a)** The city of Houston has suffered from flooding with increasing frequency due to intense downpours. **(b)** Extreme heavy precipitation events are those where the total that falls over 2 days is exceeded only once in a 5-year period. The graph compares the number of such events in each decade to the average number of such events in the United States for the base period 1901–1960, showing they have become more common in recent decades.

Adapted from (b) US Global Change Research Program (2017)

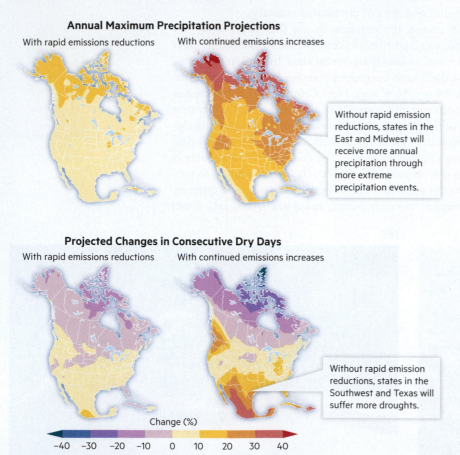

Annual Maximum Precipitation Projections

With rapid emissions reductions

With continued emissions increases

Without rapid emission reductions, states in the East and Midwest will receive more annual precipitation through more extreme precipitation events.

Projected Changes in Consecutive Dry Days

With rapid emissions reductions

With continued emissions increases

Without rapid emission reductions, states in the Southwest and Texas will suffer more droughts.

Change (%)

−40 −30 −20 −10 0 10 20 30 40

FIGURE 11.18 **Extreme Precipitation Projections** The frequency and intensity of extreme precipitation events will likely increase throughout the United States, as will the length of dry spells. The maps in this figure use the lowest and highest GHG emission scenarios to project changes to the annual amount of precipitation and the annual maximum number of consecutive dry days (projections are for the years 2070–2099 compared to the years 1971–2000).

Adapted from Melillo et al. (2014).

acidification. Although we focus on the physical effects of climate change in this section, there are significant economic effects linked to these changes too, including property damage, declining agricultural productivity, and human health problems. So the actual temperature increase is not the only challenge we will face in a warming world. Let's take a look at some of the effects of climate change.

Precipitation and the Water Cycle

In April 2016, the city of Houston received nearly 10 inches of rain in a single day, causing widespread flooding and more than $5 billion in damage. The rainstorm lasted several days and dumped an estimated 240 billion gallons of water over the city. In August 2017, Hurricane Harvey dumped more than 4 feet of rain on the Houston area— breaking the national rainfall record for a single storm and causing more than $150 billion in damage. An extreme rain event in 2021 near Houston nearly matched the record rainfall for a day—dumping more than 16 inches. Extreme rain events have become more common in Houston, which has seen a 167% increase in heavy downpours since the 1950s. The National Climate Assessment found that the frequency and intensity of heavy downpours have been increasing across most of the United States over the past three to five decades (**FIGURE 11.17**).

Changes in the amount of annual precipitation in most regions of the United States are difficult to project with confidence. However, climate models do project with a high degree of certainty that periods of extreme wetness and extreme dryness will increase in most areas of the country (**FIGURE 11.18**). In other words, the climate will become more variable and less predictable.

Increased evaporation associated with warming also produces **droughts**, which have been increasing in intensity and frequency in the Southwest and West. The National Climate Assessment expects drought frequency and intensity to increase this century to an extent that will affect agricultural yields in these regions and heighten the severity of wildfires. There has been an increase in the frequency and intensity of daily temperature extremes on a global scale since 1950, and the IPCC in its Fifth Assessment Report concluded it is very likely that the frequency and duration of heat waves will increase over the next century. The IPCC also found it is very likely that human influence on the climate has more than doubled the probability of heat waves in some regions. Extremely hot days that would have occurred only once every 20 years in the past century will occur every 2 to 3 years in this century. The World Meteorological Organization reports that between 2000 and 2010, 136,000 deaths were caused by heat waves compared to 6,000 such deaths in the decade before.

These predictions would mean that extreme droughts, such as the 2011 Texas drought mentioned in the introduction, will become more common. In 2011, Texas experienced the warmest summer recorded in any US state since 1895: the average temperature exceeded 88°F. It was also the driest summer on record in Texas: average rainfall across the state was less than 2.5 inches. More than 18,000 forest fires burned in Texas that year, engulfing a total of 3.5 million acres of land (an area the size of Connecticut), a larger swath than that burned by wildfires in Texas in the previous 5 years combined. Compounding the more than $5 billion in direct damage, agricultural losses in Texas during this time exceeded an additional $8 billion. And these types of trends are not limited to the Southwest. The National Climate Assessment predicts that the Pacific Northwest will see years of abnormally low precipitation and extended drought conditions this century. In 2021, a record heat wave hit this region, with many cities, including Portland and Seattle, reaching all-time high temperatures often several degrees above previous record highs. With temperatures as high as 116°F, roads buckled and hundreds of people died from heat-related illness, while extreme drought conditions and large wildfires continued through the summer.

Hurricanes and Severe Weather

Hurricanes are another extreme weather event affected by climate change. Warming oceans evaporate more readily and increase the amount of energy feeding the tropical storm systems that become hurricanes. Other factors besides the temperature of the ocean's surface also contribute to the occurrence of these storms, making it difficult to determine their relationship to climate change. While climate change is not an immediate cause

Frontline Communities

As world leaders convened in the early 2000s to address the human impacts affecting Earth's climate, a network of nongovernmental organizations hosted climate justice events to draw attention to the needs of *frontline communities*: those disproportionately affected by the consequences of climate change, who often experience its "first and worst" impacts. Frontline communities tend to be located in more vulnerable areas, such as floodplains and low-lying coastal areas, and are often deprived of adequate infrastructure and resources to adapt to climate risks. Because these communities frequently have low income levels and populations of people of color, Indigenous people, or younger generations, the adverse impacts of climate they experience exacerbate existing socioeconomic disparities.

Further, many frontline communities have relatively low rates of material consumption and so produce disproportionately low amounts of the greenhouse gases causing climate change. Because younger and future generations will bear the most severe consequences of climate change, there is a strong youth contingent leading actions such as the Global Climate Strike, as well as grassroots efforts to include community members in planning and decision making.

of any particular storm event, it is a background condition that increases the likelihood of more intense and frequent storm events. Although there has not been a detectable increase in the frequency of tropical storms or hurricanes that make landfall, there has been a shift toward stronger hurricanes in the Atlantic over the past 70 years, with more category 4 and 5 hurricanes and fewer category 1 and 2 hurricanes. The National Climate Assessment predicts increasing hurricane intensity in the coming years. Other types of storms, such as severe thunderstorms, tornadoes, and winter storms, may also be affected by climate change, but these events happen on small scales that are difficult to incorporate into climate models.

drought a prolonged period of low precipitation and high evaporation rates that can lead to water shortages (Chapter 7).

FIGURE 11.19 **Global Mean Sea Level Is Rising** This figure shows the change in global mean sea level (GMSL) from the GMSL in 1880 (indicated by the dashed line). GMSL changes calculated from current and historical tide gauge readings are shown by the dark blue line, while more recent readings from satellites are shown by the orange line. The light blue shading represents the range of possible values. Global mean sea level has risen steadily through the industrial age.

Adapted from US Global Change Research Program (2020)

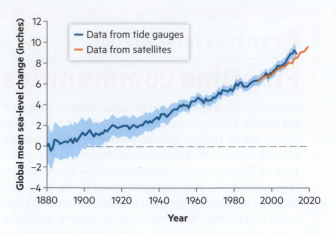

thermal expansion the expansion of something as it warms, in particular, the water in the oceans, which increases sea level.

Sea-Level Rise

Sea-level rise is a very significant outcome of climate change. Globally, sea level has risen more than 9 inches since the late 19th century (**FIGURE 11.19**). When water warms, it expands, a phenomenon known as

thermal expansion. As oceans continue to warm, the expansion will raise sea level. This process has contributed about 40% of the sea-level rise since 1980. Most of the remaining sea-level rise since that time has been due to melting ice from the land. The warmer climate has caused less water to be retained on land in the form of glaciers, including in the ice sheets at the poles, which means a larger volume of water is held in the oceans.

Climate models that incorporate melting of major ice sheets project that sea-level rise across the various emission scenarios ranges from 1 to 4 feet by 2100. Although this may not seem like much, sea-level rise of just a few feet is enough to cause major damage to low-lying coastal cities. A rise of 4 feet would likely displace about 3.7 million people in the United States, primarily those living in the lowest-lying areas of Miami, New Orleans, New York, and Boston. Should major glaciers on Greenland and Antarctica melt entirely, the effects (as shown in **FIGURE 11.20**) would be felt much farther inland.

Sea-level rise is not uniform across Earth. The sea surface itself is not flat, and some coastal areas are rising due to geologic uplift while others are sinking. This means some areas are more vulnerable to sea-level rise than others. Islands in the South Pacific are particularly at risk. Five of the more than 1,000 Solomon Islands in the South Pacific have already been fully submerged beneath the ocean because of rising sea levels. Moreover, increasingly intense storm surges can amplify the effects of a higher sea level. More than 23% of the human population now lives in coastal areas that are affected by rising sea levels, a percentage that is growing as urbanization in coastal areas expands.

Ecosystem Effects

Polar bears have become symbols of the danger that climate change poses to the survival of nonhuman species because they make their living on sea ice that is melting due to global warming. In other words, their habitat is melting. For this reason, polar bears are now listed as "threatened" under the Endangered Species Act. But climate change will affect many more types of ecosystems and species, as 25% of human-caused CO_2 emissions is absorbed into the oceans, causing **ocean acidification**. The National Oceanic and Atmospheric Administration estimates that the pH of surface ocean waters has fallen by 0.1 units since the beginning of the Industrial Revolution. While this might not sound like much, the decrease in pH translates to a 30% increase in acidity and can do significant damage to marine ecosystems. Increasing acidity interferes with the formation of calcium carbonate shells by many marine species, including shell-forming plankton that are at the base of the marine food chain. Significant loss of plankton would devastate marine life.

FIGURE 11.20 **Consequences of Sea-Level Change** Significant sea-level rise would inundate many parts of the United States. This figure shows the areas that the US Geological Survey predicts would be flooded if (1) glaciers on Greenland melt, causing a 6-meter (about 20-foot) sea-level rise, and (2) if Antarctica's glaciers also melt, causing a 70-meter (about 265-foot) rise.

Adapted from Climate Central (n.d.) and Marshak and Rauber (2017).

On land, the warming climate is altering the seasonal behaviors and geographic range of many species. The study of the seasonal timing of biological activities, such as the breeding, flowering, and migration of various species, is known as **phenology**. Recent studies have observed that in areas of the United States and Europe, many species of frogs are breeding earlier and birds are laying eggs earlier than they were four decades ago. The range of places suitable for species adapted to particular conditions changes as the temperatures, moisture levels, and other characteristics of various places on Earth are transformed over time by climate change. Consequently, climate change will alter where species are located along with behaviors such as animal migration, breeding patterns, and the timing of plant flowering. The ranges of many midlatitude species have been moving poleward and to higher elevations, where temperatures are cooler (**FIGURE 11.21**). For example, over the past 50 years, Sachem skipper butterflies have extended their range in Northern California through Oregon and into southeastern Washington. During this time, the Edith's checkerspot butterfly has also moved north and to higher elevations. One assessment showed certain species moving northward an average of 3.5 miles each decade and upslope 20 feet each decade. Species that depend on the conditions of polar regions—such as the polar bear—are experiencing the highest rates of climate warming, will not be able to extend their range, and may not be able to adapt. The Pacific walrus population was halved between 1981 and 1999 by overhunting, and its decline has been worsened by the loss of sea ice. Dependent on ice cover in the Arctic Ocean to breed and provide access to offshore food sources, walrus populations have been declining over the past four decades.

Many of the predicted outcomes of climate change are manifested in recent examples of extreme events and observed trends (**TABLE 11.2**).

ocean acidification the ongoing decrease in the pH of Earth's oceans caused by absorption of carbon dioxide (CO_2) from the atmosphere (Chapter 10).

phenology the study of the seasonal timing of biological activities, such as the breeding, flowering, and migration of various species.

FIGURE 11.21 Changing Habitats Species such as the loggerhead sea turtle, Pacific walrus, polar bear, and Edith's checkerspot butterfly are experiencing effects on their habitats because of climate change.

TABLE 11.2 Summary of the Effects of Climate Change

Predicted Outcomes of Climate Change	Recent Examples of the Outcomes
PRECIPITATION AND THE WATER CYCLE	
Extreme rain events increase in frequency	• August 2017: Hurricane Harvey dumps more than 4 feet of rain on the Houston area. • July 2021: Areas of Germany and Belgium receive over 8 inches of water in just 9 hours, and a downpour in Zhengzhou, China, dumps 24 inches of rain in a single day.
Increased intensity and number of droughts	• 2021: More than half of the United States experiences a drought, while the western states face extreme and exceptional droughts—the most severe classifications.
Longer and more frequent heat waves	• 2000–2010: 136,000 deaths are caused by heat waves compared to 6,000 such deaths from 1990 to 2000. • Summer 2021: Heat waves in the United States set record temperatures in seven states, with temperatures as high as 116°F.
Increased frequency and burn area of wildfires	• 2018: The Camp Fire, California's deadliest and most destructive fire to date, occurs. • 2020: The August Complex Fire, the largest recorded fire in California history, occurs. • 2019–2020: Bushfires in Australia burn more than 24 million acres. • 2019–2020: Unprecedented Arctic wildfires burn peatlands, releasing record levels of carbon dioxide.
HURRICANES AND SEVERE WEATHER	
Increased frequency and strength of hurricanes	• 1945–present: More category 4 and 5 hurricanes and fewer category 1 and 2 hurricanes have occurred in the Atlantic.
Larger and more frequent storm surges	• 2012: Storm surge from Superstorm Sandy damages or destroys more than 600,000 homes and businesses in New York and New Jersey. • 2018: Wind and storm surges from Hurricane Michael cause catastrophic damage in the Florida panhandle.
SEA-LEVEL RISE	
Global mean sea level increases	• Late 19th century–present: Global mean sea level has risen about 9 inches.
Thermal expansion of ocean water increases	• 1980–present: 40% of sea-level rise is attributable to thermal expansion.
Increased melting of glaciers, including continental ice sheets	• 90% of Earth's glaciers are melting rather than advancing. • 1996–present: Rate of mass loss from West Antarctic Ice Sheet increased by 75%.
Submersion of low-lying islands and coastal areas	• Five of the more than 1,000 Solomon Islands have been fully submerged. • Kiribati regularly experiences flooding, contaminated water, and crop failure due to rising sea levels. • Residents of Isle de Jean Charles, Louisiana, are being relocated as the island sinks underwater.
ECOSYSTEM EFFECTS	
Ocean acidification will occur	• Late 18th century–present: pH of surface ocean waters has decreased by 0.1 (a 30% increase in acidity).
Habitat destruction	• Polar bear sea-ice habitat is melting, decreasing populations so much that they are now listed as "threatened" under the Endangered Species Act. • Pacific walrus sea-ice habitat is melting, worsening a population decline that began in the 1980s due to overhunting.
Species ranges move poleward and to higher elevations	• 1970–present: Sachem skipper butterflies have extended the northern edge of their range from Northern California into southeastern Washington. • 1970–present: Edith's checkerspot butterflies have moved north and to higher elevations.
Seasonal behaviors move to an earlier time in the year	• 1980–present: In the United States and Europe, many species of frogs breed earlier and birds lay eggs earlier.

TAKE-HOME MESSAGE Small temperature changes alter global weather patterns. Warming oceans evaporate more readily, increasing the energy that feeds tropical storm systems. Sea level is rising because of warmer oceans and melting glacial ice. Increases in the ocean's CO_2 interfere with shell formation, including that of important plankton. Warming climate alters the range and behaviors of many land species.

11.6 What about the Future?

As we gain confidence in how global climate models represent past conditions, we can run them forward to forecast climate change in the future. But scientists modeling future climate change face significant challenges given the complexity of Earth systems.

We learned in Chapter 10 that feedbacks are loops that either amplify or counter a change to the system and that accurately simulating these feedbacks presents a significant challenge. Recall that a feedback that enhances the original change is called a reinforcing, or positive, feedback, and one that counters the original change is called a balancing, or negative, feedback. **FIGURE 11.22** illustrates some of the reinforcing feedbacks associated with a warming climate. For example, a reinforcing feedback occurs when a warming climate diminishes ice and snow and therefore reduces the reflectivity (or albedo) of Earth's surface. When the ocean and land are depleted of ice and snow, they absorb more of the Sun's energy, which results in more warming and thus more reduction of ice and snow cover. This is one of the reasons why the high-latitude polar regions are experiencing more warming than are lower latitudes. Other reinforcing climate feedbacks include warming oceans adding more water vapor to the atmosphere, where it acts as a GHG, and melting permafrost that releases the GHG methane from the ground.

Balancing feedbacks further complicate climate models. When oceans warm, algae growth increases, which leads to more absorption of CO_2. Greater evaporation rates from warming oceans lead to more cloud cover that reflects more solar radiation. How clouds are affected by climate change and the effects of cloud cover on future climate is an expanding area of inquiry where many questions remain.

Scientists are also studying potential tipping points in the climate system, where sudden changes can have a rapid and significant effect on global climate. For example, if a large land-based ice sheet—such as the West Antarctic Ice Sheet—melted and flowed into the ocean, it would raise the sea level by as much as 3 meters. Or the addition of a vast amount of fresh water from the melting ice sheet over Greenland could alter the thermohaline circulation pattern that moves warm surface water

in the Atlantic toward northern Europe, producing a sudden regional cooling. Accurately predicting the timing of such rapid changes remains a significant challenge.

Modeling Alternative Scenarios for Future Human Response

A final complicating factor is that human behavior can be a key feedback source. For example, warming in urban areas can lead to a greater use of fossil fuels

---

> **SUSTAINABILITY**
>
> # Melting Ice Sheets
>
> Melting ice sheets will be key factors in our planet's not-too-distant future. While thermal expansion of the warming oceans has thus far been the primary contributor to sea-level rise since the 19th century, melting ice sheets will become far more important as the future unfolds. The Antarctic Ice Sheet covers an area of 5.4 million square miles, about the size of the contiguous United States and Mexico combined. There has been an accelerated loss of ice in many of the glaciers that constitute the West Antarctic Ice Sheet (WAIS)—a section that contains about 10% of the total volume of Antarctic ice. In this section, the rate of mass loss increased by 75% since 1996. The base of the WAIS lies below sea level, and it is thinning as warm ocean water flows beneath and along its edges. The part of the WAIS that is losing the most ice per year is the Pine Island Glacier. The photo shows a 226-square-kilometer iceberg that broke off the glacier in 2018. Different teams of scientists studying the retreat of this ice sheet have all reached the same conclusion: the ice sheet has reached a point of instability, and melting will continue to accelerate. The pace of this process will have a big effect on how sea-level rise influences coastal areas around the world.
>
>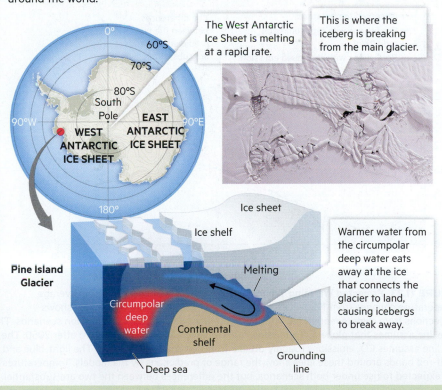

FIGURE 11.22 Reinforcing Feedback Loops Accelerate Global Warming Some important reinforcing feedback loops in Earth's climate system are shown.

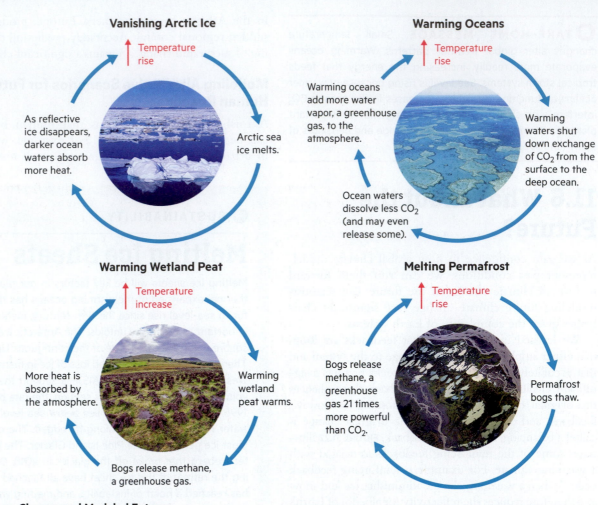

Vanishing Arctic Ice

Temperature rise

As reflective ice disappears, darker ocean waters absorb more heat.

Arctic sea ice melts.

Warming Oceans

Temperature rise

Warming oceans add more water vapor, a greenhouse gas, to the atmosphere.

Warming waters shut down exchange of CO_2 from the surface to the deep ocean.

Ocean waters dissolve less CO_2 (and may even release some).

Warming Wetland Peat

Temperature increase

More heat is absorbed by the atmosphere.

Warming wetland peat warms.

Bogs release methane, a greenhouse gas.

Melting Permafrost

Temperature rise

Bogs release methane, a greenhouse gas 21 times more powerful than CO_2.

Permafrost bogs thaw.

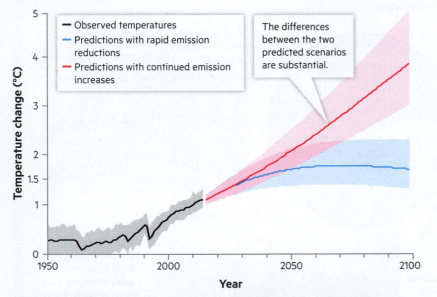

Observed Global Temperature Change and Modeled Futures

- Observed temperatures
- Predictions with rapid emission reductions
- Predictions with continued emission increases

The differences between the two predicted scenarios are substantial.

Temperature change (°C) — Year

FIGURE 11.23 Projected Warming Trends This graph shows observed global average temperature changes from 1950–2014 relative to the 1850–1900 average. It also shows projections from climate models for future years under two different CO_2 emission scenarios. The blue line assumes that global CO_2 emissions are cut severely, reaching net-zero after 2050. The red line assumes CO_2 emissions roughly double from current levels by 2100. The light-blue and light-red bands around these lines show the range of predictions by the models. Temperatures are expected to rise under both scenarios, but the differences between the two are substantial.

Adapted from IPCC (2022).

to power air-conditioning, which leads to more GHG emissions and greater warming. Or this same warming could lead to human actions that provide cooling but without the GHG emissions, helping to balance the warming. How can global climate models account for human behavior?

Climate modelers address this challenge by running simulations for different scenarios of future human-caused GHG emissions. The IPCC makes projections of four different climate trajectories on the basis of four possible future concentrations of GHGs in the atmosphere. Each trajectory makes assumptions about future economic activity, energy sources, population growth, land-use patterns, and other human factors to arrive at estimates for emissions until the year 2100. These trajectories are then used in climate models to produce different scenarios. **FIGURE 11.23** displays the projected warming trends associated with two of these scenarios.

All of the scenarios project a rise in CO_2 concentrations and some rise in mean global temperature, but the difference in temperature change between lower (about 3.6°F) and higher (about 14.4°F) emissions trajectories is about 11°F. So the uncertainty in climate science is not

Fuel Efficiency and the Carbon Footprint of Your Vehicle

In the United States, more than 25% of our human-caused GHG emissions are linked to our transportation choices. Many individual decisions to reduce our carbon footprint in transportation are fairly straightforward: for example, biking or walking instead of driving, and carpooling or using public transit instead of driving alone. But if you are driving a car to get around, there are some more complicated trade-offs linked to the kind of car you purchase.

Because each gallon of gas you burn releases about 24 pounds of CO_2, the fuel efficiency of your vehicle is quite important. When you shop for a car, fuel efficiency is typically listed in terms of miles per gallon (mpg). So in general, replacing a low-mpg vehicle with a higher-mpg vehicle ought to reduce GHG emissions. This has its biggest effect when very inefficient vehicles are replaced. For example, replacing an older sport-utility vehicle (SUV) that gets just 16 mpg with a newer model that gets 20 mpg will save 125 gallons of gas for every 10,000 miles driven (the average distance a car is driven each year). And replacing that old 16-mpg SUV with a hybrid sedan that gets 40 mpg or better would save 375 gallons of gas or more for every 10,000 miles driven—substantially reducing your carbon footprint.

But using miles per gallon as our guide can be deceiving, as increases in fuel efficiency begin to equate with less fuel savings in gallons per total miles driven. As the graph shows, replacing a vehicle that gets 16 mpg with one that gets 20 mpg saves a lot more gas (125 gallons per 10,000 miles) than does replacing a 40-mpg vehicle with a 44-mpg vehicle (27 gallons per 10,000 miles)—even though there is an improvement of 4 mpg in each case. For this reason, the Environmental Protection Agency has changed its fuel-economy stickers on new cars, so consumers can better compare the fuel trade-offs between models.

Why does this matter for the carbon footprint of the vehicle you choose? Life-cycle analysis on cars shows that about 25% to 30% of the total GHG emissions related to a vehicle comes from its manufacturing process. So if you are replacing a vehicle with very poor fuel efficiency (like that old SUV), the fuel savings of the more efficient new vehicle will offset the carbon footprint of its manufacturing process relatively quickly. However, if you are upgrading from one highly efficient vehicle to another that is slightly more efficient, the relatively small gain in fuel efficiency is not likely to offset the carbon footprint of the new vehicle's manufacturing process. In these cases, you may have a smaller carbon footprint if you keep driving the older (and still relatively fuel-efficient) vehicle.

What Would You Do?

Driving more fuel-efficient vehicles is one of the most important steps to fighting climate change. However, as soon as gas prices drop, Americans in particular have a history of switching very quickly to vehicles that get poor gas mileage, even if they tried to change to fuel-efficient vehicles when gas prices were high. If you were in the position to recommend US policy for transportation and climate change, would you take steps to address this behavior? For example, would you create incentives to modify behavior, regulate or deregulate aspects of business, or spur innovation? Or you might believe that targeting behaviors other than driving would prove more effective and favor other courses of action. Either way, describe how your decision would decrease CO_2 emissions.

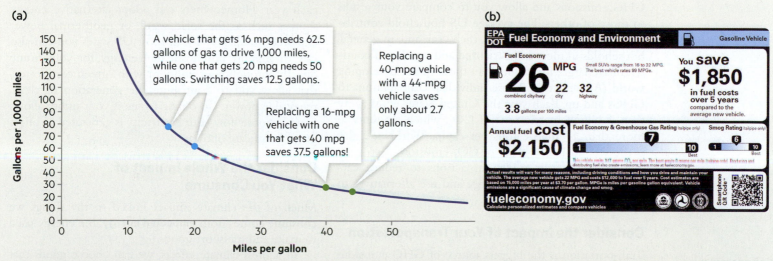

(a)

(b)

(a) Incremental gains in fuel economy are much higher when vehicles with very low miles per gallon (mpg) make efficiency improvements. This graph shows the number of gallons used per 1,000 miles depending on the average mpg of a vehicle. Notice the savings are much greater with switches from very low-mpg vehicles than with switches from higher-mpg vehicles. **(b)** Fuel-economy stickers on cars tell you the projected annual fuel cost for a vehicle, as well as the projected savings for the particular vehicle compared with average fuel cost for that vehicle class.

about whether or how the global climate is changing but about how much it will change. Even if all GHG emissions ended within the next few decades, scientists see a warmer Earth in our future because of the GHG emissions already pumped into the atmosphere. About one-third of the annual human-caused CO_2 emissions remain in the atmosphere for at least a century. Also, the energy from recent warming that is now held in the surface waters of the oceans will take many years to be transferred to the deep ocean. Because of these and other delayed effects of our recent, human-caused GHG emissions, the climate will continue to warm by about 0.5°F over the next several decades even if humans could suddenly halt all GHG emissions.

⬡ **TAKE-HOME MESSAGE** Earth's climate includes reinforcing and balancing feedback mechanisms that enhance and counter warming. Climate models attempt to simulate the way complicated feedbacks would interact over various timescales at various places on Earth. Even if humans immediately curtail GHG emissions, the delayed effects of recent emissions drive climate models to predict at least several additional decades of warming.

11.7 What Can I Do?

When we speak about climate change, it is often with broad sets of data and at the level of nations. However, even change across these groups starts with individual actions. So what are some decisions you might make or actions you can take that affect climate change?

Know Your Carbon Footprint

Online carbon footprint calculators (Appendix A) approximate the way your daily actions contribute to GHG emissions and allow you to compare your results to those of others. The average US household contributes 48.5 metric tons of CO_2 each year, and, in general, households in more industrialized and developed countries have larger footprints than those in the developing world (Appendix B). These individual contributions to GHGs add up. Each year the United States emits more than 6 billion metric tons of CO_2, which amounts to more than 17 metric tons per capita. Worldwide, this is the second-largest share of CO_2 emissions. Two general principles can guide our actions to reduce carbon emissions: conserve energy and use alternatives to GHG-emitting sources.

Consider the Impact of Your Transportation

Transportation is the biggest source of GHG emissions in the United States. Conserving or using less energy could mean driving less, carpooling, taking public transit,

biking, or walking. One of the most meaningful decisions affecting our transportation options is the choice to live in a place that is close enough to where we work and to our other daily routines so that we can walk, bike, or access public transit to get around. When we do drive, we can improve the efficiency of our vehicles and the way we drive (see this chapter's Sustainability Matters feature). However, it is also important to maintain the efficiency of a vehicle by keeping its tires properly inflated and the engine tuned-up, as a vehicle's efficiency can decrease if this routine maintenance is not performed.

Reduce Home Energy Use and Choose Green Power

The second-biggest source of GHG emissions in the United States is the energy we use to heat our homes and the fuel used to generate electricity for them. Some simple behavioral steps we can take to reduce GHG emissions include turning down the thermostat a bit in the winter and turning it up a bit in the summer. We can choose lights and appliances that are more energy efficient and save money over the long run. Light-emitting diode (LED) bulbs are 75% more energy efficient than traditional lighting, and the Department of Energy estimates that widespread adoption of LEDs in the United States could save the equivalent of the annual electrical output of 44 large electric power plants and save homeowners a total of $30 billion each year. Homeowners can adopt alternative energy sources such as the solar, wind, or geothermal power explored in Chapter 14. In most US states you can also buy or support non–fossil fuel energy for your home directly through your utility or with "green certificates" that allow you to contribute the generation of clean, renewable power. Go online to see if your utility allows you to choose how your electricity is generated or offers green certificate options.

Consider the Whole Impact of What You Consume

Much of the climate impact linked to the things we consume comes from **embodied energy**: the energy used to make and transport goods.

To address climate effects, we can choose goods that have smaller carbon footprints. For example, as **TABLE 11.3** shows, we can choose natural fibers over synthetic fibers,

embodied energy the energy used to make and transport goods.

TABLE 11.3 Embodied CO$_2$ Emissions in Textiles

KILOGRAMS OF CO$_2$ EMISSIONS PER TON OF SPUN FIBER				EMBODIED ENERGY USED IN PRODUCTION OF VARIOUS FIBERS	
Fiber	Crop Cultivation	Fiber Production	Total	Fiber	Energy Use in Megajoules per Kilogram of Fiber
Polyester (US)	0	9.52	9.52	Flax (linen)	10
Cotton, conventional (US)	4.2	1.7	5.9	Cotton	55
Hemp, conventional	1.9	2.15	4.05	Wool	63
Cotton, organic (India)	2.0	1.8	3.8	Polypropylene	115
Cotton, organic (US)	0.9	1.45	2.35	Polyester	125
				Acrylic	175
				Nylon	250

Adapted from O Ecotextiles (2019).

which are manufactured from petroleum products. We can choose organic cotton over nonorganic cotton because organic cotton does not use synthetic fertilizer: production of these fertilizers involves high energy inputs and their use produces the potent GHG methane. The foods we eat have different impacts too (see Chapter 12). In particular, as the figure here shows, conventional meat production requires a lot of embodied energy. We can also opt for more vegetables than meats in our diets and for organic produce over conventional produce.

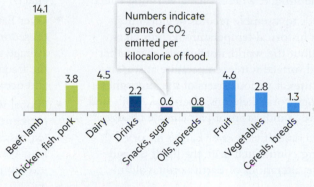

Adapted from Shrink That Footprint (n.d.).

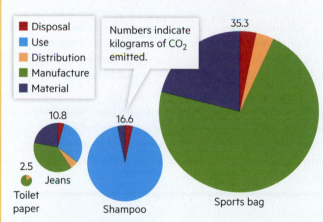

Adapted from Shrink That Footprint (n.d.).

Similarly, if feasible, we can choose to patronize or work for companies that have good environmental practices. A survey by the Yale Center for Business and the Environment of 3,711 business school students in 25 countries found that 44% would accept a lower salary to work for a company with better "green" credentials. Twenty percent said they would refuse a job at a company with bad environmental practices no matter how high the salary. Stuart DeCew, program director at the center, summarizes these findings by saying, "There is a carbon tax on talent."

Learn about Climate Justice

In Chapter 17 on environmental health and justice, we'll take a closer look at the ways in which climate impacts such as rising sea levels and extreme weather events, including hurricanes, floods, heat waves, and droughts, are disproportionately affecting low-income communities, people of color, Indigenous people, and younger generations. You can learn more about how exposure to different climate-related health risks in the United States varies across different groups by looking at Chapter 14: Human Health in the 2018 National Climate Assessment.

Make Changes Visible to Others

Especially in the age of social media, individuals and small groups can have a lot of impact on large numbers of people, businesses, and government. Famously, Greta Thunberg aggressively called out for action on climate change as a teenager. Her effective communication and powerful stances helped lead a youth climate movement. For most people, though, the solution can be as simple as making others aware of the effect that humans have on climate change. For example, environmental reporter

Julia Kumari Drapkin founded the "ISeeChange" project on Instagram in 2013, which encourages users to document observations of nature that may be related to climate change. It is now the website iseechange.org that exists as a global community with its own custom app to download. This experiment in citizen science builds awareness of climate change while providing information that may be useful to scientists gathering data on the effects of climate change. Similarly, the website Climate Generation (climategen.org) suggests creating and telling your own "Climate Story ... a personal story about you and your experience of climate change," making it known to others, and then reading other accounts.

Chapter 11 **Review**

SUMMARY

- Global climate is warmed by the greenhouse effect—a process through which GHGs such as water vapor and CO_2 reradiate heat from Earth back toward the surface, where it has a warming effect.

- Scientists use many methods to assemble records of global temperature change over time: historical temperature records taken at weather stations throughout the world; instruments that gather marine air temperature and surface water temperature at sea; satellites that detect temperatures at various layers of the atmosphere; and proxies such as tree rings, coral skeletons, glacial ice, fossils, and sedimentary strata that serve as indicators of past climate change.

- Global climate has changed dramatically over Earth's history, affected by factors including continental drift, the uplift of landmasses, the evolution of life, alterations of Earth's orbit, volcanic eruptions, and asteroid strikes.

- The water in Earth's oceans is perpetually on the move. Currents at the ocean surface are driven by Earth's rotation and shaped by the continental landmasses. Surface currents also interact with currents deep below the ocean surface, which are driven by differences in water temperature and salinity.

- Ocean currents are a major factor contributing to Earth's climate.

- Today, Earth's climate is in a period of global warming, and human activities are the dominant factor influencing this trend.

- Human-caused GHGs, such as CO_2 and CH_4 related to the combustion of fossil fuels and land use, enhance Earth's greenhouse effect.

- Over Earth's history, sudden and severe shifts in global climate have caused mass extinctions, and even a few degrees of change in global average temperature can significantly alter precipitation patterns, the frequency and intensity of extreme weather events, sea level, and ocean acidity.

- Global warming is exacerbating periods of extreme wetness and extreme dryness in many regions of the United States. Thermal expansion of warming oceans combined with meltwater from land-based glaciers being added to the oceans is causing sea level to rise.

- The ocean is becoming more acidic because of the increase in CO_2 dissolved in seawater. In particular, this interferes with the formation of calcium carbonate shells for many marine species. A warming climate on land is altering the seasonal behaviors of many species.

- Calculating your carbon footprint approximates how your daily actions contribute to GHG emissions. Some individual actions to reduce these emissions include changes in your transportation choices, home energy use, and the goods you consume.

KEY TERMS

albedo	feedback	hindcasting	proxy
Coriolis effect	geothermal gradient	instrumental period	salinity
deep current	global climate	Milankovitch cycles	surface current
drought	greenhouse effect	ocean acidification	thermal expansion
El Niño–Southern Oscillation (ENSO)	greenhouse gas	paleoclimate	thermohaline conveyor
embodied energy	gyre	phenology	upwelling

REVIEW QUESTIONS

The letters following each Review Question refer to the Chapter Objectives.

1. What is the difference between climate and weather? **(A)**

2. What is the difference between the climate of a particular region and the global climate? **(A)**

3. Identify at least three methods that scientists use to track changes in global climate over time. **(B)**

4. Describe how the greenhouse effect has a warming effect on Earth. **(A)**

5. Name at least three GHGs. **(A)**

6. Identify at least three factors that have contributed to significant changes to Earth's global climate in the past. **(C)**

7. Identify two factors affecting surface currents in the ocean and two factors affecting deep currents in the ocean. **(D)**

8. What is the most significant human-caused GHG? Is the concentration of this gas in the atmosphere decreasing, increasing, or staying constant? **(E)**

9. What are climate models, and what types do scientists use to analyze the climate? Which models do a better job of simulating the actual global climate change that has occurred: those that incorporate human factors or those that omit human factors? **(F)**

10. Explain how global warming can cause sea level to rise. **(G)**

11. What is ocean acidification? Describe how ocean acidification affects marine ecosystems. **(G)**

12. Identify at least three types of individual choices that affect your individual contribution to GHG emissions. **(H)**

FOR FURTHER THOUGHT

The letters following each item refer to the Chapter Objectives.

12. Describe the role that climate models play in helping us understand climate change and its likely effects and how these models are tested and refined. Explain the extent to which they support or contradict the idea that human activities are the dominant factor influencing the recent warming of Earth's climate. If future human behavior is an important factor influencing the likely effects of climate change, how do models incorporate something like this that is as yet unknown? **(F)**

13. If you were given the job of recommending five individual actions that one could take to reduce GHG emissions, what would they be? How would you rank them in terms of effectiveness and why? How would you rank these same actions in terms of ease of adoption and why? **(H)**

14. The Federal Emergency Management Agency (FEMA)—the agency that helps state and local governments prepare for and respond to extreme weather events such as flooding, hurricanes, and wildfires—chose to omit all mention of "climate change," "global warming," and "sea-level rise" from its strategic plan for 2018 through 2022. FEMA administrators have argued that the term *climate change* triggers negative reactions in some people and decreases their willingness to take steps to address its effects. What do you think of this argument? Does avoiding mention of climate change and related terms make it more or less likely that communities will prepare for the effects of climate change? Explain your reasoning. **(H)**

Make Your Case ?

The following exercises use real-world data and news sources. Check your understanding of the material and then practice crafting well-supported responses.

Use the News

The following article appeared on the "News Stories" website maintained by the US Department of Defense (DOD). This site is used both to promote activities of the various DOD activities and to maintain a connection with the press. This article highlights the recently released DOD Climate Adaptation Plan (CAP). Use this article to answer the questions that follow. The first two questions test your understanding of the article. Questions 3–5 are short answer, requiring you to apply what you have learned in this chapter and cite information in the article. Answers to Questions 1–2 are provided at the back of the book. Question 6 asks you to make your case and defend it.

"DOD Efforts Showcased in Climate Adaptation Companion Document," Office of the Deputy Assistant Secretary of Defense for Environment and Energy Resilience, November 4, 2021

A wide range of activities are addressed in the DOD Climate Adaptation Plan. Much of this is not new; the DOD has been working to address climate change for over a decade, and many projects have been undertaken during that time. A recently released Companion Document to the CAP showcases a selection of these projects.

The DOD has been integrating climate information into everyday decisions, ranging from the 2014 Climate Change Adaptation Roadmap and updated building codes to guidance specific to the Departments of the Army, Navy and Air Force. Work on Line of Effort 1, climate-informed decision-making, is well underway. In addition to highlighting these different resources, the CAP Companion Document provides information about the DOD Climate Assessment Tool and the DOD Regionalized Sea Level Database, which provide key information to DOD planners and engineers.

The increased frequency, intensity and extent of wildland fires are unfortunate realities of changing climate, and DOD personnel are central to firefighting operations on land the department owns and in support of firefighting efforts on public lands. Ensuring this is possible is one way the DOD is working on Line of Effort 2: train and equip a climate-ready force. The CAP Companion Document highlights wildland fire training operations across the military departments.

In fiscal year 2020, the Air Force Wildland Fire Branch taught close to 800 students from the Air Force and partner organizations and partnered with the Army Wildland Fire Program to deliver 12 additional higher-level wildland fire training courses. The U.S. 3rd Fleet, Naval Air Forces Pacific and Navy Region Southwest partner with the California's CAL FIRE Air Program on semi-annual training to ensure an immediate response capability in support of local authorities for emergency events.

While fires rage, coastal and inland flooding are becoming more common with climate change, as well. The DOD is working to protect military installations through Line of Effort 3: resilient built and natural installation infrastructure. DARPA's Reefense program is one example highlighted in the CAP Companion Document. By developing novel hybrid biological and engineered reef-mimicking structures, Reefense can mitigate wave and storm damage. Coral or oyster settlement and growth on the Reefense structures will enable them to be self-sustaining and address infrastructure-related impacts of sea level rise over time.

Climate change affects more than just military bases; impacts anywhere along the supply chain for materials and equipment could affect DOD operations. This is addressed in Line of Effort 4: supply chain resilience and innovation. Considering direct and indirect impacts is not new for the DOD. One example in the CAP Companion Document is the Air Force's demand-side optimization program, which works to optimize logistical support requirements—including water to improve resilience—and make supply lines less vulnerable to the effects of climate change and adversaries.

Resilient supply chains have good implications for the DOD's own climate impact, as well. Another example in the CAP Companion Document is the Defense Logistics Agency's Defense Supply Center Columbus Environmental Division, where the Management System includes a requirement for identifying and employing source reduction; sustainable resource use; climate change mitigation and adaptation; diversity and ecosystem protection; and pollution prevention opportunities, where practicable.

The DOD won't be able to address the impacts of climate change alone, which leads to Line of Effort 5: enhance adaptation and resilience through collaboration. One part of the collaboration highlighted in the CAP Companion Document are the grant programs administered by the DOD that support community coordination with local installations on climate change and extreme weather: the longstanding Compatible Use Plan and two pilot programs established in fiscal year 2020—Military Installation Resilience and Defense Community Infrastructure. These build on the success of the Readiness and Environmental Protection Integration Program, which partners with communities to acquire real property or other interests in land from willing sellers to promote compatible land uses, enhance military installation resilience, and preserve habitat to relieve existing or future restrictions on military activities.

The CAP provides a roadmap to ensure the department maintains the ability to operate under changing climate conditions while preserving operational capability and protecting systems essential to our success. This companion document highlights many of the ways that the DOD is already putting this roadmap into action and will continue to do so.

1. Which statement best describes how the DOD uses information about the climate in their formal decision making?
   a. The DOD integrates this information into everyday decisions the department makes.
   b. The DOD does not factor climate change into its decision making, because it does not see it as strategically important.
   c. The DOD factors climate change mostly into its long-term planning, looking several years out.
   d. The DOD factors climate change into special initiatives only, such as training people to combat wildfires.

2. What are three specific reasons the article cites for the DOD needing to address climate change?
   a. The DOD wants to avoid lawsuits from surrounding communities for failing to address climate change.
   b. Firefighting is a central operation on many DOD lands, and climate change increases its necessity.
   c. The DOD needs to protect the entire supply chain that supports bases from climate change threats.
   d. The DOD needs to protect the United States from foreign attacks where warming waters open up frozen ports.
   e. Impacts such as fires and floods from climate change present a threat to military bases.

3. The article notes "Line of Effort 5: enhance adaptation and resilience through collaboration." In a few sentences, describe what this means in terms of how the DOD addresses climate change.

4. Using the information in this article, describe what "Reefense" is and why it is important to the DOD.

5. The article says the CAP is seen as particularly important to the DOD and "essential to our success." In this context and considering this source, describe what types of "success" they hope to achieve.

6. 🌎❓ **Make Your Case** Do you agree with the DOD's plans to adapt to climate change based on their assessment of future risks? Or do you have another position? Describe your position using information in the article, in the chapter, or from other sources to support your case.

Use the Data

We learned in this chapter that climate change can affect the range of habitats and/or growing seasons for certain species. Sometimes changes like this can have implications for human health. The figures below show the current and projected probability in the year 2080 of the establishment of tick populations that cause Lyme disease, a bacterial infection that currently affects about 300,000 people in the United States each year.

Study the figures shown here, and use them to answer the questions that follow. The first three questions test your understanding of the figures. Questions 4–5 are short answer, requiring you to apply what you have learned in this chapter and cite data in the figures. Answers to Questions 1–3 are provided at the back of the book. Question 6 asks you to make your case and defend it.

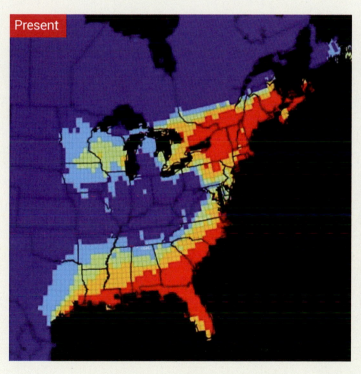

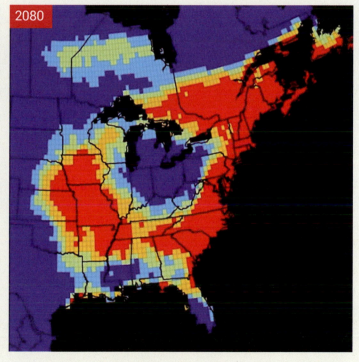

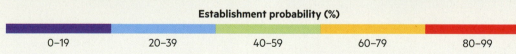

Establishment probability (%)

| 0–19 | 20–39 | 40–59 | 60–79 | 80–99 |

1. Which two of the following states will experience the largest increase in the likelihood of Lyme disease–carrying ticks establishing populations in those states between the present and 2080?
   a. Missouri
   b. Iowa
   c. Maine
   d. Vermont
   e. Florida

2. Which two of the following states that today have a significant likelihood of being infested with these ticks will very likely still have about the same risk in 2080?
   a. Missouri
   b. Iowa
   c. Maine
   d. Vermont
   e. Florida

3. On the basis of the data in the figures, name two states that would see a dramatic reduction in the establishment probability of tick populations between the present and 2080.

4. Tick populations are strongly influenced by climatic factors. In general, an earlier onset of higher temperatures in the spring and a later onset of cold and frost in the fall and winter lead to higher tick densities in suitable physical habitats. However, extremely high temperatures reduce tick populations. Describe how these climatic influences will likely affect Lyme disease contraction rates in different regions or states pictured on the two maps.

5. On the basis of the data in the figures, describe how the autumn season in the southern parts of Ontario and Quebec will likely change in 2080 (including factors such as temperature and onset of frost). Explain your reasoning.

6. **Make Your Case** Per question 3, some states are likely to experience a lower risk of ticks. In your opinion, will climate change make these areas more desirable places to live? Explain your reasoning using data from the figures and/or information from the chapter.

LEARN MORE

- National Climate Assessment, 2017: https://science2017.globalchange.gov/downloads/CSSR2017_FullReport.pdf
- IPCC Sixth Assessment Report, 2022: www.ipcc.ch/assessment-report/ar6/
- NOAA website on ocean acidification: www.noaa.gov/resource-collections/ocean-acidification
- NASA climate change website: https://climate.nasa.gov/
- Special issue of *Science* on climate change: www.sciencemag.org/site/special/climate2013/index.xhtml
- California climate strategy website: http://climatechange.ca.gov/
- Video: NASA, 2018 Fourth Hottest Year on Record: digital.wwnorton.com/environsci
- Video: NASA, Usual Suspects: digital.wwnorton.com/environsci
- Video: NOAA, Tracking an Ocean of Carbon: digital.wwnorton.com/environsci
- Video: NASA, Unusual Arctic Warming Event: digital.wwnorton.com/environsci
- Video: NASA, 22-Year Sea-Level Rise: digital.wwnorton.com/environsci
- Video: NASA, What Is Sea Ice and Why Is It Shrinking?: digital.wwnorton.com/environsci
- Video: NASA, Sea-Level Rise: digital.wwnorton.com/environsci
- Video: NPS, Climate Change in the Arctic: digital.wwnorton.com/environsci

12

Food
How Do We Feed Ourselves?

These strawberries are growing in controlled conditions on a commercial, hydroponic farm, with roots in enriched water. Are these the strawberries that are going on your cereal? Are these strawberries more or less nutritious than those grown in soil? As agriculture becomes more sophisticated, the questions about what we eat and how it is grown get more complex too.

What did you eat for breakfast this morning? Maybe you had a bowl of cereal: it is the top-selling breakfast food in the United States by far. Do you know what the ingredients of your cereal are? Where they were grown? Who harvested them? Were they genetically modified? Or maybe you're a coffee-only person who skips breakfast. Do you know how your coffee was grown or who harvested it? How did it reach your cup from the tropical places where it is usually grown? There are complex food systems behind the staples that we consume every day.

Now think about what you chose to eat over the past few days. Why did you make those choices? In addition to nourishment, you might find that taste and convenience play important roles. But there are often other factors influencing our food choices too. In most societies, food serves a range of social functions as well as basic physical needs. People can have strong opinions about food that can extend beyond nutrition and taste. Food choices often reflect and project beliefs and values, including ideas about ourselves, our obligations to others, and the wider environment.

In this chapter, we'll see that what we choose to eat affects more than just our own well-being. Our choices are linked to food systems that have major impacts on the world in which we live. Consider the work of artist, community activist, and self-proclaimed "gangsta gardener" Ron Finley. In 2010, he was concerned that residents in his neighborhood in South Central Los Angeles lacked access to fresh produce and relied largely on convenience store items and fast-food restaurants. Not only did reliance on these outlets funnel money from the community to distant fast-food franchises and the corporations that produce highly processed, prepared foods; it also contributed to an epidemic of diet-related diseases such as obesity, diabetes, and heart disease. In response, Finley converted the narrow strip of land between the road and the sidewalk in front of his house to a food forest, complete with fruit trees and vegetables. Because the city owns these small strips of land, initially Finley ran afoul of government regulations and faced fines. In response, he launched a successful grassroots campaign to petition the city to approve gardens like his. With the support of many volunteers, city council members, and a new group called Green Grounds, he created urban gardens in parking strips, vacant lots, and schoolyards and at homeless shelters. Today, Finley, like many other urban gardeners across the country, teaches

Chapter Objectives

By the end of this chapter, you should be able to . . .

A. explain how human societies came to rely on cultivated plants and domesticated animals for food.

B. identify the environmental impacts associated with modern methods of crop production.

C. describe modern changes in the way livestock and seafood are produced and identify the associated environmental impacts.

D. discuss important trends with food systems, such as the scale of farming, the diversity of food resources, and the components of our diet.

E. explain how organic and sustainable agricultural practices differ from conventional agriculture.

F. summarize how our individual food choices can link to sustainable food production.

> ### In dirt is life.
>
> —George Washington Carver

FIGURE 12.1 Grassroots Change in LA Ron Finley started a movement in his home of South Central Los Angeles to grow fresh produce in the city.

people how to grow and prepare their own food where they live, in the hopes of strengthening his community while providing it with healthier food. In Finley's words, "Growing your own food is like printing your own money" (**FIGURE 12.1**).

Finley is one of many people who have identified what they see as problems with the way we produce and consume our food and then initiated changes to the way we feed ourselves. While Finley's efforts are focused on small-scale changes at the local level, these challenges multiply when feeding a human population that now exceeds 7.5 billion people. Although overall we produce enough food to feed every person on the planet, more than 800 million people worldwide still suffer from hunger and malnourishment.

With this in mind, how can we start to approach this problem? First, let's take a closer look at what we eat, how it is produced, and the impacts of our food choices. Then, we can explore some of the background of our modern agricultural systems, how they have changed in recent decades, and the environmental impacts associated with how they have developed. Finally, we'll consider our place as consumers in the systems that produce our food and the ways that our food choices can make a difference.

12.1 How Did Our Modern Agriculture Develop?

Until about 10,000 years ago, early humans dined on what they could find, ranging from acorns to insects, and including plenty of roots, leaves, and captured animals. Societies of hunter-gatherers worked in groups of up to 50 individuals, moving from place to place to find food. To maintain a healthy population, these societies had to move constantly over a large area, a lifestyle that ultimately limited the number of humans on Earth. Agricultural economists have estimated that Earth's carrying capacity for hunter-gatherers could not exceed 30 million people—less than one-half of 1% of today's population. For most of humanity's existence, overall population stayed relatively constant at just under 10 million people.

But as humans' ability to learn, plan, and work in groups grew, we started to hunt game larger than ourselves, control cooking fires, and eventually cultivate crops,

domesticate animals, and preserve foods. Humans greatly expanded Earth's carrying capacity with **agriculture**, the process of cultivating plants and domesticating animals for consumption. As we learned in Chapter 6, a series of agricultural revolutions enabled the human population to skyrocket, greatly reducing the area of land needed to support a person from 100 hectares to just a fifth of a hectare.

Historical Agricultural Practices

Humans first cultivated plants by simply fostering the growth of the wild plants they favored while removing undesirable plants. Later, they learned to breed and create domesticated varieties better suited for agriculture. For example, wild relatives of our agricultural wheat have shattering seed heads, a trait that helps seeds disperse under even a gentle wind or contact with an animal. But this makes for an inefficient harvest of the plant for food. So plants with a mutation that kept seeds on the stalk—a disadvantage for the plant in the wild— was a desirable trait for humans, who eventually cultivated these varieties all over the world. Similarly, wild relatives of maize (corn) produced only a dozen kernels per cob. Over time, domesticated corn was bred by humans to produce 500 kernels per cob by consistently selecting and cross-breeding plants with the most kernels (**FIGURE 12.2**). Domestication of animals proceeded as well, first with dogs for hunting and then with animals for consumption: sheep, goats, cattle, chickens, and pigs. Even cats have a history of domestication linked to agriculture, having initially been bred to control rodents attracted to crops.

Crop selection developed alongside other agricultural technologies, particularly human- or animal-powered tools such as seed drills, hoes, and plows (**FIGURE 12.3**). Plowing cut and turned the ground over completely into wide furrows of loose soil, controlling weeds and

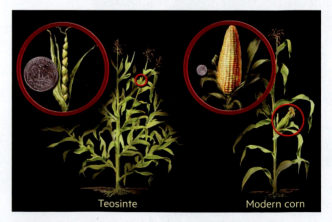

Teosinte | Modern corn

FIGURE 12.2 Ancient Corn Teosinte, a Mexican grass, is considered a historical precursor plant to modern corn. Modern crops that are familiar to us, such as corn, were carefully selected over many generations by humans who cultivated varieties with beneficial traits, such as more and larger kernels.

agriculture the process of cultivating plants and domesticating animals for consumption.

(a)

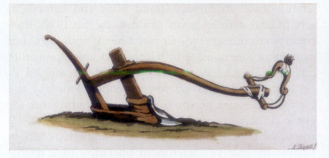

FIGURE 12.3 Historical Agricultural Technologies
(a) Ancient agricultural implements included seed drills like this one, which were used in China more than 2,000 years ago to ensure that seeds were inserted at a uniform depth and covered with soil. (b) This present-day photo illustrates the Pueblo Zuni practice of "waffle gardening." Traditionally, Zuni farmers dug these shapes into the hard clay soil to help keep precious water close to the plants.

(b)

making it easier to plant seeds. Planned harvesting began as humans first reaped (cut) ears of grain and then separated their seeds from their inedible husks (known as chaff) through threshing and winnowing. Fields needed water, so about 6,000 years ago societies in Mesopotamia and Egypt began constructing canals and dams along rivers to irrigate crops. Irrigation provides a regular water supply for crops and can boost yields as much as threefold. Using plant and animal matter—**organic matter**—to fertilize crops began in China at least 6,000 years ago.

More than 2,000 years ago, the Roman Empire spread the idea of rotating crops between fields and allowing fields to remain unplanted, or **fallow**, on a regular basis, plowing weeds back into the soil to help restore it. Native Americans along the East Coast enriched the soil with fish. They and many other societies, including the Hopi and Pueblo peoples of the Southwest, also practiced companion planting, the practice of growing multiple crops in close proximity to facilitate nutrient uptake, pollination, and pest control. In this case, beans and corn were grown together, with beans providing their nitrogen-fixing capabilities (see Chapter 10) and corn acting as a trellis on which the beans could climb. Until the 1950s, most farms raised a variety of animals as well as crops, so that they had a ready source of manure for fertilizer, muscle power for farm implements, and edible protein in the form of meat, eggs, and dairy.

Modern Agricultural Practices: The Green Revolution

In the 20th century, new technologies transformed agriculture, leading to what became known as the "Green Revolution" in the latter half of the century. As these changes took shape, they were seen as the way to improve yields and food security. Prior to 1960, it had taken humans 1,000 years to increase the average yield

of grain from 0.5 to 2 metric tons per hectare, but the Green Revolution took just 40 years to triple average yields to 6 metric tons per hectare. What were some of the most important technological developments?

Gasoline- or Diesel-Powered Farm Equipment Replacing human- or animal-powered plows and harvesting tools with gasoline- or diesel-powered equipment greatly reduced the time and effort required to cultivate farmland.

organic matter decaying plant and animal matter.

fallow an agricultural field that remains unplanted for the purpose of restoring soil health.

> **CONTEXT**

Dogs from Wolves

The wide variety of domesticated dogs we're familiar with today descended from wolves domesticated by humans. Recent research of DNA from ancient wolf bones shows that this domestication may have occurred more than 40,000 years ago. Over time, humans bred dogs to select for traits they found useful, such as hunting, herding, guarding, and companionship, leading to more than 400 distinct breeds of domesticated dogs today.

What Is Genetic Engineering?

Genetic engineering is different from selective breeding because it involves directly manipulating an organism's genome. As the illustration shows, once a desired genetic trait is identified, scientists cut and splice its gene into a small, circular DNA molecule known as a plasmid. The new DNA is then either inserted directly into a seed with a gene gun or inserted into bacteria that infect the cells of a seed. The result is a genetically modified organism, or GMO. Later in this chapter, we'll see that GMOs in widespread use for crops include those resistant to herbicide applications and those with traits toxic to certain insect pests, and we'll review controversies surrounding them.

Step 1 — Identify an organism with a trait of interest.

Search appropriate environment

OR

Screen list of chosen plants

Step 2 — Isolate the genetic trait of interest.

Comparative analysis of genomes to identify gene for trait

GTACTGCACTCT
CATGACGTGAGA

TTACAGCTCTCG
AATGTCGAGAGC

Step 3 — Insert the desired trait into the new genome.

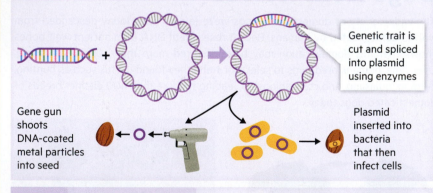

Genetic trait is cut and spliced into plasmid using enzymes

Gene gun shoots DNA-coated metal particles into seed

Plasmid inserted into bacteria that then infect cells

Step 4 — Growing the GMO.

Carefully controlled growth chambers are monitored to ensure that the new GMO grows and replicates. Ultimate growth conditions are determined at this stage.

In the early 19th century, farmers using human- and animal-drawn machinery could plant and harvest an acre of land with 58 hours of work. By the 1930s, they could finish the same job in just 3 hours with the aid of machinery. The number of tractors working the world's cropland has tripled since 1961, which has helped us cultivate vast areas with fewer people. Today in the United States, one person is employed in farm work for every 20 hectares of cropland, and less than 2% of the employed population works in agriculture. By contrast, Bhutan has just 117 tractors in operation (one tractor for every 5,000 people), six people are employed in farm work for every hectare of cropland, and more than 62% of the employed population works in agriculture.

Modern Irrigation Large-scale dams and reservoirs, electric aquifer pumps, long-distance canals and pipelines, and automated sprinkler systems (**FIGURE 12.4**) have increased the area of irrigated land in the world by about 2% a year since 1962. Now, 20% of the world's cropland is irrigated, which has made a tremendous difference. In Haiti's Artibonite Valley, irrigation infrastructure improvements enhanced 7,000 hectares of cropland, helping to double rice production. A study in Ethiopia, where rainfall is unreliable, found that the income per hectare of irrigated cropland was double that of rain-fed areas.

Synthetic Fertilizers and Pesticides First manufactured in the early 20th century, **synthetic fertilizers** (see Chapter 10) became widespread only in the latter half of the 20th century, with global use increasing from 14 million tons in 1950 to more than 177 million tons today. These industrially produced chemicals have one or more of the primary nutrients for plant growth. Similarly, **synthetic pesticides**—manufactured chemicals applied to combat insects and weeds in agriculture—also gained widespread use.

Genetic Manipulation A key to the success of the Green Revolution was that scientists started using selective breeding to alter and "improve" crops. Notably, plant geneticists used cross-breeding to develop new "high-responder" varieties of wheat, corn, and rice. These varieties were selected to thrive with more intensive irrigation and fertilization and to resist common diseases, thereby boosting yields. By 1991, the new high-yield varieties of wheat and rice composed three-quarters of the global totals for these crops.

In the 1970s, a new technique known as genetic engineering was developed that could splice genetic material from one species into the DNA of another to develop new varieties of **genetically modified organisms (GMOs)** in far less time than that needed with traditional cross-breeding. For example, in the early 1980s, work began on genetically modifying rice so that it contains beta-carotene, a necessary precursor to vitamin A and an essential nutrient for both good vision and healthy

synthetic fertilizer
an industrially produced chemical that has one or more of the primary nutrients for plant growth.

synthetic pesticide
an industrially produced chemical applied to combat insects and weeds in agriculture.

genetically modified organism (GMO) a variety developed by splicing genetic material from one species into the DNA of another.

12.2 How Does Modern Agriculture Impact the Environment?

If you look out the window of an airplane as it flies over irrigated regions of the western United States, you can often see circle after circle of green crops laid out across the landscape (**FIGURE 12.5**). The scale and organization of crop production can be particularly impressive from this vantage point. Modern agricultural practices have helped a few of us feed a lot of people, yet it was not too long ago that most of us would have spent a significant amount of time each day securing food to live at the expense of other pursuits (such as education). But gains

FIGURE 12.4 Water Diversion and Delivery for Irrigation Tapping water sources for modern irrigation often involves pumps that draw water out of aquifers, transport of water long distances (as through this canal in Canyon County, Idaho), and then distribution of water over fields by use of automated mobile sprinklers.

immune systems. This so-called *golden rice* could thus serve to address vitamin A deficiency, a health problem that affects more than 140 million children worldwide each year. The breakthrough genetic modification that created golden rice included the addition of two genes from daffodil plants and one from a bacterium. The first field trials of golden rice began in 2004, but as we will see later in this chapter, the use of GMOs remains controversial and so golden rice is still not widely consumed.

FIGURE 12.5 Irrigation Technology Central pivot irrigation creates circles of crops laid out on a vast scale across agricultural regions of the United States. This aerial view is near Greeley, Colorado.

agrobiodiversity the variety and variability of genetic material in the life-forms used by humans for food and other agricultural applications.

genetic erosion the process by which genetic diversity is lost.

monoculture an agricultural practice of growing a single crop, plant, livestock species, or breed.

from modern agricultural practices have come with significant environmental impacts, ones that have been accepted for years but are now being questioned. In this section, we will examine what is becoming a growing list of challenges.

The Loss of Agrobiodiversity

When certain crop varieties are intensively produced, other crop varieties are cultivated less. In some cases, these other varieties can even die out, reducing **agrobiodiversity**—the variety and variability of genetic material in the life-forms used by humans for food and other agricultural applications. **Genetic erosion** is the process by which genetic diversity is lost (**FIGURE 12.6**). For example, 96% of the sweet corn varieties cultivated in 1903 were lost by 1983 as higher-yield varieties became a dominant, single-crop **monoculture**. But although the newer varieties had higher yields, they were more vulnerable to pests and diseases. In 1970, 80% of the US corn crop was planted in a variety that was vulnerable to a mutant strain of fungus (*Helminthosporium maydis*).

When this fungus appeared that year, it reduced yields by 15% and caused lost revenues of $1 billion. More recently, a disease called citrus greening is threatening Florida's orange crop (which is now a monoculture), causing declining harvests and losses of more than $1 billion per year in the past five years.

The Green Revolution has had great success boosting yields of a narrow range of grains—corn, wheat, and rice—and focusing production on particular crops. However, these gains came at the expense of the cultivation of a wider range of crops and the availability of a healthier range of foods. Changes in land-use patterns also disrupted aspects of rural cultures that traditionally stored food reserves and provided community support during times of scarcity. As we will see later in this chapter, modern agricultural practices associated with these new varieties, and the greater investment needed for them, have driven the emergence of global agribusiness and a decrease in the number of small family farms.

Pesticides

As the use of synthetic fertilizers and intensive cultivation of single-species fields became widespread in the 1940s, farmers started to notice an increase in agricultural pests. It was suspected by some, and now confirmed by more recent studies, that the combination of intensive plowing and chemical fertilizer and pesticide use disrupted plant defense systems and relationships with beneficial soil microorganisms. This resulted in increased susceptibility to *pests*—those species competing with, feeding on, or causing disease in desired crops—and increased the use of pesticides to combat them.

Widespread use of pesticides had some other unexpected outcomes. Some were soon banned in the United States because of the harm they cause non-pest species. Most famously, dichlorodiphenyltrichloroethane (DDT) was banned by the United States and many other countries in the 1970s because of the way this toxin bioaccumulates in species above insects in the food chain. DDT accumulation in birds of prey such as bald eagles and peregrine falcons caused the thinning of eggshells, endangering the survival of these species (**FIGURE 12.7**).

Further, when crops were treated repeatedly with a particular pesticide, pests with traits enabling them to tolerate the poison survived and propagated. Eventually, entire populations became resistant to the pesticide. For example, the repeated application of the insecticide *malathion* on fruit trees bred resistance in fruit fly populations. Pesticides pose risks to humans too, especially when ingested, absorbed through the skin, or inhaled. The Centers for Disease Control and Prevention estimates that as many as 20,000 of the 2 million farmworkers in the United States suffer from pesticide poisoning each year. In the developing world, the poisoning rate is double that. Some research indicates that a class of pesticides called neonicotinoids that are used on most

FIGURE 12.6 Declining Agrobiodiversity
The number of varieties of plants cultivated for food has declined significantly with the rise of modern agriculture. This graphic shows the different number of seed varieties available for purchase both at the beginning and toward the end of the 20th century. The available varieties of corn seeds, for example, have declined by more than 25 times.

Adapted from Rural Advancement Foundation International (n.d.).

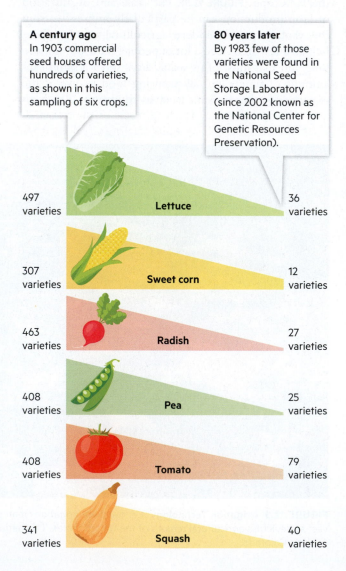

A century ago
In 1903 commercial seed houses offered hundreds of varieties, as shown in this sampling of six crops.

80 years later
By 1983 few of those varieties were found in the National Seed Storage Laboratory (since 2002 known as the National Center for Genetic Resources Preservation).

	A century ago	80 years later
Lettuce	497 varieties	36 varieties
Sweet corn	307 varieties	12 varieties
Radish	463 varieties	27 varieties
Pea	408 varieties	25 varieties
Tomato	408 varieties	79 varieties
Squash	341 varieties	40 varieties

FIGURE 12.7 **Pesticide Impacts on Wildlife** Before it was banned in the 1970s, DDT was widely used as a pesticide. However, DDT accumulated in ever higher concentrations up the food chain. So it also endangered several species of birds—including bald eagles—because it thinned their eggshells, leading to high mortality rates for their young.

conventionally grown fruit and vegetable crops may harm pollinators such as honeybees. Impacts on pollinators are also impacts on our food systems, as 70 to 100 of the top food crops rely on pollinators.

Genetically Modified Organisms

Today, the most widely used herbicide is glyphosate, a chemical that is most commonly sold under the brand name Roundup. This chemical is typically sprayed so it can be absorbed by the leaves of the plants it targets, where it inhibits the production of critical amino acids, killing the plant. It eventually breaks down in soil and water, although this can take anywhere from just a few days to more than 100 days. It has been found in a variety of foods and even in human urine. Roundup is used so widely because genetically engineered crops that are resistant to Roundup applications have also been developed. "Roundup-ready" crop varieties are GMOs that enable farmers to apply the herbicide broadly and more intensely without directly harming the productivity of their crops. This allows for easy, effective weed control. Other common GMOs include crops that have genetic material from *Bacillus thuringiensis* (Bt), a soil-dwelling bacterium, spliced into them. The Bt genetic material produces a protein toxic to the digestive system of

FIGURE 12.8 **Erosion: A Modern-Day Dust Bowl** Dust bowl conditions caused by extensive erosion driven by drought, deforestation, and rapid urban and industrial growth have plagued China, with airborne soil blowing as far away as the United States and Canada.

certain insects, and Bt varieties of corn, soybeans, and potatoes thereby gain resistance to these insects.

What does this mean for agriculture today? GMOs now constitute a significant share of US crops, including 88% of the corn acreage and 94% of the soybean acreage. Further, genetically modified organisms such as Bt and Roundup-ready crops, which are designed and marketed to combat pests, may actually make crops increasingly vulnerable over time. Farmers are now seeing pests and weeds that are resistant to these products arising through natural selection. In addition, using just a few patented crops is contributing to less diversity overall in what we grow, as we discussed earlier.

Soil Loss and Degradation

When land is cultivated intensively with farm machinery, plowed rows of single crops leave large areas of bare soil vulnerable to *erosion*, a process where water and wind redistribute soil particles across Earth's surface. Bare soil can erode more than 100 times more rapidly than soil with plant cover, and such erosion can quickly remove soil that took centuries to form (**FIGURE 12.8**).

Similarly, irrigation systems can speed erosion. Water in gravity-fed furrows can carry away bare topsoil at rates of 5–50 tons per hectare per year. Sprinkler or spray irrigation also contributes to erosion when the application rate exceeds the ability of the cultivated soil to absorb water. Intensive irrigation also speeds a process called **leaching**, where water moves mineral elements deeper, potentially beneath the reach of plant roots. It also causes **salinization**, when mineral salts build up in the soil, eventually building up to amounts damaging or lethal to plants (see also Chapter 9). Globally, 20% of irrigated croplands are affected by salinization. In the United States, where irrigation has been practiced more intensely, salinization is estimated to affect 25% to 30% of irrigated cropland. In the most extreme national

leaching a process driven by water where mineral elements move down to deeper soil layers, potentially beneath the reach of plant roots, to be carried away in groundwater or stream flow (Chapter 9).

salinization the process of mineral salts building up in the soil (Chapter 9).

desertification a process prompted by drought, extreme erosion, and soil infertility that causes land to lose more than 10% of its productivity.

subsidence the process of land settling as formerly water-filled spaces collapse under the weight of overlying rock and soil (Chapter 7).

example, almost half the irrigated cropland in Uzbekistan suffers from salinization.

Even applying fertilizers can degrade the soil, actually diminishing its fertility over time. Excessive application of nitrogen fertilizers can acidify soil and degrade its natural organic matter content as it stimulates microbial decay processes. Soil ecosystems can become dominated by bacteria that do not promote crop health. This in turn can increase the need to rely on fertilizers and pesticides.

As topsoil is lost or degraded, it becomes harder to feed ourselves. Researchers estimate that agricultural soils in the United States have lost about half their soil organic matter, on average, due to conventional farming practices. Over the course of history, about one-third of the world's potentially farmable land has become unproductive due to soil degradation and erosion. Globally, we lose approximately 360,000 hectares of land each year to **desertification**, in which drought, extreme erosion, and soil infertility cause land to lose more than 10% of its productivity.

Water Quality and Air Pollution

When soil is blown or washed away by erosion, it can cause adverse impacts where it ends up. Agricultural lands now account for much of the sediment arriving in rivers, lakes, estuaries, and oceans. The sediment reduces light penetration, impairing the ability of algae and rooted aquatic plants to photosynthesize. When these sediments finally settle, they can smother aquatic life and fill in waterways.

Synthetic fertilizer is another water-pollution source. Because synthetic fertilizers are highly concentrated and designed to dissolve quickly, water runoff transports them away from cropland in greater quantities than organic fertilizers or composts. Under conventional practices only about half of the nitrogen in synthetic fertilizer is utilized by crops in the area where it is applied; the rest is lost to the wider environment. Many agricultural areas use underground tile drainage systems to keep the soil from accumulating too much moisture during rainy seasons. These systems rapidly move water along with sediment and fertilizer pollution into nearby streams. As we have seen in Chapters 7 and 10, both nitrogen and phosphorus fertilizers in runoff can cause increased plant and algae growth, leading to eutrophication of water bodies. One of the consequences of eutrophication linked to food production upstream is the downstream harm to fish and seafood production in affected ocean environments. For example, seasonal hypoxic (low oxygen) conditions in the Gulf of Mexico have been shown to decrease the diversity and abundance of benthic (bottom-dwelling) organisms, including some species of shrimp, crabs, and oysters that are commercially harvested (**FIGURE 12.9**).

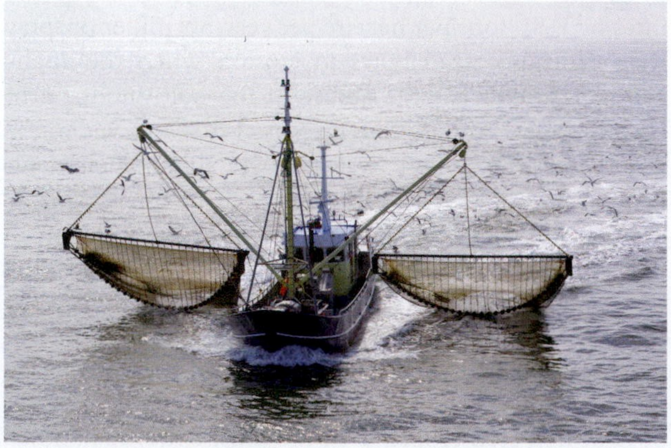

FIGURE 12.9 Runoff Affects Marine Life and Fishing in the Gulf of Mexico The influx of nutrients (such as nitrogen and phosphorus) from synthetic fertilizers in agricultural runoff causes eutrophication and hypoxic conditions when runoff carries them into the Gulf of Mexico. This environmental impact of conventional agriculture can kill marine organisms—such as shrimp—that are not able to move to higher-oxygen areas. Of course this impact also threatens the seafood industry in the Gulf, which supplies the majority of shrimp and oysters harvested in the United States.

Materials blown into the air also cause problems. Synthetic fertilizer applications are the largest source of nitrous oxide (N_2O) emissions in the United States: two-thirds of this pollutant and greenhouse gas in the atmosphere is derived from these applications. Soil microbes rapidly convert nitrogen-based fertilizers into N_2O emissions, and N_2O is 300 times more potent as a greenhouse gas than carbon dioxide. Globally, agricultural application of synthetic fertilizers contributes more than 90% of ammonia (NH_3) emissions and fine particulate matter.

Water Availability

Today, water for agriculture is the largest draw on freshwater resources and composes two-thirds of all freshwater use. In particular, drilling and pumping technologies have enabled groundwater to supply 40% of the world's irrigation water. However, groundwater supplies are slow to recharge, and if groundwater is drawn out faster than it is replenished by natural sources, the water table drops. In a well-known example, Saudi Arabia tapped deep aquifers for irrigation in the 1970s and became a self-sufficient producer of wheat. But since then, these water sources have been overpumped, crop production has fallen, and the country will soon be, once again, completely dependent on imported wheat. Also, as an aquifer dries up, the land on top settles in a process called **subsidence**, as formerly water-filled spaces collapse under the weight of overlying rock and soil (**FIGURE 12.10**). Subsidence permanently reduces the ability of an aquifer to store water.

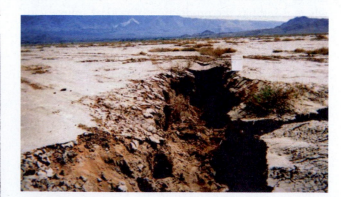

FIGURE 12.10 Subsidence Depleted aquifers drawn down from irrigation can cause the land to sink in a process called subsidence (image from Lucerne Lake area, California).

12.3 How Is Meat Production Changing What We Consume?

A popular entrée at most college cafeterias is fried chicken strips, although hamburgers, spaghetti and meatballs, beef tacos, and fish are also typically on the menu. Around the globe, people are getting more and more of their calories from meat. The Food and Agriculture Organization of the United Nations (FAO) has found that both global meat consumption per person per year and annual seafood catch per person per year have more than doubled since 1950 (**FIGURE 12.11**). As with crops, the way we raise livestock has changed radically this century, and we see this changing the form of the meat itself.

Changing Livestock Production Practices

Do you remember the song "Old MacDonald Had a Farm"? In the United States, this is still a standard in the musical repertoire of preschoolers despite the fact that

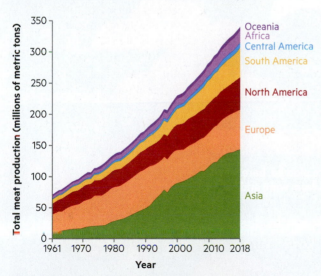

FIGURE 12.11 Total Meat Production As population and meat consumption have increased in recent decades, meat production has increased dramatically. Meat production includes beef, poultry, sheep, goat, pork, and wild game.

Adapted from Ritchie and Roser (2017) and Food and Agriculture Organization of the United Nations (2020).

Will the Water Below Us Dry Up?

The Ogallala Aquifer beneath the Great Plains in the United States and the North China Plain Aquifer are two of the largest sources of groundwater in the world, and each supports a major share of its country's agricultural production. However, their recharge rates are slow enough that they are nonrenewable resources, and the water levels in these aquifers have been dropping. Water levels in some parts of the Ogallala Aquifer in Kansas and western Texas are dropping a foot or more per year, while the water level in some parts of the North China Plain Aquifer are falling at an even faster rate. While it is difficult to know how much water is in these aquifers, with the rate of withdrawal far exceeding recharge, they cannot continue to support irrigated agriculture at current rates indefinitely.

few children live on or near a farm and few farms in the United States have the variety of farm animals mentioned in the lyrics. A few generations ago, farms raised a wide variety of **livestock**, or animals for food (such as chickens, cows, and pigs), along with the crops they grew. Horses and mules were used to plow the fields, chickens fed on crop remnants and pecked around for insects and worms, while pigs were fed kitchen "slops." Beef and dairy production was based on grazing on pastures, fallow fields, or open rangeland. Livestock manure was the primary fertilizer for the crops.

But since the 1940s, the widespread use of synthetic fertilizers in conjunction with powered farm machinery made animal labor unnecessary. Farmers shifted away from the practice of raising animals to produce manure for their fields and, spurred by government incentives, started to specialize. Some specialized in raising livestock, and over time the predominant method of raising

livestock animals raised for food.

Environmental Impacts of Conventional Agriculture

Many practices of conventional agriculture have impacts on the environment. Finding solutions to these problems while feeding a growing population is a significant challenge.

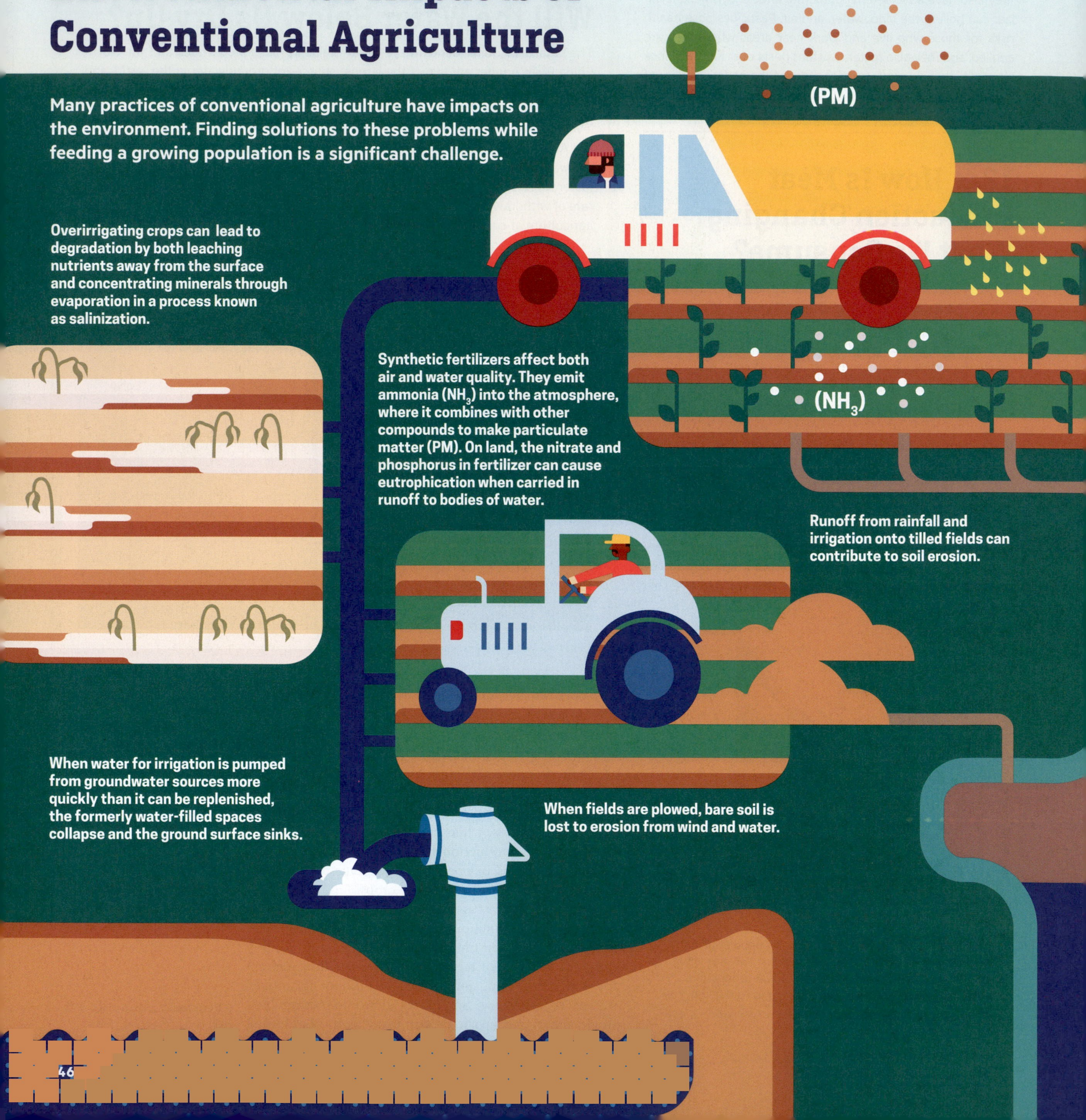

(PM)

Overirrigating crops can lead to degradation by both leaching nutrients away from the surface and concentrating minerals through evaporation in a process known as salinization.

Synthetic fertilizers affect both air and water quality. They emit ammonia (NH_3) into the atmosphere, where it combines with other compounds to make particulate matter (PM). On land, the nitrate and phosphorus in fertilizer can cause eutrophication when carried in runoff to bodies of water.

(NH_3)

Runoff from rainfall and irrigation onto tilled fields can contribute to soil erosion.

When water for irrigation is pumped from groundwater sources more quickly than it can be replenished, the formerly water-filled spaces collapse and the ground surface sinks.

When fields are plowed, bare soil is lost to erosion from wind and water.

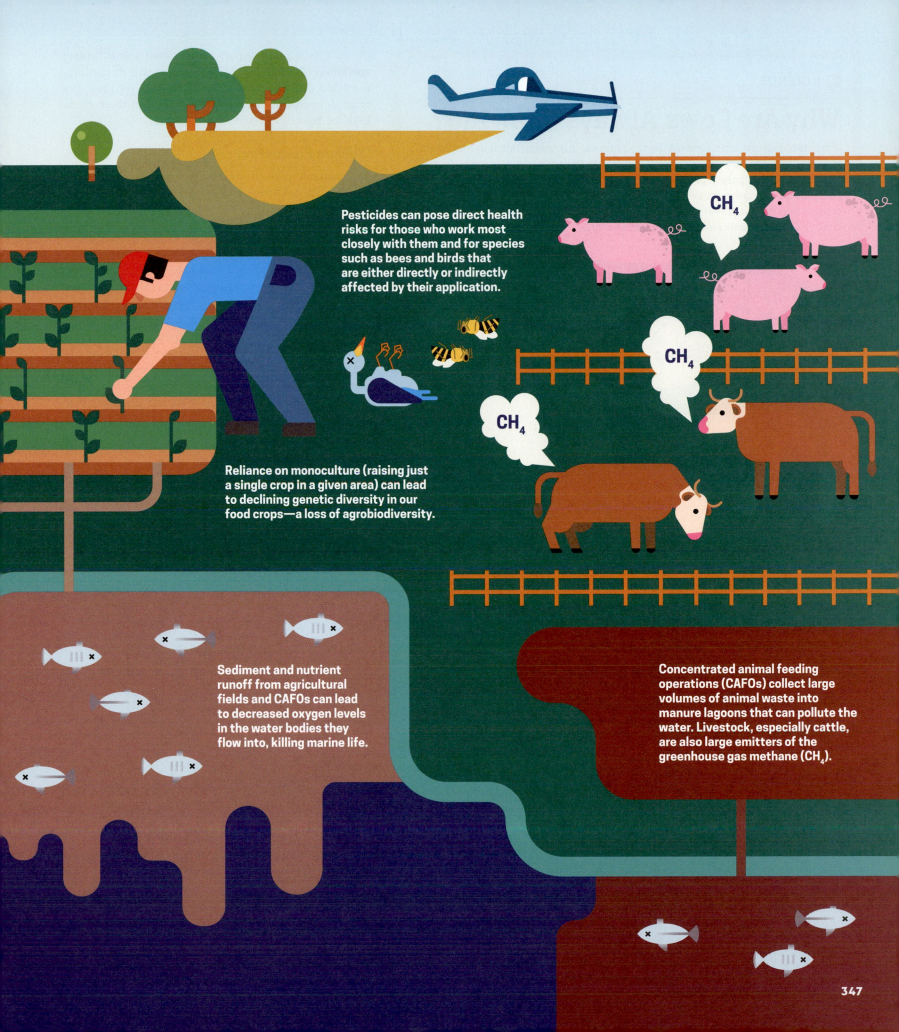

Pesticides can pose direct health risks for those who work most closely with them and for species such as bees and birds that are either directly or indirectly affected by their application.

Reliance on monoculture (raising just a single crop in a given area) can lead to declining genetic diversity in our food crops—a loss of agrobiodiversity.

Sediment and nutrient runoff from agricultural fields and CAFOs can lead to decreased oxygen levels in the water bodies they flow into, killing marine life.

Concentrated animal feeding operations (CAFOs) collect large volumes of animal waste into manure lagoons that can pollute the water. Livestock, especially cattle, are also large emitters of the greenhouse gas methane (CH_4).

CH₄

347

SCIENCE

Why Are Cows Always Chewing?

Ruminants, such as cattle, ferment the food they eat in a compartment of their stomach known as the rumen. Ruminants regurgitate their food several times (sometimes this is called "chewing the cud"), and microbial species—particularly anaerobic bacteria—in the rumen break down the cellulose in plant fibers. This fermentation process also produces methane, a potent greenhouse gas contributing to climate change, that ruminants emit in significant quantities.

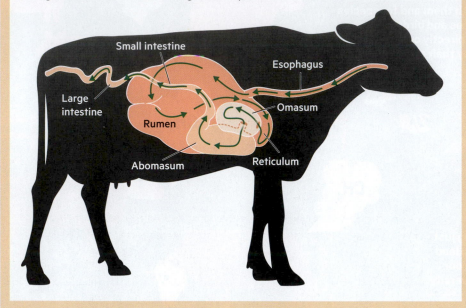

livestock became industrial **concentrated animal feeding operations (CAFOs)**. A CAFO raises animals in confinement and does not grow its own livestock feed but uses grain shipped from farms that specialize in these crops. Let's see how this started.

Chicken Production

Concentrated animal feeding operations began with chickens. As described by the National Chicken Council, since 1950 the chicken industry has evolved from "fragmented, locally oriented businesses into a highly efficient, vertically integrated, progressive success story." Feed mills, hatcheries, farms, and processors were integrated into single operations, and chickens bred for meat production, known as "broilers," became the dominant variety. The overall size of broilers increased steadily as feed was enhanced with animal protein, vitamins and minerals, and amino acids. As conditions became crowded, CAFOs used antibiotics to prevent diseases and stimulate more growth. Artificial light was also introduced to stimulate growth. In pure production terms, CAFOs' statistics show they have succeeded in growing bigger broilers more quickly, with less feed needed and fewer deaths before slaughter (**FIGURE 12.12**).

Pork Production

Pork production underwent a similar concentration, with particular attention to artificial insemination practices to develop high-producing breeds. In 1990, artificial insemination accounted for just 7% of swine breeding, but within

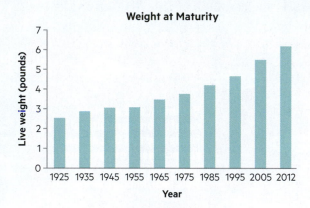

FIGURE 12.12 Bigger Chickens Raising chickens in CAFOs has changed the industry. These data from Georgia show much larger chickens are being raised more quickly, with fewer deaths and less feed.

Adapted from Bishop et al. (2015).

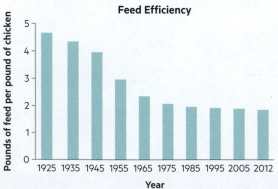

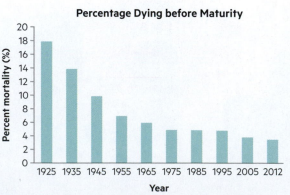

Genetically Modified Organisms

Substantial controversy surrounds the labeling of foods from genetically modified organisms (GMOs). Proponents of labeling maintain that people have a right to know what is in their food and what farming practices they are supporting with their consumer dollars. Critics of labeling maintain there is no evidence that GMOs are hazardous to consume and there is therefore no need to label them. While these are all important issues, the relationship of GMOs to sustainability is neither simple nor straightforward.

Big claims were made about GMOs when they were first introduced, notably that they would increase crop yields and decrease pesticide use. Proponents of GMOs tend to argue they are essential to feed our growing population. But while crop yields have indeed increased, a 2016 National Research Council report found that the increases in corn and soybean yields occurred at the same pace before and after adoption of GMO varieties. Other studies comparing trends in corn and soybean yields find no compelling difference between the pace of yield increases in Europe, where GMOs are not allowed, and in the United States, where GMOs are widely adopted. This implies that the increased crop yields since the adoption of GMO crops are due to factors other than the genetically modified traits. In particular, the yield increase has been attributed to advances in traditional crop breeding of new varieties that the genetically modified traits are then engineered into.

The record is mixed in regard to the claim that GMOs would reduce pesticide use. Consider the introduction of crops genetically engineered to produce Bt delta endotoxin, a protein naturally produced by the soil-dwelling bacterium *Bacillus thuringiensis*. While the bacterium is toxic to many insect pests, such as the European corn borer and corn earworms, it does not harm most other types of animals, including vertebrates. Because of this relative specificity for pests, the bacterium is commonly used as a biological pesticide in organic farming. Adoption of genetically modified Bt crops (corn and cotton) helped reduce overall insecticide use by

The most widespread genetically modified crops include those engineered to resist applications of the herbicide glyphosate (brand name Roundup) and those engineered to produce Bt delta endotoxin in order to combat certain insect pests. After years of letting states decide whether or not to label products that contain GMO crops as such, the United States instituted a labeling requirement in 2022. Brands that do not use GMOs often promote this fact with a label that certifies their products to be GMO free.

more than 25% and greatly reduced reliance on some of the most environmentally damaging broad-spectrum insecticides.

A major factor in the widespread adoption of GMOs in the United States was the development of crops resistant to the herbicide glyphosate. This delivered an easy means of effective weed control, which previously had been accomplished through labor-intensive weeding or energy-intensive plowing. Herbicide-resistant crops propelled adoption of no-till farming, which cut down on soil erosion and tillage-induced loss of soil organic matter. In short order, however, it also led to the development of herbicide-resistant weeds that now plague many North American farms.

A simple Google search will turn up wildly divergent opinions about the human health effects of glyphosate exposure. Even different branches of the United Nations have come to radically different conclusions as to whether glyphosate is a human carcinogen, with the World Health Organization concluding on the one hand that it is unlikely to pose a cancer risk in people and the International Agency for Research on Cancer concluding on the other hand that it *is* probably a cause of cancer in humans. So far, however, few studies have been done on the potential health effects of long-term exposure to glyphosate and its breakdown products in the food supply.

While we are not aware of direct evidence that GMOs themselves present an acute toxicity risk and are therefore unsafe to eat, recent studies in Europe of the effects of glyphosate on livestock (chickens, cows, and goats) showed detrimental effects at relatively low concentrations. Researchers have suggested that glyphosate may disrupt the microbiomes of host organisms and that this indirect disruption may play a role in some chronic illnesses. If this suspicion holds up to scrutiny, it may turn out that adverse health impacts of GMOs are not due to the genetically modified crops themselves but to increased use of the herbicide, which accelerated after the introduction of glyphosate-tolerant corn and soybeans.

(continued)

Regulating GMOs

Governments in Europe and the US government have taken very different approaches to regulating genetically modified crops. In Europe, the regulatory assumption is that new products need to be shown to be safe before being adopted. The burden of proof is flipped in the United States, where the regulatory assumption is that new products are safe until shown not to be. These differing regulatory perspectives made a huge difference in the adoption of genetically modified crops on different sides of the Atlantic Ocean.

Of course, there are also moral and ethical issues at play in the GMO debate. Proponents like to point out that farmers have been developing new crop varieties since the dawn of agriculture and that plant breeding also modifies the genetic makeup of organisms. For example, teosinte, the wild ancestor of maize (corn), had thumb-sized cobs before farmers began breeding for larger kernels and cobs. And if you go back more than 50 million years, rice, wheat, and corn all evolved from a common ancestor. Modern genetic studies also show that gene transfers have occurred naturally in the past between bacteria and plants and animals. So is it wrong to put bacterial or insect genes into plants and plant genes into animals? Consider a question like this: If scientists were able to splice the ability to photosynthesize into people, would it be a major achievement in human evolution or a moral and ethical debacle?

concentrated animal feeding operation (CAFO) an industrial livestock operation in which large numbers of animals are raised in confinement.

ruminant an animal with a digestive system that can turn plant-based cellulose into protein.

10 years nearly all hog farms bred their stock in this way—with as many as 40 sows artificially inseminated at the same time. Smithfield Foods, one of the major US meat producers, has used artificial insemination to propagate a single genetic line of pigs on thousands of farms. In this way, the company was able to establish a production process with standardized pigs, feed, and operations. Producers even started designing the machinery used to slaughter the pigs and butcher the meat to fit the characteristics of these carefully developed breeds. This standardization process has increased the average number of piglets per litter, as well as the amount of edible meat produced by each hog.

Pigs raised today now have 75% less fat than those raised in 1950.

Beef Production

Concentrating beef production presents special challenges. Like humans, calves take 9 months to gestate and are normally born just one at a time; they also suckle for 4 months before they can be weaned. Further, cattle are **ruminants**: animals with digestive systems that can turn plant-based cellulose into protein through compartmentalized stomachs, regurgitation, and fermentation. Humans, who cannot digest cellulose, can benefit from this process by consuming the protein in meat and dairy products. But digestive systems of ruminants are equipped to process grass rather than grain.

So cattle production must still begin on traditional grass-covered expanses of rangeland and pasture with cows raising their calves. However, calves are now typically moved to a feedlot when they are 6 months old (**FIGURE 12.13**). There the calves are given corn-based feed with soybeans, fat and protein rendered from slaughterhouses, vitamins, synthetic estrogen, antibiotics, and roughage. Growth promoters, such as zilpaterol hydrochloride (brand name Zilmax), are also added. Although this decreases the flavor and juiciness of meat, Zilmax boosts muscle growth in the final 3 weeks before slaughter, adding about 30 additional pounds of meat to each animal. The corn-based feed fattens cattle quickly but also causes severe bloating and heartburn that must be treated regularly with esophageal tubes to relieve the pressure. This diet causes acid to eat away at the rumen wall, which leaves the animal vulnerable to bacterial infections—and leads to greater use of antibiotics. In the 1950s, it took up to 3 years for a steer to grow to 830 pounds for slaughter. On a feedlot, this time is cut down to as short as 14 months. CAFOs have increased the pounds of beef per head of cattle by more than 40%.

FIGURE 12.13 CAFO for Cattle Feedlots such as this one in Nebraska concentrate large numbers of cattle where they can be fed with grain rather than dispersing livestock widely in pastures to feed on grass. In this photo the black dots on the barren lots are cattle, and the large yellow mound is the feed.

(a) Purse Seine Nets

(b) Gillnets

(c) Bottom Trawling

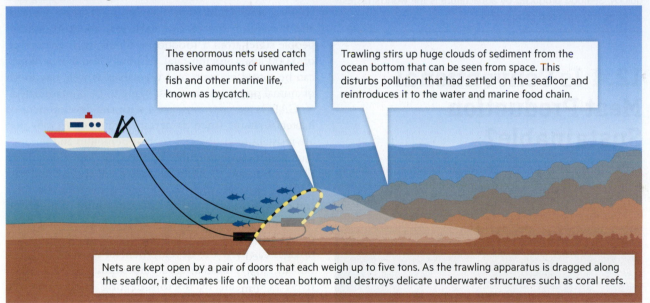

The enormous nets used catch massive amounts of unwanted fish and other marine life, known as bycatch.

Trawling stirs up huge clouds of sediment from the ocean bottom that can be seen from space. This disturbs pollution that had settled on the seafloor and reintroduces it to the water and marine food chain.

Nets are kept open by a pair of doors that each weigh up to five tons. As the trawling apparatus is dragged along the seafloor, it decimates life on the ocean bottom and destroys delicate underwater structures such as coral reefs.

FIGURE 12.14 **Fishing Technology** Modern commercial fishing gear utilizes techniques such as longline fishing, (a) purse seine nets, (b) gillnets, and (c) bottom trawling that can catch larger volumes of fish. However, these methods are indiscriminate and catch more than just commercial fish species—including other fish, birds, marine mammals, and sea turtles.

Catching and Raising Seafood

Whether it's *Moby-Dick*, *The Old Man and the Sea*, or the TV show *Deadliest Catch*, many cultures have long-held traditions of people risking everything on the open ocean. Open-water fishing is a harvest of wild creatures—like hunting on a much larger scale. There are no costs or inputs for feeding or rearing the stock; fishing boats only need to find and catch their products and bring them to market. However, technological advances have enhanced our ability to find, catch, and process large quantities of fish quickly for sale in markets around the world—increasing the scale and intensity of fishing. Since 1950, global annual oceanic fish catch has quintupled, exceeding 90 million tons. The global annual seafood catch per capita has risen from 15 to 35

pounds during this time. Fish-finding sonar, global positioning systems, 3-D mapping software, and even helicopters help fishing fleets locate their catch. The gear used to catch fish has become more sophisticated as well. The introduction of synthetic materials for nets and lines in the 1960s made this gear lighter, stronger, and less visible to fish. Fishing fleets once used large underwater "walls" of gillnets that stretched up to 40 miles across the ocean (today these are limited to lengths of 1–2 miles). However, fishing lines can still extend more than 50 miles, with baited hooks placed at regular intervals. Nets have become larger, and trawlers drag nets behind them that can catch huge volumes of fish—and other unfortunate marine animals (**FIGURE 12.14**). Large fishing vessels are now equipped with processing

aquaculture the practice of raising seafood in controlled ponds, tanks, or pens.

organic dust a dust that comes from materials such as animal feed, bedding, pesticide residues, animal wastes, and dander.

equipment and freezers so that the catch can be prepared and preserved for sale while at sea.

Seafood can also be raised and fed in ponds, tanks, or pens, much like a CAFO. This practice is called **aquaculture**. Freshwater fish are confined in ponds, shrimp in indoor tanks, and saltwater fish in floating pens at sea. They are held in high concentrations and fed a formulated combination of fish meal and fish oil derived from smaller ground-up wild fish, as well as other animal proteins, antibiotics, and additives to combat disease and enhance appearance. Commercial aquaculture—seafood farming—is the world's fastest-growing form of food production. It now accounts for more than half of the seafood consumed in the United States.

⬠ **TAKE-HOME MESSAGE** The way we raise livestock and harvest seafood has changed dramatically in recent decades. Most livestock production now occurs at large-scale CAFOs that increase the amount of meat per animal and reduce the time to reach slaughter weight. The scale and intensity of open-water fishing has also increased through our enhanced ability to find, catch, and process large quantities of fish quickly. Seafood produced by aquaculture is the world's fastest-growing form of food production.

12.4 Is Conventional Meat Production Sustainable?

Let's take a closer look at the environmental impacts of modern livestock and seafood production. For livestock, the impacts tend to be associated with waste, pollution,

(a)   (b)

FIGURE 12.15 Animal Waste and Food Poisoning (a) The bacterium *E. coli* from animal waste can contaminate food products and cause food poisoning that is sometimes fatal. (b) Manure lagoons can be a source of water pollution, especially after heavy storms. This photo shows a manure lagoon in North Carolina spilling over its containment walls after a hurricane flooded the area.

and greenhouse gas emissions. For seafood production, the primary concern is the impact fishing and aquaculture can have on wild species.

Impacts of Livestock Production

Concentrated animal feeding operations in the United States generate tremendous quantities of animal waste: more than 1.3 billion tons of manure each year. This is more than 40 times the mass of human waste processed each year at all of the publicly owned waste treatment facilities in the United States. CAFO waste is concentrated in tremendous lagoons onsite (**FIGURE 12.15**) or distributed with water over "spray farms," and pollution occurs when the effluent runs off the fields or leaks into groundwater. Animal wastes carry microorganisms, such as *Salmonella, Escherichia coli*, and *Cryptosporidium*, that can spread infectious disease. In 2000, seven people died and 2,500 people became ill in the town of Walkerton in Ontario, Canada, when *E. coli* from animal manure contaminated the well water. Pathogens can also contaminate meat during the slaughter process, which is why meats sold in the United States come with safe handling and cooking instructions. The Centers for Disease Control and Prevention reports that each year about 48 million people suffer from a foodborne illness, with an average of 3,000 cases proving fatal. Although foodborne illness can be spread by fruits and vegetables as well, raw foods of animal origin are the most likely to be contaminated.

CAFOs are now associated with the development of bacterial strains (such as certain strains of *E. coli*) that are resistant to antibiotics such as penicillin and tetracycline. Although vulnerable bacteria are wiped out, those that are resistant to antibiotic treatment survive and pass on their resistant genes to the next generation. Bacteria can even pass their traits to different strains of unrelated bacteria, further spreading resistant genes. Using these drugs only when absolutely necessary and with correct dosages can slow the development of resistant bacteria, but in CAFOs these drugs are not administered as selectively as they are by doctors treating human illnesses. Instead, at CAFOs, antibiotics are often included in the daily feed. This unmonitored delivery creates an ideal setting for rapid spread of antibacterial resistance. Resistant strains jeopardize livestock production and impact treatment options for people struggling with bacterial infections.

Decomposing animal urine and feces emit air pollutants such as ammonia and hydrogen sulfide, which can cause respiratory distress for those working in and living near CAFOs. Similarly, **organic dust**, a dust that comes from materials such as animal feed, bedding, pesticide residues, animal wastes, and dander, can cause allergic reactions in humans. Finally, the digestive process and manure from livestock account for about one-third of the anthropogenic methane emissions both in the United States and globally (**FIGURE 12.16**).

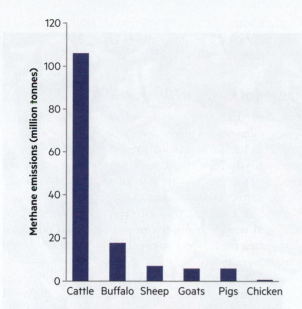

FIGURE 12.16 **Total Methane Emissions from the Global Livestock Sector by Livestock Type** Worldwide, cows produce the majority of methane (CH₄) emissions. Pigs and chickens produce far less CH₄ (a fraction of the amount that cows do), with pigs emitting slightly more than chickens.

Adapted from United Nations Environment Programme (2021).

Impacts of Seafood Production

The primary environmental concern linked to open-water fishing is simply **overfishing**. When the quantity of fish caught exceeds the productive capacity of a species, the population can decline and even collapse. For hundreds of years, the cod fishery off the coast of Newfoundland, Canada, produced hundreds of thousands of tons of fish because less than 10% of the population was removed each year—an amount the population could support. Then in the 1960s, the fishing take increased to more than 30% of the population each year, and by 1992 the population collapsed and the fishery was closed.

Open-water fishing also carries the environmental impact of **bycatch**. This is incidental capture of non-food species in fishing nets and lines. Millions of tons of sea turtles, seabirds, marine mammals, and fish are caught in this way each year (**FIGURE 12.17**). Shrimp trawling produces the highest bycatch levels, at an average global rate of 1.5 kilogram for every 1 kilogram of shrimp caught. Seafood guides (produced by conservation programs such as Seafood Watch) provide advice on purchasing seafood so that you can avoid overfished species and also choose fish that are caught using methods that reduce bycatch. For example, albacore tuna from the US and Canadian North Pacific is often recommended over other kinds because it is caught using safer methods. Tuna are caught in other places by use of longlines, which snare significant numbers of sea turtles, sharks, and seabirds.

Aquaculture avoids both overfishing and bycatch, but it can still weaken the genetic traits of wild fish if the farmed fish escape from their pens. The amount of waste produced can also affect water quality. Poorly

located operations can spread disease among wild fish and destroy habitats for other animals too; the latter has been the case with many shrimp farms developed on the site of former mangrove forests.

TAKE-HOME MESSAGE Wastes from CAFOs can carry pathogens and pollute the air and water and are significant contributors to increasing atmospheric greenhouse gases. Increases in wild seafood harvests are associated with both overfishing of certain stocks and the bycatch of non-food species. Aquaculture can impact wild stocks by polluting or destroying natural habitats and farmed fish interbreeding with wild fish.

→ SCIENCE

Antibiotic-Resistant Bacteria

Antibiotic resistance develops when disease-causing bacteria acquire genetic traits that deactivate the effectiveness of antibiotics designed to kill the bacteria. When antibiotic treatments are given, the nonresistant bacteria die, leaving the resistant bacteria to survive and multiply, passing their resistance on to subsequent generations. Over time, this resistance increases. In this way, raising animals with antibiotics to stimulate growth or combat disease can help resistant bacteria develop. These resistant bacteria can spread to humans through food, soil, or water contaminated with animal feces.

How Antibiotic Resistance Happens

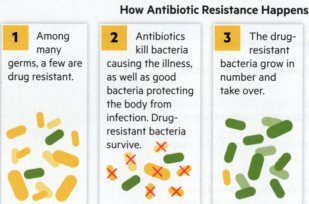
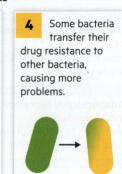

1 Among many germs, a few are drug resistant.

2 Antibiotics kill bacteria causing the illness, as well as good bacteria protecting the body from infection. Drug-resistant bacteria survive.

3 The drug-resistant bacteria grow in number and take over.

4 Some bacteria transfer their drug resistance to other bacteria, causing more problems.

12.5 How Have Our Food Systems Changed?

In many ways, agriculture can be seen as the most far-reaching human endeavor. Food production is the world's biggest industry and the largest human alteration of land cover. Together, croplands and pastures are now the largest terrestrial biome, covering almost 40% of Earth's land surface. But the impacts of crop, livestock, and seafood production are also intertwined with how

overfishing a practice that causes the quantity of fish caught to exceed the productive capacity of a species.

bycatch the incidental capture of non-food species in fishing nets and lines.

(a) Wasteful Fisheries

1 Gulf of Alaska Flatfish Trawl Fishery 35% Discarded
- More than 34 million pounds of fish were thrown overboard in one year, including 2 million pounds of halibut and 5 million pounds of cod.

2 California Set Gillnet Fishery 65% Discarded
- More than 30,000 sharks and rays as well as valuable fish were discarded as waste over three years.

3 California Drift Gillnet Fishery 63% Discarded
- About 550 marine mammals were entangled or killed over 5 years.

4 Southeast Shrimp Trawl Fishery 64% Discarded
- Thousands of turtles are killed annually.

5 Southeast Snapper–Grouper Longline Fishery 66% Discarded
- More than 400,000 sharks were captured and discarded in one year.

6 Atlantic Highly Migratory Species Longline Fishery 23% Discarded
- More than 75 percent of the wasted fish are highly valuable tuna, swordfish, and other billfish.

7 Mid-Atlantic Bottom Trawl Fishery 33% Discarded
- Almost 200 marine mammals and 350 sea turtles were captured or killed in one year.

8 New England & Mid-Atlantic Gillnet Fishery 16% Discarded
- More than 1,200 endangered sturgeon were captured in one year.
- More than 2,000 dolphins, porpoises, and seals were captured in one year.

9 Northeast Bottom Trawl Fishery 35% Discarded
- More than 50 million pounds of fish are thrown overboard each year.

(b)

FIGURE 12.17 Bycatch (a) Wasteful fisheries: In the United States, nine specific fishing activities and locations (fisheries) account for more than 50% of annual US bycatch. (b) These animals were snared as bycatch. Some species are in danger of extinction primarily due to bycatch, such as Hector's dolphin, the North Atlantic right whale, the Amsterdam albatross, and the Mediterranean monk seal.

Adapted from (a) National Marine Fisheries Service (2013).

we process them for sale in our grocery stores and with our consumption trends. Linked processes of agricultural production, food processing and distribution, and consumption constitute a **food system**. In this section, we look at some significant trends in our modern food system.

Agricultural Production: Increasing Scale and Shrinking Diversity

If you walk into a grocery store and look around, you might assume that we consume a tremendous variety of crops. But what we really have is an amazing variety of food products, because the number of crops we consume has decreased considerably. Historically and globally, a total of roughly 3,000 species of plants have been eaten by humans on a regular basis, but only 150 plants are now grown on a significant scale, and three crops—rice, maize (corn), and wheat—account for about 60% of the calories and protein humans derive from plants (**FIGURE 12.18**).

Agricultural production is now undertaken on a huge scale but is more concentrated geographically. Modern agricultural technologies increasingly cultivate great swaths of land more quickly and with less labor. The

FIGURE 12.18 Reliance on a Small Number of Grains Global grain production has increased dramatically since 1950 to keep pace with population growth, and per capita grain production has held steady. But we rely on a small variety of plants to sustain us. Rice, maize, and wheat currently account for about 60% of human caloric and protein needs. Wheat production takes up 1.5% of all continental landmass.

United States now produces half of the world's corn, and China, India, and the United States together produce more than half of the world's rice and wheat. More than half of the food production in the United States comes from just 10 states. For some foods, such as meats, production is also concentrated in the hands of a small number of companies (**FIGURE 12.19**).

These changes are affecting agrobiodiversity. Over the past 100 years, there has been a 90% decline in the varieties of all food crops cultivated, and more than half of the global varieties of food crops have gone extinct. For example, US Department of Agriculture (USDA) records indicate that 86% of the 7,100 named apple varieties grown in the United States in the 1800s are now extinct. According to the US Apple Association, red delicious, gala, golden delicious, and Granny Smith apples constitute more than half of all the apples grown in the United States. Further, the variety of domesticated animals kept for livestock has declined by 50%, and 1,600 of the 8,000 known livestock breeds are now endangered or extinct.

Food Security

This global decrease in agrobiodiversity reduces the resilience of our food systems and can threaten **food security**—the affordable access to enough nutritious food to maintain dietary needs. Reliance on a shrinking variety of crops and livestock from a decreasing amount of fertile land in the world makes the food supply vulnerable to large-scale disruptions, such as pests and diseases, decreasing soil fertility, and limited water supplies. In recent years, the percentage of US households experiencing food insecurity has held steady at just over 10%. However, the disruptions of the COVID-19 pandemic more than doubled the percentage of US households facing this challenge, with as many as one in four households reporting food insecurity in the spring of 2020. Many reported that their eating patterns had been disrupted and they were reducing their food intake. Households with children are more likely to face food insecurity than those without children, and during the pandemic, food insecurity for households with children tripled. **FIGURE 12.20** shows that many college students also grapple with food insecurity. These trends have revealed the importance of school lunches, food assistance programs, and proximity to affordable, nutritious food to ensure everyone can meet their dietary needs.

Food Consumption: Increasing Role of Meat and Processed Foods

The typical diet has also changed. On average, we now consume significantly more meat and processed foods and less fruit and vegetables. Some studies in the United States indicate that fresh fruit and vegetable consumption fell by as much as 25% over the past 50 years. In contrast, meat consumption is on the rise in both the United States and the wider global community. The FAO has reported that between 1950 and 2010, global meat consumption per person per year increased from 38 to 88 pounds. Until the 1980s, much of this growth was concentrated in the United States and Western Europe. But in 1992, China surpassed the United States in total meat consumption and now consumes twice as much total meat as the United States does.

At the same time, rather than cooking food with raw ingredients, we are increasingly relying on **processed foods** such as frozen pizzas, microwavable dinners, packaged snacks, and fast foods. Processed foods are commercially processed for ease of use, preparation, consumption, and storage. These foods are often individually wrapped in specially treated materials that can be heated and/or

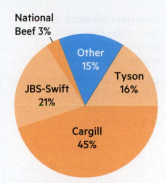

FIGURE 12.19 Beef Producers in the United States Meat production in the United States is controlled by a small number of large corporations. For beef, just four companies produce 85% of the meat. The story is similar for chicken and pork, with just three to five companies producing more than half of the meat supply.

Adapted from USDA (2019).

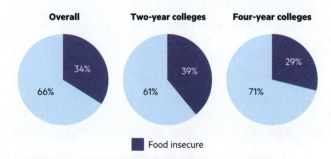

FIGURE 12.20 Percentage of Students Who Were Food Insecure in 2020 by Institution Type Many college and university students found it difficult to get enough food to eat during the COVID-19 pandemic. A 2020 survey found that more than a third of college students reported that in the past 30 days they couldn't afford to eat balanced meals, bought food that didn't last until they had money to buy more, or worried that their food would run out before they had money to buy more.

Adapted from The Hope Center for College, Community, and Justice (2021).

food system the processes that link agricultural production to food consumption.

food security the affordable access to enough nutritious food to maintain dietary needs.

processed food food that is commercially processed to optimize ease of preparation, consumption, and storage.

overnourishment excessive food intake that causes energy use to be less than the food energy consumed.

obesity a condition of being substantially overweight because of excessive fat accumulation.

can resist saturation by oil and water; this packaging is both a growing source of waste and of PFAS (per- and polyfluoroalkyl substances) contamination. PFAS are a class of "forever chemicals," called this because they are very resistant to breaking down and so persist in the environment. They are under increasing scrutiny from the Food and Drug Administration and Environmental Protection Agency for their links to certain kinds of cancer and to reduced immune responses. Highly processed foods require more production steps between agricultural producers and consumers, with processes that often remove fiber and add sweeteners, salt, fats, colors, artificial flavors, and preservatives. According to the USDA, processed foods now provide more than 70% of the caloric intake in the average US diet.

Nutritionists warn that although processed foods are tasty and convenient, when consumed in large amounts their lack of fiber and high levels of sugar, fat, and salt are bad for our health. Compared to 100 years ago, the average American now eats 60% more sugar, double the amount of fat, and 3.5 times the salt. This dramatic shift in diet is not unique to the United States. According to Euromonitor International, processed food consumption per person has risen 66% in China since 2008, making that nation the global leader in consumption of such food. The portion sizes of the food we consume are increasing too, and in some cases doubling in recent decades (**FIGURE 12.21**).

While hunger and undernourishment affect more than 800 million people worldwide, the trends identified above are associated with **overnourishment**, when energy use does not match the food energy consumed. This produces excess body fat, sometimes causing **obesity**, which is linked to diabetes, heart disease, respiratory problems, and other ailments. These diseases have replaced infectious diseases as the primary killers in the developed world. The FAO found in 2010 that for the

first time in history, more people in the world are overweight and obese than are suffering from hunger. More than 1.5 billion people currently consume more calories than they need but are deficient in essential micronutrients such as vitamins and minerals. In 2019, the Centers for Disease Control and Prevention found that more than half of US adults are overweight, and the percentage of adults who are obese has increased from 15% to 32% since 1980. The United Nations has found the percentage of overweight and obese adults to be increasing in all regions of the world.

Shifting Food Culture

Nutritionists see overnourishment as a global "nutritional transition." Humans crave sweet and fatty foods. For most of human existence, high-calorie foods such as sweets and fats were difficult to obtain, and we evolved to store up these rare high doses of energy. Our survival depended on consuming these high-energy foods whenever we could get them. But our modern food production technologies and policies have made these foods cheaper and readily available. So although these high-energy foods are no longer rare and difficult to obtain, our craving for them persists.

Cultural practices related to foods have also changed. The way in which we meet our daily needs has become increasingly fast paced and mobile (even as our physical activity level has declined), and studies show that eating outside of the home and on the go is becoming more common. One survey revealed that home-cooked meals have fallen to just one-third of US household food consumption. Convenient processed foods have played an important role in this lifestyle change, especially because most heads of US households, whether women or men, single parents or couples, now are responsible for both work outside the home and meal preparation at home. Countries such as China that are rapidly urbanizing and expanding industrial and service-sector employment are following the dietary patterns of nations such as the United States that underwent these same changes decades earlier (**FIGURE 12.22**).

The way we sell and buy or exchange food has changed as well. Public markets that fostered a direct connection between food producers and consumers have declined, while supermarkets have increased. In Latin America, the share of all food sales at supermarkets increased from 15% to 60% between 1990 and 2000 and is nearing the US share of 80%. Supermarkets typically use long-distance food supply chains that link large-scale and often foreign producers of low-price specialized crops instead of local sources. In one study, researchers tracking the distance fruits and vegetables traveled before arriving for sale at grocery stores in Chicago found that most products had been shipped more than 500 miles. The top 10 agricultural producing states supply more than half of all the food sold in the United States, with California

FIGURE 12.21 Growing Portion Sizes Average portion sizes have grown considerably over the past few decades, leading to greater caloric intake. A study by the British Heart Foundation found that portion sizes across 245 popular food products in Great Britain doubled between 1993 and 2013. Similar studies in the United States have found that portion sizes for nearly every food product increased during this time.

Adapted from National Heart, Lung, and Blood Institute (n.d.).

20 years ago	Today	Difference
333 calories	590 calories	**257** more calories
500 calories	850 calories	**350** more calories
210 calories	500 calories	**290** more calories

Where Americans Ate, 1987–2017

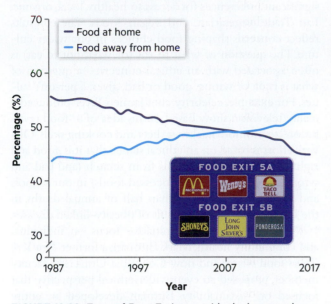

FIGURE 12.22 Eat and Run Eating outside the home and on the go is increasing the share of processed foods in our diets. In 1970, food eaten outside the home made up just about 25% of the market share for food consumption in the United States. Today, food eaten outside the home makes up the majority of the market share for food consumption.

Graph adapted from Saksena et al. (2018).

alone producing about 14% of the food sold. And a glance at the labels of produce in the grocery store reveals that much of the food for sale is produced overseas as well. Supermarkets have also greatly expanded the availability of processed foods, which are packaged and preserved to hold up over long shipping routes and lengthy storage periods in warehouses. The online grocery delivery business that ships food from regional warehouses to homes is also expanding, with grocery e-commerce in the United States more than doubling during the COVID-19 pandemic from just over 4% to more than 10% of total grocery sales.

Food Policy

Government interventions that facilitate cheaper food prices have been particularly important in shaping our current food systems. For example, government subsidies for corn production in the United States, including crop insurance, direct payments, and market loss protections, encouraged farmers to produce ever-larger harvests of corn and sell it cheaply. Between 1995 and 2014, government subsidies for corn production totaled more than $94 billion. The cheap subsidized corn supply made possible many other cheap food products, from processed foods sweetened with corn syrup to corn-fed cattle, pigs, and chickens. Economists estimate that the operating costs for CAFOs would be 7% to 10% higher if not for government crop subsidies for animal feed grains such as corn

ENVIRONMENTAL JUSTICE

Controlling the Food System

There are racial disparities on the production side of the food system. People of color make up more than 80% of US farmworkers, who bear the highest risks of pesticide exposure, but account for only 5% of farm owners. People of color also compose nearly two-thirds of the workforce in meatpacking plants. These are among the most dangerous and lowest-paid jobs in the US workforce. Today organizations such as the Earthseed Collective and the Farmers of Color Network are trying to help marginalized communities gain greater control over the food system—connecting land justice issues to food justice issues. In particular, the Farmers of Color Network provides training and financial assistance opportunities as farmers are getting started. Groups like this are helping farmers of color acquire or lease land and produce affordable healthy food.

and soybeans. Government food-assistance policies reduce the cost of food for consumers more directly; examples include free and reduced lunch programs in schools and the Supplemental Nutrition Assistance Program (SNAP) that provides benefit cards, similar to debit cards, to pay for food.

Government policies also shape public knowledge about food in important ways. Nutrition information on food packaging, nutritional guidelines established by the US Food and Drug Administration, and food labeling that certifies foods as organic are all examples of policies that influence consumer food choices. Government decisions about how to convey this information to consumers, or about which standards to use, are often hotly contested political issues because they affect not only consumer health but also the bottom line of many food businesses.

Our food system has undergone significant changes in recent decades. The scale of agricultural production has increased while the number of farms, farm jobs, and crop and livestock varieties has decreased. Agricultural operations are now concentrated in a few geographic areas and are in the hands of a few corporations. Typical diets now include more meats and processed foods, trends linked to a global rise in obesity. These trends are driven by an increasingly fast-paced and mobile lifestyle along with more global, industrialized, and large-scale systems of food exchange.

12.6 How Do Our Food Choices Link to Sustainability?

Problems linked to modern agriculture are making many people wonder if we can feed ourselves in a more sustainable way. What role can our individual food choices play in supporting sustainable agriculture and food production? If you had a chance to select the food provided at your college cafeteria or food court, what kinds of concerns would guide your decisions?

In a survey at Boston University, students expressed significant preferences for access to healthy, local, organic, Fair Trade, vegetarian, and vegan foods. These results reflect concerns shaping food choices in our wider culture. The question of what to eat (or what not to eat) is often associated with an ethical concern—a question of what is right or wrong, good or bad, given a person's values. For example, celebrity chef Jamie Oliver produced a reality television show based on his idea of a "food revolution" to promote healthier diets and cooking skills. Oliver's effort is based on an ethical stance that it is good and right to cook healthier meals from scratch (and bad and wrong to rely heavily on processed foods) in our schools and homes because more than half of annual deaths in the United States are the result of obesity-linked diseases.

While Oliver and his initiative focus on individual and community health, Mark Bittman, a former *New York Times* food writer and now Columbia University faculty member, proposed an even wider ethical perspective that focused on sustainability. Bittman developed an entire food ethic that focused on reducing meat consumption because of the environmental impacts associated with livestock production and the consequences these impacts will have on future generations. Bittman's and Oliver's initiatives represent only a few popular perspectives, and others can go much deeper into the ethics of consumption. Let's take a closer look at some of the major food choices associated with a move to a sustainable food system.

The Efficiency Argument: Eating Low on the Food Chain

In Chapter 3, when we explored how energy affects life, we saw that consumption at different levels of a *food chain* or *food web* involves the transfer of energy from one level to another. When we eat, we gain some energy from the plant or animal we consume. But you may also recall that this energy transfer is not an efficient process. Energy is lost along the way as transfers are made between trophic levels, and some argue that we should be eating lower on the food chain because less energy is lost (**FIGURE 12.23**).

Where does the food you eat appear on the food chain? Are you eating animal-based food that is higher on the food chain or plant-based food that is lower? Some people—including vegetarians and vegans—argue that eating lower on the food chain has positive environmental impacts. As we have seen, conventional production of animal-based foods generally leads to more water use and greenhouse gas emissions than does the production of plant-based food. When you consider the additional emissions of greenhouse gases from livestock and waste products, there is a strong argument for reducing consumption of CAFO-produced meat in order to reduce overall human environmental impacts (**TABLE 12.1**; see also Chapter 2, Sustainability Matters: Is Eating a Vegetarian Diet More Sustainable?).

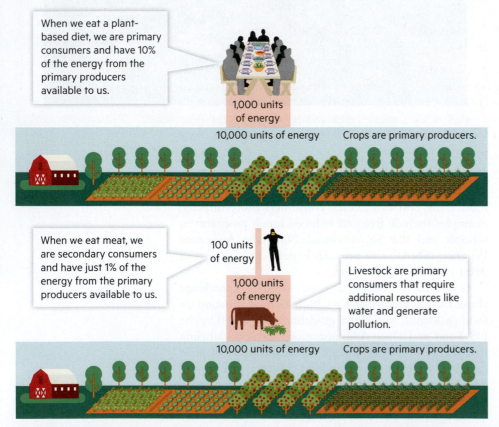

When we eat a plant-based diet, we are primary consumers and have 10% of the energy from the primary producers available to us.

1,000 units of energy

10,000 units of energy

Crops are primary producers.

When we eat meat, we are secondary consumers and have just 1% of the energy from the primary producers available to us.

100 units of energy

1,000 units of energy

Livestock are primary consumers that require additional resources like water and generate pollution.

10,000 units of energy

Crops are primary producers.

FIGURE 12.23 **Where Are You Eating on the Food Chain?** Eating lower on the food chain is more efficient because less energy is required to produce the food that is eaten.

Concern for Animal Welfare

Others may have values centered on a concern for animals. We are now slaughtering far more animals for food than at any time in our history. Because CAFOs exist to gain economic advantage of scale, they crowd animals into small spaces while simultaneously promoting their rapid growth. Confinement from modern farming can cause many problems for chickens, from heart and lung problems to musculoskeletal deficiencies that prevent them from standing. These painful conditions for animals lead ethicists such as Peter Singer to argue that we should consider more than anthropocentric (human-focused) needs. These ethical arguments are not limited to animals raised on farms but can apply to wild seafood species too, as they are threatened by overfishing or bycatch.

Organic and Sustainable Agricultural Practices

Still others focus on how food is produced, and various certifications help signal certain practices. The organic certification process in the United States requires that producers pass inspections ensuring crops are grown without synthetic fertilizers, pesticides, or sewage sludge and are not genetically modified. To meet this standard, organic farmers often employ crop rotation and cover crops, and use animal and plant waste to add nutrients and conserve the soil. While studies tend to report lower yields from organic farms than from conventional farms, this difference depends on the health of the soil (whether it is degraded) and on the specific practices used on both organic and conventional farms. One 15-year study on the productivity of corn and soybeans found that crops enriched with organic plant and animal matter not only have yields equal to those using synthetic fertilizers but also build healthier soils. Consumer demand for organic foods has led to dramatic growth in their sale and retail availability. While in the United States just over 5% of all the food sold is certified organic, it constitutes more than $55 billion in sales and is growing by about 9% each year on average (**FIGURE 12.24**).

But organic practices are not sustainable if they degrade or erode the soil through practices such as plowing for weed control. And naturally occurring but toxic chemical pesticides such as copper sulfate, which are allowed under the "organic" label, can be just as hazardous as synthetic pesticides. To combat these problems, farmers have developed a variety of more sustainable approaches. These include incorporating trees into crop and grazing systems, building self-sufficient agricultural systems modeled on natural ecosystems, and using regenerative soil-building practices (see Chapter 9).

Meat can also be certified organic if livestock are raised with organic feed, not treated with hormones or antibiotics, and have access to the outdoors. Others argue that grass-fed cattle are more sustainable than those fed

with organic grain. First, these cattle can be raised on pastures that are not suitable for cropland. Second, controlled grazing practices can actually promote the health of native grassland ecosystems and rebuild the fertility of degraded agricultural soils.

As for seafood, there is no USDA organic standard for aquaculture, though several have been proposed over the past decade. One sticking point is that many producers and consumers dispute whether fish raised or farmed organically is superior to fish caught in the wild. In addition, because fish eat other fish, it's difficult for farmers to guarantee that their fish are eating organic sources.

Concern for Global Equity and Community: Fair Trade and Buying Local

The World Food Programme of the United Nations estimates that 75% of those suffering from hunger live and work in farming communities. This leads some people to question how modern agriculture could ever become truly sustainable for those most involved in it and how economic incentives can be provided to promote more sustainable farming. The **Fair Trade** certification standard

Fair Trade a certification standard that assures customers that the price of a product provides for adequate wages and environmentally sustainable production and helps farming communities thrive.

TABLE 12.1 Meat-Based versus Plant-Based Diets

Conventional (Grain-Fed) Meat	Plants
• Requires more land, water, and energy	• Requires less land, water, and energy
• Produces about twice the greenhouse gas emissions of vegetarian diets	• Produces about half the greenhouse gas emissions of conventional meat-based diets
• Livestock are fed grains that people could be eating	• Conventional tillage causes soil erosion
• Animal wastes can pollute water bodies	• Plowing grasslands to plant crops reduces soil organic matter and releases carbon dioxide
• Ethical issues surround eating animals and confining them in feedlots	• Synthetic fertilizers and pesticides can pollute water bodies
• Overuse of antibiotics can cause antibiotic resistance	

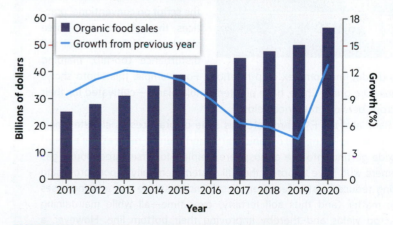

FIGURE 12.24 Total US Organic Food Sales and Growth, 2011–2020 Organic food sales have grown every year in the past decade.

Adapted from Organic Trade Association (2021).

is one response. It tries to assure customers that the price of a product provides for adequate wages and environmentally sustainable production and helps farming communities thrive. Research on coffee has found that producers following the Fair Trade standard received significantly higher prices per pound and enjoyed higher average incomes than those for conventional producers. Fair Trade coffee producers were also more likely to have implemented environmentally friendly soil and water conservation measures than were conventional producers. If Fair Trade can become a label that people will pay a premium for, increased sustainable production should follow, and recent research has demonstrated that Fair Trade is showing success as a market-based program (**FIGURE 12.25**).

Closer to home, many grocery stores and restaurants highlight items grown locally, and there are now more than 8,000 farmers' markets in over 100 cities and towns across the United States, an increase of more than 75% since 2008. What's behind the growing campaign to buy local? One rationale is that supporting local agriculture reduces the greenhouse gas emissions from transporting food over long distances. However, it's possible that a food grown a great distance away might be associated with fewer environmental impacts than one

FIGURE 12.25 Fair Trade The Fair Trade standard tries to better support local communities, such as the workers on this coffee plantation in Jimma, Ethiopia.

grown nearby. This will vary from place to place and product to product. For example, a recent study found that produce raised in energy-intensive greenhouses in New York State had a bigger climate impact than did produce raised by conventional means in California and shipped 3,000 miles to New York. But there are other arguments made for local agriculture. Author Michael Pollan encourages consumers to buy local food at farmers' markets or through consumer-supported agriculture (CSA) cooperatives. CSAs work by having consumers pay a subscription to a farm in exchange for a regular supply of food grown on that farm, and sometimes the consumer can even contribute labor on the farm as payment. Pollan argues that farmers' markets and CSAs help consumers discover and develop relationships with food producers, while ensuring that farmers are using the practices important to local consumers.

Another argument for buying local includes support for crop diversity. Consuming local foods requires eating seasonal varieties compatible with local conditions, and some local farms are engaged in seed bank programs to preserve and share local crop varieties. Supporting local agriculture can help to keep the rural areas surrounding cities viable and prevent them from being converted into sprawling developments. New York City's Greenmarket program established more than 50 farmers' markets for more than 200 local farmers and food producers. SNAP benefits can be used at Greenmarkets, and the city government gives users an additional dollar of farmers' market benefits for every $2 of SNAP benefits spent. This program has helped keep more than 20,000 acres of land near the city in agriculture.

➔ **SUSTAINABILITY**

How Can We Restore the Soil?

Over recent decades, farmers around the world have increasingly adopted agricultural practices based on restoring soil life to build soil health. This new model of agriculture, called *conservation agriculture,* flips the script on conventional practices through adopting three simple principles: minimal soil disturbance (no-till), planting cover crops, and adopting complex crop rotations. For example, this photo shows how farmers near Dongting, China, have planted canola plants—cultivated for their oil—as cover crops between their rice fields.

The World Bank and the FAO promote this suite of practices as climate-smart agriculture because it builds up soil organic matter and thereby reduces the amount of carbon dioxide getting into the atmosphere, helping to offset greenhouse gas emissions. Farmers who have adopted the full system generally report promising results, including reduced water, fertilizer, diesel, and pesticide use and increases in soil organic matter (and thus soil fertility) over time—all while maintaining or increasing crop yields and thereby improving their bottom line. However, a several-year transition period in which yields can decline somewhat can prove a disincentive to adopting soil-building practices that have lower long-term costs.

Reducing Food Waste

In 2017, the FAO released a report on the future of food and agriculture. The agency reported that future productivity gains must come mainly from improvements in efficiency, including reducing food loss and waste

Sir Albert Howard, Grandfather of the Organic Farming Movement

In 1905, English agriculturalist Sir Albert Howard accepted an offer to become the Imperial Economic Botanist to the colonial government of India. A key attraction of the new post was 75 acres of farmland on which Howard could experiment as he pleased. Thus began his journey to become the grandfather of the modern organic farming movement.

Soon after he arrived, he noticed that local peasant farmers consistently raised remarkably healthy crops without the chemicals that agronomists were experimenting with and recommending as essential to modern farming. His interest raised, Howard began to study traditional farming practices. Over the next several decades, he developed a composting process, called the Indore process after his research region, and found that he could raise healthy crops without agrochemicals. He also saw that the incidence of pests and crop diseases was increasing and spreading on British colonial fields that relied on chemicals, but there were far fewer problems on his composted fields and those of local traditional farmers. He set up large-scale tests of his composting method and found that within a few years, he could double his yields and suppress crop diseases with impressive efficiency. On the basis of his observations, he concluded that the use of chemical fertilizers and pesticides actually made it harder to grow healthy crops. The chemicals made plants more susceptible to pests and disease, which in turn increased the need for more pesticides.

Howard reasoned that the problem with modern agricultural methods lay in disrupting the role of beneficial fungi in plant nutrition and health. On his composted fields, fungi thrived, and their root-like hyphae (filaments) intertwined with plant roots. He suspected that heavy fertilizer and pesticide applications disrupted this fungal connection and were detrimental to the health of the crop. Building up soil organic matter, he thought, was the way to feed the fungal "partners" that helped maintain plant health and recycled organic matter into nutrients that plants could take up again.

Mainstream agronomists at the time dismissed Howard's ideas. They felt that organic matter didn't nourish crops, as most soil organic matter was insoluble and could not be taken up by plant roots. And they viewed compost as an open invitation for pests to savage crops.

While Howard saw soil biology as the key to sustained soil fertility, in the end he could not explain how this all worked. Without a documented mechanism to point to, skeptical scientists were quick to dismiss Howard's idea that abundant, healthy fungi were central to fertile soil and healthy crops. The large-scale composting that Howard advocated was soon eclipsed by mechanized agrochemical agriculture, fertilizer-intensive practices, and modern crop varieties, a combination that continues to define the dominant form of farming today.

Over the past several decades though, modern science has confirmed the importance of complex nutrient and chemical exchanges happening in the root zone between microbes and plants. Through photosynthesis, plants have the ability to manufacture organic matter. Yet plants don't keep it all for themselves. Many push more than a quarter of what they make out of their roots in the form of carbohydrate- and protein-rich secretions. This feeds the bacteria and fungi that inhabit their root zone. When these microbes consume this rich fare, they in turn produce substances that benefit the growth and health of the plants. In other words, plants have outsourced the manufacturing of beneficial compounds to microbes that the plants in turn feed through the bounty of root secretions. Such symbioses provide a mechanism through which soil rich in organic matter helps support the growth and health of plants. Plants fed a heavy diet of fertilizers ramp down their secretions, which reduces microbial production of substances that improve plant health and defense.

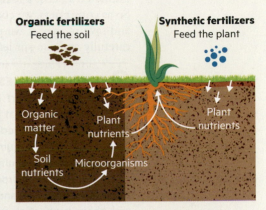

Organic fertilizers
Feed the soil

Organic matter
Soil nutrients
Plant nutrients
Microorganisms

Synthetic fertilizers
Feed the plant

Plant nutrients

Sir Albert Howard was impressed by the health of the crops raised in Indian fields with traditional farming methods. He promoted an approach that emphasizes organic matter and the role that the community of life within soil plays in soil fertility and plant growth.

What Would You Do?

When Howard proposed his ideas in the early 1900s, they were rejected in favor of other solutions, in great part because Howard could not explain how his ideas worked. Today, organic products have become a $55 billion industry; clearly lots of people are buying organic. But the USDA has also found that when the price of organic products is compared to that of their non-organic counterparts, organic products are on average about 20% more expensive, which drives most consumers to lower-price options. Are you willing to pay more for organic produce? If you were the head of dining services for your school and had a choice between buying all organic food but having higher prices in the dining halls or buying the lowest-priced foods possible and passing the savings on to customers, what choice would you make and why?

361

(which currently claims about one-third of the world's agricultural output). Food loss or waste can occur at several stages along the food supply chain from the farm to the consumer, but in the United States more than 60% of the food waste occurs because consumers buy food they do not end up eating (read more about this in Chapter 15). This represents a waste not only of food but also of the resources that went into producing it. It is also a waste of money. Food waste in the United States is estimated to cost a total of about $165 billion each year. So as consumers, we could lessen environmental impacts and save money if we plan our grocery shopping more carefully and eat our leftovers.

 TAKE-HOME MESSAGE The individual food choices of consumers are part of food production systems. The question of what to eat is often associated with an ethical concern. Concerns linked to food choices can include animal welfare, efficiency arguments, a desire to support organic and sustainable agricultural practices, and/or support for Fair Trade and local producers. Awareness of how various foods are produced and the relative impacts associated with these production methods can help inform individual food choices.

12.7 What Can I Do?

Because we all need to eat, we can all impact our food systems by making better choices about what we eat. There are several simple steps we can take to encourage sustainable food production and reduce the amount of food that needs to be produced.

Make Sustainable Choices about the Meat You Eat

While vegetarians and vegans eat lower on the food chain by avoiding meat and/or other animal products entirely, others suggest that we can still make positive steps toward these goals by simply reducing our consumption of animal-based foods. This is what food writer Mark Bittman put forth in his VB6 (vegan before 6:00 PM) cookbook: a diet that eliminates animal-based foods from breakfasts, lunches, and snacks but allows for moderate consumption of these foods at dinner. This is also the idea behind the "meatless Mondays" plan that has been promoted on several college campuses, such as Miami University of Ohio and SUNY Binghamton.

Support Sustainable Fishing Practices

There are also resources available to help improve consumer choices. The Monterey Bay Aquarium Seafood Watch program encourages consumers to be "responsible seafood lovers" by providing guides and an app that steer shoppers to "Best Choices": those species that are not overfished and for which fishermen avoid bycatch, which ensures that marine species persist at sustainable levels. Take a look at their seafood guide and assess its effectiveness at guiding your consumer choices.

Reduce Your Food Waste

There are several simple steps you can take to reduce the amount of food you waste. First, if you're picking up food in a cafeteria line or ordering food from a restaurant, try to select only what you'll eat. Choosemyplate.gov offers resources for different audiences to consider what and how much you are eating. If you're shopping for food to make at home, go to the store with a plan. Having a grocery list helps to restrict your purchases to items that you plan on making. It might help if you keep in mind how much the food costs when you buy it and visualize throwing that money away when you throw out your food.

Support a Community Garden, Farmers' Market, or Food Bank

You can take an active role in your local food system by helping plant or harvest at a community garden, shopping at a farmers' market, or volunteering at a food bank. Many colleges have community gardens on campus, and the American Community Gardening Association maintains a map that can help you find other sites near you. Similarly, the US Department of Agriculture has a searchable database of farmers' markets. Community gardens and farmers' markets are a great way to participate in, support, and enjoy local food production. By volunteering at a food bank, you can help address food insecurity in your area. FeedingAmerica.org can help you connect with the organizations near you that are providing food assistance.

Chapter 12 **Review**

SUMMARY

- Humans have been able to support a growing population by transforming the way food is grown and harvested through developments such as new crop varieties, farm machinery, advanced irrigation, synthetic fertilizers, and pesticides.

- While modern agricultural practices have increased food production, they have also had adverse environmental impacts, including depleting soil fertility and water supplies; polluting the land, air, and water; contributing to climate change; and reducing the genetic variety of food-producing crops.

- Meat consumption is increasing globally, and most livestock production now occurs at large-scale concentrated animal feeding operations (CAFOs). CAFOs rely on grain and growth-inducing additives and increase greenhouse gas emissions, pollution from animal waste, and the spread of antibiotic-resistant bacteria.

- Technologies for finding, catching, and processing large quantities of fish have increased the scale and intensity of open-water fishing.

- Overfishing and bycatch are adverse environmental impacts associated with open-water fishing.

- Seafood farmed by aquaculture is the fastest-growing form of food production globally. Pollution, habitat destruction, and genetic weakening of wild fish stocks are negative consequences of aquaculture.

- Diets have shifted globally to include more meats and processed foods, which are associated with a global rise in obesity, while the lack of affordable access to nutritious food remains a global problem.

- Food production, processing, and distribution have become more global, industrialized, and concentrated in the hands of a few large companies.

- Consumer food choices are also part of food systems. Ethical concerns linked to food choices include animal welfare, efficiency arguments, a desire to support organic and sustainable agricultural practices, and/or support for Fair Trade and local producers.

KEY TERMS

agriculture
agrobiodiversity
aquaculture
bycatch
concentrated animal feeding
  operation (CAFO)
desertification

Fair Trade
fallow
food security
food system
genetic erosion
genetically modified organism
  (GMO)

leaching
livestock
monoculture
obesity
organic dust
organic matter
overfishing

overnourishment
processed food
ruminant
salinization
subsidence
synthetic fertilizer
synthetic pesticide

REVIEW QUESTIONS

The letters following each Review Question refer to the Chapter Objectives.

1. Describe the earliest plant cultivation method practiced by humans. **(A)**

2. Name three technological developments associated with modern agriculture. **(A, B)**

3. Explain two ways through which agricultural practices contribute to topsoil loss. **(B)**

4. Describe two ways in which agricultural practices affect water resources. **(C)**

5. What effect did the use of the pesticide DDT have on bald eagles and other birds of prey? **(B)**

6. Describe two uses of genetically modified organisms for crop production. **(B, C)**

7. Explain the risks associated with declining agrobiodiversity and monocultures. **(C)**

8. How has global meat consumption per person changed in recent decades? **(C)**

9. Describe two environmental impacts associated with concentrated animal feeding operations (CAFOs). **(C)**

10. Describe two environmental impacts associated with seafood production. **(C)**

11. How has the proportion of undernourished to overnourished people changed globally in the past decade? What has driven this shift? **(D)**

12. Describe one way our cultural practices regarding food have changed in recent years. **(D, F)**

13. What are three ways that consumer food choices can lead to more sustainable outcomes? **(E, F)**

FOR FURTHER THOUGHT

The letters following each item refer to the Chapter Objectives.

14. While genetically modified organisms (GMOs) have been developed to complement the use of certain pesticides, they have also been developed to address certain human health problems linked to malnutrition, such as the "golden rice" developed to address vitamin A deficiency. In your view, how should GMOs be used, if at all? What sorts of policies (if any) should be in place to regulate their development and use? **(B, C, D)**

15. Where is the healthy food? Many areas of the country lack easy access to healthy food. Use a mapping program on your phone or computer to identify the grocery stores and farmers' markets that are closest to where you live, and report your findings. Relative to others who attend your university, do you think you have more or fewer options to obtain healthy food? **(F)**

16. Examine the menu at your local campus dining hall or a local campus restaurant. Using the material in this chapter, identify which two meals you think have the lowest environmental impacts and which two have the highest. Report and explain your choices. **(B, D)**

17. Shifts in our overall food culture have affected the way we eat, our health, and environmental impacts linked to our food consumption. What sorts of personal and societal changes could be made to shift food culture in ways that improve human health and reduce environmental impacts linked to our food choices? Are there aspects of your own personal food culture that you would like to change? Explain if there are or are not. **(D, F)**

18. There are many visions for the future of agriculture. In particular, one vision embraces new technology (e.g., GMOs and precision fertilizer applications) to help improve our agriculture. Another looks to updating practices used in the past to create regenerative soil-building systems (e.g., combining no-till, cover crops, and diverse rotations) to sustainably feed the world of tomorrow. What do you see as the potential advantages and pitfalls of each approach? **(D, E, F)**

Make Your Case ?

The following exercises use real-world data and news sources. Check your understanding of the material and then practice crafting well-supported responses.

Use the News

The following article appeared in the online publication of the *Telegram and Gazette*, a local news source in Worcester, Massachusetts. Use this article to answer the questions that follow. The first three questions test your understanding of the article. Questions 4 and 5 are short answer, requiring you to apply what you have learned in this chapter and cite information in the article. Answers to Questions 1–3 are provided at the back of the book. Question 6 asks you to make your case and defend it.

"Worcester Farmers Markets Offer Fresh Produce for All Budgets, Target Food Insecurity," Nicole Shih, *Telegram and Gazette,* October 8, 2021

WORCESTER—Yajaira Calderon of Worcester shopped at the Beaver Brook Farmers Market on Chandler Street on Friday morning for the first time for some fresh fruits and vegetables even though she's been enrolled in federal government's Supplemental Nutrition Assistance Program (SNAP) since the pandemic hit last year.

Calderon and her daughter heavily rely on the program as she tries to budget her spending on food, usually within $150 weekly.

Beaver Brook Farmers Market provides access to locally grown fresh veggies and fruits to residents at a discount and accepts SNAP, state Healthy Incentives Program (HIP) benefits and Women, Infants and Children (WIC) farmers market coupons.

The market at 316 Chandler St. is open from 9 a.m. to noon Mondays and Fridays.

"I feel like it is really helpful," Calderon said, adding that the program can also help budget her food spending. "It's nice to be able to help other farmers and them to help us, too. It's a good program."

State Reps. David LeBoeuf, Worcester City Councilor Sarai Rivera, the Regional Environmental Council and Project Bread joined the Beaver Brook Farmers Market Friday morning to raise awareness of food justice.

In early 2021, REC received a $25,000 grant from funds raised through the charitable organization Project Bread's Walk for Hunger to support 60 community gardens and 500 gardeners across Worcester, which provide foods for the local standing and mobile markets as well as other anti-hunger organizations, including Project Bread's local community partner, El Buen Samaritano.

Last week, Project Bread submitted a written testimony to the state Joint Committee on

Children, Families and Persons with Disabilities in support of an act relative to HIP.

If passed, House Bill 250 of the Massachusetts Healthy Incentives Fund would become a permanent program that would generate new jobs in the local farm industry. Farmers would be able to better protect their land and allow more than 160,000 individuals to increase their fruit and vegetable intake by one serving per day, according to a release from Project Bread.

"Since the program started, it's been subjected to the annual state budget," Erin McAleer, chief executive officer of Project Bread, said. "We wanted to make this program permanent. It's a win-win and helps low-income families afford and access healthy local foods, and it's great for our farmers across Massachusetts."

Over the last six months, according to a release from Project Bread, 18.1% of households with children in Massachusetts were worried about running out of food before having money to buy more, with Black and brown households disproportionately burdened by food insecurity.

Before the pandemic, 9% of households in the state are food-insecure, but it spiked to 18% immediately after COVID-19 hit, McAleer said. Even though the percentage of food-insecure households has lowered to around 14%, it's still higher than pre-pandemic levels and, if counting the households with children, the rate would be higher as 18%.

"The reasons why we've seen a decline in food insecurity is because of federal government programs, like expanding school meals for all and the SNAP program," McAleer said, adding that HIP benefits also serve as another opportunity for residents to easily obtain healthy fruits and vegetables.

Launched in April 2017, HIP provides monthly incentives to SNAP households to shop at local farmers markets, like Beaver Brook Farmers Market, farm stands, Community Supported Agriculture and mobile markets.

Families in SNAP have monthly incentives of up to $40 for households of one or two people, $60 for households of three to five and $80 for households of six or more.

As of July, nearly 100,000 households have participated in the program.

"We see the tremendous benefit of HIP in action daily at REC Markets," Grace Sliwoski, director of programs at REC, said in a statement. "Our customers share how HIP allows them to stretch their food budgets, increase their families' fruit and vegetable intake, and allows them to support our local farming community. HIP has also helped build a strong and reliable customer base for local farms in Central Massachusetts, strengthening our local food economy and food system."

1. Which two of the following does the Beaver Brook Farmers Market do to increase purchases of their produce?
   a. selling produce at discounted prices
   b. free delivery of produce ordered either by phone or on-line to households
   c. providing information on the nutritional value of the produce compared to fast food
   d. free counseling on food choices
   e. accepting payments and coupons from government benefits programs

2. How did local food security trends change during the pandemic?
   a. Normally 18% of households are food insecure. Government relief in the pandemic brought this down to 9% at the time of the article (October 2021).
   b. Normally 9% of households in the state are food insecure. This rose to 18% at the start of the pandemic, and then dropped to 14% at the writing of the article (October 2021).
   c. At the time the article was written (October 2021), food insecurity was a problem for 9% of all households, which is much better than the pre-pandemic level of 14%.
   d. At the time the article was written (October 2021), food insecurity had stopped being a problem in the area because of the passing of House Bill 250.

3. Which two of the following are government incentives that the article says were launched in 2017?
   a. SNAP provides coupons to families to help buy additional produce.
   b. HIP provides $40 incentives to families of one or two people at farmers markets.
   c. HIP provides direct funding to community gardens to help them grow more produce to sell at the markets.
   d. HIP provides monthly incentives to families with SNAP benefits to shop at farmers' markets.
   e. SNAP removed all restrictions on what types of food people could buy.

4. Briefly explain what the article says the Regional Environmental Council (REC) does to support local agriculture.

5. Describe steps being taken at the state level to improve food security, as noted in the article. How do these state actions supplement federal benefits like SNAP and free school meals for all?

6. Incentives to purchase items are put in place to impact the demand for something. Incentives to grow or produce goods will affect their supply. Based on this article and other information in the book or outside sources that you find, do you think governments should choose demand incentives, supply incentives, a combination of both, or possibly neither to best alleviate food insecurity in an area? Briefly explain your choice.

Use the Data

One of the most fundamental factors shaping the impact of our food systems is the amount of food that is never consumed. Food that is lost or wasted still contributes to the kinds of environmental impacts associated with agriculture and livestock production detailed in this chapter but without the benefit of providing nourishment for people. Food loss or waste can occur at several stages along the food supply chain.

Study the figure, and use it to answer the questions that follow. The first three questions test your understanding of the figure. Questions 4 and 5 are short answer, requiring you to apply what you have learned in this chapter and cite data in the figure. Answers to Questions 1–3 are provided at the back of the book. Question 6 asks you to make your case and defend it.

1. What are the units in the figure?

    a. kilograms of food lost or wasted per capita

    b. kilograms of food lost or wasted per person per year

    c. percentage of calories of food that is lost or wasted

    d. calories of food lost or wasted

2. Which region has the largest share of its total available food lost or wasted?

3. In which two regions does food wasted in consumption exceed food lost during all the other stages combined?

4. What stage in the supply chain shows the least variation across regions in terms of food lost or wasted? Cite data from the figure to support your answer. On the basis of what you have learned in this chapter, why do you think there is so little variation in this stage across regions?

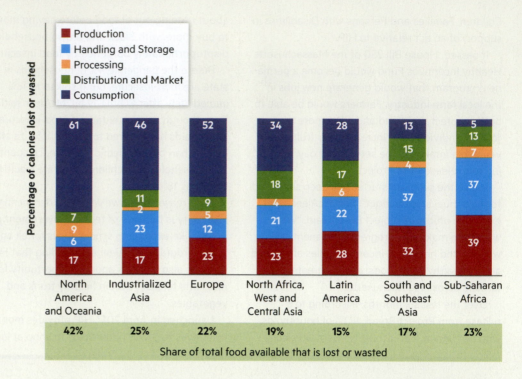

5. What stage in the supply chain shows the greatest variation across regions in terms of food lost or wasted? Cite data from the figure to support your answer. On the basis of what you have learned in this chapter, why do you think there is so much variation in this stage across regions?

6. **Make Your Case** If you were working for the Food and Agriculture Organization of the United Nations and your job was to develop programs to reduce food loss and waste in two regions, North America/Oceania and South/Southeast Asia, what stage(s) of the process would you target first and why? Propose at least one strategy to reduce food loss and waste for each region, focusing on the stage(s) you chose. Use the data in the figure and information in the chapter or other sources to make your case.

LEARN MORE

- Fair Trade Certified website: www.fairtradecertified.org/
- Food and Agriculture Organization of the United Nations (news, videos, reports, and progress on sustainable development goals): www.fao.org/home/en/
- David R. Montgomery and Anne Biklé. 2016. *The Hidden Half of Nature: The Microbial Roots of Life and Health*. New York: Norton.
- David R. Montgomery. 2017. *Growing a Revolution: Bringing Our Soil Back to Life*. New York: Norton.

- David R. Montgomery and Anne Biklé. 2022. *What Your Food Ate: How to Heal Our Land and Reclaim Our Health*. New York: Norton.
- NOAA Office of Sustainable Fisheries: www.fisheries.noaa.gov /about/office-sustainable-fisheries
- Real Food Challenge (campus activism group): www .realfoodchallenge.org/
- USDA Certified Organic Agriculture resources: www.usda.gov /topics/organic

13

Fossil Fuels
Energy of the Industrial Age

Fossil fuels, such as the oil making its way through this pipeline from the Prudhoe Bay Oil Field in Alaska, remain the dominant source of global energy consumption despite the many environmental impacts linked to their extraction, production, and combustion.

A re you energized? If you are reading this—even if you feel a little "low energy"—you are using energy in many ways. Of course, your body is using energy generated from what you consume to carry out even your most basic functions, such as your breathing and movement.

But chances are you're also drawing on energy to power your lights, phone, and the vehicle that brought you to where you are now. If you've ever experienced a power outage or been in a vehicle that has run out of gas, you can relate to a sudden awareness of just how dependent you are on energy. We rely on energy not only for our survival but also to power our electronics, build our cities and towns, and of course move us around. It is this demand that sustains the global energy industry. This industry is rapidly changing, but it remains reliant on **fossil fuels**—energy sources formed from living organisms from earlier geologic eras.

Today, nearly 80% of all energy used to power our modern lifestyle comes from fossil fuels, driving us to great lengths to obtain resources that are no longer as readily available. For example, one project drills for oil 9,500 feet (2,900 meters) below the Gulf of Mexico. Another company proposes building an artificial island 18 football fields in size in the Beaufort Sea off the coast of Alaska. Coal mining is similarly disruptive, often achieved by digging open pits, a technique known as mountaintop removal. The extraction of coal, oil, and gas pollutes the land, air, and water. Greenhouse gas emissions from fossil fuel production and use are the leading cause of climate change. Every stage of fossil fuel production—from extraction at wells and mines; to distribution by pipelines, trucks, and trains; to production at refineries and combustion in power plants, factories, and vehicles—is linked to significant environmental health impacts for those who live and work nearby. These impacts are not borne equally across society and raise serious environmental justice concerns.

While each of us is tied to some extent to a fossil fuel economy right now, this was not always the case. Fossil fuels were once emerging technologies with start-up costs like all new technologies. Their success was driven not only by innovation but also by government support of the tremendous investments necessary to

Chapter Objectives

By the end of this chapter, you should be able to . . .

A. discuss how and why humans have developed energy technologies throughout history.

B. understand how economic forces and government actions influenced the rise of fossil fuel use.

C. describe the ways in which fossil fuels are extracted and produced and the associated environmental impacts.

D. identify challenges with transitions between energy sources.

> **Our civilization runs by burning the remains of humble creatures who inhabited the Earth hundreds of millions of years before the first humans came on the scene.**
> —Carl Sagan

get started. In this chapter, we explore fossil fuels: what they are, how we use them, their environmental impacts, and how we came to rely on them in the first place. We will also consider some of the choices we now have about using them. With a world that is built around their existence and use, what would it mean to change this system? While one company wants to make an island to drill in the Arctic, another (Royal Dutch Shell) suspended its Arctic drilling in 2015. Two major oil pipeline extensions—the Keystone XL and the Atlantic Coast Pipeline—were recently canceled by the companies proposing them in the face of mounting public opposition, rising costs, and increasing concern that fossil fuel infrastructure may not be a good long-term investment. Is this a sign of change? Possibly, but first, let's set the stage with some basic background on the history of human energy use.

fossil fuel fuel formed from living organisms from earlier geologic eras.

hydrocarbon material made of strands of hydrogen and carbon molecules derived from ancient, solar energy–capturing photosynthesis.

13.1 What Are Fossil Fuels, and How Important Are They Today?

Today we rely heavily on fossil fuel energy, especially to power our transportation and our electric grid and to provide heating and cooking fuel. But where does this energy come from? First, it is important to remember that the original source of all energy in fossil fuels is the Sun. Oil and natural gas are known as **hydrocarbons** because they are made of strands of hydrogen and carbon molecules derived from ancient, solar energy–capturing photosynthesis. These fuels began to form more than 100 million years ago as microscopic organisms—most often the plankton commonly found in oceans and lakes (and not dinosaurs as sometimes thought)—drew on the Sun's energy. As these organisms died, some became buried beneath layers of sediment, failed to decompose, and were then subjected to both extreme heat and pressure deep underground (**FIGURE 13.1A**). Over millions of years, the energy stored

(a) How Oil Formed

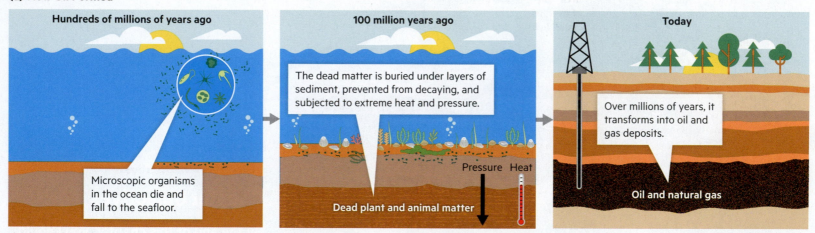

(b) How Coal Formed

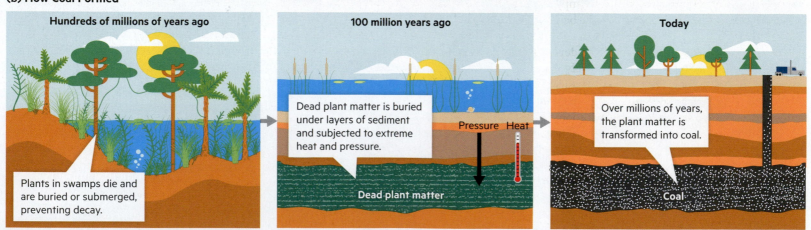

FIGURE 13.1 **The Formation of Fossil Fuels** (a) How Oil Formed (b) How Coal Formed

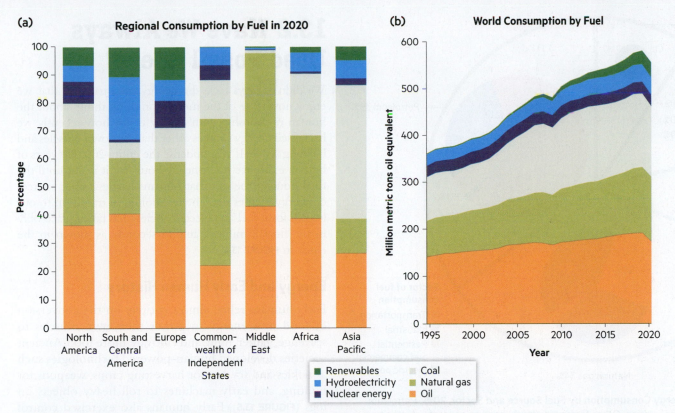

(a) Regional Consumption by Fuel in 2020

(b) World Consumption by Fuel

Legend:
- Renewables
- Hydroelectricity
- Nuclear energy
- Coal
- Natural gas
- Oil

FIGURE 13.2 Energy Consumption by Fuel Source **(a)** Both globally and in the United States, the fossil fuels petroleum (or oil), natural gas, and coal are the leading sources of our energy consumption. **(b)** Overall, global energy consumption has risen about 2% per year on average, although it actually decreased during the recession in 2009 and again in 2020 during the COVID-19 pandemic.

Adapted from BP (2021).

in the dead bodies of these organisms transformed into the liquid we know as **oil** (or crude oil) and a gaseous, methane-heavy fuel known as **natural gas**. Similarly, **coal** formed from ancient tropical swamps that stored energy from the Sun in their plant matter. If the plant matter in these swamps dies without rotting away, becomes buried, and is subjected to immense heat and pressure, it turns into this solid hydrocarbon over millions of years. Coal mostly consists of carbon molecules, and because it was once part of an entire swampy landscape, it is often found in large masses called beds or seams (**FIGURE 13.1B**).

We think of these masses of coal or deposits of oil and natural gas that can be economically accessed as **reserves**. Easily obtained oil or gas deposits are called **conventional reserves**, and deposits that are more difficult to extract are called **unconventional reserves**. But because we extract fossil fuels at a much faster rate than they form, any reserves are effectively finite (nonrenewable) resources. **FIGURE 13.2** shows the share of primary energy sources consumed in the United States and the world. Both globally and in the United States, fossil fuels account for nearly 80% of the energy we use, while all other non–fossil fuel resources account for just over 20% of the energy we use (although, as

we will see in Chapter 14, wind and solar energy production are expanding at a rapid rate). Total world energy consumption has steadily increased, averaging a gain of more than 2% each year for the past decade (see Figure 13.2).

In the United States, you are part of an energy flow that is consuming more than 300 tons of fossil fuel every second. If you take a look around you right now, you can probably identify dozens of devices performing a wide range of functions, each drawing on energy. Overall uses for primary energy sources in the United States are displayed in **FIGURE 13.3** and include transportation, industrial processes, heating and cooking in residential and commercial properties, and an ever-growing variety of electrical demands ranging from air-conditioning to power for appliances, computers, and phones.

⬡ **TAKE-HOME MESSAGE** The majority of global energy needs (80%) is provided by fossil fuels: oil, coal, and natural gas. Fossil fuels take millions of years to form from the remnants of organisms. Fossil fuels are a nonrenewable resource because we consume them far more quickly than they can form.

oil a liquid fossil fuel that formed from plants and microscopic animals that lived millions of years ago.

natural gas a gaseous fossil fuel that formed deep beneath Earth's surface and contains primarily methane.

coal a solid carbon-based fossil fuel formed from ancient tropical swamps that were buried and subjected to immense pressure over millions of years.

reserve a known resource of a fossil fuel that can be economically accessed with current technology at current prices.

conventional reserve an easily obtained deposit of fossil fuels.

unconventional reserve a difficult-to-extract deposit of fossil fuels.

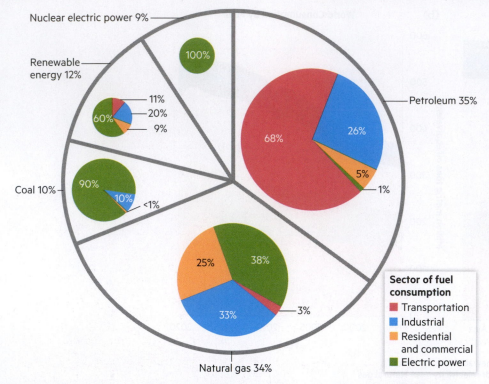

FIGURE 13.3 **US Primary Energy Consumption by Fuel Source and Sector, 2015** Petroleum (oil), is primarily used for transportation, Other forms are mostly used to generate electricity.

Adapted from US Energy Information Administration (2021).

Legend:

Sector of fuel consumption
- Transportation
- Industrial
- Residential and commercial
- Electric power

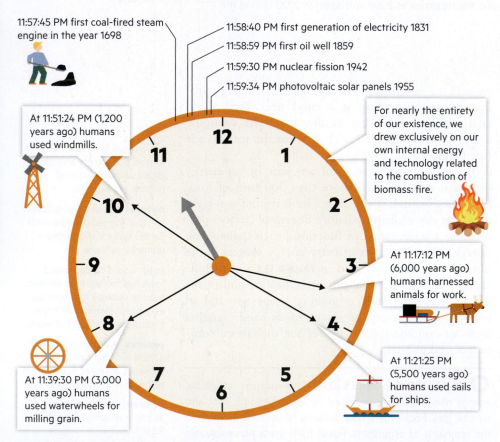

FIGURE 13.4 **Human Energy History** If 200,000 years of human history were compressed into a single 24-hour day, each year would equate to less than half a second. This clock captures key events as times in a day, with all the technological advances occurring between 11:00 PM and 12:00 AM.

Labels on clock figure:
- 11:57:45 PM first coal-fired steam engine in the year 1698
- 11:58:40 PM first generation of electricity 1831
- 11:58:59 PM first oil well 1859
- 11:59:30 PM nuclear fission 1942
- 11:59:34 PM photovoltaic solar panels 1955
- At 11:51:24 PM (1,200 years ago) humans used windmills.
- For nearly the entirety of our existence, we drew exclusively on our own internal energy and technology related to the combustion of biomass: fire.
- At 11:17:12 PM (6,000 years ago) humans harnessed animals for work.
- At 11:21:25 PM (5,500 years ago) humans used sails for ships.
- At 11:39:30 PM (3,000 years ago) humans used waterwheels for milling grain.

13.2 Have We Always Used Fossil Fuels?

We didn't always have the energy infrastructure that we enjoy now. One of the most astonishing things about the human relationship with energy is just how recently we harnessed energy sources other than our own muscles and burning wood. If we condense the entire 200,000 years of human existence into one 24-hour day, it is not until the final hour of the day that humans harness other sources of energy besides fire. The technological transition to our largely fossil fuel and electric society has taken place only in the final 2 minutes and 15 seconds, as shown in the clock in **FIGURE 13.4**.

Energy and Early Human History

Early humans relied almost entirely on the conversion of food into energy to power their own muscles to scavenge, gather, hunt, and cultivate more food. Ancient societies developed human-powered technologies such as sickles and scythes for harvesting crops, weapons for hunting, and early machines to roll heavy objects on logs (**FIGURE 13.5**). Early humans also exercised control over fire for warmth, light, and cooking by burning **biofuels**, or recently living matter or by-products of its decomposition as an energy source. Humans also cleared land for hunting or agriculture by burning forests. Plant-based biofuels were the most common fuels burned: woody materials from tree branches, roots, leaves, shrubs, and dry grasses. Animal dung and fat were also fuel sources, and various forms of plant-based fuels are still the primary energy sources in many places today. Ancient agricultural and urban cultures developed more sophisticated forms of these biofuels, including *charcoal*, a biofuel refined by cooking piles of wood in a low-oxygen environment (such as a sod- or clay-covered mound) to remove water and other compounds. Relative to other biomass, charcoal remains a cleaner-burning fuel, and it has more energy by weight than does wood.

It was not until about 5,000 years ago that humans began using draft animals like horses and cattle to pull carts, cultivate fields, move heavy objects, and power grain mills. The first mechanical energy application not provided by human or animal muscle was probably the wind-powered sails used on ships by ancient Sumerians and Egyptians at least 5,500 years ago. About 3,000 years ago, the kinetic energy of flowing water was harnessed by use of "undershot" *waterwheels* to mill grains. Three hundred years later, "overshot" waterwheels were devised to capture water falling from above, either from a waterfall or from a water supply controlled by a dam and a descending diversion channel or flume (**FIGURE 13.6**). The first windmills were developed in the Middle East about 1,200 years ago.

(a)

(b)

FIGURE 13.5 Muscle Power Ancient cultures relied on the power of their own muscles but often gained mechanical advantages by using tools such as **(a)** this sickle to harvest crops (wall painting from Thebes, Egypt) or **(b)** by elaborate log-rolling operations to move heavy objects (engraving of ancient Assyrian wall frieze).

Power and Productivity

These new energy technologies greatly expanded human access to **power**, or the rate at which work is done. Power can be measured in *joules per second*, a measure of energy flow over time, which is simplified into *watts* (W), named for the Scottish engineer James Watt, inventor of the steam engine. The power of human manual labor is roughly 100 watts per person. In comparison, cattle power is 3 times greater at about 300 watts per head, ox power is about 600 watts per head, and horse power is 700 watts per head. Machine technologies work even harder, with power measured in kilowatts (kW), or thousands of watts (1 kilowatt is equivalent to the power of 10 people). Ancient water-wheels could produce 2 kilowatts of power, or the work of 20 people, and operate continually without requiring food or rest. By the 1800s, the most advanced windmills could produce 40 kilowatts of power, while the best waterwheels could produce more than 200 kilowatts of power.

No doubt early operators of these devices would have recognized the effect of power on relative **productivity**, defined as the amount of inputs (such as labor time) required to attain a certain level of output (such as milled grain). A human can mill just over 6 pounds (2.7 kilograms) of grain an hour, a working animal can more than triple that output, and a waterwheel can mill 250 pounds (113 kilograms) of grain—a little more than 40 times as much as a human—per hour. More advanced Roman milling systems with multiple waterwheels produced as much as 28 tons of flour per day—almost 400 times the output of a single human working around the clock (**FIGURE 13.7**).

Even these early energy technologies were associated with environmental impacts. Extensive use of biofuels such as wood and peat caused air pollution and

biofuel recently living matter or by-products of its decomposition used as an energy source.

power the rate at which work is done, measured as energy flow over time.

productivity the amount of inputs required to attain a certain level of output.

FIGURE 13.6 Water and Wind Power With the development of waterwheels and windmills like these in Syria and the Greek islands, respectively, the power of water and wind were put to use for milling grain.

FIGURE 13.7 Gains in Productivity Employing new energy sources and mechanical devices led to significant gains in productivity. Replacing milling of grain by hand with animal power and later waterwheel power (either single mills or complexes) meant that flour could be produced at a much faster rate.

resulted in deforestation, erosion, and fuel shortages throughout history. Both ancient Greece and Rome were eventually forced to import wood from outlying areas hundreds of miles away. Draft animals, such as horses, require as much as 10 pounds (4.5 kilograms) of feed per day, and harnessing this resource on a large scale meant converting large areas of land from forest to agriculture. Waterwheels often used dams and other diversions that obstructed stream flow and fish migration. So how did fossil fuels become so prevalent? This will be the subject of the next section.

⬡ **TAKE-HOME MESSAGE** Humans have the ability to accomplish work with more than just our own muscles. For most of our species' existence, we relied on the controlled combustion of biomass (such as firewood) for warmth, light, cooking, and clearing land. Over the past 7,000 years, we have developed more powerful energy technologies that expand the kinds of work we can do and the rate at which we can do them. All energy technologies have environmental effects.

13.3 What Drove the Rise of Fossil Fuels?

We often overlook the fact that coal, oil, and gas were once new, "alternative" energy sources, replacing firewood for heat and horsepower for transportation. Substituting fossil fuel energy resources for existing ones took a long time, occurred inconsistently across different places, and still remains incomplete overall. After coal attained widespread adoption as a heating fuel in England in the early 1600s, it took more than 300 years for it to surpass wood as the dominant energy source. Oil, which was first commercialized in the 1850s, took more than 100 years to usurp coal's position as the leading energy source when it became the most common fuel for transportation. The use of natural gas to generate electricity began in the late 1930s, but it wasn't until 2015 that natural gas surpassed coal as the top energy source for electricity in the United States (**FIGURE 13.8**). So how did fossil fuels gain their

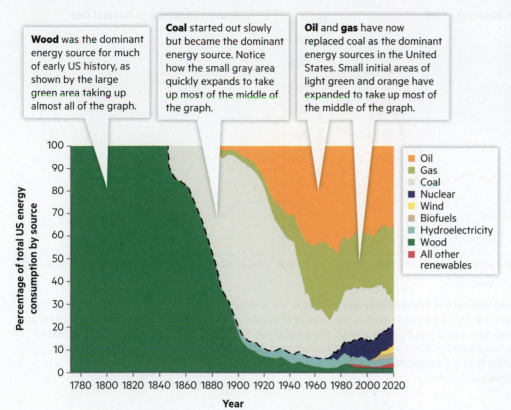

Wood was the dominant energy source for much of early US history, as shown by the large green area taking up almost all of the graph.

Coal started out slowly but became the dominant energy source. Notice how the small gray area quickly expands to take up most of the middle of the graph.

Oil and **gas** have now replaced coal as the dominant energy sources in the United States. Small initial areas of light green and orange have expanded to take up most of the middle of the graph.

FIGURE 13.8 Share of Energy Consumption by Year Energy transitions have taken significant lengths of time to occur. It took coal more than 100 years to replace wood, and oil and gas more than 50 years to replace coal as the largest share of energy consumption. Note that total energy consumption has risen steeply over this time, so the gross amount of energy provided by wood in the early 1800s was far less than that provided by other sources in more recent years. Fossil fuels are grouped above the dashed line, while nonfossil fuels are below. In the past few decades, the share of nonfossil fuels has grown dramatically, although fossil fuels still dominate.

Adapted from US Energy Information Administration (2021).

dominance? Let's take a closer look at some of the economic forces and government actions that played an important role in the rise of fossil fuels.

Economic Forces and the Rise of Fossil Fuels

The rise of fossil fuels shows how supply and demand influence the adoption of energy sources and technologies. First, scarcity inspires innovation and substitution; when demand for a good (including an energy source) rises relative to supply, the cost of that good increases. In turn, this can inspire people to develop technologies that allow more efficient use of that energy source and search for substitutes to replace it. Let's see how this worked for the adoption of coal as a heating fuel in England in the 1500s. At this time, firewood around major cities in England was becoming scarce, and the transportation costs associated with harvests farther and farther from consumers caused the price of firewood to quadruple between 1540 and 1580. This scarcity and the resulting price increase created a market opportunity for coal as a substitute. But this was not the whole story. Even though coal had a higher heat output and was cheaper than firewood, widespread adoption was slow because it was still considered in economics terms an **inferior good**: something consumed because people cannot afford what they prefer. Coal was used primarily by those who could not afford wood: most domestic fires burned in open hearths, and living with the smoke,

foul smell, and sooty residue from burning coal made it unappealing compared to wood. However, wood's scarcity inspired the development and widespread adoption of chimneys—a technological advance at the time—that enabled households to burn coal more efficiently without smoke polluting the indoors. By the 1600s, coal was widely used for home heating in England, and even the king used it to warm his palace, signaling that this was no longer an inferior good. So the substitution of coal for wood as a dominant heating fuel in England was related not only to the scarcity and rising price of wood but also to the development of a technology (widespread adoption of chimneys in this case) that made coal more appealing to consumers.

In addition, the availability of an energy resource depends on the technical factors needed to extract it. By the mid-1700s, coal prices in England were on the rise. There was plenty of coal underground, but existing mines tended to flood, making extraction expensive or even impossible. To address this problem, James Watt invented a steam engine that could power water pumps and conveyors to bring coal out of deeper depths more efficiently than could existing technologies. Thus, Watt's technology made more coal available for use. Technological advances can make resources more available and expand what people consider to be the **proven reserves** of that resource, or the amount of a resource that can be profitably accessed with current technology at current prices. Of course, the total amount of a nonrenewable resource, such as a fossil

inferior good something consumed because people cannot afford what they prefer.

proven reserve the amount of a resource that can be profitably accessed with current technology at current prices.

(a)
Proven Reserves of Oil

Billion barrels (y-axis): 0, 15, 20, 25, 30, 35, 40, 45, 50
Year (x-axis): 1979, 1984, 1989, 1994, 1999, 2004, 2009, 2014, 2019

(b)
Proven Reserves of Natural Gas

Trillion cubic feet (y-axis): 150, 200, 250, 300, 350, 400, 450, 500, 550
Year (x-axis): 1979, 1984, 1989, 1994, 1999, 2004, 2009, 2014, 2019

FIGURE 13.9 **US Crude Oil and Natural Gas Proven Reserves, 1977–2019** As technological advances help us locate and extract resources from places that were previously difficult to access, the amount of a resource we can access with current technology at current prices can increase over time. These graphs show that US proven reserves for **(a)** oil and **(b)** natural gas have increased about 100% and 150%, respectively, between 2000 and 2019. This has occurred due to advances in fracking and horizontal drilling—technologies we explore later in this chapter.

Adapted from US Energy Information Administration (2021c).

Jevons paradox a finding that efficiency gains in the use of a resource can lower the cost of that resource, which can cause consumption of the resource to rise (Chapter 2).

fuel, remains fixed and is depleted with use over time (**FIGURE 13.9**).

Watt's steam engine illustrates two additional lessons about how energy technologies develop. First, there often are up-front costs and risks associated with new energy technologies that can initially limit their adoption rate, even if those technologies are more efficient and will save money (**FIGURE 13.10**). The new steam engines would cost mine and factory owners a lot of money up front, even though these machines were more efficient and provided cost savings over the long run.

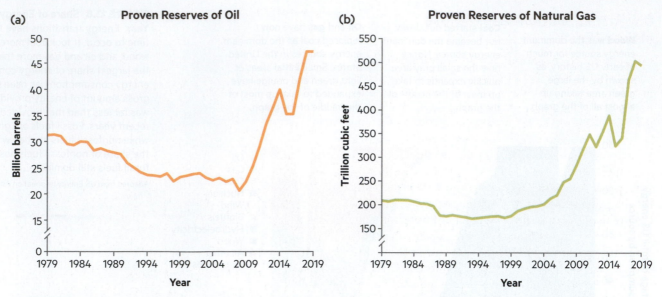

FIGURE 13.10 **Watt's Steam Engine** New technology can be expensive and can require support to become adopted. To promote adoption of his new steam engine, James Watt leased his technology, thus assuming many of the up-front costs and risks. Similar leasing strategies are used in many states now for rooftop solar energy installations.

Early adopters of the steam engine would also face the risks associated with new and unproven technologies. Would it provide the promised efficiency gains? What if it broke? Who would fix it? To address these concerns early on, Watt offered to lease his steam engines. He supplied the steam engines to mine operators and factory owners but maintained ownership of the engines himself, assuming the risk of engine malfunction and sharing the cost of repair or replacement. Moreover, instead of making money on the sale of the device, Watt was paid one-third of the estimated energy savings the client gained from the device.

Second, although Watt's steam engine enabled mines to extract more coal with less energy, these efficiency gains did not translate into energy conservation because of the **Jevons paradox** (introduced in Chapter 2). When efficiency gains raise the supply of an energy resource, this lowers the cost of consuming energy, leading to new and expanded applications for the energy and greater consumption overall. In this case, the efficiency of steam engines increased the coal supply and enabled this resource to become much more than a heating fuel. Over the course of the 1800s, coal became the dominant fuel for transportation on ships and trains, and it powered industrial processes such as those in textile mills and iron foundries. By the end of the century, coal became the dominant fuel for the production of electricity. In this manner, technologies that improve energy efficiency and lower the cost of consuming energy can actually facilitate greater energy consumption overall.

Government Actions and the Rise of Fossil Fuels

In addition to economic forces, governments have played (and still play) a powerful role in favoring fossil fuels and spurring their adoption. In the case described in this section, as England was facing wood shortages in the late 1500s, many local governments adopted regulations that banned bakers, brewers, and iron foundries from using wood for fuel. This essentially forced these large fuel consumers to rely on coal instead. At the same time, many local governments subsidized coal purchases for the poorest people in the community and invested heavily in improving the navigability of waterways used to bring coal from mines into the urban areas. These government actions provided a guaranteed market and reduced transportation costs for coal producers, both of which helped develop and expand the industry. Later, as James Watt was developing the steam engine, he benefited from government support of his workshop at the University of Glasgow and patents protecting his new technology. This support helped Watt succeed in a way that ultimately expanded the market for coal.

Governments can also intervene in the way that they tax and/or regulate competing energy sources and technologies. In the mid-1800s in the United States, whale oil fuel (not a fossil fuel) was still the primary source of household lighting. But whale populations were in decline, and whalers had to sail farther from shore to attain their catch, ultimately increasing the cost of the fuel. Then, the first oil well was drilled in Titusville, Pennsylvania, in 1859 (**FIGURE 13.11**), and kerosene derived from oil pumped from wells became a promising alternative fuel. But the leading alternative fuel to whale oil at the time was actually the much cheaper camphene, a mixture of plant-based chemicals including alcohol, turpentine, and camphor oil that sold for about 50 cents a gallon. This started to change in the 1860s as the United States adopted a tax on camphene of $2.00 a gallon, while kerosene from oil wells was taxed at just 10 cents a gallon on top of its price of 60 cents a gallon. As a result of this government policy, kerosene became much cheaper than its closest competing fuel, and it quickly became the dominant replacement for whale oil. Tax breaks help too. For example, for many years the US Congress passed tax deductions for major start-up expenses such as labor, insurance, chemicals, and equipment for new wells to encourage investment in oil and natural gas.

Finally, governments often provide **public goods** (see also Chapter 2), or things that cannot be profitably produced because it's difficult to exclude nonpaying customers from receiving the benefits. In the case of fossil fuels, governments often step in to provide key pieces of infrastructure, such as constructing roads and highways that facilitate automobile transportation. State governments established regulated monopolies with utility companies

FIGURE 13.11 **Striking Oil in Titusville, Pennsylvania** The first oil well was drilled in Titusville, Pennsylvania, in 1859. Kerosene from oil soon became the dominant fuel for lighting thanks in part to US tax policies on various alternative fuels.

in order to build transmission lines and natural gas pipelines (**FIGURE 13.12**). Funding for energy research (and the subsequent knowledge gained) is another public good the government provides. Between 1978 and 2004, the

public good a thing that cannot be profitably produced because it's difficult to exclude nonpaying customers from receiving the benefits (Chapter 2).

FIGURE 13.12 **Interstate Highways and Other Public Goods** Governments often help provide things that benefit the public but can't be profitably produced by private entrepreneurs. The Federal-Aid Highway Act of 1956, signed into law by President Dwight Eisenhower, authorized the construction of 41,000 miles (66,000 kilometers) of interstate highway. State governments granted monopolies to utility companies as incentives to make huge investments in power lines.

Is "Clean Coal" Clean?

There are ways to reduce emissions from coal power plants, often referred to as "clean coal" technology. But are these methods truly clean? A coal supply with a relatively low sulfur content can be used, like that mined in the Powder River Basin in Montana and Wyoming. This can reduce the amount of sulfur dioxide (SO_2) pollution emitted compared to that from power plants burning coal with higher sulfur content, but it does not reduce carbon dioxide (CO_2) emissions and other pollutants. Another option is designing power plants that burn coal at much higher "ultra-supercritical" temperatures (like the one pictured here in Shanghai, China), which makes the plants more efficient, producing more energy relative to the emissions they release. These plants are expensive, and there are only a handful of them currently operating in the world.

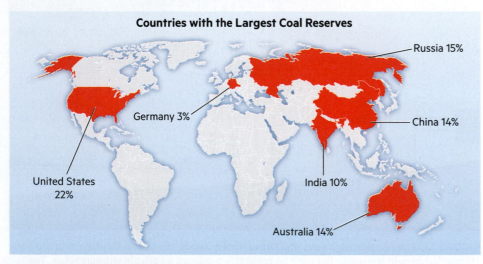

Countries with the Largest Coal Reserves

Russia 15%

Germany 3%

China 14%

United States 22%

India 10%

Australia 14%

FIGURE 13.13 **The Global Share of Coal Reserves** Together these six countries contain 80% of the global recoverable coal reserves.

Adapted from US Energy Information Administration (2021).

federal government invested more than $25 billion in fossil fuel energy research. Research in hydraulic fracturing (discussed in the next section)—a technology first used commercially in the 1980s by Texas oilman George Mitchell—was funded by the US Department of Energy in the 1970s.

⬡ **TAKE-HOME MESSAGE** Fossil fuels such as coal, oil, and natural gas became our dominant sources of energy over a long period of time. Widespread adoption of fossil fuels was aided both by economic factors such as the increasing scarcity (and cost) of the energy sources they replaced and by government actions. Government subsidies, tax policies, and the provision of public goods encouraged their development and use.

13.4 What Are the Environmental Impacts of Obtaining and Using Fossil Fuels?

Fossil fuels must be mined and burned to provide energy. In this section, we will look at ways we use all three fossil fuels and consider some of their impacts. By and large, fuel extraction, processing, and combustion result in pollution of the land, water, and air, as well as the release of large quantities of greenhouse gases.

Coal

Coal, a sedimentary rock, was the first fossil fuel widely used by humans, and it is still the fossil fuel with the most abundant reserves. The United States along with Russia, China, and Australia account for more than two-thirds of global reserves (**FIGURE 13.13**). The United States hosts the largest recoverable reserves of coal in the world (22%) and is a net exporter. If current production levels continue, the United States has enough recoverable coal to last more than 200 years. While we learned that coal started as a replacement for firewood in heating and cooking, nearly all coal energy now is used to produce electricity. In the United States, coal is the second largest source of electricity (23% in 2020). And while the United States is the third leading consumer of coal, accounting for about 6.5% of global coal consumption, China accounts for nearly half of the world's total coal consumption. After decades of expansion, global coal consumption began to fall in 2015 as natural gas and alternative energy production fueled a greater portion of global electricity generation.

(a)

(b)

FIGURE 13.14 Mining Methods Coal mining involves either **(a)** extraction underground by boring machines like this one or **(b)** mountaintop removal that blasts away the soil and rock at the surface to access the coal below (as shown here in Wise County, Virginia).

Coal is either extracted from underground mines by machines that work in tunnels or, more commonly, from surface mines that use blasting and machinery to completely remove the soil and rock above the coal. In some regions, entire mountaintops are removed to scoop out the underlying coal—a process fittingly called **mountaintop removal**—and the waste material is deposited in adjacent valleys (**FIGURE 13.14**). Coal mining has many associated hazards and risks. Underground coal mining is one of the world's most dangerous professions because of the accidents and lung diseases that lead to high death rates among miners. Mountaintop removal alters the landscape and smothers streams with rocks and dirt, destroying aquatic habitats. Water that drains from both types of mines is often acidic and contaminated with heavy metals. Mines can cause land subsidence as tunnels collapse, resulting in craters on the surface (**FIGURE 13.15**). Coal mines contribute 7% of US methane emissions because this potent greenhouse gas must be vented from the mines to prevent explosions.

Once coal is extracted, it is crushed and washed at a *beneficiation plant* and transported to power plants by trucks, railroads, ships, or pipelines. Transportation of coal requires up-front energy costs in the form of vehicle fuels, but it is also associated with hidden impacts such as diesel train, truck, and ship emissions as well as the coal dust that escapes en route, which causes air and water contamination.

Most coal–fired power plants produce electricity by blowing coal dust into a boiler, where it heats water to produce steam. The steam turns the blades of a turbine, which powers an electrical generator. In this process, the burned coal produces ash and gaseous emissions.

Though most plants use pollution-control devices to capture and bury some of these by-products (see the Stories of Discovery feature later in the chapter), emissions from coal-fired power plants include nitrogen oxides (NO_x) and particulates that cause smog and respiratory illnesses, as well as sulfur dioxide (SO_2; see Figures 8.22 and 8.23 in Chapter 8) that causes acid rain damage to land and aquatic environments. And coal plants emit significantly larger quantities of the greenhouse gas carbon dioxide (CO_2), contributing 40% more CO_2 per unit of energy than oil and 50% more

mountaintop removal
a mining process commonly used to extract coal, where entire mountaintops are removed to scoop out the underlying resource.

FIGURE 13.15 Craters from Coal Mines Subsidence above coal tunnels can create craters on the surface like this one in southwestern Indiana.

than natural gas. Coal production also consumes more than twice as much water per unit of energy as does natural gas.

Oil

In our day-to-day lives, most of us are never far from a gas station. Oil, in the refined form we know as gasoline, seems always to be available for us to refill our tanks. It is now the most consumed fuel in the world, accounting for about one-third of global energy consumption and about the same share of US energy consumption (**FIGURE 13.16**). World oil reserves are sufficient to provide a 54-year supply at current levels of global production. The Middle East has more than half (55%) of the world's proven oil reserves, led by Saudi Arabia, Iran, and Iraq. North America has the next largest share of proven oil reserves (15%).

Oil deposits form within rocks, especially porous ones such as limestone or sandstone that are surrounded by nonporous rock. Oil is extracted by drilling and pumping, and when a well is successfully drilled, underground pressure sometimes forces oil to the surface in an intense gusher (**FIGURE 13.17**). Once this flow has slowed, the oil must be pumped by processes known as secondary and tertiary extraction. In *secondary*

FIGURE 13.16 Leading Oil Producers and Consumers
(a) The 15 countries with the largest proven reserves of oil are shown. **(b)** The United States is the world's leading consumer of oil, outstripping all countries in Europe combined. However, China's and India's use is increasing sharply.

Adapted from Central Intelligence Agency (2019) and BP (2022).

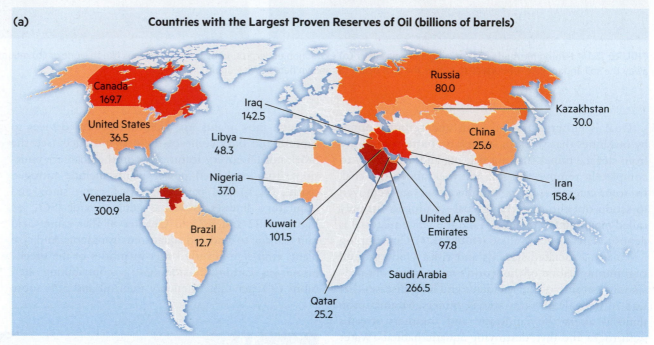

(a) **Countries with the Largest Proven Reserves of Oil (billions of barrels)**

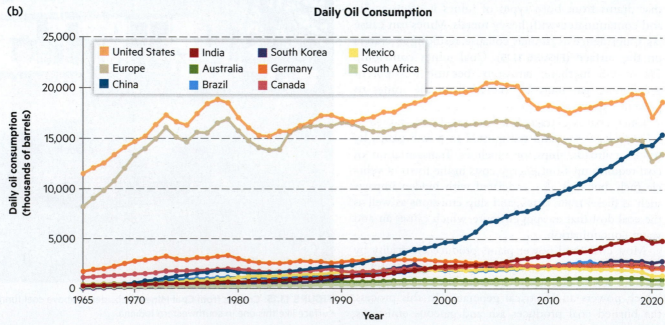

(b) **Daily Oil Consumption**

(a)

(b)

FIGURE 13.17 **Getting Oil to the Surface** Wells are drilled to access oil. **(a)** In some cases, underground pressure forces oil to the surface, causing a "gusher" as in this historical image of the Lakeview Gusher in Kern County, California, which spewed 18,000 barrels a day for 18 months in 1910 and 1911. **(b)** Oil can be pumped to the surface with a pumpjack and tank system. Water and steam can be injected into the well to increase the pressure too.

extraction, water is injected to increase pressure, forcing more oil to the surface, and in *tertiary extraction*, steam is injected to enable the oil to flow more easily from the rock into the well.

While drilling used to be only up and down (or vertical), it is now also possible to drill horizontally to access thin layers of oil and to prepare rock for **hydraulic fracturing**, or **fracking** (explained in more detail in the Sustainability Matters feature). Fracking uses a mixture of water, sand, and chemicals pumped at high pressure into the well to fracture the rock in order to release oil that is locked up inside. Fracking has allowed the extraction of oil from unconventional reserves much more easily. Dense rock layers such as the oil-bearing shale of the Bakken Formation underlying parts of North Dakota, Montana, and the Canadian provinces of Saskatchewan and Manitoba (**FIGURE 13.18**) and of the Green River Formation in Wyoming, Colorado, and Utah have made the United States and Canada the largest sources of these reserves. Fracking can also be used to extract natural gas, as we will see later in the chapter.

In the Athabasca area of Alberta, Canada, oil is found bonded with loose, fine-grained deposits known as **tar sands**. Steam and direct heat can be applied to separate this oil. The oil is then conveyed by pipelines or other means to refineries, which distill and collect the different components of crude oil. Extracting and refining oil from tar sands is a more energy- and water-intensive process than that used to produce conventional

FIGURE 13.18 **Oil Boom in the Bakken** In the early 2000s, application of hydraulic fracturing technology led to a boom in oil production from the Bakken Formation, a layer of oil-bearing shale. By 2014, North Dakota and Montana, the US states overlying the Bakken Formation, were producing 1 million barrels of oil per day and contributing 10% of US oil production.

(a)

(b)

FIGURE 13.19 Land Use for Oil Development (a) Oil wells used for extraction have a significant impact on the landscape, as these wells in Paso Robles, California, show. Each well requires a drilling pad to be cleared and a road to access it. Processing facilities are also built in close proximity to the wells. **(b)** Despite protests by Indigenous peoples and their supporters, the Dakota Access Pipeline was built less than a mile from the Standing Rock Reservation.

hydraulic fracturing (fracking) a process using a mixture of water, sand, and chemicals pumped at high pressure into an oil or gas well to fracture the rock and release the fuel locked up inside.

tar sand a type of loose-grained rock deposit bonded with oil.

oil. Tar sands oil production emits about 15% more CO_2 than does conventional oil production and consumes more than 3 times as much water.

Oil use, extraction, and transportation do not come without risk. Highly visible oil spills, such as those resulting from the *Exxon Valdez* grounding in 1989 and the *Deepwater Horizon* explosion in 2010, and cumulative damage from leaking storage tanks, pipelines, fueling stations, and vehicles illustrate the risks (see **AT A GLANCE: FOSSIL FUEL ENERGY DISASTERS**). In addition, oil extraction generally requires rigs, production facilities, and roads that impact local and regional environments, as well as pipelines that transport oil (or natural gas), which can be controversial too because of their sheer scale (**FIGURE 13.19A**). The controversial Keystone Pipeline runs about 3,000 miles (4,800 kilometers) from Alberta, Canada, through Illinois to Texas, and a proposed 1,200-mile (1,900-kilometer) extension was canceled in 2021 after years of protests and lawsuits challenging regulatory approvals. The Dakota Access Pipeline, connecting oil production facilities in North Dakota to Midwest refineries, was completed in 2017, despite intense opposition from the Standing Rock Sioux Tribe and other Indigenous peoples, as the pipeline crosses the Missouri River less than a mile from the Standing Rock Reservation (**FIGURE 13.19B**). Similarly, the Enbridge Line-3 replacement pipeline through Minnesota was completed in 2021 despite opposition from Indigenous peoples and environmental groups. These last two projects continue to face legal challenges. Overall, there are already 72,000 miles (116,000 kilometers) of crude oil pipelines in the United States.

Burning oil (or gasoline) produces many of the same polluting emissions as does burning coal, including CO_2, carbon monoxide (CO), sulfur dioxide (SO_2), nitrogen oxides (NO_x), and particulate matter. This means oil is also a source of greenhouse gas, acid rain, smog, and respiratory illness; in fact, it accounts for the largest share of energy-related CO_2 emissions in the United States (45%).

Natural Gas

Most of us have used a small Bunsen burner in a high school or middle school science class experiment, not knowing its invention was a key catalyst in the wider adoption of natural gas. Though natural gas was first used in China around 500 BCE—moved by bamboo pipes to boil seawater to separate the fresh water from its salt—its commercial use, as for other fossil fuels, is recent. In the 19th century, cities such as Baltimore started using natural gas for lighting, but the invention of the Bunsen burner showed its usefulness for heating. Over the past decade, natural gas has been the fastest-growing fossil fuel energy source produced and consumed. Natural gas, which is primarily methane (CH_4), is often found along with oil deposits, but it is also found on its own and bound up in layers of shale or other impermeable rock formations. In 1930, it accounted for just 5% of the energy market and was typically burned off as a waste product of oil extraction. But in recent years, natural gas consumption has surpassed that for coal. Two-thirds of gas consumption goes to residential, commercial, and industrial uses such as heating, cooking, and manufacturing; however, most of the gains in the use of this fuel come from the growth of natural gas turbine power plants for electricity.

Although burning natural gas creates the same types of emissions associated with other fossil fuels, it produces fewer of nearly all types of air pollutants and less CO_2 per unit of heat produced than do coal or refined petroleum products. For example, to generate the same amount of energy, natural gas produces 30% less CO_2 than oil and 50% less than coal. Current gas-fired power plants are also twice as efficient as a typical coal plant. Increasingly, natural gas is being used in more vehicles, especially newer public buses and transportation fleets in Los Angeles, New York, Chicago, and many other cities.

Historically, the downside of gas has been that it is harder to contain and capture than oil or coal. Gas requires a lot of infrastructure to distribute, including a large network of pipelines (**FIGURE 13.20**) across long distances and gas lines right into homes. This network must be carefully maintained. When gas lines leak and meet with a source of ignition, they create a fireball, like the one that instantaneously ignited a house in Salem, Pennsylvania, in 2016 and caused the evacuation of the surrounding neighborhood. The Atlantic Coast Pipeline through West Virginia, Virginia, and North Carolina was cancelled in 2020 by the company proposing it due to uncertainties and cost overruns linked to legal challenges. Not all leaks ignite, but chronic methane leaks during natural gas extraction and transmission have significant climate impacts. Studies based on satellite readings indicate that oil and gas wells may be leaking as much as 12% of the methane they are extracting.

As with oil, hydraulic fracturing has dramatically increased the amount of natural gas that can be profitably extracted from unconventional reserves. The United States sits atop many large, gas-bearing rock layers that have been brought into production in recent years, including in Texas, Wyoming, and Pennsylvania. As discussed in the Sustainability Matters feature, fracking has enabled the United States to become the world's leading producer of natural gas. Domestic production increased by more than 80% between 2000 and 2020, reaching record levels in 2019. But this transformation is not without its own environmental costs linked to the practice of fracking.

Extraction methods that utilize fracking are both water intensive and polluting, and the contaminated wastewater from this process is injected into deep wells for disposal. These wells carry the risk of hazardous spills or leaks, potentially contaminating groundwater and drinking water wells. In 2016, a former EPA scientist and Stanford researcher Dominic DiGiulio connected local fracking activity to polluted groundwater near Pavilion, Wyoming. These injection wells can also trigger earthquakes, such as those that are now regularly experienced in Oklahoma.

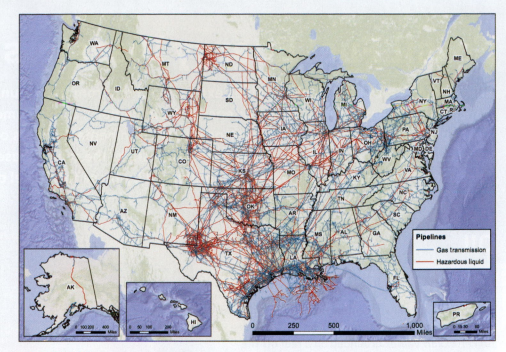

FIGURE 13.20 Natural Gas Pipelines This map, produced by the US Department of Transportation, shows the complex web of natural gas pipelines (blue lines) that distributes this fuel throughout the United States.

Adapted from US Department of Transportation (2022).

⬡ **TAKE-HOME MESSAGE** All fossil fuels when burned produce greenhouse gases and other pollutants, such as CO_2, NO_x, and SO_2. Their extraction also degrades the environment, through either the direct destruction of the land during mining or the leaks and waste products that result from extraction. Hydraulic fracturing (fracking) has permitted the increased extraction of less-polluting fuels from unconventional reserves, but it has environmental drawbacks of its own.

13.5 What Factors Will Impact the Future of Fossil Fuels?

When costs, technologies, and infrastructure evolve, change can happen rapidly. Consider pay phones, a common sight in most cities until very recently. Why have they largely disappeared? As cell phones went from brick-sized, expensive oddities to inexpensive items with global access, the need for a pay phone disappeared (one wonders if ATMs will soon see the same fate). Similarly, coal, oil, and natural gas have provided more than 80% of all the energy consumed in the United States each year for the past 100 years. To be certain, economic growth inside the

Fossil Fuel Energy Disasters

The production of fossil fuels involves extracting them from the ground, refining them, transporting them, burning them for energy (combustion), and in some cases disposing of waste products. Although the highest-profile environmental impacts of fossil fuels are related to the air pollution and climate impacts of combustion, the other phases of production also cause significant harms such as those highlighted by these fossil fuel disasters.

OIL

Tanker Spills: 1989 *Exxon Valdez*

Using oil for fuel involves transport from extraction locations to refineries, often by ship. In 1989 the tanker *Exxon Valdez* grounded on a reef off the coast of Alaska, tearing open its hull and spilling nearly 11 million gallons of crude oil, which covered 11,000 square miles of water and coated more than 1,300 miles of coastline.

Oil Platform Disasters: 2010 *Deepwater Horizon*

When oil wells "blow out," they gush oil into the surrounding area. In 2010, a blowout at the offshore *Deepwater Horizon* drilling rig in the Gulf of Mexico caused an explosion that killed 11 people and leaked 200 million gallons of oil from the seafloor. This was the largest accidental oil spill in history.

Pipeline and Rail: Enbridge Line 5 and 2013 Lac-Mégantic Accident

Other means of transporting oil besides ships also present hazards and result in spills—these can be gradual or sudden. Pinhole-sized leaks in the decaying Enbridge Line 5 pipeline between Superior, Wisconsin, and Sarnia, Ontario, Canada, have slowly spilled more than 1 million gallons of oil into the surrounding area since 1968. In 2013, an oil tanker train derailed in the town of Lac-Mégantic in the Canadian province of Quebec, causing an explosion and fire that killed 47 people and destroyed 30 buildings in the downtown area.

2010 Deepwater Horizon oil platform disaster

NATURAL GAS

Pipeline Explosions: 2010 San Bruno, California

Natural gas is distributed by vast networks of pipelines that bring it into buildings for heating and cooking. Leaks in these pipelines can cause fatal explosions. In 2010, a natural gas pipeline in San Bruno, California, exploded, killing 8 people and destroying nearly 40 buildings.

Gas Leaks: 2015 Aliso Canyon, California

The worst natural gas leak in history occurred over 4 months and began in 2015 in a well at the Aliso Canyon natural gas storage facility in Southern California. The well leaked nearly 100,000 tons of natural gas, and more than 11,000 people had to be temporarily relocated from the nearby Porter Ranch community.

2015 gas leak in Aliso Canyon, California

2008 Kingston Fossil Plant coal ash spill

COAL

Slurry Spills: 2000 Martin County Coal Slurry Spill

Toxic coal slurry is generated when raw coal is washed with water and chemicals. In 2000, about 300 million gallons of coal slurry broke free of its containment pond in Martin County, Kentucky. The spill damaged property, contaminated the drinking water supply of more than 25,000 people, and killed aquatic life in the Ohio River.

Hazardous Ash: 2008 Kingston Fossil Plant Coal Ash Spill

Once coal is burned, the leftover ash is hazardous and often held in containment ponds. In 2008, a dike in a coal ash containment pond in Roane County, Tennessee, ruptured, sending more than 1 billion gallons of coal ash slurry into the surrounding area.

Mining Disasters: 2010 Upper Big Branch Coal Mine Disaster

Coal mining is among the most hazardous professions, possessing risks including tunnel collapse, methane explosions, and poisonous gases. In 2010, the Upper Big Branch Coal Mine explosion in West Virginia killed 29 miners. A government review board found that the disaster would have been prevented had proper safety measures been in place.

Where Is It Built?

More than 17 million Americans live within one mile of an active oil or gas well where fuels are extracted. In some cases the proximity of communities to these sources of pollution was determined by racial segregation policies. In the 1930s, Los Angeles County restricted housing options for people of color to petroleum-producing areas. Today the county has the biggest urban oil field in the world, hosting over 6,000 active oil and gas wells. More than 75% of county residents currently living within a mile of these sites are people of color.

In the United States, counties scoring higher on a social vulnerability index (a measure of race, income, and other demographic factors) have significantly higher densities of gas pipelines. Indigenous peoples have protested the construction of major oil pipelines such as the Dakota Access Pipeline, and Black communities have led the recent opposition to the Atlantic Coast Pipeline. People of color make up 62% of the populations living closest to the most polluting oil and gas refineries in the United States.

FIGURE 13.21 Laying Claim to Oil and Gas China has been building artificial islands in the disputed waters of the South China Sea in order to lay claim to oil and natural gas reserves there. This has caused conflict with other countries with claims on these waters, including Malaysia, Indonesia, Vietnam, and the Philippines.

United States and expanding demand for more energy in emerging economies is causing the overall demand for fossil fuels to increase. But their collective share of the energy market has also been shrinking gradually over the past decade. Why is this happening? One reason is advancements in alternative technologies, the subject of the next chapter. A second reason is the increased costs associated with certain fossil fuels, which will be discussed in this section. As the environmental and economic costs associated with fossil fuels continue to rise, the choices we make regarding them may change quickly.

The Costs of Fossil Fuel Dominance

As a finite resource, fossil fuels have costs that intensify the more we draw on them. While technological advances have enabled us to locate and access more and more fossil fuels, extraction will continue to get more difficult as more fossil fuels are removed. Consider plans to obtain oil from the Arctic described at the beginning of the chapter. An offshore oil rig costs an average of $650 million to set up, which is 15–20 times more than the cost for a typical rig on land. The most expensive offshore rigs, like those planned to obtain oil from the Arctic, will likely cost more than $1 billion each to set up. Competition for access to scarce fossil fuels has also been a source of international conflict and thus raises national security concerns (and costs) in many countries. Because fossil fuels are in limited supply and resources are not distributed equally throughout the world, countries engage in struggles to ensure access. For example, oil and natural gas reserves are at the heart of a territorial dispute in the South China Sea between China—which is building artificial islands to lay claim to the area—and other countries bordering the sea that also have claims, such as Malaysia, Indonesia, and the Philippines. The struggle for these resources not only can spark conflict but also comes at the cost of providing military protection and support to gain and defend access (**FIGURE 13.21**).

Other costs associated with fossil fuels come in the form of harm caused by air pollution and climate change. A recent study found that as energy from wind and solar power grew from providing 2% to 7% of US electrical generation between 2007 and 2015, these technologies saved an estimated $56 billion in costs that would have been caused by fossil fuels. The social cost of carbon—an economic measure that estimates the cost of damage to property, agricultural productivity, and human health resulting from climate change—is estimated to be at least $51 per ton of greenhouse gas emissions, which when applied to US emissions would translate into more than $300 billion each year. Many economists argue that the social cost of carbon is closer to $100 per ton of greenhouse gas emissions. Public health costs from pollution and property damage from the effects of climate change are

What Is Fracking?

Fracking is a method of extracting oil and natural gas that has been in the news a lot in recent years. It has greatly expanded natural gas production in the United States and has had a significant impact on both our mix of energy resources and greenhouse gas (GHG) emissions. Also, as a relatively new technology, its environmental impacts are hotly contested.

"Fracking" is actually short for hydraulic fracturing: the use of water (along with sand and chemicals) under high pressure to fracture layers of rock that contain oil or natural gas. Historically, these unconventional reserves were not worth mining, but fracking has made their processing profitable. Let's take a closer look at how this works. The process begins with a deep vertical well that's often drilled more than 5,000 feet (1,500 meters) beneath Earth's surface through rock layers. Once the well reaches the layer of rock containing the oil and/or natural gas, the well is then drilled horizontally through the layer. The entire well is encased in steel, with tiny perforations made in the horizontal section of the well casing. Then, explosive charges, along with water, sand, and chemicals that are pumped down the well at high pressure, fracture the surrounding rock. Sand particles help to keep the cracks open as they develop. Some chemicals are used to facilitate the flow of natural gas back into the well, while others are used to reduce corrosion of the well lining. The liquid that was injected in the well then flows back to the surface, and it must be disposed of. The oil or natural gas that was held in the rock layer then begins flowing into the horizontal section of the well and back up to the surface where it can be collected.

Fracking has been used to extract natural gas since 1949, but it was not widely practiced until 2007 when higher oil and natural gas prices made it worthwhile to pursue innovative and more expensive drilling technologies. The Energy Policy Act of 2005, signed by President George W. Bush, also helped reduce regulatory costs, exempting fluids used in the practice from federal pollution protections. Vast oil and natural gas reserves (or "plays," to use an industry term) were identified and developed in layers of flaky rock, known as shale, in North Dakota, Texas, and Pennsylvania. As fracking

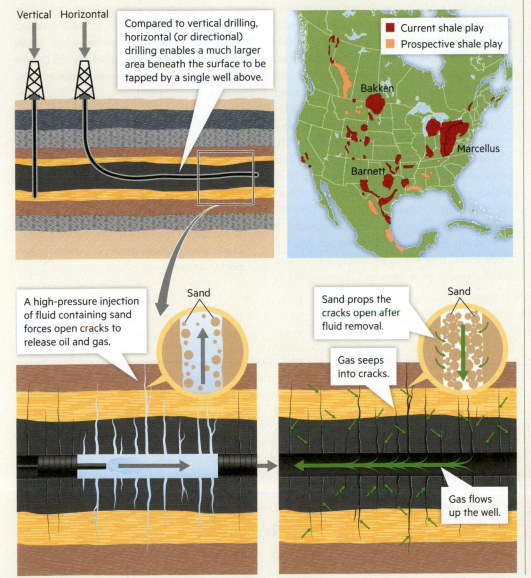

Vertical   Horizontal

Compared to vertical drilling, horizontal (or directional) drilling enables a much larger area beneath the surface to be tapped by a single well above.

Current shale play
Prospective shale play

Bakken

Marcellus

Barnett

A high-pressure injection of fluid containing sand forces open cracks to release oil and gas.

Sand

Sand props the cracks open after fluid removal.

Sand

Gas seeps into cracks.

Gas flows up the well.

Horizontal drilling and fracking greatly increased the accessibility of oil and natural gas resources. They also cause environmental impacts such as contamination of water used in the drilling process.

Art adapted from Marshak (2019).

(continued)

technologies were applied more widely, they became cheaper to employ, dramatically increasing the supply of oil and natural gas produced in the United States. Gains in natural gas production were particularly striking: the United States reached record production levels in 2013 with an annual production of 24 trillion cubic feet (680 million cubic meters) of natural gas, an increase of more than 40% over 1990 production levels.

As the supply of natural gas rose, its price declined, which caused big changes in electric utilities. Natural gas became a cheaper option than coal for power plants, and in 2015 natural gas surpassed coal as the primary fuel for electricity generation. Natural gas is, of course, a fossil fuel that emits the greenhouse gas CO_2 when it is burned in power plants. However, natural gas power plants emit less than half the CO_2 per unit of energy than do coal-fired power plants. This has led some to argue that fracking has helped reduce our environmental impact. Indeed, the boom in natural gas utilities and the related decline in coal use have played a role in declining US CO_2 emissions, which dropped by 12% between 2005 and 2016. For this reason, some have argued that by replacing coal, natural gas can serve as a "bridge" fuel that helps us reduce our GHG emissions in an affordable way.

However, critics of fracking point out that this method of extraction is responsible for other GHG emissions and a range of negative environmental effects. Natural gas is largely methane, a potent GHG, and it leaks during the extraction process. There is still a high degree of uncertainty regarding the magnitude of these leaks, but studies show a range between 3% and 12% for leakage of methane associated with fracking. And once this infrastructure is built, it will be in place for years to come.

Critics of fracking also take issue with other environmental impacts, particularly those related to water and the sand and chemicals used for extraction. Wells require up to 10,000 tons of frac sand—a special type of quartz that is primarily mined in the Great Lakes region. These sand mines pollute the water and air and pose a risk of silicosis—a respiratory disease linked to exposure to airborne silica dust from the sand. The chemicals used in the extraction process vary by company, and current regulations in the United States do not require companies to disclose their proprietary mixes. More than 2,500 different chemicals, some of which are toxic, have been used in fracking processes. For this reason, fracking wastewater is typically disposed of either by injecting it into the ground or by storing it in holding tanks. However, accidents can happen. When a fracking well blew out in Bradford County, Pennsylvania, in 2011, it leaked wastewater into a nearby creek. Two years earlier, a wastewater storage tank from a fracking operation in Dimock, Pennsylvania, leaked more than 8,000 gallons (30,000 liters) of drilling fluids into a creek, causing a fish kill. An Environmental Protection Agency study found that although contamination was not systematic or widespread, fracking did in fact contribute to drinking water contamination. Finally, fracking requires large volumes of water, with a typical well using between 2 million and 7 million gallons (7.6 million and 26.5 million liters) of water. This can be a particular concern in regions experiencing drought, where water use for fracking can compete with that for agriculture.

In the United States, the vast majority of the oil and natural gas fracking wells are on state and private land, so the regulation of this new technology is largely left to the states. In some cases, the surface rights to the land (where people live and farm) are split from the mineral rights to the oil and gas below. In many of these situations, property owners on the surface do not have the power to prevent fracking underneath their property. State policies on fracking vary widely, with some such as New York and Maryland banning the practice, while others such as Pennsylvania, Texas, and North Dakota welcoming it as an important piece of their economic development.

What Would You Do?

If you were a lawmaker in Congress, would you support or oppose a measure requiring companies to disclose the mix of chemicals they use in the fracking process? What reasons would you use to support your case?

---

enhanced oil recovery (EOR) a process where captured CO_2 is condensed and pumped underground as a way to force more oil out of depleted wells.

carbon capture and storage (CCS) a process that prevents CO_2 emissions from escaping into the atmosphere by injecting them underground.

costs that are avoided by the use of alternative energy sources. This provides another rationale for continued investments in alternative energy technologies. And the fossil fuel industry is itself vulnerable to the effects of climate change. The oil distribution infrastructure of Alaska and Russia is increasingly disrupted as pipelines sink and break under melting permafrost. Wildfires in Alberta, Canada, have disrupted extraction from tar sands mines. Many oil refineries are located on coastlines throughout the world that are increasingly vulnerable to sea level rise, storm surges, and hurricanes. Further, proven reserves and fossil fuel extraction projects will become *stranded assets*: obsolete investments that cannot yield a profit, as other forms of energy replace them. Many economists worry that this could cause losses of $1 trillion to $4 trillion from the global economy as these investments go bad. Some refer to these financial losses as the "carbon bubble," reflecting the great losses that will occur when it "bursts." While this might not seem like everyone's problem, these economists note that "only" $250 billion in losses started the financial crash in 2008. With this problem looming on the horizon, some wonder why we are still investing in fossil fuel infrastructure in the first place.

Policy Responses

The range of impacts and costs linked to fossil fuel dominance has led to policy responses from governments at a variety of levels. These include pollution regulations enforcing air-quality standards and requiring power

Capture, Use, and Storage of Carbon Emissions

More than 75 years ago, technologies were first developed to bind and separate carbon dioxide (CO_2) from the exhaust of coal- or gas-fired power plants. In this way, carbon can be "captured," diverting greenhouse gas emissions from the atmosphere. More recently, giant fans and new chemical processes that can capture CO_2 directly from the atmosphere have been developed for facilities.

But what happens to the CO_2 after it's captured? Commercial applications of carbon capture began in the 1970s as oil prices were rising. Some natural gas processing facilities and fertilizer plants started capturing and selling their CO_2 to oil and gas companies. These companies condensed and pumped it underground as a way to force more oil out of depleted wells, a practice known as **enhanced oil recovery (EOR)**. For example, the Petra Nova project near Houston, Texas, paired a CO_2 carbon capture system with a coal-fired power plant and used the diverted CO_2 for enhanced oil recovery in the region. This process of diverting

Direct air capture (DAC) fans on the roof of this garbage incinerator outside Zurich, Switzerland, can remove up to 135 kilograms of CO_2 from the air daily. The captured CO_2 is used by a nearby commercial greenhouse (in the background) to increase the crop yield and a carbonated water manufacturer.

CO_2 from exhaust and injecting it underground is now known as **carbon capture and storage (CCS)** technology. Although the CO_2 from this project was diverted from the atmosphere and remains trapped underground, the process was used to extract more oil and gas—which in turn contributed to more greenhouse gas emissions.

According to one proposed scenario, CCS use would grow by helping oil producers recover more oil from existing sites. Power companies would invest in technology to capture and sell CO_2, and oil and gas companies would invest to transport and store it. Eventually, these processes would become so cost-effective that CCS could be used to store extra CO_2 without requiring more oil production.

However, so far market forces have not driven this scenario. The success of another technology, hydraulic fracturing (fracking), has increased oil and gas supplies, and their prices have fallen dramatically. Oil and gas companies have not had an incentive to use the more expensive EOR methods in depleted oil fields. This means there has not been much of a demand for CO_2, and there is little incentive for power plants and other CO_2-emitting industries to invest in CCS technology. The Petra Nova project was idled in 2020, after less than 4 years of operation, due to falling oil prices.

In the United States, the federal government has stepped in to facilitate the development of CCS technology largely through hundreds of millions of dollars in grants to private companies. One funded project in Illinois has successfully installed a carbon capture system at an ethanol facility (ethanol is an alcohol fuel derived from plant materials that is often blended with gasoline). It simply sequesters the CO_2 underground without any accompanying oil recovery. Federal funding has also supported the development of direct air capture (DAC) technologies that can pull and capture CO_2 from the atmosphere and either use it for industrial processes or sequester it underground. Industrial processes that use CO_2 include producing carbonation for soda, concrete, and plastics. Some of these manufactured materials may

(continued)

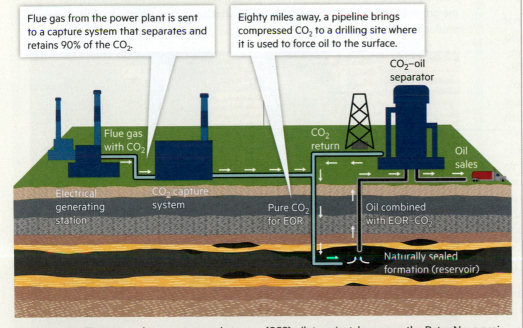

Flue gas from the power plant is sent to a capture system that separates and retains 90% of the CO_2.

Eighty miles away, a pipeline brings compressed CO_2 to a drilling site where it is used to force oil to the surface.

CO_2–oil separator

Flue gas with CO_2

CO_2 return

Oil sales

Electrical generating station

CO_2 capture system

Pure CO_2 for EOR

Oil combined with EOR–CO_2

Naturally sealed formation (reservoir)

Near Houston, Texas, a carbon capture and storage (CCS) pilot project, known as the Petra Nova project, captured CO_2 from a coal-fired plant, compressed it, and then pumped it 80 miles (130 kilometers) underground to enhance oil recovery at a depleted field. However, due to falling oil prices, the Petra Nova project was stopped after less than four years of operation.

Art adapted from *Electric Light & Power* (2017).

even "lock up" carbon for long periods of time—keeping it out of the atmosphere. However, DAC technologies are also expensive. There are legislative proposals designed to support CCS and DAC, including federal financing for larger projects and corporate tax credits that increase with each ton of CO_2 captured.

But environmental groups are divided on government support for these technologies. Opponents argue that CCS boosts greenhouse gas emissions linked to oil production in the short term and funding for DAC diverts resources away from fossil fuel alternatives. Supporters argue that CCS is the only tool available to address the CO_2 emissions of the existing coal- or gas-fired power plants that will continue to operate for the next two or three decades. They also point out that it can address the 20% of global greenhouse gas emissions that come from industrial facilities such as cement and steel plants. For now, CCS and DAC remain technologies that have long been under development but are rarely employed.

What Would You Do?

If you were in a leadership position with a prominent national environmental group, what would your position be on this issue? Would you support adding CCS technologies to the mix of greenhouse gas reduction strategies your group advocates for? Why or why not?

plants and vehicles to use emissions-reduction technologies. Other strategies use taxes or market mechanisms to increase the cost of fossil fuels, thus encouraging energy conservation and the development of alternative energy resources.

One approach is seen in a **carbon tax** levied in British Columbia, Canada, in 2008. This taxed the carbon emissions from homes, vehicles, businesses, and industry across the province. While this tax increased energy bills in the province, revenue from it was used to reduce income and corporate tax rates. This trade-off has been quite effective. Since that time, per capita gasoline consumption in the province dropped by 16% while it has risen by 3% in the rest of Canada. And thanks at least in part to the revenue from this tax, British Columbia now enjoys the lowest income tax rate in Canada.

Another approach is to set up an exchange for pollution, such as a **cap-and-trade** program (see also Chapter 2). In this system, a government sets an overall maximum allowable emissions standard (cap). The emissions under the cap are allotted to firms in polluting industries in the form of allowances. Allowances can be bought, sold, traded, or banked for the future in an open market. If a firm pollutes less than its allowance, the firm can sell the remainder to firms that pollute more. This creates both an incentive for firms to reduce pollution and a higher operating cost for those that pollute more. Since 2013, California has instituted a cap-and-trade program for utilities and large industrial polluters as well as oil and gas companies, which is credited with helping the state stay on track to meet its goal to reduce its greenhouse gas emissions to 1990 levels by 2020.

However, it is important to remember that ambitious greenhouse gas emission reduction targets and policies designed to phase out fossil fuels also displace people and communities that have been dependent on these industries. The 2018 "Yellow Vest" movement in France involved widespread protests reacting in part to new diesel taxes. Protesters claimed the government was

→ CONTEXT

Earthquakes—A Reasonable Trade-Off?

Fracking can cause numerous small to moderate earthquakes like the one that dislodged these stones from a building in Pawnee, Oklahoma. Normally, this state would be expected to experience just two tremors a year registering above a barely noticeable 2.5 on the Richter scale, which measures earthquake magnitude. However, since fracking has become widespread, the state has experienced thousands of such quakes, including record magnitude events for Oklahoma, such as the 5.8 magnitude quake in Pawnee. While this is still a moderately sized earthquake, the increase in frequency and magnitude of quakes in Oklahoma is a concern. The fracking itself is not causing the earthquakes, but the injection of wastewater into the ground puts pressure on existing faults, which can cause rocks to shift, thereby producing tremors.

A Cap-and-Trade Success Story?

California instituted a cap-and-trade system for carbon emissions in 2013. The cap is decreased a bit each year to meet overall pollution-reduction goals. The state gains about $2 billion per year in revenue from auctioning pollution allowances. This money is deposited in a Greenhouse Gas Reduction Fund that invests in alternative energy and other efforts to reduce carbon emissions. Since instituting this program, California's GHG emissions have declined while its economic growth has increased.

Greenhouse Gas Reduction Fund (GGRF) investments through fiscal year 2018–2019
At least 35% of these cumulative investments directly benefit disadvantaged communities across California.

$5.8 billion	$821 million	$1 billion	$559 million
SUSTAINABLE COMMUNITIES AND CLEAN TRANSPORTATION	**ENERGY EFFICIENCY AND CLEAN ENERGY**	**NATURAL RESOURCES AND WASTE DIVERSION**	**COMMUNITY AIR PROTECTION**
✔ High-speed rail ✔ Public and alternative transportation ✔ Affordable housing near transit ✔ Low- or zero-carbon cars, trucks, buses, and freight	✔ Weatherization and solar energy for low-income households ✔ Water and energy efficiency for agriculture ✔ Wood smoke reduction	✔ Wetland restoration ✔ Urban forests ✔ Forest fire prevention ✔ Increased composting and recycling	✔ Community-level air pollution monitoring ✔ Neighborhood emission reduction plans ✔ Accelerated technology upgrades at facilities

carbon tax a tax levied on fossil fuels.

cap and trade a system where a government sets an overall maximum allowable emissions standard (cap) and then creates a market that enables pollution allowances to be bought, sold, traded, or saved for the future (Chapter 2).

not sensitive to the burdens these taxes put on them. Similarly, in 2021, rising gas prices forced President Joe Biden on the defensive, as he criticized oil companies for not producing as much gasoline as they could to lower prices.

More broadly, some policy makers believe new energy policies must be tied with initiatives to address overall inequality. Governments can respond to economic upheaval with policies that provide workers and communities with new opportunities. Spain initially resisted a European Union effort to end government support for coal power but then adopted what it calls a "just transition" away from coal. This includes retraining coal miners for work in alternative energy industries and environmental restoration, as well as offering early retirement pensions. On a smaller level, the Coalfield Development Corporation in West Virginia retrains coal miners to work in housing construction, environmental restoration, and clean energy technologies.

🏠 **TAKE-HOME MESSAGE** Fossil fuels have been our dominant source of energy for the past century, but their share of the energy market is shrinking as we transition to alternative energy resources. The costs of fossil fuel use, especially the many environmental impacts but also international security issues, make them less appealing when compared to alternative sources of energy. Governments at various levels are enacting policy responses to aid in a shift away from fossil fuels.

❓ 13.6 What Can I Do?

What can we do as individuals to reduce our dependence on fossil fuels? We all use energy directly for transportation and electricity and indirectly in the products we consume, which is a good place to start. We can also play a role in shaping future energy policies. Here are some steps you can take.

Choose Transportation That Uses Less Fossil Fuel

Some very concrete, day-to-day decisions impact one's use of oil. First, transportation is by far the biggest consumer and driver of demand for oil products. As we have written before (and will point out again), any steps you can take that lessen or eliminate your demand or decrease the length of commutes are important. Walking, riding a bike, taking public transportation, or driving an electric vehicle are great strategies to reduce fossil fuel use. Barring this, you can save a lot of fuel (and money on fuel costs) by improving the fuel economy of gas vehicles you drive. Shifting from a vehicle that averages 16.5 miles per gallon (mpg) to one that averages 33 mpg typically saves 300 gallons (1,100 liters) of gas a year. But even switching from a car that runs at 20.0 mpg instead of 16.5 mpg saves 100 gallons (380 liters) of gas a year, so operating just a little more efficiently (and per Chapter 8, avoiding idling) can lead to big savings. Fuel-economy labels now show you how much in fuel costs you might save (or not save) with a particular make and model compared with an average new car to help you choose what kind of vehicle to buy.

Opt for Products and Services That Use Less Fossil Fuel

As we saw in Chapter 11, cutting back on energy use in the home has the next biggest impact on your pollution production after that from transportation. Steps to shift from energy powered by coal are especially important, and many local energy utilities will let you opt into other types of provider. For example, Con Edison, a major energy utility in New York, lets you actively shop and choose your energy supplier. But be prepared that cleaner sources may not have the lowest costs; as the utility's website says, "Energy is a commodity . . . so prices can fluctuate."

Oil is a component of many other things that we use, so you can make choices that reduce your oil use.

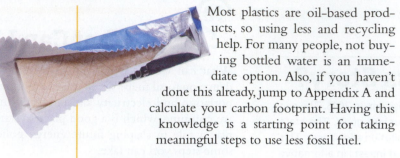

Most plastics are oil-based products, so using less and recycling help. For many people, not buying bottled water is an immediate option. Also, if you haven't done this already, jump to Appendix A and calculate your carbon footprint. Having this knowledge is a starting point for taking meaningful steps to use less fossil fuel.

Ten Surprising Products Derived from Petroleum

1. Chewing gum
2. Pantyhose
3. Cosmetics
4. Nonstick coating
5. Crayons
6. Synthetic fabrics
7. Aspirin
8. Sports equipment
9. Dentures
10. Toothpaste

Find Out How Familiar Organizations Are Committed to Reducing Fossil Fuel Use

It's important to recognize that groups and organizations—like schools, businesses, and faith communities—often can make a much greater impact on reducing reliance on fossil fuels than can individuals. For example, many groups have adopted climate commitments that include reducing fossil fuel use or investment. This can include companies that replace their vehicle fleets with cleaner vehicles or organizations that choose to remove or "divest" fossil fuel stocks from their investment portfolios. In 2021, Harvard University joined nearly 1,500 other institutions in divesting its nearly $42 billion endowment from fossil fuel companies. You most likely are a student, member, employee, or customer who is part of several groups and organizations. You can learn how your organizations are directly using fossil fuels or indirectly supporting fossil fuels with investments. You can then look for ways to suggest changes.

Be Informed about the Policies and Economics of Fossil Fuel Use

Carbon taxes, cap and trade, and subsidies of alternative energies are often criticized as unfair and operating against free market economics. However, in 2017, Martin Feldstein, the chair of the Council of Economic Advisers under President Ronald Reagan, and Greg Mankiw, the chair of the council under President George W. Bush, were part of a group proposing a broad tax on carbon, writing that "the idea of using taxes to correct a problem like pollution is an old one with wide support among economists." But accompanying this tax on carbon would be an elimination of many regulations, especially some created by the Obama administration. Who do you think is correct? Or do you feel both policies are wrong? Decisions on similar policies will likely continue to be the subject of hot debate. While this book provides some background on basic policies and economic concepts (see Chapter 2), becoming better informed about the economics of energy and the concept of externalities is an important step any individual can take. Of course, as we learned in Chapter 1, it's always important to consider your sources—and to make your voice heard by your elected representatives.

Chapter 13 Review

SUMMARY

- Fossil fuels are formed from the remains of organisms that were buried and transformed into coal, oil, or natural gas through intense heat and pressure over millions of years.

- Fossil fuels are currently the leading source of energy in the United States and the world.

- Prior to the development of fossil fuels, humans relied primarily on burning wood for energy.

- It took a long transition period, spanning centuries, for coal, oil, and natural gas to become the dominant sources of energy.

- Economic factors such as the scarcity and cost of alternative energy sources and government policies such as subsidies and taxes have played key roles in facilitating and maintaining the dominance of fossil fuels.

- The combustion of fossil fuels produces greenhouse gases that contribute to climate change along with other air pollutants.

- The extraction, processing, and transport of fossil fuels also causes significant negative environmental impacts.
- Fracking is a technology that has changed the mix of energy use in the United States, in particular, causing a move away from coal power. However, it still encourages fossil fuel use and can have other impacts, especially on water supplies and quality.
- While fossil fuel consumption remains high, its share of energy use among all energy sources is declining. Factors such as the cost of extraction, the price of alternative energies, and policy decisions will shape the extent of future reliance on fossil fuels.

KEY TERMS

biofuel	conventional reserve	Jevons paradox	proven reserve
cap and trade	enhanced oil recovery (EOR)	mountaintop removal	public good
carbon capture and storage (CCS)	fossil fuel	natural gas	reserve
	hydraulic fracturing (fracking)	oil	tar sand
carbon tax	hydrocarbon	power	unconventional reserve
coal	inferior good	productivity	

REVIEW QUESTIONS

The letters following each Review Question refer to the Chapter Objectives.

1. Explain how the Sun is ultimately the source of energy for fossil fuels. **(A)**

2. What is the difference between power and productivity? How did early energy technologies such as waterwheels and windmills affect these measures? **(A)**

3. What energy source did coal compete with and replace in England during the 16th and 17th centuries? What led to this transition? **(B)**

4. Explain how the reserves of a natural resource (such as a fossil fuel) can be expanded even though the resource itself is nonrenewable and will diminish over the long term. **(B)**

5. Explain how efficiency improvements in how we use fossil fuels can actually lead to greater overall consumption of these energy sources. **(B)**

6. Name two government actions that have facilitated the use of fossil fuels. **(B)**

7. Describe two environmental impacts caused by coal mining. **(C)**

8. What environmental impacts are common to all fossil fuels when they are burned? **(C)**

9. How has hydraulic fracturing (fracking) expanded our ability to extract oil and natural gas from the ground? What are the environmental impacts of fracking? **(C)**

10. Why have natural gas–fueled power plants overtaken coal-fueled power plants as the primary source of energy generation in the United States? **(C)**

11. Name two government actions that could be taken to reduce the use of fossil fuels. **(D)**

FOR FURTHER THOUGHT

The letters following each item refer to the Chapter Objectives.

12. While each fossil fuel energy source is linked to certain environmental impacts, it is also the source of employment for workers who extract, process, and burn the fuel to provide energy. In recent years, the coal industry has been slumping as cheaper and more efficient natural gas (and some of the alternative energy sources we'll learn about in the next chapter) increasingly replaced it. The coal industry has lobbied hard for the federal government to step in and use its regulatory power to ensure that coal power plants stay in business. Do you think this is a proper use of government authority? Explain your reasoning. **(C, D)**

13. Identify the fossil fuels that power your life. Find out from your utility company how much of your electricity is powered by coal and gas. What about your transportation and heating? What do you think the most important step you could take to reduce your fossil fuel consumption would be? What are the primary obstacles to you taking this step? **(D)**

14. Many college campuses are considering (or are in conflict over) divestment—institutional decisions to avoid financial investments in fossil fuel–related companies. Do you think divestment from fossil fuels is a good idea? Explain your position on this issue, considering both economic and ethical factors. **(D)**

15. Some local governments in states that allow fracking have passed regulations banning or restricting the practice. In response, some of these states, such as Texas, have overridden these local ordinances. What level of government should have the authority to regulate fracking? Why? **(C, D)**

Make Your Case ?

The following exercises use real-world data and news sources. Check your understanding of the material and then practice crafting well-supported responses.

Use the News

The following article appeared on the website of *USA Today*, a national digital news network. It details results on worldwide emissions of carbon dioxide from a report by the International Energy Agency, an intergovernmental agency originally founded in 1974. Use this article to answer the questions that follow. The first two questions test your understanding of the article. Questions 3 and 4 are short answer, requiring you to apply what you have learned in this chapter and cite information in the article. Answers to Questions 1–2 are provided at the back of the book. Question 5 asks you to make your case and defend it.

"Global Carbon Dioxide Emissions Reach Highest Level in History," Doyle Rice, *USA Today*, March 8, 2022

Worldwide emissions of carbon dioxide—the greenhouse gas most responsible for global warming—have rebounded to their highest level in history, experts reported Tuesday, as the world economy rebounded strongly from the COVID-19 crisis and relied heavily on coal to power that growth.

The report, which was prepared by the International Energy Agency, found that emissions of carbon dioxide rose by 6% in 2021 to 36.3 billion metric tons.

"The numbers make clear that the global economic recovery from the COVID-19 crisis has not been the sustainable recovery that IEA executive director Fatih Birol called for during the early stages of the pandemic in 2020," the IEA said in a statement.

"The world must now ensure that the global rebound in emissions in 2021 was a one-off—and that an accelerated energy transition contributes to global energy security and lower energy prices for consumers," the IEA said Tuesday.

COVID-19 restrictions in 2020 caused a massive plunge in fossil fuel use, and Birol predicted at that time that "the energy industry that emerges from this crisis will be significantly different from the one that came before." But that prediction didn't come true.

The burning of fossil fuels such as coal, oil and gas releases "greenhouse" gases such as carbon dioxide (CO_2) and methane into Earth's atmosphere and oceans. The emissions have caused the planet's temperatures to rise to levels that cannot be explained by natural factors, scientists say.

In the past 20 years, the world's temperature has risen about two-thirds of a degree Fahrenheit, the National Oceanic and Atmospheric Administration said.

Coal accounted for more than 40% of the overall growth in global CO_2 emissions in 2021, reaching an all-time high of 15.3 billion metric tons, according to the IEA. CO_2 emissions from natural gas rebounded well above their 2019 levels to 7.5 billion metric tons.

The rebound of global CO_2 emissions above pre-pandemic levels has largely been driven by China, where emissions increased by 750 million metric tons from 2019 to 2021.

China was the only major economy to experience economic growth in both 2020 and 2021, the IEA said.

Meanwhile, in the United States, CO_2 emissions in 2021 were 4% below their 2019 level.

1. In the first section of the article, what factors are given for carbon dioxide emissions reaching their highest levels in history in 2021?

   a. The economic recovery from the COVID-19 pandemic shutdowns was very strong, and nations used coal more than expected as a source of power.

   b. The economic recovery from the COVID-19 pandemic shutdowns was very strong and relied on energy from increased oil drilling and fracking.

   c. A massive increase in the use of natural gas in particular led to more pollution, trapping more carbon dioxide in the atmosphere.

   d. China and the United States had a stronger recovery than expected, and both saw increased CO_2 emissions compared to pre-COVID 2019 emissions.

2. As a percentage, how much of the additional CO_2 growth in 2021 is attributed to coal?

   a. 4%
   b. 6%
   c. 19%
   d. 40%

3. In a few sentences, describe what the executive director of the IEA predicted would happen in terms of overall energy use trends as a result of the 2020 restrictions driven by COVID. Was this prediction correct?

4. What does the article suggest was the major cause of this CO_2 growth? What specific details support your answer?

5. **Make Your Case** The article notes that CO_2 emissions in 2021 in the United States were 4% below their pre-pandemic 2019 level, but it does not provide a lot of detail about why this might have happened or whether this trend is expected to continue. Using what you read in the text, as well as other sources on the internet like the IEA's website iea.org or the US government's Energy Information website EIA.gov, make your case for what you think caused the 4% decrease, whether you think this trend will continue, and why?

Use the Data

The age of US power plants generating electricity from the fossil fuels coal, natural gas, and oil varies greatly—with some beginning operation as long ago as the 1950s. These graphs, from the Government Accountability Office report *Air Emissions and Electricity Generation at US Power Plants*, show the share of emissions produced by fossil fuel–burning plants, sorted by the age of the plant, and the emissions produced by different types of fuel per unit of electricity generated.

Study the graphs, and use them to answer the questions that follow. The first three questions test your understanding of the graphs. Questions 4 and 5 are short answer, requiring you to cite data in the graphs. Answers to Questions 1–3 are provided at the back of the book. Question 6 asks you to make your case and defend it.

(a)

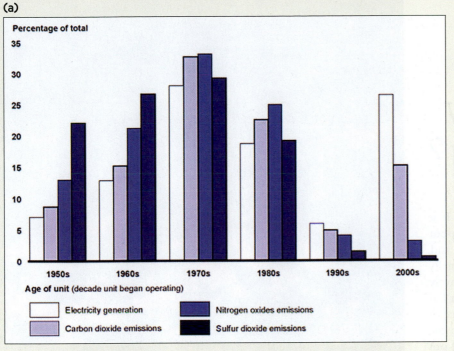

(b)

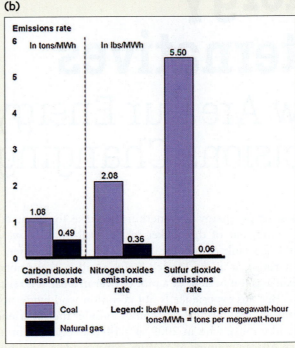

1. Which decade of power plants produces the largest share of electricity relative to their share of nitrogen oxides and sulfur dioxide emissions?

   a. 1950s   c. 1990s

   b. 1970s   d. 2000s

2. Which decade of power plants produces the largest share of nitrogen oxides and sulfur dioxide emissions relative to their share of electricity generation?

   a. 1950s   c. 1990s

   b. 1970s   d. 2000s

3. Which pollutant is emitted the most from coal-burning power plants?

   a. carbon dioxide

   b. nitrogen oxides

   c. sulfur dioxide

4. Most power plants in the United States use either coal or natural gas as fuel. On the basis of graph (b), which of these two fuels do you think is more commonly used in newer power plants? Why?

5. Under a cap-and-trade policy, plants that pollute more can purchase credits from plants that pollute less. Which plants are most likely to have credits to sell to polluters? Which plants are most likely to need to purchase credits? Explain your answers.

6. **Make Your Case** It is expensive to build pollution controls for an older power plant, but it is also expensive to buy pollution credits. In addition, newer power generation technologies are more efficient than older ones, generating more electricity per unit of fuel than do older plants. If you were the owner of an older power plant, would you rebuild your plant with newer power generation and pollution-control technologies? Or would you continue to buy pollution credits? Would your personal values about pollution factor into your decision?

LEARN MORE

- US Energy Information Administration page explaining energy: www.eia.gov/energyexplained/
- International Energy Agency page on carbon capture and storage: www.iea.org/topics/carbon-capture-and-storage/
- International Energy Agency pages on oil, coal, and gas (like this one on oil): www.iea.org/topics/oil/
- NOAA educational page on the Gulf oil spill and oil spills in general: www.noaa.gov/resource-collections/gulf-oil-spill
- US Geological Survey page on fracking: www.usgs.gov/mission-areas/water-resources/science/hydraulic-fracturing
- Video: NOAA, Black Carbon: digital.wwnorton.com/environsci

14

Energy Alternatives

How Are Our Energy Decisions Changing?

For most of us, energy is something we take for granted. It is what comes out of the socket in the wall or what we pour into the gas tank of our car. Energy seems omnipresent and, considering that in the United States a gallon of gas normally costs less than a gallon of milk or a pint of beer in a bar, it is relatively inexpensive. But this experience is not common worldwide, nor has it been common until relatively recently. Consider the experience of William Kamkwamba, who in 2002, as a 14-year-old student in the southeastern African country of Malawi, constructed a windmill out of a bicycle and a tractor fan set atop a makeshift tower of blue gum logs (**FIGURE 14.1**). Prior to constructing his windmill, he had dropped out of school because his family could no longer afford tuition. At night, while reading by the light of smoky kerosene lamps, he found descriptions of wind power in textbooks on energy and physics he had borrowed from the library. At this time in Malawi, the Electricity Supply Corporation of Malawi (ESCOM) was the only source of electricity, and less than 10% of the nation's population had access to it. Kamkwamba realized that if he perched a tractor fan high enough, the wind could turn the blades, which could in turn act like pedals to spin a bicycle tire. He could then attach a bicycle dynamo—a small machine that converts the mechanical energy of the spinning wheel into electricity—to power indoor lights, a radio, a water pump for drinking and irrigation, and even a charging station for cell phones.

Innovators like Kamkwamba have long been experimenting with ways to harness energy to help us survive and accomplish the many tasks we've incorporated into our daily lives. Kamkwamba overcame numerous challenges to successfully harness the wind. He had to teach himself principles of science and engineering and scavenge parts for his invention from discarded junk at the community scrapyard. Once his windmill was assembled, Kamkwamba had to solve technical problems, such as how to fashion homemade transformers, switches, and circuit breakers to effectively use the power he was generating.

But technical problems were not the only ones Kamkwamba faced: once news of Kamkwamba's windmill spread, some warned that he would be arrested for providing power outside the reach of ESCOM.

Alternatives to fossil fuels supply just under 20% of global energy demand, but some alternative energy sources (such as wind power being generated here in Garner, Iowa) are growing rapidly.

Chapter Objectives

By the end of this chapter, you should be able to . . .

A. describe the ways in which alternatives to fossil fuels are employed and the environmental impacts and challenges associated with these processes.

B. identify the major energy alternatives for transportation.

C. explain how energy conservation plays a role in addressing energy challenges.

D. discuss the challenges associated with a transition to alternative energy and the steps governments are taking to facilitate this change.

I'd put my money on the sun and solar energy. What a source of power! I hope we don't have to wait till oil and coal run out before we tackle that.

—Thomas Edison

FIGURE 14.1 A Wind Power Entrepreneur When he was just 14 years old, William Kamkwamba built an electricity-generating windmill out of parts he found in a scrapyard near his home in Malawi.

wind turbine a machine that uses flowing air to turn large blades that in turn power a generator and creates electricity.

Others accused him of practicing witchcraft and using wind power to blow away rain clouds and prolong a famine the area was experiencing. These cultural challenges were similar to those faced by an innovative knight in Suffolk, England, who began operating one of Europe's first windmills more than 800 years ago. At that time, the local authorities forced him to destroy it. This new energy technology threatened the profitability of other mills in the area that relied on animal power, as wind energy was free of the expenses required to care for animals.

This chapter considers alternatives to fossil fuel, such as wind, solar, hydroelectric, geothermal, and nuclear power, and the factors influencing their use. It explores the technologies that make these energy sources available to us and the environmental impacts associated with them. The popular saying "Necessity is the mother of invention" may be insightful, but widespread development, acceptance, and adoption of new energy technologies requires much more than just a need. As we saw with fossil fuels (see Chapter 13), factors such as economic forces and government actions also drive the use and development of alternative energy technologies.

14.1 Why Are Wind and Solar Power Use Growing?

There are many alternative energy sources available, and a pragmatic approach is to consider them all important contributors to a transition from fossil fuels. Currently, all the alternatives to fossil fuels supply less than one-fifth of global energy consumption. While most of the alternative share is now provided by hydroelectric power and nuclear power (which we explore later in the chapter), the most rapidly growing alternatives are wind and solar power. Some of this growth occurs because of government subsidies that have made these options attractive. However, even as the subsidies are phased out, businesses predict the opportunities for growth will get even better. Depending on where you live, you may be witnessing this. It's not uncommon to see large wind turbines in Texas and across states in the Midwest, and rooftop solar panels have been widely adopted in California and Arizona, as well as many other areas of the country. In fact, in each year since 2008, wind and solar have been competing with natural gas for the top spot among sources of newly added electricity generation in the United States. In this section, we will explore how these rapidly expanding sources of alternative energy work and their benefits and limitations.

Wind Power

On October 19, 2017, Amazon CEO Jeff Bezos smashed a champagne bottle on the top of a 300-foot-high (90-meter-high) wind turbine on a newly opened wind farm in Texas, one of more than 35 Amazon wind projects in operation or development (**FIGURE 14.2**). The company is betting on this technology as a major part of its sustainable energy strategy. This reflects a growing trend; investors such as Goldman Sachs forecast that wind energy will see $3 trillion in investments in 20 years. So how does wind power work?

As we learned in Chapter 8, wind flow is caused by uneven heating of the air by the Sun and by features on Earth's surface, such as bodies of water and terrain. **Wind turbines** use the flowing air to turn large blades on a

FIGURE 14.2 Amazon and Wind This wind farm in Indiana has 65 turbines.

shaft that in turn powers a generator as the shaft spins, creating electricity. Massive utility-scale turbines provide electricity to the distribution grid for use by homes and businesses just like power plants do. Turbines at this scale range from 300 to 600 feet (90 to 180 meters) in height (taller than the Washington Monument), with rotor diameters (the width of the circle traced around the sweeping blades) more than 410 feet (125 meters) across. Each turbine is capable of powering more than 2,000 homes when operating at maximum capacity. Wind farms concentrate hundreds or sometimes even thousands of turbines in areas with favorable wind resources, either on relatively large areas of open land or offshore like the turbines off the coast of England in the North Sea (**FIGURE 14.3A**).

Using wind as an energy source has many advantages. First, it is freely accessible—so there are no fuel bills—and nondepletable, so using wind today has no effect on the availability of more wind in the future. In addition, there is no need to extract or transport fuel because the wind is moving to the turbine. Once construction of the turbine is complete, wind power has no air or water emissions because it doesn't rely on steam turbines to generate electricity and it doesn't require a supply of water for heating or cooling (**FIGURE 14.3B**). Finally, while most new power plants take years to complete, wind farm construction is typically completed in less than a year.

Although wind power supplies just over 2% of global energy consumption, it has experienced rapid growth, with consumption tripling between 2010 and 2020. Wind power exceeds a 10% share of electricity generation in several countries, including Denmark (more than 48%) and Spain, Portugal, and Ireland (20% to more than 30%). In the United States, which has more than 67,000 turbines on more than 1,500 wind farms across 44 states, wind power provides more than 8% of total electricity generated. Among US states, Iowa uses wind power for the largest share (more than 37%) of its in-state electricity generation, while Texas is the largest producer of wind power overall.

But wind power has several associated challenges, a key one being **variable generation**: its generating capacity changes according to the time of day, weather conditions, or other factors. Turbines don't always run at maximum capacity: they vary with the wind. Most only reach maximum capacity when wind speed is in the range of 27–55 miles per hour (43–88 kilometers per hour). When wind speeds are slower, they produce less power, and turbines typically generate only 10% to 40% of their maximum potential output over the course of a year. Productive wind power depends on sites with strong, steady winds, and the highest-quality wind resource areas in the United States are primarily found in the wide-open, windswept areas of the Midwest. This creates a new challenge: because favorable wind resources are often in remote places, new transmission lines are frequently required to connect turbines to a larger grid.

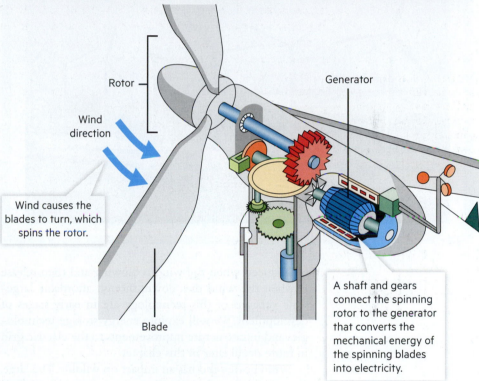

FIGURE 14.3 Modern Wind Turbines (a) Huge utility-scale wind turbines can be constructed either on land or offshore (like this one off the English coast in the North Sea). If these turbines were in New York Harbor they would tower above the Statue of Liberty. (b) The basic components of a wind turbine.

Adapted from US Department of Energy (n.d.).

And wind farms must be balanced with other sources of power, such as conventional power plants, that can ramp up production when the wind dies down and phase out when the wind turbines are operating at peak capacity. Natural gas and coal power plants frequently serve this function because they are plentiful and can be adjusted to provide more or less power.

Using stored energy is another solution to wind variability. Various systems can temporarily hold energy that

variable generation a power source with generating capacity that changes according to the time of day, weather conditions, or other factors.

Harvesting Wind

In the United States, some of the best wind resource areas overlap with agricultural land. It is increasingly common for farmers and ranchers to receive payments from utilities that lease rural land for between $3,000 and $10,000 per turbine per year. Rural landowners in the United States earned more than $250 million in 2020 from the placement of wind turbines on their properties. Although wind farms require large areas of land covering many square miles, the footprint of the actual turbines makes up only about 1% of the total area. So there generally is space to continue agricultural production around the turbines.

is produced when the wind is blowing and then release it when the wind dies down, though affordable large-scale versions of this technology are in early stages of development. We will explore energy storage technologies and infrastructure improvements to the electric grid in more detail later in this chapter.

Wind power also has an impact on wildlife. The large, spinning blades of wind turbines are responsible for bird and bat kills; however, strategies such as placing units out of the path of predictable flight patterns can reduce the harm to birds, and adjusting the speed at which the blades turn can reduce bat kills.

photovoltaic (PV) solar panel a panel that produces an electric charge when it is exposed to sunlight.

concentrated solar thermal (CST) plant a solar power technology that captures heat using huge mirrors to focus sunlight toward liquid-filled pipes or a centrally located "power tower."

Solar Power

Some see solar power as the most logical long-term solution to our energy needs. Inventor Elon Musk (founder of Tesla Motors and SpaceX) notes that providing the entire energy needs of the United States through solar power would require a 100-mile-by-100-mile area (10,000 square miles, or 26,000 square kilometers) somewhere in the Southwest. More realistic estimates by the National Renewable Energy

Laboratory have forecast about twice that area—about 14 million acres (22,000 square miles, or 57,000 square kilometers) of solar panels, approximately the size of Lake Michigan. The most common solar energy technology by far is the **photovoltaic (PV) solar panel**. PV solar panels use sunlight to directly produce an electric current. They are typically made of specially treated silicon (the same material used for the electronic circuitry in our computers and phones) that produces an electric charge when it is exposed to sunlight. PV solar panels can be applied at many scales, from small ones designed to recharge a battery-powered device—like a phone—to an array of one or two dozen panels on a rooftop, to hundreds or thousands of panels covering many acres on a solar farm (**FIGURE 14.4**). A second technology, **concentrated solar thermal (CST) plants**, captures heat from the Sun by using huge mirrors to focus sunlight toward liquid-filled pipes or a "power tower" that holds a large container of liquid. The liquids heated by the Sun are then used to heat water, which generates the steam needed to turn turbines for generation of electricity.

Of course, solar power can be used to provide basic heating functions too. "Passive" solar buildings are designed so that their exposure angle to the Sun, placement of windows, air circulation, and insulation will slowly collect heat during the day and release it at night. And solar energy technology for home use pumps water or another heat-exchanging fluid through a solar collector on the roof to heat up during the day and then stores it in a tank for later use (**FIGURE 14.5**).

Solar power for electricity shares many of the advantages and limitations of wind power. The fuel (sunlight) is freely accessible and nondepletable, and there are no air or water emissions; there are no direct costs or environmental impacts associated with fuel extraction, transportation, or pollution. Solar energy systems can be installed relatively quickly—a year or less for large projects and a matter of weeks for a rooftop installation—and in a wide variety of settings: urban rooftops, parking lots, and farm fields, to name a few. But all require reliable access to direct sunlight. This does not mean that the effectiveness of solar panels is limited to places such as Southern California or Arizona, which have mainly sunny days and high solar intensity. Germany, which has a climate similar to the notoriously cloudy and rainy Pacific Northwest in the United States, has been a world leader in installed PV solar panel capacity since 2005. However, it does mean that each installation site must be assessed to determine the extent to which it is shaded by trees or buildings and how much sunlight it receives on average throughout the year. Like wind power, solar power is a variable energy resource. Every day it stops generating power when the Sun sets and must be balanced with other power sources and/or coupled with energy storage to provide a reliable power supply. On a small scale, homeowners can store solar power with

FIGURE 14.4 Solar Power Both light and heat from the Sun can be used to generate electricity.

(a) Solar panels are composed of many photovoltaic (PV) cells made of specially treated material (typically silicon).

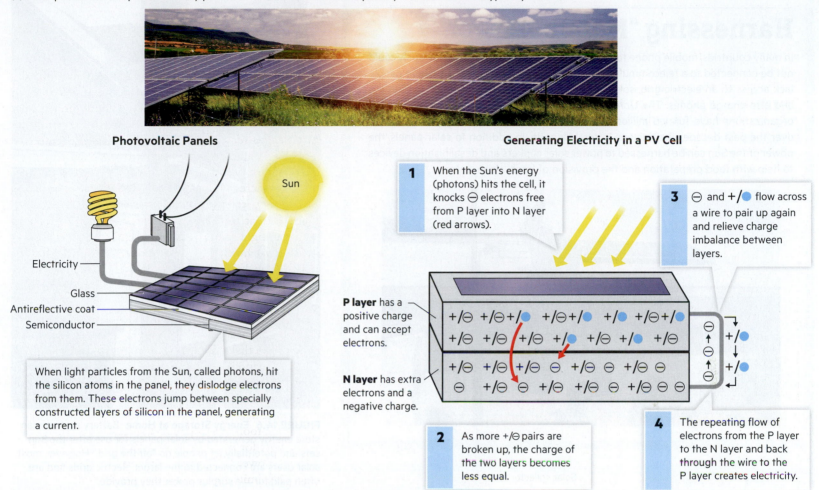

Photovoltaic Panels

Sun

Electricity

Glass
Antireflective coat
Semiconductor

When light particles from the Sun, called photons, hit the silicon atoms in the panel, they dislodge electrons from them. These electrons jump between specially constructed layers of silicon in the panel, generating a current.

Generating Electricity in a PV Cell

1 When the Sun's energy (photons) hits the cell, it knocks ⊖ electrons free from P layer into N layer (red arrows).

3 ⊖ and +/● flow across a wire to pair up again and relieve charge imbalance between layers.

P layer has a positive charge and can accept electrons.

N layer has extra electrons and a negative charge.

2 As more +/⊖ pairs are broken up, the charge of the two layers becomes less equal.

4 The repeating flow of electrons from the P layer to the N layer and back through the wire to the P layer creates electricity.

(b) Concentrated solar thermal (CST) plants focus sunlight toward a receiving point on a tall tower.

Concentrated Solar Thermal Plant

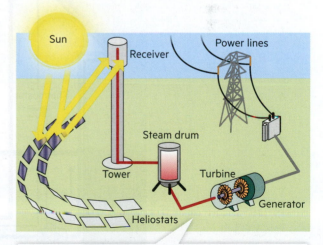

Sun
Receiver
Power lines
Steam drum
Tower
Turbine
Generator
Heliostats

In power tower CST systems, numerous large, flat, Sun-tracking mirrors, known as heliostats, focus sunlight onto a receiver at the top of a tall tower. A heat-transfer fluid heated in the receiver is used to generate steam, which in turn is used in a conventional turbine generator to produce electricity.

Art adapted from (a) US Department of Energy (2014) and Australian Academy of Sciences (2015) and (b) A Student's Guide to Global Climate Change (2016b).

Harnessing "Mobile" Power

In many countries, mobile phone technology connected people who previously could not be connected to a telecommunication network. Similarly, in places where people lack access to an electric grid, solar power can replace kerosene lamps for lighting and also charge phones. The United Nations, World Bank, and other international organizations have funded millions of solar home systems throughout the world over the past decade, like this one in rural Uganda. In addition to solar panels, the power of the Sun can be harnessed to power solar cookers and desalinization devices to help with food preparation and the provision of fresh, clean water.

FIGURE 14.6 Energy Storage at Home Battery systems can store energy generated by solar panels for use after the Sun sets and potentially let people go "off the grid." However, most solar users are connected to the larger electric grids and are often paid for the surplus power they provide.

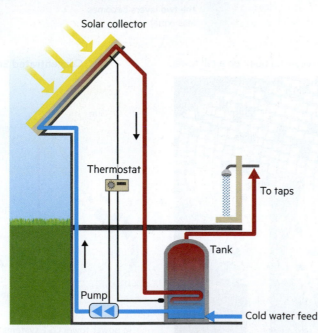

FIGURE 14.5 Using the Sun for Hot Water Sunlight can heat a home and provide hot water. A rooftop solar collector can generate hot water by circulating fluid through tubes that are heated by the Sun. The heated fluid then moves through a coil inside a water-filled tank, providing a supply of hot water.

Adapted from NC Sustainable Energy Association (n.d.).

specialized batteries (**FIGURE 14.6**). But more often, solar panel users rely on their connection to the larger electric grid so that they can send surplus power out when it's sunny (often getting paid in the process) and draw on power from other sources when it's not.

There are, however, some indirect costs and environmental impacts associated with the production and disposal of solar panels. The technology uses hazardous materials such as corrosive acids, as well as some heavy metals, and requires workplace protections and disposal precautions to protect human health and the environment. PV solar panels also rely on the use of rare earth elements, the mining of which (discussed in Chapter 9) has adverse human and environmental impacts. In the future, limits on the supply of these minerals may also affect the availability of PV technology. PV solar panels are relatively durable because they have no moving parts, but their efficiency declines over time, requiring most to be replaced after 30 years. And CST power plants use steam turbines, so they require a water supply—although once used, the water can be condensed and recirculated. Further, large solar farms (and certainly Musk's theoretical

100-by-100-mile one) require the conversion of large areas of land, often in sensitive desert habitats where they can have negative impacts on biodiversity. The planned construction of a 7,000-acre (2,800-hectare) PV farm outside of Las Vegas, Nevada, will likely negatively impact the threatened desert tortoise that makes that area home.

Solar power supplies just over 1% of global energy consumption, but it is increasing rapidly. The United States also uses solar to meet about 1% of its electricity consumption, and this is expanding quickly too. Notably, the number of rooftop PV systems more than doubled since 2016 and now exceeds 2 million. The United States also hosts more than 20 CST plants and more than 2,500 utility-scale PV farms.

⬡ **TAKE-HOME MESSAGE** Both wind and solar power can be nondepletable, nonpolluting sources of electricity. Although they currently supply a relatively small share of global energy consumption, both wind and solar power are expanding at a rapid rate in the United States and globally. Wind and solar power provide variable energy production that changes in response to weather conditions or the time of day. For this reason, they must be balanced with other power sources or coupled with energy storage and shared across regions to provide a reliable source of power.

14.2 What Are Other Energy Alternatives?

Wind and solar are not the only energy alternatives to fossil fuels. Power also can be drawn from moving water (hydropower), splitting atoms (nuclear power), and tapping into the heat below Earth's surface (geothermal power). These technologies have seen heavy development in the past and have larger shares of current energy generation than do solar and wind power. However, for different reasons, including their impacts, trade-offs, and scarcity of the power generation source, they are not growing as quickly as wind and solar. For example, as water supplies have been threatened or challenged, the environmental organization Pacific Institute estimates that nearly half of the hydropower plants in the western United States may see decreases in power generation. Both hydropower and nuclear power have had more significant negative environmental impacts than wind and solar power. In the rest of this section, we'll explore the advantages and limitations of each.

Hydropower

Hydropower uses water to spin turbines and generate electricity. It is the source of about 6% of global energy consumption and just under 3% of the energy consumed in the United States—which makes it the current leading

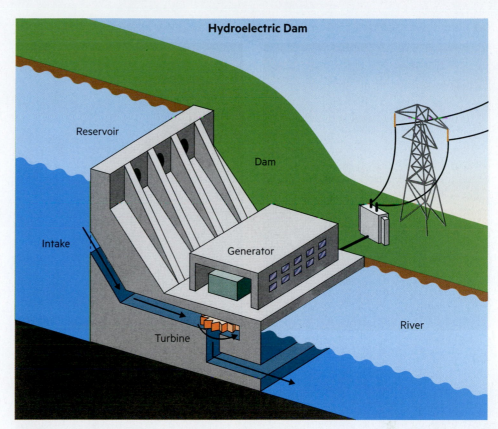

FIGURE 14.7 Hydropower Hydropower uses the power of flowing water to turn a turbine and generate electricity. A system using a dam that creates a reservoir is pictured here.

fossil fuel alternative. Although new devices exist to harness the mechanical energy of waves and tides, at this time nearly all hydropower is generated by two methods. The first involves constructing dams and reservoirs (**FIGURE 14.7**) to retain water and provide a reliable source of stored energy behind the dam. By adjusting the flow of water through the turbines of the dam, hydropower can be ramped up or down to quickly meet fluctuations in energy demand or variations in other energy sources contributing to the electric grid (such as wind and solar power). A second form called a run-of-the-river system does not hold back or store water but diverts some of the water flowing downstream through a pipe or canal that leads to a turbine. This method requires less investment and does not completely obstruct the river system, but it does generate less power.

Compared to other forms of power, hydroelectric has some obvious advantages. The energy source is the flow of water itself, which requires no extraction and produces none of the environmental effects related to mining and none of the emissions associated with fossil fuels. However, because many dams flood large areas to create reservoirs, they do disrupt the local ecosystem and can sometimes displace large numbers of people as well. China's Three Gorges Dam on the Yangtze River, the world's largest hydropower facility, displaced more than 1 million people, altered entire ecosystems, and endangered more than 50 local plant species and local fish

hydropower power generated when water is used to spin turbines and generate electricity.

FIGURE 14.8 China's Three Gorges Dam The Three Gorges Dam on the Yangtze River in China is the world's largest hydropower station. **(a)** These aerial photos show the river before and after dam construction. Construction of the dam and the resulting reservoir displaced more than a million people living in towns and villages along the river—such as the city of Kaixian **(b)** that was leveled to make way for the reservoir.

nuclear power power generated when the nuclei of atoms are split, releasing a large amount of energy.

fission a process where the nuclei of atoms are split, releasing a very large amount of energy.

populations (**FIGURE 14.8**). And recent research has indicated that hydropower facilities still contribute to greenhouse gas (GHG) emissions. When rivers are dammed to create reservoirs, they often flood large vegetated areas; over time this vegetation decomposes, releasing carbon dioxide (CO_2) and methane (CH_4). Further, dams are often made from concrete, the production of which generates large amounts of CO_2. Nevertheless, hydropower still contributes far less GHG emissions than do fossil fuel plants. Run-of-the-river dams built on rivers with high enough minimum flows generally avoid the

impacts associated with reservoirs; however, by diverting some water away from the river, they can still alter the water level, flow rate, and temperature in ways that adversely affect aquatic organisms. Another issue with hydropower is that many of the best sites for generating hydropower have already been developed, which has slowed the rate of new hydroelectric development in recent decades. Plentiful and powerful rivers with adequate sites for construction of dams and reservoirs enable Norway, Brazil, Venezuela, and Canada to get the majority of their electricity from hydropower. Similar conditions in the Pacific Northwest of the United States allow hydropower to provide more than half of the electricity to the region. But not all rivers have fast enough currents or enough of an elevation drop to reliably power large-scale commercial generating facilities, and some are seeing less flow because of climate change.

There are, however, untapped hydropower resources in Earth's oceans. China, Great Britain, Canada, and the United States are among the nations experimenting with tidal machines—essentially large floats with submerged turbines. Other efforts are attempting to harness the energy of ocean waves. Despite promising developments, tidal and wave hydropower are not yet commercially viable technologies for providing a significant share of electricity.

Nuclear Power

Say the words *nuclear power* and negative reactions often come to mind, yet this power source generates little air pollution and virtually no CO_2 emissions. Nuclear energy accounts for about 4% of the total energy consumed worldwide. The United States has more nuclear energy capacity (94 reactors) and generates more electric power from nuclear reactors than any other nation. These reactors provide 9% of the overall US energy consumption and 20% of the electricity generated in the country. However, several countries rely on nuclear power to an even greater extent, particularly France, Belgium, Hungary, and Slovakia, who each use nuclear power to meet more than half of their domestic electricity needs.

So what is **nuclear power**? In one sense, nuclear power plants operate very much like coal or natural gas power plants, using heat to convert water to steam that turns turbines, thereby generating electricity. However, nuclear power utilizes a completely different sort of energy conversion to generate heat. When we burn fossil fuels for energy, the chemical energy of the electrons in atoms produces heat. But nuclear reactions harness the power of **fission**, a process where the nuclei of atoms are split. This releases a very large amount of energy (in the form of heat) as well as neutrons, which collide with and split other unstable atoms, thereby creating and sustaining a powerful chain reaction. The most common atoms used in nuclear power plants are unstable uranium isotopes (**FIGURE 14.9**).

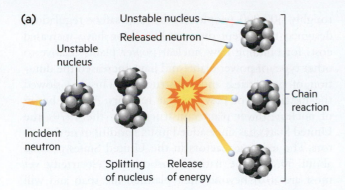

(a)

Unstable nucleus

Released neutron

Unstable nucleus

Incident neutron

Splitting of nucleus

Release of energy

Chain reaction

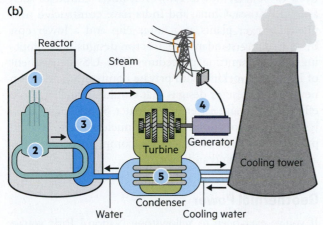

(b)

Reactor

Steam

1

2

3

Turbine

4

Generator

5

Condenser

Cooling tower

Water

Cooling water

FIGURE 14.9 Nuclear Fission (a) In nuclear fission, a neutron collides with and splits the nucleus of an atom. This releases energy and more neutrons that can in turn split the nuclei of additional atoms in a chain reaction. **(b)** 1: In the nuclear reactor of a nuclear power plant, fuel rods full of uranium pellets are placed in water. 2: Inside the fuel rods, uranium atoms split, releasing energy. 3: This energy heats water, creating steam. 4: The steam moves through a turbine, which turns a generator to create electricity. 5: The steam cools back into water, which can then be used over again. At some nuclear power plants, extra heat is released from a cooling tower.

Adapted from (b) A Student's Guide to Global Climate Change (2016a).

One advantage of nuclear fission is that it utilizes a much larger share of each atom's potential energy than does the chemical reaction used for fossil fuel energy sources (**FIGURE 14.10**). This makes nuclear power a much more efficient energy source: 1 kilogram (2.2 pounds) of coal fuel in a power plant could power a 100-watt lightbulb for 4 days. The same weight of natural gas could power the lightbulb for 6 days, while 1 kilogram of nuclear fuel could power the lightbulb for 140 years. The volume of overall waste associated with nuclear power is also many thousands of times less than that stemming from fossil fuel sources. Like coal and natural gas, nuclear power can provide a reliable, consistent source of electricity that can be adjusted to provide more or less power as needed. But unlike fossil fuels, nuclear power produces no ash, carbon dioxide, carbon monoxide, nitrogen oxide, sulfur dioxide, or particulates. Even when the GHG emissions associated with the industrial processes of mining and processing uranium for nuclear

→ **CONTEXT**

Trade-Offs

The Onkalo Spent Nuclear Fuel Repository in Finland (at right) is expected to be the world's first permanent disposal facility for high-level waste from nuclear power plants. It is a series of tunnels dug more than 1,700 feet (520 meters) beneath the surface. Construction of a similar type of facility in the United States at Yucca Mountain, Nevada (below), has been controversial and has been on hold since 2010. While nuclear energy producers have generally favored the site, others, including environmental groups and the local gambling industry, have opposed it, pointing to the downsides of nuclear power, as well as the potential dangers of transporting the fuel past lucrative casinos.

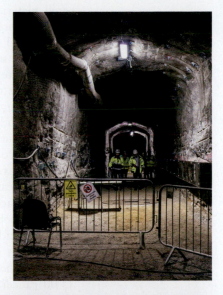

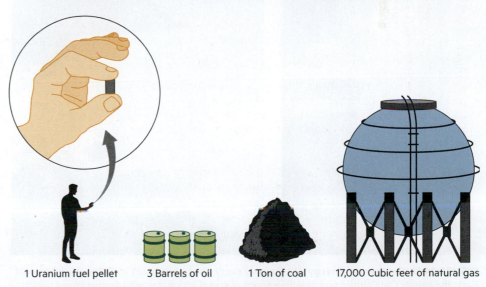

1 Uranium fuel pellet   3 Barrels of oil   1 Ton of coal   17,000 Cubic feet of natural gas

FIGURE 14.10 Energy Production by Fuel Weight Pound for pound, nuclear power plants produce much more energy by fuel weight than do fossil fuels.

Adapted from Center for Nuclear Science and Technology Information (n.d.).

geothermal power power generated by heat from below Earth's surface.

fuel are factored in, nuclear power has a significantly smaller CO_2 footprint than those of fossil fuel–based energy sources.

Yet nuclear power has significant downsides too. First, the spent nuclear fuel and other contaminated materials from nuclear reactors remain radioactive for thousands of years and must be isolated from humans and other species to prevent harm. Spent fuel is currently stored in temporary cooling pools or steel-and-concrete casks at the reactor sites. At this point, no country has developed a long-term storage facility for spent fuel from nuclear energy production, although some are under construction.

While nuclear power has an impressive safety record overall, uncontrolled nuclear reactions resulting from accidents have led to the radioactive contamination of large areas of land, air, and water. The two most significant accidents at nuclear power plants were the 1986 Chernobyl disaster in Ukraine and the 2011 tsunami-related accident at the Fukushima Daiichi nuclear power plant in Japan. The release of radioactive debris from the Chernobyl meltdown is related to an estimated 4,000 cancer deaths, and it forced 350,000 people to relocate. The Fukushima meltdown will eventually cause an estimated 100 to 1,500 deaths and the relocation of

roughly 300,000 people. Health and safety regulations designed to prevent disasters like these have increased costs for building new nuclear power plants relative to other types of power plants and have increased the duration of permitting and construction. This has slowed the installation of new nuclear reactors. After a boom of nuclear power plant construction in the 1970s, the United States has since added just a handful of new reactors. The existing reactors in the United States provide about 50% of the nation's carbon-free electricity, yet most are at or beyond their planned life span and will need to be shut down. However, other countries such as South Korea, China, and India have constructed new nuclear power plants at a faster clip and a lower cost by employing standardized reactor designs and adopting faster permitting procedures. The US Department of Energy is working with private companies to develop new nuclear reactor designs that would be smaller, more efficient, safer, and easier to construct. Some energy analysts and investors think that new nuclear reactors could provide consistent electricity to complement the variable power of wind and solar.

Geothermal Power

If you've ever been to Yellowstone National Park, you've probably seen the spectacular geysers like Old Faithful and various steaming hot springs, some with temperatures exceeding 450°F (232°C) (**FIGURE 14.11A**). This is geothermal activity—heat from below Earth's surface. Under the right conditions it can be harnessed to generate electricity or piped throughout buildings to provide heat. To generate electricity, **geothermal power** plants rely on wells that bring hot water and gases to the surface to turn a steam turbine. Or if the water temperature is slightly below the boiling point (100°C/212°F), special plants use it to heat up and vaporize a different fluid with a lower boiling temperature. The vapor from this fluid powers the turbine. Geothermal power has very low emissions when compared to those of fossil fuels, although geothermal gases can contain toxic hydrogen sulfide, which must be treated before it is released.

Like hydropower, the potential for geothermal energy production depends on location. This is why it supplies only a fraction of a percent of global energy consumption. These plants often have large start-up costs, and energy sources can cool down and no longer be useful. However, in certain locations it is a major energy source. Iceland uses geothermal power for a larger share of its electricity production (29%) than does any other country. This is because Iceland straddles the Mid-Atlantic Ridge where the Eurasian and North American tectonic plates are spreading apart. Here, within just a mile of the surface, underground temperatures exceed 250°C (480°F) (**FIGURE 14.11B**). Although geothermal power represents a tiny share of US energy production, the United States has more installed geothermal capacity than any

(a)

(b)

FIGURE 14.11 Geothermal Energy (a) Roughly every one to two hours in Yellowstone National Park, the geyser Old Faithful erupts with a spout of steam and water, with temperatures over 350°F (177°C). (b) Heat from beneath Earth's surface can be used on a large scale to power a steam turbine and generate electricity. Iceland taps geothermal energy for 90% of its space heating in buildings and nearly 30% of its electricity.

other country and hosts the largest collection of geothermal generating plants in the world at the Geysers project north of San Francisco.

The steam from geothermal wells can also directly provide heat for buildings. A network of pipes circulates the steam throughout neighborhoods, business complexes, or towns. In Iceland, about 90% of the space heating for buildings is provided by geothermal resources. In the United States, a system like this provides heat for the city of Boise, Idaho, including the state capitol building. And the heat beneath your feet need not be extreme for geothermal resources to play a role in heating a building. Simple household heat pumps take advantage of the constant subsurface temperature of 15°C (59°F) in shallow underground wells, using the temperature differential between the surface and underground to provide heat in the winter and cooling in the summer (**FIGURE 14.12**).

🏠 **TAKE-HOME MESSAGE** Alternatives to fossil fuel use include nuclear and hydropower, as well as geothermal energy. Each of these sources can provide a constant, reliable supply of electricity, unlike wind and solar, and do so on a large scale. However, limitations of these alternative energy sources include safety concerns about radioactive waste from nuclear power plants, habitat impacts in rivers with hydroelectric power plants, and the limited availability of suitable sites for large hydroelectric and geothermal energy plants.

14.3 How Is Alternative Energy Used for Transportation?

When we drive, get on a bus, or board a flight, we are depending on some source of energy to move us faster than we could move ourselves. More than 25% of energy consumption globally and in the United States goes toward transportation, and more than 90% of energy for transportation is provided by petroleum products such as gasoline and diesel fuel. A simple way to reduce the energy we use to get around is to increase the use of human-powered transportation (like biking or walking) and public transportation. Another is to drive more fuel-efficient vehicles such as hybrids that use both gas and electric power to achieve savings in fuel use of 20% to 30% over conventional gas-powered cars. Still another way is to change what is going in the tank, if there is a tank at all. This section will focus on different fuel choices that might replace what we now find at the pump.

Ethanol and Other Biofuels

Oil captures the energy from matter that was living millions of years ago, but there is a way to do the same thing without waiting so long. **Biofuels** use recently living matter, or by-products of their decomposition, as an

biofuel recently living matter or by-products of its decomposition used as an energy source (Chapter 13).

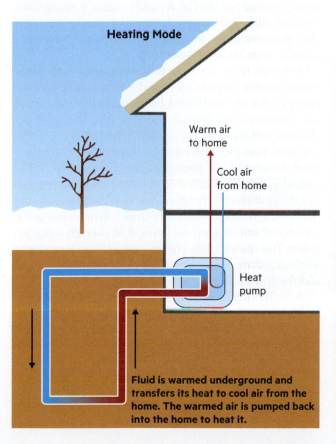

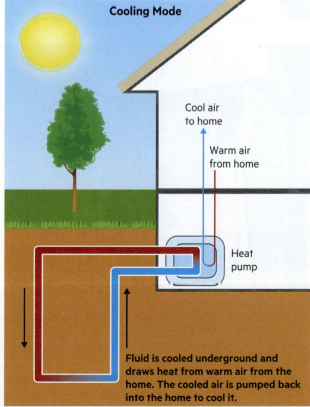

FIGURE 14.12 Geothermal at Home On a small scale, household heat pump systems take advantage of the fact that the temperature 20–30 feet (6–9 meters) down is almost always 15°C (59°F). Fluid is circulated through a loop buried underground, where it is warmed or cooled by natural temperatures beneath the surface. This fluid then warms or cools air from inside the home through the heat pump.

Adapted from US Environmental Protection Agency (2016).

ethanol a type of biofuel made of alcohol and produced by fermenting sugars in plants such as corn or sugarcane; it is often blended with gasoline.

electric vehicle (EV) a vehicle powered by an electric motor using a magnetic field to generate motion.

energy source. In fact, you might already be using some biofuel in your vehicle, as **ethanol**—an alcohol produced by fermenting sugars in plants such as corn or sugarcane—is often blended with gasoline. More than 40% of the corn grown in the United States is used for ethanol, and most gasoline in the United States now contains 10% ethanol. If you own a flexible-fuel vehicle in the United States, you can safely put gasoline blends with up to 85% ethanol in your tank (**FIGURE 14.13**). Currently, biofuel (most of it ethanol from corn) is a source of just under 5% of the fuel consumed for transportation in the United States, but its production has increased more than 200-fold since 2001 to more than 2 billion gallons (7.5 billions liters). Biofuels have the advantage of functioning as "drop-in" fuels because they do not require significant modifications to vehicle engines to use them. Biodiesel, which is made from vegetable oils and animal fats, already fuels about 1% of the US transportation sector. It primarily powers large trucks and heavy equipment, generating less particulate-matter air pollution than diesel from fossil fuels.

Any environmental advantages of ethanol are weakened when resource-intensive crop production is used and the crops are grown with farming practices that degrade the soil. In some cases in Brazil, forested land has been cleared, often through the use of fire, to grow biofuel crops, disrupting ecosystems and reducing biodiversity. In addition, deforesting land to plant biofuel crops causes a loss of the climate benefit trees provide by absorbing and storing CO_2. When biofuels are refined from food crops like corn, soybeans, or palm oil, this new demand can reduce the overall supply and drive up food prices. In recent years, demand for biodiesel has caused scarcity and price spikes for both soybean and palm oil.

With these impacts in mind, a great deal of research is focused on finding better sources of biofuels. Some see existing waste streams such as landfills, sewage treatment plants, and even discarded vegetable oils from restaurants as promising biofuel sources. Others look for new sources. For example, switchgrass and algae are two promising sources of biofuel because they grow quickly with relatively few inputs, are normally harvested on short rotations, sequester more carbon, and consume far less area for production. Government agencies and oil companies alike have made sizable investments in new techniques to grow and extract oil from algae. These include growing algae in salt water (to reduce the impact on freshwater supplies) and devising methods to extract a higher percentage of the algal oil with fewer energy inputs.

Electric Vehicles

Unlike biofuels, which substitute a different fuel in the engine of a vehicle, **electric vehicles (EVs)** are powered by an electric motor. This gives them many advantages. The engines in typical vehicles are internal combustion engines. These engines essentially generate motion by producing many controlled explosions to move a multitude of metal parts that are lubricated with oil. In contrast, electric motors generate motion much more directly with a magnetic field. This means EVs waste less energy as heat, so they're typically about 4 times more efficient than gas-powered cars and accelerate faster. EVs also have fewer moving parts, so they're easier and cheaper to maintain (**FIGURE 14.14**).

Although EVs are very promising, they have some environmental impacts to consider. Since most EVs use batteries that are recharged by plugging into an outlet, one environmental impact is tied to the source of the electricity leading to the outlet. While EVs are GHG emission–free, they still contribute to your carbon footprint if the source of your electricity is primarily coal or natural gas. In the United States, depending on the mix of sources of electricity where you live, the fuel efficiency of EVs ranges from 50 to more than 130 miles per gallon equivalent, with a national average of over 80 miles per gallon. But if alternative energy continues to generate an increasing share of our electricity, these numbers will rise, and the emissions related to EV use will fall.

Another impact is connected to the materials used in the batteries themselves. Lithium is the favored material for EV batteries (known as lithium-ion batteries) because it can store a large amount of energy for its weight and can be charged repeatedly without losing much energy storage capacity. But raw lithium is typically extracted from open-pit mines that use sulfuric acid to separate the lithium from other materials, or it is pumped out of underground saltwater deposits that

FIGURE 14.13 Ethanol as a Biofuel There are more than 19 million "flex-fuel" vehicles operating in the United States that can use a blend of gasoline and up to 85% ethanol (a biofuel typically made from corn).

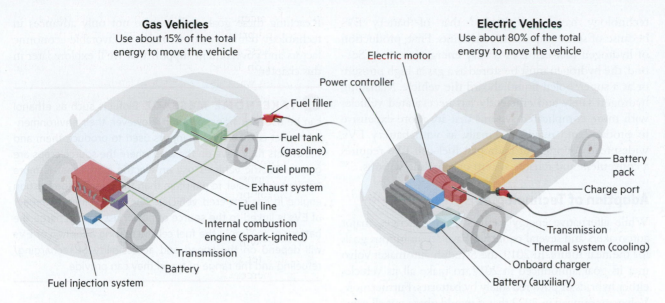

Gas Vehicles
Use about 15% of the total energy to move the vehicle

- Fuel filler
- Fuel tank (gasoline)
- Fuel pump
- Exhaust system
- Fuel line
- Internal combustion engine (spark-ignited)
- Transmission
- Battery
- Fuel injection system

Electric Vehicles
Use about 80% of the total energy to move the vehicle

- Electric motor
- Power controller
- Battery pack
- Charge port
- Transmission
- Thermal system (cooling)
- Onboard charger
- Battery (auxiliary)

FIGURE 14.14 Gas versus Electric Cars Much of the energy in gas vehicles is wasted as heat emitted from the engine and out the tailpipe. The electric current generated by large batteries in electric vehicles is efficiently sent to an electric motor as needed by the power controller.

Adapted from US Department of Energy (n.d.).

are evaporated at the surface. Both mining processes use a lot of energy and water, while producing solid and liquid waste that contains arsenic and other toxic materials. Most of the world's raw lithium is now mined in South America and Australia, although new mines have been proposed in the southwestern United States. Other minerals like cobalt and rare earth elements are also important in the batteries and electronics in these vehicles, and these resources also come from mines that have adverse environmental impacts (see Chapter 9).

Generally, however, GHG reductions throughout the life span of an EV more than compensate for GHG emissions during manufacturing and disposal. Recent life-cycle studies show that the environmental benefits of EVs far outweigh the costs when compared to internal combustion vehicles. In addition, many auto manufacturers, like Ford and BMW, are now employing environmental firms that vet mining operations to minimize environmental harms.

There are now more than 1.3 million EVs on the road in the United States and more than 7 million globally. But EVs still make up only about 1% of the more than 250 million passenger cars registered in the United States. Important constraints are the lack of availability of easy and quick recharging stations, the range (distance) EVs can travel on a single charge, and a battery life that meets or exceeds the typical useful life of a vehicle. Currently, the fastest-charging EVs can power up to 80% in about 15 minutes. While many charging stations are being installed, they are not nearly as ubiquitous as gas stations. The longer that EV batteries can last, the greater the range they can hold for each charge, and the more convenient charging stations become, the better EVs will compete with gas-powered vehicles.

Fuel Cells

Fuel cells are like batteries, but they do not need to be recharged. Instead, they can continue to generate electricity as long as a fuel (typically hydrogen) is supplied.

Fuel-cell systems work by stripping hydrogen molecules of their electrons as the hydrogen molecules enter the system, which creates a flow of electricity. These electrons and the positively charged hydrogen ions from which they were stripped combine with oxygen to form water as they leave the fuel-cell system. So hydrogen fuel-cell vehicles emit only water. Refueling is quicker than charging a battery, and for large vehicles, fuel-cell EVs can achieve better ranges than EVs powered with batteries. The shipping company UPS is now testing a fleet of electric trucks powered by hydrogen fuel cells, and several car companies including BMW and Honda have prototype fuel-cell cars (**FIGURE 14.15**). But hydrogen fuel-cell

fuel cell a device that creates an electric current by stripping electrons from hydrogen molecules.

FIGURE 14.15 Hydrogen Fuel-Cell Technology Water is the only emission generated directly by these vehicles. The delivery company UPS makes extensive use of fuel cells in their vans and is now testing the technology in semi-trailer trucks.

cogeneration a principle where systems capture waste heat from power plants and use it as an additional source of heating or cooling.

technology has lagged behind that of battery EVs because of some important constraints. First, production of hydrogen fuel-cell EVs is very energy intensive. Second, the hydrogen must be stored as a gas at high pressure or as a supercooled liquid aboard the vehicle. Managing hydrogen safely and efficiently has necessitated vehicles with more complicated systems that are more expensive to produce and maintain. Finally, as with battery EVs, widespread adoption of hydrogen fuel-cell EVs requires a convenient and reliable system of refueling stations.

Adoption of Technologies

While alternative energy sources face challenges, major automakers and several countries have set ambitious goals for the near future. In 2019, the Swedish automaker Volvo met its goal, announced in 2017, to make all its vehicles either hybrids or powered solely by batteries. Furthermore, Volvo announced in 2022 that it would phase out all vehicles with internal combustion engines, including hybrids, and have an electric-only fleet by 2030. General Motors has pledged that all of its cars and trucks will be all-electric by 2035. Norway, a country in which more than 25% of the cars are already battery EVs, has a goal to be free of gas and diesel cars by 2025. Germany and India have a similar goal set for 2030, and England and France have set 2040 as the target date. Both California and China have committed to phasing out gas and diesel vehicle sales by 2035. These are signs that a transition is under way.

Reaching these goals will require not only advances in technology but also a combination of favorable economic factors and government support that we'll explore later in this chapter.

⬠ **TAKE-HOME MESSAGE** Biofuels such as ethanol can be used for transportation. However, their environmental impact is linked to the inputs used to produce them and the effects they have on the land where the source crops are grown. Electric vehicles use an electric motor that is more efficient and easier to maintain than the internal combustion engine in gas-powered vehicles. The environmental impact of EVs is linked to the source of power used to recharge the battery or to supply the fuel cell. Increasing adoption of EVs will depend on the speed and accessibility of recharging/refueling and the range of travel they can provide.

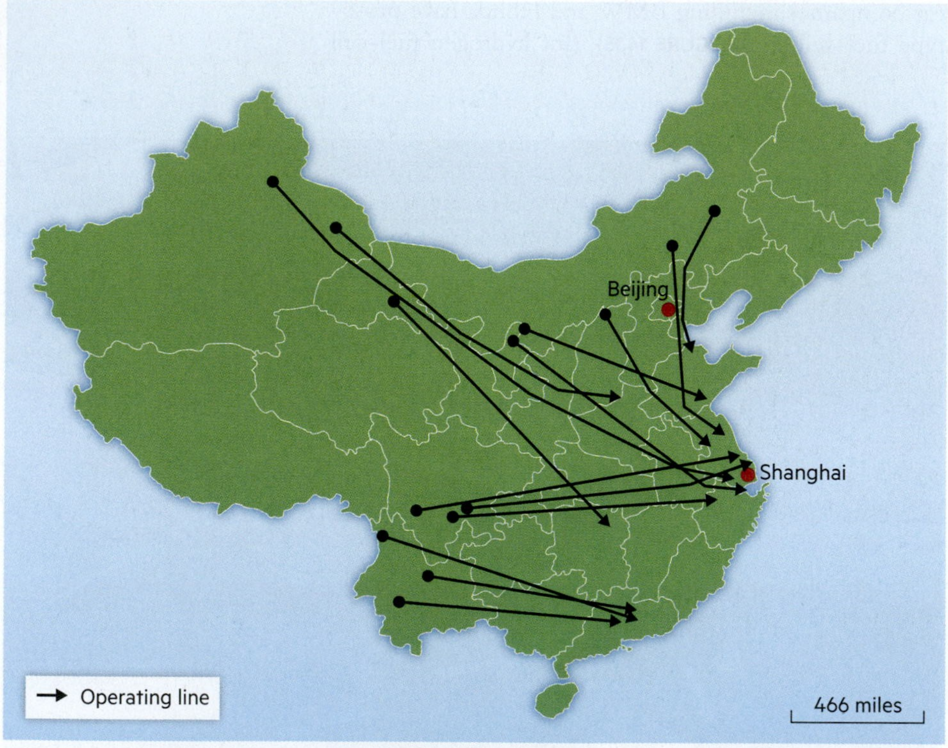

14.4 What Role Does Energy Conservation Play in Energy Transitions?

At the start of the chapter, we discussed how most of us have come to take energy for granted. But what might happen if we decided not to? Conserving and making better use of energy is part of formal *energy transition plans* that many businesses, communities, and schools create to change how they use energy. For example, in 2009, Michigan State University developed its Energy Transition Plan. While the overall goal is to "create transformational change," the first steps emphasize conservation: "Avoid wasteful energy- and carbon-intensive practices" and "improve efficiency."

In fact, the lowest-hanging fruit for energy conservation is to avoid waste. When we waste energy, we are consuming it without putting it to any productive use—like when we leave the lights on in a room we're not using or run the furnace to heat a room while leaving the windows open. Energy audits estimate that the United States could reduce energy consumption by 23% and save more than $1 trillion annually simply by wasting less energy. Moreover, the high rate of growth in energy use among developing nations could be cut in half by cutting waste. And about 10% of the electricity produced at power plants is lost through its transmission over long transmission lines: in the United States this amounts to enough electricity to power a city the size of New York for about 6 months. New, higher-voltage transmission lines have very low energy losses and would suffer less from power outages and blackouts. China has recently developed the world's first ultrahigh-voltage transmission grid: it connects high-consumption population centers with distant power generators and has less than half the energy losses of conventional transmission lines (**FIGURE 14.16**).

FIGURE 14.16 Ultrahigh-Voltage Transmission in China China now has ultrahigh-voltage transmission systems that enable long-distance energy transmission with significantly reduced energy losses. These lines bring power from areas of low population to densely packed urban areas.

Adapted from *The Economist* (2017).

The Solutions Project

Have you ever set a goal for yourself that will take decades to achieve? What do you want to accomplish by the year 2050? We're used to setting goals and planning what we'll do day to day, week to week, and even for a few years for bigger accomplishments such as earning a college degree. But thinking forward across decades is so full of uncertainty and so distant from our decisions in the present that individuals and organizations rarely plan over such time frames. Yet long-term planning is often required to achieve really ambitious societal goals—such as shifting energy use to entirely renewable sources by the year 2050.

One thing that can help in the planning process is to develop a more detailed, concrete vision of what achieving the goal would look like. A team of civil and environmental engineering professors and graduate students at Stanford University and the University of California, Berkeley, sought to do just that for the goal of 100% renewable energy. Calling themselves the Solutions Project, they developed detailed technological road maps for how each of the 50 US states could reach this goal.

For each state, these researchers estimated future demand for electricity, transportation, heating and cooling, and industry. They then analyzed current and future installed capacity for wind turbines, CST plants, solar PV installations, hydroelectric power plants, geothermal power plants, and tidal and wave-energy power. Determining the best available renewable energy resources in each state involved identifying the generating capacity of each power source, estimating their cost-effectiveness and overall feasibility in terms of land use and other concerns, and considering how variable generating capacity could be balanced to provide a reliable energy supply. The researchers also factored in energy conservation upgrades in buildings, as well as the impact of shifting to electric vehicles, increasing reliance on public transportation, and renewable energy for heating and cooling such as through use of heat pumps. All of this results in a mix of renewable energy technologies unique to each state.

Next, the team recognized that one of the greatest concerns associated with a large-scale shift to renewable energy is the variable nature of

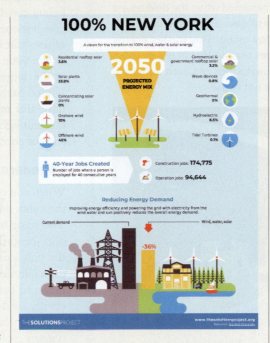

wind and solar power. So they turned their attention to identifying the most cost-effective, current energy storage technologies to help balance out the variable generating capacity of wind and solar. They ran simulations of various energy resource arrangements combined with different energy storage technologies to identify the most cost-effective way to ensure reliable energy supplies.

Finally, the research team highlighted the economic, health, and environmental benefits associated with a transition to 100% renewable energy in each state, including jobs produced, cost savings compared to fossil fuels, air pollution–related deaths avoided, and greenhouse gas reduction. The hope is that this work will help the public and policy makers see not just the value in pursuing such an ambitious goal but also how it might be realized. To that end, the Solutions Project is now an organization that includes not just scientists but also business leaders, activists, and even

celebrities such as Mark Ruffalo and Leonardo DiCaprio. They are dedicated to raising public awareness about these technological pathways to 100% renewable energy and are encouraging policy makers to start taking steps in this direction. In recent years, the states of California, New York, and Illinois have set long-term renewable energy goals informed by the Solutions Project. For example, part of New York has launched an initiative to expand offshore wind power—a key part of the energy mix for this state outlined in the Solutions Project. And the engineers in the Solutions Project have now published research identifying 100% renewable plans for 139 countries to support the global community's goal of keeping global warming to less than 2°C (3.6°F) above preindustrial levels under the Paris Climate Agreement. Setting a goal is important, but making a road map that shows in detail how to reach it can make it more real and achievable.

🔎 What Would You Do?

Look up the closest city to you at the Solutions Project website (https://thesolutionsproject.org/), and read about the benefits associated with a transition to 100% renewable energy for your area. If you were a community leader trying to encourage a transition to renewable energy, which of these benefits do you think would resonate the most with your neighbors? How would you try to convince your community to adopt more renewable energy resources?

Vampire Power

According to the US Department of Energy, about 25% of the energy used by home electronics, such as TVs, computers, and printers, is consumed when the devices are turned off. Why? As long as they are plugged in, these devices are still drawing a small amount of power to remain in a kind of standby mode. This wastes energy and money, but it can be solved by unplugging devices that are not in use or using a power strip that can easily be switched off. Some power strips can even automatically stop the flow of electricity to electronics that are not in use.

We can also develop and adopt technologies that use energy more efficiently. There are a lot of conservation opportunities in the average vehicle, as less than 13% of the energy in a gallon of gasoline is used to move an average car. Sixty percent of that energy is lost in waste heat from the engine itself, as well as from braking, idling, and powering accessories. A simple way to conserve is to have lighter cars. Lighter cars (or planes, ships, trains, etc.) simply take less energy to move. Reducing the rolling resistance of tires can improve fuel economy by 50%: remember to inflate your tires! Homes and office buildings are another prime target for energy conservation. Buildings in the United States account for 42% of our energy consumption, and adding insulation, better windows, and more efficient lighting, heating, cooling, and appliances can save a lot of energy, with opportunities to go much further.

Cogeneration

When you turn on the heat in a car and hot air blasts out of the vents, do you wonder where the hot air comes from? The waste heat generated by a car's engine as it burns gasoline heats the air around it, and this heated air is pumped through a car's vents to provide heat for the interior. This principle lies at the heart of **cogeneration**, or combined heat and power (CHP) systems. About two-thirds of the energy used to make electricity in fossil fuel power plants is typically lost as heat to the atmosphere, but cogeneration systems capture this waste heat and use it as an additional source of heating or cooling. On a large scale, Denmark has been able to provide more than half of its household heating with cogeneration technology (**FIGURE 14.17**). But more common are new local projects adopted by businesses. In 2006, the University of New Hampshire built a cogeneration plant where the "extra" heat is used to heat buildings. However, a 2017 project at Duke University was met with controversy, as groups questioned the wisdom of building a plant that was still fueled by natural gas. The project was put on hold indefinitely in 2018.

⬡ **TAKE-HOME MESSAGE** Although conservation measures are not a source of energy production, they can still contribute to a reduction in the use of fossil fuels. Conserving energy can include simply avoiding waste, but it can also include efficiency measures in how we use, produce, and transmit energy. Harnessing waste heat is another way to make the most of fossil fuels.

14.5 What Could Drive a Transition away from Fossil Fuels?

What factors would drive a large-scale switch from fossil fuels to energy alternatives? Likely this would involve a combination of market forces, innovations,

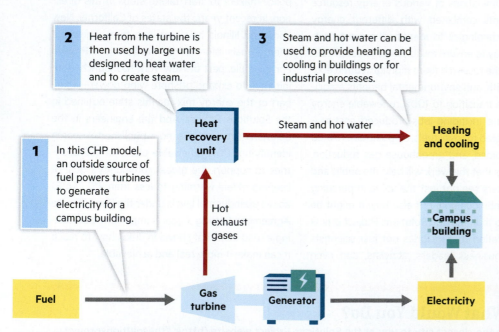

FIGURE 14.17 How Cogeneration Works In cogeneration, wasted thermal energy (heat) from generating electricity is used for industrial processes or building heating and cooling. These are known as combined heat and power (CHP) systems.

Adapted from US Environmental Protection Agency (2018).

infrastructure improvements, government policy, and individual actions. But first let's consider why an energy transition like this is so challenging.

In the last chapter, we learned that coal, oil, and gas were once new "alternative" energy sources, and it took centuries for these fossil fuels to gain dominance over firewood and horse-powered transportation. Energy transitions take time. This transition also teaches us that even as new energy resources gain traction, global energy use continues to increase. So we are seeing that even as the share of energy consumption supplied by coal has declined relative to that supplied by oil and gas, in absolute terms annual coal production continues to rise. This means that a shift to alternative energy sources requires these sources to meet both current and growing demand (barring an unforeseen shift to conservation). Because of these hurdles, it's best not to think of a transition to alternative energy as an immediate or complete replacement of fossil fuels. However, many people see the start of this transition happening now. Let's look at some of the key factors driving this shift to alternative energy.

Economic Forces and Employment

In economic terms, alternative energy sources—particularly wind and solar power—are poised to be a substitute and eventual replacement for fossil fuels. But can they ever compete on price? At certain points in our recent history, alternative energy has benefited from scarcity and rising prices of fossil fuels. Big technological gains in solar and wind power and in vehicle efficiency were made during the 1970s and early 2000s when the price of oil was rising rapidly. However, rising fossil fuel prices also led to advances in fossil fuel extraction technologies and increased the supply and demand for these fuels, especially as prices have declined again. But in the midst of these boom-and-bust cycles for fossil fuels, the cost of wind and solar has fallen steadily. In 2009, expanded production and technological advances in solar panel and wind turbine design began to drive costs down. As **FIGURE 14.18** shows, between 2009 and 2020 the cost of wind and solar power declined by almost 71% and 90%, respectively.

While natural gas is still dominant, wind and solar are now the cheapest forms of energy in many regions in the United States—and coal is not competing very

FIGURE 14.18 Declining Production Costs for Wind and Solar The levelized cost of energy (LCOE) measures the lifetime costs of an energy source relative to the amount of energy it produces. It shows the value of an energy source over the long term and is summarized as cost in dollars per unit energy in megawatt-hours (MWh). Between 2009 and 2020, the LCOE of wind power and especially photovoltaic solar power declined dramatically, with solar power becoming even less expensive than wind in 2019.

Adapted from Lazard (2020).

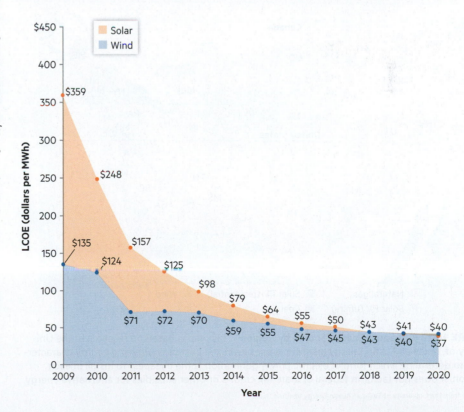

When a Greenhouse Is Not a "Greenhouse"

The Phipps Conservatory and Botanical Gardens in Pittsburgh was designed to be one of the most energy-efficient buildings of its type in the world. It uses three emission-free energy sources—solar panels, geothermal wells, and a wind turbine—along with energy conservation strategies to generate more energy than it uses. It is also host to a tropical forest conservatory/greenhouse. Ordinarily, a structure like this would become uncomfortably hot in the summer months because of the heat-trapping aspects of the greenhouse. However, the innovative shape of the greenhouse sections at the Phipps Conservatory mitigates the consequences of this "greenhouse effect" and allows it to stay cooler inside than out during the summer months while using very little electricity for cooling.

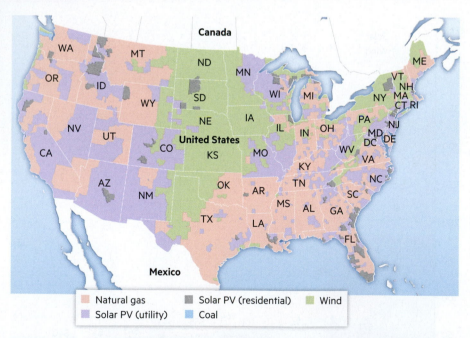

FIGURE 14.19 Cheapest Energy Source by Region This map is based on research at the University of Texas at Austin Energy Institute. It uses levelized cost of energy and county characteristics to determine which form of electric power generation is the cheapest to install in each location. It does not factor in federal or state tax credits or other subsidies for alternative energy.

Adapted from the University of Texas at Austin Energy Institute (n.d.).

well at all (**FIGURE 14.19**). As costs for wind and solar have fallen, the share of global electricity supplied by these sources has grown rapidly. Both are projected to continue their rapid expansion (**FIGURE 14.20**).

Transitioning to alternative energy and sources like wind and solar provides new job opportunities. As we see in **AT A GLANCE: AN ENERGY TRANSITION IS UNDER WAY**, the solar and wind industries now employ a larger share of the electric power generation workforce than all fossil fuel industries combined. Many of these jobs are related to the installation of solar arrays or construction of new wind turbines. However, there are plenty of other jobs, including energy transmission engineers and technicians; heating, HVAC, and energy conservation specialists; designers, engineers and manufacturing professionals working on electric vehicles; and sales and business services professionals.

Innovation and Infrastructure for Energy Storage and Distribution

The variable nature of wind and solar energy will limit the growth of these energy sources as both require innovations in energy storage and distribution. Engineers and utility analysts estimate that as variable sources of power, such as wind and solar, approach 25% of the supply of our electricity consumption, they will threaten the reliability of our current distribution system. This is because the larger the share of variable power on the existing electric grid, the more difficult it is to balance with other sources when there is a mismatch between energy demand and energy supply. Under our current electric distribution system, there is almost no capacity to store energy. Once it is produced, it is moved around on the grid to meet demand, but for the most part it cannot be saved for later use. This means that energy production must meet consumption in real time or else the system becomes unreliable and fails to provide the power needed. A great deal of energy research and development is now focused on devising energy storage solutions and better ways to distribute power.

The most common technology for large-scale energy storage is called pumped hydro. It uses electricity (typically from a hydroelectric dam) to pump water uphill to a reservoir when energy is plentiful, which allows water to be run back downhill through hydropower turbines to generate electricity when energy is scarce (**FIGURE 14.21**). Another existing technology uses electricity to pump compressed air into underground caverns when energy is plentiful, and this air can then be released to turn a turbine when energy is needed. The potential capacity for both of these storage methods depends on location-specific features: pumped hydro requires steep topography and water, while compressed air requires large caverns. And while pumped hydro can store the energy for long periods of time in the reservoir, the energy held as compressed air dissipates within 24 hours as the air expands and escapes. Yet another storage technology uses energy to heat salt to temperatures in excess of 1,000°F (540°C). Molten salt cools slowly, and at these temperatures it can be used to heat water for steam turbines. Molten salt is hot enough to generate power for about 24 hours before it needs to be reheated.

Large batteries are an option, but they cannot yet provide storage for very long periods of time. The electric grid in Alaska is connected to a 1,300-ton nickel–cadmium battery that can provide 40 megawatts (enough to power about 40,000 homes) for about 14 minutes. The world's largest battery (actually an assembly of more than 18,000 lithium–ion batteries) is under development in Long Beach, California. When completed, it will be able to deliver 100 megawatts (enough to power about 100,000 homes) for about 4 hours. But batteries can work on a smaller scale too. Pairing batteries with wind turbines can store some of the power the turbines create. Similarly, pairing rooftop solar panels with batteries can store energy for use after the Sun sets each day. And auto manufacturers are working on vehicle-to-grid-enabled (V2G) electric vehicles that could make the power stored in their batteries available to the grid when they are not in use. As electric vehicles become more common, V2G technology could provide considerable storage capacity.

Replacing gasoline-powered cars is a central and challenging aspect of any energy transition away from

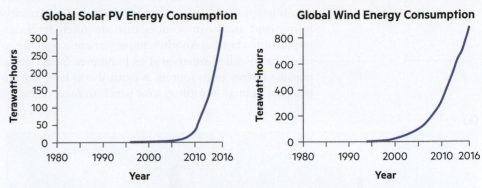

Global Solar PV Energy Consumption

Global Wind Energy Consumption

FIGURE 14.20 Rapid Growth of Wind and Solar Power The share of global electricity supplied by photovoltaic (PV) solar and wind power has grown quickly, doubling several times since 2000. Adapted from Ritchie and Roser (2019).

fossil fuels. Petroleum supplies the largest share of energy consumption both in the United States and globally, and most of this oil is used for transportation. To transition away from direct fossil fuel power to indirect power from stored electricity, drivers will need convenient and reliable infrastructure and charging stations along major roads and highways. Drivers need confidence that the range of their vehicles will meet their transportation needs and that they can recharge their batteries quickly with the same convenience they have come to expect from gas stations.

Finally, the current electric grid would need to change. Improvements might include better high-voltage

FIGURE 14.21 Energy Storage The United States currently has storage capacity for just 2% of the energy it generates. More than 90% of this is stored via pumped hydro, as at the Ludington Pumped Storage Plant in Michigan, pictured here. Pumped hydro uses electricity to pump water up an incline to a reservoir so that it can be run back down through a turbine when it is needed.

transmission lines that efficiently transmit electricity from wind and solar sources that are often far from population centers. Another improvement could be a move to so-called smart-grid technologies. Smart grids manage energy from sources as expansive as large power plants to as small as rooftop solar panels to meet demand.

Or microgrids—smaller systems designed for hundreds or thousands of energy users—can exist independently or connect to the larger grid as needed. In this way, many microgrids can be combined to form a larger grid, which provides resiliency: the ability to adapt to disruptions in the system (**FIGURE 14.22**). Should one part of a grid experience an outage, it does not ripple through the entire system, and other parts can provide the additional energy needed.

Government Actions

Government policies historically have played a big role in shaping energy transitions. This includes government support for research and development, tax advantages for consumers, regulating the energy industry, and funding infrastructure. Policies like these remain in place. The US government undertakes research directly or in partnership with private firms through institutions such as the US Department of Energy National Laboratories, and national and state governments provide grant money to encourage and support research at colleges and universities. The US government has also offered assistance to alternative energy companies either with direct start-up funding through stimulus grants or with guaranteed loans and tax advantages. At the consumer level, there are numerous tax incentives for rooftop solar installations, energy conservation improvements in the home, and

(a)

(b)

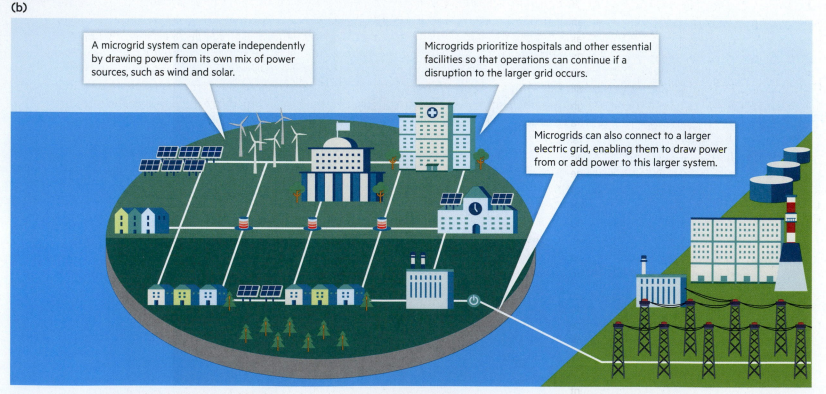

A microgrid system can operate independently by drawing power from its own mix of power sources, such as wind and solar.

Microgrids prioritize hospitals and other essential facilities so that operations can continue if a disruption to the larger grid occurs.

Microgrids can also connect to a larger electric grid, enabling them to draw power from or add power to this larger system.

FIGURE 14.22 An Energy Grid within a Grid (a) The Brooklyn Microgrid Startup lets neighbors produce, consume, and purchase power using solar panels they have installed. **(b)** A microgrid operates like a self-sufficient "island" that can power a concentrated area. It is connected to the main electric grid to both balance and exchange electricity. Microgrids can keep power on during blackouts, storms, and other disasters.

Self-Driving Cars

Imagine getting into a car and sitting in the driver's seat but letting the car do the driving while you safely respond to e-mails, scan the news headlines, and send texts to friends. Most major auto companies, some tech companies such as Google, and ride-sharing companies such as Uber are racing to develop viable autonomous vehicles: self-driving cars and trucks that will navigate, adapt to changing traffic patterns, drive, and park more safely and efficiently than can human drivers. Waymo, Google's autonomous vehicle company, has logged hundreds of thousands of miles in test drives throughout California, and other companies have begun road tests in Nevada, Michigan, and Florida. Could the widespread use of self-driving vehicles lead us to consume less energy for our transportation and help to reduce greenhouse gas emissions? The answer depends on both the way the technology is developed and how the cars are used.

As with any vehicle, self-driving cars can significantly reduce their climate impact if they are powered by electricity rather than gas. Electric motors are more efficient than gas engines, and if the share of alternative energy powering the electric grid continues to grow, widespread use of electric vehicles (EVs) will help cut greenhouse gas emissions. Although self-driving technology is not dependent on a particular fuel, most prototypes are in fact EVs. Like all EVs, widespread adoption of these vehicles depends on improvements in their range and a convenient and reliable infrastructure of charging stations that are relatively quick and easy to use. Self-driving EVs in particular would benefit from wireless charging, which requires only that the car park or slowly drive over a charging pad in the ground rather than relying on a person to plug the car into an outlet.

Other aspects of self-driving cars might also significantly boost energy efficiency and reduce environmental impact. The ability of these vehicles to rapidly process information from other vehicles on the road and the surrounding environment can help them to choose more efficient routes, adjust to congestion, practice optimal driving speed and braking responses, and quickly locate parking—all of which save energy and improve safety. Moreover, some have speculated that if driverless cars excel at accident prevention, this may allow them to be built with lighter-weight materials than those used in today's vehicles, making them even more energy efficient.

But perhaps the most important factors are related to vehicle ownership. Will they be primarily owned by individuals or used as part of a vehicle-sharing service in which members pay per use or by a monthly subscription? Vehicle sharing has several environmental advantages. For example, with vehicle sharing you could select a vehicle size for each trip that suits your needs rather than having to buy a bigger one-size-fits-all vehicle that suits all of your possible needs. If vehicle sharing reduces vehicle ownership, it also reduces the overall number of vehicles that need to be produced and maintained. Unlike individually owned vehicles that remain parked most of the day, shared vehicles will be driven throughout the day, trip after trip. Some vehicle-sharing plans could also facilitate carpooling if multiple members are heading to nearby destinations, thus reducing the number of individual vehicle miles traveled. For this reason, vehicle sharing has the potential to reduce congestion and the number of parking spaces needed.

However, if self-driving vehicles—whether privately owned or as part of a vehicle-sharing program—become a substitute for public transportation such as buses and trains, there could be negative environmental impacts. This is because public transportation moves large numbers of people very efficiently through busy transportation corridors. Moving this same number of people via self-driving vehicles as individuals or even small groups of passengers would likely consume more energy and lead to greater congestion. For this reason, urban planners have recommended that regional land-use and transportation systems integrate self-driving vehicle use in ways that complement (rather than replace) public transportation—that is, as a means to get to and from major public transit access points.

What Would You Do?

All of us have primarily been passengers or drivers in human-controlled vehicles our entire lives. Would you be comfortable riding in a self-driving car? For you, do the convenience and potential environmental benefits outweigh the unfamiliarity and potential environmental impacts of self-driving cars? Explain your reasoning.

An Energy Transition Is Under Way

Although history tells us that transitions to new sources of energy occur incrementally over long periods of time, current energy trends also show that we are undergoing a transition right now from fossil fuels to alternative sources of energy such as wind and solar.

The combined use of wind and solar to generate power is growing.

This map shows the location, type, and size of plants generating electric power in the United States at the end of 2018. While natural gas plants are the most prevalent, power from wind and solar is growing, especially in the middle of the country and the eastern and western coasts. The 2021 US government forecasts predict that by 2030, renewables will collectively surpass natural gas to be the predominant source of electricity in the U.S.

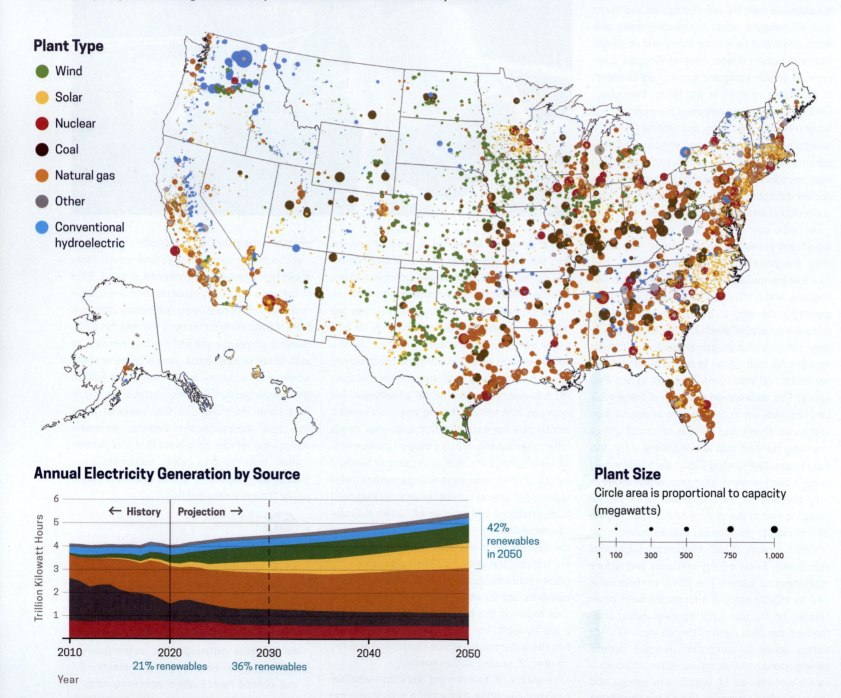

Plant Type

- Wind
- Solar
- Nuclear
- Coal
- Natural gas
- Other
- Conventional hydroelectric

Annual Electricity Generation by Source

← History | Projection →

42% renewables in 2050

21% renewables   36% renewables

Trillion Kilowatt Hours

Year

Plant Size

Circle area is proportional to capacity (megawatts)

1 100 300 500 750 1,000

418

Workers by Industry

In 2019, solar power jobs greatly outnumbered other power generation jobs.

Almost 250,000 people—31% of the entire electric power generation workforce—worked in solar power, mostly in constructing new solar power-generating technologies. Workers in all fossil fuel generation industries combined (coal, oil, and natural gas) made up only 27% of the workforce.

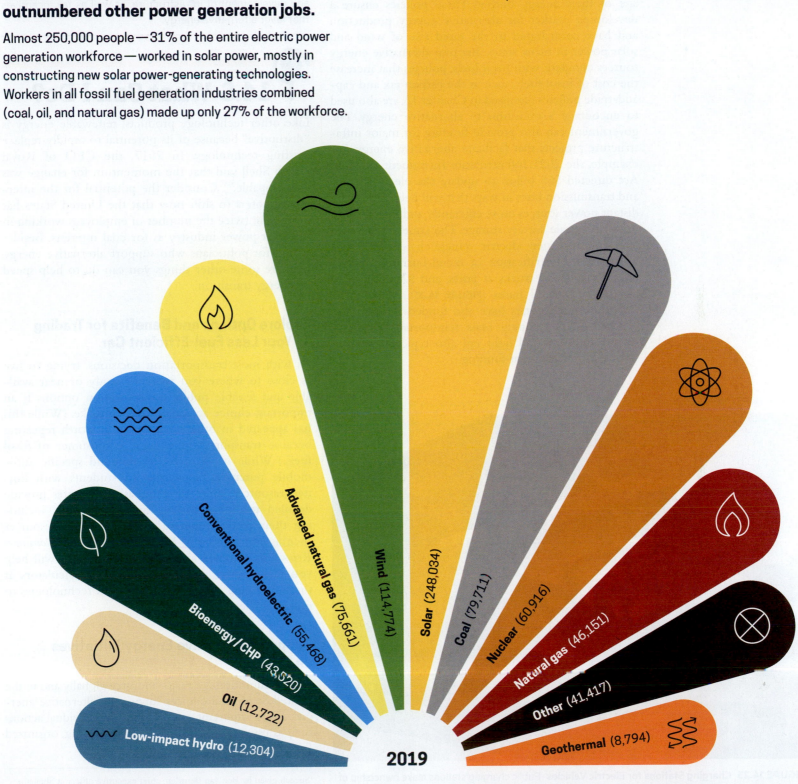

Wind (114,774)

Advanced natural gas (75,661)

Conventional hydroelectric (55,468)

Bioenergy / CHP (43,520)

Oil (12,722)

Low-impact hydro (12,304)

Solar (248,034)

Coal (79,711)

Nuclear (60,916)

Natural gas (46,151)

Other (41,417)

Geothermal (8,794)

2019

Adapted from NASEO (2020) and US Energy Information Administration (2021a, 2021b).

low- or zero-emission vehicles. This is one way to help consumers afford the significant initial costs of adopting many of these technologies.

Regulatory policies on energy production include state-level energy portfolio standards (now in place in more than 30 US states) requiring local energy utilities to use alternative energy for a specific percentage of their energy output. These policies ensure a developing market for alternative energy production and have contributed to the rapid rise of wind and solar power in many states. Because alternative energy sources compete with fossil fuels, policies that increase the cost of fossil fuels, such as the carbon tax and cap-and-trade policies discussed in Chapter 13, are also used to encourage a transition to alternative energy. The government can also provide funding for major infrastructure projects that facilitate alternative energy. For example, the 2021 Infrastructure Investment and Jobs Act directed $65 billon to update the electrical grid and transmission lines in ways that will help coordinate distant power sources more efficiently and $6 billion to fund large-scale battery storage. The same law devoted $7.5 billion to new electric vehicle chargers, $17 billion to the electrification of diesel-burning vehicles like tugboats and trucks at ports, and $2.5 billion to electrifying school buses (**FIGURE 14.23**). Many states and some major cities have also funded similar types of projects. Developing mass transportation systems such as subways and light rail also typically requires significant government funding.

FIGURE 14.23 Charging Stations for Electric Vehicles Public charging stations make ownership of electric vehicles more convenient. Infrastructure projects like this often require government funding.

⬠ **TAKE-HOME MESSAGE** A transition to alternative energy entails facing the challenge of increasing global demand for energy. This will require not only favorable market conditions but also innovation and improvements in energy technologies and infrastructure. Government efforts to promote alternative energy include support for research and development of infrastructure, as well as tax incentives that favor alternative energy.

14.6 What Can I Do?

Like other technology products, renewable energy is "disruptive" because of its potential to rapidly replace existing technology. In 2017, the CEO of Royal Dutch Shell said that the momentum for change was "unstoppable."[1] Consider the potential for the interests of voters to shift now that the United States has more than twice the number of employees working in the solar power industry as for coal interests. Besides voting for politicians who support alternative energy, here are some other things you can do to help speed an energy transition.

Explore Options and Benefits for Trading in Your Less Fuel-Efficient Car

As with most transportation decisions, trying to live as close to where you work or study or near available and feasible public transportation options is an important choice an individual can make. (While this has appeared in other chapters, it is worth repeating because transportation is such a consumer of fossil fuels.) While it's hard to recommend specific automobile purchases, especially to students with limited resources, most car-purchasing websites provide detailed recommendations on electric cars, including their cost to operate per mile. Using resources such as edmunds.com, *Kelley Blue Book*, or *Consumer Reports* to compare and understand pricing will help you make better decisions. And often calculators at these sites help compare fuel-efficient technologies to older vehicles.

Support Alternative Energy Initiatives on Campus

Higher education is a major industry globally and in the United States, and choices supporting alternative energies create important markets. Simple individual actions such as using more efficient electric lighting, organized

[1] Speech given by Ben van Beurden, chief executive officer at Shell, for the Aurora Spring Forum in Oxford, England, on March 22, 2017.

DORM ELECTRICITY COMPETITION

Piper ↓ 43.51%	↑ 12.87%	Sturtevant
GoHo ↓ 42.32%	↑ 1.65%	MaryLow
Drummond ↓ 36.34%	↑ 0.20%	Pierce
Treworgy ↓ 33.73%	↓ 0.31%	Marriner
PeWi ↓ 24.93%	↓ 1.09%	Coburn

Overall: April. 19th ↓ 14.05%

across campus, will have an impact. One way to promote this is to implement or organize a competition. In Maine, Bowdoin College and Colby College organize a competition every year to see which school and which dorm in either school can conserve the most energy. In the first year of the competition, Colby lowered its usage by 7% and Bowdoin by 8.7%. Bentley College has an annual "Blackout Challenge" between residence halls, which compete to see which one can save the most energy. On a broader scale, the Campus Conservation Nationals (CCN), supported by the National Wildlife Federation, US Green Building Council, and Alliance to Save Energy, supports both individual competitions on a campus and competitions with other similar organizations.

In addition, many campuses of all sizes are making new choices to power the campus. Another part of the energy transition plan at Michigan State University was to stop burning coal and shift to a natural gas–powered cogeneration plant in 2015. This school built a facility that uses campus meal waste and other biomass from on and off campus to produce additional electricity as part of its sustainability initiatives. In 2021 the University of Michigan responded to student demand for more ambitious energy goals by pledging to become carbon neutral by 2040. This will include an electric campus bus fleet, energy-efficient heating and cooling, and purchasing electricity from renewable sources. Are any initiatives like these either ongoing or being explored at your school? Learning more about them, deciding if you support them, and making your thoughts known is an important action anyone can take.

Join in the Debate about Local Alternative Energy Projects

Offshore wind farms of multiple and massive wind turbines are being operated, built, or planned globally. Recently, new "floating turbines" have been released off the coast of Scotland that extend 175 meters high (almost 575 feet) and another 78 meters (255 feet) below the surface. Similarly, the Icebreaker Wind Project located in Lake Erie about 8 miles (13 kilometers) north of Cleveland is being planned and debated. Supporters include local unions valuing the employment opportunities and interests saying these are investments in the technological future. However, opponents noted that this site and its huge blades would be a very large threat to hundreds of millions of birds and bats that migrate every year. These concerns had blocked construction at a similar project on the shore of Lake Erie. Opinions about this project were presented to local judges in Cleveland for their recommendation. If you were arguing this case, what side would you emphasize?

Learn More about Global Energy Policies

Germany is ranked third in economic output among industrialized nations, behind the United States and Japan, and is regularly cited as a competitor in the manufacturing industry. The German government promotes itself as "leaving the age of fossil and nuclear energy clearly behind it and heading fast for a future that hinges on sustainable energy sources." It goes on to describe its strategy as one that "pays off twofold, because on the one hand the impact on the environment and climate declines, while on the other new fields of business and jobs are created." Do you feel that we should be competing with Germany for these businesses and jobs, or do you feel that Germany is making the wrong decisions? Or do you feel like this competitive mindset is the wrong stance to take? Learn more about the alternative energy policies of major countries such as China, Germany, Japan, and India by searching for their department of energy websites (most major countries maintain an English-language version). Becoming informed about major global energy policies can give you an idea about which policies might be successful where you live.

Chapter 14 **Review**

SUMMARY

- The use of wind and solar power for electricity is growing rapidly in the United States and the world.

- Wind and solar power are nondepletable resources, but their energy production is variable, so they each must be balanced with other energy sources or paired with energy storage technologies.

- The most established and highest-producing alternative energy sources to fossil fuels are hydropower and nuclear power.

- Both hydropower and nuclear power can produce a steady and reliable (nonvariable) source of power on a large scale, but each has environmental impacts and siting restrictions that increase their cost and limit expanded production.

- Geothermal power can be applied on a large scale in places where extreme heat below Earth's surface can produce steam or on a household scale in any location by using heat pumps that take advantage of the moderate temperature differential between the surface and subsurface.

- Biofuels from corn or other plants are a viable source of energy for transportation, but the environmental impacts from their production can be significant.

- Electric vehicles are much more efficient than internal combustion engines, and their environmental advantages are linked to how much the source of electricity used for recharging draws on alternatives to fossil fuels.

- The widespread adoption of electric vehicles will depend on the range of travel they can provide before recharging and the speed and accessibility of the recharging process.

- Fuel-cell vehicles emit only water but rely on the energy-intensive process required to supply hydrogen.

- Governments can take steps to speed a transition to fossil fuel alternatives by supporting research and development of new energy technologies, funding the infrastructure needed to support alternative energy, and adopting tax policies with incentives that favor the adoption and use of fossil fuel alternatives.

- A transition to fossil fuel alternatives will also require improvements in energy conservation that address the challenge of rising global energy demand.

KEY TERMS

biofuel

cogeneration

concentrated solar thermal (CST) plant

electric vehicle (EV)

ethanol

fission

fuel cell

geothermal power

hydropower

nuclear power

photovoltaic (PV) solar panel

variable generation

wind turbine

REVIEW QUESTIONS

The letters following each Review Question refer to the Chapter Objectives.

1. Which energy alternative to fossil fuels provides the largest share of global power? **(A)**

2. Which two forms of alternative energy are growing at the most rapid rate? **(A)**

3. Why don't wind turbines consistently reach maximum capacity? How much of their maximum capacity do they typically generate? **(A)**

4. Name two new developments that will help variable sources of power such as wind and solar provide a larger share of energy. **(A, D)**

5. What are the primary technologies used to generate electricity from solar energy? **(A)**

6. What environmental impact is common to both wind and solar power? What environmental impacts are unique to each type of power? **(A)**

7. How does hydropower, which uses a non–fossil fuel, still contribute to greenhouse gas emissions? **(A)**

8. How and where do we currently store spent nuclear fuel (nuclear waste) from power plants in the United States? **(A)**

9. Define what a biofuel is. **(B)**

10. When are the environmental advantages of biofuel the greatest? **(B)**

11. Why does the carbon footprint and fuel efficiency of an electric vehicle vary depending on where you are driving it? **(B, D)**

12. Why are energy conservation strategies part of most energy transition plans? Describe a few common energy conservation strategies. **(C)**

13. What is cogeneration, and why is it not a true alternative to fossil fuels? **(C)**

FOR FURTHER THOUGHT

The letters following each item refer to the Chapter Objectives.

14. Which of the alternative energy sources described in this chapter is most feasible where you live? Why? **(A)**

15. Imagine you are in the market for a new vehicle. Identify the kinds of things you are looking for in a vehicle according to your current and projected future uses. Then do some comparison shopping for a vehicle online, and select three candidates for purchase. But here's the catch: one of your three candidates must be an electric vehicle (EV). How does the EV compare to your other candidates? What sorts of factors would affect your final selection? What sorts of factors make (or would make) the EV more competitive? **(B)**

16. Revisit the various types of government actions that could support the transition to alternative energy. Which two government actions do you think would be most effective in this regard? Why? What obstacles would the policy makers likely face in taking these actions? **(D)**

17. Nuclear power has been a controversial source of power for decades, such that few new power plants have been built since 1980. Explain your position on nuclear power with detailed reasoning as to why you would support or oppose expansion of this energy source. **(A)**

Make Your Case?

The following exercises use real-world data and news sources. Check your understanding of the material and then practice crafting well-supported responses.

Use the News

In 2010, solar power made up a negligible portion of electricity generation in the United States, but by 2020 it grew to 6.4%. The following article from the *Wall Street Journal* looks at factors driving its growth in Georgia. Use this article to answer the questions that follow. The first three questions test your understanding of the article.

Questions 4 and 5 are short answer, requiring you to apply what you have learned in this chapter and cite information in the article. Answers to Questions 1–3 are provided at the back of the book. Question 6 asks you to make your case and defend it.

"Solar Power Booms in Georgia, Where It Isn't Mandated," Elena Shao, *The Wall Street Journal*, August 22, 2021

Georgia has no mandates requiring power companies to add renewable energy and hasn't made climate change a political priority. Solar power is booming there anyway.

The state went from having virtually no solar industry a decade ago to ranking ninth nationwide in installed solar capacity this year, according to the Solar Energy Industries Association. Solar has flourished in Georgia as tech companies such as Facebook look to locate facilities near cheap renewable-energy sources and rural communities turn to solar farms to create tax revenues and jobs.

Much of the initial build-out of solar and wind power in the U.S. over the past three decades was driven by mandates in states such as Iowa, California, Colorado and New York that required utilities to source a certain amount of renewables. But wind and solar are now gaining market share even in states with no such requirements, as Georgia's experience shows.

Republican regulators have pushed the state's major utility, Southern Co. subsidiary Georgia Power, to invest in solar, saying that economic factors make it an attractive energy source beyond its carbon-free characteristics.

"Don't come into my office talking about climate change or the environment," said Tim Echols, who has served for the past decade on the elected, all-Republican public-service commission that regulates the state's investor-owned utilities. "Talk about new jobs, talk about low-cost energy, talk about reduction of transmission lines," he said. "Learn to speak Republican here." '

Solar installation prices in Georgia have fallen 43% over the past five years, according to data from SEIA. Similar declines in price are behind the solar-industry growth in states across the

Southeast, many of which also lack renewable-energy mandates.

"I oppose any renewable portfolio standard—it's not necessary." said Lauren "Bubba" McDonald, who serves with Mr. Echols on the public-service commission.

Georgia Power, which serves 2.6 million customers across the state, has added more than 570 solar projects totaling close to 2,000 megawatts to its energy portfolio in the past decade, according to company spokesman Jeff Wilson.

While Georgia Power has led the state in solar additions, rural Georgia's smaller electric membership cooperatives, or EMCs, have leaned into solar power as well, catching the eye of large companies like Facebook.

The tech giant has partnered with Walton EMC in northern Georgia, which will buy energy on Facebook's behalf through a long-term power purchase agreement with a solar developer, Tennessee-based Silicon Ranch Corp.

Silicon Ranch will in turn develop, own and operate the six projects, totaling 434 megawatts, that will help power Facebook's data center in Newton County. The solar developer said it has invested more than $1 billion in rural communities in Georgia, making it the largest taxpayer in multiple jurisdictions.

"We are committed to bringing renewable energy close to where we operate," said Urvi Parekh, who leads Facebook's renewable-energy team. Data centers like the one in Newton County represent more than 90% of the electricity that Facebook consumes, she added.

The bump in investment from the solar-power projects ends up having a significant impact on the local communities where the projects are sited, said Jeff Pratt, president of Green Power EMC, an entity that secures renewable resources on behalf of the state's 41 electric cooperatives. Those EMCs serve roughly 4.4 million of the state's 10 million residents in mostly rural areas, which encompass some of the poorest counties in the country.

But, while solar projects can add jobs, a lot of the new employment growth can be temporary, with opportunities petering out after construction is completed, said John Howard, mayor of Monroe in Walton County. Coal power plants take twice as many workers to generate a megawatt hour of electricity as do solar farms, according to estimates from BW Research Partnership, an economic and workforce consulting firm.

The projects can be unpopular with farmers, since plots of agricultural land usually need to be cleared for the installation of solar panels.

"You'll find that in a rural, farming community, folks don't like seeing that farmland being taken away and used for anything other than what they think it should be intended for," Mr. Howard said.

Integrating the solar projects with agricultural practices has helped ease some of those concerns, said Matt Beasley, Silicon Ranch's chief commercial officer. On several projects, the solar developer partners with local farmers to move in flocks of sheep and cattle to naturally fertilize the land and graze on the grass, which prevents erosion and manages the length of the grasses so they don't block the panels.

Mr. Beasley has started to see the growth of solar power changing the politics of renewable energy in the state. People are beginning to associate solar plants with energy resiliency, job creation and economic growth without much disturbance to the rural landscape, he said.

Solar power has also gained the support of conservative coalitions such as the Conservatives for Clean Energy Georgia and the Atlanta Tea Party Patriots, which are also advocating for the expansion of rooftop solar in the state.

"The pendulum is swinging back, and climate change is starting to become a part of the discussion now," Mr. Beasley said. "The economics are such that you can have your cake and eat it too."

---

1. According to the article, why are state regulators in Georgia pushing for its major power company to invest in solar?
   a. Georgia will be at a competitive disadvantage with other states and countries without this power source.
   b. Regulators think it is important to invest in renewable energy to combat global warming and climate change.
   c. The economic factors of solar power make it an attractive source of energy for the state.
   d. Adding solar plants to unused land actually prevents erosion if the companies are required to plant cover crops too.

2. How many solar power projects had Georgia Power completed in the ten years before the article was written?
   a. 41
   b. 434
   c. 570
   d. 2000

3. Which three of the following are factors that are cited as driving the growth of solar in Georgia?
   a. state regulations and incentives
   b. integration of solar projects with local agriculture instead of replacing it
   c. creation of new jobs in the state

   d. construction of bigger and stronger transmission lines to carry the energy across the state
   e. tech companies partnering with local energy providers to power their facilities

4. Does what is described in this article match or differ from the trends described for solar power in the chapter? Support your answer with at least two points from the article and how they compare or contrast with the text.

5. The article notes that most initial solar power growth came from states with mandates to build it. Given what you read in this article, what do you think its author feels about the usefulness of mandates? Do you agree or disagree with this stance and why?

6. **Make Your Case** In the article, a member of the state public service commission is quoted saying "I oppose any renewable portfolio standard—it's not necessary." First, using the article, explain what they mean in terms of solar power in the state. Then, using the article, what you have read in the text, and outside sources if necessary, briefly explain why you agree or disagree with this statement.

Use the Data

This table from the US Department of Energy shows the number of "light duty" alternative fuel vehicle models (think trucks, cars, and SUVs that consumers buy) offered by manufacturers (US and foreign) from 2006 through 2018. Alternative fuel vehicles are classified into four major categories: alternative fuel vehicles (AFVs) powered by ethanol (E85), compressed natural gas (CNG), propane, or hydrogen; hybrid electric vehicles (HEVs); electric vehicles (EVs); and diesel models.

Study the table, and use it to answer the questions that follow. The first three questions test your understanding of the table. Questions 4 and 5 are short answer, requiring you to apply what you have learned in this chapter and cite data in the figure. Answers to Questions 1–3 are provided at the back of the book. Question 6 asks you to make your case and defend it.

Alternative Fuel Vehicle Light Duty Model Offerings by Fuel Type, 2006–2018

Fuel Type	2006	2007	2008	2009	2010	2011	2012	2013	2014	2015	2016	2017	2018	Total per Type
Ethanol	22	31	31	36	34	72	62	84	90	84	66	45	53	710
CNG	5	1	1	1	1	1	6	11	19	17	12	9	9	93
Diesel	6	7	6	12	14	16	17	22	35	39	29	21	38	262
EV	0	0	1	1	1	2	6	15	16	27	29	51	57	206
Hybrid	8	11	16	19	20	29	31	38	43	46	31	44	43	379
Propane	0	0	1	1	0	0	1	6	14	10	5	8	7	53
Hydrogen	0	0	0	0	0	0	1	1	2	3	3	2	2	14
Total per year	**41**	**50**	**56**	**70**	**70**	**120**	**124**	**177**	**219**	**226**	**175**	**180**	**209**	**1,717**

Adapted from US Department of Energy (2019).

1. Using the information in this table, which four vehicle types have had the most total models from 2006–2018? List them in order from highest to lowest.

2. How many total vehicles of all types were for sale in the year 2006? How many were for sale in the year 2018?

3. In the year 2018, which type of vehicle had the most models available, and how many models were available? Which type of vehicle had the second most models available, and how many models were available for this type?

4. Looking at the data on ethanol vehicles, describe the trend for the number of available models between 2006 and 2014, and then after 2014. What do you think explains the changing trend since 2014? Use information in this table and/or information in the chapter to support your answer.

5. Looking at the data on electric vehicles, what is the trend in the number of models over the past 5 years? Briefly, what do you think is causing this trend? Use information in this table and/or information in the chapter to support your answer.

6. **Make Your Case** In 2016, there was a large drop in the total number of model types, from 226 in 2015 to 175 in 2016. However, after this low point, the total number of models rises again, eventually nearing 2015 levels in 2018. Using the data in the table and information in this text or from other sources, briefly describe what you think is driving this trend. In your answer, describe at least one factor, though you may choose to describe more.

LEARN MORE

- Energy Information Administration, background and stats on renewable and alternative fuels: www.eia.gov/renewable/
- Energy Information Administration, background on nuclear power: www.eia.gov/nuclear/
- Nuclear Regulatory Commission, information about Fukushima: www.nrc.gov/reactors/operating/ops-experience/post-fukushima-safety-enhancements.html
- California Energy Commission, information about renewable energy: www.energy.ca.gov/renewables/
- International Energy Agency, "Tracking Clean Energy Progress": www.iea.org/tcep/
- Video: NSF, QESST (Quantum Energy and Sustainable Solar Technologies) for Solar Power to Feed an Energy-Hungry World: digital.wwnorton.com/environsci
- Video: NSF, A "Clear Path" to Solar Power: digital.wwnorton.com/environsci

15

Waste

What Happens to All the Stuff We Use?

Have you ever tried to discard an empty plastic grocery bag, only to watch it fall to the ground and then be swept out of reach by the wind? Or maybe you've seen plastic bottles, bags, and other garbage littering the shoulder of the highway as you speed by. Where do these bits of waste end up? Many of them find their way to the ocean. How do these types of plastic trash end up in the sea? They could be blown by the wind, but in most cases they are carried by water into storm drains that discharge into rivers and streams that ultimately empty into the ocean. In this way, even trash from areas far inland can pollute oceans thousands of miles away.

In 1997, Captain Charles Moore was about 1,000 miles (1,600 kilometers) from land, sailing between Hawaii and California, when he discovered the "Great Pacific Garbage Patch": a thin but pervasive soup of suspended tiny plastic flakes intermixed with bobbing nets, bags, floats, tarps, bottles, and innumerable other plastic items. The plastics are caught in the North Pacific Gyre—a swirling 20-million-square-mile area stretching across the Pacific Ocean from California to Japan (**FIGURE 15.1**). This is a recent form of pollution. Moore recalled crossing an unlittered ocean "wilderness" in the North Pacific Gyre in the 1960s, when plastic was a newly developed material used in specialty products. Now it is a nearly universal packaging material produced at a rate of more than 200 million tons per year. Although the fishing and shipping industries are contributing nets, floats, buoys, bottles, and other ocean debris to this area, researchers have found that most of the plastics in the ocean are coming from land.

And while the amount of plastic in the ocean is increasing, the average size of these pieces of plastic is decreasing. Though it may take hundreds or even thousands of years to completely decompose, plastic can rapidly break down into tiny fragments of microplastics when exposed to sunlight, water, and certain microorganisms. Microplastics mimic both zooplankton and phytoplankton, which are food for filter feeders such as krill and mussels as well as small fish. In some samples of ocean water, microplastics outweigh zooplankton by as much as six to one. When marine animals ingest microplastics and process them in their guts, they absorb and in some cases accumulate toxic plastic additives, such as phthalates (added to plastic for flexibility), bisphenol A (BPA; added for hardness), and polybrominated diphenyl ethers (PBDEs; added for flame resistance). Research

Plastic waste (such as bags, bottles, and straws) that ends up accumulating in the ocean is a growing threat to marine organisms and ecosystems.

Chapter Objectives

By the end of this chapter, you should be able to . . .

A. explain what the waste stream is and what the concepts upstream and downstream mean.

B. describe some of the environmental and socioeconomic impacts of different forms of dumping.

C. describe how combustion, isolation, and conversion processes are used as modern waste management strategies and the environmental effects of each.

D. compare and contrast the advantages and limitations of recycling.

E. explain the advantages of reducing consumption on the waste stream and aspects of culture that make this strategy challenging.

F. give historical examples of ways consumers reduced waste and of new strategies developed by consumer groups, businesses, governments, and communities.

has even shown that some filter feeders incorporate microplastics into their cells. Animals higher on the food chain, such as tuna and salmon, ingest the additives and microplastics when they consume animals at lower trophic levels.

A new generation of young environmental entrepreneurs have taken on the challenge of cleaning up plastic in the ocean. Boyan Slat, a Dutch inventor who while still a teenager began work on a floating mechanism that could gather plastic, has successfully tested plastic harvesting vessels in the North Pacific. Fionn Ferreira, a young Irish inventor, recently won a Google Science Fair competition with his plan to collect microplastics from water. His plan uses a mixture of vegetable oil and magnetite powder (an iron oxide) that enables magnets to collect microplastics. Though early in their development, such technologies are promising.

But finding solutions to our dependence on plastics may be even more challenging than figuring out how to clean them up. Plastic has material properties that serve functions we would find difficult to forgo or to replace. Medical treatments are carried out with sanitary plastic gloves, syringes, intravenous bags and tubing, lab equipment, prosthetics, and even plastic pill coatings for time-released medication. Plastic food packaging provides a sanitary barrier, and cars include plastic safety features such as airbags, shatter-resistant windshield coatings, and impact-absorbing foam. Our houses are encased, insulated, plumbed, and wired with durable, energy-efficient, corrosion- and fire-resistant plastic products. Our clothing is often made from insulating and water-resistant plastic fibers. All of this plastic is lightweight, which makes it easy to handle and transport—and unfortunately, easy to be swept out to sea.

As you can infer from our dependence on plastics, thoroughly exploring our relationship to waste requires that we understand how and why we consume the multitude of material items that pass through our lives. We'll learn in this chapter that the most common materials in residential waste are paper, plastic, food, and yard waste—so chances are you threw some of these

(a)

(b)

North Pacific Gyre

(c)

FIGURE 15.1 Great Pacific Garbage Patch (a) Captain Charles Moore holding some of the plastic debris from the North Pacific Gyre. (b) The debris ranges in size from large, abandoned fishing nets to tiny microplastics less than 5 millimeters (0.2 inches) in diameter. This debris is concentrated in several locations across the gyre. (c) Inventions by Boyan Slat use computer models to locate concentrations of plastic in the ocean and then employ floating barriers that are 3 meters (10 feet) deep to capture the plastic from the ocean surface.

Adapted from (b) National Ocean Service (2018).

FIGURE 15.2 The Waste Stream All of our stuff comes from somewhere and goes somewhere once we're through with it. We can think of the waste stream as a flow of materials starting "upstream" with the extraction of raw materials (through mining and harvesting) and ending "downstream" with the disposal or recycling of materials once consumers are done using them.

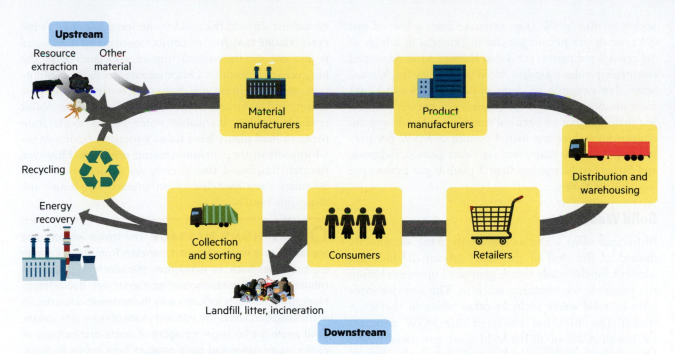

materials away very recently. When you used them, did you consider how they'll be disposed of or whether they can be recycled or reused? Or whether you really needed to consume them in the first place? This chapter engages these questions by considering the environmental effects associated with waste and how we try to manage it. The chapter also explores the cultural context that influences our consumption. To understand how we can manage waste differently, we must first ask questions about how we define waste, what we do with it, and why we generate it.

15.1 What Is Waste?

Broadly speaking, we use the terms *waste* or *garbage* to define things as unusable or unwanted. This is the stuff we throw away. Of course, in Chapter 3 we learned that stuff—matter—continues to cycle through the Earth system. We can't destroy it, so it has to go somewhere. In

this section, we'll look at the **waste stream**: the way in which our consumption is linked to the steady flow of materials from "upstream" processes (such as extraction, production, and distribution) and to their disposal "downstream" from us (**FIGURE 15.2**).

Municipal Solid Waste

Consumers in the United States generate a lot of waste. The Environmental Protection Agency (EPA) estimates that American consumers *directly* contribute about 4.9 pounds of garbage to the waste stream per person per day—nearly 1,800 pounds of garbage per year. This is the waste we drag to the curb each week from our households and commercial businesses, known as **municipal solid waste (MSW)**. Per capita MSW generation in the United States is now nearly double what it was when the government began collecting data in 1960 (**FIGURE 15.3A**).

On a per capita basis, the United States is one of the largest producers of MSW. Different countries vary

waste stream the steady flow of materials from "upstream" processes, such as extraction, production, and distribution, to their disposal, which is "downstream."

municipal solid waste (MSW) the waste consumers dispose of from their households and businesses.

FIGURE 15.3 Municipal Solid Waste Rates (a) The blue line shows the total municipal solid waste (MSW) generated in the United States in millions of tons per year, and the orange line shows the pounds generated per person per day. (b) The United States is one of the largest producers of MSW per person per day, even when you consider only developed countries.

Adapted from (a) US EPA (2018c) and (b) OECD (2022).

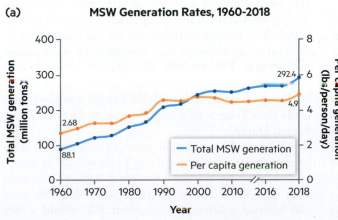

(a) MSW Generation Rates, 1960-2018

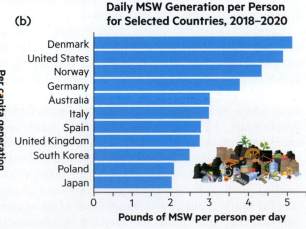

(b) Daily MSW Generation per Person for Selected Countries, 2018–2020

Creating
plastic resin
13%

14%
Manufacturers
Distribution

Disposal and
recycling processes

73%

Other impacts include

Air
pollution

Greenhouse gas
emissions

FIGURE 15.4 **Solid Waste Associated with a Plastic Water Bottle** This graphic summarizes the solid waste generation associated with the plastic material in a 12-ounce polyethylene terephthalate (PET) water bottle.

Adapted from Franklin Associates (2007).

solid waste all discarded material in solid, liquid, semisolid, or contained gaseous form.

life-cycle assessment an evaluation of the environmental impacts of all the steps involved in making, distributing, using, and ultimately disposing of a product.

open dump unregulated waste disposal in uncovered and openly accessible piles.

widely in the MSW they generate from a low of just 0.4 pounds per person per day in Ethiopia to a high of 6.2 pounds per person per day in Bahrain. Most developed countries produce far less MSW than does the United States. For example, residents of Japan and South Korea each produce less than 2.4 pounds per person per day: less than half that of a typical American. Most European countries produce less than 4 pounds of MSW per person per day, with many such as Great Britain, Belgium, and Spain producing less than 3 pounds per person per day (**FIGURE 15.3B**).

Solid Waste

Municipal solid waste includes only what we directly discard at the end of the waste stream. It does not account for the waste that is generated upstream before the products we consume reach us. The broader category of **solid waste** includes other waste in the waste stream. The EPA has estimated that MSW accounts for less than 2% of all the solid waste generated in the United States and that each year the products a typical American consumes are linked to a staggering 1 million pounds of waste.

To better understand this, let's consider the waste associated with the production of the 30 or so silicon and metal microchips in a typical computer. If you were to throw away your computer after it has reached the end of its useful life, these microchips would contribute just 1.5 ounces to the waste stream, less than one-tenth of a pound. However, the hundreds of steps that it takes to extract and refine the silicon and metals and ultimately to manufacture the chips generates nearly 90 pounds of waste.

Life-Cycle Assessment

Accounting for the environmental impacts of all the steps involved in making and ultimately disposing of a product is called **life-cycle assessment** (see the one for jeans in **AT A GLANCE**). Applying the metaphor of a life span to a product, "from cradle to grave" emphasizes the importance of considering each step from the extraction and refinement of raw materials through production, distribution, consumer use, and disposal.

Consider disposable plastic water bottles. While it's true that most of the solid waste associated with this product comes from the empty bottles that we throw in the trash or recycle after we've consumed the water, that's only part of the story. Creating the plastic resin itself, which in a typical bottle is made of polyethylene terephthalate (PET), generates 13% of the solid waste associated with the bottle. Fabricating the bottle and distributing it contribute another 14%, mostly related to the fuel used to manufacture and transport the materials and finished product. Then, recycling and disposal processes after the bottle is out of your hands (analysts assume that only 23% of these bottles are recycled by consumers)

contribute 73% of the solid waste for this product's life cycle (**FIGURE 15.4**). And remember, generating solid waste is just one environmental impact associated with the life cycle of a product. Other impacts include air pollution and greenhouse gas emissions, water use and pollution, and environmental health effects. What about refillable bottles? Refillable water bottles made from metal or hard plastic have more associated environmental impacts in their manufacturing processes. However, research has found that if they are reused more than 50 times, they have less of an impact than single-use disposable bottles.

🏠 **TAKE-HOME MESSAGE** The things we consume are linked to a waste stream. Upstream from the consumer are processes such as extraction, manufacturing, and distribution, while waste disposal processes are downstream. Municipal solid waste includes only the materials downstream of consumption and is the smallest share of the waste stream. Solid waste is a far larger category of waste that includes all of the waste generated for a product from cradle to grave. Life-cycle assessment is a tool that measures environmental impacts from the processes that create a product through to its ultimate disposal.

15.2 What Happens When Waste Is Dumped?

Why should we be concerned about solid waste? For a better understanding of its impacts, let's begin by exploring what happens when waste is simply dumped in an area that is open to the elements. For most of human history, trash was not neatly placed in bins to be systematically collected, transported, and managed at regulated disposal facilities. Prior to 1900, most US cities lacked a municipally run system for disposing of garbage. And in most of the world today, the norm is still for residents and trash collectors to heap garbage in uncovered, openly accessible, and uncontrolled piles known as **open dumps**. In major metropolitan areas, these dumps operate on a very large scale. For example, until recently, waste from the city of Manila in the Philippines (a city of 2 million people in a greater metropolitan area of more than 25 million) accumulated at truly monumental open dumps (**FIGURE 15.5**). The Smokey Mountain Garbage Dump, for example, operated for more than 50 years. When it closed in 1995, thousands of Manila residents who made their living scavenging there moved to the nearby Payatas Dump.

Closer to home, if you were to collect and heap your trash for a month in a small pile outside, it would exhibit many of the environmental impacts associated with large-scale dumps. Some of it would blow away and become a litter problem. Your pile would attract

FIGURE 15.5 Smokey Mountain Garbage Dump Over five decades, Smokey Mountain in Manila accumulated more than 2 million metric tons of garbage and hosted a resident population of more than 30,000 people who made their living scavenging the waste for food and materials to sell or trade. When it closed, residents moved to the nearby Payatas Dump (shown in both photos), which itself was closed in 2017.

vermin—species such as rats associated with the spread of disease. Even on this small scale, your waste pile would likely drain pollutants into the water system and emit air pollution as it decomposed. If your pile included such items as car batteries, nail polish, used motor oil, or other products containing substances the EPA classifies as hazardous, the waste would pose a danger to human health and the surrounding environment.

Uncontrolled dumping on a larger scale amplifies these effects. Open dumps like Smokey Mountain obviously create an undesirable and even hazardous environment for those who live on or near them. But decomposing waste causes environmental problems with broader impacts, particularly from polluting gas and liquids that can leak and spread from disposal sites. Let's take a closer look at these impacts.

The Life Cycle of One Pair of Jeans

The life cycles of even common items can be very resource-intensive, involving many people and processes, with different by-products at each step. Here we show what typically goes into making a single pair of jeans, from seed to store, and also what is discarded as waste at each stage.

1

From Fiber to Fabric

Your favorite pair of jeans likely contains cotton and another fiber like elastane for stretch. Over 60 percent of the world's cotton is produced by an estimated 40 million small farmers, 99 percent of them in developing countries. The elastane fibers are made by manufacturers that spin petroleum-based polymers into thread.

🗑 **Pesticides and chemical fertilizers during farming; waste fiber as cotton is spun into thread; chemicals in wastewater from the dyeing process**

2

Design

Independent companies often work on behalf of brands to create samples. Once the sample is approved, an entire production run is created, modeled after the sample.

🗑 **Scraps during fabric cutting**

Reprinted and adapted with permission from Luu (2018).

5 Shipping and Retail

Because garment factories are often located in developing countries, jeans are usually shipped by plane or cargo ship to retail stores. Store employees stock the jeans on the shelves and sell them to consumers.

🗑 **Labels, tags, and waste from shipping containers; end-of-life waste when the jeans themselves are discarded**

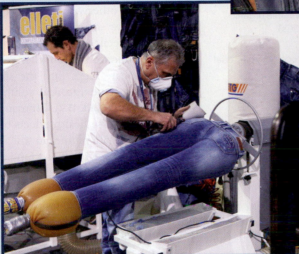

4 Finishing

Jeans can undergo a "wet process" to add design effects like whiskers. Sometimes jeans are finished by hand or with sanders to "distress" them, or create artificial wear.

Last, the jeans are inspected before being tagged and packed for distribution.

🗑 **Silica, pumice, sandpaper, and other abrasives used to distress jeans**

3 Manufacturing

At the mills, the yarns are dyed and woven together. A precise formula of dye and solvents helps the dye stick to the fibers. Mills typically hire denim specialists, who usually have a background in chemistry, to oversee the denim finishing processes.

By far, the people who sew clothes make up the biggest group of workers in a cut-and-sew factory. Garment factories in Vietnam and India range in size from fewer than 10 workers to more than 5,000. After the jeans are assembled, they are sent to the trim-and-notions department, where garment workers sew on buttons, snaps, and other small articles.

🗑 **Chemical detergents during fabric conditioning; scraps during fabric cutting**

Bacterial decomposition: Near the surface, in the presence of oxygen, bacteria decompose waste and release carbon dioxide.

CO_2 emissions

Volatile organic compound (VOC) emissions

Methane emissions

Volatilization: Beneath the surface, in the absence of oxygen, anaerobic microbes decompose waste and release methane.

Chemical reactions: Certain chemicals either on their own or in reaction with carbon dioxide, methane, or other compounds can vaporize, releasing hazardous chemicals and volatile organic compounds (VOCs).

FIGURE 15.6 **Landfill Gas Emissions** Landfill gas is produced by three processes: bacterial decomposition, volatilization, and chemical reactions.

Polluting Gases

In most open dumps, two gases are created as waste breaks down and decomposes: methane and carbon dioxide (**FIGURE 15.6**). As we learned in Chapter 11, both of these compounds are greenhouse gases that contribute to climate change. Methane, or natural gas, is also explosive. Methane explosions ignited Smokey Mountain several times (giving the dump its nickname), injuring and killing scores of people and releasing air

leachate a foul-smelling, soupy liquid that forms when rainwater or groundwater mixes with the decomposing waste at the dump.

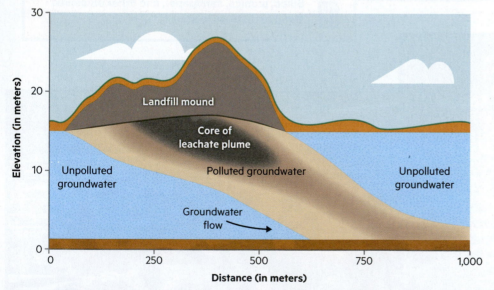

FIGURE 15.7 **Leachate** When rainwater or groundwater mixes with decomposing waste, it forms leachate, which can create a contaminated plume as it seeps into the surrounding soil and water. This sketch of a landfill in Norman, Oklahoma, shows how a leachate plume (the entire beige and gray area) is polluting a confined aquifer.

Adapted from USGS (n.d.).

pollutants such as dioxins and fine particulate matter associated with asthma and lung cancer. Methane is produced in open dumps when the piles of waste provide a moist *anoxic environment*—one with a very low oxygen concentration—for anaerobic microbes. Carbon dioxide and methane can also strip other dangerous compounds from coatings, paints, and adhesives on manufactured goods. This process releases small but dangerous quantities of hazardous materials, such as mercury, benzene, and chloroform, as well as volatile organic compounds that form smog when exposed to sunlight.

Leachate

Dumps also produce a foul-smelling black, yellow, or orange soupy liquid called **leachate**, which forms when rainwater or groundwater mixes with the decomposing waste at the dump. Besides its noxious look and smell, leachate can be particularly dangerous. Leachate can dissolve or otherwise incorporate harmful substances from the waste stream, including acids, ammonia, methane, heavy metals, and dioxins. Leachate can also carry pathogenic microorganisms that grow in the waste itself. Then, as leachate seeps into the surrounding soil and water, it creates underground columns of tainted liquid called plumes (**FIGURE 15.7**). Humans and other organisms can suffer adverse health effects by coming in direct contact with leachate plumes in soil and water (including drinking water) or by consuming plants and other organisms lower on the food chain that have taken up leachate contamination. Plumes can have lasting environmental effects. More than 20% of the hazardous waste sites prioritized for cleanup in the United States are former municipal open dumps. Some dumps of the ancient Roman Empire are still producing leachate plumes.

Ocean Dumping

Depositing of waste in the ocean also forms a type of open dump, whether the waste comes from simple solids dumped by individuals or from larger-scale, city-based waste dumps. Two common forms of city-based waste are sewage sludge (the waste remaining after sewage treatment) and dredge spoils (sediment removed from bodies of water to improve navigation). Sewage sludge is organic waste and can cause *eutrophication* (see Chapter 7), leading to anoxic dead zones in shallow coastal waters. Sediments in dredge spoils often contain heavy metals (dense metals such as mercury or lead that are potentially toxic) and industrial wastes that can accumulate in fish and other sea life in the marine food chain, which may in turn be consumed by humans. Commercial sites can supply other ocean waste, notably construction and demolition debris as well as industrial wastes such as acids and refinery ash. For example, in 1986 a waste management company in Philadelphia dumped 14,000 tons of ash from waste incinerators into the Atlantic Ocean after the state of New Jersey and the countries of the Dominican Republic, Honduras, Panama, Senegal, Morocco, and Sri Lanka all refused to accept it. There is even a history of national governments dumping chemical and radioactive weapons wastes at sea (**FIGURE 15.8**).

Waste Trade

In 1987, a barge hauling more than 3,000 tons of municipal solid waste (known as the "garbarge") from Islip, New York drew national attention as it sought a site for disposal along the Eastern Seaboard (**FIGURE 15.9**). After the state of North Carolina rejected it, followed by other states bordering the Gulf of Mexico and the countries of Mexico and Belize, the barge eventually returned to New York where the waste was ultimately incinerated. Although in this case the waste shipment ended up relatively close to its source, the story brought attention to the fact that there is a national and international trade in waste that can send trash far from where it was produced, often to developing nations. Two years after the gar-barge incident, 121 countries signed the **Basel Convention** to restrict transboundary movements of household waste and hazardous waste from more-developed to less-developed countries (the United States has not ratified this agreement). The convention requires the notice and consent of waste-importing countries, and it prohibits waste export when the exporting country has reason to believe that the wastes will not be handled in an environmentally sound manner.

Some of the most egregious examples of the international waste trade involve **e-waste** from our old televisions, computers, cell phones, and other electronic devices. These devices contain valuable materials such as gold and copper, along with hazardous materials such as lead, cadmium, and toxic flame retardants. While many e-waste collection centers and processors are legitimate, some have shipped these materials to developing countries where they are "de-manufactured" at great risk to human health and the surrounding environment. People may burn or pick apart the electronics on open fires before the remnants are heaped in open dumps (**FIGURE 15.10**).

⬡ **TAKE-HOME MESSAGE** Uncontrolled or open dumps were the dominant waste disposal method for most of human history and are still common throughout the world. Open dumps pollute the air and water, creating undesirable and hazardous environments. The waste arriving at open dumps sometimes comes from far away, as producers in developed countries ship their waste to developing countries.

15.3 How Do We Manage Waste?

Most of us have little understanding of how the waste generated upstream of our consumption is handled, nor do we have much idea about where it goes after we drop

FIGURE 15.8 Dumping at Sea From the 1940s through the 1960s, the US government dumped hazardous and radioactive waste from chemical and nuclear weapon production into the sea. This photo shows the US Army dumping one-ton mustard gas containers into the Atlantic in 1964.

our trash into a trash can or a garbage chute. Although open dumps are becoming less prevalent throughout the world, disposal strategies to address the environmental impacts of solid waste are still limited. Waste can be isolated in a variety of ways or burned (combusted), and both strategies can generate electricity for human use (conversion). In some cases, biological agents can convert pollutants or hazardous materials into more benign components. In this section, we will see that each

Basel Convention an international agreement negotiated to restrict movements of hazardous waste from more-developed to less-developed countries.

e-waste waste created from and by electronic devices.

FIGURE 15.9 Garbage in Search of a Home The infamous "gar-barge," carrying more than 3,000 tons of municipal solid waste from New York, was rejected at numerous ports as it sought disposal for its cargo before it finally returned to New York where the waste was incinerated.

Illegal International Dumping

Despite the Basel Convention, many waste shipments from developed countries have found their way to open dumps in the developing world. For example, British waste management companies have been repeatedly caught shipping containers filled with municipal waste to dumps in China and Indonesia, in violation of European law. A report by European Union waste inspectors estimates that 20% of the shipping containers labeled as holding recyclable paper for shipment abroad are disguising significant amounts of household waste bound for open dumps in Asia and Africa. This practice allows the waste management company to avoid the costs of regulated waste disposal in their home countries. Interpol, the international police organization, has even found that organized crime groups profit from smuggling plastic waste across borders. In 2020 the Implementation and Compliance Committee of the Basel Convention issued new guidance and support for countries to address this problem. This included better training and technology for customs and trade personnel to monitor, detect, and investigate waste shipments.

FIGURE 15.10 E-Waste Dump in Accra, Ghana E-waste includes the electronics we discard—such as computers, TVs, and cell phones. When these products are de-manufactured to strip out valuable materials such as gold and copper (as seen here), they can cause harm to the people working with them and to the surrounding environment.

environments such as wetlands and floodplains, which facilitated the flow of leachate plumes into the surrounding area. Often, this land-use strategy was adopted to use the garbage as fill, creating "new" land to develop. Sites like these were also chosen simply because they were close to the waste-producing population and had a low property cost. Today, more stringent regulations require considering environmental factors that influence how well a facility can isolate waste. These factors include a location's topography, hydrology, geology, and normal precipitation levels.

In modern landfills, as each day's waste is spread on top, it is covered with a layer of dirt to prevent it from blowing away and to keep vermin from it. As we see in **FIGURE 15.11**, important engineered aspects of landfill design include subsurface layers of impermeable clay or plastic liners; drainage, collection, and treatment systems for leachate; venting and collection systems for methane gas; and monitoring wells on the outskirts of the landfill to detect escaped contaminants. Landfills devoted to hazardous waste disposal have a similar structure with more restrictive site-selection criteria, more liners beneath the surface, more monitoring requirements, and waste deposited within sealed containers.

Another strategy for isolation is **geologic disposal**, injecting or placing waste in the rock and other natural formations beneath Earth's surface. For example, liquid wastes from industry, mining, and oil and gas production (much of it hazardous) are injected through deep wells into dry, porous rock beneath the impermeable layers of rock that surround an aquifer. The United States has hundreds of thousands of these disposal wells. The US government also disposes of a class of radioactive waste from nuclear weapons production (known as transuranic waste) in a deep geologic repository carved out of impermeable salt deposits located more than 2,000 feet

strategy has advantages as well as its own environmental effects (**TABLE 15.1**).

Isolation

isolation the segregation of waste from significant contact with humans or the wider environment, normally belowground but sometimes in buildings aboveground.

sanitary landfill a disposal site that isolates and contains waste, manages its contents, and treats liquid and gas releases.

geologic disposal injecting or placing waste beneath Earth's surface.

The dominant disposal strategy in the United States is **isolation**—segregating waste from significant contact with humans or the wider environment, normally below the surface but sometimes in buildings aboveground.

The isolation technology most familiar to us is the controlled or **sanitary landfill**, used for more common forms of waste. Sanitary landfills improve on the problems associated with open dumps by attempting to isolate and contain the waste, better manage its contents, and treat liquid and gas releases. Perhaps the most important advance in landfill design is location selection. As late as the 1960s, municipal dumps were located in wet

TABLE 15.1 Modern Waste Management Strategies

Waste Management Strategy	Benefits	Impacts
Isolation • Sanitary landfills • Geologic disposal • Containment buildings	Improvement over dumps Caps and liners prevent waste from escaping into the environment Some strategies allow waste to be moved and treated at a later date	Does not reduce the volume of waste Requires a large area of land to be converted for this use Caps and liners often leak Can produce the greenhouse gas methane from decomposition Risk of water pollution due to failure of caps and liners
Incineration	Reduces waste volume Does not require large areas of land Less risk of direct water pollution	Requires high energy inputs to attain extremely high temperature for combustion Produces a wide range of air pollution and greenhouse gases Harmful materials can remain in residual ash
Conversion	Waste-to-energy conversion leads to some benefit from disposal, such as electricity or heat Remediation technologies can remove certain toxins from the waste	Still uses landfills so has their associated challenges Requires precautions to prevent removed toxins from contaminating the environment

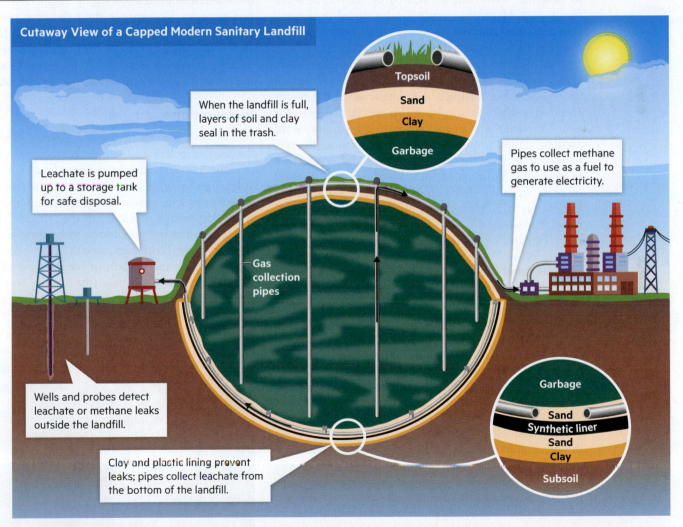

FIGURE 15.11 Landfill Design Modern sanitary landfills are designed both to protect the environment and to recover and use the methane gas produced in them. This graphic shows a capped landfill, which has been sealed off from further garbage inputs. Multiple layers protect both the surface water and the groundwater below the landfill. Pipes sunk into the landfill capture the escaping gas, which can be used to generate electricity.

Adapted from US EPA (2018b).

FIGURE 15.12 Geologic Disposal The Waste Isolation Pilot Plant near Carlsbad, New Mexico, stores nuclear waste canisters in tunnels carved into deep geologic salt deposits.

containment building a structure that isolates the waste from the surrounding area while allowing constant monitoring and retrieval of it.

combustion in waste disposal, the strategy of burning waste.

(600 meters) beneath the Chihuahuan Desert outside of Carlsbad, New Mexico (**FIGURE 15.12**).

Still other disposal strategies offer less isolation from the environment while providing greater retrievability so that the waste can be treated or moved at some future date. Hazardous waste can be stored in secured concrete **containment buildings** that have air-lock doors, liquid collection drains, and negative air-pressure and dust-control systems. These isolate the waste from the surrounding area while allowing constant monitoring and relatively easy retrieval of the waste. Hazardous waste is also stored in metal, fiberglass, or concrete tanks either above- or belowground or in earthen impoundment ponds; these options use systems that can convey the waste into and out of the storage area. As of yet, there is no permanent disposal strategy in place for radioactive waste from spent nuclear fuel at nuclear power plants. Power plants must store this waste on-site using water-filled pools and dry metal and concrete casks.

Challenges of Isolation While modern isolation technologies are a dramatic improvement over open dumps, complete and permanent isolation is impossible, and the potential for waste-related environmental impacts persists. More than half of all the municipal solid waste in the United States is taken to sanitary landfills for disposal. Yet the EPA and waste consultants have found that leachate will eventually escape from these facilities as synthetic liners degrade: landfills in Virginia with several feet of clay and synthetic fabric linings have contaminated groundwater after operating for only a decade. Even though municipal solid waste is not defined as hazardous waste, the leachate and gas from landfills can become hazardous as pollutants such as heavy metals and volatile organic compounds are released. Leaks from disposal tanks, containment buildings, or impoundment ponds are another potential problem. Landfills also produce significant greenhouse gas emissions and worldwide are the largest human-generated source of methane gas (37%).

The growing practice of deep-well injections of waste is also associated with thousands of failures. These failures include leaks from drilling accidents, cracked well linings, and injections that shatter the rock meant to contain the waste. Deep-well injection can even destabilize geologic faults, causing earthquakes. Such geologic hazards have exposed the vulnerabilities of even highly engineered waste-isolation strategies, for example, the radioactive waste cooling ponds at the Fukushima nuclear power plant in Japan. These were compromised during a 2011 earthquake and tsunami, releasing radioactive material into the ground, water, and air that remains a containment challenge (**FIGURE 15.13**).

Incineration

Incineration is a waste disposal strategy with roots in the ancient practice of simple **combustion**, or burning, of waste. Humans have long burned waste to reduce its volume and provide heat, but often at the cost of uncontrolled fires and serious air pollution. Though simple waste combustion still takes place around the globe—from small fires in open dumps to burning yard waste—some developed countries, such as Japan and Switzerland, use a more advanced form of combustion called incineration to manage the majority of their waste disposal. In the United States, less than 12% of the waste stream is managed by incinerators. Strong public opposition

FIGURE 15.13 The Risk of On-site Radioactive Waste Storage In 2011, a tsunami engulfed the Fukushima nuclear power plant in Japan. This caused a breach in the storage tanks holding water contaminated with nuclear waste (pictured here).

Who Bears the Impacts of Waste?

When you throw something in the trash can, where does it go? In the United States, most of our municipal solid waste—our garbage—goes to a landfill, and a smaller percentage is burned in incinerators. And where are these facilities located? While the final destination of your garbage may not be in your neighborhood, it may be near the place other people call home. For more than four decades, the city of Chapel Hill, home of the University of North Carolina, has been sending its waste to a multisite landfill complex in Afton, North Carolina, a town several counties away. As we have learned in this chapter, waste disposal carries with it a range of environmental impacts. Is it fair for the environmental impacts from the actions of people in one place to be concentrated in and borne by the people living in another place?

The environmental justice movement (explored in more detail in Chapter 17) is often considered to have begun in the town of Afton. The movement focuses on the belief that no group of people should have to bear a disproportionate share of the negative environmental consequences generated by industry, government, or society at large. In the 1970s, the government of Afton agreed to allow Chapel Hill to develop and manage a new landfill in the town in exchange for new water lines, sewage treatment plants, sidewalks, and a community center. The landfill was built and even expanded, but Chapel Hill did not provide the promised municipal services. In the 1980s, the state of North Carolina built another landfill, this one for hazardous waste, in Afton over the opposition of local government, and hundreds of residents conducted civil disobedience

Environmental justice protesters in Warren County, North Carolina, practice civil disobedience by lying down in the road leading to the construction site of a hazardous waste facility.

protests by blocking access to the site. The protesters drew on the tradition of the Civil Rights Movement and pointed to the fact that the area chosen for the site had a much larger population of African American and lower-income residents than did other potential host communities in the region.

Although the protests failed to block construction of the hazardous waste site, they garnered national attention. In 1987, widely publicized statistical research found that throughout the United States, a community's racial makeup was more strongly correlated with the location of hazardous waste facilities than any other single factor. Moreover, more than 40% of the nation's total volume of hazardous waste was going to communities with predominantly African American or Hispanic populations. When the research was conducted again in 2007, the results were similar, and the report corroborated that most of the people living closest to hazardous waste facilities in the United States are African American or Latino.

In 2016, the US Commission on Civil Rights found that although the EPA has had the power since 1993 to address environmental justice concerns, it has never made a formal finding of discrimination, citing the difficulty in linking disparate environmental impacts to discriminatory intent. However, this report also singled out the town of Afton as one positive example of EPA efforts to address environmental justice. In 2019, after decades of citizen struggle and numerous complaints filed with the EPA, the Chapel Hill government finally provided the infrastructure improvements it had long promised the residents of Afton.

What Would You Do?

How would you react to the siting of a waste disposal facility in your neighborhood? What would you do to inform yourself about the situation? What steps might you take to influence the local decision-making process?

---

from prospective host communities has thwarted most newly proposed incinerators. **Incineration** makes use of controlled combustion with pollution controls in place. It occurs at temperatures high enough (1,000°C [1,830°F]) to reduce the volume of waste by 80% to 90% (**FIGURE 15.14**). About three-quarters of all incinerators are **mass burn** technologies that combust unprocessed waste on a moving grate system. Mass burn first performs primary combustion of the waste and then secondary combustion of the resulting gases. These systems use filters to capture fine particles, liquid spray to neutralize acid gases, and emission-control devices to capture some (but not all) pollutants. Prompted by federal regulations, incinerator technology has become

incineration in waste disposal, a technology of controlled combustion at high temperatures with pollution controls in place.

mass burn a technology that combusts solid waste first and then performs secondary combustion of the resulting gases.

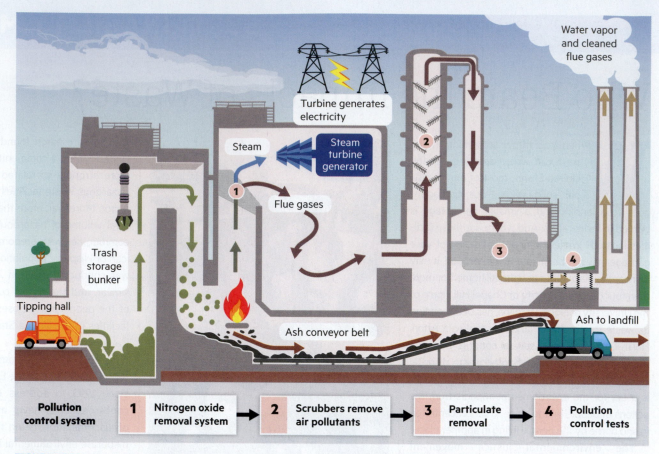

FIGURE 15.14 Incinerators Waste-to-energy incinerators work by using the heat from the burning trash to generate steam that turns turbines to generate electricity. This system still yields solid ash that must be disposed of in a landfill.

Adapted from US EPA (2016).

In the diagram labels:

Water vapor and cleaned flue gases

Turbine generates electricity

Steam

Steam turbine generator

Flue gases

Trash storage bunker

Tipping hall

Ash conveyor belt

Ash to landfill

Pollution control system

1 Nitrogen oxide removal system → **2** Scrubbers remove air pollutants → **3** Particulate removal → **4** Pollution control tests

---

increasingly effective at limiting the pollution produced. In the United States, new emission controls introduced in 1987 reduced dioxin releases from incinerators by more than 99%.

Despite these advanced technologies, incineration still releases significant amounts of carbon dioxide. Concentrations of heavy metals and harmful compounds such as dioxins and furans are also still present in both the residual ash and the emissions. Some incinerators achieve greater efficiency and pollution control by sorting waste prior to incineration. These facilities use *refuse-derived fuel* in the form of dehydrated waste pellets (created in a separate process before burning) in which noncombustibles such as metal and glass have been removed. Some municipalities are also sorting out batteries and plastics from their fuel to further reduce pollution.

Conversion

A third strategy is **conversion**, where waste is changed to something else, most often by harnessing energy from combustion and isolation to generate electricity. For example, incineration generates heat, which can produce steam to heat buildings and/or turn a turbine to generate electricity. Globally, there are more than 1,000

of these **waste-to-energy facilities**, which are especially popular in Japan, Brazil, and Western Europe. There are more than 70 such facilities in the United States, including the Covanta waste-to-energy facility in Tulsa, Oklahoma, that provides electricity for nearby industrial facilities. Landfills can also be paired with conversion technologies to filter out the methane from landfill gases and burn it to produce steam. Landfill gas can even be converted into vehicle fuel.

Other new technologies augment and/or speed natural decomposition processes. In the traditional sanitary landfill described earlier in this chapter, the buried waste is kept as dry as possible, or "dry tombed," to minimize leachate. However, a new strategy of wet **bioreactor landfills** injects additional water and air through horizontal wells or trenches beneath the surface while recirculating leachate with sprayers. This accelerates waste decomposition by as much as 50% and reduces the volume of waste at the site. Because it is aerated, this type of landfill also reduces the amount of methane produced by as much as 90%. (Recall that methane is created only in an oxygen-free environment.)

The newest strategies for disposing of hazardous waste convert it to less hazardous substances. These **remediation** processes use microorganisms and enzymes (*bioremediation*),

conversion a disposal strategy where waste is converted to something else useful.

waste-to-energy facility a place where the heat produced by incineration is used to power a steam turbine that generates electricity.

bioreactor landfill a disposal strategy using injected water and air to accelerate decomposition and reduce the volume of waste.

remediation the process that converts hazardous waste to less-hazardous substances.

fungi (*mycoremediation*), or plants (*phytoremediation*) in conditions that control temperature, pH, moisture, oxygen, and nutrients. These strategies have seen some success at contaminated sites and treatment facilities. Both bacteria and fungi have also successfully reduced the toxic components in petroleum, and other bioremediation successes tackled pesticides and polychlorinated biphenyls (PCBs). Scientists have even identified bacteria and fungi that can break down the polyethylene terephthalate (PET) that is used for many plastic bottles and other products. Some plants, such as poplar trees and mustard, tobacco, and sunflower plants, can effectively absorb and remove heavy metals from contaminated areas. In all these cases, the treated environment must be carefully controlled to prevent the toxins from accumulating in the food chain. Contaminated plants may need to be incinerated or buried as hazardous waste themselves.

⬡ **TAKE-HOME MESSAGE** Modern waste management strategies that were developed to address the environmental effects of uncontrolled open dumps involve isolation, incineration, or conversion. Each of these strategies has environmental impacts.

15.4 How Do We Recycle and Reuse Waste?

Do you have leftovers in your fridge? Most of us do, and much of this food ends up as waste. A recent study showed that 40% of the food produced in the United States is uneaten and wasted. This amounts to $165 billion per year of wasted food, primarily perishable items, including fresh produce (22%), dairy (19%), and meat, poultry, and eggs (18%) (**FIGURE 15.15**). As we learned in Chapter 12, food is wasted all along the production chain, from the farm to the supermarket to those leftovers in your fridge. But it is also possible to divert this food from the waste stream. First, you could eat the leftovers on their own or as an ingredient in a different recipe. This is an example of reuse, and restaurants do it all the time by using the main ingredients from one night's entree in the next night's soup. Some college cafeterias, such as those of New York University and the University of Maryland, even have programs to make use of leftover food each day by sharing it with area food banks. If the food goes uneaten, recycling is another option. Composting is a form of recycling that decomposes and converts food waste into organic material that can then be used to enrich soil to grow more crops. Individual households can compost their food waste with cheap do-it-yourself methods or by purchasing more expensive high-tech compost bins that require less work and accelerate the decomposition process by grinding,

Am I Standing on a Landfill?

In the first half of the 20th century, municipal landfills were often located in wetlands and coastal areas that were drained and filled with garbage so they could be developed. This was the case in the Meadowlands area of New Jersey where MetLife Stadium, home to the New York Jets and Giants football teams, now stands. The site was contaminated with a variety of hazardous wastes from years of dumping in the area, so it had to be remediated before construction could begin. It was also replacing an older stadium nearby. The new stadium used 40,000 tons of recycled steel and 30,000 tons of recycled concrete from the demolition of the old stadium.

FIGURE 15.15 Food Waste and Compost This supermarket food waste is destined to be processed into compost.

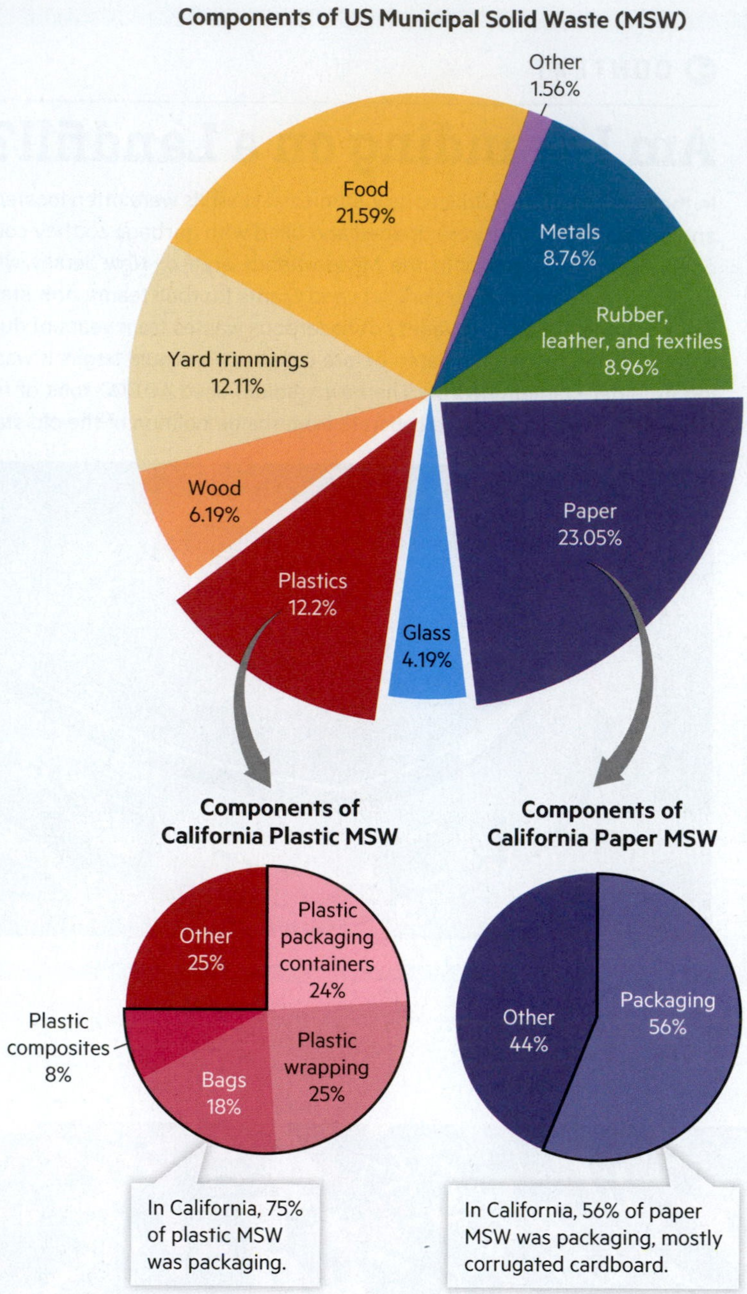

Components of US Municipal Solid Waste (MSW)

Other 1.56%
Food 21.59%
Metals 8.76%
Rubber, leather, and textiles 8.96%
Yard trimmings 12.11%
Wood 6.19%
Plastics 12.2%
Glass 4.19%
Paper 23.05%

Components of California Plastic MSW

Other 25%
Plastic packaging containers 24%
Plastic composites 8%
Bags 18%
Plastic wrapping 25%

In California, 75% of plastic MSW was packaging.

Components of California Paper MSW

Other 44%
Packaging 56%

In California, 56% of paper MSW was packaging, mostly corrugated cardboard.

FIGURE 15.16 **Municipal Solid Waste Generation by Material** The annual municipal solid waste generated in the United States weighs 292.4 million tons. Paper, food, yard trimmings, and plastics are the largest component materials of this waste stream. Waste characterization studies that examine the types of products composing the waste stream show that packaging makes up the majority of both the plastic and paper waste.

Adapted from US EPA (2018a) and CalRecycle (2020).

recycling a strategy of redefining "waste" as "resources" for new products, thus diverting materials from the waste stream.

primary recycling (closed-loop recycling) a process that converts waste materials into the same sort of product from which they came.

secondary recycling (open-loop recycling) a process that converts waste material from a product into a different sort of product.

---

cities in the United States now collect and compost their food waste. In this section, we take a closer look at the potential for waste diversion strategies such as recycling and reuse for the materials we discard.

Recycling

The latest EPA report on municipal solid waste shows that the four biggest components by weight of this waste stream are paper and paperboard, food scraps, yard trimmings, and plastics (**FIGURE 15.16A**). A statewide study in California revealed that packaging materials made up 75% of the plastic waste and the largest share of the paper waste (56%) (**FIGURE 15.16B**).

Analyses like these show that most components of our household waste are recyclable. Formally, we define **recycling** as redefining "waste" as "resources" for new products, thus diverting materials from the waste stream. By definition, we should not confuse recycling with reuse: for example, clothes that are donated or "handed down" and reused by others. Recycling helps to save energy and reduce the impact associated with the creation of products from scratch, but it also requires some processing. This means that recycling still involves some energy inputs and environmental impacts. Some materials can be processed into the same sort of product from which they came by **primary recycling** (or **closed-loop recycling**). Nearly half of the aluminum cans on today's grocery store shelves will be recycled into another can within 2 months. Other materials can undergo **secondary recycling** (or **open-loop recycling**) by conversion to different products. For instance, plastic water bottles made of PET are sometimes recycled into polar fleece products such as jackets and blankets (**FIGURE 15.17**). Composting food waste is also a form of secondary recycling because it converts high-value food into a useful but lower-value product to enrich soil for gardening and farming. While primary recycling tends to be most efficient, secondary recycling is often referred to as "downcycling" because it requires additional energy and materials while reducing the quality of the source material. For example, recycling plastic often results in a mix of plastic materials that is a lower-quality material overall, has more limited uses, and is more difficult to recycle in the future than newly created plastic.

The percentage of municipal solid waste that gets recycled—that is, reprocessed into a new product—is known as the *recycling rate* (**FIGURE 15.18**). In the United States, the recycling rate has grown from 6% in 1960 to more than 32% in 2018. The Netherlands, Germany, Austria, Switzerland, Taiwan, and South Korea have met or exceeded a recycling rate of 50%. Differences in national recycling rates are tied to variations in waste management policies and to different cultural approaches to consumption and disposal. How do the recycling rates of various materials compare? In the United States, more than two-thirds of paper products in municipal solid waste is recycled, more

---

drying, and churning the waste. A growing number of big cities and smaller municipalities, including New York City, Denver, Seattle, San Francisco, and Davis, California, now collect household food waste in curbside bins and then send it to commercial-scale composting facilities that sell the finished product to landscaping companies, farmers, and home gardeners. Many individual households and an increasing number of campuses and

(a)

(b)

FIGURE 15.17 **Different Types of Recycling** (a) Recycling aluminum cans into new cans is an example of closed-loop recycling. (b) Using plastic bottles as a material source for fleece jackets is an example of open-loop recycling.

than one-third of metals, about one-quarter of glass, and less than one-tenth of plastics. Additionally, nearly two-thirds of yard trimmings in municipal solid waste is composted. About one-third of food waste is diverted from the waste stream either by composting or other management strategies like use for animal feed.

➡ **SUSTAINABILITY**

Can We Recycle Human Waste?

Human waste—sewage—is a major cause of water pollution in much of the world. Many developed countries have addressed this problem using sewage treatment plants that separate out the solids (called sludge or biosolids) and disinfect waste-water. While the disinfected water is released for reuse, can anything be done with the biosolids? If this material is suitably treated (heated, microbially digested, and composted) to eliminate disease-causing microbes, it can be converted into fertilizer. The city of Boston converts biosolids into dehydrated fertilizer pellets. Tacoma, Washington, mixes its biosolids with composted yard waste to make topsoil and mulch products it calls Tagro, which it sells to area gardeners and landscapers.

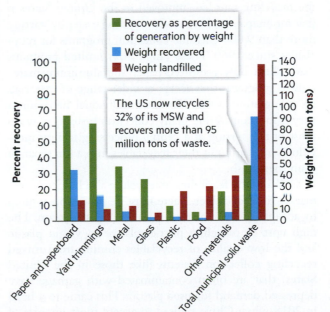

FIGURE 15.18 **US Municipal Solid Waste Recycling Rate and Materials Recycled**

Adapted from US EPA (2018c).

- ■ Recovery as percentage of generation by weight
- ■ Weight recovered
- ■ Weight landfilled

The US now recycles 32% of its MSW and recovers more than 95 million tons of waste.

Percent recovery (y-axis: 0–100)
Weight (million tons) (y-axis: 0–140)

Categories: Paper and paperboard, Yard trimmings, Metal, Glass, Plastic, Food, Other materials, Total municipal solid waste

The Economics of Recycling

The economics of recycling vary across different materials. Recycling paper is a straightforward process involving sorting, removing foreign materials (such as staples), converting it back into pulp, and removing glue before pressing and rolling it back into paper. Producing recycled paper consumes 30% to 64% less energy than does paper production from trees, providing considerable cost savings for the producers. Recycled aluminum consumes 95% less energy than does producing new aluminum, which requires refining mined bauxite ore. Using magnetic fields at recycling centers also makes aluminum the easiest packaging material to separate, as these fields separate the aluminum out. This makes aluminum scrap a valuable commodity, recently fluctuating between $0.40 and $1.00 per pound.

In contrast, recycling plastic presents serious challenges because of sorting difficulties. Plastic products can be made from a variety of polymer types, which are reflected in the resin identification codes visible on soda bottles and other plastic products (**FIGURE 15.19**). These various types of plastics differ in their chemical makeup and cannot be processed together. Products often laminate multiple types of plastics together or integrate them with non-plastic materials, making the sorting process complex and contamination by foreign materials common. Plastics also contain many additives to provide color, temperature stability, flexibility, and strength, which must be removed during the recycling process. The extensive processing required for plastics recycling has kept the cost of recycled plastic as much as 40% above the cost of virgin plastic.

The success of recycling hinges on more than just the materials and processes involved. Collection systems, the behavior of individual consumers, and product design play significant roles too. One of the reasons the recycling rate has improved in the United States is that municipalities have made recycling easier by starting more than 9,000 curbside collection programs for recyclables since 1960. These range from limited programs that collect only presorted paper and aluminum materials to programs that accept a wider range of materials including food waste. However, individual behavior is very important. While many collection systems now do most of the sorting at a central processing facility, individuals must still put their recyclables into a bin separate from their trash.

Globally, the feasibility of recycling also hinges on the market for recyclables. There must be processors willing to accept plastics diverted from the waste stream. The high cost of recycling plastic relative to virgin plastic and the low quality of recyclables coming from mixed recycling collection systems (like those in the United States) that are often contaminated with garbage have depressed demand for used plastics. This came to a head in 2018 when China refused to accept most imports of

Code	Generally accepted
1 PET	**Polyethylene terephthalate** Soda/water bottles, food containers
2 HDPE	**High-density polyethylene** Milk jugs and detergent bottles
3 PVC	**Polyvinyl chloride** Shampoo and window cleaner bottles
4 LDPE	**Low-density polyethylene** Squeeze bottles
5 PP	**Polypropylene** Yogurt containers and ketchup bottles
6 PS	**Polystyrene** Not generally accepted
7 OTHER	**Other** Not generally accepted

FIGURE 15.19 Plastic Resin Codes Plastic packaging is made from a variety of resins, making sorting and recycling challenging.

Adapted from Johnston (n.d.).

recyclables under its National Sword policy. Prior to this shift, 70 percent of the plastics collected through recycling programs in the United States and 95% of those collected in Europe were sold and shipped to processors in China. But the processors found that they were unable to use much of this material and instead ended up burning or burying millions of tons of waste plastics. The United States and other nations that relied on exporting recyclables have not been able to meet the new 99.5% purity standard China placed on imports and have been forced to look for other solutions. In the short term, many US towns and cities halted their recycling programs or restricted the types of material collected—leaving residents to put more materials in the trash. In

Great Britain, the amount of plastics sent to incinerators increased. Over the longer term, recycling industry analysts hope that the estimated 111 million metric tons of plastic that are displaced by the Chinese policy over the next decade will lead to dramatic changes in the design, use, and recovery of plastics. Industry groups have developed **design for recovery** guidelines to standardize the manufacture of materials and products in order to facilitate efficient recycling (**FIGURE 15.20**). For example, the Association of Plastic Recyclers has issued recommendations on inks, adhesives, and colorants used in plastics that help to improve recycling operations. There has been growth in government policies throughout the world that ban certain plastic products and force producers to help pay for waste management related to their products. Many governments are also investing in public education campaigns to help change consumption habits, reduce waste production, and encourage better sorting of recyclables.

The Advantages and Limitations of Recycling

In general, there are tremendous environmental advantages to recycling. Recycling alleviates the environmental effects of landfills and incineration. Recycled materials reduce the consumption of raw materials and also reduce the many environmental effects from extracting and processing these materials. Recycled metal, paper, and plastic reduce the need for mining, logging, and oil extraction and thereby reduce their accompanying land, air, and water pollution as well as associated habitat and climate impacts. Processing recyclables typically saves energy and lowers greenhouse gas emissions compared to processing raw materials.

However, this does not mean that all recycling is a net gain for the environment. First, it's important to make sure that recycled products are safe. When the products undergoing recycling are *hazardous*, the new products may carry human health risks and produce additional pollution. For example, under US law, refinery and incinerator wastes contaminated with heavy metals can be recycled into additives for fertilizers and soil amendments. However, these processes can add toxins to the food chain. Similarly, hazardous wastes reprocessed into concrete and asphalt products for road construction or road de-icers can contaminate water.

When a product is labeled "recyclable" or "recycled," we should all consider what this really means. These terms can sow confusion for consumers. Labeling a product as "recyclable" merely suggests a possibility rather than an assurance or even a high probability that the product will be diverted from the waste stream and converted to a useful product. Technologies exist to recycle nearly any material, yet it is often not economically advantageous to adopt these technologies. And even collecting recyclables does not ensure recycling. The market for recyclables fluctuates, and the demand for some materials,

FIGURE 15.20 Packaging Design and Recycling Some common items are very hard to recycle. The smaller the size of the yogurt cup, the worse the economics are to recycle it. Most curbside recycling programs in the United States do not accept plastic bags or films because they clog up the machinery at recycling facilities, although separate collection centers for this type of plastic can sometimes be found. Toothpaste tubes are almost impossible to recycle.

such as glass, has fallen off so drastically that municipal collection services now routinely send these recyclables to landfills. Even when a product is labeled "recycled," it may not be diverting materials from the waste stream. For example, unless a package of "recycled paper" gives a percentage of postconsumer content on the label, scraps from the virgin paper-making process that paper mills have always used could be what is "recycled."

Reuse, Refurbish, Repurpose, and Upcycle

Used materials do not need to be recycled if we can simply reuse, refurbish, or repurpose them. This was once a common practice. Until the 1970s, glass bottles in the United States were commonly collected and returned to producers for cleaning and reuse. In many developing countries, glass bottles and other containers of various materials are not treated as disposables because bottling companies are willing to pay for their return. They are washed, refilled, and reused at a much higher rate than occurs in developed countries. This contributes to a far different municipal solid-waste stream composition: even though these countries lack the extensive curbside recycling collection infrastructure of more developed countries, individual consumers have a strong economic incentive to safeguard and return glass containers for reuse.

Some industries have devised new ways to use materials from their production processes. Eco-industrial parks facilitate reuse and repurposing of materials. The best-known example is in Kalundborg, Denmark, where a coal-fired power plant provides steam heat to nearby factories; gypsum captured from the pollution-control devices of the power plant provides raw material for a neighboring wallboard manufacturer; and ash from burned coal is used for cement production. Each of these industries uses the waste from the power plant, reducing the need for new extraction of raw materials.

Local businesses that resell durable goods, such as cars, furniture, electronics, and clothing, are already common in most communities, and some "recycled" goods, such as clothing and furniture, attain "vintage" status and can net a considerable profit. "Freecycling" and "Buy Nothing" groups and websites or other online markets such

design for recovery
a process where guidelines are established to standardize the manufacture of materials and products in order to facilitate efficient recycling.

The Challenge of Handling Biodegradable Plastics

Biodegradable plastics are typically made from polylactic acid (PLA) resin rather than from the petroleum-based polyethylene terephthalate (PET) resin. PLA plastics are made from renewable plant materials (typically corn), and under the right conditions they can break down and be assimilated back into the environment. But PLA plastics pose a problem for recycling systems. They are hard to compost with food waste because they do not degrade at the same rate as other materials in the compost. It is also hard to distinguish PLA from PET plastics. PET plastics must be removed from food waste recycling, while PLA is hard to recycle with other plastic because it is not as strong as PET plastic (and therefore breaks down more rapidly). Until PLA containers are easier to sort from PET containers, they will remain a problematic challenge for recycling systems.

100% Biodegradable

consumption the ways we use and dispose of material things.

as eBay, Craigslist, and Facebook connect people with used stuff to those who want that stuff. Other goods can be repurposed or creatively "upcycled" for new uses that bring value above and beyond that of the original products: glasses made from bottles, handbags from reclaimed plastic, and furniture from wooden pallets are just a few examples (**FIGURE 15.21**).

⬡ **TAKE-HOME MESSAGE** Most components of US household waste are recyclable. Recycling involves processing waste into a new product. The recycling loop is not complete until used materials are converted into useful products that are purchased by consumers. The environmental effects associated with making new products and recycling old ones are avoided when materials are reused, refurbished, or repurposed.

FIGURE 15.21 Upcycling Remaking goods into creative new products makes them more valuable than the original product. For example, this tiny house is built out of old shipping pallets.

15.5 How Does Our Culture Affect Consumption?

At its root, solid waste is linked to the goods we consume. The most fundamental way to address the problems associated with waste is to reduce the waste source by reducing consumption. The longtime slogan "reduce, reuse, recycle" is not meant to give equal weight to each of those waste management strategies: it is a prioritized list, and "reduce" is given top billing. Because our consumption is linked to a far bigger waste stream than we can see directly in the goods we purchase, each choice we make that reduces our consumption reduces waste associated with extraction, production, and distribution as well.

Reducing consumption is easier said than done. Think about the things you've purchased in the past few weeks. How many do you really need? This is not a simple question. At a basic level, consumption is a biological function. We have bodily responses that signal us to meet our needs, and the environment shapes what is available for us to consume. Over the past few weeks, you've consumed things related to the provision of food, water, warmth, and shelter that you need to survive. But you've probably purchased other things recently—maybe a phone or certain clothes—that aren't directly linked to your basic survival, even though you might have a hard time doing without them. Humans are social animals, so our **consumption** also plays an important cultural role in our lives.

To better understand the role consumption plays, anthropologists have spent months living with and observing individuals as they engaged in everyday tasks and shopping. These researchers have found that individual consumption patterns have social functions extending beyond mere survival and comfort. Personal electronics and clothing are often discarded and replaced not because they are broken or worn out but because they are perceived as unfashionable. This finding matches

FIGURE 15.22 Conspicuous Consumption Conspicuous consumption of goods is consumption intended to be seen—like goods produced by expensive designer brands.

the concept of **conspicuous consumption** (consumption intended to be seen) that economist Thorstein Veblen developed in 1899 to describe how people purchase certain goods to project particular identities within society (**FIGURE 15.22**). Anthropologists have also found that people buy and exchange various products with each other to establish and express relationships. In other words, consumption has a cultural context, where **culture** is the systematic, learned, and shared understandings and behaviors of a particular group. Let's look in more detail at how culture influences consumption in the United States.

Culture and Consumption in the United States

We all know that different people think and act differently toward the environment. Exploring these differences across different historical periods and cultures can be very revealing.

Historians have found that prior to the 1950s, household manuals, magazines, and newspapers often defined the success of households (and, at the time, "housewives") by how little they "wasted." In other words, people often compared themselves on the basis of thrift. At that time, it was not uncommon for food scraps to be fed to animals or tilled into the ground as fertilizer; corn cobs were dipped in a reserve of tar to use for kindling; worn bedsheets were refurbished by cutting them down the middle and sewing the outer edges together; and even leftover soap suds and ashes were used to enrich the soil for young plants. As long as these and other items had a use, they were not "waste."

Similarly, in the first half of the 20th century, repurposing food packaging was a widely shared cultural phenomenon. One of the most popular transformations occurred with old cloth sacks that held flour or sugar that were later sewn into quilts, clothing, towels,

and tablecloths (**FIGURE 15.23**). Food manufacturers provided sewing instructions and packaged their products in attractive patterned fabrics, while retailers saw value in buying used sacks back from customers who did not care to sew and then selling them to others as fabric or finished cloth products. Even government officials gave the practice a boost by posing in flour-bag clothing to promote thrift.

But a growing economy and increasing prosperity after World War II gave consumers more buying power. New technologies and materials inspired an ever-expanding array of consumer goods. Today, most Americans do not value food packaging for anything beyond its first use, so it is one of the largest components of household waste in the United States. Much of this waste is plastic, one of the new materials that gained widespread use after World War II. By the 1980s, technological advances produced plastic containers, bags, and packaging film that were as much as 30% lighter and 40% thinner than they had been in the 1950s. Lightweight plastic lowered both the production cost (by using less raw materials per item) and the transportation cost (by decreasing weight). And an increasing demand for portable, single-serving prepared foods by a culture of consumers spending more time in their cars and working outside the home

conspicuous consumption the act of purchasing certain goods to project particular identities or images within society.

culture the systematic, learned, and shared understandings and behaviors of a particular group.

FIGURE 15.23 A Culture Facilitating Waste Reduction In the early 20th century, flour companies packaged their product in cloth with attractive patterns so that it could be sewn into dresses. The companies even provided sewing instructions.

provided a growing market for cheap, disposable plastic packaging.

Archaeological research using core samples drilled out of landfills reveals that each time a new technology (such as plastic food packaging) gains widespread cultural adoption, it makes a significant impact on the waste stream (see **STORIES OF DISCOVERY** for more on the archaeology of garbage).

Of course, individual consumption is also set within our economic and political systems. Businesses that sell material goods are striving to raise profits by boosting consumption of their products. Economists use measures of consumption, such as retail sales, food sales, and house construction, as indicators of economic growth. Politically, different strategies are often used to regulate behavior. Prior to 1866, it was legal and common to simply dump garbage out the window or back door or even in the streets of American cities. Now cities have detailed laws governing waste disposal, and most provide collection services and disposal facilities to accommodate our increasing consumption and consequent waste production.

⬡ **TAKE-HOME MESSAGE** The most fundamental way to address the problems associated with waste is to reduce consumption. However, waste reduction is a challenging goal to meet because consumption patterns are nested within our cultural context. Culture consists of the learned and shared understandings and behaviors of particular groups.

15.6 Can We Reduce Our Waste?

Reducing what we consume is often challenging because it can require significant lifestyle changes. One strategy for reduction is to target aspects of our society and culture that drive consumption. Let's explore some of these strategies.

Business and Consumer Strategies

Some of the most promising waste reduction strategies target businesses and consumers. For example, waste reduction certification and marketing campaigns recognize some consumers' desire to identify and purchase less wasteful products (and possibly identify themselves as environmentally conscious) and businesses' desire to attract these consumers. The US Green Building Council offers the TRUE Zero Waste certification to facilities that meet stringent waste reduction criteria. Facilities that earn the TRUE certification are encouraged to publicize their status and use the TRUE logo in their press releases. The *dose certa* certification program in the European

Union is extended to restaurants that meet waste reduction targets and submit to monitoring of their waste stream. British retailer Marks and Spencer is marketing itself as the world's most sustainable major retailer with an ambitious set of environmental goals known as "Plan A." Since 2007, the company has reduced its waste by 30% and, more impressive, sent zero waste to landfills (**FIGURE 15.24**).

There are also examples of consumer groups pressuring businesses to adopt less wasteful practices. In 1987, the Citizens Clearinghouse for Hazardous Waste launched a successful series of protests and boycotts to convince McDonald's restaurants to eliminate Styrofoam food packaging. The group's "send-it-back" campaigns flooded McDonald's corporate headquarters with Styrofoam containers collected from trash cans. McDonald's responded by reducing the volume of its solid waste by 70% to 90% with a change to paper food wrapping.

More recently, businesses and institutions that purchase large volumes of products have influenced their suppliers to reduce waste (and shipping costs) as a condition of sale. Walmart implemented a very successful waste reduction plan that encourages manufacturers to redesign packaging (and reduce shipping costs). The slight redesign of a single cooking oil bottle reduced plastic

FIGURE 15.24 Business Support for Waste Reduction Marks and Spencer's Plan A environmental goals include waste reduction initiatives such as hanger recycling.

What Can We Learn from Our Garbage? The Archaeology of Us

Imagine yourself on an archaeological dig, sifting through dirt and rock to find bits of broken pots, vases, jewelry, tools, and building materials left long ago by another civilization. Archaeologists will be the first to tell you that what you are actually doing as you dig and sift is going through someone else's trash: discarded objects from the distant past. When discarded objects fade from living memory (after a period of about 60 years), they often take on special significance as artifacts—objects made by humans that reveal something about the cultures that created and used them.

But why shouldn't we also examine our more recent waste as artifacts with the systematic care of an archaeologist? What would we learn about our own society if we were to conduct an archaeological dig through the decades of waste accumulated at our municipal landfills? These questions led pioneering anthropologist William Rathje to found the Garbage Project at the University of Arizona. Rathje began his career examining pottery sherds and other discarded objects related to exchange systems and trade routes of the ancient Mayan culture, but this work led to questions about our own waste.

Rathje called his work "the archaeology of us." He and his students carefully sorted, weighed, and categorized more than 250,000 pounds of waste, either intercepted from city garbage collection or retrieved from the depths of more than a dozen different landfills in the United States and Canada. They also conducted interviews and surveys on consumption patterns. Other anthropologists have taken up this work too, and the new field of study is often called "garbology."

One of the earliest findings in the 1980s was simply that paper, which composed nearly half of municipal solid waste, did not biodegrade in the landfills as waste managers expected it to. The archaeological digs found newspapers from the 1950s that were still readable. This finding contributed to the development of more ambitious paper-recycling efforts. In the late 1990s, garbology studies found that

William Rathje pioneered "garbology"—the archaeological study of modern landfills. Other approaches to waste characterization studies count and categorize the materials in a waste stream before they are sent to a disposal facility. These students at Williams College are conducting a waste audit of garbage from 16 residence halls on campus.

technological advances that produced lighter and stronger plastics using less resin did not reduce plastic waste but instead facilitated the use of plastic in a lot more disposable products and packaging.

Studies of curbside garbage pickup also found that certain kinds of hazardous wastes were closely associated with household income. For example, motor oil was most often disposed of in low-income neighborhoods where people were most likely to change their own oil, while pesticides were most often disposed of in middle-income neighborhoods where people were most likely to do their own gardening. Information like this has helped many cities design targeted education and hazardous-waste collection programs that are effective.

Finally, researchers combined curbside waste analysis with household surveys of actual waste producers. They found that people tend to drastically underestimate and/or underreport how much waste they generate and the extent to which they consume prepared and packaged food. In other words, this research reveals a disconnect between how we think of and portray our own behaviors and what we actually do.

What Would You Do?

If archaeologists of the future were excavating waste from your generation, what sorts of artifacts would set it apart from previous generations? What cultural factors drive the consumption and discard of these waste artifacts? With this in mind, what do you propose we can do to reduce these aspects of the waste stream?

How Much Is a Bag Worth to You?

Since 2015, nearly half of US states have considered bills regulating plastic bags in retail stores. In 2014, California became the first state to ban plastic bags; this law also levies a 10-cent fee on the use of recycled paper bags. Many states, including Connecticut, New York, and Oregon, followed California's lead. In other places, the regulation is done at the local level. Most counties in Hawaii ban the use of plastic bags and paper bags with less than 40% recycled content. Still other states, including Florida, Michigan, and Texas, have passed laws preventing local governments from instituting plastic bag bans.

producers of various kinds of plastic products to cover the cost of waste management, recycling, data gathering, and awareness campaigns for the public. Many US states have passed producer responsibility laws for e-waste, requiring that manufacturers of products and packaging also take responsibility for collection, recycling, reuse, or disposal. The state of Washington passed a law requiring manufacturers of consumer electronic products to pay an administrative fee to support an e-waste collection and recycling program. The fee is based on the number of products a manufacturer sells within the state and the presence of those products in the waste stream. This means that although state and local governments are administering the management of e-waste, the cost is covered directly by the manufacturers and indirectly by the consumers of electronics rather than by the taxpayers at large. Since 1991, Germany has implemented a policy of *extended producer responsibility* across all product packaging, which makes manufacturers responsible for collecting, recycling, and final disposal of packaging waste. German manufacturers pay a fee on the basis of the type and volume of packaging material, which is used to operate a recycling collection program.

Community Initiatives and Individual Actions

Other waste reduction strategies are taking place at the community level by making it easier to share goods. Food co-ops and other neighborhood food-sharing organizations are ways that communities can help localize food distribution and reduce the amount wasted. Bike-sharing programs that rent out bikes for short trips have been popular in Europe since the 1970s and have enjoyed widespread adoption in US cities in recent years, including in Boston, New York, Fort Worth, Chicago, and San Francisco.

While "car culture" is still very strong around the world, many are predicting changes that could radically alter the consumption and waste created by transportation. Some forecasts predict far greater sharing or multiple-consumer use of cars in our future (**FIGURE 15.25**). In one study, Barclays Bank estimated that by 2040, household car ownership will drop from 2.1 cars to 1.2 cars. Uber, Lyft, the Chinese DiDi, and businesses such as Zipcar are all built on the idea that in the future far more people will not own a car and instead will rely on hiring or renting one. However, it is also true that the pandemic slowed down interest in ride sharing, and recent surveys have shown a shifting trend back to buying cars. Still, in 2021, 46% of people in Gen Z agreed with the statement "Having transportation is necessary, but ownership is not." There is a good chance pre-pandemic trends will resume.

packaging by 415 tons a year, the redesign of shoeboxes reduced paper waste by 692 tons, and new toy packaging eliminated 1 billion feet of wire ties. In 2011, Tacoma, Washington, became the first "life-cycle city" when the government adopted a policy of making operations and purchasing decisions on the basis of life-cycle assessment of products and services to reduce waste and other environmental impacts as well as cost. Many colleges and universities have also developed green purchasing criteria to encourage suppliers to reduce waste.

Government Action

Governments have also designed policies to reduce certain components of the waste stream. For example, more than 500 US cities and counties have some form of legislation discouraging or banning the distribution of disposable plastic bags at the checkout of large retailers. These policies encourage shoppers to bring reusable bags to the store and thus reduce the overall consumption of plastic bags. In 2021 the European Union banned many other single-use plastic products, including straws, cutlery, and food containers. It also established **producer responsibility laws** that force the

Of course, waste reduction initiatives rely on the support and participation of individuals making decisions to reduce their waste. As consumers, we can support businesses taking steps to reduce waste, and as citizens, we can

producer responsibility law
a law requiring manufacturers of products and packaging also to take responsibility for collection, recycling, reuse, or disposal.

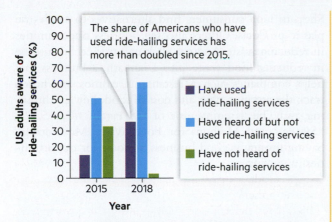

The share of Americans who have used ride-hailing services has more than doubled since 2015.

Legend:
- Have used ride-hailing services
- Have heard of but not used ride-hailing services
- Have not heard of ride-hailing services

(Y-axis: US adults aware of ride-hailing services (%); X-axis: Year — 2015, 2018)

FIGURE 15.25 Ride Sharing Is Growing Rapidly Programs and businesses that facilitate sharing certain products and reduce individual ownership can play a role in reducing waste. The increasing popularity of ride-hailing services may reduce the prevalence of individual car ownership.

Adapted from Jiang (2019).

support government waste reduction policies. But more direct actions that get to the heart of our consumption are also needed. We must ask ourselves if we really need a new product before buying it, if there are ways to get more use out of the things we already have, and how we can recycle materials rather than dispose of them.

⬡ **TAKE-HOME MESSAGE** Many strategies have been developed to reduce waste by working with the cultural factors that influence our consumption choices. These include targeting the behaviors of businesses and consumers as well as creating new government and community-level initiatives.

15.7 What Can I Do?

Each one of us occupies a place in the waste stream through our consumption and disposal decisions and actions. Here are some steps we can take to reduce the waste we generate and to play a role in reducing the overall waste stream.

Reduce Your Food Waste

Some estimates hold that Americans throw away 35 million tons of food each year, 50% more than the amount thrown away in 1990 (and 3 times more than that in 1960). This adds more than $2,000 to the average family's annual grocery bill. An easy suggestion to help you buy less is to make a shopping list, which helps eliminate impulse purchases. Another is to do what the best chefs do and try to use everything in the kitchen. At home, you can do this by eating your leftovers before they spoil and using as much of an ingredient as possible. For example, you can use bones and vegetable scraps to make stock, pickle edible rinds, and eat the edible greens of root vegetables.

Stop Hitting the "Buy" Button So Often

During the pandemic, most Americans had less to do during the week, and the US government found that people were able to save more. Much of this saving turned into a pattern of buying. With the inability to spend money on such activities as going out, eating out, and travel, people shifted to buying more things online. Shortages initially blamed on shipping problems were later shown by a *New York Times* study as often being caused by people just buying a lot more. As we move away from this difficult time, it will be important for us to reconsider how and what we are purchasing. How many purchases were necessary and which ones were not? How many could be made by a quick stop at a store on the way home from work or class instead of being shipped in a box?

Track How Much Waste You Produce

You can raise your understanding of your role in the waste stream by keeping track of what you discard week to week and trying to reduce it. The average person throws out 4.3 pounds of stuff a day. See how low you can go. A woman named Bea Johnson and her family decided to adopt a zero-waste lifestyle in 2008, and over 2 years they were able to capture all their waste in a small jar. She says, "Refuse, Reduce, Reuse, Recycle, Rot (and only in that order) is my family's secret to living waste-free since 2008!" Her blog contains many suggestions, though they start with the simple instruction to "arm yourself with a reusable water bottle, a couple grocery totes, a few cloth bags and reusable jars and bottles."

Take Action against Wasteful Practices

This chapter already described how a campaign to end the use of Styrofoam by McDonald's caused the company to change their packaging materials. Are there similar wasteful practices going on today, either by a company or at your university? If you don't know, maybe your school should conduct a waste audit for its buildings or campus. The Post Landfill Action Network (PLAN) has a manual you can download to help you get started. This is a big

WASTE AUDIT MANUAL

project to initiate, of course, but you might consider joining with some friends and classmates to research the possibilities. Or consider starting small and doing a waste audit of just one department, office, or building on campus.

Help Others Use Less

Lauren Singer, a former environmental studies major, opened a store and company in 2017 called Package Free Shop to help consumers find alternatives to single-use plastic products. Others also see business opportunities in reducing waste. Andrew Shakman turned his interest in reducing food waste into Leanpath, a business that helps companies such as restaurants, casinos, and hotels track the amount, type, and cost of food they are throwing away. He is coinventor of US Patent #7,415,375, "Systems and Methods for Food Waste Monitoring." Saving money is good business, so look for opportunities in your area.

Chapter 15 Review

SUMMARY

- The waste linked to the things we consume includes not only the materials we put in the trash, the recycling bin, or the compost pile (downstream), but also waste from the processes that make and distribute the goods we consume (upstream).

- Life-cycle assessment is a way of measuring environmental impacts of a product from its creation through its disposal.

- For most of human history, and still today in many places, waste is simply dumped or burned in an unregulated way that causes a range of environmental and human health harms.

- There is an international trade in waste from developed to developing countries. An international agreement known as the Basel Convention addresses inequities in the environmental harms resulting from these shipments.

- The three modern waste management strategies are isolation (often in sanitary landfills), combustion (often in mass burn incinerators), and conversion (through either waste-to-energy facilities or remediation technologies).

- Recycling can be a way to reduce the environmental effects associated with the creation of new products, although recycling itself is an industrial process that produces pollution.

- The largest components of the household waste stream in the United States are recyclable, although the recycling loop is not complete until materials are diverted from the waste stream, collected, and converted into new products that are purchased by consumers.

- Reusing, refurbishing, or repurposing materials has a greater environmental benefit than does recycling.

- The cultural context of consumption plays a major role in waste production.

- Reducing consumption is the most direct and beneficial strategy for reducing waste.

KEY TERMS

Basel Convention
bioreactor landfill
combustion
conspicuous consumption
consumption
containment building
conversion

culture
design for recovery
e-waste
geologic disposal
incineration
isolation
leachate

life-cycle assessment
mass burn
municipal solid waste (MSW)
open dump
primary recycling (closed-loop recycling)
producer responsibility law
recycling

remediation
sanitary landfill
secondary recycling (open-loop recycling)
solid waste
waste stream
waste-to-energy facility

REVIEW QUESTIONS

The letters following each Review Question refer to the Chapter Objectives.

1. What is the difference between solid waste and municipal solid waste? Give an example of waste that is not included as part of municipal solid waste. **(A)**

2. What does life-cycle assessment of a product analyze? **(A)**

3. Name two environmental impacts associated with open (uncontrolled) dumping on land and at sea. **(B)**

4. List and define the three modern strategies for managing waste disposal. **(C)**

5. Which waste disposal technology is most commonly used in the United States? **(C)**

6. What types of waste are sometimes managed with geologic disposal? Why are containment buildings sometimes used to isolate this type of waste, despite being less secure? **(C)**

7. Describe two examples of waste conversion. **(C, D)**

8. What materials are the four biggest components by weight in the US municipal solid waste stream? **(A, E)**

9. Describe the difference between closed-loop and open-loop recycling, and identify the type that is associated with downcycling. **(D)**

10. In the United States, which material has the lowest recycling rate? What economic factors lead to the low rate of recycling of this material? **(D)**

11. Explain why the waste management slogan "reduce, reuse, recycle" lists these strategies in that order. **(D, E, F)**

12. How did US cultural attitudes about conservation and reuse shift after World War II and why? **(F)**

13. How can businesses influence their suppliers to reduce waste? **(F)**

FOR FURTHER THOUGHT

The letters following each item refer to the Chapter Objectives.

14. Select a material item that you use or consume, and describe where your usage falls in the life cycle of this item. What contributions has it made to the waste stream both upstream and downstream of you? **(A, C)**

15. Earlier in this chapter, we identified the phenomenon of conspicuous consumption as an example of a way in which culture affects our consumption and waste patterns. Identify and describe an example of conspicuous consumption in a social setting familiar to you. Propose a waste reduction strategy directed toward the type of consumption you are describing. **(E, F)**

16. The Stories of Discovery feature in this chapter focuses on the ways in which archaeologists have gained insights about culture through careful analysis of garbage. Find some garbage near you that you can safely sort through (perhaps the trash bin in your room), and identify, categorize, and count the contents. What inferences can you make about our culture and consumption from this garbage? **(E)**

17. In the United States, the cost of disposing of or recycling waste from our homes and businesses is primarily borne by local tax-payers through municipal governments, such as those of cities and counties. However, the policy strategy of producer responsibility seeks to shift the cost to the manufacturers of products and packaging that end up in the waste stream and to the consumers that buy these goods. Would you support a policy shift toward producer responsibility? Explain your reasoning in detail. **(E, F)**

18. Finding a location for a new waste disposal facility is often a contentious process. Imagine you are a government official overseeing such a process in your hometown. Describe how you would determine which waste disposal method to choose and where to locate the facility. What conflicts would likely emerge, and how would you handle them? **(C)**

19. In this chapter, we learned that there is an international trade in hazardous waste. The Basel Convention is an international agreement to ensure that the receiving country (typically a developing nation) consents to the import of waste and that the waste will be managed in an environmentally sound manner. Although 121 countries have signed this agreement, the United States has not yet ratified it. Take a position on the international trade in hazardous waste and the Basel Convention, and explain your reasoning. **(C, F)**

20. Identify a material that you think constitutes a significant component of the waste stream on your campus, and propose a strategy that aims to reduce, reuse, and recycle this material at higher rates. **(D, E, F)**

Make Your Case

The following exercises use real-world data and news sources. Check your understanding of the material and then practice crafting well-supported responses.

Use the News

The following article from the Canadian Broadcasting Corporation (CBC) website discusses food waste in Canada. Use this article to answer the questions that follow. The first three questions test your understanding of the article. Questions 4 and 5 are short answer, requiring you to apply what you have learned in this chapter and cite information in the article. Answers to Questions 1–3 are provided at the back of the book. Question 6 asks you to make your case and defend it.

"Grocers, Innovators Work to Save $31B in Food from Being Trashed in Canada Each Year," Maryse Zeidler, *CBC News*, November 10, 2018

Lori Nikkel first discovered how much food gets wasted in Canada when she was a single mother to three hungry boys in the Toronto suburb of Etobicoke. Nikkel and a group of other low-income mothers had convinced their local Loblaws grocery store to donate food to their children's school for a student nutrition program. Often, Nikkel found out, the store was overstocked because it ordered too much of certain items. The excess food went to Nikkel and the school—otherwise, it would have gone to waste. "I remember once getting 500 pineapples," Nikkel said. "It was great." Nikkel is now the CEO of Toronto-based charity Second Harvest, which connects suppliers and distributors with non-profit organizations across Ontario to distribute excess food so it doesn't end up in landfills. This week she was on a panel at the Zero Waste Conference in Vancouver.

Canada wastes an estimated $31 billion worth of food each year—about 40 per cent of all the food produced. More than half of that occurs before it gets to the dinner table, when it's discarded by farmers, food manufacturers and distributors. Nikkel says food gets wasted across the supply chain, from farmers who can't sell blemished apples to manufacturers that slightly mess up a batch of bread, to grocery stores that end up with too many items about to reach their best-before date. "People just consider it the cost of doing business," she said. "Where food is getting made, there is loss across the chain."

Food waste has become a salient topic, not only because edible food gets tossed out while millions of people go hungry. Organic waste also produces methane when it decomposes in landfills, making it one of the biggest contributors of greenhouse gas emissions. But people like Nikkel are trying to change that. And businesses that want to root out inefficiencies to protect their bottom line are listening.

Overproduction of Food

Second Harvest collects food that would otherwise go to waste and delivers it to more than 300 non-profit organization [*sic*] across Ontario. The organization focuses on fresh food, like produce, meat and dairy. "There is an overproduction of food globally and it actually leapfrogs over people and goes into landfills, creating a huge environmental challenge," she said during a lunch break at the conference on Thursday. Other groups at the conference dedicated to preventing waste included Vancouver-based tech start-up FoodMesh, which also connects surplus foods to charities, and Provision Coalition, which works directly with food and drink manufacturers to help them become more efficient.

One of the other people on the panel with Nikkel was Sam Wankowski, Walmart Canada's senior vice-president of operations for Western Canada. Last April, Walmart committed to have zero food waste across its organization by 2025. "It just makes good business sense," Wankowski said. "If we can . . . improve the efficiencies of our processes and our infrastructure, that just increases value for everyone."

"We Can Do a Lot More"

Walmart came under a lot of criticism in 2016, when a CBC Marketplace investigation found garbage bins full of produce, bottled water, frozen foods, meat and dairy products that appeared to still be fresh and safe for consumption. But Wankowski says the company's new policy didn't stem from that criticism. He says Walmart's food waste reduction initiatives were already underway by then, and the company has decreased food waste by 23 per cent over the last few years. "We know we can do better and we can do a lot more," he said.

Walmart has implemented several tactics to decrease food waste, Wankowski said. First and foremost it's trying to sell the food it carries—this means more precise tools to forecast demand so it can order accordingly—and dropping the prices of soon-to-be expired items.

And through its philanthropic arm, the Walmart Foundation, the company is giving away $19 million to organizations that help reduce food waste in Canada.

1. As described in the article, what is the total value and estimated percentage of food produced that is wasted annually in Canada?

   a. $5 billion, about 40% of the total
   b. $19 million, about 23% of the total
   c. $31 billion, about 40% of the total
   d. $40 billion, about 31% of the total

2. Which three of the following reasons does the article give for why the issue of food waste is important in Canada?

   a. Wasted food could be used to feed the hungry in Canada.
   b. Wasted food in dumpsters is often a biological health hazard.
   c. Wasted food is an important alternative energy source in Canada.
   d. Wasted food in landfills produces methane, which is a greenhouse gas.
   e. Food waste is inefficient for businesses in this industry and causes them to lose money.
   f. Food waste is causing Canada's agriculture sector to be less competitive with other countries.

3. What does the charity Second Harvest do, as described in this article?

   a. It connects food suppliers and distributors with charities that would use their excess food.
   b. It helps food suppliers obtain more food by harvesting fields for a second time.

   c. It buys food from distributors and gives it away to local grocery stores.
   d. It gets free food from local suppliers and stores it in warehouses until it is needed by charities.

4. What principle in the "reduce, reuse, recycle" waste management slogan is Lori Nikkel focusing on through her efforts with Second Harvest? Explain your answer using information from the article and the definitions of these waste management principles from the chapter.

5. The article notes that food is wasted in many places in its supply chain and that waste is often considered "a cost of doing business." What attitudes toward food are driving these costs? Per this chapter, do you think these same attitudes were prevalent in the general population 100 years ago? Why or why not?

6. **Make Your Case** The article notes that Walmart is decreasing food waste because it is good for business. However, like most large stores, its business is based on consumption and encouraging increased consumption. Do you believe that Walmart's actions will ultimately decrease its overall waste production, or will this simply substitute food waste at its stores with another type of waste?

Use the Data

The two graphs shown below illustrate recycling rates by type of product and by year and were produced by the Environmental Protection Agency (EPA).

Study the graphs and use them to answer the questions that follow. The first three questions test your understanding of the graphs. Questions 4 and 5 are short answer, requiring you to apply what you have learned in this chapter and cite data in the graphs. Answers to Questions 1–3 are provided at the back of the book. Question 6 asks you to make your case and defend it.

(a)

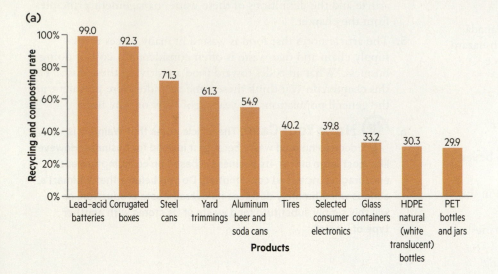

(b)

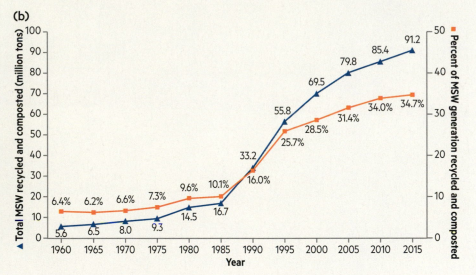

1. In graph (a), what type of material makes up the two products with the lowest recycling rate?

    a. aluminum
    b. glass
    c. paper
    d. plastic

2. According to graph (b), between what years did the sharpest increase in recycling and composting rates of municipal solid waste (MSW) by US households occur?

    a. 1960–1965
    b. 1980–1985
    c. 1990–1995
    d. 2005–2010

3. In graph (b), the orange and blue lines represent different values. Match each line to the value that it represents (not all values will be matched).

    Orange line   Percentage of MSW that is recycled and composted

    Blue line   Percentage of all waste that is recycled and composted

          Total weight of MSW that is recycled and composted

          Total weight of all waste that is recycled and composted

4. Use graph (b) to calculate the following: If US households recycled and composted 91.2 million tons of MSW in 2015, how many tons of MSW were generated in total? How many tons of material remained in the waste stream for disposal? How does the number of tons of material left in the waste stream in 2015 compare to the total MSW that was recycled and composted in 1990? Using information in the chapter, explain why recycling and composting rates in 1990 and 2015 are different.

5. Of the products included in the bar chart in graph (a), what is the recovery rate for lead–acid batteries, the type of battery that you find in vehicles? What is the rate for white translucent bottles? Which rate is lower? On the basis of what you have learned in this chapter, what factors do you think contribute to the different recovery rates of these two products?

6. **Make Your Case** Using the data from graph (a), if you wanted to improve the tons of MSW recycled, what two materials would you target first and why? Use the data in the graph and any information in the chapter about recycling to help support your choices. What would you do or change to drive these improvements, including policies, incentives, or other initiatives?

LEARN MORE

- Basel Action Network (activism on e-waste and hazardous waste trade and environmental justice): www.ban.org/about-us/
- EPA "Reduce, Reuse, Recycle" page: www.epa.gov/recycle
- EPA "Facts and Figures about Materials, Waste and Recycling" page: www.epa.gov/facts-and-figures-about-materials-waste-and -recycling/advancing-sustainable-materials-management-0
- NOAA "What Are Microplastics?" page (microplastics in the ocean): https://oceanservice.noaa.gov/facts/microplastics.html
- Story of Stuff Project: https://storyofstuff.org/
- Video: Ocean Dead Zones: digital.wwnorton.com/environsci
- Video: National Park Service, Waste in the Ocean: digital .wwnorton.com/environsci

16

Urbanization
Why Are Cities Growing?

Stuck in traffic again! It seems inescapable. Whether you are commuting to school or work in and out of a city each day or attempting a weekend getaway, everybody seems to be on the road at the same time. Even when the number of lanes in nearby roads is expanded, the congestion continues to increase. In the 1980s, the average driver in the United States wasted about 18 hours stuck in traffic each year. Now that number has surpassed 40 hours, and American drivers overall waste nearly 7 billion hours and more than 3 billion gallons of gasoline in traffic jams each year. Today, most of the world's people live in urban areas, and traffic congestion is just one common problem they face.

The most rapid urban growth is occurring in high-density cities of the developing world. Cities such as Dhaka, Bangladesh; Kinshasa, Democratic Republic of Congo; and Delhi, India, each added over 8 million people between 2000 and 2020. These cities often contain areas that struggle to provide basic water supplies, sanitation, safe dwellings, and social services to keep pace with growing populations. Closer to home, urbanization is occurring too, but in the form of lower-density urbanization in suburbs. Today, more Americans live, work, and shop in suburban areas than in the central cities. Compared to urban areas, the challenge of the suburbs can be one of **infrastructure**—physical structures such as roads, buildings, and parking lots—extending across the landscape and needing more energy and being less resource efficient than higher-density forms of urbanization.

Recently though, the COVID-19 pandemic led many people to rethink and sometimes rearrange where and how they lived, worked, and carried out their daily activities. For some, opportunities for remote work led them away from city centers and allowed them to avoid the traffic of their former commutes. As city residents and leaders attempted to respond to the pandemic and population declines, they also thought about how best to "future-proof" their cities. They're imagining more resilient cities that can better respond not only to future pandemics but also to threats like extreme weather events and sea-level rise linked to climate change. At the same time, others suggested that more community participation in city development during this time might address housing shortages and other

Urban populations continue to grow rapidly throughout the world. To accommodate more people, cities are often built upward, with taller buildings, and outward, like this new residential development in New Delhi, India.

Chapter Objectives

By the end of this chapter, you should be able to . . .

A. describe how recent and rapid demographic trends have made us a predominantly urban species.

B. understand that cities function as networks connecting a multitude of people, things, and information.

C. identify environmental and human health effects associated with urbanization and the opportunities cities provide for more sustainable and resilient living.

D. explain the characteristics, causes, and challenges of slums and suburban sprawl.

E. discuss how planning, participation, and design can develop more livable, sustainable, and resilient cities.

> **We shape our cities and then our cities shape us.**
>
> —Andres Duany, Elizabeth Plater-Zyberk, and Jeff Speck (*Suburban Nation*)

ronmental impacts, many experts believe that urbanization also provides our greatest opportunities for improvement. As climate activists such as Xiye Bastida have written, meeting current challenges begins with imagining what an ideal city would look like—using new ways of building and providing energy and transportation that better meet the needs of its residents while reducing environmental impacts.

16.1 How Are Urban Areas Changing?

infrastructure physical structures and facilities such as roads, buildings, and parking lots.

urban transition term referring to the trend of populations becoming more urban and less rural.

urban area a place where large numbers of people live together in relatively dense settlements and work in nonagricultural jobs.

rural area a place where populations are more spread out geographically and more reliant on agricultural employment.

Guess who is coming to town? Nearly everyone. Globally, we are in the midst of a rapid and momentous rural to **urban transition.** In 2008, for the first time in human history, more people lived in urban areas than in rural areas. What are **urban areas**? While formal criteria vary by country, they generally are places where large numbers of people live together in relatively dense settlements and work in nonagricultural jobs. **Rural areas,** by contrast, have populations that are more spread out geographically and more reliant on agricultural employment. According to the United Nations (UN), the world's cities collectively grow by more than 1 million people per week. Some of the growing urban population is due to natural increase, where the number of in-city births outpaces the number of deaths. But more than

issues of inequality often seen in large cities.

In this chapter, we will see that while urban and suburban areas may generate some of the most serious challenges to our health and our most significant envi-

half of the growth is due to migration from rural areas. In fact, the world's rural population is beginning to shrink. Even though there will be more than 9 billion people on Earth in 2050, forecasts predict fewer rural inhabitants at that time than there are today. The UN projects that two-thirds of the global population will live in urban areas. Much of this migration and expansion is taking place in developing nations, home to three-quarters of the world's urban population.

Putting the Urban Transition into Historical Perspective

Most of us have a hard time picturing a transition this large and rapid. Look at the aerial photo of New York City, the most populous city in the United States, home to more than 8 million people (**FIGURE 16.1**). Now consider that urban populations are expanding globally at a rate equal to six New York–sized cities each year. The scale of this growth is unprecedented in human history. There are now more people living in metropolitan Tokyo alone than the total population of Earth 6,000 years ago, when cities were only starting to form.

In ancient times, some of the largest cities were the Chinese cities of Chang'an and Hangzhou and the Kingdom of Angkor Wat in present-day Cambodia. These cities were home to hundreds of thousands of people about 800 years ago. Then, as now, cities tended to form around trade and transportation hubs and/or centers of political power (**FIGURE 16.2A**). The first city to have more than 1 million inhabitants was London: it reached that milestone in 1800, though at that time only one out of every 50 people on Earth lived in an urban area. Even as late as 1900, no single country was predominantly urban. Now, not only are urban areas home to the majority of humans, but nearly 25% of the global population lives in a city with at least 1 million inhabitants. Today, most countries are predominantly urban, and the populations of some countries are entirely urban (**FIGURE 16.2B**). As remarkable as this urban transition has been, the world's urban populace is projected to double by 2050. Between now and then we will have to develop additional settlements and services for as many people as there are today in all of the world's cities.

Urbanization and Environmental Impacts

Cities are now the epicenters of humanity's impact on the wider environment, transforming landscapes by occupying land that had previously been used less intensely. Consider an urban area you know well. What was that place like several hundred years ago? Chances are that forest, prairie, wetland, and/or shoreline habitat was fragmented or destroyed by its development. These changes can be very obvious. The settlement of Ciudad Nezahualcóyotl adjoining Mexico City is located on the drained remnants of Lake Texcoco and neighboring salt marshes and is now home to more than 1 million

FIGURE 16.1 New York City With a population of more than 8 million people, New York is the most populous city in the United States. Globally, the urban population is growing at a rate of six New York–sized cities each year.

(a)

(b)

FIGURE 16.2 Cities Then and Now (a) Angkor Wat is a temple complex in present-day Cambodia. In the middle of the 12th century, it was at the center of several large cities that were connected by a network of canals and roads and that had a total population in the hundreds of thousands. (b) China's special administrative region of Hong Kong (pictured), as well as Singapore and Monaco, are 100% urban. Other countries in which more than 95% of the population lives in urban settings include Belgium, Kuwait, and Uruguay.

(a)

(b)

FIGURE 16.3 Cities and Habitat Loss (a) Ciudad Nezahualcóyotl adjoining Mexico City is located on top of the drained remnants of Lake Texcoco and neighboring salt marshes. This image shows the few remaining marshes on the right, bordering the city's open dump. (b) In the United States, California has lost most of its coastal wetlands to urbanization; marsh areas that remain tend to be small and fragmented like this area in Newport Beach.

people. On a larger scale, urban development has altered or eliminated more than half of the once common wetlands in Florida and more than 90% of coastal wetlands in California (**FIGURE 16.3**).

In addition, the inhabitants of the world's cities are drawing on resources from and discharging waste into an area that extends far beyond urban borders. International trade networks enable urban populations to consume resources grown or extracted from far away. Solid waste is most often hauled beyond city boundaries for disposal, and air and water pollution flow out of the city as well. Attempts to measure the resources required to maintain a city yield ecological footprints (see Chapter 6) that are very high. The San Francisco–Oakland–Hayward Metropolitan Statistical Area, which has been ranked as one of the "greenest" urban areas to live in the United States, has an ecological footprint of 7.1 global hectares per person (GHP), considerably higher than the estimated biocapacity of the world of 1.8 GHP (**FIGURE 16.4**).

FIGURE 16.4 Cities Draw on Outside Resources Cities rely on resources from throughout the world to support their populations. The ecological footprint of the San Francisco–Oakland–Hayward Metropolitan Statistical Area relies on resources that would require land about 48 times larger than the actual size of this metropolitan area—or nearly three-quarters of the entire state of California.

The Urban Transition

Today, the world is more urban than rural. Globally, more than half of people now live in cities. This trend is occurring in every region of the world.

Percentage of Population Living in Urban Areas

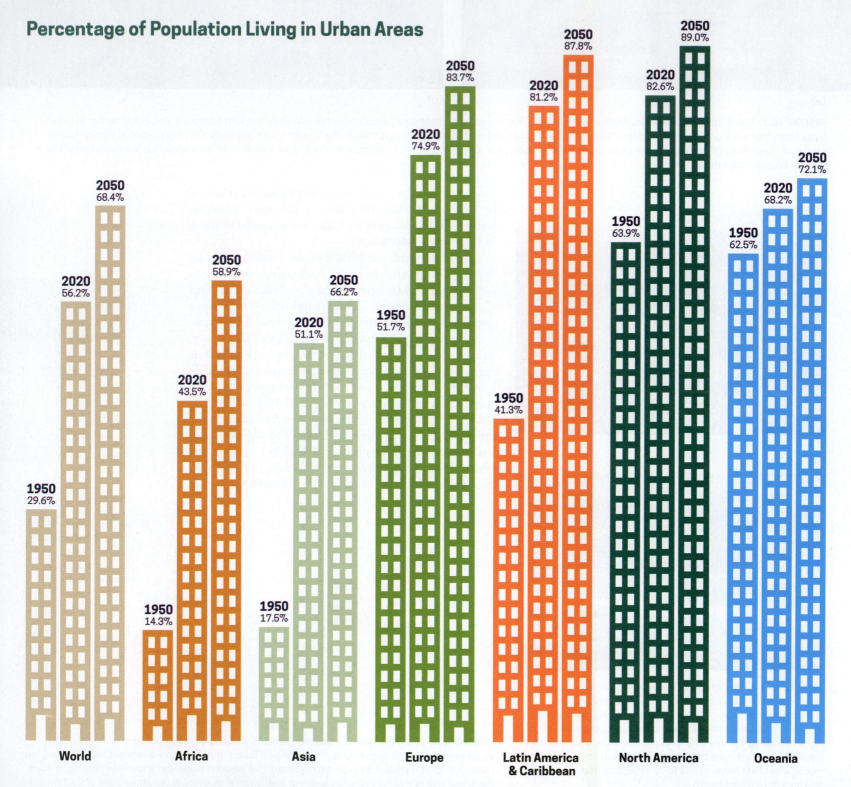

World
- 1950: 29.6%
- 2020: 56.2%
- 2050: 68.4%

Africa
- 1950: 14.3%
- 2020: 43.5%
- 2050: 58.9%

Asia
- 1950: 17.5%
- 2020: 51.1%
- 2050: 66.2%

Europe
- 1950: 51.7%
- 2020: 74.9%
- 2050: 83.7%

Latin America & Caribbean
- 1950: 41.3%
- 2020: 81.2%
- 2050: 87.8%

North America
- 1950: 63.9%
- 2020: 82.6%
- 2050: 89.0%

Oceania
- 1950: 62.5%
- 2020: 68.2%
- 2050: 72.1%

Information on this two page spread is adapted from UN, DESA (2015), WorldAtlas (2020), and Hoornweg and Pope (2014).

Urbanization by Income Category

The urban transition in developing countries is happening at a much faster rate than it did in the United States and Europe. In 1980 China was just 20% urban. Now more than half of its population lives in cities. Demographers predict that China will be 70% urban in 2035 after undergoing the largest urban transition in human history.

Percentage of Population Living in Urban Areas by Income Category

$$$ High-income countries

$$ Middle-income countries

$ Low-income countries

The World's Most Populous Cities

How familiar are you with the list of the world's most populous cities in 1950 compared to now? More than 400 cities now have populations over 1 million—two-thirds of these cities are in the developing world. Twenty-one of the 26 cities with populations exceeding 10 million people are in the developing world.

Most Populous Cities in 1950
(Population in millions)

1. New York—Newark, US (12.4)
2. Tokyo, Japan (11.3)
3. London, UK (8.4)
4. Osaka, Japan (7.0)
5. Paris, France (6.3)
6. Moscow, Russia (5.4)
7. Buenos Aires, Argentina (5.1)
8. Chicago, US (5.0)
9. Kolkata, India (4.5)
10. Shanghai, China (4.3)

Most Populous Cities in 2020
(Population in millions)

1. Tokyo, Japan (37.4)
2. Delhi, India (30.3)
3. Shanghai, China (27.1)
4. São Paulo, Brazil (22.0)
5. Mexico City, Mexico (21.8)
6. Dhaka, Bangladesh (21.0)
7. Cairo, Egypt (20.9)
8. Beijing, China (20.5)
9. Mumbai, India (20.4)
10. Osaka, Japan (19.2)

Predicted Most Populous Cities in 2050
(Population in millions)

1. Mumbai, India (42.4)
2. Delhi, India (36.2)
3. Dhaka, Bangladesh (35.2)
4. Kinshasa, Democratic Republic of the Congo (35.0)
5. Kolkata, India (33.0)
6. Lagos, Nigeria (32.6)
7. Tokyo, Japan (32.6)
8. Karachi, Pakistan (31.7)
9. New York, US (24.8)
10. Mexico City, Mexico (24.3)

urban penalty the human health problems associated with urban living.

urban density the number of residents per unit of space.

economy of scale a situation in which the per capita cost for services decreases as a result of the increasing scale of operations.

Historically, cities have been regarded as a threat to human health and well-being—and not without reason. Horrendous living conditions characterized industrial cities in Europe and North America during the 19th and early 20th centuries. Cholera and other infectious disease epidemics were so common that until 1890, death rates and infant mortality in industrial cities were higher than those in rural areas. The human health problems associated with urban living are sometimes referred to as an **urban penalty**. Crowding, congestion, pollution, crime, disease, and poor sanitary conditions are all real and pressing health threats in many cities around the world. Though improvements in sanitation and public health have reduced global infant mortality rates in urban areas to less than those in rural areas, many rapidly growing areas remain plagued by pneumonia, diarrhea, malaria, and other diseases.

The Opportunity of Urban Areas

So why would anyone live in a city? Researchers at the Santa Fe Institute found that on average, as city population increases, per capita measures such as wages and wealth increase. In fact, when the population of a city doubles, they discovered that all types of economic output increase by about 15% per person; in other words, if a person moves to a city with twice the number of people, that person produces 15% more than they used to. And social indicators that measure education, cultural activities, and innovation similarly increase. Using available data on both the material and social aspects of cities, the researchers observed that these improvements result from the way cities function as *networks*—connections that allow a multitude of people, things, and information to mix. Their research found that urbanization attracts, concentrates, and accelerates social interactions

that often lead to more opportunity. These researchers also learned that as city populations grow, the cities use resources more efficiently. For example, when a city population doubles, it typically requires only about an 85% increase in most aspects of its physical infrastructure, such as total road surface, length of electrical cables, and number of water and gas lines. They found an increase in the **urban density**—the number of residents per unit of space—achieves benefits resulting from an **economy of scale**, a situation in which the per capita cost for services decreases as a result of the increasing scale of operations.

Further, negative environmental impacts of cities can be counterbalanced by opportunities for improvement (**FIGURE 16.5**). As anyone stuck in a traffic jam knows, more people using shared resources such as city roads can stress the infrastructure and resources of cities: roads, sewers, green space, water supply, and air quality. But this shared use also presents opportunities to solve environmental problems. One way is through efficiency gains. How urban infrastructure is designed can influence as much as 75% of our daily ecological footprint. For example, heating and cooling costs per person can decline as people in multi-dwelling buildings share space. As parking becomes more expensive, public transportation becomes more viable. Per capita energy use and greenhouse gas emissions may be reduced as urban density increases. In fact, most cities have lower per capita greenhouse gas emissions than the per capita national average of their respective countries (**FIGURE 16.6**).

While urbanization presents many opportunities to reduce environmental impacts, the effects will vary considerably depending on the conditions present in a particular urban area. In the next several sections, we will study various types of urbanization.

(a)

(b)

(c)

FIGURE 16.5 City Efficiencies A well-designed city can actually reduce its environmental impacts. As cities are designed and built to accommodate greater population density, the space allotted per capita shrinks, and infrastructure is used more intensely. **(a)** Multifamily dwellings such as apartments and condos use energy and water more efficiently per capita than do single-family dwellings. **(b)** Denser, walkable neighborhoods with public transportation reduce the use of single-occupancy vehicles, lowering per capita air pollution and greenhouse gas emissions. **(c)** Concentrating populations in urban areas can reduce the pressure to develop agricultural land, forests, and other green spaces in outlying areas.

We are undergoing a rapid global urban transition. Cities concentrate our impacts on the environment, and inadequate living conditions in many cities are a threat to human health and well-being. Yet cities also reduce the per capita use of physical infrastructure, energy consumption, and pollution while increasing opportunities for education, cultural activities, and innovation. These urban advantages are realized when increasing population density is matched by the effective provision of urban infrastructure and services.

16.2 What Are Slums?

In 2009, the Academy Award for Best Picture went to *Slumdog Millionaire*, a film set in the settlement of Dharavi, which sits on the margins of the city of Mumbai, India. More than 600,000 people live there in a maze of shacks packed into 1 square mile that sits on a former mangrove swamp (**FIGURE 16.7**). The population density of this area is about 25 times higher than that of New York City. Most Dharavi residents lack access to indoor plumbing, and the city is polluted by open sewer ditches. Dharavi is just one example of what the United Nations Human Settlements Programme defines as a **slum**, or *informal settlement*, an urban area characterized by substandard housing, a lack of formal property ownership arrangements, inadequate urban services (such as sewage systems, trash collection, and water supplies), and high rates of poverty. Similar places are known as shantytowns

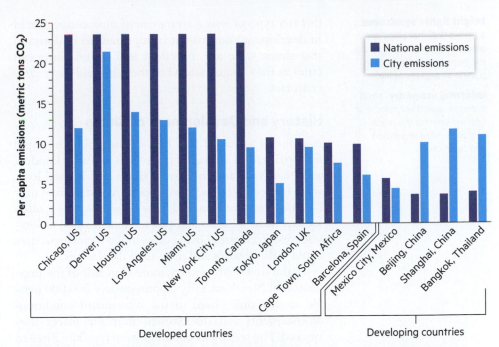

FIGURE 16.6 **City and National Per Capita Emissions (CO$_2$)** On a per capita basis, cities are often (though as this graph shows, not always) lower emitters of greenhouse gases than their host countries.

Adapted from World Bank (2010).

in South Africa, *favelas* in Brazil, *barriadas* in Latin America, among others.

Globally, slums are home to nearly 1 billion people, a number expected to double by 2030. This means the population in informal settlements alone will equal the global population of a century earlier. And while

slum an urban area characterized by substandard housing, a lack of formal property ownership arrangements, inadequate urban services, and high rates of poverty.

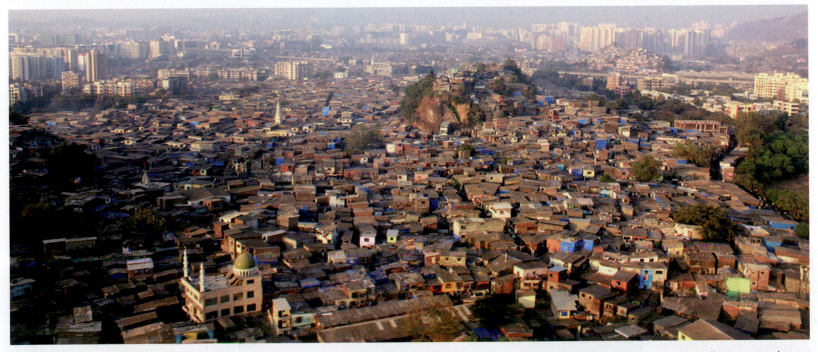

FIGURE 16.7 **Dharavi Settlement in Mumbai, India** The settlement of Dharavi (the setting for the movie *Slumdog Millionaire*) is just 1 square mile in area yet home to more than 600,000 people.

this is a type of living arrangement now concentrated in developing countries, it is important to understand that slums were also prevalent in US and European cities as they industrialized in the 19th and early 20th centuries.

History and Development of Slums

A defining factor of the 19th century was the rapid industrialization of major cities such as Paris, London, and New York and the rise of slums that housed low-wage earners. Why did this happen? As cities grew, new factories clustered around new power sources, often located near rail hubs and ports. Slums formed as low-wage workers started living close to the factories so they could walk to work and meet their daily needs.

In the mid-19th century, more than half of the population of New York City (approximately 500,000 people at that time) lived in the substandard conditions of tenements and shantytowns near the places they worked. The tenement buildings were typically three to five stories high, had apartments of just over 200 square feet, and were leased or informally sublet to families of five or more. These buildings had poor ventilation and often lacked indoor toilets and running water. The New York City shantytowns, with names such as Slab City and Tinkersville, housed tens of thousands of residents who built homes out of scavenged lumber and tin on marshy land near the waterfront to which they had no legal title (**FIGURE 16.8**). A lack of policies governing living conditions and industrial practices in these areas led to many of the environmental and health threats we now associate with slums. For example, in the absence of sewers, waste was discharged into open ditches and the streets themselves. Smoke from the factories was released close to ground level, and foul-smelling air pollution and soot deposits were ever-present. While the slums of New York are now gone, settlements like these still exist in such major cities as Cape Town, Nairobi, and Mexico City, where lack of clean water and sanitation, as well as crowded substandard buildings, pose daily challenges for residents.

FIGURE 16.8 Slums of New York The city of New York was once host to shantytowns like these Depression-era ones where people made homes out of any materials they could find.

Opportunities in Slums and Cities in the Developing World

Why do people live in these conditions? Whether we are looking historically at cities in the United States and Europe or at present-day slums in developing countries, the answer to this question includes factors that both push and pull people into cities. Slums function as a kind of staging ground for many people who move to the city. These residents are often moving away from undesirable conditions somewhere else, and in this sense they are pushed into slums. For example, refugees from war, political repression, famine, natural disasters, or economic hardships in other areas often migrate to cities in search of work and cheap housing. Sometimes the economic hardships sending people into such living conditions are related to technological changes in agriculture. Mechanized field preparation and harvesting, along with fertilizers, pesticides, and high-yield crop varieties, have greatly reduced the number of farm laborers needed for production, which forces them to move to cities in search of new opportunities.

The perception of the city as a place of opportunity, where people can find a better job, greater economic security, more excitement, and freedom to pursue all aspects of life—often called the **bright lights syndrome**—is another factor pulling people into cities. Migrating to a city, despite the problems associated with slums, does tend to offer greater employment and education opportunities than are typically available in rural areas, particularly for women. Many find work in the **informal economy**—entrepreneurial ventures such as small manufacturers, service providers, and street vendors that operate beyond the reach of government regulation and taxation and don't have formally trained personnel or access to credit from banks. UN and World Bank studies on slums have found that in developing cities, the informal sector provides most of the food products, services, and construction in the area.

Despite the scarcity of formal services, slums can be communities with tight and robust networks of support that are based on family and other forms of social organization. Dharavi, for example, may lack many government-provided services, but it hosts citizen-organized schools, clubs, and organizations, as well as nearly every service you would expect in a city, from hairdressers and health clinics to hardware stores and internet cafés. It is a site of cultural mixing and innovation (**FIGURE 16.9**). Residents meet most of their daily needs within a very small area or by commuting into Mumbai via one of the two train lines bordering the settlement. Others work in one of the 10,000 workshops amid and within the shacks of Dharavi that produce food products, ceramics, leather and textile goods, and many other items.

FIGURE 16.9 Slums Have Thriving Social and Economic Networks Despite their often-challenging living conditions, Dharavi (shown here) and other slums have thriving social and economic networks.

Most experts agree that the challenge for urbanization is improving living and working conditions without disrupting or destroying the positive effects of social density. Often, governments have attempted to solve the challenges of slums by clearing these areas and relocating the residents to newly constructed apartment blocks elsewhere. This may improve the physical infrastructure, but such policies can detach people from familiar livelihoods. For example, in the late 1990s the government of the Philippines undertook a policy to relocate slum dwellers on Manila's Pasig River to new apartment blocks on the outskirts of the city. Most of the relocated people have since moved away from these apartments to different slums closer to the jobs and social networks from which they were relocated. The UN Human Settlements Programme now recommends upgrading the existing infrastructure of slums rather than relocating residents to new sites.

Development experts have also found that improving slum conditions coincides with economic growth, rising wages, and health and safety regulations for buildings and workplaces, as well as the extension of land and property rights. These are the factors that transformed historical slums in New York City, London, and Paris, and they have been the key to improvements more recently in cities across Asia and Latin America.

⬡ **TAKE-HOME MESSAGE** Slums are characterized by substandard housing, a lack of formal property ownership, inadequate urban services, and high rates of poverty. These conditions develop when the provision of affordable infrastructure and services in cities fails to keep pace with increasing population. Yet slums can also provide opportunities to residents through well-developed social networks and highly productive and innovative informal economies.

16.3 What Is Suburban Sprawl?

In a well-known joke about New Jersey that started on *Saturday Night Live* in the 1980s, one New Jersey resident meets another for the first time and, instead of asking what town or city the person is from, asks, "What exit?" The joke gets to the heart of **suburban sprawl**, a relatively new form of urbanization that dominates the United States, Canada, Australia, and some European countries. Suburban sprawl is defined as the spread of urban populations away from the center of cities to widely dispersed areas on the outskirts with relatively low population densities. Sprawl also separates where residents live, work, and shop into isolated zones and provides roads and highways for commutes between zones. Sprawl occurs when policies increase the supply of land for development while reducing the transportation costs associated with commuting. Although suburban sprawl is often associated with moving from a central city out to its suburbs, it is not a form of de-urbanization. Instead, suburban sprawl is a less dense and more resource-intensive form of urbanization that has its own environmental and social challenges (**FIGURE 16.10**).

suburban sprawl
the spread of urban populations away from the centers of cities to widely dispersed areas that have relatively low population densities.

FIGURE 16.10 Suburban Sprawl Suburban sprawl, shown here near Las Vegas, is a form of low-density urbanization that spreads horizontally across the landscape and tends to separate the functions of daily life—such as working, shopping, school, and home life—into isolated zones that require a car to move from place to place.

How Street Design Shapes Transportation

The patterns used to lay out new streets have a big impact on the range of transportation available. The illustration shows how roads in the United States evolved away from grid patterns toward more disconnected "loops" during the past century as cars became the dominant form of transportation. Typical suburban street loop patterns are curved and disconnected. The lack of a direct route from housing to shopping or school makes it much less pedestrian friendly. A grid pattern is a much older street design form that allows more direct routes and a larger variety of routes from point to point in the neighborhood.

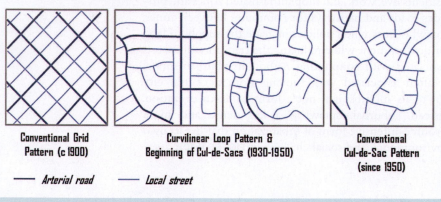

Conventional Grid Pattern (c 1900)

Curvilinear Loop Pattern & Beginning of Cul-de-Sacs (1930-1950)

Conventional Cul-de-Sac Pattern (since 1950)

—— Arterial road   —— Local street

metropolitan area (metropolis) a cluster of densely populated suburbs and cities.

megalopolis a chain of roughly adjacent metropolitan areas, such as the region from Boston to Washington, DC.

The 2010 census revealed that a majority of the US population now lives in suburbs. Though we tend to think of suburbs as residential areas, suburbs are now the primary locations for economic and commercial activity in the United States. More than 75% of the working people who live in suburbs find their employment in suburban areas. Since 1980, the typical US work commute has been not from suburb to city but from one suburban area to another, and there are twice as many manufacturing jobs in suburbs as there are in central cities. Suburban areas also have 90% of the office space in the United States: for example, the suburbs around Atlanta have twice the office space of downtown Atlanta, and northern New Jersey alone has more office space than that of Manhattan. Further, in the 1970s, the downtown areas of central cities lost their dominance over retail sales to suburban shopping centers and malls. This trend has continued, with more than 80% of new retail space in the United States coming in the form of "big box" superstores located outside of central cities.

Because of these changes, demographers and urban planners now use the term **metropolitan area**, or **metropolis**, to identify these clusters of suburbs and cities where most Americans now live and work. Metropolitan areas are frequently identified with the names of the multiple boundaries that they span, such as Dallas–Fort Worth and Washington–Baltimore. They are more or less continuous human settlements sprawling across vast areas of space. Every county in the states of Rhode Island, New Jersey, and Delaware—even those that are less populated—are considered part of a metropolitan area. In fact, these states are part of a vast corridor of urbanization stretching from Boston to Washington, DC. When metropolitan areas like this creep closer to each other and integrate across broad corridors, they are called a **megalopolis**, a chain of roughly adjacent metropolitan areas (**FIGURE 16.11**). A metropolitan area can grow even as the central city or cities within it are shrinking. For example, the population of the city of Baltimore declined by almost one-third between 1970 and 2010, while the population of the Baltimore–Washington metropolitan area, including areas in northern Virginia, grew by more than two-thirds during this time.

Characteristics of Sprawl

In suburbs, the places where we live, work, shop, and engage with civic institutions tend to be segregated into distinct zones for each use, or **one-use zones**. So suburbs include geographically distinct residential subdivisions, shopping centers, office parks, and civic spaces such as schools and government buildings. The buildings across these various zones tend to be low-rise, extending outward rather than upward. The most prevalent example is large-lot, single-family housing, which spreads residential use over a relatively large land area compared to that of multifamily dwellings in multistory buildings. One way to grasp variations in residential development is to calculate the dwelling units per acre (DUs/acre), as seen in **FIGURE 16.12**. Single-family homes are typically set 3–4 DUs/acre, but townhouses, condos, and apartments can achieve many more DUs/acre.

Another measurement that illustrates differences in building design is the floor-to-area ratio (FAR). This

Boston
New York City
Philadelphia
Baltimore
Washington, DC

FIGURE 16.11 Megalopolis The cities and suburbs from Boston, Massachusetts, to Washington, DC, form an unbroken chain of urbanization known as a megalopolis. Its lights are visible from space at night.

Dwelling Units per Acre

one-use zone a distinct zone, often in suburbs, that serves one particular purpose.

3–4 dwelling units per acre (single-family detached homes)

30 dwelling units per acre (low-rise apartment complex)

30 dwelling units per acre (mid- to high-rise apartment buildings)

FIGURE 16.12 Measuring Land Use A good way to visualize different levels of urban density is to measure the dwelling units per acre. Higher-density forms can often provide for a lot of shared open space.

Adapted from Chakrabarti (2013).

ratio compares the total floor space of a building with the area of land on which it is built. A one-story building that covers an entire lot would have an FAR of 1.0. A low-rise building surrounded by a vast area of parking will have an FAR of less than 1. So while a typical shopping mall has an FAR of 0.25 and the average suburban office park has an FAR of 0.29, a higher-rise building can have a much greater FAR. The typical US suburb has an FAR of less than 0.2, while an FAR of 1.5 is understood to be the minimum threshold associated with walkable, car-free living (**FIGURE 16.13**).

Because of its low population density, suburban sprawl is heavily dependent on automotive infrastructure to move individuals. Sprawl is said to be *built for cars*, with vast acreage of pavement devoted to streets and highways as well as driveways and parking lots at each individual destination. The street networks in such areas are often curved and disconnected.

Because uses are separated, these areas lack urban centers or efficient transportation hubs for mixed-use urban activity. Newly developed areas also tend to "leapfrog" over undeveloped areas in a scattered pattern, and major roads take commuters long distances past undeveloped land on their way to the zone they aim to reach.

What Are the Problems of Suburban Sprawl?

Sprawl occupies land that had been used less intensively, such as farmland or formerly broad stretches of forests, wetlands, and shorelines (**FIGURE 16.14**). Sprawl is paved with impermeable surfaces for roads, parking lots, and driveways. The vast distances covered by sprawl make any form of transportation infrastructure

(a)

Three Examples of FAR = 1

1 story (covers all of lot)

2 stories (covers half of lot)

3 stories (covers one-third of lot)

(b)

FIGURE 16.13 Floor-to-Area Ratio (FAR) (a) The floor-to-area ratio (FAR) shows the difference between building out versus building up. All three examples have an FAR of 1.0, but the three-story building leaves more open space on its lot. **(b)** High-rises have a much higher FAR. The Comcast Center in Philadelphia, which opened in 2008, has an FAR of 18.0!

Adapted from (a) Lindeke (2016).

FIGURE 16.14 Encroaching Sprawl Sprawl consumes previously undeveloped land. Contrast the development on the left and its abundance of paved roads and driveways with the less developed farmland on the right. Researchers who have developed a sprawl index to compare cities have found that sprawl is associated with reduced pedestrian travel and public transit use, as well as increased average vehicle ownership, daily vehicle miles traveled per capita, and annual traffic fatality rates.

less energy efficient than one serving a higher-density urban area. Public transportation is much more difficult to implement successfully because its users are coming from and going to places that are spread out from each other. In suburban areas, walking- and biking-friendly infrastructure such as sidewalks and bike paths or lanes rarely link all of the places one needs to go to meet daily needs. This in turn leads to less physical activity for residents.

Comparing a sprawling city such as Atlanta, Georgia, and a denser, centralized city such as Barcelona, Spain, offers a revealing contrast (**TABLE 16.1**). These two metropolitan areas have similar overall populations, but Barcelona's residents live in a much smaller area. Atlanta is often in noncompliance with federal air-quality standards, and Atlanta residents have the longest average daily commute in the United States at 36.5 miles (59 kilometers). Further, in 2018, data company INRIX found that Atlantans spent an average of 108 hours stuck in traffic jams each year. (In the United States, Boston led the 2018 INRIX rankings, with Bostonians spending more than 164 hours in traffic each year. Note that congestion dropped dramatically in cities during the pandemic, but researchers such as the Texas A&M Transportation Institute saw levels return to normal in 2021.) And yet suburban sprawl continues in the United States. To better understand why, we'll explore how this trend in land use and urbanization developed in the United States over the past several decades.

TABLE 16.1 Different Densities: Atlanta, Georgia versus Barcelona, Spain

Urban Area	Atlanta	Barcelona
Population	5.58 million	4.81 million
Built-up Urban Area	2,817 square miles	415 square miles
Population Density	1,700 people per square mile	11,600 people per square mile
Average Annual Gas Consumption	782 gallons per person	64 gallons per person
Public Transportation as a Percentage of Annual Motor Transportation	1%	35%

Adapted from Demographia World Urban Areas (2019) and Newman et al. (2009).

16.4 How Did Suburban Sprawl Develop in the United States?

What happened to bring about the building boom that led to sprawl? Let's start with the economic dynamics of the housing market in the 1950s. After World War II, the United States experienced a marriage boom and a baby boom. Ten million new households were created between 1945 and 1955, all needing places to live. Many of these people were initially living with relatives or renting small apartments or rooming houses in a city. Yet since at least the 1920s, a widely shared cultural perspective in America held out single-family home ownership as the ideal.

In the 1920s, before the Great Depression, there had been a boom in single-family home construction in areas adjacent to cities. These places were often incorporated into city boundaries and connected by streetcar lines. At this time, technological improvements in automobile manufacturing and road design also made it possible for upper middle-class people living in these places to drive into the city, while back at home they were enjoying many new electrical appliances. "Moving up" in terms of one's class status also meant "moving out" of the central city and started to symbolize the American Dream (**FIGURE 16.15**). Moving out to the suburbs could also mean moving away from the diversity of social groups in the city. Zoning, a regulation at the local level initially designed to separate residential areas from the pollution of heavy industry, was applied to reinforce aspects of social organization that separated people by class, race, ethnicity, and religion.

Building Policies

Suburban sprawl could not have expanded as widely or quickly as it did without federal government policies designed to expand home ownership. In 1920, living in the suburbs lay beyond the means of most households since mortgages typically required a 40% down payment and a 10-year payoff period. This changed with federal policies in the 1930s and 1940s administered by the Federal Housing Administration (FHA) and Veterans

Administration (VA) that guaranteed loans made by private banks to home owners. If a home owner defaulted on a loan, the federal agencies would protect the bank against losses. This incentivized banks to take more risks and to offer mortgages with much more favorable interest rates, longer payoff periods, and small down payments.

⊘ ENVIRONMENTAL JUSTICE

Restrictive Covenants and Redlining

The development of many residential neighborhoods in the United States was rooted in discrimination against various minority groups that was enforced by two primary mechanisms: restrictive covenants and redlining. Racially restrictive covenants barred certain racial or ethnic groups from purchasing, renting, or occupying a property. Legally, a covenant stays with a property's deed in perpetuity even as ownership changes hands over time, effectively blocking entire groups of people from home ownership for generations. If an owner violated the terms of the covenant, the property would be forfeited. For example, 95% of single-family residences in 1920s Los Angeles were off-limits to Black and Asian people. The government reinforced discrimination through the Federal Housing Administration and the Veterans Administration, which produced manuals for determining which government-guaranteed mortgages to approve and deny. The manuals denied mortgages for neighborhoods with certain racial and ethnic characteristics, drawing red lines around these areas. The practice became known as "redlining," and it too denied home ownership to minority groups and prevented their investment in existing urban areas. Current disparities in wealth and education, proximity to environmental harms, and access to environmental benefits reflect the lines drawn on these historical maps in many US cities, demonstrating that the effects of discriminatory policies and practices persist today.

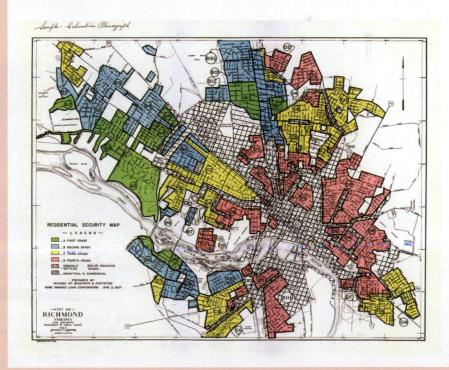

FIGURE 16.15 **Suburbs and the American Dream** This ad from the 1940s touts prefabricated homes in one of the first mass-produced suburban housing tracts ever built. It reflects how moving to suburbs became a cultural phenomenon associated with higher status, freedom, and "the good life" in the United States.

Combined with a tax policy that enabled the deduction of home mortgage interest from income tax payments, this made home ownership much more affordable. However, these loans did come with a big catch. The government would only back loans for homes that met certain criteria, and it favored the newly constructed single-family homes with relatively large lots in suburban areas.

The government loan guarantees also encouraged larger-scale real estate development. Prior to this time, suburban home construction was done primarily by small local businesses. One study of the pre–World War II homes in a suburban Boston area found that design and construction was done by more than 9,000 individual builders. Most of these builders constructed only two or three homes in their careers, and 70% lived in the towns where they worked. But by the late 1940s, construction companies had emerged that could develop large tracts of land with many houses at once. The most famous of these was Levitt and Sons, which constructed 17,500 houses at a rate of more than 30 houses per day on 4,000 acres (1,600 hectares) of farmland on Long Island, New York. They made all of the lots one-eighth of an acre and standardized the land by clearing trees, leveling hills, and filling valleys. The company owned lumber mills and centralized warehouses where they organized kits of pre-cut building materials and fixtures for each of the six housing designs they offered. The kits were delivered to the construction sites, where workers could assemble a house quickly without measuring or cutting (**FIGURE 16.16**). They also sold these homes at an astonishing rate of 1,600 per month, in part because their home designs were preapproved for FHA and VA mortgages. By 1950, *Time* magazine reported that four of every five new houses built each year in the United States were

(a)

(b)

FIGURE 16.16 **Mass-Produced Homes** Large suburban developments by single contractors—like **(a)** Levittown, New York, built by Levitt and Sons—were constructed at a rate of 30 per day with **(b)** ready-made kits of building materials and appliances in areas that had been farmland and forest. These houses were preapproved for government mortgages, and the contractors streamlined the loan application process so that home buyers could complete all of the paperwork while visiting the model house at the development site.

Transportation Infrastructure

The federal government also encouraged sprawl by passing the Federal-Aid Highway Act of 1956. It provided more than $26 billion in funding ($230 billion today adjusted for inflation) for more than 41,000 miles (66,000 kilometers) of roadways throughout the country. The new highways established rapid automobile access to and from the new suburban areas, while also making older urban areas less accessible and less attractive. New roads often intentionally bypassed urban areas or went through the middle of neighborhoods or along waterfronts in ways that diminished the quality of life in these places (**FIGURE 16.17**). During the last half of the 20th century, the federal government spent 8 times as much money on roads as it did on mass transit. It also facilitated new commercial development in the suburbs by offering tax incentives when businesses erected buildings on undeveloped land. This helped fuel the construction of shopping centers and malls, which increased from just 10 nationwide in 1953 to more than 7,000 by 1964.

Sprawl and Consumption

Consumer culture also drove sprawl, as consumers wanted more space to live in—and more space for amenities and material goods. When Levitt and Sons and others built the suburbs of the 1950s, their average home provided 983 square feet of floor space and typically included two bedrooms and one bathroom on a 5,000-square-foot lot. This was at a time when the average US population per household was a little more than 3.5 people. Today, the population per household has fallen to about 2.5 people, yet the average home size now stands at 2,349 square feet on a 12,910-square-foot lot (**FIGURE 16.18**). Among new homes built in 2012, 61% had more than two bathrooms, and 87% had more than two bedrooms, a form of sprawl at the

FIGURE 16.17 Fragmented Cities These photos of central Kansas City, Missouri, in 1955 and today show the way in which highways can fragment the grid network of cities.

constructed by large companies using mass-production methods in newly developed suburbs.

The lack of policies or planning for land use at the regional and metropolitan levels also facilitated sprawl. During the postwar boom, local governments competed to attract residential and commercial developments, enticing builders with tax reductions and offers to create infrastructure and roads. Developers built their projects in places where they could get the best deal, without a wider geographic plan that considered transportation and other community needs. This was and remains in contrast to practices in many other countries, which have long had centralized regulations that plan and provide for various land uses while curtailing sprawl. For example, the Town and Country Planning Act of 1947 in England was established to halt residential development in the rural areas around cities. Urban limit lines were drawn around cities, and development was directed into the existing areas, while undeveloped greenbelts were protected in the surrounding areas.

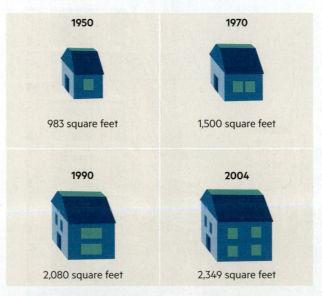

FIGURE 16.18 Growing Average Home Size The average square footage of a new single-family home, 1950–2004.

Adapted from Beach (n.d.) and National Association of Home Builders (n.d.).

household level: use of more space to meet the needs of fewer people. Expanding homes are also places to fill with more goods, purchased at bigger and bigger retail stores. Americans have built almost twice as much retail space per citizen as any other country in the world: more than 19 square feet per person, most of which is in low-rise malls and big box stores.

⬡ **TAKE-HOME MESSAGE** The majority of the US population now lives in suburbs. This was facilitated by a high demand for new housing after World War II and federal policies that made home mortgages more affordable, subsidized highway construction, and encouraged commercial development in suburban areas. Unlike many other countries, the United States lacks centralized land-use planning that would curtail sprawling developments.

➔ **SUSTAINABILITY**

Urban Design That Encourages Community Interactions

Consider the trees, grass, pavement, and structures that make up New York City's High Line, a public park built on a former elevated freight rail line above the streets on Manhattan's West Side. Urban planners designed this space to facilitate certain community interactions that would make it meaningful and useful. On the High Line, people walk, sit, people-watch, eat at food and drink carts, enjoy musical and theater performances, view and experience art installations, participate in community programs, stop and look over the city streets below them, and interact in a variety of other ways. All this happens while also serving as an urban walkway linking more than 20 blocks of Manhattan above street level.

16.5 What Is Urban Planning?

The challenges of slums and sprawl remind us that for growing cities to be livable, authorities must plan and provide for essential infrastructure. In the remaining sections of this chapter, we'll explore how cities can be designed to foster more livable and sustainable urban areas with resilient characteristics: physical and social systems that can adapt to changes and crises such as rising fuel prices, resource shortages, and even natural disasters.

Urban Planning

Urban planning is the development of guidelines to shape the future of an urban area. It can be applied at different scales: for a particular neighborhood or shopping area within a city, an entire metropolitan area, or the major population centers of an entire country. Planning like this involves assessing the current conditions of a place, anticipating likely changes, setting goals for the future of the place, and choosing and implementing strategies to meet these goals. Urban planners often think of themselves as making careful decisions about places that set the stage for certain kinds of human activities. This involves working with both the natural and human-built aspects of the physical environment, such as landscapes, infrastructure, and transportation.

Challenges of Planning

Designing plans that actually transform urban life is no easy task. A major reason for this difficulty is that good plans need to work not only for local individuals and communities but also for broader regions. Planning at the city level is often ineffective without regional planning because daily activities send residents into the surrounding area. However, if regional plans fail to engage local communities, they may fail to address the needs of the residents they are supposed to serve, and local opposition may thwart implementation. For example, in the 1980s the World Bank funded a regional plan in Mumbai to upgrade slum conditions for more than 100,000 dwellings. Regional authorities determined the size, location, and features of the new dwellings as well as the rent and form of community organization that would govern these new neighborhoods, but they made these plans without participation from the settlement dwellers themselves. The World Bank found that residents never fully embraced the new living arrangements because they did not feel invested in the decision-making process. The project was discontinued after 10 years of failed implementation. Urban planners sometimes refer to this need to operate at both a local and a regional scale as the **planner's dilemma**.

But urban planners also point to successes. The UN Human Settlements Programme successfully organized

How Can Cities Respond When Disaster Strikes?

When Hurricane Katrina struck New Orleans in 2005, the category 5 storm flooded more than 80% of the city. As a result, more than 1,800 people died; 250,000 of the city's residents were displaced; and property damage exceeded $100 billion. It was one of the costliest and deadliest hurricanes in US history. We often call extreme weather events like this "natural disasters," but in reality the effects of such an event also depend on human decisions and actions leading up to it.

Much of the criticism leveled at government officials after the hurricane centered on cost-cutting measures undertaken during earlier construction of levees and floodwalls. These decisions contributed to more than 50 breaches in a system that was supposed to protect the city from flooding. But many more decisions also contributed to the human costs from this disaster. After the levee system was built in 1965, residential development was allowed to expand in low-lying areas that were then assumed to be safe, adding more than 170,000 households in high-risk areas.

In New Orleans, creating new residential areas required pumping groundwater away from the lower-lying areas to make it less marshy. This groundwater removal caused the land surface to sink (a process known as subsidence) about 10 feet (3 meters) below sea level in the years leading up to Katrina. Before these housing developments were built, natural subsidence in these areas was offset by sediment deposits from the Mississippi River and organic material accumulating from the plants and forests nearby. But the levee and canal systems along the river interrupted the sediment supply. These structures, along with oil and gas development along the coast, also led to the loss of sprawling wetlands and barrier islands that once helped absorb storm surges.

In addition to these infrastructure and development decisions, as the storm approached, New Orleans found itself ill-equipped to deal with an evacuation and emergency response. While many people with cars were able to evacuate, those without cars were left stranded because of a shortage of bus drivers. Tens of thousands of people remained in flooded areas, often without adequate

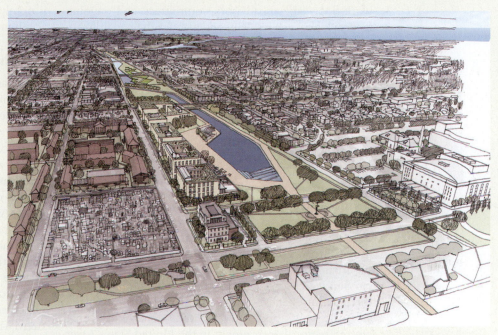

Plans to make the city of New Orleans more resilient include using landscaped green spaces and restored wetlands to manage stormwater.

food, water, or shelter, while various government agencies struggled to communicate with each other and coordinate rescue and relief efforts.

Today, New Orleans is one of more than 100 cities with a chief resilience officer, an official whose job is to anticipate the challenges that a city could face, identify and reduce vulnerabilities to those challenges, and plan effective responses and recovery actions before the city finds itself in the midst of a crisis. These challenges can be acute, such as from a hurricane, earthquake, flood, or prolonged heat wave and drought, but they can also be chronic, such as from rising unemployment and poverty. Often, climate change is the impetus for a city's resilience planning, anticipating shocks from sea-level rise, increasing frequency of extreme weather events, and food and water shortages.

In New Orleans, resilience efforts include infrastructure and environmental improvements, managing stormwater with landscaped spaces rather than pumped drainage systems, and restoring coastal wetlands. They also include developing energy and transportation systems with redundancies that enable one part of a system to take over if another part is

lost during an extreme event. Finally, in reflecting on Katrina, leaders in New Orleans found that those who died or suffered the most in the hurricane and its aftermath tended to be the most economically vulnerable and isolated residents. To address this problem, much of the city's resilience plan is also devoted to human systems, with a focus on reducing economic inequality, boosting social cohesion, and facilitating education and communication across government agencies and social organizations.

What Would You Do?

If you assumed the role of chief resilience officer for your college or university, what challenges would you anticipate your campus or parts of your campus could face in the future from problems such as floods, droughts, or storms? What research would you do to identify vulnerable areas on campus? What are some of the changes you would propose to make your campus more resilient?

475

transportation and land use in eight cities of the Lake Victoria Basin in Uganda, Kenya, and Tanzania. These three nations coordinated with each other as well as with the city governments and community members to sensibly distribute urban infrastructure and services. This UN agency also leads the Participatory Slum Upgrading Programme, which funds projects that strengthen the capacity of informal settlement residents to participate in the design and implementation of urban improvements. Agency officials have found that community design workshops, citizens' advisory committees, and ongoing feedback between residents and planners helped improve design decisions while also building community.

In the United States, Portland, Oregon, is often held up as an example of successful regional planning and public participation. In 1973, Oregon passed a state law that required each city and county to develop comprehensive land-use plans and empowered a new state agency to ensure that these plans conformed to statewide goals. At the same time, dozens of workshops were held throughout the state, which drew participation from tens of thousands of residents to help prioritize land-use goals. The goals included preserving green space and farmland, providing affordable housing, developing regional mass transit systems, and establishing urban growth boundaries that would constrain suburban sprawl. Oregon also created the Metropolitan Service District (Metro) to govern land use for the three counties and 24 cities that make up the Portland area, the state's population center. Metro is the only elected regional government in the United States. Over several decades, in accordance with statewide and Metro goals, Portland built an extensive light-rail system, dedicated streets for bus travel, and provided bikeways and pedestrian paths while boosting urbanization within the urban growth boundary and keeping it from spreading to nearby forest and farmland.

🏠 **TAKE-HOME MESSAGE** Careful planning is required to address the challenges of slums and suburban sprawl and match urban infrastructure to population needs. The planner's dilemma describes the way in which plans can fail if they do not effectively operate at both a regional and a local scale. Engaging local communities can help ensure that plans meet the actual needs of residents, while a regional authority can provide the resources and coordination necessary.

16.6 Why Is Transportation Important?

In the late 1990s, Governor Christine Todd Whitman of New Jersey and Governor Parris Glendening of Maryland were grappling with the challenges of urbanization.

Although they represented different political parties—Whitman was a Republican and Glendening was a Democrat—they agreed on the need for urban design that accommodated more modes of transportation than just private automobiles. The two governors said, "If you design communities for automobiles, you get more automobiles. If you design them for people, you get walkable, livable communities." These two governors had effectively intuited a core tenet of urban development and planning: where and how we develop our urban areas is tightly linked to the modes of transportation that we accommodate, encourage, and provide.

The Challenge of Induced Traffic

Whitman and Glendening recognized that as their transportation departments added lanes to existing roads and built more highways to lessen traffic congestion, traffic actually increased. Transportation planners call this phenomenon **induced traffic** (FIGURE 16.19). Initially, the boost in highway capacity enables and encourages people to drive faster, farther, and more frequently. But as more people utilize and depend on the expanded capacity of highway systems, congestion once again becomes a problem. A 20-year study of California's highway system found that each 10% increase in road capacity was met with a 9% increase in traffic within 4 years. If the increase in traffic is met with another round of road construction, the pattern will repeat itself. A study of US cities over the past 30 years conducted by the Texas A&M Transportation Institute found no significant difference between the congestion levels of cities that had invested heavily in road construction and those that did not. Even as technological improvements reduce automobile fuel consumption and emissions, these gains will diminish if the cycle of road construction and increased traffic continues.

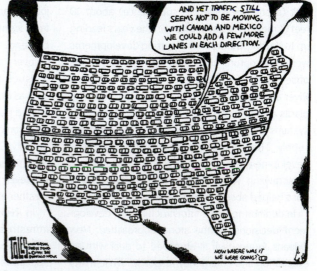

FIGURE 16.19 Induced Traffic Building more lanes on highways tends to increase traffic volumes.

This information is starting to raise questions about land-use and building patterns and laws that facilitate car travel. The typical layout of a suburban subdivision has wide, cul-de-sac style road systems linked by only a single collector road to a larger highway. To better accommodate cars, roads in suburbs are widened, sidewalks are narrowed or eliminated, and even buildings represent what some have called "car-chitecture." Homes are set back on bigger lots to accommodate driveways and built-in garages. Further, suburban building ordinances often require businesses, schools, and government offices to provide a certain quantity of on-site parking, which is most often done with large parking lots in front of the buildings. As a result, Houston has enough parking area to accommodate 30 car spaces per resident, and one study found that more than one-third of the land area in America's cities is devoted to either driving or parking cars.

Transportation Alternatives

The obvious alternative to more cars is more mass transit options, such as bus, streetcar, and subway systems. There is an efficiency associated with the infrastructure of mass transit: a two-track rail line can carry the same number of passengers as 16 highway lanes full of automobiles, and an average bus at full capacity gets 163 passenger miles per gallon, while a car carrying just the driver gets only 25 passenger miles per gallon. In addition, many cities use natural gas to fuel buses and electricity to power rail transit. These are less-polluting fuel sources than the gasoline powering most cars. Walking and biking are options that come with the added health benefits of increased activity levels.

The term **modal split** describes the percentage of travelers in an area using a particular type of transportation to get to work. The denser the city, the higher the modal split for mass transit ridership, for reasons we will see next.

The Relationship between Transit and Density

Any advantage of mass transit depends on high levels of use. A relatively empty bus or train car can be even less efficient and more polluting per passenger than transportation by car. For example, although full buses can often carry as many as 70 seated and standing passengers, the average bus in the United States carries only 10 passengers at a time, just 14% of capacity. Streetcars and subway systems do better, averaging 46% and 24% of capacity, respectively.

Urban planners have found that the advantages of mass transit depend on having large numbers of people living and working near transit stops. Hourly bus service requires at least four houses per acre in a residential area or 5 million square feet of retail space in a commercial area to draw enough use that it is self-supporting. Streetcars and subways require even denser urban and commercial areas to work well, and more frequent service of mass transit requires higher levels of density. Researchers

have also found that the average person will not walk more than 8–10 minutes to get to a transit stop.

⬡ **TAKE-HOME MESSAGE** Our transportation systems and land-use patterns influence each other. When we add lanes to existing highways and build new ones, we induce more traffic. Transportation alternatives include various forms of mass transit that can be associated with environmental, health, and lifestyle benefits. However, mass transit success requires close proximity to significant population density and a diverse range of daily uses.

modal split the percentage of travelers in an area using a particular type of transportation to get to work.

➡ SCIENCE

Modal Split for Cities

The modal split measures the share of commuters who use various types of transportation. In the United States, New York City has achieved the highest modal split for mass transit ridership in an urban area at 56%. Only 22% of New Yorkers use a private vehicle to get to work, and only about half of New Yorkers even own a car. In fact, the New York metropolitan area generates about 40% of all the mass transit trips in the entire United States. The average person in New York City uses less than half the amount of gasoline per year as a person in Atlanta, where only 13% of residents report using mass transit to get to work.

City (top 15 by number of commuters)	Car Driven Alone (%)	Carpool (%)	Transit (%)	Bike (%)	Walk (%)	Other (%)	Home (%)
New York, NY	22	4	56	1	10	2	5
Los Angeles, CA	70	9	9	1	3	2	6
Chicago, IL	49	8	28	2	67	2	5
Houston, TX	78	10	4	0	1	3	4
Phoenix, AZ	74	13	3	1	2	1	5
San Diego, CA	74	9	4	1	4	1	7
San Antonio, TX	78	12	3	0	2	1	3
Philadelphia, PA	50	7	26	2	9	2	4
Dallas, TX	77	11	4	0.2	2	1	5
Austin, TX	75	8	3	1	3	2	8
San Jose, CA	76	12	4	1	2	1	4
San Francisco, CA	30	9	34	4	13	4	6
Columbus, OH	78	10	3	1	3	1	4
Charlotte, NC	75	10	3	0	2	2	8
Seattle, WA	44	7	23	4	12	2	8

Adapted from Freemark (n.d.).

16.7 How Are Cities Changing?

As urban population increases and cities are faced with new challenges, many cities in the United States and worldwide have implemented new planning efforts guided by the principles of **smart growth** (FIGURE 16.20). Smart-growth planning uses inclusive community processes to determine how to accommodate urban growth to meet the needs of residents and avoid the negative consequences of sprawl and many environmental impacts. One of the primary smart-growth strategies is the development of **mixed-use areas**, in which a combination of housing, shops, restaurants, grocery stores, offices, and even public amenities, such as parks, museums, and event venues, are all within a half-mile of public transportation. For example, the PS1200 development on the Southside of Fort Worth, Texas, combines apartments, office space, shops, and restaurants with a public park. It also makes use of green building techniques by using recycled steel and building methods that keep the inside cool, while promoting alternative energy sources through solar power and providing electric vehicle charging stations.

Planners are also trying to make mass transit more appealing to users. For example, research shows that the transit stops themselves facilitate use when they are safe, sheltered, and integrated with shops and cafés where riders can comfortably wait for their ride. A study of mass transit in 84 world cities found that rail transit was the fastest and most reliable option, traveling an average of 12 miles per hour faster than bus service, the latter often being slowed by traffic and indirect routes. However, when a bus system is given dedicated lanes and traffic signal priority, it can provide service that is faster than cars and nearly as fast as rail transit for less cost. This is known as *bus rapid transit*, and it has worked on a large scale in the cities of Curitiba, Brazil; Bogotá, Colombia; and the central Manhattan section of New York City.

Walkability

Cities are also doing more to promote walking. Urban planners have found that people feel most comfortable walking when they are protected from car traffic and are close to buildings with interesting things to look at, such as shop windows. Trees and curbside parking can serve as a buffer between the sidewalk and fast-moving vehicles, giving pedestrians a safe sense of space in which to walk. Building mixed-use commercial developments, such as retail stores, restaurants, and offices at ground level up to the edge of the sidewalk rather than set back behind big parking lots, also facilitates foot traffic. Short city blocks and relatively narrow streets in a grid pattern help to give pedestrians a continuous network of routes to various destinations, calmer traffic, and shorter and safer street crossings. In the United States, many urban areas that are tourist destinations—such as Charleston, South Carolina; Santa Fe, New Mexico; and the Georgetown neighborhood of Washington, DC—share these characteristics (FIGURE 16.21).

Transit-Oriented and Pedestrian-Oriented Developments

Urban planners often use the terms **transit-oriented development (TOD)** and **pedestrian-oriented development (POD)** to describe design strategies to integrate travel and land-use patterns in more sustainable and resilient ways. TOD strategies emphasize public transportation, while POD strategies emphasize pedestrian travel. The Kronsberg District in Hanover, Germany, illustrates how

(a)

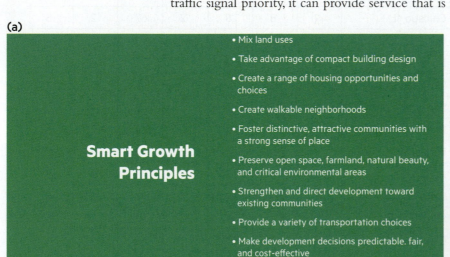

Smart Growth Principles

- Mix land uses
- Take advantage of compact building design
- Create a range of housing opportunities and choices
- Create walkable neighborhoods
- Foster distinctive, attractive communities with a strong sense of place
- Preserve open space, farmland, natural beauty, and critical environmental areas
- Strengthen and direct development toward existing communities
- Provide a variety of transportation choices
- Make development decisions predictable. fair, and cost-effective
- Encourage community and stakeholder collaboration in development decisions

(b)

FIGURE 16.20 Smart Growth (a) These principles are increasingly used to create thoughtful projects with community input. **(b)** PS1200 in Fort Worth, Texas, provides a new example for mixed-used development.

Adapted from (a) Smart Growth Network (2006).

Walk Score® ChoiceMaps™

Learn more about Walk Score data

People can walk to:
5.89 restaurants in
5 minutes

5 |———————————————| 20
minutes

Map type: Choice

Category: Restaurants

■ = 122+ choices

Restaurants: 8,547
Population: 2,686,606
Routes Analyzed: 14,837,868

More Chicago information

© 2014 Walk Score

Walk Score
78   Chicago is Very Walkable
Most errands can be accomplished on foot.

Walk Score Map

FIGURE 16.21 **Walkability** How walkable is your neighborhood? If you enter your address in the website walkscore.com, it will calculate a walkability score that is based on the characteristics of your neighborhood.

TODs and PODs can be developed from scratch. This high-density development was built in 1998 along a tram line that connects it to the larger city of Hanover (**FIGURE 16.22**). The project's residential density is 47 units per acre. No resident lives more than one-third of a mile (0.5 kilometers) from a transit stop.

Other cities have decided to facilitate additional modes of transportation. For example, the Complete Streets policy of Buffalo, New York, results in at least 10 miles (16 kilometers) of new bike lanes laid out each year on city streets, and in Amsterdam more than 40% of commuters travel by bicycle because of the extensive network of bikeways throughout the city and region. Many cities now have bike- and car-sharing programs (**FIGURE 16.23**). Some cities have even altered their land use and policies to discourage driving cars: Seoul, South Korea; Portland, Oregon; and Providence, Rhode Island, all decided to remove or relocate major sections of urban freeways splitting areas of the city. London, Stockholm, and Singapore have instituted congestion taxes on those who drive cars within the city. Drivers are assessed taxes by automatic gates, depending on where they drive, paying higher taxes in congested parts of the city at busy times. These cities have seen significant increases in the share of commuters using mass transit.

Finally, another strategy is to enable and encourage workers to commute less. Traffic in and around US cities declined significantly in 2020 as many employees worked online. Although many companies that halted in-person work during the early months of the pandemic have brought their employees back to the office for at least a portion of the work week, remote work will likely continue for many workplaces. Office-sharing resources close to where people live and more reliable internet infrastructure can help to make remote work more accessible.

FIGURE 16.22 **Transit-Oriented and Pedestrian-Oriented Design** The Kronsberg District in Hanover, Germany, has several features that facilitate transit-oriented and pedestrian-oriented travel, including a central bike path, a grid pattern of streets, and proximity to a tram line.

FIGURE 16.23 Reducing Reliance on Car Ownership Many cities, such as Spartanburg, South Carolina, pictured here, now have bike-sharing and even car-sharing programs.

Green Buildings and Infrastructure

The buildings where people live and work have a major impact on a city's environmental impact. Because of this, some governments have adopted policies to mandate green building performance standards. For example, Barcelona, Spain, requires that new construction and remodels get a minimum of 60% of their hot water needs from solar power. Sydney, Australia, passed an ordinance that requires new homes to produce at least 40% less greenhouse gas emissions and consume 40% less water than a standard house. Some US states, including California and Washington, have incorporated green building standards on energy and water conservation into their permitting processes for construction.

Voluntary rating and certification systems, such as the Leadership in Energy and Environmental Design (LEED) standards in the United States, can also provide incentives for builders and owners to pursue sustainable features. Many college campuses have adopted LEED standards for buildings constructed on their campuses (**FIGURE 16.24**). There are a range of ways buildings can accumulate points for LEED certification, including through energy and water conservation, the materials used in construction, the way waste is managed during construction, and the inclusivity of the planning process. The most advanced green buildings can serve as positive examples and as publicity for green building designs. On an even larger scale, the Bosco Verticale high-rise complex in Milan, Italy, hosts hundreds of trees and thousands of shrubs that grow along the sides of the building and provide shade, habitat, and carbon dioxide storage.

The term **green infrastructure** captures the many ways a wide variety of green spaces can be incorporated into urban life. Many cities are well known for the parks and green spaces that are part of their urban designs, such as New York's Central Park and Boston's Emerald Necklace. More recently, green roofs atop office buildings in Chicago have been made into park-like retreats for workers while also improving the heating and cooling efficiency of the buildings. Bioswales and rain gardens absorb stormwater while providing small green zones that reduce the area of impermeable surfaces in cities. The Glenwood Park development 2 miles

green infrastructure the wide variety of green spaces that can be incorporated into urban life.

(a)

(b)

FIGURE 16.24 Green Buildings (a) LEED-certified buildings, such as the (b) Paul L. Foster Campus for Business and Innovation at Baylor University, incorporate features that reduce a building's environmental impact. This building has energy- and water-efficiency features including a solar reflective roof, adjustable daylighting in most rooms, and water-saving landscaping and irrigation systems. (c) The Bosco Verticale building in Milan, Italy, was designed to be a "vertical forest" promoting sustainable development.

(c)

The Bullitt Center—Meeting the Living Building Challenge

Consider the buildings you inhabit. These structures provide shelter, but the resources that go into them and the way they influence our actions make buildings a primary source of our environmental impacts. Some builders are taking a closer look at the environmental impacts associated with their structures and attempting to minimize these impacts with new design features. The Living Building Challenge sets the most ambitious green building standards, such as self-sufficiency. Living buildings must produce more energy than they use and collect and treat all water on-site. And these buildings must be designed to operate for at least 100 years. What would living buildings look like? Are they even possible?

The architects and engineers at the firm Miller Hull who designed the Bullitt Center set out to exceed the Living Building Challenge and create the world's "greenest" commercial building, a "model for 21st century sustainability." The completed six-story building, which opened in 2013, provides 52,000 square feet of office space in Seattle. It is designed to last 250 years, far longer than other commercial buildings and well beyond the lifetimes of the people who built it. The building materials were chosen not only for their durability but also for the environmental performance of their harvesting and manufacturing processes and from producers operating within a 600-mile (965-kilometer) radius of the construction site. All of the water used in this building is collected

rainwater from the roof that is filtered and held in a 56,000-gallon (212,000-liter) cistern. Special toilets use just two tablespoons of water per flush, and the waste is composted on-site for use as fertilizer. Wastewater is cleaned in a constructed wetland on-site and returned back into the local water table. The solar panels on the roof and the many energy conservation features inside enable this building to produce more energy than it uses.

So how is this living building performing so far? Seattle is famous for its rain, so perhaps it's not surprising that this building does capture and store all the water it needs on-site. But rainy Seattle is one of the cloudiest cities in the United States. How would it fare generating all of the electricity it needs from the Sun? Once it was open and occupied, the building actually proved to generate more energy than it used. The solar panels performed as expected, but what was surprising even to the designers was how much the building's energy conservation measures lowered its energy demand. The Bullitt Center uses approximately 75% less energy than comparable Seattle office buildings. Reaching for a "stretch" goal like the Living Building Challenge—a standard that many think is slightly out of reach—can result in innovations, like those in the Bullitt Center, that redefine what is possible.

The Bullitt Center in Seattle is part of the Living Building Challenge. It produces more energy than it uses through solar panels installed on the roof, and it collects and treats all of the water used in the facility.

🔍 What Would You Do?

If you were on the team designing a new building tasked with reducing environmental impacts, what building attributes would you assign the highest priority? Assume you are told by the building owners you are much more likely to get extra funding for the first three items on your list. Your current list includes building materials used, source of building materials, height of the building, energy source used for heating, energy source for cooling, energy source for electricity, water supply, garbage and waste disposal, and longevity of building before repairs. Which items would you list first and why?

Green Gentrification

Today, sustainable planning and building features are often highly desirable. Especially for younger buyers or renters, homes near work, parks, and other "urban amenities" are hard to come by and typically make properties cost more. For example, a study by the real estate company Redfin found that each point increase in the "walk score" of a neighborhood increases the price of a home by an average of $3,250. New mixed-use development proposals are often met with concerns about **gentrification**—a process of displacement that occurs when changes are made to a neighborhood's character, leading to more expensive housing, higher taxes, and an influx of wealthier people. "Green gentrification" can occur when green buildings go up and even small improvements such as community gardens are added. Such changes can make these communities less affordable and cause longtime residents to move.

Many planners today feel it's important to implement a range of affordable housing strategies and supports to ensure that current residents will not be forced out as the environmental benefits come to fruition. Another idea is to build more housing everywhere and increase housing density—an idea known as YIMBY, for "Yes, in my backyard." Groups like YIMBYAction.org seek to address restrictive zoning practices, especially in areas that are already desirable but not open to high-density developments. Opening up these areas makes communities more accessible to more people, while also beginning to break down the legacy of redlining.

Protestors rally against re-development of long-standing row houses in an area of Boston long populated by Chinese immigrants.

gentrification a process of displacement that occurs when changes are made to a neighborhood's character, leading to more expensive housing, higher taxes, and an influx of wealthier people.

(3.2 kilometers) from downtown Atlanta, with more than 300 dwelling units around a central park, has a stormwater system directing rainfall to residential landscaping and a rain garden in the park area. This has reduced runoff by two-thirds and eliminated the use of potable water for irrigation.

⬡ **TAKE-HOME MESSAGE** Modern urban planning efforts have been focused on mixed-use areas, integrating all aspects of daily life with close proximity to public transportation. Transit-oriented and pedestrian-oriented developments foster use of mass transit and other alternatives to cars and improve community walkability. Urban areas provide many opportunities for lowering per capita environmental impacts through green building practices and green infrastructure.

❓ 16.8 What Can I Do?

At times it may seem like changes that foster sustainability in the places we live take a long time. This is because the infrastructure we build often lasts decades or longer. And planning, design, and new construction require substantial investments and significant public support. Though infrastructure may last a long time, it is not static: it requires maintenance and improvements. If you look around your campus, city, or town, you will notice that it is always changing. Maybe a new building is going up here, an old building is being refurbished there, a park is getting a landscaping improvement, the waterfront is being cleaned up and converted to a park, bus routes are added, or a new bike path is being established. Or maybe none of these things are happening and you wish they were. What's your role in the midst of all this change?

What Is Transportation Like Where You Live?

You can visit walkscore.com to see how walkable your community is. You can also get a commute report to see what options you have for getting around by car, bus, mass transit, and bike. Once you get a feel for the transportation options where you live, compare it to some other places. What sorts of factors make a place more walkable or ease the commute? If you have the chance to travel to some other cities here or abroad, make some comparisons of the way people get around in their daily lives.

Learn More about Your Campus Plan

Often colleges and universities have long-term planning processes to guide changes to their infrastructure, and many of these plans are developed by committees that include student representation. Research the planning process at your school. Who is involved? Is there a focus on sustainability and green design? Maybe you too could get involved in the process. Campuses often seek student feedback on a wide range of planning topics. For example, the University of Connecticut's Department of

Transportation recruited students to participate in surveys, focus groups, and strategy meetings to help determine optimal bus schedules and routes for student needs. The resulting changes led to increased student ridership of mass transit on and around campus.

Attend Planning Board Meetings in Your City or Town

Cities, counties, and towns typically have planning commissions that make decisions on where and how land will be developed. What government body makes these decisions where you live? What are some of the recent decisions they've made, and what is on their agenda now? Most municipalities are required to hold public meetings

when major changes to infrastructure are proposed, so you can make your voice heard. In 2018, students at the University of California, Berkeley, began organizing groups to participate in a city planning meeting to push for more affordable and "car-free" housing developments near the campus. Several students were also approved to serve on various city planning commissions.

Chapter 16 Review

SUMMARY

- Globally, urban populations are growing at a rapid rate, driven by both birth rates and the migration of people from rural areas to cities.

- The extent to which effective infrastructure and services are provided in rapidly urbanizing environments shapes the environmental and human health impacts of these growing cities.

- When the provision of affordable infrastructure and services in cities fails to keep pace with increasing population, it produces dense areas of substandard housing, poor sanitation, and high rates of poverty.

- Low-density, resource-intensive suburban sprawl is characterized by zones separating daily activities, low-rise buildings, and a reliance on long-distance automobile transportation.

- In the United States, suburban sprawl is the predominant form of urbanization, supported in part by federal policies that made home

mortgages more affordable, subsidized highway construction, and encouraged commercial development in suburban areas.

- Planning efforts for urban improvements can fail if they do not engage local needs and concerns and if they do not coordinate with important regional authorities that influence transportation networks and other factors affecting land-use patterns.

- Public mass transportation alternatives, such as buses, trains, and subways, require close proximity to significant population density and a diverse range of daily uses to be successful.

- Urban improvement strategies often focus on cultivating mixed-use areas that integrate many aspects of daily life, while also favoring transit-oriented and pedestrian-oriented developments and green building practices.

KEY TERMS

bright lights syndrome
economy of scale
gentrification
green infrastructure
induced traffic
informal economy
infrastructure

megalopolis
metropolitan area (metropolis)
mixed-use area
modal split
one-use zone
pedestrian-oriented development (POD)

planner's dilemma
rural area
slum
smart growth
suburban sprawl
transit-oriented development (TOD)

urban area
urban density
urban penalty
urban planning
urban transition

REVIEW QUESTIONS

The letters following each Review Question refer to the Chapter Objectives.

1. Compare and contrast rural areas and urban areas. **(A)**

2. If less than half the growth in the world's urban populations is due to natural increase (births outpacing deaths in cities), what accounts for the rest of the population growth? **(A, B)**

3. How do the environmental impacts of cities extend beyond their borders? **(C)**

4. Provide an example of a way in which increasing population density and urban infrastructure can reduce the per capita use of natural resources. **(C, E)**

5. What characteristics define urban slums? **(D)**

6. Despite the disadvantages, why might people migrate from a rural area to an urban slum? **(B, C)**

7. Describe the characteristics of suburban sprawl and how they are associated with less efficient use of natural resources. **(C, D)**

8. Which government policies facilitated the development of suburban sprawl in the United States? How did consumption play a role? **(D)**

9. How can urban planning help address the negative characteristics of slums and suburban sprawl? **(D, E)**

10. Explain how mass transit and urban density are related and dependent on each other. **(D, E)**

11. Name the factors that can make an urban area more walkable. **(E)**

12. What is green infrastructure? Provide an example. **(E)**

FOR FURTHER THOUGHT

The letters following each item refer to the Chapter Objectives.

13. Although urban planning can help to solve many of the problems associated with slums and suburban sprawl, earlier in the chapter we presented the challenge of the "planner's dilemma" that makes it difficult to fulfill regional goals while satisfying local interests. Consider a metropolitan region near you. Identify and list the local and state governments that would potentially be involved in a new transportation plan covering the region. If you were a planner, what steps could you take to ensure that the new transportation plan was well integrated across the region and also gained enough support across the various local affected communities? **(D, E)**

14. First, list what forms of transportation you use each day and the factors that drive these choices. Then research two alternative forms of transportation that you do not typically use. As best you can, how long do you estimate your commute times would be if you used the alternatives (compared to your typical commute times), and what would they cost? **(E)**

15. Describe at least three ways that the design and infrastructure of the place where you live affects quality of life and environmental impacts. If you could advocate for two changes to the infrastructure near where you live, what would they be and why? **(B, C, D, E)**

Make Your Case

The following exercises use real-world data and news sources. Check your understanding of the material and then practice crafting well-supported responses.

Use the News

The following article appears in a news website run by the World Economic Forum and is written by economists from the World Bank and the World Data Lab. The article describes two global "megatrends" largely in terms of potential for economic growth. Use this article to answer the questions that follow. The first three questions test your understanding of the article. Questions 4 and 5 are short answer, requiring you to apply what you have learned in this chapter and cite information in the article. Answers to Questions 1–3 are provided at the back of the book. Question 6 asks you to make your case and defend it.

"The Two Population Megatrends Set to Shape the Next Decade," Wolfgang Fengler and Max Heinze, December 23, 2021

It is shopping time. Many people, especially in the Northern Hemisphere, participated in Black Friday and Cyber Monday while getting ready for Christmas. For many, the 2021 shopping experience also reflects a return to normality at a time of never-ending waves of COVID-19 infections and lockdowns.

Many large stores are in cities, which reflect two megatrends that are reshaping the global economy: the rise of the global consumer class and urbanization. Both megatrends experienced tipping points in this century. In 2008, the world became majority urban and in 2019, World Data Lab projected that half the world would be middle class or wealthier. By the end of 2021, there will be 4 billion people in the global consumer class, and, in the absence of another major economic crisis, the global consumer class will reach 5.2 billion people by 2030.

Urbanization will also continue steadily throughout this decade. People are moving from villages to cities to find better education, health care, and jobs. Additionally, in emerging markets, former villages are contributing to urbanization as they grow rapidly to the point that they represent a peri-urban segment. These are often the new suburbs of the world's megacities such as Jakarta, Mexico City, Mumbai, São Paolo, or Lagos.

Urban areas are more prosperous than rural areas. Proximity breeds innovation and allows for economies of scale. In "Triumph of the City," Edward Glaeser showed that cities are healthier, wealthier, and better for the environment because higher population density makes it possible to produce new goods and services at scale and thus at a lower cost for everyone. This is also why most of the global consumer class is now living in cities.

To estimate the urban consumer class, we use urbanization projections from the U.N. along with education and income measures and investigate how personal spending depends on each of these factors, especially urbanization. Projections for India are derived from survey microdata for the country.

World Data Lab projects that out of today's 3.8 billion consumers, almost 3.2 billion (80 percent) are living in urban areas (see Table 1).

However, the split is uneven across countries as many emerging markets are still

TABLE 1. Two-thirds of the global consumer class are urban

	Consumer class	Poor and vulnerable	TOTAL
Urban	3.2 billion	1.6 billion	4.6 billion
Rural	0.6 billion	2.6 billion	3.2 billion
Toal	3.8 billion	4.0 billion	7.8 billion

Consumer class = greater than $11/per day. Poor and vulnerable = less than $11/per day.

Image: MarketPro, World Data Lab projections.

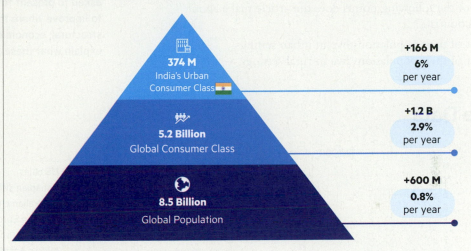

The number of consumers in India and globally.

Image: World Data Lab projections.

predominantly rural. In OECD [Organization for Economic Cooperation and Development] countries, almost everyone is part of the consumer class. Since OECD economies are highly urbanized, approximately 80 percent of the OECD consumer class is urban. The same is true for emerging markets and developing countries but for different reasons. Poorer countries are still majority rural, but only a very small portion of rural folk are part of the consumer class. This means that among poorer economies also approximately 80 percent of consumers are in urban areas. In Asia where urbanization has been advancing rapidly, the urban consumer class dominates with 1.7 billion people, representing approximately 54 percent of the global urban consumer class. Asia's consumer class is not only rising in China and India. There are also several emerging markets that are below the radar, especially Indonesia, Bangladesh, Pakistan, the Philippines, and Vietnam.

But India remains the most dynamic market for urban consumers because it remains relatively rural even though it is urbanizing rapidly. In India both megatrends converge on scale. Today India has an estimated 208 million urban consumers, which will rise to an estimated 374 million. This represents an annual growth of 6 percent. The growth of India's urban consumer class is six times larger than global population growth, almost double growth of the global consumer class, and even substantially larger than the average of urban consumer class growth and Asia's consumer class growth (see Table 1).

Many of us will enjoy the festive season of the year looking back at two difficult years dominated by the COVID-19 pandemic. Despite the massive economic impact of COVID-19 in 2020, its impact has only been temporary. The fundamental forces of global megatrends will shape this decade and they will be most visible in Asia.

1. What two "megatrends" does the article identify as occurring now? Choose the two correct answers:

   a. growth of the global consumer class
   b. improved health outcomes for the global population
   c. gentrification of inner-city communities
   d. urbanization of the population
   e. an increase in poverty tied to more urbanization

2. How is the "consumer class" defined in the article?

   a. people who make more than $11/day
   b. people who have access to shopping within walking distance
   c. wealthier members of cities who have a lot of disposable income
   d. anyone who is able to purchase food for themselves instead of growing it

3. Which of the following points does the article make about poorer countries?

   a. Most of their populations live in urban settings.
   b. Most of their populations live in rural settings.

   c. Their populations are stagnant.
   d. Their populations are in decline.

4. Using the article and material in this chapter, identify at least two likely benefits of these megatrends. Then identify two environmental problems they will likely cause.

5. This article was written by two economists. Do you think its conclusions would be different if it had been written by two primary care doctors? Explain if you think doctors would have emphasized the same points or if they would have emphasized different points regarding the two megatrends.

6. **Make Your Case** Imagine you are an experienced city planner who is helping to develop an area close to a rapidly growing city center in a poor but quickly developing country. You are asked to present to city officials the first two things you will focus on to improve where these people live and work. These could be social, structural, economic, health, or any other issues. Briefly identify and explain what these two priorities will be and why you chose them.

Use the Data

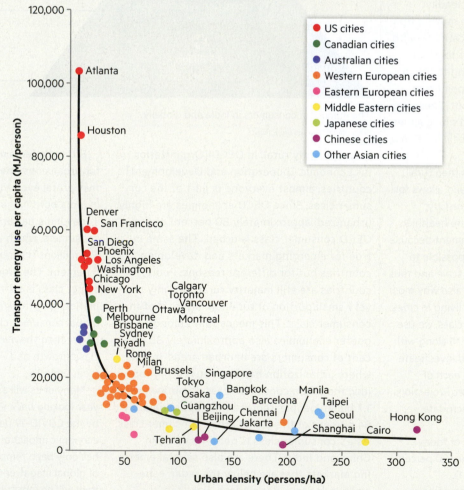

Adapted from World Health Organization (2011).

The United Nations' World Health Organization (WHO) included the graph shown here in a 2011 report titled *Health in the Green Economy* that explores how energy-efficient developments can also benefit public health. The graph compares transport energy per capita (measured in megajoules [MJ] per person) with urban density (measured in persons per hectare [ha]) and comes from research on land-use patterns and transportation in 32 countries.

Study the graph and use it to answer the questions that follow. The first three questions test your understanding of the graph. Questions 4 and 5 are short answer, requiring you to apply what you have learned in this chapter and cite data in the graph. Answers to Questions 1–3 are provided at the back of the book. Question 6 asks you to make your case and defend it.

1. Using the key on the graph, identify the country or global region that hosts the cities using the most transport energy per capita.

   a. Western Europe
   b. United States
   c. Canada
   d. China

2. Roughly how much more transport energy per capita does Denver use than Brussels?

   a. 2 times as much
   b. 3 times as much
   c. 4 times as much
   d. 5 times as much

3. Which of the following statements best summarizes the relationship in the graph?

   a. The lower the urban density, the higher the transport energy use per capita.
   b. The lower the energy use per capita, the lower the urban density.

   c. The higher the urban density, the higher the transport energy use per capita.
   d. There is no strong relationship between urban density and transport energy use per capita.

4. Use the graph and the material in this chapter to identify at least two reasons for the relationship between urban density and transport energy use.

5. Why might the WHO be interested in these data for its report on the connection between human health and urban planning? Use the graph and the material in this chapter to explain your answer.

6. **Make Your Case** A friend who is a native of Atlanta defends its high per capita transport energy use by citing its low population density and argues that this is the best energy use it can achieve for its low population density. Do you agree with your friend's assessment? Explain your reasoning using data from the graph and/or information from the chapter.

LEARN MORE

- Brookings Institution Metropolitan Policy Program: www.brookings.edu/program/metropolitan-policy-program/
- Rockefeller Foundation 100 Resilient Cities program: https://resilientcitiesnetwork.org/about/
- Lincoln Institute of Land Policy (climate change resources): www.lincolninst.edu/key-issues/climate-change

- UN Habitat—Human Settlements Programme (info on sustainability, slums, and more): www.unhabitat.org
- World Green Building Council: www.worldgbc.org/what-green-building
- US Environmental Protection Agency (resources on green infrastructure): www.epa.gov/green-infrastructure/what-green-infrastructure

17

Environmental Health and Justice

How Do Environmental Factors Affect the Places People Live, Work, and Play?

I f you go to the doctor for a routine physical exam, you might expect to step on a scale, have your temperature taken, and even have some blood drawn. You might anticipate answering questions about personal habits—such as smoking, drinking, and exercise—and your family history of disease. But you might not expect the doctor to ask you for your ZIP code or about your workplace. Yet doctors are increasingly asking these questions. Why is this? Many factors affecting our health are determined by our environments. For example, a child's ZIP code might reveal that he lives in a neighborhood with older houses and a high likelihood of lead exposure from old paint and water fixtures. This could help explain symptoms such as abdominal pain, headaches, and learning disabilities. Or a doctor might better understand what's ailing someone with respiratory problems if she learns that the patient works in a nail salon with repeated exposure to chemical hazards.

Where a person lives can put their health and safety at risk in much more obvious and immediate ways—as in the case of communities that face the threat of hurricanes each year. Although we tend to call these storms "natural disasters," scientists are learning that human-caused climate change may be increasing their severity and making extreme events more likely. Regardless, choices to locate development in areas at high risk for hurricane damage and flooding can result in substantial health, housing, and infrastructure effects.

Many factors affecting our health are shaped by the environments in which we live, work, and play. This includes the water supply. In 2016 a federal state of emergency was declared in the city of Flint, Michigan, due to lead-contaminated water.

Chapter Objectives

By the end of this chapter, you should be able to . . .

A. define and discuss environmental justice issues.

B. explain the factors that facilitate the spread of major biological hazards.

C. identify the routes of exposure affecting risks that communities face from chemical hazards.

D. describe how values play a role in risk management for environmental hazards.

E. explain how human factors can influence harms from physical hazards.

F. identify ways you can prepare for and address environmental health challenges in your community.

489

> **Environmental justice means that no community should be saddled with more environmental burdens and less environmental benefits than any other.**
>
> —Majora Carter, activist

This was the case in 2017 during Hurricane Harvey in Houston, where many buildings, roads, and parking lots were constructed on wetlands and other green spaces that formerly helped absorb floodwaters (**FIGURE 17.1A**). In the same year, damage caused by Hurricane Maria in Puerto Rico resulted in widespread power outages, failed water-delivery systems, and downed cell-phone service. Poor infrastructure maintenance before the hurricane and slow or nonexistent emergency response afterward caused these problems to linger for months. In this case, risks from and exposures to harms were caused by how some communities were prioritized over others (**FIGURE 17.1B**).

The decisions made before and after Hurricanes Harvey and Maria illustrate how issues of environmental health are not equally felt across the population. In Houston, as in many US cities, low-income communities of color have historically been situated in the areas most at risk of flooding and in closest proximity to chemical plants and other polluting facilities (**FIGURE 17.2**). When Hurricane Harvey struck, it unleashed a swirl of interrelated environmental threats, and researchers found that these underserved communities felt their impacts most intensely. The physical damage of the high winds was followed by record rainfall and flooding. The floods cut off power to and inundated a chemical plant, which exploded and released toxic substances. The storm also caused the release of raw sewage from treatment plants and septic tanks and promoted mold growth in flood-damaged buildings. In Puerto Rico after Hurricane Maria, slow federal emergency response times and less disaster relief than for similar disasters in the continental United States raised questions about whether equal treatment is provided to all US citizens (**FIGURE 17.3**).

In this chapter, we will explore how environmental factors affect our health and how human decisions frequently influence environmental health. We will explore what happens when humans are exposed to **environmental hazards**: biological, chemical, and physical factors in the environment that threaten the health of humans and

(a)

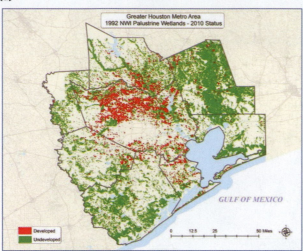

(b)

FIGURE 17.1 Human Factors in Natural Disasters (a) Extensive wetland development in Houston has reduced the land's ability to absorb floodwaters, which increased the damage caused by Hurricane Harvey. (b) Poor infrastructure maintenance and delayed response times slowed recovery from Hurricane Maria in Puerto Rico.

FIGURE 17.2 Protesting Environmental Injustice Marginalized communities in Houston have a long history of protests against hazardous facilities in their neighborhoods—such as this 1979 march against a landfill in Northwood Manor, a predominantly Black middle-class neighborhood.

environmental hazards biological, chemical, and physical factors in the environment that threaten the health of humans and living things.

(a)

(b)

(c)

FIGURE 17.3 Interrelated Environmental Threats Hurricane Harvey exposed the city of Houston not only to **(a)** physical damage from record flooding but also to **(b)** chemical hazards from industrial sites damaged by the storm (such as the Arkema chemical plant, which exploded), as well as biological hazards from raw sewage that contained disease-causing bacteria. **(c)** One year after Hurricane Maria ripped the roofs off their homes, tens of thousands of residents in Puerto Rico were still living under the blue tarps installed as a temporary fix by the Federal Emergency Management Agency.

living things. In doing so, we will pay close attention to the concept of environmental justice—the way that some hazards affect certain groups more than others and the range of ways that we can address environmental health challenges. We will also consider how climate change may affect issues related to environmental health in the future and how we can meet these changes as they arise.

17.1 How Are Environmental Benefits and Burdens Distributed?

In the fall of 2021, while the COVID-19 pandemic was affecting everyone in the United States, hospitalization and death rates for people of color were significantly higher than those for white people. For Sacoby Wilson (**FIGURE 17.4**), an environmental health scientist at the University of Maryland, this was a trend he had seen before. Researchers like Wilson and his team study the way that hazards in the environment accumulate and affect human well-being—often leading to increased vulnerability to disease. Wilson and his colleagues had previously studied cancer risk disparities between different communities in South Carolina. The research team found that the risk disparities were closely associated with the prevalence of hazardous air pollutants measured by the National Air Toxics Assessment of the Environmental Protection Agency (EPA). Because polluting facilities and busy roadways in South Carolina were more often located in or near communities with predominantly non-white populations, the overall cancer risk for communities of color in the state was much higher.

FIGURE 17.4 Sacoby Wilson An environmental health scientist at the University of Maryland, Dr. Wilson studies the links between environmental hazards, vulnerability to health problems, and racial disparities.

During the pandemic Wilson highlighted a similar connection. He felt that environmental factors associated with where people lived were likely affecting their vulnerability to COVID-19. In fact, statistics have shown that people experiencing higher levels of air pollution also had a higher COVID-19 death rate. These people were disproportionately from communities of color where higher pollution rates have long been a fact of life. What Wilson saw was that COVID-19 was following the trend that links environmental harms and disease to racial disparities.

FIGURE 17.5 Uranium Mine Hazards Hundreds of Navajo uranium miners contracted lung cancer because they were not adequately warned of or protected from exposure to radon gas.

environmental justice
the principle that no community should bear more environmental burdens or enjoy fewer environmental benefits than others.

This relationship is at the center of the idea of environmental justice. As we learned in Chapter 1, **environmental justice** centers on the idea that any given community should not bear more environmental burdens or have access to fewer environmental benefits than any other community. Ultimately, these environmental burdens and benefits shape human health and well-being. Environmental injustice occurs when some groups experience more burdens or benefits than others.

Problems in this area have been well documented. In 1983, a General Accounting Office study found that 75% of hazardous waste sites in eight southwestern states were located in low-income areas and communities of color. Further, a 1987 landmark report titled *Toxic Wastes and Race in the United States* found that race was the most significant variable associated with the location of hazardous waste sites. It also concluded that siting these facilities in communities of color was the deliberate result of local, state, and federal land-use policies. Workplace exposures to environmental hazards also raise justice concerns. For decades, Navajo uranium mine workers in the southwestern United States faced unsafe working conditions. Despite research establishing a link between exposure to radon gas in uranium mines and lung cancer, miners were not informed of these risks, nor were proven health and safety measures taken to protect them. More than 600 Navajo miners died from cancer linked to this workplace exposure. It was not until 1990, after decades of advocacy by the Navajo

Nation, that Congress passed the Radiation Exposure Compensation Act to provide compensation to the miners and their families (**FIGURE 17.5**).

In more recent years, the American Lung Association (ALA) has shown that proximity of a community to highways, industrial facilities, or other sources of air pollution is linked to an elevated incidence of asthma, other types of respiratory disease, heart disease, and certain kinds of cancer. In 2021, the ALA found that although air pollution in the United States had decreased in general over several decades, people of color still had higher exposures to the most harmful air pollutants than did white people across all 50 states and Washington, DC, even when researchers accounted for differences in income.

Similarly, the federal government supports the identification and replacement of lead water lines because lead exposure is known to increase the risk of damage to the brain, kidneys, and blood. An analysis of the EPA's lead service line inventory in Illinois—the US state with the most reported lead lines—found that people of color are twice as likely as white people in the state to live in communities with the highest exposure to this health hazard.

Further, many issues of environmental injustice are tied to discriminatory policies. Redlining (see Chapter 16) has driven racial and ethnic disparities in community proximity to sources of air pollution, hazardous waste facilities, and potentially harmful industrial sites, as well as to parks, greenspaces, and even tree cover in major US cities (**FIGURE 17.6**). Redlined communities often face cumulative impacts as individuals interact with multiple environmental health hazards where they live and work. The risks of multiple sources of air pollution near a residential community, for example, might be compounded by industrial sources of air pollution where many of the residents work.

Responding to Issues of Environmental Justice

In 1991, more than 300 leaders representing Black, Latino, Native American, and Asian American communities convened in Washington, DC, to draft 17 principles of environmental justice. These principles included the right to protection from hazardous waste and other forms of pollution to the air, land, and water; the right to a safe work environment; and the right to participate in environmental decision making. They expanded the range of concerns that had animated the environmental movement up to that point, extending it beyond simply conservation and wilderness preservation to equal protection for the places people live, work, and play. At the 1992 United Nations Rio Earth Summit, the environmental justice concept was applied to disparities among high- and low-income nations as well.

That same year the EPA established the Office of Environmental Justice and guidelines clarifying two important aspects of environmental justice for all people "regardless of race, color, national origin, or income."

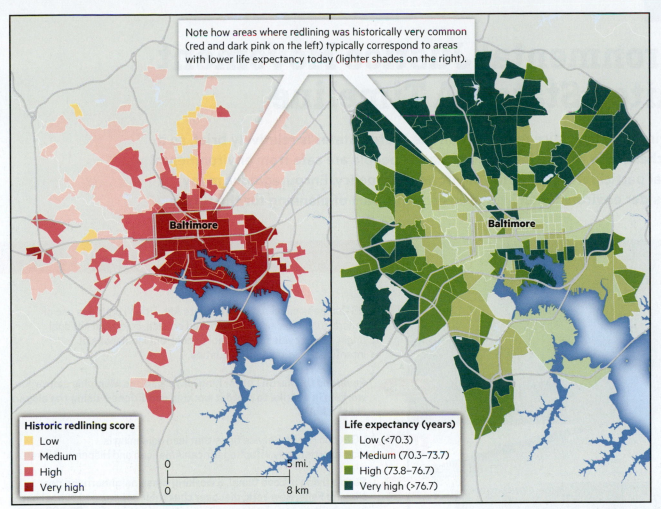

Note how areas where redlining was historically very common (red and dark pink on the left) typically correspond to areas with lower life expectancy today (lighter shades on the right).

Historic redlining score
- Low
- Medium
- High
- Very high

Life expectancy (years)
- Low (<70.3)
- Medium (70.3–73.7)
- High (73.8–76.7)
- Very high (>76.7)

0 — 5 mi.
0 — 8 km

FIGURE 17.6 Life Expectancy and Redlined Areas in Baltimore, MD The effects of redlining—discriminatory policies excluding people of color from desirable neighborhoods—are far-reaching. Decades after redlining was banned in Baltimore, life expectancy in the most heavily redlined neighborhoods remains significantly lower than in other parts of the city.

Adapted from National Community Reinvestment Coalition (2020).

The first aspect is "fair treatment," which the EPA says will be achieved "when everyone enjoys the same protection from environmental and health hazards." The EPA also emphasized the importance of "meaningful involvement," which means that everyone should have "an opportunity to participate in decisions about activities that may affect their environment and/or health." Marginalized communities often lack the resources and political power to fight for environmental benefits and to fight against unwanted environmental hazards.

One important environmental justice tool that can empower community participation is the ability to document and map environmental concerns. Sacoby Wilson, the environmental health scientist introduced earlier in this section, has helped develop and refine public participatory geographic information systems that help residents learn and share information about environmental hazards and health disparities with mapping software. In Charleston, South Carolina, the program is called EJ Radar, and it is designed to empower residents to become more engaged in environmental decisions affecting their communities.

The federal government and many states now include environmental justice in their long-term planning. New Jersey and Washington State now require environmental justice analysis as part of the permitting process for private and public projects. A 2020 law in Connecticut also requires an "environmental justice public participation plan" to ensure meaningful involvement of affected communities. California and Michigan each have mechanisms to direct revenue gained from environmental taxes, fees, or penalties toward communities that have faced environmental justice challenges. The federal government requires agencies to consider environmental justice impacts, and infrastructure funding plans have directed spending for the replacement of lead service lines and the cleanup of hazardous waste sites in overburdened communities. In 2022, the Department of Justice established a new Office of Environmental Justice to enforce environmental justice plans and protect underserved communities.

⬡ **TAKE-HOME MESSAGE** Environmental burdens and benefits shape human health and well-being and are not equally distributed across society. Often disparities in exposure to environmental burdens and access to environmental benefits are rooted in discriminatory policies. Environmental justice promotes the fair treatment and meaningful involvement of all people, regardless of background, in developing, implementing, and enforcing environmental laws, regulations, and policies.

The Environmental Justice Movement in the United States: A Timeline

The history of the environmental justice movement in the United States is relatively brief, so it is not surprising that challenges remain in ensuring that all people are safe from environmental hazards and have equal participation in setting environmental policy. Encouragingly, many governments now consider environmental justice an essential part of planning for the future.

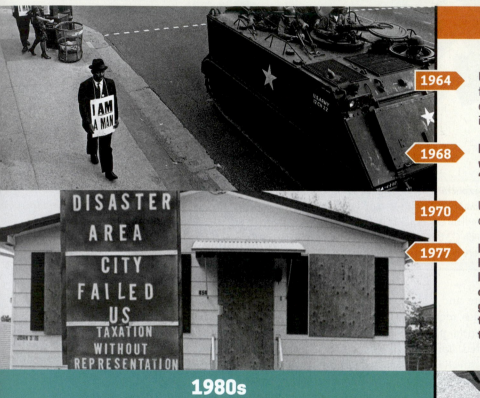

1960s and 1970s

1964 US Congress passes the Civil Rights Act. Title VI prohibits use of federal funds to discriminate based on race, color, and national origin. This provides a legal tool to confront incidences of intentional environmental injustice.

1968 Reverend Dr. Martin Luther King, Jr., organizes Memphis sanitation workers in a strike to protest working conditions—using the slogan "I AM A MAN."

1970 US Public Health Service finds that lead poisoning is disproportionately affecting African American and Hispanic children.

1977 People living in Love Canal, a working-class neighborhood near Niagara Falls, New York, discover that their houses and school had been built on top of a dump site containing more than 20,000 tons of hazardous waste from area chemical factories. Protests lead the government to fund the relocation of more than 800 families from the area and inspire a new federal law known as "Superfund" to clean up contaminated sites.

1980s

Residents of Warren County, North Carolina, protest the siting of a hazardous waste landfill in their community using civil disobedience tactics. The protests are widely seen as the catalyst for the environmental justice movement.

1982

The US General Accounting Office publishes Siting of Hazardous Waste Landfills and Their Correlation with Racial and Economic Status of Surrounding Communities, which finds that 75% of the commercial hazardous waste facilities in the southeastern United States are located in predominantly African American communities.

1983

West Harlem Environmental Action is formed. New York's first environmental justice group protests pollution from the North River Sewage Treatment Plant. In response, the city provides pollution control upgrades to the plant and funds for community amenities.

1988

1990s

1990
The Indigenous Environmental Network is founded to protect the sacred sites, land, water, air, and other natural resources of indigenous communities.

1991
The first National People of Color Environmental Leadership Summit is held in Washington, DC, outlining 17 principles of environmental justice.

1994
President Bill Clinton signs Executive Order 12898, Federal Actions to Address Environmental Justice in Minority Populations and Low-Income Populations. This provides administrative tools for confronting agency decisions that lead to disparate environmental impacts for these populations and funds to remedy them.

1999
America's Parks, America's People Conference is held in San Francisco, California, to draw attention to access to parks and open space in the United States for minority and low-income populations.

2000s to 2020s

Warren County, North Carolina, receives state and federal money to remediate the hazardous waste site located there. Cleanup is completed in 2003.
2001

EJSCREEN, an environmental justice mapping and analysis tool, is released by the EPA to aid communities and decision makers in identifying and addressing environmental justice concerns.
2015

EPA releases EJ 2020 Action Agenda, a strategic plan to improve the health and environment of overburdened communities and demonstrate progress on environmental justice challenges.
2016

Plans to build the Dakota Access oil pipeline less than a mile from the Standing Rock Sioux Reservation catalyze the "Water is Life" movement of Indigenous peoples acting to protect their ancestral lands and waters. Their struggle inspires opposition to other pipeline projects throughout the country.
2016

The Union Hill, Virgina community successfully halts the Atlantic Coast Pipeline Project, a proposed 600-mile natural gas pipeline with a massive compressor station sited less than a mile from the predominantly Black community. The 4th Circuit Court of appeals found that the environmental justice analysis was lacking.
2020

toxin a poisonous substance that can cause illness or death.

environmental health the assessment and control of the biological, chemical, and physical factors in the environment that affect our well-being.

epidemiology the study of the distribution and determinants of health-related states and events in specific populations.

toxicology the study of the negative health effects of substances on an organism.

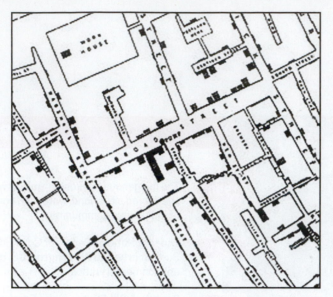

FIGURE 17.7 Epidemiologic Troubleshooting In a famous example of epidemiologic investigation, physician John Snow in the mid-1800s carefully mapped the incidence of cholera in London neighborhoods and discovered that disease outbreaks were spread through contaminated water supplies. This portion of the map focuses on the contaminated Broad Street water pump, which is in the center of the picture. The black bars represent incidences of cholera.

THE ART OF HAT-MAKING.

Engrav'd for the Universal Magazine according to Act of Parliament 1750 for IHinton at the Kings Arms in S.t Pauls Church Yard London.

FIGURE 17.8 Mad as a Hatter Toxicologists working in the 19th century discovered that mercury compounds used to condition animal fur in hat making caused neurologic problems for workers, which led to the phrase "mad as a hatter." This engraving shows workers making hats in a London factory.

17.2 What Is the Study of Environmental Health?

More than 2,400 years ago, the Greek philosopher Hippocrates, known as the father of medicine, recognized the poisonous properties of metals such as lead and sulfur. He understood that interacting with these materials, which we have come to call **toxins**, could make people sick. From this beginning, we now have the field of **environmental health**, one the World Health Organization defines as the assessment and control of the biological, chemical, and physical factors that affect our well-being. It has three main branches: epidemiology, toxicology, and public health.

Epidemiology

In the mid–1800s, cholera ran rampant through London. Cholera is a diarrheal disease that can be fatal if untreated, and in the 19th century several pandemics (global outbreaks) of the disease killed more than 30 million people. Cholera outbreaks were widely believed to be caused by *miasmas*, or poisonous vapors in the air. However, in 1854, British doctor John Snow carefully mapped the incidence of cholera in the Broad Street neighborhood of London and found that incidences were much more prevalent near certain public pumps that households used for drinking water (**FIGURE 17.7**). Although every pump used water from the Thames River, those associated with cholera clusters were contaminated by sewage, while others drew water from cleaner intakes. Snow's discovery initiated a movement for sanitation and clean drinking water that helped reduce the environmental hazard. This launched the field of **epidemiology**: the who, what, when, and where of diseases in populations, or per the Centers for Disease Control and Prevention (CDC), the study of what causes diseases, how they occur and are distributed, and the overall health of specific populations.

Toxicology

A second branch of the environmental health field, **toxicology**, studies the negative health effects of substances on an organism. One classic case gave rise to the phrase "mad as a hatter," popularized by the Mad Hatter character in Lewis Carroll's 1865 book *Alice's Adventures in Wonderland*. At the time, many hatmakers in Europe and the United States suffered from erethism, a neurologic disorder with symptoms including tremors, delirium, personality changes, and memory loss (**FIGURE 17.8**). During the 1860s, scientists in both France and the United States linked the condition to a mercury compound used to treat animal fur in the hat-making process. Although some European countries adopted policies to protect hatmakers from mercury exposure, the United States did not take such measures until 1941.

Alice Hamilton

No one embodied the application of epidemiology and toxicology in pursuit of public health more than Alice Hamilton, a medical and environmental justice pioneer of the early 20th century. After graduating from medical school at the University of Michigan in 1893, she traveled to Germany for training in pathology: the study of the causes and effects of diseases. Women were not admitted to German universities at the time, so she attended classes informally on the understanding that she remain "invisible" to the male students. She put her training to use in Chicago, where she taught

at the Northwestern University medical school and in 1897 joined the settlement movement at Hull House. The settlement movement was an effort to bring college-educated professionals into the poorest, most ethnically diverse areas of cities to provide social services such as education and health care. Hamilton initially used her expertise to provide a clinic for infants and mothers, but she was soon called to address a typhoid epidemic in the neighborhood. Her epidemiologic research eventually found that the bacterial disease was being spread by broken water mains that had been contaminated with sewage, a public services breakdown that the Chicago Board of Health had tried to conceal.

But Hamilton also observed that many of the ailments affecting immigrant populations in Chicago were due to workplace hazards, so she began to study industrial toxins. She researched hospital records to identify incidences of illness, interviewed workers to understand their symptoms and jobs, and visited factories to learn about industrial processes and chemical applications. Her findings of high mortality rates associated with the use of lead convinced her to identify trades and industrial processes that put workers at risk of exposure to this toxin. She documented more than 70 industrial processes that were causing lead poisoning. This included notoriously toxic trades such as paint and ceramic manufacturing

but also lesser-known cases of exposure, such as of workers who rolled products in "tin foil" that was actually made with lead. Hamilton worked to convince factory owners to find alternative materials, install lead-dust-protection devices, hire doctors to inspect factory processes, and provide medical examinations of workers.

In 1910, she was hired by the state of Illinois to survey industrial-related illnesses in the state. Her report documenting the connection between various occupations, diseases, and certain toxins led the state to pass the country's first comprehensive workplace safety regulations in 1911. In subsequent years, the US Department of Commerce asked her to conduct studies of a wide range of toxins in the workplace, including mercury, benzene, and carbon monoxide. By 1919, she was the country's foremost authority on industrial toxicology, and she was invited to join the faculty of Harvard Medical School. There she was the first female professor in a program that still had yet to allow women to enroll as students. In 1925, she published *Industrial Poisons in the United States*, the first textbook on toxicology in the workplace. Hamilton continued to study workplace hazards and push for government regulation until she died at the age of 101 in 1970—just 3 months before Congress passed the landmark Occupational Safety and Health Act. This act sought "to assure so far as possible every working man and woman in the Nation safe and healthful working conditions."

Alice Hamilton was a pioneering doctor and environmental health professional in the early 20th century. She identified workplace hazards in numerous industrial facilities and partnered with state agencies to establish some of the first environmental health regulations in the United States for what she called "the dangerous trades." One of these hazards was the luminescent paint used by watch-dial painters (pictured), which could cause radiation poisoning.

What Would You Do?

Alice Hamilton's work often took her into factories where owners both denied that the workplace was causing health problems and argued that the health problems of their workers were not their responsibility. Yet prior to health and safety regulations, Hamilton often relied on owners to make voluntary changes to reduce the risks to their workers. If it was your job to convince factory owners to voluntarily improve workplace conditions, how would you make the case to them?

FIGURE 17.9 Public Health Initiatives Mass vaccination centers established throughout the United States during the COVID-19 pandemic are just one way that public health agencies help protect the well-being of a community.

Public Health

Epidemiologists and toxicologists are two types of trained experts contributing to a team of public health specialists. **Public health**, often administered by a city, county, or state agency, is the science and practice of protecting and improving the health of people where they live and work. In addition to epidemiologists and toxicologists, public health departments might include doctors, nurses, social workers, educators, and outreach specialists. Public health professionals are devoted not only to diagnosing and treating health problems in a community but also to reducing and preventing health problems while improving overall health. This process involves doing research to identify health challenges, crafting educational programs that encourage healthy behaviors, making policy recommendations, delivering preventative treatments, and making safety inspections at workplaces and public facilities. All of this requires careful risk analysis. Because public perceptions of risk often differ from the real and most pressing health risks facing a community, public health experts must help individuals and leaders prioritize actions taken.

The COVID-19 pandemic highlighted the important role that public health agencies play. Steps taken in most communities included creating testing sites, distributing vaccines, and tracing contacts to determine how the disease spread from person to person (**FIGURE 17.9**). Other environmental health responses in recent years have included establishing cooling centers during heat waves and testing drinking water for lead and other contaminants.

🏠 **TAKE-HOME MESSAGE** The study of environmental health seeks to understand and control biological, chemical, and physical factors in the environment that cause health

public health the science and practice of protecting and improving the health of people where they live and work.

pathogen a microorganism that causes illness or infection when it takes up residence in our bodies.

biological hazard an organic substance that poses a threat to the health of living organisms

respiratory infection an infectious disease affecting the lungs and airways.

diarrheal disease a disease caused by pathogens that affect the digestive tract.

problems. It includes the fields of epidemiology, toxicology, and public health. Epidemiologists study the who, what, when, and where of diseases in populations, while toxicologists study the adverse health effects of substances that harm living things. Public health agencies coordinate teams of experts to protect and improve the health of people where they live and work.

17.3 How Do Microorganisms Make Us Sick?

Trillions of microorganisms live on and in our bodies; each of us contains as many or more resident microorganisms as we do human cells. Many of these organisms are beneficial, aiding in digestion, boosting immunity, and helping with other bodily functions. But other microorganisms called **pathogens** cause illness or infections when they take up residence in our bodies. Infectious diseases account for about one in four human deaths worldwide. In environmental health terms, these are considered **biological hazards**.

What Are the Major Types of Biological Hazards?

Whether through personal experience or through outbreaks that often make the news, the major categories of biological hazards are familiar to most of us. **Respiratory infections** that affect the lungs and airways are something we all have experienced, so it's probably not surprising to learn that even before the COVID-19 pandemic they were the leading cause of sickness and death worldwide. These infections spread by pathogens in an infected person's cough or sneeze that can be inhaled by those nearby or picked up through contact with a contaminated surface (such as a doorknob). While most people overcome the most common respiratory diseases—the common cold, flu (influenza), and bronchitis—each causes hundreds of thousands of deaths per year, mainly among the very young and very old. And all three of these common respiratory infections can lead to pneumonia: an inflammation of the lungs that causes their small air sacs to fill with fluid. Pneumonia kills more than 1 million people each year and is the leading infectious cause of death for children (**FIGURE 17.10**). Similarly, tuberculosis, a bacterial infection that primarily affects the lungs, also kills more than 1 million people each year. By June of 2022, the COVID-19 pandemic had directly claimed more than 6 million lives globally.

Diarrheal diseases affect the digestive tract and are the second leading cause of death for children under 5 years of age. The pathogens that cause these diseases are typically transmitted through water used for drinking,

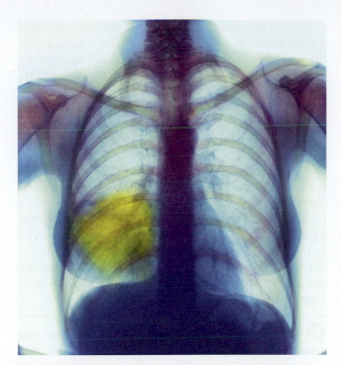

FIGURE 17.10 Pneumonia Pneumonia is a respiratory disease that is a leading cause of death for all age groups globally. It tends to result from a viral or bacterial infection that moves from the nose or throat into the lungs when a person's immune system is weakened from another illness. The yellow area in this colored X-ray highlights the part of the lung affected by pneumonia in this patient.

cooking, or cleaning that is contaminated with human or animal feces. Diarrhea can be fatal if not treated because it causes the loss of fluids, salts, and other nutrients essential to survival.

Many other infectious diseases are caused by **blood-borne pathogens**. For example, HIV is a virus that attacks the immune system and is transmitted person to person by blood and certain bodily fluids through sexual contact or by sharing needles. Those who have AIDS, the final stage of an HIV infection, experience a weakened immune system and are particularly susceptible to other infectious diseases discussed earlier, such as diarrhea and pneumonia. About 1 million people die from HIV/AIDS each year.

While less familiar in developed countries, malaria is a blood-borne disease that is rampant in tropical regions, spreading by bites from female *Anopheles* mosquitoes (**FIGURE 17.11**). Malaria causes fevers, seizures, and organ failure and is responsible for more than 400,000 deaths each year in tropical climates worldwide, where mosquito populations thrive. Other, less common blood-borne diseases include the West Nile virus and the Zika virus (which are both carried by mosquitoes) and Lyme disease (which is carried by ticks). The 2014 outbreak of the Ebola virus in West Africa that killed more than 11,000 people was caused by human contact with the infected raw meat of forest animals such as fruit bats,

monkeys, and porcupines. Once infected, people can spread this disease through sexual transmission or other contact with bodily fluids (**FIGURE 17.12**).

Where Do Pathogens Strike?

Infectious diseases can strike anywhere. As the COVID-19 pandemic revealed, countries worldwide, regardless of income, suffered from this respiratory disease. But other infectious diseases—particularly those affecting young children—are far more prevalent in low-income countries, as **FIGURE 17.13** illustrates. The prevalence of childhood death from diarrhea and pneumonia is often facilitated by crowded, unsanitary conditions and pollution. Today, nearly 800 million people worldwide still lack access to clean drinking water, and more than 2.5 billion people lack access to sanitation systems such as flush toilets that effectively prevent human contact with waste. Air pollution—especially particulate matter—indoors and outdoors increases vulnerability to respiratory diseases including pneumonia. Malnutrition and hunger also make children vulnerable to infection, so providing nutritious food is essential in disease prevention.

The major blood-borne illnesses, malaria and HIV/AIDS, are among the top 10 causes of death in low-income countries. Most fatalities are concentrated on the continent of Africa, where 92% of worldwide deaths from malaria and 70% of deaths from HIV/AIDS occur. Transmission of malaria is greatest in tropical climates where the mosquitoes' life span is longest, and bed nets treated with insecticide and insecticide sprays are

blood-borne pathogen
an infectious microorganism in human blood that can cause disease in humans.

FIGURE 17.11 Mosquito Bites Spread Malaria Malaria is an example of a blood-borne disease that is spread through mosquito bites. Since 2014, global malaria deaths have hovered around 550,000 each year, but in 2020 deaths climbed to over 600,000 due to disruptions to malaria control efforts by the COVID-19 pandemic.

FIGURE 17.12 Ebola Treatment Center, Sierra Leone, 2014 Ebola is an example of a disease that has caused a relatively small number of fatalities but has raised significant concern because it spreads rapidly and is fatal for more than 50% of those who contract it. Outbreaks must be identified and treated quickly to prevent a pandemic—the global spread of a disease.

the most widespread prevention measures in use today. There are also vaccines and preventative drugs that are partially effective. HIV/AIDS deaths can be reduced through testing, education campaigns that encourage safe sex practices (such as the use of condoms), and drugs that suppress viral replication and can reduce the risk of transmitting the virus.

Future Risks

In recent years, improvements in public health, education, and sanitation have led to significant progress in reducing global mortality from infectious diseases. The annual number of deaths from infectious diseases fell by about 30% between 1990 and 2015 even as the world population grew by more than 2 billion. Impressive gains were made in combating malaria, HIV/AIDS, and diarrheal diseases, each of which saw mortality declines of more than 30% over this time. Still, more than 10 million people die from these diseases each year, and developing countries bear a disproportionate share of this burden.

But health professionals see larger risks on the horizon. Pathogens can also develop resistance to the drugs used to treat them. For example, bacteria can develop resistance to antibiotics. As these drugs eliminate the members of a bacterial population that are not resistant, those that have developed immunity are left to survive and multiply, greatly increasing their share of the population. The development of resistant strains of bacteria is hastened by the overprescription of antibiotics for ailments they can't cure, such as the common cold and the flu, which are viral infections. Another cause is the widespread use of antibiotics in the livestock industry (see Chapter 12) to prevent infection and boost animal growth rates. As the effectiveness of antibiotics declines, the risk of death from pneumonia and other bacterial infections will increase.

The global spread of a deadly infectious disease on the scale of the COVID-19 pandemic had long been feared by infectious disease experts. A virus with a high transmission and high mortality rate can spread rapidly in our very interconnected world. And even as testing and vaccines for COVID-19 were developed with historic speed, mutations produced rapidly transmissible variants like Delta and Omicron that outpaced vaccinations.

⬡ **TAKE-HOME MESSAGE** The deadliest biological hazards are pathogens that cause infectious diseases. The most common infectious diseases are respiratory diseases (spread through the air), diarrheal diseases (spread through contaminated water), and blood-borne diseases (spread through person-to-person contact or animals). Low-income countries bear a bigger burden of the world's fatalities from infectious diseases, but most are preventable with improvements in sanitation, access to clean water, and access to medical care.

Region or country	Percentage of deaths under age 5 attributed to diarrhea	Percentage of deaths under age 5 attributed to pneumonia
West and Central Africa	10	17
Sub-Saharan Africa	10	17
Eastern and Southern Africa	10	17
South Asia	10	15
Middle East and North Africa	7	14
East Asia and the Pacific	5	15
Latin America and the Caribbean	4	12
United States	1–4	<5
Canada	<1	<5
Italy, Germany, UK, Australia, Chile	<1	<5

FIGURE 17.13 Disparity in Percentage of Childhood Deaths from Pneumonia and Diarrhea This data shows a disparity among nations in the percentage of childhood deaths from pneumonia and diarrhea in each country. Worldwide, these diseases account for one out of every four deaths of children under 5 years of age. However, more than 90% of deaths from these diseases occur in low-income countries, even though these countries are home to just 60% of the world's population of children younger than 5. These diseases are largely preventable and treatable by addressing water and air pollution, improving sanitation, and increasing access to medical care.

United Nations Children's Fund (2016).

DDT: Trading a Biological Hazard for a Chemical Hazard?

Sometimes addressing a public health challenge involves choosing among a range of treatments that raise separate environmental concerns. Many diseases, including malaria, yellow fever, and Zika, are caused by pathogens that are spread by mosquitoes. In the mid-20th century, the insecticide dichlorodiphenyltrichloroethane (DDT) was used to combat the spread of malaria by controlling the mosquito population in areas affected by the disease. DDT was sprayed from planes over large areas, and it was also sprayed indoors and even applied directly on people's clothing and skin. When used in combination with antimalarial drugs, DDT was a relatively cheap and effective way of combating the disease. Paul Müller, the Swiss scientist who discovered DDT's insecticidal properties, was awarded the Nobel Prize in Physiology or Medicine in 1948. By the 1950s, DDT was the cornerstone of the World Health Organization plan for global eradication of malaria.

However, in 1962, a marine biologist named Rachel Carson published *Silent Spring*, which detailed the harm this pesticide was doing to the environment. This book was incredibly influential and is considered by many to mark the beginning of the environmental movement (see the Stories of Discovery feature in Chapter 20). It became clear that DDT had adverse environmental impacts as it persists in the environment and moves up the food chain, killing not only insects but also birds and other species. In response, the United States banned the use of DDT as did many other countries. Research over the ensuing decades found that DDT accumulates in the fatty tissue of humans as well. There is some evidence that it may disrupt reproductive and hormonal processes, and it has been identified by the International Agency for Research on Cancer as a possible carcinogen. In other words, fighting malaria with DDT represented use of a chemical hazard to combat a

The insecticide dichlorodiphenyltrichloroethane (DDT), which was once widely used to control disease-causing mosquitoes, was later banned because of the adverse environmental effects it had on birds and other species.

biological hazard. Today, malaria remains one of the top 10 causes of death globally, and banning DDT to control malaria is thought to have contributed to a surge in the disease in some areas.

The trade-offs between these two environmental hazards were highlighted in 1995 during negotiations on an international agreement to ban DDT and other pollutants. Arguments against the continued use of DDT included ecological and health concerns, along with the fact that several species of mosquitoes had already become resistant to the chemical. Increasing resistance over time would decrease DDT's effectiveness, while the chemical would remain in the environment for decades. Critics also noted alternative methods to fight mosquito-borne diseases. These included special insecticides that break down in sunlight after 1 or 2 days, civil engineering projects that eliminate pools of water where mosquitoes breed, and construction of well-screened housing to protect people from bites when mosquitoes feed in the evening.

Arguments in favor of using DDT include the assertion that DDT is a relatively cheap and easy-to-use mosquito-control method.

Alternative insecticides can cost as much as 20 times more than DDT and carry their own ecological risks. And historically, reducing mosquito-to-breeding waters has involved draining and filling wetlands (as was done in the southeastern United States)—actions that carry adverse ecological impacts. Further, the research on DDT's toxicity to humans is still uncertain at present, while fatalities from malaria are known and occur at a high rate.

In the end, the 2001 international agreement known as the Stockholm Convention on Persistent Organic Pollutants banned DDT for most uses. However, it did allow its use to control mosquitoes in disease-prone areas when safe, effective, and affordable alternatives are not available.

What Would You Do?

The Stockholm Convention allowed use of DDT to control mosquitoes. Do you think the right decision was made? If you had to vote on this agreement, how would you have voted and why?

Herd Immunity

Public vaccination efforts play a vital role in environmental health. Much of the reduced mortality from infectious diseases has been achieved through public vaccination programs, and broader immunization is an important part of the United Nations' third sustainable development goal (see all the United Nations' sustainable development goals in Chapter 1). Why is this? When vaccination is widespread, even the small number of people who are unable to receive vaccinations—because they are too young or have compromised immune systems—are unlikely to become infected because most of the population is inoculated against the pathogen. This is known as **herd immunity**: when a large enough proportion of the population is immune to a pathogen, it indirectly protects people who are not immune. When herd immunity breaks down, a disease that was once largely suppressed can begin to spread among the nonimmunized population.

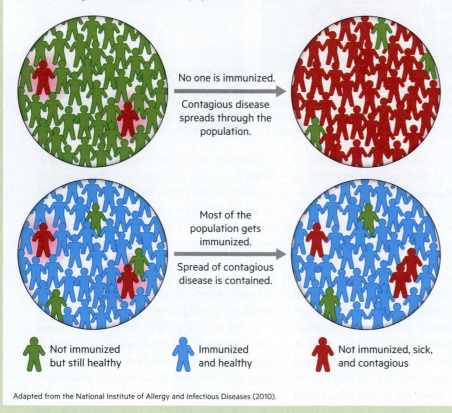

No one is immunized.

Contagious disease spreads through the population.

Most of the population gets immunized.

Spread of contagious disease is contained.

Not immunized but still healthy

Immunized and healthy

Not immunized, sick, and contagious

Adapted from the National Institute of Allergy and Infectious Diseases (2010).

17.4 What Does It Mean When Something Is Toxic?

herd immunity when members of a population are unlikely to become infected by a pathogen because most of the population is either vaccinated or has been previously infected.

chemical hazard a chemical linked to immediate or delayed health effects after exposure

Chemicals make up everything around us, and we ourselves are made up of them. But some chemicals are toxic: they have the capacity to cause injury or death. Known as toxins, these **chemical hazards** can be naturally occurring, like the elements lead and arsenic, or can be compounds that have been synthesized by humans.

Toxins have a variety of harmful effects on the body, depending on the substance and the exposure. **FIGURE 17.14** shows that these effects range from tissue and organ damage to disruption of particular biological functions.

Further, toxins can be *additive* (where the toxic impact is the sum of the effects of interacting chemicals) or *synergistic* (where the toxic impact is greater than the sum of the effects of interacting chemicals). A common example of the synergistic effect occurs when the cleaning products ammonia and bleach are combined. The result is chlorine gas, a toxin that is far more deadly than either ammonia or bleach alone.

Risk Factors

Just because a substance is toxic does not necessarily mean that there is a high risk that the harms associated with it will be realized. In fact, any substance has the potential to be toxic, but a range of factors shapes the likelihood of whether it will cause harm. For example, we typically think of oxygen as essential to life and table salt as common seasoning for our food. While we comfortably breathe air with just over 20% oxygen content, air that is 50% oxygen would kill us. And salt is fine in moderation, but eating a half cup or more a day would eventually cause kidney failure. In other words, the dose of a substance affects its toxicity.

Toxicologists often measure risks in terms of the percent mortality of the exposed population, identifying the dose required to kill half the individuals. This threshold can vary considerably by chemical, from those for dioxin and botulinum toxin (two of the most toxic substances), which are lethal to 50% of the exposed population in extremely tiny doses, to those for salt, vitamin C, and water, which require massive doses to provoke a toxic response.

Toxins do not affect everyone in the same way. A familiar example is the way individuals respond to alcohol consumption, with lighter-weight people typically feeling the effects after fewer drinks than would be the case for heavier people. Genetics is another way the effects of a toxin can vary by individual. Vulnerability to many types of cancer is determined not only by exposure to a toxin that initiates or promotes the growth of cancer cells but also by the genetic susceptibility of individuals to certain types of cell damage. Age, sex, and overall health are other factors that shape an individual's vulnerability to particular toxins. For example, many toxins have a pronounced effect on infants and children because the toxins alter growth and development.

⬡ **TAKE-HOME MESSAGE** Toxic chemicals have the capacity to cause injury or death. Many different factors determine the risk for harm that these chemical hazards possess, including the dose and the characteristics of the individuals exposed to them.

Neurotoxins
Damage the nervous system.
Example: Lead. The presence of lead compounds in paints, gasoline, and pipes used in plumbing has caused developmental disabilities and other neurologic problems in those affected by exposure to this metal.

Corrosive toxins
Damage human tissue when they come into contact with skin, eyes, or the respiratory tract.
Example: Cleaning products such as ammonia, bleach, and acids.

Asphyxiants
Restrict the body's ability to absorb oxygen.
Example: Carbon monoxide, a common air pollutant from the incomplete combustion of fossil fuels, inhibits the ability of hemoglobin in the blood to absorb oxygen.

Carcinogens
Damage cell DNA and initiate or promote uncontrolled growth of tumors.
Example: Cigarette smoke is a carcinogen and the leading cause of lung cancer.

Reprotoxins
Disrupt reproductive functions. This includes chemicals that can cause miscarriages and birth defects as well as those that affect fertility.
Example: Exposures to the metals lead, mercury, arsenic, and cadmium are all linked to miscarriage, low birth weight, and various birth defects.

Hepatotoxins (target liver) and nephrotoxins (target kidneys)
Target the liver or kidneys, organs that process toxins.
Example: Excessive alcohol consumption can cause liver damage over time. Heavy metals such as lead, mercury, and cadmium can cause damage to both the liver and the kidneys.

FIGURE 17.14 The Six Major Types of Toxins and Examples of Their Effects

volatile organic compound (VOC) a gas released through the evaporation or incomplete combustion of fossil fuels and other organic compounds (Chapter 8).

17.5 How Can I Be Exposed to Something Toxic?

Unfortunately, toxic exposure normally happens through the most basic activities of breathing, drinking, and eating. The fastest way for a substance to be absorbed into the body is through inhaling it, and the term *volatility* describes the tendency of a chemical to evaporate in the air. Benzene, formaldehyde, and toluene are examples of toxins known as **volatile organic compounds (VOCs)**, which are used in products such as paints, wood preservatives, and nail polish. VOCs are easily inhaled and absorbed through the lungs (**FIGURE 17.15**).

Another important characteristic is whether a chemical is water soluble or fat soluble. *Water-soluble* substances such as cyanide dissolve easily in water and move quickly into the bloodstream when ingested. There they can have an immediate effect before being filtered through the kidneys and excreted in urine or other fluids. *Fat-soluble* toxins such as DDT and mercury accumulate in the body fat of animals and are released within the body over time as the fat is burned to make energy. But if an organism absorbs the toxin at a faster rate than they metabolize or excrete it, the toxin will build up in the tissues of that organism. When these fats are then eaten by other animals, they accumulate these toxins too. Animals higher up on the food chain absorb the toxins eaten by animals lower in the food chain.

→ **SCIENCE**

The Dose Makes the Poison

This classic principle of toxicology was coined by the alchemist Paracelsus in the 1500s, and its truth is quite evident in the case of botulinum toxin, which is a powerful neurotoxin that is produced by *Clostridium botulinum* bacteria and causes paralysis. Just a tiny amount can be lethal: as little as 1.3 billionths of a gram (1.3 nanograms) per kilogram of body weight, or about 100 nanograms for a 75-kilogram (165-pound) person. However, in even tinier amounts, the toxin (marketed most commonly as Botox) is used to treat muscle spasms and applied in cosmetic procedures to reduce facial wrinkles. The typical cosmetic dose is about 0.15 nanogram (1/667 of the lethal dose), which when injected directly into a facial muscle, leads to smooth skin rather than deadly paralysis.

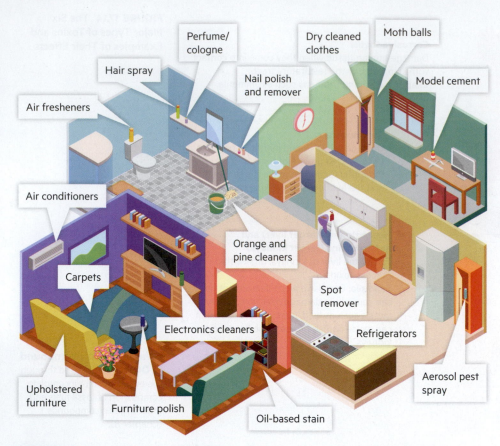

FIGURE 17.15 Volatile Organic Compounds Volatile organic compounds (VOCs) readily evaporate into the air. The EPA has found that they are a common cause of indoor air pollution because they are in many household products. Exposure to VOCs can cause respiratory ailments and headaches, and some VOCs are known carcinogens.

Adapted from New York State Department of Health (n.d.).

Labels in figure: Air fresheners; Hair spray; Perfume/cologne; Nail polish and remover; Dry cleaned clothes; Moth balls; Model cement; Air conditioners; Carpets; Orange and pine cleaners; Spot remover; Electronics cleaners; Refrigerators; Upholstered furniture; Furniture polish; Oil-based stain; Aerosol pest spray

The **persistence** of a toxin, or the extent to which a chemical resists being broken down, affects its risk to human health. The longer it takes for a chemical to degrade from its toxic form, the more available it is for exposure and accumulation. A particularly persistent group of chemicals known as **persistent organic pollutants (POPs)**—DDT is one example (see the Sustainability Matters feature earlier)—resists breakdown through chemical reactions, biological processes, or exposure to sunlight. So although the use of these and many other POPs has been curtailed or banned in the United States and other countries, they remain in the environment because they persist in the soil and can be passed up the food chain as organisms consume one another (**FIGURE 17.16**).

Where Exposure Occurs

Exposure to chemical hazards most commonly occurs in the neighborhoods, homes, schools, and workplaces where people carry out their daily lives. Research has also shown that exposure to toxins is an environmental justice concern. An EPA report summarizing two decades of research found that low-income communities and communities of color are most vulnerable to the health impacts associated with these hazards.

persistence the extent to which a chemical resists being broken down.

persistent organic pollutant (POP) a chemical that is resistant to breakdown through chemical reactions, biological processes, or exposure to sunlight.

Lead exposure provides one example. The most common sources of lead exposure are old paint, pipes, and plumbing fixtures from buildings constructed before 1980 and lead in the soil deposited by industrial sources, such as metal-smelting plants. Studies of more than 1 million children living in Chicago between 1995 and 2013 found that those living in areas with older, dilapidated houses and schools and in proximity to smelting plants tended to have higher lead levels in their blood than those of children living elsewhere. The studies also revealed that the city's Black and Latino neighborhoods tended to be in these areas and exhibited a disproportionate share of blood tests indicating lead poisoning. Similarly, in Flint, Michigan, a predominantly Black city with one-third of its residents living below the poverty line, more than 100,000 people were exposed to high levels of lead when the city switched to a new water source (the Flint River) that corroded lead pipes, releasing lead into the drinking water (see also Chapter 7). It took more than 3 years for lead levels to decrease to the acceptable levels set by the federal government, and replacement of the lead-containing pipes in the city was not completed until late 2021, 7 years after the initial crisis (**FIGURE 17.17**).

Food consumption patterns are important too. In 1982, more than 1,000 residents of the predominantly Black community of Triana, Alabama, were found to have some of the highest levels of DDT ever recorded in humans (more than 40 times the federal limit). Residents relied on subsistence fishing in the Alabama River, where a pesticide manufacturer had been discharging DDT-contaminated wastewater for several decades.

The workplace is another place for exposure, and studies have shown that risks are disproportionately high for lower-wage workers and people of color. In the United States, agricultural workers face unique risks because they often both work and live where toxins from agrochemicals can be inhaled, ingested, and absorbed through the skin. The EPA has found that 10,000–20,000 incidences of pesticide poisoning are diagnosed by doctors each year in the United States among farmworkers and their families. Nail salons, where workers spend long hours in confined indoor spaces with chemicals such as dibutyl phthalate (DBP), toluene, and formaldehyde, are another common problem area. Internationally, the recycling and disposal of electronic equipment discarded from high-income countries is often done in poor communities with few if any safety precautions, exposing the workers and those living in the communities to toxic chemicals (**FIGURE 17.18**).

⬡ **TAKE-HOME MESSAGE** A chemical's health risk depends on how people are exposed to it. Chemical characteristics that influence routes of exposure include volatility, the extent to which a material is soluble in water or fat, and its persistence in the environment. Marginalized populations are commonly exposed disproportionately to certain chemical hazards.

The Most Common POPs (Persistent Organic Pollutants)

POP	Global historical use/source
Aldrin	Insecticides used on crops such as corn and cotton; also used for termite control.
Dieldrin	
Chlordane	Insecticide used on crops, including vegetables, small grains, potatoes, sugarcane, sugar beets, fruits, nuts, citrus, and cotton. Used on home lawn and garden pests. Also used extensively to control termites.
DDT	Insecticide used on agricultural crops, primarily cotton, and insects that carry diseases such as malaria and typhus.
Endrin	Insecticide used on crops such as cotton and grains; also used to control rodents.
Mirex	Insecticide used to combat fire ants, termites, and mealybugs. Also used as a fire retardant in plastics, rubber, and electrical products.
Heptachlor	Insecticide used primarily against soil insects and termites. Also used against some crop pests and to combat malaria.
Hexachlorobenzene	Fungicide used for seed treatment. Also an industrial chemical used to make fireworks, ammunition, synthetic rubber, and other substances. Also unintentionally produced during combustion and the manufacture of certain chemicals. Also an impurity in certain pesticides.
Polychlorinated biphenyls (PCBs)	Used for a variety of industrial processes and purposes, including in electrical transformers and capacitors, as heat-exchange fluids, as paint additives, in carbonless copy paper, and in plastics. Also unintentionally produced during combustion.
Toxaphene	Insecticide used to control pests on crops and livestock, and to kill unwanted fish in lakes.
Dioxins	Unintentionally produced during most forms of combustion, including burning of municipal and medical wastes, backyard burning of trash, and industrial processes. Also can be found as trace contaminants in certain herbicides, wood preservatives, and PCB mixtures.
Furans	

The POPs aldrin, chlordane, dieldrin, toxaphene, and endrin were widely used as insecticides in crop production.

DDT is a POP that was used as an insecticide to control mosquito populations. Its current use is restricted.

The POPs mirex and heptachlor were used as insecticides to control termites, ants, and soil insect populations.

The POP hexachlorobenzene was widely used as a fungicide to protect wheat crops from a fungal blight known as Karnal bunt.

Polychlorinated biphenyls (PCBs) were POPs used as heat-exchange fluids in electrical transformers.

Dioxins and furans are POPs produced by combustion in industrial and other processes.

FIGURE 17.16 Common Persistent Organic Pollutants (POPs) Although certain POPs once used as insecticides or fungicides have had their use curtailed or banned, the toxins persist in the environment. Dioxins and furans, which are primarily the result of combustion in industrial processes, are harder to control, and emission of these toxins remains a concern.

Adapted from US EPA (2017).

FIGURE 17.17 Flint Water Crisis In 2014, the city of Flint, Michigan, switched its water source from the Detroit water supply to the Flint River to save money, which led to elevated lead exposure for its predominantly low-income and minority residents. The residents relied on bottled water for more than 3 years until lead levels decreased to acceptable levels.

17.6 How Do We Manage Risks Associated with Toxins?

We often assess risk based on how much harm a society feels is acceptable. Generally, we use the number of people potentially affected by a toxin to express its potential harm. For example, toxicologists have determined that drinking water contaminated with arsenic at concentrations of 50 parts per billion causes a lifetime cancer risk of 15 per every 1,000 people. If the arsenic concentration is reduced to 10 parts per billion, the lifetime cancer risk drops to 3 per every 1,000 people.

With this understanding, risks are then managed through government regulations, though these can differ

physical hazard an event or phenomenon that causes harm to humans through physical damage.

from one region to the next. In the United States and many other countries, regulatory agencies typically control the approval of new chemical compounds, establish labeling requirements for hazardous products, and set and enforce workplace safety rules. For example, the EPA regulates new chemicals under the Toxic Substances Control Act and sets standards for toxins in the air and drinking water under the Clean Air Act and the Safe Drinking Water Act.

In creating regulations, an important consideration is often where the "burden of proof" lies in determining the safety of a particular substance. Should the burden be on the manufacturer to prove that the use of a particular substance is safe, or should the burden be on a government agency to prove that it is hazardous? The European Union uses what is known as the "precautionary principle" in the way it manages chemicals under a regulation known as REACH, or "Registration, Evaluation, Authorisation and Restriction of Chemicals" (**FIGURE 17.19**). This is widely regarded as the world's most stringent law governing potential toxins. It requires all companies manufacturing or importing chemicals in Europe to register each substance, which includes providing a chemical safety assessment. If the European Chemicals Agency determines that there is credible evidence of danger to human or environmental health, it takes protective actions—such as a ban or limitations on the use of the substance—even if there is continuing scientific uncertainty regarding the harm. This law emphasizes avoiding harms over any benefits that might flow from the use of the substance in question.

In contrast, the United States takes the opposite approach. The burden to conduct safety assessments on a new chemical is the responsibility of the EPA, and the agency must clearly establish that a harm exists within a relatively short amount of time to prevent a substance from going to market. If there is uncertainty, the benefit of the doubt typically goes to the manufacturer of the substance in question. As a result, many chemicals—most notably chemicals such as bisphenol A (BPA) and phthalates—are more stringently regulated in Europe than in the United States.

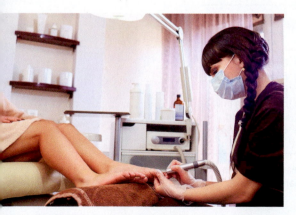

FIGURE 17.18 Environmental Injustice in the Workplace When particular populations are concentrated in certain jobs, workplace hazards can be a vehicle for disproportionate impacts on environmental health.

Which standard should be used to regulate drinking water? In 2000, both the European Union and the United States established rules to reduce the allowable arsenic in drinking water from 50 to 10 parts per billion. However, in 2001, George W. Bush's incoming administration suspended this rule, arguing that the cost of installing improved water-treatment systems to remove arsenic might not be worth the decreased risk of cancer. In response to public pressure, the administration eventually reversed course and lowered the standard to 10 parts per billion.

Addressing Disproportionate Impacts

In 1994, President Clinton issued Executive Order 12898, which directed federal agencies to identify and address "disproportionately high and adverse human health or environmental effects" on marginalized communities. Although this order does not establish new rights or avenues of redress for affected communities, it was the first act to consider impacts related to environmental justice concerns. Increasingly, efforts in environmental health are trying to address global issues of environmental justice. For example, a 2004 international agreement on organic pollutants included financial support to lower-income countries for cleanup and elimination of these toxic materials. Similarly, the Minamata Convention on Mercury manages the mining of this metal as well as its use in products, its emissions from various industries, and its disposal, mandating that lower-income nations do not bear the disproportionate burden of disposal or pollution.

⬡ **TAKE-HOME MESSAGE** Regulating chemical hazards engages the science of toxicology as well as value determinations regarding the acceptable level of risk. The precautionary principle is the idea that protective measures ought to be taken to prevent harm even when scientific certainty regarding the harm is lacking.

17.7 What Are the Most Common Physical Hazards in the Environment?

Each year in the United States, tens of thousands of wildfires burn through shrublands and forests, consuming millions of acres. The 2017 wildfires that swept through Napa and Sonoma Counties in Northern California destroyed more than 9,000 structures, killed 47 people (making it one of the deadliest fires in state history), and caused more than $12 billion in damage. Smoke from the fires caused the worst air pollution ever recorded

in the San Francisco Bay Area, producing as much particulate matter pollution in 2 days as all the vehicles in the state produce over a year. Although wildfires can be naturally ignited by lightning, a 2016 study at the University of Colorado's Earth Lab using decades of US government records found that humans ignited 84% of the wildfires studied and were responsible for more than half of the area burned.

Wildfires are one example of a **physical hazard**, which is an event or phenomenon that causes harm to humans through physical damage. These include disastrous events—such as tornadoes, hurricanes, floods, and other extreme weather events (covered more in Chapters 8 and 11), as well as earthquakes, tsunamis, landslides, and other geologic hazards. Many physical hazards fall under the category of "natural disasters" that can cause great damage. But increasingly, humans are playing a role in these hazards, as human activities increase their likelihood and the extent of the damage inflicted on affected communities. It is also important to note that physical hazards can also affect human health indirectly, by setting the stage for other types of hazards. For example, physical hazards such as earthquakes, tsunamis, hurricanes, and floods often damage infrastructure in ways that lead to

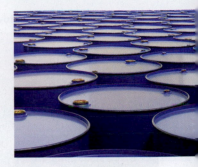

FIGURE 17.19 Regulating Toxins in the European Union The European Union takes a precautionary approach to approving new chemicals in the marketplace. The Registration, Evaluation, Authorisation and Restriction of Chemicals (REACH) program requires manufacturers to submit a chemical safety assessment for review before new products are cleared for sale.

⮕ **CONTEXT**

Dumping Toxic Burdens on Developing Countries

In 1984, an accident at a chemical plant in Bhopal, India, exposed more than 500,000 nearby residents to poisonous gas. Government officials in India found that more than 2,200 people died immediately as a result of the exposure, and more than 10,000 people died from the exposure in subsequent weeks, months, and years. Many more people were sickened, and the area remains contaminated to this day. The plant was operated by the US company Union Carbide to produce pesticides. The company agreed to accept moral responsibility for the disaster and pay $554 million in compensation to the survivors at an average payout of $2,200 per person.

(a) Landslide Outside Freetown, Sierra Leone, 2017

(b) Buildings Destroyed by Earthquake, Port au Prince, Haiti, 2010

(c) Tsunami Damage near Minamisanriku, Japan, 2011

(d) Flowing Lava from Kilauea Volcano, Hawaii, 2018

FIGURE 17.20 **Geologic Hazards** Although geologic hazards are caused by movement in Earth's crust, land-use decisions at the surface play a role in determining the scope of the harm human populations suffer from these hazards.

contaminated water supplies or the release of hazardous materials into the environment.

Geologic hazards are large-scale events that can cause tremendous damage (**FIGURE 17.20**). **Landslides**, for example, occur when rock or other debris detaches from a slope and slides downhill. Ultimately, landslides are caused by sloping areas becoming unstable, and human activities such as deforestation and mining can destabilize slopes. In 2017 a landslide in Freetown, Sierra Leone, killed more than 1,000 people and destroyed 2,000 homes in its path (**FIGURE 17.20A**). Although extreme rains were the immediate cause of the slide, experts also blamed the tragedy on deforestation and the construction of substandard housing on dangerous slopes. And forest fires, such as those in California mentioned earlier, can leave slopes bare and vulnerable to slides should heavy rains occur.

Earthquakes are a powerful geologic hazard typically caused when parts of Earth's crust shift along faults (see Chapter 9). The impact of an earthquake is shaped not only by its strength and location but also by the way in which structures are built and by the emergency response to the event. In 2010, more than 200,000 people died as a result of an earthquake in Haiti, making it one of the deadliest in human history (**FIGURE 17.20B**). Most of the fatalities were attributed to the lack of building codes and substandard housing in the country. Many structures were built without foundations on steep, deforested slopes that were prone to collapse. After the quake and its aftershocks, additional hazards such as fire and lack of access to clean water and medical care compounded the impact. In 2015, a powerful earthquake in Nepal killed nearly 9,000 people and upended the lives of more than 8 million people who scrambled to find clean water, food, shelter, and medical

geologic hazard a large-scale event that can cause tremendous damage.

landslide an event that occurs when rock or other debris detaches from a slope and slides downhill.

earthquake a powerful geologic hazard typically caused when parts of Earth's crust shift along faults.

care. Political conflicts and poor management slowed recovery efforts so much that 3 years after the earthquake, only half of the disaster victims had been able to rebuild their homes.

Earthquakes can also trigger **tsunamis**, powerful waves that can flood coastal areas, destroying structures and unleashing additional hazards. The meltdown of the Fukushima nuclear power plant was caused by a 40-foot tsunami that breached the 30-foot seawall protecting the plant, flooding the facility and disabling emergency backup generators, ultimately leading to a loss of cooling for the nuclear reactors. That tsunami caused tremendous damage along 1,242 miles (2,000 km) of Japan's Pacific coast (**FIGURE 17.20C**).

Volcanoes can also pose physical hazards from lava, hot mud flows (called lahars), and ash. Lava flows from Kilauea on the Big Island of Hawaii have engulfed entire neighborhoods (**FIGURE 17.20D**).

Radiation Hazards

The nuclear disaster in Japan triggered by the tsunami exposed the population to radiation—a different type of physical hazard that inflicts damage at the cellular level. **Radiation** is a form of energy that travels through space and penetrates various materials. It is all around us, and everyday examples include the light we see and the microwaves we use to heat up food. But two kinds of radiation pose particular hazards to human health: ionizing radiation and ultraviolet radiation.

Ionizing radiation can remove electrons from an atom, and when it penetrates and reacts with our bodies, it can damage cells. A large, short-lived dose of ionizing radiation can cause immediate, widespread tissue damage, leading to acute radiation sickness. Chronic, long-term exposure to smaller doses of ionizing radiation can lead to cancer. Most of us receive a dose of this type of radiation every year. Natural sources such as cosmic rays and radioactive elements in Earth's crust account for more than 50% of the annual dose people typically receive. Medical use of ionizing radiation, including X-rays and radiation therapy for cancer, is the largest nonnatural source. These radiation doses are an obvious trade-off for the life-saving and health benefits that come with medical treatment, and physical shielding greatly reduces the risk of exposure to patients and medical staff.

But other human-made sources of radiation exposure are far more detrimental. Creating and using nuclear power is one example. The two most serious accidents at nuclear power plants, rated at level 7 (the most dangerous type of event on the International Nuclear Event Scale), occurred at Chernobyl, Ukraine, and Fukushima, Japan (see Chapter 14). More than two dozen first responders died of acute radiation sickness at Chernobyl. The number of cancer deaths caused by long-term radiation exposure is projected to reach 4,000 at Chernobyl and up to 1,500 at Fukushima.

Though less common today than during the heyday of nuclear weapons development in the 1950s and 1960s, the production and obviously the use of nuclear weapons is another example. When these weapons are tested aboveground, they release large amounts of radioactive materials into the atmosphere. Producing these weapons can also lead to radioactive contamination in and around the facilities where they are made (**FIGURE 17.21**).

A more common concern for most people is **radon**, a naturally occurring radioactive gas found in rock, soil, and groundwater. Radon gas is odorless and colorless, and it can seep into the basements of buildings. As discussed earlier in the chapter, radon caused cancer in Navajo uranium miners, and in the general population the US surgeon general has found that radon is second only to smoking as a cause of lung cancer. While the risk of lung cancer from radon is far lower than from smoking, especially for nonsmokers, we can avoid radon exposure by installing systems that vent the gas to the outside so that it does not get trapped and accumulate in indoor air. According to the EPA, about 1 out of every 15 homes in the United States has elevated radon levels, and radon gas causes several thousand deaths each year. Kits are available to test for it in the home.

While emergencies such as nuclear power plant meltdowns highlight the deadliness of ionizing radiation, far more fatalities are caused by **ultraviolet (UV) radiation**. UV radiation is composed of invisible rays that come from the Sun. Two forms of it, known as UVA and UVB, reach Earth's surface and are the leading causes of skin cancer. The amount of UV exposure

tsunami a powerful wave that can flood coastal areas.

radiation a form of energy that travels through space and penetrates various materials.

ionizing radiation a form of energy that can remove electrons from an atom.

radon a naturally occurring radioactive gas found in rock, soil, and groundwater.

ultraviolet (UV) radiation invisible rays that are part of the energy that comes from the Sun. Ultraviolet radiation that reaches Earth's surface is made up of two types of rays, called UVA and UVB.

FIGURE 17.21 Radioactive Contamination Between 1952 and 1992, the Rocky Flats Plant near Denver, Colorado, produced plutonium triggers for hydrogen bombs. The FBI and EPA raided the plant in 1989 because of environmental health and safety violations. The site was decommissioned and demolished, and cleanup of residual plutonium at the site continued until 2005. Today, it is the protected Rocky Flats National Wildlife Refuge, home to many native prairie species.

flood an event that occurs when water inundates land that is normally dry, especially through intense rainfall and rapid snowmelt.

a person receives is dependent on many factors, such as the time of day, the time of year, and location on the planet (**FIGURE 17.22**). Risk factors that increase the incidence of skin cancer include cumulative time spent in the Sun without the protection of clothing or sunscreen, having a fair complexion, the number of sunburns experienced, and residing closer to the equator where the Sun's rays are more direct and intense. But the thinning of Earth's ozone layer due to human-made, ozone-depleting substances (see Chapter 8) has also made humans more vulnerable to UV radiation. This is particularly true in the Southern Hemisphere, where thinning of the ozone layer has been very pronounced. The World Health Organization estimates that more than 65,000 people die each year from melanoma, the deadliest form of skin cancer (**FIGURE 17.23**).

⬡ **TAKE-HOME MESSAGE** Physical environmental hazards include wildfires, geologic events, radiation hazards, extreme weather events, and harm caused by the built environment. Although many of these hazards are considered natural disasters, human actions often serve as a direct or background condition shaping the harms they cause.

Skin Cancer Rate

Rank	Country	Age-standardized rate per 100,000
1	Australia	33.6
2	New Zealand	33.3
3	Norway	29.6
4	Denmark	27.6
5	Netherlands	25.7
6	Sweden	24.7
7	Germany	21.6
8	Switzerland	21.3
9	Belgium	19.9
10	Slovenia	18.6
11	Luxembourg	16.5
12	Ireland	16.3
13	Finland	15.8
14	United Kingdom	15.0
15	Austria	13.6
16	France (metropolitan)	13.6
17	United States	12.7
18	Czech Republic	12.6
19	Canada	12.4
20	Italy	12.4

FIGURE 17.23 Cancer Risk and a Thinning Ozone Layer The highest rates of skin cancer occur in Australia and New Zealand, which are located in the Southern Hemisphere. In these countries, a thinning ozone layer reduces atmospheric protection from UV radiation, and the fair complexions of large shares of both populations predispose them to Sun damage.

Adapted from World Cancer Research Fund International (n.d.).

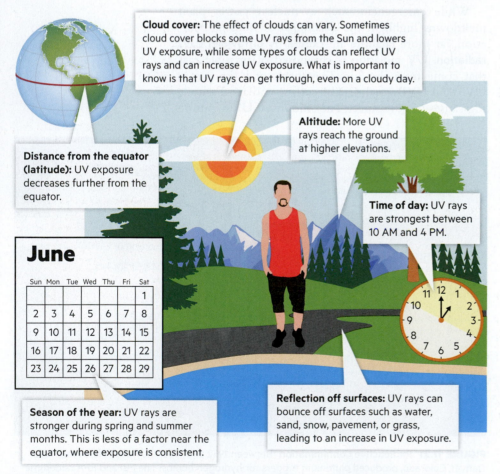

Cloud cover: The effect of clouds can vary. Sometimes cloud cover blocks some UV rays from the Sun and lowers UV exposure, while some types of clouds can reflect UV rays and can increase UV exposure. What is important to know is that UV rays can get through, even on a cloudy day.

Altitude: More UV rays reach the ground at higher elevations.

Distance from the equator (latitude): UV exposure decreases further from the equator.

Time of day: UV rays are strongest between 10 AM and 4 PM.

Season of the year: UV rays are stronger during spring and summer months. This is less of a factor near the equator, where exposure is consistent.

Reflection off surfaces: UV rays can bounce off surfaces such as water, sand, snow, pavement, or grass, leading to an increase in UV exposure.

FIGURE 17.22 Reducing Exposure to Ultraviolet Radiation An individual's UV exposure depends on many variables.

Adapted from American Cancer Society (2017).

17.8 How Is Climate Change Affecting Environmental Health Issues?

Increasingly, scientists are focusing on climate change as a threat to our health. As we saw in Chapter 11, scientists are gradually accumulating evidence to attribute the increased intensity of extreme weather events to climate change. Warming temperatures are also extending the ranges of disease-carrying organisms, such as ticks and mosquitoes, exposing more people to these diseases.

Extreme Weather

Extreme weather events include individual events such as hurricanes, as well as persistent trends that last months or even years such as extreme droughts. But while images of neighborhoods flattened by tornadoes and hurricanes make us aware of the destructive force of high winds, on average, floods and heat waves result in more fatalities. **Floods** occur when water inundates land that is normally dry, especially through intense rainfall and rapid snowmelt. In coastal areas, **storm surges** can also cause flooding as high winds push extremely large waves of seawater ashore (**FIGURE 17.24**). Hurricanes can bring both storm surges and intense precipitation to an area at the same time. This is what happened after Hurricane Harvey made landfall in Houston, Texas, in 2017. A storm surge as high as 6 feet (1.8 meters) occurred in some places along the coast, while as much as 60 inches (152 centimeters) of rain fell during the week of the storm over parts of the metropolitan area. Floods can also occur in many natural floodplain areas around rivers and streams. In Myanmar, the Irrawaddy River regularly floods during the July monsoon season, which produces weeks of heavy rain. These floods routinely displace tens of thousands of residents, yet without other options to make a living in this impoverished country, most residents move back when the waters recede.

Human land-use decisions directly influence flood damage. The loss of wetlands in floodplains and coastal areas eliminates buffers that can disperse the destructive energy of floodwaters and storm surges. Areas paved with impervious surfaces prevent water from infiltrating the ground and facilitate the rapid movement of water along paved surfaces. Even structures such as levees and reservoirs along rivers and seawalls that are designed to protect against floods can contribute to catastrophic events if they encourage human development in flood-prone areas. Because of the failure of more than 50 levees and seawalls, flooding in the wake of Hurricane Katrina in 2005 swept through 80% of the metropolitan

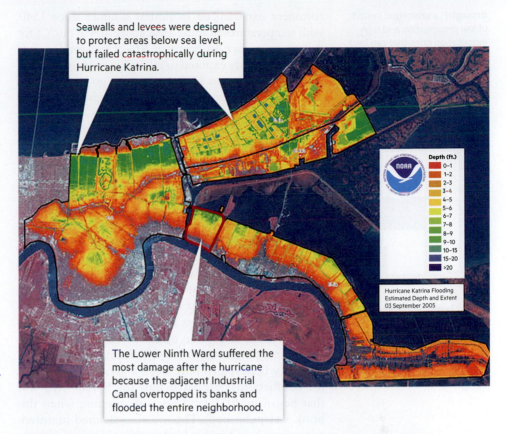

Seawalls and levees were designed to protect areas below sea level, but failed catastrophically during Hurricane Katrina.

The Lower Ninth Ward suffered the most damage after the hurricane because the adjacent Industrial Canal overtopped its banks and flooded the entire neighborhood.

Hurricane Katrina Flooding Estimated Depth and Extent 03 September 2005

Depth (ft.): 0–1, 1–2, 2–3, 3–4, 4–5, 5–6, 6–7, 7–8, 8–9, 9–10, 10–15, 15–20, >20

FIGURE 17.25 Flood Risk and Development Levees and seawalls are meant to protect developments from flooding, but they can also encourage populations to build and settle in areas that are at high risk of flooding should these protections fail. This satellite map shows the depth of flooding in New Orleans after Hurricane Katrina in 2005. The hardest-hit areas—those located well below sea level—are shown in orange and red. These areas were flooded when stormwater flowed over or broke the seawalls and levees that were meant to protect them.

area of New Orleans, much of which lay below sea level (**FIGURE 17.25**).

Heat waves are periods of extremely and unusually high temperatures that last days or weeks and can be deadly. A heat wave in Europe in 2003 killed more than 25,000 people across 16 countries as the

storm surge an abnormal rise of marine waters generated by a storm, over and above the predicted tide.

heat wave a period of extremely and unusually high temperatures that lasts days or weeks and can be deadly.

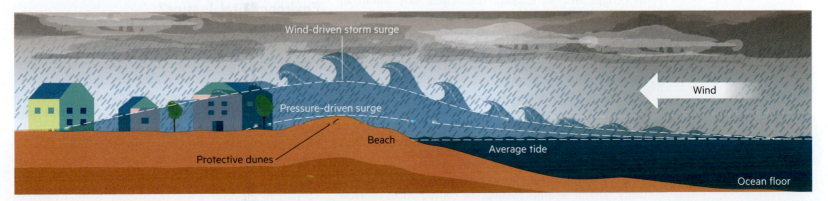

FIGURE 17.24 Storm Surge Although we often think of floods as the result of large volumes of water flowing downstream, they can also be caused by a rise in sea level that forces water inland and upstream during a storm surge.

Adapted from Marshak and Rauber (2017).

drought a prolonged period of low precipitation and high evaporation rates that can lead to water shortages (Chapters 7 and 11).

continent experienced its hottest summer since 1540, with temperatures in some areas above 104°F (40°C) for 8 consecutive days. An estimated 30% of the world's population is currently exposed to 20 days or more of extreme heat events each year. Heat waves often go hand in hand with water shortages or **droughts**. Droughts occur over months or years in places affected by below-average precipitation and/or above-average evaporation due to high temperatures. For example, when large areas of Texas and Oklahoma experienced more than 100 days of temperatures exceeding 100°F (38°C) in the summer of 2011, the rate of water loss from evaporation in each state was double the long-term average. Sometimes droughts are not just a summer phenomenon. From 2011 to 2017, California endured its driest period since record-keeping began.

In extreme cases, water scarcity and agricultural losses due to drought spark international conflict and crises. A severe drought affecting Syria from 2006 to 2009 caused crop failures and restrictive water policies that led to the migration of more than 1 million people from rural areas into cities. This internal migration is widely believed to have been a precipitating factor for conflicts that escalated into the Syrian civil war that began in 2011 (and was still ongoing when this book was published). This war has claimed hundreds of thousands of lives and led to the migration of more than 5 million people—a refugee crisis second only to those displaced by the 2022 war in Ukraine. Research has shown a link between climate change and increasing aridity in the eastern Mediterranean, and the relationship between drought and conflict is one of the reasons that the US Department of Defense considers climate change an emerging security threat.

Expanded Ranges for Diseases

Ticks and mosquitoes need the right temperatures to live because they cannot control their body temperature. In some locations, warmer weather is encouraging the populations of these disease carriers to increase. In certain areas of Canada, researchers have found an increase in Lyme disease due to an expansion of tick populations in the region, and a 2016 study found that in the United States, the list of counties hosting potential breeding areas for ticks has expanded by 400 since 1996. The CDC believes that the incidence of West Nile virus may rise as climate change produces hotter and wetter weather that increase the population of mosquitoes after storms hit.

⬡ **TAKE-HOME MESSAGE** Climate change can worsen the effects of biological and physical hazards, particularly extreme weather events. The consequences of these disasters can be severe, in some cases causing armed conflict and eroding environmental justice.

17.9 Where Are Vulnerable Communities Located?

The Lower Ninth Ward in New Orleans is a predominantly Black community that was hit hardest by the effects of Hurricane Katrina in 2005. This is because it was adjacent to the Industrial Canal, an artificial waterway that flooded the neighborhood with 12 feet (3.5 meters) of water when its banks gave way during flooding. The area was also the last place in New Orleans to be pumped dry and have its electricity and water restored after the storm. This is just one example of how where a person lives affects his or her health and shows how environmental justice issues can involve physical hazards.

Much of the danger associated with physical hazards is determined by where communities are located and how they are built. Often, the areas most vulnerable to these hazards are home to low-income and minority communities. In some cases, the vulnerability is due strictly to proximity to the hazard. For example, a 2015 analysis of populations located in coastal and inland flood-risk zones in the Miami metropolitan area found that these places were disproportionately occupied by marginalized populations unless the area in question was also the site of a valuable amenity, such as an ocean view or beach access.

But other factors, such as the way the land is developed (as seen in the Ninth Ward) and the way buildings are constructed, contribute to the risk. Research on the deadly potential of earthquakes across the world's major cities found that nearly all of the highest-risk places were in low-income countries, with potential fatalities estimated to be in the thousands. The primary reason was "building frailty": the highest death tolls from earthquakes occur in places where buildings are not designed to withstand seismic activity, as in Haiti in 2010 and Nepal in 2015 (**FIGURE 17.26**).

What Actions Promote Environmental Justice?

How we design the places we live influences the day-to-day risks that affect public health. Studies have shown that lower-income communities suffer more traffic-related injuries and fatalities than do higher-income communities. One study of ambulance records in Montreal found that this disparity holds across injuries to pedestrians, bicyclists, and drivers and is linked to road design. The study also found that lower-income neighborhoods had twice as many intersections crossing major thoroughfares, fewer sidewalks, and a higher proportion of four-way stops (which confused drivers and pedestrians) than did wealthier neighborhoods. All these factors are known to increase the likelihood of traffic accidents. Even access to parks can be an

environmental justice concern. For example, a study in Los Angeles found that while predominantly white neighborhoods enjoyed more than 31 acres of parks per 1,000 residents, predominantly Black and Latino neighborhoods had access to fewer than 2 acres per 1,000 people.

As we learned in Chapter 16, thoughtful urban planning can ensure that building codes suit the physical hazards of an area, can make streets more accessible and safer for pedestrians, and can ensure that amenities such as parks are available. Increasingly, urban planning efforts center on the concept of *resilience* (see Chapter 1)—ensuring that both human and infrastructure systems can endure or recover from future challenges. Although well-prepared emergency responses to natural disasters such as wildfires, earthquakes, hurricanes, and floods do of course save lives, preventative measures can also keep more people out of harm's way. Most often this touches on land-use and development decisions. For example, restricting development in the areas that exist between unoccupied forest and shrubland and locations of residential development can reduce the number of people and human structures facing the highest risks of wildfire.

Another land-use strategy that promotes community well-being (as well as environmental justice) is buyout programs offered in flood zones by the Federal Emergency Management Agency and many state and local agencies. For example, New Jersey's Blue Acres program purchases homes from owners of properties in flood-prone areas. Once purchased, the properties are demolished, and the areas are designated as open spaces. Programs like this can have the added benefit of restoring wetland and sand dune ecosystems while providing buffers to absorb and disperse floodwaters and their force (**FIGURE 17.27**).

On a broader scale, the executive order on environmental justice signed by President Clinton in 1994 has had far-reaching effects. In addition to ensuring that federal environmental policies and practices do not harm marginalized communities, it mandates that a portion of federal funding for environmental projects go to federally determined environmental justice areas. In this way these communities are both protected from harm and given environmental benefits that they have historically been denied.

⬡ **TAKE-HOME MESSAGE** Much of the danger associated with physical hazards is determined by where communities are located and how they are built. The way we plan for and respond to physical hazards can limit the scope of their damage and promote environmental justice.

FIGURE 17.27 Relocating Communities in Flood-Prone Areas This house in Sayreville, New Jersey, is being torn down after it was purchased by the government as part of New Jersey's Blue Acres program, which started in the aftermath of Superstorm Sandy.

ENVIRONMENTAL JUSTICE

Climate Justice

Countries producing the most emissions are wealthier and thus better able to protect themselves against the harms of climate change and to reduce greenhouse gas emissions going forward. The United States is responsible for 25% of global greenhouse gas emissions since the dawn of the industrial age in 1751, whereas the entire continents of South America and Africa have been responsible for just 3% each and contain a large proportion of the nations most threatened by climate change.

At the 2021 UN Climate Change Conference (COP26) in Glasgow, Scotland, then British prime minister Boris Johnson raised the issue of climate justice in his opening remarks, stating that "the countries most responsible for historic and present-day emissions are not yet doing their fair share of the work." Although wealthier nations have been paying into a global fund since 2009 to help less wealthy nations adapt to climate change, the richer countries have yet to fully honor their commitments. It is estimated that the global cost of loss and damage due to climate change may average as much as $500 billion per year by 2030. A new "Loss and Damage" fund was initiated at COP26, with Scotland making the initial contribution.

Follow Good Sanitation and Hygiene Practices

While COVID-19 has been the biological hazard of greatest global concern for the past few years, other pathogens continue to pose a risk to human health. To protect yourself against the flu, another of the most common biological hazards, the CDC recommends getting the flu vaccine each year, washing your hands with soap and water, and avoiding touching your eyes, nose, and mouth (this advice might sound familiar, as it is the general advice for all respiratory diseases, including COVID-19). If you do get sick, seeking medical attention when the illness persists can prevent it from developing into something more serious such as pneumonia. To prevent spreading your sickness to others, you should stay away from work or school for at least 24 hours after your last fever has ended. And in general, good sanitation practices in the restroom and kitchen can prevent many other pathogens from causing illness. So washing up, cleaning sinks and countertops, washing produce, and thoroughly cooking meats are all important practices.

Avoid Chemical and Radiation Hazards When Possible

Because air pollution contains chemical hazards that can increase vulnerability to respiratory illnesses, it's also important to avoid environments such as rooms filled with cigarette smoke or busy traffic areas where smog and other pollutants accumulate, if it all possible. Other steps to avoid chemical hazards include selecting nontoxic paints, cleaning products, and gardening supplies and making sure you have a working carbon monoxide detector in your home. You can also test where you live or work for radon gas to see if you need to take steps to reduce this risk. Many local and state health departments provide these detectors for free, and there are other places to obtain them inexpensively (such as the National Radon Program Services at Kansas State University). And to protect against UV radiation, you can wear hats, sunglasses, and more protective clothing or apply sunscreen to reduce exposure outdoors, and you can also avoid indoor tanning beds.

Prepare a Personal and Community Disaster Plan

Having a personal disaster plan can help prepare you for the physical hazards affecting your region, but you can also get involved in the disaster plan for your larger

17.10 What Can I Do?

There are many individual actions you can take to prevent or address health issues caused by environmental factors. But despite these individual actions, addressing environmental health challenges is really a community effort. Many of the hazards we face where we live and work are shaped by more than individual behaviors. And in these cases, it is important to become aware of local environmental risks and be an active and engaged citizen in the decision-making processes that influence how the risks are managed. Let's look at some of the individual and community-based actions you can take to improve your health and that of the community.

community. For example, various county and city planning boards typically identify areas that are at risk of landslides, floods, and other physical hazards; determine what sorts of construction are permitted in these areas; and put emergency procedures in place. By attending public hearings and even participating in the planning process, you can learn about the hazards affecting your area and provide feedback on how your community manages these risks.

Become Aware of Environmental Injustices in Your Area

You can learn a lot about your neighborhood with EJSCREEN, a mapping tool developed by the EPA to help both the EPA and local communities identify environmental justice concerns that can be addressed by government enforcement, permitting, and planning activities. This tool allows you to see 11 different environmental indicators, including several measures of air quality; traffic volume; the locations of hazardous waste treatment, storage, and disposal facilities; estimates of lead paint exposure; and the locations of facilities that are permitted to use or discharge various kinds of hazardous or polluting materials. It also maps various demographic variables and allows you to compare your neighborhood to those around you and to state and national averages for various parameters to give you a sense of whether your community is bearing a disproportionate impact.

Understand Your Rights to a Safe Workplace

Under federal law, employees have the right to request and receive information about chemicals used in the workplace, injuries and illnesses that have happened there, and the results of tests conducted at the workplace regarding various hazards. Employees also have the right to receive training on safety and health hazards in the workplace. If employees have concerns about workplace conditions, they can file complaints with and even request inspections by the Occupational Safety and Health Administration (OSHA). Often, the implementation of safe and healthy workplace practices requires active participation on the part of employees, which can take some courage. That is why it is also against federal law to discriminate against employees who speak up about hazards in the workplace.

AS A WORKER
you have the right to:
- a safe and healthful workplace
- tools and equipment needed to do your job safely
- training in a language you understand
- and more...

If you think your job is unsafe and you have questions, call OSHA. It's confidential. We can help!

1-800-321-OSHA (6742)

OSHA 3385-4N-10

Chapter 17 Review

SUMMARY

- The goal of environmental justice is that no community should bear more environmental burdens or enjoy fewer environmental benefits than others.

- Environmental justice concerns include disproportionate exposure to chemical hazards, lack of access to sanitation and health-care resources that reduce biological hazards, and the increased risk of harm from physical hazards that is caused by where communities are located and how the structures they live in are built.

- Biological, chemical, and physical factors in the environment can cause human health problems.

- Epidemiology, toxicology, and public health are the three main fields of environmental health. Epidemiology seeks to understand the spread of diseases in populations, while toxicology investigates the adverse health impacts of substances on organisms. Public health is concerned with protecting and improving the health of people where they live and work.

- Pathogens are microorganisms that cause illness or infections, and they are the deadliest type of biological hazard. Most pathogen-caused diseases are preventable with sanitation improvement and access to clean water and medical care.

- Infectious diseases spread through the air or water or through person-to-person or person-to-animal contact.

- Chemicals that have the capacity to cause injury or death are considered toxic. The risk of harm from a toxin is determined by a range of factors including the dose as well as the characteristics of the individual exposed to the toxin.

- Chemical characteristics affecting exposure to toxins include volatility, solubility in water or fat, and persistence in the environment.

- Determining an acceptable level of risk for exposure to chemical hazards involves the application of values.

- Environmental health hazards can also come from the physical environment, including extreme weather events, geologic events, wildfires, radiation, and harms caused by human-built structures.

- Even the harm from physical hazards categorized as natural disasters tends to have human actions as direct or indirect factors shaping the extent of the harm.

- Climate change is worsening the environmental health effects of certain biological and physical hazards.

KEY TERMS

biological hazard
blood-borne pathogen
chemical hazard
diarrheal disease
drought
earthquake
environmental hazard
environmental health

environmental justice
epidemiology
flood
geologic hazard
heat wave
herd immunity
ionizing radiation
landslide

pathogen
persistence
persistent organic pollutant (POP)
physical hazard
public health
radiation
radon

respiratory infection
storm surge
toxicology
toxin
tsunami
ultraviolet (UV) radiation
volatile organic compound (VOC)

REVIEW QUESTIONS

The letters following each Review Question refer to the Chapter Objectives.

1. What is the term the EPA uses to capture the idea that all people regardless of background are entitled to fair treatment and meaningful involvement in the development, implementation, and enforcement of environmental laws, regulations, and policies? **(A)**

2. Which field of environmental health research studies the spread of diseases in populations? **(B)**

3. What type of infection is the leading cause of sickness and death worldwide? **(B)**

4. Name at least two ways the pathogens that cause infectious diseases can spread in populations. **(B)**

5. Name at least two factors that shape the risk of a toxin harming an individual. **(C)**

6. What is the name of the principle that places the "burden of proof" on proving the safety of a particular substance before it is sold commercially? **(D)**

7. Describe at least two ways in which human actions can increase the harms caused by forest fires. **(E)**

8. Describe at least two ways that the location and construction practices of a community place it at increased risk of physical hazards. **(E)**

9. Identify one natural and one human-caused source of radiation. **(E)**

10. What is the leading cause of skin cancer? **(E)**

11. Identify two types of extreme weather events that show increasing intensity linked to climate change. **(E)**

12. How might climate change expand the range of certain infectious diseases? **(B, E)**

13. What are three urban planning and/or land-use strategies that promote environmental justice in traditionally marginalized communities? **(F)**

FOR FURTHER THOUGHT

The letters following each item refer to the Chapter Objectives.

14. Low-income countries suffer disproportionately from infectious diseases. Identify the factors that are responsible for this disparity, and develop a proposal to address them. **(A, B)**

15. Imagine you are a city planner responsible for identifying areas for residential development and areas prone to physical hazards. Describe at least three types of locations where you would recommend against siting a residential development and why. **(E)**

16. Imagine you are a toxicologist working for an agency charged with deciding when a new chemical is allowed to be used in commercial products. Describe at least three kinds of risk factors you would examine to assess the toxicity of the new substance. Would you prefer that your agency use the precautionary principle to make its final determination or not? Explain your answer. **(C)**

17. The ideal of environmental justice is that no community should bear more environmental burdens or enjoy fewer environmental benefits than others. Identify and describe places in the region where you live that fall short of this ideal. What are their environmental health concerns, and what could be done to improve the situation? **(A)**

Make Your Case

The following exercises use real-world data and news sources. Check your understanding of the material and then practice crafting well-supported responses.

Use the News

We have seen in other chapters that our clothing choices can have major environmental impacts. This article describes a study of the impacts of a popular retail category known as "fast fashion." Use this article to answer the questions that follow. The first three questions test your understanding of the article. Questions 4 and 5 are short answer, requiring you to apply what you have learned in this chapter and cite information in the article. Answers to Questions 1–3 are provided at the back of the book. Question 6 asks you to make your case and defend it.

"How Fast Fashion Hurts Environment, Workers, Society: Brown School Research Examines Environmental, Social Costs of Ready-Made, Inexpensive Clothing," Neil Schoenherr, *The Source*, January 9, 2019

The overabundance of fast fashion—readily available, inexpensively made clothing—has created an environmental and social justice crisis, claims a new paper from an expert on environmental health at Washington University in St. Louis.

"From the growth of water-intensive cotton, to the release of untreated dyes into local water sources, to worker's low wages and poor working conditions, the environmental and social costs involved in textile manufacturing are widespread," said Christine Ekenga, assistant professor at the Brown School and co-author of the paper "The Global Environmental Injustice of Fast Fashion," published in the journal *Environmental Health*.

"This is a massive problem," Ekenga said. "The disproportionate environmental and social impacts of fast fashion warrant its classification as an issue of global environmental injustice."

In the paper, Ekenga and her co-authors—Rachel Bick, MPH '18, and Erika Halsey, MPH '18—assert that negative consequences at each step of the fast-fashion supply chain have created a global environmental justice dilemma.

"While fast fashion offers consumers an opportunity to buy more clothes for less, those who work in or live near textile manufacturing facilities bear a disproportionate burden of environmental health hazards," the authors wrote.

"Furthermore, increased consumption patterns have created millions of tons of textile waste in landfills and unregulated settings. This is particularly applicable to low- and middle-income countries (LMICs) as much of this waste ends up in second-hand clothing markets. These LMICs often lack the supports and resources necessary to develop and enforce environmental and occupational safeguards to protect human health."

In the paper, the researchers discuss the environmental and occupational hazards during textile production, particularly for those in LMICs, and the issue of textile waste.

They also address a number of potential solutions, including sustainable fibers, corporate sustainability, trade policy, and the role of the consumer.

Globally, 80 billion pieces of new clothing are purchased each year, translating to $1.2 trillion annually for the global fashion industry. The majority of these products are assembled in China and Bangladesh, while the United States consumes more clothing and textiles than any other nation in the world.

Approximately 85 percent of the clothing Americans consume, nearly 3.8 billion pounds annually, is sent to landfills as solid waste, amounting to nearly 80 pounds per American per year.

---

1. According to the article, why should fast fashion be classified as a global environmental justice issue?
   a. It offers consumers the opportunity to buy more clothing for less money.
   b. Its environmental and social impacts fall disproportionately on certain populations.
   c. The United States consumes more clothing and textiles than any other country.
   d. It is readily available, inexpensively made clothing sold on the global fashion market.

2. How does the article quantify the global scale and size of the fast-fashion industry?
   a. 80 billion pieces of clothing purchased each year, worth $1.2 trillion
   b. 8 billion pieces of clothing purchased each year, creating 1.2 million jobs

   c. 3.8 billion pounds of clothing purchased each year, or 80 pounds per person
   d. $80 billion invested in production each year, creating 1.2 trillion pieces of clothing

3. What does the article identify as some of the environmental and social costs of fast fashion? Choose the correct responses from the following list.
   a. low wages and poor working conditions
   b. high incidence of disease
   c. water-intensive cotton crops
   d. unfair political processes
   e. pesticides leaching into the water
   f. release of untreated dyes into water
   g. millions of tons of textile waste

4. What do the letters LMIC refer to in the article? Why are LMICs more likely to be affected by the negative consequences of fast fashion? Use information from the article and the chapter to support your answer.

5. Do you shop at popular fast-fashion stores or buy similar brands sold at large discount or department stores? Do you feel this article is suggesting that you or others should not do so? Provide an answer and explain your reasoning in a few sentences.

Use the Data

The World Health Organization (WHO) keeps data on fatalities to track global death rates. The graphs shown here compare the top 10 causes of death in 2019 in low-income countries with those in high-income countries.

1. Which health issue is a higher-ranked cause of death in low-income countries than in high-income countries?
   a. lower respiratory infections
   b. stroke
   c. ischemic heart disease
   d. Alzheimer's disease

2. Which cause of death is more highly ranked in high-income countries than in low-income countries?
   a. lower respiratory infections
   b. stroke
   c. HIV/AIDS
   d. road injuries

3. Most of the top 10 causes of death in low-income countries fall into which cause group?
   a. communicable
   b. noncommunicable
   c. injuries

4. Identify at least two causes of death that are not on the low-income countries graph but are present on the high-income countries graph. On the basis of what you have learned in this chapter, what might explain these differences?

5. Identify two causes of death that are more highly ranked in higher-income countries than in lower-income countries. On the basis of what you have learned in this chapter, what might explain these differences?

6. **Make Your Case** Imagine you run a major foundation dedicated to addressing environmental health. Explain which of the causes of death listed in the graphs you would identify as highest priority and why. What sorts of programs would you fund to address them?

6. **Make Your Case** Imagine you are a fashion designer who has been given the opportunity by a large company to start your own clothing brand. If you do not take this opportunity, you know that many others will jump at the chance. However, you have read reports that this company produces clothing only in low-income countries with low wages relative to those in the United States and often with lax environmental standards. Would you decide to work with this company? In a paragraph or two, explain your choice and any actions you would take.

Study the graphs and use them to answer the questions that follow. The first three questions test your understanding of the graphs. Questions 4 and 5 are short answer, requiring you to apply what you have learned in this chapter and cite data in the graphs. Answers to Questions 1–3 are provided at the back of the book. Question 6 asks you to make your case and defend it.

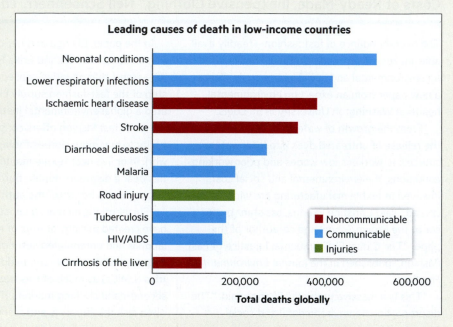

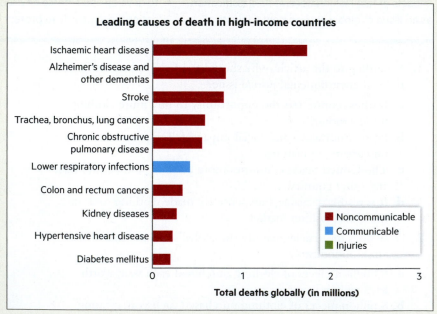

LEARN MORE

- Centers for Disease Control and Prevention (CDC), "Global Health" page: www.cdc.gov/globalhealth/index.html
- United Nations, "Sustainable Development Goal No. 3" ("ensure healthy lives and promote well-being for all at all ages") progress and information: https://sustainabledevelopment.un.org/sdg3
- US EPA, "Environmental Justice" page (definition, policies, programs, and resources): www.epa.gov/environmentaljustice
- California Department of Forestry & Fire Prevention page: fire.ca.gov
- US National Climate Assessment. 2018. https://nca2018.globalchange.gov/chapter/1/

- Video: Science 360, Flint, Michigan, Water Crisis: digital.wwnorton.com/environsci
- Video: NSF, Impact of Nanomaterials on Living Things: digital.wwnorton.com/environsci
- Video: Science 360, A Future with Fewer Forests: digital.wwnorton.com/environsci
- Video: Science NSF, Low-Cost Environmental and Pollution Sensors: digital.wwnorton.com/environsci

18

Decision Making
Why Do Our Choices Matter?

This wall is made of disposable plastic bags used once and discarded. How do we change people's behavior so they stop using plastic bags, when even many environmentally committed customers do so out of convenience?

As you approach the checkout clerk at a grocery store, you know the question is coming: "Do you need a bag?" Did you bring your own reusable bag or request a disposable paper or plastic bag from the store? In a growing number of places, you now have to pay for a disposable bag. So did you decide that the fee was worth it? Grocery store chains are interested in the behavior of consumers like you and have hired market research firms to conduct research on this issue. They have found that a majority of US consumers who consider themselves frequent grocery shoppers prefer reusable bags, primarily because they believe that reusable bags are better for the environment than disposable paper or plastic bags. But there is another issue: surveys and observational studies have shown that even when people express personal concern about the environment, are aware of the environmental benefits associated with reusable bags, and are already in possession of several reusable bags, they still use disposable bags about 4 out of every 10 times they go shopping. Why is this?

Apparently, basic convenience and forgetfulness play an important role in this decision even for environmentally motivated consumers. Edelman Berland, a market research firm, has shown that about one-third of these environmentally committed shoppers still use disposable bags because they think that reusable bags are inconvenient, and the other two-thirds have reusable bags in their cars and simply forget to bring them into the store. We see a range of strategies designed to influence behaviors such as these. Some efforts include **incentives**: positive or negative signals that pull us toward or push us away from certain choices or behaviors. Incentives for reusable bags include taxes and fees on disposable bags, discounts and refunds for reusable bags, signs to help people remember to bring the bags into the store, and even bans on the use of disposable plastic bags in grocery stores (**FIGURE 18.1**).

So what else can drive our decisions and those of others? In this chapter we will explore the science of decision making. We will see that actions are influenced by many factors, including not only what we know and care about but also competing needs and desires,

Chapter Objectives

By the end of this chapter, you should be able to . . .

A. describe research on how experiences, beliefs, and group dynamics influence decisions.

B. discuss several ways that individuals prioritize their needs and desires.

C. recognize how attitudes and automatic thought processes drive decisions.

D. explain how effective strategies work to influence behaviors.

E. understand how positive and negative incentives motivate decisions.

F. understand the ways in which words and images are used to communicate about environmental issues and influence attitudes and behaviors.

> **Thinking is easy, acting difficult, and to put one's thoughts into action, the most difficult thing in the world.**
>
> —Goethe

social pressures, biases, and a variety of emotional responses. When we think about how people decide to reduce their environmental impact, we often imagine a simple recipe: people develop an awareness of and concern for an issue, learn what they can do, and take action. But the shopping-bag example demonstrates that human behavior tends to be more complicated than this—even for motivated people engaging in relatively simple behaviors. Let's take a closer look at what's going through our minds as we make decisions that impact the environment.

incentive a positive or negative signal that pulls us toward or pushes us away from a certain choice or behavior (Chapter 1).

hierarchy of needs a model of the way people prioritize some needs and desires over others.

automatic thinking an instantaneous cognitive system distinct from our slower, conscious, and more reflective ways of thinking.

status quo the current situation; the existing state of affairs.

loss averse a bias whereby people prefer avoiding the loss of something they already have more than they prefer acquiring an equivalent amount of that same thing.

18.1 What Are the Key Factors Influencing Our Decisions?

As with the choice between paper and plastic, the choice to drive or take public transportation is one many of us make every day as we consider factors like travel time, convenience, and flexibility. And in one study, researchers found that when people have the option to drive, they are reluctant to take public transportation when they perceive bus or train terminals as unsafe places. Because the need to feel safe is a high priority for most people in most situations, this basic need has a very strong effect on choices individuals make. We will prioritize our need for safety over many other needs. But our need or desire for safety is just one factor influencing our decisions.

FIGURE 18.1 Encouraging the Use of Reusable Bags There are many ways to encourage shoppers to use reusable bags at the grocery store, such as signs that remind shoppers to bring their bags into the store.

Most often when we are faced with a choice, we have multiple needs and desires, some of which will win out over others and influence the action we ultimately take. This presents a challenge when addressing environmental impacts, which almost always involves changing our own and others' way of doing something.

Hierarchy of Needs

In 1943, the psychologist Abraham Maslow set out to model the way people prioritized some needs and desires over others and developed his **hierarchy of needs** pyramid (**FIGURE 18.2A**). In this model, physiologic needs linked to our survival, such as the need to satisfy hunger and thirst, are prioritized, followed by needs linked to our feelings of safety, belongingness, self-esteem, and finally self-actualization (a feeling that we are realizing the potential of our talents and skills). Although Maslow's original research was based on biographical analysis of a relatively small number of historical and contemporary figures, including Abraham Lincoln and Albert Einstein, subsequent research in this area has been conducted on large numbers of people in many different countries. Researchers at the University of Illinois collected data on more than 60,000 participants in studies across 123 countries. The results demonstrated that people do tend to act to meet their various needs roughly according to Maslow's hierarchy. While Maslow's model often matches the ways we are influenced by various kinds of needs, individuals do not always prioritize needs according to his pyramid. For example, people sometimes sacrifice their basic physiologic needs because they place a higher priority on being part of a group or sticking to their personal convictions. In this way, teenagers often engage in risky behaviors such as binge drinking to fit in with a particular group of peers. Similarly, protesters might go on a hunger strike or sacrifice their own personal safety in other ways to further a cause they think is important (**FIGURE 18.2B**). The main thing to remember is that not all needs carry the same weight in decision making, and understanding this can help explain certain choices and behaviors.

Automatic Thinking

Think about how you might react if you were presented with a stinking pile of rotting food—a flash of disgust might appear on your face as you quickly pull away from it. Similarly, people with common phobias, such as fear of spiders or snakes, exhibit reflexive stress responses to photos or even just the mention of these critters. Often, the thinking that drives decisions is more reflexive than deliberate. When the mental processes guiding our behavior are operating quickly and effortlessly like this, we are engaging in **automatic thinking**: an instantaneous thought process distinct from our slower, conscious, and more reflective ways of thinking. Automatic thinking is

Maslow's Hierarchy of Needs

Self-actualization
Morality, creativity, spontaneity, acceptance, experience purpose, meaning and inner potential

Self-esteem
Confidence, acheivement, respect of others, the need to be a unique individual

Love and belonging
Friendship, family, intimacy, sense of connection

Safety and security
Health, employment, property, family and social stability

Physiologic needs
Breathing, food, water, shelter, clothing, sleep

(a)

(b)

FIGURE 18.2 Hierarchy of Needs **(a)** In a 1943 article, Abraham Maslow first proposed the hierarchy of needs pyramid as a "theory of human motivation." **(b)** The hierarchy of needs does not always match human behavior. These protesters are sacrificing their physical needs and risking their own physical safety to express themselves on the issue of logging old-growth forests.

Adapted from (a) Maslow (1943).

not limited to sensory reactions. Knowledge and skills can also become part of our automatic thinking when these have been repeated or practiced many times. This is why we can quickly understand simple sentences in our native language, know that $2 + 2 = 4$, and carry out routine physical activities such as riding a bicycle.

But automatic thinking can also be a source of bias. In Chapter 1, we learned that *bias* is an unreasonable weighting of our thinking that leads to misunderstandings. In 1974, psychologists Daniel Kahneman and Amos Tversky published a groundbreaking article in the journal *Science* that identified and described more than a dozen biases in the automatic thinking of their experimental subjects. One of the most significant patterns of bias is our tendency to prefer our current situation, or **status quo**. This pattern can happen simply because we don't examine the things we already do as critically as we examine possible changes to our behavior. People also tend to be **loss averse**, preferring to avoid the loss of something they already have more than they prefer acquiring an equivalent amount of that same thing. We tend to be twice as unhappy about losing something we already have as we are happy about gaining the same amount of that thing. **FIGURE 18.3** illustrates an experiment conducted on college students by Kahneman and others that demonstrates loss aversion through the random distribution of coffee mugs to half the class. In this experiment, the students given coffee mugs became very reluctant to sell them. Owning one of these mugs became the status quo for those who received them, and,

The Mug Experiment

1 Students are divided into two groups of "buyers" and "sellers." "Sellers" are given a coffee mug as a gift.

$2.25? $4.50

2 Sellers were willing to sell the mugs for a price much higher than the buyers were willing to pay.

3 This is an example of loss aversion. People appreciate things that they own more than things they do not.

FIGURE 18.3 Loss Aversion Because people are loss averse, they tend to value an object they possess more than one they don't.

Implicit Bias

Automatic thinking can also lead to implicit bias against particular groups of people. Psychologists test for unconscious attitudes about race, gender, and other social categories with implicit association tests (IATs). In IATs, participants rapidly match images or words with evaluations or qualities like good and bad, forcing them to make associations more quickly than the conscious brain can function. While IATs are not reliable predictors of any one participant's bias on a single test, in the aggregate across many participants, they do reveal that most people hold unconscious racial and gender biases. Although implicit bias operates at the individual level, researchers have found that they reflect biases in broader social and cultural contexts too.

Additional studies have found that because implicit bias operates at the unconscious level, it is highly resistant to change. So rather than pursuing strategies to eliminate implicit bias in individuals, actions to counter it are recommended instead. These include documenting disparities (e.g., workforce composition in an organization) and creating processes to remedy these disparities that are based on objective and inclusive criteria.

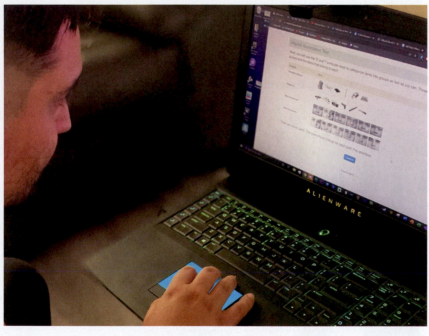

IATs test for unconscious bias against different groups of people.

emotional defense mechanism a quick and automatic thought process that relieves us of bad feelings associated with a distressing situation.

denial an emotional defense mechanism by which people refuse to acknowledge a problem's existence.

as the illustration shows, they in fact valued them about twice as much as those who did not receive them.

In other words, we often overvalue what we already possess or the way we do things now relative to what we might gain from a change. So we have a strong tendency to stick with the status quo because we fear the losses that could be associated with a change. When benefits associated with a change are far in the future and diffused over the wider community rather than just focused on ourselves, they have an even weaker influence on our behaviors. Our bias against losses makes us fear the potential losses more than we desire the potential gains associated with the behaviors.

Emotional Defense Mechanisms

While many of us enjoy a challenge, few people will say that they like problems. In fact, people react to the existence of problems with a range of **emotional defense mechanisms**: quick and automatic thought processes that relieve us of bad feelings associated with a distressing situation. In 1977, psychiatrist George Vaillant established widely accepted categories for defense mechanisms based on his research of people confronting addiction and various mental health conditions. Two of the most common defense mechanisms are **denial** (refusing to acknowledge the existence of a problem in the first place) and **emotional distancing** (when a person refuses to feel emotions associated with the problem). Emotional distancing often pushes issues that are not immediately affecting us out of mind. Research into how people process information about climate change reveals that many feel initial concern but perceive the problem as abstract and distant from themselves. This can cause them to assign it a very low priority of care.

Similarly, **resignation** is a feeling of helplessness that leaves a person believing he or she cannot change the situation. **Delegation** is a defense mechanism that absolves the individual of responsibility. It puts the blame for a problem on others or deems it someone else's responsibility. People tend to prioritize actions they feel responsible for. If we don't feel responsible, we tend to wait for others to act on the problem.

Social Acceptance and Conformity

Studies of the evolution of our behavior have found that people across different cultures are strongly motivated by the need to feel accepted by a wider group of people. We are social animals. Social bonds within groups facilitate cooperation, something that enhanced our ancestors' ability to survive by hunting, foraging, and defending the group from threats. Our need to belong often pushes us to adjust our behavior to conform to the groups of which we are a part. **Social norms** are expectations or standards for behavior that are widely shared within a group, and we generally act to avoid rejection and make favorable impressions, adjusting our behavior to fit what we perceive the norms to be.

An obvious place to see and feel the influence of our need to belong is in the way people dress (**FIGURE 18.4**). Most people base their decisions about what to wear not only on their own personal comfort but also partly on the impression they think they will make on people who see them. Even as many people worked from home during the COVID-19 pandemic and adopted a more casual wardrobe, most still made clothing choices based

FIGURE 18.4 The Need for Social Acceptance Shapes Behavior The human need for social acceptance can be seen in the way the clothes we choose to wear conform to shared expectations in certain social contexts. If we chose our clothing on the basis of function alone, our wardrobes would likely be far more limited.

in part on how coworkers would perceive them while connecting via Zoom or other video conferencing platforms—at least from the waist up. This fits a normal pattern of human behavior. We place a high value on our social interactions, and all of these interactions are influenced by the impressions we make on each other, including the way we look. We may no longer be socially dependent on groups for hunting, gathering, and defending ourselves, but the impressions we make on others still have a significant effect on many of the

things we value in life, such as friendships, romantic relationships, and career opportunities.

Advertisements for consumer goods and influencers on social media often play on our need for belonging, attempting to convince us that a new product will help us gain more social acceptance. They also communicate that by sticking with an older style, we might risk rejection; so we might choose to discard and replace a still-functioning product because we no longer perceive it as meeting our social needs. The power of social acceptance and norms is frequently used in campaigns to address public health problems and foster positive behaviors. For example, many college campuses have attempted to address the binge-drinking problem by publicizing statistics showing that the vast majority of students do not engage in this practice.

What Do You Believe?

People have a strong tendency to accept or reject the information they receive—even scientific information—in a way that conforms with the views dominant in social, political, and religious groups that are important to them. Further, once people have formed beliefs on an issue, they tend to cling to them tightly even when they are presented with evidence to the contrary. This human tendency is called **belief perseverance**. This does not mean that one's beliefs cannot change, but numerous studies have shown that when presented with new information, people tend to contort the new evidence in order to stick to their current belief.

Another human tendency is **confirmation bias**. People tend to selectively "cherry pick" information that confirms their existing beliefs. They also attempt to assimilate or "fit" any information they encounter into their existing way of thinking. Experimental subjects have been shown to trust and have a higher regard for those who present information with which they agree. For example, in one of many experiments conducted by Dan Kahan's Cultural Cognition Project at Yale, participants were given different quotes on climate change attributed to fictitious experts with identical qualifications. The participants had a strong tendency to rank the expert who confirmed their existing beliefs on climate change as the most qualified on the topic.

⬡ **TAKE-HOME MESSAGE** People's needs and desires form a hierarchy of importance from basic physiologic needs to needs for safety, acceptance, and maximizing one's potential. Automatic thinking plays a significant role in our decision making and is associated with biases such as a preference for the status quo. There is great social pressure to conform, and once we form beliefs, they tend to persevere because we engage in confirmation bias.

emotional distancing an emotional defense mechanism by which people prevent themselves from feeling emotions associated with a problem.

resignation a feeling of helplessness that leaves a person believing he or she cannot change the situation.

delegation an emotional defense mechanism by which individuals absolve themselves of responsibility for a problem by blaming others or deeming it someone else's responsibility.

social norm an expectation or standard for behavior that is widely shared within a group.

belief perseverance a bias that occurs when people cling to existing beliefs on an issue even when they are presented with evidence to the contrary.

confirmation bias the tendency for people to select information that confirms their existing beliefs and "fit" information they are given into their existing way of thinking.

Pressure to Conform

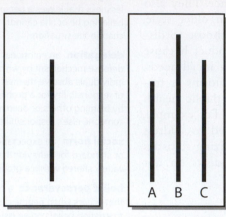

Which of the lines on the right is the same size as the one on the left? In 1955, a famous study of conformity known as the Asch experiment asked a volunteer participant to join two others at a table and observe various diagrams of different-length lines like those shown here. The correct answer was straightforward, but the two others were secret collaborators with the researchers and were instructed to agree on an answer that was clearly wrong. More than one-third of the time, volunteer participants were willing to conform to the collaborators' opinion.

This and other conformity experiments have shown that certain conditions strengthen the push we feel to conform. Conformity is strongest when the behavior in question can be observed directly by the members of a group, the group consists of at least three people, and the group members are all adopting the same behavior. Research has shown that any resistance we might have to conforming with a group behavior is significantly weakened if we greatly admire or value the status of the group members and if we feel incompetent or insecure about ourselves.

18.2 What Are Some Successful Strategies to Influence Behavior?

prompt a conspicuous reminder that targets behaviors at the moment people are deciding how to act.

Influencing behavior often requires more than simply educating people about an issue; there is often a gap between what people know and care about and what they actually do. For example, in the 1970s a social

psychologist at Smith College set up individual interviews with 500 students about their views on litter. Just outside the interview location, the researcher placed several pieces of litter in plain sight on the path interviewees used to depart from the interview. An observer was stationed nearby to see whether each interviewee picked up and disposed of the litter. Nearly every interviewee (94%) identified litter as a concern and indicated that individuals had a responsibility to pick up litter when they encountered it. Yet nearly all of the interviewees walked right past the litter in their path as they left the interview site. Only 2% of the interviewees picked anything up.

In another well-known experiment, 40 participants attended a 3-hour workshop on residential energy conservation. The researcher measured each participant's knowledge of and attitude toward energy conservation before and after the workshop. He found that each participant demonstrated both an increased knowledge of residential energy conservation strategies and an intent to implement one or more of the energy conservation strategies after completing the workshop. The researcher also distributed free low-flow showerheads to participants to help with water conservation. When the researcher visited the homes of each workshop participant months later, he found that just one participant had adopted an energy conservation measure that was learned at the workshop. Only eight of the participants had installed the low-flow showerheads they were given.

So what can be done to bring our behaviors more in line with our knowledge and concerns? In the rest of this chapter, we will consider strategies that have been shown to influence our individual behaviors.

Prompts

When you get in the car and turn on the ignition without buckling your seat belt, normally you hear a constant ringing sound and see a light blinking on the dashboard. This is an example of a **prompt**: a conspicuous reminder that targets behaviors at the moment people are deciding how to act (**FIGURE 18.5**). Prompts are based on the recognition that the effect of an attitude (such as concern for the environment) on behavior can be strengthened if the favorable attitude (e.g., "I ought to use a reusable bag") is brought to mind quickly as we decide to take an action. People are more likely to remember to take reusable bags to the grocery store if they store the bags in their cars, and they are more likely to remember to bring them from their cars into the store if there is a prompt prominently posted outside. Similarly, studies show that recycling increases when recycling and trash bins are placed next to each other and recycling bins have openings that clearly signal what types of items go in each place, such as a round hole for beverage containers and a long narrow opening for papers.

FIGURE 18.5 Prompts Prompts such as stickers and signs can encourage behaviors that facilitate water and energy conservation. Making recycling containers more noticeable and easier to understand makes recycling more likely.

Addressing Product Obsolescence: How Can We Extend the Useful Life of Our Stuff?

Do you love your phone? Chances are if you bought it in the past couple of months, you still feel a buzz of excitement about its new features and even its aesthetics: the way it looks and feels. On the other hand, if your phone is more than 2 years old, you are probably ambivalent about it or even frustrated with it. Its processing speed may have slowed down, its battery life isn't what it once was, and it no longer looks new and cool. Although the average life span of a cell phone is now about 5 years, Americans upgrade to a new cell phone about every 3 years on average. This means we keep our phones about as long as we keep a T-shirt or a pair of flip flops—a practice that causes significant environmental impacts. It takes a lot of energy and resources to manufacture new products, and making and disposing of these products pollutes the environment.

Our phones illustrate three important ways in which products become obsolete. First, all products break down eventually and reach the end of their useful lives. The useful life of your phone can end after several years as the battery power diminishes or all of a sudden if you break it or submerge it in water. Second, products can become functionally obsolete as new products with better features outcompete older models. Finally, in some cases our social context can play an important role if we don't think an older product is as fashionable or socially desirable as newer models.

So what can be done to reduce the speed at which products become obsolete? Consumer advocates and environmental groups often point the finger at businesses for practicing "planned obsolescence" by intentionally producing lower-quality goods that will break quickly, designing products that are difficult to fix or modify, or releasing new products with minor functional changes that are incompatible with older models. Cell phone companies have been the target of charges such as these because phones break relatively easily and have casings that make it difficult or impossible to

Producer responsibility or product stewardship laws require companies to assume a significant share of the cost of dealing with their discarded products. This provides producers with incentives to design products in ways that reduce waste.

replace batteries or switch out components. Software upgrades for cell phones are often incompatible with or inhibit the functionality of older models.

Some countries have adopted policies to encourage companies to make more durable products. For example, France adopted legislation that requires companies to communicate the expected life span of their products to consumers. The European Union also has producer responsibility or product stewardship laws that require companies to assume a significant share of the cost of discarding their products. Other policy ideas include levying higher sales tax rates on products with short life spans, establishing minimum standards of durability and upgradability for certain types of products, and mandating longer product warranties. Bills have been considered recently at the federal and state levels that would establish a "right to repair" certain products for consumers.

However, many businesses and economists point out that obsolescence is often a consequence of useful innovation and a reaction to the changing tastes and needs of their customers. Further, they point to solutions that are based on business model innovations and consumer actions. For example, computer and cell phone companies are now establishing buy-back programs that credit customers for returning used products that can then be refurbished and resold. The electric vehicle company Tesla has designed car batteries that can be easily refurbished into home energy storage systems.

Another business model is for companies to rent or lease products to consumers. Under this arrangement, consumers pay a monthly subscription fee to use the product, and the company takes responsibility for upgrading or replacing it for the length of the contract. This can give the company a greater incentive to ensure product durability. Apple has moved in this direction with its iPhone upgrade program.

Finally, consumers can take actions to address obsolescence by purchasing goods that are likely to last a long time and remain in style. Consumer advocates suggest that this can be done by supporting companies with generous return policies or lifetime warranties and by resisting immediate adoption of the trendiest new products.

What Would You Do?

Can you think of a product you own that has become or is becoming functionally or fashionably obsolete? Who do you think should bear the primary responsibility for addressing the environmental impacts of rapid product obsolescence: consumers or manufacturers? Why? What do you think is the most promising strategy to deal with this challenge?

FIGURE 18.6 **Visual Feedback** Devices like this that give visual real-time feedback about energy consumption are effective ways to encourage energy conservation.

feedback a prompt that contains information specific to the impact of an individual's behavior.

commitment an action that is taken to express a general attitude on an issue.

cognitive dissonance an unpleasant tension or stress we feel when we are made aware that our attitudes and actions do not match.

Feedback

A prompt that contains information specific to the impact of an individual's behavior is called **feedback**. Many cars now have dashboard displays of fuel efficiency that provide continual feedback on how one's driving behavior affects fuel use.

If you accelerate quickly, you can see the fuel economy dip, and research shows that such devices do lead people to adjust their driving behavior to conserve fuel. Studies have also shown that when utility bills compare energy use for the current month with that of the same month in previous years, people will make modest reductions in their energy use. The more timely and specific the feedback, the more influence it has on individual behavior. One way to get continual and timely feedback is by installing a thermostat that provides real-time information about energy consumption for home heating and cooling (**FIGURE 18.6**). Studies have shown

FIGURE 18.7 **Environmental Commitments** Research shows that people who make commitments are more likely to adopt the behaviors. People who make their commitments public, such as by posting photos of themselves holding environmental pledge signs, reinforce their intentions. Musician Jack Johnson even established the website allatonce.org to provide a forum for his fans to post their photos holding environmental pledge signs.

that households with devices like this have reduced their energy use significantly.

Commitments

In Toronto, drivers sitting in idling cars at some school areas were approached by volunteers who distributed an informational pamphlet and asked the drivers to commit to stop idling. Drivers who made the commitment received a sticker that read "For Our Air: I turn my engine off when parked." Observational studies were done to compare driver behavior before and after the campaign was conducted at these sites and at control areas where no commitment was sought. The sites where drivers were approached to make commitments saw a reduction in idling that was 72% greater than that observed at the control sites. This is not a unique example, as researchers have found that making what are known as **commitments**—actions that are taken to express a general attitude on an issue—increases the adoption of environmentally beneficial behaviors. They often come in the form of a pledge that we affirm with a signature or oral agreement. Many times, commitments are made public with buttons, stickers, petitions, or postings online to add more weight to our intention (**FIGURE 18.7**).

Several experiments have shown that the effectiveness of the commitment strategy is strengthened when people are also made mindful of their past failures to match their concern with their behavior. In one experiment, some members of a university swim team who identified themselves as supporting water-conservation efforts were first asked if they *always* turned off the shower while soaping up or shampooing and if they *always* made their showers as short as possible to conserve water. The purpose of the question was to bring to mind the swimmers' past failures to behave in ways consistent with their concern about water conservation. These swimmers were then asked to write down their commitment to water-conservation practices on a flyer. Throughout the course of the study, these swimmers took significantly shorter showers than those who were not questioned but wrote a commitment and those who identified themselves as supporters of water conservation but were neither questioned nor wrote a commitment.

The effectiveness of this approach is due to a mental phenomenon known as **cognitive dissonance**: an unpleasant tension or stress we feel when we are made aware that our attitudes and actions do not match. In other experiments on energy and water conservation behavior, survey respondents were sent mailings that juxtaposed their statements of support for conservation in a survey and their actual household water and energy use information. In the months following the experiment, these people consumed significantly less water and energy than those who were not reminded of their survey responses but only provided with information on their water and energy use. However, other studies have

Applying a Nudge to Reduce Food Waste

If you've been through a buffet line at a cafeteria or an all-you-can-eat restaurant, you can probably relate to the saying "your eyes are bigger than your stomach." When given the option to serve themselves, people tend to take more food than they can comfortably eat. This leads to wasted food. Studies have shown that about one-third of the food we produce never gets eaten. Why should we care? Wasted food is wasted money, energy, and resources. If we wasted less and consumed only what we needed, we would spend less while reducing the environmental impacts associated with food production and waste, including air and water pollution, soil degradation, and emission of greenhouse gases that contribute to climate change.

Food waste happens for many reasons, and much of it is generated in our households, but restaurants (and buffet lines in particular) exemplify how our individual behavior and the choices we are presented with can play a role in the problem and the solution. About two-thirds of the food waste from restaurants comes from the discarded leftovers on customers' plates. We might know we're supposed to take only what we need and clean our plates, but this rarely happens in practice. In buffet lines, it would seem that you have total control over how much food you choose to serve yourself. So it might be surprising to learn that the size of the plate you use has a big effect on how much food you end up taking. Behavioral research has shown that a larger plate makes adequate portion sizes look too small to us, so we tend to fill the plate up with more food than we need. Even studies conducted on nutrition professionals have shown that they tend to overserve themselves when given larger plates.

A European consumer research group known as GreeNudge ran an experiment to see what would happen to food waste at buffet restaurants in Scandinavia's biggest hotel chain when the restaurants shrank the diameter of the plates they were using by a little over an inch. In addition to switching out the plates, the restaurants also provided prompts reminding customers that they could return to the line more than once to refill their plates. The kitchen staff at the restaurants weighed and recorded food waste both before and after this experimental treatment. The results showed that the restaurants wasted nearly 20% less food on average after switching to the slightly smaller plates. Customer surveys conducted before and after the experimental treatment did not reveal any significant decrease in satisfaction.

The authors of the study noted that if these results could be achieved throughout the European food service industry, food waste could be reduced by 2.5 million metric tons per year, leading to billions of dollars in savings and a reduction of food production–related carbon dioxide emissions by 6 million metric tons per year (about the amount the state of Vermont produces each year).

This experiment is an example of a "nudge": a subtle, often imperceptible change in the way options are presented that helps guide us toward certain desired outcomes. GreeNudge conducts research to help businesses and governments design nudges that reduce environmental impacts. In Great Britain, the government set up a Behavioral Insights Team known as the "nudge unit," whose research includes experiments on energy conservation and waste reduction behavior. This promising area of research shows that minor adjustments to the context in which we make our individual decisions can have major collective impacts.

24 centimeters (9.45 inches)

21 centimeters (8.27 inches)

More than 1 billion tons of food are wasted each year at various stages from production through consumption. One solution to food waste in buffet-style restaurants is to provide slightly smaller plates, an example of a "nudge" that alters our behavior.

What Would You Do?

Can you think of an example on your campus where a "nudge" could reduce your school's overall environmental impact? Describe why you think the nudge would work and how you would make the case to implement the nudge to your administration.

Feedback and Fuel Economy

Last 100 mi

Best
36.8
mpg

Avg
32.4
mpg

Fuel Economy

14

Fuel economy driver interface (FEDI) devices like this one give feedback to drivers on how their driving is affecting the fuel efficiency of their vehicles. Research by the National Highway Traffic Safety Administration measured the fuel economy of drivers who were asked to "drive as you would normally" in stop-and-go traffic to get a baseline of the drivers' fuel economy. Then the drivers were asked to "drive fuel efficiently" in the same kind of traffic, although they were given no instruction on driving behaviors that would improve efficiency. During this time, two groups of drivers were provided with FEDI devices (one displayed average fuel-efficiency information and the other displayed both average and instantaneous fuel-efficiency information) and one group was not. While all of the drivers increased their fuel economy compared to the baseline when they were asked to "drive fuel efficiently," those with the FEDI devices that displayed both average and instantaneous fuel-efficiency feedback drove more "smoothly" with less abrupt acceleration and braking. Their fuel economy was several miles per gallon greater than the fuel economy of drivers with FEDI devices that only provided average fuel efficiency and of drivers without any FEDI device.

---

motivation the drive we feel toward the behaviors we pursue.

shown that this strategy can backfire if the participant's attitude of concern is not strongly held. This is because people can also resolve cognitive dissonance by bringing their attitudes in line with their behavior: justifying their actions by reducing or eliminating their attitude of concern to eliminate discomfort with the idea that they were hypocritical.

⌂ **TAKE-HOME MESSAGE** Strategies that bring attitudes to mind at the point at which decisions are being made can influence behavior. This can be done with prompts that help people link an attitude to a behavior at the moment they are deciding how to act. It can also be done by providing feedback on an individual's impact while they are considering or carrying out a behavior. Gaining written or oral commitments from people to carry out certain behaviors increases the likelihood that they will follow through with the prescribed actions.

18.3 How Can Incentives Motivate Behavior?

Any time you are poised to make a decision, you are likely considering various incentives. Imagine you are at an outdoor summer music festival with three friends and you are walking between venues. You haven't had anything to drink all day, and you are feeling thirsty when you see an advertisement directing you toward "ICE COLD, REFRESHING WATER AHEAD." The sign leads you to a vendor selling bottled water for $5, and you have exactly $5 in your pocket. Each of your friends buys a bottle of water and stops to socialize, but are you motivated enough to buy a bottle?

If we examine how this situation works, we can see both internal and external factors acting on us. Our **motivation** is the drive we feel toward the behaviors we pursue, and it is the product of an interplay between these factors. For example, water is a fundamental human need, and our physiologic response to dehydration is a growing feeling of thirst. This is a push we feel from within: an internal factor motivating us to buy the water. Other internal factors operating here might be the need or desire to conform to our friends' behaviors and our attitude about bottled water and packaging waste. But external factors also play a role. As important as internal factors such as attitudes, needs, and desires are to individual behavior, it is important to remember that external factors such as incentives also play a big role in shaping our decisions and actions. There are incentives at work here, such as the advertisement for the water, the price, and what our friends are doing. In this section, we will study some of their effects on us.

How Do Incentives Work?

Incentives heighten our motivation to act a certain way. When we make decisions, we are pulled by the consequences and outcomes we anticipate. In everyday conversation, incentives are typically spoken of only as positive motivators, but incentives can either lure us toward or repel us from certain behaviors. In the bottled water example, we might feel our need for water in our thirst, but our motivation to meet that need by buying bottled water is enhanced by the advertisement. The advertisement is a positive incentive luring us to buy the water. The price of the bottled water is also an incentive. If you thought $5 for a bottle of water was a bargain—a better deal than you would find elsewhere at this concert—the price would be functioning as a positive incentive. However, most people consider $5 relatively expensive for a bottle of water, and some might decide not to buy the water at that price. You might even be motivated to bring your own reusable water bottle to your next concert. In these cases, the cost would be a negative incentive.

(a)

(b)

PAY AS YOU THROW

because waste is just a resource out of place

(c)

FIGURE 18.8 **Recycling Incentives** **(a)** Many US states have bottle deposit fees that consumers can regain if they return cans or bottles for recycling. **(b)** In some cities, passengers can even pay for their public transportation with recyclables. This vending machine in Istanbul accepts plastic bottles and aluminum cans and adds their value to a commuter's fare card. **(c)** Yale University tested a Pay As You Throw disposal system on campus. Customers were charged for the amount of garbage they threw away. They were charged at a lower rate for recycling and composting.

We often think of incentives in terms of economic costs and benefits. Although we are powerfully motivated by economics, humans are also motivated by many other needs and desires. It might be more useful to think of incentives as the pros and cons you would consider when making a major decision such as choosing where to go to college. For most people, financial cost is a major factor they consider, but other things would likely be tallied as pros and cons as well: location, availability of classes and online programs, whether the school offers the academic programs and extracurricular activities that interest you, whether you like the campus, and maybe even how others will regard your decision. Similarly, in most large cities, if the decision to drive a car or take public transit were based on economic costs alone, public transportation would be a much more attractive option. However, other influential pros and cons for these transportation alternatives match up with other needs and desires, such as prestige, travel time, flexibility, and social comfort. Many people are willing to pay a little bit more to drive because they are attracted to the other types of benefits it provides. But when the cost of driving is increased by substantial amounts because of increasing gas prices, traffic congestion, or lack of parking, public transportation becomes more attractive.

Incentives as a Strategy to Influence Choice

Frequently, we see incentives being used to promote certain choices (**FIGURE 18.8**). Positive incentives will reward people for adopting a particular behavior. For example, to encourage recycling, many states charge a deposit on beverages and will repay the deposit when empty containers are returned. Nationally, the government incentivizes purchasing homes by letting home buyers deduct all the interest on their mortgage from their income. Negative incentives, or disincentives, raise the cost associated with particular behaviors in an attempt to reduce or eliminate them. State taxes on cigarettes and alcohol are a negative incentive, sometimes called "sin taxes." Some cities now charge a fee for garbage collection that goes up relative to the volume of the bin used by residents. In order for a financial incentive like this to influence our choices, we need to be aware of it as we are poised to make a decision. Some grocery stores attempt to discourage customers from using disposable plastic bags by either charging a fee for plastic bag use or granting a credit for use of a reusable bag. But if the fee or credit is simply incorporated into the bill and the amount is too small for us to notice, we may not even be aware of the incentive, and it is unlikely to influence our behavior.

Social Incentives

A survey of California residents revealed that people often perceive financial and altruistic motivations (such as benefits for the environment or future generations) as having a strong influence on their behavior, but they consistently rank social norms (conformity with how others are behaving) as having the weakest influence. But are we really so independent? Probably not. Earlier in this chapter, we learned that humans have a fundamental need for acceptance and often conform to the social norms of the people with whom they associate. Although we might like to think of ourselves as immune to peer pressure, research shows that our perception of social norms has a strong influence on our behavior. Simply receiving a message that "everybody's doing it" tends to motivate people to conform their behavior to the norm and perform the action. Consider signs you may have seen in hotel rooms asking you to hang your towels back on the racks for reuse in order to save water. The same research team that

Strategies for Influencing Behavior

Even when we are concerned and knowledgeable about an issue, it sometimes takes an extra nudge to get us to change our behavior. Here are some successful strategies that help us align our actions with our ideals.

STUDENT CENTER

Commitments

Commitments elicit pledges from people to carry out certain behaviors, which increases the likelihood that they will follow through.

I PROMISE TO REDUCE WASTE
...by bringing my reusable mug to the café.

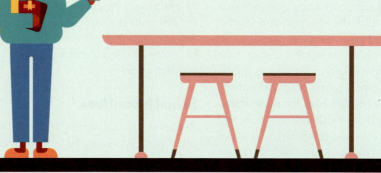

Reuse Your Cup
SAVE 20%

Incentives

Incentives are positive or negative signals that influence our behaviors.

100% organic

PLEASE RECYCLE

Feedback

Feedback provides information specific to the impact of a particular behavior.

CAMPUS-WIDE ENERGY USAGE:

Social Influence

Social influence can be used to encourage or discourage certain behaviors by making individuals aware that other people around them are behaving in a certain way.

ENERGY- SAVING REMINDER
Remember to turn your computer off at the end of the day.

Prompts

Prompts can be signs or physical reminders that call a person's attention to a desired behavior at the moment the person is deciding how to act.

PLEASE RECYCLE

Labels and Guides

Labels and guides can help consumers compare the environmental impacts of various products.

SUSTAINABLE SHOPPING GUIDES

administered the survey to California residents conducted experiments in hotel rooms using these signs, giving some rooms signs with added information stating that 75% of the guests at this hotel participated in this water-conservation effort. Guests in rooms given information about the behavior of other guests reused their towels at a significantly higher rate than guests in rooms with the standard signs. When the signs were modified further to report that 75% of guests staying in this room (with the percentage and room number handwritten on the sign) chose to reuse their towels, reuse rates climbed even higher relative to those of the other rooms in the hotel.

What about the California residents who thought social norms only had a weak effect on their energy conservation behavior? A follow-up experiment showed that they may have felt this way, but their actions were different. This survey divided hundreds of households in a California neighborhood into four groups. Each household in a group received a door hanger that described energy conservation actions and included an appeal specific for that group. The door hanger for the first group made an appeal based on environmental concerns, the appeal to the second group was based on social responsibility, the appeal to the third group was based on financial savings, and the appeal to the fourth group was based on a statement about what others in the neighborhood were doing. The message for this last group displayed survey results from the community indicating that a majority of residents in the neighborhood were taking certain conservation actions. After the study, the households that received the last message ranked the effectiveness of the door hanger significantly lower than did any of the other groups. Clearly, they did not think that social norms had much influence on their actions. Yet their energy consumption tells a different story. Researchers studied energy consumption across all of the door hanger groups before and after the door hangers were distributed. The households who were given the message about their neighbors' energy conservation actions reduced their energy consumption significantly more than any of the other groups in the study (**FIGURE 18.9A**).

Social media can be used as a tool to deliver social incentives and rapidly multiply the number of individuals taking particular actions. In recent years the movement known as Fridays for Future, or Youth for Climate, has used social media to organize students to strike in an effort to raise climate awareness. Climate activist Greta Thunberg famously staged a protest outside the Swedish parliament that was widely shared, and more than 1 million youth across 125 countries joined to strike in March 2019 (**FIGURE 18.9B**).

⬡ **TAKE-HOME MESSAGE** Incentives can be positive or negative stimuli that either lure us toward or repel us from certain behaviors. Incentives are often put in terms of costs and benefits, but they can involve many of our needs and desires, including prestige and comfort. Social incentives work on the basis of the human need for acceptance and our tendency to conform with our peers.

18.4 How Important Are the Words We Use?

Often, the words we use to describe things have different connotations. Consider the words *swamp* and *wetland*. Does one of these words have a more "positive" meaning for you when you hear it? Although the two words can describe the same physical feature, most people associate more positive meanings with the word *wetland*. Wetlands tend to be understood as natural areas and habitat for migratory birds, while swamps tend to be associated with less desirable attributes such as mud, muck, and mosquitoes. These connotations matter because they can lead us toward certain policies. It's common for people to associate wetlands with policies that protect those areas, while swamps are associated with policies designed to remove them (think of the phrase "drain the swamp").

Each day, most of us are exposed to hundreds of different messages designed to influence our decisions,

FIGURE 18.9 Social Norms as Incentives
(a) Showing consumers how their energy use compares to the energy use of those around them can motivate behaviors that conserve energy.
(b) The original protest of Greta Thunberg (in the yellow jacket) outside the Swedish parliament was widely shared across social media and kicked off the global Youth for Climate movement.

(a)

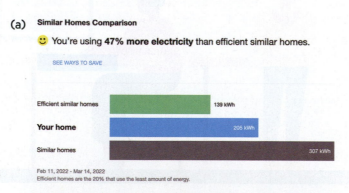

Similar Homes Comparison

☺ You're using **47% more electricity** than efficient similar homes.

SEE WAYS TO SAVE

Efficient similar homes — 139 kWh
Your home — 205 kWh
Similar homes — 307 kWh

Feb 11, 2022 - Mar 14, 2022
Efficient homes are the 20% that use the least amount of energy.

(b)

mostly advertisements on the internet, television, radio, print media, billboards, and other outlets we encounter. If you examine these messages closely, you will see that they are not simply describing products or services. Advertisers attempt to capture our attention and influence our attitudes and behaviors by use of marketing research and carefully crafted messages. Most of these messages use what are called **frames**, or mental shortcuts that use word associations and images to help people quickly apply meaning to new information. As we grow, develop, and observe the world around us and interact with others, our brains build neural circuitry for frames, which act as shortcuts to help us judge, decide, and act on complex information. When you hear a word, one or more frames come to mind, giving you a simple and limited interpretation.

To emphasize just how quickly frames evoke meaning for us, in a famous example, the linguistics professor George Lakoff would ask his students to think about what happens when someone tells you, "Don't think of an elephant." When we receive this command, our unconscious minds quickly conjure up a mental image of an elephant before our reflective mental functions can interpret the order not to think of one. Taking this example a step further, what comes to mind when you read the words "baby elephant"? Chances are these two words evoke more than just an image of a small (and probably cute) elephant: you probably also have an emotional response. We might feel a sense of affection for the baby elephant and even an ethical duty to protect it—all in a split second—because we are associating the baby elephant with ready-made frames stimulating our emotions as well as ideas of good and bad or right and wrong. Frames can even evoke sensory reactions such as pleasant (or disgusting) smells, tastes, or sounds; think of a food you love (or hate), and you can experience this right now.

Although individuals have many frames in common, people sometimes differ in the associations they make with certain words because of variations in cultural background and personal experiences. This is why the chocolate candy bar Plopp is quite popular in Sweden but has never really caught on in English-speaking countries where the sound "plop" carries negative connotations we would rather not have associated with our food (**FIGURE 18.10**). But let's return to elephants. What does the image of the elephant in your mind look like? What experiences provide the basis for your image? Is it a cartoon elephant, a circus elephant, an elephant

in the wild? An individual who grew up on a farm in Botswana where elephants threatened lives and livelihoods by trampling crops would likely have a different frame for the word *elephant* than someone whose most significant associations come from animated movies or elephant conservation campaign pamphlets. The farmer's frame might be associated with fear and the ethical intuition to protect people from elephants rather than the other way around.

But frames can change. For example, until the late 19th century, the word *wilderness* had a negative and fearful connotation: something to be conquered. It was also sometimes referred to as *wasteland* because it was regarded as something underutilized that ought to be cultivated. Since that time, the dominant frame for *wilderness* has come to connote special places deserving of protection where human use is limited. This is codified in the Wilderness Act of 1964, which defines wilderness as "an area where the earth and its community of life are untrammeled by man, where man himself is a visitor who does not remain." Similarly, in the late 19th and early 20th centuries, cityscapes with billowing smokestacks were symbols of economic pride and described as "silvery banners of progress" in city promotional materials (**FIGURE 18.11**). By the 1960s, the dominant frame associated with smokestacks became pollution and threats to human health.

Using Frames to Design Narratives

While frames may provide only simple interpretations that fail to capture the complex realities behind various ideas, our reliance on them makes them very powerful. Influential communication is built around messages that

frame a mental shortcut that uses word associations and images to help people quickly apply meaning to new information.

FIGURE 18.10 Words Matter Words in one culture can have a completely different frame in another. This brand of Swedish candy might not do so well in an English-speaking country.

FIGURE 18.11 Frames Can Change over Time In the early 20th century, the image of smokestacks actually carried a positive connotation of economic and technological progress and prosperity. This image is from a 1910 travel postcard.

use frames that drive certain attitudes and motivate certain behaviors. We often see this in practice in political campaigns where competing narratives try to win over public opinion and support. **FIGURE 18.12** shows competing narratives addressing the issue of gray wolf reintroduction and protection in the southwestern United States. Note how both the billboard and the protesters use images and a small number of words to quickly convey strikingly different narratives. The billboard identifies wolves as something to be outright eliminated, with an image of one simply crossed out and the words "NO, NO, NO." The implication is clear that wolves are threats who have no place in the landscape. On the other hand, the protesters' signs include "I stand for wolves," and one protester is even dressed in a friendly wolf costume. They say wolves should be protected because "they were here before anyone" and so are a natural part of the ecosystem. Note that neither the billboard nor the protest devotes space to facts or figures on the wolf issue. Each is a not-so-subtle attempt to link the wolf issue to a particular narrative in the hopes that it will influence how people interpret the rest of the information they encounter on the issue.

Labels and Guides

The way products present information either on their packaging or in a shopping guide can activate frames that have a significant impact on consumer choices. In experiments where some consumers were given detailed information on the hazardous and solid waste impacts associated with products, these consumers were much more likely to purchase products with the lowest waste impact than consumers not provided with this information. There are several ways that information strategies like this can be widely implemented. One way is to provide warnings or disclosure statements for products that have certain environmental impacts—much like the government-mandated health

warnings that are posted on cigarettes and alcoholic beverages. We often see these warnings on household products such as cleaning supplies, paint, and pesticides that contain toxins requiring disposal as hazardous waste. This messaging helps consumers choose products that are made without toxic materials. Similar warnings have worked on products that contribute significantly to long-term environmental damage. Prior to their phasing out in the 1990s, products with chemicals that depleted Earth's ozone layer required labels with this information.

Labels or special certifications can also identify products that meet or exceed the highest government environmental standards. The US Department of Agriculture certifies crops as organic if they are produced without synthetic fertilizers, prohibited pesticides, genetically modified organisms, irradiation, or sewage sludge. The Energy Star label, used in the United States and several other countries, identifies energy-efficient products such as computers and appliances that go beyond government energy-efficiency requirements, typically using 20% to 30% less energy than competing products (**FIGURE 18.13**). Nongovernmental organizations also create certifications for products meeting certain environmental standards. For example, as we saw in Chapter 9, the Forest Stewardship Council certifies wood products such as lumber and paper that are sourced from trees harvested with approved forest management practices.

Even with labels like these, consumers sometimes find that making comparisons while shopping is difficult. To help, more informative guides have been created. The Environmental Protection Agency (EPA) has developed Energy Guide labels for appliances: these are designed to help shoppers compare the estimated annual electricity use and the average annual cost of various brands and models (**FIGURE 18.14**). The EPA produces a similar label to help consumers compare the fuel-economy estimates for different kinds of cars.

FIGURE 18.12 Competing Narratives On contentious issues, proponents and opponents attempt to influence the public by assembling powerful frames into compelling narratives, such as the ones pictured here advocating and protesting the reintroduction of gray wolves in New Mexico.

(a)

(b)

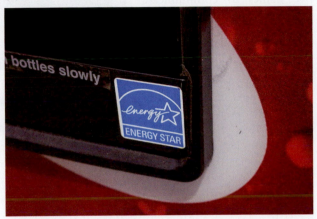

FIGURE 18.13 Using Labels to Convey Environmental Information Product labeling is a communication strategy that can be used to convey information about **(a)** environmental harms or **(b)** environmental benefits.

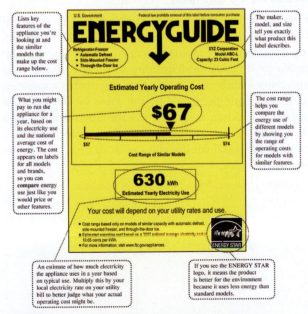

FIGURE 18.14 Using Guides to Help Consumers Make Comparisons Energy Guide labels, like this one for water heaters, can help consumers compare products.

➔ CONTEXT

How Do Frames Work?

The word *frame* comes from the way in which photographers choose to represent the reality they see by composing photos in different ways. Photographers can set boundaries around the images they produce, either by zooming in or out before taking the picture or by cropping the photo afterward, and they use other techniques to draw the attention of viewers toward certain focal points in a photo. Notice how these two photos differ. The photo on the right is cropped to include only the zebras and does not allow the viewer to see the larger context of the city of Nairobi, Kenya, in the background. These two images can lead the viewer to understand the image in different ways. The image on the left invites the viewer to consider the impact of a city so close to the zebras' habitat. A similar thing happens when words are used to frame a situation in particular ways. Consider the way words for similar things are associated with different thoughts and feelings, like *dirt* versus *soil* or *swamp* versus *wetland*. Consider each word pair, and think of the associations that come to mind.

These labels identify not only the estimated annual fuel economy and cost but also how well the vehicle compares to other vehicles in its class. Some environmental groups have developed shopper's guides to help consumers purchase products with the least environmental impact. The Monterey Bay Aquarium Seafood Watch program identifies particular kinds of seafood as a best choice, good alternative, or food to avoid. This program started as a wallet-sized pamphlet but is now available as an app that consumers can use on their phones by scanning products while they shop. Similarly, the Joro app allows shoppers to see the carbon footprint of each purchase they make.

⬡ **TAKE-HOME MESSAGE** The way in which information is presented can influence our choices and actions. Our brains make sense of information we receive through patterns of meaning known as frames. The choice of images and words one uses to communicate activates different frames. Multiple frames can be assembled into narratives designed to influence attitudes and behavior.

18.5 What Can I Do?

At the start of this chapter, we talked about deciding on the kind of bag to use at the grocery store. Bringing a reusable bag to the store has a lower environmental impact than that of disposable paper or plastic bags after just a couple of months of weekly use. While this may seem like one of the easiest behaviors we can adopt to benefit the environment, we can now see that it is not so simple. The individual choices we make are influenced by a range of factors, including the way we prioritize our needs and desires, common biases in our automatic thinking, and the pressure to conform to those around us. But we have also seen examples of strategies that can be employed to encourage beneficial behaviors such as using a reusable bag. Let's look at a few ways that you can use decision-making strategies such as prompts, commitments, financial and social incentives, and communication strategies to motivate yourself and others to act sustainably.

Make Commitments, Use Prompts, and Use Tools to Keep Track of Your Progress

The EcoChallenge Platform is an online tool hosted by EcoChallenge.org (formerly the Northwest Earth Institute) that is designed to promote the accumulation of individual actions into big collective impacts. You and/or your team representing a workplace, group of friends, or college class make commitments to adopt certain environmentally beneficial behaviors,

EcoChallenge.Org

such as taking shorter showers, carpooling, or eating a certain number of meatless meals each week. As you log progress on your goals, the website gives you feedback on your environmental impact and your progress relative to other individuals and teams taking part in the challenge. As individuals are spurred toward their goals, they adopt environmentally beneficial habits that add up to significant impacts across all of the participants. Since the program began in late 2016, almost 221,000 people have participated and have collectively taken and recorded more than 1.9 million actions to reduce their environmental impacts. Since participants are

part of teams, social incentives also play a role. For example, teams spur each other toward eating more sustainably by swapping recipes and shopping and food preparation tips. This has resulted in hundreds of thousands of plant-based and zero-waste meals consumed in recent years. Some of the other accumulated benefits of the actions include more than 623,000 miles (more than 1 million kilometers) traveled by foot, bike, public transportation, or carpool, and more than 6.7 million gallons (25.3 million liters) of water conserved.

Communicate More Effectively to Influence Changes You Want to See

Consider using the most effective frames to communicate your message. In a recent article in *Nonprofit Quarterly*, Julie Sweetland and Rob Shore, experts at FrameWorks Institute (a nonprofit communications think tank), suggest several ideas. They cite research that negative "crisis" messaging becomes less effective than messaging that provides solutions. Use this concept to write more positive social media posts to influence sustainable behavior. Similarly, they suggest using more educational posts to build interest and credibility and provide information to your readers.

Find Places Where Prompts Might Influence Behavior on Campus

Prompts are a proven strategy in advertising. Especially if your university has a business school, consider partnering with someone in marketing or advertising to help create prompts around campus. Or learn where your university stands in a ranking of sustainable campuses, such as the one published by the *Princeton Review*. Find out your ranking, identify areas for improvement, and then start brainstorming some ways your institution could make sustainability gains with strategies for influencing behavior.

Chapter 18 **Review**

SUMMARY

- People do not weigh all of their needs and desires equally. Instead, we are generally guided by an internal hierarchy that privileges needs linked to survival and social acceptance.

- Our decision-making processes are not wholly based on deliberate, conscious, and reflective thinking. Automatic thinking also plays an important role.

- All people are prone to patterns of bias—errors due to automatic thinking that lead to misunderstandings and errors in judgment.

- Humans are social animals, and the desire for social acceptance has a significant influence on behavior.

- Successful strategies to influence behavior often work by helping people link an attitude to a behavior at the moment the decision to act is being made.

- Incentives are stimuli that either lure or repel us from certain behaviors, and they can involve money or material goods as well as intangible goods such as prestige.

- We interpret information we receive with the help of mental shortcuts known as frames.

- Influential communication is built around images, words, and narratives that use frames to drive certain attitudes and motivate certain behaviors.

KEY TERMS

automatic thinking	delegation	frame	prompt
belief perseverance	denial	hierarchy of needs	resignation
cognitive dissonance	emotional defense mechanism	incentive	social norm
commitment	emotional distancing	loss averse	status quo
confirmation bias	feedback	motivation	

REVIEW QUESTIONS

The letters following each Review Question refer to the Chapter Objectives.

1. Which need is most important (ranks highest) on Maslow's hierarchy of needs? Which ranks lowest? **(B)**

2. Define and describe what *automatic thinking* is and provide an example of how it is a common source of bias. **(C)**

3. Name three types of emotional defense mechanisms. **(B, C)**

4. What is the evolutionary root of humans' tendency toward conformity? **(A)**

5. Define and describe *confirmation bias*, and provide an example. **(C)**

6. What are prompts, and how can they be used to influence behavior? **(D)**

7. What differentiates feedback from other types of prompts? Provide an example of how feedback can be used to promote more sustainable behaviors. **(D)**

8. What psychological phenomenon explains why acknowledging failures to live up to certain environmental commitments seems to promote future adherence to these environmental commitments? **(D)**

9. What are incentives, and how do they influence our motivation to act a certain way? Provide an example of a positive incentive and of a negative incentive. **(E)**

10. Describe how our psychological need for acceptance and desire to conform can be used to influence our behavior. Provide an example. **(A, D)**

11. What are frames? How can they be used to influence our decision making? Provide an example of framing on an environmental issue. **(F)**

12. Compare and contrast labels and guides. **(D, F)**

FOR FURTHER THOUGHT

The letters following each item refer to the Chapter Objectives.

13. In this chapter, we learned about key psychological factors influencing our decisions. Choose one of the following factors, and describe an instance in which it influenced your decisions affecting the environment: hierarchy of needs, bias, emotional defense mechanisms, social acceptance, conformity. **(A, B, C)**

14. Identify a common behavior on your campus or where you live that you think could be changed or improved to reduce an environmental impact. Describe why you think the behavior is occurring and how you think it should be improved, and use a combination of the strategies to influence behavior covered in this chapter to design a plan to achieve these improvements. **(A, B, C, D)**

15. Find an example in which certain incentives are preventing widespread adoption of a sustainable behavior. How could the incentives be changed to promote this behavior? **(E)**

16. Choose a controversial environmental issue that is important to you. Find information provided by interests on various sides of the issue, and provide analysis of the way each side is framing the issue. Explain how the sides differ in their representation of the issue and why you think they are choosing specific words and/or images. **(F)**

Make Your Case

The following exercises use real-world data and news sources. Check your understanding of the material and then practice crafting well-supported responses.

Use the News

We have seen in other chapters that food waste is a large problem in the United States and Canada. This article describes a study suggesting that food label dates are a factor contributing to this problem. Use this article to answer the questions that follow. The first three questions test your understanding of the article. Question 4 is short answer, requiring you to apply what you have learned in this chapter and cite information in the article. Answers to Questions 1–3 are provided at the back of the book. Question 5 asks you to make your case and defend it.

"Widespread Confusion about Food Safety Labels Leads to Food Waste, Survey Finds," Natalie Wood-Wright, *The Hub*, March 14, 2019

A recent survey examining US consumer attitudes and behaviors related to food date labels found widespread confusion, leading to unnecessary discards, increased waste, and food safety risks. The survey analysis was led by researchers at the Johns Hopkins Center for a Livable Future, which is based at the Johns Hopkins Bloomberg School of Public Health.

The study, published online in the journal *Waste Management*, comes at a time of heightened awareness of food waste and food safety among both consumers and policymakers. The US Department of Agriculture estimates that 31 percent of food may be wasted at the retail and consumer levels.

This study calls attention to the issue that food may be discarded unnecessarily based on

food safety concerns, though relatively few food items are likely to become unsafe before becoming unpalatable. Clear and consistent date label information is designed to help consumers understand when they should and should not worry.

Among survey participants, the research found that 84 percent discarded food near the package date "at least occasionally," and 37 percent reported that they "always" or "usually" discard food near the package date. Notably, participants between the ages of 18 and 34 were particularly likely to rely on label dates to discard food. More than half of participants incorrectly thought date labeling was federally regulated or reported being unsure. In addition, the study found that those perceiving labels as reflecting safety and

those who thought labels were federally regulated were more willing to discard food.

New voluntary industry standards for date labeling were recently adopted. Under this system, "Best if used by" labels denote dates after which quality may decline but the products may still be consumed, while "Use by" labels are restricted to the relatively few foods where safety is a concern and the food should be discarded after the date. Previously, all labels reflected quality and there was no safety label. The researchers found that among labels assessed, "Best if used by" was most frequently perceived as communicating quality, while "use by" was one of the top two perceived as communicating safety. But many had different interpretations.

"The voluntary standard is an important step forward. Given the diverse interpretations, our study underlines the need for a concerted effort to communicate the meanings of the new labels," says lead author Roni Neff, who directs the Food System Sustainability Program with the CLF and is an assistant professor with the Bloomberg School's Department of Environmental Health and Engineering. "We are doing further work to understand how best to message about the terms."

Using an online survey tool, Neff and colleagues from Harvard University and the National Consumers League assessed the frequency of discards based on date labels by food type, interpretation of label language, and knowledge of whether date labels are regulated by the federal government. The survey was conducted with a national sample of 1,029 adults ages 18 to 65 and older in April 2016. Recognizing that labels are perceived differently on different foods, the questions covered nine food types including bagged spinach, deli meats, and canned foods.

When consumers perceived a date label as an indication of food safety, they were more likely to discard the food by the provided date. In addition, participants were more likely to discard perishable foods based on labels than nonperishables.

Raw chicken was most frequently discarded based on labels, with 69 percent of participants reporting they "always" or "most of the time" discard by the listed date. When it came to prepared foods, 62 percent reported discards by the date label and 61 percent reported discards of deli meats. Soft cheeses were near the bottom of the list with only 49 percent reporting discards by the date label, followed by 47 percent reporting discards of canned goods and breakfast cereals.

Among foods included in the survey, prepared foods, deli meats, and soft cheeses are particularly at risk of contamination with listeria, which can proliferate in refrigerated conditions. Despite concerns of listeria, soft cheeses were rarely discarded by the labeled date. On the other hand, raw chicken was frequently discarded even though it will be cooked prior to consuming and is not considered as big of a risk. Unopened canned goods and breakfast cereal pose the least concern based on time since packaging, but were still discarded by just under half of respondents.

"Foodborne illness is misery—or worse," says Neff. "As date labeling becomes standardized, this research underlines the need for a strong communications campaign and highlights a particular need for education among those ages 18 to 34."

---

1. According to the article, what percentage of food at the retail and consumer levels is wasted, as estimated by the US Department of Agriculture?

   a. 31%
   b. 47%
   c. 69%
   d. 84%

2. Which group of individuals does the article indicate is particularly likely to rely on food label dates when making a decision to discard food?

   a. 18–34 years old
   b. 35–54 years old
   c. 55–64 years old
   d. people 65 and older

3. The article discusses the confusion caused by current labels. Which two of the following statements are supported by information in the article?

   a. People discarded perishable and nonperishable food at the same rates, interpreting the dates on the labels in the same way for both categories of food.
   b. Many people believe the federal government regulates labeling, and those who have this belief discard more food than others.

   c. People should be holding onto refrigerated foods for far longer than they do, because there is little to no risk of these foods being contaminated.
   d. Most confusion over labeling is due to federal government labeling requirements and overregulation of the food industry.
   e. If a consumer believes a date on the label reflects the safety of the food, that individual is more likely to discard the food by that date.

4. Referring back to the chapter, describe in a few sentences the problems with these labels in terms of *frames* and *framing*. Suggest one way that the idea of frames might be used to make a change on a food label to influence consumer behavior.

5. **Make Your Case** The article describes how the behavior of people 18–34 years old is an issue of particular interest to the researchers. Imagine you have been hired by the government to design a new label that will go on cereal boxes. They want you to use design elements such as color and graphics to better inform customers about the manufacturer recommendations for the date by which a store should sell the cereal to guarantee freshness and for how long a customer can keep an unopened box before it needs to be thrown away. Describe at least two ways you would do this, and make your case for why these strategies will be effective, using ideas from the chapter about how to influence the behavior of individuals.

Use the Data

The way in which data are presented on product labels can influence consumer decisions. The Environmental Protection Agency has used different labels on automobiles over the years.

Study the two labels shown here, and use them to answer the questions that follow. The first three questions test your understanding of the labels. Questions 4 and 5 are short answer, requiring you to apply what you have learned in this chapter and cite data in the labels. Answers to Questions 1–3 are provided at the back of the book. Question 6 asks you to make your case and defend it.

For Model Years 1995–2008

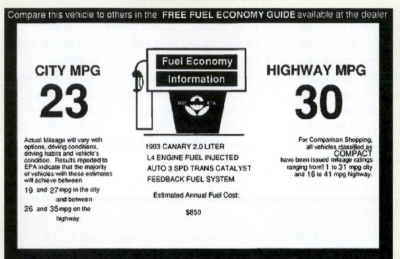

For Model Years 2013–Present

1. How much more is the annual fuel cost of the vehicle in the newer label compared to that of the vehicle in the older label?

   a. $850
   b. $1,300
   c. $2,150
   d. $3,000

2. What are three pieces of information about the vehicles that are given on both labels? Choose the three correct responses from the following list.

   a. fuel cost savings over 5 years
   b. gallons used per 100 miles
   c. estimated annual fuel cost
   d. city miles per gallon
   e. highway miles per gallon
   f. fuel cost per mile driven
   g. smog rating
   h. vehicle cost

3. What are three pieces of information that are only provided on the newer label? Choose the three correct responses from the following list.

   a. fuel cost savings over 5 years
   b. gallons used per 100 miles
   c. estimated annual fuel cost
   d. city miles per gallon
   e. highway miles per gallon
   f. fuel cost per mile driven
   g. smog rating
   h. vehicle cost

4. Using what you have learned from the chapter, name two strategies employed by these labels to influence consumer choices. Explain in a few sentences how, in a general sense, differences in the way the data are presented on the labels might influence consumer choices.

5. Describe two specific changes in how the data are presented from the older to the newer label, and explain how you think the changes would influence consumer choices.

6. **Make Your Case** Imagine you are hired to redesign the most recent version of the label. Applying what you have learned in the chapter, describe at least three design changes you would make to the most recent label and how these changes might influence consumer choices. Explain why you feel these changes are important.

LEARN MORE

- Fostering Sustainable Behavior (community-based social marketing): www.cbsm.com
- Yale Cultural Cognition Project: www.culturalcognition.net/
- Yale Program on Climate Change Communication: https://climatecommunication.yale.edu

- The Behavioural Insights Team ("nudge unit" in the United Kingdom): www.bi.team
- UC Davis Center for Environmental Policy & Behavior: http://environmentalpolicy.ucdavis.edu/

19

Groups and Organizations

How Do We Work Together for Sustainability?

Imagine yourself reading the following newspaper headline: "Millions of Wild Animals Have Already Disappeared from Africa This Century." You continue reading and learn about species on that continent that have recently gone extinct and many more that are on the brink of extinction because of habitat loss and poaching. The author of the article argues that something must be done to address this crisis and ends with a plea for readers to take action. How would you respond? How could you make this change happen?

In 1960 Victor Stolan, a Czech immigrant to England with no experience in biology, conservation work, or politics, read that headline in his Sunday newspaper and decided to organize a response to the crisis. Stolan reached out to the article's author, Julian Huxley, and pitched an idea to form an international organization to protect endangered animals by raising public awareness and gathering funds for conservation efforts. Huxley, who was a highly regarded biologist in England's Royal Society and director of the London Zoo, connected Stolan to Max Nicholson, the director of a British government research council on conservation. Then Nicholson worked with advertising executive Guy Mountfort to formalize Stolan's idea through a proposal in 1961 called "How to Save the World's Wildlife." They invited Peter Scott, a well-known ornithologist, illustrator, and host of nature shows on radio and television, to serve as the first chair of their new group: the World Wildlife Fund (WWF). Scott accepted the invitation and sketched an image of a panda that still serves as the group's logo (**FIGURE 19.1**). Today, the group is the world's largest conservation organization, with more than 5 million members across more than 100 countries.

After reading this story, you might think about the incredible work of this particular individual. But think a little more about all the connections Stolan forged to make his idea a reality. Harnessing broad networks can be very powerful. For example, in the 1990s the

The roofs of "big-box" retailers such as IKEA are great locations for solar panels. Expanding commitments to alternative energy is one way that businesses and other organizations are incorporating sustainability into their operations.

Chapter Objectives

By the end of this chapter, you should be able to . . .

A. understand the role that social networks play in influencing individuals and groups.

B. identify how an organization's purpose can influence its environmental impacts.

C. discuss the ways an organization can adjust its strategy, structure, and culture to match its mission.

D. describe the ways organizations change by developing, adopting, and implementing new ideas.

E. summarize how people within organizations and forces on the outside initiate organizational change.

Many ideas are more easily changed by aiming at a group than by aiming at an individual.

—Josephine Klein

WWF found that nearly all environmental degradation in the places throughout the world they targeted for conservation was associated with the production of the 16 commodities shown in **FIGURE 19.2**. They also discovered that 300–500 companies effectively control the trade of these commodities by playing a dominant role in how the commodities are produced, processed, consumed, or financed. For example, IKEA, the world's largest furniture retailer, uses 1% of the world's commercially produced wood to manufacture its products each year. Through its purchasing power, one major company like this has much more influence on forest practices than that of a multitude of individual consumers.

The World Wildlife Fund realized that just as its organization can exert more influence than can single individuals, businesses are organizations too and exert substantial influence. While WWF continues its education and awareness campaigns to influence consumers, it now also partners directly with large companies such as IKEA to facilitate business practices that reduce environmental impacts. IKEA, in partnership with WWF, met a 2020 goal to source more than 98% of its wood either from producers who adhere to sustainable forest management practices certified by the Forest Stewardship Council (FSC) or from recycled wood. This shift at IKEA has directly contributed to an increase of more than 35 million hectares (86 million acres) of FSC-certified forests throughout the world—an area the size of Germany.

In this chapter, we will study how different types of private organizations such as environmental groups and businesses can impact sustainability (governments and other public organizations are the subject of Chapter 20). We will see that individuals expand their influence through connections and collaborations within social networks. And organizations link networks of individuals together for shared purposes. The efforts of individuals within an organization often accomplish more than the individuals could achieve on their own, and the organizations can persist long after individual members are gone. While the collective impact of businesses and other types of organizations can be the source of adverse environmental impacts, the example in this introduction shows that organizations can also provide opportunities to address these problems.

19.1 How Are We Connected?

If you weren't reading this book and you had some free time, who are the people you would consider spending it with? If you wanted to talk about something important to you, who are the people you would contact? Researchers have found that most respondents to these questions come up with a small list of four to six friends, family members, or coworkers. A study by social scientists at Harvard on the overlap between people's social contacts found that in the United States, there is a more than 50% chance that any two of your social contacts also know each other. We tend to share the interactions in life that are most important to us with a relatively small and closed group of people. Even in the virtual social world of Facebook where users average more than 300 "friends," once researchers sort the closeness of these contacts by analyzing users' photo-posting behaviors, the results show that the average Facebook user has just 6.6 close "friends." Researchers say the people closest to you are part of your **social context**: the collection of people and groups you have contact with as you go about your daily life. What are some of these groups?

Social Groups and Organizations

Some of the people you identified earlier as your closest connections probably share membership in a group with you, as in a family or a group of friends who hang out together. We have regular interactions with people in other groups as well, even though some of the members may not be our closest contacts: for example, a group of people who regularly meet to play basketball or carpool to work. **Social groups** are just collections of people who regularly interact with each other. However, if you have a job or are enrolled in school, you are also part of an organization. An **organization** is a larger, more formal, and less personal group, such as a business or a school. Organizations are not strictly social. They are set up with specific objectives in mind—such as making a profit and/or providing certain services—and they tend to have rules, processes, and even physical structures designed to achieve these objectives. For example, most 2- and 4-year colleges and universities have a mission that is supported by thousands of people in various roles,

World Wildlife Fund Priority Commodities

Tuna
Whitefish
Farmed salmon
Farmed shrimp
Tropical shrimp
Fish meal and fish oil
Beef
Dairy
Palm oil
Soy
Sugarcane
Cotton
Pulp and paper
Timber

FIGURE 19.2 Priority Commodities to Influence Business The World Wildlife Fund prioritizes its conservation efforts by identifying the commodities that have the most adverse impacts on biodiversity and then working to influence the companies and consumers linked to producing and consuming those commodities.

Adapted from World Wildlife Fund (n.d.).

FIGURE 19.1 Forming an Environmental Organization Peter Scott, who designed the World Wildlife Fund's now famous panda logo, later said, "We wanted an animal that is beautiful, is endangered, and one loved by many people in the world for its appealing qualities. We also wanted an animal that had an impact in black and white to save money on printing costs."

from the president to the deans and various staff members, and of course the faculty and students. The people in these positions are governed by rules and less formal norms as well. When you are part of an organization, you play some role in meeting the objectives the organization is designed to accomplish. While some of your friends may be part of the organization—for instance, a classmate or a coworker—organizations also put you in association with people you know little about other than their membership in that organization.

Social Networks

Social connections extend well beyond the people closest to us and the groups and organizations of which we are members. Each of us is enmeshed in a multitude of interconnected webs known as **social networks** (FIGURE 19.3). At their core, social networks are systems of people and their connections that allow certain kinds of things—money, information, influence, and even disease—to flow between people. Social network analysis has shown that many beliefs, values, behaviors, and emotions such as happiness (or sadness) can spread from person to person to person even in the absence of conversations explicitly communicating these things. The way we think, act, and feel is often influenced by people with whom we are not directly interacting, in much the same way that a contagious illness spreads through a community.

Scholars measure social connections in **degrees of separation**, a measure of social distance between people. A direct interaction you have with another person, such as your friend or a waiter at a restaurant, is one degree

FIGURE 19.3 Social Networks Imagine that you are in the middle of this figure, and think about all the other people with whom you connect. These people, in turn, are in the center of their own networks with another web of links.

How Many Friends Do You Really Have?

Although it is common for people to have hundreds (or in some cases thousands) of "friends" on Facebook, research on the users of this social media site shows that the number of close friends is only about six people. This finding on virtual social networks tracks closely with Oxford psychologist Robin Dunbar's work on social networks within various communities. Dunbar found that the average number of friends considered "dependable" was just 4.1, and the number of friends who expressed sympathy during an emotional crisis was just 13.6.

of separation from you. The interactions those people have with others that indirectly affect you represent two degrees of separation from you. When you start thinking in these terms, you begin to realize that our social connections are quite extensive. The folk wisdom that people are connected to one another by an average of six degrees of separation is supported by experiments measuring how many steps it takes on average for a message to reach a target individual unknown to the original message sender. For example, in a recent version of this experiment, 18 target individuals from a variety of professions were identified in countries throughout the world. Then 98,000 participants from more than 150 countries were randomly assigned a target and asked to e-mail someone they knew who could help get the message closer to the target. It took about six steps on average for the e-mail messages to reach the targets.

Structure and Influence in Social Networks

You may have heard people talk about professional meetings or events as opportunities to "network" or "make connections." Often, these professionals are trying to enhance their position in the structure of social networks. Because every connection is a potential point of influence in a social network, the more connections you have with other people and the more extensive their connections, the more central you are in the network. Some networks operate within groups that have few connections to individuals on the outside. Individuals within groups like this tend to be very susceptible to whatever is flowing within their tightly knit social network, but they are fairly protected from outside influence. Other networks are more integrated, with many connections linking various groups to each other. Many of us see these structures in how social media platforms

social context the collection of people and groups you have contact with as you go about your daily life.

social group a collection of people who regularly interact with each other.

organization a larger, more formal, and less personal group, such as a business or a school, set up with specific objectives in mind.

social network a system of people and their connections that allows certain kinds of things—money, information, influence, and even disease— to flow between people.

degree of separation a measure of social distance between people.

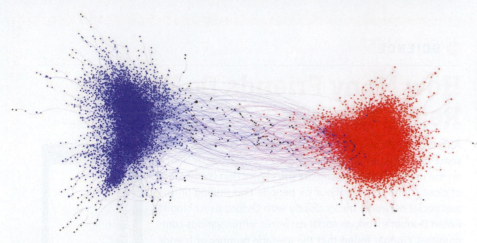

FIGURE 19.4 Polarized Networks This figure shows a network analysis of a selection of topics on Twitter. Notice the polarized nature of these networks (red represents conservative users and blue is liberal users). The lines represent retweets of another user's message, and most connections are made between similar ideological perspectives. Both the liberal and conservative users are relatively tight-knit networks that are highly susceptible to the ideas flowing through them but nearly immune to outside perspectives.

such as Instagram, TikTok, Twitter, and Facebook have connected us to individuals and groups outside of our closest associates. However, even these networks can become insular if we remain open only to those with similar views (**FIGURE 19.4**).

More-integrated groups offer their members more points of contact, and the individuals who serve as connections between groups tend to be particularly

FIGURE 19.5 Influencers Microsoft and National Geographic use an influencer marketing strategy in their "Make What's Next" campaign to encourage girls to study science, technology, engineering, and mathematics during International Women's Day. In 2017, they featured Jenny Adler (pictured), a conservation photographer and ecologist with over 16,000 Instagram followers in their campaign.

influential people. Network analysis has shown that these people are held in high regard within their communities and tend to be the ones sought after for information and advice. Marketing researchers call such people "influencers" and try to monitor what attitudes and consumer preferences are likely to spread through their social networks. These marketers sometimes harness the power of these influential individuals to help sell their products or promote their ideas. For example, rather than using traditional advertising, the watch company Daniel Wellington provides free watches to celebrities and other influencers who have large numbers of followers on social media; the company then benefits when these people feature its product in their posts.

Organizations can be influencers too. Year after year National Geographic is rated a top brand on social media by industry experts on the basis of number of followers and their engagement. To take advantage of National Geographic's social media reach, Microsoft partners with National Geographic to promote International Women's Day through Microsoft's "Make What's Next" campaign. They post photos featuring prominent female scientists and adventurers across five of National Geographic's Instagram channels. With a combined total following of 91 million users across these five channels, the Make What's Next campaign can draw more than 3.5 million total likes in a single day (**FIGURE 19.5**).

⬡ **TAKE-HOME MESSAGE** We are enmeshed within social networks that extend from those closest to us to people two or more steps removed from our direct interactions. Social networks play a role in shaping the way individuals think, feel, and act. The structure and function of certain organizations and the location of certain individuals within social networks makes these organizations and individuals particularly influential.

19.2 Why Are Organizations Important for Change?

Organizations can be particularly influential not only because of the size and reach of their social networks but also because of their scale and the particular functions they serve. We rely on organizations for our employment, schooling, health care, and physical infrastructure and for the provision of countless goods and services. Often, a shift in a business or manufacturing process can have a tremendous impact simply because of the global reach of a company. For example, the Flyknit sneakers designed by Nike use computers to conserve material during manufacturing and rely primarily on recycled plastic. The success of this model has reduced the waste generated by conventional sneaker manufacturing by as much as 80%, avoiding more than 2 million pounds (900,000 kilograms) of fabric waste a year. With changes like this, Nike is now taking a more active role in sustainability initiatives and receiving praise for its efforts. But over the years, Nike has also been the target of critics who have pointed to poor working and environmental conditions in their factories. It was in fact different sorts of organizations—colleges and universities—that pressured Nike to pay more attention to the social and environmental impacts of its manufacturing practices. In the late 1990s and early 2000s, student protests led academic institutions such as Duke University, the University of Arizona, and the University of Michigan to require companies (such as Nike) licensing the school name and logo to provide certain information. This included disclosing the locations of their manufacturing facilities for independent monitoring of wages, working conditions, and environmental impacts. In this section, we look at several different kinds of organizations, including for-profit businesses (such as Nike), nonprofits (such as universities), and benefit corporations (a hybrid of the two) to get a better sense of how they are run. Understanding what each type of organization is set up to do can help to explain its potential as a change agent for sustainability.

For-Profit Businesses

For-profit businesses largely direct their activities toward earning money. This may not be their only purpose, but it is usually the primary purpose. With few exceptions, businesses cannot succeed without a steady flow of income. As we will see later in this chapter, a business may also have a mission to provide a quality service, meet a social need, and even reduce human impacts on the environment, but it earns its revenue and maintains its existence primarily by selling goods and services. No matter what other goals this type of organization may have, if it loses money for an extended period, it will likely cease to exist.

The ownership structure of a business can have a big influence on organizational decision making. The most common type of business is **privately (or closely) held**. This means that a single individual or a relatively small group of people holds all of the ownership shares for the business. Privately held businesses can range in size from a local corner store or family-run restaurant to very large businesses such as Dell, Bloomberg, Cargill, or the candy company Mars. Sometimes these types of businesses are able to take more risks and make decisions on the basis of longer-term goals because of their ownership structure; owners need only answer to others in the ownership group. IKEA is an example of a closely held business that has undertaken ambitious sustainability initiatives such as the sustainable forest management practices mentioned at the beginning of the chapter.

In contrast, the ownership shares of **publicly traded** businesses can be bought and sold on a stock exchange

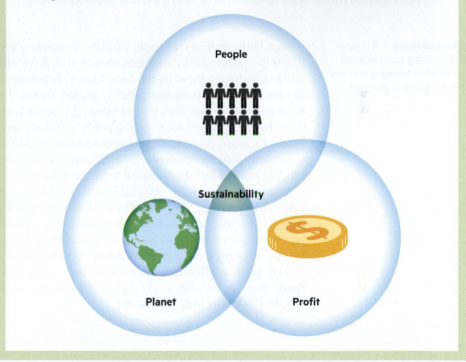

⊙ SUSTAINABILITY

Triple Bottom Line Accounting

Although for-profit businesses are established to prioritize profit above other concerns, there are models that incorporate other priorities. The **triple bottom line (TBL)** framework is one such approach. In addition to profit, TBL measures a company's impact on social equity and the environment. These three metrics are sometimes called "people, planet, and profit." TBL frameworks operate with the belief that all three should be priorities within a company working to accomplish its goals. The world's largest logistics/delivery company, DHL, is one major organization that uses the TBL approach. This has inspired a shift in DHL's operations to favor non–fossil fuel vehicles in an effort to reduce greenhouse gas emissions.

People

Sustainability

Planet

Profit

(e.g., NYSE, NASDAQ, United States; LSE, United Kingdom; FWB, Germany; TSE, Japan; SEHK, Hong Kong). Individuals or groups of shareholders such as mutual funds or pension funds typically buy publicly traded businesses in the hopes that their shares will increase in value or the company will provide regular payments (dividends) based on the number of shares (**FIGURE 19.6**). People tend to be very familiar with publicly traded businesses, as their share prices and transactions are actively reported, and they range from the most highly valued and profitable companies in the world (e.g., Apple, Google, Amazon) to smaller companies (e.g., Peet's Coffee, Krispy Kreme). Maximizing shareholder

privately (or closely) held the type of business in which a single individual or a relatively small group of people holds all of the ownership shares.

publicly traded the type of business that has ownership shares that can be bought and sold on a stock exchange.

triple bottom line (TBL) an accounting method for business operations that values social and environmental concerns as well as profit.

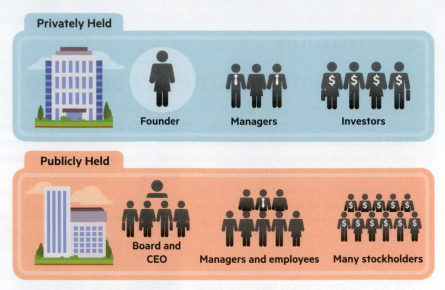

Privately Held

Founder

Managers

Investors

Publicly Held

Board and CEO

Managers and employees

Many stockholders

FIGURE 19.6 For-Profit Businesses Businesses are for-profit organizations that can be either privately held (an individual or small group holds all of the ownership shares) or publicly traded (shares of the company can be bought and sold on a stock exchange).

greenwashing the process of advertising environmental values while engaging in unsustainable business practices.

wealth is the primary purpose of these businesses, and this will almost always take precedence and drive any initiatives the business might take. Court decisions in shareholder lawsuits against publicly traded companies regularly uphold this principle. However, in recent years there has been an emphasis among publicly traded companies to adopt corporate social responsibility (CSR) as a company goal. For example, L'Oréal USA—a skin-care products company—has adopted a goal to source 100% of its raw materials from sustainable sources that do not cause deforestation. Similarly, BMW of North America aims to run its plants only on renewable energy and is pursuing one effort to use methane from a landfill near its manufacturing plant in Spartanburg, South Carolina, to help provide electricity. It's even possible to invest in funds that are focused solely on companies that meet certain CSR standards (**FIGURE 19.7**).

FIGURE 19.7 Socially and Environmentally Responsible Investing One attempt to increase investments made to generate positive impacts is the nonprofit Global Impact Investing Network (GIIN). It attempts to promote investments with the "intention to generate social and environmental impact alongside a financial return."

There are many critics of CSR, most famously the economist Milton Friedman, who wrote, "There is one and only one social responsibility of business—to use its resources and engage in activities designed to increase its profits so long as it stays within the rules of the game." Others criticize CSR efforts as more about marketing and less about solving problems—a shortcoming that is sometimes called **greenwashing**. This term was coined by the environmentalist Jay Westerveld in 1986 to criticize companies that used advertising to extol environmental values while their actual business practices were unsustainable. A famous example was a 1980s advertising campaign by Chevron that used ads promoting the company's small investments to preserve animals such as butterflies, while the company itself was violating environmental laws such as the Clean Air Act. Nevertheless, consumers are increasingly holding companies to higher social and environmental standards. The research firm Nielsen has reported that more than 80% of millennial and Gen Z consumers expect companies whose products they buy to act sustainably and ethically.

Nonprofit Organizations

Nonprofit organizations are fundamentally different from for-profit businesses in that they do not have owners or shareholders who can profit from the revenue they generate. This type of organization also tends to rely on a different mix of financial resources than do businesses, including charitable donations, membership dues, and grants from government agencies or private foundations. While nonprofits may gain revenue by selling something, their primary missions are larger public goods. For example, nonprofit colleges, universities, and community colleges charge tuition, but their primary objective is providing higher education, and they often gain large amounts of revenue and support from state funding and from donations and grants contributed by groups and individuals. Similarly, the Sierra Club, an environmental nonprofit group, sells books, clothing, gear, and adventure tours on its website, but its primary purpose is advocacy for environmental protection. The group could not function if it relied solely on the goods and services it sells. Instead, it gains most of its revenue from donations. And although they are called "nonprofits," these organizations can accumulate surplus revenue. A university's endowment is a good example: some of the wealthiest schools have billions of dollars invested to generate future revenue.

Nonprofits do not have owners or shareholders and are typically run by executives and a board of directors or trustees. These individuals must ensure that the organization pursues its mission and maintains its financial health (**FIGURE 19.8**). Many nonprofits share this responsibility with a broader group of members who elect the board and sometimes have the power to shape the organizational purpose and policies. Although nonprofits are free from the need to maximize profits for shareholders, they still need to maintain financial stability, which can come into conflict with pursuit of their purpose.

Benefit Corporations

A type of organization less familiar to most of us is a **benefit corporation**. Benefit corporations enable businesses to identify and pursue purposes beyond profit. In many US states, *benefit corporation* is an official legal status for business organizations that want to pursue more socially or environmentally driven objectives. Under these arrangements, shareholders' interests are no longer treated as the primary purpose of the business. Warby Parker is an eyewear company chartered as a benefit corporation in New York State, and for every pair of glasses purchased, the business donates the money to produce an additional pair of glasses. These glasses can then be sold at local prices in developing countries. Warby Parker has also conducted a greenhouse gas emission audit on their production, shipping, warehousing, and office operations and purchased carbon offsets for these emissions.

A similar though less formal designation is *B Corp*, a certification awarded by a company called B Lab to companies that have goals similar to those of benefit corporations. B Corp designations protect businesses from lawsuits charging that their actions were not adequately prioritizing shareholder profits, and in some cases these designations enable legal challenges if the business fails to adequately pursue the social benefits identified in its charter. There are now over 5,000 businesses certified as B Corps. They include Method Products, a manufacturer of cleaning products that is based in California. Method subjects nearly all of its products to "cradle-to-cradle" certification—an independent testing process that ensures the human health and environmental safety of all materials used. The company also uses 100% recycled plastic in its packaging and has developed a program to recover and incorporate plastic waste from the ocean into some of its bottles.

🔴 **TAKE-HOME MESSAGE** Organizations coordinate the ideas and actions of many individuals to accomplish objectives that individuals could not accomplish on their own. An organization's purpose influences its actions and can dramatically affect its environmental impact. The primary purpose of for-profit businesses is to earn revenue by selling goods and services, while that of nonprofit organizations is to provide a social benefit. Benefit corporations combine aspects of both by being businesses that pursue social and environmental benefits as well as profit.

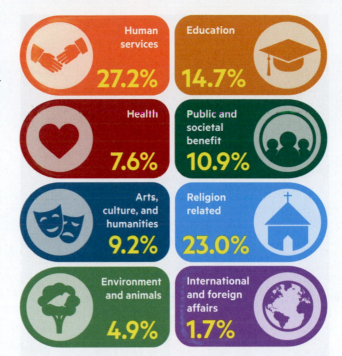

FIGURE 19.8 Nonprofit Organizations Nonprofits typically are created with specific purposes in mind. In the United States, about 4.9% of nonprofits were created with specifically environmental goals.

Adapted from National Council of Nonprofits (2019).

19.3 How Do Organizations Integrate the Ideal of Sustainability?

You might think that college has always been, well, college. Though individuals have come and gone, the overall purpose of the school and its social network has been the same. But if you look back several decades at the history of whatever institution you are in, you'd find that things have probably changed quite a bit. All sorts of things from admissions standards and processes to the range and type of academic programs offered and even the food available in the cafeteria has changed at colleges and universities over the decades. Like your college, organizations are dynamic: they often adopt new ideas and behaviors in response to perceived problems and opportunities. But why change at all? At a fundamental level, organizational change begins with a perceived need to do things differently.

Sometimes ideals can come from within. In the case of King Arthur Flour described in the **Stories of Discovery** feature, that organization wanted to take the steps to become a B Corp. In many other cases, the perceived need for change also has roots in pressure from outside the organization. For businesses, forces that threaten profits can play a big role in sparking change. For example, employees at furniture manufacturer Herman Miller recognized a threat to the production of one of their best-selling products, an upholstered rosewood

benefit corporation
an official legal status for business organizations that want to pursue more socially or environmentally driven objectives.

FIGURE 19.9 Climbing "Mount Sustainability" at Interface, Inc. CEO Ray Anderson changed his organization to focus on sustainable goals.

mission statement a brief and general value statement presenting an ideal vision of what an organization hopes to achieve.

FIGURE 19.10 Sustainable Carpet These interchangeable and recyclable "cool carpet" tiles are also third-party verified as carbon neutral.

ottoman. The employees recognized that rosewood was an endangered tropical hardwood. They convinced their managers that if the company continued to use this wood, it risked increasing production costs because of the scarcity of the resource and losing customers because of rising public concern over logging in tropical rain forests. Because of this perceived need for change, Herman Miller stopped using rosewood in its products. Rising environmental awareness linked to this small product change eventually led the company to adopt the TBL approach to accounting described earlier in this chapter.

Interface, Inc., a leading manufacturer of commercial carpeting, provides another example of pressure from outside a business sparking change. In this case, customers began to question the company's environmental practices, which led founder and CEO Ray Anderson to learn more about environmental impacts linked to his manufacturing processes and about life-cycle assessment (an approach we explored in Chapter 15). Anderson outlined a new set of goals for Interface in 1994 that is nicknamed "Mount Sustainability" (**FIGURE 19.9**). These goals included seven broad objectives:

1. eliminating waste from their business,

2. eliminating their use of toxic materials,

3. using 100% renewable energy,

4. changing processes to use recovered and natural materials,

5. using transportation that eliminates waste and emissions,

6. having a business culture that promotes sustainability across its business network, and

7. creating a business model that supports sustainable goals.

The performance of the company toward these newly established goals helped to define a series of necessary changes. The changes included adopting non-petroleum-based materials and a strategic shift toward leasing carpet that the company can later collect from the customer and recycle into new products (**FIGURE 19.10**). These goals continue to drive Interface toward more sustainable production. The company now ensures that all of its carpet tile is carbon neutral across the entire product life cycle.

Mission Statement and Strategy

Most organizations adopt overarching **mission statements** or other official goals: brief and general value statements presenting an ideal vision of what the organization hopes to achieve. Mission statements are typically published in employee manuals, on websites, and in annual reports to foster shared values within the organization and build legitimacy with people outside the organization who identify with the values in the statement

B Corp: Inventing a New Kind of Corporation

If you want to increase your influence, joining with others in an organization can be one way to do it. But what kind of organization would have the biggest impact? That was the question friends and business partners Jay Coen Gilbert, Bart Houlahan, and Andrew Kassoy wrestled with in 2006. Gilbert and Houlahan had just sold their successful basketball shoe and apparel company AND1, while Kassoy was a successful private equity investor on Wall Street. All three wanted to build on their business success, while also providing some social and environmental benefits—that is, making the world a better place.

Their first instinct was to start a business. But they concluded that even if they could build the rare company that was able to succeed while also benefiting society and the planet, it would still be just one company with a relatively small impact. So they instead considered starting a socially and environmentally responsible investment fund. In this way, they could help fund and support dozens of companies that had aims beyond just making a profit. But the more they learned about the challenges of succeeding as a socially and environmentally responsible business, the more they recognized a need for a new marketing and legal infrastructure to encourage this work. The result was B Lab, a nonprofit organization dedicated to promoting social benefit corporations, or B Corps. The three reasoned that this project might take longer to have an effect, but it could create an entirely new sector of business.

The first step was simply to familiarize the public with the concept of B Corps and to establish standards for a certification program. This would let companies pursuing social and environmental goals easily make themselves known to consumers and investors. The B Lab team began by creating the B Impact Assessment, a rating system for socially and environmentally responsible performance that B Lab could use

King Arthur Flour was the first company to obtain B Corp status—a certification that holds the company accountable for attaining workplace, social, and environmental standards. There are now thousands of B Corp companies operating in more than 60 nations throughout the world.
Adapted from Bové et al. (2017).

to offer independent certifications for businesses that met the highest standards. In 2007, King Arthur Flour became the first company to put the B Corp certification on its product. Many other companies soon followed, such as Method Products and the New Belgium Brewing Company. Now there are more than 5,000 companies certified as B Corp organizations.

But the more ambitious project for the B Lab team was changing corporate law, which varies state by state, to allow for an expansion of a company's duties beyond just maximizing shareholder value. A new corporate form—the benefit corporation—had to be created by law, state by state. On the one hand, this legislation would protect these companies from shareholder lawsuits contending that pursuit of social and environmental goals was emphasized at the expense of profits. On the other hand, it would allow shareholders legal recourse if the company was not adequately upholding the social and environmental commitments written into its charter.

Establishing these laws required extensive legal research and analysis as well as education and lobbying campaigns for policy makers. In 2010, Maryland became the first state to approve benefit corporation legislation. Now more than 30 states have passed similar legislation, and the movement even has a foothold in Europe, as Italy became the first country outside of the United States to establish legal status for benefit corporations. Now there are certified B Corp companies in more than 60 nations, and the B Lab team has thus realized an even wider sphere of influence for its organization.

What Would You Do?

Imagine you are starting your own company and going through the legal steps to create a corporation. Describe what kind of business you have in mind, and choose whether you would register as a conventional corporation or a benefit corporation. Identify the factors that influenced your choice.

Why Are Organizations Addressing Sustainability?

The global consulting firm McKinsey & Company regularly surveys global executives on the most significant issues they face. One of these issues is implementing sustainability programs in their businesses. These surveys show that the importance of this goal is growing, driven by the opinions of their customers and employees.

Top reasons why organizations are addressing sustainability topics
(by percentage of respondents)

- ● Improve operational efficiency

- ● Build, maintain, or improve reputation

- ● Align with our goals, mission, or values

- ● Respond to regulatory requirements

- ● Develop new growth opportunities

- ● Ensure our ability to grow
  in 2012 and 2014 this was phrased differently: "Strengthen competitive positioning"

- ● Meet consumers' expectations

- ● Make tangible, positive impact

- ● Attract, motivate, or retain employees

- ● Meet industry norms or standards

- ● Meet investors' expectations
  *this was not offered as an answer choice in 2012 or 2014

Adapted from Bové et al. (2017).

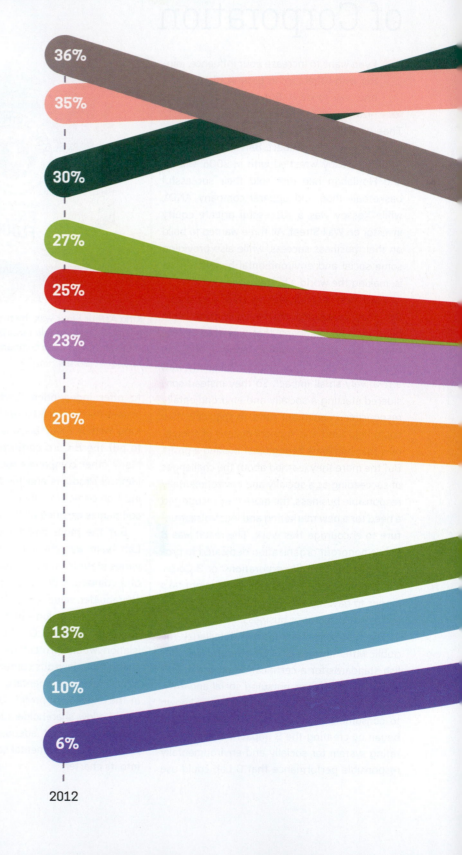

36%
35%
30%
27%
25%
23%
20%
13%
10%
6%

2012

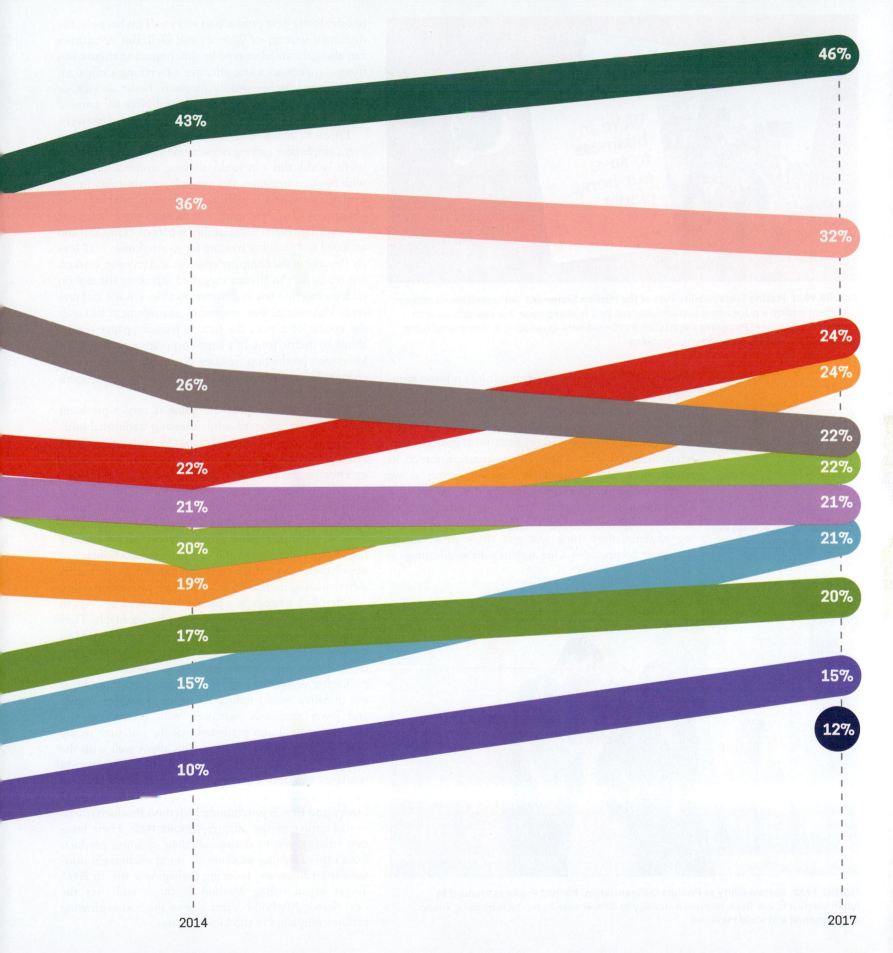

46%

43%

36%

32%

26%

24%

24%

22%

22%

22%

22%

21%

21%

21%

20%

20%

19%

17%

15%

15%

12%

10%

2014

2017

FIGURE 19.11 Making Sustainability Part of the Mission Statement An organization's mission statement is often a place where sustainability can be a featured value. For example, outdoor clothing manufacturer Patagonia emphasizes its commitment to reducing environmental harm and helping to address environmental problems.

strategy the way in which an organization directs its resources toward its goals.

(**FIGURE 19.11**). In this way, an organization's mission statement can be a powerful tool to influence the internal and external social networks to which it connects.

Whereas incorporating sustainability into a mission statement establishes it is an organizational goal, **strategy** refers to the way in which an organization directs its resources toward its goals. In business terms, as we learned earlier in this chapter, for-profit businesses have an overarching purpose to generate a profit for their owners and investors. They attempt to do this by adopting strategies they think will give them an advantage over their competitors. One way to gain an advantage is

to offer lower-cost products or services. This has been the dominant strategy of Walmart and Dell. But companies can also gain an advantage by differentiating themselves from competitors using effective advertising campaigns and offering higher-quality products, better service, or distinctive product features. Apple and Nike are familiar examples of companies using these strategies effectively.

Efforts to integrate sustainability often work best when they complement existing organization strategies. In the for-profit world, this can mean aligning sustainability goals with business strategies. For example, sustainability initiatives can often (but not always) align with reducing costs. Unilever is a large company that includes many brands. It also has emphasized sustainability as part of its mission and included sustainability training for its employees at all levels. Because of the company's mission and training, workers at a tea factory in Britain suggested shrinking the seal on each tea bag by a few millimeters to reduce waste and save costs. This change was adopted by management and over the course of a year, the factory reduced paper use by about 10 metric tons. In a large corporation such as Unilever, with production facilities all over the world, simple changes like these can add up to big sustainability gains and significant cost savings as well.

Similarly, the sustainability gains at carpet producer Interface were achieved while meeting traditional business performance metrics. In the 25 years after establishing its sustainability goals, Interface reduced its net greenhouse gas emissions by 96%, decreased water use by 89%, reduced the waste sent to landfills by 92%, and drew 99% of its energy from renewable sources. Interface achieved these environmental gains while doubling its profits and increasing its sales by two-thirds because the changes met a need: the company took advantage of opportunities to reduce costs and expand its market by differentiating itself as an industry leader in sustainability.

Interface's strategy is a pioneering model of ways in which sustainability measures could be profitable. They have since been scaled up by many larger companies. Most notably, in 2006 Walmart began working with more than 60,000 of its suppliers throughout the world to reduce packaging waste. Although the company announced that the initiative would reduce the waste going to landfills and lower emissions associated with production and transportation, it also predicted that the initiative would save the company $3.4 billion. This aligns well with the company's low-cost business strategy. Differentiation strategies can also align with environmental and social measures. In 2000, two former college roommates, Adam Lowry and Eric Ryan, founded Method Products, mentioned earlier in the chapter (**FIGURE 19.12**). Their business strategy was to distinguish their cleaning products from other cleaning products by using exclusively nontoxic ingredients and reducing packaging waste. In 2002, Target began selling Method Products, and over the next decade Method became one of the fastest-growing private companies in the United States.

FIGURE 19.12 Sustainability as Product Differentiation Method Products, founded by Adam Lowry and Eric Ryan, pursued a strategy to differentiate its products by using strong environmental and social measures.

Organizational Culture and Structure

The actions an organization takes are significantly influenced by the culture and structure of the organization. **Organizational culture** refers to the widely shared assumptions and patterns of behavior that group members learn and share. This includes shared expectations of loyalty, trust, personal development, and other aspects of interpersonal relations. **Organizational structure** refers to the way in which working relationships are arranged and authority is exercised among the members. This includes the degree of specialization: the extent to which tasks are delegated to particular departments and individuals or shared more widely across the organization.

Sociologists who study organizations have found that change is often facilitated by certain structures and cultures, especially when employees can interact with people who have different skills and backgrounds from their own and when they are granted time, freedom, and support to research and try out new ways of doing things. For example, the initial concern that sparked environmental change at the Herman Miller furniture company only came to the attention of executives because the company had an open governance structure that provided employees with opportunities to voice concerns to upper managers. At Interface, Inc., new ideas to improve environmental outcomes were facilitated through a program called QUEST—Quality Utilizing Employee Suggestions and Teamwork—that gave employees the freedom to float new ideas to management and collaborate with people across a wide range of tasks within the company. Similarly, the Collins Companies, a manufacturer of wood products, reorganized its employees to encourage new waste reduction goals. Employees were formed into teams on the basis of production processes. This reorganization allowed people with different tasks and training within the company to share what they knew about how products were made. Each team was encouraged to critically examine production processes and develop new ways to reduce waste. This support for new ideas generated more than 100 new projects, some of which led to production changes that resulted in about $1 million in savings per year through reductions in energy use, water consumption, and waste disposal.

⬡ **TAKE-HOME MESSAGE** Organizations adopt new ideas and behaviors in response to perceived problems and opportunities. Organizations direct their resources toward their goals with various strategies. When an organization adopts a new goal, it is sometimes necessary to change certain aspects of its culture or structure to facilitate work toward the new objective. Organizations can facilitate new ideas by giving employees opportunities to interact with people who have different skills and backgrounds and by encouraging workers to develop and try new ways of doing things.

19.4 How Can Members within an Organization Facilitate Change?

Many colleges and universities have made changes in recent decades to make their campuses more environmentally sustainable. In 1990, Tufts University was one of the first to adopt an environmental mission statement and strategy, largely due to the efforts of President Jean Mayer. Mayer appointed a committee of faculty, staff, vice presidents, and deans to develop an environmental policy for the institution and then hosted university presidents from around the world to sign the Talloires Declaration: a 10-point commitment to integrate sustainability and environmental literacy into teaching, research, operations, and outreach at higher-education institutions. To date, presidents from nearly 500 institutions across more than 50 nations have signed this commitment.

This is an example of change initiated at the highest levels of an organization. Yet under the right conditions, change can begin at any level. In fact, individuals lower down in the hierarchy of an organization are often those best positioned to see problems, opportunities, and new ideas because they tend to have the most direct interactions with customers, suppliers, and production processes. In the 1990s, Kevin Lyons, then the senior procurement staff member at Rutgers University, took it upon himself to establish environmental objectives for the university's purchasing contracts. Similarly, Ciscoe Morris, the head of groundskeeping at Seattle University, initiated a successful effort to make the campus pesticide-free in 1986. This was the first step in many other sustainability initiatives and innovations at the school. In 1996 at Oberlin College, Professor David Orr's studies of energy, water, and materials use at college campuses inspired him to lead the effort to design and build the first substantially green building at a US college (**FIGURE 19.13**). At Stanford University, it was an engineering student who initiated a task force on sustainable building and ultimately convinced his university's administration to develop guidelines for sustainable buildings.

Of course, it is important to recognize that change is hard. Many of you may have experienced this already when working in teams on class assignments or within clubs and organizations. Those advocating for change, whether from the top or bottom of an organization, can face significant hurdles. Even highly motivated CEOs can see their environmental initiatives fail because of managers and staff who are slow to change the way they carry out standard operating procedures. And of course, unsupportive leaders or executives can easily block innovative approaches generated by employees

organizational culture the widely shared assumptions and patterns of behavior that group members learn and share.

organizational structure the way working relationships are arranged and authority is exercised among the members of a group.

FIGURE 19.13 People Driving Change (a) Professor David Orr led the effort to design and build the Adam Joseph Lewis Center at Oberlin College (completed in 2000), which ushered in a new era of green buildings on college campuses. (b) Professor Kevin Lyons established more sustainable purchasing guidelines in the 1990s for Rutgers University when he was the senior procurement officer. (c) Seattle University's former head of groundskeeping, Ciscoe Morris, eliminated pesticides from campus in 1986. (d) Debbie Mielewski spearheaded the development of plant-based car parts at Ford Motor Company beginning in the 1990s despite a lack of support from company executives.

(a)

(b)

(c)

(d)

and managers lower in the organizational hierarchy. Lasting change typically requires support from all areas of influence within the organization because change requires adoption by both those with the power to make decisions and those who work in areas effecting the change.

Building this broad base of support requires both considerable effort and patience. Consider the experience of Debbie Mielewski, the technical leader of the plastics research division for Ford Motor Company. She perceived a need to find plant-based alternatives to petroleum-based plastic car components. Even while company executives repeatedly rejected this idea throughout the 1990s, she broadened support for her idea among a growing team of researchers who devoted some of their efforts toward using materials such as soybeans, wheat straw, coconuts, tomato skins, and cellulose from tree fiber to create various car parts. But the technologies they developed only gained widespread adoption in Ford vehicles in 2008, after new executives had vowed to make Ford an environmental leader in the industry and rising oil prices made alternatives to plastic more cost-effective. Today plant-based seat cushions,

backs, and headrests are used on every Ford vehicle, and these materials are estimated to have reduced greenhouse gas emissions by over 228 million pounds (more than 100 million kilograms).

Mielewski's research on alternatives to plastics gained more support as the company established sustainability leaders in each unit of the company. These leaders worked with employees to identify environmental performance targets, and each salaried employee developed performance goals with his or her supervisor that reflected overall company goals, including sustainability. Regular reviews of progress toward performance goals enabled communication to flow up and down the company hierarchy.

⬡ **TAKE-HOME MESSAGE** Under the right conditions, change can begin at any level in an organization, but lasting change requires adoption and implementation at all levels of influence within the organization. When organizations integrate change teams and leaders throughout their hierarchy, they improve both communication and their chances for successful implementation.

Should Organizations Divest from Fossil Fuel Companies?

When environmental activists ask individual consumers to rally behind a cause, they often encourage us to "vote with our dollars" by directing our purchases away from products with negative environmental impacts and toward more environmentally friendly alternatives. What if a "vote with your dollars" campaign was extended to investors as well, by directing money toward or away from stock holdings of various companies on the basis of their environmental impact? As an individual you might not have significant stock holdings yourself, but you might be a part of an institution that does. Some of the largest investors are actually nonprofit organizations such as agencies that manage employee retirement plans and universities that manage endowment funds for their institutions. The investments of these groups are often valued in billions of dollars.

In recent years, students at more than 100 colleges and universities have campaigned for divestment from stocks in the fossil fuel industry because of the impact these companies have on global climate change. Divestment is the opposite of investment. It is a decision to remove a particular kind of asset from an investment portfolio. Proponents of this strategy argue that it is a way to send a powerful signal both to the companies they are divesting from and to the wider public. Divestment supporters also argue that it is a way to free up and redirect investment dollars toward companies pursuing alternative energy technologies such as wind and solar, while clearly establishing the environmental values of their college or university. In the United States, several colleges and universities have fully or partially divested from fossil fuel companies, including Harvard, Stanford, the New School, and the University of Dayton. Other institutions such as churches, cities, and charitable organizations have also committed to this strategy.

But many colleges and universities are hesitant to take this step. Opponents to divestment

Endowment Funds of the 20 Colleges and Universities with the Largest Endowments, by Rank Order: Fiscal Year 2018

Institution	Rank Order	Market Value of Endowment (in billions of dollars)
Harvard University (MA)	1	39.2
University of Texas System	2	30.6
Yale University (CT)	3	29.4
Stanford University (CA)	4	26.5
Princeton University (NJ)	5	25.4
Massachusetts Institute of Technology	6	16.4
University of Pennsylvania	7	13.8
Texas A&M University, College Station	8	12.7
University of Michigan, Ann Arbor	9	11.7
University of Notre Dame (IN)	10	11.0
Columbia University (NY)	11	10.9
University of California System	12	10.8
Duke University (NC)	13	8.5
Northwestern University (IL)	14	8.4
Emory University (GA)	15	8.0
Washington University in St. Louis (MO)	16	7.7
University of Chicago (IL)	17	7.0
Cornell University (NY)	18	6.9
University of Virginia, Main Campus	19	6.9
Rice University (TX)	20	6.2

As of the latest US Department of Education report, these schools had the highest overall endowment funds. Students at many campuses are questioning some of the investment decisions made by these funds.

Table adapted from US Department of Education, National Center for Education Statistics (2019).

(continued)

question the impact it actually has on the target companies. When an institution divests, it sells its shares in a company to other investors. If there are plenty of investors willing to buy those shares, then the stock value of the company is unlikely to fall, and the company itself may not feel much of an impact from the divestment. Evidence from past divestment campaigns targeting companies linked to human rights abuses in South Africa and Sudan show that although the campaigns did much to raise awareness, they had little effect on the share prices of the targeted companies.

Moreover, many opponents worry that divesting will lead to a decrease in investment returns. One controversial economic study sponsored by the Independent Petroleum Association used past returns to estimate what the investment performance of universities with the largest endowments would have been over the past 50 years if they had been divested from fossil fuels. The results showed that the portfolios of these universities would have been on average 23% lower had they divested. Opponents argue that such results suggest divesting means universities would have significantly less money in their endowments to support student scholarships and other needs central to their mission.

But divestment supporters question this analysis because it is based on past performance of fossil fuel stocks. They argue that it may be risky to keep assuming fossil fuel stocks will continue to perform well as less-polluting energy sources gain a larger share of the energy market. And in part because of divestment campaigns, there is now a growing market for "low carbon" and "fossil free" stock indexes and funds in which to invest. We will not know for sure what the comparative returns of fossil fuel versus fossil fuel–free investment strategies will be until they are borne out in the coming years. Finally, proponents of divestment emphasize the moral reasoning behind their strategy. They want to clearly signal their opposition to the fossil fuel industry and its associated environmental impacts by withdrawing their institution's financial support to these enterprises.

What Would You Do?

Do you think your college or university should divest from fossil fuel companies? Explain why or why not.

19.5 How Can Those Outside an Organization Facilitate Change?

Wherever the internal initiative for change is coming from—whether from those at the top or those working further down in an organization—it typically begins as a response to forces acting outside the organization. Each organization is a social network that is connected to and at least somewhat responsive to a wider network of individuals and groups on the outside. For-profit organizations are directly linked to a range of outside influencers through their contact with customers, suppliers, distributors, competitors, regulators, and labor groups. Surveys of company executives have found that nearly half of the respondents' companies were taking environmental actions to align with their mission or values and about a third were taking such actions to enhance their reputation and not only profits (see **At a Glance: why Are Organizations Addressing Sustainability?** earlier in the chapter). This means there is a role for outsiders to play in influencing these organizations.

Environmental Nonprofits Exerting Influence on Businesses

Environmental groups have long tailored their strategies to influence business practices by encouraging customers to boycott certain products or by creating bad publicity for companies through protests. But more recently, many environmental groups have also begun to help businesses find ways to improve their environmental performance to align with company strategies and missions.

Consider the strategies used by Greenpeace, an organization well known for its use of protests and boycotts. In 1989, Greenpeace activists scaled a 185-foot water tower at a New Jersey DuPont plant that produced Freon—the primary ozone-depleting refrigerant chemical in use at the time—and hung a 65-foot blue ribbon that read "#1 Ozone Destruction." The protest and the arrest of the activists at the event were covered in the national news, and executives within DuPont felt compelled to respond. Greenpeace also pursued a political strategy by lobbying for international agreements and national policies to prevent companies such as DuPont from producing ozone-depleting chlorofluorocarbons (CFCs).

Yet Greenpeace also pursued a range of strategies to influence business practices on this issue. In 1992, Greenpeace set out to facilitate change within the industries using CFCs by creating a business opportunity. They did this by organizing a group of inventors to develop alternative refrigerant technologies and then hiring a manufacturer to build what they called "Greenfreeze" refrigerators, using these technological advances. The environmental group aggressively marketed and sold the Greenfreeze refrigerators in Europe and made their product specifications open source so any manufacturer could use the new technology. Large companies such as Panasonic, Siemens, LG, and Whirlpool adopted the Greenfreeze technology, and over the next two decades, more than 650 million Greenfreeze refrigerators were sold worldwide (**FIGURE 19.14**). In this case, Greenpeace influenced business practices by assuming the cost of research and development and making it freely available to the industry. Greenpeace also helped to boost the market for the new technology by convincing Coca-Cola, PepsiCo, and Unilever to adopt this and other less-polluting refrigerant technologies into their global operations.

Many environmental groups now employ a range of strategies like this. WWF and other environmental groups have put considerable pressure on Coca-Cola to reduce its impact on water resources. Coca-Cola (and the company's many beverages such as Dasani bottled water and Vitamin Water) have been implicated in the creation of severe water shortages and in limiting access to water near its bottling plants in several countries. While groups such as the WWF have supported legal challenges and publicity campaigns that have attempted to hold Coca-Cola to account for its environmental impacts, they have also worked directly with the company on water-conservation measures. WWF helped Coca-Cola adopt a more comprehensive view of the company's "water footprint" by expanding its water use and watershed impact analysis to include its suppliers of raw ingredients as well. This analysis has led to a number of initiatives designed to improve agricultural practices in strategic watersheds. For example, in Guangxi Zhuang Autonomous Region, China, where 60% of China's sugar is produced and more than 20 million people are employed in sugarcane agriculture, Coca-Cola has worked with the United Nations and the Chinese government to help sugar-processing plants develop facilities to treat water from the factory so that it can be returned to irrigate the sugarcane fields (**FIGURE 19.15**).

The World Wildlife Fund has entered into partnerships with many large companies since that time. However, a strategy like this is only effective if the environmental group is willing to hold a company accountable when it fails to make adequate progress toward environmental goals. This was what happened when the Asia Pulp & Paper (APP) company failed to meet the impact reductions on tropical rain forests it had promised in a partnership agreement. WWF not only withdrew its partnership with the company but also convinced the Forest Stewardship Council (FSC) to prevent APP from using the FSC label. With that change in place, WWF then successfully campaigned to have large paper sellers such as Staples and Office Depot refuse to carry APP's products.

Forces Changing Environmental Groups

Some large environmental organizations have themselves been challenged by external forces. In 1991, the United Church of Christ Commission for Racial Justice convened a National People of Color Environmental Leadership Summit in Washington, DC, that put pressure on national environmental groups such as the Sierra Club, Environmental Defense Fund, National Wildlife Federation, and Nature Conservancy to address the health, safety, and equity concerns of urban communities where they "live, work and play." Several hundred delegates at the event drafted 17 principles of environmental justice that have worked their way into the mission and strategy of many environmental groups (see Chapter 17). Challenges like this have sometimes caused internal disagreements, leading to new groups. Disagreements within the

(a)

(b)

FIGURE 19.14 **Activist Strategies** **(a)** Greenpeace activists are well known for their protest actions, like this one supporting stricter ozone laws in Germany, **(b)** but they have also used research and development of new environmental technologies—such as these Greenfreeze refrigerators made without ozone-depleting substances—to combat the growing hole in Earth's ozone layer.

Sierra Club in the late 1960s regarding negotiations and compromises between the club and utility companies led to the formation of Friends of the Earth. Similarly, in the late 1970s, the Sea Shepherd Conservation Society, known for its risky and sometimes violent confrontations with whaling vessels and seal hunters, was born out of a

FIGURE 19.15 **Partnerships for Change** In Guangxi, China, Coca-Cola formed a public–private partnership with the United Nations Development Programme to improve the treatment and recycling of water used by sugar-processing factories. While sprinkling irrigation systems like the one shown here were installed during the early stages of this partnership, more recent projects are promoting an even more efficient drip irrigation system. This is particularly important as this sugarcane-farming region suffers from regular droughts and flooding.

Green 2.0: Increasing Diversity in Environmental Organizations

In 2014, the fledgling advocacy group Green 2.0 commissioned environmental justice scholar Dorceta Taylor to research the state of diversity in more than 300 environmental organizations. She found that many of these organizations had significantly improved gender diversity over the years, with women making up more than half of the new hires and leadership positions in conservation and preservation organizations. However, men were still more likely than women to occupy executive positions, and the percentage of people of color in these organizations was less than 16%, far lower than in the population as a whole.

Spurred by these results, Green 2.0 has served as a watchdog group for inequality in the environmental sector since 2014, providing tools and best practices to help environmental groups recruit, retain, and promote people of color in their organizations while providing a supportive workplace culture. They have also continued to track demographic data and issue "transparency report cards" on diversity within environmental groups. In 2021, the data showed improvements in the percentage of people of color occupying full-time staff, senior staff, and board positions in environmental organizations, with a 10% improvement over the previous 5 years across the overall staff. The majority of the organizations studied had also adopted several best practices for recruiting and retaining more diverse workforces. However, although the staff composition of the environmental sector now includes 30% people of color, this is still about 10% less than their representation in the population at large.

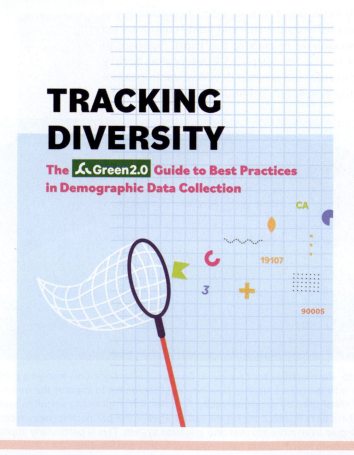

Media and Social Media as a Force for Change

Consumers become aware of the environmental impacts of various businesses through stories in the media. The global furniture manufacturer IKEA adopted its sustainability strategy in response to two crises in the early 1990s that were uncovered by media organizations. In 1992, a German newspaper found that IKEA's popular "Billy" bookshelves were off-gassing formaldehyde emissions in excess of the amount mandated by European environmental regulations. The company spent several million dollars recalling the products. Two years later, IKEA suffered more bad publicity when a Swedish television reporter exposed child labor practices by the company's suppliers in Pakistan.

Of course, media platforms such as TikTok, Facebook, and Twitter can spread information about a company much faster than traditional media outlets. For example, many people point to a video of a straw in a live turtle's nose that was posted in 2015 by a marine biologist as driving renewed interest in eliminating the widespread use of disposable straws. A #StopSucking social media campaign and related Kickstarter initiatives to develop collapsible, reusable straws contributed too, especially after gaining support on celebrity media feeds. These efforts have led to companies such as Starbucks discontinuing their use of disposable plastic straws and many communities considering laws to discontinue their use as well (**FIGURE 19.16**). While critics of this effort have noted that it may be misguided, it does seem to be driving home the need for broader change. For example, the vice president of government affairs at the Plastics Industry Association defended their product by noting that "plastic straws have legitimate and important uses,

split within Greenpeace over protest tactics and strategies of engagement with government and industry.

FIGURE 19.16 Moving Beyond Disposable Straws In response to widespread public campaigns against disposable plastic straws and the litter they generate, many communities and companies are discontinuing their use.

and banning them gives a false sense of accomplishment, which is more harmful in the long run." But he was careful to add, "Our goal is to make sure that every product, no matter how small, can be properly recovered."

Partnerships and Economic Incentives

When businesses see opportunity in sustainable strategies, some often surprising partnerships have formed. In 2005, Walmart invited the Environmental Defense Fund (EDF) to assess the chemical composition of all the cleaning and bath supplies it sells. EDF and Walmart agreed on targeting 10 chemicals of concern and then began working with suppliers such as Colgate-Palmolive and Procter & Gamble to reformulate products containing the chemicals of concern.

Similarly, in 2008 Walmart and Patagonia began partnering to set shared standards for the apparel industry. Walmart CEO Lee Scott reportedly approached Patagonia for advice on how to improve his company's environmental and social responsibility. A market survey had revealed that a majority of US consumers had a negative impression of the store and an estimated 14 million people may have stopped shopping there because of its declining reputation. The unlikely pair of companies convened the Sustainable Apparel Coalition, consisting of 16 of the largest global clothing companies, to develop open-source metrics and assessment tools for sustainability that will enable companies and consumers to rate suppliers and finished products. This effort was based on Patagonia's leadership of the Outdoor Industry Association's Eco Index, a shared set of social and environmental benchmarks and assessment tools agreed on by roughly 1,500 manufacturers, suppliers, and retailers of gear for outdoor recreation (**FIGURE 19.17**).

Finally, the Leadership in Energy and Environmental Design (LEED) certification system (a voluntary

FIGURE 19.17 Sustainability in Retail Apparel company Patagonia and big-box store chain Walmart are very different, but Patagonia has worked directly with Walmart to help the latter's sustainability efforts. Patagonia's CEO, Rose Marcario, has announced the company wants to be 100% carbon neutral by 2025.

→ CONTEXT

Outsized Influence

A large manufacturer and retailer such as IKEA can exercise considerable influence over its suppliers. IKEA purchases products from nearly 1,500 suppliers in more than 50 countries. A purchaser this large has considerable leverage when negotiating prices, quality, and style. The same leverage can be applied to environmental impacts and labor conditions. For example, in just 2 years, IKEA raised the amount of sustainably grown cotton used in its products from 74% to its target of 100%. IKEA now requires its suppliers to adopt its code-of-conduct guidelines and then uses a life-cycle assessment tool to gauge each supplier's environmental and social performance. While IKEA is still a business driven by profit, its sustainable actions will drive the actions of other businesses.

system to encourage green building that we discussed in Chapter 16) was developed in 1993 in Kansas City when a small group of architects partnered with a developer and the executive of a heating and cooling equipment company to form the US Green Building Council (USGBC). Although the USGBC began with just 10 members, this group gradually extended its social networks into professional associations representing real estate brokers, engineers, and home finance professionals. USGBC now has more than 120,000 individual members and more than 5,000 organizational members in 80 countries. And the LEED certification system it developed has become the industry standard for commercial buildings in the United States. There are now more than 100,000 LEED-certified buildings, including projects in 167 different countries.

⬣ **TAKE-HOME MESSAGE** People who initiate change within an organization are typically responding to external forces in their wider network of outside individuals and groups. Environmental groups often coordinate external forces in an attempt to change the behavior of businesses. This can be done with a range of strategies from protests, boycotts, and publicity campaigns to partnerships with businesses to develop new standards and practices. Trade and professional groups can also influence organizational behavior.

19.6 What Can I Do?

In this chapter, we've examined various types of groups and organizations and the ways they change over time. We are all linked to social networks. Our connections with others are avenues for influence that can have impacts far greater than those of our individual actions. Let's look at some of the ways you can use your networks to enact change.

Examine Your Sphere of Influence

We are all members of various informal and formal groups. Identify some of the groups you are in and the role you play in each group. If you're on social media, this is a good place to start. Where do you think your greatest potential for influence is? What sorts of changes could you help bring about?

Get Involved with What's Happening on Your Campus

Many colleges and universities have committees or task forces in place to work on sustainability initiatives. There are also environmental groups that have student chapters. Learn what groups are working in this area on your campus. Is there a role for students like you to play? How could you get involved?

Bring Sustainability Groups Together on Campus

At the University of Virginia, student government leaders discovered there were 43 groups on campus involved in sustainability. These groups decided to meet once a month to develop a community of "Green Leaders" with the idea of increasing communication, cooperating, and decreasing the number of overlapping initiatives. Check if there is a similar alliance on your campus or consider forming one of your own.

Connect to Other Campuses

If you are looking for ideas for campus sustainability projects, check out the "For Students" tab at the Association for the Advancement of Sustainability in Higher Education (AASHE) website (www.aashe.org/resources/for-students/). The annual Sustainable Campus Index describes what top-performing schools are doing, and the Campus Sustainability Hub has tools and case studies for campus sustainability projects. At the website you can also find webinars, conferences, and other opportunities to connect with students working on sustainability.

Learn More about Financial Factors That Drive Business Decisions

Whether it be large national and international companies like the ones mentioned in this chapter (IKEA, Apple, Ford Motor Company) or local businesses in your city or town that work with your university, creating profits and avoiding losses are normal goals. So if your group is trying to convince a business to change its practices, it helps to be able to understand these financial motiva-  tions. You could take a class at your school to learn the basics or search the internet or your library for good sources on how to read a profit-and-loss or income statement. Even watching shows such as *Shark Tank* will give you a good idea about the factors and terms businesspeople use. Maybe you can craft a campaign to influence a business that has an impact on the environment where you live (in 2018, *Shark Tank* teamed with Discovery Channel's *Shark Week* for a special on shark conservation).

Chapter 19 **Review**

SUMMARY

- Individuals act within social networks that extend from close relations to people several steps removed from direct interactions. These networks play a role in shaping the way people think, feel, and act.

- Some people are more influential than others according to their place in social networks and organizations.

- An organization's purpose is important because it directs individuals within the organization to coordinate ideas and actions to accomplish the shared purpose.

- Identifying environmental values as part of an organization's purpose can have a big impact on that organization's actions.

- While for-profit businesses have the purpose of earning revenue and nonprofit organizations have the purpose of providing a social benefit, benefit corporations combine aspects of both.

- Organizational culture and structure are factors that shape how an organization sets out to achieve its purpose and how it responds to problems and opportunities.

- Change can begin at any level within an organization, but for change to be lasting it must be adopted and implemented at all levels.

- Individuals and groups outside organizations can also play a role in transforming organizational behavior. Some common strategies include protests, boycotts, and publicity campaigns.

KEY TERMS

benefit corporation	organization	publicly traded	social network
degree of separation	organizational culture	social context	strategy
greenwashing	organizational structure	social group	triple bottom line (TBL)
mission statement	privately (or closely) held		

REVIEW QUESTIONS

The letters following each Review Question refer to the Chapter Objectives.

1. Describe the difference between a social group and an organization. Provide an example of each. **(A)**

2. Explain how an individual's place within a social network affects that person's influence on others and susceptibility to the influence of others. **(A)**

3. What is the difference between for-profit businesses that are publicly traded and those that are privately held? **(B)**

4. What distinguishes benefit corporations from other types of for-profit businesses? **(B)**

5. How do the financial resources of nonprofit organizations differ from those of for-profit businesses? **(B)**

6. Describe two factors that can lead organizations to integrate the ideal of sustainability, and provide examples. **(C, D)**

7. Name two ways that individuals or groups outside an organization can facilitate change in that organization, and provide examples. **(E)**

8. Identify one force for change that influences the mission and strategy of many environmental groups, and describe its impact. **(D, E)**

9. Describe how media can influence business practices. **(E)**

10. Provide an example of how two or more organizations can create a partnership for change. **(E)**

The letters following each item refer to the Chapter Objectives.

11. Choose two organizations you belong to or otherwise interact with. What types of organizations are they? What are their primary functions? How do their actions influence other people and the environment? In what ways are they influenced by their members and by those outside the organization? **(B, E)**

12. Choose an environment-related cause that is important to you. What steps would you take to create an organization to address this issue? What type of organization would it be (consider the three types outlined in this chapter)? Where would the funding come from? What organizations could you partner with? **(D, E)**

13. Choose an organization that you think is not doing enough to fulfill the ideal of sustainability. If you were a member of this organization, what steps could you take to facilitate change toward this end? If you were a person outside this organization, what steps could you take to facilitate change toward this end? Explain your reasoning in each case. **(D, E)**

Make Your Case

The following exercises use real-world data and news sources. Check your understanding of the material and then practice crafting well-supported responses.

Use the News

Publicly traded companies such as Apple have annual shareholder meetings that can provide an avenue for feedback and influence between the company's leadership and its investors. The article below describes a dramatic dialogue about the company's priorities for environmental and social sustainability between Apple CEO Tim Cook and a group called the National Center for Public Policy Research (NCPPR) that owns shares of Apple's stock. Use this article to answer the questions that follow. The first three questions test your understanding of the article. Questions 4 and 5 are short answer, requiring you to apply what you have learned in this chapter and cite information in the article. Answers to Questions 1–3 are provided at the back of the book. Question 6 asks you to make your case and defend it.

"Tim Cook Soundly Rejects Politics of the NCPPR, Suggests Group Sell Apple's Stock," Bryan Chaffin, *The Mac Observer*, February 28, 2014

In an emotional response to the National Center for Public Policy Research (NCPPR), Apple CEO Tim Cook soundly rejected the politics of the group and suggested it stop investing in Apple if it doesn't like his approach to sustainability and other issues. Mr. Cook's comments came during the question and answer session of Apple's annual shareholder meeting, which the NCPPR attended as shareholder. The self-described conservative think tank was pushing a shareholder proposal that would have required Apple to disclose the costs of its sustainability programs and to be more transparent about its participation in "certain trade associations and business organizations promoting the amorphous concept of environmental sustainability."

As I covered in depth yesterday, the proposal was politically-based, and rooted in the premise that humanity plays no role in climate change. Other language in the proposal advanced the idea that profits should be the only thing corporations consider. That shareholder proposal was rejected by Apple's shareholders, receiving just 2.95 percent of the vote. During the question and answer session, however, the NCPPR representative asked Mr. Cook two questions, both of which were in line with the principles espoused in the group's proposal.

The first question challenged an assertion from Mr. Cook that Apple's sustainability programs and goals—Apple plans on having 100 percent of its power come from green sources—are good for the bottom line. The representative asked Mr. Cook if that was the case only because of government subsidies on green energy. Mr. Cook didn't directly answer that question, but instead focused on the second question: the NCPPR representative asked Mr. Cook to commit right then and there to doing only those things that were profitable.

What ensued was the only time I can recall seeing Tim Cook angry, and he categorically rejected the worldview behind the NCPPR's advocacy. He said that there are many things Apple does because they are right and just, and that a return on investment (ROI) was not the primary consideration on such issues. "When we work on making our devices accessible by the blind," he said, "I don't consider the bloody ROI." He said the same thing about environmental issues, worker safety, and other areas where Apple is a leader.

As evidenced by the use of "bloody" in his response—the closest thing to public profanity I've ever seen from Mr. Cook—it was clear that he was quite angry. His body language changed, his face contracted, and he spoke in rapid fire sentences compared to the usual metered and controlled way he speaks. He didn't stop there, however, as he looked directly at the NCPPR representative and said, "If you want me to do things only for ROI reasons, you should get out of this stock." To me, it was a clear rejection of the group's politics, especially the anything-for-the-sake-of-profits mentality the NCPPR was asking him to embrace. It was also an unequivocal message that Apple would continue to invest in sustainable energy and related areas.

1. What would the National Center for Public Policy Research (NCPPR) shareholder proposal have required Apple to do?
   a. divest from sustainability-related investments and improvements
   b. disclose the costs of its sustainability programs
   c. invest more in environmentally friendly measures such as renewable energy
   d. refrain from lobbying on environmental issues

2. Which of the following statements is true?
   a. The NCPPR proposal passed.
   b. The NCPPR proposal was narrowly defeated.
   c. The NCPPR proposal was rejected by the vast majority of shareholders.
   d. The NCPPR proposal was tabled until next year's shareholder meeting.

3. Which statement most accurately reflects Apple CEO Tim Cook's view on return on investment (ROI)?
   a. ROI should be the only consideration for company decisions.
   b. ROI should not be the only factor driving company decisions.
   c. Environmental and social improvement goals should be the primary drivers of company decisions.
   d. Government subsidies on green energy are the main factor driving the company's sustainability programs.

4. Use the material in the chapter and this article to identify at least one way this situation would be different if Apple was a privately held company rather than a publicly traded one.

5. Use the material in the chapter and this article to identify at least one way this situation would be different if Apple was a benefit corporation.

6. **Make Your Case** If you were an Apple investor at this shareholder meeting and you could address attendees about the NCPPR shareholder proposal, what case would you make? Would you urge people to vote yes or no, and what are your reasons?

Use the Data

The global consulting firm McKinsey & Company regularly surveys global business executives on different aspects of business, typically receiving several thousand responses to a survey. In 2012 and 2017, the company asked business leaders about how they used different sustainability technologies in their businesses.

Study the graphs shown here and use them to answer the questions that follow. The first three questions test your understanding of the graphs. Question 4 is short answer, requiring you to apply what you have learned in this chapter and cite data in the graphs. Answers to Questions 1–3 are provided at the back of the book. Question 5 asks you to make your case and defend it.

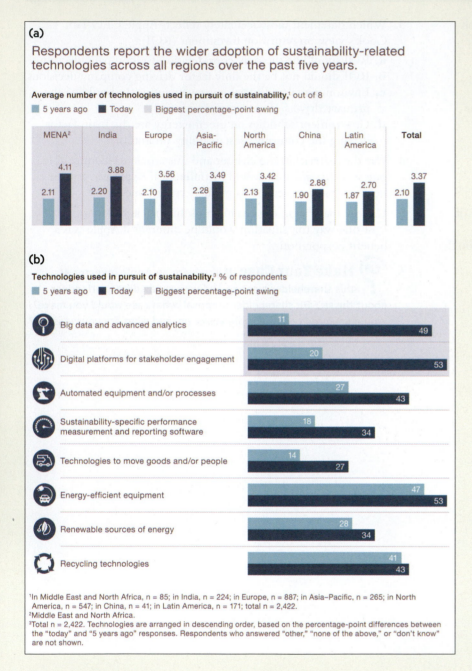

(a)

Respondents report the wider adoption of sustainability-related technologies across all regions over the past five years.

Average number of technologies used in pursuit of sustainability,[1] out of 8

■ 5 years ago ■ Today ■ Biggest percentage-point swing

	MENA[2]	India	Europe	Asia-Pacific	North America	China	Latin America	Total
5 years ago	2.11	2.20	2.10	2.28	2.13	1.90	1.87	2.10
Today	4.11	3.88	3.56	3.49	3.42	2.88	2.70	3.37

(b)

Technologies used in pursuit of sustainability,[3] % of respondents

■ 5 years ago ■ Today ■ Biggest percentage-point swing

Technology	5 years ago	Today
Big data and advanced analytics	11	49
Digital platforms for stakeholder engagement	20	53
Automated equipment and/or processes	27	43
Sustainability-specific performance measurement and reporting software	18	34
Technologies to move goods and/or people	14	27
Energy-efficient equipment	47	53
Renewable sources of energy	28	34
Recycling technologies	41	43

[1]In Middle East and North Africa, n = 85; in India, n = 224; in Europe, n = 887; in Asia–Pacific, n = 265; in North America, n = 547; in China, n = 41; in Latin America, n = 171; total n = 2,422.
[2]Middle East and North Africa.
[3]Total n = 2,422. Technologies are arranged in descending order, based on the percentage-point differences between the "today" and "5 years ago" responses. Respondents who answered "other," "none of the above," or "don't know" are not shown.

1. According to graph (b), which of the following two sustainability technologies have seen the biggest increase in use by respondents between 5 years ago and today?

    a. automated equipment and/or processes
    b. big data and advanced analytics
    c. digital platforms for stakeholder engagement
    d. energy-efficient equipment
    e. recycling technologies
    f. renewable sources of energy
    g. sustainability-specific performance measurement and reporting software
    h. technologies to move goods and/or people

2. According to graph (a), businesses in which region of the world today use the most sustainability technologies on average?

    a. China
    b. Europe
    c. India
    d. Middle East and North Africa (MENA)
    e. North America

3. Of the technologies listed in graph (b), which technology was most widely used both 5 years ago and today? Which technology was used the least 5 years ago? Which is used the least today?

4. Given the data in the graphs, how would you describe the overall trend in using sustainability technologies over the past 5 years? In your answer, describe the trend in the regions of the world as well as in the types of technologies used, citing data in the graphs in your answer. How does the information in At a Glance: Why Are Organizations Addressing Sustainability? earlier in this chapter help explain the trends seen in the graphs here?

5. **Make Your Case** Imagine you were hired as the chief sustainability officer for a new transportation company based in Latin America. Using the information in these graphs, how many technology solutions would you recommend your new company put in place, and why? If you were required to implement only two technologies, which of these would you choose, and why? In your answer, use trends seen in the data and/or information you have learned in this chapter to support your position.

- The Natural Step (planning framework for organizational change toward sustainability): https://thenaturalstep.org/
- Global Reporting Initiative (sustainability reporting): www .globalreporting.org/information/about–gri/Pages/default.aspx
- Association for the Advancement of Sustainability in Higher Education (AASHE): www.aashe.org/

20

Government
How Can Policy Influence Sustainability?

As you approach a recycling bin to deposit some papers, you notice that the bin is already "contaminated" with food waste and nonrecyclables. You realize that this bin will now need to be disposed of as trash, and you think, "Somebody should do something about this." This is a common response when we are brought face to face with an environmental problem, and the "somebody" we have in mind is often the government. But the "somebody" taking action could also be us, both through our individual behaviors and by working to bring government attention to problems we feel are important. How can we do this?

When he was 25 years old, in 1970, a graduate student named Denis Hayes coordinated the first Earth Day. Hayes grew up in Camas, Washington, a small town on the Columbia River that was established to support a paper mill, where his father worked. The mill is historically so central to the economy and identity of Camas that the school teams are named the "Papermakers." As a teenager, Hayes recalls that he was depressed by the problems of the world around him: the clear-cut forests in the region, the sight of thousands of dead fish in a creek near the mill, and the rotten-egg smell of sulfurous gases from the mill's smokestacks that left him with a sore throat most mornings. Prior to the 1970s, there were no national policies governing air and water pollution. National forests were primarily used for timber production to support local mills, and there were no protections for endangered species.

In college, Hayes became active in **politics**, the processes by which decisions are made for and applied to a group of people. After serving an internship with Senator Gaylord Nelson of Wisconsin, Hayes was hired by the senator to orchestrate a demonstration of public support for national environmental policies that would spur Congress and President Richard Nixon to act. Hayes recalls coming up with the name "Earth Day" over pizza and beer with other student organizers. Tapping into student networks across the country, they organized a national event that at the time was the largest demonstration in US history, with more than 20 million people in cities across the United States participating in marches and protests. About 1,500 colleges and universities and more than 10,000 elementary and high schools

Climate activist Greta Thunberg speaks outside the 26th UN Climate Change Conference in Glasgow, Scotland, known as COP26. Thunberg has mobilized the political power of students and youth worldwide to enact lasting and meaningful action on climate change.

Chapter Objectives

By the end of this chapter, you should be able to . . .

A. discuss how governments exercise authority.

B. compare how governments are organized and the role that elections and parties play in democratic systems.

C. explain how policies are made and the ways in which they are designed to influence behavior.

D. describe some major policies governing pollution control, land use, and natural resource management in the United States.

E. understand how countries attempt to work together to address international environmental issues.

> ## In the absence of strong environmental laws and institutions, we will be unable to protect and restore our planet.
> —United Nations Environment Programme

politics the process by which decisions are made for and applied to a group of people.

policy authoritative decisions such as laws, regulations, and court rulings guiding the behavior of people and organizations subject to a government (Chapter 2).

organized teach-ins to raise awareness of environmental problems. Earth Day helped to spark a decade of bipartisan environmental policy changes in the United States, including creation of the Environmental Protection Agency and passage of landmark laws such as the Clean Air Act, Clean Water Act, and Endangered Species Act (**FIGURE 20.1**).

Policies are decisions adopted by authorities to influence behavior. Policies affecting the environment continue to change from year to year—particularly as elections alter the balance of power between the major governing parties. We are reminded of this each time a

new president replaces the previous administration from another political party. When President Barack Obama (a Democrat) replaced President George W. Bush (a Republican), he established rules governing a range of environmental practices and actively participated in the Paris Agreement on Climate Change. Likewise, when President Donald Trump (a Republican) replaced President Obama, the new administration made it a priority to roll back many of the rules established over the previous 8 years, including pulling the United States out of the Paris Agreement. In 2021, when President Joe Biden (a Democrat) was inaugurated, he recommitted the United States to the Paris Agreement and helped lead international climate negotiations at the UN Climate Change Conference in Glasgow, Scotland, known as COP26.

But ordinary citizens—particularly students—also continue to play an active role in the political process. In 2021 hundreds of young people between the ages of 15 and 29 representing 186 different countries convened as youth delegates to COP26, as world leaders convened during the COP26 climate summit. 350.org, a group started in 2008 by college students to link global activists and organizations concerned about climate change, continues to organize online campaigns and mass public protests (**FIGURE 20.2**). Students in this and other groups coordinated a walkout at hundreds of college campuses in January 2017 to protest the US government's diminishing commitment to address climate change.

Simply exercising your right to vote is important too. Consider the results of some recent local elections for legislative seats that often control everyday events in your life. In 2017, an election for a seat in the Virginia state legislature ended as a tie after a total of 23,000-plus votes were cast. The tie was broken by randomly drawing the name of one of the two candidates, which gave

FIGURE 20.1 The First Earth Day Denis Hayes (top) turned his environmental concern into political activism and organized the first Earth Day in 1970. Pictured here, tens of thousands gathered on Independence Mall in Philadelphia to support the movement.

FIGURE 20.2 Student Activism The student-initiated group 350.org pushes for policies to address climate change. The name comes from 350 parts per million, the concentration of carbon dioxide that is deemed safe for our atmosphere—a threshold we have passed.

the Republicans control of the statehouse by one seat. Similarly, in Iowa in 2020, a US House of Representatives race was decided by six votes of a total of nearly 197,000 votes cast.

In this chapter, we will learn more about government authority and the ways in which environmental policy making involves political struggles among competing interests and parties. There are decision-making processes under way right now from the local to the international level, from town land-use planning commissions to international treaty negotiations. Let's see how this works.

20.1 How Are We All Subject to Government Decisions?

We often think of ourselves as independent, and throughout this book we emphasize choices you can make as an individual to reduce your environmental impact. But the truth is that the range of choices you can make and the incentives affecting your decisions are deeply influenced by government policies. Think about your own transportation choices. Your decision on whether to drive, walk, bike, or take public transportation is influenced by local decisions governing the way streets, sidewalks, and bike paths are laid out; state and federal funding for public transportation; and even national trade, regulatory, and tax policies that affect the cost of gasoline. So we are all subject to government decisions to some extent.

To govern is to rule. It involves the exercise of authority: the power to make rules, give orders, and enforce obedience for a group. On a small scale, parents have authority over their children, and on a larger scale, company executives might exercise authority over employees. Broader still, **governments** consist of the governing bodies and procedures that have authority over the people residing within a particular area. If you think about which governments you are subject to where you are now, you might find yourself within the boundaries of several governments, including a local government such as a city or town, a state government, and the federal government (**FIGURE 20.3**).

How Do Governments Maintain Authority?

Governments depend on a certain level of compliance from the governed. Most maintain their authority by cultivating a sense of **legitimacy**: a widely shared belief that government authority is appropriately held. Long-established governments often hold a degree of legitimacy because of traditions and identities widely shared among those they govern and simply because

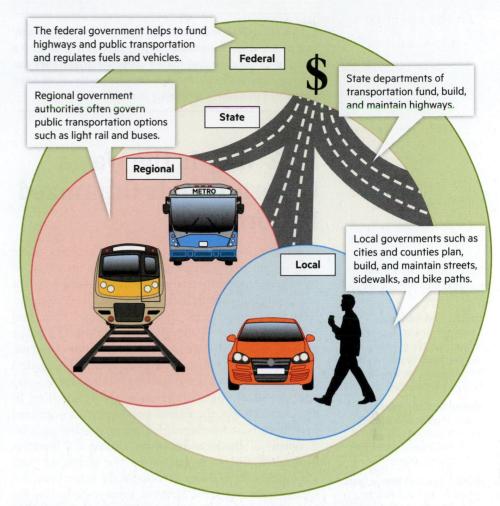

The federal government helps to fund highways and public transportation and regulates fuels and vehicles.

Federal

State

Regional government authorities often govern public transportation options such as light rail and buses.

Regional

State departments of transportation fund, build, and maintain highways.

Local

Local governments such as cities and counties plan, build, and maintain streets, sidewalks, and bike paths.

FIGURE 20.3 Levels of Government There are many levels of government. Which ones have authority over you?

people have become accustomed to obeying particular authorities. But legitimacy can also come from a certain degree of responsiveness to the governed and through adherence to certain widely supported decision-making procedures. For example, stable democracies depend on public confidence in election procedures. However, another way that governments, including democracies, maintain authority is through **coercion**—the use or threat of physical force. Such governments can include those that stifle criticism or protest by punishing dissenting activists. But they can also include police enforcement of ordinances that govern more day-to-day activities. If you can think of a time when you made a conscious choice to obey a particular law, such as a speed limit, because of fear of police action against you, then you have responded to coercion. But too much reliance on coercion can undermine government authority. Government leaders are outnumbered by those they govern, and governments can be changed. This has been the case in revolutions throughout the world, including the American Revolution that established independence from British rule.

government the governing bodies and procedures that have authority over the people residing within a particular area.

legitimacy a widely shared belief that government authority is appropriately held.

coercion the use or threat of physical force to maintain authority.

20.2 Why Is the Way a Government Is Organized Important?

Governing large numbers of people requires more than simply establishing rules for people to follow. To be effective, governments establish **institutions**—formal and enduring organizations, rules, and processes—to carry out the laws, settle disagreements over them, and enforce compliance. Political institutions we are familiar with include the presidency, Congress, and the Supreme Court. To impact a government, you need to understand how its institutions are organized, though there is no one "standard" form of government organization. In the United States, the federal government is organized into three branches: legislative, executive, and judicial. A system of **checks and balances** among the branches exists such that each branch has the means to exert some influence over the activities of the other branches. Other countries have different arrangements. In the United Kingdom, the legislative branch elects one of its own members to serve as prime minister and lead the executive branch. Some countries such as France and South Korea have two executives with different powers: a prime minister elected by the legislative branch and a president elected by the public at large.

The level of public involvement in government institutions varies among countries as well (**FIGURE 20.4**). In **authoritarian systems**, the public plays no significant role, and governments tend to severely limit individual liberties such as freedom of speech and assembly. Although in some cases authoritarian systems concentrate power in a single leader, such as a king or dictator, these systems more commonly manifest as rule by a relatively small group of elites who occupy several government institutions.

In **democratic systems**, the public plays a significant role in government decision making. In Switzerland, in some local government units known as cantons, citizens vote directly on policy questions. More often, democratic systems take the form of a representative government, like in the United States, where citizens indirectly control policy by electing decision makers. Democratic governments also tend to have either informal or formal constitutions that limit government authority and protect individual freedoms.

Finally, power can be distributed geographically. The United States, along with other countries such as Canada, Mexico, Brazil, and India, has a **federal governing system** in which states or provinces have some powers that are independent of the central government. The Tenth Amendment to the US Constitution "reserves" powers for state governments that are not specifically given to the federal government or prohibited to states. This means that states retain some important authority, including developing and enforcing their own criminal codes and regulating public health and safety and many business and labor practices.

In other countries, the central government maintains more authority. The government of China is designed so that the government is replicated in miniature within 27 provinces, thousands of counties and townships, and nearly 1 million villages. But every level of government exists to carry out the policy directives from above. This is one version of a **unitary governing system** whereby the central government alone directs policy. Although China is an example of an authoritarian government with a unitary system, most democracies throughout the world also use unitary systems. France and Japan are two examples. The regional and local governing institutions in these countries do not have powers independent of the national government.

How the Type of Government Impacts Environmental Policy

The way a government is organized can greatly affect environmental policy. Consider two very different governments: those of China and the United States. While both nations face environmental challenges, their policy responses are shaped by important differences.

Let's start with China. Its government is considered authoritarian because power is concentrated in the Chinese Communist Party (CCP), which is dominated by one paramount leader and about three dozen influential political elites. This can lead to what seems from the outside to be rapid decision making. For example, once

---

institution a formal and enduring organization, rule, and process.

checks and balances a system that ensures a part or branch of the government has the means to exert some influence over the activities of the other branches.

authoritarian system a form of government in which power is concentrated in a single individual or among a small group of elites.

democratic system a form of government in which the public plays a significant role in government decision making.

federal governing system a form of government in which states or provinces have some powers that are independent of the central government.

unitary governing system a form of government in which the central government alone directs policy.

Pure Democracy	Representative Democracy		Authoritarian Government
Citizens vote directly on policy	Citizens elect officials to make policy		Rule by one or a few without meaningful elections

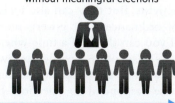

High public participation in government ←———————————→ Low public participation in government

FIGURE 20.4 Governing Systems and Public Participation We can think of various types of governing systems as a continuum stretching from low to high public participation, with authoritarian systems at the low end and democratic systems at the high end.

the party leadership became concerned about population growth, the government implemented a decades-long policy allowing only one child per family before shifting to a two-child limit in 2016 and a three-child limit in 2021 (see Chapter 6). Similarly, as China's government prioritized quick economic growth over environmental regulation, many of its cities developed severe air pollution. Yet in recent years China has begun to rapidly shift toward alternative energy and conservation measures. In 2004, the CCP charged a science and policy think tank with drafting renewable energy legislation, and in less than 1 year China's national legislature passed the law without amendments. With few checks and balances among its governing institutions and a unitary government across geographic regions, China's governmental agencies simply implemented the new directives. Just 1 year after passage, the percentage of China's total energy consumption from renewables tripled from 2.5% to 7.5%.

Of course, authoritarian governments often restrict the rights of their people, and Human Rights Watch noted in 2021 that "governments, civil society groups, and United Nations officials expressed growing concern over the Chinese government's human rights violations," including the silencing of journalists and activists along with increasing restrictions on the internet. Even in its quest to improve air quality, the Chinese government significantly restricted individual behavior. For example, Chinese car owners were required to leave their cars at home at least one day a week, and individuals were no longer allowed to use elevators for the first three floors of public buildings.

In the United States, the public has greater involvement in the policy-making process, power is shared among the three branches of government, and the states often challenge the actions of the federal government. But this means that policies often develop incrementally over longer periods of time and frequently change course as various interests move into and out of power.

For example, in the early 2000s, the Republican administration of George W. Bush declined to regulate greenhouse gases under the Clean Air Act. This decision was challenged in court by several states, cities, and environmental groups such as the Sierra Club. Eventually, a 2009 US Supreme Court case called *Massachusetts v. Environmental Protection Agency* found that the executive branch had not fully upheld the Clean Air Act and called on the EPA to reexamine the regulation of greenhouse gases. Then, in that year, the newly elected Democratic administration of Barack Obama tried and failed to adopt climate legislation over Republican opposition in Congress. The Obama administration further attempted to address climate change in 2015 through a new EPA rule known as the Clean Power Plan, which would require states to meet carbon dioxide (CO_2) emission-reduction standards at power plants by 2030 (**FIGURE 20.5**). However, before the new rule could take effect, it was challenged in federal court by 27 states and several coal and energy companies.

FIGURE 20.5 Changing Politics and Changing Policy In the United States, policy direction often changes course in response to the party in power. For example, the EPA declined to regulate greenhouse gas emissions for 8 years under President George W. Bush, and then President Barack Obama directed the EPA to develop a Clean Power Plan to regulate these emissions. However, before these regulations took force, President Donald Trump directed the EPA to rescind them. In 2021 President Joe Biden proposed his own rules regulating emissions and a series of alternative energy initiatives.

Before the court had a chance to rule, the executive branch shifted back to Republican control with the 2016 election of Donald Trump. The Trump administration promptly set out to repeal the Clean Power Plan in 2017 and establish its own rule. In 2021, a federal court struck down the Trump administration's climate rule, and the newly elected Biden administration began drafting its own rule. In 2022, the US Supreme Court, now with several new justices, limited the authority of the EPA to issue sweeping rules regulating the energy sector in *West Virginia v. EPA*. But later that year, the Biden administration successfully passed the Inflation Reduction Act, which provided record levels of government funding for alternative energy and conservation.

Certainly, this political back-and-forth can seem tedious and frustrating, although the US system of government does more to protect individual rights and the prerogatives of states than does a more authoritarian and unitary approach. And it also allows for some unique solutions. For example, because of the federal system of government, states are often free to pursue their own policies. We learned in Chapter 2 that California has become a world leader in adopting policies to reduce greenhouse gas emissions through energy conservation measures, a CO_2 cap-and-trade policy, and tighter emissions standards for vehicles. And sometimes compromises emerge among competing interests on an environmental issue. For example, as Congress strengthened the regulations on sulfur dioxide emissions from coal

Rapid Change in China

China now owns two of the world's top five wind turbine manufacturers, the six largest photovoltaic solar panel manufacturers, and the largest lithium-ion battery manufacturer. It also subsidized the creation of about 80 electric-car makers. However, in 2019 the government rapidly slashed these automobile subsidies, in what many assumed was a move to encourage growth in some companies over others. While these changes occurred quickly, they were made almost unilaterally without input from the public.

plants in 1990, it also funded job-training partnerships and unemployment support for states most dependent on coal mining.

⬡ **TAKE-HOME MESSAGE** Governments differ in the way they organize authority among various institutions and different geographic areas, spanning a continuum that stretches from authoritarian systems to direct democracies. The role the public plays in government affairs affects how environmental policies are established and implemented. These differences determine who is involved in environmental policy making and the level of coordination required.

20.3 How Do Elections Shape Government?

The idea of an election is simple, right? You hold the election and count the votes to determine the winner. However, there are many ways to organize elections. Who gets to vote? How many votes does it take to win?

single-member-district plurality (SMDP) system an electoral system in which each geographic district chooses a single representative by selecting the candidate who receives the largest number of votes.

proportional representation (PR) system an electoral system that elects multiple members from each district and distributes representation in proportion to each party's percentage of the vote.

What role do political parties play? Is it a winner-take-all election, or is representation granted in proportion to the percentage of votes? The answers to questions such as these make a big difference in determining the winners and losers of elections in democratic systems.

One way to see the important effect that electoral systems have is by considering the role that environmental issues play in different countries. In most of the world's democracies, environmental activists have formed so-called Green Parties dedicated to addressing environmental issues and fostering sustainability. Yet in the United States, no Green Party candidate has held a national office. Simply put, none of their candidates have received enough votes to beat the other candidates in our winner-take-all system. However, this is not the case in other countries. For example, since the mid-1990s, the German Green Party has averaged more than 50 seats in Germany's 600-plus seat Bundestag (federal parliament). Green Parties in Finland, Belgium, and the Netherlands have also gained some influence in parliament. While these success rates reflect support for Green Party policies, in this section we will see how these countries' electoral systems also play an important role in effecting change.

Organizing Election Systems

Most democracies employ one of two electoral systems: the **single-member-district plurality (SMDP) system** or the **proportional representation (PR) system**. The United States, Canada, and the United Kingdom are among the countries that use an SMDP electoral system. In this system, the area to be governed is divided up into geographic districts, and each elects a representative in government. It is important to note that it is a *plurality* system rather than a *majority* system; if more than two candidates are running, the candidate with the most votes may have fewer than 50% of the votes. It is a winner-take-all system because there is only one representative per district. The US House of Representatives follows this model.

On the other hand, most of the world's democracies use a PR electoral system that elects multiple members from each district and distributes representation in proportion to each party's percentage of the vote. Typically, voters in each district are presented with a list of candidates from each party. Each party can put forward a number of candidates equal to the number of representatives that will be elected from the district. Voters indicate their preference for a party on the ballot. On the basis of the proportion of votes received in the election, each party is awarded a number of seats in the legislature. Green Parties have had some success in these systems. For example, if a district has 10 seats and the Green Party receives 20% of the vote, it will receive two seats for that district in the legislature. The top two people on that party's list in the district will serve in the legislature. Notice that this is not a winner-take-all system. A small

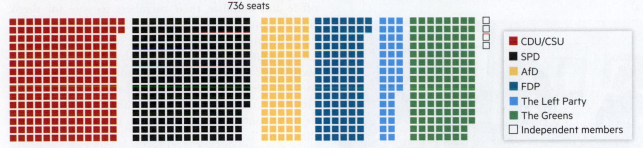

736 seats

■	CDU/CSU
■	SPD
■	AfD
■	FDP
■	The Left Party
■	The Greens
□	Independent members

FIGURE 20.6 **The 2021 Distribution of Party Seats in the German Parliament by Party** Because of proportional representation, the Green Party has a significant number of seats in the Bundestag (the national parliament of Germany) and plays a major role in governing when one of the two major parties requires its numbers to attain a majority.

Adapted from Deutscher Bundestag (2021).

number of candidates from parties receiving a relatively small share of votes in the district can still serve in the legislature (**FIGURE 20.6**).

Political Parties

The type of electoral system a democracy employs has a big effect on **political parties**, organizations that attempt to influence government by working to get their members elected to government offices. Political parties recruit and nominate candidates, mobilize voters who support their candidates, and coordinate decision making among party members. In the United States, most elected officials come from just two parties: Democrats and Republicans. Why? Voters tend to act strategically. In a winner-take-all situation, many voters are not so much choosing the candidate they most prefer as they are casting a ballot for the least objectionable candidate among those with a significant chance of winning. This often works against candidates campaigning with an environmental message. If they are not part of a major party, voters might not "waste a vote" on a candidate who, in their view, does not have a serious chance of winning. In this way, support coalesces around candidates from a small number of parties. Over time, these parties can amass resources and, once in office, establish rules that favor large parties over small parties.

In contrast, PR systems favor the existence of many small parties. Each party receives legislative seats in proportion to its share of votes, so even those that garner a very small share of the votes may have an opportunity to serve in office. The impact of electoral systems on political parties affects both the distribution of power and the strategies used to influence government. In a PR system, a group with new ideas and approaches might have success creating its own party, gaining a foothold in parliament, and maybe even playing a role in a governing coalition of parties. But in an electoral system dominated by two major parties, a similar group is more likely to follow a strategy of changing one of the major parties from within to develop policies more to its liking. For example, in recent years, environmental activists within the Democratic Party who wanted a more comprehen-

sive climate change policy than their party had offered began advocating a Green New Deal—a plan to tackle greenhouse gas emissions from the transportation and energy sectors while also addressing issues of economic opportunity and equity.

⬡ **TAKE-HOME MESSAGE** In democracies, electoral systems and political parties play an important role in determining government leadership. When the electoral system is designed so that a single winner is determined by plurality vote, there is a tendency for a small number of large parties to dominate government. Proportional representation systems allow for smaller parties to play a role in government.

20.4 How Are Environmental Policies Made?

Political scientists—people who study systems of government and political behavior—often describe their work as the study of "who gets what, when, and how." Politics pits people with different perspectives against one another in power struggles. In this context, **power** is the capacity of individuals and groups to make their concerns or interests count in decision-making processes. The competitive nature of politics can make it difficult or impossible for both sides to "win." For example, if one group of citizens wants to protect a forested area as a nature preserve and another group wants to use the area for residential development, it is not likely that a decision will fully satisfy the desires of both groups. If a decision on the issue is reached, it is likely that either one group's desires win out over those of the other or a compromise is achieved.

To increase their overall power, people and organizations with shared interests often form **interest groups** in order to influence government. They attempt to persuade a government decision maker through a practice known as **lobbying**. They support certain candidates for

political party an organization that attempts to influence government by working to get its members elected to government offices.

power the capacity of individuals and groups to make their concerns or interests count in decision-making processes.

interest group a group formed by people and organizations with shared interests in order to influence government.

lobbying a practice in which groups or individuals attempt to persuade a government decision maker.

(a)

(b)

(c)

FIGURE 20.7 Interest Groups Interest groups do many things including **(a)** directly lobbying elected officials for their preferred policies; **(b)** supporting candidates for office by campaigning and fundraising; and **(c)** organizing media, information, and/or protest campaigns to influence public opinion.

office and mobilize their members to vote for them. Interest groups may also try to influence decision makers indirectly by mobilizing public opinion behind their cause (**FIGURE 20.7**). This can involve media and public awareness campaigns as well as protest events designed to focus attention on a particular issue. Each step in the process is an opportunity for influence and is likely to be hotly contested by groups with competing views on the issue at hand.

Politics and the Arctic National Wildlife Refuge

Let's take a closer look at policy making by studying the issue of drilling for oil on the coastal plain of Alaska's Arctic National Wildlife Refuge (ANWR) (**FIGURE 20.8**). Prior to 1960, the nearly 20 million acres of land (about the size of South Carolina) that now compose ANWR were owned by the federal government but not designated for a specific purpose. They were available for various kinds of development. But in 1960, the US Department of the Interior set aside much of the area as "national interest" lands for a future national park, national forest, wildlife refuge, or wilderness status. This situation created a dilemma: what sort of policy would the government establish for this land? Furthermore, until these lands attained formal protected status, they could be open to development. A struggle ensued, pitting wilderness conservation advocates such as the Wilderness Society and Friends of the Earth against advocates for developing oil and gas resources in the area, such as ARCO, British Petroleum (BP), and the state government of Alaska. Oil and gas interests advocated drilling, and wilderness advocates opposed provisions allowing any development. Each side lobbied decision makers and launched media campaigns in an attempt to influence the outcome.

Problem Identification and Agenda Setting In the ANWR case, each side first tried to define the agenda in its own terms. Until an issue is clearly defined as a problem to solve, it will not make it onto a government's agenda. Further, there are practical limits to the number of issues a government has the time and resources to address and that can capture public concern. For this reason, problems compete for attention, with each interest group trying to rally support for its issue of choice.

We see this struggle in the ANWR case. Wilderness advocates defined the problem in terms of the threat that development would pose to wildlife: the caribou, peregrine falcons, snow geese, tundra swans, wolves, polar bears, and other species for which ANWR is critical habitat. Development advocates argued that oil and gas exploration and extraction could be done in ways that would not pose a problem for these species. Throughout the struggle, commercial interests argued that drilling in

ANWR could help boost economic growth, decreasing reliance on foreign oil and contributing revenue to Alaska and the federal government.

Policy Formulation and Legitimation Over the years, competing policies varied in the level of protection given to different areas of ANWR. In 1980, these policies found their way into bills considered by the US Congress. The focal point of the bills was the coastal plain, an area that wilderness advocates often called "the ecological heart" of ANWR and that development interests coveted for oil and gas resources. The US House of Representatives favored wilderness designation—the highest level of protection—for the area, but the US Senate favored allowing more exploration for oil and gas development. In 1980, Congress passed and President Jimmy Carter signed a compromise measure called the Alaska National Interest Lands Conservation Act. The law required the Department of the Interior to study the potential for resource development on the coastal plain. It also empowered Congress to authorize future oil and gas production. So ANWR was added to the National Wildlife Refuge System, but it lacked the protection of wilderness status.

This example shows not only that government policies are often compromises among competing interests but also that the passage of a policy does not halt political activity. Over the ensuing decades, oil and gas interests have continually tried to persuade Congress to allow drilling in the coastal plain, while environmental groups have worked to maintain the area's protected status. New policies simply set the stage for future political struggles.

Policy Implementation and Evaluation One typical area of struggle involves choosing the policies that are actually implemented and then ultimately evaluated as successes or failures. When Congress passes laws, it also empowers agencies to administer the laws. These agencies draft rules, regulations, and management plans clarifying how a law will be implemented. Political interests attempt to influence specific rules and regulations and challenge the way in which they are interpreted and administered. Consider recent updates to the US Fish and Wildlife Service's management of ANWR. In 2015, this agency identified several wilderness study areas within ANWR, including the coastal plain, which meant that these areas were to be managed as wilderness. However, this plan was short lived, as the ANWR issue made it onto the congressional agenda in 2017. This time, the changes proposed were designed to generate revenue in a Republican-backed tax-cut bill in Congress. When President Trump signed this bill into law, it began the process of opening ANWR to oil and gas exploration and development. But before these projects could begin, in 2021, the Biden administration suspended oil and gas leases in ANWR—a move that the state of Alaska and fossil fuel companies challenged in court.

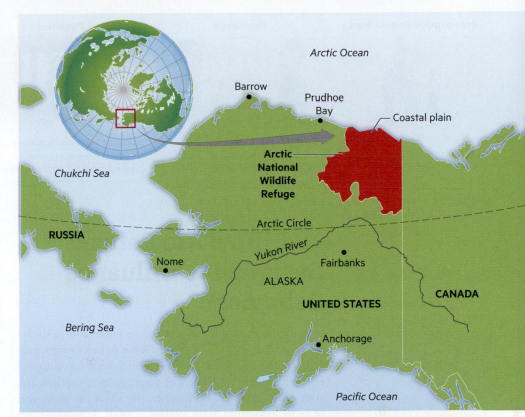

FIGURE 20.8 Arctic National Wildlife Refuge The Arctic National Wildlife Refuge in Alaska was the product of a contentious policy-making process. This involved significant compromises between those who wanted to preserve the area and those who wanted to open it up to oil and gas development.

⬡ **TAKE-HOME MESSAGE** In politics, it is difficult or impossible for all interested groups to get exactly what they want. Because of this, policy making involves power struggles between competing groups as they try to influence how problems are defined, which issues get on the government agenda, and how policy is formulated, approved, implemented, and evaluated.

Prescriptive Regulations

Establish mandates about what can and can't be done.

Payments

Offer positive incentives like subsidies, grants, and tax benefits to encourage certain behaviors.

Penalties

Impose negative incentives such as fines or criminal punishment for noncompliance.

Property Rights

Create property rights to achieve policy goals.

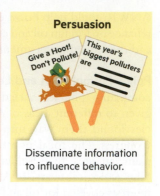

Persuasion

Disseminate information to influence behavior.

FIGURE 20.9 The "5 Ps" of Policy Design

20.5 How Are Policies Designed to Influence Behavior?

Imagine you are part of a group interested in promoting more fuel-efficient vehicles. What policies might you promote to affect the purchasing behavior of consumers? Whatever the specifics of your group's proposals are, your policies would be designed to affect the costs and benefits associated with certain actions. Policies create incentives by utilizing some combination of five policy tools: the "5 Ps" of prescriptive regulations, payments, penalties, property rights, and persuasion (**FIGURE 20.9**). In this section, we will look at each of these tools.

Prescriptive Regulations

prescriptive regulation a mandate about what can and cannot be done.

payment in policy terms, a positive incentive designed to encourage certain actions.

subsidy a direct payment from the government to encourage a certain activity.

Most environmental policies apply the tool of **prescriptive regulation**: that is, mandates about what can and cannot be done. For example, Congress passed the Energy Policy and Conservation Act of 1975 empowering the US Department of Transportation to develop corporate average fuel economy (CAFE) standards for vehicles, which determine the minimum fuel economy of the vehicles available. Manufacturers must meet these minimum standards or pay a fine. Other policies using this tool include laws and rules that establish water-quality standards as well as various land-use designations—such as wilderness areas and wildlife refuges (see Chapter 5), which forbid certain kinds of activities and uses.

Payments

What if the government passed a policy lowering the cost of a more fuel-efficient vehicle? This would be a kind of **payment**—a positive incentive designed to encourage certain actions. Payments can come in many forms.

>> **Subsidies** are direct payments from the government to encourage certain activities. They often come in the form of cash payments or other items of value such as tax incentives, grants, and low-interest loans. When automaker General Motors (GM) was facing bankruptcy in 2009, the US government bought a 61% stake in the company for $50 billion to keep it afloat. Among the many changes GM made as a condition of the bailout was a greatly expanded production of hybrid vehicles. At the same time, the US government subsidized electric vehicles by offering consumers who purchased them a tax credit of $7,500.

>> Governments also use *research grants* to encourage developments in science and technology. Government agencies such as the National Science Foundation provide general support for scientific progress by grants that fund promising work in academic fields. Other government agencies such as the US Department of Energy, National Institutes of Health, US Department of Agriculture, and EPA provide grants for research on targeted problems (**FIGURE 20.10**).

>> Payments sometimes come in the form of *contracts*—agreements to pay for the provision of certain goods or services. Because they are such large buyers, governments can demand certain kinds of business practices from their suppliers. For example, the US

FIGURE 20.10 Grants as Policies of Payment Each year the Environmental Protection Agency encourages teams from colleges and universities to pursue sustainability research projects at the National Sustainable Design Expo. Award-winning projects receive research grants.

Department of Defense has an annual budget in excess of half a trillion dollars. In 2004, it adopted a procurement policy to favor providers of "environmentally preferable products and services" such as recycled content products, energy- and water-efficient products, alternative-fuel vehicles, and alternatives to hazardous chemicals.

>> **Permits** are a form of payment by which the government authorizes certain actors to engage in behaviors that would otherwise be illegal. For example, the High Seas Fishing Compliance Act of 1995 empowers the US Department of Commerce to issue fishing permits to vessels that adhere to certain international conservation measures (**FIGURE 20.11**). Permits are thus privileges granted by the government in exchange for compliance with mandated procedures.

Penalties

Prescriptive regulations are often accompanied by **penalties** for noncompliance. In policy terms, a penalty is a disincentive designed to discourage certain actions. *Criminal penalties* can mean imprisonment but may also take the form of major fines. *Civil penalties* do not include imprisonment but extract payments or service to punish and provide restitution for illegal activities or negligence. Environmental laws make use of both criminal and civil penalties. After the explosion of BP's *Deepwater Horizon* drilling rig in the Gulf of Mexico in 2010, BP pled guilty to criminal manslaughter charges for the 11 workers killed by the explosion. In addition, under the Clean Water Act, BP pled guilty to knowingly and directly discharging pollution into the water and not accurately reporting the amount. The company also faced criminal charges under the Migratory Bird Treaty Act for the birds killed by the oil spill. In total, BP paid $4 billion in criminal fines and penalties, the largest criminal resolution of an environmental crime in US history, and $62 billion in civil penalties.

Governments can also use taxes as a negative incentive to discourage and diminish certain behaviors by raising the cost of products or services. The United States has a "gas guzzler" tax on new cars that do not meet the federal fuel-efficiency standards of 22.5 miles per gallon. The amount of the tax—ranging from $1,000 to $7,700—is displayed on a sticker, along with the fuel efficiency of the car (**FIGURE 20.12**). The government also applies penalties for noncompliance by *withholding payments*. This can be a powerful motivation for compliance because government subsidies, licenses, and contracts are valuable, and many individuals and groups depend on them to conduct their business. Both the Clean Water Act and Clean Air Act empower the EPA to exclude businesses violating these acts from entering into contracts with the federal government.

FIGURE 20.11 Permits Commercial fishing harvests, like this net full of yellowfin tuna, are often regulated by policies that make it illegal to fish without government-issued permits. These permits are a kind of privilege granted in exchange for abiding by certain rules.

Property Rights

We have seen with ANWR an example of the government having the power of property rights. By changing the property rights of an area, the government can drive a policy goal. This has been going on for some time. In the 19th century, the US government turned millions of acres of public land into private land to spur economic growth through farming, mining, and railroad construction. In another example, sulfur dioxide pollution from coal-fired power plants was turned into a property right by the Clean Air Act. This created a market for pollution allowances (effectively the right to pollute a certain amount) as part of a cap-and-trade system. The allowances could be bought and sold as property on a commodities market (see also Chapter 8).

Persuasion

A final policy tool is simple persuasion, normally initiated by producing and disseminating information to influence behavior. The Emergency Planning and Community Right-to-Know Act of 1986 requires facilities that manufacture, process, or use any one of 650 chemicals identified as toxic to report the quantity of their annual releases of these chemicals into the air, land, and water. The EPA collects and disseminates this information to the public through the Toxics Release Inventory (TRI). Information-disclosure requirements enable environmentally concerned groups and media sources to identify and put pressure on the worst polluters to improve their operations. Several studies have found that TRI reporting requirements have facilitated significant reductions in toxic releases.

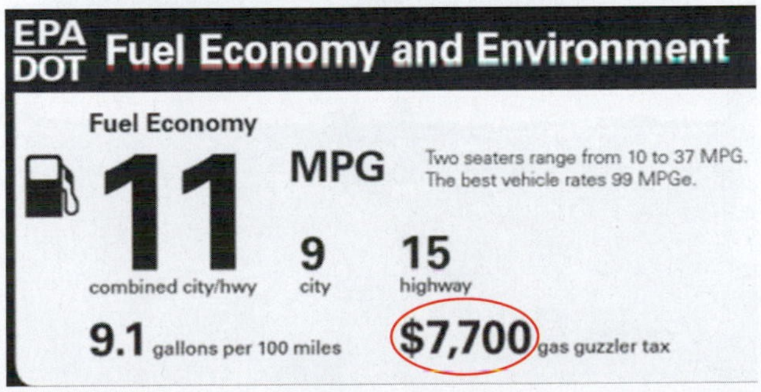

permit government-authorized permission to engage in behavior that would otherwise be illegal.

penalty in policy terms, a disincentive designed to discourage certain actions.

EPA DOT Fuel Economy and Environment

Fuel Economy

11 MPG
combined city/hwy

Two seaters range from 10 to 37 MPG. The best vehicle rates 99 MPGe.

9 city   **15** highway

9.1 gallons per 100 miles   **$7,700** gas guzzler tax

FIGURE 20.12 Taxes as Penalties Cars that do not meet federal fuel-efficiency standards are subject to the gas guzzler tax—in this case $7,700.

TAKE-HOME MESSAGE Policies influence behavior by using some combination of policy tools: prescriptive regulations, payments, penalties, property rights, and persuasion. Payments provide positive incentives to encourage certain behaviors. Penalties are negative incentives for those who fail to comply with a policy. Governments also attempt to achieve policy goals by seizing or granting private property rights and by persuasion through information campaigns.

20.6 How Did Important Environmental Policies in the United States Develop?

Policies affecting the environment in the United States are often divided into two broad issue areas: those that regulate pollution and those that manage natural resource impacts. Each policy area has its own history linked to environmental problems and policy responses.

Pollution Control, the EPA, and *Silent Spring*

After World War II, the rapid development and use of synthetic fertilizers and pesticides amplified the impact of pollution in many areas of the country. Until the 1970s, pollution was regulated largely by state and local governments. But state and local laws were often poorly enforced, contained exemptions for certain influential industries, and could not control pollution sources outside their boundaries.

Then a series of air- and water-pollution events began to focus media attention and influence public opinion (**FIGURE 20.13**). But perhaps the single most powerful factor influencing a rise in public concern over pollution was the 1962 publication of the book *Silent Spring* by a US Fish and Wildlife Service marine biologist named Rachel Carson (see the **Stories of Discovery** feature). Her best-selling book detailed her research on the health and environmental effects of the pesticide dichlorodiphenyltrichloroethane (DDT), and as a result, interest groups devoted to banning DDT and passing national pollution-control legislation, such as the Environmental Defense Fund, were established in the 1960s.

By the 1970s, elected officials felt the need to respond to a public demand for national pollution-control laws. They created the Clean Water Act to govern water pollution,

(a)

(b)

(c)

(d)

FIGURE 20.13 Environmental Events Influencing Public Opinion **(a)** Donora, Pennsylvania, 1948: Sulfuric acid, nitrogen dioxide, fluorine, and other poisonous gases emitted from US Steel's Zinc Works in Donora killed 20 people and sickened nearly half the city. **(b)** Los Angeles, California, 1950s: Smog in Los Angeles became a common and highly visible problem. **(c)** Santa Barbara, California, 1969: An offshore well leaked 20,000 gallons (76,000 liters) of oil a day for several weeks off the coast of Santa Barbara. **(d)** Cleveland, Ohio, 1952: A fire that started in an oil slick in Cleveland's Cuyahoga River caused $1.5 million in damage. A similar fire in 1969 spurred people to demand greater regulation of water pollution.

Rachel Carson: Speaking Out about *Silent Spring*

Rachel Carson is widely recognized as a pioneering leader of the modern environmental movement. Her 1962 book, *Silent Spring*, is often credited with sparking the concern over pollution and endangered species in the United States that led to landmark environmental laws in the late 1960s and early 1970s.

As an undergraduate student, Carson had an interest in both English and biology. In graduate school she focused on marine biology, and her first job was with the US Bureau of Fisheries in the 1930s, where she analyzed data on fish populations and wrote educational brochures for the public. She often expanded on her research and writing tasks, turning them into nature-themed articles for national magazines such as the *Atlantic* and the *New Yorker*. By the 1950s, Carson was the editor-in-chief of publications for her government agency and a best-selling author, with books about marine life to her credit; one of her books was even made into a documentary. Although her interest in the environmental impacts of synthetic pesticides began in the 1940s, it was only when she became a well-known writer that she was able to resign from her government job in 1952 and dig into this topic full time.

When residents in Nassau County, New York, lost a court challenge in 1957 opposing the gypsy moth eradication program that involved aerial spraying of dichlorodiphenyltrichloroethane (DDT) over their properties, the issue gained some national attention. Carson wrote to the editor of the *New Yorker* and suggested that he write an article about the topic, but he replied that she should write the article herself. So over the next 4 years, Carson interviewed scientists working in the area, pored over the scientific literature, and probed documents produced by government agencies in the United States and Canada.

Carson assembled the research into *Silent Spring*, which was first serialized in the *New Yorker* and then published in book form a few months later. It described the food chain and human health effects of several pesticides (including DDT) and the development of pesticide resistance in insects. As a scientist, she was able to find, evaluate, and integrate many strains of scientific research on the topic and then, as a gifted writer, communicate the information in a way the general public and policy makers could understand. In the introduction, Carson asked her readers to imagine a springtime without birdsong—a "silent spring"—and the cascading effects through the food web, including effects on agricultural production and human well-being. As she completed the research and writing of the later chapters of the book on cancer-causing agents and rising rates of cancer, she herself was stricken with breast cancer.

When the book was finally published in 1962, it was a best seller that gained considerable national distribution as a Book-of-the-Month Club selection and through serialization in the *New Yorker* and newspapers around the country. The chemical industry launched fierce criticism against Carson and the book, which she defended in congressional hearings and on numerous news programs and in debates, even while she was undergoing radiation treatment for cancer. Many of these critiques were sexist in nature, labeling her a "hysterical spinster." Carson responded, "I'm not interested in things done by women or by men but in things done by people." In 1963, the President's Science Advisory Committee backed Carson's claims in the book.

Although Carson succumbed to cancer and died in 1964, her call for more restricted and managed use of pesticides and greater understanding of the ecological effects of widespread chemical applications had a big impact on federally funded research and legislation. DDT was ultimately banned in the United States in 1972. In 2012, the American Chemical Society, which had once been a fierce critic of Carson, deemed *Silent Spring* a National Historic Chemical Landmark for its role in the development of the modern environmental movement. The group credited Carson's book for inducing a paradigm shift in the discipline of chemistry by focusing attention on the environmental and human health impacts of chemical technologies.

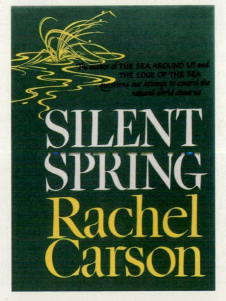

Rachel Carson's best-selling book *Silent Spring* helped draw public attention to the ecological impacts of pesticide use and to environmental issues in general.

What Would You Do?

Rachel Carson took complicated scientific discussions of an environmental problem and presented them in a way that was accurate, compelling, and easy to understand. Choose an issue that's important to you. Describe what you would do and what tools (social media, publicity, writing, etc.) you would use to publicize the problem more widely.

TABLE 20.1 Major US Pollution-Control Policies

Policy	Substance or Standard Regulated	Federal Responsibilities	State and Local Responsibilities
Clean Air Act (CAA) of 1970 (see Chapter 8)	Air quality and emissions	• EPA sets standards for overall air quality and emissions from polluting sources such as utilities, factories, and vehicles • Act was amended in 1990 to create tradable emission allowances for sulfur dioxide from coal-burning power plants	• State environmental agencies often issue permits and carry out enforcement
Clean Water Act (CWA) of 1972 (see Chapter 7)	Toxic pollutants in surface water	• EPA sets allowable pollution levels for each industry or polluter category on the basis of available pollution-control technology	• State environmental agencies determine water-quality goals for respective state waterways • Often issue permits and carry out enforcement
Resource Conservation and Recovery Act (RCRA) of 1976	Solid and hazardous waste buried in the ground	• EPA sets standards for solid and hazardous waste management and cleanup • EPA implements a hazardous waste program if a state does not have one	• Implement solid waste programs • Implement hazardous waste programs if authorized by EPA • Issue permits and carry out enforcement
Comprehensive Environmental Response, Compensation, and Liability Act (CERCLA, also known as "Superfund") of 1980	Preexisting hazardous waste sites that threaten public health and/or the environment	• EPA identifies potentially responsible parties (PRPs) for the contamination and assigns liability • Act establishes a trust fund to pay for cleanup when PRPs cannot be identified • EPA implements cleanup plan if a state cannot	• Implement cleanup plan if authorized by EPA • Pay part of cleanup costs depending on the state's role in generating hazardous waste

sovereignty the ability to make policies and carry out actions within a certain territory.

United Nations (UN) an intergovernmental organization with 193 member states.

the Clean Air Act of 1970 to govern air pollution, and the Resource Conservation and Recovery Act to govern solid and hazardous waste disposal, all of which limited and prohibited polluting activities before they caused harm. Further, the EPA was created to implement and enforce laws like these by setting minimum national standards for pollution levels, governing polluting activities, and working with states for enforcement and compliance. **TABLE 20.1** provides details of some of the major US pollution-control policies.

Natural Resource Policy

The federal government owns 28% of the landmass of the United States and controls the use of natural resources on these lands. However, this leaves 72% of the landmass under the control of states and means that many policy decisions affecting land and water use are left to state governments, which often delegate authority to local governments. **TABLE 20.2** lists some of the major responsibilities divided between federal and state authorities.

🛑 **TAKE-HOME MESSAGE** In the 1970s, the federal government gained more authority over pollution-control policy. In general, the EPA sets standards and works with states to issue and enforce permits governing polluting activities. Authority over natural resource policy is shared among different levels of government in the United States. Policies governing federally owned land are carried out by land-use agencies that make and enforce rules for particular uses on the lands they administer. State and local governments still have authority over most land- and water-use decisions.

20.7 How Are Environmental Issues Addressed on the International Scale?

Impacts on the environment are not constrained by national borders. The Colorado River and its branches flow across seven US states and then into Mexico. The United States, China, and Europe produce more than half of the world's greenhouse gas (GHG) emissions, contributing to sea-level rise that threatens 52 small island nations. Although every country has **sovereignty**—the ability to make policies and carry out actions within its territory—this authority ends at the national border. In this section, we will explore the daunting challenge facing sovereign nations in their attempt to address global environmental issues through intergovernmental organizations and agreements.

Intergovernmental Organizations

Intergovernmental organizations (IGOs) are formed by member countries to facilitate cooperation. With 193 member states, the **United Nations (UN)** is the largest IGO, and it works in many ways to influence change, such as by facilitating negotiations on international agreements.

Two important current initiatives related to the environment are the sustainable development goals (SDGs;

TABLE 20.2 Federal and State Natural Resource Policies

Resource Area	Federal Responsibilities	State Responsibilities
Land	• Designate protected areas on federal lands and waters (e.g., national parks and monuments, wilderness areas, marine protected areas, etc.) • Set prescriptive regulations for these areas, with penalties for disallowed uses • Issue permits for regulated use of federal lands, such as grazing, timber harvests, or mining (see Chapter 5)	• Set policies governing extractive resources like forestry and mining that take place on state and private land
Plants and animals	• Establish species-protection laws (e.g., Marine Mammal Protection Act, Endangered Species Act) • Prohibit or strictly limit the hunting or harvest of certain species • Require development of recovery plans, designation of critical habitat, and consideration of human impacts on species (see Chapter 5)	• Regulate hunting, fishing, and harvesting of plants unless a species is listed under the Endangered Species Act • Many states also set their own species-protection laws
Environmental impact and development	• National Environmental Policy Act requires federal agencies to consider alternatives and complete environmental impact statements prior to taking actions that could have significant environmental impacts (see Chapter 2)	• Local governments pass zoning ordinances and building codes that regulate where and how land within their boundaries is used • Some states have growth-management laws that require municipalities to protect the environment and preserve farmland in their urban development plans
Water	• The Bureau of Reclamation and the US Army Corps of Engineers both construct and operate large water-management projects such as dams, levees, reservoirs, and canals • The Army Corps of Engineers also regulates and issues permits for dredging and filling of waterways and for draining and filling of wetlands	• Set policies for distribution of water • Water-management agencies administer water rights • State courts resolve disputes between various users • Significant variation in water-distribution policies across the country

see also Chapter 1) and the Intergovernmental Panel on Climate Change (IPCC). The SDGs are a set of 17 goals on climate change, poverty, and inequality intended to be a blueprint for the future. Adopted by UN member states in 2015, these detailed goals are designed to motivate and measure progress and to influence through publicity and persuasion (**FIGURE 20.14**). Similarly, through

the IPCC the UN convenes thousands of experts to assess the scientific literature on climate change for policy makers. Their reports are produced and updated regularly, providing the data to observe and make decisions about the impacts of climate change.

Other influential IGOs are financial organizations such as the World Bank and the International Monetary

FIGURE 20.14 Sustainable Development Goals The United Nations has created 17 goals for a sustainable future. These goals relate to many global efforts to promote equality and prosperity, including projects pictured here in Papua New Guinea, Argentina, and Peru.

ENVIRONMENTAL JUSTICE

Policies to Address Cumulative Impacts

Cumulative impacts are incremental environmental harms—like various forms of air and water pollution—that accumulate and worsen over time. In recent years there has been an effort at the national and state levels to design policies to address cumulative impacts. In 2021 President Joe Biden issued an executive order that incorporated a "whole of government" approach to environmental justice, bringing together department heads and task forces that work across many government agencies to try to reduce the cumulative impacts on communities that are marginalized, underserved, and overburdened by pollution.

Some state governments are now requiring agencies to define, identify, map, and monitor pollutants in overburdened communities in ways that will direct agency actions. For example, New Jersey's environmental justice law requires the state Department of Environmental Protection to deny permits for polluting facilities in overburdened communities if the project will add to the cumulative impacts those communities already face. Washington State's Healthy Environment for All (HEAL) Act increases pollution monitoring in overburdened communities and requires agencies to focus 40% of environmental expenditures on reducing harm and creating environmental benefits in communities that have faced a disproportionate share of cumulative impacts.

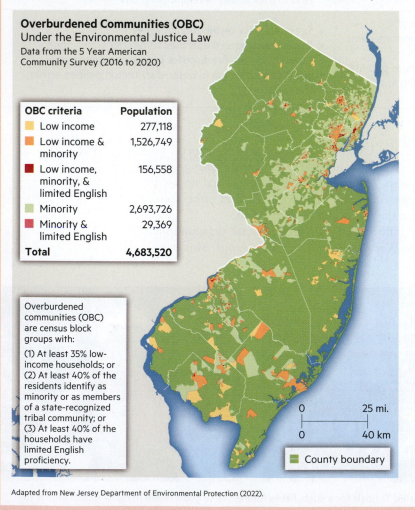

Overburdened Communities (OBC)
Under the Environmental Justice Law
Data from the 5 Year American Community Survey (2016 to 2020)

OBC criteria	Population
Low income	277,118
Low income & minority	1,526,749
Low income, minority, & limited English	156,558
Minority	2,693,726
Minority & limited English	29,369
Total	**4,683,520**

Overburdened communities (OBC) are census block groups with:

(1) At least 35% low-income households; or (2) At least 40% of the residents identify as minority or as members of a state-recognized tribal community; or (3) At least 40% of the households have limited English proficiency.

■ County boundary

0   25 mi.
0   40 km

Adapted from New Jersey Department of Environmental Protection (2022).

Fund (IMF), which gather funds from member countries to provide economic assistance. Historically, the World Bank has funded projects in developing countries, such as large dams and oil pipelines, and it has promoted logging in tropical rain forests. However, since the 1990s, these organizations have instituted environmental best-practice guidelines for the projects they fund. Most recently, the World Bank and the IMF worked together to provide guidance for countries on tax and regulatory policies that could help address climate change.

There are also influential regional IGOs. The most powerful of these is the European Union (EU). The EU includes 27 countries and is the only regional organization with the authority to make policy decisions binding on its members. It is unique in that it has some authority to negotiate international agreements on behalf of its member countries. However, like all international agreements, countries may decide to withdraw their commitment and membership, and in 2020 the United Kingdom voted to leave the EU.

International Agreements

There are many ways for two or more countries to make agreements that commit them to take various actions or abide by common rules. These can go by a variety of names including *treaties*, *protocols*, *conventions*, *accords*, or simply *agreements*, which vary in their level of specificity and commitment. **FIGURE 20.15** recaps some of the international agreements presented in Chapter 2. Since the 1970s international agreements have been reached on a range of environmental issues.

The Montreal Protocol on Substances That Deplete the Ozone Layer

An example of a very successful international agreement is the Montreal Protocol of 1987, which was responsible for dramatic reductions in the global emissions of chlorofluorocarbons (CFCs). The agreement was eventually

1970s
1972 United Nations Conference on the Human Environment ("Stockholm Conference")
1973 Convention on International Trade in Endangered Species of Wild Fauna and Flora (CITES)

1980s
1987 Montreal Protocol on Substances That Deplete the Ozone Layer
1989 Basel Convention on the Control of Transboundary Movements of Hazardous Wastes and Their Disposal

1990s
1992 United Nations Conference on Environment and Development ("Earth Summit")
1992 United Nations Convention on Biological Diversity
1997 Kyoto Protocol to the United Nations Framework Convention on Climate Change

2010s
2016 Paris Agreement Under the United Nations Framework Convention on Climate Change

FIGURE 20.15 Timeline of Major International Agreements

ratified by 196 countries and the European Union, making it the first universally ratified treaty in modern history. The process started in 1974, when the chemists Mario Molina and F. Sherwood Rowland published research hypothesizing that CFCs in the stratosphere would cause a series of reactions depleting the ozone layer over time (see Chapter 8). This was viewed as a serious threat because the ozone layer shields all life on the planet from the harmful effects of ultraviolet solar radiation.

In 1977, the governments of the United States, Canada, Norway, Sweden, and 28 other countries developed a *World Plan of Action on the Ozone Layer* under the auspices of the United Nations Environment Programme (UNEP). Their first step was a research plan that provided support for monitoring ozone and solar radiation and assessing the effects of ozone depletion on human health and ecosystems. By 1985, this research demonstrated that CFCs in the atmosphere had doubled since 1975 and that an ozone hole was forming over Antarctica. In 1985, UNEP facilitated a framework convention—an agreement on broad goals that leaves specific targets or actions to future negotiations—called the Vienna Convention for the Protection of the Ozone Layer, which established agreement on the findings regarding CFC emissions and a shared but nonbinding goal of taking "appropriate measures" to reduce CFC emissions. The convention established a plan to negotiate a binding protocol for action in the future. A period of intense bargaining and negotiations followed, leading to the 1987 Montreal Protocol, which established CFC reduction targets along with assessment mechanisms.

However, this agreement was driven by more than just scientific findings on the issue. Another critical factor was that DuPont, the single largest manufacturer of CFCs, had developed substitutes for them and thus stood to profit from a CFC phaseout. This factor led to the United States taking a strong role, steering the effort and persuading other countries to ratify the protocol. Finally, several key compromises were negotiated. European nations were persuaded to join once the CFC phaseout requirements were moderated, and many developing countries signed on once a special international fund to aid in the transition was established and a 10-year postponement was allowed. In the decades since this basic framework was established, the Montreal Protocol has been used by the international community to phase out other ozone depleting substances besides CFCs—including those that are greenhouse gases.

The UN Framework Convention on Climate Change

Especially in the United States, attempts to adopt an agreement about climate change have been contentious. Negotiations began in 1988 when UN member

states organized the IPCC to summarize scientific research on this new concern. In 1990, the IPCC issued its first report that global warming was a serious threat. In response, the UN General Assembly initiated a Framework Convention on Climate Change (UNFCCC) in 1991. It was finalized and took force in 1994, although it had no specific targets to reduce emissions because of US opposition. In 1997, negotiations continued in Kyoto, Japan, where a central tension existed between developed countries (such as the United States) and developing countries (such as China and India). The developing countries resisted limits on their GHG emissions, feeling that they were an unavoidable by-product of the industries that would improve living standards.

In 1997, the Kyoto Protocol to the UNFCCC was established, and it differentiated between developed (Annex I) and developing (Annex II) countries. Annex I countries agreed to various reduction targets, while Annex II countries had no commitment to reduce emissions. It also created an emissions trading program incentivizing emissions reduction.

When the United States, which was the largest emitter of GHGs, withdrew from its commitment to the Kyoto Protocol in 2001, it became very hard for the agreement to hit its overall targets. In 2007, the heads of government of several developed countries (such as the United States, Japan, and Germany) and developing countries (such as China, India, and Mexico) started negotiating a new agreement that would apply

FIGURE 20.16 Paris Agreement In September 2016, the United States ratified the Paris Agreement. Less than 9 months later, President Donald Trump began the process of withdrawing the United States from the agreement. In 2021, the United States recommitted to the Paris Agreement under President Joe Biden.

GHG reduction targets to both developed and developing nations.

This new initiative finally took the form of the Paris Agreement under the UNFCCC, whose text was finalized late in 2015, which nearly every country in the world, including the United States, has signed (**FIGURE 20.16**). The GHG emission-reduction goals of the Paris Agreement were intended to ensure that the average global temperature does not increase more than 2°C (3.6°F). Initially, the United States and China, the two largest emitters of GHGs, supported the agreement, and it was adopted in the fall of 2016. In the spring of 2017 during the beginning of the Trump administration, the United States began the process of withdrawing from the Paris Agreement. In 2021, the United States recommitted to the Paris Agreement under President Biden, and the United States and China agreed to significant new GHG reduction targets. The effects of the wavering US climate commitments across presidential administrations are not yet clear, but other cities have promised to redouble their commitment to the agreement, as have many states and cities within the United States.

US Participation in International Agreements

In the United States, an international treaty is difficult to sign, as it requires approval by a two-thirds vote in the Senate. For this reason, the United States largely negotiates and signs *executive congressional agreements*, which can be adopted with presidential support and simple majorities in the House of Representatives and Senate. The North American Free Trade Agreement (NAFTA), which included environmental side agreements between the United States, Canada, and Mexico, was adopted in this way in 1994 and renegotiated in 2018 as the US–Mexico–Canada Agreement. The president can also enter into *executive agreements* that do not require any

congressional approval and that future presidents can reverse or change. Some executive agreements, such as the 1991 Air Quality Agreement between the United States and Canada, have had long-lasting effects. This agreement signed by President George H. W. Bush and the government of Canada is still in force today, and it has influenced measures taken to combat acid rain, smog, and fine particle pollution. More recently, in 2014 President Obama used executive agreements with China to lay the groundwork for the Paris Agreement. As mentioned previously, when Donald Trump became president, he withdrew from the Paris Agreement and the climate-related commitments Obama had made as president. In 2021, President Biden reentered into these agreements.

TAKE-HOME MESSAGE Addressing human impacts on the environment extends beyond national borders, requiring the formation of intergovernmental organizations and international agreements. Intergovernmental organizations such as the United Nations and its agencies help facilitate international agreements and can provide an authoritative and independent source of information on environmental issues. International agreements emerge after lengthy negotiations that begin with broad agreements and work toward specific goals, commitments, and rules.

20.8 What Can I Do?

There is a saying that "all politics is local." In other words, political change always starts with groups of individuals interacting on a relatively small scale—often where they live—and sometimes these local associations build to effect policy changes on a much larger scale. The decision of the United States to withdraw from the Paris Agreement was tied to the result of a particular presidential election, which was determined by the individuals who took the initiative to vote. However, less than 60% of registered voters decided to cast a ballot in the 2016 presidential election. Fewer still participated actively in other ways, such as by campaigning, fundraising, attending political events, or voting in primaries. Four years later, the 2020 presidential election had the highest voter turnout on record with nearly 67% of registered voters casting a ballot. While not everyone who takes an active role in politics realizes their goals, people who decide not to participate will likely have no influence at all. So it's important to understand not only how various levels of government make policies but also the opportunities that each of us has to shape that government and the policies it will make in the future. Here are some things you can do to take an active role in the political process.

Can the World's Nations Work Together to Confront Climate Change?

The Paris Agreement is a commitment shared by nearly 200 nations to keep global temperature rise this century from exceeding a 2°C (3.6°F) increase above preindustrial levels. Two degrees doesn't sound like much, so why set the threshold here? From a scientific perspective, this is the amount of temperature increase that would lead to catastrophic changes, such as the flooding of coastal cities and low-lying countries and more pervasive and frequent extreme weather events. From an economics perspective, a 2-degree increase is the threshold beyond which we can expect severe effects on both the global economy and basic infrastructure. The insurance industry has also pegged a 2-degree increase as the point beyond which climate-related damages will exceed the finances of the industry—that is, its capacity to insure.

However, global temperature has already risen about 1.1°C (2°F) above preindustrial levels. In order to stay below a 2°C rise by 2100, the countries that are party to the agreement will have to figure out how to reduce human-caused greenhouse gas (GHG) emissions to less than 40 gigatons (Gt) by 2030. To this end, under the Paris Agreement each country has agreed to establish an Intended Nationally Determined Contribution (INDC): a pledge to reduce GHG emissions. More than 150 countries have submitted GHG reduction commitments. These countries represent more than 90% of global economic activity and more than 90% of global energy-related CO_2 emissions.

Despite the relative success of the Paris Agreement, its implementation still entails enormous challenges. First, each country is free to establish its own INDC pledge, which has led to a variety of methods and approaches that makes coordination of the overall goal difficult. Second, current INDC pledges do not yet add up to avoiding a 2-degree global temperature increase. The UN Environment Programme "emissions gap analysis" found that

UNEP Emissions Gap Analysis (2030 projection)

Neither current climate policies (red bar) nor fully implemented emissions-reduction pledges from nations that committed to the Paris Agreement (blue bar) will decrease CO_2 emissions enough to limit global warming to 2°C (3.6°F). More expansive emissions-reduction commitments by the international community will be required to reach the goals of the Paris Agreement. Gt = gigatons.

Graph adapted from Levin and Tirpak (2018) and United Nations Environment Programme (2018).

initial pledges would still likely lead to a rise in global temperature of 3.2°C (5.8°F) by 2100 because projected emissions in 2030 will be 15–18 Gt greater than the 40 Gt target. This led to renewed efforts by world leaders to set more ambitious pledges at the COP26 climate summit in 2021—including many net-zero pledges from large emitters like the United States (by 2050), China (by 2060), and India (by 2070). Third, many developing nations are dependent on substantial funding and technical assistance that is paid for out of the Green Climate Fund, which is financed by developed nations and private industry.

But perhaps the most significant challenge for the international community is that the Paris Agreement is not binding. Its success hinges on the extent to which each country will follow through with its commitments. As political dynamics shift within a country's government, its commitment to this agreement may also change. If the country is a major emitter of GHGs or a major donor to the Green Climate Fund, this could jeopardize global progress toward emission-reduction goals. This

weakness was exposed when President Trump began the process of withdrawing the United States from the agreement in the first months of his presidency. Although the INDC pledges are not binding under this agreement, in many cases government leaders are still subject to pressures from governments of other nations that can hold them to their commitment. These pressures might be in the form of economic and trade relationships or a country's relative status as a leader or trusted member of an international alliance.

What Would You Do?

If you were the leader of another large country (such as China) that has signed on to the Paris Agreement, how might the shifting of the US commitment back and forth across different administrations affect your actions regarding climate change? Why? What if you were the leader of a historical ally such as the United Kingdom or Canada?

Opportunities to Influence Policy Making

We all have the right to make our voices heard and influence the policies made for our communities. There are many ways to become involved in policy making at local, regional, and national levels.

Voting

Voting matters. If you don't vote, your voice is not represented in an election. Sometimes even a few votes can turn an election.

Running for Office

Become a decision maker yourself by running for elected office or being appointed to a board, commission, or task force.

Donating/Contributing

Make a monetary contribution. Campaigns cost money and candidates rely on small donors as well as large ones.

Lobbying

Write letters to, call, and lobby your representatives directly on issues you care about.

Campaigning

Campaign for a candidate you support. Think of it as multiplying your vote by influencing others.

Challenging Decisions in Court

Policy decisions are often challenged in court by ordinary citizens like you. For example, a group of children and teens is challenging the US government's lack of climate change policy in the lawsuit *Juliana v. United States*.

Speaking at a Public Hearing

Many environmental decisions require public input. Speak up at public hearings and comment on proposed decisions.

Protesting

Protests and other public demonstrations draw attention to issues.

Get Registered and Vote

A relatively easy and direct way to take part in politics is to vote. There are elections at all levels of government, from school boards to state offices and Congress, but before you can cast your ballot, you have to make sure you are registered. In the United States, registration rules and procedures vary by state, but there are many websites that can simplify this process for you. Check out Vote.org and Rockthevote.org to see if you are registered or learn how to register in your state.

Support a Campaign

There are also many ways to support those running for office who share your views or to promote a policy measure that is important to you. First, campaigns always need financial support and campaign volunteers. Although you may not be a big donor, even donations less than $20 can add up. And if you have some free time, campaigns always need volunteers to mobilize voters and support through door-to-door visits, telephone calls, texts, mailers, and petition drives. Grassroots work like this makes a big difference in elections and is often the first step in gaining valuable political experience, which can lead to roles in policy issues and political strategy.

Work with a Group You Support

Political groups work year-round to influence policy. The group 350.org described at the beginning of the chapter is just one example. Getting connected to a group like this can give you a chance to organize or join in protests, take part in lobbying trips to influence public officials, and craft public relations campaigns to raise awareness of issues important to you.

Tune In to Social Media

Follow local, state, and federal policy makers on social media to see where they stand on issues that are important to you. Use social media platforms to express your views and influence others to take action. For example, you can join or even form a group of your own on social media to promote causes that are important to you.

Chapter 20 **Review**

SUMMARY

- Governments exercise authority over people using policies designed to influence behavior.

- Governments vary in organizational structure, geographic breadth, degree of centralized control, and extent of public involvement in decision making.

- In democracies, the rules of the electoral system affect who wins and loses and the overall distribution of power in government.

- Groups with different interests compete to influence policy making.

- Policy making often entails power struggles over how problems are defined, which issues get on the government agenda, and how policy is formulated, approved, implemented, and evaluated.

- Policies influence behavior by using some combination of policy tools: prescriptive regulations, payments, penalties, property rights, or persuasion.

- Pollution-control policy in the United States is shared between the federal and state governments, with the federal government setting standards and the states working under federal guidelines to issue and enforce pollution permits.

- Authority over land use and natural resources in the United States is shared among different levels of government, with executive land-use agencies enforcing rules on federal land and the state and local authorities regulating their own lands and private property.

- International agreements are required to address environmental problems that extend beyond national borders.

- International organizations such as the United Nations and its agencies help facilitate international agreements and can provide an authoritative and independent source of information on environmental issues.

KEY TERMS

authoritarian system	interest group	political party	single-member-district plurality
checks and balances	legitimacy	politics	(SMDP) system
coercion	lobbying	power	sovereignty
democratic system	payment	prescriptive regulation	subsidy
federal governing system	penalty	proportional representation	unitary governing system
government	permit	(PR) system	United Nations (UN)
institution	policy		

REVIEW QUESTIONS

The letters following each Review Question refer to the Chapter Objectives.

1. Name two ways that governments maintain a degree of legitimacy. **(A)**

2. Describe the institutional organization of the US government into three branches, and provide an example of how a "check and balance" between the branches works. **(B)**

3. Explain the difference between an authoritarian versus a democratic system of government, and provide an example of each. **(A)**

4. Explain the difference between federal and unitary systems of government and how this affects environmental policy. **(B)**

5. Identify and describe two types of electoral systems and the effect each system has on the role of political parties in a political system. **(B)**

6. Name two strategies that interest groups use to influence government. **(B)**

7. Identify the stages in the policy process in which groups and interests compete for influence. **(C)**

8. Give two examples of ways that payments can be used as a policy strategy to influence behavior. **(C)**

9. Give two examples of ways that penalties can be used as a policy strategy to influence behavior. **(C)**

10. How has the relationship between state governments and the federal government changed over time with regard to pollution-control policies? **(D)**

11. Name three federal agencies that play a major role administering federal lands. **(D)**

12. Name two types of protected areas established by federal land designations and one example of a species-protection law. **(D)**

13. Compare and contrast the ways that the federal government and state and local governments influence the management of natural resources and land use. **(D)**

14. Name two types of intergovernmental organizations, and describe what they do. **(E)**

15. Identify two important international environmental agreements, and describe what they do. **(E)**

FOR FURTHER THOUGHT

The letters following each item refer to the Chapter Objectives.

16. Without legitimacy, a government cannot maintain authority over environmental policy or any other aspect of governing. Describe how the US government uses the three methods of maintaining legitimacy described in this chapter (traditions and identities, responsiveness to the governed, and coercion), and provide an example of each. **(A)**

17. In this chapter, we learned that the level of public involvement varies from authoritarian to democratic systems, and these differences affect the way and the speed in which environmental policies are implemented. Explain why you think the US system of government has too much or too little public involvement in the policy-making process. **(A, B)**

18. Even local governments play a significant role in policies that affect the environment. Do some research to identify and

describe one way in which your local government makes decisions affecting the environment. What avenues are there for you to get involved in these decision-making processes? **(D)**

19. In this chapter, we used the Arctic National Wildlife Refuge case to illustrate the way that conflicting interests engage in the policy-making process. How would you define the problem at the heart of the policy issue? Given what you have learned, what policies do you support and why? **(C)**

20. Elections determine the balance of power in our governing system, and they have an impact on environmental policies. Provide an example of a way in which past election results had an impact on environmental policies. Provide another example of a way in which future election results could have an impact on environmental policies. **(B, C, D)**

Make Your Case?

Use the News

This chapter has discussed how the political process can work and be affected in different ways and at different levels of government. This article describes a snapshot of the political process involved in a 2019 decision in Florida to ban hydraulic fracturing, or fracking. Use this article to answer the questions that follow. The first three questions test your understanding of the article. Question 4 is short answer, requiring you to apply what you have learned in this chapter and cite information in the article. Answers to Questions 1–3 are provided at the back of the book. Question 5 asks you to make your case and defend it.

"Lawmakers Say Partial Fracking Ban Is Better Than Nothing: Environmentalists Disagree," Samantha J. Gross, *Herald/Times* Tallahassee Bureau, March 26, 2019

Senate and House committees Tuesday passed bills aimed at banning fracking, an oil-drilling technique widely criticized for its negative environmental impact.

While the bills have faced significant pushback from environmentalists across the state, no significant changes have been made to address their concerns. Both bills have loopholes that would allow use of another type of drilling technique called matrix acidizing, a process that involves injecting an acid solution at a pressure low enough to not be considered hydraulic fracturing.

Lawmakers who support the proposals say that even if the bills aren't exactly what environmentalists want, it does more than doing nothing at all.

"Is it completely perfect? Is it everything we want? No, it's not," said Rep. Kristin Jacobs, D-Coconut Creek. "We have to do something. The idea that we're not going to act because it isn't perfect, I reject."

Sen. Lizbeth Benaquisto [*sic*] echoed the sentiment as she questioned an environmentalist who asked that the senators vote no on the bill.

"Wouldn't it be prudent to take the wins where you can get the wins?" the Fort Myers Republican asked.

Tuesday morning, the House Appropriations subcommittee on agriculture and natural resources voted 10-2 for HB 7029, filed by Key Largo Republican Holly Raschein.

Later Tuesday afternoon, the Senate Innovation, Industry and Technology Committee voted 6-4 along party lines for SB 7064, a bill filed on behalf of the Senate Agriculture Committee by Sen. Ben Albritton, R-Wauchula. Albritton also proposed an amendment to protect the Everglades, which was adopted in its last committee stop.

Environmentalists have repeatedly said matrix acidizing puts Florida's underground water supply at risk of pollution. Groups have argued that the state's porous geology of limestone rock makes fracking a serious threat, especially because most of the state's drinking water comes from underground.

David Cullen, a lobbyist for the Sierra Club, said because Florida is so porous, matrix acidizing is used disproportionately compared to other forms of fracking that fall under the bill.

"What they're saying is 'you're getting two-thirds. Don't be greedy,'" he said. "What you're really getting is 3 percent and the industry gets 97 percent. It's not a bargain for anybody in the state of Florida except the oil and gas industry."

Kim Ross of ReThink Energy Florida said it's not a fracking ban until matrix acidizing is included.

"Two of the three forms of fracking being banned leaves open the same risks," she said. "If only two-thirds of this process is banned, you still have 100 percent of the risk."

Well drillers and lobbyists from the petroleum industry argue that matrix acidizing is mainly used to remediate damage and maintain wells that get clogged or damaged from drilling operations, and therefore should not be included in the fracking ban. However in the Senate committee Tuesday, the executive director of the Florida Petroleum Council said those in the industry often use matrix acidizing to extract oil as well.

"We do both," David Mica said. "The term is a scary term, isn't it? Matrix acidizing? That chemical process is much like what we do when we're cleaning a water well."

A similar fracking ban bill filed by Sen. Bill Montford, D-Tallahassee, includes the matrix acidizing provision but will likely not be heard this session. Environmentalists call his the "good bill."

Albritton said critics who say his bill has a loophole are misinformed.

"I'm not sure, quite frankly, what they're talking about as a loophole," he said Tuesday. "[Matrix acidizing] is a term that has popped up. This bill bans fracking and gives definitions; hydraulic fracking. I'm certain they're incorrect."

According to a staff analysis of the Senate bill, oil and natural gas production in Florida peaked at 47 million barrels in 1978. In 2017, just 2 million barrels were produced and as of 2018, only 57 active wells were left. Rather than hydraulic fracturing, well operators in the state have generally preferred alternatives to fracking to recover oil and gas resources.

In 1990, the state enacted a drilling ban for state water. Earlier this year, Gov. Ron DeSantis unveiled sweeping measures to protect the state's aquifer and clean up the state's water supply. He then went further and announced fracking bans as a priority.

Vermont, New York, and Maryland all prohibit hydraulic fracturing.

If a fracking ban passes this legislative session, Florida's Department of Environmental Protection will have to revise existing rules to implement the ban.

1. According to the article, which of the following statements about the legislation described is correct?

   a. Review committees in the Florida House and Senate passed two versions of similar bills on fracking; the House passed HB 7029 and the Senate passed SB 7064.

   b. Review committees in the Florida House and Senate passed two different bills. The House passed HB 7029, which *allows* fracking, and the Senate passed SB 7064, which *bans* fracking.

   c. Review committees in the Florida House and Senate passed two different bills. The House passed HB 7029, which *bans* fracking, and the Senate passed SB 7064, which *allows* fracking.

   d. Review committees in the Florida House and Senate passed similar bills that met with the approval of environmental groups such as the Sierra Club.

2. The article says that the petroleum industry and the Sierra Club used a type of policy specialist to help influence legislators. What is this type of specialist called?

   a. executive director
   b. governor
   c. legal aide
   d. lobbyist
   e. representative

3. Which statement best describes the status of these bills at the time the article was written?

   a. While the bills have been approved by committees and the full House and Senate, the governor has not signed them.

   b. While the governor has approved the bills along with the full House and Senate, they cannot be approved until there is a state vote by Florida citizens.

   c. While the two committees have approved the bills, they are being stopped by environmental regulations.

   d. While the bills have been approved by committees, the full House and Senate have not voted on them.

4. Read the last paragraph of the article again. What does the article imply will be the next political battle regarding any passed legislation? Based on the descriptions of the policy-making process included in this chapter, does this seem like a typical step in the policy process? Explain your answer.

5. **Make Your Case** The article notes that a supporter of the legislation says to its opponents, "Wouldn't it be prudent to take wins where you can get wins?" But an opponent of the bills is quoted as saying, "Two of the three forms of fracking being banned leaves open the same risks. . . . If only two-thirds of this process is banned, you still have 100 percent of the risk." Given these two statements, and any other information in the article, do you believe that the political process is working in this situation? In other words, is this policy being crafted according to the usual policy-making steps described in the chapter? Make your case, using at least two statements in the article and information from the chapter to support your argument.

Use the Data

Public opinion plays an important role in environmental politics in the United States, as does party identification. The graphs shown here are from a periodic Pew Research Center poll that is based on an online survey of a randomly recruited panel of just over 10,000 adults aged 18 and older who are drawn from all 50 states and Washington, DC. The respondents were asked if policies aimed at reducing the effects of global climate change generally help or hurt the US economy, and if they do more good than harm to the environment. The left side of the graphs summarize the responses as a whole by year, while the middle and right parts of the graphs summarize the responses by the respondents' party identifications (Rep = Republican, Dem = Democrat) and year.

1. In what years was there a higher percentage of US adults who said that climate change policies help the economy than those who said these policies hurt the economy? Select all that apply.
   a. 2018 c. 2022
   b. 2019 d. none

2. In what year were there more Republicans who thought climate change policies do more good than harm than Republicans who thought these policies did more harm than good?
   a. 2018 c. 2022
   b. 2019 d. none

Study the graphs, and use them to answer the questions that follow. The first three questions test your understanding of the graphs. Questions 4 and 5 are short answer, requiring you to apply what you have learned in this chapter and cite data in the graphs. Answers to Questions 1–3 are provided at the back of the book. Question 6 asks you to make your case and defend it.

3. Which year has the biggest difference between how many Democrats and Republicans think that climate change policies help the US economy?
   a. 2018 b. 2019
   c. 2022

4. The graphs show that the percentage of Republicans who think climate change policies do more good than harm *decreases* between 2019 and 2022, while the percentage of Democrats who think these policies do more good than harm *increases* by a larger amount during the same period. During the same period, the percentage of Republicans who think climate change policies *hurt* the economy increases, while the percentage of Democrats who think these policies *help* the economy increases. Use information from the chapter to explain the partisan trends behind these numbers.

5. Explain how the difference in how many Democrats and Republicans think that climate change policies help the economy and do more good than harm for the environment could affect the environmental policies that members of these two parties support.

6. **Make Your Case** Imagine you have been hired as a political consultant to advise a congressional candidate in your home district and of a party of your choosing. What environmental policy positions would you encourage the candidate to take on the basis of these data?

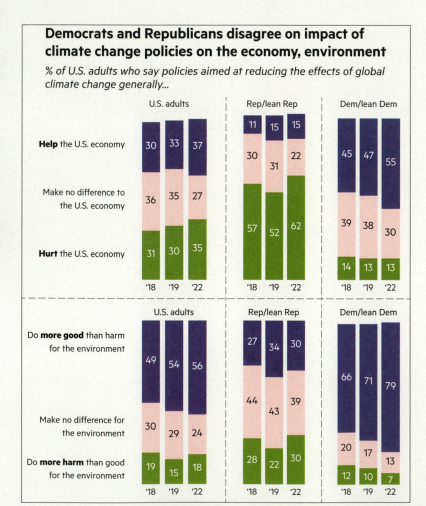

Democrats and Republicans disagree on impact of climate change policies on the economy, environment

% of U.S. adults who say policies aimed at reducing the effects of global climate change generally...

LEARN MORE

- Nora Gallagher and Lisa Meyers (eds.). 2016. *Tools for Grassroots Activists*. Patagonia

- Council on Environmental Quality. 2007. *A Citizen's Guide to the NEPA: Having Your Voice Heard*. www.energy.gov/nepa/downloads /citizens-guide-nepa-having-your-voice-heard-ceq-2007

- Robinson Meyer. 2017. "How the US Protects the Environment, from Nixon to Trump." *The Atlantic*, March 29, 2017. www .theatlantic.com/science/archive/2017/03/how-the-epa-and-us -environmental-law-works-a-civics-guide-pruitt-trump/521001/

- Video: USGS, Eastern US Biological Carbon Storage Potential Assessment: digital.wwnorton.com/environsci

- Video: House of Representatives Arctic National Wildlife Refuge (ANWR) Amendment: digital.wwnorton.com/environsci

- USA.gov. 2019. "Branches of the US Government." www.usa.gov /branches-of-government

- United Nations Environment Programme: www.unenvironment .org/

- Osita Nwaneyu. 2017. "If You Want to Run for Office (and You Should), Start with These Resources." *Slate*, January 28, 2017. www .slate.com/articles/news_and_politics/politics/2017/01/a_guide _to_resources_for_running_for_office.html

Appendix A

Footprint Calculators

Knowing your impact on the planet is an important step in your decision making. Several chapters in this text suggest different footprint calculators that you can find online to get a better sense of your impact. The most common footprint calculators help you calculate your carbon and ecological impacts, but you can also find calculators for nutrients such as nitrogen and resources such as water, forests, or the impact of the car you drive.

Calculators ask you a series of questions related to your daily, and often yearly, activities and take about 5 to 15 minutes to complete. Although calculators often yield specific numbers based on your inputs, it's often best to think of these as comparison tools: they give you a sense of what activities lead to a higher number and what activities will give you a lower number. If you have the time, you might try a tool several times, the first time with your most honest answers possible, but then again, changing certain inputs you could change to see how they affect your performance.

The following list provides a variety of calculators, with some of the more popular ones listed first. Because weblinks change, if you can't find a site using the link, try searching for the page name.

Global Footprint Network—Footprint Calculator

www.footprintcalculator.org

Available in eight languages and calculates your personal Earth overshoot day.

Carbon Footprint Ltd

www.carbonfootprint.com

A UK-based business with a free carbon footprint calculator. This calculator has options for both individuals and small businesses.

The Nature Conservancy—Calculate Your Carbon Footprint

www.nature.org/en-us/get-involved/how-to-help/consider -your-impact/carbon-calculator

A comprehensive calculator from a nonprofit group focused on initiatives driven by scientific research.

Conservation International—Calculate Your Carbon Footprint and Reduce Your Impact

www.conservation.org/carbon-footprint-calculator#

The free calculator at this site is provided with an emphasis on purchasing carbon offsets.

WWF—How Big Is Your Environmental Footprint?

https://footprint.wwf.org.uk/#

A multistep calculator from this global conservation agency. This organization also has a downloadable My Footprint App

Water Footprint Calculator

www.watercalculator.org

This tool from a nonprofit group helps you estimate your total water use, including embedded water.

Foodprint

https://foodprint.org/what-is-foodprint/

A tool from a nonprofit agency that helps you better assess the environmental impact of the food you eat.

Nitrogen Footprint

www.n-print.org/YourNFootprint

A calculator that helps assesses nitrogen impact at a variety of scales (including the individual).

US Department of Energy—Save Money

www.fueleconomy.gov/feg/savemoney.jsp

A US government calculator that helps you assess and compare the fuel economy and associated costs of different vehicles.

Appendix B

Global Footprints by Country

The following table provides data from Global Footprint Network's assessment of country ecological footprints in 2018. It is ordered from highest to lowest Ecological Footprint per person. This table also reports the overall biocapacity per person of the country, and how much the footprint is above or below this. The information is available from http://data.footprintnetwork.org in their open data platform.

Country	Life Exectancy	Per Capita GDP	2018 Ecological Footprint [gha per capita]	2018 biocapacity [gha per capita]	2018 biocapacity (Deficit) or Reserve [gha per capita]
Qatar	80.10	$63,990	14.27	0.92	−13.35
Luxembourg	82.30	$111,381	12.95	1.21	−11.74
Bahrain	77.16	$21,478	8.17	0.47	−7.71
United States of America	78.64	$54,724	8.12	3.39	−4.74
United Arab Emirates	77.81	$41,325	8.10	0.52	−7.58
Canada	82.05	$51,518	8.07	14.74	6.67
Estonia	78.24	$19,919	8.01	9.33	1.32
Kuwait	75.40	$33,113	7.90	0.50	−7.40
Belize	74.50	$4,217	7.87	3.51	−4.36
Trinidad and Tobago	73.38	$15,161	7.44	1.50	−5.95
Mongolia	69.69	$4,211	7.28	14.14	6.87
Australia	82.75	$57,032	7.09	11.47	4.39
Belgium	81.60	$46,768	6.87	0.76	−6.10
Denmark	80.95	$64,672	6.65	3.62	−3.04
Finland	81.73	$48,890	6.44	12.14	5.69
Latvia	74.78	$16,252	6.41	8.16	1.75
Korea, Republic of	82.63	$28,336	6.32	0.64	−5.67
Oman	77.63	$15,797	6.29	1.43	−4.86
Sweden	82.56	$59,068	6.28	8.96	2.68
Austria	81.69	$49,711	6.06	2.68	−3.39
Lithuania	75.68	$17,744	6.00	4.75	−1.25

Country	Life Exectancy	Per Capita GDP	2018 Ecological Footprint [gha per capita]	2018 biocapacity [gha per capita]	2018 biocapacity (Deficit) or Reserve [gha per capita]
Singapore	83.30	$57,838	5.93	0.06	−5.87
Czech Republic	79.03	$23,721	5.72	2.33	−3.39
Netherlands	81.81	$55,369	5.69	0.78	−4.91
Norway	82.76	$91,608	5.67	6.91	1.24
Malta	82.45	$31,604	5.45	0.49	−4.97
Brunei Darussalam	75.72	$31,437	5.42	2.71	−2.70
Slovenia	81.38	$26,710	5.37	2.20	−3.17
Russian Federation	72.66	$11,940	5.31	6.72	1.40
New Zealand	81.86	$39,215	5.30	8.84	3.54
Ireland	82.26	$77,407	5.25	3.07	−2.18
Bhutan	71.46	$3,129	5.16	4.98	−0.18
Saudi Arabia	75.00	$20,829	4.96	0.41	−4.55
Kazakhstan	73.15	$11,136	4.95	3.57	−1.38
Turkmenistan	68.07	$7,648	4.90	2.29	−2.61
Poland	77.60	$16,666	4.75	1.88	−2.87
Slovakia	77.27	$20,539	4.73	2.79	−1.94
Germany	80.89	$47,158	4.67	1.49	−3.18
Israel	82.80	$36,867	4.66	0.18	−4.48
Antigua and Barbuda	76.89	$15,135	4.66	0.86	−3.80
Japan	84.21	$48,506	4.61	0.59	−4.02
Réunion			4.59	0.17	−4.42
Portugal	81.32	$24,180	4.55	1.29	−3.26
France	82.72	$45,057	4.42	2.38	−2.04
Belarus	74.18	$6,608	4.40	2.86	−1.55
Spain	83.43	$32,981	4.39	1.54	−2.86
Switzerland	83.75	$79,173	4.35	0.99	−3.35
Italy	83.35	$35,277	4.31	0.82	−3.49

(continued)

Country	Life Exectancy	Per Capita GDP	2018 Ecological Footprint [gha per capita]	2018 biocapacity [gha per capita]	2018 biocapacity (Deficit) or Reserve [gha per capita]
Chile	80.04	$15,110	4.29	3.34	−0.96
Bahamas	73.75	$28,846	4.28	9.18	4.89
Malaysia	76.00	$12,116	4.26	2.16	−2.10
Montenegro	76.77	$8,180	4.23	2.51	−1.72
United Kingdom	81.26	$42,894	4.18	1.03	−3.15
Greece	81.79	$24,044	4.10	1.17	−2.93
French Polynesia	77.46	–	4.00	1.31	−2.69
Barbados	79.08	$16,137	3.97	0.19	−3.78
Bosnia and Herzegovina	77.26	$6,097	3.89	2.00	−1.89
Croatia	78.07	$15,708	3.88	2.83	−1.05
Hungary	76.07	$16,894	3.87	2.57	−1.29
Cyprus	80.83	$23,395	3.85	0.21	−3.64
South Africa		$7,440	3.80	1.00	−2.80
China	76.70	$7,469	3.80	0.92	−2.87
Bulgaria	74.96	$8,642	3.62	3.28	−0.34
Guadeloupe			3.62	0.51	−3.10
Romania	75.36	$11,535	3.58	3.18	−0.41
Martinique			3.49	0.41	−3.09
Guyana	69.77	$5,825	3.41	72.70	69.30
Turkey	77.44	$15,181	3.35	1.30	−2.04
Argentina	76.52	$10,076	3.31	6.18	2.87
Libyan Arab Jamahiriya	72.72	$8,039	3.28	0.60	−2.68
Iran, Islamic Republic of	76.48	$6,443	3.26	0.73	−2.53
Mauritius	74.42	$10,561	3.22	0.67	−2.55
Bolivia	71.24	$2,560	3.11	14.25	11.14
Republic of North Macedonia	75.69	$5,452	3.10	1.49	−1.60
Serbia	75.89	$5,472	3.07	1.68	−1.39
Paraguay	74.13	$5,380	3.07	10.06	6.99

Country	Life Exectancy	Per Capita GDP	2018 Ecological Footprint [gha per capita]	2018 biocapacity [gha per capita]	2018 biocapacity (Deficit) or Reserve [gha per capita]
Suriname	71.57	$8,100	3.06	80.37	77.31
Lebanon	78.88	$6,205	2.96	0.25	−2.71
Panama	78.33	$11,755	2.92	2.51	−0.41
Tonga	70.80	$4,378	2.92	1.08	−1.84
Fiji	67.34	$4,795	2.67	2.34	−0.33
Samoa	73.19	$3,754	2.65	1.74	−0.91
Brazil	75.67	$11,076	2.59	8.61	6.02
Grenada	72.38	$9,092	2.54	2.10	−0.44
Mauritania	64.70	$1,704	2.52	3.92	1.40
Namibia	63.37	$5,942	2.51	6.32	3.81
Eswatini	59.40	$4,762	2.46	0.96	−1.51
Dominica	0.00	$6,697	2.46	1.14	−1.32
Costa Rica	80.10	$9,937	2.45	1.50	−0.94
Djibouti	66.58	–	2.44	0.71	−1.72
Ukraine	71.58	$2,961	2.42	2.74	0.32
Venezuela, Bolivarian Republic of	72.13	–	2.39	2.76	0.37
Mexico	74.99	$10,381	2.38	1.15	−1.23
Botswana	69.28	$8,033	2.37	3.30	0.93
Peru	76.52	$6,440	2.36	3.60	1.24
Thailand	76.93	$6,366	2.35	1.27	−1.08
Algeria	76.69	$4,760	2.34	0.57	−1.77
Georgia	73.60	$4,407	2.32	1.26	−1.06
Viet Nam	75.32	$1,968	2.27	0.94	−1.33
Saint Lucia	76.06	$9,237	2.26	0.32	−1.94
Azerbaijan	72.86	$5,795	2.06	0.79	−1.27
Uzbekistan	71.57	$2,409	2.04	0.66	−1.38
El Salvador	73.10	$3,507	2.03	0.59	−1.44
Tunisia	76.51	$4,408	2.02	0.69	−1.34

(continued)

Country	Life Exectancy	Per Capita GDP	2018 Ecological Footprint [gha per capita]	2018 biocapacity [gha per capita]	2018 biocapacity (Deficit) or Reserve [gha per capita]
Equatorial Guinea	58.40	$10,135	2.02	3.36	1.33
Ghana	63.78	$1,808	2.00	1.30	−0.70
Gabon	66.19	$9,051	1.95	18.92	16.97
Armenia	74.95	$4,407	1.90	0.62	−1.28
Albania	78.46	$5,046	1.89	0.99	−0.89
French Guiana			1.88	92.14	90.26
Republic of Moldova	71.81	$2,356	1.87	1.25	−0.62
Colombia	77.11	$7,692	1.86	3.44	1.58
Guinea	61.19	$897	1.84	2.06	0.21
Lao People's Democratic Republic	67.61	$1,786	1.84	1.70	−0.14
Egypt	71.83	$2,906	1.84	0.37	−1.47
Cabo Verde	72.78	$3,740	1.83	0.47	−1.36
Guatemala	74.06	$3,164	1.82	0.96	−0.86
Solomon Islands	72.84	$1,774	1.77	3.63	1.86
Dominican Republic	73.89	$7,698	1.77	0.67	−1.10
Iraq	70.45	$5,464	1.76	0.16	−1.61
Cuba	78.73	$6,817	1.76	0.77	−0.99
Jordan	74.41	$3,307	1.76	0.15	−1.61
Morocco	76.45	$3,414	1.75	0.79	−0.97
Papua New Guinea	64.26	$2,398	1.72	3.43	1.70
Indonesia	71.51	$4,296	1.72	1.24	−0.48
Ecuador	76.80	$5,181	1.70	1.89	0.18
Myanmar	66.87	$1,573	1.66	1.86	0.20
Chad	53.98	$812	1.65	1.70	0.05
Jamaica	74.37	$4,855	1.63	0.42	−1.22
Mali	58.89	$779	1.63	1.88	0.25
South Sudan	57.60	–	1.61	1.73	0.12
Benin	61.47	$1,211	1.61	0.92	−0.69

Country	Life Exectancy	Per Capita GDP	2018 Ecological Footprint [gha per capita]	2018 biocapacity [gha per capita]	2018 biocapacity (Deficit) or Reserve [gha per capita]
Honduras	75.09	$2,219	1.57	1.74	0.16
Lesotho	53.71	$1,375	1.54	0.80	−0.75
Niger	62.02	$552	1.54	1.30	−0.24
Kyrgyzstan	71.40	$1,094	1.54	1.20	−0.34
Sri Lanka	76.81	$4,028	1.51	0.45	−1.06
Cambodia	69.57	$1,203	1.50	1.20	−0.31
Sudan	65.10	$1,590	1.48	1.36	−0.11
Philippines	71.10	$3,188	1.47	0.51	−0.95
Guinea-Bissau	58.00	$623	1.47	2.67	1.20
Nicaragua	74.28	$1,857	1.43	2.19	0.77
Senegal	67.67	$1,547	1.41	1.11	−0.30
Côte d'Ivoire	57.42	$1,668	1.40	1.79	0.39
Syrian Arab Republic	71.78	–	1.34	0.49	−0.85
Sao Tome and Principe	70.17	$1,297	1.28	0.77	−0.51
Zambia		$1,678	1.28	1.76	0.48
Burkina Faso	61.17	$800	1.26	1.00	−0.26
Uruguay	77.77	$14,618	1.26	9.32	8.06
Liberia	63.73	$541	1.24	2.99	1.75
Zimbabwe		$1,306	1.22	0.62	−0.59
Cameroon	58.92	$1,502	1.21	1.59	0.38
India	69.42	$2,085	1.21	0.45	−0.76
Central African Republic	52.81	$379	1.20	7.51	6.30
Comoros	64.12	$1,403	1.19	0.35	−0.84
Nepal	70.48	$818	1.19	0.62	−0.57
Tanzania, United Republic of	65.02	$931	1.17	0.98	−0.19
Korea, Democratic People's Republic of	72.10	-	1.17	0.56	−0.61
State of Palestine	73.90	$2,805	1.14	0.10	−1.05

(continued)

Country	Life Exectancy	Per Capita GDP	2018 Ecological Footprint [gha per capita]	2018 biocapacity [gha per capita]	2018 biocapacity (Deficit) or Reserve [gha per capita]
Sierra Leone	54.31	$473	1.13	1.00	−0.14
Tajikistan	70.88	$1,073	1.13	0.48	−0.65
Nigeria	54.33	$2,384	1.09	0.69	−0.40
Uganda	62.97	$934	1.06	0.47	−0.59
Togo	60.76	$677	1.04	0.59	−0.45
Gambia	61.74	$791	1.04	0.52	−0.52
Congo	64.29	$2,304	1.03	8.78	7.75
Ethiopia	66.24	$571	1.02	0.56	−0.46
Somalia	57.07	–	1.02	0.82	−0.20
Kenya	66.34	$1,201	1.00	0.47	−0.53
Madagascar	66.68	$490	0.95	2.21	1.26
Bangladesh	72.32	$1,202	0.90	0.41	−0.49
Malawi	63.80	$515	0.88	0.68	−0.21
Angola	60.78	$3,234	0.86	1.84	0.97
Burundi	61.25	$211	0.83	0.55	−0.29
Eritrea	65.94	–	0.81	2.25	1.44
Mozambique	60.16	$593	0.80	1.57	0.77
Congo, Democratic Republic of	60.37	$419	0.78	2.16	1.38
Pakistan	67.11	$1,197	0.77	0.33	−0.44
Rwanda	68.70	$845	0.76	0.39	−0.37
Afghanistan	64.49	$565	0.68	0.34	−0.34
Haiti	63.66	$1,282	0.64	0.25	−0.39
Timor-Leste	69.26	$840	0.59	1.67	1.07
Yemen	66.10	$633	0.51	0.37	−0.13
Vanuatu	70.32	$2,865	0.00	0.00	0.00

Scale: Ecological footprint in global hectares per person.

Note: While detailed feedback is given for most countries, this is not the case for all of them. According to the website, this is "probably because the results weren't up to Global Footprint Network quality standards. National Footprint Accounts use internationally available data from multiple datasets for all countries, mostly provided by the United Nations and the UN Food and Agriculture Organization. These datasets are official, widely obtainable, and are available in a consistent format across countries. However, data is limited, unavailable, or contains apparent errors for some countries."

Adapted from York University Ecological Footprint Initiative & Global Footprint Network. National Footprint and Biocapacity Accounts, 2022 edition. Produced for the Footprint Data Foundation and distributed by Global Footprint Network. Available online at: https://data.footprintnetwork.org.

Selected Answers

Chapter 1

USE THE NEWS

1. b
2. a
3. d

USE THE DATA

1. d
2. d
3. a

Chapter 2

USE THE NEWS

1. c
2. b and e
3. c and e

USE THE DATA

1. c
2. (i) a; (ii) b
3. d

Chapter 3

USE THE NEWS

1. b
2. c

USE THE DATA

1. d
2. a, $1.49/15,000 hours = $0.000099/hour
3. (a) chemical; (b) electrical, radiant; (c) radiant, thermal; (d) electrical

Chapter 4

USE THE NEWS

1. b
2. d
3. c

USE THE DATA

1. Terbang Utara: 20
   Kisar: 38
   Damar: 50
   Atauro: 70
   Roti: 100
   Timor: 170
2. less than 800 km; more than 3,000 km

3. Atauro is closer. Although it is about the same size as Sermata, it has more species, so it is more likely closer to the large island that is the source of colonizing species.

Chapter 5

USE THE NEWS

1. a
2. c
3. b

USE THE DATA

1. b
2. d
3. c, 63 − (7 + 4) = 52 acres more

Chapter 6

USE THE NEWS

1. (i) c; (ii) a, b, d
2. a
3. b

USE THE DATA

1. Country C
2. Country A
3. Women

Chapter 7

USE THE NEWS

1. d
2. c
3. a, d
4. d

USE THE DATA

1. a
2. Wisconsin: 3.3
   Lake Michigan: 1.5
   Indiana: 1.7
   Upper Mississippi: 34
   Upper Ohio: 95
3. b

Chapter 8

USE THE NEWS

1. c
2. e
3. c

USE THE DATA

1. a
2. d
3. c

Chapter 9

USE THE NEWS

1. a
2. d
3. b

USE THE DATA

1. b
2. b
3. a

Chapter 10

WHAT CAN I DO?

Inflow: faucet

Outflow: drain

Some ways that soil carbon can be increased: increased sequestration of carbon in soil, decreased microbial respiration/decomposition of soil carbon, increased production of terrestrial biomass that leads to increased soil carbon storage

Some ways that atmospheric carbon (CO_2) can be decreased: increased absorption of CO_2 by surface ocean, decreased release of CO_2 from surface ocean (these first two are both gas exchange processes between the atmosphere and surface ocean), decreased combustion of fossil fuels, decreased microbial respiration/decomposition, increased photosynthesis, decreased respiration/combustion of terrestrial biomass

If rate of photosynthesis doubled, some ways that carbon in terrestrial biomass could stay the same: respiration/combustion of terrestrial biomass increases by same amount, soil carbon storage increases by the same amount, or CO_2 in the atmosphere decreases enough to decrease the rate of photosynthesis to original levels

USE THE NEWS

1. c
2. d
3. c
4. d

USE THE DATA

1. d
2. c
3. Restoring and constructing wetlands; $1.50 a pound

Chapter 11

USE THE NEWS

1. c
2. b, c, e

USE THE DATA

1. a and b
2. c and d
3. Florida, Texas, Louisiana, Alabama, Mississippi, and Georgia will all see a decrease in establishment probability.

Chapter 12

USE THE NEWS

1. a, e
2. b
3. b, d

USE THE DATA

1. c
2. North America and Oceania (42%)
3. North America and Oceania (61%), and Europe (52%)

Chapter 13

USE THE NEWS

1. a
2. d

USE THE DATA

1. d
2. a
3. c

Chapter 14

USE THE NEWS

1. c
2. c
3. b, c, e

USE THE DATA

1. Ethanol, hybrid, diesel, EV
2. 2006, 41; 2018, 209
3. EV, 57 models; ethanol, 53 models

Chapter 15

USE THE NEWS

1. c
2. a, d, e
3. a

USE THE DATA

1. d
2. c
3. Orange line: Percentage of MSW that is recycled and composted
   Blue line: Total weight of MSW that is recycled and composted

Chapter 16

USE THE NEWS

1. a, d
2. a
3. b

USE THE DATA

1. b
2. b
3. a

Chapter 17

USE THE NEWS

1. b
2. a
3. a, c, f, g

USE THE DATA

1. a
2. b
3. a

Chapter 18

USE THE NEWS

1. a
2. a
3. b, e

USE THE DATA

1. b
2. c, d, e
3. a, b, g

Chapter 19

USE THE NEWS

1. b
2. c
3. b

USE THE DATA

1. b, c
2. d
3. Most widely used 5 years ago and today, energy-efficient equipment; least used 5 years ago, big data and advanced analytics; least used today, technology to move goods and/or people

Chapter 20

USE THE NEWS

1. a
2. d
3. d

USE THE DATA

1. b, c
2. b
3. c

Selected Sources

CHAPTER 1

American Lung Association. 2022. "State of the Air 2022." https://www.lung.org/research/sota.

APM Research Lab Staff. 2021. "The Color of Coronavirus: Covid-19 Deaths By Race and Ethnicity in the U.S. (march 5, 2021)." https://www.apmresearchlab.org/covid/deaths-by-race-03-05-21

Beardsley, Eleanor. 2021. "Trees from the Forest of Villefermoy Will Help Rebuild Notre Dame." NPR, *Morning Edition*, April 15, 2021. https://www.npr.org/2021/04/15/987552098/trees-from-the-forest-of-villefermoy-will-help-rebuild-notre-dame?ft=nprml&f=987552098.

Berman, Robby. 2021. "Online Game Teaches Players How Misinformation Works." Medical News Today, May 15, 2021. https://www.medicalnewstoday.com/articles/the-go-viral-game-makes-a-person-better-at-spotting-misinformation.

Berndt, Brooks. 2021. "30th Anniversary: The First National People of Color Environmental Leadership Summit." United Church of Christ, October 26, 2021. https://www.ucc.org/30th-anniversary-the-first-national-people-of-color-environmental-leadership-summit/.

Bick, Rachel, Erika Halsey, and Christine C. Ekenga. 2018. "The Global Environmental Injustice of Fast Fashion—Environmental Health." *Environmental Health* 17 (December 27, 2018). https://ehjournal.biomedcentral.com/articles/10.1186/s12940-018-0433-7.

Bison Range Restoration. 2022. Confederated Salish and Kootenai Tribes. Accessed May 25, 2022. https://bisonrange.org/.

Brand, Stewart. 2010. "Oak Beams, New College Oxford." Atlas Obscura, March 23, 2010. https://www.atlasobscura.com/places/oak-beams-new-college-oxford.

Bruce, Dylan. 2021. "Analysis: State Laws Are Codifying Environmental Justice." Bloomberg Law, March 9, 2021. https://news.bloomberglaw.com/bloomberg-law-analysis/analysis-state-laws-are-codifying-environmental-justice.

Centers for Disease Control and Prevention. 2013. "Minority Health: CDC Health Disparities & Inequalities Report (CHDIR)." November 26, 2013. https://www.cdc.gov/minorityhealth/CHDIReport.html.

Chipperfield, Martyn P., and Vitali E. Fioletov (lead authors). 2007. "Global Ozone: Past and Present." In *Scientific Assessment of Ozone Depletion: 2006*, Chapter 3. World Meteorological Organization Global Ozone Research and Monitoring Project, report no. 50. www.esrl.noaa.gov/csd/assessments/2006/report.html

Derissen, Sandra, Martin F. Quaas, and Stefan Baumgärtner. 2011. "The Relationship between Resilience and Sustainability of Ecological-Economic Systems." *Ecological Economics* 70, no. 6 (April 15, 2011): 1121–28. https://www.sciencedirect.com/science/article/abs/pii/S0921800911000103.

Diamond, Jared. 2008. "What's Your Consumption Factor?" *New York Times*, January 2, 2008. https://www.nytimes.com/2008/01/02/opinion/02diamond.html.

Dunne, Daisy. 2018. "Severe Coral Reef Bleaching Now 'Five Times More Frequent' Than 40 Years Ago." Carbon Brief, January 4, 2018. https://www.carbonbrief.org/severe-coral-reef-bleaching-now-five-times-more-frequent-than-40-years-ago.

Ebbs, Stephanie, and Devin Dwyer. 2020. "America's National Parks Face Existential Crisis over Race." ABC News, July 1, 2020. https://abcnews.go.com/Politics/americas-national-parks-face-existential-crisis-race/story?id=71528972.

Fleischman, Lesley, Marcus Franklin, and Sarah Uhl. 2021. "Fumes across the Fence-Line: The Health Impacts of Air Pollution from Oil & Gas Facilities on African American Communities." Edited by Katherine Taylor. NAACP, May 20, 2021. https://naacp.org/resources/fumes-across-fence-line-health-impacts-air-pollution-oil-gas-facilities-african-american.

Fleming, Nic. 2020. "Coronavirus Misinformation, and How Scientists Can Help to Fight It." Nature News, June 17, 2020. https://www.nature.com/articles/d41586-020-01834-3.

Friedman, M. S., K. E. Powell, L. Hutwagner, L. M. Graham, and W. G. Teague. 2001. "Impact of Changes in Transportation and Commuting Behaviors during the 1996 Summer Olympic Games in Atlanta on Air Quality and Childhood Asthma." *Journal of the American Medical Association* 285, no. 7 (February 21, 2001): 897–905.

Global Footprint Network National Footprint and Biocapacity Accounts, 2022 Edition Downloaded 7/29/22 from https://data.footprintnetwork.org.

Halpern, Benjamin S., Shaun Walbridge, Kimberly A. Selkoe, Carrie V. Kappel, Fiorenza Micheli, Caterina D'agrosa, John F. Bruno, et al. 2008. "A Global Map of Human Impact on Marine Ecosystems." *Science* 319, no. 5865 (February 15, 2008): 948–52.

Hoekstra, J. M., J. L. Molnar, M. Jennings, C. Revenga, M. D. Spalding, T. M. Boucher, J. C. Robertson, T. J. Heibel, and K. Ellison. 2010. *The Atlas of Global Conservation: Changes, Challenges, and Opportunities to Make a Difference.* Edited by J. L. Molnar. Berkeley: University of California Press.

Kuntsman, Ben, Eric Schaffer, and Alexandra Shaykevich. 2021. "Environmental Justice and Refinery Pollution: Benzene Monitoring around Oil Refineries Showed More Communities at Risk in 2020." Environmental Integrity Project, April 28, 2021. https://environmentalintegrity.org/wp-content/uploads/2021/04/Benzene-report-4.28.21.pdf.

Mau, Dhani. 2019. "Toms Shifts Away from One for One, the Giving Model It Originated." Fashionista, November 20, 2019. https://fashionista.com/2019/11/toms-evolves-one-for-one-model.

McDonald, Kendrick. 2021. "Special Report: 2020 Engagement Analysis Unreliable News Sites More Than Doubled Their Share of Social Media Engagement in 2020." NewsGuard, September 16, 2021. https://www.newsguardtech.com/special-report-2020-engagement-analysis/.

Meriam Library, California State University Chico. 2010. "Evaluating Information – Applying the CRAAP Test." https://library.csuchico.edu/sites/default/files/craap-test.pdf.

Mikati, Ihab, Adam F. Benson, Thomas J. Luben, Jason D. Sacks, and Jennifer Richmond-Bryant. 2018. "Disparities in Distribution of Particulate Matter Emission Sources by Race and Poverty Status." *American Journal of Public Health* 108, no. 4 (March 7, 2018): 480–85.. https://www.ncbi.nlm.nih.gov/pmc/articles/PMC5844406/.

Minkler, Meredith, Victoria Breckwich Vásquez, Mansoureh Tajik, and Dana Petersen. 2006. "Promoting Environmental Justice through Community-Based Participatory Research: The Role of Community and Partnership Capacity." *Health Education and Behavior* 35, no. 1 (May 31, 2006). https://journals.sagepub.com/doi/10.1177/1090198106287692.

Nesbitt, Lorien, Michael J. Meitner, Cynthia Girling, Stephen R. J. Sheppard, and Yuhao Lu. 2019. "WHO Has Access to Urban Vegetation? A Spatial Analysis of Distributional Green Equity in 10 US Cities." *Landscape and Urban Planning* 181 (January 2019): 51–79. https://www.sciencedirect.com/science/article/abs/pii/S0169204618307710.

Newman, Paul A., and Markus Rex (lead authors). 2007. "Polar Ozone: Past and Present." In *Scientific Assessment of Ozone Depletion: 2006*, Chapter 4. World Meteorological Organization Global Ozone Research and Monitoring Project, report no. 50. www.esrl.noaa.gov/csd/assessments/2006/report.html.

Office of Legacy Management. 2017. "Environmental Justice History." Energy.gov, January 13, 2017. https://www.energy.gov/lm/services/environmental-justice/environmental-justice-history.

Pauly, Daniel, and Villy Christensen. 1995. "Primary Production Required to Sustain Global Fisheries." *Nature* 374, no. 6519 (March 1995): 255–57.

Robbins, Jim. 2021. "How Returning Lands to Native Tribes Is Helping Protect Nature." Yale Environment 360, Yale School of the Environment, June 3, 2021. https://e360.yale.edu/features/how-returning-lands-to-native-tribes-is-helping-protect-nature.

Rodriguez-Ramirez, Alberto, Manuel González-Rivero, Oscar Beijbom, Christophe Bailhache, Pim Bongaerts, Kristen T. Brown, Dominic E. P. Bryant, et al. 2020. "A Contemporary Baseline Record of the World's Coral Reefs." *Nature* 7, no. 355 (October 20, 2020). https://www.nature.com/articles/s41597-020-00698-6.

Rowland-Shea, Jenny. 2020. "The Nature Gap Confronting Racial and Economic Disparities in the Destruction and Protection of Nature in America." Center for American Progress, July 21, 2020. https://www.americanprogress.org/article/the-nature-gap/.

Sanderson, Eric W., Malanding Jaiteh, Marc A. Levy, Kent H. Redford, Antoinette V. Wannebo, and Gillian Woolmer. 2002. "The Human Footprint and the Last of the Wild: The Human Footprint Is a Global Map of Human Influence on the Land Surface, Which Suggests That Human Beings Are Stewards of Nature, Whether We Like It or Not." *BioScience* 52, no. 10 (October 2002): 891–904.

Shearer, Elisa, and Elizabeth Grieco. 2019. "Americans Are Wary of the Role Social Media Sites Play in Delivering the News." Pew Research Center, October 2, 2019. https://www.journalism.org/2019/10/02/americans-are-wary-of-the-role-social-media-sites-play-in-delivering-the-news/.

Sherman, Daniel J. 2008. "Sustainability: What's the Big Idea? A Strategy for Transforming the Higher Education Curriculum." *Sustainability* 1, no. 3 (June 2008): 188–95.

Simon – Kucher & Partners. 202. "Global Sustainability Study 2021. Consumers are key players for a sustainable future." https://www.simon-kucher.com/sites/default/files/studies/Simon-Kucher_Global_Sustainability_Study_2021.pdf.

Tessum, Christopher, Joshua Apte, Andrew L. Goodkind, Nicholas Z. Muller, Kimberley A. Mullins, David A. Paolella, Stephen Polasky, et al. 2019. "Inequity in Consumption of Goods and Services Adds to Racial–Ethnic Disparities in Air Pollution Exposure," *Proceedings of the Natural Academy of Sciences of the United States of America* 116, no. 13 (March 11, 2019): 6001–6. https://www.pnas.org/doi/10.1073/pnas.1818859116.

Tovey, Catherine, and Siet Meijer. 2014. *The Bangladesh Responsible Sourcing Initiative: A New Model for Green Growth.* Washington, DC: World Bank, South Asia Environment and Water Resources Unit, April 2014. https://documents1.worldbank.org/curated/en/614901468768707543/pdf/922610WP0P11950DEL0FOR0GREEN0GROWTH.pdf.

Understanding Evolution. 2019. *Gathering Evidence to Study Mass Extinctions.* www.evolution.berkeley.edu/evolibrary/article/0_0_0/massextinct_07.

United Nations Environment Programme. 2012. *Green Economy in a Blue World.* Nairobi: UNEP. https://www.unep.org/resources/report/green-economy-blue-world-full-report-0.

United Nations General Assembly. 2015. "Transforming Our World: The 2030 Agenda for Sustainable Development." Resolution A/RES/70/1. www.un.org/en/development/desa/population/migration/generalassembly/docs/globalcompact/A_RES_70_1_E.pdf.

US Environmental Protection Agency. 2021. "Learn about Environmental Justice." Accessed September 22, 2021. https://www.epa.gov/environmentaljustice/learn-about-environmental-justice.

US Environmental Protection Agency. 2022. "Environmental Justice." Accessed March 23, 2022. https://www.epa.gov/environmentaljustice#:~:text=Environmental%20justice%20is%20the%20fair,laws%2C%20regulations%2C%20and%20policies.

Woods Hole Oceanographic Institution. 2014. "Mercury in the Global Ocean." August 6, 2014. whoi.edu/press-room/news-release/mercury-in-global-ocean.

World Wildlife Fund for Nature. 2016. "Living Planet Report 2016." Gland, Switzerland: WWF. https://www.worldwildlife.org/pages/living-planet-report-2016.

Wu, X., R. C. Nethery, M. B. Sabath, D. Braun, and F. Dominici. 2020. Air Pollution and COVID-19 Mortality in the United States: Strengths and Limitations of an Ecological Regression Analysis. Science advances 6, no. 45 (November 4, 2020). https://pubmed.ncbi.nlm.nih.gov/33148655/.

CHAPTER 2

Backman, Melvin. 2015. "High Gas Prices Will Weigh on Bond Investors Tied to SUVs." *Quartz*, May 19, 2015. https://qz.com/407765/high-gas-prices-will-weigh-on-bond-investors-tied-to-suvs/.

Barclay, Eliza. 2012. "A Nation of Meat Eaters: See How It All Adds Up." Iowa Public Radio, June 27, 2012. https://www.iowapublicradio.org/2012-06-27/a-nation-of-meat-eaters-see-how-it-all-adds-up.

Brune, Michael. 2020. "Pulling Down Our Monuments." Sierra Club, December 18, 2020. https://www.sierraclub.org/michael-brune/2020/07/john-muir-early-history-sierra-club.

Capper, J. L. 2011. "The Environmental Impact of Beef Production in the United States: 1977 Compared with 2007." *Journal of Animal Science* 89, no. 12): 4249–61. doi:10.2527/jas.2010-3784.

Center for Biological Diversity. n.d. "Texas Wild Rice." Accessed October 1, 2019. www.biologicaldiversity.org/campaigns/esa_works/profile_pages/TexasWildRice.html.

Earth Charter Initiative. n.d. "The Earth Charter." Accessed October 16, 2018. https://earthcharter.org/read-the-earth-charter/.

Epstein, Paul R., Jonathan J. Buonocore, Kevin Eckerle, Michael Hendryx, Benjamin M. Stout III, Richard Heinberg, Richard W. Clapp, Beverly May, Nancy L. Reinhart, Melissa M. Ahern, Samir K. Doshi, and Leslie Glustrom. 2011. "Full Cost Accounting for the Life Cycle of Coal." *Annals of the New York Academy of Sciences* 1219 (February 2011): 73–98.

Gannett, Henry, United States National Conservation Commission, and Theodore Roosevelt. 1909. Report of the National Conservation Commission, February 1909: Special Message from the President of the United States Transmitting a Report of the National Conservation Commission, with Accompanying Papers. 3 vols., vol. 1. 60th Congress, 2d Session, Document No. 676, February 1909. Washington, DC: US Government Printing Office. https://books.google.com/books?id=y6YyAQAAMAAJ&printsec=frontcover&source=gbs_ge_summary_r&cad=0#v=onepage&q&f=false.

Harris, James. 2021. "A Brief History: 1911–1986." Sierra Club Angeles Chapter, January 25, 2021. https://angeles.sierraclub.org/about/chapter_history/brief_history_1911_1986.

Horn, Paul. 2015. "The Downside to Lower Gas Prices" [graphic]. Inside Climate News, April 13, 2015. https://insideclimatenews.org/content/downside-lower-gas-prices. Jevons, William Stanley. 1865. *The Coal Question.* London: MacMillan.

Johnson, Eric Michael. 2014. "How John Muir's Brand of Conservation Led to the Decline of Yosemite." *Scientific American*, August 13, 2014. https://blogs.scientificamerican.com/primate-diaries/how-john-muir-s-brand-of-conservation-led-to-the-decline-of-yosemite/.

Kaufman, Frederik A. 2003. *Foundations of Environmental Philosophy: A Text with Readings.* New York: McGraw-Hill.

Kollipara, Puneet. 2016. "Updated: United States Adopts Major Chemical Safety Overhaul." ScienceMag.org, June 8, 2016. www.sciencemag.org/news/2016/06/updated-united-states-adopts-major-chemical-safety-overhaul.

Li, Guangdong, and Chuanglin Fang. 2014. "Global Mapping and Estimation of Ecosystem Services Values and Gross Domestic Product: A Spatially Explicit Integration of National 'Green GDP' Accounting." *Ecological Indicators* 46 (November 2014): 293–314.

Marshak, Stephen. 2016. *Essentials of Geology*, 5th ed. New York: Norton.

Nowak, David J., and Daniel E. Crane. 2002. "Carbon Storage and Sequestration by Urban Trees in the USA." *Environmental Pollution* 116, no. 3 (March 2002): 381–89.

Pelletier, Nathan, Rich Pirog, and Rebecca Rasmussen. 2010. "Comparative Life Cycle Environmental Impacts of Three Beef Production Strategies in the Upper Midwestern United States." *Agricultural Systems* 103, no. 6 (July 2010): 380–89.

Pinchot, Gifford. 1911. "The Conservation of the Forests." *Scientific American*, August 12, 1911. https://www.scientificamerican.com/article/the-conservation-of-the-forests/.

Purdy, Jedediah. 2015. "Environmentalism's Racist History." *New Yorker*, August 13, 2015. https://www.newyorker.com/news/news-desk/environmentalisms-racist-history.

Rational Walk. 2010. "Brazil Pushes Technical Limits in Deepwater Oil Exploration." July 20, 2010. www.rationalwalk.com/?p=8301.

Ravenscraft, Eric. 2016. "The Best Ways to Contact Your Congresspeople, from a Former Staffer." Lifehacker, November 15, 2016. https://lifehacker.com/the-best-ways-to-contact-your-congress-people-from-a-f-1788990839.

Shapiro, Craig. 2016. "Thousands of Animals Saved with a Pen Stroke." PETA, May 23, 2016. www.peta.org/blog/new-law-will-spare-thousands-animals-agony-cruel-chemical-tests/.

Shindell, Drew T. 2015. "The Social Cost of Atmospheric Release." *Climatic Change* 130, no. 2 (February 25, 2015): 313–26.

Solnit, Rebecca. 2021. "John Muir in Native America." Sierra Club Angeles Chapter, March 2, 2021. https://www.sierraclub.org/sierra/2021-2-march-april/feature/john-muir-native-america.

Stillman, Amy. 2010. "Brazil's Fledgling Deep-Sea Oil Industry Faces Up to Risks." *Financial Times*, July 19, 2010.

Tavernise, S. 2012. "F.D.A. Bans BPA from Baby Bottles and Sippy Cups." *New York Times*, July 17, 2012. www.nytimes.com/2012/07/18/science/fda-bans-bpa-from-baby-bottles-and-sippy-cups.html.

Taylor, Dorceta E. 2014. "The State of Diversity in Environmental Organizations." Prepared for Green 2.0, July 2014. https://vaipl.org/wp-content/uploads/2014/10/ExecutiveSummary-Diverse-Green.pdf.

US Bureau of Economic Analysis. 2021. "Light Weight Vehicle Sales: Autos and Light Trucks (ALTSALES)." FRED, Federal Reserve Bank of St. Louis. https://fred.stlouisfed.org/series/ALTSALES.

US Bureau of Labor Statistics. 2021. "Average Price: Gasoline, Unleaded Regular (Cost per Gallon/3.785 Liters) in U.S. City Average (APU000074714)." FRED, Federal Reserve Bank of St. Louis. https://fred.stlouisfed.org/series/APU000074714.

US Energy Administration. 2021. "Crude Oil Prices: West Texas Intermediate (WTI)." FRED, Federal Reserve Bank of St. Louis. https://fred.stlouisfed.org/series/DCOILWTICO.

US Energy Information Administration. 2021. "US Regular All Formulations Gas Price (GASREGW)." FRED, Federal Reserve Bank of St. Louis. https://fred.stlouisfed.org/series/GASREGW.

US Energy Information Administration. n.d. "Crude Oil Prices: West Texas Intermediate (WTI)—Cushing, Oklahoma [DCOILWTICO]." FRED, Federal Reserve Bank of St. Louis. Accessed October 16, 2018.

US Federal Highway Administration. 2021. "Moving 12-Month Total Vehicle Miles Traveled (M12MTVUSM227NFWA)." FRED, Federal Reserve Bank of St. Louis. https://fred.stlouisfed.org/series/M12MTVUSM227NFWA.

USA.gov. 2019. "How to Contact Your Elected Officials." www.usa.gov/elected-officials.

Woolworth, Charles. 2010. "Conservation and Eugenics." *Orion Magazine*, June 2010. https://orionmagazine.org/article/conservation-and-eugenics/.

CHAPTER 3

American Iron and Steel Institute. n.d. "Recycling." Accessed October 1, 2019. www.steel.org/sustainability/recycling.

Atkins, Peter. 2010. *The Laws of Thermodynamics: A Very Short Introduction*. Oxford: Oxford University Press.

BBC News. 2015. "What Causes Southeast Asia's Haze?" October 26, 2015.

Bretz, Stacey Lowery, Geoffrey Davies, Natalie Foster, Thomas Gilbert, and Rein V. Kirss. 2017. *Chemistry: The Science in Context*, 5th ed. New York: Norton.

Department of Ecology, State of Washington. n.d. "Addressing Priority Toxic Chemicals." https://ecology.wa.gov/Waste-Toxics/Reducing-toxic-chemicals/Addressing-priority-toxic-chemicals.

Hazen, Robert, and James Trefil. 2009. *Science Matters: Achieving Scientific Literacy*. New York: Anchor.

Rieland, Randy. 2016. "Meet Eight Young Energy Innovators with Ingenious Ideas." *Smithsonian Magazine*, July 18, 2016. www.smithsonianmag.com/innovation/meet-eight-young-energy-innovators-with-ingenious-ideas-180959761/?page=6.

Shepard, Peggy, and Jana Walker. 2002. "Fish Consumption and Environmental Justice." National Environmental Justice Advisory Council (NEJAC), revised November 2022. https://www.epa.gov/sites/default/files/2015-02/documents/fish-consump-report_1102.pdf.

Smil, Vaclav. 2012. *Energy: A Beginner's Guide*. Oxford: Oneworld.

Sussman, Art. 2006. *Dr. Art's Guide to Science: Connecting Atoms, Galaxies, and Everything in Between*. San Francisco: WestEd/Jossey-Bass.

University of Southampton. 2018. "News: Global Analysis Reveals How Sharks Travel the Oceans to Find Food." January 18, 2018. www.southampton.ac.uk/news/2018/01/shark-feeding-habits.page.

US Environmental Protection Agency. 2014. "Estimated Fish Consumption Rates for the U.S. Population and Selected Subpopulations (NHANES 2003–2010): Final Report." April 2014. https://www.epa.gov/sites/default/files/2015-01/documents/fish-consumption-rates-2014.pdf.

CHAPTER 4

Biocyclopedia. n.d. "General Zoology: Competition and Character Displacement." Accessed October 1, 2019. https://biocyclopedia.com/index/general_zoology/competition_and_character_displacement.php.

BiodiversityMapping.org. 2012. [Biodiversity maps based on data from BirdLife International and the International Union for Conservation of Nature.] https://biodiversitymapping.org/.

Brockington, Daniel, and James Igoe. "Eviction for Conservation: A Global Overview." *Conservation and Society* 4, no. 3 (July–September 2006): 424–70. http://www.jstor.org/stable/26396619.

Charlesworth, Brian, and Deborah Charlesworth. 2003. *Evolution: A Very Short Introduction*. Oxford: Oxford University Press.

Cincotta, Richard P., Jennifer Wisnewski, and Robert Engelman. 2000. "Human Population in the Biodiversity Hotspots." *Nature* 404 (April 27, 2000): 990–92. https://www.nature.com/articles/35010105.

Cooke, Lacy. 2016. "Millions of Genetically Altered Mosquitoes Are Being Released in the Cayman Islands to Fight Zika." Inhabitat, May 10, 2016. https://inhabitat.com/millions-of-genetically-altered-mosquitoes-are-being-released-to-fight-zika/.

Cornell CALS. "Pollinator Network." 2021. Cornell College of Agriculture and Life Sciences, June 23, 2021. https://pollinator.cals.cornell.edu/threats-wild-and-managed-bees/pesticides/neonicotinoids/.

Critical Ecosystem Partnership Fund. n.d. "Explore the Biodiversity Hotspots." Accessed July 6, 2019. www.cepf.net/our-work/biodiversity-hotspots.

DiStasio, Cat. 2016. "Brazil Unleashes Millions of Genetically Modified Mosquitoes to Combat Zika." Inhabitat, October 31, 2016. https://inhabitat.com/brazilian-scientists-unleash-millions-of-genetically-modified-mosquitoes-to-combat-zika/.

DiStasio, Cat. 2016. "World's Largest 'Mosquito Factory' in China to Release 20 Million Bugs a Week." Inhabitat, March 20, 2016. https://inhabitat.com/worlds-largest-mosquito-factory-in-china-to-release-20-million-bugs-a-week/.

Durant, Sarah M., Nicholas Mitchell, Rosemary Groom, Nathalie Pettorelli, Audrey Ipavec, Andrew P. Jacobson, Rosie Woodroffe, et al. 2016. "The Global Decline of Cheetah *Acinonyx jubatus* and What It Means for Conservation." *Proceedings of the National Academy of Sciences USA* 114, no. 3 (December 27, 2016): 528–33. www.pnas.org/content/early/2016/12/20/1611122114.

Fisher, Brendan, and Treg Christopher. 2007. "Poverty and Biodiversity: Measuring the Overlap of Human Poverty and the Biodiversity Hotspots." *Ecological Economics* 62, no. 1 (April 1, 2007): 93–101.

Gorenflo, L. J., Suzanne Romaine, Russell A. Mittermeier, and Kristen Walker-Painemilla. 2012. "Co-occurrence of Linguistic and Biological Diversity in Biodiversity Hotspots and High Biodiversity Wilderness Areas." *Proceedings of the National Academy of Sciences USA* 109, no. 21 (May 7, 2012): 8032–37.

Harris, William. 2010. "What Happens When a Keystone Species Goes Extinct?" How Stuff Works. https://science.howstuffworks.com/environmental/conservation/issues/keystone-species-extinct.htm.

Hirsch Tim, Kieran Mooney, and David Cooper. 2020. *Global Biodiversity Outlook 5*. Montreal: Secretariat of the Convention on Biological Diversity. https://www.cbd.int/gbo/gbo5/publication/gbo-5-en.pdf.

Houtman, Anne, Megan Scudenari, Cindy Malone, and Anu Singh-Cundy. 2015. *Biology Now*. New York: Norton.

International Union for Conservation of Nature, Red List. "Table 1A: Number of Species Evaluated in Relation to the Overall Number of Described Species, and Numbers of Threatened Species by Major Groups of Organisms." n.d.

Jones, Rhett. 2016. "Scientists Hope to Eradicate Disease with Massive Mosquito Orgy." Gizmodo, October 30, 2016. https://gizmodo.com/scientists-hope-to-eradicate-disease-with-massive-mosqu-1788379798.

Klein, D. 1968. "The Introduction, Increase, and Crash of Reindeer on St. Matthew Island." *Journal of Wildlife Management* 32, no. 2 (April 1968): 350–67.

Labour Voices. 2021. "Caring for the Land, Caring for People." New Zealand Labour Party, June 5, 2021. https://www.labour.org.nz/news-world_environment_day_2021.

Larsen, Brendan B., Elizabeth C. Miller, Matthew K. Rhodes, and John J. Wiens. 2017. "Inordinate Fondness Multiplied and Redistributed: The Number of Species on Earth and the New Pie of Life." *Quarterly Review of Biology* 92, no. 3 (September 2017): 229–65.

MacArthur, Robert H. 1958. "Population Ecology of Some Warblers of Northeastern Coniferous Forests." *Ecology* 39, no. 4 (October 1958): 599–619. www.jstor.org/stable/1931600.

MacArthur, Robert, and Edward O. Wilson. 1963. "An Equilibrium Theory of Insular Zoogeography." *Evolution* 17, no. 4 (December 1963): 373–87.

Main, Douglas. 2015. "The 21 Most Excellent New Animal Species of 2015." *Newsweek*, December 30, 2015. www.newsweek.com/21-coolest-new-animal-species-2015-410008.

Millennium Ecosystem Assessment. 2005. "Current State and Trends Assessment." www.millenniumassessment.org/en/Condition.html.

Mora, C., D. P. Tittensor, S. Adl, A. G. B. Simpson, and B. Worm. 2011. "How Many Species Are There on Earth and in the Ocean?" *PLOS Biology* 9, no. 8 (August 23, 2011): e1001127. doi:10.1371/journal.pbio.1001127.

Muth, F., and A. S. Leonard. 2019. "A Neonicotinoid Pesticide Impairs Foraging, but Not Learning, in Free-Flying Bumblebees." *Scientific Reports* 9, no. 4764 (March 18, 2019). https://www.nature.com/articles/s41598-019-39701-5.

Myers, Norman, Russell A. Mittermeier, Cristina G. Mittermeier, Gustavo A. B. da Fonseca, and Jennifer Kent. 2000. "Biodiversity Hotspots for Conservation Priorities." *Nature* 403 (February 24, 2000): 853–58.

NOAA Fisheries. n.d. "Adult Male Fur Seal Counts, 1911–2004, St. Paul Island." Accessed July 6, 2019.

Novak, Sara. 2018. "Scientists Race to Kill Mosquitoes before They Kill Us." *Sierra*, January 9, 2018. www.sierraclub.org/sierra/scientists-race-kill-mosquitoes-they-kill-us.

Oldekop, J. A., G. Holmes, W. E. Harris, and K. L. Evans. 2015. "A Global Assessment of the Social and Conservation Outcomes of Protected Areas." *Conservation Biology* 30, no. 1 (June 10, 2015): 133–41.

Schoolbaginfo. n.d. "The Living World." Accessed October 1, 2019. https://schoolbag.info/biology/living/286.html.

Trainor, Colin, and Pedro Leitão. 2007. "Further Significant Bird Records from Atauro Island, Timor-Leste (East Timor)." *Forktail* 23 (August 2007): 155–58.

Whittaker, Robert Harding. 1975. *Communities and Ecosystems*. New York: Macmillan.

Zukerman, Wendy. 2009. "Australia's Battle with the Bunny." ABC Science, April 8, 2009. www.abc.net.au/science/articles/2009/04/08/2538860.htm.

CHAPTER 5

Ceballos, Gerardo, Paul R. Ehrlich, Anthony D. Barnosky, Andrés García, Robert M. Pringle, and Todd M. Palmer. 2015. "Accelerated Modern Human–Induced Species Losses: Entering the Sixth Mass Extinction." *Science Advances* 1, no. 5 (June 19, 2015): e1400253.

Crochetiere, Heather. 2015. "Did You Know? 10 Amazing Things about Salmon." WWF, March 19, 2015. https://blog.wwf.ca/blog/2015/03/19/did-you-know-10-amazing-things-about-salmon/.

Daily, Gretchen C., ed. 1997. *Nature's Services: Societal Dependence on Natural Ecosystems*. Washington, DC: Island Press.

Dinerstein, E., C. Vynne, E. Sala, A. R. Joshi, S. Fernando, T. E. Lovejoy, J. Mayorga, et al. 2019. "A Global Deal for Nature: Guiding Principles, Milestones, and Targets." *Science Advances* 5, no. 4 (April 19, 2019).

Ecotourism World. 2021. "Eco Friendly Travel against Shortage of Water." February 23, 2021. https://ecotourism-world.com/eco-friendly-travel-against-shortage-of-water/.

Elbein, Saul. 2015. "Grim Prospects for Sustainable Miners in Peru." Al Jazeera America, September 21, 2015. http://america.aljazeera.com/multimedia/2015/9/Peru-mining.html.

Food and Agriculture Organization of the United Nations. 2010. *Global Forest Resources Assessment 2010*. Rome: FAO. www.fao.org/forestry/fra/62219/en/irl/.

Forest Inventory and Analysis National Program, USDA Forest Service. n.d. "FIA Library." https://www.fia.fs.usda.gov/

Friends of Times Beach Nature Preserve. n.d. "Passenger Pigeon." Accessed October 1, 2019. www.friendsoftimesbeachnp.org/passenger-pigeon.html.

Grebenstein, Emily. 2013. "Escape of the Invasives: Top Six Invasive Plant Species in the United States." Smithsonian Sparks, April 19, 2013. https://insider.si.edu/2013/04/top-six-invasive-plant-species-in-the-united-states/.

Haaland, Deb, Thomas J. Vilsack, Gina M. Raimondo, and Brenda Mallory. 2021. "Conserving and Restoring America the Beautiful 2021." Washington, DC: US Departments of the Interior, Agriculture, and Commerce, and Council on Environmental Quality.

Hardy Vincent, Carol, Laura A. Hanson, and Carla N. Argueta. 2017. *Federal Land Ownership: Overview and Data*. Homeland Security Digital Library, pp. 9, 11. Congressional Research Service Report for Congress, no. R42346. www.hsdl.org/?abstract&did=799426.

Hedgelink. n.d. "About Hedgerows." Accessed October 1, 2019. www.hedgelink.org.uk/index.php?page=16.

Hoekstra, Jonathan, Timothy Boucher, Taylor Ricketts, and Carter Roberts. 2004. "Confronting a Biome Crisis: Global Disparities of Habitat Loss and Protection." *Ecology Letters* 8 (December 3, 2004): 23–29. doi: 10.1111/j.1461-0248.2004.00686.x.

ICCA Consortium. 2021. *Territories of Life: 2021 Report*. https://report.territoriesoflife.org/wp-content/uploads/2021/09/ICCA-Territories-of-Life-2021-Report-FULL-150dpi-ENG.pdf.

IKEA. 2022. "Being Forest Positive." https://about.ikea.com/en/sustainability/responsible-sourcing/being-forest-positive.

Intergovernmental Science-Policy Platform on Biodiversity and Ecosystem Services. 2019. *Summary for Policymakers of the Global Assessment Report on Biodiversity and Ecosystem Services of the Intergovernmental Science-Policy Platform on Biodiversity and Ecosystem Services*. May 29, 2019. www.ipbes.net/system/tdf/ipbes_7_10_add.1_en_1.pdf?file=1&type=node&id=35329.

IPBES. 2019. *Global Assessment Report on Biodiversity and Ecosystem Services of the Intergovernmental Science-Policy Platform on Biodiversity and Ecosystem Services*. Edited by Eduardo Brondizio, Sandra Diaz, and Hien T. Ngo. Bonn, Germany: IPBES Secretariat. https://ipbes.net/global-assessment.

Kareiva, Peter, and Michelle Marvier. 2011. *Conservation Science: Balancing the Needs of People and Nature*. Greenwood Village, CO: Roberts.

Kolbert, Elizabeth. 2014. *The Sixth Extinction: An Unnatural History*. New York: Henry Holt.

LoSchiavo, A. J., R. G. Best, R. E. Burns, S. Gray, M. C. Harwell, E. B. Hines, A. R. McLean, T. St. Clair, S. Traxler, and J. W. Vearil. 2013. "Lessons Learned from the First Decade of Adaptive Management in Comprehensive Everglades Restoration." *Ecology and Society* 18, no. 4: 70.

MacArthur, Robert H., and Edward O. Wilson. 2016. *The Theory of Island Biogeography*. Princeton Landmarks in Biology. Princeton, NJ: Princeton University Press.

Montgomery, David. R. 2003. *King of Fish: The Thousand-Year Run of Salmon*. Boulder, CO: Westview.

Nagy, R. Chelsea, Emily J. Fusco, Jennifer K. Balch, John T. Finn, Adam Mahood, Jenica M. Allen, and Bethany A. Bradley. 2020. "A Synthesis of the Effects of Cheatgrass Invasion on US Great Basin Carbon Storage." *Journal of Applied Ecology* 58, no. 2 (September 26, 2020): 327–37.

National Geographic Society. n.d. "BioBlitz Program." Accessed October 1, 2019. www.nationalgeographic.org/projects/bioblitz/.

National Ocean Service. 2018. "What Is Coral Bleaching?" NOAA. https://oceanservice.noaa.gov/facts/coral_bleach.html.

New York State Adirondack Park Agency, Geographic Information Services. 2017. "125 Years: New York State Adirondack Park Forest Preserve Lands, 1892 and 2017" [graphic]. October 2017. https://apa.ny.gov/gis/_assets/ForestPreserve1892-2017.png.

Ontiri, Enoch M., Martin Odino, Antony Kasanga, Paula Kahumbu, Lance W. Robinson, Tom Currie, and Dave J. Hodgson. 2019. "Maasai Pastoralists Kill Lions in Retaliation for Depredation of Livestock by Lions." *People and Nature* 1, no. 1 (February 25, 2019): 59–69. https://doi.org/10.1002/pan3.10.

Rutten, Marcel. 2002. "Amplifying Local Voices, Striving for Environmental Justice." 80–97.

Solutions Project. 2022. "Bezos Earth Fund." https://thesolutionsproject.org/bezos-earth-fund/.

Taylor, Michael. 2021. "Analysis-G7 Brightens Outlook for New Nature Pact but Pandemic Threatens Deadline." Reuters, June 15, 2021. https://www.reuters.com/article/us-global-nature-climate-politics-analys/analysis-g7-brightens-outlook-for-new-nature-pact-but-pandemic-threatens-deadline-idUSKCN2DR1L5.

Tilman, D., P. B. Reich, J. Knops, D. Wedin, T. Mielke, and C. Lehman. 2001. "Diversity and Productivity in a Long-Term Grassland Experiment." *Science* 294, no. 5543 (October 26, 2001): 843–45.

Tizard, Ian R. 2020. "A Brief History of Veterinary Vaccines." In *Vaccines for Veterinarians*, 1–12. St. Louis: Elsevier.

US Department of Agriculture, Farm Service Agency. 2022. "Conservation Reserve Program." https://www.fsa.usda.gov/programs-and-services/conservation-programs/conservation-reserve-program/.

US Geological Survey. 2016. "National Gap Analysis Project." https://gapanalysis.usgs.gov/padus/.

Vogdrup-Schmidt, Mathias, Niels Strange, Søren B. Olsen, and Bo Jellesmark Thorsen. 2017. "Trade-off Analysis of Ecosystem Service Provision in Nature Networks." *Ecosystem Services* 23 (February 2017): 165–73.

Wenzel, Lauren, and Mimi D'Iorio. 2011. "Definition & Classification System for U.S. Marine Protected Areas." National Marine Protected Areas Center, March 2011. https://nmsmarineprotectedareas.blob.core.windows.net/marineprotectedareas-prod/media/archive/pdf/helpful-resources/factsheets/mpa_classification_may2011.pdf.

Wilson, Kerrie A., Emma C. Underwood, Scott A. Morrison, Kirk R. Klausmeyer, William W. Murdoch, Belinda Reyers, Grant Wardell-Johnson, et al. 2007. "Conserving Biodiversity Efficiently: What to Do, Where, and When." *PLOS Biology* 5, no. 9 (August 21, 2007): e223.

CHAPTER 6

Atlas of Economic Complexity. 2022. "China, Vietnam, and Indonesia among Fastest-Growing Countries for Coming Decade in New Harvard Growth Lab Projections." July 26, 2022. https://atlas.cid.harvard.edu/growth-projections.

Berquó, Elza, and Suzana Cavenaghi. 2006. "Fertility in Decline: A Brief Note on the Decrease in the Number of Births in Brazil." *Novos Estudos-CEBRAP* 2 (March 2006): 11–15.

Bongaarts, John, and Ann K. Blanc. 2015. "Estimating the Current Mean Age of Mothers at the Birth of Their First Child from Household Surveys." *Population Health Metrics* 13, no. 1 (September 14, 2015). https://doi.org/10.1186/s12963-015-0058-9.

Brazilian Institute of Geography and Statistics. n.d. "IBGE Demographic Censuses of 1940 to 2004, Graph 1, Total Fertility Rates: Brazil, 1940 to 2004." Rio de Janeiro: IBGE.

Camill, Philip. 2010. "Global Change: An Overview." NatureEducation Knowledge Project. www.nature.com/scitable/knowledge/library/global-change-an-overview-13255365.

CBC News. 2015. "5 Things to Know about China's 1-Child Policy." October 29, 2015. https://www.cbc.ca/news/world/5-things-to-know-about-china-s-1-child-policy-1.3294335.

Colby, Sandra L., and Jennifer M. Ortman. 2014. *The Baby Boom Cohort in the United States: 2012 to 2060*. Current Population Reports, P25-1141. Washington, DC: US Census Bureau. www.census.gov/prod/2014pubs/p25-1141.pdf.

Degroot, Dagomar, Kevin Anchukaitis, Martin Bauch, Jakob Burnham, Fred Carnegy, Jianxin Cui, Kathryn de Luna, et al. 2021. "Towards a Rigorous Understanding of Societal Responses to Climate Change." *Nature* 591, no. 7851 (March 24, 2021): 539–50. https://doi.org/10.1038/s41586-021-03190-2.

Demographic Dividend. n.d. "Rwanda." Accessed June 14, 2022. https://demographicdividend.org/country_highlights/rwanda/.

EFA Global Monitoring Report Team. 2014. *Teaching and Learning, Achieving Equality for All*. EFA Global Monitoring Report 2013/4, p. 194. Paris: UNESCO. https://unesdoc.unesco.org/ark:/48223/pf0000225660.

Ehrlich, Paul R., Anne H. Ehrlich, and Gretchen C. Daily. 1995. *The Stork and the Plow: The Equity Answer to the Human Dilemma*. New Haven, CT: Yale University Press.

Ezeh, Alex. 2018. "Empowering Women Lies at the Centre of Controlling Population Growth in Africa." Conversation, September 20, 2018. https://theconversation.com/empowering-women-lies-at-the-centre-of-controlling-population-growth-in-africa-103539.

Global Footprint Network National Footprint and Biocapacity Accounts, 2022 Edition Downloaded 7/29/22 from https://data.footprintnetwork.org.

Global Partnership for Education. n.d. "Girls' Education and Gender Equality." Accessed October 1, 2019. www.globalpartnership.org/focus-areas/girls-education.

Kassam, Ashifa, Rosie Scammell, Kate Connolly, Richard Orange, Kim Willsher, and Rebecca Ratcliffe. 2015. "Europe Needs Many More Babies to Avert a Population Disaster." *Guardian*, August 22, 2015. www.theguardian.com/world/2015/aug/23/baby-crisis-europe-brink-depopulation-disaster.

La Ferrara, Eliana, Alberto Chong, and Suzanne Duryea. 2012. "Soap Operas and Fertility: Evidence from Brazil." *American Economic Journal: Applied Economics* 4, no. 4 (October 2012): 1–31.

Lancaster, Henry Oliver. 2012. *Expectations of Life: A Study in the Demography, Statistics, and History of World Mortality*. New York: Springer Science & Business Media.

Liu, Daphne H., and Adrian E. Raftery. 2020. "How Do Education and Family Planning Accelerate Fertility Decline?" *Population and Development Review* 46, no. 3 (July 23, 2020): 409–41.

Lutz, Wolfgang, and Anne Goujon. 2004. "Literate Life Expectancy: Charting the Progress in Human Development." In *The End of World Population Growth in the 21st Century: New Challenges for Human Capital Formation and Sustainable Development*, edited by Wolfgang Lutz, Warren C. Sanderson, and Sergei Scherbov, chap. 5. London: Earthscan.

Malthus, Thomas Robert. 1798. *An Essay on the Principle of Population*. London: J. Johnson.

Minow, Newton N., and Craig L. LaMay. 2008. *Inside the Presidential Debates: Their Improbable Past and Promising Future*, 24–25 (p. 170, "sustainability"). Chicago: University of Chicago Press, 2008.

Ndahindwa, Vedaste, Collins Kamanzi, Muhammed Semakula, François Abalikumwe, Bethany Hedt-Gauthier, and Dana R Thomson. 2014. "Determinants of Fertility in Rwanda in the Context of a Fertility Transition: A Secondary Analysis of the 2010 Demographic and Health Survey." *Reproductive Health* 11, no. 1 (December 13, 2014). https://doi.org/10.1186/1742-4755-11-87.

Pearce, Fred. 2010. *The Coming Population Crash and Our Planet's Surprising Future*. Boston: Beacon.

Population Reference Bureau. n.d. "Data Sheets." Accessed October 1, 2019. www.prb.org/datasheets/.

Rizvi, Najma. 2014. "Successful Family Planning in Bangladesh." D+C, March 19, 2014. www.dandc.eu/en/article/successful-family-planning-bangladesh-holistic-approach-leads-lower-fertility-rates-rates.

Roser, Max. 2018. "Fertility Rate." OurWorldInData.org .https://ourworldindata.org/fertility-rate.

Track20 Project. "Technical Note: Developing scenarios for long-term population growth in SSA." http://www.track20.org/download/pdf/Track20_Technical_Note_on_SSA_Fertility_Models.pdf.

UN News. 2015. "UN Projects World Population to Reach 8.5 Billion by 2030, Driven by Growth in Developing Countries." UN Sustainable Development Goals, July 29, 2015. https://news.un.org/en/story/2015/07/505352-un-projects-world-population-reach-85-billion-2030-driven-growth-developing#.WR34Q7zytTY.

United Nations, Department of Economic and Social Affairs, Population Division. 2017. *World Population Prospects: The 2017 Revision*, DVD ed. New York: UN Population Division.

United Nations, Department of Economic and Social Affairs, Population Division. 2021. *World Population Prospects: The 2021 Revision*. New York: UN Population Division.

United Nations, Department of Economic and Social Affairs. 2021. "Sustainable Development Goals—Goal 5: Achieve Gender Equality and Empower All Women and Girls." https://sdgs.un.org/goals/goal5.

United Nations Population Fund. 2019. "Twenty-Five Years Ago, Leaders Promised to Advance Women's Health and Rights: Have They Delivered?" July 10, 2019. https://www.unfpa.org/news/twenty-five-years-ago-leaders-promised-advance-womens-health-and-rights-have-they-delivered.

United Nations—Population Division (2019 revision). 2021. "Women's Educational Attainment vs. Number of Children per Woman." Our World in Data, Global Change Data Lab. https://ourworldindata.org/grapher/womens-educational-attainment-vs-fertility?tab=table&time=1950..latest&country=KWT~TUN~GTM~KOR~GAB~MDV~IDN~CHN~BHR~CUB~MNG~DZA.

Upadhyay, Ushma D., Jessica D. Gipson, Mellissa Withers, Shayna Lewis, Erica J. Ciaraldi, Ashley Fraser, Megan J. Huchko, and Ndola Prata. 2014. "Women's Empowerment and Fertility: A Review of the Literature." *Social Science & Medicine* 115 (August 2014): 111–20.

US Census Bureau. n.d. "U.S. and World Population Clock." Accessed Sept 1, 2022. www.census.gov/popclock/.

US Census Bureau. 2013. "A Century of Population Change in the Age and Sex Composition of the Nation." September 12, 2013. www.census.gov/dataviz/visualizations/055/.

US Census Bureau. 2020. "Southern and Western Regions Experienced Rapid Growth This Decade." May 21, 2020. https://www.census.gov/newsroom/press-releases/2020/south-west-fastest-growing.html.

US Census Bureau. 2021. "2020 Census: Percent Change in County Population: 2010 to 2020." August 12, 2021. https://www.census.gov/library/visualizations/2021/dec/percent-change-county-population.html.

Wackernagel, Mathis, and William Rees. 1996. *Our Ecological Footprint: Reducing Human Impact on Earth*. Gabriola Island, BC: New Society.

Wee, Sui-Lee. 2021. "China Says It Will Allow Couples to Have 3 Children, up from 2." *New York Times*, May 13, 2021. https://www.nytimes.com/2021/05/31/world/asia/china-three-child-policy.html.

Westoff, Charles F. 2018. *Trends in Reproductive Behavior in Rwanda: Further Analysis of the 2014–15 Demographic and Health Survey*. DHS Further Analysis Reports No. 107. Rockville, MD: ICF. https://dhsprogram.com/pubs/pdf/FA107/FA107.pdf.

Wikipedia. n.d. "Total Fertility Rate." Accessed October 1, 2019. https://en.wikipedia.org/wiki/Total_fertility_rate#/media/File:Countriesbyfertilityrate.svg.

World Bank. 2021a. "School Enrollment, Secondary (% Gross)." UNESCO Institute for Statistics, September 2021. https://data.worldbank.org/indicator/SE.SEC.ENRR.

World Bank. 2021b. "School Enrollment, Tertiary, Female (% Gross)." UNESCO Institute for Statistics, September 2021.

World Wildlife Federation. 2016. *Living Planet Report 2016: Risk and Resilience in a New Era*. Gland, Switzerland: WWF International.

CHAPTER 7

American Water Works Association. 2020. "Survey Shows High Confidence in U.S. Tap Water, Lower Satisfaction among Black, Hispanic Residents." July 15, 2020. https://www.awwa.org/AWWA-Articles/survey-shows-high-confidence-in-us-tap-water-lower-satisfaction-among-black-hispanic-respondents.

Baehler, Karen, Carley Weted, and Theo Affonso Laguna. 2020. "Lead Pipes and Environmental Justice: A Study of Lead Pipe Replacement in Washington, DC." Washington, DC: American University, School of Public Affairs, and Environmental Defense Fund. https://www.edf.org/sites/default/files/u4296/LeadPipe_EnvironJustice_AU%20and%20EDF%20Report.pdf.

Black, Maggie. 2004. *The No-Nonsense Guide to Water*. Oxford: New Internationalist.

Brenan, Megan. 2021. "Water Pollution Remains Top Environmental Concern in U.S." Gallup, April 19, 2021. https://news.gallup.com/poll/347735/water-pollution-remains-top-environmental-concern.aspx.

California Department of Food and Agriculture. 2012–2013. *California Agricultural Statistics Review*. Sacramento: Office of Public Affairs.

Carle, David. 2009. *Introduction to Water in California*. Berkeley: University of California Press.

Cave, Eloise J., and Stephen M. Kajiura. 2018. "Effect of *Deepwater Horizon* Crude Oil Water Accommodated Fraction on Olfactory Function in the Atlantic Stingray, *Hypanus sabinus*." *Scientific Reports* 8, no. 15786 (October 25, 2018).

FAO. 2018. "Global Trends in the State of the World's Marine Fish Stocks, 1974-2015" (see Figure 14). In *The State of World Fisheries and Aquaculture 2018—Meeting the Sustainable Development Goals*. Rome: Food and Agriculture Organization of the United Nations. www.fao.org/3/i9540en/I9540EN.pdf.

Felton, Ryan. 2019. "Should We Break Our Bottled Water Habit?" *Consumer Reports*, October 9, 2019. https://www.consumerreports.org/bottled-water/should-we-break-our-bottled-water-habit/.

Grandoni, Dino. 2021. "Biden Hopes to Dig up the Nation's Lead Water Pipes: This City Wants His Help." *Washington Post*, May 15, 2021. https://www.washingtonpost.com/climate-environment/2021/05/15/biden-hopes-dig-up-nations-lead-water-pipes-this-city-wants-his-help/.

Hutchinson, Adam. 2017. "Transforming Wastewater to Drinking Water: How Two Agencies Collaborated to Build the World's Largest Indirect Potable Reuse Project." Fountain Valley, CA: Orange County Water District.

Illinois State Water Survey. n.d. "About the Illinois State Water Survey." https://www.isws.illinois.edu/.

International Bottled Water Association. 2021a. "Bottled Water Consumption Shift." October 14, 2021. https://bottledwater.org/bottled-water-consumption-shift/.

International Bottled Water Association. 2021b. "Consumer Choice and Availability." July 23, 2021. https://bottledwater.org/consumer-choice-and-availability/.

Lakhani, Nina. 2020. "Millions of Americans Can't Afford Water, as Bills Rise 80% in a Decade." *Consumer Reports*, July 10, 2020. https://www.consumerreports.org/personal-finance/millions-of-americans-cant-afford-water-as-bills-rise-80-percent-in-a-decade/.

Liles, M., S. Thomas, and T. Sovich. 2001. "Saltwater Intrusion in Orange County, California: Planning for the Future." First International Conference on Saltwater Intrusion and Coastal Aquifers—Monitoring, Modeling, and Management, Essaouira, Morocco, April 23–25, 2001. https://docslib.org/doc/12099931/saltwater-intrusion-in-orange-county-california-planning-for-the-future.

Maupin, Molly A. 2018. "Summary of Estimated Water Use in the United States in 2015." US Geological Survey, June 2018. https://pubs.usgs.gov/fs/2018/3035/fs20183035.pdf.

McGuire, V. L. 2004. "Water-Level Changes in the High Plains Aquifer, Predevelopment to 2002, 1980 to 2002, and 2001 to 2002." US Geological Survey, Fact Sheet 2004-3026. https://pubs.usgs.gov/fs/2004/3026/pdf/fs04-3026.pdf.

Mekonnen, M. M., and A. Y. Hoekstra. 2010. "The Green, Blue and Grey Water Footprint of Crops and Derived Crop Products." Value of Water Research Report Series, no. 47. Delft, Netherlands: UNESCO-IHE. https://waterfootprint.org/media/downloads/Mekonnen-Hoekstra-2011-WaterFootprintCrops.pdf.

Mintz, Joel A. 2015. "Animas River Spill: Root Causes and Continuing Threats." The Hill, September 2, 2015. https://thehill.com/blogs/congress-blog/energy-environment/252426-animas-river-spill-root-causes-and-continuing-threats.

National Ocean Service. 2018. "What Is a Gyre?" NOAA https://oceanservice.noaa.gov/facts/gyre.html.

Oleson, Tim. 2014. "Virtual Water, Tracking the Unseen Water in Goods and Resources." Earthmagazine.org, August 28, 2014. www.earthmagazine.org/article/virtual-water-tracking-unseen-water-goods-and-resources.

Orange County Water District, Orange County Sanitation District. 2020. "Groundwater Replenishment System." Fountain Valley, CA: OCWD, OCSD. https://www.ocwd.com/media/8861/ocwd-technicalbrochure_web-2020.pdf.

Orange County Water District. 2022. "Frequently Asked Questions." https://www.ocwd.com/gwrs/frequently-asked-questions/.

Park, Alex, and Julia Lurie. 2014. "It Takes HOW Much Water to Make Greek Yogurt?!" MotherJones, March 10, 2014. www.motherjones.com/environment/2014/03/california-water-suck/.

Roper Center for Public Opinion Research. n.d. "Water Crisis: Worry and a Lack of Trust." Cornell University. Accessed June 14, 2022. http://ropercenter.cornell.edu/water-crisis-worry-and-lack-trust.

Thornton, Stuart. 2022. "From Toilet to Tap." National Geographic, June 2, 2022. https://education.nationalgeographic.org/resource/toilettotap.

UNDP India. n.d. "Conserving Water the Traditional Way." Accessed October 1, 2019. www.in.undp.org/content/india/en/home/sustainable-development/successstories/conserving_waterthetraditionalway.html.

US Drought Monitor. n.d. Map of United States [graphic]. Accessed July 6, 2019. https://droughtmonitor.unl.edu.

US Environmental Protection Agency. 2001. Action Plan for Reducing, Mitigating, and Controlling Hypoxia in the Northern Gulf of Mexico. Mississippi River/Gulf of Mexico Watershed Nutrient Task Force. Washington, DC: Environmental Protection Agency. www.epa.gov/sites/production/files/2015-03/documents/2001_04_04_msbasin_actionplan2001.pdf.

US Environmental Protection Agency. 2016a. "Climate Change Indicators: Glaciers." www.epa.gov/climate-indicators/climate-change-indicators-glaciers.

US Environmental Protection Agency. 2016b. "Lead and Copper Rule Revisions White Paper." Washington, DC: US Environmental Protection Agency, Office of Water, October 2016. https://www.epa.gov/sites/default/files/2016-10/documents/508_lcr_revisions_white_paper_final_10.26.16.pdf.

US Environmental Protection Agency. 2017. "Find Your Local CCR." May 19, 2017. https://ofmpub.epa.gov/apex/safewater/f?p=ccr_wyl%3A102.

US Geological Survey. n.d. "Groundwater Use in the United States." Accessed October 1, 2019. www.usgs.gov/special-topic/water-science-school/science/groundwater-use-united-states?qt-science_center_objects=0#qt-science_center_objects.

WGMS (World Glacier Monitoring Service). 2015. Global Glacier Change Bulletin No. 1 (2012–2013), edited by M. Zemp, I. Gärtner-Roer, S. U. Nussbaumer, F. Hüsler, H. Machguth, N. Mölg, F. Paul, and M. Hoelzle. ICSU (WDS)/IUGG (IACS)/UNEP/UNESCO/WMO. Zurich: World Glacier Monitoring Service. https://wgms.ch/data_databaseversions/

WGMS (World Glacier Monitoring Service). 2016. Update to 2015 Global Glacier Change Bulletin, no. 1 (2012–2013), edited by M. Zemp, I. Gärtner-Roer, S. U. Nussbaumer, F. Hüsler, H. Machguth, N. Mölg, F. Paul, and M. Hoelzle. ICSU (WDS)/IUGG (IACS)/UNEP/UNESCO/WMO. Zurich: World Glacier Monitoring Service.

Williams, A. Park, E. R. Cook, J. E. Smerdon, B. I. Cook, J. T. Abatzoglou, K. Bolles, S. H. Baek, A. M. Badger, and B. Livneh. 2020. "Large Contribution from Anthropogenic Warming to an Emerging North American Megadrought." Science 368, no. 6488 (April 17, 2020): 314–18. https://doi.org/10.1126/science.aaz9600.

Williams, Justin, and Tara Jagadeesh. 2020. "Data Points: The Environmental Injustice of Lead Lines in Illinois." Metropolitan Planning Council (Chicago), November 10, 2020. https://www.metroplanning.org/news/9960/Data-Points-the-environmental-injustice-of-lead-lines-in-Illinois.

CHAPTER 8

American Lung Association. n.d. "Most Polluted Cities: State of the Air." Accessed October 1, 2019. www.lung.org/our-initiatives/healthy-air/sota/city-rankings/most-polluted-cities.html.

Brainard, Jean. 2012. "Atmospheric Pressure." CK12. https://www.ck12.org/c/physics/

Carle, David. 2006. Introduction to Air in California. Berkeley: University of California Press.

Colmer, Jonathan, Ian Hardman, Jay Shimshack, and John Voorheis. 2020. "Disparities in $PM_{2.5}$ Air Pollution in the United States." Science 369, no. 6503 (July 31, 2020): 575–78. https://doi.org/10.1126/science.aaz9353.

Congressional Research Service. 2021. "Vehicle Fuel Economy and Greenhouse Gas Standards." https://sgp.fas.org/crs/misc/IF10871.pdf

Diesel Technology Forum. n.d. "What Is Selective Catalytic Reduction?" Accessed October 1, 2019. www.dieselforum.org/about-clean-diesel/what-is-scr.

Dreyfuss, Emily. 2015. "Volkswagen Says Emissions Deception Actually Affects 11 Million Cars." Wired, September 22, 2015. www.wired.com/2015/09/vw-emissions-deception-11-million-cars/.

Kirby, Jen. 2017. "The Environmental Fallout of Hurricane Harvey." New York Magazine, September 1, 2017. http://nymag.com/intelligencer/2017/09/the-environmental-fallout-of-hurricane-harvey.html?gtm=top.

Kooser, Amanda. 2021. "Scientists and Satellites Watch as Air Pollution Climbs Back to Pre-COVID Levels." CNET, March 15, 2021. https://www.cnet.com/news/scientists-warn-air-pollution-is-climbing-back-to-pre-covid-levels/.

Lefohn, Allen S. 2022. "State of the Air 2022." Chicago: American Lung Association. https://www.lung.org/getmedia/74b3d3d3-88d1-4335-95d8-c4e47d0282c1/sota-2022.pdf

Lehmann, Christopher M. B., Van C. Bowersox, Robert S. Larson, and Susan M. Larson. 2007. "Monitoring Long-Term Trends in Sulfate and Ammonium in US Precipitation: Results from the National Atmospheric Deposition Program/National Trends Network." In Acid Rain—Deposition to Recovery, edited by P. Brimblecombe, H. Hara, D. Houle, and M. Novak, 59–66. Dordrecht, Netherlands: Springer.

Marshak, Stephen, and Robert Rauber. 2017. Earth Science, 1st ed. (See Figure 17.8; Figure Bx18.1.) New York: Norton.

Mass, Cliff. 2008. The Weather of the Pacific Northwest. Seattle: University of Washington Press.

New World Climate. 2017. "How to Read a Weather Map: Beginners Guide." February 14, 2017. www.nwclimate.org

Niiler, Eric. 2015. "VW Could Fool the EPA, but It Couldn't Trick Chemistry." Wired, September 22, 2015. www.wired.com/2015/09/vw-fool-epa-couldnt-trick-chemistry/.

Pierce, David. 2015. "'Ridiculous Ridge' May Be Back to Prolong California Drought [storm track graphic]." KQED. www.kqed.org/science/26293/ridiculous-ridge-may-be-back-to-prolong-california-drought.

Tessum, Christopher W., David A. Paolella, Sarah E. Chambliss, Joshua S. Apte, Jason D. Hill, and Julian D. Marshall. 2021. "$PM_{2.5}$ Polluters Disproportionately and Systemically Affect People of Color in the United States." Science Advances 7, no. 18 (April 28, 2021). https://doi.org/10.1126/sciadv.abf4491.

Trimble, David, and Frank Rusco. 2012. Air Emissions and Electricity Generation at U.S. Power Plants. Washington, DC: US Government Accountability Office.

U.S. Department of Energy - Alternative Fuels Data Center 2021. "Maps and Data – Vehicle Fuel Efficiency (CAFÉ) Requirements by Year." https://afdc.energy.gov/data/

US Energy Information Administration. 2012. "Fuel Economy Standards Have Affected Vehicle Efficiency." EIA.gov, Today in Energy, August 3, 2012. www.eia.gov/todayinenergy/detail.php?id=7390.

US Environmental Protection Agency. 2018a. "Air Quality—National Summary." www.epa.gov/air-trends/air-quality-national-summary.

US Environmental Protection Agency. 2018b. "Health and Environmental Effects of Particulate Matter (PM)." www.epa.gov/pm-pollution/health-and-environmental-effects-particulate-matter-pm.

US Environmental Protection Agenct. 2018c. "Particulate Matter (PM) Basics." www.epa.gov/pm-pollution/particulate-matter-pm-basics#PM.

US Environmental Protection Agency. 2020. "Our Nation's Air." https://gispub.epa.gov/air/trendsreport/2020/#home.

US Environmental Protection Agency. 2021. "Particulate Matter (PM2.5) Trends." June 1, 2021. https://www.epa.gov/air-trends/particulate-matter-pm25-trends.

World Health Organization. 2021. "Ambient (Outdoor) Air Pollution." September 22, 2021. https://www.who.int/news-room/fact-sheets/detail/ambient-(outdoor)-air-quality-and-health.

Yin, Peng, Michael Brauer, Aaron J. Cohen, Haidong Wang, Jie Li, Richard T. Burnett, Jeffrey D. Stanaway, et al. 2020. "The Effect of Air Pollution on Deaths, Disease Burden, and Life Expectancy across China and Its Provinces, 1990–2017: An Analysis for the Global Burden of Disease Study 2017." Lancet Planetary Health 4, no. 9 (August 17, 2020): e386–98. https://doi.org/10.1016/s2542-5196(20)30161-3.

CHAPTER 9

Bargout, Remy N., and Manish N. Raizada. 2013. "Soil Nutrient Management in Haiti, Pre-Columbus to the Present Day: Lessons for Future Agricultural Interventions." *Agriculture & Food Security* 2, no. 11 (July 29, 2013). https://doi.org/10.1186/2048-7010-2-11.

Bierman, Paul R., and David R. Montgomery. 2013. *Key Concepts in Geomorphology*. New York: W. H. Freeman.

Britannica. 1998. "Crop Rotation." www.britannica.com/topic/crop-rotation.

Brouwer, C., A. Goffeau, and M. Heibloem. 1986. "Chapter 7—Salty Soils." In *Irrigation Water Management: Training Manual No. 1—Introduction to Irrigation*. Rome: FAO. www.fao.org/3/r4082e/r4082e08.htm.

CK-12. n.d. "Avoiding Soil Loss." Accessed July 6, 2019. www.ck12.org/earth-science/avoiding-soil-loss/lesson/Avoiding-Soil-Loss-MS-ES/.

Claassen, Roger, Maria Bowman, Jonathan McFadden, David Smith, and Steven Wallander. 2018. *Tillage Intensity and Conservation Cropping in the United States*. EIB-197. US Department of Agriculture, Economic Research Service. https://www.ers.usda.gov/webdocs/publications/90201/eib-197.pdf?v=7027.1.

Feller, Christian, George G. Brown, Eric Blanchart, Pierre Deleporte, and Sergey S. Chernyanskii. 2003. "Charles Darwin, Earthworms and the Natural Sciences: Various Lessons from Past to Future." *Agriculture, Ecosystems & Environment* 99, no. 1-3 (October 2003): 29–49. .

Glaser, B., L. Haumaier, G. Guggenberger, and W. Zech. 2001. "The Terra Preta Phenomenon: A Model for Sustainable Agriculture in the Humid Tropics." *Naturwissenschaften* 88: 37–41.

Haxel, Gordon B., James B. Hedrick, and Greta J. Orris. 2005. *Rare Earth Elements—Critical Resources for High Technology*. US Geological Survey, Fact Sheet 087-02. https://pubs.usgs.gov/fs/2002/fs087-02/.

King, Hobart M. n.d. "REE—Rare Earth Elements and Their Uses." Geology.com. Accessed July 6, 2019. https://geology.com/articles/rare-earth-elements.

Marshak, Stephen. 2016. *Essentials of Geology*, 5th ed. New York: Norton.

Marshak, Stephen. 2018. *Earth: Portrait of a Planet*, 6th ed. New York: Norton.

Montgomery, David R. 2007. *Dirt: The Erosion of Civilizations*. Berkeley: University of California Press.

National Minerals Information Center. 2022. "Rare Earths Statistics and Information." US Geological Survey. https://www.usgs.gov/centers/nmic/rare-earths-statistics-and-information.

Natural Resources Conservation Service. n.d. "Natural Resources Inventory." Accessed October 1, 2019. www.nrcs.usda.gov/wps/portal/nrcs/main/national/technical/nra/nri/.

Natural Resources Conservation Service. n.d. "Soil Erosion on Cropland 2007." Accessed October 1, 2019. www.nrcs.usda.gov/wps/portal/nrcs/detail/national/technical/?cid=stelprdb1041887.

Philadelphia Water Department. n.d. "Stormwater Plan Review." Accessed July 6, 2019. www.pwdplanreview.org/manual/introduction.

Poole, Robert M. 2011. "What Became of the Taíno?" *Smithsonian Magazine*, October 2011. https://www.smithsonianmag.com/travel/what-became-of-the-taino-73824867/.

Qadir, Manzoor, Emmanuelle Quillérou, Vinay Nangia, Ghulam Murtaza, Murari Singh, Richard J. Thomas, Pay Drechsel, and Andrew D. Noble. 2014. "Economics of Salt-Induced Land Degradation and Restoration." *Natural Resources Forum* 38, no. 4 (October 28, 2014): 282–95.

US Department of Agriculture. 2015. *Summary Report: 2012 National Resources Inventory*. Washington, DC: Natural Resources Conservation Service; Ames: Center for Survey Statistics and Methodology, Iowa State University. www.nrcs.usda.gov/Internet/FSE_DOCUMENTS/nrcseprd396218.pdf.

US Department of Agriculture. 2016. *RCA Report, Erosion Rates on Cropland, 1982–2012, by Farm Region*. Washington, DC: Natural Resources Conservation Service.

US Forest Service. n.d. "Monarch Butterfly Migration and Overwintering." https://www.fs.usda.gov/wildflowers/pollinators/Monarch_Butterfly/migration/index.shtml.

US Government Accountability Office. 2014. *Overview of GAO's Past Work on the National Flood Insurance Program*. GAO-14-297R. Washington, DC: US GAO. www.gao.gov/assets/670/662438.pdf.

Van Gosen, Bradley S., Philip L. Verplanck, and Poul Emsbo. 2019. "Rare Earth Element Mineral Deposits in the United States." US Geological Survey, April 11, 2019. https://pubs.er.usgs.gov/publication/cir1454.

Williams, Megan Mansell. 2004. "Magnetic Maps Guide Migrating Turtles." ScienceMag.org, April 28, 2004. www.sciencemag.org/news/2004/04/magnetic-maps-guide-migrating-turtles.

Zielinski, Sarah. 2014. "Earth's Soil Is Getting Too Salty for Crops to Grow." Smithsonian, October 28, 2014. www.smithsonianmag.com/science-nature/earths-soil-getting-too-salty-crops-grow-180953163/.

CHAPTER 10

Belva, Keeley. 2017. "Gulf of Mexico 'Dead Zone' Is the Largest Ever Measured." National Oceanic and Atmospheric Administration, August 2, 2017. www.noaa.gov/media-release/gulf-of-mexico-dead-zone-is-largest-ever-measured.

Cavallaro, N., G. Shresta, R. Birdsey, M. A. Mayes, R. G. Najjar, S. C. Reed, P. Romero-Lankao, and Z. Zhu. 2018. "Second State of the Carbon Cycle Report (SOCCR2): A Sustained Assessment Report." US Global Change Research Program. https://doi.org/10.7930/soccr2.2018.

Chesapeake Bay Foundation. n.d. "Cost of Nitrogen Pollution Reduction by Sector and Practice (per Pound)." Accessed February 16, 2019. www.cbf.org/assets/images/old-site-body-content/sotb-2014-pollution-reduction.jpg.

Chesapeake Bay Foundation. 2012. "Nitrogen Pollution to the Chesapeake Bay by Sector." www.cbf.org/assets/images/graphs-graphics/pollution-pie-march2013.jpg.

Cowling, E., J. Galloway, C. Furiness, and J. W. Erisman, eds. 2002. *Optimizing Nitrogen Management in Food and Energy Production and Environmental Protection*. Report from the Second International Nitrogen Conference, Potomac, MD, October 14–18, 2001. Washington, DC: Ecological Society of America.

Ellen MacArthur Foundation. n.d. "Closing the Nutrient Loop." Accessed October 1, 2019. www.ellenmacarthurfoundation.org/case-studies/closing-the-nutrient-loop.

Fowler, David, Mhairi Coyle, Ute Skiba, Mark A. Sutton, J. Neil Cape, Stefan Reis, Lucy J. Sheppard, et al. 2013. "The Global Nitrogen Cycle in the Twenty-First Century." *Philosophical Transactions of the Royal Society B: Biological Sciences* 368, no. 1621 (July 5, 2013): 20130164.

Galloway, J. N., F. J. Dentener, D. G. Capogne, E. W. Boyer, R. W. Howarth, S. P. Seitzinger, G. P. Asner, C. Cleveland, P. Green, E. Holland, D. M. Karl, A. F. Michaels, J. H. Porter, A. R. Townsend, and C. Vörösmarty. 2004. "Nitrogen Cycles: Past, Present, and Future." *Biochemistry* 70 (September 2004): 153–226.

Global Carbon Project. 2020. "Global Carbon Budget." US Carbon Cycle Science Program. December 11, 2020. https://www.globalcarbonproject.org/carbonbudget/21/infographics.htm.

IPCC, 2022: Climate Change 2022: Mitigation of Climate Change. Contribution of Working Group III to the Sixth Assessment Report of the Intergovernmental Panel on Climate Change [P.R. Shukla, J. Skea, R. Slade, A. Al Khourdajie, R. van Diemen, D. McCollum, M. Pathak, S. Some, P. Vyas, R. Fradera, M. Belkacemi, A. Hasija, G. Lisboa, S. Luz, J. Malley, (eds.)]. Cambridge University Press, Cambridge, UK and New York, NY, USA. doi: 10.1017/9781009157926

IPCC, 2022: Summary for Policymakers. In: Climate Change 2022: Mitigation of Climate Change. Contribution of Working Group III to the Sixth Assessment Report of the Intergovernmental Panel on Climate Change [P.R. Shukla, J. Skea, R. Slade, A. Al Khourdajie, R. van Diemen, D. McCollum, M. Pathak, S. Some, P. Vyas, R. Fradera, M. Belkacemi, A. Hasija, G. Lisboa, S. Luz, J. Malley, (eds.)]. Cambridge University Press, Cambridge, UK and New York, NY, USA. doi: 10.1017/9781009157926.00

MacCallum, Taber, Jane Poynter, and David Bearden. 2004. "Lessons Learned from Biosphere 2: When Viewed as a Ground Simulation/Analog for Long Duration Human Space Exploration and Settlement." SAE Technical Paper 2004-01-2473. https://pdfs.semanticscholar.org/f99e/cc0dc23d873964c2baa4efdec477fa6f8599.pdf.

Meadows, Donella H. 2008. *Thinking in Systems: A Primer*. White River Junction, VT: Chelsea Green.

Nielsen, R. 2005. "Can We Feed the World? The Nitrogen Limit of Food Production." http://home.iprimus.com.au/nielsens/nitrogen.html.

Nitrogen Footprint. n.d. "Your Nitrogen Footprint." Accessed October 1, 2019. www.n-print.org/YourNFootprint.

NOAA. 2021. "Larger-than-average Gulf of Mexico 'dead zone' measured." https://www.noaa.gov/news-release/larger-than-average-gulf-of-mexico-dead-zone-measured

Northwest Earth Institute. n.d. "A Systems Thinking Model: The Iceberg." Accessed February 16, 2019. www.nwei.org/iceberg/.

Patoway, Kaushik. 2015. "Nauru: An Island Country Destroyed by Phosphate Mining." Amusing Planet, June 13, 2015. www.amusingplanet.com/2015/06/nauru-island-country-destroyed-by.html.

Rosen, Julia. 2021. "Humanity Is Flushing Away One of Life's Essential Elements." *Atlantic*, February 8, 2021. https://www.theatlantic.com/science/archive/2021/02/phosphorus-pollution-fertilizer/617937/.

Slagter, Martin. 2017. "University of Michigan Adds 'Pee-Cycling' to Recycling Efforts." Mlive.com, February 21, 2017. www.mlive.com/news/ann-arbor/2017/02/researchers_hoping_pee-cycling.html.

Shibata, Hideaki, James N. Galloway, Allison M. Leach, Lia R. Cattaneo, Laura Cattell Noll, Jan Willem Erisman, Baojing Gu, et al. 2017. "Nitrogen Footprints: Regional Realities and Options to Reduce Nitrogen Loss to the Environment." *Ambio* 46, no. 2 (March 2017): 129–42.

"Tool G: Systems Thinking & the Iceberg Model." 2020. Seattle: Just Lead Washington.

United Nations Environment Programme (UNEP) and GRID-Arendal. 2005. "Vital Climate Change Graphics Update." Accessed December 3, 2019. www.grida.no/resources/6903.

US Carbon Cycle Science Program. 2019. "What Is the Carbon Cycle? What Is the Science behind It?" July 1, 2019. https://www.carboncyclescience.us/what-is-carbon-cycle#CarbonCycle.

Volk, Tyler. 1998. *Gaia's Body: Toward a Physiology of Earth.* Cambridge, MA: MIT Press.

CHAPTER 11

Briggs, Helen. 2021. "Carbon: How Calls for Climate Justice Are Shaking the World." BBC News, May 3, 2021. https://www.bbc.com/news/science-environment-56941979.

British Antarctic Survey. 2018. "Role of Oceanic Forcing in West Antarctic Ice Sheet Retreat." www.bas.ac.uk/project/role-of-oceanic-forcing-in-west-antarctic-ice-sheet-retreat/.

Capucci, Matthew. 2021. "Flash Flood Emergencies in Louisiana While Double-Digit Rainfall Deluges Texas." *Washington Post,* May 18, 2021. https://www.washingtonpost.com/weather/2021/05/18/flash-flooding-louisiana-texas/.

Church, John A. 2007. "A Change in Circulation?" *Science* 317, no. 5840 (August 17, 2007): 908–9.

Climate Central. n.d. "Surging Seas: 2 °C Warming and Sea Level Rise." Accessed November 11, 2019. https://ss6m.climatecentral.org/#5/10.898/-66.401.

Climate Justice Alliance. 2022. "About Climate Justice Alliance." Accessed March 21, 2022. https://climatejusticealliance.org/about/.

Climate Science Investigations. 2016. "Global Wind Patterns." www.ces.fau.edu/nasa/content/resources/global-wind-patterns.php.

Fernholz, Tim. 2021. "This Is How the Federal Reserve Will Confront Climate Change." *Quartz,* November 2, 2021. https://qz.com/2069774/this-is-how-the-federal-reserve-will-confront-climate-change/.

Fields, Samantha. 2021. "California Orders Insurers Not to Drop Homeowners in Wildfire-Prone Areas." Marketplace, September 21, 2021. https://www.marketplace.org/2021/09/21/california-orders-insurers-not-to-drop-homeowners-in-wildfire-prone-areas/.

Ganthier, Tiffany, Lisa Anne Hamilton, Annie Bennett, Katherine McCormick, Anne Perrault, Joel Smith, Sara Hoverter, Jennifer Li, and Jessica Grannis. 2020. "Equitable Adaptation Legal & Policy Toolkit." Georgetown Climate Center. https://www.georgetownclimate.org/adaptation/toolkits/equitable-adaptation-toolkit/introduction.html?full.

Gleck, P. H. 1996. *Encyclopedia of Climate and Weather.* New York: Oxford University Press.

Hayhoe, Katharine, Donald J. Wuebbles, David R. Easterling, David W. Fahey, Sarah Doherty, James P. Kossin, William V. Sweet, Russell S. Vose, and Michael F. Wehner. 2018. "Chapter 2: Our Changing Climate." In *The Fourth National Climate Assessment,* vol. 2. Edited by D. R. Reidmiller, C. W. Avery, D. R. Easterling, K. E. Kunkel, K. L.M. Lewis, T. K. Maycock, and B. C. Stewart. Washington, DC: US Global Change Research Program. https://doi.org/10.7930/nca4.2018.ch2.

Insurance Information Institute. n.d. "Facts + Statistics: Hurricanes." www.iii.org/fact-statistic/facts-statistics-hurricanes#Top 10 Costliest Hurricanes In The United States (1).

Intergovernmental Panel on Climate Change. 2007. "TS.3.1.1 Global Average Temperatures" (AR4 WGI Technical Summary). www.ipcc.ch/publications_and_data/ar4/wg1/en/tssts-3-1-1.html.

IPCC, 2022: Climate Change 2022: Impacts, Adaptation, and Vulnerability. Contribution of Working Group II to the Sixth Assessment Report of the Intergovernmental Panel on Climate Change [H.-O. Pörtner, D.C. Roberts, M. Tignor, E.S. Poloczanska, K. Mintenbeck, A. Alegría, M. Craig, S. Langsdorf, S. Löschke, V. Möller, A. Okem, B. Rama (eds.)]. Cambridge University Press. Cambridge University Press, Cambridge, UK and New York, NY, USA, 3056 pp., doi:10.1017/9781009325844.

Jackson, Rob. 2020. "Understanding and Predicting Wetland Methane Emissions." John Wesley Powell Center for Analysis and Synthesis, Fort Collins, CO, February 27 2020. https://www.usgs.gov/news/understanding-and-predicting-wetland-methane-emissions?qt-news_science_products=7#qt-news_science_products.

Jay, A., D. R. Reidmiller, C. W. Avery, Dan Barrie, B. J. DeAngelo, A. Dave, M. Dzaugis, et al. 2018. "Chapter 1: Overview." In *The Fourth National Climate Assessment,* vol. 2. Edited by D. R. Reidmiller, C. W. Avery, D. R. Easterling, K. E. Kunkel, K. L.M. Lewis, T. K. Maycock, and B. C. Stewart. Washington, DC: US Global Change Research Program. https://doi.org/10.7930/nca4.2018.ch1.

Johnson, Raymond N. 2010. "Carbon Dioxide: The 800-Pound Gorilla That We Have to Talk About." Institute of Climate Studies, USA, November 2010. www.icsusa.org/pages/articles/2010-icsusa-articles/november-2010—-carbon-dioxide-the-800-pound-gorilla-that-we-have-to-talk-about.php#.XRzF9i2ZOi4.

J. J. Kennedy, P. W. Thorne, T. C. Peterson, R. A. Ruedy, P. A. Stott, D. E. Parker, S. A. Good, H. A. Titchner, and K. M. Willett, 2010: [in "State of the Climate in 2009"]. Bull. Amer. Meteor. Soc., 91 (7), S79-106.

Lindsey, Rebecca. 2018. "Climate Change: Atmospheric Carbon Dioxide" [graphic]. Climate.gov, June 23, 2022. www.climate.gov/news-features/understanding-climate/climate-change-atmospheric-carbon-dioxide.

Lüthi, Dieter, Martine Le Floch, Bernhard Bereiter, Thomas Blunier, Jean-Marc Barnola, Urs Siegenthaler, Dominique Raynaud, Jean Jouzel, Hubertus Fischer, Kenji Kawamura, and Thomas F. Stocker. 2008. "High-Resolution Carbon Dioxide Concentration Record 650,000–800,000 Years before Present." *Nature* 453 (May 15, 2008): 379–82.

Mandel, Kyla. 2017. "America's Climate Refugees Have Been Abandoned by Trump." *Mother Jones,* October 17, 2017. www.motherjones.com/environment/2017/10/climate-refugees-trump-hud/.

Maldonado, Samantha. 2021. "Flood Insurance Hikes Haunt Homeowners Still Recovering from Ida and Henri." *The City,* October 12, 2021. https://www.thecity.nyc/2021/10/12/22723591/flood-insurance-hikes-haunt-homeowners.

Marshak, Stephen, and Robert Rauber. 2017. *Earth Science,* 1st ed. New York: Norton.

Melillo, Jerry M., Terese (T. C.) Richmond, and Gary W. Yohe, eds. 2014. *Climate Change Impacts in the United States: The Third National Climate Assessment.* Washington, DC: US Global Change Research Program. doi: 10.7930/J0Z31WJ2.

Menezes, Viviane V., Alison M. Macdonald, and Courtney Schatzman. 2017. "Accelerated Freshening of Antarctic Bottom Water over the Last Decade in the Southern Indian Ocean." *Science Advances* 3, no. 1 (January 25, 2017): e1601426.

Miettinen, Dylan. 2021. "The Climate Crisis Is Here: Are Insurance Companies Keeping Up?" Marketplace, August 6, 2021. https://www.marketplace.org/2021/08/06/the-climate-crisis-is-here-are-insurance-companies-keeping-up/.

Morice, C. P., J. J. Kennedy, N. A. Rayner, and P. D. Jones. 2012. "Quantifying Uncertainties in Global and Regional Temperature Change Using an Ensemble of Observational Estimates: The HadCRUT4 Dataset." *Journal of Geophysical Research* 117 (April 17, 2012), D08101.

NASA. 2009. "The Thermohaline Circulation—the Great Ocean Conveyor Belt" [video]. Scientific Visualization Studio, October 8, 2009. http://svs.gsfc.nasa.gov/vis/a000000/a003600/a003658/.

National Centers for Environmental Information (NCEI). n.d. "Paleoclimatology Data Map: GIS Maps." Accessed October 1, 2019. https://gis.ncdc.noaa.gov/maps/ncei/paleo?layers=1.

National Climatic Data Center, n.d. "Climate at a Glance." https://www.ncdc.noaa.gov/cag/.

National Climate Assessment. 2014a. "Climate Science Supplement." https://nca2014.globalchange.gov/report/appendices/climate-science-supplement#intro-section-2.

National Climate Assessment. 2014b. "Future Climate Change." https://nca2014.globalchange.gov/report/our-changing-climate/future-climate-change.

National Climate Assessment. 2014c. "Human Health." https://nca2014.globalchange.gov/report/sectors/human-health#tab2-images.

Nielsen_Gammon, John W., Jay L. Banner, Benjamin I. Cook, Darrel M. Tremaine, Corinne I. Wong, Robert E. Mace, Huilin Gao, et al. 2020. "Unprecedented Drought Challenges for Texas Water Resources in a Changing Climate: What Do Researchers and Stakeholders Need to Know?" *Earth's Future* 8, no. 8 (June 29, 2020). https://doi.org/10.1029/2020ef001552.

NOAA National Centers for Environmental Information, State of the Climate: Monthly Global Climate Report for August 2022, published online September 2022, https://www.ncei.noaa.gov/access/monitoring/monthly-report/global/202208.

O Ecotextiles. 2019. "Textile Industry and Climate Change." https://oecotextiles.wordpress.com/category/embodied-energy-in-textiles/.

Porter, Eduardo. 2013. "For Insurers, No Doubts on Climate Change." *New York Times,* May 14, 2013. www.nytimes.com/2013/05/15/business/insurers-stray-from-the-conservative-line-on-climate-change.html?_r=1&.

Riebeek, Holly. 2010. "Global Warming." Earth Observatory, NASA, June 3, 2010. https://earthobservatory.nasa.gov/features/GlobalWarming/page3.php.

Rohde, Robert. 2021. Global Temperature Report for 2020. Berkeley Earth, January 14, 2021. http://berkeleyearth.org/global-temperature-report-for-2020/.

Rosoff, Stephanie, and Jessica Yager. 2017. "Data Brief: Housing in the U.S. Floodplains." New York: NYU Furman Center, May 2017.

Scarborough, Peter, Paul N. Appleby, Anja Mizdrak, Adam D. M. Briggs, Ruth C. Travis, Kathryn E. Bradbury, and Timothy J. Key. 2014. "Dietary Greenhouse Gas

Emissions of Meat-Eaters, Fish-Eaters, Vegetarians and Vegans in the UK." *Climatic Change* 125, no. 2 (June 11, 2014): 179–92.

Scism, Leslie. 2021. "Ida Storm Damage Expected to Cost Insurers at Least $31 Billion." *Wall Street Journal*, September 22, 2021. https://www.wsj.com/articles/ida-storm-damage-expected-to-cost-insurers-atleast-31-billion-11632303002.

Shrinkthatfootprint.com. n.d. "The Carbon Footprints of Products." Accessed October 13, 2018. shrinkthatfootprint.com/wp-content/uploads/2009/08/productLCA.gif [with pcf-project, and mistral data sources].

Simmons, Daisy. 2020. "What Is 'Climate Justice'?" Yale Climate Connections, July 29, 2020. https://yaleclimateconnections.org/2020/07/what-is-climate-justice/.

Stein, Theo. 2021. "Carbon Dioxide Peaks Near 420 Parts per Million at Mauna Loa Observatory." NOAA Research News, June 7, 2021. https://research.noaa.gov/article/ArtMID/587/ArticleID/2764/Coronavirus-response-barely-slows-rising-carbon-dioxide.

Strauss, Ben, Claudia Tebaldi, and Remik Ziemlinski. 2012. "Sea Level Rise, Storms, and Global Warming's Threat to the US Coast." Climate Central.

US Department of Commerce, National Oceanic and Atmospheric Administration. "What Is a Gyre?" NOAA's National Ocean Service, March 12, 2018. https://oceanservice.noaa.gov/facts/gyre.html.

US Department of Energy. n.d. "Gasoline Vehicles: Learn More about the Label." Fueleconomy.gov. Accessed October 21, 2019. www.fueleconomy.gov/feg/label/learn-more-gasoline-label.shtml.

US Department of Energy. 2021. "Buying Clean Electricity." Energy.gov. https://www.energy.gov/energysaver/buying-clean-electricity.

US Geological Survey. n.d. "Coral Reef Ecosystem Studies (CREST)." Accessed October 1, 2019. www.usgs.gov/centers/spcmsc/science/coral-reef-ecosystem-studies-crest?qt-science_center_objects=0#qt-science_center_objects.

US Global Change Research Program/Climate Change Science Program. 2009. *Climate Literacy: The Essential Principles of Climate Sciences.* Washington, DC: US Global Change Research Program/Climate Change Science Program.

US Global Change Research Program. n.d. "Sea Level Rise." GlobalChange.gov. Accessed June 17, 2022. https://www.globalchange.gov/browse/indicators/global-sea-level-rise.

Vergun, David. 2021. "Action Team Leads DOD Efforts to Adapt to Climate Change Effects." US Department of Defense, April 22, 2021. https://www.defense.gov/Explore/News/Article/Article/2577354/action-team-leads-dod-efforts-to-adapt-to-climate-change-effects/.

CHAPTER 12

Bastasch, Michael. 2012. "UN: More Obese in the World Than Suffering from Hunger." *Daily Caller*, August 10, 2012. https://dailycaller.com/2012/08/10/un-more-obese-in-the-world-than-suffering-from-hunger/.

Beam, Christopher. 2009. "Which Animal Did We Domesticate First?" *Slate*, March 6, 2009. https://slate.com/news-and-politics/2009/03/which-animal-did-we-domesticate-first.html.

Bishop, Sanford D. Jr., Nathaniel L. Tablante, Michael Reed, Namrata Kollad, and Maxwell Gigle. 2015. "Georgia's

Poultry Industry and Its Impact on the Local Economy and Global Trade." *International Food and Agribusiness Management Review* 18, Special Issue A (January 2015): 91–98. https://bishop.house.gov/media-center/op-eds/georgia-s-poultry-industry-and-its-impact-on-the-local-economy-and-global-trade.

Blagg, Kristin, Diane Whitmore-Schanzenbach, Craig Gundersen, and James Ziliak. 2017. "Assessing Food Insecurity on Campus." Urban Institute. www.urban.org/research/publication/assessing-food-insecurity-campus.

Brown, Lester R. 2012. *Full Planet, Empty Plates: The New Geopolitics of Food Scarcity.* New York: Norton.

Centers for Disease Control and Prevention. 2011. "NIOSH Pesticide Poisoning Monitoring Program Protects Farmworkers." www.cdc.gov/niosh/docs/2012-108/.

Centers for Disease Control and Prevention. 2013. *Antibiotic Resistance Threats in the United States, 2013*, p. 14. www.cdc.gov/drugresistance/threat-report-2013/pdf/ar-threats-2013-508.pdf.

Clift, Joseph. 2013. "Portion Distortion Report 2013." British Heart Foundation. www.bhf.org.uk/informationsupport/publications/policy-documents/portion-distortion-report-2013.

"Disease Threatens Florida Citrus Industry." 2013. *Tampa Bay Times*, May 11, 2013. www.tampabay.com/news/business/economicdevelopment/disease-threatens-florida-citrus-industry/2120337.

Dunham, Will. 2015. "Dog Domestication Much Older Than Previously Known." *Scientific American*, May 21, 2015. www.scientificamerican.com/article/dog-domestication-much-older-than-previously-known/.

Earthseed Land Collective. 2020. "Our Collective." September 14, 2020. https://earthseedlandcoop.org/about/.

Ebner, Paul. 2007. "CAFOs and Public Health: Pathogens and Manure." Concentrated Animal Feeding Operations ID-356-W. Purdue Extension, Purdue University. www.extension.purdue.edu/extmedia/id/cafo/id-356.pdf.

Economic Research Service. n.d. "Key Statistics & Graphics." US Department of Agriculture. https://www.ers.usda.gov/topics/food-nutrition-assistance/food-security-in-the-us/key-statistics-graphics.aspx.

Economic Research Service. 2022. "Cash Receipts by Commodity State Ranking." US Department of Agriculture. https://data.ers.usda.gov/reports.aspx?ID=17844.

Food and Agriculture Organization. 2003. *World Agriculture: Towards 2015/2030—An FAO Perspective.* Edited by Jelle Bruinsma. London: Earthscan. www.fao.org/3/y4252e/y4252e05b.htm.

Food and Agriculture Organization of the United Nations (FAO) (2020). http://www.fao.org/faostat/en/#data And Hannah Ritchie and Max Roser (2017) – "Meat and Dairy Production". Published online at OurWorldInData.org. Retrieved from: 'https://ourworldindata.org/meat-production' [Online Resource]

Gardner, Gary T., and Brian Halweil. 2000. *Underfed and Overfed: The Global Epidemic of Malnutrition.* Washington, DC: Worldwatch Institute.

Gerber, Pierre J., Benjamin Henderson, and Harinder P. S. Makkar. 2013. *Mitigation of Greenhouse Gas Emissions in Livestock Production: A Review of Technical Options for Non-CO2 Emissions.* No. 177. Rome: Food and Agriculture Organization of the United Nations.

Gerber, P. J., H. Steinfeld, B. Henderson, A. Mottet, C. Opio, J. Dijkman, A. Falcucci, and G. Tempio. 2013. *Tackling Climate Change through Livestock—A Global Assessment of Emissions and Mitigation Opportunities*, p. 16. Rome: Food and Agriculture Organization of the United Nations. www.fao.org/3/a-i3437e.pdf.

Golmohammadi, Golmar, Ramesh Rudra, Shiv Prasher, Ali Madani, Mohamed Youssef, Pradeep Goel, and Kourosh Mohammadi. 2017. "Impact of Tile Drainage on Water Budget and Spatial Distribution of Sediment Generating Areas in an Agricultural Watershed." *Agricultural Water Management* 184 (April): 124–34. https://doi.org/10.1016/j.agwat.2017.02.001.

Hill, Holly. 2008. "Food Miles: Background and Marketing." Butte, MT: NCAT, ATTRA Sustainable Agriculture. https://attra.ncat.org/product/food-miles-background-and-marketing/.

History World. n.d. "History of the Domestication of Animals." Accessed October 2, 2019. www.historyworld.net/wrldhis/plaintexthistories.asp?historyid=ab57.

Hope Center for College, Community, and Justice. 2021. "#RealCollege 2021: Basic Needs Insecurity During the Ongoing Pandemic." Philadelphia.

Klein, Ezra, and Susannah Locke. 2014. "40 Maps That Explain Food in America." *Vox*, June 9, 2014. https://www.vox.com/a/explain-food-america.

Kuo, Lily. 2013. "By 2015, China Will Be the World's Largest Consumer of Processed Food." *Quartz*, September 23, 2013. https://qz.com/127235/by-2015-china-will-be-the-worlds-largest-consumer-of-processed-food/.

Lakhani, Nina, Aliya Uteuova, and Alvin Chang. 2021. "Revealed: The True Extent of America's Food Monopolies, and Who Pays the Price." *Guardian*, July 14, 2021. https://www.theguardian.com/environment/ng-interactive/2021/jul/14/food-monopoly-meals-profits-data-investigation.

Leopold Center for Sustainable Agriculture. 2001. "How Far Do Your Fruits and Vegetables Travel?" (Leopold Letter), Iowa State University, Ames, April 1, 2001. https://dr.lib.iastate.edu/entities/publication/e2f64054-b5cd-4a2f-b52a-6d8b8c93d2a1.

Lipinski, Brian, Craig Hanson, James Lomax, Lisa Kitinoja, Richard Waite, and Tim Searchinger. 2013. "Reducing Food Loss and Waste." Working paper, installment 2 of Creating a Sustainable Food Future. Washington, DC: World Resources Institute. https://www.wri.org/research/reducing-food-loss-and-waste.

Lowe, Marcy, and Gary Gereffi. 2009. *A Value Chain Analysis of the U.S. Beef and Dairy Industries.* Report prepared for Environmental Defense Fund. Durham, NC: Center on Globalization, Governance & Competitiveness, Duke University. https://gvcc.duke.edu/cggclisting/a-value-chain-analysis-of-the-u-s-beef-dairy-industries/.

McGuire, Virginia L. 2011. *Water-Level Changes in the High Plains Aquifer, Predevelopment to 2009, 2007–08, and 2008–09, and Change in Water in Storage, Predevelopment to 2009.* US Geological Survey Scientific Investigations Report 2011-5089. https://pubs.er.usgs.gov/publication/sir20115089.

McGuire, Virginia L. 2017. *Water-Level and Recoverable Water in Storage Changes, High Plains Aquifer, Predevelopment to 2015 and 2013–15.* US Geological Survey Scientific Investigations Report 2017-5040. https://pubs.usgs.gov/sir/2017/5040/sir20175040.pdf.

Meador, Ron. 2017. "Agriculture Is Depleting World Aquifers, New Satellite Measurements Show." *MinnPost*, April 7, 2017. www.minnpost.com/earth-journal/2017/04/agriculture-depleting-world-aquifers-new-satellite-measurements-show/.

Minnesota Pollution Control Agency. 2021. "PFAS 101." March 1, 2021. https://www.pca.state.mn.us/waste/pfas-101.

Montgomery, David R. 2017. *Growing a Revolution: Bringing Our Soil Back to Life*. New York: Norton.

Montgomery, D. R., and A. Biklé. 2016. *The Hidden Half of Nature: The Microbial Roots of Life and Health*. New York: Norton.

National Heart, Lung, and Blood Institute. n.d. "Portion Distortion." Accessed October 2, 2019. www.nhlbi.nih.gov/health/educational/wecan/eat-right/portion-distortion.htm.

National Marine Fisheries Service. 2013. *U.S. National Bycatch Report,* 1st ed., Update 1. Edited by L. R. Benaka, C. Rilling, E. E. Seney, and H. Winarsoo. US Department of Commerce.

Nicole, Wendee. 2013. "Cafos and Environmental Justice: The Case of North Carolina." *Environmental Health Perspectives* 121, no. 6 (June 1, 2013). https://doi.org/10.1289/ehp.121-a182.

Organic Trade Association. 2021. "U.S. organic sales soar to new high of nearly $62 billion in 2020." https://ota.com/news/press-releases/21755.

Patel, Raj. 2012. *Stuffed and Starved: The Hidden Battle for the World Food System*. Brooklyn, NY: Melville House.

Pimentel, David, and Marcia H. Pimentel. 2008. *Food, Energy, and Society*, 3rd ed. Boca Raton, FL: CRC Press.

Powell, Chelsea, and Anna Maurer [illustrator]. 2015. "How to Make a GMO." Science in the News, Harvard University, August 9, 2015. http://sitn.hms.harvard.edu/flash/2015/how-to-make-a-gmo/.

RAFI (Rural Advancement Foundation International). 2022. "Farmers of Color Network." May 19, 2022. https://www.rafiusa.org/programs/farmers-of-color-network/.

Ravishankara, A. R., Johan C. I. Kuylenstierna, Eleni Michalopoulou, Lena Höglund-Isaksson, Yuqiang Zhang, Karl Seltzer, Muye Ru, et al. 2021. *Global Methane Assessment: Benefits and Costs of Mitigating Methane Emissions*. UN Environment Programme, May 6, 2021. https://www.unep.org/resources/report/global-methane-assessment-benefits-and-costs-mitigating-methane-emissions.

Redman, Russell. 2020. "Online Grocery to More Than Double Market Share by 2025." Supermarket News, September 18, 2020. https://www.supermarketnews.com/online-retail/online-grocery-more-double-market-share-2025.

Ritchie, Hannah, and Max Roser. 2017. "Meat and Dairy Production." Our World in Data, August 25, 2017. https://ourworldindata.org/meat-production#citation.

Ritchie, Hannah, and Max Roser. 2019. "Meat and Seafood Production & Consumption." Our World in Data, last updated November 2019. https://ourworldindata.org/meat-and-seafood-production-consumption.

Ron Finley Project. 2022. https://ronfinley.com/.

Rural Advancement Foundation International. n.d. "Seeds and Breeds for 21st Century Agriculture." Accessed February 27, 2019. https://rafiusa.org/issues/seeds/.

Saksena, Michelle J., Abigail M. Okrent, Tobenna D. Anekwe, Clare Cho, Christopher Dicken, Anne Effland, Howard Elitzak, Joanne Guthrie, Karen S. Hamrick, Jeffrey Hyman, Young Jo, Biing-Hwan Lin, Lisa Mancino, Patrick W. McLaughlin, Ilya Rahkovsky, Katherine Ralston, Travis A. Smith, Hayden Stewart, Jessica Todd, and Charlotte Tuttle. 2018. *America's Eating Habits: Food away from Home*. Economic Information Bulletin, no. 196. US Department of Agriculture, Economic Research Service. www.ers.usda.gov/webdocs/publications/90228/eib-196.pdf?v=1045.6.

Shahbandeh, M. 2022. "Organic Food Sales in the U.S. 2021." Statista, February 2022. https://www.statista.com/statistics/196952/organic-food-sales-in-the-us-since-2000/.

Shih, Nicole. 2021. "Worcester Farmers Markets Offer Fresh Produce for All Budgets, Target Food Insecurity." *Worcester Telegram & Gazette*, October 9, 2021. https://www.telegram.com/story/news/2021/10/08/worcester-farmers-markets-food-insecurity-snap-assistance-programs/6008274001/.

Silberner, Joanne. 2008. "A Not-So-Sweet Lesson from Brazil's Cocoa Farms." NPR, June 14, 2008. www.npr.org/templates/story/story.php?storyId=91479835.

Silva, Christianna. 2020. "Food Insecurity in the U.S. by the Numbers." NPR, September 27, 2020. https://www.npr.org/2020/09/27/912486921/food-insecurity-in-the-u-s-by-the-numbers.

Steward, Karen. 2020. "Harmful PFAS Only Come from Old Food Packaging, Right? Wrong!" Technology Networks, Applied Sciences, April 15, 2020. https://www.technologynetworks.com/applied-sciences/articles/harmful-pfas-only-come-from-old-food-packaging-right-wrong-333426.

Tosi, Simone, James C. Nieh, Fabio Sgolastra, Riccardo Cabbri, and Piotr Medrzycki. 2017. "Neonicotinoid Pesticides and Nutritional Stress Synergistically Reduce Survival in Honey Bees." *Proceedings of the Royal Society B: Biological Sciences* 284, no. 1869 (December 20, 2017).

United Nations Environment Programme. 2012. "A Glass Half Empty: Regions at Risk Due to Groundwater Depletion." Reproduced from United Nations Environment Programme Global Environmental Alert Service. *Environmental Development* 2 (January 2012): 117–27.

United Nations Environment Programme and Climate and Clean Air Coalition (2021). Global Methane Assessment: Benefits and Costs of Mitigating Methane Emissions. Nairobi: United Nations Environment Programme.

University of Minnesota Extension. 2018. "The Ruminant Digestive System." https://extension.umn.edu/dairy-nutrition/ruminant-digestive-system.

US Department of Agriculture. 2019. "Food Consumption & Demand." Economic Research Service. www.ers.usda.gov/topics/food-choices-health/food-consumption-demand/.

US Department of Agriculture. 2020. "Packers and Stockyards Division: Annual Report 2019." USDA Agricultural Marketing Service.

US Environmental Protection Agency. 2022. "Overview of Greenhouse Gases." EPA.gov, May 16, 2022. https://www.epa.gov/ghgemissions/overview-greenhouse-gases#CH4-reference.

US Geological Survey. 2017. "USGS: High Plains Aquifer Groundwater Levels Continue to Decline." June 16, 2017. www.usgs.gov/news/usgs-high-plains-aquifer-groundwater-levels-continue-decline.

CHAPTER 13

American Lung Association. 2022. "State of the Air: Key Findings." https://www.lung.org/research/sota/key-findings.

American Petroleum Institute. 2021. "Where Are the Pipelines?" https://www.api.org/oil-and-natural-gas/wells-to-consumer/transporting-oil-natural-gas/pipeline/where-are-the-pipelines.

Bowen, Lauren. 2017. "10 Products You Won't Believe Are Derived from Petroleum." Care2.

BP. 2021. *Statistical Review of World Energy*, 70th ed. British Petroleum. https://www.bp.com/content/dam/bp/business-sites/en/global/corporate/pdfs/energy-economics/statistical-review/bp-stats-review-2021-full-report.pdf.

Central Intelligence Agency. 2021. *The World Factbook*. www.cia.gov/library/publications/resources/the-world-factbook/.

Comstock, Owen. 2021. "Nonfossil Fuel Sources Accounted for 21% of U.S. Energy Consumption in 2020." US Energy Information Administration, July 1, 2021. https://www.eia.gov/todayinenergy/detail.php?id=48576.

Con Edison. n.d. "How to Choose Your Energy Supplier." Accessed October 2, 2019. www.coned.com/en/save-money/shop-for-energy-service-companies/how-to-choose-your-energy-supplier.

Cumming, Daniel G. 2018. "Black Gold, White Power: Mapping Oil, Real Estate, and Racial Segregation in the Los Angeles Basin, 1900–1939." *Engaging Science, Technology, and Society* 4 (March): 85–110. https://doi.org/10.17351/ests2018.212.

Cusick, Marie, Reid Frazier, and Susan Phillips. 2016. "Man Injured after Pipeline Explodes Near His Home in Western Pa.." NPR, State Impact Pennsylvania, April 29, 2016. https://stateimpact.npr.org/pennsylvania/2016/04/29/1-injured-after-gas-pipeline-explosion-in-western-pa/.

Czolowski, Eliza D., Renee L. Santoro, Tanja Srebotnjak, and Seth B. C. Shonkoff. 2017. "Toward Consistent Methodology to Quantify Populations in Proximity to Oil and Gas Development: A National Spatial Analysis and Review." *Environmental Health Perspectives* 125, no. 8 (August 23, 2017). https://doi.org/10.1289/ehp1535.

Dewan, Shaila. 2008. "Coal Ash Spill Revives Issue of Its Hazards." *New York Times*, December 25, 2008. www.nytimes.com/2008/12/25/us/25sludge.html?mtrref=undefined&gwh=78A505F23C4AA90838D9141F3E9E705F&gwt=pay.

Earthworks. 2017. "Frac Sand Mining." December 11, 2017. https://earthworks.org/issues/frac_sand_mining/.

Editors of *Electric Light & Power*. 2017. "Mitsubishi Completes CO_2 Capture System for Petra Nova Carbon Capture Project." UtilityProducts.com, January 11, 2017. https://www.utilityproducts.com/line-construction-maintenance/article/16014888/mitsubishi-completes-co2-capture-system-for-petra-nova-carbon-capture-project.

Eilperin, Juliet, and Steven Mufson. 2015. "Royal Dutch Shell Suspends Arctic Drilling Indefinitely." *Washington Post*, September 28, 2015. www.washingtonpost.com/news/energy-environment/wp/2015/09/28/royal-dutch-shell-suspends-arctic-drilling-indefinitely/?utm_term=.9245c30dced9.

Emanuel, Ryan E., Martina A. Caretta, Louie Rivers, and Pavithra Vasudevan. 2021. "Natural Gas Gathering and Transmission Pipelines and Social Vulnerability in the United States." GeoHealth 5, no. 6 (May 18, 2021). https://doi.org/10.1029/2021gh000442.

Feldstein, Martin S., Ted Halstead, and N. Gregory Mankiw. 2017. "A Conservative Case for Climate Action." *New York Times*, February 8, 2017. www.nytimes.com/2017/02/08/opinion/a-conservative-case-for-climate-action.html.

Gertner, Jon. 2013. "George Mitchell, Father of Fracking." *New York Times*, December 21, 2013. www.nytimes.com/news/the-lives-they-lived/2013/12/21/george-mitchell/.

Gillis, Justin. "The Era of the Oil Gusher." *New York Times*, June 21, 2010. https://green.blogs.nytimes.com/2010/06/21/the-era-of-the-oil-gusher/?_r=0&mtrref=undefined&gwh=E1F9791C248CD357A9097138BCD37B4C&gwt=pay.

Goodman, Jasper G., and Kelsey J. Griffin. 2021. "Harvard Will Move to Divest Its Endowment from Fossil Fuels." *Harvard Crimson*, September 10, 2021. https://www.thecrimson.com/article/2021/9/10/divest-declares-victory/.

Havlin, Laura. 2017. "Portraits of Protest: The Women of Standing Rock." Magnum Photos, June 26, 2017. https://www.magnumphotos.com/newsroom/environment/alessandra-sanguinetti-women-anti-dakota-acess-pipeline-demonstrations/.

Hepburn, Cameron, Ella Adlen, John Beddington, Emily A. Carter, Sabine Fuss, Niall Mac Dowell, Jan C. Minx, Pete Smith, and Charlotte K. Williams. 2019. "The Technological and Economic Prospects for CO_2 Utilization and Removal." *Nature* 575 (November): 87–97. https://doi.org/10.1038/s41586-019-1681-6.

Joling, Dan. 2017. "Hilcorp Seeks to Build Island to Drill in the Arctic." *Houston Chronicle*, October 20, 2017. www.houstonchronicle.com/business/article/Hilcorp-seeks-to-build-island-to-drill-in-the-12295448.php.

Joselow, Maxine. 2021. "'Seriously Flawed': Experts Clash over Social Cost of Carbon." *E&E News*, August 24, 2021. https://www.eenews.net/articles/seriously-flawed-experts-clash-over-social-cost-of-carbon/.

Kim, Jed. 2014. "3 Years after San Bruno Gas Explosion, Is California Safer?" Southern California Public Radio, March 13, 2014. www.scpr.org/news/2014/03/13/42787/3-years-after-san-bruno-gas-explosion-is-californi/.

Konkel, Lindsey. 2017. "In the Neighborhood of 18 Million: Estimating How Many People Live Near Oil and Gas Wells." *Environmental Health Perspectives* 125, no. 12 (December 7, 2017). https://doi.org/10.1289/ehp2553.

Kroepsch, Adrianne C., Peter T. Maniloff, John L. Adgate, Lisa M. McKenzie, and Katherine L. Dickinson. 2019. "Environmental Justice in Unconventional Oil and Natural Gas Drilling and Production: A Critical Review and Research Agenda." *Environmental Science & Technology* 53, no. 12 (May 22, 2019): 6601–15. https://doi.org/10.1021/acs.est.9b00209.

Lee, Jinjoo. 2021. "Big Oil Is Vulnerable to Climate Change. Literally." *Wall Street Journal*, July 30, 2021. https://www.wsj.com/articles/big-oil-is-vulnerable-to-climate-change-literally-11627637435.

Makower, Joel. 2019. "The Growing Concern over Stranded Assets." Greenbiz, September 10, 2019. https://www.greenbiz.com/article/growing-concern-over-stranded-assets.

Marshak, Stephen. 2019. *Essentials of Geology*, 6th ed. New York: Norton.

Martinez, Kimiko, and Kate Kiely. 2014. "Fracking Report: 5.4 Million Californians Now Live within a Mile of Oil or Gas Wells, Majority Are People of Color." Natural Resource Defense Council, October 22, 2014. https://www.nrdc.org/media/2014/141022.

Minnesota Department of Natural Resources. 2013. "DNR and Silica Sand. https://www.lrl.mn.gov/guides/guides?issue=fracsands

O'Connor, Peter A., and Cutler J. Cleveland. 2014. "U.S. Energy Transitions 1780–2010." *Energies* 7 (November 27, 2014): 7955–93. doi:10.3390/en7127955.

Reilly, Michael. 2016. "A Mississippi Power Plant Highlights All That's Wrong with 'Clean Coal.'" *MIT Technology Review*, July 5, 2016. www.technologyreview.com/s/601840/a-mississippi-power-plant-highlights-all-thats-wrong-with-clean-coal/.

Rice, Doyle. 2022. "Global Carbon Dioxide Emissions Reach Highest Level in History." *USA Today*, March 8, 2022. https://www.usatoday.com/story/news/world/2022/03/08/global-carbon-dioxide-emissions-soar/9429433002/.

Roberts, David. 2019. "These Uses of CO_2 Could Cut Emissions—and Make Trillions of Dollars." *Vox*, November 27, 2019. https://www.vox.com/energy-and-environment/2019/11/13/20839531/climate-change-industry-co2-carbon-capture-utilization-storage-ccu.

Roedner Sutter, Katelyn. n.d. "California's Cap-and-Trade Program Step by Step." Sacramento: Environmental Defense Fund.

Sierra Club, Wisconsin Chapter. 2019. "Frac Sand Mining." December 2, 2019. https://www.sierraclub.org/wisconsin/frac-sand-mining.

Smil, Vaclav. 2017. *Energy and Civilization: A History.* Cambridge, MA: MIT Press.

Trimble, David, and Frank Rusco. 2012. *Air Emissions and Electricity Generation at U.S. Power Plants.* Washington, DC: US Government Accountability Office.

US Department of Transportation. 2022. "Gas Transmission and Hazardous Liquid Pipelines" (map: pipeline data as of 08/26/2022). https://www.npms.phmsa.dot.gov/Documents/NPMS_Pipelines_Map.pdf.

US Energy Information Administration. n.d. "Energy Explained: Your Guide to Understanding Energy." Accessed October 2, 2019. www.eia.gov/energyexplained/index.php.

US Energy Information Administration. 2007/2008. "About U.S. Natural Gas Pipelines." www.eia.gov/naturalgas/archive/analysis_publications/ngpipeline/develop.html.

US Energy Information Administration. 2018a. "Natural Gas Explained: Natural Gas Pipelines." www.eia.gov/energyexplained/index.php?page=natural_gas_pipelines.

US Energy Information Administration. 2018b. *Monthly Energy Review*, Tables 1.3, 1.4a, 1.4b, and 2.1–2.6. www.eia.gov/energyexplained/?page=us_energy_home.

US Energy Information Administration. 2021. "Coal Explained: How Much Coal Is Left?" October 19, 2021. https://www.eia.gov/energyexplained/coal/how-much-coal-is-left.php.

US Energy Information Administration. 2021. "U.S. Energy Facts explained." https://www.eia.gov/energyexplained/us-energy-facts/

US Energy Information Administration. 2021c. *US Crude Oil and Natural Gas, Proved Reserves, Year-End 2020.* Washington, DC: US Energy Information Administration.

US Energy Information Association. 2021. "Nonfossil fuel sources accounted for 21% of U.S. energy consumption in 2020." https://www.eia.gov/todayinenergy/detail.php?id=48576# https://www.eia.gov/naturalgas/crudeoilreserves/.

US Government Accountability Office. 2012. *Air Emissions and Electricity Generation at U.S. Power Plants.* GAO-12-545R. Washington, DC: US Government Accountability Office. www.gao.gov/assets/600/590188.pdf.

Vaughan, Adam. 2020. "Fracking Wells in the US Are Leaking Loads of Planet-Warming Methane." *New Scientist*, April 22, 2020. https://www.newscientist.com/article/2241347-fracking-wells-in-the-us-are-leaking-loads-of-planet-warming-methane/.

Vogelsong, Sarah. 2020. "What Sank the Atlantic Coast Pipeline? It Wasn't Just Environmentalism." *Virginia Mercury*, July 8, 2020. https://www.virginiamercury.com/2020/07/08/what-sank-the-atlantic-coast-pipeline-it-wasnt-just-environmentalism/.

Wethe, David, Jennifer A. Dlouhy, and Ari Natter. 2021. "Who's to Blame for the Pain at the Pump? It's Complicated." Bloomberg, November 20, 2021. https://www.bloomberg.com/news/articles/2021-11-30/fact-checking-the-finger-pointing-on-high-gasoline-prices.

Wikipedia. n.d. "Lac-Mégantic Rail Disaster." Accessed October 2, 2019. https://en.wikipedia.org/wiki/Lac-Mégantic_rail_disaster.

Wines, Michael. 2016. "Geologist Sees Clues, and Further Dangers, in Puzzle of Oklahoma's Earthquakes." *New York Times*, September 7, 2016. www.nytimes.com/2016/09/07/us/in-puzzle-of-oklahomas-earthquakes-new-data-may-provide-clues.html?_r=0.

WORC (Western Organization of Resource Councils). 2021. "Split Estate Leaves People Living Near Mineral Deposits at the Mercy of Energy Companies." January 21, 2021. http://www.worc.org/split-estate-leaves-home-and-landowners-at-the-mercy-of-energy-companies/.

Yeomans, Matthew. 2004. *Oil: A Concise Guide to the Most Important Product on Earth.* New York: New Press.

CHAPTER 14

A Student's Guide to Global Climate Change. 2016a. "Nuclear Energy."

A Student's Guide to Global Climate Change. 2016b. "Solar Energy."

Alternative Fuels Data Center. 2020. "AFVs by Fuel Type." 2020. Washington, DC.

Australian Academy of Science. 2015. "From Sunlight to Electricity." November 7, 2015. https://www.science.org.au/curious/technology-future/solar-pv.

BP. 2021. *Statistical Review of World Energy*, 70th ed. British Petroleum. https://www.bp.com/content/dam/bp/business-sites/en/global/corporate/pdfs/energy-economics/statistical-review/bp-stats-review-2021-full-report.pdf.

Bureau of Labor Statistics. 2019. "May 2018 National Industry-Specific Occupational Employment and Wage Estimates." www.bls.gov

Campus Conservation Nationals. 2017. "How It Works." https://www.nwf.org/Home/EcoLeaders/Campus-Conservation-Nationals.

Center for Nuclear Science and Technology Information. n.d. "Source Energy Equivalents." Accessed July 6, 2019.

Choi-Granade, H., J. Creyts, A. Derkach, P. Farese, S. Nyquist, and K. Ostrowski. 2009. *McKinsey Global Energy and Materials: Unlocking Energy Efficiency in the US Economy.* New York: McKinsey & Company.

"Daily Chart: China Powers Ahead with a New Direct-Current Infrastructure." 2017. *Economist*, January 16, 2017. www.economist.com/graphic-detail/2017/01/16/china-powers-ahead-with-a-new-direct-current-infrastructure.

Deutschland.de. 2012. "A Pioneer in Climate Policy." August 13, 2012. www.deutschland.de/en/topic/environment/earth-climate/a-pioneer-in-climate-policy.

Edwards, Chris. 2020. "The EV Lifecycle Conundrum." E&T, Engineering and Technology, July 10, 2020. https://eandt.theiet.org/content/articles/2020/07/the-ev-lifecycle-conundrum/.

Energy.gov. n.d. "Fuel Cell Animation (Text Version)." Accessed October 2, 2019. www.energy.gov/eere/fuelcells/fuel-cell-animation-text-version.

Engineering.com. n.d. "Passive Solar Systems & Solar Hot Water." Accessed July 6, 2019. www.engineering.com/SustainableEngineering/RenewableEnergyEngineering/SolarEnergyEngineering/PassiveSolarSystemsSolarHotWater/.

Fourwinds10.com. 2011. "World Bank Bringing Solar Power to Over 1 Million Homes, Shops in Rural Bangladesh." November 21, 2011. http://fourwinds10.com/siterun_data/science_technology/energy_free_energy/news.php?q=1321979049.

Goldman Sachs. 2017. "The Wind and Solar Boom." July 21, 2017. www.goldmansachs.com/our-thinking/pages/alberto-gandolfi-wind-and-solar-boom.html?mediaIndex=1&autoPlay=true&cid=sch-pd-google-windandsolarboom-searchad-201707—&mkwid=wMvTgASo.

Green Age. n.d. "Iceland Geothermal Power." Accessed October 2, 2019. www.thegreenage.co.uk/cos/iceland-geothermal-power/.

Hawken, Paul, ed. 2017. Drawdown: The Most Comprehensive Plan Ever Proposed to Reverse Global Warming. London: Penguin.

Holthaus, Eric. 2015. "Hot Dam." Slate, June 1, 2015. www.slate.com/articles/business/moneybox/2015/06/the_future_of_hydroelectricity_it_s_not_good.html.

"How Many Solar Panels Would You Need to Power the USA?" 2022. https://atlantickeyenergy.com/how-many-solar-panels-would-you-need-to-power-the-unitedstates/#:~:text=That's%20how%20much%20power%20solar,to%20power%20the%20whole%20USA

Kamkwamba, William, and Bryan Mealer. 2009. The Boy Who Harnessed the Wind: Creating Currents of Electricity and Hope. New York: William Morrow.

Lambert, Fred. 2016. "White House Unlocks $4.5 Billion for Electric Vehicle Infrastructure and Announces New EV Programs." Electrek, July 21, 2016. https://electrek.co/2016/07/21/white-house-unlocks-4-5-billion-for-electric-vehicle-infrastructure-and-announces-new-ev-programs/.

Lazard. 2020. "Lazard's Levelized Cost of Energy Analysis: Version 14.0." October 2020. https://www.lazard.com/media/451419/lazards-levelized-cost-of-energy-version-140.pdf.

LeVine, Steve. 2021. "America Isn't Ready for the Electric-Vehicle Revolution." New York Times, November 10, 2021. https://www.nytimes.com/2021/11/10/opinion/electric-vehicle-climate-battery.html.

Lucchesi, Nick. 2017. "Here's Elon Musk's Plan to Power the U.S. on Solar Energy." Inverse, July 16, 2017. www.inverse.com/article/34239-how-many-solar-panels-to-power-the-usa.

MacDonald, James. 2016. "The Unexpected Problem with Wind Power." Jstor Daily, Novembeer 17, 2016. https://daily.jstor.org/the-unexpected-problem-with-wind-power/.

Matheny, Keith. 2021. "University of Michigan Pledges to Make Its Campuses Carbon-Neutral in Climate Change Fight." Detroit Free Press, May 20, 2021. https://www.freep.com/story/news/local/michigan/2021/05/20/u-m-carbon-neutral-campus-climate-change/5185944001/.

McFadden, Christopher. 2021. "The Paradox of 'Clean' EVs and the 'Dirty' Lithium Mining Business." Interesting Engineering, April 10, 2021. https://interestingengineering.com/clean-evs-and-dirty-lithium-mining-business.

Mey, Alex. 2019. "Most U.S. Utility-Scale Solar Photovoltaic Power Plants Are 5 Megawatts or Smaller." US Energy Information Administration, February 7, 2019. https://www.eia.gov/todayinenergy/detail.php?id=38272.

Millstein, Dev, Ryan Wiser, Mark Bolinger, and Galen Barbose. 2017. "The Climate and Air-Quality Benefits of Wind and Solar Power in the United States." Nature Energy 2, no. 9 (August 14, 2017): 17134.

NASEO (National Association of State Energy Officials). 2020. 2020 U.S. Energy and Employment Report. https://www.naseo.org/data/sites/1/documents/publications/USEER-2020-US-Energy-Employment-Report.pdf.

NC Sustainable Energy Association. n.d. "Solar-thermal-energy-1." Accessed November 18, 2019. https://energync.org/solar/solar-thermal-energy-1/.

Office of Energy Efficiency and Renewable Energy. 2022. "About the Weatherization Assistance Program." https://www.energy.gov/eere/wap/about-weatherization-assistance-program.

Ong, Sean, Clinton Campbell, Paul Denholm, Robert Margolis, and Garvin Heath. 2013. "Land-Use Requirements for Solar Power Plants in the United States." National Renewable Energy Laboratory (NREL), June 2013. https://www.nrel.gov/docs/fy13osti/56290.pdf.

Penn, Ivan, and Eric Lipton. 2021. "The Lithium Gold Rush: Inside the Race to Power Electric Vehicles." New York Times, May 6, 2021. https://www.nytimes.com/2021/05/06/business/lithium-mining-race.html.

Plumer, Brad. 2017. "This Neat Interactive Map Shows Why Renewables and Natural Gas Are Taking Over the US." Vox, April 21, 2017. www.vox.com/energy-and-environment/2016/12/12/13914942/interactive-map-cheapest-power-plant.

Plumer, Brad, Nadja Popovich, and Blacki Migliozzi. 2022. "Electric Cars Are Coming: How Long Until They Rule the Road?" New York Times, March 10, 2022. https://www.nytimes.com/interactive/2021/03/10/climate/electric-vehicle-fleet-turnover.html.

Randall, Tom. 2016. "Wind and Solar Are Crushing Fossil Fuels." Bloomberg, April 6, 2016. www.bloomberg.com/news/articles/2016-04-06/wind-and-solar-are-crushing-fossil-fuels.

Reichmuth, David. 2020. "Are Electric Vehicles Really Better for the Climate? Yes. Here's Why." Union of Concerned Scientists, February 11, 2020. https://blog.ucsusa.org/dave-reichmuth/are-electric-vehicles-really-better-for-the-climate-yes-heres-why/.

Ritchie, Hannah, and Max Roser. 2019. "Renewable Energy." Our World in Data. https://ourworldindata.org/renewable-energy.

Roberts, David. 2017. "2 Remarkable Facts That Illustrate Solar Power's Declining Cost." Vox, February 3, 2017. www.vox.com/science-and-health/2016/12/22/14022114/solar-year-two-remarkable-facts.

Shiff, Blair. 2017. "Amazon CEO Commemorates Opening of Giant Wind Farm in a Smashing Way." ABC News, October 20, 2017. http://abcnews.go.com/Business/amazon-ceo-commemorates-opening-giant-wind-farm-smashing/story?id=50608858.

Sieminski, Adam. 2014. "International Energy Outlook 2014," p. 18. US Energy Information Administration, September 22, 2014. www.eia.gov/pressroom/presentations/sieminski_09222014_columbia.pdf.

Smil, Vaclav. 2017. Energy Transitions: Global and National Perspectives, 2nd ed. Santa Barbara, CA: Praeger.

Solar Energy Industries Association. 2019. "United States Surpasses 2 Million Solar Installations." May 9, 2019. https://www.seia.org/news/united-states-surpasses-2-million-solar-installations.

n.d. "Our 100% Clean Energy Vision." Accessed October 2, 2019. http://thesolutionsproject.org/why-clean-energy/#/map/cities/.

University of Texas at Austin Energy Institute. n.d. "Levelized Cost of Electricity." Accessed July 6, 2019. http://calculators.energy.utexas.edu/lcoe_map/#/county/tech.

US Department of Energy. n.d. "How Do Gasoline Cars Work." Alternative Fuels Data Center. Accessed July 6, 2019. https://afdc.energy.gov/vehicles/how-do-gasoline-cars-work.

US Department of Energy. 2014. "The Year of Concentrating Solar Power." May 2014. www.energy.gov/sites/prod/files/2014/05/f15/2014_csp_report.pdf.

US Department of Energy. 2019. "Light Duty Model AFV, HEV, and Diesel Model Offerings, by Technology/Fuel," 1991–2019." Alternative Fuels Data Center. https://afdc.energy.gov/data/10303.

US Department of Energy. 2022a. "Wind Turbines: The Bigger, the Better." Office of Energy Efficiency and Renewable Energy, August 16, 2022. https://www.energy.gov/eere/articles/wind-turbines-bigger-better.

US Department of Energy. 2022b. "Solar Energy in the United States." Office of Energy Efficiency and Renewable Energy. https://www.energy.gov/eere/solar/solar-energy-united-states.

US Energy Information Administration. 2021a. "EIA projects renewables share of U.S. electricity generation mix will double by 2050." Today in Energy, February 8, 2021

US Energy Information Administration. 2021b. "Preliminary Monthly Electric Generator Inventory (Based on Form EIA-860M as a Supplement to Form EIA-860)."

US Environmental Protection Agency. 2016. "Geothermal Heating and Cooling Technologies." www.epa.gov/rhc/geothermal-heating-and-cooling-technologies.

US Environmental Protection Agency. 2018. "What Is CHP?" www.epa.gov/chp/what-chp.

CHAPTER 15

Brooks, Amy L., Shunli Wang, and Jenna R. Jambeck. 2018. "The Chinese Import Ban and Its Impact on Global Plastic Waste Trade." Science Advances 4, no. 6 (June 20, 2018). https://doi.org/10.1126/sciadv.aat0131.

CalRecycle. 2020. 2018 Disposal-Facility-Based Characterization of Solid Waste in California. California Department of Resources Recycling and Recovery, May 15, 2020. https://www2.calrecycle.ca.gov/Publications/Details/1666.

Comolli, Virginia. 2021. "Plastic for Profit: Tracing Illicit Plastic Waste Flows, Supply Chains and Actors." Global Initiative, October 31, 2021. https://globalinitiative.net/analysis/illicit-trade-plastic-waste/.

Covanta. 2022. "Tulsa." https://www.covanta.com/where-we-are/our-facilities/tulsa.

Cox Automotive. 2021. "Evolution of Mobility 2.0." August 2021. https://www.coxautoinc.com/wp-content/uploads/2021/09/Full-Report_2021-Evolution-of-Mobility-2.0.pdf.

Durning, Alan Thein, and Northwest Environment Watch. 1997. *Stuff: The Secret Lives of Everyday Things*. Seattle: Northwest Environment Watch.

Foulkes, Imogen, Matthew Davis, Thomas Buch-Andersen, Tristina Moore, Richard Galpin, Jeremy Bowen, and Hamadou Tidiane Sy. 2005. "Recycling around the World." BBC News, June 25, 2005. http://news.bbc.co.uk/2/hi/europe/4620041.stm.

Franklin Associates. 2007. *Final Report: LCI Summary for PLA and PET 12-Ounce Water Bottles*. Report prepared for PET Resin Association. Prairie Village, KS: Franklin Associates. http://petresin.org/pdf/FranklinPETPLAlifecycleanal_12-oz.pdf.

Gamio, Lazaro, and Peter S. Goodman. 2021. "How the Supply Chain Crisis Unfolded." *New York Times*, December 5, 2021. https://www.nytimes.com/interactive/2021/12/05/business/economy/supply-chain.html.

Haro, Alexander. 2021. "The Ocean Cleanup Crew Just Proved That System 002 Works." Inertia, November 3, 2021. https://www.theinertia.com/environment/the-ocean-cleanup-crew-just-proved-that-system-002-two-works/.

Jiang, Jingjing. 2019. "More Americans Are Using Ride-Hailing Apps." Pew Research Center, January 4, 2019. www.pewresearch.org/fact-tank/2019/01/04/more-americans-are-using-ride-hailing-apps/.

Johnston Food Service and Cleaning Solutions. n.d. "Crunching the Numbers: Which Plastics Are Actually Recyclable?" Accessed April 9, 2019. www.johnston.biz/which_plastics_are_recyclable/.

Kaza, Silpa, Lisa C. Yao, Perinaz Bhada-Tata, and Frank Van Woerden. 2018. *What a Waste 2.0: A Global Snapshot of Solid Waste Management to 2050*. Open Knowledge Repository, "Urban Development." Washington, DC: World Bank Group. https://openknowledge.worldbank.org/handle/10986/30317.

Luu, Paula. 2018. "The Hundred People behind Your Favorite Pair of Jeans." University of Michigan Erb Institute, April 23, 2018. https://erb.umich.edu/2018/04/23/the-hundred-people-behind-your-favorite-pair-of-jeans/.

Moore, Charles, with Cassandra Phillips. 2011. *Plastic Ocean*. New York: Avery.

Moore, Patty. 2015. "Recycling Is Not Dead." *Resource Recycling News*, July 1, 2015. https://resource-recycling.com/recycling/2015/07/01/recycling-is-not-dead/.

National Ocean Service. 2018. "What Is a Gyre?" https://oceanservice.noaa.gov/facts/gyre.html.

Organisation for Economic Co-operation and Development (OECD). 2019a. "Environment at a Glance—OECD Indicators." https://www.oecd.org/environment/environment-at-a-glance/.

OECD. 2019b. "Municipal Waste" [indicator].

OECD. 2022. "Circular Economy—Waste and Materials." Environment at a Glance Indicators. https://doi.org/10.1787/f5670a8d-en.

Pearson, Bryan. 2018. "Food Waste Is the New Sales Driver: 4 Ways Kroger, Walmart Are Changing Shopper Thinking." *Forbes*, November 5, 2018. www.forbes.com/sites/bryanpearson/2018/11/05/food-waste-is-the-new-sales-driver-4-ways-kroger-walmart-are-changing-shopper-thinking/#5292e5df7c09.

Phillips, Anna M. 2021. "Trafficking of Plastic Waste Is on the Rise and Criminal Groups Are Profiting, Report Says." *Los Angeles Times*, November 8, 2021. https://www.latimes.com/environment/story/2021-11-08/report-trafficking-of-plastic-waste-is-on-the-rise-and-criminal-groups-are-profiting.

Rathje, William, and Cullen Murphy. 2001. *Rubbish! The Archaeology of Garbage*. Tucson: University of Arizona Press.

Royte, Elizabeth. 2017. "The Compost King of New York." *New York Times*, February 15, 2017. www.nytimes.com/2017/02/15/magazine/the-compost-king-of-new-york.html.

Smith, Carl. 2021. "After China's Recyclable Ban, Municipalities Shift Gears." Governing, August 27, 2021. https://www.governing.com/now/after-chinas-recyclable-ban-municipalities-shift-gears.

Strasser, Susan. 1999. *Waste and Want: A Social History of Trash*. New York: Henry Holt.

US Environmental Protection Agency. 2016. "Wastes—Non-Hazardous Waste—Municipal Solid Waste." https://archive.epa.gov/epawaste/nonhaz/municipal/web/html/basic.html.

US Environmental Protection Agency. 2018a. *Advancing Sustainable Materials Management: 2015 Fact Sheet*, p. 7. www.epa.gov/sites/production/files/2018-07/documents/2015_smm_msw_factsheet_07242018_fnl_508_002.pdf.

US Environmental Protection Agency. 2018b. "Municipal Solid Waste Landfills." www.epa.gov/landfills/municipal-solid-waste-landfills.

US Environmental Protection Agency. 2018c. "National Overview: Facts and Figures on Materials, Wastes and Recycling." https://www.epa.gov/facts-and-figures-about-materials-waste-and-recycling/national-overview-facts-and-figures-materials

US Environmental Protection Agency. 2019. "U.S. State and Local Waste and Materials Characterization Reports." www.epa.gov/facts-and-figures-about-materials-waste-and-recycling/advancing-sustainable-materials-management-0.

US Environmental Protection Agency. 2022. "National Overview: Facts and Figures on Materials, Wastes, and Recycling." See Figure "Municipal Solid Waste Management: 1960–2018." Updated July 31, 2022. https://epa.gov/facts-and-figures-about-materials-waste-and-recycling/national-overview-facts-and-figures-materials.

US Geological Survey. n.d. Toxic Substances Hydrology Program—Normal Landfill Project. "Cross-Sectional View of Normal Landfill." Accessed October 21, 2019.

Vidal, John. 2005. "UK Firms Caught in Illegal Waste Dumping." *Guardian*, March 27, 2005. www.theguardian.com/society/2005/mar/28/environment.uknews.

Woollacott, Emma. 2021. "The Fungus and Bacteria Tackling Plastic Waste." BBC, July 30, 2021. https://www.bbc.com/news/business-57733178.

Zeidler, Maryse. 2018. "Grocers, Innovators Work to Save $31B in Food from Being Trashed in Canada Each Year." CBC News, November 10, 2018. www.cbc.ca/news/canada/british-columbia/grocers-innovators-work-to-save-31b-in-food-from-being-trashed-in-canada-each-year-1.4898285.

CHAPTER 16

Adler, Margot. 2006. "Behind the Ever-Expanding American Dream House." NPR, July 4, 2006. www.npr.org/templates/story/story.php?storyId=5525283.

Al, Stefan, ed. 2016. *Mall City: Hong Kong's Dreamworlds of Consumption*. Pokfulam: Hong Kong University Press.

Al, Stefan. 2017. "The Mall Isn't Dead—It's Just Changing." *Conversation*, May 14, 2017. https://theconversation.com/the-mall-isnt-dead-its-just-changing-72237.

Bastida, Xiye, Elli Stuhler, and Space 10. 2021. "Afterward." In *The Ideal City: Exploring Urban Futures*. Edited by Robert Klanten and gestalten. Berlin: Gestalten Verlag.

Beach, Doug. n.d. "From Modest to McMansion." National Public Radio. Accessed May 4, 2019.

Bettencourt, Luís M. A. 2013. "The Origins of Scaling in Cities." *Science* 340, no. 6139 (June 21, 2013): 1438–41. https://science.sciencemag.org/content/340/6139/1438.

Bigelow, Pete. 2021. "Average Time Spent in U.S. Traffic Jams Cut by 73 Hours during Pandemic, Study Finds." *Automotive News*, March 11, 2021. https://www.autonews.com/mobility-report-newsletter/average-time-spent-us-traffic-jams-cut-73-hours-during-pandemic-study.

Boeri, Stefano. 2018. "Vertical Forest." Stefano Boeri Architetti. https://www.stefanoboeriarchitetti.net/en/project/vertical-forest/.

Bokhari, Sheharyar. 2020. "How Much Is a Point of Walk Score Worth?" Redfin Real Estate News, October 6, 2020. https://www.redfin.com/news/how-much-is-a-point-of-walk-score-worth/.

Busch, Chris, and Hallie Kennan. 2013. "Urbanization Can Actually Reduce Greenhouse Gas Emissions." *LiveScience*, November 21, 2013. www.livescience.com/41396-urbanization-can-reduce-carbon-emissions.html.

Cervero, Robert. 2017. *Suburban Gridlock*. Abingdon, UK: Routledge.

Chakrabarti, Vishaan. 2013. "Building Hyperdensity and Civic Delight." *Places Journal*, June 2013.

Demographia.com. 2019. *Demographia World Urban Areas*. 15th annual ed. http://demographia.com/db-worldua.pdf.

Duany, Andres, Elizabeth Plater-Zyberk, and Jeff Speck. 2000. *Suburban Nation: The Rise of Sprawl and the Decline of the American Dream*. New York: North Point.

Duany, Andres, and Jeff Speck with Mike Lydon. 2010. *The Smart Growth Manual*. New York: McGraw-Hill.

"Environment: A Pall over the Suburban Mall." 1978. *Time*, November 13, 1978. http://content.time.com/time/magazine/article/0,9171,946182,00.html.

Freemark, Yonah. n.d. "Travel to Work by City in 2014: 15 Largest Cities and All Cities with More than 30,000 Commuters." The Transport Politic. Accessed May 4, 2019. www.thetransportpolitic.com/databook/travel-mode-shares-in-the-u-s/.

Green, Jared. n.d. "Is Urban Revitalization without Gentrification Possible?" Smart Cities Dive. https://www.smartcitiesdive.com/ex/sustainablecitiescollective/urban-revitalization-without-gentrification-possible/614391/.

Hoornweg, Daniel, and Kevin Pope. 2014. "Socioeconomic Pathways and Regional Distribution of the World's 101 Largest Cities." Working paper no. 04. Toronto: Global Cities Institute.

Hu, Winnie, Patrick McGeehan, and Nate Schweber. 2021. "As Traffic Roars Back, Neighborhoods outside Manhattan Feel the Pain." *New York Times*, December 28, 2021. https://www.nytimes.com/2021/12/28/nyregion/nyc-traffic-today.html.

INRIX. n.d. "INRIX 2021 Global Traffic Scorecard." http://inrix.com/scorecard/.

International Living Future Institute. 2021. "The Kendeda Building for Innovative Sustainable Design."

Kent, John. 2021. "Green Mixed-Use Development Underway in Fort Worth." GreenSource DFW, August 4, 2021. https://greensourcedfw.org/articles/green-mixed-use-development-underway-fort-worth.

Lehrer, Jonah. 2010. "A Physicist Solves the City." New York Times Magazine, December 19, 2010. www.nytimes.com/2010/12/19/magazine/19Urban_West-t.html.

Lindeke, Bill. 2016. "Floor Area Ratio 101: This Obscure but Useful Planning Tool Shapes the City." MinnPost, October 3, 2016. www.minnpost.com/cityscape/2016/10/floor-area-ratio-101-obscure-useful-planning-tool-shapes-city/.

Maantay, Juliana A., and Andrew R. Maroko. 2018. "Brownfields to Greenfields: Environmental Justice versus Environmental Gentrification." International Journal of Environmental Research and Public Health 15, no. 10 (October 12, 2018). https://doi.org/10.3390/ijerph15102233.

Manila Today Staff. 2017. "In Photos: Why Are You Destroying Our Homes?" Manila Today, October 18, 2017.

Momin, Kashaf. 2019. "Confronting Environmental Gentrification: The Case of the Anacostia." Environmental Law Institute, June 3, 2019. https://www.eli.org/vibrant-environment-blog/confronting-environmental-gentrification-case-anacostia.

Moore, David. 2011. Ecological Footprint Analysis San Francisco-Oakland-Fremont, CA. Oakland: Global Footprint Network.

MTA (Metropolitan Transit Authority, New York City). n.d. "Introduction to Subway Ridership." Accessed October 2, 2019. http://web.mta.info/nyct/facts/ridership/.

National Association of Home Builders. n.d. "Housing Facts, Figures, and Trends for March 2006." Washington, DC: National Association of Home Builders.

Newman, Peter, Timothy Beatley, and Heather Boyer. 2009. Resilient Cities: Responding to Peak Oil and Climate Change. Washington, DC: Island.

Office of Sustainable Communities, Dan Emerine, Christine Shenot, Mary Kay Bailey, Lee Sobel, and Megan Susman. 2014. This Is Smart Growth. Edited by Jane Cotnoir. Smart Growth Network. https://www.epa.gov/sites/default/files/2014-04/documents/this-is-smart-growth.pdf.

Owen, A., D. M. Levinson, and B. Murphy. 2017. Access across America: Transit 2015 Data [data set]. Data Repository for the University of Minnesota (DRUM). doi:10.13020/D63G6F.

Satterthwaite, David. 2020. "The World's Fastest Growing Cities." International Institute for Environment and Development, March 24, 2020. https://www.iied.org/worlds-fastest-growing-cities.

Southworth, Michael. 2005. "Designing the Walkable City." Journal of Urban Planning and Development 131, no. 4 (December 1, 2005): 246–57.

Stromberg, Joseph. 2015. "The Real Reason American Public Transportation Is Such a Disaster." Vox, August 10, 2015. www.vox.com/2015/8/10/9118199/public-transportation-subway-buses.

Texas A&M Transportation Institute, Mobility Division. 2021. "COVID-19 Gave Us an Historic Traffic Hiatus, but the Pause Didn't Last Long." https://mobility.tamu.edu/umr/media-information/press-release/.

Uberti, David. "The Death of the American Mall." Guardian, June 19, 2014. www.theguardian.com/cities/2014/jun/19/-sp-death-of-the-american-shopping-mall.

United Nations Department of Economics and Social Affairs, Population Division. 2015. World Urbanization Prospects: The 2014 Revision (ST/ESA/SER.A/366). New York: United Nations. https://population.un.org/wup/publications/files/wup2014-report.pdf.

United Nations Department of Economics and Social Affairs, Population Division. 2018. The World's Cities in 2018—Data Booklet (ST/ESA/SER.A/417). New York: United Nations.

Valmero, Anna. 2016. "Cleaning Up Manila's Pasig River, One Tributary at a Time." CityLab, June 27, 2016. www.citylab.com/environment/2016/06/cleaning-up-manilas-pasig-river-one-tributary-at-a-time/488885/.

World Bank. 2010. Cities and Climate Change: An Urgent Agenda. Urban Development Series; Knowledge Papers, no. 10. Washington, DC: World Bank. https://openknowledge.worldbank.org/handle/10986/17381.

World Health Organization. 2012. Health in the Green Economy: Health Co-benefits of Climate Change Mitigation—Transport Sector. Geneva: World Health Organization.

Zimmermann, Kim Ann. 2015. "Hurricane Katrina: Facts, Damage & Aftermath." LiveScience, August 27, 2015. www.livescience.com/22522-hurricane-katrina-facts.html.

CHAPTER 17

American Cancer Society. 2017. "What is Ultraviolet (UV) Radiation?" www.cancer.org/cancer/skin-cancer/prevention-and-early-detection/what-is-uv-radiation.html.

Amouzou, A., L. Carvajal Velez, H. Tarekegn, and M. Young. 2016. One Is Too Many: Ending Child Deaths from Pneumonia and Diarrhea. New York: UNICEF. https://data.unicef.org/resources/one-many-ending-child-deaths-pneumonia-diarrhoea/.

Bagley, Katherine. 2020. "Connecting the Dots between Environmental Injustice and the Coronavirus." YaleEnvironment360, May 7, 2020. https://e360.yale.edu/features/connecting-the-dots-between-environmental-injustice-and-the-coronavirus.

Centers for Disease Control and Prevention. n.d. "Fight Flu." Accessed October 21, 2019. www.cdc.gov/flu/pdf/freeresources/take3-fight-flu-infographic.pdf.

Centers for Disease Control and Prevention. 2018. "Malaria." www.cdc.gov/malaria/about/biology/index.html.

Chakraborty, J., T. W. Collins, and S. E. Grineski. 2019. "Exploring the Environmental Justice Implications of Hurricane Harvey Flooding in Greater Houston, Texas." American Journal of Public Health 109, no. 2 (February 2019): 244–50.

Fears, Darryl. 2022. "Redlining Means 45 Million Americans Are Breathing Dirtier Air, 50 Years after It Ended." Washington Post, March 9, 2022. https://www.washingtonpost.com/climate-environment/2022/03/09/redlining-pollution-environmental-justice/.

Frevert, Jürgen. 2010. "Content of Botulinum Neurotoxin in Botox®/Vistabel®, Dysport®/Azzalure®, and Xeomin®/Bocouture®." Drugs in R&D 10, no. 2 (November 27, 2010): 67–73.

Frumkin, Howard, ed. 2010. Environmental Health from Global to Local. San Francisco: Jossey-Bass.

Gesch, Dean. n.d. "Topography-Based Analysis of Hurricane Katrina Inundation of New Orleans." US Department of the Interior, US Geological Survey. Accessed May 6, 2019. https://pubs.usgs.gov/circ/1306/pdf/c1306_ch3_g.pdf.

Gill, Peter. 2018. "Nepal's Desperate Race to Rebuild." Diplomat, April 25, 2018. https://thediplomat.com/2018/04/nepals-desperate-race-to-rebuild/.

Gottlieb, Robert. 1993. Forcing the Spring: The Transformation of the American Environmental Movement. Washington, DC: Island.

Gunderman, Richard M., and Angela S. Gonda. 2015. "Radium Girls." Radiology 274 (January 27, 2015): 314–18. https://pubs.rsna.org/doi/pdf/10.1148/radiol.14141352.

Helme, Ned, Stacey Davis, Suzanne Reed, Nancy Ginn Helme, Michelle Levinson, and David Wooley. 2017. Advancing Environmental Justice: A New State Regulatory Framework to Abate Community-Level Air Pollution Hotspots and Improve Health Outcomes. Berkeley: Goldman School of Public Policy, University of California, Berkeley.

Insurance Information Institute. n.d. "Facts + Statistics: Wildfires." Accessed October 2, 2019. www.iii.org/fact-statistic/facts-statistics-wildfires.

Jacob, John S., Kirana Pandian, Ricardo Lopez, and Heather Biggs. n.d. Houston-Area Freshwater Wetland Loss, 1992–2010. TAMU-SG-14-303. College Station, TX: Texas A&M AgriLife Extension Service.

Johns Hopkins University and Medicine, Coronavirus Resource Center. "Mortality Analyses." 2022. June 23, 2022. https://coronavirus.jhu.edu/data/mortality.

Lane, Haley M., Rachel Morello-Frosch, Julian D. Marshall, and Joshua S. Apte. 2022. "Historical Redlining Is Associated with Present-Day Air Pollution Disparities in U.S. Cities." Environmental Science & Technology Letters 9, no. 4 (March 9, 2022): 345–50. https://doi.org/10.1021/acs.estlett.1c01012.

Los Angeles County Department of Public Health. 2016. Parks and Public Health in Los Angeles County: A Cities and Communities Report. May 2016. http://publichealth.lacounty.gov/chronic/docs/Parks%20Report%202016-rev_051816.pdf.

Liu, Jiawen, Lara P. Clark, Matthew J. Bechle, Anjum Hajat, Sun-Young Kim, Allen L. Robinson, Lianne Sheppard, Adam A. Szpiro, and Julian D. Marshall. 2021. "Disparities in Air Pollution Exposure in the United States by Race/Ethnicity and Income, 1990–2010." Environmental Health Perspectives 129, no. 12 (December 15, 2021). https://doi.org/10.1289/ehp8584.

Marshak, Stephen, and Robert Rauber. 2017. Earth Science. New York: Norton.

Morency, P., J. Archambault, M. Cloutier, M. Tremblay, and C. Plante. 2015. "Major Urban Road Characteristics and Injured Pedestrians: A Representative Survey of Intersections in Montréal, Quebec." Canadian Journal of Public Health/Revue Canadienne de Santé Publique 106, no. 6 (September–October 2015): E388–E394. www.jstor.org/stable/90005915.

National Community Reinvestment Coalition

National Institute of Allergy and Infectious Diseases. 2010. "Community Immunity ('Herd' Immunity)."

National Weather Service.. www.weather.gov/hazstat/.

New York State Department of Health. n.d. "Volatile Organic Compounds (VOCs) in Commonly Used Products." Accessed May 6, 2019. www.health.ny.gov/publications/6513.pdf.

Petley, Dave. 2011. "Mining Related Landslides in the Philippines." AGU, Landslide Blog, April 28, 2011. https://blogs.agu.org/landslideblog/2011/04/28/mining-related-landslides-in-the-philippines/.

Pieper, Kelsey J., Min Tang, and Marc A. Edwards. 2017. "Flint Water Crisis Caused by Interrupted Corrosion Control: Investigating 'Ground Zero' Home." *Environmental Science & Technology* 51, no. 4 (February 1, 2017): 2007–14.

Richardson, Jason, Bruce C. Mitchell, Helen C. S. Meier, Emily Lynch, and Jad Edlebi. 2020. "Redlining and Neighborhood Health." NCRC. https://ncrc.org/holc-health/.

Ritchie, Hannah. 2019. "Who Has Contributed Most to Global CO$_2$ Emissions?" Our World in Data, October 1, 2019. https://ourworldindata.org/contributed-most-global-co2.

Rogers, Paul, Matthias Gafni, and George Avalos. 2017. "PG&E Power Lines Linked to Wine Country Fires." *Mercury News*, October 13, 2017. www.mercurynews.com/2017/10/10/pge-power-lines-linked-to-wine-country-fires/.

"Sacoby Wilson." 2022. University of Maryland, School of Public Health. https://sph.umd.edu/people/sacoby-wilson.

Schoenherr, Neil. 2019. "How Fast Fashion Hurts Environment, Workers, Society: Brown School Research Examines Environmental, Social Costs of Ready-Made, Inexpensive Clothing." *Source*, January 9, 2019. https://source.wustl.edu/2019/01/how-fast-fashion-hurts-environment-workers-society/.

Simmons, Ann M. 2017. "Sierra Leone Faces Long Slog to Recovery after Devastating Mudslide." *Los Angeles Times*, September 7, 2017. www.latimes.com/world/africa/la-fg-global-sierra-leone-20170906-story.html.

Taylor, Alan. 2014. "Bhopal: The World's Worst Industrial Disaster, 30 Years Later." *Atlantic*, December 2, 2014. www.theatlantic.com/photo/2014/12/bhopal-the-worlds-worst-industrial-disaster-30-years-later/100864/.

United Nations Children's Fund. 2016. *One Is Too Many: Ending Child Deaths from Pneumonia and Diarrhea.* New York: UNICEF.

US Department of Energy. 2005. "DOE Certifies Rocky Flats Cleanup 'Complete.'" December 8, 2005. www.energy.gov/articles/doe-certifies-rocky-flats-cleanup-complete.

US Environmental Protection Agency. 2017. "Persistent Organic Pollutants: A Global Issue, a Global Response" (see table "The 'Dirty Dozen'").

Walker, Peter, Rowena Mason, and Fiona Harvey. 2021. "Cop26 Summit at Serious Risk of Failure, Says Boris Johnson." *Guardian*, October 31, 2021. https://www.theguardian.com/environment/2021/oct/31/cop26-summit-at-serious-risk-of-failure-says-boris-johnson.

WE ACT for Environmental Justice. n.d. "History." Accessed October 2, 2019. https://community.weact.org/history.

WHO and the Maternal and Child Epidemiology Estimation (MCEE) Group. 2015. Provisional Estimates. World Health Organization.

Williams, Justin, and Tara Jagadeesh. 2020. "Data Points: The Environmental Injustice of Lead Lines in Illinois." Metropolitan Planning Council (Chicago), November 10, 2020. https://www.metroplanning.org/news/9960/Data-Points-the-environmental-injustice-of-lead-lines-in-Illinois.

Wilson, Sacoby, Kristen Burwell-Naney, Chengsheng Jiang, Hongmei Zhang, Ashok Samantapudi, Rianna Murray, Laura Dalemarre, LaShanta Rice, and Edith Williams. 2015. "Assessment of Sociodemographic and Geographic Disparities in Cancer Risk from Air Toxics in South Carolina." *Environmental Research* 140 (July): 562–68. https://doi.org/10.1016/j.envres.2015.05.016.

World Cancer Research Fund International. n.d. "Skin Cancer Statistics." Accessed May 6, 2019. www.wcrf.org/dietandcancer/cancer-trends/skin-cancer-statistics.

World Health Organization. 2020. "The Top 10 Causes of Death." December 9, 2020. https://www.who.int/news-room/fact-sheets/detail/the-top-10-causes-of-death.

Wu, X., R. C. Nethery, M. B. Sabath, D. Braun, and F. Dominici. 2020. "Air Pollution and Covid-19 Mortality in the United States: Strengths and Limitations of an Ecological Regression Analysis." *Science Advances* 6, no. 45 (November 4, 2020). https://doi.org/10.1126/sciadv.abd4049.

Zhong, Raymond, and Nadja Popovich. 2022. "How Air Pollution across America Reflects Racist Policy from the 1930s." *New York Times*, March 11, 2022. https://www.nytimes.com/2022/03/09/climate/redlining-racism-air-pollution.html.

CHAPTER 18

Asch, S. E. 1951. "Effects of Group Pressure upon the Modification and Distortion of Judgment." In *Groups, Leadership and Men: Research in Human Relations*, 177–90. Edited by H. Guetzkow. Pittsburgh, PA: Carnegie Press.

Asch, S. E. 1952. "Group Forces in the Modification and Distortion of Judgments." In *Social Psychology*, 450–51. Edited by S. E. Asch. Englewood Cliffs, NJ: Prentice Hall.

Asch, S. E. 1956. "Studies of Independence and Conformity: I. A Minority of One against a Unanimous Majority." *Psychological Monographs: General and Applied* 70, no. 9: 1–70.

Bertrand, Marianne, and Sendhil Mullainathan. 2004. "Are Emily and Greg More Employable Than Lakisha and Jamal? A Field Experiment on Labor Market Discrimination." American Economic Review 94, no. 4 (September 2004): 991–1013. https://doi.org/10.3386/w9873.

Brown, Eileen. 2020. "Office Depot Reveals Habits of US Tech Device Users." ZDNet, January 28, 2020. https://www.zdnet.com/article/office-depot-reveals-usage-habits-of-us-tech-device-users/#:~:text=It%20wanted%20to%20explore%20consumer's,their%20laptops%20every%204.8%20years.

Cialdini, Robert, and Wesley Schulz. 2004. *Understanding and Motivating Energy Conservation Via Social Norms.* Final report prepared for the William and Flora Hewlett Foundation.

Dickerson, Chris Ann, Ruth Thibodeau, Elliot Aronson, and Dayna Miller. 1992. "Using Cognitive Dissonance to Encourage Water Conservation." *Journal of Applied Social Psychology* 22, no. 11 (June 1992): 841–54.

Dunbar, Robin. 1993. "Coevolution of Neocortex Size, Group Size, and Language in Humans." *Behavioral and Brain Sciences* 16: 681–735.

Ecochallenge. 2022. "The Power of Collective Impact." https://ecochallenge.org/collective-impact/.

Fine, Eve, Jennifer Sheridan, Molly Carnes, Jo Handelsman, Christine Pribbenow, Julia Savoy, and Amy Wendt. 2014. "Minimizing the Influence of Gender Bias on the Faculty Search Process." *Gender Transformation in the Academy (Advances in Gender Research)* 19 (October): 267–89. https://doi.org/10.1108/s1529-212620140000019012.

Forscher, Patrick S., Calvin K. Lai, Jordan R. Axt, Charles R. Ebersole, Michelle Herman, Patricia G. Devine, and Brian A. Nosek. 2019. "A Meta-Analysis of Procedures to Change Implicit Measures." *Journal of Personality and Social Psychology* 117, no. 3: 522–59. https://doi.org/10.1037/pspa0000160.

Geller, Scott E. 1981. "Evaluating Energy Conservation Programs: Is Verbal Report Enough?" *Journal of Consumer Research* 8, no. 3 (December 1, 1981): 331–35. doi:10.1086/208872.

Goldstein, Noah, Robert Cialdini, and Vladas Griskevicius. 2008. "A Room with a Viewpoint: Using Social Norms to Motivate Environmental Conservation in Hotels." *Journal of Consumer Research* 35, no. 3 (March 3, 2008): 472–82.

Joro. 2021. "How It Works." https://www.joro.app/how-it-works.

Kahan, Dan M., Hank Jenkins-Smith, and Donald Braman. 2011. "Cultural Cognition of Scientific Consensus." *Journal of Risk Research* 14, no. 2 (September 10, 2011): 147–74.

Kahneman, Daniel, Jack L. Knetsch, and Richard H. Thaler. 1990. "Experimental Tests of the Endowment Effect and the Coase Theorem." *Journal of Political Economy* 98, no. 6: 1325–48.

Kallbekken, Steffen, and Håkon Sælen. 2013. "'Nudging' Hotel Guests to Reduce Food Waste as a Win–Win Environmental Measure." *Economics Letters* 119, no. 3 (June 2013): 325–27. doi:10.1016/j.econlet.2013.03.019.

Kolbert, Elizabeth. 2017. "Why Facts Don't Change Our Minds." *New Yorker*, February 19, 2017. www.newyorker.com/magazine/2017/02/27/why-facts-dont-change-our-minds.

Kollmuss, Anja, and Julian Agyeman. 2002. "Mind the Gap: Why Do People Act Environmentally and What Are the Barriers to Pro-Environmental Behavior?" *Environmental Education Research* 8, no. 3: 239–60. https://doi.org/10.1080/13504620220145401.

Lakoff, George. 2004. *Don't Think of an Elephant: Know Your Values and Frame the Debate.* White River Junction, VT: Chelsea Green.

Lokhorst, Anne Marike, Carol Werner, Henk Staats, Eric van Dijk, and Jeff L. Gale. 2013. "Commitment and Behavior Change: A Meta-Analysis and Critical Review of Commitment-Making Strategies in Environmental Research." *Environment and Behavior* 45, no. 1 (June 9, 2013): 3–34. https://doi.org/10.1177/0013916511411477.

Manser, Michael P., Michael Rakauskas, Justin Graving, and James W. Jenness. 2010. *Fuel Economy Driver Interfaces: Develop Interface Recommendations—Report on Task 3.* DOT HS 811 319. US Department of Transportation, May 2010. www.nhtsa.gov/sites/nhtsa.dot.gov/files/documents/811319.pdf.

Maslow, A. H. 1943. "A Theory of Human Motivation." *Psychological Review* 50, no. 4: 370–96.

Sweetland, Julie, and Rob Shore. 2014. "Reframing Issues in the Digital Age: Using Social Media Strategically." *Nonprofit Quarterly*, August 29, 2014. https://nonprofitquarterly.org/reframing-issues-in-the-digital-age-using-social-media-strategically/.

Taylor, Dorceta E. 2010. *The State of Diversity in Environmental Organizations.* https://www.researchgate.net/publication/323518991_The_State_of_Diversity_in_Environmental_Organizations_Mainstream_NGOs_Foundations_Government_Agencies.

Vaillant, George. 1977. *Adaptation to Life.* Boston: Little, Brown.

Wood-Wright, Natalie. 2019. "Widespread Confusion about Food Safety Labels Leads to Food Waste, Survey Finds." *Hub*, March 14, 2019. https://hub.jhu.edu/2019/03/14/misunderstanding-food-labels-waste/.

Yeginsu, Ceylan. 2018. "Istanbul Vending Machines Offer Subway Credit for Recycled Bottles and Cans." *New York Times*, October 17, 2018. www.nytimes .com/2018/10/17/world/europe/istanbul-vending-machines-recycling-subway.html.

CHAPTER 19

Association for the Advancement of Sustainability in Higher Education. n.d. "For Students." Accessed October 2, 2019. www.aashe.org/resources/for-students/.

Banks, Matthew C. 2010. "World Wildlife Fund." In *Good Cop Bad Cop: Environmental NGOs and Their Strategies toward Business*, 171–83. Edited by Thomas P. Lyon. Washington, DC: RFF.

Bonini, Sheila, and Stephan Görner. 2011. "The Business of Sustainability." McKinsey & Company. www.mckinsey .com/business-functions/sustainability/our-insights/the-business-of-sustainability-mckinsey-global-survey-results.

Bové, Anne-Titia, Dorothee D'Herde, and Steven Swartz. 2017. "Sustainability's Deepening Imprint." McKinsey & Company. www.mckinsey.com/business-functions/sustainability/our-insights/sustainabilitys-deepening-imprint.

Caron, Christina. "Starbucks to Stop Using Disposable Straws by 2020." *New York Times*, July 9, 2018. www. nytimes.com/2018/07/09/business/starbucks-plastic-straws.html.

Chaffin, Bryan. 2014. "Tim Cook Soundly Rejects Politics of the NCPPR, Suggests Group Sell Apple's Stock." *Mac Observer*, February 28, 2014. www.macobserver .com/tmo/article/tim-cook-soundly-rejects-politics-of-the-ncppr-suggests-group-sell-apples-s.

Christakis, Nicholas, and James H. Fowler. 2009. *Connected: The Surprising Power of Our Social Networks and How They Shape Our Lives*. Boston: Little, Brown.

Creighton, Sarah Hammond. 1999. *Greening the Ivory Tower: Improving the Environmental Track Record of Universities, Colleges, and Other Institutions*. Cambridge, MA: MIT Press.

Davies, Kert. 2010. "Greenpeace." In *Good Cop Bad Cop: Environmental NGOs and Their Strategies toward Business*, 195–207. Edited by Thomas P. Lyon. Washington, DC: RFF.

"Debbie Mielewski." 2014. Ford Motor Company, Media Center. 2014. https://media.ford.com/content/fordmedia/fna/us/en/people/debbie-mielewski.html.

Dodds, P. S., Duncan Watts, and Roby Muhamad. 2003. "An Experimental Study of Search in Global Social Networks." *Science* 301 (August 8, 2003): 827–29.

Doppelt, Bob. 2003. *Leading Change toward Sustainability: A Change-Management Guide for Business, Government and Civil Society*. Sheffield, UK: Greenleaf.

Elkington, John. 1997. *Cannibals with Forks: The Triple Bottom Line of Twenty-First Century Business*. Oxford: Capstone.

Fischel, Daniel R. n.d. *Fossil Fuel Divestment: A Costly and Ineffective Investment Strategy*. Compass Lexecon. Accessed October 2, 2019. http://divestmentfacts.com/pdf/Fischel_Report.pdf.

Fischetti, Mark, and Jen Christiansen. 2018. "Only 150 of Your Facebook Contacts Are Real Friends." *Scientific American*, September 1, 2018. https://www .scientificamerican.com/article/only-150-of-your-facebook-contacts-are-real-friends/.

Gilliland, Nikki. 2017. "11 Impressive Influencer Marketing Campaigns." Econsultancy, June 22, 2017. https:// econsultancy.com/impressive-influencer-marketing-campaigns/.

Granskog, Anna, Eric Hannon, Solveigh Hieronimus, Marie Klaeyle, and Angela Winkle. 2021. "How Companies Capture the Value of Sustainability: Survey Findings." McKinsey & Company, April 28, 2021. https://www .mckinsey.com/business-functions/sustainability/our-insights/how-companies-capture-the-value-of-sustainability-survey-findings.

Hazan, Louise, Yossi Cadan, Richard Brooks, Alex Rafalowicz, and Brett Fleishman. n.d. *1000 Divestment Commitments and Counting*. Go Fossil Free. Accessed October 2, 2019. https://gofossilfree.org/wp-content/uploads/2018/12/1000divest-WEB-.pdf.

IKEA. n.d. *Inter IKEA Sustainability Summary Report*. Accessed October 21, 2019. https://preview. thenewsmarket.com/Previews/IKEA/DocumentAssets/502623.pdf.

"Interface Carbon Neutral Floors Program Adds Nora Rubber Flooring." 2019. Facility Executive, February 26, 2019. https://facilityexecutive.com/2019/02/interface-carbon-neutral-floors-program-adds-nora-rubber-flooring/.

Interface. n.d. "A Look Back: Interface's Sustainability Journey."

Kulkarni, Smruti, and Arnaud LeFevebvre. 2018. "How Can Sustainability Enhance Your Value Proposition?" Nielsen Company.

LaCroix, Kevin. 2014. "Environmental Liability and D&O Exposure." *D&O Diary*, May 30, 2014. www .dandodiary.com/2014/05/articles/environmental-liability/environmental-liability-and-do-exposure/.

Malangone, Kathy. 2012. "Green Building Outlook Strong for Both Non-Residential & Residential Sectors." Dodge Data & Analytics. .

McCarthy, Stephen, and Luca Tacconi. 2011. "The Political Economy of Tropical Deforestation: Assessing Models and Motives." *Environmental Politics* 20, no. 1 (January 18, 2011): 115–32.

McGregor, Jena. 2015. "What Etsy, Patagonia and Warby Parker Have in Common." *Washington Post*, April 20, 2015. www.washingtonpost.com/news/on-leadership/wp/2015/04/20/what-etsy-patagonia-and-warby-parker-have-in-common/?utm_term=. c13348ea67a7.

National Center for Education Statistics. 2021. "Endowment Funds of the 20 Degree-Granting Postsecondary Institutions with the Largest Endowments, by Rank Order: Fiscal Year 2020" [data table]. Digest of Education Statistics. https://nces.ed.gov/fastfacts/display.asp?id=73.

National Council of Nonprofits. n.d. "What Is a Nonprofit?" Accessed June 1, 2019. www.councilofnonprofits.org/what-is-a-nonprofit.

New Belgium Brewing Company. n.d. *Sustainability* [brochure]. Fort Collins, CO: New Belgium Brewing.

Polman, Paul, and C. B. Bhattacharya. 2016. "Engaging Employees to Create a Sustainable Business." *Stanford Social Innovation Review*, Fall 2016. https://ssir.org/articles/entry/engaging_employees_to_create_a_sustainable_business.

Senge, Peter, Bryan Smith, Nina Kruschwitz, Joe Laur, and Sara Schley. 2008. *The Necessary Revolution: How Individuals and Organizations Are Working Together to Create a Sustainable World*. New York: Doubleday.

Silvestri, Luciana, and Ranjay Gulati. 2015. "From Periphery to Core: A Process Model for Embracing Sustainability." In *Leading Sustainable Change: An Organizational Perspective*, 81–110. Edited by Rebecca Henderson, Ranjay Gulati, and Michael Tushman. Oxford: Oxford University Press.

Taylor, Dorceta E. 2014. *The State of Diversity in Environmental Organizations*. Prepared for Green 2.0, July 2014. https://vaipl.org/wp-content/uploads/2014/10/ExecutiveSummary-Diverse-Green.pdf.

US Department of Education, National Center for Education Statistics. 2019. *Digest of Education Statistics (NCES 2017-2014)*. Washington, DC: US Department of Education.

US Green Building Council. n.d. "Green Building Leadership Is LEED." Accessed October 21, 2019. https://new.usgbc.org/leed.

Watson, Bruce. 2016. "The Troubling Evolution of Corporate Greenwashing." *Guardian*, August 20, 2016. www.theguardian.com/sustainable-business/2016/aug/20/greenwashing-environmentalism-lies-companies.

Watson, Stephen, Michael Mulet Solon, Wouter-Jan Schouten, Sharon Hesp, Alessandro Runci, and Marije Willems. 2016. *Slow Road to Sustainability*. Gland, Switzerland: World Wildlife Fund for Nature.

World Wildlife Fund. n.d. "Priority Commodities." Accessed May 31, 2019. www.worldwildlife.org/initiatives/transforming-business.

CHAPTER 20

350.org. n.d. "About 350." Accessed October 2, 2019. https://350.org/about/.

Andrews, Richard N. L. 2006. *Managing the Environment, Managing Ourselves: A History of American Environmental Policy*. New Haven, CT: Yale University Press.

Bomey, Nathan. 2016. "BP's *Deepwater Horizon* Costs Total $62B." *USA Today*, July 14, 2016. www.usatoday.com/story/money/2016/07/14/bp-deepwater-horizon-costs/87087056/

Brenan, Megan. 2018. "Polarized Americans Rate Environment Worst since 2009." Gallup, March 29, 2018. https:// news.gallup.com/poll/231971/polarized-americans-rate-environment-worst-2009.aspx.

Carey, Bjorn. 2016. "Environmental Conservation Efforts in China Are Making a Positive Impact, Stanford Scientists Say." *Stanford News*, June 16, 2016. https:// news.stanford.edu/2016/06/16/chinas-environmental-conservation-efforts-making-positive-impact-stanford-scientists-say/.

Deutscher Bundestag. 2021. "Sitzverteilung des 20. Deutschen Bundestages." October 26, 2021. https://www.bundestag.de/parlament/plenum/sitzverteilung_20wp.

Ginsberg, Benjamin, Theodore Lowi, Margaret Weir, Caroline Tolbert, and Robert J. Spitzer. 2015. *We the People*. New York: Norton.

Grad, Shelby. 2017. "The Environmental Disaster That Changed California—and Started the Movement against Offshore Oil Drilling." *Los Angeles Times*, April 28, 2017. www.latimes.com/local/lanow/la-me-santa-barbara-spill-20170428-htmlstory.html.

Gross, Samantha J. 2019. "Lawmakers Say Partial Fracking Ban Is Better Than Nothing: Environmentalists Disagree." *Herald/Times*, Tallahassee Bureau, March 26, 2019. www.miamiherald.com/news/politics-government/state-politics/article228430179.html.

Human Rights Watch. n.d. "China: Events of 2016." In *World Report 2017*. Accessed October 21, 2019. www .hrw.org/world-report/2017/country-chapters/china- and-tibet.

Human Rights Watch. 2021. *World Report 2021: Events of 2020*. https://www.hrw.org/sites/default/files/ media_2021/01/2021_hrw_world_report.pdf.

Klyza, Christopher McGrory, and David J. Sousa. 2013. *American Environmental Policy beyond Gridlock*. Cambridge, MA: MIT Press.

Krogstad, Jens Manuel, and Mark Hugo Lopez. 2017. "Black Voter Turnout Fell in 2016, Even as a Record Number of Americans Cast Ballots." Pew Research Center. www.pewresearch.org/fact-tank/2017/05/12/ black-voter-turnout-fell-in-2016-even-as-a-record- number-of-americans-cast-ballots/.

Layzer, Judith A. 2006. *The Environmental Case: Translating Values into Policy*. Washington, DC: CQ Press.

Levin, Kelly, and Dennis Tirpak. 2018. "2018: A Year of Climate Extremes." World Resources Institute. www .wri.org/blog/2018/12/2018-year-climate-extremes.

Martinson, Erica. 2017. "Opening ANWR to Oil Drilling Is Priority in Trump's Proposed Budget." *Anchorage Daily News*, May 23, 2017. www.adn.com/ politics/2017/05/23/opening-anwr-to-oil-drilling-is- priority-in-trumps-proposed-budget/.

New Jersey Department of Environmental Protection. n.d. "Environmental Justice." https://www.nj.gov/dep/ej/ communities.html.

Oklahoma Cooperative Extension Service. 2017. *Whose Water Is It Anyway? Comparing the Water Rights Frameworks of Arkansas, Oklahoma, Texas, New Mexico, Georgia, Alabama, and Florida*. OSU Extension Fact Sheets, April 2017. http://factsheets.okstate.edu/documents/e-1030- whose-water-is-it-anyway-comparing-the-water-rights- frameworks-of-arkansas-oklahoma-texas-new-mexico- georgia-alabama-and-florida/.

O'Neil, Patrick. 2017. *Essentials of Comparative Politics*. New York: Norton.

Salzman, James, and Barton H. Thompson Jr. 2003. *Environmental Law and Policy*. St. Paul, MN: Foundation Press.

Speth, James Gustave, and Peter M. Haas. 2007. *Global Environmental Governance*. Noida: Pearson Education India.

United Nations Environment Programme. 2018. "Executive Summary." In *Emissions Gap Report 2018*. November 27, 2018. www.unenvironment.org/ resources/emissions-gap-report-2018.

US Environmental Protection Agency. n.d. "Resource Conservation and Recovery Act (RCRA) Overview." Accessed October 21, 2019. www.epa.gov/rcra/ resource-conservation-and-recovery-act-rcra- overview.

US Environmental Protection Agency. 1989. "State and Local Involvement in the Superfund Program." Publication no. 9375.5-01/FS. Fall 1989. https:// semspub.epa.gov/work/HQ/174431.pdf.

Glossary

A

3Es (environment, economy, and equity) a framework that considers the environmental effects as well as the economic and social effects of human actions. (Ch. 1)

acid a compound that yields positively charged hydrogen ions (H^+) when dissolved in water. (Ch. 3)

acid deposition a phenomenon where air pollutants react with water making an acid that precipitates from the sky, also known as "acid rain." (Ch. 8)

adaptive radiation an evolutionary process where over many generations an ancestral species diversifies, producing a variety of new species adapted to specific sets of conditions, or *niches*, of an ecosystem. (Ch. 4)

agriculture the process of cultivating plants and domesticating animals for consumption. (Ch. 12)

agrobiodiversity the variety and variability of genetic material in the life-forms used by humans for food and other agricultural applications. (Ch. 12)

air mass a large volume of air typically several kilometers thick and a thousand or more kilometers wide that has relatively uniform temperature and humidity. (Ch. 8)

air pollution what occurs when the addition of materials into the atmosphere adversely affects the health of humans and/or other organisms. (Ch. 8)

albedo a measure of the reflectivity of a surface. (Ch. 7, 11)

allele alternate form of a gene. (Ch. 4)

ammonification a process where bacteria and fungi break down nitrogen compounds in animal waste products and in dead plant and animal matter and release ammonia. (Ch. 10)

Anthropocene a new, current epoch of Earth history suggested by many scientists, which is marked by conspicuous human effects on the planet. (Ch. 1)

anthropocentrism an approach to ethics where moral concerns are focused on the interests of humans. (Ch. 2)

aquaculture the practice of raising seafood in controlled ponds, tanks, or pens. (Ch. 12)

aquifer a subsurface area of rock or sediment where water can accumulate or pass through. (Ch. 7)

asthenosphere an area of Earth's mantle that is relatively pliable and is situated below the more rigid lithosphere. (Ch. 9)

atmosphere a dynamic envelope of gases extending up from Earth's surface that clings to our planet because of gravitational pull. (Ch. 8)

atom the smallest unit of an element that retains all the characteristics of that element. (Ch. 3)

authoritarian system a form of government in which power is concentrated in a single individual or among a small group of elites. (Ch. 20)

automatic thinking an instantaneous cognitive system distinct from our slower, conscious, and more reflective ways of thinking. (Ch. 18)

B

balancing feedback in a system, a type of feedback that counteracts the direction in which a particular stock is changing. (Ch. 10)

barometric pressure a measure of the pressure exerted by Earth's atmosphere at any given point. (Ch. 8)

base a compound that yields negatively charged hydroxide ions (OH^-) when dissolved in water. (Ch. 3)

Basel Convention an international agreement negotiated to restrict movements of hazardous waste from more-developed to less-developed countries. (Ch. 15)

belief perseverance a bias that occurs when people cling to existing beliefs on an issue even when they are presented with evidence to the contrary. (Ch. 18)

benefit corporation an official legal status for business organizations that want to pursue more socially or environmentally driven objectives. (Ch. 19)

bias an unreasonable weighting, inclination, or prejudice of one's thinking that leads to misunderstandings. (Ch. 1)

biocentrism an approach to ethics where the interests of all living things are considered. (Ch. 2)

biodiversity the variety of species and life in the world or in a particular ecosystem. (Ch. 4)

biodiversity hot spot place on Earth where a large number of species are particularly vulnerable to extinction. (Ch. 4)

biofuel recently living matter or by-products of its decomposition used as an energy source. (Ch. 13, 14)

biogeochemical cycle a path that shows how matter on Earth flows through different parts of the environment. (Ch. 10)

biological hazard an organic substance that poses a threat to the health of living organisms. (Ch. 17)

biome a particular region of Earth that has a distinctive type of climate, organisms, vegetation, and overall ecosystem. (Ch. 4)

bioreactor landfill a disposal strategy using injected water and air to accelerate decomposition and reduce the volume of waste. (Ch. 15)

biosphere the entirety of the regions of Earth occupied by living things. (Ch. 4)

blood-borne pathogen an infectious microorganism in human blood that can cause disease in humans. (Ch. 17)

boycott the act of withholding your buying power (as an individual or group) from a company to motivate change. (Ch. 2)

bright lights syndrome a perception that cities are places of greater opportunity, excitement, and freedom to pursue all aspects of life. (Ch. 16)

bycatch the incidental capture of non-food species in fishing nets and lines. (Ch. 12)

C

Cairo Consensus an agreement saying that demographic and development goals can only be met when the rights and opportunities of men and women are balanced. (Ch. 6)

cap and trade a system where a government sets an overall maximum allowable emissions standard (cap) and then creates a market that enables pollution allowances to be bought, sold, traded, or saved for the future. (Ch. 2, 13)

captive breeding the process in which surviving individuals are removed from the wild so that they can be raised and reproduce in a controlled environment. (Ch. 5)

carbon capture and storage (CCS) a process that prevents CO_2 emissions from escaping into the atmosphere by injecting them underground. (Ch. 13)

carbon tax a tax levied on fossil fuels. (Ch. 13)

carrying capacity the maximum number of individuals of a species that a habitat can sustainably support. (Ch. 4, 6)

checks and balances a system that ensures a part or branch of the government has the means to exert some influence over the activities of the other branches. (Ch. 20)

chemical bond the forces that hold the atoms in molecules together. (Ch. 3)

chemical hazard a chemical linked to immediate or delayed health effects after exposure. (Ch. 17)

chemical reaction the process by which one or more substances are converted into different substances. (Ch. 3)

Clean Air Act of 1970 a law that established two important policy strategies: air quality standards and regulations on the source of pollutants. (Ch. 8)

Clean Water Act a 1972 federal law that regulates the discharge of pollution in water. (Ch. 7)

climate the long-term average of weather conditions for a given region. (Ch. 8)

climate change long-term change in climate conditions, such as temperature and precipitation (Ch. 1).

closed system a system that is self-contained, neither receiving inputs nor sending outputs beyond the system's borders. (Ch. 10)

coal a solid carbon-based fossil fuel formed from ancient tropical swamps that were buried and subjected to immense pressure over millions of years. (Ch. 13)

coercion the use or threat of physical force to maintain authority. (Ch. 20)

coevolution where two species evolve together, and the adaptations of one species cause a second species to adapt too. (Ch. 4)

cogeneration a principle where systems capture waste heat from power plants and use it as an additional source of heating or cooling. (Ch. 14)

cognitive dissonance an unpleasant tension or stress we feel when we are made aware that our attitudes and actions do not match. (Ch. 18)

cold front a cold air mass replacing a warm air mass. (Ch. 8)

combustion in waste disposal, the strategy of burning waste. (Ch. 15)

commitment an action that is taken to express a general attitude on an issue. (Ch. 18)

communication strategy verbal or nonverbal ways of connecting with others to influence or communicate. (Ch. 1)

community in biology, populations of different species living and interacting in a particular place. (Ch. 4)

competition in biology, what occurs when one individual reduces for others the availability of a resource—such as food, water, or potential mates. (Ch. 4)

compound a substance made of atoms of two or more elements bonded to each other. (Ch. 3)

concentrated animal feeding operation (CAFO) an industrial livestock operation in which large numbers of animals are raised in confinement. (Ch. 12)

concentrated solar thermal (CST) plant a solar power technology that captures heat using huge mirrors to focus sunlight toward liquid-filled pipes or a centrally located "power tower." (Ch. 14)

confirmation bias the tendency for people to select information that confirms their existing beliefs and "fit" information they are given into their existing way of thinking. (Ch. 18)

conspicuous consumption the act of purchasing certain goods to project particular identities or images within society. (Ch. 15)

consumption the ways we use and dispose of material things. (Ch. 15)

containment building a structure that isolates the waste from the surrounding area while allowing constant monitoring and retrieval of it. (Ch. 15)

contraception technology, such as condoms or birth-control pills, that greatly reduces the probability of impregnation. (Ch. 6)

controlled experiment a test in which researchers intentionally manipulate some specific aspect of a system to see how this change affects the outcome. (Ch. 1)

Convention on Biological Diversity an agreement committing countries to pass laws to expand protected areas, restore degraded ecosystems, and ensure the sustainable and equitable human use of ecosystem services. (Ch. 5)

Convention on International Trade in Endangered Species of Wild Fauna and Flora (CITES) an agreement banning the hunting, capture, and selling of endangered and threatened species. (Ch. 5)

conventional reserve an easily obtained deposit of fossil fuels. (Ch. 13)

convergence when unrelated species resemble one another because of evolution under similar environmental conditions. (Ch. 4)

conversion a disposal strategy where waste is converted to something else useful. (Ch. 15)

core the center of Earth, with a solid inner section and a liquid outer section both made primarily of iron and nickel. (Ch. 9)

Coriolis effect a force driven by Earth's rotation that deflects objects, winds, and currents on the surface of Earth and in the ocean or the atmosphere. (Ch. 11)

crude death rate the total number of deaths per year per 1,000 people. (Ch. 6)

crust Earth's outermost layer. (Ch. 9)

cryosphere the frozen part of Earth's surface and crust. (Ch. 7)

cultural services benefits to humans from the environment that may not be essential to survival but that greatly enhance our quality of life. (Ch. 5)

culture the systematic, learned, and shared understandings and behaviors of a particular group. (Ch. 15)

D

deep current the flow of ocean water below the surface caused by variations in density, temperature, and salinity. (Ch. 11)

deforestation the clearing of large areas of forested land. (Ch. 5)

degree of separation a measure of social distance between people. (Ch. 19)

delegation an emotional defense mechanism by which individuals absolve themselves of responsibility for a problem by blaming others or deeming it someone else's responsibility. (Ch. 18)

delta a landform created where the river deposits its sediment load as it flows into the ocean. (Ch. 9)

demand in economics, a concept reflecting how much someone (the consumer) desires and is willing to pay for a good. (Ch. 2)

democratic system a form of government in which the public plays a significant role in government decision making. (Ch. 20)

demographer a social scientist who studies the characteristics and consequences of human population growth. (Ch. 6)

demographic transition decrease in the birth and death rates of a population linked to improvements in basic human living conditions, the availability of modern birth-control technologies, and economic growth. (Ch. 6)

demographic window a time when a country's population is dominated by people of working age; the share of the population younger than 15 years is less than 30%, and the share older than 64 years is less than 15%. (Ch. 6)

denial an emotional defense mechanism by which people refuse to acknowledge a problem's existence. (Ch. 18)

denitrification a process in which microorganisms use nitrogen compounds for respiration and create N_2 gas, returning it to the atmosphere. (Ch. 10)

deontological ethics an approach to ethics that establishes duties that ought to be upheld and rights that ought to be protected. (Ch. 2)

desalination a process of removing salt from seawater to create fresh water. (Ch. 7)

desertification a process prompted by drought, extreme erosion, and soil infertility that causes land to lose more than 10% of its productivity. (Ch. 12)

design for recovery a process where guidelines are established to standardize the manufacture of materials and products in order to facilitate efficient recycling. (Ch. 15)

diarrheal disease a disease caused by pathogens that affect the digestive tract. (Ch. 17)

diazotroph a bacterium with an enzyme that uses free nitrogen from the air along with hydrogen to produce ammonia. (Ch. 10)

disinformation deliberately misleading, incorrect, or false information. (Ch. 1)

dominance in biology, the opposite of diversity; when a single or small number of species is far more abundant than other species. (Ch. 4)

drought a prolonged period of low precipitation and high evaporation rates that can lead to water shortages. (Ch. 7, 11, 17)

E

Earth Charter a United Nations initiative completed in 2000 that outlines duties for the people of Earth, including protecting and restoring Earth's ecological systems. (Ch. 2)

earthquake a powerful geologic hazard typically caused when parts of Earth's crust shift along faults. (Ch. 17)

ecocentrism an approach to ethics concerned with all the living and non-living components of ecosystems. (Ch. 2)

ecological footprint analysis an analysis that tries to tally the area of land (and water) required for each category of consumption and waste discharge to make human consumption impacts more visible. (Ch. 6)

ecological island effect negative effects on a population when its protected habitat is isolated amid wider unprotected areas. (Ch. 5)

ecological niche a particular role or position of a species in a community. (Ch. 4)

ecological resilience the ability of an ecosystem to recover from damage suffered in a disturbance and return to its pre-disturbed state. (Ch. 1)

ecological resistance the ability of an ecosystem to remain unchanged in the face of a disturbance. (Ch. 1)

ecological restoration the process of assisting the recovery of an entire ecosystem. (Ch. 5)

economic system a chain of exchange that helps shape the production, distribution, and consumption of things we use. (Ch. 2)

economy of scale a situation in which the per capita cost for services decreases as a result of the increasing scale of operations. (Ch. 16)

ecosystem a community of life and the physical environment with which it interacts. (Ch. 1, 4)

ecosystem services a concept that assigns a value to human benefits derived from naturally functioning ecosystems. (Ch. 1, 2, 5)

edge effect the distinctive environmental conditions and community of life that occur in a place where two or more types of habitats come together. (Ch. 5)

El Niño–Southern Oscillation (ENSO) a change from normal ocean currents that significantly weakens or even shifts the direction of trade winds and ocean currents in the southern Pacific. (Ch. 11)

electric vehicle (EV) a vehicle powered by an electric motor using a magnetic field to generate motion. (Ch. 14)

electron a particle within atoms that carries a negative charge. (Ch. 3)

element a substance that cannot be broken down into other substances. (Ch. 3)

embedded water a concept that accounts for the water that is used to produce goods that we consume. (Ch. 7)

embodied energy the energy used to make and transport goods. (Ch. 11)

emergent property an outcome arising from the function of the system as a whole. (Ch. 1, 10)

emotional defense mechanism a quick and automatic thought process that relieves us of bad feelings associated with a distressing situation. (Ch. 18)

emotional distancing an emotional defense mechanism by which people prevent themselves from feeling emotions associated with a problem. (Ch. 18)

Endangered Species Act an act making it illegal to directly and intentionally harm endangered species through activities such as hunting, commercial development, and trapping. (Ch. 5)

energy the capacity to do work. (Ch. 3)

Energy Policy and Conservation Act a 1975 act that set minimum miles per gallon (mpg) fuel-economy standards for cars and light trucks. (Ch. 8)

enhanced oil recovery (EOR) a process where captured CO_2 is condensed and pumped underground as a way to force more oil out of depleted wells. (Ch. 13)

entropy a measure of the degree of disorder in a system. (Ch. 3)

environment the multitude of living and nonliving things on Earth that sustain life, including our own. (Ch. 1)

environmental hazards biological, chemical, and physical factors in the environment that threaten the health of humans and living things. (Ch. 17)

environmental health the assessment and control of the biological, chemical, and physical factors in the environment that affect our well-being. (Ch. 17)

environmental justice the principle that no community should bear more environmental burdens or enjoy fewer environmental benefits than others. (Ch. 1, 17)

epidemiology the study of the distribution and determinants of health-related states and events in specific populations. (Ch. 17)

erosion a process where natural forces—such as wind, water, ice, and gravity—move weathered rock particles. (Ch. 9)

estuary a wetland where fresh water meets salt water. (Ch. 7)

ethanol a type of biofuel made of alcohol and produced by fermenting sugars in plants such as corn or sugarcane; it is often blended with gasoline. (Ch. 14)

ethics a set of moral principles that provides guidelines for our behavior. (Ch. 2)

eutrophication a process where marine and freshwater environments are enriched with nutrients, such as phosphorus or nitrogen from agricultural runoff or other sources, causing rapid growth, death, and decomposition of algae and phytoplankton. The decomposition process consumes and depletes oxygen levels in the water, harming oxygen-dependent organisms. (Ch. 7)

evapotranspiration the combined water vapor released from Earth's land and water surfaces along with transpiration from plants. (Ch. 7)

evolution the process of genetic change in populations over generations. (Ch. 4)

e-waste waste created from and by electronic devices. (Ch. 15)

exponential growth rapid population growth that occurs when environmental conditions allow for a constant ratio in the increase of individuals over time, resulting in population growth proportional to population size. (Ch. 4)

externality failures of markets to account for all the costs and benefits of goods and services. (Ch. 2)

extinction a complete and permanent loss of that particular species. (Ch. 4)

F

Fair Trade a certification standard that assures customers that the price of a product provides for adequate wages and environmentally sustainable production and helps farming communities thrive. (Ch. 12)

fallow an agricultural field that remains unplanted for the purpose of restoring soil health. (Ch. 12)

fault a fracture in Earth's crust, along which one body of rock slides past another. (Ch. 9)

federal governing system a form of government in which states or provinces have some powers that are independent of the central government. (Ch. 20)

feedback a loop in a system that responds to and produces changes in levels of stocks that either amplify or counter a change. (Ch. 10, 11)

feedback (behavior) a prompt that contains information specific to the impact of an individual's behavior. (Ch. 18)

Ferrel cell an air circulation pattern between latitudes 30° N and 60° N and between latitudes 30° S and 60° S. (Ch. 8)

fertility transition a period when the population growth slows because the birth rate decreases, even though the death rate remains low. (Ch. 6)

first law of thermodynamics the physical law that energy is conserved: it cannot be created or destroyed; it merely changes form. (Ch. 3)

fission a process where the nuclei of atoms are split, releasing a very large amount of energy. (Ch. 14)

flood an event that occurs when water inundates land that is normally dry, especially through intense rainfall and rapid snowmelt. (Ch. 17)

floodplain a place where floods frequently send water over the banks of a river or stream channel and deposit sediment. (Ch. 9)

flow the mechanism and rate by which a stock in a system changes over time. (Ch. 10)

food security the affordable access to enough nutritious food to maintain dietary needs. (Ch. 12)

food system the processes that link agricultural production to food consumption. (Ch. 12)

footprint analysis a method to understand the magnitude of the impact of choices and actions individuals make, both over time and collectively. (Ch. 1)

fossil fuel fuel formed from living organisms from earlier geologic eras. (Ch. 13)

frame a mental shortcut that uses word associations and images to help people quickly apply meaning to new information. (Ch. 18)

fraud an attempt to deceive people by communicating findings that are simply false. (Ch. 1)

fuel cell a device that creates an electric current by stripping electrons from hydrogen molecules. (Ch. 14)

G

gene an organism's basic unit of inheritance between a parent and its offspring. (Ch. 4)

gene flow a genetic transfer process where genetic differences between populations decrease. (Ch. 4)

genetic diversity the number of different kinds of genetic characteristics present within a population or species. (Ch. 4)

genetic drift a change in frequency of a trait within the gene pool of a population caused by chance. (Ch. 4)

genetic erosion the process by which genetic diversity is lost. (Ch. 12)

genetically modified organism (GMO) a variety developed by splicing genetic material from one species into the DNA of another. (Ch. 12)

genotype the genetic makeup of an individual. (Ch. 4)

gentrification a process of displacement that occurs when changes are made to a neighborhood's character, leading to more expensive housing, higher taxes, and an influx of wealthier people. (Ch. 16)

geologic disposal injecting or placing waste beneath Earth's surface. (Ch. 15)

geologic hazard a large-scale event that can cause tremendous damage. (Ch. 17)

geothermal gradient how temperature increases with depth in Earth's crust due to decaying radioactive elements. (Ch. 11)

geothermal power power generated by heat from below Earth's surface. (Ch. 14)

glacier a mass of ice that flows over land surfaces and lasts all year. (Ch. 7, 9)

global climate the average weather conditions over a significant period of time for the planet as a whole. (Ch. 11)

government the governing bodies and procedures that have authority over the people residing within a particular area. (Ch. 20)

green infrastructure the wide variety of green spaces that can be incorporated into urban life. (Ch. 16)

greenhouse effect any system where a barrier causes the inflow of energy to outpace the outflow in a way that warms the interior. Specifically, the warming effect on Earth due to atmospheric greenhouse gases that prevent radiant heat emitted from the surface from escaping into space. (Ch. 11)

greenhouse gas a gas in the atmosphere that redirects heat rising from Earth's surface back toward the surface, causing a warming effect. (Ch. 11)

greenwashing the process of advertising environmental values while engaging in unsustainable business practices. (Ch. 19)

groundwater-dependent ecosystem a community of organisms that requires groundwater to meet at least some of its water needs. (Ch. 7)

gyre a large circular ocean current. (Ch. 11)

H

habitat in ecology, the place(s) an organism inhabits. (Ch. 4)

Hadley cell a looping weather pattern forming circulation systems between the equator and latitude 30° N and between the equator and latitude 30° S. (Ch. 8)

heat wave a period of extremely and unusually high temperatures that lasts days or weeks and can be deadly. (Ch. 17)

herd immunity when members of a population are unlikely to become infected by a pathogen because most of the population is either vaccinated or has been previously infected. (Ch. 17)

hierarchy of needs a model of the way people prioritize some needs and desires over others. (Ch. 18)

high-pressure system a system of air formed by cooling air that becomes denser and heavier and then sinks to form an area of high pressure. It is usually associated with clear, dry conditions. (Ch. 8)

hindcasting a scientific technique where simulations are run on a particular model to see how its predictions match up to actual or historical real-world observations. (Ch. 11)

humidity the amount of water in a given volume of air. (Ch. 8)

humus in soil, a complex, dark, sticky organic material that can remain relatively stable over time. (Ch. 9)

hydraulic fracturing (fracking) a process using a mixture of water, sand, and chemicals pumped at high pressure into an oil or gas well to fracture the rock and release the fuel locked up inside. (Ch. 13)

hydrocarbon material made of strands of hydrogen and carbon molecules derived from ancient, solar energy–capturing photosynthesis. (Ch. 13)

hydropower power generated when water is used to spin turbines and generate electricity. (Ch. 14)

hydrosphere all the places that hold water on Earth, including surface water, groundwater, and water in the atmosphere. (Ch. 7)

hypothesis a proposed explanation for a phenomenon or answer to a scientific question. (Ch. 1)

I

I=PAT shorthand for the idea that environmental impact (I) is a function of not only human population (P) but also affluence (A) and technology (T). (Ch. 6)

ice age a time when the climate was much colder, and glaciers at times covered larger areas of continents, and mountain glaciers grew. (Ch. 7)

igneous rock rock formed by the cooling of molten rock. (Ch. 9)

impervious something that does not allow water to pass through it. (Ch. 9)

inbreeding breeding between closely related individuals, often occurring when populations are small. (Ch. 4)

incentive a positive or negative signal that pulls us toward or pushes us away from a certain choice or behavior. (Ch. 1, 18)

incineration in waste disposal, a technology of controlled combustion at high temperatures with pollution controls in place. (Ch. 15)

induced traffic increased traffic and congestion caused by adding lanes to existing roads and highways. (Ch. 16)

inferior good something consumed because people cannot afford what they prefer. (Ch. 13)

infiltration in science, a process where water (or other liquid) moves down into the soil. (Ch. 9)

informal economy small entrepreneurial ventures that operate beyond the reach of government regulation and taxation. (Ch. 16)

infrastructure physical structures and facilities such as roads, buildings, and parking lots. (Ch. 16)

institution a formal and enduring organization, rule, and process. (Ch. 20)

instrumental period the current era in which we have access to temperature readings taken directly with instruments such as thermometers. (Ch. 11)

instrumental value the usefulness of particular species for human purposes. (Ch. 5)

interest group a group formed by people and organizations with shared interests in order to influence government. (Ch. 20)

intrinsic value the value of something in and of itself and apart from its usefulness to others. (Ch. 2, 5)

ionizing radiation a form of energy that can remove electrons from an atom. (Ch. 17)

island biogeography an ecological pattern in which large isolated areas tend to have more species richness than smaller isolated areas of the same habitat type. (Ch. 5)

isolation the segregation of waste from significant contact with humans or the wider environment, normally belowground but sometimes in buildings aboveground. (Ch. 15)

isotope an atom of an element with a particular number of neutrons. (Ch. 3)

J

jet stream a prevailing wind pattern near the top of the troposphere. (Ch. 8)

Jevons paradox a finding that efficiency gains in the use of a resource can lower the cost of that resource, which can cause consumption of the resource to rise. (Ch. 2, 13)

K

keystone species species that exhibit a particularly strong influence over the abundance and diversity of other organisms in their ecosystem. (Ch. 4)

kinetic energy the energy embodied in something due to its being in motion. (Ch. 3)

L

landform a surface feature of the landscape. (Ch. 9)

landslide an event that occurs when rock or other debris detaches from a slope and slides downhill. (Ch. 17)

law of conservation of mass the physical law that matter cannot be created or destroyed, and that the mass of the constituent parts in a chemical reaction remains unchanged even as the atoms involved in the reaction are rearranged. (Ch. 3)

leachate a foul-smelling, soupy liquid that forms when rainwater or groundwater mixes with the decomposing waste at the dump. (Ch. 15)

leaching a process driven by water where mineral elements move down to deeper soil layers, potentially beneath the reach of plant roots, to be carried away in groundwater or stream flow. (Ch. 9, 12)

legitimacy a widely shared belief that government authority is appropriately held. (Ch. 20)

lentic ecosystem an ecosystem in a lake or pond or other relatively still water. (Ch. 7)

life-cycle assessment an evaluation of the environmental impacts of all the steps involved in making, distributing, using, and ultimately disposing of a product. (Ch. 15)

literate life expectancy the average number of years in one's life a person has the ability to read and write. (Ch. 6)

lithosphere the rigid outer portion of Earth, consisting of the crust and the very top of the mantle. (Ch. 9)

livestock animals raised for food. (Ch. 12)

lobbying a practice in which groups or individuals attempt to persuade a government decision maker. (Ch. 20)

loss averse a bias whereby people prefer avoiding the loss of something they already have more than they prefer acquiring an equivalent amount of that same thing. (Ch. 18)

lotic ecosystem an ecosystem in flowing water such as a river or stream. (Ch. 7)

low-pressure system a system of air in which warm, rising air cools as it expands into higher and colder altitudes. The cooling can cause the water vapor in the air mass to condense into clouds that may eventually produce precipitation. (Ch. 8)

M

macronutrient an element that organisms use in large amounts. (Ch. 10)

magnetosphere an area of space around Earth that shields the planet from charged particles emitted from the Sun. (Ch. 9)

mantle the rock section of Earth above the core and below the crust. (Ch. 9)

Marine Mammal Protection Act law protecting marine mammal species such as whales, dolphins, seals, and manatees. (Ch. 5)

market in economics, a system that brings buyers and sellers together to exchange goods. (Ch. 2)

mass a measure of the amount of matter in something. (Ch. 3)

mass burn a technology that combusts solid waste first and then performs secondary combustion of the resulting gases. (Ch. 15)

matter anything in the universe that takes up space. (Ch. 3)

megalopolis a chain of roughly adjacent metropolitan areas, such as the region from Boston to Washington, DC. (Ch. 16)

mesosphere the layer of atmosphere above the stratosphere. It has Earth's coldest temperatures and is where most meteors burn up. (Ch. 8)

metamorphic rock rock produced when one type of rock changes to another because of extreme heat and pressure. (Ch. 9)

methanogen a methane-producing microorganism that thrives in oxygen-poor environments. (Ch. 10)

metropolitan area (metropolis) a cluster of densely populated suburbs and cities. (Ch. 16)

Migratory Bird Treaty Act a treaty between the United States and Canada protecting more than 800 types of birds, including the bald eagle. (Ch. 5)

Milankovitch cycles regular patterns of variation in the shape of Earth's orbit and the tilt and direction of Earth's rotational axis. These alter the amount of the Sun's radiation and energy that reach Earth and its atmosphere. (Ch. 11)

mineral a natural solid from Earth's crust that forms rocks. Minerals also provide key materials for many practical human endeavors. (Ch. 9)

misinformation false or incorrect information that may be spread intentionally or unintentionally. (Ch. 1)

mission statement a brief and general value statement presenting an ideal vision of what an organization hopes to achieve. (Ch. 19)

mixed-use area an area where a mixture of housing, shops, restaurants, grocery stores, offices, and public amenities are all within a half-mile of public transportation. (Ch. 16)

modal split the percentage of travelers in an area using a particular type of transportation to get to work. (Ch. 16)

model a simplified concept or representation of a complex process that is designed to help understand interactions among different factors. (Ch. 1, 10)

molecule what is formed when two or more atoms (either the same or different) bond together. (Ch. 3)

monoculture an agricultural practice of growing a single crop, plant, livestock species, or breed. (Ch. 12)

mortality transition a period that occurs as access to food, clean water, and medical care improves and the country's death rate declines. (Ch. 6)

motivation the drive we feel toward the behaviors we pursue. (Ch. 18)

mountaintop removal a mining process commonly used to extract coal, where entire mountaintops are removed to scoop out the underlying resource. (Ch. 13)

municipal solid waste (MSW) the waste consumers dispose of from their households and businesses. (Ch. 15)

mutation a random change to DNA that can produce an altered trait in an organism. (Ch. 4)

N

national forest lands owned and managed by the US Forest Service for multiple purposes including timber harvests, recreation, and fish and wildlife conservation. (Ch. 5)

national park an area of scenic, historical, or scientific importance protected and maintained by the federal government through the National Park Service. (Ch. 5)

national recreation area an area that conserves and provides recreation, typically around the reservoirs created by large dams. (Ch. 5)

National Wildlife Refuge System a system of public lands and waters set aside and managed by the US Fish and Wildlife Service to conserve America's fish, wildlife, and plants. (Ch. 5)

natural gas a gaseous fossil fuel that formed deep beneath Earth's surface and contains primarily methane. (Ch. 13)

natural increase a process in which population growth due to birth rates exceeds death rates. (Ch. 6)

Natural Resource Damage Assessment a legal process that fines polluters who create oil spills or hazardous waste sites and then uses the revenue for restoration projects. (Ch. 5)

natural selection the process where organisms better adapted to their environment survive and tend to produce more offspring. (Ch. 4)

neutron a particle within atoms that has no electric charge. (Ch. 3)

nitrification a process in which microorganisms convert ammonia to nitrogen compounds. (Ch. 10)

nitrogen fixation a process where free nitrogen—inert nitrogen gas (N_2)—in the air is converted to soluble ammonium (NH_4^+) and nitrate (NO_3^-) ions that plants, algae, and bacteria can take in and use. (Ch. 10)

nonpoint source a broad or diffuse source of pollution, such as agriculture or residential runoff. (Ch. 7)

nuclear power power generated when the nuclei of atoms are split, releasing a large amount of energy. (Ch. 14)

O

obesity a condition of being substantially overweight because of excessive fat accumulation. (Ch. 12)

ocean acidification the ongoing decrease in the pH of Earth's oceans caused by absorption of carbon dioxide (CO_2) from the atmosphere. (Ch. 10, 11)

oil a liquid fossil fuel that formed from plants and microscopic animals that lived millions of years ago. (Ch. 13)

one-use zone a distinct zone, often in suburbs, that serves one particular purpose. (Ch. 16)

open dump unregulated waste disposal in uncovered and openly accessible piles. (Ch. 15)

open system a system that is affected by outside influences. (Ch. 10)

organic dust a dust that comes from materials such as animal feed, bedding, pesticide residues, animal wastes, and dander. (Ch. 12)

organic matter decaying plant and animal matter. (Ch. 12)

organization a larger, more formal, and less personal group, such as a business or a school, set up with specific objectives in mind. (Ch. 19)

organizational culture the widely shared assumptions and patterns of behavior that group members learn and share. (Ch. 19)

organizational structure the way working relationships are arranged and authority is exercised among the members of a group. (Ch. 19)

overfishing a practice that causes the quantity of fish caught to exceed the productive capacity of a species. (Ch. 12)

overnourishment excessive food intake that causes energy use to be less than the food energy consumed. (Ch. 12)

P

paleoclimate ancient climate conditions understood through use of proxies. (Ch. 11)

parasitic relationship an interaction where one organism (known as a parasite) lives off a host organism without immediately killing it. (Ch. 4)

particulate matter (PM) tiny particles and droplets less than 10 micrometers in size—about one-seventh the diameter of a strand of human hair—that are suspended in the air we breathe. (Ch. 8)

pathogen a microorganism that causes illness or infection when it takes up residence in our bodies. (Ch. 17)

payment in policy terms, a positive incentive designed to encourage certain actions. (Ch. 20)

pedestrian-oriented development (POD) use of design strategies that integrate pedestrian travel and land-use patterns in more sustainable and resilient ways. Also see *transit-oriented development*. (Ch. 16)

peer review a process for refining research design and ensuring that conclusions can indeed be drawn from the evidence by subjecting work to assessment by experts in the field of study in question. (Ch. 1)

penalty in policy terms, a disincentive designed to discourage certain actions. (Ch. 20)

permafrost water perpetually frozen in soil or rock. (Ch. 7)

permit government-authorized permission to engage in behavior that would otherwise be illegal. (Ch. 20)

persistence the extent to which a chemical resists being broken down. (Ch. 17)

persistent organic pollutant (POP) a chemical that is resistant to breakdown through chemical reactions, biological processes, or exposure to sunlight. (Ch. 17)

phase change a change in matter from one state (solid or liquid or gas) to another without a change in its chemical composition. (Ch. 3)

phenology the study of the seasonal timing of biological activities, such as the breeding, flowering, and migration of various species. (Ch. 11)

phenotype an individual's observable characteristics, or traits. (Ch. 4)

photosynthesis a series of chemical reactions involving water and CO_2 through which plants and some other organisms store the Sun's energy in simple sugars. (Ch. 3, 10)

photovoltaic (PV) solar panel a panel that produces an electric charge when it is exposed to sunlight. (Ch. 14)

physical hazard an event or phenomenon that causes harm to humans through physical damage. (Ch. 17)

planner's dilemma the need (and difficulty) for a project to operate at both a local and a regional scale. (Ch. 16)

point source a clearly identifiable source of pollution, such as a drainpipe, channel, or ditch. (Ch. 7)

polarity an imbalance of positive and negative charge within an atom. (Ch. 3)

policy authoritative decisions such as laws, regulations, and court rulings guiding the behavior of people and organizations subject to a government. (Ch. 2, 20)

political party an organization that attempts to influence government by working to get its members elected to government offices. (Ch. 20)

politics the process by which decisions are made for and applied to a group of people. (Ch. 20)

population in ecology, a collection of the same species living in a given area. (Ch. 4)

potential energy energy in something that has yet to be released. (Ch. 3)

power (energy) the rate at which work is done, measured as energy flow over time. (Ch. 13)

power (politics) the capacity of individuals and groups to make their concerns or interests count in decision-making processes. (Ch. 20)

predation an interaction where one organism, called the predator, feeds on another, known as the prey. (Ch. 4)

prescriptive regulation a mandate about what can and cannot be done. (Ch. 20)

pressure the force exerted on or acting against something. (Ch. 3, 8)

primary consumer an organism that consumes plants directly. (Ch. 3)

primary pollutant a substance that is harmful in its directly emitted form. (Ch. 8)

primary producer an organism that transforms sunlight into chemical energy held within the molecules of its structure. (Ch. 3)

primary recycling (closed-loop recycling) a process that converts waste materials into the same sort of product from which they came. (Ch. 15)

privately (or closely) held the type of business in which a single individual or a relatively small group of people holds all of the ownership shares. (Ch. 19)

processed food food that is commercially processed to optimize ease of preparation, consumption, and storage. (Ch. 12)

producer responsibility law a law requiring manufacturers of products and packaging also to take responsibility for collection, recycling, reuse, or disposal. (Ch. 15)

productivity the amount of inputs required to attain a certain level of output. (Ch. 13)

prompt a conspicuous reminder that targets behaviors at the moment people are deciding how to act. (Ch. 18)

proportional representation (PR) system an electoral system that elects multiple members from each district and distributes representation in proportion to each party's percentage of the vote. (Ch. 20)

proton a particle within atoms that has a positive charge. (Ch. 3)

proven reserve the amount of a resource that can be profitably accessed with current technology at current prices. (Ch. 13)

provisioning services goods humans consume that are directly provided by ecosystems. (Ch. 5)

proxy an observable and measurable phenomenon that serves as an indirect indicator of changes in climate. (Ch. 11)

pseudoscience claims that are not the result of scientific inquiry or are derived by a process that is not open to scientific scrutiny. (Ch. 1)

public good a thing that cannot be profitably produced because it is difficult to exclude nonpaying customers from receiving the benefits. (Ch. 2, 13)

public health the science and practice of protecting and improving the health of people where they live and work. (Ch. 17)

publicly traded the type of business that has ownership shares that can be bought and sold on a stock exchange. (Ch. 19)

R

radiation a form of energy that travels through space and penetrates various materials. (Ch. 17)

radon a naturally occurring radioactive gas found in rock, soil, and groundwater. (Ch. 17)

recharge the process of adding water to a groundwater system. (Ch. 7)

recycling a strategy of redefining "waste" as "resources" for new products, thus diverting materials from the waste stream. (Ch. 15)

regulating services natural processes of ecosystems that provide favorable conditions for humans. (Ch. 5)

reinforcing feedback a loop that responds to the direction of change in the stock by enhancing that same direction of change. (Ch. 10)

relative humidity the amount of water in the air as a percentage of the maximum amount of water that the air can hold at a given temperature. (Ch. 8)

remediation the process that converts hazardous waste to less-hazardous substances. (Ch. 15)

replacement cost in economics, an estimate of how much it would cost to replace something with a substitute good. (Ch. 5)

replacement fertility a total fertility rate of 2.1 children, which is the rate at which the population does not grow or decline. (Ch. 6)

reproductive isolation the inability of populations to successfully interbreed, due to factors such as geographic isolation; this can lead to evolutionary divergence into distinct species. (Ch. 4)

reserve a known resource of a fossil fuel that can be economically accessed with current technology at current prices. (Ch. 13)

reservoir a part of Earth where a material (such as water) remains for a period of time; also a term used for an artificial water body behind a dam. (Ch. 7)

residence time the time a molecule of a nutrient (like water) spends in a particular reservoir. (Ch. 7)

resignation a feeling of helplessness that leaves a person believing he or she cannot change the situation. (Ch. 18)

respiratory infection an infectious disease affecting the lungs and airways. (Ch. 17)

rock cycle the geologic process by which Earth recycles and renews its surface. (Ch. 9)

ruminant an animal with a digestive system that can turn plant-based cellulose into protein. (Ch. 12)

rural area a place where populations are more spread out geographically and more reliant on agricultural employment. (Ch. 16)

S

Safe Drinking Water Act a federal law that sets the maximum allowable levels of contaminants in drinking water. (Ch. 7)

salinity the concentration of salt in water, generally measured in parts per thousand. (Ch. 11)

salinization the process of mineral salts building up in the soil. (Ch. 9, 12)

sanitary landfill a disposal site that isolates and contains waste, manages its contents, and treats liquid and gas releases. (Ch. 15)

science our way of asking and answering questions and testing ideas about the natural world by using evidence gathered from the natural world. (Ch. 1)

scientific method a formal process of inquiry designed to test problems and ideas. (Ch. 1)

sea ice frozen ocean water that is usually seasonal. (Ch. 7)

second law of thermodynamics the physical law that with each transformation or transfer of energy, some energy is degraded or wasted, and that the tendency of any isolated system is to disorder. (Ch. 3)

secondary pollutant a pollutant that is the product of reactions occurring in the atmosphere. (Ch. 8)

secondary recycling (open-loop recycling) a process that converts waste material from a product into a different sort of product. (Ch. 15)

sediment eroded material that is transported and accumulates in different places. (Ch. 7, 9)

sedimentary rock rock formed when sediment buried under many layers of material cements together due to high pressures and temperatures. (Ch. 9)

single-member-district plurality (SMDP) system an electoral system in which each geographic district chooses a single representative by selecting the candidate who receives the largest number of votes. (Ch. 20)

sink a place where matter accumulates and is held for a long period of time. (Ch. 10)

slum an urban area characterized by substandard housing, a lack of formal property ownership arrangements, inadequate urban services, and high rates of poverty. (Ch. 16)

smart growth principles of urban planning that emphasize inclusive community processes and aim to reduce suburban sprawl and other negative environmental impacts. (Ch. 16)

smog a secondary pollutant that forms when chemicals from the combustion of fossil fuels interact in the presence of sunlight at warm temperatures. (Ch. 8)

social context the collection of people and groups you have contact with as you go about your daily life. (Ch. 19)

social group a collection of people who regularly interact with each other. (Ch. 19)

social network a system of people and their connections that allows certain kinds of things—money, information, influence, and even disease—to flow between people. (Ch. 19)

social norm an expectation or standard for behavior that is widely shared within a group. (Ch. 18)

soil a complex mixture of weathered rock and mineral particles (sediment), dead and decaying plant and animal matter, and the multitude of organisms that live within these materials. (Ch. 9)

soil horizon a layer in soil created by the action of living (biotic) and non-living (abiotic) factors. (Ch. 9)

soil profile the collection of soil horizons at a location. (Ch. 9)

solid waste all discarded material in solid, liquid, semisolid, or contained gaseous form. (Ch. 15)

sovereignty the ability to make policies and carry out actions within a certain territory. (Ch. 20)

speciation the process by which subsets of a population diverge enough genetically to no longer produce fertile offspring when they interbreed. (Ch. 4)

species a group of organisms that are closely related to each other and are usually able to breed with each other to produce viable offspring. (Ch. 4)

species evenness a measure of the relative abundance of each species in a certain area. (Ch. 4)

species richness the number of different kinds of species in an area. (Ch. 4)

stability transition a period when low birth rates match low death rates. (Ch. 6)

status quo the current situation; the existing state of affairs. (Ch. 18)

stock a supply of something that we want to observe and measure over time. (Ch. 10)

storm surge an abnormal rise of marine waters generated by a storm, over and above the predicted tide. (Ch. 17)

strategy the way in which an organization directs its resources toward its goals. (Ch. 19)

stratosphere the layer of the atmosphere above the troposphere. It holds a layer of ozone that protects and warms Earth. (Ch. 8)

subsidence the process of land settling as formerly water-filled spaces collapse under the weight of overlying rock and soil. (Ch. 7, 12)

subsidy a direct payment from the government to encourage a certain activity. (Ch. 20)

suburban sprawl the spread of urban populations away from the centers of cities to widely dispersed areas that have relatively low population densities. (Ch. 16)

supply in economics, the total quantity of a good or service that is available. (Ch. 2)

supporting services the fundamental natural conditions on which many other ecosystem services depend. (Ch. 5)

surface current an ocean current affecting the top 400 meters of water that starts because of air blowing across the water surface. (Ch. 11)

surface water freshwater sources visible to us, such as rivers, streams, lakes, and wetlands. (Ch. 7)

sustainability the management of natural resources in ways that do not diminish or degrade Earth's ability to provide them in the future. (Ch. 1)

sustainable development development that meets the needs of the present without compromising the ability of future generations to meet their own needs. (Ch. 1)

sustainable forest management a strategy to manage forests not only for harvest but also as ecosystems that maintain their biodiversity. (Ch. 5)

synthetic fertilizer an industrially produced chemical that has one or more of the primary nutrients for plant growth. (Ch. 12)

synthetic pesticide an industrially produced chemical applied to combat insects and weeds in agriculture. (Ch. 12)

system a collection of components interacting with each other to produce outcomes that each component could not achieve on its own. (Ch. 1, 10)

T

tar sand a type of loose-grained rock deposit bonded with oil. (Ch. 13)

tectonic plate a section of Earth's crust that rides on top of denser material below; interactions among these plates shape Earth's surface. (Ch. 9)

temperature in common terms, a measurement of the hotness or coldness of something; formally, a measure of how vigorously the atoms in a material are moving and colliding with each other. (Ch. 3, 8)

thermal expansion the expansion of something as it warms, in particular, the water in the oceans, which increases sea level. (Ch. 11)

thermohaline conveyor a large-scale ocean circulation driven by surface and deepwater ocean currents and changes in water temperature and salinity (density). (Ch. 11)

thermosphere the top layer of the atmosphere. It is very thin and where most man-made satellites orbit. (Ch. 8)

tillage preparing the soil for planting by breaking it up and turning it over with a plow. (Ch. 9)

tilth the soil's overall structure and conditions that facilitate plant growth. (Ch. 9)

topography the shape of the land. (Ch. 9)

topsoil typically the first 2–6 inches of soil, encompassing the top two soil horizons. (Ch. 9)

total fertility rate (TFR) the average number of children a woman would have in her reproductive years in a given population. (Ch. 6)

toxicology the study of the negative health effects of substances on an organism. (Ch. 17)

toxin a poisonous substance that can cause illness or death. (Ch. 17)

tradable emission allowance a strategy where the EPA sets the total allowable annual emissions for a pollutant from utilities, then divides this total into tradable units called allowances that are bought and sold. (Ch. 8)

trade-offs pros and cons, benefits and costs of alternative courses of action. (Ch. 1)

tragedy of the commons a situation where a shared resource is overexploited and degraded by many individual users because the short-term incentives felt by each user discourage conservation. (Ch. 2)

transit-oriented development (TOD) use of design strategies that integrate public transportation and land-use patterns in more sustainable and resilient ways. Also see *pedestrian-oriented development*. (Ch. 16)

triple bottom line (TBL) an accounting method for business operations that values social and environmental concerns as well as profit. (Ch. 19)

trophic level an organism's position in a food chain indicating its number of steps away from primary producers. (Ch. 3)

troposphere the lowest area of the atmosphere from 5 to 10 miles above the surface where Earth's weather occurs. (Ch. 8)

true cost accounting a bookkeeping method that incorporates direct and indirect costs (including externalities) throughout the life cycle of a product. (Ch. 2)

tsunami a powerful wave that can flood coastal areas. (Ch. 17)

U

ultraviolet (UV) radiation invisible rays that are part of the energy that comes from the Sun. Ultraviolet radiation that reaches Earth's surface is made up of two types of rays, called UVA and UVB. (Ch. 17)

unconventional reserve a difficult-to-extract deposit of fossil fuels. (Ch. 13)

unitary governing system a form of government in which the central government alone directs policy. (Ch. 20)

United Nations (UN) an intergovernmental organization with 193 member states. (Ch. 20)

upwelling a place where ocean currents draw up colder water from the deep. (Ch. 11)

urban area a place where large numbers of people live together in relatively dense settlements and work in nonagricultural jobs. (Ch. 16)

urban density the number of residents per unit of space. (Ch. 16)

urban penalty the human health problems associated with urban living. (Ch. 16)

urban planning the development of guidelines to shape the future of an urban area. (Ch. 16)

urban transition term referring to the trend of populations becoming more urban and less rural. (Ch. 16)

utilitarianism an approach to ethics that defines what is right by determining what actions would bring as much good (or as little harm) as possible. (Ch. 2)

V

value chain economic assessment that integrates the market value of a good with the market value of things linked to its production. (Ch. 5)

values reflections of our understanding of how we want things to be—what we desire, aim for, or demand. (Ch. 1)

variable generation a power source with generating capacity that changes according to the time of day, weather conditions, or other factors. (Ch. 14)

volatile organic compound (VOC) a gas released through the evaporation or incomplete combustion of fossil fuels and other organic chemicals. (Ch. 8, 17)

W

warm front a warm air mass replacing a cold air mass. (Ch. 8)

waste stream the steady flow of materials from "upstream" processes, such as extraction, production, and distribution, to their disposal, which is "downstream." (Ch. 15)

waste-to-energy facility a place where the heat produced by incineration is used to power a steam turbine that generates electricity. (Ch. 15)

water mining withdrawals from water reservoirs that are not renewable. (Ch. 7)

water recycling a strategy of taking residential water, treating it, and then using it typically for other agricultural or industrial uses. (Ch. 7)

water security reliable access to adequate safe and affordable water. (Ch. 7)

watershed an area of land that drains to a particular point along a river or stream. (Ch. 9)

weather short-term variations in conditions such as temperature, moisture, and wind in a specific place. (Ch. 8)

weathering physical and chemical processes that reduce rocks to smaller particles (such as gravel, sand, or silt) and alter minerals. (Ch. 9)

wetland an area where the ground is seasonally or permanently saturated with water. (Ch. 7)

Wilderness Act of 1964 an act creating areas where road construction and motorized transportation are prohibited, as are permanent structures such as cabins, bathrooms, and picnic shelters. (Ch. 5)

wildlife corridor a protected strip of land that enables migration from one habitat to another. (Ch. 5)

wind air in motion as it flows from high-pressure to low-pressure areas. (Ch. 8)

wind turbine a machine that uses flowing air to turn large blades that in turn power a generator and creates electricity. (Ch. 14)

work in science, the act of applying force to an object over some distance. (Ch. 3)

World Heritage site an area with "outstanding value to humanity" established by a national government and recognized by the United Nations Educational, Scientific and Cultural Organization (UNESCO). (Ch. 5)

Y

youth bulge a prolonged period of low mortality with high fertility, leading to explosive population growth and a very large population of young people. (Ch. 6)

Z

zoning ordinance a regulation that mandates the types of development, land uses, and human activities that are allowed (and disallowed) in particular places. (Ch. 5)

Credits

FRONT MATTER

Photos: pp. ii–iii: Mike Harrington/Getty Images; p. v: Christina Sherman; p. v: Photo by Cooper Reid; p. xxxiii: United States Department of Agriculture; p. vii: Jenson/Shutterstock; p. viii: Cultura Creative (RF)/Alamy Stock Photo; p. ix: Crystite RF/Alamy Stock Photo; p. x: OlgaKok/Shutterstock; p. xi: Sekar Balasubramanian/Alamy Stock Photo; p. xii: Sirtravelalot/Shutterstock; p. xiii: Makieni/Shutterstock; p. xiv: Tetra Images/Alamy Stock Photo; p. xv: Steven J Taylor/Shutterstock; p. xvi: Ian Leonard/Alamy Stock Photo; p. xvii: Torresigner/Getty Images; p. xviii: Stephanie Jackson - Agriculture/Alamy Stock Photo; p. xix: BP via Getty Images; p. xx: JG Photography/Alamy Stock Photo; p. xxi: Rich Carey/Shutterstock; p. xxii: AsiaDreamPhoto/Alamy Stock Photo; p. xxiii: Chip Somodevilla/Getty Images; p. xxiv: Andrea Matone/Alamy Stock Photo; p. xxv: A.P.S. (UK)/Alamy Stock Photo; p. xxvi: Jeff J Mitchell/Getty Images

CHAPTER 1

Photos: pp. 2–3: Jenson/Shutterstock; 1.1 p. 4 (top left): STR/AFP/Getty Images; 1.1 p. 4 (bottom left): Charles Stirling (Travel)/Alamy Stock Photo; 1.2 p. 4 (right): Astrid Stawiarz/Getty Images for Zady; 1.3 p. 5 (left): Melinda Lee Patelli; 1.4 p. 5 (right): Tony Linck/The LIFE Picture Collection/Shutterstock; 1.5 p. 6: VeryBigAlex/Getty Images; 1.6 p. 7: NASA Earth Observatory/NOAA; 1.8 (a) p. 8 (left): Thomas Kline/Newscom; 1.8 (b) p. 8 (center): Gunter Marx/NO/Alamy Stock Photo; 1.8 (c) p. 8 (right): Kari Neumeyer/Northwest Indian Fisheries Commission; 1.9 p. 9: Mauritius Images GmbH/Alamy Stock Photo; 1.10 p. 12 (top): Monticello/Shutterstock; 1.11 p. 12 (bottom): Russell Kord/Alamy Stock Photo; 1.12 (a) p. 13 (left): AP Photo/Steve Helber; 1.12 (a) p. 13 (right): AP Photo/Greg Gibson; 1.15 p. 15: Jill Odice Photography; 1.16 (a) p. 16 (top): Kyodo News via Getty Images; 1.16 (b) p. 16 (bottom): Joe Raedle/Getty Images; p. 17: Science Photo Library/Alamy Stock Photo; p. 18: AP Photo/Margot Ingoldsby; 1.18 19 (top left): Brian Snyder/Reuters/Newscom; 1.18 p. 19 (top right): AP Photo/Michael Felberbaum; 1.19 19 (bottom): Clarence Holmes/age fotostock/Superstock; 1.20 p. 20 (icon): Studiomango/Shutterstock; p. 21: NASA; 1.21 p. 22 (top): Sharkshock/Shutterstock; p. 22 (bottom): Taylor Spicer; 1.22 p. 23: Global Footprint Network, www.footcalculator.org; p. 24: Indiaforte/Alamy Stock Photo; p. 25: AP Photo/Peter Dejong; p. 26: Yonhap News/YNA/Newscom; p. 27: fairelectionscenter.org

Drawn art/text: 1.13: Christopher W. Tessum et al., "Inequity in consumption of goods and services adds to racial–ethnic disparities in air pollution exposure," Proceedings of the National Academy of Sciences (March 2019). Used by permission of the author; 1.20: "CRAAP Test." Created by The Meriam Library, California State University, Chico. CC-BY-4.0 https://creativecommons.org/licenses/by/4.0/

CHAPTER 2

Photos: pp. 32–33: Cultura Creative (RF)/Alamy Stock Photo; 2.1 (b) p. 34: U.S. Coast Guard; 2.2 p. 35 (left): BSIP/Newscom; 2.3 p. 35 (right): American Presidency Project; 2.4 p. 36: The Earth Charter Initiative; 2.5 p. 37: John Sylvester/Alamy Stock Photo; 2.6 p. 38: Courtesy of the Aldo Leopold Foundation and University of Wisconsin-Madison Archives; p. 39: Coprid/Shutterstock; 2.9 p. 42 (left): World History Archive/Alamy Stock Photo; 2.9 p. 42 (center left): World History Archive/Alamy Stock Photo; 2.9 p. 42 (center right): Gado Images/Alamy Stock Photo; 2.9 p. 42 (right): Ashley Cooper pics/Alamy Stock Photo; 2.10 p. 43 (top): Howard C/Getty Images; 2.10 p. 43 (bottom): Mike Goldwater/Alamy Stock Photo; 2.12 p. 45 (top): Ross Anania/Getty Images; 2.13 p. 45 (bottom): TT News Agency/Alamy Stock Photo; 2.14 p. 46: Jim West/Science Source; 2.16 p. 48 (top left): Brandon Klein Video/Shutterstock; 2.16 p. 48 (top center): All Canada Photos/Alamy Stock Photo; 2.16 p. 48 (top right): Nikita Sursin/Stocksy; 2.16 p. 48 (bottom left): VegterFoto/Stocksy; 2.16 p. 48 (bottom center): Bonninstudio/Stocksy; 2.16 p. 48 (bottom right): Per Swantesson/Stocksy; p. 49: U.S. Department of Energy; p. 50 (top): Courtesy of Society of American Foresters Archive; p. 50 (center): Library of Congress; p. 50 (bottom): U.S. National Park Service; p. 51: WE ACT for Environmental Justice; 2.18 (a) p. 52 (top): Maniglia Romano/Pacific Press/LightRocket/Getty Images; 2.18 (b) p. 52 (bottom): Issac Lawrence/AFP/Getty Images; p. 53 (left): Charles Stirling/Alamy Stock Photo; p. 53 (right): League of Women Voters Education Fund

Drawn art/text: 2.4: Principles of the Earth Charter Initiative from Earth Charter Commission, The Earth Charter, 2000, http://earthcharter.org/discover/the-earth-charter/. Reprinted with permission from Earth Charter International; (p39): J.L. Capper, Journal of Animal Science, December, 2011 via Eliza Barclay, Jessica Stoller-Conrad, and Kevin Uhrmacher/NPR.

CHAPTER 3

Photos: pp. 62–063: Crystite RF/Alamy Stock Photo; 3.1 p. 64 (top): Yenni Safana/Anadolu Agency/Getty Images; p. 64 (bottom): NYPL/Science Source; 3.2 p. 65: Jeff Dai/Science Source; 3.5 p. 67: Djavitch/Shutterstock; 3.7 (a) p. 68 (left): Photo by Peggy Greb, courtesy of USDA; 3.7 (b) p. 68 (center): Demarcomedia/Shutterstock; 3.7 (c) p.68 (right): Vitaly Raduntsev/Shutterstock; p. 69 (left): AP Photo/San Francisco Chronicle, Frederic Larson; p. 69 (right): ZUMA Press, Inc./Alamy Stock Photo; p. 73: TCD/Prod.DB/Alamy Stock Photo; 3.9 p. 74 (top): Ivan Kuzkin/Shutterstock; 3.9 p. 74 (bottom): Karl R. Martin/Shutterstock; 3.10 p. 75 (top left): Gresei/Shutterstock; 3.10 p. 75 (top right): Sergii Koval/Alamy Stock Photo; p. 75 (center): Marek Slusarczyk/Alamy Stock Photo; p. 75 (bottom): Doug Wilson/Alamy Stock Photo; 3.11 p. 76: KeithSzafranski/iStock/Getty Images; 3.13 p. 78: Ian Hubball/Alamy Stock Photo; 3.14 (b) p. 79: Robbert Lalisang/EyeEm/Getty Images; p. 80: Riau Images/Barcroft Images/Barcroft Media via Getty Images; p. 81: Hennell/Alamy Stock Photo

CHAPTER 4

Photos: pp. 86–087: OlgaKok/Shutterstock; 4.1 p. 88 (top left): Diego Zapata; 4.1 p. 88 (bottom left): The Xerces Society, Inc.; 4.2 (a) p. 88 (top right): Alexander Wild; 4.2 (b) p. 88 (bottom right): Jeff Rotman/Science Source; 4.3 p. 89 (top): Photononstop/Alamy Stock Photo; 4.4 p. 89 (center left): Mark Windom/Stocksy; 4.4 p. 89 (center): Tom Uhlman/Alamy Stock Photo; 4.4 p. 89 (center right): Bildagentur Zoonar GmbH/Shutterstock; 4.4 p. 89 (bottom left): FOTOimage Montreal/Shutterstock; 4.4 p. 89 (bottom center): Science Photo Library/Alamy Stock Photo; 4.4 p. 89 (bottom right): Cultura Creative (RF)/Alamy Stock Photo; 4.5 p. 90 (left): Classic Image/Alamy Stock Photo; 4.5 p. 90 (top right): MichaelGrant/Alamy Stock Photo; 4.5 p. 90 (top center): David Kennedy/Alamy Stock Photo; 4.5 p. 90 (bottom center): Michelle Gilders/Alamy Stock Photo; 4.5 p. 90 (bottom): Rich Reid/Alamy Stock Photo; 4.6 (b) p. 91 (top): cgwp.co.uk/Alamy Stock Photo; 4.6 (b) p. 91 (center): Peter Ptschelinzew/Alamy Stock Photo; 4.6 (b) p. 91 (bottom): The Photo Works/Science Source; 4.7 p. 92: Everett Collection Historical/Alamy Stock Photo; p. 94: 2C2C/Shutterstock; 4.10 p. 96 (top left): Robert Harding / Alamy Stock Photo; 4.10 p. 96 (top right): B.A.E. Inc./Alamy Stock Photo; 4.10 p. 96 (bottom left): Pascale Gueret/Shutterstock; 4.10 p. 96 (bottom right): Felix Choo/Alamy Stock Photo; p. 98: Steve Gschmeissner/Science Source; 4.13 (c) p. 99: Igor Tichonow/Alamy Stock Photo; p. 101 (top left): Saraporn/Shutterstock; p. 101 (top center): GoodMoodPhoto/Shutterstock; p. 101 (top right): Inigo Cia Da Riva/Stocksy; p. 101 (center left): Juan Vilata/Alamy Stock Photo; p. 101 (center): Christian Gideon/Stocksy; p. 101 (center right): Ventdusud/Shutterstock; p. 101 (bottom left): Molly Steele/Stocksy; p. 101 (bottom center): STILLFX/Shutterstock; p. 101 (bottom right): Gabriel Ozon/Stocksy; 4.15 p. 102 (top): Krys Bailey/Alamy Stock Photo; 4.16 (a) p. 102 (bottom left): Design Pics Inc/Alamy Stock Photo; 4.16 (a) p. 102 (bottom center): Norman Pogson/Alamy Stock Photo; 4.16 (a) p. 102 (bottom right): Design Pics Inc/Alamy Stock Photo; 4.18 104 (top): Gabbro/Alamy Stock Photo; 4.19 p. 104 (bottom left): Bianca Pontes/Getty Images; 4.19 104 (bottom center): Erik A/Getty Images; 4.19 p. 104 (bottom right): Dennis W Donohue/Shutterstock; p. 105 (top left): Ian Macrae Young/Alamy Stock Photo; p. 105 (bottom left): DJ40/Shutterstock; p. 105 (center): Greg Vaughn/Alamy Stock Photo; p. 105 (right): Arian Wallach; 4.21 p. 107: Dmytro Pylypenko/Alamy Stock Photo; 4.23 p. 108 (left): LJ Wilson-Knight/Alamy Stock Photo; 4.23 p. 108 (right): ImageBROKER/Alamy Stock Photo; 4.24 p. 109 (left): Alexey U/Shutterstock; 4.24 p. 109 (right): ArCaLu/Shutterstock; p. 111 (top): Edo Schmidt/Alamy Stock Photo; p. 111 (bottom): Predator Free NZ; p. 112 (left): Hemis/Alamy Stock Photo; 4.27 p. 112 (right): Ferenc Cegledi/Alamy Stock Photo; 4.28 p. 113 (top left): Accent Alaska.com/Alamy Stock Photo; 4.28 p. 113 (top center): National Geographic Image Collection/Alamy Stock Photo; 4.28 p. 113 (top right): Robert McGouey/Wildlife/Alamy Stock Photo; p. 113 (bottom left): Enrique Mendez/Alamy Stock Photo; p. 113 (bottom right): © Leave No Trace Center for Outdoor Ethics; p. 115: Henrik Larsson/Shutterstock

Drawn art/text: 4.11a: IUCN. 2021. Summary Statistics Table 1a. The IUCN Red List of Threatened Species. Version 2021-3. https://www.iucnredlist.org. Accessed on 11.07.22; 4.13a (p. 88): Map adapted from https://biodiversitymapping.org/wordpress/index

.php/home/. Data from IUCN RedList and BirdLife International: http://datazone.birdlife.org/species/requestdis. Reprinted by permission of Clinton N. Jenkins, Instituto de Pesquisas Ecológicas and by permission of BirdLife International and IUCN.; 4.16: From "The Population Biology of Isle Royale Wolves and Moose: An Overview" by John A. Vucetich. Reprinted by permission of John A. Vucetich; 4.21b: Reprinted with permission from John Wiley & Sons, Inc. Adapted from Klein, D (1968) The "Introduction, Increase, and Crash of Reindeer on St. Matthew Island," *Journal of Wildlife Management*, 32(2), 350-367. © 1968 John Wiley & Sons, Inc. Permission conveyed through Copyright Clearance Center, Inc.; 4.26: Biodiversity Hot Spots. © Conservation International. Reprinted with permission; Use the Data, a (p. 117): Figure reprinted from Colin R. Trainor and Pedro J. Leitão, "Further significant bird records from Atauro Island, Timor-Leste (East Timor)." Forktail 23: 155-158 Reprinted with permission; Use the Data, b (p. 117): Figure 45.13 from David M. Hillis et. al., *Principles of Life*. © 2012 Sinauer Associates. Reproduced with permission of the Licensor through PLSclear.

CHAPTER 5

Photos: pp. 118-119: Sekar Balasubramanian/Alamy Stock Photo; p. 121: North Wind Picture Archives/Alamy Stock Photo; 5.4 (a) p. 122: NASA/METI/AIST/Japan Space Systems, and U.S./Japan ASTER Science Team; 5.5 p. 123 (top left): Jerry Whaley/Media Bakery; 5.5 p. 123 (top center): Irra/Shutterstock; 5.5 p. 123 (top right): Rafael Ben-Ari/Alamy Stock Photo; 5.6 p. 123 (bottom): NOAA; 5.7 (a) p. 124 (left): Wildlife GmbH/Alamy Stock Photo; 5.7 (b) p. 124 (center): Inga spence/Alamy Stock Photo; 5.7 (c) p. 124 (right): Onoky/Superstock; 5.8 p. 125 (top left): Pecold/Shutterstock; 5.8 p. 125 (top right): Niday Picture Library/Alamy Stock Photo; 5.9 p. 125 (bottom): Design Pics Inc/Alamy Stock Photo; 5.10 p. 126: Paul Souders/Alamy Stock Photo; p. 127: Jim Richardson; 5.11 p. 130 (top): Paul Tessier/Stocksy; 5.11 p. 130 (bottom): Peter Wey/Stocksy; 5.12 p. 131: April Bahren, CBNERRVA/National Oceanic and Atmospheric Administration/Department of Commerce; p. 132 (left): The White House; 5.13 p. 132 (right): Barry O'Neill/National Park Service; p. 133 (top): Ramon Carretero/Alamy Stock Photo; 5.14 p. 133 (bottom): Rudmer Zwerver/Shutterstock; 5.15 p. 134 (bottom): Jeremy Sutton-Hibbert/Alamy Stock Photo; p. 134 (top): Dante Piaggio/El Comercio de Peru/Newscom; 5.16 p. 135 (top): Lindsay Fendt/Alamy Stock Photo; 5.17 p. 135 (bottom): George Grantham Bain Collection, Library of Congress; p. 136 (top): Brian Koprowski/Stocksy; p. 136 (top center): Save the Redwoods League/Humboldt County Historical Society; p. 136 (bottom center): David Nelson/Alamy Stock Photo; p. 136 (bottom): Shujaa_777/Shutterstock; p. 137 (top): John Norman/Alamy Stock Photo; p. 137 (bottom): Keith Homan/Shutterstock; 5.18 p. 139 (top left): Helmut Corneli/Alamy Stock Photo; 5.18 p. 139 (top right): Jan-Dirk Hansen/Alamy Stock Photo; 5.18 p. 139 (bottom left): Design Pics Inc/Alamy StockPhoto; 5.18 p. 139 (bottom right): Steve Bloom Images/Alamy Stock Photo; 5.20 (a) p. 140 (left): Sanit Fuangnakhon/Shutterstock; 5.20 (b) p. 140 (right): Blickwinkel/Alamy Stock Photo; 5.21 (a) p. 141 (top): Morley Read/Alamy Stock Photo; 5.21 (b) p. 141 (bottom): Tony Eveling/Alamy Stock Photo; 5.22 (a) p. 142 (left): National Park Service; 5.22 (b) p. 142 (right): Natural History Collection/Alamy Stock Photo; 5.24; p. 143: Lynnette Peizer/Alamy Stock Photo; p. 144 (left): Rawpixel.com/Shutterstock; p. 144 (right): Razvan Cornel Constantin/Alamy Stock Photo

Drawn art/text: 5.3: Figure 3B from IPBES. 2019. Summary for policymakers of the global assessment report on biodiversity and ecosystem services of the Intergovernmental Science-Policy Platform on Biodiversity and Ecosystem Services. S. Díaz, et. al. (eds.). IPBES secretariat, Bonn, Germany. Reprinted by permission of IPBES; 5.19: Food and Agriculture Organization of the United Nations, 2012, Figure 5: Annual Change in forest area by country , 2005-2010 from Global Forest Resources Assessment 2010, http://foris.fao.org/static/data/fra2010/fig0.5.jpg. Reproduced with permission.

CHAPTER 6

Photos: pp. 150-151: Sirtravelalot/Shutterstock; 6.2 p. 153 (top): GL Archive/Alamy Stock Photo; 6.3 (a) p. 153 (bottom left): Dave And Les Jacobs/Stocksy; 6.3 (b) p. 153 (bottom center): Nagel Photography/Shutterstock; 6.3 (c) p. 153 (bottom right): Realimage/Alamy Stock Photo; p. 154 (left): The Print Collector/Alamy Stock Photo; p. 154 (center): Eye35.pix/Alamy Stock Photo; p. 154 (right): Monkey Business Images/Shutterstock; p. 155 (left): Leungchopan/Shutterstock; p. 155 (center): Blvdone/Shutterstock; p. 155 (right): ImageBroker/Alamy Stock Photo; p. 156: Archive PL/Alamy Stock Photo; 6.5 p. 158: Ahturner/Shutterstock; 6.6 p. 159 (top): David Pollack/Corbis via Getty Images; p. 159 (bottom): Muhammad Mostafigur Rahman/Alamy Stock Photo; p. 160 (top): Found Image Holdings/Corbis via Getty Images; 6.7 p. 160 (bottom): Roslan Rahman/AFP/Getty Images; p. 161 (left): Courtesy of Bocconi University; p. 161 (right): AP Photo/Felipe Dana; 6.11 p. 164: Jiji Press Photo/Morio Taga/Newscom; p. 166: Ahmad Gharabli/

AFP/Getty Images; 6.13 p. 167 (top): Film 4/Celador Films/Pathe International/Kobal/Shutterstock; 6.13 p. 167 (bottom): Globo/Kobal/Shutterstock; 6.15 p. 169: Sean Sprague/Alamy Stock Photo; p. 174 (top): Global Footprint Network; p. 174 (bottom): DeVisu/Shutterstock

Drawn art/text: (p. 161): Figure 3 from "Soap Operas and Fertility: Evidence from Brazil," Eliana La Ferrara, Alberto Chong, Suzanne Duryea, *American Economic Journal: Applied Economics*, vol. 4, no. 4, October 2012 (pp. 1-31). © American Economic Association; reproduced with permission of the *American Economic Journal: Applied Economics*; 6.9: Graph adapted from the World Bank. CC-BY-4.0 https://creativecommons.org/licenses/by/4.0/; 6.14: Graph adapted from "Developing scenarios for long-term population growth in SSA," by Track20. Reprinted by Permission; (p. 168): Figure from p. 183, Teaching and learning: achieving quality for all; EFA global monitoring report, 2013-2014, UNESCO. Director-General, 2009-2017 (Bokova, I.G.). writer of foreword (Paris: UNESCO, 2014). Reprinted by permission of UNESCO.

CHAPTER 7

Photos: pp. 180-181: Makieni/Shutterstock; 7.1 p. 162: Mike Hutchings/Reuters/Newscom; 7.5 (a) p. 186 (top): Chris Proctor; 7.5 (b) p. 186 (bottom): All Canada Photos/Alamy Stock Photo; 7.6 (a) p. 187 (top): NASA/Alamy Stock Photo; 7.6 (b) p. 187 (bottom left): Michela Ravasio/Stocksy; 7.6 (c) p. 187 (bottom right): FLPA/Alamy Stock Photo; 7.8 p. 188 (top left): Sompol/Shutterstock; 7.8 p. 188 (top right): Valerii_M/Shutterstock; 7.8 p. 188 (center left): Ba8389/Alamy Stock Photo; 7.8 p. 188 (center right): Rosanne Tackaberry/Alamy Stock Photo; 7.8 p. 188 (bottom left): Martyn Williams/Alamy Stock Photo; 7.8 p. 188 (bottom right): Peter Martin Rhind/Alamy Stock Photo; 7.10 (b) p. 191 (top): Justin Brandt/USGS; 7.11 (a) p. 191 (bottom left): Chris Geszvain/Shutterstock; 7.11 (b) p. 191 (bottom right): National Geographic Image Collection/Alamy Stock Photo; 7.11 (c) p. 191 (bottom center): Mark Conlin/Alamy Stock Photo; 7.12 p. 192 (top left): Kevin Fleming/Getty Images; 7.12 p. 192 (bottom left): US Army Photo/Alamy Stock Photo; 7.13 p. 192 (right): NASA Earth Observatory; 7.14 (a) p. 193 (top left): Kevin Murch/Alamy Stock Photo; 7.14 (b) p. 193 (top right): TMI/Alamy Stock Photo; 7.15 (b) p. 193 (bottom): Geogphotos/Alamy Stock Photo; 7.16 p. 195 (left): AP Photo/Nasser Nasser; 7.17 p. 195 (right): Paul Jeffrey/Alamy Stock Photo; p. 196: Steven Coutts/Alamy Stock Photo; 7.18 p. 197: The U.S. Drought Monitor is jointly produced by the National Drought Mitigation Center at the University of Nebraska-Lincoln, the United States Department of Agriculture, and the National Oceanic and Atmospheric Administration. Map courtesy of NDMC; 7.19 p. 198: Niklas Hallen/UNDP India; 7.21 p. 200: © S. Augustin, Watercone.com; p. 201: PhotoStock-Israel/Alamy Stock Photo; 7.22 p. 202 (left): Hans Engbers/Alamy Stock Photo; 7.23 (a) p. 202 (top right): WC Alden/USGS Photo Library; 7.23 (a) p. 202 (bottom right): Kevin Jacks/USGS; p. 203 (top): U.S. Fish and Wildlife Service; 7.24 p. 203 (bottom): Elementix/Alamy Stock Photo; 7.25 (b) p. 205 (left): Steve Downer/Science Source; 7.25 (c) p. 205 (center): George Ward/Alamy Stock Photo; 7.25 (d) p. 205 (right): F.Bettex/Alamy Stock Photo; p. 206 (row 1): Olenalavrova/Shutterstock; p. 206 (row 2): Design Pics/Jeff Schultz/Newscom; p. 206 (row 3): NASA; p. 206 (row 4): Ihi/Shutterstock; p. 206 (row 5): Alexander Semenov/Science Source; p. 206 (row 6): Jean-Paul Ferrero/Pantheon/Superstock; p. 207: Kathryn Roach/Shutterstock

Drawn art/text: (p. 189): Illustration adapted with from MLive.com. © 2015 MLive Media Group. All rights reserved. Reprinted with permission; 7.16: Map adapted from Running the Sahara. Reprinted by permission of NEHST Studios; 7.18: The U.S. Drought Monitor is jointly produced by the National Drought Mitigation Center at the University of Nebraska-Lincoln, the United States Department of Agriculture, and the National Oceanic and Atmospheric Administration. Map courtesy of NDMC; Table 7.1, Overfishing: Food and Agriculture Organization of the United Nations, Figure 14 from 2018, *The State of World Fisheries and Aquaculture 2018 - Meeting the sustainable development goals*, http://www.fao.org/3/i9540en/i9540en.pdf. Reproduced with permission; (p. 211): Provided courtesy of the Illinois State Water Survey, University of Illinois Urbana-Champaign.

CHAPTER 8

Photos: pp. 212-213: Tetra Images/Alamy Stock Photo; 8.1 (a) p. 214 (top left): Chad Ehlers/Alamy Stock Photo; 8.1 (b) p. 214 (bottom left): Hans Blossey/Alamy Stock Photo; 8.1 (c) p. 214 (right): Walter Stein/AP Photo; 8.3 (a) p. 215 (left): Jes2ufoto/Alamy Stock Photo; 8.3 (b) p. 215 (center): Turgay Koca/Alamy Stock Photo; 8.3 (c) p. 215 (right): Meoita/Shutterstock; 8.4 (a) p. 216: Photoshot/Newscom; 8.6 p. 218 (top): Ojal/Shutterstock; p. 218 (bottom): Archistoric/Alamy Stock Photo; 8.8 p. 219: Chang W. Lee/The New York Times/Redux Pictures; 8.10 (b) p. 220: NOAA; 8.11 (b) p. 221: NASA/NOAA GOES Project; p. 223 (left): Kimberly Prather; p. 223 (right): NASA; 8.14 p. 224 (left): Tatsuya

Kanabe/Stocksy; 8.14 p. 224 (right): Zack Frank/Shutterstock; 8.15 (a) p. 225 (top): David Wall/Alamy Stock Photo; 8.15 (b) p. 225 (bottom): James Anderson; 8.16 p. 226 (top): SPL/Science Source; p. 226 (bottom): Ed Rooney/Alamy Stock Photo; 8.17 p. 227 (left): EPA; 8.18 (a) p. 227 (top right): Dennis MacDonald/Alamy Stock Photo; 8.18 (b) p. 227 (bottom right): Dennis MacDonald/Alamy Stock Photo; 8.19 (a) p. 228 (top): NASA/GSFC/METI/ERSDAC/JAROS, and U.S./Japan ASTER Science Team; 8.19 (b) p. 228 (bottom): NASA/GSFC/METI/ERSDAC/JAROS, and U.S./Japan ASTER Science Team; p. 230: Extarz/Shutterstock; 8.21 p. 231: NASA Ozone Watch; p. 234 (left): NASA Earth Observatory; 8.23 (a) p. 234 (top right): Photofusion Picture Library/Alamy Stock Photo; 8.23 (b) p. 234 (bottom right): Vandathai/Shutterstock; p. 235 (top left): View Stock/Alamy Stock Photo; p. 235 (bottom left): Testing/Shutterstock; p. 235 (right): Kevin Frayer/Getty Images; p. 237 (top): Melinda Lee Patelli; p. 237 (bottom): Melinda Lee Patelli; p. 238: CRM/Shutterstock; p. 242 (left): NASA; p. 242 (right): NASA

Drawn art/text: 8.9a: From The Met Office, UK. © Crown copyright, Met Office and for outside the UK © British Crown copyright, Met Office.

CHAPTER 9

Photos: pp. 244–245: Steven J Taylor/Shutterstock; 9.1 (a) p. 246 (top left): ShakeOut .org; 9.1 (b) p. 246 (bottom left): Hideo Kurihara/Alamy Stock Photo; p. 246 (right): JHVEPhoto/Alamy Stock Photo; 9.2 p. 247 (left) Special Collections & Archives, UC San Diego, La Jolla, 92093-0175 (https://lib.ucsd.edu/sca); 9.4 p. 247 (top right): Capture That-Landscapes/Alamy Stock Photo; 9.4 p. 247 (bottom right): James Stone/Alamy Stock Photo; 9.8 p. 249 (left): Martin Harvey/Alamy Stock Photo; 9.8 p. 249 (center): Jeffrey Isaac Greenberg 14+/Alamy Stock Photo; 9.8 p. 249 (right): Kip Evans/Alamy Stock Photo; 9.10A p. 251 (top left): Incamerastock/Alamy Stock Photo; 9.10B p. 251 (top right): S6Wigj/Shutterstock; 9.11 p. 251 (bottom left): Inge Johnsson/Alamy Stock Photo; 9.11 p. 251 (bottom right): Andreas Werth/Alamy Stock Photo; 9.12 p. 252 (top): MPI/Getty Images; p. 252 (bottom): Julio Etchart/Alamy Stock Photo; 9.13 p. 253 (top): Napat/Shutterstock; p. 253 (bottom): Ed Jones/AFP via Getty Images; 9.15 p. 255: The Metropolitan Museum of Art, New York/Thomas Cole (American, Lancashire 1801–1848 Catskill, New York)/The Metropolitan Museum of Art, Gift of Mrs. Russell Sage, 1908 (08.228) www .metmuseum.org; 9.16 p. 256 (top left): Ken Schulze/Shutterstock; 9.16 p. 256 (top right): Nickolay Khoroshkov/Alamy Stock Photo; 9.16 p. 256 (center): Timothy Mulholland/Alamy Stock Photo; 9.16 p. 256 (bottom left): Paul Brady Photography/Shutterstock; 9.16 p. 256 (bottom center left): Zachary Frank/Alamy Stock Photo; 9.16 p. 256 (bottom center right): ZUMA Press, Inc./Alamy Stock Photo; 9.16 p. 256 (bottom right): Anton Foltin/Shutterstock; 9.17 (a) p. 257 (top left): NASA; 9.17 (b) p. 257 (bottom left): NASA; 9.18 p. 257 (right): US Army Photo/Alamy Stock Photo; 9.19 (a) p. 258 (left): Peter Wey/Stocksy; 9.19 (b) p. 258 (center): PureStock/Alamy Stock Photo; 9.19 (c) p. 258 (right): Claudia Weinmann/Alamy Stock Photo; 9.23 (a) p. 260 (left): Blickwinkel/Alamy Stock Photo; 9.23 (b) p. 260 (center left): Joe Blossom/Alamy Stock Photo; 9.24 p. 260 (center right): Dr. Bruno Glaser; 9.24 p. 260 (right): Dr. Bruno Glaser; p. 261 (top): The Picture Art Collection/Alamy Stock Photo; p. 261 (bottom): Paolo Trovo/Shutterstock; 9.26 p. 263: Eduardo Pucheta/Alamy Stock Photo; 9.27 p. 264 (top left): Ebubekir Olcok/Shutterstock; 9.27 p. 264 (top center): Fotokostic/Shutterstock; 9.27 p. 264 (top right): David Litman/Shutterstock; 9.28 p. 264 (center left): Thuong Tran/Alamy Stock Photo; 9.28 p. 264 (center right): Helga_foto/Shutterstock; 9.28 p. 264 (bottom left) Paul Gregg Travel NZ/Alamy Stock Photo; 9.28 p. 264 (bottom right): Edwin Remsberg/Alamy Stock Photo; p. 265: Lanmas/Alamy Stock Photo; 9.30 p. 266: Rich Carey/Shutterstock; 9.31 p. 267: Jim Hargan/Alamy Stock Photo; p. 268 (top): Mario Tama/Getty Images; p. 268 (center): James MacDonald/Bloomberg via Getty Images; p. 268 (bottom): Roberto Cornacchia/Alamy Stock Photo; p. 269 (top): Terrence Antonio James/MCT/Newscom; p. 269 (center): Robert Harding/Alamy Stock Photo; p. 269 (bottom): Artur Szymczyk/Alamy Stock Photo; p. 270 (left): age fotostock/Alamy Stock Photo; p. 270 (right): Robert K. Chin/Alamy Stock Photo

CHAPTER 10

Photos: pp. 276–277: Ian Leonard/Alamy Stock Photo; 10.1: p. 278: ClassicStock/Alamy Stock Photo; 10.9 p. 282: Noppharat Studio 969/Shutterstock; 10.10 p. 283 (top): Steve Cukrov/Shutterstock; p. 283 (bottom): LeStudio/Shutterstock; 10.12 p. 285: Tom Jastram/Shutterstock; 10.14 p. 286: Jim Damaske/Tampa Bay Times/ZUMA Wire/Alamy Live News; p. 287 (left): Oticki/Shutterstock; p. 287 (right): Criniger Kolio/Shutterstock; 10.16 (a) p. 288 (left): Dark Moon Pictures/Shutterstock; 10.16 (b) p. 288 (center left): Dr Jeremy Burgess/Science Source; 10.16 (c) p. 288 (center right): Mycelium/Shutterstock; 10.16 (d) p. 288 (right): Kevin Britland/Alamy Stock Photo; p. 289 (top): Embrapa Agrobiologia; p. 289 (bottom left): Wildlife GmbH/Alamy Stock Photo; p. 289 (bottom right): Dr Jeremy Burgess/Science Source; 10.17 p. 290: Courtesy of N. Rabalais, LSU/LUMCON/NOAA; p. 291

(left): Lynn Betts, USDA Natural Resources Conservation Service; p. 291 (right): Lynn Betts, USDA Natural Resources Conservation Service; 10.21 p. 293: Sunshine Seeds/Shutterstock; p. 299: AP Photo/Carlos Osorio; p. 301: Chesapeake Bay Foundation

Drawn art/text: 10.4a: Illustration by Alexandra Rose. Reprinted with permission; 10.18: Figure 3 from Nielsen, R. 2005, 'Can We Feed the World? The nitrogen limit of food production', http://home.iprimus.com.au/nielsens/; 10.21a (p. 282): Figure SPM.1 IPCC, 2014: Climate Change 2014: Synthesis Report. Contribution of Working Groups I, II and III to the Fifth Assessment Report of the Intergovernmental Panel on Climate Change [Core Writing Team, R.K. Pachauri and L.A. Meyer (eds.)]. IPCC, Geneva, Switzerland.

CHAPTER 11

Photos: pp. 302–303: Torresigner/Getty Images; 11.1 (a) p. 304: Spencer Platt/Getty Images; 11.2 p. 305 (top): AP Photo/San Angelo Standard-Times, Patrick Dove; 11.2 p. 305 (center): David A Litman/Shutterstock; 11.3 (a) p. 305 (bottom left): Photo © Christie's Images/Bridgeman Images; 11.3 (b) p. 305 (bottom right): NOAA/NASA; 11.5 (a) p. 306 (left): NOAA; 11.5 (b) p. 306 (center): Courtesy of the U.S. Geological Survey; 11.5 (c) p. 306 (right): Robert Mulvaney/British Antarctic Survey; p. 307: Chris Butler/Science Source; 11.7 (a) p. 308 (left): Westend61/Getty Images; 11.7 (b) p. 308 (right): lnzyx/iStock/Getty Images; 11.9 (a) p. 310 (top): Science Photo Library/Alamy Stock Photo; 11.9 (b) p. 310 (bottom): Merlin74/Shutterstock; 11.13 p. 313: Josh Haner/The New York Times/Redux; p. 317: Courtesy of Hiroshi Arakawa via UCLA; 11.16 (a) p. 319 (left): Mohamed Shareef/Shutterstock; 11.16 (b) p. 319 (right): AP Photo/Mohammed Seeneen; 11.17 (a) p. 320: Jim Olive/Polaris/Newscom; p. 321: AP Photo/Craig Ruttle; 11.21 p. 323 (top left): U.S. Department of the Interior; 11.21 p. 323 (top right): Captain Budd Christman, NOAA Corps/National Oceanic and Atmospheric Administration/Department of Commerce; 11.21 p. 323 (bottom left): Gecko1968/Shutterstock; 11.21 p. 323 (bottom right): David Byron Keener/iStock/Getty Images; p. 324: Brooke Medley/NASA; 11.22 p. 326 (top left): Lphoto/Alamy Stock Photo; 11.22 p. 326 (bottom left): John and Penny/Shutterstock; 11.22 p. 326 (top right): Atiger/Shutterstock; 11.22 p. 326 (bottom right): Bernhard Edmaier/Science Source; p. 327: EPA; p. 328 (top): Powerbeephoto/iStock/Getty Images; p. 328 (bottom): AP Photo/Paul Sakuma; p. 333 (left): GlobalChange.gov; p. 333 (right): GlobalChange.gov

Drawn art/text: Fig. 11.18: Figure from *Climate Change Impacts in the United States: The Third National Climate Assessment, 2014.* NOAA NCDC / CICS-NC.

CHAPTER 12

Photos: pp. 336–337: Stephanie Jackson - Agriculture/Alamy Stock Photo; 12.1 p. 338 (top): Todd Williamson/Getty Images; 12.2 p. 338 (bottom): Nicolle Rager Fuller, National Science Foundation; 12.3 (a) p. 339 (top left): World History Archive/Alamy Stock Photo; 12.3 (b) p. 339 (top right): ZUMA Press Inc/Alamy Stock Photo; p. 339 (bottom right): Bildagentur Zoonar GmbH/Shutterstock; p. 339 (bottom left): Derek Meijer/Alamy Stock Photo; 12.4 p. 341 (top left): David R. Frazier Photolibrary, Inc./Alamy Stock Photo; 12.4 p. 341 (bottom left): Igor Stevanovic/Alamy Stock Photo; 12.5 p. 341 (right): Design Pics Inc/Alamy Stock Photo; 12.7 p. 343 (left): Old Paper Studios/Alamy Stock Photo; 12.8 p. 343 (right): Lou Linwei/Alamy Stock Photo; 12.9 p. 344 (left): Split Second Stock/Shutterstock; 12.10 p. 344 (right): U.S. Geological Survey; p. 349 (top): Martin Bond/Alamy Stock Photo; p. 349 (center): USDA; p. 349 (bottom): B Christopher/Alamy Stock Photo; 12.13 p. 350: Aerial Archives/Alamy Stock Photo; 12.14 p. 351 (left): Design Pics Inc/Alamy Stock Photo; 12.14 p. 351 (right): Mark A. Johnson/Alamy Stock Photo; 12.15 (a) p. 352 (left): Scimat/Science Source; 12.15 (b) p. 352 (right): Rodrigo Gutierrez/Reuters/Newscom; 12.17 (b) p. 354: Images & Stories/Alamy Stock Photo; 12.18 p. 355: Melinda Lee Patelli; 12.22 p. 357 (left): Joseph Sohm/Shutterstock; p. 357 (right): Jade Wilson; 12.25 p. 360 (top): Eric Lafforgue/age fotostock/Superstock; p. 360 (bottom): Edward Parker/Alamy Stock Photo; p. 361: Jay Stocker/Shutterstock; p. 362 (top): Melinda Lee Patelli; p. 362 (bottom): U.S. Department of Agriculture

Drawn art/text: (p. 340): Figure by Anna Maurer. Reprinted by permission of SITN Boston. http://sitn.hms.harvard.edu/flash/2015/how-to-make-a-gmo/; 12.11: Graph: Meat Production from Hannah Ritchie and Max Roser (2019) – "Meat and Seafood Production & Consumption". Published online at OurWorldInData.org. CC-BY-4.0 https://creativecommons.org/licenses/by/4.0/; 12.12: "Research and Innovation Have Led to Dramatic Progress in Broiler Performance." Data from National Chicken Council. Reprinted by permission of the Georgia Poultry Federation/Brian Fairchild, PhD; (p. 348): Illustration of Cow's Digestive Tract reprinted by permission of the University of Minnesota, Dairy Extension; 12.16: From "United Nations Environment Programme and Climate and Clean Air Coalition Global Methane Assessment: Benefits and Costs of Mitigating Methane

Emissions. Nairobi." Reprinted by permission of the United Nations Environment Programme; (p. 361): Organic Fertilizers vs. Synthetic Fertilizers graphic, Milwaukee Metropolitan Sewerage District. Reprinted by permission of Milorganite

CHAPTER 13

Photos: pp. 368–369: BP via Getty Images; 13.5 (a) p. 373 (top left): Robert Harding/Alamy Stock Photo; 13.5 (b) p. 373 (top right): Sarin Images/GRANGER; 13.6 p. 373 (center): Ivan Vdovin/Alamy Stock Photo; 13.6 p. 373 (bottom): Carolyn Clarke/Alamy Stock Photo; 13.7 p. 374 (top left): Natalya Erofeeva/Alamy Stock Photo; 13.7 p. 374 (bottom left): Oleksiy Maksymenko Photography/Alamy Stock Photo; 13.7 p. 374 (right): CathyRL/Shutterstock; 13.10 p. 376: Universal History Archive/UIG/Bridgeman Images; 13.11 p. 377 (top): Science History Images/Alamy Stock Photo; 13.12 p. 377 (bottom): Shirley Kilpatrick/Alamy Stock Photo; p. 378: IEA Clean Coal Centre; 13.14 (a) p. 379 (top left): Chris Howes/Wild Places Photography/Alamy Stock Photo; 13.14 (b) p. 379 (top right): Alan Gignoux/Alamy Stock Photo; 13.15 p. 379 (bottom): Indiana Department of Natural Resources; 13.17 (a) p. 381 (top left): Mark Smith/Shutterstock; 13.17 (b) p. 381 (top right): Gado Images/Alamy Stock Photo; 13.18 p. 381 (bottom): Seth Haines, USGS; 13.19 (a) p. 382 (left): David Wall/Alamy Stock Photo; 13.19 (b) p. 382 (right): Alessandra Sanguinetti/Magnum Photos; p. 384: Tom Atkeson/MCT/Newscom; p. 385 (top): ZUMA Press Inc/Alamy Stock Photo; p. 385 (bottom): AP Photo/Wade Payne; p. 386 (top): UPI/Alamy Stock Photo; 13.21 p. 386 (bottom): Gallo Images/Getty Images; p. 387: Mladen Antonov/AFP/Getty Images; p. 389: Orjan Ellingvag/Alamy Stock Photo; p. 390: REUTERS/Lenzy Krehbiel-Burton/Newscom; p. 391 (left): Environmental Defense Fund, based on data from California Air Resources Board and the California State Legislature; p. 391 (right): Flipser/Shutterstock; p. 392: Maximilian Weinzierl/Alamy Stock Photo; p. 395 (left): United States Government Accountability Office; p. 395 (right): United States Government Accountability Office

Drawn art/text: Fig. 13.2: Graphs reprinted from *BP Statistical Review of World Energy*, p. 10. © 2021. Reprinted by permission of BP p.l.c.; Fig. 13.16b: Figure: Oil consumption by country, Fossil Fuels from Hannah Ritchie and Max Roser (2019) - "Fossil Fuels". Published online at OurWorldInData.org. Retrieved from: 'https://ourworldindata.org /fossil-fuels.' CC-BY-4.0 https://creativecommons.org/licenses/by/4.0/

CHAPTER 14

Photos: pp. 396–397: JG Photography/Alamy Stock Photo; 14.1 p. 398 (top): Lucas Oleniuk/ZUMApress/Newscom; 14.2 p. 398 (bottom): Luke Sharrett/Bloomberg/Getty Images; 14.3 p. 399 (top): Chris Laurens/Alamy Stock Photo; 14.3B p. 399 (bottom): National Renewable Energy Laboratory, U.S. Department of Energy; p. 400: Brian A Jackson/Shutterstock; 14.4 (a) p. 401 (top): Diyana Dimitrova/Shutterstock; 14.4(b) p. 401 (bottom): Hans Blossey/Alamy Stock Photo; p. 402 (left): Agence FOB/Alamy Stock Photo; 14.6 p. 402 (right): Adrian Arbib/Alamy Stock Photo; 14.8 (a) p. 404 (top): NASA; 14.8 (b) p. 404 (bottom): Cyril Young/Alamy Stock Photo; p. 405 (top): Antti Yrjonen/NurPhoto/Getty Images; p. 405 (bottom): Everett Collection Historical/Alamy Stock Photo; 14.11(a) p. 406 (left): Jacob Lowenstern, Yellowstone Volcano Observatory; 14.11(b) p. 406 (right): Daniel S. Ragnarsson/Alamy Stock Photo; 14.13 p. 408 (top): GIPhotoStock Z/Alamy Stock Photo; 14.13 p. 408 (bottom): Inga Spence/Alamy Stock Photo; 14.15 p. 409: Jim West/Alamy Stock Photo; p. 411: The Solutions Project; p. 412: Radius Images/Alamy Stock Photo; p. 413: Photo courtesy of WAPTAC; p. 414: Randy Duchaine/Alamy Stock Photo; 14.21 p. 415: Consumers Energy; 14.22(a) p. 416: AP Photo/Mark Lennihan; p. 417: Metamorworks/Shutterstock; 14.23 p. 420: Gado Images/Alamy Stock Photo; p. 421 (left): Colby College Office of Sustainability; p. 421 (right): Ecco/Shutterstock

Drawn art/text: 14.5: Solar Collector Diagram. Samantha Radford. © NC Sustainable Energy Association. Reprinted with permission; 14.18: Figures reprinted from *Lazard's Levelized Cost of Energy Analysis Version 14.*0. Copyright © 2021 Lazard. Reprinted by permission of Lazard. 14.19: Map adapted from Levelized Cost of Electricity Map, University of Texas at Austin, Energy Institute, http://calculators.energy.utexas.edu/lcoe_map/#/county/tech. Reprinted by permission of Joshua D. Rhodes, PhD.

CHAPTER 15

Photos: pp. 426–427: Rich Carey/Shutterstock; 15.1(a) p. 428 (top): Citizen of the Planet/Alamy Stock Photo; 15.1(c) p. 428 (bottom): Ray Chavez/East Bay Times/Getty Images; 15.5 p. 431 (top): Christina Simons/Alamy Stock Photo; 15.5 p. 431 (bottom): Peter Charlesworth/LightRocket via Getty Images; p. 432 (top): Joerg Boethling/Alamy Stock

Photo; p. 432 (bottom): Paula Luu, University of Michigan; p. 432 (bottom, inset): Paula Luu, University of Michigan; p. 433 (top): Alice Musbach/Alamy Stock Photo; p. 433 (center): ImageBROKER/Alamy Stock Photo; p. 433 (bottom): Paula Luu, University of Michigan; 15.6 p. 434: Muellek Josef/Shutterstock; 15.8 p. 435 (top): AP Photo/U.S. Army via Daily Press; 15.9 p. 435 (bottom): AP Photo/David Bookstaver; p. 436 (left): Adi Weda/EPA-EFE/Shutterstock; 15.10 p. 436 (right): Friedrich Stark/Alamy Stock Photo; 15.12 p. 438 (top left): Joe Raedle/Getty Images; 15.12 p. 438 (top right): Science History Images/Alamy Stock Photo; 15.13 p. 438 (bottom): The Asahi Shimbun/Getty Images; p. 439: AP Photo/Steve Helber; p. 441 (top): Richard Cavalleri/Shutterstock; 15.15 p. 441 (bottom): Norma Jean Gargasz/Alamy Stock Photo; 15.17(a) p. 443 (top left): Tetra Images/Alamy Stock Photo; 15.17(b) p. 443 (center left): Teerasak Ladnongkhun/Shutterstock; 15.17(b) p. 443 (bottom left): Andrey Burstein/Shutterstock; p. 443 (right): Dan Sherman; 15.20 p. 445 (left): Africa Studio/Shutterstock; 15.20 p. 445 (center): Picture Partners/Alamy Stock Photo; 15.20 p. 445 (right): Chuck Rausin/Shutterstock; p. 446 (left): Tim Gainey/Alamy Stock Photo; 15.21 p. 446 (right): I-BEAM DESIGN; 15.22 p. 447 (top): Richard Levine/Alamy Stock Photo; 15.23 p. 447 (bottom): Margaret Bourke-White/The LIFE Picture Collection/Shutterstock; 15.24 p. 448: Marks and Spencer; p. 449: Williams College, Zilka Center for Environmental Initiatives; p. 450: Image courtesy of the City of Santa Barbara; p. 451 (left): Tim Hill/Alamy Stock Photo; p. 451 (right): Philippe Huguen/AFP via Getty Images; p. 452: Post-Landfill Action Network

Drawn art/text: 15.19: Plastic Resin Codes Chart. Reprinted by permission of Recycling Resource Systems, Inc. (RRS)

CHAPTER 16

Photos: pp. 458–459: AsiaDreamPhoto/Alamy Stock Photo; 16.1 p. 460: TierneyMJ/Shutterstock; 16.2(a) p. 461 (top left): Karnnapus/Shutterstock; 16.2(b) p. 461 (top right): Bisual Studio/Stocksy; 16.3(a) p. 461 (center): Aerial Archives/Alamy Stock Photo; 16.3(b) p. 461 (bottom): ZUMA Press, Inc./Alamy Stock Photo; 16.5(a) p. 464 (left): Melinda Lee Patelli; 16.5(b) p. 464 (center): Catherine Zibo/Shutterstock; 16.5(c) p. 464 (right): A.P.S. (UK)/Alamy Stock Photo; 16.7 p. 465: FotoFlirt/Alamy Stock Photo; 16.8 p. 466: GRANGER; 16.9 p. 467 (top): Knyazevfoto/Shutterstock; 16.10 p. 467 (bottom): Janis Miglavs/Newscom; p. 468 (top): Copyright 2017 from The Geography of Transport Systems by Jean-Paul Rodrigue. Reproduced by permission of Taylor and Francis, a division of Informa PLC.; 16.11 p. 468 (bottom): NASA Earth Observatory/NOAA NGDC; 16.13(b) p. 469: William Ju/Alamy Stock Photo; 16.14 p. 470: Keith Richard Mountain; p. 471: Hum Historical/Alamy Stock Photo; 16.15 p. 472 (top): Todd Strand/Alamy Stock Photo; 16.16(a) p. 472 (bottom left): Ewing Galloway/Alamy Stock Photo; 16.16(b) p. 472 (bottom right): Tony Linck/The LIFE Picture Collection/Shutterstock; 16.17 p. 473 (top): U.S. Geological Survey; 16.17 p. 473 (bottom): U.S. Geological Survey; p. 474: Batchelder/Alamy Stock Photo; p. 475: Waggonner & Ball Architecture; 16.19 p. 476: Toles © The Washington Post. Reprinted with permission of ANDREWS MCMEEL SYNDICATION. All rights reserved.; 16.20 (a) p. 478 (left): Environmental Protection Agency; 16.20 (b) p. 478 (right): Courtesy of Marlon Blackwell Architects; 16.21 p. 479 (top left): Walk Score; 16.21 p. 479 (top right): Walk Score; 16.22 p. 479 (bottom): Look/Alamy Stock Photo; 16.23 p. 480 (top left): Jim West/Alamy Stock Photo; 16.24 (a) p. 480 (center left): Michael Vi/Alamy Stock Photo; 16.24 (b) p. 480 (bottom left): VIVA Railings, LLC; 16.24 (c) p. 480 (bottom right): Ilyas Ayub/Alamy Stock Photo; p. 481 (top left): Nic Lehoux-VIEW/Alamy Stock Photo; p. 481 (top right): Living Building Challenge; p. 481 (bottom): Nic Lehoux-VIEW/Alamy Stock Photo; p. 482 (left): John Tlumacki/The Boston Globe via Getty Images; p. 482 (right): Walk Score; p. 483: Jim West/Alamy Stock Photo; p. 486: Adapted from World Health Organization (2011)

Drawn art/text: (pp. 485–486) Table 1 & 'Use the Data' Graph: Adapted from "The Rapid Rise of the Urban Consumer Class," by Wolfgang Fengler and Max Heinze. *Brookings.edu* December 15, 2021. Reprinted by permission of World Data Lab.

CHAPTER 17

Photos: pp. 488–489: Chip Somodevilla/Getty Images; 17.1 (a) p. 490 (top): Texas A&M AgriLife Extension Service; 17.1 (b) p. 490 (bottom left): Alessandro Pietri/Shutterstock; 17.2 p. 490 (bottom right): Dr. Robert Bullard; 17.3 (a) p. 491 (top left): Joe Raedle/Getty Images; 17.3 (b) p. 491 (top center): The Washington Post/Getty Images; 17.3 (c) p. 491 (top right): CaseSensitiveFilms/Shutterstock; 17.4 p. 491 (bottom): Courtesy of Sacoby Wilson; 17.5 p. 492: AP Photo; p. 494 (top left): Barney Sellers/ZUMApress/Newscom; p. 494 (bottom left): Bettmann/Getty Images; p. 494 (top right): AP Photo/Steve Helber; p. 494 (bottom right): WE ACT for Environmental Justice; p. 495 (top): Clinton Presidential Library; p. 495 (bottom): UPI/Alamy Stock Photo; 17.7 p. 496 (top): Map of the book "On the Mode of Communication of Cholera" by John Snow, originally published in 1854 by

C.F. Cheffins, Lith, Southhampton Buildings, London, England; 17.8 p. 496 (bottom): World History Archive/Alamy Stock Photo; p. 497 (top): Library of Congress, Bain News Service; p. 497 (bottom): Rutgers University Library; 17.9 p. 498: Christianthiel.net/Shutterstock; 17.10 p. 499 (top left): SPL/Science Source; 17.11 p. 499 (top right): Ian Dagnall Computing/Alamy Stock Photo; 17.12 p. 499 (bottom): Tommy E Trenchard/Alamy Stock Photo; 17.13 p. 500: Sipa USA/Alamy Stock Photo; p. 501 (left): AP Photo/Horst Faas; p. 501 (right): AP Photo; p. 503: Wavebreak Media/Alamy Stock Photo; 17.17 p. 506 (top): Jim West/Alamy Stock Photo; 17.18 p. 506 (bottom left): Roman Lacheev/Alamy Stock Photo; 17.18 p. 506 (bottom center): ImageBROKER/Alamy Stock Photo; 17.18 p. 506 (bottom right): Jeremy Sutton-Hibbert/Alamy Stock Photo; 17.19 p. 507 (top): Photomaster/Shutterstock; p. 507 (bottom): Dinodia Photos/Alamy Stock Photo; 17.20 (a) p. 508 (top left): AP Photo/Manika Kamara; 17.20 (b) p. 508 (top right): Robert Harding/Alamy Stock Photo; 17.20 (c) p. 508 (bottom left): Robert Gilhooly/Alamy Stock Photo; 17.20 (d) p. 508 (bottom right): Science History Images/Alamy Stock Photo; 17.21 p. 509: Jim West/Alamy Stock Photo; 17.25 p. 511: NOAA/NASA. Adapted from Gesch (n.d.) https://pubs.usgs.gov/circ/1306/pdf/c1306_ch3_g.pdf; 17.26 p. 513 (top): Think4photop/Shutterstock; 17.27 p. 513 (bottom): Rosanna Arias/FEMA; p. 514 (left): Reuters/Alamy Stock Photo; p. 514 (right): Center for Disease Control; p. 515: Occupational Safety and Health Administration; p. 518 (top): World Health Organization; p. 518 (bottom): World Health Organization

Drawn art/text: (p. 518): Graphs: Reproduced from "WHO Fact sheet: The top 10 causes of death," © 2020 World Health Organization. Reprinted by permission.

CHAPTER 18

Photos: pp. 520–521: Andrea Matone/Alamy Stock Photo; 18.1 p. 522: Moravka Images/Alamy Stock Photo; 18.2 (b) p. 523: Sarah Northrop and Emerald Media Group; p. 524: Melinda Lee Patelli; p. 525: Maxx-Studio/Shutterstock; 18.4 p. 526 (top): Noriko Cooper/Alamy Stock Photo; 18.4 p. 526 (center): Flamingo Images/Shutterstock; 18.4 p. 526 (bottom): Michael Jung/Shutterstock; 18.5 p. 527 (left): Southern Illinois University Carbondale; 18.5 p. 527 (right): Sweeann/Shutterstock; 18.6 p. 528 (top): W.W. Norton & Company; 18.7 p. 528 (bottom): W.W. Norton & Company; p. 529: Victoria Shapiro/Shutterstock; p. 530: Melinda Lee Patelli; 18.8 (a) p. 531 (left): Melinda Lee Patelli; 18.8 (b) p. 531 (center): Bulent Kilic/AFP/Getty Images; 18.8 (c) p. 531 (right): Yale Office of Sustainability; 18.9 (a) p. 534 (top): Melinda Lee Patelli; 18.9 (b) p. 534 (bottom): REUTERS/Alamy Stock Photo; 18.10 p. 535 (left): Urbanbuzz/Alamy Stock Photo; 18.11 p. 535 (right): Photo © CCI/Bridgeman Images; 18.12 p. 536 (left): Jim Thompson/ZUMA Wire/Alamy Stock Photo; 18.12 p. 536 (right): Jim West/Alamy Stock Photo; 18.13 p. 537 (top left): Stonel/Shutterstock; 18.13 p. 537 (center left) David Tonelson/Shutterstock; 18.14 p. 537 (bottom left): U.S. Consumer Federal Trade Commission; 18.13 p. 537 (right): Huang Jenhung/Shutterstock; p. 538 (left): EcoChallenge.org; p. 538 (right): JaCrispy/Alamy Stock Photo; p. 542 (left): Environmental Protection Agency; p. 542 (right): Environmental Protection Agency

Drawn art/text: 18.2a: Figure from *Theory and Treatment Planning in Counseling and Psychotherapy*, 2e © 2016 South-Western, a part of Cengage, Inc. Reproduced by permission. www.cengage.com/permissions.

CHAPTER 19

Photos: pp. 544–545: A.P.S. (UK)/Alamy Stock Photo; 19.1 p. 546: Ricochet64/Shutterstock; p. 547 (top): Melinda Lee Patelli; 19.3 p. 547 (bottom): Designer Things/Shutterstock; 19.4 p. 548 (top): Moral contagion in social networks by William J. Brady, Julian A. Wills, John T. Jost, Joshua A. Tucker, Jay J. Van Bavel. Proceedings of the National Academy of Sciences Jun 2017, 201618923; DOI:10.1073/pnas.1618923114; 19.5 p. 548 (bottom): Bob Croslin/National Geographic; p. 549: Dpa Picture Alliance/Alamy Stock Photo; 19.7 p. 550: Global Impact Investing Network; 19.9 p. 552 (top): Green Stock Media/Alamy Stock Photo; 19.10 p. 552 (bottom): Richard Gardner/Alamy Stock Photo; p. 553 (left): Warren Price Photography/Shutterstock; p. 553 (right): B Lab; 19.11 p. 556 (top): PATAGONIA; 19.12 p. 556 (bottom): Gabriela Hasbun/Redux; 19.13 (a) p. 558 (top left): Daderot; 19.13 (b) p. 558 (top right): Dr. Kevin Lyons, Ph.D; 19.13 (c) p. 558 (bottom left): Mary Flewelling Morris; 19.13 (d) p. 558 (bottom right): Ford Motor Company; p. 559: Reza A. Marvashti/The Free Lance-Star/AP Photo; 19.14 (a) p. 561 (top left): Interfoto/Alamy Stock Photo; 19.14 (b) p. 561 (top right): Grzegorz Czapski/Shutterstock; 19.15 p. 561 (bottom): UNDP China; p. 562 (left): Green 2.0; 19.16 p. 562 (right): Rich Carey/Shutterstock; 19.17 p. 563 (left): Laure Joliet/The New York Times/Redux; p. 563 (right): Nigel Cattlin/Alamy Stock Photo; p. 564 (left): GaudiLab/Shutterstock; p. 564 (right): MediaPunch/Shutterstock; p. 568: McKinsey & Company

CHAPTER 20

Photos: pp. 570–571: Jeff J Mitchell/Getty Images; 20.1 p. 572 (top): AP Photo; 20.1 p. 572 (bottom left): AP Photo; 20.2 p. 572 (bottom right): CulturalEyes - AusGS2/Alamy Stock Photo; 20.5 p. 575: REUTERS/Alamy Stock Photo; p. 576: Novarc Images/Alamy Stock Photo; 20.7 (a) p. 578 (top): Jim West/Alamy Stock Photo; 20.7 (b) p. 578 (center): B. Leighty/Photri Images/Alamy Stock Photo; 20.7 (c) p. 578 (bottom): David Grossman/Alamy Stock Photo; p. 579: John Schwieder/Alamy Stock Photo; 20.10 p. 580: Environmental Protection Agency; 20.11 p. 581 (top): Nature Picture Library/Alamy Stock Photo; 20.12 p. 581 (bottom): Environmental Protection Agency/Department of Transportation; 20.13 (a) p. 582 (top left): Walter Stein/AP Photo; 20.13 (b) p. 582 (top right): Everett Collection Historical/Alamy Stock Photo; 20.13 (c) p. 582 (bottom left): Santa Barbara News-Press/ZUMApress.com/Alamy Stock Photo; 20.13 (d) p. 582 (bottom right): Bettmann/Getty Images; p. 583 (left): Science History Images/Alamy Stock Photo; p. 583 (right): Universal Art Archive/Alamy Stock Photo; 20.14 p. 585 (left): ZUMA Press, Inc./Alamy Stoack Photo; 20.14 p. 585 (center): UNDP Papua New Guinea; 20.14 p. 585 (right): UNDP Peru/Anette Andresen; p. 587: Alterov/Shutterstock; 20.16 p. 588: David Grossman/Alamy Stock Photo; p. 589: REUTERS/Alamy Stock Photo; p. 592: Mak/Stockimo/Alamy Stock Photo; p. 596: Pew Research Center

Drawn art/text: (p. 589): Figure: "Emissions Gap in 2030" reprinted by permission of the World Resources Institute.

USE THE NEWS

Drawn art/text

Ch1: "Go Viral Game 'pre-bunks' coronavirus conspiracies," by Fred Lewsey. Reprinted by permission of University of Cambridge and Fred Lewsey.

Ch2: Reprinted with permission from AAAS. "Updated: United States adopts major chemical safety overhaul," by Puneet Kollipara. *Science* June 8, 2016. © 2016 AAAS. Permission conveyed through Copyright Clearance Center, Inc.

Ch3: "Global analysis reveals how sharks travel the oceans to find food," University of Southampton News, 18 January 2018. Reprinted by permission of University of Southampton Media Relations.

Ch4: "Scientists Race to Kill Mosquitoes Before They Kill Us," by Sara Novak. *Sierra*, January 9, 2018. Reprinted by permission of the Sierra Club.

Ch5: "Sea Lion-Salmon Cycle Starts Again in Columbia, This Time with Expanded Ability to Kill," by Karina Brown. *Courthouse News Service*, March 25, 2021.

Ch6: Republished with permission of United Nations Publications. "9.7 billion on Earth by 2050, but growth rate slowing, says new UN population report," by UN News, 29 July 2015. © United Nations 2015. Permission conveyed through Copyright Clearance Center, Inc.

Ch7: "Animas River spill: Root causes and continuing threats," by Joel A. Mintz, The Hill, September 2, 2015. Reprinted by permission of The Hill.

Ch8: "VW Could Fool the EPA but It Couldn't Trick Chemistry," by Eric Niiler. Wired, September 22, 2015. © Conde Nast. Reprinted with permission.

Ch9: "Earth's Soil is Getting Too Salty for Crops to Grow," by Sarah Zielenski. Smithsonianmag.com, October 28, 2014. Copyright 2014 Smithsonian Institution. Reprinted with permission from Smithsonian Enterprises. All rights reserved. Reproduction in any medium is strictly prohibited without permission from Smithsonian Institution.

Ch10: "University of Michigan adds 'pee-cycling' to recycling efforts," by Martin Slagter, MLive.com, February 21, 2017. © 2017 MLive Media Group. All rights reserved. Reprinted with permission.

Ch12: "Worcester farmers markets offer fresh produce for all budgets, target food insecurity," by Nicole Shih. *Telegram & Gazette*, October 8, 2021. © Nicole Shih– USA TODAY NETWORK.

Ch13: "Global carbon dioxide emissions reach highest level in history," by Doyle Rice. *USA Today*, March 8, 2022. © Doyle Rice– USA TODAY NETWORK.

Ch14: Reprinted with permission from Dow Jones & Company, Inc. "Solar Power Booms in Georgia, Where It Isn't Mandated," by Elena Shao. *Wall Street Journal* August 22, 2021. © 2021 Dow Jones & Company, Inc. Permission conveyed through Copyright Clearance Center, Inc.

Ch15: "Grocers, innovators work to save $31B in food from being trashed in Canada each year," by Maryse Zeidler, CBC News, November 10, 2018. Reprinted by permission of CBC Licensing.

Index

Note: *Material in figures and tables is indicated by italic page numbers.*

A

acidification, oceanic, 205, *206*, 296, 322
acidity, oceanic, 16
acid rain (acid deposition), 230, *231*, 234, *234*
acids and bases, 72, *72*
actions. *See* individual choices and actions
adaptive radiation, 90, *91*
additive toxins, 502
Afton, North Carolina, 439
agriculture
  agricultural practices and conservation, 141–42, *141*
  agrobiodiversity, 342, *342*
  burning of forests for agricultural development, 62, 64, *64*
  conservation agriculture, 360
  consumer-supported agriculture (CSA), 360
  defined, 338
  desertification, 344
  development and human population, 157–58, *158*
  Dust Bowl, 252, *252*
  and environmental decision-making, 24
  fallow fields, 339
  farm equipment, 339–40
  genetically modified organisms (GMOs), 340–41, 343, 349–50
  genetic erosion, 342
  grazing practices, 265–66
  Green Revolution, 339–41, 342
  historical agricultural technologies, 338–39, *338, 339*
  hydroponic farming, *336–37*
  impacts on ocean, *206*
  irrigation practices, 340, *341*
  monoculture, 342
  seed drills, 338–39, *339*
  soil fertility, 127, 361
  soil loss and degradation, 263–66, *264, 266,* 343–44, *343*
  subsidence, 344
  synthetic fertilizers and pesticides, 340, 341, 342–43, *343*
  tilling practices, 264–65, *264, 266,* 273–74, 287, *287*
  water availability, 344, *344*
  wind and water erosion of cropland soils, 264–65, *264*
  and wind power, 400
  See also food
agrobiodiversity, 342, *342*
air
  mass, 220, *220*
  pressure, 215–16, *215, 216*
  See also air pollution; atmosphere; climate; weather
air pollution, 232–33
  acid deposition (acid rain), 230, *231*, 234, *234*
  air quality regulations, 236–37, *236*

automobile use and frequency of asthma-related events, 16, 30–31, *30*
Black Americans and, 13
CAFE standards, 236, *236*
carbon monoxide, 226, *226*
chlorofluorocarbons (CFCs), 229–30, *229, 230, 231*
in cities, 214, *214,* 236
Clean Air Act of 1970, 234, 236
defined, 225
Donora, Pennsylvania, inversion, 214, *214, 582*
Energy Policy and Conservation Act, 234, 236
imbalance of consumption and exposure, 13, *14*
individual choices and actions, 237–38
and lead in the environment, 14–15, *15*
nitrous oxide emissions, 318, *318,* 344
ozone layer, 229–30, *229, 231*
particulate matter (PM), 226–27, *227,* 230
primary pollutants, *226,* 226–27, *227,* 232
public health research, 16
secondary pollutants, 227–28, *227, 228,* 233
smog, 227–28, *227, 582*
smoking bans, 226
soot, 225, *225,* 226
thermal inversion, 214, *214,* 218, 548
tradeable emission allowances, 234, 236
Use the Data, 242
Use the News, 240–41
vehicle exhaust, 212
volatile organic compounds (VOCs), 226, 227, *227*
  See also atmosphere
Alam, Mohammed-Reza, 49
albedo, 202, 203, 308, 309
Albritton, Ben, 594
aldrin, *505*
Alice in Wonderland (Carroll), 496
alleles, 91, *91*
almond milk, 199
alternative energy
  biofuels, 372, 373–74, 407–8, *408*
  energy transition plans, 410, 412–13, 414–16, *415, 416,* 420, *420*
  ethanol, 408, *408,* 425
  fuel cells, 409–10, *409*
  geothermal power, 76, 406–7, *406, 407*
  hydropower, 403–4, *403, 404,* 414, *415*
  individual choices and actions, 420–21
  nuclear power, 76, 404–6, *405*
  ocean waves, 49
  solar power, 400, *401,* 402–3, *402,* 413–14, *414, 415, 418,* 423–24, 544–45
  Solutions Project, 411
  wind power, 373, *373,* 396, 396–97, 398–400, *399, 398,* 413–14, *413, 414, 415, 418,* 421, 576
altitude and air pressure, 216, *216*
Alvarez, Luis, 17

Alvarez, Walter, 17
ammonification, 290
Amster-Olszewski, David, 81
Ancestral Puebloan people, 156
Anderson, Ray, 552, *552*
animal welfare, 359
Animas River pollution, 209–10
anoxic environment, defined, 434
Antarctic Ice Sheet, 325
Anthropocene epoch, 6
anthropocentrism, *37,* 37–38
antibiotic-resistant bacteria, 352, 353, 500
Antiquities Act of 1906, 130
aphotic zone, 186, *187,* 204
Apple, 566–67
aquaculture, 352, 359
aquatic and marine protected areas, 131, *131*
aquifers, 183, *183*
Arakawa, Akio, 317, *317*
Arctic National Wildlife Refuge (ANWR), 578–79, *579*
asphyxiants, *503*
Association for the Advancement of Sustainability in Higher Education (AASHE), 22, 26
asthenosphere, 248, *248*
asthma, automobile use and frequency of asthma related events, 16, *30,* 30–31, *30*
atmosphere
  acid rain, 230, *231,* 234, *234*
  air circulation cells, 222, *222,* 224
  air quality regulations, 236–37, *236*
  carbon cycle, 291–96, *292, 293*
  chlorofluorocarbons (CFCs), 21, *21,* 25, 229–30, *229, 230, 318,* 586–87
  circulation, 222, *222,* 224–25, *224, 225*
  clouds, 223
  components and structure, 215–17, *215, 216, 217*
  defined, 212, 214
  dust transport, 224–25, *224, 225*
  equatorial low-pressure belt, 222, *222*
  Ferrel cells, *222,* 224
  Hadley cells, 222, *222*
  jet streams, 224, *224*
  mesosphere, 216, 217, *217*
  nitrogen cycle, 286, 288–90, *288, 290, 291*
  oxygenating extinction, 283
  oxygen cycle, 283–84, *284, 285*
  stratosphere, 216, 217, *217*
  subtropical high-pressure belt, 222, *222*
  thermal inversion, 214, *214,* 228, *228*
  thermosphere, 217, *217*
  troposphere, 216, *217*
  water cycle, 184–85
  See also air pollution; climate change

atomic number, 65, *65*
atoms, 65–68, *66, 67, 68*
authoritarian systems, 574, *574*
automatic thinking, 522–24, *523*

B

bacteria, antibiotic-resistant, 352, 353, 500
Bakken Formation, 41, 381, *381*
balancing feedback, 282–83, *283*
Baotou, Inner Mongolia, 253, *253*
barometric pressure, 219–21, *220*
Basel Convention, 435
bases, acids and, 72, *72*
Basol, Melisa, 29
Bastida, Xiye, 460
battery storage for solar power, 400, 402, *402,* 415
Baumgartner, Felix, *216*
beavers, *113*
beef production, 348, 350, *350,* 355, *355*
bees
    bee-safe plants, 144
    flowering plants and pollination, 86–88, *88*
belief perseverance, 525
Benaquisto, Lizbeth, 594
beneficiation plant, 379
benefit corporations, 551, 553
benthic zone, 204, *205*
Beschta, Robert, 105, 106
Bezos, Jeff, 398
Bhopal, India, 507
bias, 19, *19*
Biddle, Mike, 69, *69*
Biden, Joe, 391, 572, 586, 588
BioBlitz events, 144
biocentrism, *37,* 38
biodegradable plastics, 446, *446*
biodiversity
    agrobiodiversity, 342, *342*
    biomes, 90, 98, *99,* 100–101, 262, *262*
    biosphere, 90
    biotic potential, 108
    carrying capacity, 106, 107, *108*
    coevolution, 103, *103*
    commensalism, 103–4, *104*
    community, 89
    community interactions, 98–99, 102–6, *104*
    competition, 99, 102, *102*
    convergence, 98, *99*
    defined, 88
    dominance, 97, *97*
    ecological niches, 90, 102, 103, *104*
    ecosystem, 89–90
    ecosystem diversity and biogeography, 98, *99*
    edge effects, 140
    endemic species, 110–11
    environmental conditions as determining force, 95, *95,* 98, *99*
    exponential population growth, 106–7, *107, 108*
    extinction rates, 121, *121*
    flowering plants and pollination, 86–88, *88*
    functional extinction, 109–10

habitat, 90
habitat alteration, 121, *122*
hot spots, 110, *110,* 111, 112, 138
human impacts, 121–23, *121, 122, 123*
individual choices and actions, 112–13
interspecific competition, 103
invasive species, 122, *123*
keystone species, 111–12, *112, 113*
K-strategists, 109, *109*
laws for protecting, 135–36, 138–39, *139*
logistic (S-shaped) growth, 107, *108*
loss, 109–12
mass extinction events, 109, *110,* 283
mutualism, 104, 106
parasitic relationships, 102–3, *103*
population, 89
population crash, 107, *107*
population density, 107–8, *108*
predation, 102–3, *103,* 105–6
Predator Free New Zealand, 111
reproductive strategies, 108–9, *109*
resource overexploitation, 122
r-strategists, 108, *109*
shared characteristics of living things, 88–90, *89*
soil fertility and agricultural productivity, 127
of species, 88
species evenness, 96–97, *97*
species richness, 95–96, *97,* 116–17
symbiotic interactions, 103–4, *104,* 106
Use the Data, 116–17
Use the News, 115–16
See also climate change; conservation; evolution; genetics; life
biofuels, 372, 373–74, 407–8, *408,* 425
biogeochemical cycles, 283
biological hazards
    antibiotic-resistant bacteria, 352, 353, 500
    blood-borne pathogens, 499–500, *499, 500*
    childhood deaths from pneumonia and diarrhea, 499, *500*
    defined, 498
    diarrheal diseases, 498–99, 500
    pathogens, 498, 499–500, *499*
    respiratory infections, 498, *499, 500*
bioluminescence, 204
biomes, 90, 98, *99,* 100–101
bioreactor landfills, 440
biosphere
    biosphere composition and climate change, 310
    carbon cycle, 291–96, *292, 293*
    defined, 90
Biosphere 2, *276–77, 278, 278, 281, 281, 282, 282*
biotic potential, 108
Bird, Christopher, 83
Birol, Fatih, 394
birth control, access to, 168–69
Bittman, Mark, 358, 362
Blue Acres program, 513, *513*
Bohr, Niels, 66
Bohr atomic model, 66, *66*
bonds, chemical, 66, 67–68, *68*
Booderee National Park, 134

Booker, Cory, 58
Botox, 503
bottom trawling, *351*
Boxer, Barbara, 58
boycotts, 55
BPA-free plastics, 35, *35*
Braverman, Inna, 81
breathing system, human, 279–80, *279*
Brexit and EU environmental policy, 587
bright lights syndrome, 466
Brooklyn Bridge Park, 12, *12*
Brown, Jerry, 196
bubonic plague, 153
Buckland, William, 286
Bullitt Center Living Building Challenge, 481
Bureau of Land Management (BLM), 53, 131
burning of forests for agricultural development, 62, 64, *64*
Bush, George H.W., 588
Bush, George W., 387, 392, 507, 572, 575
bycatch, 353, *354, 362*

C

CAFE (corporate average fuel economy) standards, 236, *236,* 352, 357
Cairo Consensus, 170
California condors, 142, *142*
Camp Fire, 302
CAP (DOD Climate Adaptation Plan), 331–32
cap-and-trade exchange system, 44, *44,* 390, 391
captive breeding programs, 142, *142*
carbon capture and storage, 388, 389–90
carbon cycle, 291–96, *292, 293*
carbon dioxide
    from burning of forests for agricultural development, 62, 64, *64*
    carbon capture and storage, 388, 389–90
    chemical formula, 67
    and "clean coal" technology, 378
    Clean Power Plan, 575
    and ocean acidity, 16
    from oil burning, 382
    Paris Agreement, 572, 588, *588,* 589
    photosynthesis, 76, 77
    soil fertility and agricultural productivity, 127
    soil formation, 260
    temperature, 308, 394
    textiles and carbon dioxide emissions, 328–29, *329*
    vehicle fuel efficiency and carbon footprints, 327, 328
    See also climate change
carbon isotopes and shark food webs, 83–84
carbon monoxide, 226, *226*
carbon taxes, 390, 391, 392
carcinogens, *503*
Carroll, Lewis, 496
carrying capacity, 106, 107, *108,* 158
Carson, Rachel, 501, 582, 583, *583*
Carter, Jimmy, 579
castings, 260
cement production, 316

Chaco Canyon, *153, 156*
charcoal, 372
checks and balances, 574
cheetahs and genetic fitness, 94
chemical bonds, 66, 67–68, *68*
chemical energy, 76
chemical formulas, 66, *67*
chemical hazards, 502, 514
chemical reactions
    chemical weathering, 251, *251*
    oxidation–reduction (redox) reactions, 74
    and phase changes, 72, 74–75, *74, 75*
    polymerization, 74–75, *75*
chicken production, 348, *348*
Chicxulub Crater, 17
China
    air pollution and, 236
    environmental policy in, 574–75, 576
    one-child policy, *159,* 159–60
    Three Gorges Dam, 403–4, *404*
chlordane, *505*
chlorofluorocarbons (CFCs), 21, *21,* 25, 229–30,
    229, 230, 231, 318, 586–87
choices. *See* decision making; individual choices and
    actions
cholera, 496, *496*
circulation, atmospheric, 222, *222,* 224–25, *224,
    225*
cities. *See* urbanization
civil unrest and youth bulge, 165, 166
Clean Air Act of 1970, 50, 52, 234, 236, 572, 581,
    584, *584*
Clean by Design, 4
"clean coal" technology, 378
Clean Power Plan, 575
Clean Water Act, 50, 53, 200, 202
Clean Water Act of 1972, 52, 200, 202, 584, *584*
Clements, Shaun, 147
climate
    defined, 217–18
    hindcasting climate models, 17–18
    oceanic influence on, 205, *206*
climate change, 9
    albedo, 308, 309
    atmospheric carbon dioxide concentrations, 316,
        316
    biosphere composition, 310
    carbon dioxide and temperature, 308
    continental drift, 309–10
    coral bleaching and, 123, *123*
    crop failure and livestock loss, 304
    DOD Climate Adaptation Plan (CAP), 331–32
    and drought, 321
    ecosystem effects, *322–23, 323, 324*
    environmental justice and, 329, 510–12, *511,* 514
    eruptions and asteroids, 310, *310*
    extreme weather, 511–12, *511*
    feedback loops and acceleration of, 325, *326*
    geoengineering, 73
    geothermal gradients, 306
    global climate, 305
    global climate models, 317, 318–19

global indicators, 314–15
greenhouse effect, 308–9, *308, 309*
greenhouse gas emissions, 293, *293,* 296, 316, 318,
    318, 327
greenhouse gases, 308–9
hindcasting, 317, 318–19
historical, 307–10
hurricanes and severe weather, 321, *324*
ice sheet melting, 325
individual choices and actions, 328–30
instrumental period, 305
Milankovitch cycles, 310
ocean currents, 311–13, *311, 312, 313*
ocean ecosystems and, *206,* 207
paleoclimates, *306,* 307
precipitation and extreme rain events, 319–21,
    320, 324
projected global warming, 326, *326,* 328
proxy indicators, 306–7, *306*
risk and insurance, 302, 304
sea-level rise, 304, 313, *313,* 315, 319, *319,* 322,
    322, 324, 325
storm damage costs, 304, *304*
temperature records, 305–7, *305, 306, 307*
thermal expansion, 322, *322*
Use the Data, 333–34
vehicle fuel efficiency and carbon footprints, 327,
    328
Clinton, Bill, 507, 513
CLORPT (climate, organisms, relief, parent material,
    and time), 262
closed system, 278
Clostridium botulinum, 503
clothing
    environmental effects of production, 2, 4, *4*
    individual choices and sustainability, 26
clouds, 223
coal
    environmental impacts of, 378, 379–80, *379,* 385,
        394
    extraction and consumption of, 378–79, *379*
    formation of, 370–71, *370*
    global reserves, 378, *378*
    Jevons paradox, 42, *42,* 376
    true cost of, 43–44, *43, 44*
    US and global energy sources and consumption,
        371, *371, 372, 375*
coastal resilience, 12, *12*
Coca Cola, 561, *561*
coercion, 573
coevolution, 102, 103, *103*
cogeneration, 410, 412, *412*
cognitive biases, "gambler's fallacy," 19
cognitive dissonance, 528, 530
cold front, 221
cold fusion, 18
Cole, Jerry, 147
Colorado River, 192, *192,* 194
combustion
    law of conservation of mass, 66, *67*
    of waste, 438–39, *440*
commensalism, 103–4, *104*

commitments, 528, *528,* 530, 532
communication strategies and environmental
    decision-making, 22
community
    community interactions, 98–99, 102–6, *104*
    defined, 89
competition, 99, 102, *102*
composting, 441–42, *441*
compounds, molecules and, 66–68, *67, 68,* 72, *72*
Comprehensive Environmental Response,
    Compensation, and Liability Act (CERCLA),
    50, 52, *584*
Comprehensive Everglades Restoration Program
    (CERP), 142–43, *143*
concentrated animal feeding operations (CAFOs),
    348, *350,* 352
concentrated solar thermal (CST) plants, 400, *401,*
    402–3
confirmation bias, 525
conformity, 524, *525*
conservation
    agricultural practices, 141–42, *141*
    alternative protection strategies, 135, *136*
    Convention on Biological Diversity, 139
    Convention on International Trade in
        Endangered Species of Wild Fauna and
        Flora (CITES), 139
    deforestation, 140–41, *140*
    easements, *136*
    ecological island effect, 132
    ecological restoration, 142–44, *143*
    ecotourism, 135, *135*
    Endangered Species Act (ESA), 136, 138, *139*
    enforcement and continued human use, 134–35,
        134
    forest land ownership and management, 148
    grazing and grassland management, 141
    human impacts on biodiversity, 121–23, *121, 122,
        123*
    individual choices and actions, 144
    instrumental and intrinsic values, 123–24
    laws for biodiversity protection, 135–36, 138–39,
        139
    lynx repopulation in Spain, 133
    Marine Mammal Protection Act, 135–36
    Migratory Bird Treaty Act, 135, *135*
    Natural Resource Damage Assessment process,
        143–44, *143*
    Pacific salmon, 118–20, *118*
    passenger pigeon extinction, 121
    predator controls, 146–47
    prioritization of efforts, 137–38
    protected areas, 126, *126,* 128–35, *130, 131, 132,
        133, 134*
    Sierra del Divisor National Park, 134
    species reintroduction, 142, *142*
    sustainable forest management, 141
    urbanization and land-use planning, 142
    Use the Data, 148
    Use the News, 146–47
    water, 196–97, 207–8
    wildlife corridors, 132, 133, *133*
    See also ecosystems

consumer-supported agriculture (CSA), 360
consumption
    conspicuous consumption, 447, *447*
    culture, 447–48, *447*
    defined, 446
    embodied energy, 328–29, *329*
    and exposure to pollution, imbalance of, 13, *14*
    fast fashion, 24
    food system and consumption, 353–57, *355, 356, 357*
    product obsolescence, 527
    rising consumption and human population, 170–71, *171, 172,* 173–74
    suburban sprawl, 473–74, *473*
    and waste, 24, 446–48, *447*
continental drift, 309–10
contraception, 159, 161, 168–69
control group, 16
controlled experiments, 15–16, *16*
Convention on Biological Diversity, 139
Convention on International Trade in Endangered Species of Wild Fauna and Flora (CITES), 139
convergence, 98, *99*
convergent plate boundary, 249, *249*
conversion of waste, *437,* 440–41
Cook, Tim, 566–67
Copernicus, Nicolaus, 16
coral bleaching, 16, *16,* 123
coral ecosystems, 6, 204
coral skeletons, 306, *306*
core of Earth, 247, *247*
Coriolis effect, 224, *224,* 311
corn
    monocultures, 342
    teosinte (ancient corn), 338, *338*
corrosive toxins, *503*
countercurrents, 312–13, *313*
covalent bonds, 68, *68*
cover crops, *264,* 265
COVID-19, 13, *13,* 19, 29–30, 242, 357, 394, 458, 491, 498, *498,* 500, 514, 524–25
"CRAAP" guide, 19–20, *20*
crude death rate, 157
crust of Earth, 247, *247,* 248, *248*
Crutzen, Paul, 73
cryosphere, 202, *203*
Cullen, David, 594
cultural services, 124–25, *125*
currents
    countercurrents, 312–13, *313*
    deep currents, 311–12
    El Niño–Southern Oscillation (ENSO), 312, *313*
    gyres, 205
    ocean currents and climate change, 311–13, *311, 312, 313*
    surface currents, 311
    thermohaline conveyor, 312, *312*
    upwelling, 312, *313*
cycles. *See* systems and cycles

D

Darwin, Charles, 90, 261
dead zone, 290, *290*
decision making
    automatic thinking, 522–24, *523*
    belief perseverance, 525
    cognitive dissonance, 528, 530
    commitments, 528, *528,* 530, 532
    confirmation bias, 525
    delegation, 524
    denial, 524
    designing competing narratives, 535–36, *536*
    effect of word choice on, 534–35, *535*
    emotional defense mechanisms, 524
    emotional distancing, 524
    feedback, 528, *528,* 533
    food waste reduction, 529, 540–41
    frames, 535–36, *535, 536,* 537, 538
    fuel economy driver interface (FEDI), 530
    fuel economy labels, 542
    hierarchy of needs, 522, *523*
    incentives, 520, 522, 530–31, *531,* 532, 534, *534*
    labels and guides, 534, 536–37, *537,* 542
    loss aversion, 522, 523–24, *523*
    motivation, 530
    product obsolescence, 527
    prompts, 526, *526,* 534
    resignation, 524
    social acceptance and conformity, 524–25, *525,* 526
    social norms, 524, 531, *534*
    status quo, 522, 523
    use of disposable *versus* reusable plastic bags, 520, *520–21, 522*
    Use the News, 540–41
    See also ethics; individual choices and actions
decomposers, 79
deep currents, 311–12
Deepwater Horizon oil spill, 32, *34,* 205, 382, 384, 581
deforestation, 140–41, *140,* 252, 266–67, *267*
delegation, 524
delta, river, 257, *258*
demand, supply and, 38, 40, *40*
democratic systems, 574
Democritus, 64
demographers, 152
demographic transition and global population dynamics, 163–66, *163, 164, 165*
demographic window, 164, *164*
denial, 524
denitrification, 290
deontological ethics, 35, 38
depolymerization, 75, *75*
deposition, 290, *291*
desalination, 198, *200*
DeSantis, Ron, 594
desertification, 344
desert soil, 262, *262*
design for recovery, 445, *445*
detritus, 259
development

energy-efficient development and public health, 486–87
environmental effects of Hurricane Harvey, 490, *490, 491*
in floodplains, 255, 511, 512
government building and loan policies, 471–73, *472*
Leadership in Energy and Environmental Design (LEED) certification, 563
low-impact, water quality, 200, *202*
redlining, 471
Sustainable Development Goals (SDGs), 9, *12,* 584–85, *585*
3Es (environment, economy, and equity) of sustainable development, 9
See also land use
Dharavi settlement, 465, *465, 466, 467*
diazotrophs, 288–90
dichlorodiphenyltrichloroethane (DDT), 342, *343,* 501, 503, *505,* 582, 583
dieldrin, *505*
DiGiulio, Dominic, 383
dioxins, *505*
disinformation, 19–20, *20*
divergent plate boundary, 249, *249*
DNA (deoxyribonucleic acid), 91
Döbereiner, Johanna, 289, *289*
DOD Climate Adaptation Plan (CAP), 331–32
dogs, humans and, 339
dominance, 97, *97*
Donana National Park, 133
dracula ants, 88, *88*
Drapkin, Julia Kumari, 330
drought
    and climate change, 321, 512
    crop failure and livestock loss, 304
    defined, 196
    reservoir levels, *305*
    and water shortages, 180, 182, *182,* 195–96, *195, 197,* 207–8
Dunbar, Robin, 547
Dust Bowl, 252, *252*
dust transport, 224–25, *224, 225*
dwelling units per acre (DU/acre), 468, *469*

E

Earth
    asthenosphere, 248, *248*
    carbon cycle, 291–96, *292, 293*
    core, 247, *247*
    crust, 247, *247,* 248, *248*
    earthquake awareness and tsunami damage, 244, 246, *246*
    erosion, 250–52, *250, 251, 252, 253*
    faults, 248, *248*
    human impacts on, 244, 246, *246*
    igneous rocks, 250, *250, 251*
    lithification, 252, *253*
    lithosphere, 248, *248*
    magnetic field and animal migrations, 246
    magnetosphere, 247, *247*
    mantle, *247,* 248

metamorphic rock, *250*, 252, *253*
Milankovitch cycles, 310
minerals, 253–54, *254*, 270
nitrogen cycle, 286, 288–90, *288, 290, 291*
oxygen cycle, 283–84, *284, 285*
plate tectonics, 248–49, *248, 249*
rock cycle, 250, *250*
sedimentary rock, *250*, 252, *253*
sedimentation, *250*, 252, *253*
snowball Earth, 307
tectonic plates, 248–49, *248, 249*
water cycle and land surface of, 184–85
weathering of rocks, 250–52, *250, 251, 252, 253*
See also landforms and topography; land use; soil;
    systems and cycles
Earth Charter, 35, *36*
Earth Day, 570, 572, *572*
earthquakes, 244, 246, *246*, 390, 508–9, *508*
Easter Island, 153, *153*
Ebola virus, 499, *499*
ecocentrism, 38, *38*
ecological footprint analysis, 23, *23*, 170–71, *172,*
    174
ecological footprint of cities, 461, *461*
ecological island effect, 132
ecological niches, 90, 102, 103, *104*
ecological resilience, 9, 12, *12*
ecological restoration, 142–44, *143*
economics
  boycotts, 55
  cap-and-trade exchange system, 44, *44*, 390, 391
  carbon taxes, 390, 391, 392
  economic systems, 38
  efficiency gains and increased consumption (Jevons
    paradox), 42, *42*
  energy transition plans, 410, 412–13, 414–16, *415,*
    416, 420, *420*
  externalities, 43–47, *43, 44*
  global waste trade, 435, *435, 436*
  individual choices and actions, 54–55
  inferior goods, 375
  informal economy, 466
  Jevons paradox, 42, *42*, 376
  markets and externalities, 47
  markets and the government, 40–41
  negative externalities, 43–45, *43, 44, 45*
  oil prices and, 38, 40, *40,* 41–42, *41*
  organizational divestment from fossil fuel
    investments, 559–60
  proven reserves, 375–76, *376*
  public goods, 46, *46*, 377–78, *378*
  recycling, 444–45, *444, 445*
  revealed preferences, 125
  rise of fossil fuel use, 375–76, *376*
  scarcity, 41–42, *41*
  stated preferences, 125
  stranded assets, 388
  supply and demand, 38, 40, *40*
  tragedy of the commons, 44–45, *45*
  triple bottom line (TBL) accounting, 549, *549*
economies of scale, 464, *464, 465*
economy, 3Es (environment, economy, and equity), 9

ecosystems
  about, 6
  climate change effects, 322–23, *323, 324*
  coral, 6, 204
  cultural services, 124–25, *125*
  defined, 6, 89–90
  diversity and biogeography, 98, *99*
  ecosystem services, 6, 46, *46*, 123–26, *124, 125*
  forests, 6, *6*
  groundwater-dependent, organisms, 186, *186*
  lentic, 186, *187*
  lotic, 186–87, *188*
  oceanic support for life, 204, *205*
  phenology, 323
  provisioning services, 124, *124*
  regulating services, 124
  replacement costs, 125–26
  supporting services, 124
  valuation of ecosystem services, 125–26
ecosystem services, 6, 46, *46*, 123–26, *124, 125*
ecotourism, 135, *135*
Eden Project, 266, *266*
edge effects, 140
education
  for girls and women, 168, 169, *169*
  health and, 167, *167*
efficiency gains and increased consumption
  (Jevons paradox), 42, *42*
Ehrlich, Paul R., 156, 157, 158, 170
Ekenga, Christine, 517
elections, 576–77
electrical energy, 76
electric grid, 415–16, *416*
electric vehicles (EVs), 408–9, *409,* 415, 420, *420,* 425
electrons, 65–66, *66*
elements, 64–65, *65*
El Niño–Southern Oscillation (ENSO), 312, *313*
embedded water, 182, *182,* 196
embodied energy, 328–29, *329*
emergent properties, 6, 280
eminent domain, *136*
emotional defense mechanisms, 524
emotional distancing, 524
Empedocles, 64
Empire State Building, 19, *19*
Endangered Species Act (ESA), 50, 53, 120, 136, 138, *139*
endemic species, 110–11
endrin, *505*
energy
  biofuels, 372, 373–74, 407–8, *408,* 425
  chemical, 76
  cogeneration, 410, 412, *412*
  conversions and inefficiency, 78, *78*
  defined, 74, 75
  electrical, 76
  electric grid, 415–16, *416*
  embodied energy, 328–29, *329*
  energy-efficient development and public health,
    486–87
  energy transition plans, 410, 412–13, 414–16, *415,*
    416, 418–20, *420*
  entropy, 78, *79*

environmental decision-making, 24
ethanol, 408, *408,* 425
first law of thermodynamics, 77–78
forms of, 76–77, *77*
fuel cells, 409–10, *409*
geothermal power, 76, 406–7, *406, 407*
global policies, 421
government actions regarding, 416, 420
home electronics, 412
home energy use, 328
human energy history, 372–74, *372, 373, 374*
hydropower, 403–4, *403, 404,* 414, *415*
individual choices and actions, 420–21
kinetic, 75
light bulbs and energy use, 84–85
mechanical, 76
milling technologies, 373–74, *373, 374*
nuclear power, 76, 404–6, *405*
from ocean waves, 49
photosynthesis, 76, 77
potential, 75–76, *76*
power, 373
productivity, 373, *374*
radiant, 76
second law of thermodynamics, 78
self-driving cars, 417
solar power, 400, *401,* 402–3, *402,* 413–14, *414,*
    415, 418, 423–24
Solutions Project, 411
storage, 400, 402, *402,* 414–16, *415*
thermal, 70
trophic levels, 78–79, *80*
ultrahigh-voltage transmission, 410, *410*
Use the Data, 425
Use the News, 423–24
variable generation, 399–400, 414
water power, 372, *373*
Watt's steam engine, 376, *376*
wind power, 373, *373,* 396, *396–97,* 398–400,
    398, 399, 413–14, *413, 414, 415, 418,* 421
work, 75
See also fossil fuels
Energy Policy and Conservation Act, 234, 236
Engle, Charlie, 194–95, *195*
enhanced oil recovery (EOR), 388, 389
ENSO (El Niño–Southern Oscillation), 312, *313*
entropy, 78, *79*
environment
  about, 5–6
  defined, 5
  environmental conditions and biodiversity, 95, *95,*
    98, *99*
  environmental decision-making, 20, 22–24, *22, 23*
  3Es (environment, economy, and equity), 9
  terrestrial habitat loss and impact on oceans, 7, *7*
  United States at night, 6–7, *7*
Environmental Defense Fund (EDF), 563
environmental effects
  ecological footprint analysis, 23, *23,* 170–71, *172, 174*
  ecological footprint of cities, 461, *461*
  Hurricane Harvey, 490, *490, 491*
  urban economies of scale, 464, *464, 465*

environmental hazards
  avoidance of, 514
  defined, 490–91
  distribution of, 491–93, *492, 493*
  earthquakes, 508–9, *508,* 512, *513*
  geological hazards, 508, *508*
  ionizing radiation, 509
  landslides, 508, *508*
  lava, *508,* 509
  radiation, 509–10, *509, 510*
  radon, 509
  tsunamis, *508,* 509
  ultraviolet (UV) radiation, 509–10, *510*
environmental justice
  about, 12–14
  Bhopal, India, 507
  Blue Acres program, 513, *513*
  and climate change, 329, 510–12, *511,* 514
  cumulative impacts, policies to address, 586
  defined, 12–13, 492
  environmental effects of Hurricane Harvey, 490, *490, 491*
  environmental health, 491–93, 496–98
  epidemiology, 496, *496,* 497
  Flint, Michigan drinking water supply, 189, 193, *488–89,* 504, *506*
  frontline communities, 321
  individual choices and actions, 514–15
  lead exposure, 194
  location of vulnerable communities, 512–13, *513*
  Love Canal, 494
  movement, timeline, 494–95
  Navajo uranium mines, 492, *492*
  outside influences on organizational change, 561–62
  protests, 13, *13, 488–89, 490*
  risk management and regulation of toxins, 506–7, *507*
  Use the Data, 518
  Use the News, 517
  Warren County, North Carolina, 13
  waste, 436, 439
  working for, 14
  workplace exposure to toxins, 497, 504, *506*
  See also biological hazards; environmental hazards; public health; toxins
environmental policy
  Arctic National Wildlife Refuge (ANWR), 578–79, *579*
  behavioral influences of, 580–81, *580, 581*
  Brexit and EU environmental policy, 587
  in China, 574–75, 576
  global environmental politics and, 51, *51,* 54, *54*
  governmental effects on, 572, 574–76, *575*
  individual choices and actions, 588, 592
  intergovernmental organizations (IGOs), 584–86, *585*
  international agreements, 54, *54,* 586–88, *586,* 589
  Montreal Protocol, 586–87
  National Environmental Policy Act (NEPA), 35, *35*
  need for, 47, *47*
  Paris Agreement, 572, 588, *588,* 589
  payments, 580–81, *580, 581*

  penalties, *580,* 581, *581*
  persuasion, *580,* 581
  policy formation process, 577–79, *578, 579*
  prescriptive regulation, 580, *580*
  process of making, 47–48, *48*
  property rights, *580,* 581
  5 Ps of policy design, 580–81, *580, 581*
  Sustainable Development Goals (SDGs), 9, *12,* 584–85, *585*
  Toxic Substances Control Act (TSCA), 57–58
  UN Framework Convention on Climate Change (UNFCCC), 587–88
  in the United States, 48, 50–51, 52–53, 572, 574–76, *575,* 582–84, *582, 584, 585*
  Use the Data, 596
  Use the News, 594
Environmental Protection Agency (EPA), 13, 14, 50, 51, 58, 116, 202, 209–10, 236–37, 429, 430, 431, 456, 491, 492–93, 506, 536, 572, 575, 584
epidemiology, 496, *496,* 497
equatorial low-pressure belt, 222, *222*
equity, 3Es (environment, economy, and equity), 9
Ereen Mine, 266
erosion
  and rock cycle, 250–52, *250, 251, 252, 253*
  soil loss and degradation, 263–66, *264, 266,* 343–44, *343*
estuaries, *188,* 189
ethanol, 408, *408,* 425
ethics
  about, 32, 34
  anthropocentrism, 37–38, *37*
  biocentrism, *37,* 38
  BPA-free plastics, 35, *35*
  and decision-making, 35–38, *37*
  Deepwater Horizon oil spill, 32, *34*
  defined, 35
  deontological, 35, 38
  ecocentrism, 38, *38*
  in individual choices and actions, 54–55
  intrinsic value, 35
  National Environmental Policy Act (NEPA), 35, *35*
  UN Earth Charter, 35, *36*
  Use the Data, 60
  Use the News, 57–58
  utilitarianism, 37
eugenics movement, 160
Eustace, Alan, *216*
eutrophication, 193, *193,* 286, 290, *290,* 344, 434
evapotranspiration, 186
Everglades, Comprehensive Everglades Restoration Program (CERP), 142–43, *143*
evolution
  adaptive radiation and, 90, 91
  alleles and, 91, *91,* 92
  coevolution, 102, 103, *103*
  Darwin's finches, 90, *90*
  defined, 90
  genes, 91, *91*
  misconceptions regarding, 92–93
  mutations, selection, and extinction, 91–92, *92*
  natural selection, 90, *90*

  e-waste, 435, *436,* 450
  exponential population growth, 106–7, *107, 108*
  externalities, economic, 43–47, *43, 44*
  extinction, 92
  extinction rates, 121, *121*
  extractive reserves, 134

F

Fair Trade standard, 359–60, *360*
fallow fields, 339
family planning, 159–60, *159*
"fast fashion," 24, 517
fat-soluble substances, 503
faults, crustal, 248, *248*
Federal-Aid Highway Act of 1956, 473
Federal Emergency Management Agency (FEMA), 255
federal governing systems, 574
Federal Housing Administration (FHA), 471
feedback
  decision making, 528, *528,* 533
  systems and cycles, 281–83, *282, 283*
Feldstein, Martin, 392
Ferreira, Fionn, 428
Ferrel cells, *222,* 224
fertility
  global total fertility rate (TFR), 163, *163*
  replacement, 157
  soap operas and fertility, 161
  total fertility rate (TFR), 156–57, *157,* 160, 165
  transition and demographic window, 164, *164*
  women's empowerment and, 168, *169*
fertilizers, synthetic fertilizers and pesticides, 340, 341, 342–43, *343*
Finley, Ron, 336, 338, *338*
first law of thermodynamics, 77–78
fishing
  aquaculture, 352, 359
  bycatch, 353, *354,* 362
  fishing technology, 351–52, *351*
  Native peoples and sustainability, 8, *8*
  overfishing, 204, *206,* 207, 353
  PBTs (persistent bioaccumulative toxins) and, 75
  sustainable practices, 362
fission, nuclear, 76, 404–5, *405*
Fleischmann, Martin, 18
Flint, Michigan drinking water supply, 189, 193, *488–89,* 504, *506*
floodplains, 255, 256–57, *257*
floods, 511, *511,* 512
floor-to-area ratio (FAR), 468–69, *469*
flows, systems and cycles, 280, *280*
food
  animal welfare, 359
  antibiotic-resistant bacteria, 352, 353
  aquaculture, 352, 359
  beef production, 348, 350, *350,* 355, *355*
  buying locally, 360
  chain, 78–79, *80,* 358, *358, 359*
  chicken production, 348, *348*
  choices and sustainability, 358–60, *358, 359, 360,* 362

composting, 441–42, *441*

concentrated animal feeding operations (CAFOs), 348, *350,* 352, 357

culture and consumption practices, 356–57, *357*

Fair Trade standard, 359–60, *360*

food composting, 441–42, *441*

food losses and waste, 360, 362, 366, 441–42, *441,* 451, 454, 529, 540–41

food policy, 357

food security, 355, *355,* 363–64

food system and consumption, 353–57, *355, 356, 357*

genetically modified organisms (GMOs), 340–41, 343, 349–50

global grain production, 354–55, *355*

individual choices and actions, 362

livestock, 345

livestock production impacts, 352, *352, 353*

meat production, 345, *345,* 348, *348,* 350, *350*

obesity, 356

organic and sustainable foods, 359, *359*

organic dust, 352

organic matter, 339

overnourishment, 356

PBTs (persistent bioaccumulative toxins), 75

pork production, 348, 350

portion sizes, 356, *356*

processed foods, 355–56

seafood production impacts, 353, *354*

seafood production practices, 351–52, *351*

synthetic fertilizers and pesticides, 340, 341

teosinte (ancient corn), 338, *338*

urban gardens, 336, 338, *338*

Use the Data, 366

Use the News, 363–64

vegetarian diets and sustainability, 39

See also agriculture

Food and Drug Administration (FDA), 35, *35,* 357

footprint analysis, 23, *23*

forest land ownership and management, 148

forests

deforestation, 140–41, *140,* 266–67, *267*

ecosystem, 6, *6*

edge effects, 140

forestry practices, 266–67, *267*

industrial, 140, *140*

land ownership and management, 148

sustainable management, 141, 267

Forest Stewardship Council (FSC), 113, 141, 546, 561

for-profit businesses, 549–50, *550*

fossil fuels

cap-and-trade exchange system, 390, 391

carbon taxes, 390, 391, 392

conventional reserves, 371

costs of, 386, *386,* 388, 391, 394

defined, 368, 370

economics and rise of fossil fuel use, 375–76, *376*

energy disasters, 384–85

formation of, 370–71, *370*

fracking, 381, *381, 382,* 387–88, 390, 594

human energy history, 372–74, *372, 373, 374*

hydrocarbons, 370

individual choices and actions, 391–92

pollution, 204–5, *206*

power plant emissions, 395

proven reserves, 375–76, *376*

reserves, 371

unconventional reserves, 371

US and global energy sources and consumption, 371, *371, 372, 375*

Use the Data, 395

Use the News, 384

See also coal; natural gas; oil

fracking, 41, *41,* 381, *381,* 382, 387–88, 390, 594

frames, 535–36, *535, 536,* 537, 538

fraud, scientific, 18–19, *19*

freshwater access improvement, 194–96, *195*

Friedman, Milton, 550

frontline communities, 321

fuel cells, 409–10, *409*

fuel economy driver interface (FEDI), 530

functional extinction, 109–10

fungicides, *505*

furans, *505*

G

Galton, Francis, 160

"gambler's fallacy," 19

Gandhi, Indira, 25

"gar-barge," 435, *435*

"garbology," 449

gases, 74

air pressure, 215–16, *215, 216*

properties of, 288, *288*

gas prices, 41, 60

gene flow, 93–94

genes, 91, *91*

genetically modified organisms (GMOs), 340–41, 343, 349–50

genetic diversity, 94–95

genetic drift, 95, *95*

genetics

cheetahs and genetic fitness, 94

extinction and, 92

gene flow, 93–94

genes, 91, *91*

genetic diversity, 94–95

genetic drift, 95, *95*

genetic erosion, 342

genotype, 92

inbreeding, 94, 95

mutation, 92, *92*

outbreeding, 94

phenotype, 92

reproductive isolation, 93–94

speciation, 93, *93*

genotype, 92

gentrification, 482

geoengineering and climate change, 73

geological hazards, 508, *508*

geologic disposal of waste, 436, 438, *438*

geologic process. *See* Earth

geothermal gradients, 306

geothermal power, 76, 406–7, *406, 407*

German Parliament, 576, 577, *577*

Gilbert, Jay Coen, 553

girls, empowerment of women and, 168–70, *169*

glaciers

defined, 186

Grinnell Glacier, 203, *203*

landforms and topography, 257, *258*

retreat, 202–3, *202, 203,* 314

Glendening, Parris, 476

global death rates, 518

global total fertility rate (TFR), 163, *163*

Good on You app, 5, *5*

government

authoritarian systems, 574, *574*

checks and balances, 574

coercion, 573

democratic systems, 574

elections, 576–77

federal governing systems, 574

government organization and authority, 573–76, *573, 574, 575*

individual choices and actions, 54–55, 570, 572, *572, 574, 574*

institutions, 574

interest groups, 577–78, *578*

legitimacy of, 573

lobbying, 577–78, *578,* 590

markets and, 40–41, 42

policies, 47–48, *48,* 572, 586

policy creation, 47–48, *48,* 577–79, *578, 579,* 590–91

political parties, 577, *577*

politics, 570, 572, *572,* 574–76, *575,* 596

power, 577

proportional representation (PR) electoral system, 576–77

single-member-district plurality (SMDP) electoral system, 576–77

sovereignty, 584

unitary governing systems, 574

United Nations (UN), 584–85, *585*

Go Viral! game, 29

grazing and grassland management, 141

Great Pacific Garbage Patch, 426, *428*

Greenfreeze refrigerators, 560, *561*

greenhouse effect, 308–9, *308, 309*

greenhouse gases

defined, 309, 310

emissions, 293, *293,* 296, 316, 318, *318,* 327, 394, 464, *465,* 514

and hydraulic fracturing (fracking), 381, *381,* 382, 387–88, 390

green infrastructure, 480, *480,* 482

Greenland ice loss, 203, *203,* 325

Greenpeace, 560, *561*

greenwashing, 550

Greer, Linda, 2, 4, *4*

Grinnell Glacier, *202*

grizzly bears, *113*

groundwater
  dependent ecosystems and organisms, 186, *186*
  formation and water cycle, 183, *183,* 185, 186
  withdrawals, 190–91, *190, 191*
groups and organizations
  Apple, 566–67
  benefit corporations, 551, 553
  Coca Cola, 561, *561*
  degrees of separation, 547
  diversity in, increasing, 562
  Environmental Defense Fund (EDF), 563
  Forest Stewardship Council (FSC), 546, 561
  for-profit businesses, 549–50, *550*
  Greenpeace, 560, *561*
  greenwashing, 550
  Herman Miller, 551–52, 557
  IKEA, 141, *544–45,* 546, 549, 562, 563, 564
  individual choices and actions, 564
  influencers, 548, *548*
  Interface, Inc., 552, *552,* 557
  internal organizational change, 557–58, *558*
  King Arthur Flour, 551, 553
  Leadership in Energy and Environmental Design
    (LEED) certification, 563
  Method Products, 551, 553, 556, *556*
  mission statements and strategy, 552, 556, *556,* 557
  National Center for Public Policy Research
    (NCPPR), 566–67
  National Geographic, 548, *548*
  Nike, 549
  nonprofit organizations, 550–51, *551*
  organizational culture and structure, 557
  organizational divestment from fossil fuel
    investments, 559–60
  organizational influence on change, 548–51
  organizations, 546–47
  outside influences on organizational change,
    560–63, *561, 562, 563*
  Patagonia, 563, *563*
  polarized networks, 548, *548*
  privately (or closely) held business, 549
  product differentiation, 556, *556*
  publicly traded business, 549–50, *550,* 566–67
  social context, 546, 547
  social groups, 546, 547
  socially and environmentally responsible investing,
    550, *550*
  social networks, 547–48, *547, 548*
  and sustainability, 551–52, 554–57, 566–67, 568
  triple bottom line (TBL) accounting, 549, *549*
  Walmart, 563, *563*
  World Wildlife Fund (WWF), 544, 546, *546,* 561
Growth Management Act, 142
gyres, 205, 311, *311*

H

habitat
  alteration, 121, *122*
  defined, 90
  urbanization and habitat loss, 460–61, *461*
Hadley cells, 222, *222*
Haiti, restoring land traditions of, 252

Halsey, Erika, 517
Hamilton, Alice, 497, *497*
Hardin, Garrett, *45,* 173
Hayes, Denis, 570, *572*
hazardous waste disposal and remediation, 438, *438,*
  440–41
health
  education and, 167, *167*
  health advances and human population, 158
  See also environmental justice
heat waves, 511–12
Hector, Mary-Pat, 55
hedgerows, *264,* 265
hepatotoxins, *503*
heptachlor, *505*
herd immunity, 502
Herman Miller, 551–52, 557
Hess, Harry, 248
hexachlorobenzene, *505*
hierarchy of needs, 522, *523*
high-pressure systems, 220, *220*
Hiltner, Lorenz, 127
hindcasting, 17–18, 317, 318–19
HIV/AIDS, 499–500
Holdren, John, 170
home energy use, 328
hot spots, 110, *110,* 111, 112, 138
Houlahan, Bart, 553
Howard, Albert, 361
Hudson, Chuck, 147
Hudson River School and landscape art, 255, *255*
human body systems, 279–80, *279*
human population
  agricultural development and technology, 157–58, *158*
  Cairo Consensus, 170
  carrying capacity, 158
  contraception, 159, 161
  crude death rate, 157
  demographic transition and global population
    dynamics, 163–66, *163, 164, 165*
  ecological footprint analysis, 170–71, *172,* 174
  education and health, 167, *167*
  empowerment of girls and women, 168–70, *169*
  eugenics movement, 160
  family planning, 159–60, *159*
  fertility transition and demographic window, 164, *164*
  global growth, 150, 152, *152, 153,* 154–55,
    176–77
  global total fertility rate (TFR), 163, *163*
  health advances, 158
  history of change and crashes, 142–53, *153,* 156
  individual choices and actions, 174–75
  infant mortality rates, 166
  literate life expectancy (LLE), 167, *167*
  low birth rate consequences, 165–66
  mortality transition, 163, *163*
  natural increase, 167
  population-control policies, 159–60, *159, 160*
  population dynamics, 150, 152, *153*
  population pyramids and growth projections, 165, *178*
  replacement fertility, 157
  and rising consumption, 170–71, *171, 172,* 173–74

soap operas and fertility, 161
social stratification and inequality, 160, 162
stability transition, 164, *164, 165*
total fertility rate (TFR), 156–57, *157,* 165
urbanization, 166–67, *167*
US demographic transition (baby boom), 164, *165*
Use the Data, 178
Use the News, 176–77
US regional population dynamics, 162–63, *162*
youth bulge, 165, 166
humans and dogs, 339
Human Settlements Programme, 474, 476
human waste recycling, 443
humidity, 219
humus, 260, *260*
hurricane, 221, *222, 304*
Hurricane Harvey, 490, *490, 491*
Hurricane Katrina, 475, 511, *511,* 512
Hurricane Maria, 490, *490, 491*
hurricanes and severe weather, 321, *324*
Huxley, Julian, 544
hydraulic fracturing (fracking), 41, *41,* 381, *381,*
  382, 387–88, 390, 594
hydrocarbons, 370
hydroelectric dams, 76, *76*
hydrofluorocarbons (HFCs), 25, *318*
hydrogen bonds, 68, *68*
hydroponic farming, *336–37*
hydropower, 403–4, *403, 404,* 414, *415*
hydrosphere, 183, *183*
hypothesis
  asteroid strikes and predicting past events, 17
  defined, 14
  and scientific method, 14, *15*

I

ice
  Antarctic Ice Sheet melting, 325
  Greenland ice loss, 203, *203,* 325
  water cycle, 184–85
ice age, 202, 203
iceberg model, 297, *297*
ice cores, *306,* 307
Iceland and geothermal power, 406, *406*
igneous rocks, 250, *250, 251*
IKEA, 141, *544–45,* 546, 549, 562, 563, 564
immunizations, 502
impervious surfaces, 258, *258*
implicit association tests (IATs), 524
implicit bias, 524
inbreeding, 94, 95
incentives, 22–23, 520, 522, 530–31, *531,* 532, 534, *534*
incineration of waste, 436, *437,* 438–40, *440*
Indigenous peoples, 30 by 30 initiative and, 132
individual choices and actions
  air pollution, 237–38
  boycotts, 55
  carbon and nitrogen footprint calculations, 296
  climate change, 328–30
  conservation projects, 144
  education and rights issues, 174

engagement with elected officials, 55
environmental policy, 587–88
food, 362
fossil fuels, 391–92
government, 54–55, 570, 572, *572*, 574, *574*
innovation opportunities, 81
personal ecological footprint calculation, 174
product selection, 144
recycling and waste stream awareness, 81
reduction of negative environmental effects, 113
sustainability and purchasing decisions, 26
sustainability groups, 26
sustainable production and development, 80–81
systems thinking, 296–97
urbanization, 482–83
voting and public policy, 55
waste, 451–52
water use and conservation, 207–8
wildlife-friendly habitat creation, 113
wildlife impact minimization, 113
Indonesian species richness, 116–17
induced traffic, 476–77, *476*
industrial forests, 140, *140*
infant mortality rates, 166
infiltration, 259
influencers, 548, *548*
infrastructure, 458, 460, 475, 480, *480*, 482
insecticides, *505*
instrumental and intrinsic values, 123–24
instrumental period, 305
interest groups, 577–78, *578*
Interface, Inc., 552, *552*, 557
intergovernmental organizations (IGOs), 54, 584–86, *585*
Intergovernmental Panel on Climate Change (IPCC), 318
international agreements, 138–39, 586–88, *586*, 589
interspecific competition, 103
interstate highways, 377, *377*
intertidal zone, 204, *205*
intrinsic value, 35
invasive species, 122, *123*
ionic bonds, 67–68, *68*
ionizing radiation, 509
ions, 66
I=PAT (Impact = Population x Affluence x Technology), 170, *171*
iron fertilization, 73
irrigation practices, 340, *341*
island biogeography, 132–33
isolation of waste, 436, *437*, 438, *438*
isotopes
  defined, 66
  shark food webs, 83–84
isotope tracking, 67, 83

J

Jacobs, Kristin, 594
Japanese population, 164, *164*, 165
Jemez Pueblo, 125, *125*
jet streams, 224, *224*

Jevons, William, 42
Jevons paradox, 42, *42*, 376
Jian River, *4*
Johnson, Bea, 451
Johnson, Boris, 514
justice. *See* environmental justice

K

Kahneman, Daniel, 523
Kamkwamba, William, 396, 398, *398*
Kassoy, Andrew, 553
Keystone Pipeline, 382
keystone species, 111–12, *112*, *113*
kinetic energy, 74, *75*
King Arthur Flour, 551, 553
Kissimmee River, 192, *192*
Kodjak, Drew, 241
Kola Superdeep Borehole, 247, *247*
K-strategists, 109, *109*
Kyoto Protocol, 587

L

labels and guides, 534, 536–37, *537*, 542
La Ferrara, Eliana, 161, *161*
Lagrange, Joseph-Louis, 64
lake and land effects on weather, 221–22, *222*
lakes and ponds, 186, *187*
Lakoff, George, 535
landfills
  bioreactor landfills, 440
  gas emissions, 434, *434*
  isolation of waste, 436, *437*, 438, *438*
  leachate, 434, *434*, 438
  Meadowlands MetLife stadium, 441, *441*
  open dumps, 430–31, *431*, 434, *434*
  sanitary landfills, 436, *437*
landforms and topography
  defined, 255
  delta, 257, *258*
  depositional environments, 257–58, *258*
  floodplains, 255, 256–57, *257*
  glaciers, 257, *258*
  Hudson River School and landscape art, 255, *255*
  impervious surfaces, 258, *258*
  landslides, 257, *257*
  Mississippi River drainage basin, 255–56, *256*
  Mississippi River floodplain, 256–57, *257*
  Oso landslide, 257, *257*
  overland flow, 258
  pervious surfaces, 258, *258*
  sediment transport, 257–58, *258*
  watersheds, 255–56, *256*
landslides, 257, *257*, 508, *508*
land trusts, *136*
land use
  agriculture and grazing practices, 264–66, *264*, *266*
  dwelling units per acre (DU/acre), 468, *469*
  floor-to-area ratio (FAR), 468–69, *469*
  forestry practices, 266–67, *267*
  individual choices and actions, 270
  mine reclamation, 266, *266*

New Orleans, Louisiana, 475
oil development, 382, *382*
one-use zones, 468, *469*
planner's dilemma, 474
Portland, Oregon, 476
redlining, 471
and resilience, 513
and urbanization, 24, 142, 267
wind and water erosion of cropland soils, 264–65, *264*
See also suburban sprawl; urbanization
Las Vegas, *467*
Lautenberg, Frank, 57
lava, *508*, 509
Lavoisier, Antoine, 62, 64
law of conservation of mass, 66, *67*
laws for biodiversity protection, 135–36, 138–39, *139*
leaching, 263, 343
Leadership in Energy and Environmental Design (LEED) certification, 563
lead in the environment, 14–15, *15*, 194, 492
LeBoeuf, David, 364
Lehmann, Marcus, 49
lentic ecosystems, 186, *187*
Leopold, Aldo, 38, *38*
Levittown, New York, 472, *472*
life
  food chain, 78–79, *80*, 358, *358*, *359*
  oceanic support for ecosystems, 204, *205*
  primary consumers, 79, *80*
  primary producers, 79, *80*
  scales of life on Earth, 89–90, *89*
  secondary consumers, 79, *80*
  shared characteristics of living things, 88–90, *89*
  10% law, 79–80, *80*
  tertiary consumers, 79, *80*
  trophic levels, 78–79, *80*
  See also biodiversity
life-cycle assessment, 430, *430*
light bulbs and energy use, 84–85
lightning strikes, 19, *19*
Lin, Kevin, 194–95, *195*
liquids, 74
literate life expectancy (LLE), 167, *167*
lithification, 252, *253*
lithosphere, 248, *248*, 285, *285*, 291, 293
littoral zone, 186, *187*
livestock
  concentrated animal feeding operations (CAFOs), 348, *350*, 352
  defined, 345
  production impacts, 352, *352*, *353*
  production practices, 345, *345*, 348, *348*, 350, *350*
Living Building Challenge, 481
loam, 259
lobbying, 577–78, *578*, 590
logging,
  deforestation, 140–41, *140*
  forest land ownership and management, 148
  sustainable forest management, 141, 267
logistic (S-shaped) growth, 107, *108*
loss aversion, 522, 523–24, *523*
lotic ecosystems, 186–87, *188*

Love Canal, 494
Lovelock, James, 21, 229
low-impact development, 200, *202*
low-pressure systems, *220*, 221
Lowry, Adam, 556, *556*
Lummi Nation First Salmon ceremony, *8*
Lyme disease, 333–34, 512
lynx repopulation in Spain, 133
Lyons, Kevin, 557, *558*

M

Maasai peoples, 134
MacArthur, Robert, 103
macronutrients, 283
Maertens, Rakoen, 29
magnetic field and animal migrations, 246
magnetosphere, 247, *247*
Mahoney, Stephen, 116
majority electoral systems, 576
"Make What's Next" campaign, 548, *548*
malaria, 471, 499, *499*, 501
Malthus, Thomas, 150, 152, 153, *153*
Mankiw, Greg, 392
mantle of Earth, *247*, 248
Marcario, Rose, *563*
Marine Mammal Protection Act, 52, 135–36
marine protected areas (MPAs), 131, *131*
markets
    defined, 40
    and externalities, 47
    and the government, 40–41, 42
Martian, The (film), 276, 278
Maslow, Abraham, 522, *523*
mass
    defined, 65
    elements and atoms, 64–66, *65, 66*
    law of conservation, 66, *67*
mass extinction events, 109, *110*, 283
matter
    acids and bases, 72, *72*
    Bohr atomic model, 66, *66*
    burning of forests for agricultural development,
        62, 64, *64*
    chemical bonds, 66, 67–68, *68*
    defined, 65
    elements and atoms, 64–66, *65, 66*
    geoengineering and climate change, 73
    individual choices and actions, 80–81
    isotopes, 66
    isotope tracking, 67
    law of conservation of mass, 66, *67*
    mass, 65
    molecules and compounds, 66–68, *67, 68*, 72, *72*
    periodic table of elements, *65*
    phase changes and chemical reactions, 72, 74–75,
        74, 75
    plastics recycling, 69
    sustainability, 62, 64
    Use the Data, 84–85
    Use the News, 83–84
    See also energy

Mayer, Jean, 557
MBA Polymers, 69, 81
McAleer, Erin, 365
McCarthy, Gina, 210
McDonald, Lauren "Bubba," 424
McGrath, Shaun, 210
Meadowlands MetLife stadium, 441, *441*
meat production, 345, *345*, 348, *348*, 350, *350*
mechanical energy, 76
megalopolis, 468, *468*
mesosphere, 216, 217, *217*
metallic bonds, 68, *68*
metamorphic rock, *250*, 252, *253*
methane emissions, 316, 318, *318*, 352, *353*, 379
methanogens, 292–93, *292*
Method Products, 551, 553, 556, *556*
metropolitan area/metropolis, 468
Mica, David, 594
microorganisms, 498–500, *499, 500*
Mielewski, Debbie, 558, *558*
Migratory Bird Treaty Act, 135, *135*, 581
Milankovitch cycles, 310
milk products and water usage, 199
Millennium Drought, 207–8
Millennium Ecosystem Assessment (2005), 124
milling technologies, 373–74, *373, 374*
minerals, 253–54, *254*, 268–69, 270
mining
    Animas River pollution, 209–10
    coal, 379, *379*
    minerals, 253–54, *254*, 268–69
    mine reclamation, 266, *266*
    Navajo uranium mines, 492, *492*
    phosphorus, 286, *286*
    water, 190
mirex, *505*
misinformation, 19–20, *19, 20*
mission statements and strategy, 552, 556, *556*, 557
Mississippi River drainage basin, 255–56, *256*
Mississippi River floodplain, 256–57, *257*
Mitchell, George, 378
Miwok people, 54
mixed-use areas, 478
modal split, 477
models
    defined, 16–18, *17*, 280
    global climate models, 317, 318–19
    hindcasting climate models, 17–18, 317, 318–19
    iceberg model, 297, *297*
    stock and flow diagrams, 280–81, *281*, 296, *296, 297*
Mojave Desert, 224, *224*
molecules and compounds, 66–68, *67, 68*, 72, *72*
Molina, Mario J., 21, 229, 586
monoculture, 342
Montford, Bill, 594
Montreal Protocol, 25, 586–87
Moore, Charles, 426, 428, *428*
Morris, Ciscoe, 557, *558*
Morris, John, 58
mortality transition, 163, *163*
mosquito-borne diseases, 115–16, 499, *499*, 512
motivation, 530

mountaintop removal, 379, *379*
Mountfort, Guy, 544
Muir, John, 50, *50*, 54, 125
Muneer, Sanwal, 81
municipal solid waste (MSW)
    generation, 429, *429*
    recycling, 442–43, *442, 443*, 456
mutation, 92, *92*
mutualism, 104, 106

N

National Center for Public Policy Research
    (NCPPR), 566–67
National Environmental Policy Act (NEPA), 35, *35*,
    50, 53
national forests, 131
National Geographic, 548, *548*
national monuments, 53, 130, *130*
national parks, 50, 126, *126*, 130, *130*
National Park Service, 13, 50, 53, 126, 130
national recreation areas, 130
National Weather Service models, 19, *19*
National Wildlife Refuge System, 53, 130–31
Native peoples and sustainability, 8, *8*
natural gas
    environmental impacts of, 382, 383, 385
    formation of, 370–71, *370*
    fracking, 381, *381*, 382, 387–88, 390, 594
    pipelines, 383, *383*
    production and consumption of, 382–83, *383*,
        386, 388
    US and global energy sources and consumption,
        371, *371, 372, 375*
natural increase, 167
Natural Resource Damage Assessment process,
    143–44, *143*
natural selection, 90, *90, 91*
Navajo uranium mines, 492, *492*
Nazi Germany, 160, 162
Neff, Roni, 541
negative externalities, 43–45, *43, 44, 45*
Nelson, Gaylord, 570
nephrotoxins, *503*
neritic zone, 204, *205*
neurotoxins, *503*
neutrons, 66, *66*
New College, Oxford dining hall ceiling beams, 8, *9*
Newmark, William, 132
New Orleans, Louisiana, 475
New York City, 460, *460*, 466, *466*
New Zealand, 111, 122
Nicholson, Max, 544
Nike, 4–5, 549
Nikkel, Lori, 454
ninja lanternsharks, 88, *88*
Niphargus aquilex, 186, *186*
Nisqually River estuary restoration, 143–44, *143*
nitrification, 290
nitrogen cycle, 286, 288–90, *288, 290, 291*
nitrogen fixation, 288–90, *288, 290, 291*
nitrogen isotope tracking, 67
nitrogen pollution, 291, 300, 344

nitrous oxide emissions, 318, *318, 344*
Nixon, Richard, *35*
nonpoint-source pollution, 193–94, *193,* 202
nonprofit organizations, 550–51, *551*
North American Free Trade Agreement (NAFTA), 588
North China Plain Aquifer, 345
no-till agriculture, *264,* 265, *266,* 273–74
Notre Dame cathedral, 8
nuclear energy, 76
nuclear fission, 404–5, *405*
nuclear fusion, 76
nuclear power, 76, 404–6, *405*
nuclear waste, 405, 406, 436, 438, *438*
nutrient recycling, 287, 299–300

O

Obama, Barack, 572, 575, *575,* 588
obesity, 356
observations and controlled experiments, 15–16, *16*
oceans
    acidification, 205, *206,* 296, 322
    acidity, 16
    coral ecosystems, 204
    dumping of waste in, 434–35, *435*
    gyres and, 205, 311, *311*
    human impacts on life in, 10–11, 204–5, *206,* 207
    influence on climate, 205, *206*
    ocean currents and climate change, 311–13, *311, 312, 313*
    sea ice, 202, *203*
    support for life and ecosystems, 204, *205*
    and water cycle, 184–85
Ogallala Aquifer, 190, *190,* 345
oil
    enhanced oil recovery (EOR), 388, 389
    environmental impacts, 382, *382,* 384, 394
    formation of, 370–71, *370*
    fracking, 381, *381,* 382, 387–88, 390, 594
    oil pipelines, *368–69*
    prices, 38, 40, *40,* 41–42, *41*
    production and consumption, 380–82, *380, 381, 382,* 386, *386,* 388
    spills, 32, *34, 206,* 382, 384, *582*
    tar sands, 381–82
    Titusville oil well, 377, *377*
    US and global energy sources and consumption, 371, *371, 375*
Old Faithful geyser, 406, *406*
Old Farmer's Almanac, 18–19, *19*
Oliver, Jamie, 358
one-use zones, 468, *469*
open dumps, 430–31, *431, 434, 434*
open system, 278
organic and sustainable foods, 359, *359*
organic dust, 352
organic fertilizers and soil fertility, 361
organic matter, 339
organizations. *See* groups and organizations
Orr, David, 557, *558*
Oso landslide, 257, *257*

Ostrom, Elinor, 45, *45*
outbreeding, 94
overfishing, *206, 207,* 353
overland flow, 258
overnourishment, 356
overpopulation. *See* human population
oxidation–reduction (redox) reactions, 74
oxygenating extinction, 283
oxygen cycle, 283–84, *284, 285*
ozone layer, 229–30, *229, 231*
    and cancer risk, 510, *510*
    and chlorofluorocarbons (CFCs), 21, *21,* 229–30, *229, 230*
    Greenfreeze refrigerators, 560, *561*

P

Pacific salmon
    conservation, 118–20, *118,* 142
    life stages of, *120*
    Native peoples and sustainability, 8, *8*
paleoclimates, *306,* 307
Pallone, Frank, Jr., 58
Paracelsus, 503
parasitic relationships, 102–3, *103*
parent material, 259
Paris Agreement, 572, 588, *588,* 589
particulate matter (PM), 226–27, *227,* 230, 290
passenger pigeon extinction, 121
passive solar buildings, 400, *402*
Patagonia, 563, *563*
Patterson, Clair, 14–15, *15*
payments for ecosystem services (PES), *136*
PBTs (persistent bioaccumulative toxins), 75
PCBs (polychlorinated biphenyls), 13, *505*
peat, 64, *326*
peat moss, 189
pedestrian-oriented development (POD), 478–79, *479*
peer review, 20
pelagic zone, 186, *187*
periodic table of elements, 65
permafrost, 185, 202, *203, 326*
persistent bioaccumulative toxins (PBTs), 75
persistent organic pollutants (POPs), 504, *505*
pervious surfaces, 258, *258*
pesticides, synthetic fertilizers and pesticides, 340, 341, 342–43, *343*
phase changes and chemical reactions, 72, 74–75, *74, 75*
phenology, 323
phenotype, 92
Phipps Conservatory and Botanical Gardens, 414
phosphorus cycle, 284–86, *285, 286,* 287
photic zone, 186, *187*
photosynthesis, 76, 77, 281, 284, *284*
photovoltaic (PV) solar panels, 400, *401,* 402–3
pH scale, *72*
physical hazard, 506, 507–10, *508, 509, 510, 513*
physical weathering, 250–51
phytoplankton, 204
Pinchot, Gifford, 50, *50*
Pisaster ochraceus (sea stars), 111–12, *112*

planner's dilemma, 448
plastics
    disposable straws, 562–63, *562*
    pollution, 205, *206,* 207
    recycling, 69, 442, *443,* 444–45, *444, 445*
    use of disposable plastic *versus* reusable bags, 520, *520–21, 522*
    waste, 426, *426–27,* 428–29, *428,* 430, *430,* 450
plate tectonics, 248–49, *248, 249*
plurality electoral systems, 576
pneumonia, 499, *500*
point-source pollution, 193–94, *193*
polarity, 68
polarized networks, 548, *548*
policy
    defined, 47
    process of making, 47–48, *48*
    See also environmental policy; government
politics, 47
Pollan, Michael, 362
pollination, bees and, 86–88, *88*
pollution
    air. *See* air pollution
    fine particulate matter, 290
    fossil fuel, 204–5, *206*
    human impacts on ocean, 204–5, *206,* 207
    imbalance of consumption and exposure, 13, *14*
    nitrogen pollution, 291, 300, 344
    nonpoint-source, 193–94, *193,* 202
    plastic, 205, *206,* 207
    point-source, 193–94, *193*
    rare earth minerals extraction, 253
    water, 2, 4, *4,* 193–94, *193,* 209–10, 268
polychlorinated biphenyls (PCBs), 13, *505*
polyethylene terephthalate (PET) resins, 446
polylactic acid (PLA) resins, 446
polymerization, 74–75, *75*
polyvinyl chloride (PVC), 67
ponds, lakes and ponds, 186, *187*
Pons, Stanley, 18
Population Bomb, The (Ehrlich), 156
populations
    defined, 88, 89
    and environmental decision-making, 24
    exponential population growth, 106–7, *107, 108*
    population crash, 107, *107*
    population density, 107–8, *108,* 112
    reproductive isolation, 93–94
    reproductive strategies, 108–9, *109*
    See also human population
pork production, 348, 350
Portland, Oregon, 476
positive feedback, 282
positive selection, 92
potential energy, 74, 75–76, *76*
Prather, Kimberly, 223, *223*
Pratt, Jeff, 424
precipitation, 221–22, *222*
precipitation and extreme rain events, 319–21, *320, 324*
predation, 102–3, *103,* 105–6
predator controls, 146–47
Predator Free New Zealand, 111

pressure
  air, 215–16, *215, 216*
  defined, 215
  of gases, 74
Price, James, 18
primary consumers, 79, *80*
primary pollutants, 226–27, *226, 227,* 232
primary producers, 79, *80*
primary (closed-loop) recycling, 442, *443*
privately (or closely) held business, 549
processed foods, 355–56
producer responsibility law, 450
product differentiation, 556, *556*
product obsolescence, 527
"Project Mohole," 247, *247*
prompts, 526, *526,* 534
proportional representation (PR) electoral system, 576–77
protected areas, 126, *126,* 128–135, *130, 131, 132, 133, 134*
protons, 65–66, *66*
provisioning services, 124, *124*
proxy indicators, 306–7, *306*
pseudoscience, 18–19, *19*
5 Ps of policy design, 580–81, *580, 581*
public goods, 46, *46*
public health, 498
  air pollution and public health research, 16
  childhood deaths from pneumonia and diarrhea, 499, *500*
  DDT and malaria, 501
  defined, 498
  energy-efficient development and public health, 486–87
  herd immunity, 502
  See also environmental justice
publicly traded business, 549–50, *550,* 566–67
Pueblo Zuni "waffle gardening," 339, *339*
pulmonary veins and arteries, 279, *279*
pumped hydro energy storage, 414–15, *415*

Q

Qadir, Manzoor, 272

R

radiant energy, 76
radiation, 509–10, *509, 510,* 514
radon, 509
Ramirez, Jose Luis, 115, 116
rare earth minerals, 253, 254, *254,* 270
Raschein, Holly, 594
Rathje, William, 449
Reagan, Ronald, 392
recharge, 190
recycling
  design for recovery, 445, *445*
  economics, 444–45, *444, 445*
  food recycling and composting, 441–42, *441*
  human waste, 443
  municipal solid waste (MSW), 442–43, *442, 443,* 456
  nutrient recycling, 287, 299–300

paper, *442,* 449
plastics, 69, 442, *443, 444,* 444–45, *445*
primary (closed-loop), 442, *443*
secondary (open-loop), 442, *443*
upcycling, 445–46, *446*
Use the Data, 456
waste stream awareness, 81
water, 197, 201
red dye, 4, *4*
redlining, 471, 492, *493*
Redondo Peak, 125, *125*
Registration, Evaluation, Authorisation and Restriction of Chemicals (REACH), 506, *507*
regulating services, 124
reindeer, 107, *107*
reinforcing feedback, 282, *282*
relative humidity, 219, *219*
remediation, 440–41
replacement costs, 125–26
replacement fertility, 157
reproductive isolation, 93–94
reproductive strategies, 108–9, *109*
reprotoxins, *503*
reservoirs, 183, *183*
residence time, 183
resignation, 524
resilience, 9, 12, *12,* 513
Resource Conservation and Recovery Act (RCRA) of 1976, 50, 52, *584*
resource overexploitation, 122
revealed preferences, 125–26
Richardson, Lewis Fry, 317
ride-sharing, 450, *451*
Ripple, Bill, 105, 106
risk
  and insurance, 302, 304
  risk management and regulation of toxins, 506–7, *507*
Rivera, Sarai, 364
rivers
  streams and, 186–87, *188*
  wild and scenic, 131
Road to Survival (Vogt), 156
rocks
  igneous, 250, *250, 251*
  lithification, 252, *253*
  metamorphic, 240–241, *250*
  rock cycle, 250, *250*
  sedimentary rock, *250,* 252, *253*
  weathering, 250–52, *250, 251, 252, 253*
Rocky Flats Plant, *509*
Roosevelt, Theodore, 54, 130
Roozenbeek, Jon, 29
Ross, Kim, 594
Roundtable on Sustainable Palm Oil (RSPO) standard, 80–81
Rowland, F. Sherwood, 21, 229, 586
r-strategists, 108, 109, *109*
ruminants, 350
runaway loops, 282
rural areas, 460
Ryan, Eric, 556, *556*

S

Sabido, Miguel, 161
Safe Drinking Water Act, 52, 200, 207
salinity, 311–12, *312*
salinization, 263, *263,* 272–73, 343
salt, ionic bonds, 68, *68*
saltwater intrusion, 190, *191*
sanitary landfills, 436, *437*
Save Your Wardrobe app, *5*
scarcity, 24, 41–42, *41*
science
  about, 14–16
  bias and misinformation, 19–20, *19, 20*
  challenges to, 18–20
  controlled experiments, 15–16, *16*
  defined, 14
  fraud and pseudoscience, 18–19, *19*
  models, 16–18, *17*
  observation and testing methods, 15–16
  peer review, 20
  and predicting past events, 17
scientific method, 14–15, *15*
Scott, Peter, 544
seafood
  production impacts, 353, *354*
  production practices, 351–52, *351*
sea ice, 202, 203
sea-level rise, 304, 313, *313,* 315, 319, *319,* 322, *322, 324, 325*
sea lions, 147
sea otters, *113*
sea stars, 111–12, *112*
secondary consumers, 79, *80*
secondary education for women, 168, 169, *169*
secondary pollutants, 227–28, *227, 228,* 233
secondary (open-loop) recycling, 442, *443*
second law of thermodynamics, 78, *80*
sediment, 192, 252
sedimentary rock, *250,* 252, *253*
sedimentation, *250,* 252, *253*
sediment transport, 257–58, *258*
seed drills, 338–39, *339*
self-driving cars, 417
Shakman, Andrew, 452
sharks
  food webs, 83–84
  ninja lanternsharks, 88, *88*
Shimkus, John, 58
Sierra Club, 54, 116, 125
Sierra del Divisor National Park, 134
Silent Spring (Carson), 501, 582, 583, *583*
Singer, Lauren, 452
Singer, Peter, 359
single-member-district plurality (SMDP) electoral system, 576–77
sinks, systems and cycles, 284
Slat, Boyan, 428
Sliwoski, Grace, 365
slums, 465–67, *465, 466, 467,* 474
smart growth of cities, 478, *478*
Smith, Adam, 40

smog, 227–28, *227*
Smokey Mountain Garbage Dump, 430–31, *431*
smoking bans, 226
SNAP (Supplemental Nutrition Assistance Program), 357, 360, 364–65
Snow, John, 496, *496*
snowballing loops, 282
Snowpiercer (film), 73
soap operas and fertility, 161
social acceptance and conformity, 524–25, *525,* 526
social media, 19–20, *20,* 592
social norms, 524, 531, *534*
social stratification and inequality, 160, 162
soil
    aeration, 259
    castings, 260
    CLORPT (climate, organisms, relief, parent material, and time), 262
    conservation agriculture, 360
    desertification, 344
    detritus, 259
    formation, 259–60, *259*
    humus, 260, *260*
    infiltration, 259
    leaching, 263, 343
    loam, 259
    organic matter in, 260, *260, 261*
    parent material, 259
    salinization, 263, *263,* 272–73, 343
    soil fertility and agriculture, 127, 361
    soil horizons, 251, 259–60, *259*
    soil loss and degradation, 263–66, *264, 266,* 343–44, *343*
    soil orders, 262, *262*
    soil pH, 263
    soil profiles, 259, *259*
    structure and organic matter content, 260, *260,* 262–63, *262, 263*
    terra preta, *260,* 265
    texture, 259, *259*
    tilth, 263
    topsoil, 259, 262–63, *263,* 270
    wind and water erosion of cropland soils, 264–65, *264*
    worms and soil formation, 261
solar power, 400, *401,* 402–3, *402,* 413–14, *414, 415, 418,* 423–24, *544–45*
solids, 74
solid waste, 430
soot, 225, *225,* 226
sovereignty, 584
speciation, 93
species
    defined, 88
    endemic, 110–11
    evenness, 96–97, *97*
    interspecific competition, 103
    invasive, 122, *123*
    keystone, 111–12, *112*
    reintroduction and conservation, 142, *142*
    richness, 95–96, *97,* 116–17
St. Matthew Island, Alaska, 107, *107*
stability transition, 164, *164,* 165

Standing Rock Reservation, 382, *382*
stated preferences, 125
state of matter, 74
status quo, 522, 523
Stefanopolou, Anna, 240
stock and flow diagrams, 280–81, *281,* 296, *296, 297*
stocks, systems and cycles, 280, *280*
Stolan, Victor, 544
storm damage costs, 304, *304*
storm surges, 511, *511*
strategy, defined, 556
stratosphere, 216, 217, *217*
streams, rivers and, 186–87, *188*
street design and transportation, 468, 469–70
stygobites, 186
subsidence, 190–91, *191,* 344
subsidies, 580
subtropical high-pressure belt, 222, *222*
suburban sprawl, 142
    and consumption, 473–74, *473*
    covenants, 471
    defined, 467
    dwelling units per acre (DU/acre), 468, *469*
    effects, 469–71, *470*
    floor-to-area ratio (FAR), 468–69, *469*
    Las Vegas, *467*
    megalopolis, 468, *468*
    metropolitan area/metropolis, 468
    one-use zones, 468, *469*
    redlining, 471
    street design and transportation, 468, 469–70
    in the United States, 471–74, *472, 473*
sulfates, 73
sulfur dioxide, 378, 379, 382
"Superfund" (CERCLA), *584*
Superstorm Sandy, 12, *12,* 302, 304
Supplemental Nutrition Assistance Program (SNAP), 357, 360, 364–65
supply and demand, 38, 40, *40*
supporting services, 124
surface currents, defined, 311
surface water, 185, 186, 191–92, *191, 192*
sustainability
    academic programs and groups, 22
    campus groups, 564
    defined, 8–9
    described, 7–8
    and environmental decision-making, 10–24, *22, 23,* 26
    environmental effects of clothing production on, 2, 4–5
    footprint analysis, 23, *23*
    forest management and, 8
    and groups and organizations, 551–52, 554–57, 566–67, 568
    groups and organizations focused on, 24
    individual choices and actions, 26
    and matter and energy, 62, 64
    and Native peoples, 8, *8*
    New College, Oxford dining hall ceiling beams, 8, *9*
    Roundtable on Sustainable Palm Oil (RSPO) standard, 80–81

trade-offs and incentives, 22–23
    values, 20
    of vegetarian diets, 39
    See also environmental policy; government; groups and organizations; individual choices and actions
Sustainable Apparel Coalition, 4–5
sustainable forest management, 141, 267
symbiotic interactions, 103–4, *104,* 106
synergistic toxins, 502
Syrian drought and civil war, 512
systems and cycles
    about, 5, 5–6
    ammonification, 290
    balancing feedback, 282–83, *283*
    biogeochemical cycles, 283
    Biosphere 2, 278, *278,* 281, *281,* 282, *282*
    breathing, 279–80, *279*
    carbon cycle, 291–96, *292, 293*
    closed system, 278
    dead zones, 290, *290*
    denitrification, 290
    diazotrophs, 288–90
    emergent properties, 280
    eutrophication, 286, 290, *290,* 344
    feedback, 281–83, *282, 283*
    flows, 280, *280*
    greenhouse gas emissions, 293, *293,* 296
    human body, 279–80, *279*
    iceberg model, 297, *297*
    individual choices and actions, 296–97
    macronutrients, 283
    methanogens, 292–93, *292*
    models, 280
    nitrification, 290
    nitrogen cycle, 286, 288–90, *288, 290, 291*
    nitrogen fixation, 288–90, *288, 290, 291*
    nitrogen pollution, 291, 300
    nutrient recycling, 287, 299–300
    ocean acidification, 296
    open system, 278
    oxygen cycle, 283–84, *284, 285*
    parts and interconnections, 280
    phosphorus cycle, 284–86, *285, 286,* 287
    reinforcing feedback, 282, *282*
    sinks, 284
    stock and flow diagrams, 280–81, *281,* 296, *296, 297*
    stocks, 280, *280*
systems definition, 6, 276, 278

T

tar sands, 381–82
Taxol, 126
tectonic plates, 248–49, *248, 249*
temperate soil, *262*
temperature, 218–19, *219*
    and carbon dioxide, 308, 394
    defined, 76
    temperature records and climate change, 305–7, *305, 306, 307*

10% law, 79–80, *80*

teosinte (ancient corn), 338, *338*

terracing, *264,* 265

terra preta, *260,* 265

tertiary consumers, 79, *80*

test group, 16

testing, controlled experiments, 15–16, *16*

textiles
    and carbon dioxide emissions, 328–29, *329*
    environmental effects of textile industry, 2, 4, *4*
    textile waste and "fast fashion," 517

thermal energy, 76

thermal inversion, 214, *214,* 228, *228*

thermodynamics
    entropy, 78, *79*
    first law of, 77–78
    second law of, 78, *80*

thermohaline conveyor, 312, *312*

thermosphere, 217, *217*

thermostats, 282–83, *283*

30 by 30 initiative and, 132

3Es (environment, economy, and equity) and sustainable development, 9

350.org group, 572, *572*

Three Gorges Dam, 403–4, *404*

Thunberg, Greta, 330, *534, 570–71*

tick populations, 333–34

tilling practices, 264–65, *264, 266,* 273–74, 287, *287*

Tilman, David, 127

tilth, 263

Titusville, Pennsylvania oil well, 377, *377*

Tolba, Mostafa, 25, *25*

topography. *See* landforms and topography

topsoil, 259, 262–63, *263,* 270

total fertility rate (TFR), 156–57, *157,* 160, 165, 169, *169*

toxaphene, *505*

toxicology, 496, *496,* 497

Toxic Substances Control Act (TSCA), 52, 57–58

toxins
    asphyxiants, *503*
    carcinogens, *503*
    corrosive, *503*
    defined, 496
    dichlorodiphenyltrichloroethane (DDT), 342, *343,* 501, 503, *505,* 582, 583
    effects of, 502, *503*
    exposure to, 503–4, *504, 505, 506*
    hepatotoxins, *503*
    nephrotoxins, *503*
    neurotoxins, *503*
    persistence, 504
    persistent organic pollutants (POPs), 504, *505*
    reprotoxins, *503*
    risk management and regulation of, 506–7, *507*
    risk of harm from, 502
    volatile organic compounds (VOCs), 503, *504*
    workplace exposure to, 504, *506,* 515

tradeable development rights (TDR), *136*

tradeable emission allowances, 234, 236

trade-offs and environmental decision-making, 22–23

tragedy of the commons, 44–45, *45*

transform plate boundary, 249, *249*

transportation
    choices and environmental policy, 47, *48*
    Federal-Aid Highway Act of 1956, 473
    induced traffic, 476–77, *476*
    modal split, 477
    street design, 468, 469–70, 473, *473*
    transit-oriented development (TOD), 478–79, *479*
    urbanization, 476–77, *476,* 482

trawling, bottom, *351*

tree rings, 306, *306*

triple bottom line (TBL) accounting, 549, *549*

trophic levels, 78–79, *80*

tropical soil, *262*

troposphere, 216, *217*

true cost accounting, 44

Trueman, Clive, 83

Trump, Donald, 572, 575, 579, 588

tsunamis, 244, *246, 508,* 509

Tversky, Amos, 523

U

ultraviolet (UV) radiation, 229, 230, 509–10, *510*

unitary governing systems, 574

United Nations (UN)
    Conference on the Human Environment, 25
    Decade of Ocean Science for Sustainable Development, 207
    Earth Charter, 35, *36*
    Environment Programme (UNEP), 131, 586
    Framework Convention on Climate Change (UNFCCC), 587–88
    Millennium Ecosystem Assessment (2005), 124
    Sustainable Development Goals, 9, *12,* 24, 584–85, *585*

upcycling, 445–46, *446*

upwelling, 312, *313*

urbanization
    bright lights syndrome, 466
    ecological footprint of cities, 461, *461*
    economies of scale, 464, *464, 465*
    gentrification, 482
    greenhouse gas emissions from cities, 464, *465*
    green infrastructure, 480, *480,* 482
    habitat loss, 460–61, *461*
    human population, 166–67, *167*
    individual choices and actions, 482–83
    induced traffic, 476–77, *476*
    informal economy, 441, 466
    infrastructure, 458, 460, 475, 480, *480,* 482
    Living Building Challenge, 481
    mixed-use areas, 478
    New York City, 460, *460,* 466, *466*
    pedestrian-oriented development (POD), 478–79, *479*
    planner's dilemma, 474
    rural areas, 460
    slums, 465–67, *465, 466, 467,* 474
    smart growth, 478, *478*
    street design, 468, 469–70
    and sustainable land use, 267
    transit-oriented development (TOD), 478–79, *479*

transportation, 476–77, *476,* 482

urban areas, 460, *460, 461*

urban density, 464

urban penalty, 464

urban planning, 474–76

urban transition, 460, *460,* 462–63

Use the Data, 486–87

Use the News, 484–85

walkability, 478, *479*

    See also development; land use; suburban sprawl

urban planning. *See* land use

US demographic transition (baby boom), 164, *165*

US Fish and Wildlife Service, 50, 130, 136, 142

US Forest Service, 50, 125, 130, 131, 148

US regional population dynamics, 162–63, *162*

utilitarianism, 37

V

vaccination and herd immunity, 502

Vaillant, George, 524

values
    defined, 20
    instrumental and intrinsic, 123–24
    valuation of ecosystem services, 125–26
    value chains, 125

van der Linden, Sander, 29

variable generation, 399–400, 414

vegetarian diets and sustainability, 39

vehicles
    electric vehicles (EVs), 408–9, *409,* 415, 420, *420,* 425
    fuel economy driver interface (FEDI), 530
    fuel economy labels, 542
    fuel efficiency and carbon footprints, 327, 328
    ride-sharing, 450, *451*
    self-driving cars, 417

Veterans Administration (VA) t, 471

vicious/virtuous circles, 282

Viking colonies in Greenland, *153,* 156

Vogt, William, 156, 157, 158

volatile organic compounds (VOCs), 226, 227, *227,* 503, *504*

volatility, 503

Volkswagen emissions testing, 240–41

voting, 26–27, 55, 590, 592

W

"waffle gardening," *339*

walkability, 478, *479*

Wallace, Alfred Russel, 95

Walmart, 454, 556, 563, *563*

Wankowski, Sam, 454

warm front, 221

Warren County, North Carolina, 13

Washington, Warren, 317, *317*

waste
    Basel Convention, 435
    bioreactor landfills, 440
    combustion, 438–39, *440*
    and consumption, 24, 446–48, *447*
    containment buildings, 438
    conversion, *437,* 440–41

and environmental justice, 436, 439
e-waste, 435, *436*, 450
food composting, 441–42, *441*
food losses and waste, 360, 362, 366, 441–42, *441*, 451, 454
"gar-barge," 435, *435*
"garbology," 449
geologic disposal, 436, 438, *438*
global waste trade, 435, *435, 436*
government actions regarding, 450
Great Pacific Garbage Patch, 426, *428*
hazardous waste disposal and remediation, 438, *438*, 440–41
human waste recycling, 443
incineration, 436, *437*, 438–40, *440*
individual choices and actions on, 451–52
isolation, 436, *437*, 438, *438*
landfill gas emissions, 434, *434*
landfill leachate, 434, *434*, 438
life-cycle assessment, 430, *430*
mass burn technology, 439–40
municipal solid waste (MSW), 429–30, *429*
nuclear waste, 405, 406, 436, 438, *438*
ocean dumping, 434–35, *435*
open dumps, 430–31, *431*, 434, *434*
plastics, 426, *426–27*, 428–29, *428*, 430, *430*, 450
producer responsibility law, 450
reduction, *447*, 448, *448*, 450–51, *451*
sanitary landfills, 436, *437*
Smokey Mountain Garbage Dump, 430–31, *431*
solid waste, 430
Use the News, 454–55
waste stream, 429, *429*
waste-to-energy facilities, 440
See also recycling
wastewater treatment, 200, *200*, 201, 287, 299–300
water
    albedo, 202, 203
    aquifers, 183, *183*
    Clean Water Act, 200, 202
    conservation, 196–97, 207–8
    covalent bonds, 68, *68*
    cryosphere, 202, 203
    cycle, 184–85
    desalination, 198, *200*
    drought and water shortages, 180, 182, *182, 195*, 195–96, *197*, 207–8
    embedded, 182, *182*, 196
    estuaries, *188*, 189
    Flint drinking water supply, 189, 193
    freshwater access improvement, 194–96, *195*
    glaciers, 186, *202*, 203, *203*
    groundwater, 183, *183*, 185, 186
    groundwater-dependent ecosystems and organisms, 186, *186*

groundwater withdrawals, 190–91, *190, 191*
harvesting, 198, *198*
human impacts on, 122, 180, 182, *182,* 189–94
hydrosphere, 183, *183*
ice age, 202, 203
importance of, 70–71
individual choices and actions, 207–8
lakes and ponds, 186, *187*
lentic ecosystems, 186, *187*
lotic ecosystems, 186–87, *188*
milk products and water usage, 199
mining, 190
nonpoint-source pollution, 193–94, *193,* 202
Ogallala Aquifer, 190, *190*
permafrost, 185, 202, 203
point-source pollution, *193,* 193–94
pollution, 193–94, *193,* 209–10, 268
quality, 198, 200, 202, 344, *344*
recharge, 190
recycling, 197, 201
reservoir, 183, *183*
residence time, 183
rivers and streams, 186–87, *188*
Safe Drinking Water Act, 200
saltwater intrusion, 190, *191*
scarcity, 197–98, *198*
sea ice, 202, 203
security, 195
sediment, 192
subsidence, 190–91, *191*
surface water, 185, 186
surface water system disruptions, 191–92, *191, 192*
Use the Data, 210–11
Use the News, 209–10
vapor and weather, 219, *219*
wastewater treatment, 200, *200,* 201
water power, 372, *373*
wetlands, 187, *188,* 189
See also oceans
watersheds, 255–56, *256*
water-soluble substances, 503
Watt, James, 42, 375–76, 377
Watt's steam engine, 376, *376*
weather
    air masses, 220, *220*
    barometric pressure, 219–21, *220*
    defined, 217–18
    environmental consequences of severe weather, 218
    floods, 511, *511,* 512
    heat waves, 511–12
    high-pressure systems, 220, *220*
    humidity, 219
    hurricanes and severe weather, 321, *324*
    lake and land effects, 221–22, *222*

low-pressure systems, *220,* 221
precipitation, 221–22, *222*
precipitation and extreme rain events, 319–21, *320, 324*
relative humidity, 219, *219*
storm damage costs, 304, *304*
storm surges, 511, *511*
temperature, 218–19, *219*
water vapor, 219, *219*
winds, fronts, and storms, 221, *221*
weather forecasting
    information sources, 19, *19*
    models, 16–18, *17*
    reports, 217–18, *218*
weathering of rocks, 250–52, *250, 251, 252, 253*
Wegener, Alfred, 248
Westerveld, Jay, 550
West Nile virus, 115–16, 499, 512
wetlands, 187, *188,* 189
Whitman, Christine Todd, 476
wild and scenic rivers, 131
Wilderness Act of 1964, 50, 53, 131, 535
wildlife corridors, 132, 133, *133*
Wilson, E. O., 133
Wilson, Sacoby, 491, *491,* 493
Wilson, Stiv, 81
windbreaks, *264,* 265
wind power, 373, *373,* 396, *396–97,* 398–400, *398, 399,* 413–14, *413, 414, 415, 418,* 421, 576
winds, fronts, and storms
    atmospheric pressure and prevailing winds, 222, *222,* 224, *224*
    described, 221, *221*
wind turbines, 398–99, *398, 399*
wolves, 89, *89,* 94, 102, 103, *103,* 105–6, 132–33
women, empowerment of girls and, 168–70, *169*
work, 74, 75
workplace exposure to toxins, 497, 504, *506,* 515
World Health Organization, 159, 194, 198, 214, 235
World Heritage Sites, 131
World Wildlife Fund (WWF), 544, 546, *546,* 561

Y

Yellowstone National Park, 50, 105–6, 126, *126,* 130, 132, 134
Yosemite National Park, 54, 125
youth bulge, 165, 166

Z

Zahab, Ray, 194–95, *195*
Zhenmin, Liu, 177
Zika virus, 499
zoning ordinances, 142